BLE OF INTEGRALS

sic Forms

1. $\int u^n\,du = \dfrac{1}{n+1}u^{n+1} + C, \ n \neq -1$

2 $\int \dfrac{du}{u} = \ln|u| + C$

3. $\int e^u\,du = e^u + C$

4. $\int a^u\,du = \dfrac{1}{\ln a}a^u + C$

5 $\quad du = -\cos u + C$

6. $\int \cos u\,du = \sin u + C$

7. $\int \sec^2 u\,du = \tan u + C$

8. $\int \csc^2 u\,du = -\cot u + C$

9. $\int \sec u \tan u\,du = \sec u + C$

10. $\int \csc u \cot u\,du = -\csc u + C$

1. $\int \tan u\,du = \ln|\sec u| + C$

12. $\int \cot u\,du = \ln|\sin u| + C$

13. $\int \sec u\,du = \ln|\sec u + \tan u| + C$

4. $\int \csc u\,du = \ln|\csc u - \cot u| + C$

15. $\int \dfrac{du}{\sqrt{a^2 - u^2}} = \sin^{-1}\dfrac{u}{a} + C$

6. $\int \dfrac{du}{a^2 + u^2} = \dfrac{1}{a}\tan^{-1}\dfrac{u}{a} + C$

17. $\int \dfrac{du}{u\sqrt{u^2 - a^2}} = \dfrac{1}{a}\sec^{-1}\dfrac{u}{a} + C$

tional Forms

8. $\int \dfrac{du}{a+bu} = \dfrac{1}{b}\ln|a + bu| + C$

19. $\int \dfrac{u\,du}{a+bu} = \dfrac{1}{b^2}(a + bu - a\ln|a + bu|) + C$

0. $\int \dfrac{u\,du}{(a+bu)^2} = \dfrac{a}{b^2(a+bu)} + \dfrac{1}{b^2}\ln|a+bu| + C$

21. $\int \dfrac{u^2\,du}{(a+bu)^2} = \dfrac{1}{b^3}\left(a + bu - \dfrac{a^2}{a+bu} - 2a\ln|a+bu|\right) + C$

2. $\int \dfrac{du}{u(a+bu)^2} = \dfrac{1}{a(a+bu)} - \dfrac{1}{a^2}\ln\left|\dfrac{a+bu}{u}\right| + C$

23. $\int \dfrac{du}{u^2(a+bu)} = -\dfrac{1}{au} + \dfrac{b}{a^2}\ln\left|\dfrac{a+bu}{u}\right| + C$

4. $\int u(a+bu)^n\,du = \dfrac{(a+bu)^{n+1}}{b^2}\left(\dfrac{a+bu}{n+2} - \dfrac{a}{n+1}\right) + C, \ n \neq -1, -2$

25. $\int \dfrac{du}{a^2 - u^2} = \dfrac{1}{2a}\ln\left|\dfrac{u+a}{u-a}\right| + C$

5. $\int \dfrac{du}{(a+bu)(c+du)} = \dfrac{1}{ad-bc}\ln\left|\dfrac{c+du}{a+bu}\right| + C, \ ad - bc \neq 0$

7. $\int \dfrac{u\,du}{(a+bu)(c+du)} = \dfrac{1}{ad-bc}\left(\dfrac{a}{b}\ln|a+bu| - \dfrac{c}{d}\ln|c+du|\right) + C, \ ad - bc \neq 0$

ms Containing $\sqrt{a+bu}$

8. $\int u\sqrt{a+bu}\,du = \dfrac{2}{15b^2}(3bu - 2a)(a+bu)^{3/2} + C$

9. $\int u^n\sqrt{a+bu}\,du = \dfrac{2}{b(2n+3)}\left[u^n(a+bu)^{3/2} - na\int u^{n-1}\sqrt{a+bu}\,du\right]$

0. $\int \dfrac{u\,du}{\sqrt{a+bu}} = \dfrac{2}{3b^2}(bu - 2a)\sqrt{a+bu} + C$

31. $\int \dfrac{u^2\,du}{\sqrt{a+bu}} = \dfrac{2}{15b^3}(8a^2 + 3b^2u^2 - 4abu)\sqrt{a+bu} + C$

2. $\int \dfrac{u^n\,du}{\sqrt{a+bu}} = \dfrac{2u^n\sqrt{a+bu}}{b(2n+1)} - \dfrac{2na}{b(2n+1)}\int \dfrac{u^{n-1}\,du}{\sqrt{a+bu}}$

33. $\int \dfrac{du}{u\sqrt{a+bu}} = \begin{cases} \dfrac{1}{\sqrt{a}}\ln\left|\dfrac{\sqrt{a+bu}-\sqrt{a}}{\sqrt{a+bu}+\sqrt{a}}\right| + C, \ a > 0 \\ \dfrac{2}{\sqrt{|a|}}\tan^{-1}\sqrt{\dfrac{a+bu}{|a|}} + C, \ a < 0 \end{cases}$

4. $\int \dfrac{du}{u^n\sqrt{a+bu}} = -\dfrac{\sqrt{a+bu}}{a(n-1)u^{n-1}} - \dfrac{b(2n-3)}{2a(n-1)}\int \dfrac{du}{u^{n-1}\sqrt{a+bu}}$

35. $\int \dfrac{\sqrt{a+bu}}{u}\,du = 2\sqrt{a+bu} + a\int \dfrac{du}{u\sqrt{a+bu}}$

5. $\int \dfrac{\sqrt{a+bu}}{u^2}\,du = -\dfrac{\sqrt{a+bu}}{u} + \dfrac{b}{2}\int \dfrac{du}{u\sqrt{a+bu}}$

ms Containing $\sqrt{u^2 \pm a^2}$

7. $\int \sqrt{u^2 \pm a^2}\,du = \dfrac{u}{2}\sqrt{u^2 \pm a^2} \pm \dfrac{a^2}{2}\ln\left|u + \sqrt{u^2 \pm a^2}\right| + C$

38. $\int u\sqrt{u^2 \pm a^2}\,du = \dfrac{1}{3}(u^2 \pm a^2)^{3/2} + C$

9. $\int u^2\sqrt{u^2 \pm a^2}\,du = \dfrac{u}{8}(2u^2 \pm a^2)\sqrt{u^2 \pm a^2} - \dfrac{a^4}{8}\ln\left|u + \sqrt{u^2 \pm a^2}\right| + C$

40. $\int \dfrac{\sqrt{u^2 + a^2}}{u}\,du = \sqrt{u^2 + a^2} - a\ln\left|\dfrac{a + \sqrt{u^2 + a^2}}{u}\right| + C$

2

SINGLE VARIABLE CALCULUS

EJA

EDITION

2

SINGLE VARIABLE CALCULUS

Leonard I. Holder
Gettysburg College

James DeFranza
St. Lawrence University

Jay M. Pasachoff
Williams College

Brooks/Cole Publishing Company
Pacific Grove, California

ITP™ The trademark ITP is used under license.

To

Jean, Bill, and Steve
Regan, Sara, and David
Naomi, Eloise, and Deborah

for their love and support

Brooks/Cole Publishing Company
A Division of Wadsworth, Inc.

Printed in the United States of America

10 9 8 7 6 5 4 3 2 1

Library of Congress Cataloging in Publication Data
Holder, Leonard Irvin. [date]–
    Single variable calculus / Leonard I. Holder, James DeFranza, Jay
M. Pasachoff.
        p.   cm.
    Includes index.
    ISBN 0-534-18990-3
    1. Calculus.   I. DeFranza, James, [date]–   .   II. Pasachoff, Jay
M.   III. Title.
QA303.H578   1993
2nd ed.
515—dc20   93-40949
                                                               CIP

Sponsoring Editor: *Jeremy Hayhurst*
Editorial Assistant: *Elizabeth Barelli Rammel*
Production Coordinator: *Marlene Thom*
Manuscript Editor: *Janet Tilden*
Interior and Cover Design: *Sharon L. Kinghan*
Cover Photo: *Orrery, Adler Planetarium; Satellite, European Space Agency/Photo Researchers Inc.*
Art Coordinator: *Lisa Torri*
Interior Illustration: *MacArt Design; Tech-Graphics*
Photo Editor: *Diana Mara Henry*
Photo Researcher: *Joan Meyers*
Typesetting: *Electronic Technical Publishing*
Cover Printing: *Phoenix Color Corporation*
Printing and Binding: *Arcata Graphics/Hawkins*

PREFACE

The cover depicts a working mechanical model of the solar system, called an *orrery* after the fourth Earl of Orrery, who commissioned the second such machine ever built, in about 1713. The earliest orreries showed only the movement of the Earth and Moon relative to the Sun, but later ones incorporated other planets such as Mercury and Venus, all of which were set in motion by a hand crank. This particular orrery includes Saturn and was built in the Netherlands in 1735. A well-crafted orrery was an elegant mechanical model used as both a calendar and a clock, and it furthered popular understanding of the physical and mathematical nature of the solar system.

The cover also portrays a currently deployed European Remote Sensing satellite (ERS1) because the artificial satellite is in large part an extension of the underlying physics and mathematics used in constructing the orrery. We have chosen the orrery and the satellite images to represent both the very beginning and the latest manifestation of the modern scientific age, which has been profoundly influenced by mathematics—especially calculus. They also serve to remind us of the remarkable duality of mathematics: its precision and truthfulness are unchanging; yet its application is amazingly flexible and adaptable to both the physical and mathematical universes.

The study of the orbits of artificial and natural satellites (planets) owes a great deal to the scientific insights made possible by calculus. Isaac Newton used calculus to show that the planets travel in elliptical orbits around the Sun. He thereby logically confirmed Johannes Kepler's earlier postulates based on years of astronomical observations by Kepler's mentor Tycho Brahe. Whereas mechanical orreries gave relatively crude approximations of the orbits of the planets, modern software-based digital orreries are vastly more accurate and are used to make predictions about the long-term (and perhaps ultimate) behavior of the solar system and the universe.

In the early eighteenth century, about the time that the first orrery was built, Bernard de Fontenelle said that the recent invention of the calculus was a revolution in mathematics. Fontenelle's enthusiasm was an understandable reaction to the astonishing success of calculus in enabling rapid progress on many previously intractable problems. Calculus was immediately seen to be effective and remarkably powerful in describing and predicting both change and motion. The complementary concepts of derivative and integral and their associated foundational problems—that of tangent (or velocity) and area—clearly set the stage for a profound reevaluation of the importance and very nature of mathematics. It was evident, even in Fontenelle's day, that calculus was a singular development in the history of mathematics, but in what sense was it a revolution? Many of the insights that calculus provided are worthy of that description, but many mathematical philosophers would deny that there have ever been revolutions in mathematics. After all, mathematics is unique among the sciences in that its truths, when based on common axioms, are unchanging once proven. Mathematics does not appear to be prone to revolutions in the sense of sudden and dramatic changes; but there is another connotation for the word *revolution*. It was first used as a technical term in the exact sciences, where it literally meant to return again, through cyclic succession, as in the changing seasons, or to ebb and flow, like the movement of tides. In this sense Fontenelle was on firmer ground. Perhaps it is natural to expect a kind of cyclic progression whereby calculus, and mathematics itself, is periodically reexamined and reexpressed.

CONTEMPORARY CALCULUS

One could argue that the teaching of calculus is currently undergoing a kind of revolution caused by widespread dissatisfaction with the current national calculus curriculum. Broadly stated, there is an awakened interest in teaching core concepts in a modern context as opposed to overwhelming students with interesting but secondary details. Although it is probably characteristic of revolutionary movements that—even among their supporters—they mean different things to different people, there are common themes in reform discussions. The principal goals seem to be: a focus on conceptual understanding rather than rote manipulation; increased relevance and applicability to other disciplines; an emphasis on visualization as a counterpoint to abstraction; and the use of relevant technology to streamline computation. The discussions seem to hold out the promise of a new and creative attitude to mathematics instruction.

We wholeheartedly approve of many of the goals of reform, but we have consciously not "thrown the baby out with the bathwater." For instance, despite calls for streamlining topics, we have retained a thorough treatment of limits and techniques of integration. We see attempts to all but eliminate their discussion as ill advised. After all, the limit is the single most important notion in calculus, which is the mathematics of limits; and techniques of integration have important theoretical implications in later mathematics courses. The same is true of computational skills. Though we want students to appreciate that mathematics is more than computation, nothing is gained by claiming that these skills are somehow suddenly less important in understanding calculus. We considered many options and weighed them carefully before revising this text. We have ultimately taken what might be called a modern but conservative approach. Our response to criticism of the calculus curriculum is evolutionary rather than revolutionary.

Some of the reform principles exhibited here are as follows: a thorough grounding in functions and their graphs; early introduction of differential equations; the proper emphasis on numerical procedures; the use of graphing calculators and computer algebra systems; discussion of broad-based and pedagogically realistic applications; a more streamlined discussion of integration techniques in favor of using CAS and tables; and the introduction of infinite series through the vehicle of polynomial approximation of functions. We emphasize the practicality and utility of calculus and its visualization through graphics and images, both computer-generated and photographic, whenever they can make an otherwise dry and abstract discussion more concrete. The occasional use of color photographs encourages visualization in relevant physical contexts and adds interest and visual appeal. A new design enables us to highlight important results and to distinguish important elements of graphs, logical connectives, and related concepts that need to be distinguished.

Computer algebra subsections also encourage efforts to visualize using appropriate technology. These subsections are intended to be useful reading for students even if they do not have access to a computer algebra system. When facilities are available, we recommend "computer-assisted algebra"—not thoughtless button pushing—and stress that using the technology is not a substitute for understanding the mathematics. Quite the contrary—one usually needs to understand the mathematics in order to verify and apply the results of the technology.

We have revised this edition with special attention to making the material clearer and more concrete as well as more readable and interesting, while retaining the appropriate level of rigor. The organization, the explanations and examples, the exercises, and the text layout have all been substantially improved. We have also added several new features to the text that we believe will enhance students' perception and understanding of the subject.

ORGANIZATION

- The material on functions and graphs has been expanded and now comprises the entire first chapter. Additional precalculus review material appears in the free Supplementary Appendices.
- Limits and continuity form a new second chapter. The last section on the theory of limits is optional and can be omitted.
- The transcendental functions have been consolidated in this edition—that is, exponential and logarithmic functions have been combined with inverse trigonometric and hyperbolic functions in one chapter. Trigonometric functions are reviewed in the first chapter. We have chosen to introduce the logarithm as an integral and present the exponential function as its inverse to reinforce the concept of the integral.
- The chapter on parametric equations and polar coordinates now precedes that on sequences and series. A brief review of conic sections is followed by the polar form of the conics. The former separate chapter on conic sections has been placed in the Supplementary Appendices so it can be presented with near-total flexibility.
- Discussion of numerical methods is treated where the material occurs naturally in the context of other topics, not in a separate chapter. Throughout the text numerical methods are given their proper due as an irreplaceable adjunct to symbolic and graphical techniques.

NEW FEATURES

- In addition to an ample number of drill exercises, applied problems, and more theoretical exercises, we included many problems calling for the use of graphing calculators or computer algebra systems. These latter problems are marked with a special icon so that instructors can easily skip over or emphasize them, depending on local preferences and resources. The number of examples and figures has more than doubled.

- Selected applied problems have been added to the end of each chapter in special exercise sets called Applying Calculus. Some of these problems will be appropriate for class projects and, at the request of some reviewers, answers to these exercises are not provided in the text. Solutions are provided in the Complete Solutions Manual, Volumes 1 and 2.

- Preceding each chapter is a "Personal View" interview or essay "about calculus." People interviewed here include representatives of science, history, economics, physics, engineering, and computer science, as well as mathematicians. The purposes are severalfold. We hope to give a glimpse of the diversity of opinion and varied personalities of scientists and mathematicians, to neutralize some myths about mathematics, and to show that mathematics is not a static discipline. We also want students to understand that calculus is widely useful in different professions, and that teaching and mathematical research are not without controversy. We hope these "personal views" will provide a little extra motivation and perhaps even inspiration for those students who ask "where will I ever use this stuff?" We hope more students can be led to perceive that the course has utility and significance beyond the immediate grade they achieve.

- Each chapter concludes with a Concept Quiz. Answers to these quizzes are provided in the Study Guides, but students should be encouraged to attempt these quizzes and verify the answers on their own.

PEDAGOGICAL CONSIDERATIONS

To succeed in any quantitative discipline, students must "get off on the right foot" in mathematics. A calculus course is critically important in this respect alone. Most students learn best by proceeding from the concrete to the more abstract, and our consistent approach is to discuss new concepts intuitively before stating formal definitions or theorems. Our style is conversational, speaking directly to the student. We have tried to explain the mathematics in simple language. We do not "talk down" to students and we do not avoid rigorous argument where it is needed. (A number of the more technical proofs are given in the Supplementary Appendices.)

We begin each chapter with a summary of the chapter goals so that students will have some idea of where the material is leading. We try to motivate new ideas via concrete problems and often ask students to make their own conjectures before we state the results. We hope that with the right tone in the exposition, many students can be persuaded to read the text before they come to class. We even dare to hope that more of them will truly enjoy calculus and decide to major in mathematics or in a mathematically related field.

ACKNOWLEDGMENTS

First, we give our sincere thanks to the contributors of the Personal Views: Paul R. Halmos, Owen Gingerich, Judith V. Grabiner, Mark J.T. Smith, Anna J. Schwartz, Phillip A. Griffiths, Keith Geddes, Deborah Tepper Haimo, Sheldon Glashow, and Benoit Mandelbrot. The efforts of supplement authors and accuracy checkers Raymond Southworth, Ben Brown, John Banks, David Royster, Marian Hoyle, and Terri Bittner are also gratefully acknowledged. Special thanks are due to Raymond Southworth, who labored long and hard to ensure accurate answers and solutions to the exercises. Any remaining errors are our responsibility.

Jay Pasachoff thanks his daughters, Eloise and Deborah, for inspiring him through *their* studies of calculus to work on this book.

In addition, we thank the following reviewers and colleagues, who have also contributed a great deal to the text: Nancy Blachman, Variable Symbols Inc.; Jon Breitenbucher, Ohio State University; Branko Curgus, Western Washington University; Lance Nielsen, University of Nebraska–Omaha; Roger Waggoner, University of Southwestern Louisiana; Richard Bonnano, Deerfield Academy; Deborah Frantz, Kutztown University; Donna Bailey, Northeast Missouri State University; Barbara Bohannon, Hofstra University; Fred Brauer, University Wisconsin–Madison; Mervin Childers; Roger Cooper, Canada College; Anthony Ferzola, University of Scranton; Dante Giarusso, St. Lawrence University; Nathaniel Martin, University of Virginia; Montie Monzingo, Southern Methodist University; John Randolph, West Virginia University; Robert Stanton, St. John's University; Richard Tucker, North Carolina A&T State University; and Andrei Verona, California State University–Los Angeles.

We thank the indexer, Nancy Kutner, for her expert work.

Thanks also to the Brooks/Cole staff: Elizabeth Barelli Rammel, editorial assistant, who coordinated volumes of handwritten, verbal, and electronic communication and resolved numerous details essential to the book's development; Sharon Kinghan, senior designer; Lisa Torri, senior art coordinator; Margaret Parks, promotions manager; Patrick Farrant, marketing manager; and Diana Henry, photo editor. Special thanks go to Marlene Thom, production coordinator, for the superb job she did guiding the book through the complex all-electronic production process. She did so with skill, and somehow maintained her good humor, despite what was often a demanding schedule. It was a delight to work with her. Finally, we wish to thank our editor, Jeremy Hayhurst, for his support and guidance. His ideas, his vision, and his encouragement were critical in making this book possible.

Leonard I. Holder
James DeFranza
Jay M. Pasachoff

PROLOGUE FOR STUDENTS

The word *calculus* comes from the Latin term for a pebble used as a token in counting and calculating. The development of calculus has a long and rich history. Although rudiments of the subject can be traced back to the ancient Greek mathematicians, Archimedes in particular, the primary impetus came in the seventeenth century from mathematicians such as Caliveri in Italy, Fermat and Descartes in France, and Wallis and Barrow in England. The invention of analytic geometry by Descartes and Fermat was an especially important prerequisite for the development of calculus. However, it was Isaac Newton and Gottfried Wilhelm Leibniz who found the crucial mathematical relationships and provided the notational tools needed to bring the emerging ideas together into a coherent calculation system—that is, a calculus—and this remains an intellectual feat of enormous importance. (See the *Personal Views* before Chapters 2 and 3 for more discussion of the history of calculus.)

Calculus can be described as the mathematics of change and motion. Since change and motion are implicit in all aspects of the physical world, the methods of calculus are useful in all the physical, natural, and social sciences, including economics. Here are only a few of the things calculus enables us to do.

- forecast the size of a growing population after a given period of time
- maximize profit and minimize costs in a business
- model ocean and air currents and their effect on weather
- design turbines, boat hulls, and electric circuits
- model the spread of a disease, rumor, or fashion
- determine how to place a satellite in geosynchronous (stationary) orbit
- predict voting behavior

Given the incredible diversity of the applications of calculus, it is all the more surprising that it evolved from two seemingly unrelated geometric problems: finding the tangent line to a curve and finding the area bounded by a curve. In part (a) of the figure in the margin we show the tangent line to a curve at point P, but how do we know exactly which line it is? In other words, given the

(a)

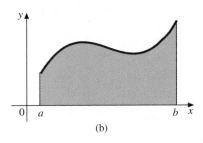

(b)

curve and the point P, how do we find the tangent line? In part (b) we show a region under a portion of a curve and above the x-axis. How do we find the area of the region?

Solving the tangent problem leads to the concept of derivative. That part of calculus that deals with derivatives is called *differential calculus*. Solving the area problem leads to the integral, which is the basis of *integral calculus*. How the derivative and the integral are related is one of the most fundamental insights in all of calculus. The key to solving both the tangent and area problems is the concept of limit. The limit is the single most important notion in calculus, and it is what separates calculus in a fundamental way from all of the mathematics that preceded it. Much of this course will deal with limits in one form or another and all of Chapter 2 is devoted to studying limits of functions.

Although the derivative has its origins in the solution of the tangent problem, it can be interpreted more broadly as a measure of the instantaneous rate of change of one quantity with respect to another. It is this rate-of-change interpretation that makes the derivative such a powerful tool. It is the derivative, for example, that enables us to find such things as the velocity and acceleration of a moving object, the rate at which blood is flowing through an artery at a given instant, the rate at which the profit on some item is increasing or decreasing, the rate at which a culture of bacteria is growing, and the rate of decay of a radioactive substance.

Although the integral is founded on a geometric problem involving an area, it can be more broadly interpreted as the total change in some quantity whose rate of change is known. For example, the integral can be used to determine the total distance traveled by an object even when its velocity changes, the work done by a variable force, the total force against a dam, the total length of a curve, and the total volume enclosed by a surface.

COMPUTERS, CALCULATORS, AND CALCULUS

During your course you will undoubtedly encounter some form of technology. Both graphing calculators and computer software can be valuable aids in studying calculus, if they are used properly; however, they are not substitutes for learning the basic mathematical concepts. In this text we discuss ways in which calculators and currently available computer software can be used to best advantage in learning calculus. We are trying to give you a general sense of where these systems are most helpful and appropriate, even if you don't have a computer or a calculator at hand. We have included discussion of software in the text and graphing calculators are discussed in the Supplementary Appendices.

With the software programs and mathematical languages known as *computer algebra systems* (abbreviated "CAS"), almost all calculus operations and computations can be done quickly and efficiently. The best-known of these systems are Maple, Mathematica, and DERIVE. The term *computer algebra system* is slightly misleading, because although these systems do algebra, they do much more, and they are forcing a reevaluation of what it is that we emphasize when teaching mathematics. CAS are designed to do symbolic manipulation, solve numerical problems, and graph functions, and they can be programmed as well. Other software programs, called *spreadsheets*, are optimized for numerical calculations. They are widely used in the business world and can be useful in calculus too, especially in work with numerical methods. Spreadsheets also have the ability to graph functions. Some widely used spreadsheets are Microsoft's Excel, Lotus's 1-2-3, and Borland's Quattro Pro.

Do not let yourself be intimidated by this type of technology. It is a servant in the course, not a master. No one is encouraging you to compete with or to try to outcompute an algebra system. It's important to understand that in many cases the software is using the same algorithm (correct mathematical procedure) that you use when you solve the problem "by hand," but in other cases it is using an algorithm that is only practical to use within the software. You will sometimes have to ask yourself how the computer or calculator is getting its results, and occasionally you will have to use the theory of calculus to verify that your results are correct. Technology changes rapidly, and there will undoubtedly be newer devices or releases of these programs available when you take this course. It is important to remember that the mathematics you are learning has not changed, and the problem-solving skills that you develop in this course will never be obsolete.

The text includes "computer algebra subsections" on a yellow screened background at the end of certain sections of the text. These subsections give guidance on carrying out calculus operations using one of the three common CAS. Not all the details are shown, but you should be able to follow the instructions as given. Don't panic if the commands don't work as indicated on your version of the software. You may have a different release that uses different commands. Also, your computer may have some other CAS, in which case you can adapt the instructions to your particular system. You may need to consult the program documentation carefully.

Basic scientific calculators have many uses in calculus just as they do in precalculus courses. The more advanced programmable calculators, especially those with graphing capabilities, are especially well suited to calculus. These advanced calculators share many features with computers. They have smaller viewing screens and do numerical calculations rather than the more general symbolic operations carried out by CAS. Although the calculators are more limited, they are more compact and portable and are cheaper than software and a computer. However, if you already have a computer, consider getting a CAS program. They rarely cost more than a good graphing calculator and will take you much further than a calculator will. Other advantages of the CAS are that you can program and update them easily as well as use them with a word processor.

Of course, it may not be necessary for you to choose between a computer and a calculator. If you have access to both, you will sometimes want to use one and sometimes the other; in most cases, we do not specify which device or program to use. We do not show instructions for using calculators in the text. (See the Supplementary Appendices for graphing calculator examples.) There are specially marked exercises for graphing calculators and CAS where technology is all but required, but we have deliberately not marked all the exercises where graphing calculators or CAS could be used.

APPROPRIATE USE OF TECHNOLOGY

What should you learn to do by hand and what should you have a computer or calculator do for you? This is analogous to asking whether grade-school students should learn fundamentals of multiplication and multiplication tables, or simply learn to push keys on calculators. Most mathematicians would argue that it is always better to understand the principles before you push the buttons. But perhaps there is more in common here than is first apparent. Using a multiplication or log table, memorized or not, and pushing buttons on a calculator both involve the use of an algorithm for carrying out a mathematical operation. In a sense, both you and the calculating device are simply employing an algorithm. The key is in understanding what the algorithm means and why it works. As your experience and confidence increase, you'll develop a sense of when the calculator or software will be most helpful. In many cases you will find that it is calculus itself, not the technology, that is the most useful tool.

In using technology to find derivatives, for example, it is essential that you first learn what a derivative is and what information it conveys. Only then is it meaningful to calculate derivatives using software, and only for more complicated derivatives (the simple ones can be found more easily and even faster by hand). If you rely solely on software to find all derivatives you encounter, you will have learned nothing about derivatives. Also, if you place too heavy a reliance on a computer or a calculator, you might draw incorrect conclusions from the output unless you have some idea of what to expect. When plotting a graph, for example, if you don't chose the x-scale and y-scale properly, you may not see all the important features of the graph. Without a knowledge of calculus, you would not know what the important features are and could easily come to incorrect conclusions. With a basic understanding of calculus, you can have total confidence in the correctness of your results.

Calculators and computers with CAS or spreadsheets are powerful devices and can be real assets in a calculus course, but they do not eliminate the need to learn the fundamental mathematical concepts. If your instructor prefers to overlook the CAS sections or the exercises marked for calculators or computers, you will not be disadvantaged in the course. We have tried to show how to take advantage of the availability of these devices, but the course is not dependent on them. If your instructor permits and encourages it, take advantage of what technology has to offer, but use it to learn your calculus. Once you truly understand what a derivative is, your instructor is unlikely to criticize you for using software to calculate one. If you don't understand what a derivative is, no degree of proficiency with software will compensate for this deficiency.

SOME SUGGESTIONS FOR STUDYING CALCULUS

Although a calculus course provides challenging goals, your success will have a great deal to do with how you study from your lecture notes and from this text. First, you will understand far more if you read the relevant section of the text before your lecture and before you attempt to work the exercises you are assigned. Don't treat this text as a novel. You need to read actively, with a pencil and paper at hand. Try to fill in any missing details you encounter in the algebraic steps. Second, do your assignments when they are assigned. Letting homework accumulate is a prescription for disaster in this or any other mathematics course. To learn mathematics, you must do mathematics. Watching your instructor do it is not enough. Don't give up too quickly if you have difficulty with a problem. If your instructor permits it, you might find it helpful to study with some of your classmates. Group study has proved to be more effective for some students than individual study.

As a student of calculus, you have much to look forward to as this rich and engaging subject unfolds before you. Read the chapter-opening Personal Views, not to learn calculus, but to learn about calculus and mathematics and their influence on the modern world. Recognize that calculus contains powerful ideas that are very general and applicable to a huge variety of problems, whatever your chosen discipline. Use whatever technology you have at your disposal to explore and investigate the mathematical ideas you encounter. This course should help you think about mathematics in a more logical fashion and to understand that definitions are not theorems, that theorems require proofs, and that abstract mathematical arguments are of great importance in efficiently describing the phenomena we see in the "real world." It is impossible to say what calculus is and to explain what all it can do in a brief introduction, because that, of course, is what the rest of the book is about. We hope that these preliminary remarks have served to whet your appetite and that you are ready now for the "main course." If you give calculus the attention it deserves, you will not regret the effort devoted to it in your future courses and career.

CONTENTS

2

SINGLE VARIABLE CALCULUS

Paul Halmos

What do you remember about your first mathematics course in college?

Most of what I learned in my first year of college was not taught in courses.... Calculus was considered too sophisticated for freshmen in those days. The mathematics I was taught in my freshman year at Illinois was algebra, trigonometry, and analytic geometry (a revelation to me). No one seemed to notice that virtually the only part of analytic geometry that calculus uses is the idea of a graph, and most of us knew about that before we entered college.

How did you do in your first calculus course?

It wasn't a big part of my life then. It was just a chore. I couldn't for the life of me understand it. B's were the best I could do. I could differentiate and integrate, but I had no understanding of things like the "four-step rule" in the text. I now know that it means "apply the definition to find the derivative."

Students sometimes have trouble in calculus because they don't know algebra and trigonometry. Why is this so important?

It's an old saying among teachers that students often trip up on the algebra—not the calculus itself. Mathematics is a hierarchical discipline. If you don't know enough algebra you're going to have trouble understanding the extensions of those ideas in calculus. My advice is to review as much precalculus as you can before you start calculus.

So what did you learn in your calculus class?

What I *did not* learn was how to study, or for that matter, what studying meant. I was just barely smart enough to know that it did not mean memorizing—but I did think that to have learned something really meant to remember it. Everything I had ever been taught was done by "here is how it goes—now you do it." That's how I was taught to count, to write, and to solve algebra problems. I didn't understand what it meant to understand something, and what one should do to get there.

Do you mean that much of what's taught in a mathematics course doesn't need to be memorized?

Not exactly. You should try to learn how to think about mathematics in a way that works for you. Don't rely on memory, but memory can't be ruled out as an asset in mathematical success. There's a story told about a mathematician named Tamarkin. On a Ph.D. oral exam, he asked the student about the convergence properties of certain series. "I don't remember," said the student, "but I can always look it up if I need it." Tamarkin was not pleased by this answer. "That doesn't seem to be true," he said, "because you sure need it now."

What would you say to a student just beginning a calculus course?

Here you sit, an undergraduate with a calculus book open in front of you.... How do you learn this, how do you study, how do you penetrate the darkness? All I can tell you is what I do, but I suspect that the same sort of thing works for everyone. It's been said before and often, but it cannot be overemphasized: study *actively*. Don't just read it; fight it! Ask your own questions, look for your own examples, discover your own proofs. Is the hypothesis necessary? Is the converse true? Why is this true? You'll learn much more this way.

What should a student concentrate on while attending a lecture?

Lectures are a standard way of learning something—one of the worst ways. Too passive, that's the trouble. *Standard recommendation*: take notes. *Counterargument*: yes, to be sure, taking notes is an activity. If you do it, you'll have something solid to refer back to afterward, but you are likely to miss the delicate details of the presentation as well as the big picture—you are too busy scribbling to pay attention. *Counter-counter argument*: if you don't take notes, you won't remember what happened, in what order it came, and chances are your attention will flag part of the time, you'll daydream, and you might even nod off. My own solution is a compromise: I take very skimpy notes, and then, whenever possible, I transcribe them in greater detail, as soon afterward as possible. By skimpy notes I mean one or two words per minute, plus, possibly a crucial formula or two and a crucial picture or two—just enough to fix the order of events, and incidentally, to keep me awake. The act of expanding your own notes is very helpful in learning what you did or didn't understand from the lecture.

We often hear that mathematics is in part a language. Is learning mathematics like learning a language?

Learning a language is different from learning a mathematical subject; in the first case the problem is to acquire a habit, in the other to understand a structure. The difference has some important implications. In learning a language from a textbook, you might as well go through the book as it stands and work the exercises in it; what matters is to keep practicing the use of the language. If, however, you want to learn calculus, it might not be a good idea to open the book on page 1 and read it, working all the problems in order.... The material is arranged so that its linear reading is logically defensible, to be sure, but we readers are human ... and each one of us is likely to find something difficult that is easy for someone else, like the author of the text. My advice is to read until you come to a definition that is new to you, and then stop and try to think of examples, and non-examples, or till you come to a theorem new to you, and then stop and try to understand it and maybe try to prove it for yourself. Read with a pencil in hand to sketch a solution or a diagram. Here's the key: when you come to an obstacle, a mysterious passage, an unsolvable problem, just skip it. Jump ahead, try the next problem, turn the page, go to the next chapter, or if it gets really bad, abandon the book and start another one.

Books may be linearly ordered, but our minds are not.

What mathematics is really all about is solving concrete problems. The biggest fault of even good students is that although they may be able to spout correct theorems and remember correct proofs they cannot give examples, or construct counterexamples, or solve special problems. So I try to start every class with a problem. A famous dictum of George Polya, the guru of problem solving, is: "if you can't solve a problem, then there is an easier problem that you can't solve. Find it!"

What about teaching calculus? I know you're an advocate of the Moore method.

Some people call it that. The "Moore method" is named for a famous professor, R. L. Moore, who had a lot of rules that he made students follow on their honor. He challenged his students to learn by asking themselves and each other questions, and getting them to present results to each other. He challenged students to learn and they did. Challenge is the best teaching tool there is, for arithmetic as well as for functional analysis. I'm convinced that it's the best way to teach anything and everything. There's an old Chinese proverb that I learned from Moore himself: "I hear, I forget; I see, I remember; I do, I understand." There's another saying I like: "The best way to learn is to do; the worst way to teach is to talk."

Do you have any advice for instructors who are new to teaching calculus?

Prepare. To really understand a subject, you must know more than the subject; to teach a course, you must know much more about the subject than you can put into the course. The first time I taught calculus at Chicago, I knew that I knew more calculus than I was expected to teach, but even so it was important to prepare for the course. Students are receptive and if you say something wrong, they'll remember it and it will do them damage. I try hard not to say something wrong on purpose and never, ever, write something deliberately wrong on the board. My teacher at Illinois once told me that he was making the point that "$0^0 = 1$" is a definition, not a theorem, and to underscore the point he said he might just as well have defined 0^0 to be 7, which he wrote on the board. Years later he got a letter from a student saying "I have in my notes that $0^0 = 7$, but I forgot how you proved it?" Obviously, this student missed the point, but at least was taking notes.

Mistakes are another matter. If I make a mistake, I admit it and if I can't answer a question I usually say: "Let me think about it." Although we try to avoid mistakes and don't like baffling questions, it's well known that they have educational value. A teacher who comes off as omniscient helps perpetuate the myth that mathematics is a rigid body of facts and perfect techniques. To see an expert admit ignorance and then reason the way to the truth can be a real eye-opener for students. The lesson is that mathematical thinking can be used to learn something that you didn't learn before. "Mathematics is not a spectator sport."

Does humor work in the classroom?

Sometimes. You can even be corny if it helps make a point. Here is a bit of innocent fun on the topic of integration. Partly as an integration drill and partly to make a point about the use of "dummy variables," I might call on several students one after another to tell me what is $\int \frac{dx}{x}$, $\int \frac{du}{u}$, $\int \frac{dz}{z}$, $\int \frac{da}{a}$, and then, as the clincher, I'll write $\int \frac{d(\text{cabin})}{(\text{cabin})}$. Some students shout out right away "log cabin." They're then surprised when I tell them that I can't agree, but I might accept the answer "houseboat" or "log cabin on the beach"—If they're still puzzled I emphasize that the answer is really "log cabin plus sea." [This riddle is explained in chapters 5 & 7.]

But let's get back to teaching by challenging. An intrinsic aspect of the method at all levels, elementary or advanced, is to concentrate attention on the definite, the concrete, the specific. Once a student understands, really and truly understands, why 3×5 is the same as 5×3, then he or she quickly gets the automatic and obvious but nevertheless exciting unshakable conviction that "it goes the same way" for all other numbers. We all have an innate ability to generalize; the teacher's function is to call attention to a concrete special case that hides (and, we hope, ultimately reveals) the germ of the student's conceptual difficulty.

You seem to have fun talking about mathematics. What would make teaching it more satisfying?

I dream of the ideal university, full of students who are brimming with intellectual curiosity. The subset of them who take a mathematics course do so because they want to learn mathematics. They may be future doctors or chemists, or executives in a shirt factory, or none of the above, but if, for whatever reason, they really want to find out what this mathematics stuff is all about, and they come to me free willing and ask me to teach it to them, I'd jump at the chance. If that happened more often, joy! But the appeal of mathematics is subtle. It takes some exposure and training to understand what makes it all so interesting. Be patient, and ask questions.

Paul R. Halmos is Professor of Mathematics at Santa Clara University. He has spent the greater part of his professional life at the University of Chicago and the University of Michigan, as well as five years at the Institute for Advanced Study in Princeton. Professor Halmos has held a Guggenheim Fellowship and was the recipient of a Chauvenet Prize for mathematical exposition from the Mathematical Association of America. He is the author of numerous books and research articles in ergodic theory, algebraic logic, and operator theory.

1

FUNCTIONS AND GRAPHS

Functions play a central role in calculus as well as in other branches of mathematics. It is hard to overestimate how important functions are both in mathematics and the "real world." The concept of a function is one of the two most basic ideas on which calculus is built. (The second is the concept of a *limit*, which we will introduce in Chapter 2.) In this chapter we will first introduce the notion of a function in general. Then we will consider ways of creating new functions based on known ones. We then explore some specific types of functions that we will use throughout the text: polynomial, rational, rational power, and trigonometric functions. We defer our study of other basic functions (such as exponential and logarithmic functions) to Chapter 7, when we can use some of the tools of calculus to analyze them.

1.1 THE FUNCTION CONCEPT

Weight w in oz.	Cost in dollars
$0 < w \leq 1$	0.29
$1 < w \leq 2$	0.52
$2 < w \leq 3$	0.75
$3 < w \leq 4$	0.98
$4 < w \leq 5$	1.21
$5 < w \leq 6$	1.44
$6 < w \leq 7$	1.67
$7 < w \leq 8$	1.90
$8 < w \leq 9$	2.13
$9 < w \leq 10$	2.36
$10 < w \leq 11$	2.59
$11 < w \leq 12$	2.82

(For $w > 12$, priority rates apply.)

Stating that one quantity *is a function of* another means that the first quantity depends on the second in some specific way. For example, the cost of mailing a letter first-class is a function of its weight. The Postal Service has set the exact form of this functional relationship; after a letter has been weighed, a Postal Service employee knows what to charge for postage. As another example, the well-known formula $C = 2\pi r$ expresses the circumference C of a circle as a function of its radius r.

In the case of circumference, the functional relationship is expressed clearly and concisely by a formula. For first-class postage, a table such as the one shown in the margin expresses the postal cost as a function of weight in an easily understood way. A formula in this case is more difficult to determine. (See Exercise 23 of Exercise Set 1.1 for a formula of this function.)

Another way of specifying a function is to use a graph. For example, the graph in Figure 1.1 shows the concentration of carbon dioxide in the atmosphere as a function of time for the years 1958 through 1992.

FIGURE 1.1

Atmospheric carbon dioxide concentrations as measured at the Mauna Loa observatory in Hawaii. This function is important for assessing the greenhouse effect and global warming.

Whether a function is specified by a formula (either in words or mathematical symbols), a table, or a graph, the basic idea is that *for each allowable value of one quantity, a value of another quantity is uniquely determined.* We make this idea more explicit in the following definition.

Definition 1.1 Function	A **function** from a set A to a set B is a rule that assigns to each element of A a unique element of B.

Note: Sometimes rules might appear to be different but yield identical results. For example, $x^2 - 1$ and $(x + 1)(x - 1)$ do not look alike, but for every value substituted for x, the two expressions yield the same value. When two rules assign the same value to each element of the set A in the definition, we will consider the rules to be the same.

The set A in Definition 1.1 is called the **domain** of the function, and the set B is called the **codomain**. If x is an element of the domain A, then the element y of B assigned by the function to x is called the **image** of x. The totality of images of points of A is called the **range** of the function (see Figure 1.2). So the range is a subset of B, but it may or may not include all of B. For example, in the case of first-class postage, the domain is the set A of all possible weights of letters between 0 and 12 oz. That is, A is the set of real numbers for which $0 < w \leq 12$. The set B could be taken as the set of all positive real numbers, but the range consists of just certain elements of B, namely, the cost for a letter weighing up to 1 oz., the cost for a weight between 1 and 2 oz., and so on, up to the cost for a weight between 11 and 12 oz.

Although the sets A and B in Definition 1.1 could consist of any sort of entities, we will assume they consist of real numbers unless otherwise specified.

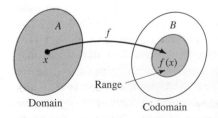

FIGURE 1.2

The set of all real numbers is denoted by $\mathbb{R}$. A function whose domain and range are subsets of $\mathbb{R}$ is called a **real-valued function of a real variable**. A function of this type is said to be *from $\mathbb{R}$ to $\mathbb{R}$*.

REMARK ───────────────────────────────

■ You can think of the set $\mathbb{R}$ of real numbers as being the set of all decimal quantities. Decimals that terminate (such as $5/4 = 1.25$) and those that repeat (such as $5/3 = 1.666\ldots$) are *rational*, meaning that they can be expressed as the *ratio* of two integers. All other real numbers are *irrational*. Examples are $\sqrt{2} = 1.4142136237\ldots$ and $\pi = 3.1415926535\ldots$. Another way of characterizing real numbers is by associating them with points on a number line. Each real number corresponds to a unique point on the line, and each point on the line corresponds to a unique real number. We discuss real numbers in more detail in Appendix 1.

─────────────────────────────────────

A function is usually designated by a single letter. The letter f is used most often, but other letters can be used. The image of an element x in A is then designated $f(x)$, read "f of x." Because an element x in A can be chosen arbitrarily, it is referred to as an **independent variable**. The image $f(x)$ is often designated by the letter y. Since its value depends on x and the function f, it is called the **dependent variable**.

It is important to distinguish between a single letter, such as f, that names a function, and the value of the function at x, symbolized by $f(x)$. So if we write, for example,

$$f(x) = x^2 + 1$$

we are in effect giving the defining rule of the function f. We often abbreviate a statement such as "Consider the function f defined by $f(x) = x^2 + 1$" by writing "Consider the function $f(x) = x^2 + 1$." Alternatively, for this same function we might write its rule by the equation

$$y = x^2 + 1$$

In this case the function is implied, but no letter such as f is given to name it.

Graphs of Functions

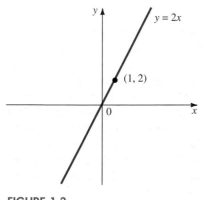

$y = 2x$

$(1, 2)$

0

FIGURE 1.3

By the **graph of a function** f from $\mathbb{R}$ to $\mathbb{R}$, we mean the set of all points (x, y) in the plane such that x is in the domain of f and $y = f(x)$. It is usually impossible to plot all points of the graph, since there are usually infinitely many of them. In some cases we can plot a few well-chosen points and connect them with a smooth curve, but this procedure has some possible pitfalls. Calculus provides the necessary tools for removing uncertainty about the shape of the graph of a function. The simple function $f(x) = 2x$, whose graph is shown in Figure 1.3, can, however, be analyzed without calculus. Letting $y = f(x)$, we see that for each increase in x, y increases twice as much. Therefore, the graph is a straight line. Since the points $(0, 0)$ and $(1, 2)$ are on the line, we can draw the line through these two points.

Since the definition of a function requires that to each x in the domain a unique value of y in the range be assigned, we can use the following test to determine whether or not a curve is the graph of a function.

Vertical Line Test

A curve in the plane is the graph of a function if and only if each vertical line intersects the curve in at most one point.

Figure 1.4 illustrates this test.

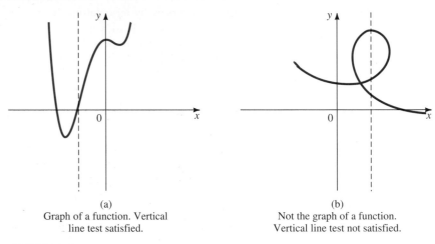

(a)
Graph of a function. Vertical
line test satisfied.

(b)
Not the graph of a function.
Vertical line test not satisfied.

FIGURE 1.4

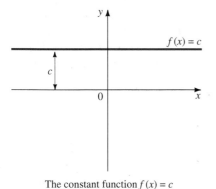

The constant function $f(x) = c$

FIGURE 1.5

A particularly simple type of function is one that gives the same value of the dependent variable for all values of the independent variable. Such a function is called a **constant function** and has the form $f(x) = c$. For example, $f(x) = 2$ is the constant function that has the value 2 for all values of x. The graph of a typical constant function is shown in Figure 1.5. It is a horizontal line and clearly passes the vertical line test. Note, however, that a vertical line does not pass this test and thus cannot be the graph of such a function.

It is sometimes helpful to think of a function as a machine. As illustrated in Figure 1.6, an element x is taken from the domain A and fed into the "f machine." The machine operates on x according to the rule for f, and out comes $f(x)$. This machine analogy suggests calling x the **input** value and $f(x)$ the **output** value. One of the best examples of such a function machine is a calculator. For example, you can input a number—say, 3—punch the $\boxed{x^2}$ key to activate the squaring function $[f(x) = x^2]$, and the output 9 appears on the display screen. As a matter of fact, a calculator combines many different "function machines" into one unit.

FIGURE 1.6

Different letters can be used to represent functions and variables. For example, in the first-class postage function we might let the independent variable (the weight) be denoted by w and the dependent variable (the cost of postage) be denoted by P. Then we could write

$$P = f(w)$$

with the rule for f determined by the Postal Service. Similarly, if we let g be the function giving the circumference of a circle in terms of its radius r, we have

$$C = g(r)$$

In this case we know the rule for g and can write

$$g(r) = 2\pi r$$

Then, for example, $g(2) = 4\pi$ expresses the fact that when $r = 2$, the circumference is 4π.

In many applications the independent variable is *time*, so it is natural to use the letter t to represent it. For example, in the function giving the concentration of carbon dioxide in the atmosphere (Figure 1.1), we might write $c = h(t)$, where t is the time in years. Then we could read from the graph to get values of $h(t)$. For example, $h(1970) \approx 325$, telling us that at the beginning of 1970 the concentration of carbon dioxide in the atmosphere was approximately 325 parts per million.

For a function to be fully defined, the rule as well as the domain and the codomain must be given. We will adopt the convention, however, that *when the domain of a function is not specified, we will understand it to be the largest subset of $\mathbb{R}$ for which the values of the function are real*. Also, *we will always assume the codomain is the set $\mathbb{R}$ of all real numbers* unless otherwise specified. This assumption means that the range of each of the functions we consider is a subset of $\mathbb{R}$. For example, the domain of

$$f(x) = \frac{1}{x - 2}$$

is the set of all real numbers except $x = 2$. We have to exclude $x = 2$, because division by 0 is not defined. Similarly, the domain of $f(x) = \sqrt{x - 2}$ is the set of all real numbers greater than or equal to 2. We can indicate this set symbolically by $\{x : x \geq 2\}$.* Values of x less than 2 result in square roots of negative numbers, which are not real.

As we have seen, functions are usually given in one of three ways: by a formula, by a table, or by a graph. Let us illustrate each way with the squaring function, that is, the function whose output is the square of the input value. We have just described the function in words, but it is more economical and less confusing to give the formula in mathematical symbols, such as $f(x) = x^2$. For this function there is no limitation on x, so we understand the domain of f to be all of $\mathbb{R}$. The three ways of describing this function are as follows:

---

*A symbol of the form $\{x : \ldots\}$ is read "the set of all x values (understood to be real) such that ..."

Graph

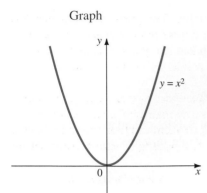

Formula	Table
$f(x) = x^2$	0 | 0
±1 | 1
±2 | 4
±3 | 9

FIGURE 1.7

Of the three ways, the table is usually least satisfactory, since it is generally impossible to show all input and output values. Based on our table alone, we have no way of knowing that when $x = \frac{1}{2}$, $f(x) = \frac{1}{4}$, or even that when $x = 5$, $f(x) = 25$. Nevertheless, tables are useful, if for no other reason, to aid in drawing graphs. More important, data from experiments or from observations often are presented in tabular form, and the function may be difficult or impossible to describe by a formula. (Also, with spreadsheets on computers, tables can be used to represent functions with a high degree of accuracy.)

Graphing, too, has its limitations. For example, just looking at the graph in Figure 1.7 does not tell us that it is the graph of $f(x) = x^2$. Since it is impossible to show the entire domain on the portion of the x-axis we draw, we have to assume that the graph continues in the manner shown. This limitation is particularly important in using a graphing calculator (or a computer algebra system). When using one, you must select an x range and a y range in such a way that you can be confident you are seeing all the important features of the graph. (You really need calculus to be certain, however.) Figure 1.8 illustrates some inadequate graphs of a function resulting from poor choices of the y range, along with the graph as it should be when an appropriate y range is used.

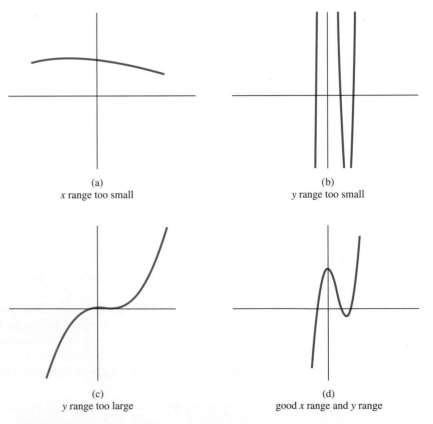

(a)
x range too small

(b)
y range too small

(c)
y range too large

(d)
good x range and y range

FIGURE 1.8

Graphs are helpful, despite their limitations. The graph of a function lets us see at a glance where the function is increasing or decreasing, whether it is bending up or bending down, where its high and low points are, and so on. All these behaviors are important in studying mathematical models of real-world phenomena, as we will see later.

A combination of ways of representing functions is usually best, especially a formula and a graph. It is possible, however, if enough tabular values are known, not only to draw the graph but also to find a formula that at least approximates the function. This "curve fitting" procedure is often used to construct mathematical models.

The following examples illustrate the main ideas introduced so far.

EXAMPLE 1.1 Let $f(x) = \sqrt{x - 1}$. What are the domain and range of f? Find $f(1)$, $f(2)$, $f(5)$, and $f(10)$. Draw the graph of f.

Solution For the values of this function to be real numbers, the expression under the radical sign must be positive or zero. That is, we can choose only those x values for which $x - 1 \geq 0$, or, equivalently, $x \geq 1$. So the domain of f is the set of x values for which $x \geq 1$. By choosing x in this domain, we can make $\sqrt{x - 1}$ equal to any nonnegative real number, so the range of f is the set of all y values for which $y \geq 0$. By substituting, in turn, $x = 1$, $x = 2$, $x = 5$, and $x = 10$, we find

$$f(1) = 0, \qquad f(2) = 1, \qquad f(5) = 2, \qquad f(10) = 3$$

To draw the graph (Figure 1.9), we set $y = f(x)$ and plot the points (x, y) we have found, connecting them with a smooth curve. ∎

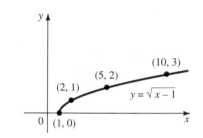

FIGURE 1.9

EXAMPLE 1.2 Let $f(t) = \dfrac{1}{2t - 1}$. Find the following:

(a) $f(1)$

(b) $f\left(\frac{1}{4}\right)$

(c) $f\left(\dfrac{1}{t}\right)$

(d) $\dfrac{1}{f(t)}$

(e) $f(t + h)$

Solution

(a) $f(1) = \dfrac{1}{2(1) - 1} = \dfrac{1}{1} = 1$

(b) $f\left(\dfrac{1}{4}\right) = \dfrac{1}{2\left(\frac{1}{4}\right) - 1} = \dfrac{1}{-\frac{1}{2}} = -2$

(c) We substitute $1/t$ for t and simplify:

$$f\left(\frac{1}{t}\right) = \frac{1}{2\left(\frac{1}{t}\right) - 1} = \frac{1}{\frac{2}{t} - 1} = \frac{1}{\frac{2}{t} - 1} \cdot \frac{t}{t} = \frac{t}{2 - t}$$

(d) $\dfrac{1}{f(t)} = \dfrac{1}{\frac{1}{2t-1}} = 2t - 1$ Note the difference between parts (c) and (d).

(e) $f(t + h) = \dfrac{1}{2(t + h) - 1} = \dfrac{1}{2t + 2h - 1}$ Replace t with $t + h$. ∎

As illustrated in Example 1.2, we evaluate a function at any admissible value by substituting that value for the independent variable wherever it occurs. For example, suppose $f(x) = x^3$. Then each of the following statements is valid:

$$f(2g) = (2g)^3 = 8g^3$$

$$f(3a + 4b) = (3a + 4b)^3$$

$$f(x + 1) = (x + 1)^3$$

$$f(x^2yz) = (x^2yz)^3 = x^6y^3z^3$$

When you substitute some expression for the independent variable, it is a good idea to place it in parentheses and then simplify. For example, with $f(x) = x^3$, we first wrote $f(2g) = (2g)^3$ and then simplified to get $8g^3$. This practice avoids the pitfall of failing to cube the 2 as well as the g.

EXAMPLE 1.3   Let $g(x) = \sqrt{x}$. Evaluate the expression

$$\frac{g(4 + h) - g(4)}{h}, \qquad h \neq 0$$

Then rationalize the numerator and simplify.

Note: We will see in Chapter 3 that the type of quotient appearing in this example forms the basis for defining what is called the *derivative*.

Solution   Since $g(4 + h) = \sqrt{4 + h}$ and $g(4) = \sqrt{4} = 2$, we have

$$\frac{g(4 + h) - g(4)}{h} = \frac{\sqrt{4 + h} - 2}{h}$$

To "rationalize the numerator" means to perform some valid operation on the function that results in a numerator that has only rational expressions—that is, no radicals (square roots, cube roots, or the like). Here, we accomplish this result by multiplying numerator *and* denominator by the factor $(\sqrt{4 + h} + 2)$. In doing this operation we rely on two facts from algebra: first, a fraction is unchanged if both numerator and denominator are multiplied by the same nonzero quantity (in effect, we are multiplying the fraction by 1); and second, the difference of two quantities times their sum is equal to the difference of their squares. Thus, we have

$$(a - b)(a + b)$$
$$= a^2 + ab - ab - b^2$$
$$= a^2 - b^2$$

Middle terms cancel.

$$\frac{\sqrt{4 + h} - 2}{h} = \frac{\sqrt{4 + h} - 2}{h} \cdot \frac{\sqrt{4 + h} + 2}{\sqrt{4 + h} + 2} = \frac{(\sqrt{4 + h})^2 - (2)^2}{h(\sqrt{4 + h} + 2)}$$

$$= \frac{(4 + h) - 4}{h(\sqrt{4 + h} + 2)} = \frac{h}{h(\sqrt{4 + h} + 2)} = \frac{1}{\sqrt{4 + h} + 2}$$

since $h \neq 0$.   ■

The domain and range of a function frequently are subsets of $\mathbb{R}$ called **intervals**. The following notation and terminology are useful in describing types of intervals.

closed interval $[a, b]$

open interval (a, b)

FIGURE 1.10

Symbol	Name	Meaning
$[a, b]$	closed interval	$\{x : a \leq x \leq b\}$
(a, b)	open interval	$\{x : a < x < b\}$

Intervals of these two types can be shown graphically as in Figure 1.10.

A set of the form $\{x: a < x \leq b\}$ is designated $(a, b]$ and is called a **half-open** (or **half-closed**) interval, as is $[a, b)$, meaning $\{x: a \leq x < b\}$. It is also convenient to use the symbol ∞, read "infinity," in interval notation as follows:

$$[a, \infty) = \{x: x \geq a\}$$

$$(a, \infty) = \{x: x > a\}$$

$$(-\infty, a] = \{x: x \leq a\}$$

$$(-\infty, a) = \{x: x < a\}$$

$$(-\infty, \infty) = \text{the set of all real numbers}$$

EXAMPLE 1.4   Find the domain of the function F defined by the rule

$$F(x) = \sqrt{2 - x - x^2}$$

Solution   Because the square root of a negative number is not real, the domain consists of those x values for which $2 - x - x^2 \geq 0$. Factoring the left-hand side of this inequality gives

$$(1 - x)(2 + x) \geq 0$$

To see where the product is greater than zero, let us first see where it is equal to zero. The numbers $x = 1$ and $x = -2$ cause the product on the left to be 0. They divide the number line into three intervals, as shown in Figure 1.11. By testing the sign of the product in each interval, we obtain the results indicated in the figure. One way to test the signs is to use a *test value* in each interval. For example, we could use $x = 2$, $x = 0$, and $x = -3$ as test values. By substituting these values, in turn, for x in $(1 - x)(2 + x)$, we find the signs as indicated. Since we want the product to be positive *or* zero, we conclude that the domain of F is the closed interval $[-2, 1]$. ∎

FIGURE 1.11

REMARK ───
■ A figure such as Figure 1.11 showing the sign of an expression for various intervals on the real line is called a **sign graph** for the expression.
───

If the actual graph of the expression whose signs we want to determine is known, an alternative to a sign graph is to observe where it is positive (above the x-axis) and where it is negative (below the x-axis). For example, to find where $(1 - x)(2 + x) \geq 0$, we could sketch the graph of $y = (1 - x)(2 + x)$, as in Figure 1.12. (For more complicated graphs, a graphing calculator or computer algebra system is useful here.) We see from the graph that $y \geq 0$ when $-2 \leq x \leq 1$, just as we found in Example 1.4.

Applications often require finding the form of a functional relationship based on known information. The next example illustrates one such situation.

y is positive or zero when $-2 \leq x \leq 1$

FIGURE 1.12

EXAMPLE 1.5   Given an isosceles triangle inscribed in a circle of radius 2, find the area A of the triangle as a function of its base b.

Solution   A typical triangle is pictured in Figure 1.13. We denote the perpendicular distance from the center of the circle to the base of the triangle by y, as shown. Then the altitude of the triangle is $y + 2$ and the area is

$$A = \frac{1}{2}b(y + 2)$$

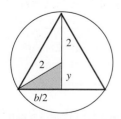

FIGURE 1.13

Pythagorean Theorem: In a right triangle the sum of the squares of the lengths of the legs equals the square of the length of the hypotenuse.

Now we must find y in terms of b. Using the Pythagorean Theorem on the shaded right triangle with legs $b/2$ and y and hypotenuse 2, we get

$$y^2 + \left(\frac{b}{2}\right)^2 = 4$$

Solving for y gives

$$y = \sqrt{4 - \frac{b^2}{4}} = \frac{\sqrt{16 - b^2}}{2}$$

Thus,

$$A = \frac{1}{2}b\left(\frac{\sqrt{16 - b^2}}{2} + 2\right) = \frac{b}{4}\left(\sqrt{16 - b^2} + 4\right) \qquad \blacksquare$$

Piecewise-Defined Functions

Some functions are defined by different rules on different parts of their domains. We call a function of this type a **piecewise-defined function**. The next example provides an illustration.

EXAMPLE 1.6   Draw the graph of the function h defined by

$$h(x) = \begin{cases} -1 & \text{if } x < 0 \\ 1 & \text{if } x = 0 \\ x + 2 & \text{if } x > 0 \end{cases}$$

Determine its domain and range.

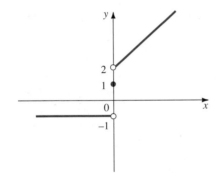

FIGURE 1.14

Solution   The graph is shown in Figure 1.14. Notice that the point $(0, 1)$ is on the graph, as indicated by the solid circle there, but $(0, 2)$ and $(0, -1)$ are not, as shown by the open circles.

Although h is defined in a piecewise fashion, its domain is all of $\mathbb{R}$. Its range is the set of y values taken on by the function, namely, -1, 1, and all numbers greater than 2. We could indicate the range as $\{-1, 1\} \cup (2, \infty)$. The symbol $\cup$ is read "union," and $A \cup B$ represents the set of all elements in A or in B (or in both). $\qquad \blacksquare$

Even and Odd Functions; Symmetry

Suppose that for every x in the domain of a function f, $-x$ is also in the domain. It may happen that for each such x, $f(x)$ and $f(-x)$ are equal, as in the case of $f(x) = x^2$, whose graph is shown in Figure 1.15. On the other hand, $f(-x)$ may always be opposite in sign to $f(x)$, as in the case of $f(x) = x^3$, shown in Figure 1.16. These special types of functions are given names in the following definition.

Definition 1.2
Even and Odd Functions

A function f is said to be **even** if $f(-x) = f(x)$ for every x in the domain of f, and f is said to be **odd** if $f(-x) = -f(x)$ for every x in the domain of f.

For example, $f(x) = x^2$ is even and $f(x) = x^3$ is odd, since $(-x)^2 = x^2$ and $(-x)^3 = -x^3$.

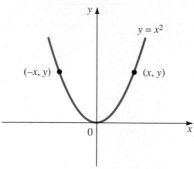

Even function: $f(-x) = f(x)$

FIGURE 1.15

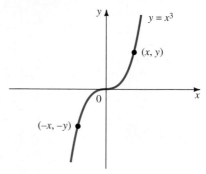

Odd function: $f(-x) = -f(x)$

FIGURE 1.16

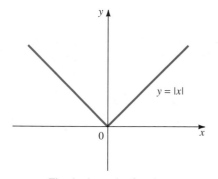

The absolute value function

FIGURE 1.17

Another important even function is the **absolute value function**, defined by $f(x) = |x|$, where

$$|x| = \begin{cases} x & \text{if } x \geq 0 \\ -x & \text{if } x < 0 \end{cases}$$

We can see that f is even, since the absolute value of any real number and of its negative are the same. For example, $|2| = 2$ and $|-2| = -(-2) = 2$. Thus, $f(2) = f(-2)$. The graph of f is shown in Figure 1.17.

REMARK ———————————————————————————

■ It should be emphasized that even and odd functions are special types. Most functions are neither even nor odd. For example, if $f(x) = x^3 + x^2 - 1$, then $f(-x) = -x^3 + x^2 - 1$, which does not equal either $f(x)$ or $-f(x)$. So this function is neither even nor odd.

For each point (x, y) on the graph of an even function, $(-x, y)$ is also on the graph, so the graph of an even function is **symmetric with respect to the y-axis**, as shown in Figure 1.18(a). When (x, y) is on the graph of an odd function, then $(-x, -y)$ is on the graph also, and the graph of an odd function is **symmetric with respect to the origin**, as shown in Figure 1.18(b).

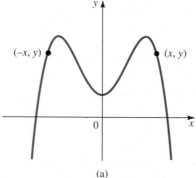

(a)
Graph of an even function
(symmetric with respect to the y-axis)

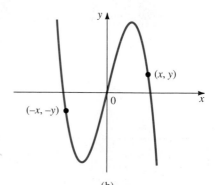

(b)
Graph of an odd function
(symmetric with respect to the origin)

FIGURE 1.18

CHART 1.1

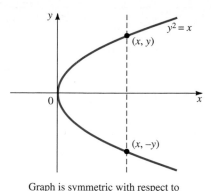

Graph is symmetric with respect to the x-axis; not a function.

FIGURE 1.19

A graph is **symmetric with respect to the x-axis** if for each point (x, y) on the graph, the point $(x, -y)$ also is on the graph. Except for the x-axis itself, in which case $f(x) = 0$ for all x, such a graph could not be the graph of $y = f(x)$ for any function f, since the graph fails the vertical line test. The graph of $y^2 = x$ (Figure 1.19), for example, is symmetric with respect to the x-axis, but this equation does not define y as a function of x. In fact, for each $x > 0$ there are two corresponding y values, namely, $y = \sqrt{x}$ and $y = -\sqrt{x}$. (*Note:* The graph of an *equation*, such as $y^2 = x$, is the set of all points (x, y) whose coordinates satisfy the equation.)

We conclude this section by showing for reference the graphs of some commonly occurring functions (Chart 1.1). We will study each of these in succeeding sections.

GRAPHING CURVES USING COMPUTER ALGEBRA SYSTEMS

Using geometric or graphical representations of mathematical concepts or problems can greatly enhance our problem-solving abilities. Whether hand-drawn or computer-generated, visualization of important concepts plays a key role in our understanding of mathematics. In selected sections throughout the text, we will consider three computer algebra systems (CAS)—Maple, Mathematica, and DERIVE—as aids to computation and visualization. But in fact, as we will demonstrate, computer algebra systems can be used for much more.

You can follow these steps on your computer with Maple, Mathematica, or DERIVE. But you do not have to do so; you can just follow along in the text, since the syntax (command logic) of these computer algebra systems is close to what we would write in mathematical notation.

In this section, we will consider how Maple, Mathematica, and DERIVE can be used to sketch curves in the plane and to obtain information about functions. For consistency we will use Maple primarily and comment on the other CAS where necessary.

CAS 1

Plot the function $f(x) = x^2$ on the interval $[-2, 2]$.

In Maple, we type the command:

plot(x^2,x=−2..2);

FIGURE 1.1.1
$y = f(x) = x^2$

Mathematica:[†]

Plot[x^2,{x,−2,2}]

DERIVE:

(At the □ symbol, go to the next step.)
a (author) □ x^2 □ p (plot window) □ p (plot) □ z (zoom out)

_____

[†]In Mathematica, the braces { } surround information showing that we are plotting x on the horizontal axis from $x = -2$ to $x = 2$. In Maple this information is not inside braces, but is written x=−2..2, with two dots. Also note that all Maple commands end with a semicolon and all Mathematica commands must begin with a capital letter. In DERIVE, zoom out to change the scale to display the desired portion of the graph. In all three CAS, the symbol ^ (found over the 6 on the keyboard) means "raise to a power."

CAS 2

Plot the function $f(x) = x^2 - 2x - 3$ on the interval $[-4, 4]$.

Maple:

plot(x^2–2*x–3,x=–4..4);

Mathematica:

Plot[x^2–2*x – 3,{x,–4,4}]

DERIVE:

(At the □ symbol, go to the next step.)
a (author) □ x^2–2x–3 □ p (plot window) □ p (plot) □ z (zoom out)

FIGURE 1.1.2
$y = f(x) = x^2 - 2x - 3$

CAS 3

Plot the function $f(x) = \frac{1}{3}x^3 + \frac{1}{2}x^2 - 2x + 3$ on the interval $[-4, 3]$. Find the point of intersection of the curve with the x-axis.

Maple:

plot(x^3/3+x^2/2–2*x+3, x=–4..3);

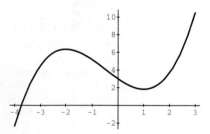

FIGURE 1.1.3
$y = f(x) = \frac{1}{3}x^3 + \frac{1}{2}x^2 - 2x + 3$
on $[-4, 3]$

Mathematica:

Plot[x^3/3+x^2/2–2*x+3, {x,–4,3}]

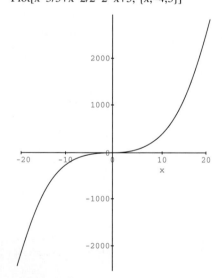

DERIVE:

(At the □ symbol, go to the next step.)
a (author) □ x^3/3+x^2/2–2x+3 □ p (plot window) □ p (plot) □ z (zoom out)

FIGURE 1.1.4
$y = f(x) = \frac{1}{3}x^3 + \frac{1}{2}x^2 - 2x + 3$
on $[-20, 20]$

The choice for the limits on x is important when generating a plot. CAS Figure 1.1.4 shows the Maple plot generated by the command

plot(x^3/3+x^2/2–2*x+3,x=–20 . . 20);

Although the domain contains the important features of the graph, the scale hides the true shape of the curve.

To find the point of intersection of the curve with the x-axis, we must find the roots of the equation $f(x) = \frac{1}{3}x^3 + \frac{1}{2}x^2 - 2x + 3$. That is, we need to solve $\frac{1}{3}x^3 + \frac{1}{2}x^2 - 2x + 3 = 0$. We can execute the following commands:

Maple:

```
f:=x^3/3+x^2/2–2*x+3;
fsolve(f=0,x);
Output: –3.744348422.
```

Mathematica:[†]

```
f=x^3/3+x^2/2–2*x+3
NRoots[f == 0,x]
```

DERIVE:

(At the □ symbol, go to the next step.)
a (author) □ x^3/3+x^2/2−2x+3 □ 1
(solve) □ [choose expression to solve]
□ x (approximate) □ [choose expression to approximate]

CAS 4

Plot the function $f(x) = \dfrac{(x+2)}{(x-1)}$ on the interval $[-10, 10]$.

Maple:

```
plot((x+2)/(x–1),x=–10..10,y=–5..5);
```

FIGURE 1.1.5
$y = f(x) = (x+2)/(x-1)$

Mathematica:

```
Plot[(x+2)/(x–1),{x,–10,10}]
```

DERIVE:

(At the □ symbol, go to the next step.)
a (author) □ (x+2)/(x−1) □ p (plot window) □ p (plot) □ z (zoom out) □ z (zoom out) □

The graph shows that f goes to infinity near $x = 1$; that is, it is asymptotic near $x = 1$. We will discuss this type of behavior in Section 1.5.

CAS 5

Use Maple to sketch the graph of $f(x) = x^2 \sin(1/x)$. This function involves the trigonometric sine function, which we will consider in Section 1.6. We show several sketches of the curve using Maple commands, starting with

```
plot(x^2*sin(1/x),x=–5..5);
```

See Figure 1.1.6.

Something interesting appears to be happening near the origin. We zoom in closer.

```
plot(x^2*sin(1/x),x=–1..1);              A little better
plot(x^2*sin(1/x),x=–0.25..0.25);        Even better
plot(x^2*sin(1/x),x=–0.025..0.025,numpoints=500);
```

[†] In Mathematica, the double equal sign is used to show that we are dealing with an equation. In both Maple and Mathematica, f is defined as the equation on the right of := and =, respectively.

This last plot is convincing as to the behavior of the function near the origin. The option, numpoints=500, plots additional points, adding to the accuracy of the plot. We see that care must be taken when interpreting computer-generated output.

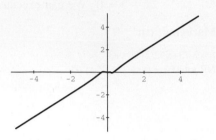

FIGURE 1.1.6
$y = f(x) = x^2 \sin(1/x)$

FIGURE 1.1.7
$y = f(x) = x^2 \sin(1/x)$

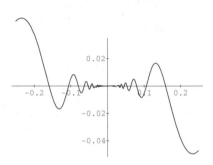

FIGURE 1.1.8
$y = f(x) = x^2 \sin(1/x)$

FIGURE 1.1.9
$y = f(x) = x^2 \sin(1/x)$

Maple:

plot(x^2*sin(1/x),x=–5..5);
plot(x^2*sin(1/x),x=–1..1);
plot(x^2*sin(1/x),x=–0.25..0.25);
plot(x^2*sin(1/x),x=–0.025..0.025);

Mathematica:‡

Plot[x^2*Sin[1/x],{x,–5,5}]
Plot[x^2*Sin[1/x],
  {x,–1,1}]
Plot[x^2*Sin[1/x],
  {x,–0.25,0.25}]
Plot[x^2*Sin[1/x],
  {x,–0.025,0.025}]

DERIVE

(At the □ symbol, go to the next step.)
a (author) □ x^2sin(1/x) □ p (plot win–dow) □ p (plot) □ z (zoom in) □ z (zoom in) □ z (zoom in) □

---

‡ In Mathematica, the sine function is Sin and must start with a capital letter. The argument for the function is in square braces, [ ]. In DERIVE, notice this time that we zoom in on the graph, rather than zoom out.

Exercise Set 1.1

1. If $f(x) = \dfrac{x^2 - 1}{x + 2}$, find
   (a) $f(1)$
   (b) $f(-1)$
   (c) $f(0)$
   (d) $f(a)$
   (e) $f(x + 1)$

2. If $g(x) = \dfrac{2x - 3}{3x + 4}$, find
   (a) $g(-2)$
   (b) $g\left(\frac{5}{6}\right)$
   (c) $g\left(-\frac{2}{3}\right)$
   (d) $g(2x)$
   (e) $g(1 - x)$

3. If $F(t) = |t - 3|$, find
   (a) $F(4)$
   (b) $F(1)$
   (c) $F(0)$
   (d) $F(t + 3)$
   (e) $F(3 - x^2)$

In Exercises 4 and 5, determine from the graph the domain and range of each function.

4.

(a)   (b)

5.

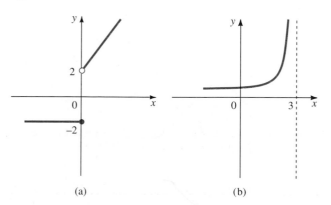

(a)   (b)

In Exercises 6–11, find the domain and range of f.

6. $f(x) = 2x + 3$

7. $f(x) = x^2 - 4$

8. $f(x) = \dfrac{1}{x}$

9. $f(t) = \sqrt{1 - t}$

10. $f(u) = \dfrac{u + 1}{u - 2}$

11. $f(s) = s^3 + 1$

In Exercises 12–21, draw the graph of the given function.

12. $f(x) = 3x - 4$

13. $g(x) = x^2 + 1$

14. $h(x) = \dfrac{2}{x}$

15. $F(x) = 4x - x^2$

16. $f(x) = \sqrt{x + 9}$

17. $G(x) = |x + 1|$

18. $h(x) = \begin{cases} x^2 & \text{if } x \geq 1 \\ 1 - x & \text{if } x < 0 \end{cases}$

19. $f(x) = \begin{cases} 2x & \text{if } x \geq 1 \\ 2 & \text{if } 0 \leq x < 1 \\ x + 2 & \text{if } x < 0 \end{cases}$

20. $f(x) = \begin{cases} \sqrt{x} & \text{if } x > 0 \\ 1 & \text{if } x = 0 \\ -1 & \text{if } x < 0 \end{cases}$

21. $g(x) = |4 - x^2|$

22. **Greatest Integer Function.** The function whose value at x is the greatest integer that does not exceed x is called the *greatest integer function*. Its value is symbolized by $[\![x]\!]$. Thus, if $k \leq x < k + 1$, where k is an integer, then $[\![x]\!] = k$.
   (a) Give the value of each of the following: $[\![3.2]\!]$, $[\![5]\!]$, $[\![-3/2]\!]$, $[\![1.99]\!]$.
   (b) Draw the graph of $y = [\![x]\!]$.

23. Verify that the first-class postage function we discussed at the beginning of this section is defined by

$$P(w) = \begin{cases} 0.29 + 0.23[\![w]\!] & \text{if } w \text{ is not an integer} \\ 0.29 + 0.23(w - 1) & \text{if } w \text{ is an integer} \end{cases}$$

   Here w is the weight in ounces ($0 < w \leq 12$) and $P(w)$ is the cost of postage in dollars. (See Exercise 22.)

24. You are driving along a straight road at a constant speed between cities A and B. Along the way you pass through city C. Draw a graph that indicates your distance from city C as a function of time, measured from the time you left city A.

25. If $f(x) = \dfrac{1}{x}$, find $\dfrac{f(x) - f(2)}{x - 2}$ for $x \neq 2$ and $x \neq 0$. Simplify the result.

26. If $g(x) = \sqrt{x}$, show that for $h \neq 0$,

$$\frac{g(x + h) - g(x)}{h} = \frac{1}{\sqrt{x + h} + \sqrt{x}}$$

27. Let $\phi(x) = \dfrac{x}{x - 3}$. Find

$$\frac{\phi(4 + h) - \phi(4)}{h} \qquad (h \neq 0)$$

and simplify the result.

28. Let $f(x) = \dfrac{x + 2}{2x + 1}$. Show that if $x \neq 0$,

$$f\left(\frac{1}{x}\right) = \frac{1}{f(x)}$$

(*Hint:* Simplify each side and show that the results are the same.)

29. For what positive integers n is $f(x) = x^n$ an even function? For what values of n is it odd?

30. Determine in each of the following whether the function is even, odd, or neither:
(a) $f(x) = \sqrt{x^2 + 1}$ (b) $f(x) = x^3 - 2x$
(c) $f(x) = x^2 + 2x$ (d) $g(x) = x - 1$
(e) $h(x) = 2x^2 - 3$

31. Follow the instructions for Exercise 30 for the following:
(a) $f(x) = \dfrac{x}{x^2 + 1}$ (b) $g(x) = \dfrac{1}{x^3 - 2x^5}$
(c) $h(t) = \dfrac{t + 2}{t - 1}$ (d) $\phi(x) = \dfrac{x^2 - 1}{x^2 + 4}$
(e) $F(t) = |t|$

32. (a) Is the sum of two even functions even? odd?
(b) What can be said about the product of two even functions? two odd functions? an even function and an odd function?
(c) What can be said about the sum of an even function and an odd function?
(d) Can a function be both even and odd? Explain.

33. Classify each of the graphs shown as being graphs of functions that are even, odd, or neither.

(a) (b)

(c) (d)

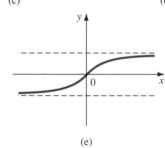

(e)

34. The graph of a function f is given for $x \geq 0$. Extend the graph for $x < 0$ if
(a) f is even (b) f is odd

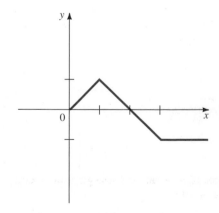

35. Repeat Exercise 34 for the graph shown here.

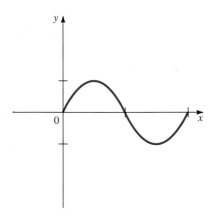

36. A piece of equipment is purchased for $200,000. For tax purposes, the value is decreased by $40,000 times the number of years x since the equipment was purchased. Let $f(x)$ = value of equipment x years after purchase.
(a) Write the rule for the function f.
(b) Find the domain of f if the value of the equipment cannot be negative.

37. If a price of x dollars per unit is charged for a certain product, the number of units sold will be $1000 - 25x$. Let $f(x)$ be the revenue when the price is x, where revenue is the price per unit times the number of units sold.
(a) Write the rule for the function f.
(b) Find the domain of f if the revenue is nonnegative.

38. The length of a rectangle is 3 units more than twice its width. Express each of the following as a function of the width: (a) perimeter and (b) area.

39. A can is in the form of a right circular cylinder with its height equal to twice the radius of the base. Express each of the following as a function of the radius of the base: (a) volume, (b) lateral surface area, and (c) total surface area.

40. Express the area of an equilateral triangle as a function of the length of a side.

41. A rectangle is inscribed in an isosceles triangle of height 4 cm and base 6 cm, as shown in the figure. Express the area A of the rectangle as a function of its base b.

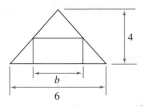

In Exercises 42–48, find the domain of the given function.

42. $f(x) = \sqrt{x^2 - 4}$ **43.** $f(x) = \sqrt{3 + 2x - x^2}$

44. $g(x) = \sqrt{\dfrac{x-1}{x+2}}$

45. $g(x) = \dfrac{\sqrt{x-1}}{\sqrt{x+2}}$. Compare with Exercise 44.

46. $g(x) = \dfrac{2}{\sqrt{1-x^2}}$ **47.** $h(x) = \dfrac{x-1}{\sqrt{6-x-x^2}}$

48. $h(x) = \dfrac{\sqrt{x^2 - 2x - 8}}{x+5}$

49. A right circular cylinder is inscribed in a sphere of radius a. Express the volume of the cylinder as a function of its base radius r.

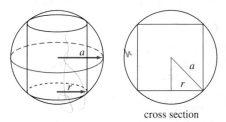

cross section

50. A right circular cone is inscribed in a sphere of radius a. Express the volume of the cone as a function of its base radius r.

51. A pyramid has a square base 4 m on a side, and its height is 10 m. Express the area of a cross section parallel to the base and h m above the base as a function of h.

52. A storage tank is in the form of a cylinder x feet long and y feet in diameter, with hemispheres at both ends (see the figure).

(a) Express the volume of the tank as a function of x and y.
(b) Express the surface area of the tank as a function of x and y.
(c) If $x = 4y$, express the volume and surface area as a function of y only.

53. An organization that issues credit cards charges cardholders interest of $1\frac{1}{2}\%$ per month for the first $1000 of unpaid balance after 30 days and 1% per month on the unpaid balance above $1000. Let $f(x)$ be the interest charged per month on an unpaid balance of x dollars, and find the rule for f.

54. A company deducts 2% of gross salary from each employee's pay as the employee's contribution to the pension fund. However, no employee pays more than $400 into the fund in a single year. Let $f(x)$ be the contribution from an employee with a gross annual salary of x dollars. Write the rule for f.

55. Prove that the range of the function f for which $f(x) = 2x/(x^2 + 1)$ is the set $\{y: |y| \leq 1\}$.

56. Redraw the graph shown, and on the same set of axes draw the graph of
(a) $y = -f(x)$ (b) $y = f(-x)$
In each case describe in words the effect of introducing the minus sign.

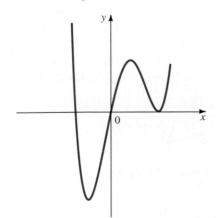

In Exercises 57–59, use a graphing calculator or CAS.

57. Straight Lines. Let $y = ax + b$, where $a \in \mathbb{R}$ and $b \in \mathbb{R}$. Describe the effect of a and b on the graph. Use both positive and negative values.

58. Parabolas. Let $y = (x - a)^2 + b$, where $a \in \mathbb{R}$ and $b \in \mathbb{R}$. Describe the effect of a and b on the graph. Consider the separate cases

$a = 0,\ b \in \mathbb{R}$	Start with $b = 0$.
$a > 0,\ b \in \mathbb{R}$	Start with $b = 0$.
$a < 0,\ b \in \mathbb{R}$	Start with $b = 0$.

59. Let $f(x) = x(x - 2)^2$. Use a graphing calculator or CAS to obtain the graph of $y = f(x)$. Then find the graphs of the following, for $a = 1, 2, -1, -2$.
(a) $y = f(x) + a$ (b) $y = f(x + a)$
(c) $y = af(x)$ (d) $y = f(ax)$
Compare each graph with that of $y = f(x)$. How would you generalize your findings?

In Exercises 60–63, first show that the given value of a is not in the domain of f, and then use a calculator to find $f(a + h)$ for each of the following values of h: ± 0.1, ± 0.01, ± 0.001, ± 0.0001. Conjecture how the function behaves as x comes arbitrarily close to a from each side.

60. $a = 3;\ \ f(x) = \dfrac{x^2 - 9}{x - 3}$

61. $a = -\dfrac{3}{2};\ \ f(x) = \dfrac{6x^2 + 17x + 12}{2x^2 + x - 3}$

62. $a = 2;\ \ f(x) = \dfrac{3x^3 - 4x^2 - 8}{3x^2 + x - 14}$

63. $a = 8;\ \ f(x) = \dfrac{x - 8}{2 - \sqrt[3]{x}}$

64. Sketch the graph of the function $f(x) = |x - 1| + |2x + 1|$.

1.2 NEW FUNCTIONS FROM OLD

If we begin with one or more known functions, we can obtain new functions from them in a variety of ways, some of which we discuss in this section. After we have introduced some calculus concepts, we will see even more ways of creating new functions from old. One of the benefits of learning the techniques in this section is that we can frequently use them to go in reverse. That is, given a new function, we may be able to recognize it as having come from some known function or functions in a way we have studied. We will then have important knowledge of the new function as a result of known information about the original function(s).

We begin by adding a constant to the independent or the dependent variable. Then we multiply each of these variables by a constant and discuss the result.

Vertical and Horizontal Translations

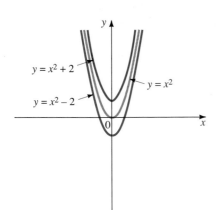

Suppose we know the graph of $y = f(x)$. (We will call f the *basic function* in the following discussion.) Let c denote a positive constant. We want to see how the graph of each of the following is related to the graph of the basic function:

$$y = f(x) + c \qquad y = f(x + c)$$
$$y = f(x) - c \qquad y = f(x - c)$$

These four new functions are special types of **transformations** of the basic function known as **translations**. We consider them in pairs.

1. $y = f(x) + c$ and $y = f(x) - c$

    Each y value of the graph of the basic function is shifted upward by c units in $y = f(x) + c$ and downward by c units in $y = f(x) - c$. These two transformations are called **vertical translations**. We illustrate them in Figure 1.20 with $y = x^2$ as the basic function and the vertical translations $y = x^2 + 2$ and $y = x^2 - 2$.

FIGURE 1.20

2. $y = f(x + c)$ and $y = f(x - c)$

    Suppose (a, b) is a point on the graph of the basic function; that is, $b = f(a)$. Then direct substitution shows that $(a - c, b)$ is on the graph of $y = f(x + c)$ and $(a + c, b)$ is on the graph of $y = f(x - c)$. So $y = f(x + c)$ shifts all points on the basic graph c units to the *left*, and $y = f(x - c)$ shifts all points on the basic graph c units to the *right*. These two transformations are called **horizontal translations**. We illustrate them in Figure 1.21, again using $y = x^2$ as the basic function, with horizontal translations $y = (x + 2)^2$ and $y = (x - 2)^2$. Note carefully the direction of each shift.

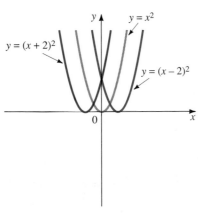

FIGURE 1.21

Stretching, Shrinking, and Reflecting

Next we consider the following four transformations of the basic function $y = f(x)$.

$$y = cf(x) \qquad y = f(cx)$$
$$y = -cf(x) \qquad y = f(-cx)$$

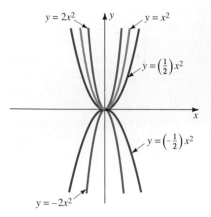

FIGURE 1.22

We again consider the transformations in pairs.

3. $y = cf(x)$ and $y = -cf(x)$

    In $y = cf(x)$, each y value of the basic function is multiplied by c, so the graph is stretched vertically if $c > 1$ and shrunk if $0 < c < 1$. For $c > 1$, we call the transformation a **vertical elongation** and for $0 < c < 1$, a **vertical compression**. In $y = -cf(x)$, each y value of $y = cf(x)$ is replaced by its negative, so the graph is **reflected about the x-axis**. Thus, $y = -cf(x)$ is a vertical elongation (if $c > 1$) or compression (if $0 < c < 1$), followed by a reflection about the x-axis. In Figure 1.22 we show the basic function $y = x^2$ along with $y = 2x^2$, $y = \frac{1}{2}x^2$, $y = -2x^2$, and $y = -\frac{1}{2}x^2$.

4. $y = f(cx)$ and $y = f(-cx)$

    If the point (a, b) is on the graph of the basic function, so that $b = f(a)$, then direct substitution shows that $(a/c, b)$ is on the graph of $y = f(cx)$. Thus, if $c > 1$, the point (a, b) on the basic graph is replaced by $(a/c, b)$, which is closer to the y-axis. Thus, the transformation $y = f(cx)$ results in a shrinking, called a **horizontal compression**. If $0 < c < 1$, then $1/c > 1$, so that $(a/c, b)$ is farther from the y-axis than (a, b). The result is that $y = f(cx)$ stretches the basic curve horizontally. We refer to the transformation in this case as a **horizontal elongation**. For $y = f(-cx)$, each x-coordinate of $y = f(cx)$ is replaced by its negative, resulting in a **reflection about the y-axis**. So $y = f(-cx)$ is a horizontal compression (if $c > 1$) or elongation (if $0 < c < 1$), followed by a reflection about the y-axis. To illustrate these transformations we use $y = (x - 1)^2$ as the basic function and show graphs of $y = (2x - 1)^2$, $y = (\frac{1}{2}x - 1)^2$, $y = (-2x - 1)^2$, and $y = (-\frac{1}{2}x - 1)^2$ (Figure 1.23).

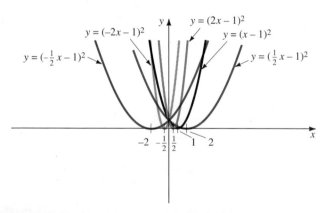

FIGURE 1.23

In Table 1.1 we summarize the eight transformations we have discussed. The basic graph we use is meant to represent an arbitrary function. For simplicity, in $y = -cf(x)$ and $y = f(cx)$ we take $c = 1$.

TABLE 1.1

Transformation	Effect	Illustration
$y = f(x) + c$	Vertical translation c units upward	
$y = f(x) - c$	Vertical translation c units downward	
$y = f(x + c)$	Horizontal translation c units to the left	
$y = f(x - c)$	Horizontal translation c units to the right	
$y = cf(x)$	Vertical elongation if $c > 1$, compression if $0 < c < 1$	

TABLE 1.1 (*continued*)

Transformation	Effect	Illustration
$y = f(cx)$	Horizontal compression if $c > 1$, elongation if $0 < c < 1$	
$y = -f(x)$	Reflection about the x-axis	
$y = f(-x)$	Reflection about the y-axis	

As the next example shows, several of these transformations can be combined to create one new function.

EXAMPLE 1.7   Discuss the function defined by

$$y = 1 - 2(x - 3)^2$$

and draw its graph.

Solution   If we rewrite the equation as

$$y = -2(x - 3)^2 + 1$$

we can show the function as a sequence of transformations of $y = x^2$ in the following diagram. (Other sequences are also possible.)

$$x^2 \longrightarrow (x - 3)^2 \longrightarrow 2(x - 3)^2 \longrightarrow -2(x - 3)^2 \longrightarrow -2(x - 3)^2 + 1$$

| shift 3 units to the right | stretch vertically by factor of 2 | reflect about x-axis | shift 1 unit upward |

The resulting graph is shown in Figure 1.24.

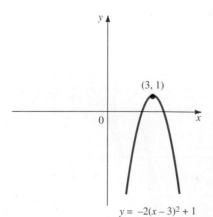

$y = -2(x - 3)^2 + 1$

FIGURE 1.24

EXAMPLE 1.8   In each of the following, obtain the graph of $y = f(x)$ by considering transformations of an appropriate basic function.

(a) $f(x) = 2\sqrt{x - 3}$ (b) $f(x) = |x + 2| - 3$

Solution

(a) Starting with the basic function $F(x) = \sqrt{x}$, we first have the vertical stretching $y = 2F(x)$, followed by a horizontal translation 3 units to the right, $y = 2F(x - 3)$. In Figure 1.25 we show both the graph of the basic function and the result of the transformations.

(b) Here we begin with the basic absolute value function $F(x) = |x|$, shift it 2 units to the left and then 3 units down, as shown in Figure 1.26.   ■

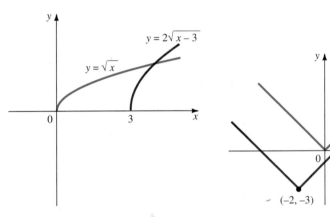

FIGURE 1.25   **FIGURE 1.26**

Sums, Differences, Products, and Quotients

Given two functions with a common domain, we can obtain new functions by adding them, subtracting them, multiplying them, or dividing them (provided the denominator is nonzero). The next definition gives precise meaning to these new functions.

Definition 1.3 Sums, Differences, Products, and Quotients of Functions	Let f and g be any two functions whose domains have nonempty intersection D. The functions $f + g$, $f - g$, $f \cdot g$, and f/g are defined as follows:

$$(f + g)(x) = f(x) + g(x) \qquad (f \cdot g)(x) = f(x)g(x)$$

$$(f - g)(x) = f(x) - g(x) \qquad \left(\frac{f}{g}\right)(x) = \frac{f(x)}{g(x)} \quad \text{if } g(x) \neq 0$$

The domain of each of the functions $f + g$, $f - g$, and $f \cdot g$ is all of D. The domain of f/g is the set of all points x in D for which $g(x) \neq 0$.

The next three examples illustrate how to form these new functions when f and g are given by formulas, tables, and graphs.

EXAMPLE 1.9   Let $f(x) = 2x - 3$ and $g(x) = x^2 - 1$. Find $(f + g)(x)$, $(f - g)(x)$, $(f \cdot g)(x)$, and $(f/g)(x)$, and give the domain of each.

Solution

$$(f + g)(x) = (2x - 3) + (x^2 - 1) = x^2 + 2x - 4; \quad \text{domain} = \mathbb{R}$$

$$(f - g)(x) = (2x - 3) - (x^2 - 1) = -x^2 + 2x - 2; \quad \text{domain} = \mathbb{R}$$

$$(f \cdot g)(x) = (2x - 3)(x^2 - 1) = 2x^3 - 3x^2 - 2x + 3; \quad \text{domain} = \mathbb{R}$$

$$\left(\frac{f}{g}\right)(x) = \frac{2x - 3}{x^2 - 1}; \quad \text{domain} = \{x : x \neq \pm 1\}$$

■

EXAMPLE 1.10   Let f and g be defined by the following table.

x	$f(x)$	$g(x)$
-1	-5	4
0	-2	5
1	1	4
2	4	1
3	7	-4
4	10	-11
5	13	-20

Find $(f + g)(x)$, $(f - g)(x)$, $(f \cdot g)(x)$, and $(f/g)(x)$.

Solution   The required values are shown in the following table.

x	$(f + g)(x)$	$(f - g)(x)$	$(f \cdot g)(x)$	$(f/g)(x)$
-1	-1	-9	-20	$-5/4$
0	3	-7	-10	$-2/5$
1	5	-3	4	$1/4$
2	5	3	4	4
3	3	11	-28	$-7/4$
4	-1	21	-110	$-10/11$
5	-7	33	-260	$-13/20$

■

EXAMPLE 1.11   The graphs of two functions f and g are shown in Figure 1.27. Find $f + g$ and $f \cdot g$ graphically.

Solution   At selected values of x, we estimate $f(x)$ and $g(x)$ from their graphs. Then we add the values to get the value of $(f + g)(x)$ and multiply them to get the value of $(f \cdot g)(x)$. Finally, we connect the points obtained to get the graphs of the functions $f + g$ and $f \cdot g$. The results are shown in Figure 1.28.

■

FIGURE 1.27

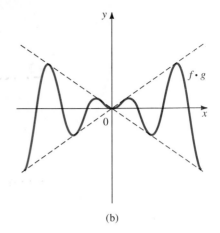

(a)   (b)

FIGURE 1.28

Exercise Set 1.2

In Exercises 1–6, the graph of a function $y = f(x)$ is given.
From it, obtain the graph of each of the following:
(a) $y = f(x) + 2$,   (b) $y = f(x - 3)$,   (c) $y = 2f(x)$,
(d) $y = f(2x)$,   (e) $y = -f(x)$,   (f) $y = f(-x)$.

1.

2.

3.

4.

5.

6.

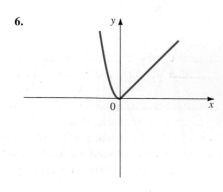

7. Draw the graph of $y = x^3$ and from it obtain the graph of $y = \frac{1}{2}(x-1)^3 + 2$.

8. Draw the graph of $y = \sqrt{x}$ and from it obtain the graph of $y = \sqrt{2-x}$.

9. The graph of a function f is given. From it obtain the graph of g, where $g(x) = 3 - 2f(\frac{1}{2}x - 1)$.

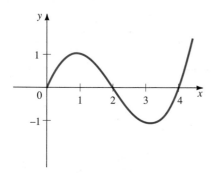

In Exercises 10–17, find $f + g$, $f - g$, $f \cdot g$, and f/g and give the domain of each.

10. $f(x) = 3x - 5$; $g(x) = 2x + 3$

11. $f(x) = 1 - x$; $g(x) = 1 + x$

12. $f(x) = \dfrac{1}{x-1}$; $g(x) = \dfrac{1}{x+1}$

13. $f(x) = \sqrt{x+4}$; $g(x) = \sqrt{4-x}$

14. $f(x) = x$; $g(x) = \sqrt{x-1}$

15. $f(x) = \dfrac{1}{x}$; $g(x) = \dfrac{x}{x-3}$

16. $f(x) = \begin{cases} -1 & \text{if } x < 0 \\ 1 & \text{if } x \geq 0 \end{cases}$; $g(x) = \begin{cases} 1 & \text{if } x < 0 \\ 0 & \text{if } x \geq 0 \end{cases}$

17. $f(x) = \begin{cases} x & \text{if } x \geq 0 \\ 0 & \text{if } x < 0 \end{cases}$; $g(x) = \begin{cases} 0 & \text{if } x \geq 0 \\ -x & \text{if } x < 0 \end{cases}$

In Exercises 18 and 19, the graphs of functions f and g are given. From these obtain the graphs of $f + g$ and $f \cdot g$.

18.

19.

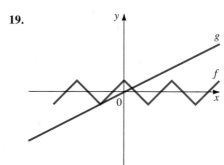

20. Refer to the greatest integer function, $f(x) = [\![x]\!]$, introduced in Exercise 22 of Exercise Set 1.1. Draw the graph of each of the following.
  (a) $y = [\![x + 1]\!]$ (b) $y = [\![x - 2]\!]$
  (c) $y = x - [\![x]\!]$ (d) $y = \sqrt{x - [\![x]\!]}$
  (e) $y = [\![x]\!] + \sqrt{x - [\![x]\!]}$
  (f) $y = (x - [\![x]\!])^2$

21. For an arbitrary function f whose domain includes $-x$ whenever it includes x, define g and h as follows:

$$g(x) = \frac{1}{2}\big[f(x) + f(-x)\big] \qquad h(x) = \frac{1}{2}\big[f(x) - f(-x)\big]$$

  (a) Show that g is even and h is odd.
  (b) Show that $f = g + h$. (This proves that every such function f can be expressed as the sum of an even function and an odd function.)

1.3 COMPOSITE FUNCTIONS AND INVERSE FUNCTIONS

In the last section we saw how two functions can be combined by addition, subtraction, multiplication, and division. Another very important way of combining two given functions to get a new function is called **composition**. In this method of combining, the output of one function becomes the input of another. For example, let $f(x) = x^3 + 1$ and $g(x) = 2x + 3$. If we input x into the g function, we get the output $2x + 3$. If we then input $2x + 3$ into the f function, we get $f(2x + 3) = (2x + 3)^3 + 1$. The result is the value of the function known as the **composition of f with g**, denoted by $f \circ g$. So for these functions we have

$$(f \circ g)(x) = f(g(x)) = f(2x + 3) = (2x + 3)^3 + 1$$

More generally, we have the following definition.

Definition 1.4 Composition	For two functions f and g, the **composition of f with g**, denoted by $f \circ g$, is defined by $$(f \circ g)(x) = f(g(x))$$ The domain of $f \circ g$ is the set of all x in the domain of g for which $g(x)$ is in the domain of f.

Figure 1.29 is a diagram of f, g, and $f \circ g$.

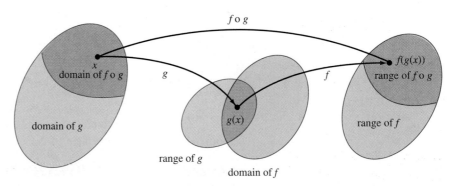

FIGURE 1.29

One reason for studying composition of functions is that if a complicated function can be recognized as the composition of two (or more) simpler functions, our knowledge of the simpler constituent functions might tell us much about the composite function. This viewpoint is similar to that taken in disciplines such as physics, where scientists study the basic constituents of matter (quarks, electrons, and other elementary particles) to understand the more complicated structures formed by grouping these constituents.

EXAMPLE 1.12   Let $f(x) = x^2$ and $g(x) = \sqrt{x - 4}$. Find $f \circ g$ and $g \circ f$, and give the domain of each.

Solution

$$(f \circ g)(x) = f(g(x)) = f(\sqrt{x-4}) = (\sqrt{x-4})^2 = x - 4$$

$$(g \circ f)(x) = g(f(x)) = g(x^2) = \sqrt{x^2 - 4}$$

The domain of f is all of $\mathbb{R}$, and the domain of g is the set $\{x : x \geq 4\}$. To determine the domain of $f \circ g$, we must find that subset of the domain of g for which $g(x)$ lies in the domain of f. But since the domain of f is all of $\mathbb{R}$, the entire range of g lies in the domain of f in this case. That is, the domain of $f \circ g$ is identical with the domain of g, namely, $\{x : x \geq 4\}$, which can also be written as the interval $[4, \infty)$. Note carefully that although $(f \circ g)(x) = x - 4$ and $x - 4$ has meaning for all of $\mathbb{R}$, the domain of $f \circ g$ is nevertheless limited to x values for which $g(x)$ is defined.

$\{x : x^2 \geq 4\} = \{x : x^2 - 4 \geq 0\}$

$x^2 - 4 = (x+2)(x-2)$

For $g \circ f$, the domain consists of all x values for which x^2 is in the domain of g, namely $\{x : x^2 \geq 4\}$. This set can be written in the equivalent form $\{x : |x| \geq 2\}$, or in interval notation as $(-\infty, -2] \cup [2, \infty)$. ∎

REMARK ───────────────────────────────────────

■ As the preceding example shows, the functions $f \circ g$ and $g \circ f$ are usually not the same.

───────────────────────────────────────

Returning to our machine analogy, the function $f \circ g$ can be thought of as being composed of the g and f machines operating in sequence, as indicated in Figure 1.30.

FIGURE 1.30

Sometimes it is important to express a function as the composition of two simpler functions, as illustrated in the next example.

EXAMPLE 1.13   Let $F(x) = (2x + 3)^5$. Find two functions f and g such that $F = f \circ g$.

Solution   To compute the value of F at x, we first compute $2x + 3$ and then raise the result to the fifth power. Thus, we have separated the evaluation of F into two operations that can be written as

$$g(x) = 2x + 3 \qquad \text{and} \qquad f(x) = x^5$$

So it is natural to express F as the composition $f \circ g$, since

$$F(x) = (2x + 3)^5 = [g(x)]^5 = f(g(x))$$

Note: There are other possible choices of the functions f and g, but the choice we made is the most natural. ∎

x	$p(x)$	$q(x)$
-2	1	5
-1	0	-8
0	-1	4
1	2	3
2	-2	-12

x	$(q \circ p)(x)$
-2	3
-1	4
0	-8
1	-12
2	5

The next example illustrates how the values of a composite function can be found when the two functions in the composition are given by a table.

EXAMPLE 1.14   Values of $p(x)$ and $q(x)$ are shown in the table. Find the values of $(q \circ p)(x)$.

Solution   The solution is given in the table in the margin. To illustrate how the values are obtained, consider the first entry. When $x = -2$, we know that $p(x) = 1$. Also, $q(1) = 3$. Thus, $(q \circ p)(-2) = q(p(-2)) = q(1) = 3$. ∎

Inverses

If we are given a function, we can sometimes obtain a new function from it by reversing the roles of the dependent and independent variables. Consider the following example.

EXAMPLE 1.15   The formula

$$C = \frac{5}{9}(F - 32)$$

expresses the Celsius temperature C as a function of the Fahrenheit temperature F. For example, when $F = 32$, $C = 0$, and when $F = 212$, $C = 100$. Find F as a function of C.

Solution   We solve for F as follows.

$$C = \frac{5}{9}(F - 32)$$

$$\frac{9}{5}C = F - 32$$

$$F = \frac{9}{5}C + 32$$

This last equation expresses F as a function of C. ∎

In Example 1.15 we began with a function of F. Suppose we call it $g(F)$. Then $g(F) = \frac{5}{9}(F-32)$. After reversing the roles of the dependent and independent variables, we obtained a function of C, say $h(C)$, where $h(C) = \frac{9}{5}C + 32$. This function h is called the **inverse** of the function g.

Unfortunately, this procedure of reversing the roles of the dependent and independent variables does not always result in a function. Consider $f(x) = x^2$, for example. Let y be the dependent variable, so that $y = x^2$. When we solve for x in terms of y, we get $x = \pm\sqrt{y}$. But this equation does not define x as a function of y, since for a given positive value of y (now being treated as the independent variable) there are two values of x, which violates the requirement that to each admissible value of the independent variable of a function there must correspond a *unique* value of the dependent variable. Figure 1.31 clearly shows the problem here. The graph of $y = x^2$ passes the vertical line test, so the equation $y = x^2$ does define y as a function of x. But if we reverse the roles of dependent and independent variables, we see that x is not a function of y. For example, if $y = 4$ we get *two* values of x, namely $x = \sqrt{4} = 2$ and $x = -\sqrt{4} = -2$. So even though $2 \neq -2$, we have $f(2) = f(-2)$. That is, two

FIGURE 1.31

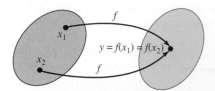

FIGURE 1.32

different x values correspond to the *same y* value. Figure 1.32 provides another way of viewing the problem. There we show two *x* values in *A* corresponding under the function to one *y* value in *B*. When we try to go in reverse, the value of *y* in *B* does not correspond to a unique *x* in *A*.

The example of $f(x) = x^2$ should make clear that only if we have a function for which no two different *x* values correspond to the same *y* value can we create a new function by interchanging independent and dependent variables. Such a function is said to be *one-to-one*.

Definition 1.5 One-to-One Function	A function f is said to be **one-to-one**, abbreviated 1–1, provided that when $x_1 \neq x_2$, then $f(x_1) \neq f(x_2)$.

Equivalently, f is 1–1 if whenever $f(x_1) = f(x_2)$, then $x_1 = x_2$. That is, each *y* value in the range of f corresponds to exactly one *x* value in the domain. We know that *all* functions must satisfy the vertical line test. By what we have just said, all 1–1 functions also must satisfy the horizontal line test.

Horizontal Line Test

A function is 1–1 if and only if every horizontal line intersects the graph of the function in at most one point.

Figure 1.33 illustrates a function that passes this test and one that does not.

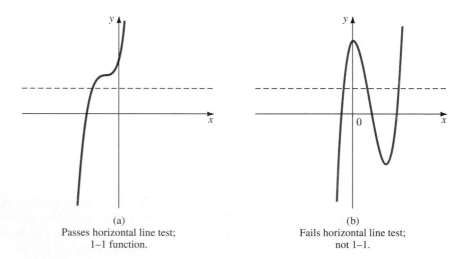

(a)
Passes horizontal line test;
1–1 function.

(b)
Fails horizontal line test;
not 1–1.

FIGURE 1.33

EXAMPLE 1.16   Let

$$f(x) = \sqrt{2x - 3} \qquad \text{and} \qquad g(x) = 2 + |x - 3|$$

Show by means of Definition 1.5 that f is 1–1 on its domain but g is not 1–1 on its domain. Confirm the results by the horizontal line test.

Solution   To show that f is 1–1 we suppose that $f(x_1) = f(x_2)$ and show that it necessarily follows that $x_1 = x_2$. Thus, we assume that

$$\sqrt{2x_1 - 3} = \sqrt{2x_2 - 3}$$

Squaring both sides gives

$$2x_1 - 3 = 2x_2 - 3$$
$$2x_1 = 2x_2$$
$$x_1 = x_2$$

So f is 1–1.

We can show that g is not 1–1 by exhibiting two different values of x that yield the same function value. For example, we can take $x_1 = 2$ and $x_2 = 4$. Then

$$g(x_1) = g(2) = 2 + |2 - 3| = 2 + |-1| = 2 + 1 = 3$$

and

$$g(x_2) = g(4) = 2 + |4 - 3| = 2 + |1| = 2 + 1 = 3$$

We show the graph of f in Figure 1.34 and the graph of g in Figure 1.35. Clearly f satisfies the horizontal line test and g does not.   ■

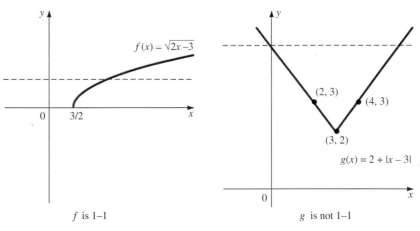

f is 1–1 g is not 1–1

FIGURE 1.34   **FIGURE 1.35**

When a function is 1–1, the new function obtained from it by interchanging independent and dependent variables is called the **inverse** of the original function. In effect, the inverse undoes what the original function does. If the given function operates on a to give b, the inverse function operates on b to give a.

Suppose y is given as a function of x by a table. For example, consider the following table.

x	0	1	2	-1	-2
y	2	5	9	0	3

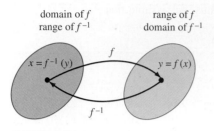

domain of f
range of f^{-1}

range of f
domain of f^{-1}

FIGURE 1.36

To obtain the inverse, we use the y values as the values of the independent variable and the x values as the values of the dependent variable. Since it is customary to use x for independent and y for dependent variables, we can show the table for the inverse function by replacing x with y and y with x:

x	2	5	9	0	3
y	0	1	2	-1	-2

If f is a 1–1 function, then it has a unique inverse. We customarily use the symbol f^{-1}, read "f inverse," for the inverse of f.

If f makes x correspond to y, then f^{-1} makes y correspond to x. (See Figure 1.36.) That is, if $y = f(x)$, then $f^{-1}(y) = x$, or equivalently

$$f^{-1}(f(x)) = x$$

The -1 in the notation f^{-1} is not to be treated as an exponent. So $f^{-1} \neq \frac{1}{f}$. This use of the superscript -1 is reserved for functions only and always indicates the inverse function.

We can now give the formal definition of the inverse of a 1–1 function.

Definition 1.6
Inverse Functions

Let f be a 1–1 function with domain A and range B. The function f^{-1} with domain B and range A satisfying

$$(f^{-1} \circ f)(x) = x \quad \text{for all } x \text{ in } A$$

and

$$(f \circ f^{-1})(y) = y \quad \text{for all } y \text{ in } B$$

is called the **inverse of f**.

REMARK ───────────────────────────────
■ If f^{-1} is the inverse of f, then f is the inverse of f^{-1}. That is,

$$(f^{-1})^{-1} = f$$
───

The function that maps each element of a given set onto itself is called the **identity mapping function** and is often denoted by I. Thus, $I(x) = x$. Since $(f^{-1} \circ f)(x) = x$ for all x in the domain of f, we see that

$$f^{-1} \circ f = I$$

on this domain. This equation bears a striking similarity to the equation

$$a^{-1} \cdot a = 1$$

for a nonzero real number a. So the identity mapping function I plays a role similar to the multiplicative identity element 1 for real numbers.

As a consequence of Definition 1.6 we can conclude the following:

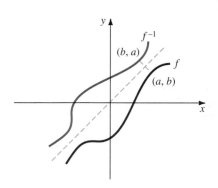

$$y = f(x) \quad \text{if and only if} \quad f^{-1}(y) = x$$

FIGURE 1.37

We have seen how to obtain a table for f^{-1} when f is given in tabular form. Graphically, it is also easy to obtain f^{-1} from f. When a point (a, b) is on the graph of f, so that $f(a) = b$, then, by Definition 1.6, $f^{-1}(b) = a$, so the point (b, a) is on the graph of f^{-1}. It follows that the graphs of f and f^{-1} are symmetrical with respect to the line $y = x$, as illustrated in Figure 1.37. So when you know the graph of a 1–1 function, to get the graph of f^{-1} simply reflect the graph of the given function about the line $y = x$.

Procedure for Finding the Inverse

Suppose now that a 1–1 function f is defined by a formula, and we set $y = f(x)$. To get f^{-1}, using x as the independent variable, we interchange x and y, getting $x = f(y)$. Then we solve this equation (if possible) for y in terms of x. To see that the result is $f^{-1}(x)$, apply f^{-1} to both sides of $x = f(y)$:

$$f^{-1}(x) = f^{-1}(f(y)) = y$$

To illustrate this procedure, consider the function f defined by

$$f(x) = \frac{2x - 3}{4}$$

Let $y = f(x)$, getting

$$y = \frac{2x - 3}{4}$$

Interchange x and y to get the inverse:

$$x = \frac{2y - 3}{4}$$

Solve for y:

$$2y - 3 = 4x$$
$$y = \frac{4x + 3}{2}$$

This last equation is in the form $y = f^{-1}(x)$; that is,

$$f^{-1}(x) = \frac{4x + 3}{2}$$

We summarize this procedure below.

Procedure for Finding the Inverse of a Function

If f is a 1–1 function, follow these steps to find f^{-1}:

1. Set $y = f(x)$.
2. Interchange x and y.
3. Solve for y in terms of x.

The resulting equation is in the form $y = f^{-1}(x)$.

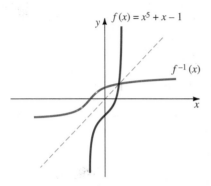

$f(x) = x^5 + x - 1$

$f^{-1}(x)$

FIGURE 1.38

There is one possible problem with this procedure that comes in step 3. It is not always possible to solve for y. Suppose, for example, that after step 2 we have the equation

$$x = y^5 + y - 1$$

Now try to solve for y! The original function in this case is given by the formula $f(x) = x^5 + x - 1$. It is possible, by means you will learn later, to show that f^{-1} does exist, but trying to find a formula for it is hopeless. We could, however, find its graph, as we show in Figure 1.38. We used a computer algebra system to get the graph of f. Notice that f does appear to be 1–1, since by what we see of its graph, the horizontal line test is satisfied. We obtained the graph f^{-1} by reflecting the graph of f about the 45°-line $y = x$.

EXAMPLE 1.17   Show that $f(x) = \sqrt{x-1}$ has an inverse on the domain $[1, \infty)$, and find the inverse. Draw the graphs of f and f^{-1}.

Solution   To show f is 1–1, suppose that x_1 and x_2 have the same image. That is, suppose

$$\sqrt{x_1 - 1} = \sqrt{x_2 - 1}$$

Squaring both sides gives

$$x_1 - 1 = x_2 - 1$$

$$x_1 = x_2$$

Thus, the only way that $f(x_1)$ and $f(x_2)$ can be the same is for x_1 to be the same as x_2. Therefore f is 1–1, and so f^{-1} exists. Following our three-step procedure, we write

$$y = \sqrt{x - 1}$$

$f^{-1}(x) = x^2 + 1, \; x \geq 0$

$f(x) = \sqrt{x - 1}$

FIGURE 1.39

and interchange x and y to get the inverse:

$$x = \sqrt{y - 1}$$

$$x^2 = y - 1$$

$$y = x^2 + 1$$

Thus, $f^{-1}(x) = x^2 + 1$. But we must state its domain. The domain of f^{-1} is the range of f, namely the set $[0, \infty)$, since $\sqrt{x-1}$ ranges over the set of all nonnegative real numbers as x takes on all values greater than or equal to 1. So the domain of f^{-1} is the set $[0, \infty)$. We show the graphs of f and f^{-1} in Figure 1.39. ∎

FIGURE 1.40

An Important Inverse Pair

We conclude this section with a brief mention of the function $f(x) = \ln x$, called the **natural logarithm function**, and its inverse, $f^{-1}(x) = e^x$, called the **natural exponential function**. These two functions are especially important in applications; they may even be the most important function-inverse pair in mathematics and science applications. We will study them in detail in Chapter 7. The letter e is used to designate a certain irrational number approximately equal to 2.71828. Your calculator has a $\boxed{\ln x}$ key. Some calculators also have an $\boxed{e^x}$ key. On others, e^x is in the $\boxed{\text{2nd}}$ register above $\boxed{\ln x}$. By plotting points, or by using a graphing calculator or computer algebra system, you can obtain the graphs shown in Figure 1.40.

FINDING INVERSE FUNCTIONS USING A COMPUTER ALGEBRA SYSTEM

The procedure for finding an inverse of a one-to-one function can be carried out easily using a CAS such as Maple, Mathematica, or DERIVE. We describe several examples here, where we assume it is known that the functions are one-to-one.

CAS 6

Find the inverse for the function $f(x) = \sqrt{x - 1}$ defined on the interval $[1, \infty]$ (the domain of f). Draw the graphs of f, f^{-1}, and $y = x$.

Following the procedure described earlier, we interchange x and y, and then solve for y in terms of x.

In Maple:

```
solve(x=sqrt(y–1),y);
```

The output is: $1 + x^2$

To plot all three functions on the same set of axes, we enter the Maple plot command:

```
plot({[x,sqrt(x–1),x=1..3],1+x^2,x},x=0..3,y=0..3);
```

The plot is on the interval $[0, 3]$ to give us a good view of the relationship between the three graphs. Notice that to plot more than one graph on the same set of axes, the expressions must be enclosed in braces, { }, and that individual expressions are separated by commas. The same is true in Mathematica.

Maple:

```
solve[x=sqrt(y–1),y);
Output: 1+x^2
plot([x,sqrt(x–1),x=1..3],1+x^2,x},
     x=0..3,y=0..3);
```

Mathematica:

```
Solve[x == Sqrt[y–1],y]
Output: {{y → 1 + x^2}}
Plot[{Sqrt[x–1],x^2+1,x},
{x,0,3},PlotRange–>{0,3}]
```

FIGURE 1.3.1
$y = f(x) = \sqrt{x - 1}$, $y = f^{-1}(x) = x^2 + 1$, $y = x$

DERIVE:

(At the □ symbol, go to the next step.)
a (author) □ x=sqrt(y−1) □ l (solve) □
[choose expression to solve for y] □ y
(solve for y)
Output: y=IF(x≥ 0,x^2+1)
a (author) □ sqrt(x−1) □ p (plot window) □ p (plot) □ a (algebra) □ a (author) □ x^2x+1 □ p (plot window) □ p (plot) □ a (algebra) □ a (author) □ x □ p (plot window) □ p (plot)

CAS 7

Find the inverse for the function $f(x) = \dfrac{(1 + 5x)}{(5 - 3x)}$ defined on the interval $[1, \infty]$. Draw the graphs of f, f^{-1}, and $y = x$.

In Maple, we enter:

```
solve(x=(1+5*y)/(5–3*y),y);
```

The output is: $-\dfrac{-5x + 1}{3x + 5}$

To get the curve shown in Figure 1.3.2 we used the plot command:

```
plot({(1+5*x)/(5–3*x),–(–5*x+1)/(3*x+5),x},x=–5..5,y=–10..10);
```

The specification y=-10..10 restricts the size of the y-axis. If, for example, we use the Maple command

```
plot({(1+5*x)/(5–3*x),–(–5*x+1)/(3*x+5),x},x=–10..10);
```

we get the curve shown in Figure 1.3.3. Again we need to be careful of the output generated by computer algebra systems. Try several modifications in the specifications of the x- and y-axes to see what pictures are generated.

Maple:

```
solve (x=(1+5*y)/(5–3*y),y);
```

Output: $-\dfrac{-5x + 1}{3x + 5}$

```
plot({(1+5*x)/(5–3*x),–(–5*x+1)/
(3*x+5),x},x=–5..5,y=–10..10);
```

Mathematica:[†]

```
Solve[x == (1+5*y)/(5–3*y),y]
```

Output: $\left\{\left\{y \to -\left(\dfrac{1-5x}{5 + 3x}\right)\right\}\right\}$

```
Plot[{(1+5*x)/(5–3*x),
    –(1–5*x/(5+3*x),x},
{x,–5,5},
PlotRange–>{–10,10}]
```

DERIVE:

(At the □ symbol, go the next step.)
a (author) □ x=(1+5y)/(5−3y) □ l (solve) □ [choose the expression to solve for y]

Output: $y = \dfrac{5x - 1}{3x + 5}$

a (author) □ (1+5x)/(5−3x) □ p (plot window) □ p (plot) □ a (algebra) □ a (author) □ (5x−1)/(3x+5) □ p (plot window) □ p (plot) □ a (algebra) □ a (author) □ x □ p (plot window) □ p (plot) □ z (zoom-out)

[†] Using the * for multiplication is optional in Mathematica.

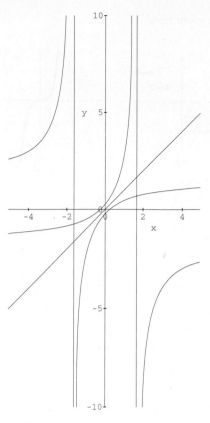

FIGURE 1.3.2
$y = f(x) = (1 + 5x)/(5 - 3x),$
$y = f^{-1}(x) = (5x - 1)/(3x + 5),\ y = x$

FIGURE 1.3.3

Exercise Set 1.3

In Exercises 1–4, find $f \circ g$ and $g \circ f$, and give the domain of each.

1. $f(x) = 2x - 1;\ g(x) = x + 2$

2. $f(x) = x^2 + 1;\ g(x) = 2x - 3$

3. $f(x) = \dfrac{x}{x - 1};\ g(x) = \dfrac{3}{x}$

4. $f(x) = \sqrt{x - 1};\ g(x) = x^2 - 3$

In Exercises 5–8, find functions f and g such that $F = f \circ g$.

5. $F(x) = (4 - 3x^2)^6$ **6.** $F(x) = \sqrt{2x + 3}$

7. $F(x) = \sqrt{(x^2 - 1)^3}$ **8.** $F(x) = 3(1 - x)^{-4}$

9. Complete the following table so that g is even and h is odd, and fill in the values for $f \circ g$ and $g \circ f$.

t	$g(t)$	$h(t)$	$(g \circ h)(t)$	$(h \circ g)(t)$
-3	2	2		
-2	-1	-1		
-1	3	3		
0	0	0		
1				
2				
3				

10. Is it ever true that $f \circ g = g \circ f$? Is it always true? Give examples or explanations to support your conclusions.

11. Show that $f \circ (g \circ h) = (f \circ g) \circ h$ whenever both sides are defined.

12. Which of the functions whose graphs are given have inverses, and which do not?

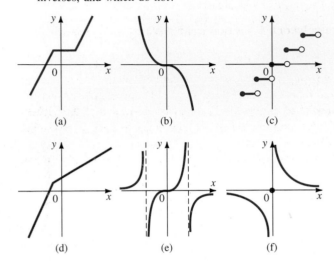

(a)   (b)   (c)

(d)   (e)   (f)

In Exercises 13–18, show that the given function has an inverse, and find it. Give the domain of the inverse.

13. $f(x) = \dfrac{5x - 2}{3}$

14. $g(x) = \dfrac{4 - 7x}{2}$

15. $F(x) = \sqrt{2x - 3}$

16. $G(x) = \dfrac{1}{\sqrt{x}}$

17. $\varphi(t) = t^2 - 1, t \geq 0$

18. $\psi(z) = \sqrt{z^2 - 1}, z \geq 1$

In Exercises 19 and 20, graphs of 1–1 functions are given. Copy each graph and on the same set of axes sketch the graph of its inverse.

19.

(a)   (b)

20.

(a)   (b)

In Exercises 21–24, show that the given function is not 1–1. Then determine a subset of its domain on which the function is 1–1, and find its inverse on this restricted domain. Give the domain of the inverse.

21. $f(x) = \dfrac{1}{x^2 + 2}$

22. $g(x) = \dfrac{1}{\sqrt{4 - x^2}}$

23. $F(x) = |3 - 2x|$

24. $h(x) = 2x - x^2$ (*Hint:* Complete the square to obtain $h(x) = 1 - (x - 1)^2$.)

25. Let $f(x) = (x + a)/(x + b)$, where $a \neq b$. Show that f^{-1} exists, and find it. Give the domain and range of both f and f^{-1}. Verify that $(f^{-1} \circ f)(x) = x$ for all x in the domain of f and that $(f \circ f^{-1})(x) = x$ for all x in the domain of f^{-1}.

26. Let $f(x) = 1/x$. Prove that for functions g and h,

$$f \circ \left(\frac{g}{h}\right) = \frac{f \circ g}{f \circ h}$$

What restrictions must be placed on the images of g and h?

In Exercises 27 and 28, use a graphing calculator or CAS to obtain the graph of the given function. Verify from the graph that the function is 1–1 and draw the graph of its inverse.

27. $f(x) = 2x^3 + x - 5$

28. $g(x) = 3 - 4x^3 - x^5$

1.4 POLYNOMIAL FUNCTIONS

The simplest functions are the **polynomial functions**, yet they are among the most important in applications. Almost all nonpolynomial functions encountered in engineering and science are evaluated numerically by polynomial approximations. Here are some examples of polynomials:

$$f(x) = 2x + 3 \qquad p(x) = 3x^2 - 4x + 1 \qquad q(x) = x^5 + x - 1$$

Every special case of a polynomial function can be written in the general form

$$p(x) = a_n x^n + a_{n-1} x^{n-1} + \cdots + a_0 \tag{1.1}$$

where n is a nonnegative integer and $a_0, a_1, \ldots, a_n$ are constants, which we will assume to be real unless otherwise specified. If $a_n \neq 0$, the expression on the right is called a **polynomial of degree** n. The coefficient $a_n \neq 0$ of the highest power of x is called the **leading coefficient**. The domain of every polynomial function is the set of all real numbers.

Functions defined by polynomials of degree 0, 1, 2, and 3, respectively, are given names as shown below.

Degree	Name	Example
0	Constant function	$p(x) = 2$
1	Linear function	$p(x) = 2x + 3$
2	Quadratic function	$p(x) = x^2 - 2x + 7$
3	Cubic function	$p(x) = 4x^3 + 2x - 8$

The names *quartic* (degree 4) and *quintic* (degree 5) are sometimes used, but polynomial functions of degree higher than 5 are seldom named.

Constant Functions and Linear Functions

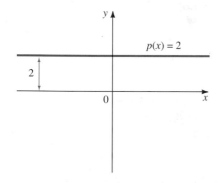

FIGURE 1.41

The graph of a constant function is a horizontal line. For example, the graph of the constant function $p(x) = 2$ consists of all points (x, y) with $y = 2$. So for every x, the value of y is 2. The graph of this function is shown in Figure 1.41.

Linear functions also have straight lines as graphs, as the name suggests. From Equation 1.1, a linear function is of the form $p(x) = a_1 x + a_0$, where $a_1 \neq 0$. It is customary to use the letters m and b for linear functions, rather than a_1 and a_0. So we write

$$p(x) = mx + b \qquad (m \neq 0)$$

for a typical linear function. By setting $x = 0$, we get $p(0) = b$. Thus, the graph crosses the y-axis at the point $(0, b)$. The number b is called the y-*intercept*. If (x_1, y_1) and (x_2, y_2) are any two different points on the line, we have $y_1 = mx_1 + b$ and $y_2 = mx_2 + b$. If we subtract the corresponding sides of the first equation from those of the second, we get $y_2 - y_1 = m(x_2 - x_1)$. Dividing by $(x_2 - x_1)$ gives

$$m = \frac{y_2 - y_1}{x_2 - x_1}$$

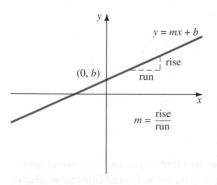

FIGURE 1.42

This number m is called the *slope* of the line. Its value is the same for *any* two distinct points (x_1, y_1) and (x_2, y_2) on the line. The numerator $y_2 - y_1$ is the change in y values corresponding to the change in x values $x_2 - x_1$ in the denominator. That is, m is the change in the vertical direction divided by the change in the horizontal direction, often described as "rise over run." Figure 1.42 illustrates a typical situation. When m is positive, the line slopes upward to the right, as in Figure 1.43(a), and when m is negative, the line slopes downward to the right, as in Figure 1.43(b).

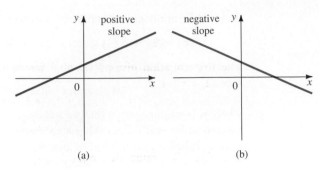

FIGURE 1.43

The next two examples illustrate how linear functions can arise in applications.

EXAMPLE 1.18   A car rental agency charges $32 a day, plus 20¢ per mile, for renting a mid-size car. Write the formula for a cost function $C(x)$ for one day's rental, where x is the number of miles driven. Find the cost of renting a car of this type and driving a distance of 215 miles.

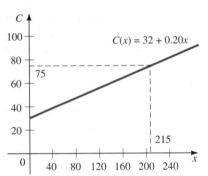

FIGURE 1.44

Solution   This situation is typical of many cost functions in that it involves both a **fixed cost** and a **variable cost**. In this example, the fixed cost is $32 and the variable cost is $0.20x$ (the cost per mile, in dollars, times the number of miles driven). So the total cost $C(x)$, in dollars, is given by the formula

$$C(x) = 32 + 0.20x$$

Notice that $C(x)$ is a linear function with slope 0.20 and y-intercept 32. Its graph is shown in Figure 1.44. The cost for driving 215 miles is

$$C(215) = 32 + 0.20(215) = \$75 \qquad \blacksquare$$

EXAMPLE 1.19   For income tax purposes, equipment is often *depreciated linearly* over a given period of time. Assume that a Mercedes-Benz originally valued at $60,000 is depreciated linearly. After five years its value is $40,000. Find a formula for the value $V(t)$ of the car t years after purchase, where $0 \le t \le 15$. What is its value eight years after purchase?

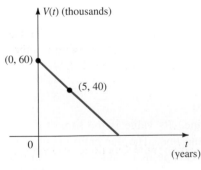

FIGURE 1.45

Solution   The graph in Figure 1.45 shows the situation. We are given that $V(0) = 60$ (in thousands) and $V(5) = 40$. So the points $(0, 60)$ and $(5, 40)$ are on the graph. Since $V(t)$ is linear, it must be of the form

$$V(t) = mt + b$$

There are two ways to find m and b. One way is to observe that since b is the y-intercept, $b = 60$. Then calculate m using the two points $(0, 60)$ and $(5, 40)$:

$$m = \frac{40 - 60}{5 - 0} = \frac{-20}{5} = -4$$

Thus,

$$V(t) = -4t + 60$$

The second method is to substitute the two known values of t and corresponding values of $V(t)$ into the equation $V(t) = mt + b$, and solve for m and b:

$$60 = m(0) + b \qquad \text{We substituted } t = 0, \; V(t) = 60.$$

$$40 = m(5) + b \qquad \text{We substituted } t = 5, \; V(t) = 40.$$

The first equation gives $b = 60$, and on substituting this value for b in the second, we get

$$40 = m(5) + 60$$

$$m = -\frac{20}{5} = -4$$

as before. The value after eight years is

$$V(8) = -4(8) + 60 = -32 + 60 = 28$$

That is, the car is worth \$28,000. ∎

When data for a function are given in tabular form, we can test for linearity by seeing if for each pair (x_1, y_1) and (x_2, y_2) in the table, the ratio $(y_2 - y_1)/(x_2 - x_1)$ remains constant. If so, this constant value is the slope. Testing for a constant slope is particularly easy when the step between consecutive x values does not change. Then, all you have to do is to see if there is also a fixed change between consecutive y values. (We are using x to mean the independent variable and y to mean the dependent variable; clearly, other letters can be used.) The next example illustrates such a situation.

EXAMPLE 1.20   Show that the data in the accompanying table indicate a linear functional relationship, and find a formula for the function.

Solution   The change in x from one entry to the next is 2 in every case. The corresponding change in y is 6 in every case. So we expect that the data come from a linear function $f(x) = mx + b$ with

$$m = \frac{\text{change in } y}{\text{change in } x} = \frac{6}{2} = 3$$

x	y
-3	-14
-1	-8
1	-2
3	4
5	10

To find b we can substitute any pair of values (x, y) into $y = mx + b$. Using $(1, -2)$, we get

$$-2 = 3(1) + b$$

$$b = -5$$

So $f(x) = 3x - 5$.

It should be made clear that although the tabular values give strong evidence that the function we have found is correct, we have no way of knowing whether points would continue to fall on this line if the table were extended. ∎

Quadratic Functions

We turn next to quadratic functions, that is, polynomial functions of degree 2. The simplest case is

$$f(x) = x^2$$

The range of f is the set $[0, \infty)$ of all nonnegative real numbers, and $f(x) = 0$ only when $x = 0$. Since $f(-x) = f(x)$, f is an even function, so its graph is symmetric with respect to the y-axis. The graph, shown in Figure 1.46, is a

$y = x^2$

FIGURE 1.46

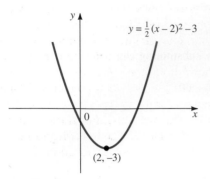

$$y = \frac{1}{2}(x-2)^2 - 3$$

$(2, -3)$

FIGURE 1.47

parabola with **vertex** at the origin and **axis** coinciding with the y-axis. From this basic parabola we can obtain the graph of every other quadratic function, using the transformations we considered in Section 1.2. For example, the graph of

$$y = a(x - h)^2 + k \qquad (1.2)$$

is obtained from that of $y = x^2$ by a vertical elongation (if $|a| > 1$) or compression (if $0 < |a| < 1$), as well as a reflection about the x-axis if $a < 0$, and then a translation so that the vertex is at the point (h, k). Equation 1.2 is called the **standard form** of the equation of a parabola. To illustrate, we show the graph of

$$y = \frac{1}{2}(x - 2)^2 - 3$$

in Figure 1.47.

The general quadratic function

$$f(x) = ax^2 + bx + c \qquad (a \neq 0)$$

can be put in the form (1.2), where $y = f(x)$, by *completing the square*, as illustrated in the next example.

EXAMPLE 1.21   Analyze and draw the graph of the function defined by

$$y = 1 - 4x - 2x^2$$

For completing the square the leading coefficient must be 1:

$$ax^2 + bx + c$$

$$= a\left(x^2 + \frac{b}{a}x\right) + c$$

Inside the parentheses, the coefficient of x^2 is 1.

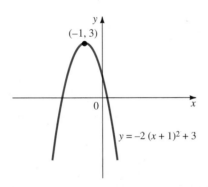

$(-1, 3)$

$$y = -2(x + 1)^2 + 3$$

FIGURE 1.48

Solution   First, rearrange terms and factor out the coefficient of x^2 from the terms involving x:

$$y = -2(x^2 + 2x \qquad) + 1$$

We have left a space inside the parentheses for adding a term to complete the square. The term we need is 1 (the square of half the coefficient of x). Adding 1 inside the parentheses has the effect of adding -2 to the right-hand side, because of the coefficient preceding the parentheses. To compensate for this change, we must add $+2$. So we have

$$y = -2(x^2 + 2x + 1) + 1 + 2$$

or

$$y = -2(x + 1)^2 + 3$$

The equation is now in the form of Equation 1.2. Its graph is that of the basic parabola $y = x^2$, stretched so that each y value is doubled, reflected about the x-axis, and translated so that its vertex is at the point $(-1, 3)$. We show the graph in Figure 1.48.   ∎

If we go through steps similar to those in Example 1.21 with the general quadratic function $f(x) = ax^2 + bx + c$, we get the result

$$y = a(x - h)^2 + k$$

where

$$h = -\frac{b}{2a} \qquad \text{and} \qquad k = \frac{4ac - b^2}{4a}$$

You will be asked to carry out the derivation in the exercises. Rather than memorizing these formulas for h and k, it is probably better to carry out the steps in each individual case, as we did in Example 1.21. Note, however, that $k = f(h)$, so it would not be too difficult to learn the formula $h = -b/2a$ and then find k by substituting h into the given function.

Applications often require finding the largest or smallest value of a given function, possibly subject to certain limitations on the independent variable. In business, we might want to maximize profit, for example, or minimize cost. In physics, we might want to determine how high an object will rise if projected at a given angle and initial velocity or how close to a planet a spacecraft can be safely sent. Problems of this type are called *optimization* problems. If the function in question happens to be quadratic, we can find the largest or smallest value by locating the vertex of the parabola. For most other functions we need the tools of calculus that we will introduce in Chapter 3. The following example illustrates the technique we can use in the case of a quadratic function.

EXAMPLE 1.22   The average midnight attendance at a "Rocky Horror Picture Show" is 100 when the ticket price is $5.00. The theater owner estimates that for each 10¢ decrease in the ticket price, the average attendance will increase by 10. What price should the owner charge to produce the maximum revenue, and what is the maximum revenue? What would the average attendance be with this new ticket price?

Solution   Let x denote the number of 10¢ reductions. Then

$$5 - 0.10x = \text{the price of each ticket (in dollars)}$$

and

$$100 + 10x = \text{the average nightly attendance}$$

The revenue, say $R(x)$, is found by multiplying the price per ticket by the number of tickets sold. That is,

$$R(x) = (5 - 0.1x)(100 + 10x)$$
$$= -x^2 + 40x + 500$$

To put this equation in the standard form of a parabola, we can complete the square, as in Example 1.21, obtaining

$$R(x) = -(x^2 - 40x + 400) + 500 + 400$$
$$= -(x - 20)^2 + 900$$

The graph of $R(x)$ is therefore a downward-opening parabola with vertex $(20, 900)$. Thus, the maximum value of $R(x)$ is $900, occurring when $x = 20$. (See Figure 1.49.) The ticket price when there are 20 reductions of 10¢ each is

$$5 - 0.1(20) = \$3.00$$

and the average attendance is

$$100 + 10(20) = 300 \qquad \blacksquare$$

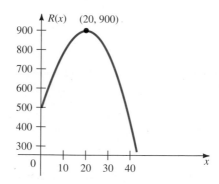

FIGURE 1.49

Polynomial Functions of Higher Degree

We will give only a brief discussion of polynomial functions of degree 3 or higher here. As we will see, calculus is needed to understand fully the nature of such functions. Figure 1.50 shows some computer-generated graphs of cubic

FIGURE 1.50

functions. These graphs are typical of cubics. If the leading coefficient is positive, as in parts (a) and (b), the function is positive for large positive values of x and negative for large negative values. When the leading coefficient is negative, as in parts (c) and (d), the opposite is true. The point $(-1, 3)$ that we have labeled in part (a) is called a **local maximum point**, because it is higher than all other points on the graph in its immediate vicinity. In part (c) the point $(1, 3)$ is a local maximum point. Similarly, the point $(1, -1)$ in (a) and $(-1, -1)$ in (c) are **local minimum points**. Parts (b) and (d) have no local maximum or minimum points.

Refer again to the graph in Figure 1.50(a). For $x < -1$, as we move from left to right on the x-axis, the curve rises, and we say the function is **increasing**. Between $x = -1$ and $x = 1$ as we move from left to right on the x-axis, the curve goes downward, and we say the function is **decreasing**. For $x > 1$, it again increases. By observing the graphs, you should verify the following results for the functions in parts (b), (c), and (d).

$$\begin{cases} \text{part (b): function is increasing for all } x \\ \text{part (c): function is decreasing for } x < -1 \text{ and for } x > 1 \\ \qquad\qquad \text{and increasing for } -1 < x < 1 \\ \text{part (d): function is decreasing for all } x \end{cases}$$

Besides knowing where a function is increasing or decreasing and knowing whether it has any local maxima (plural of maximum) or minima (plural of minimum), it is important to know if the function is bending up, as in Figure 1.51(a), or bending down, as in Figure 1.51(b). In the first instance we say the graph is **concave up** and in the second, **concave down**. For example, the graphs in Figure 1.50(a) and (b) are both concave down for $x < 0$ and concave

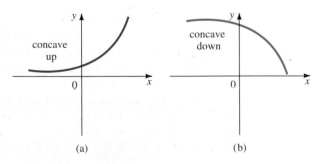

FIGURE 1.51

up for $x > 0$. Similarly, the graphs in parts (c) and (d) are both concave up for $x < 0$ and concave down for $x > 0$. Note that whether a function is increasing or decreasing on an interval has nothing to do with concavity.

We emphasize that the observations we have made about the graphs of the cubic functions in Figure 1.50 are based on computer-generated graphs. As we have seen, pictures can sometimes be deceiving. Maybe something unexpected happens beyond the range we are plotting. We need some means of determining exact locations of local maxima and minima, intervals over which a function is increasing and those over which it is decreasing, and intervals over which the graph is concave up and those over which it is concave down. As you might guess, it is calculus that provides the key, and we will return to these questions when we have introduced the appropriate calculus concepts.

Graphs of polynomial functions of degree 4 can have at most three local maxima and minima (two maxima and one minimum, or the reverse). There may be only one maximum or minimum, however. The graphs in Figure 1.52 illustrate typical possibilities in which the leading coefficient is positive. Notice that in each case the function is positive both for large positive and large negative values of x. After all, for these large values, the fourth power of x is larger than all lower powers combined. If the leading coefficient is negative, the graph always goes downward for large x (either positive or negative). Notice that the graph in part (c) looks almost like a parabola. So you cannot always tell the degree of a polynomial function just by looking at its graph.

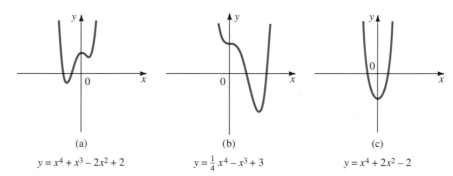

(a)   (b)   (c)

$y = x^4 + x^3 - 2x^2 + 2$ $y = \frac{1}{4}x^4 - x^3 + 3$ $y = x^4 + 2x^2 - 2$

FIGURE 1.52

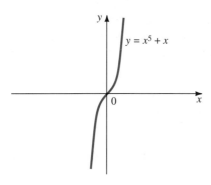

$y = x^5 + x$

FIGURE 1.53

For each increase of one in the degree of a polynomial function, the graph *may* add one local maximum or minimum point. Thus, a polynomial function of degree n, with $n \geq 2$, has *at most* $n - 1$ local maxima or minima. It is possible, however, that an *odd*-degree polynomial function will have no maxima or minima. Consider, for example, the graph of $f(x) = x^5 + x$, which we show in Figure 1.53. Here again, you could not tell the degree just by looking at the graph. In fact, one might guess that this is the graph of a cubic polynomial. It is useful to note that the range of every odd-degree polynomial function is all of $\mathbb{R}$, since for large positive and large negative values of x, the function "goes to infinity" in opposite directions.

A value of x that causes a function to be zero is called a **zero** of the function. The zeros are the x-intercepts of the graph. Not all functions have real zeros. For example, $f(x) = x^2 + 1$ is always positive. If a polynomial can be factored, however, the zeros (if any) can be determined easily. For example, consider

$$f(x) = x^3 - 3x^2 - 4x + 12$$

$$= x^2(x - 3) - 4(x - 3)$$

$$= (x - 3)(x^2 - 4)$$

$$= (x - 3)(x + 2)(x - 2)$$

The only zeros are 3, 2, and -2, since if any of the factors is zero, the product is zero, and no other value of x can cause this product to be zero. If a polynomial cannot be factored, or is hard to factor, a graphing calculator or computer algebra system can be used to approximate the zeros. (We will give a method using calculus for approximating zeros in Chapter 4.)

Finding a Polynomial Function Given Its Zeros and Another Value

The next two examples illustrate how the degree of a polynomial function, its zeros, and one additional point uniquely determine the function.

EXAMPLE 1.23   The zeros of a cubic polynomial function f are -1, 2, and 4, and $f(0) = 3$. Find $f(x)$.

Solution   For $f(x)$ to be 0 when $x = -1$, 2, or 4, the polynomial must contain the factors $(x + 1)$, $(x - 2)$, and $(x - 4)$. Since f is of degree 3, there can be no other nonconstant factors. Thus,

$$f(x) = a(x + 1)(x - 2)(x - 4)$$

where a is the unknown leading coefficient. We find a using the fact that $f(0) = 3$:

$$3 = a(1)(-2)(-4)$$

$$3 = 8a$$

$$a = \tfrac{3}{8}$$

So

$$f(x) = \tfrac{3}{8}(x + 1)(x - 2)(x - 4)$$

We show the graph of f in Figure 1.54.   ■

EXAMPLE 1.24   Find the formula for the function whose graph is shown in Figure 1.55 if it is known to be a polynomial function of lowest possible degree.

Solution   Since there are two local maximum points and one local minimum point, the lowest degree possible is 4. Reading from the graph, we get the zeros -4, -2, 1, and 5. So, as in Example 1.23, we have

$$f(x) = a(x + 4)(x + 2)(x - 1)(x - 5)$$

Since (2, 3) is on the graph, it follows that $f(2) = 3$. So, on substituting $x = 2$ we get

$$3 = a(2 + 4)(2 + 2)(2 - 1)(2 - 5)$$

$$3 = a(6)(4)(1)(-3)$$

$$a = -\frac{1}{24}$$

If $x = r$ is a zero of a polynomial function, then $x - r$ is a factor, and conversely.

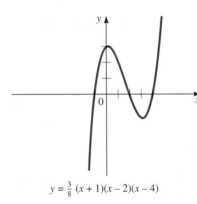

$y = \tfrac{3}{8}(x + 1)(x - 2)(x - 4)$

FIGURE 1.54

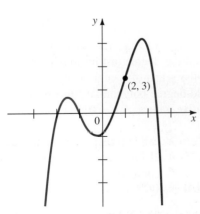

FIGURE 1.55

Thus,

$$f(x) = -\frac{1}{24}(x+4)(x+2)(x-1)(x-5)$$

We could have anticipated that the leading coefficient would be negative, since the graph is negative both for large positive and large negative values of x. The fact that $(2, 3)$ is on the graph forces a to be negative. ∎

Dominant Term of a Polynomial

A technician inspects parabolic mirrors used in military optical systems.

The highest-degree term of a polynomial is the *dominant term*. For example, it has been found that the cost of building a large telescope increases approximately with the cube of the diameter of the main mirror. The cost may go up with the square of the number of instrument ducts and linearly with the number of staff. However, the diameter of the mirror determines the major cost of a large telescope. (Recent ideas of new ways of building telescopes have lowered the degree of the cost function below a cubic.) The dominant term determines the behavior of a polynomial function when x is large in absolute value. For example, consider the cubic function

$$f(x) = 2x^3 + 10x^2 - 5x + 4$$

We can demonstrate why the highest-degree term dominates by the following trick. By factoring out x^3, we can write $f(x)$ in the form

$$f(x) = x^3\left(2 + \frac{10}{x} - \frac{5}{x^2} + \frac{4}{x^3}\right) \qquad (x \neq 0)$$

If x is large, either positively or negatively, each of the terms

$$\frac{10}{x}, \qquad -\frac{5}{x^2}, \qquad \text{and} \qquad \frac{4}{x^3}$$

is close to 0. Each can be made arbitrarily small by taking $|x|$ sufficiently large. Thus, $f(x)$ is approximately equal to $2x^3$, which is indicated by writing

$$f(x) \approx 2x^3 \qquad \text{for } |x| \text{ large}$$

More generally, if

$$f(x) = a_n x^n + a_{n-1} x^{n-1} + \cdots + a_0 \qquad (a_n \neq 0)$$

then we can factor out x^n to get

$$f(x) = x^n\left(a_n + \frac{a_{n-1}}{x} + \cdots + \frac{a_0}{x^n}\right) \qquad (x \neq 0)$$

Each term inside the parentheses except the first can be made as close to 0 as we please by taking $|x|$ sufficiently large. Thus,

$$f(x) \approx a_n x^n \qquad \text{for } |x| \text{ large}$$

So the highest-degree term, $a_n x^n$, dominates all other terms.

ESTIMATING ROOTS OF EQUATIONS USING A COMPUTER ALGEBRA SYSTEM

Finding the roots of an equation is typically a very hard problem, unless the function can be easily manipulated algebraically. There are techniques for approximating roots (see Section 4.5). Here we will use a computer algebra system to estimate roots of an equation.

CAS 8

Estimate the roots of the equation $\frac{2}{3} - \frac{x}{3} - \frac{5}{3}x^2 + x^3 = 0$.

Maple:[*]

Eq:=2/3–x/3–5/3*x^2+x^3;
solve(Eq=0,x);

Output:

$$\frac{2}{3}, \frac{1}{2} + \frac{1}{2}\sqrt{5}, \frac{1}{2} - \frac{1}{2}\sqrt{5}$$

fsolve(Eq=0,x);

Output:

–.6180339877, .6666666667, 1.618033989

Mathematica:

Eq = 2/3–x/3–5/3*x^2+x^3

[defines a name for the equation]

solve(Eq == 0,x);

Output:

$$\left\{ \left\{ x \to \frac{2}{3} \right\}, \left\{ x \to \frac{1 + \text{Sqrt}[5]}{2} \right\}, \left\{ x \to \frac{1 - \text{Sqrt}[5]}{2} \right\} \right\}$$

N[%]

[numerical roots of previous output]
Output:
{{x → 0.666667}, {x → 1.61803}, {x → −0.618034}}

DERIVE:

(At the □ symbol, go to the next step.)
a (author) □ 2/3−x/3−5/3x^2+x^3 □ 1
(solve)
Output:

$$\frac{2}{3}$$

$$\frac{1}{2} - \frac{\sqrt{5}}{2}$$

$$\frac{\sqrt{5}}{2} + \frac{1}{2}$$

x (approximate) □
[choose expression]

Output: 0.666666

x (approximate) □
[choose expression]

Output: −0.618033

x (approximate) □
[choose expression]

Output: 1.61803

*Note that in Maple `fsolve` gives numerical answers instead of the radicals that result from `solve`. In Mathematica, N[%] simplifies the previous output. In general, % refers to the previous output in Mathematica. In Maple, the double quote "refers to the previous output.

Exercise Set 1.4

1. For a certain linear function f, it is known that $f(-1) = 3$ and $f(7) = -5$. Find $f(x)$. What is $f(10)$?

2. The graph of a function g is a straight line passing through the points $(2, -1)$ and $(5, 3)$. Find a formula for $g(x)$.

3. The daily car rental rate for agency A is $21, plus 21¢ per mile. For agency B the rate is $32, plus 18¢ per mile. A driver plans to rent from one of these agencies and take a 320-mile round trip. From which agency should the driver rent to get the better rate? What is the minimum number of miles beyond which agency B's cost is less than that of agency A?

4. In depreciating equipment it is reasonable to assume that certain items will not depreciate to 0 but will have some residual value, at least as junk. Suppose a car originally valued at $15,000 depreciates linearly to a residual value of $200 in 20 years. Find its value $V(t)$, t years after purchase, where $0 \le t \le 20$.

5. (a) Find a linear function f such that $(f \circ f)(x) = 9x + 16$.
 (b) For which linear functions $cx + d$ is there a function f such that $(f \circ f)(x) = cx + d$? Under what conditions is f unique?

6. One of the functions f or g in the accompanying table is linear and one is quadratic, of the form at^2. Determine which is which and find formulas for $f(t)$ and $g(t)$.

t	$f(t)$	$g(t)$
-2	2	2
2	2	14
6	18	26
10	50	38

7. Draw the graph of each of the following, using your knowledge of the basic quadratic function and the transformations studied in Section 1.2.
 (a) $f(x) = \frac{1}{4}x^2$ (b) $f(x) = -\frac{1}{2}x^2$
 (c) $f(x) = 2x^2 - 3$ (d) $f(x) = (x - 2)^2$
 (e) $f(x) = (x + 1)^2$
 (f) $f(x) = 2(x + 1)^2 - 3$

8. By completing the square, write the equation $y = 2x^2 - 4x + 5$ in the form $y = a(x - h)^2 + k$ and draw its graph.

9. Repeat Exercise 8 with the equation

$$y = 1 - 2x - \frac{x^2}{2}$$

10. For a small manufacturing firm the unit cost $C(x)$ in dollars of producing x items per day is given by

$$C(x) = x^2 - 120x + 4000$$

How many items should be produced per day to minimize the unit cost? What is the minimum unit cost?

11. A company that produces computer terminals analyzes production and finds that the profit function $P(x)$, in

dollars, for selling x terminals per month is given by

$$P(x) = 160x - 0.1x^2 - 20,000$$

How many terminals should be sold per month to produce the maximum profit? What is the maximum profit?

12. In a contest, a baseball is thrown upward from the ground with an initial velocity of 57.6 meters per second. If air resistance is neglected, its distance $s(t)$, in meters, above the ground after t seconds is given by $s(t) = -4.8t^2 + 57.6t$. Find the maximum height the baseball will rise.

13. A restaurant has a fixed price of $18 for a complete dinner. The average number of customers per evening is 200. The owner estimates that for each 50¢ increase in the price of the dinner, there will be four fewer customers per evening on average. What price for the dinner will produce the maximum revenue?

14. From each of the data sets given, sketch the graph of f and determine a possible formula for $f(x)$.

(a) x	$f(x)$	(b) x	$f(x)$
0	-26	-3	7
1	-6	-2	5
2	6	-1	3
3	10	0	1
4	6	1	-1
5	-6	2	-3
6	-26	3	-5

15. (a) Is the composition of two linear functions a linear function? Prove your answer.
 (b) Is the composition of two quadratic functions quadratic? Prove your answer.
 (c) What is the degree of $p \circ q$ if p is a polynomial function of degree m and q is a polynomial function of degree n?

16. Let $f(x) = 100x^2$ and $g(x) = 0.1x^3$. For what values of x is $f(x) \ge g(x)$, and for what values is $g(x) \ge f(x)$?

 17. Consider the family of functions

$$f_n(x) = x^n$$

for n a positive integer. On one set of axes draw the graph of f_n for $n = 2$, 4, and 6. On another set of axes draw f_n for $n = 1$, 3, and 5. Discuss your results. How can you generalize them? For a given positive integer k, compare the sizes of $f_k(x)$ and $f_{k+1}(x)$ on each of the intervals $0 < x < 1$ and $x > 1$.

18. The zeros of a certain cubic function g are 1, 2, and -1, and $g(0) = -4$. Find $g(x)$, and draw its graph.

19. A cubic function f has zeros -1 and 2, with 2 a double zero, indicating that one of the factors of $f(x)$ is $(x-2)^2$. If $f(1) = 6$, find $f(x)$ and draw its graph.

20. The function h whose graph is shown is a polynomial function of degree 4. Find $h(x)$.

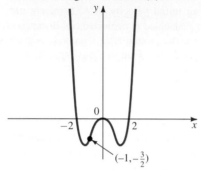

21. The function whose graph is shown is a polynomial function of degree n. What is the lowest possible value of n? Give the intervals over which the function is (a) increasing, (b) decreasing, and the intervals over which the graph is (c) concave up, and (d) concave down.

22. From the graph of the function f shown, draw the graphs of $y = |f(x)|$ and $y = f(|x|)$. Compare the graphs of $y = |f(x)|$ and $y = f(|x|)$ with the graph of f.

$y = f(x)$

23. Draw the graph of each of the following.

(a) $y = |2x - 3|$ (b) $y = |x^2 + x - 2|$
(c) $y = |2x - x^3|$

24. Let $f(x) = x^2 - \frac{3}{2}x$ and $g(x) = x^2 - 2$. Find the domains of $\sqrt{f(x)}$ and $\sqrt{g(x)}$, graphically and algebraically.

25. If $f(x) = ax^2 + bx + c$, what do you know about the values of a, b, and c if
(a) $(1, 1)$ is on the graph of f?
(b) $(1, 1)$ is the vertex?
(c) the y-intercept is 6?
(d) conditions (a), (b), and (c) are all satisfied?
What are the domain and range of the resulting function?

26. The rate R at which a population in a confined space increases is proportional to the product of the current population P and the difference between the maximum population the environment can sustain and the current population.
(a) Write R as a function of P. (Denote the proportionality constant by k.)
(b) For a fixed $k > 0$ sketch the graph of the function in part (a).

27. Cubics. In this exercise we show how the graph of every cubic function is obtainable from a basic cubic curve of the form $y = x^3 + kx$, through a series of transformations. We begin with the general cubic function

$$f(x) = ax^3 + bx^2 + cx + d \qquad (a \neq 0)$$

(a) Rewrite $f(x)$ as $af_1(x)$, where

$$f_1(x) = x^3 + b_1x^2 + c_1x + d_1$$

and specify b_1, c_1, and d_1 in terms of a, b, and c. How is the graph of f related to the graph of f_1?

(b) Let $f_2(x) = f_1\left(x - \dfrac{b_1}{3}\right)$. Show that

$$f_2(x) = x^3 + c_2x + d_2$$

and specify c_2 and d_2 in terms of b_1, c_1, and d_1. How is the graph of f_1 related to the graph of f_2?

(c) Let $f_3(x) = f_2(x) - d_2$. Then

$$f_3(x) = x^3 + c_2x$$

How is the graph of f_2 related to the graph of f_3?

(d) Let $k = c_2$, and write

$$g(x) = x^3 + kx$$

By collecting the results of parts (a), (b), and (c), explain how the graph of f can be obtained from the graph of g.

(e) Use a graphing calculator or CAS to graph g for various values of k. Try to describe the possible shapes of the graph.

In Exercises 28–30, obtain the graph from a graphing calculator or CAS and estimate each of the following: (a) local maximum and minimum points, (b) intervals where the function is increasing and those where it is decreasing, and (c) intervals where the graph is concave up and those where it is concave down.

28. $y = x^4 - 4x^3 + 5$

29. $y = \dfrac{x^5}{5} - \dfrac{10x^3}{3} + 9x$

30. $y = \dfrac{x^4}{2} - \dfrac{x^3}{3} - 4x^2 + 4x + 7$

1.5 RATIONAL FUNCTIONS AND RATIONAL POWER FUNCTIONS

A **rational function** is a function that can be expressed in the form

$$f(x) = \frac{p(x)}{q(x)}$$

where p and q are polynomial functions. The domain of f is the set of all real numbers x for which the denominator $q(x) \neq 0$. Here are some examples:

$$f(x) = \frac{x+1}{x-1} \qquad g(x) = \frac{x^2 - 2x + 3}{3x^3 + 4x - 1} \qquad h(x) = \frac{1}{x}$$

A polynomial function p is also a rational function, since we can take the denominator $q(x)$ to be 1 for all x; that is, $p(x) = p(x)/1$.

If $p(x)$ and $q(x)$ have a nonconstant factor in common, then we divide by that common factor, ruling out of the domain any value of x causing this factor to be 0. For example, let

$$f(x) = \frac{x^2 - 4}{x - 2}$$

The domain of f consists of all real numbers except $x = 2$. For all such x, we have

$$f(x) = \frac{(x-2)(x+2)}{x-2} = x + 2 \qquad (x \neq 2)$$

The graph of f, shown in Figure 1.56, is the line $y = x + 2$ with the single point $(2, 4)$ missing (indicated by the open circle). *Throughout the remainder of this section, unless otherwise specified, we will assume that all common factors have been divided out, so that $p(x)/q(x)$ is in lowest terms (in which case p and q are said to be **relatively prime**).*

The zeros of a rational function are the zeros of its numerator, $p(x)$, since a quotient is zero if and only if the numerator is 0 (when, as we are assuming, the denominator is nonzero). For example, the zeros of

$$f(x) = \frac{x^2 - 1}{x^2 - 4}$$

are 1 and -1.

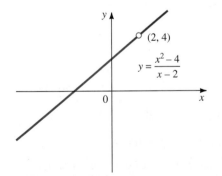

FIGURE 1.56

Vertical Asymptotes

The zeros of the denominator of a rational function are not in its domain, but investigating the behavior of f as x approaches each zero of the denominator is particularly useful in drawing its graph. Consider, for example, the function

(x > 1)		(x < 1)	
x	$f(x)$	x	$f(x)$
1.1	21	0.9	−19
1.01	201	0.99	−199
1.001	2001	0.999	−1999
1.0001	20001	0.9999	−19999

FIGURE 1.57

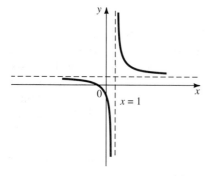

Graph of $f(x) = (x^2 - 1)/(x - 1)^2$. Same as $f(x) = (x + 1)/(x - 1)$. Line $x = 1$ is a vertical asymptote.

FIGURE 1.58

Graph of $f(x) = (x^2 - 1)/[(x + 3)(x - 1)]$. Same as $f(x) = (x + 1)/(x + 3)$, minus the point $(1, \frac{1}{2})$. Line $x = -3$ is a vertical asymptote, but $x = 1$ is not.

FIGURE 1.59

f defined by

$$f(x) = \frac{x + 1}{x - 1}$$

In the margin we show two tables, one for values of x greater than 1 but close to 1 and one for x less than 1 but close to 1. These tables provide strong evidence that when x approaches 1 through values greater than 1 (when x approaches 1 from the right), $f(x)$ gets arbitrarily large, and when x approaches 1 through values less than 1 (when x approaches 1 from the left), $f(x)$ gets arbitrarily large in absolute value but is negative. Symbolically, we can indicate this behavior as follows:

$$\text{As } x \to 1^+, \ f(x) \to \infty.$$

$$\text{As } x \to 1^-, \ f(x) \to -\infty.$$

Figure 1.57 illustrates how the graph of f looks in the vicinity of $x = 1$. The line $x = 1$ is called a **vertical asymptote** to the graph.

As this example illustrates, if $q(a) = 0$, then the line $x = a$ is a vertical asymptote to the graph of $f(x) = p(x)/q(x)$ when p and q have no common factor.

REMARK ———————————————————

■ If p and q have a factor $(x - a)$ in common, then $x = a$ *may* be a vertical asymptote. For example, if

$$f(x) = \frac{x^2 - 1}{(x - 1)^2}$$

we can rewrite $f(x)$ in the form

$$f(x) = \frac{(x + 1)(x - 1)}{(x - 1)(x - 1)} = \frac{x + 1}{x - 1} \qquad \text{if } x \neq 1$$

and conclude, as above, that $x = 1$ is a vertical asymptote (see Figure 1.58). However, if

$$f(x) = \frac{x^2 - 1}{x^2 + 2x - 3}$$

then, by factoring, we get

$$f(x) = \frac{(x + 1)(x - 1)}{(x + 3)(x - 1)} = \frac{x + 1}{x + 3} \qquad \text{if } x \neq 1$$

In this case, then, $x = 1$ is not a vertical asymptote even though the denominator of the original fraction is 0 when $x = 1$. In fact, as x approaches 1 from either side, $f(x) \to \frac{1}{2}$ (but note that the point $(1, \frac{1}{2})$ is not on the graph; see Figure 1.59). The point to remember is that when checking for vertical asymptotes, be sure to divide out any common factors first.

————————————————————————————

EXAMPLE 1.25   Find all vertical asymptotes to the graph of

$$f(x) = \frac{3x + 2}{x^2 - x - 2}$$

Solution   We factor the denominator to obtain

$$f(x) = \frac{3x + 2}{(x - 2)(x + 1)}$$

There is no common factor between numerator and denominator, so we conclude from the preceding discussion that the lines $x = 2$ and $x = -1$ are vertical asymptotes. By letting x approach 2 and -1, respectively, from the right and left, and examining the signs of all of the factors in each case, we can determine the following. (You should verify these results.)

$$\text{As } x \to 2^+, \ f(x) \to \infty$$

$$\text{As } x \to 2^-, \ f(x) \to -\infty$$

$$\text{As } x \to -1^+, \ f(x) \to \infty$$

$$\text{As } x \to -1^-, \ f(x) \to -\infty$$

We will show the graph of this function in Example 1.27, after we have discussed horizontal asymptotes.   ■

Horizontal Asymptotes

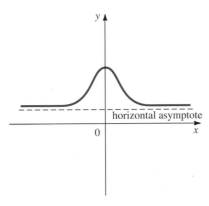

FIGURE 1.60

A line $y = b$ is a **horizontal asymptote** to the graph of a function f if $f(x) \to b$ as x becomes arbitrarily large, either positively or negatively, that is, as $x \to \infty$ or as $x \to -\infty$. (See Figure 1.60.) To determine if there is a horizontal asymptote to the graph of a rational function $f(x) = p(x)/q(x)$, we can use the fact that for $|x|$ large, both $p(x)$ and $q(x)$ are approximated by their dominant terms (their highest-degree terms), as we showed in Section 1.3. We illustrate this procedure in the next example.

EXAMPLE 1.26   For each of the following functions determine whether the graph has a horizontal asymptote, and if so, what the asymptote is.

(a) $f(x) = \dfrac{2x^2 - 3}{5x^3 + 4x - 7}$ (b) $g(x) = \dfrac{3x^2 - 1}{2x^2 + 5}$ (c) $h(x) = \dfrac{1 - x^3}{x^2 - 5}$

Solution

(a) For x large in absolute value, the numerator behaves approximately like its dominant term, $2x^2$, and the denominator like $5x^3$. So we can say that for $|x|$ large,

$$f(x) \approx \frac{2x^2}{5x^3} = \frac{2}{5x}$$

and since $2/5x \to 0$ as $|x| \to \infty$, we conclude that the line $y = 0$—that is, the x-axis—is a horizontal asymptote.

(b) Reasoning as in part (a), we see that for $|x|$ large,

$$g(x) \approx \frac{3x^2}{2x^2} = \frac{3}{2}$$

So $y = \frac{3}{2}$ is a horizontal asymptote.

(c) Again, for $|x|$ large,

$$h(x) \approx \frac{-x^3}{x^2} = -x$$

Since $-x$ does not approach a finite limiting value as x becomes arbitrarily large in absolute value, there is no horizontal asymptote.   ■

The preceding example illustrates the following general result, which can be proved using the idea of a dominant term.
Let

$$f(x) = \frac{a_m x^m + a_{m-1} x^{m-1} + \cdots + a_0}{b_n x^n + b_{n-1} x^{n-1} + \cdots + b_0} \qquad (m, n \geq 1)$$

where $a_m \neq 0$ and $b_n \neq 0$. Then

(i) If $m < n$, the line $y = 0$ is a horizontal asymptote.

(ii) If $m = n$, the line $y = a_m/b_n$ is a horizontal asymptote.

(iii) If $m > n$, there is no horizontal asymptote.

EXAMPLE 1.27   For the function in Example 1.25, find the horizontal asymptote. Also find the x- and y-intercepts and use this information, together with the results of Example 1.25, to draw the graph.

Solution   The function is

$$f(x) = \frac{3x + 2}{x^2 - x - 2}$$

Since the degree of the denominator is greater than the degree of the numerator (case (i) above), we conclude that $y = 0$ is a horizontal asymptote. The x-intercept is $-\frac{2}{3}$, found by setting the numerator $3x + 2$ equal to 0. The y-intercept is -1, found by setting $x = 0$.

In Example 1.25 we found that the lines $x = 2$ and $x = -1$ are vertical asymptotes, and we saw how the graph approaches these asymptotes from each side. When we combine all this information, we can draw the graph as shown in Figure 1.61 with reasonable confidence. We should note, however, that we need the tools of calculus to be certain of the shape. In particular, using calculus we will be able to verify that the middle section of the graph does change from concave up to concave down and to determine exactly where the concavity changes.   ■

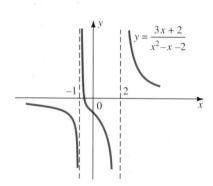

FIGURE 1.61

Rational Power Functions

We now consider functions of the form

$$f(x) = x^{m/n}$$

where m and n are positive integers. A function of this type is called a **rational power function**, since the exponent (power) is a rational number. For example, $f(x) = x^{2/3}$ is such a function. The meaning given to the expression $x^{m/n}$ is

$$x^{m/n} = (\sqrt[n]{x})^m$$

When m/n is in lowest terms, we also can write

$$x^{m/n} = \sqrt[n]{x^m}$$

If n is even, we must restrict x to be nonnegative when m/n is in lowest terms, since even roots ($\sqrt{N}$, $\sqrt[4]{N}, \ldots$) of negative numbers do not exist within the real number system.

FIGURE 1.62

FIGURE 1.63

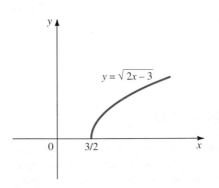

FIGURE 1.64

If the exponent is a negative rational number, say $-m/n$, then we define

$$x^{-m/n} = \frac{1}{x^{m/n}}$$

and in this case, we must also restrict x to be nonzero.

The simplest of the rational power functions is the **square root function**

$$f(x) = x^{1/2}$$

or equivalently, $f(x) = \sqrt{x}$. The domain, as well as the range, is the set of all nonnegative real numbers. The graph is shown in Figure 1.62. This curve looks as though it might be the upper half of a parabola with a horizontal axis. To show that this is the case, consider the function defined by $y = x^2$ with $x \geq 0$. This curve is the half-parabola shown in Figure 1.63. By the horizontal line test, this curve is the graph of a 1–1 function, so the function has an inverse. To find the inverse, we interchange x and y, then solve for y:

$$x = y^2$$
$$y = \sqrt{x}$$

(We used the positive square root only, since both x and y are nonnegative.) Thus, we have shown that the square root function $f(x) = \sqrt{x}$ is the inverse of the squaring function $y = x^2$ in which x is restricted to be nonnegative.

Using transformations of the types considered in Section 1.2, we can now analyze and graph all functions of the form

$$g(x) = \sqrt{ax + b}$$

The next example illustrates such a function.

EXAMPLE 1.28   Identify and draw the graph of the function

$$g(x) = \sqrt{2x - 3}$$

Solution   First we write $g(x)$ in the form

$$g(x) = \sqrt{2\left(x - \frac{3}{2}\right)}$$

We can view $g(x)$ as having come from $f(x) = \sqrt{x}$ by the horizontal compression $f(2x)$, followed by a horizontal translation 3/2 units to the right. That is,

$$g(x) = f\left(2\left(x - \frac{3}{2}\right)\right)$$

So the graph of g is the half-parabola shown in Figure 1.64.   ∎

REMARK
■ The horizontal compression $f(2x) = \sqrt{2x}$ also can be thought of as the vertical stretching $\sqrt{2}f(x) = \sqrt{2}\sqrt{x}$, since $\sqrt{2x} = \sqrt{2}\sqrt{x}$. The result is the same.

The square root function is one of the family of nth root functions

$$h_n(x) = x^{1/n}$$

You will be asked to discuss general characteristics of this family for n even and n odd in the exercises.

Another important root function is obtained from the composition of the square root function and the polynomial $a^2 - x^2$, with $a > 0$ and $-a \le x \le a$, giving

$$f(x) = \sqrt{a^2 - x^2}$$

If we set $y = f(x)$ and square both sides, we get $y^2 = a^2 - x^2$, or equivalently

$$x^2 + y^2 = a^2 \qquad (1.3)$$

Points (x, y) whose coordinates satisfy this equation are all a units from the origin, as can be seen by the Pythagorean Theorem (see Figure 1.65). Thus, the graph of Equation 1.3 is a circle of radius a, centered at the origin. (See also Appendix 1.)

All points on the graph of $y = \sqrt{a^2 - x^2}$ lie on the circle given by Equation 1.3, but since $\sqrt{a^2 - x^2}$ is nonnegative, its graph is the upper half of the circle. That is, the graph of the function $f(x) = \sqrt{a^2 - x^2}$ is the upper semi-circle shown in Figure 1.66.

The circle $x^2 + y^2 = a^2$ is not the graph of a function (it fails the vertical line test), but it can be thought of as combining the function $y = \sqrt{a^2 - x^2}$ and its negative, the function $y = -\sqrt{a^2 - x^2}$. These two functions represent the upper and lower semicircles, respectively (see Figure 1.67). We will have occasion in the next section to use the equation of a circle.

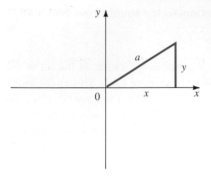

FIGURE 1.65
$x^2 + y^2 = a^2$

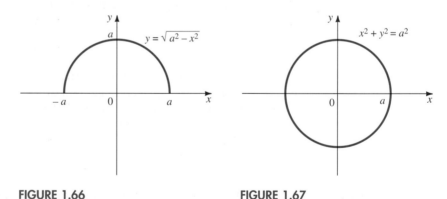

FIGURE 1.66

FIGURE 1.67

The next example illustrates a rational power function $y = x^{m/n}$ in which m is even and n is odd.

EXAMPLE 1.29   Discuss and sketch the graph of

$$y = x^{2/3}$$

Solution   Because cube roots of all real numbers exist, the domain is all of $\mathbb{R}$. Writing $x^{2/3} = \sqrt[3]{x^2}$, we see that replacing x with $-x$ leaves y unchanged, so that the graph is symmetric with respect to the y-axis. The range is contained in the set of all nonnegative real numbers, since $\sqrt[3]{x^2} \ge 0$ for all x. In fact,

x	y
1	1
2	1.59
3	2.08
4	2.52
5	2.92

for $k \geq 0$ we can take $x = k^{3/2}$ and get $(k^{3/2})^{2/3} = k$, which shows that the range is the set of all nonnegative real numbers. With the aid of the points shown in the table, and making use of symmetry, we can draw the graph in Figure 1.68.

After we have introduced some calculus concepts, we will be able to confirm that the graph is tangent to the y-axis at the origin (it is said to form a **cusp** there) and that the graph is always concave down. ■

As our final example in this section, we consider the composition of the square root function and a rational function.

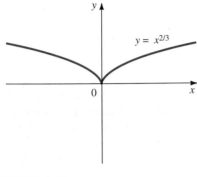

FIGURE 1.68

EXAMPLE 1.30   Discuss and sketch the graph of

$$y = \sqrt{\frac{x-1}{x+2}}$$

Solution   The domain consists of all real numbers x for which

$$\frac{x-1}{x+2} \geq 0$$

FIGURE 1.69

One way to determine the x values satisfying this inequality is to use the sign graph shown in Figure 1.69, where, after marking the zeros 1 and -2 of the numerator and denominator, respectively, we determined the sign of the fraction $(x-1)/(x+2)$ in each of the three intervals $(-\infty, -2)$, $(-2, 1)$, and $(1, \infty)$ into which these zeros divide the line. Using test values, such as $x = -3$, $x = 0$, and $x = 2$, shows that the signs are as given above the intervals in Figure 1.69. We conclude that $(x-1)/(x+2)$ is positive in $(-\infty, -2)$ and in $(1, \infty)$, and since the fraction is zero at $x = 1$ and undefined at $x = -2$, the domain of f is

$$(-\infty, -2) \cup [1, \infty)$$

Since $f(1) = 0$, the x-intercept of the graph is 1. There is no y-intercept because $x = 0$ is not in the domain.

To find asymptotes, we proceed as with rational functions, taking care to stay within the domain. The line $x = -2$ is a vertical asymptote, but to stay in the domain, x can approach -2 only from the left. More explicitly, we have,

$$\text{As } x \to -2^-, \ y \to \infty$$

Using the dominant term approximation for numerator and denominator, we see that for $|x|$ large,

$$y \approx \sqrt{\frac{x}{x}} = \sqrt{1} = 1$$

So $y = 1$ is a horizontal asymptote. In this case, the domain "extends to infinity" in both directions, so that the graph is asymptotic to the line $y = 1$ on the left and on the right. In general, you must check to see if points arbitrarily far out to the left and to the right are in the domain. Some graphs have a horizontal asymptote on one side only, and some have no horizontal asymptote at all.

Finally, note that $y \geq 0$ for all x in the domain. With the information we now have, we get the graph in Figure 1.70. ■

FIGURE 1.70

Algebraic and Transcendental Functions

All the functions we have considered so far fall into a broad category known as **algebraic functions**. These are functions that can be obtained from polynomial functions by any finite combination of the following operations: addition, subtraction, multiplication, division, raising to integral powers, and extracting integral roots. Some examples that we have not yet discussed are

$$f(x) = (5x^4 - 2x^2 + 3)^{7/3} \quad \text{and} \quad g(x) = \frac{\sqrt{x - \sqrt[3]{x^2 - 7}}}{(x^3 - 1)^4}$$

You might well ask whether there are any functions that are not algebraic. The answer is that there are *many* others. In fact, there are more nonalgebraic functions than there are algebraic functions (a "higher order of infinity"). These nonalgebraic functions are called **transcendental**, and you have already encountered many of them. The most familiar ones are the trigonometric functions that we will discuss in the next section. Others include the exponential, logarithmic, inverse trigonometric, and hyperbolic functions, all of which we will study in some detail in Chapter 7, when we have calculus to aid in our analysis.

Exercise Set 1.5

In Exercises 1 and 2, find all vertical and horizontal asymptotes.

1. (a) $y = \dfrac{2x + 1}{x - 2}$ (b) $y = \dfrac{x}{x^2 - 4}$

2. (a) $y = \dfrac{2x^2 - 3x + 1}{3x^2 + 5x - 2}$ (b) $y = \dfrac{x^3 + 2x - 4}{x^2 + x - 2}$

In Exercises 3–6, find x- and y-intercepts, vertical and horizontal asymptotes, and sketch the graph. Make use of symmetry where applicable.

3. $f(x) = \dfrac{1}{x - 3}$ 4. $f(x) = \dfrac{x - 1}{x + 2}$

5. $f(x) = \dfrac{x^2 - 4}{x^2 - 1}$ 6. $f(x) = \dfrac{x}{x^2 + 1}$

7. Sketch a graph that satisfies each of the following conditions:
   (a) passes through $(0, 0)$
   (b) is concave down for all $x \leq 0$
   (c) is concave up for $0 < x < 2$
   (d) has a vertical asymptote at $x = 2$
   (e) is concave down for all $x \geq 2$
   (f) has a horizontal asymptote at $y = 4$

 8. Consider the family of functions

$$g_n(x) = \frac{1}{x^n}$$

where $x \neq 0$ and n is a positive integer. On one set of axes draw the graph of g_n for $n = 2, 4$, and 6. On another set of axes draw g_n for $n = 1, 3$, and 5. Discuss your results. How can you generalize from them? For a given positive integer k, compare the sizes of $g_k(x)$ and $g_{k+1}(x)$ on each of the intervals $0 < x < 1$ and $x > 1$.

9. Show that if $f(x) = p(x)/q(x)$ is a rational function in lowest terms with the degree of p *exactly one greater* than the degree of q, then by long division $f(x)$ can be written in the form

$$f(x) = mx + b + \frac{r(x)}{q(x)}$$

where the degree of r is less than the degree of q. Explain why the graph of $y = f(x)$ approaches the line $y = mx + b$ as $x \to \infty$. (This line is called an **oblique asymptote**.)

In Exercises 10 and 11, use the idea of Exercise 9 to find the oblique asymptote.

10. (a) $f(x) = \dfrac{x^2 - 4}{x + 1}$ (b) $f(x) = \dfrac{x^2 - 3x + 4}{x - 2}$

11. (a) $f(x) = \dfrac{x^3 - x}{x^2 + 1}$ (b) $f(x) = \dfrac{x^3 - 2x^2 + 3x - 4}{x^2 + 2x - 1}$

12. Let $h_n(x) = x^{1/n}$ for n a positive integer. On one set of axes draw the graph of h_n for $n = 2, 4,$ and 6 and on another draw h_n for $n = 1, 3,$ and 5. Pay particular attention to the domain for n even and for n odd. How can you generalize your results? For a given positive integer k, compare $h_k(x)$ and $h_{k+1}(x)$ on each of the intervals $0 < x < 1$ and $x > 1$.

In Exercises 13–15, explain how the graph can be obtained from the graph of $y = \sqrt{x}$, and draw it.

13. (a) $y = \sqrt{x + 2}$ (b) $y = \sqrt{x - 1}$

14. (a) $y = \sqrt{1 - x}$ (b) $y = \sqrt{2x + 1}$

15. (a) $y = \sqrt{4 - 3x} - 2$ (b) $y = 1 + 3\sqrt{x}$

16. Identify and draw the graph of each of the following.
(a) $f(x) = \sqrt{4 - x^2}$ (b) $g(x) = -\sqrt{1 - x^2}$

17. Draw the graph of each of the following, using a transformation of the semicircle $y = \sqrt{a^2 - x^2}$ for the appropriate value of a. Identify the resulting curve.
(a) $y = \sqrt{9 - 4x^2}$ (b) $y = \frac{3}{2}\sqrt{4 - x^2}$

18. Analyze and draw the graph of each of the following. Can you identify the curve in each case? Can you express in words why such a seemingly minor change from part (a) to part (b) has such a profound result?
(a) $y = \sqrt{x^2 + 4}$ (b) $y = \sqrt{x^2 - 4}$

19. Analyze and draw the graph of each of the following functions.
(a) $f(x) = x^{3/2}$ (b) $g(x) = (x - 2)^{2/3}$

In Exercises 20–24, discuss the domain, intercepts, symmetry, and asymptotes, and draw the graph.

20. $y = \sqrt{\dfrac{x + 2}{x - 1}}$ **21.** $y = \sqrt{\dfrac{2 - x}{2 + x}}$

22. $y = \sqrt{\dfrac{x^2 - 1}{x^2 + 1}}$ **23.** $y = \dfrac{x}{\sqrt{4 - x^2}}$

24. $y = \dfrac{2\sqrt{x^2 - 4}}{x - 1}$

25. In Exercises 25–29, use a graphing calculator or a CAS to obtain the graph of each of the functions in Exercises 20–24.

1.6 THE TRIGONOMETRIC FUNCTIONS

In this section we give a brief review of the trigonometric functions, emphasizing how they can be defined with real-number domains, just as is the case with the other functions we have studied. This approach is in contrast to the one in which these functions are defined in terms of *angles*, although the two are closely related, as we shall see.

You may have wondered why two definitions of trigonometric functions are needed. Whether we choose to use the definitions in terms of real numbers or in terms of angles often depends on how we intend to apply the trigonometric functions. For example, the angle definition is natural when we are considering applications that involve finding sides or angles of triangles. Architects, surveyors, and navigators, for example, use trigonometric functions of angles. The real-number definition is more natural when we work with harmonic motion and with more complicated functions. Thus, physicists and engineers working with electronics, optics, and so on, tend to use trigonometric functions of real numbers. In Appendix 2 we give a more extensive review of both approaches.

Sine and Cosine Functions

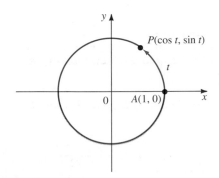

FIGURE 1.71

We begin with a circle of radius 1, called a **unit circle**, centered at the origin, as shown in Figure 1.71. Let A denote the point $(1, 0)$ and let t be any real number. If $t > 0$, measure t units of arc on the circle starting at A and going in a counterclockwise direction. If $t < 0$, measure $|t|$ units from A around the circle and in a clockwise direction. Let P denote the point on the circle arrived at in this way. In Figure 1.71 we have illustrated such a point P for $t > 0$. Since the circumference of the circle is $2\pi \approx 6.28$ ($C = 2\pi r$, and $r = 1$), if t exceeds 2π in absolute value, it is necessary to go around the circle more than once to arrive at P. The values of the **sine** and **cosine** functions at the real number t, abbreviated $\sin t$ and $\cos t$, are defined as follows:

$$\sin t = \text{the } y\text{-coordinate of } P$$

$$\cos t = \text{the } x\text{-coordinate of } P$$

Recall that in Section 1.5 we found the equation of a circle of radius a, centered at the origin, to be $x^2 + y^2 = a^2$. Since we have defined $\cos t$ and $\sin t$ as the x- and y-coordinates of a point P on the unit circle, we can conclude that for all real numbers t, $(\cos t)^2 + (\sin t)^2 = 1$. It is customary to write $\sin^2 t$ to mean $(\sin t)^2$, and similarly to write $\cos^2 t$ (we use the same type of notation for higher powers also). So we have

$$\sin^2 t + \cos^2 t = 1 \qquad (1.4)$$

Equation (1.4) is an example of a **trigonometric identity** since it is identically true for all t. Other identities are derived in Appendix 2.

Up to this point we have made no mention of angles. For example, $\sin 2$ means the sine function evaluated at the *number* 2. Theoretically, at least, we know how to find its value. We measure 2 units from A on the unit circle in a counterclockwise direction, arriving at the point P, as shown in Figure 1.72. Note that P is somewhere in the second quadrant since 2 is more than one-quarter but less than one-half of the entire circumference (approximately 6.28). The value of $\sin 2$ is the y-coordinate of P, as shown. Of course, this method is not a practical way of finding $\sin 2$, since we cannot measure distances around the circle very accurately. In practice you would find $\sin 2$ with high accuracy using a calculator by entering 2 and punching the $\boxed{\sin}$ key (or on some calculators doing this sequence in reverse). You might well ask how anyone is able to program a calculator to give values of the sine and cosine. At this stage we cannot answer this question, but in Chapter 10 you will see a way of evaluating $\sin t$ and $\cos t$ to any desired degree of accuracy, using what are known as infinite series. Calculators and computers do use infinite series (or approximations of them) in evaluating the trigonometric functions.

From Figure 1.71 we *can* obtain exact values of $\sin t$ and $\cos t$ when P falls on one of the coordinate axes. In particular, if $t = 0$, then P coincides with A and has coordinates $(1, 0)$. So we have

$$\sin 0 = 0 \qquad \text{and} \qquad \cos 0 = 1$$

One-quarter of the circumference is $\pi/2$, half is π, and three-quarters is $3\pi/2$. So we can construct Table 1.2, as follows.

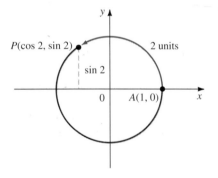

FIGURE 1.72

TABLE 1.2

t	P	$\sin t$	$\cos t$
0	$(1, 0)$	0	1
$\pi/2$	$(0, 1)$	1	0
π	$(-1, 0)$	0	-1
$3\pi/2$	$(0, -1)$	-1	0

By using certain facts from geometry, we can also find exact values of $\sin t$ and $\cos t$ for $t = \pi/6$, $t = \pi/4$, and $t = \pi/3$. We carry out the details in Appendix 2. The results are shown in Table 1.3 for reference.

TABLE 1.3

t	$\sin t$	$\cos t$
$\pi/6$	$1/2$	$\sqrt{3}/2$
$\pi/4$	$\sqrt{2}/2$	$\sqrt{2}/2$
$\pi/3$	$\sqrt{3}/2$	$1/2$

Now, we can get additional exact values, making use of the symmetries exhibited in Figure 1.73. There the point P_1 is t units from A, where $0 < t < \pi/2$. The points P_2, P_3, and P_4 are, respectively, $\pi - t$, $\pi + t$, and $2\pi - t$ units from A in a counterclockwise direction. As we have labeled P_1, we see that $\sin t = b$ and $\cos t = a$. Thus, from the relationship of the coordinates of P_2, P_3, and P_4 to those of P_1, we can conclude the following:

$$\sin(\pi - t) = \sin t \quad \sin(\pi + t) = -\sin t \quad \sin(2\pi - t) = -\sin t$$
$$\cos(\pi - t) = -\cos t \quad \cos(\pi + t) = -\cos t \quad \cos(2\pi - t) = \cos t \tag{1.5}$$

FIGURE 1.73

REMARK ——————————————————————————————

■ The requirement $0 < t < \pi/2$ was made only to simplify the picture in Figure 1.73. Equations 1.5 are true for all values of t.

To illustrate how we can use Equations 1.5, let $t = \pi/3$, for example. Then $\pi - t = \pi - \pi/3 = 2\pi/3$. So from the first and fourth equations in (1.5),

$$\sin \frac{2\pi}{3} = \sin \frac{\pi}{3}$$

$$\cos \frac{2\pi}{3} = -\cos \frac{\pi}{3}$$

and since we know $\sin \pi/3 = \sqrt{3}/2$ and $\cos \pi/3 = 1/2$, we now also know that $\sin 2\pi/3 = \sqrt{3}/2$ and $\cos 2\pi/3 = -1/2$. In the exercises you will be asked to extend Tables 1.2 and 1.3 of known exact values of the sine and cosine using each of Equations 1.5 with $t = \pi/6$, $\pi/4$, and $\pi/3$.

There is another bit of useful information we can get from Figure 1.73. To arrive at P_4, instead of going $2\pi - t$ units in the positive (counterclockwise) direction, we could go t units in the negative (clockwise) direction. That is, we can view the coordinates of P_4 as $(\cos(-t), \sin(-t))$. Comparing these with the coordinates of P_1, which are $(\cos t, \sin t)$, we conclude that

$$\sin(-t) = -\sin t$$

and

$$\cos(-t) = \cos t$$

That is, the **sine function is odd, and the cosine function is even**.

As we have seen, the exact values of $\sin t$ and $\cos t$ can be found for certain special values of t. But for most values of t there is no way to obtain exact values, and you will probably use a calculator to approximate them. (Before scientific calculators were invented, extensive printed tables were used to get values of the trigonometric functions. Such tables still exist, but they are seldom used now.) When using a calculator to get the sine or cosine of a number t, it is essential that the calculator be set in *radian mode* (as opposed to degree mode). To explain why, we will have to say a few words about angle measurement.

In our definition of $\sin t$ and $\cos t$ using Figure 1.71, think of P as initially coinciding with A and then moving around the circle until it comes to its final position t units from A. As it does so, the line segment OP turns through an angle at the center of the circle. The measure of this angle, in radians, is defined to be the number t, that is, the directed length of the arc from A to P. This definition is illustrated in Figure 1.74. For each 1 unit of arc on the circle, there is an angle of 1 radian at the center. In one complete revolution there are 2π units of arc. So there are 2π radians in a complete revolution. Since there also are $360°$ in one revolution, we see that 2π radians $= 360$ degrees, or

$$\pi \text{ radians} = 180°$$

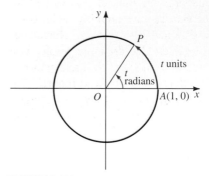

FIGURE 1.74

Dividing by π, we find that 1 radian $= (180/\pi)° \approx 57.3°$.

Now suppose again that you want to find the sine of the real number 2 on a calculator. The calculator is programmed to give trigonometric functions of *angles*, either in degrees, radians, or grads (100 grad $= 90°$). Since the sine of an angle of radian measure 2 is the same as the sine of the number 2, we find $\sin 2$ by setting the calculator in radian mode to get the answer:

$$\sin 2 \approx 0.9093$$

One of the most common mistakes students make in using a calculator with trigonometric functions is failing to use the right mode. For example, if your calculator is in degree mode and you ask for $\sin 2$, meaning the sine of 2 radians, the calculator interprets what you want as $\sin 2°$ and gives the answer as 0.034899, which is incorrect.

REMARK ————————————————————————————

■ To convert from degree measure to radian measure, or vice versa, you can use the relationships $1° = \pi/180$ radians and 1 radian $= (180/\pi)°$. So, for example,

$$2 \text{ radians} = 2\left(\frac{180}{\pi}\right)° = \left(\frac{360}{\pi}\right)°$$

and you could use this result to get $\sin 2$ even with the calculator in degree mode, but there is no advantage to doing so.

————————————————————————————————————

Let us now return to the definitions of $\sin t$ and $\cos t$ for an arbitrary real number t, using Figure 1.71. Since the circumference of the unit circle is 2π, for any value of t corresponding to the point P, increasing t by 2π units brings us back to the same point. Thus,

$$\sin(t + 2\pi) = \sin t$$

$$\cos(t + 2\pi) = \cos t$$

Any function f for which $f(t + k) = f(t)$ for some constant k and all t in the domain is said to be **periodic**, with **period** k. The smallest positive value of the period k is called the **fundamental period**. So both the sine and cosine functions are periodic with (fundamental) period 2π.

To obtain the graph of $f(t) = \sin t$, visualize how the y-coordinate of P in Figure 1.71 varies as P moves around the circle. You might imagine yourself at the point P on a ferris wheel of radius 1 unit, and visualize stationary horizontal and vertical axes through the center, in the plane of the ferris wheel. As the wheel rotates, your vertical displacement from the horizontal axis varies like a sine function, and your horizontal distance from the vertical axis varies like a cosine function. Visualizing these displacements, you should be able to get the general shape of the graphs of $f(t) = \sin t$ and $g(t) = \cos t$. Remember that t is a measure of the arc length along the circle. To draw the graphs with greater accuracy, you can use the values in Tables 1.2 and 1.3, and you can get additional points using a calculator. The graphs are shown in Figure 1.75. Notice that the two graphs have the same basic shape but are shifted horizontally from each other. In fact, if the cosine curve were shifted $\pi/2$ units to the right, it would coincide with the sine curve. That is,

$$\cos\left(t - \frac{\pi}{2}\right) = \sin t$$

Similarly,

$$\sin\left(t + \frac{\pi}{2}\right) = \cos t$$

as can be seen by shifting the sine curve $\pi/2$ units to the left.

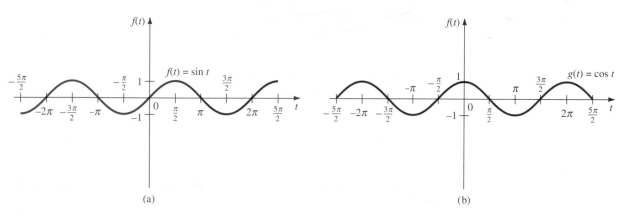

(a)   (b)

FIGURE 1.75

Since the radius of the unit circle is 1, the maximum height attained by either $\sin t$ or $\cos t$ is 1 and the minimum is -1. The maximum displacement from the t-axis is called the **amplitude**. So for both $f(t) = \sin t$ and $g(t) = \cos t$, the amplitude is 1. (See Figure 1.76.)

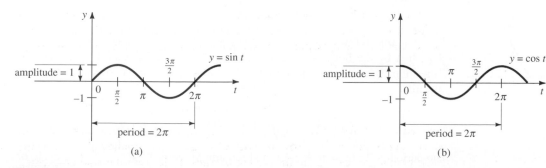

(a)   (b)

FIGURE 1.76

REMARK

■ Now that we have the graphs of the sine and cosine, there is no need to continue using the letter t to designate the independent variable. We can use the familiar x for this purpose (even though in the definition using the unit circle, x had a different meaning). In the future, then, we will often write $y = \sin x$ or $y = \cos x$.

The next two examples show the effects of various transformations (studied in Section 1.2) on the basic sine and cosine curves.

EXAMPLE 1.31   Discuss and sketch the graph of

$$y = 2 \sin 3x$$

Solution   If we let $f(x) = \sin x$, then the function we want is $2f(3x)$. We know from Section 1.2 that the factor 2 causes a vertical stretching, multiplying each y-coordinate of the basic function by 2. So the amplitude of the new function is 2. The factor 3 causes a horizontal compression by dividing each x-coordinate of the basic curve by 3. For the basic function $f(x) = \sin x$, as x goes through one period from 0 to 2π, the graph is said to complete one cycle. For the new function, one cycle is completed when x goes from 0 to $2\pi/3$. That is, the new period is $2\pi/3$. One cycle of the graph is shown in Figure 1.77.   ■

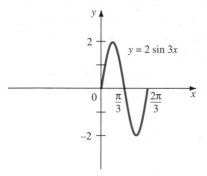

FIGURE 1.77

As Example 1.31 illustrates, for a function of the form

$$y = a \sin bx \qquad (a > 0, \ b > 0)$$

the amplitude is a and the period is $2\pi/b$. The same results hold true for the cosine function

$$y = a \cos bx \qquad (a > 0, \ b > 0)$$

EXAMPLE 1.32   Discuss and sketch the graph of

$$y = \cos(2x - 3) + 1$$

Solution   First we factor out the coefficient of x, getting

$$y = \cos 2 \left(x - \frac{3}{2} \right) + 1$$

This equation represents a transformation of the basic cosine function $g(x) = \cos x$ first by the horizontal compression $g(2x) = \cos 2x$, then by a horizontal translation $3/2$ units to the right, and finally by a vertical translation 1 unit upward:

$$g \left[2 \left(x - \frac{3}{2} \right) \right] + 1 = \cos 2 \left(x - \frac{3}{2} \right) + 1$$

The period is $2\pi/2 = \pi \approx 3.14$, and the amplitude is 1. We show one cycle of the graph in Figure 1.78.   ■

FIGURE 1.78

The sine and cosine functions are especially important in modeling periodic phenomena, such as ocean tides, brain waves, vibrations, and electric currents. A weighted spring set in motion vertically and moved horizontally (to take a multiflash exposure) exhibits a sinusoidal path. Often we are given data whose graph appears to be sinusoidal (i.e., in the shape of a sine curve) and we want to find the form of the sine (or cosine) function that seems to fit the graph. The next example illustrates how to work backward from the graph to the equation.

EXAMPLE 1.33   Find a function that seems to fit the graph in Figure 1.79.

Solution   There is no unique answer, since the curve could be the graph of either a shifted sine curve or a cosine curve. Let us attempt to find a sine function whose graph appears to match the given curve.

To determine the amplitude, think of a horizontal axis 3 units up from the x-axis. The maximum displacement from this horizontal axis is 2 units, since the point $(\frac{3}{2}, 1)$ is 2 units below this axis. So the amplitude is 2. The horizontal distance between the points $(\frac{1}{2}, 3)$ and $(\frac{3}{2}, 1)$ is 1, which is one-fourth of the period. So the period is 4. Before translation (and reflection), the equation would be of the form $y = a \sin bx$, with $a = 2$, and $2\pi/b = 4$. Thus, $b = \pi/2$, and therefore the equation would be

$$y = 2 \sin \frac{\pi}{2} x$$

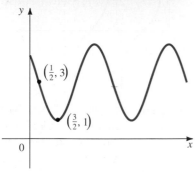

FIGURE 1.79

Next, we note that as we move to the right from $(\frac{1}{2}, 3)$, the curve goes downward, suggesting that our graph is reflected about the x-axis, giving

$$y = -2 \sin \frac{\pi}{2} x$$

Finally, the curve is shifted one-half unit to the right and 3 units upward, giving the result

$$y = -2 \sin \frac{\pi}{2} \left(x - \frac{1}{2} \right) + 3$$

You might wish to fit a cosine curve to the given graph. A correct answer (again, not unique) is

$$y = 2 \cos \frac{\pi}{2} \left(x + \frac{1}{2} \right) + 3 \qquad \blacksquare$$

The Tangent, Cotangent, Secant, and Cosecant Functions

The sine and cosine are the basic two trigonometric functions. The other four are the **tangent** (tan), **cotangent** (cot), **secant** (sec), and **cosecant** (csc) and are defined as follows:

$$\tan x = \frac{\sin x}{\cos x} \qquad\qquad \sec x = \frac{1}{\cos x}$$

$$\cot x = \frac{\cos x}{\sin x} \qquad\qquad \csc x = \frac{1}{\sin x}$$

For $\tan x$ and $\sec x$ the domain includes all real numbers x except those for which $\cos x = 0$, namely $x = \frac{\pi}{2} + n\pi$, $n = 0, \pm 1, \pm 2, \ldots$ For $\cot x$ and $\csc x$ the domain excludes x values for which $\sin x = 0$, namely $x = n\pi$, $n = 0, \pm 1, \pm 2, \ldots$ Each of the four functions has a vertical asymptote at each x value for which the denominator is 0. The graphs are shown in Figures 1.80–1.83.

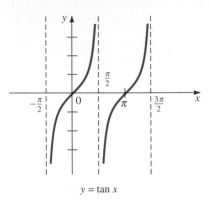

$y = \tan x$

FIGURE 1.80

$y = \cot x$

FIGURE 1.81

FIGURE 1.82

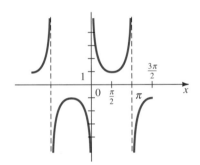

FIGURE 1.83

REMARK

■ On a calculator you will not find keys for $\cot x$, $\sec x$, or $\csc x$. To obtain these values you can get reciprocals of $\tan x$, $\cos x$, and $\sin x$, respectively. For example, if you want $\sec 3$, which is the same as $1/\cos 3$, you can use the keystroke sequence (if the calculator is in radian mode) 3 $\boxed{\cos}$ $\boxed{1/x}$. The answer displayed will be approximately 1.00137.

INVESTIGATING TRIGONOMETRIC FUNCTIONS USING A COMPUTER ALGEBRA SYSTEM

The names used by Maple, Mathematica, and DERIVE for the trigonometric functions are the standard names given in the previous section.

CAS 9

Graph, on the same set of axes, the curves

$$y = \cos x$$

$$y = \cos 2x$$

$$y = 2\cos 3x + 2$$

for $-10 \le x \le 10$.

Maple:

```
plot({cos(x),cos(2*x),
   2*cos(3*x)+2},
   x=-10..10,numpoints=500);
```

Mathematica:*

```
Plot[{Cos[x],Cos[2*x],
   2*Cos[3*x]+3},
   {x,-10,10}]
```

DERIVE:

(At the □ symbol, go to the next step.)
a (author) □ cos(x) □ p (plot window)
□ p (plot) □ a (algebra) □ a (author) □
cos(2x) □ p (plot window) □ p (plot) □
a (algebra) □ a (author) □ 2cos(3x)+3
□ p (plot window) □ p (plot)

FIGURE 1.6.1
$y = \cos x$, $y = \cos(2x)$,
$y = 2\cos(3x) + 2$

CAS 10

(a) Graph $y = \tan(x)$, $-\frac{\pi}{2} < x < \frac{\pi}{2}$. (Figure 1.6.2)

(b) Graph $y = \sec(x)$ and $y = \cos(x)$ for $x = -10$ to 10 on the same set of axes. (Figure 1.6.3)

Maple:

```
(a) plot(tan(x),x=-Pi/2..Pi/2,
y=-5..5);
(b) plot ({sec(x), cos(x)},
x=-10..10,y=-5..5);
```

Mathematica:

```
(a) Plot[Tan[x],
   {x,-Pi/2,Pi/2},
   PlotRange->{-5,5}]
(b) Plot[{Sec[x],Cos[x]},
   {x,-10,10},PlotRange->{-5,5}]
```

DERIVE:

(At the □ symbol, go to the next step.)
(a) a (author) □ tan(x) □ p (plot window) □ p (plot)
(b) a (author) □ sec(x) □ p (plot window) □ p (plot) □ a (algebra) □ a (author) □ cos(x) □ p (plot window) □ p (plot) □ z (zoom-out)

---

*Recall that functions in Mathematica begin with capital letters.

FIGURE 1.6.2

$y = \tan x, \ -\frac{\pi}{2} < x < \frac{\pi}{2}$

FIGURE 1.6.3

$y = \sec x, \ y = \cos x$

CAS 11

Solve the equation $x \sin x - x^2 = 0$.

Maple:

Eq := x*sin(x)–x^2;
solve(Eq=0,x);

Output:

0, 0

(Maple gives two solutions)
If we rewrite the equation as $x(\sin x - x) = 0$, we see one solution is at $x = 0$ and the other is where $\sin x = x$. A graph of $y = \sin x$ and $y = x$ on the same axes shows the second result is also 0. To see this graph enter

plot({x,sin(x)},x=–10..10,y=–4..4);

Mathematica:[†]

Solve[x*Sin[x]–x^2 == 0,x]

Output: **None**

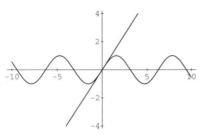

FIGURE 1.6.4

$y = \sin x, \ y = x$

DERIVE:

(At the □ symbol, go to the next step.)
a (author) □ xsin(x)−x^2 □ l (solve) □ [choose expression]

Output:

x=0
sin(x)–x=0

[†]Our version of Mathematica could not solve this equation. DERIVE was able to find $x = 0$ but also could not solve $x - \sin(x) = 0$.

Exercise Set 1.6

1. Construct a table giving values of $\sin x$, $\cos x$, $\tan x$, $\cot x$, $\sec x$, and $\csc x$ at each of the following values of x for which the function is defined: $x = 0$, $\pi/6$, $\pi/4$, $\pi/3$, $\pi/2$, $2\pi/3$, $3\pi/4$, $5\pi/6$, π, $7\pi/6$, $5\pi/4$, $4\pi/3$, $3\pi/2$, $5\pi/3$, $7\pi/4$, and $11\pi/6$. For a value of x for which a function is undefined, write DNE (for "does not exist").

In Exercises 2–13, give the amplitude and the period, and draw the graph.

2. $y = \cos 2x$

3. $y = 2\cos x$

4. $y = 2\sin 3x$

5. $y = 3\sin \frac{x}{2}$

6. $y = 1 - \cos x$

7. $y = 2 + \sin \pi x$

8. $y = 2\cos \pi x - 1$

9. $y = \frac{1}{2}\sin 2x + \frac{\pi}{2}$

10. $y = \sin(3x + \pi)$

11. $y = 3\cos(2x - \frac{\pi}{2})$

12. $y = 2 - \cos(x - 1)$

13. $y = -\frac{3}{2}\sin(2x + 1) + 2$

In Exercises 14–17, find a sine function or a cosine function whose graph matches the given curve.

14.

15.

16.

17.

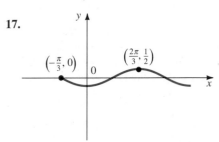

18. Fit a sine curve to the given data.

x	0	0.5	1	1.5	2	−0.5	−1	−1.5	−2
y	1	2.4	3	2.4	0	−0.4	−1	−0.4	0

19. Fit a cosine curve to the given data.

x	0	0.5	1	1.5	2	2.5	3	3.5	4
y	1.5	1.13	1	1.13	1.5	2	2.5	2.87	3

20. Let a, b, and c be positive constants. Describe what happens to the graph of $y = a\cos(bx + c)$ if
(a) a and b are fixed and c is doubled
(b) a and c are fixed and b is doubled
(c) b and c are fixed and a is doubled

21. Find the smallest positive value of x for which

$$f(x) = -2\sin\left(2x + \frac{\pi}{3}\right)$$

(a) has the value 0
(b) attains a maximum value
(c) attains a minimum value

22. Write each of the following as a composition of two functions f and g.
(a) $h(x) = 2\sin(3x + 2)$
(b) $h(x) = \sqrt{\cos x}$
(c) $h(x) = -5\tan^2 x$

Draw the graph of each function in Exercises 23–26.

23. $y = 2 \tan 3x$ **24.** $y = \frac{1}{2} \sec 2x$

25. $y = \cot \pi x + 1$ **26.** $y = -3 \csc(2x - \frac{\pi}{3})$

 27. Discuss the family of functions

$$y = a \tan(bx + c)$$

Where are the zeros? What are the asymptotes?

 28. Discuss the family of functions

$$y = a \sec(bx + c)$$

Where do the local minima and maxima occur? What are the asymptotes?

29. Draw the graph of $y = \sin x + \cos x$ by adding the ordinates (y-values) of $y_1 = \sin x$ and $y_2 = \cos x$ for selected values of x. How would you describe the resulting curve? Find its equation as a single sine function.

30. Repeat Exercise 29, with $y = \sqrt{3} \cos x - \sin x$ and $y_1 = \sqrt{3} \cos x$, $y_2 = -\sin x$.

 31. Use a graphing calculator or CAS to obtain the graph in each of the following. Discuss the main features of the graph in each case.

(a) $y = \dfrac{\sin x}{x}$ $(x \neq 0)$ (b) $y = \sin \dfrac{1}{x}$ $(x \neq 0)$

(c) $y = \sin^2 x$ (d) $y = |\sin x|$

(e) $y = \sin x^2$ (f) $y = x \sin \dfrac{1}{x}$ $(x \neq 0)$

 32. Use a graphing calculator or CAS to obtain graphs of

$$y = x \sin \frac{1}{x} \quad \text{and} \quad y = x^2 \sin \frac{1}{x}$$

By zooming in repeatedly on the curves near the origin, observe and discuss the differences in their behaviors near the origin.

33. (a) Find all points where $y = x \sin \dfrac{1}{x}$ and $y = x$ intersect.

(b) Find all points where $y = x \sin \dfrac{1}{x}$ and $y = -x$ intersect.

 34. Use a graphing calculator or CAS to graph $y = \sec x$ and $y = \cos x$ on the same axes, for $-2\pi \leq x \leq 2\pi$. Discuss the relationship between the two graphs.

 35. Repeat Exercise 34 with $y = \csc x$ and $y = \sin x$, for $-\dfrac{3\pi}{2} \leq x \leq \dfrac{5\pi}{2}$.

36. Use a graphing calculator or CAS to find a third-degree polynomial that "closely" approximates $y = \sin x$ on $[\pi, \pi]$.

Chapter 1 Review Exercises

1. Give the domain of each of the following functions.
(a) $f(x) = \sqrt{3x^2 - 8x - 35}$
(b) $g(x) = \sqrt{\dfrac{x^2 - 1}{x^2 - 2x - 3}}$

2. A circle is inscribed in an equilateral triangle, each side of which has length s. Express the area of the circle as a function of s.

3. Let $f(t) = t^2 - 3$ and $g(t) = \sqrt{1 - t}$. Find $(f \circ g)(t)$ and $(g \circ f)(t)$ and give the domain of each.

4. At temperature x the thermal conductivity of a wall is given by $f(x) = k(1 + ax)$, where k and a are constants. Write the rule for the thermal resistance $g \circ f$, given that $g(u) = T/(Au)$, where T is the thickness of the wall and A is its cross-sectional area.

5. Let

$$f(x) = \begin{cases} 1 & \text{if } x > 2 \\ \dfrac{1}{|x - 2|} & \text{if } 0 \leq x < 2 \\ -x & \text{if } x < 0 \end{cases}$$

Find each of the following, and use your results to aid in drawing the graph of f.
(a) $f(-3)$ (b) $f(0)$
(c) $f(1)$ (d) $f(1.8)$
(e) all vertical and horizontal asymptotes

6. A manufacturer of solar energy cells finds that the cost, in dollars, of producing x cells per month is $30x + 1500$, and he sets the selling price per unit at $120 - 0.1x$ dollars. Find the profit function $P(x)$ and draw its graph. How many units should be sold in order for $P(x)$ to be greatest? What is the maximum profit?

7. Let $g(t) = 1/(t - 1)$ and $h(t) = 1/(t + 2)$. Find $g \circ h$ and $h \circ g$, and give the domain of each.

8. Let $f(x) = x^2 - 2x - 3$, $g(x) = x^2 + 5x + 4$, and $h = \dfrac{f}{g}$. Find all vertical and horizontal asymptotes to the graph of $y = h(x)$.

9. Let $f(x) = 1 + x^2$ and $g(x) = \tan \dfrac{\pi x}{4}$. Find each of the following:

  (a) $(f - g)(3)$ (b) $\left(\dfrac{f}{g}\right)(-1)$

  (c) $(f \circ g)(x)$

10. Find two functions f and g so that $F = f \circ g$, where
  (a) $F(x) = 1 + \sqrt{(2x - 3)^3}$
  (b) $F(x) = \dfrac{3}{|x^2 - 4|}$
  (c) $F(x) = \sin^2\left(\dfrac{1}{2x - 1}\right)$

11. For the function f whose graph is given, draw the graph of each of the following.
  (a) $y = f(x) + 1$ (b) $y = f(x + 1)$
  (c) $y = 2f(x)$ (d) $y = f(-x)$
  (e) $y = f\left(\dfrac{x}{2}\right)$

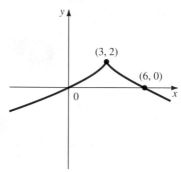

12. Draw the graphs of $f(x) = \sin 2x$ and $g(x) = \cos 3x$, and use your results to obtain the graph of $h(x) = \sin 2x + \cos 3x$. Give the period of each of the functions f, g, and h.

13. Show that f^{-1} exists and find it.
  (a) $f(x) = 1 - \dfrac{1}{\sqrt{x}}$ (b) $f(x) = \dfrac{x + 2}{x - 3}$

14. Discuss intercepts, symmetry, and asymptotes, and draw the graph of
$$y = \frac{4x}{1 - x^2}$$

15. Find the zeros of
$$f(x) = -\frac{1}{4}(x^4 - 2x^3 - 8x^2)$$

and draw its graph. Determine the intervals on which the function f is (a) increasing and (b) decreasing, and the intervals on which the graph is (c) concave up and (d) concave down.

16. Find the formula for the function f whose graph is shown if it is known that f is a polynomial function of degree 4 or less.

17. Discuss and draw the graph of each of the following:

  (a) $y = (2x - 1)^{2/3}$ (b) $y = \sqrt{\dfrac{1 + x}{1 - x}}$

18. Find the period and amplitude, and draw the graph.
  (a) $y = 1 - 2\sin(3x - \pi)$
  (b) $y = 3\cos \frac{\pi}{2}(x + 1) - 2$

19. Find a sine or cosine function whose graph matches the given graph.
  (a)

  (b)

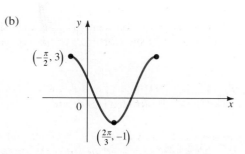

20. Prove the identity $\sec^2 t = 1 + \tan^2 t$, if $t \neq \frac{\pi}{2} + n\pi$, $n = 0, \pm 1, \pm 2, \ldots$ (*Hint:* Begin with the equation $x^2 + y^2 = 1$ of the unit circle.)

Chapter 1 Concept Quiz

1. Define each of the following. Try to do so in your own words before referring to the text.
   (a) a function from A to B
   (b) an even function
   (c) an odd function
   (d) the composite function $f \circ g$
   (e) a one-to-one function
   (f) the inverse of a function
   (g) a rational function
   (h) $\sin t$, where t is a real number
   (i) $\cos t$, where t is a real number
   (j) $\tan t$, where t is a real number

2. State the following in your own words:
   (a) the vertical line test
   (b) a necessary and sufficient condition for a function to have an inverse
   (c) a procedure for finding the inverse of a function
   (d) the horizontal line test
   (e) a way of determining vertical and horizontal asymptotes for a rational function

3. Describe the relationship between the graph of $y = f(x)$ and the graph of each of the following, where c is a positive constant:
   (a) $y = f(x + c)$ (b) $y = f(x) + c$
   (c) $y = f(cx)$ (d) $y = cf(x)$
   (e) $y = f(x) - c$ (f) $y = f(x - c)$
   (g) $y = -cf(x)$ (h) $y = f(-cx)$

4. Give the period and amplitude of each of the following, and explain how the graph is related to that of $y = \sin x$ or $y = \cos x$:
   (a) $y = 3 \sin 2x$ (b) $y = 2 \cos(\frac{\pi}{2}x)$
   (c) $y = -2 \sin(x + 1)$
   (d) $y = 1 - 3 \cos(2x - 3)$

5. Give the domain and range of $\sin x$, $\cos x$, $\tan x$, $\cot x$, $\sec x$, and $\csc x$.

6. Fill in the blanks:
   (a) The domain of $f \circ g$ is the largest subset of the domain of _____ such that _____ is in the domain of _____.
   (b) The sum of two odd functions is _____ and their product is _____ .
   (c) If f is a function from A to B and g is a function from B to A such that $(f \circ g)(x) = x$ for each x in _____ and $(g \circ f)(x) = x$ for each x in _____, then $g =$ _____.
   (d) The sine of a real number x is the same as the sine of x _____.
   (e) The graph of a polynomial function of degree n can have at most _____ maxima and minima.

7. Which of the following statements are true? Which are false?
   (a) The range of a function may be a single number.
   (b) The domain of the quotient function f/g is the intersection of the domains of f and g.
   (c) If f is 1–1, then $f^{-1}(x) = \dfrac{1}{f(x)}$.
   (d) If $f(x_1) < f(x_2)$ whenever $x_1 < x_2$, then f is an increasing function.
   (e) If $q(a) = 0$, the line $x = a$ is always a vertical asymptote for the rational function $p(x)/q(x)$.

APPLYING FUNCTIONS

The importance of this type of problem stems from the need to choose a function $f(x)$ which meets a particular requirement, in this case A_{max}. It differs from those problems in which we analyze functions derived from particular situations or scientific laws. This type of problem is a simple example of what's now known as the calculus of variations. After Queen Dido there was little progress in this area until the work of the Bernoullis nearly 900 years later.

1. *Queen Dido's Problem*   Legend tells us that a woman named Dido, from the ancient Phoenician city of Tyre, founded the African city-state of Carthage in 814 A.D. and became its queen. When Dido arrived as a refugee in North Africa she had very little, but bargained for some land by agreeing to pay a fixed price for all the property "lying within the bounds of the hide of an ox." Naturally she paid very little, but Queen Dido's ingenuity and knowledge of geometry made her purchase much better than it sounds. She cut the ox-hide into very narrow strips, sewed the ends together, and laid them out to obtain a much larger piece of property than those making the sale had in mind. Use her reasoning to solve the more general problem.

(a)   Given a fixed perimeter (the sewn together strips of ox-hide) determine whether she chose a circle, square, equilateral triangle, or a rectangle in which the length is twice the width. Express the area A of each of the four figures as a function of the perimeter P.

(b)   Suppose she were able to get 500 m of ox-hide strips. What would be the maximum area of her ox-hide territory?

(c)   If the intact ox-hide measured 2.5 m^2, by what factor had Queen Dido's ingenuity increased her intended real estate holdings? (This was probably the end of ox-hide measurements of real estate.) Express it as a percentage. Neglecting the length of material needed to fasten the strips, what was the average width of her original strips?

(d)   The legend goes that Dido's even better idea was to use the strips to form the boundary for a region along the coast (where no hide strips were needed). It was the land enclosed by this boundary that eventually became the city of Carthage. If she used her 500 m of strips to maximize her land this way, what would be the shape of the city limits of Carthage? How large would the original city be?

2.   *A baseball diamond is actually a square with sides 90 feet long.*
A (fast) player runs around the bases at a constant speed of 30 ft/sec, starting from home plate at time $t = 0$. Express his distance D from home plate as a function of time t.

3.   *The intensity of illumination (sound) E at a point P which is d units from a light (sound) source varies directly as the strength I of the source and inversely as the square of the distance from the source.*
(a)   Determine an equation that expresses the intensity of illumination (sound) in terms of the strength of the source and the distance from the source.

(b)   Suppose two light sources are 100 meters apart and one is 10 times brighter than the other. Find an equation for the set of all points $P(x, y)$ at which the amount of light from each will be the same.

(c)   Two sound speakers are in the center of a large room. They are 20 feet apart. Find an equation describing where the person should stand to receive an equal amount of sound from each speaker. Graph the equation.

4.   The accompanying table gives statistics for marriages and divorces in the United States between the years 1980 and 1991 (*Information Please Almanac Atlas and Yearbook 1993*, Houghton Mifflin). The divorces include annulments and the rate is per 1,000 population.
(a)   Plot the marriage and divorce rates on a graph.

(b) Fit a straight line, as "best" as you can, through the data.

(c) Based on the function determined in part (a), predict the marriage and divorce rate in the year 2000.

(d) From the trends, do you expect the rate of divorce to exceed the rate of marriage? If so, in what year?

Marriages and divorces in the United States

Year	Marriages Number	Rate	Divorces Number	Rate
1980	2,406,708	10.6	1,182,000	5.2
1981	2,438,000	10.6	1,219,000	5.3
1982	2,495,000	10.8	1,180,000	5.1
1983	2,444,000	10.5	3,179,000	5.0
1984	2,487,000	10.5	1,155,000	4.9
1985	2,425,000	10.2	1,187,000	5.0
1986	2,400,000	10.0	1,159,000	4.8
1987	2,421,000	9.9	1,157,000	4.8
1988	2,389,000	9.7	1,183,000	4.8
1989	2,404,000	9.7	1,163,000	4.7
1990	2,448,000	9.8	1,175,000	4.7
1991	2,371,000	9.4	1,187,000	4.7

 5. The accompanying table shows the public debt in the United States for selected years from 1800 to 1991 (*Information Please Almanac Atlas and Yearbook 1993*, Houghton Mifflin).

(a) Plot the year versus debt and fit a sixth degree polynomial to the data. Plot the data points together with the polynomial determined.

(b) Estimate the error between the actual values and the values on the curve used to fit the data.

(c) Plot the error together with the data and the curve.

(d) Determine the years in which the public debt increased by more than 50% over the previously recorded amount or increased by more than 50 billion dollars. (Assume a constant rate of increase in years that are not listed in the table.)

The public debt

Year	Amount (in millions)	Year	Amount (in millions)
1800	83	1950	256,097
1860	65	1955	272,807
1865	2,678	1960	284,093
1900	1,263	1965	313,819
1920	24,299	1970	370,094
1925	20,516	1975	533,189
1930	16,185	1980	907,701
1935	28,701	1985	1,823,103
1940	42,968	1990	3,233,313
1945	258,682	1991	3,665,303

6. *Monkey Problem* A monkey that escaped from the zoo is found in a tree in the local park by the zoo keeper. Unable to entice the monkey out of the tree the zoo keeper decides to try and shoot the monkey with a tranquilizer dart. The zoo keeper points the gun directly at the monkey, but the instant the dart leaves the gun barrel the monkey drops from the tree to escape. The problem is to show that the dart will always hit the monkey regardless of the gun's muzzle velocity (provided it gets to the monkey before he hits the ground). Placing the end of the gun barrel at the origin of an xy-axis system, if the dart is shot with an initial velocity of v_0 and at an initial angle of θ_0, then the x and y coordinates of the position of the dart (on its trajectory) are given by

$$x = (v_0 \cos \theta_0)t$$

and

$$y = (v_0 \sin \theta_0)t - \frac{1}{2}gt^2$$

(Notice that x and y are both functions of time t.)

Since the monkey drops vertically downward, the x-coordinate of his position at time t is always the distance from the zoo keeper, d, and the y-coordinate is given by the equation for a freely falling body,

$$y = d \tan \theta_0 - \frac{1}{2}gt^2$$

where $d \tan \theta_0$ is the initial position of the monkey and g the constant acceleration due to gravity.

Show the monkey always gets hit with the tranquilizer dart.

7.   Agricultural research on a certain variety of peach has shown that if thirty trees are planted per acre, then each tree will yield an average of 300 pounds of peaches per season. For each additional tree planted per acre, the average yield per tree is reduced by 5 pounds.

(a)   Determine an equation expressing Y, the yield of a 1 acre plot, as a function of the number of trees planted, and draw its graph.

(b)   How many trees should be planted per acre to produce the maximum yield? What is the maximum yield?

The Founders of Calculus

The invention of calculus is one of the landmark achievements in the history of human thought. Its impact on scientific development, and therefore on society in general, has been profound. Major credit for the introduction of calculus as an organized body of knowledge goes to two great geniuses of the seventeenth century, Gottfried Wilhelm Leibniz (1646–1716) and Sir Isaac Newton (1642–1727).

In their formulation of calculus, Newton and Leibniz worked independently, and although the published work of Leibniz preceded that of Newton, it is now known that Newton obtained his results before Leibniz. At the time, their supporters engaged in an infamous and bitter argument over whose work came first. This unfortunate affair forced mathematicians to take sides, and the lack of communication significantly inhibited mathematical progress, especially in England, for close to a hundred years.

Much of the controversy resulted from Newton's reluctance to publish his results, and therefore it was Leibniz who first revealed calculus's general capabilities in 1684. Newton had only hinted at his work on calculus in 1687 and didn't discuss it thoroughly until he released his own papers on the subject in 1704 and, more completely, in 1711. Although they approached their ideas from different directions, Newton and Leibniz achieved essentially identical results. (See the following biographies and *A Personal View* on pages 82 and 137.)

Leibniz was known all over Europe and corresponded with many of the greatest figures of his day, including Newton. It was partly the known correspondence between the two that fanned the flames of controversy later on. Leibniz has been called a "universal genius" for his command of virtually all branches of knowledge at the time, something that would be impossible today. He made lifelong contributions to mathematics, history, general philosophy, as well as "natural philosophy," especially physics and geology. He developed his own versions of the laws of motion, although his reputation as a physicist was overshadowed by that of Newton.

Newton is among the greatest mathematicians in history and is certainly the greatest applied mathematician. Probably no other individual has had so great an impact on the development of science. He is equally famous as a physicist because of his remarkable accomplishments in physics and astronomy. In particular, he made huge contributions to the new science of optics and is credited with the invention of the reflecting telescope. It was his insight into gravitation, which he said was inspired by an apple falling from an undisturbed tree, that led him to the quantitative description of the laws of universal gravitation. Our understanding of the "Newtonian Universe" was not seriously improved on until the work of Einstein in the twentieth century. Newton described his goals in his masterwork, the *Principia Mathematica*: "For I have deigned only to give a mathematical notion of the forces, without considering their physical causes. . . ."

Newton and his supporters never accepted that Leibniz had done his work independently and persisted in believing that Leibniz had plagiarized Newton. Long after Leibniz's death, Newton was still attacking him in his written works. Leibniz was more charitable about Newton, saying: "From the beginning of the world until Newton, what he did is much the better half."

Isaac Newton

A page from Newton's manuscript on optics and color and the prism he used in his experiments.

Gottfried Wilhelm Leibniz

Isaac Newton (1642–1727)

Newton was born prematurely on Christmas day, 1642, into a farming family in Woolsthorpe, England. His father died before Newton was born and his mother despaired of making him into a farmer. At the recommendation of an uncle he was sent off to Trinity College, Cambridge, where his early interests focused on chemistry and alchemy. At Cambridge he also studied Euclid, Descartes, and Galileo, and became familiar with Copernican astronomy.

One of the most famous periods in the history of science and mathematics was the 18 months after June 1665, when Cambridge University was closed because of danger from the bubonic plague. Newton went home to Woolsthorpe, and during this period of sustained isolation and concentration he discovered his version of the calculus. However, it was nearly 20 years before he published any real account of his calculus methods. Newton's reluctance to publish even extended to the *Principia Mathematica*, and he had to be persuaded to do so by his friend Edmond Halley, who was supportive enough to pay for its publication. The development of the laws of motion and gravity stem from this monumental work. Prior to the *Principia*, Newton had written three separate accounts of his calculus methods, but his avoidance of controversy meant that only his friends knew of Newton's progress.

The Trinity library in Cambridge has Newton's copy of a book by Descartes, with "error" noted here and there in Newton's handwriting. He later wrote: "In the beginning of the year 1665, I found the Method of approximating series & the Rule for reducing any [power] of any Binomial into such a series," which we now call the Binomial Theorem. Newton continued his work on series and corresponded with many people about them, including Leibniz. In 1665 he began the study of rates of change, which he called *fluxions*, and this study led to the modern concept of the *derivative*, a term invented by Leibniz. Newton originally used the letters p, q, and r for the derivatives of x, y, and z with respect to time. Later he introduced a more cumbersome notation that was adopted by British mathematicians but was eventually displaced by that of Leibniz.

Newton's practical and political talent led to later public work as Master of the Royal Mint and as President of the Royal Society. He led English science for many years and on his death was buried with great fanfare in Westminster Abbey. On his tomb are engraved the words: "Mortals, congratulate yourselves that so great a man has lived for the honor of the human race." The famous poet Alexander Pope added in tribute:
Nature and Nature's law lay hid in night;
God said: 'Let Newton be', and all was light.

Gottfried Wilhelm Leibniz (1646–1716)

Leibniz was born in Hanover, Germany, in 1646. Although he loved mathematics from an early age, he had little opportunity to study it formally either in high school or at the University of Leipzig, which he entered at the age of 15. Here he studied law, philosophy, and logic and read the work of great scientists and philosophers such as Galileo, Hobbes, and Descartes. However, he was eventually denied his law degree from Leipzig simply because he was so young.

Leibniz not only made mathematical inventions such as the binary system of computation, but he also invented many practical devices. In the 1670s he invented a calculating machine, called the Leibniz Wheel, that could add, subtract, multiply, and divide. He offered it as a labor-saving device and wrote: "It is unworthy of excellent men to lose hours like slaves in the labor of computation." Leibniz exhibited his machine in Paris and specifically traveled to London to demonstrate it to the Royal Society, the most prestigious scientific society of the era. His visit came six years after the Royal Society had published Newton's *Principia*.

In Paris in 1672, Leibniz met the great Dutch physicist and mathematician Christiaan Huygens and persuaded Huygens to tutor him in mathematics. Within a year he was developing insights into the calculus. According to his own writings, it was by reading Pascal's work that he arrived at his first important insight, using an infinite series to find a circular quadrant, in 1673. We now know that Leibniz developed his version of calculus in the years 1673–1676 and published his research in 1684 and 1686. Between 1674 and 1676 he formulated the notation we now use for differential and integral calculus. In particular, the notations for derivative and integral were much better designed than those offered by Newton. By 1684, he had worked out much of differential calculus, which he published in a book entitled *New Method for the Greatest and the Least*. It was this book that brought calculus to the world's attention, before Isaac Newton's earlier, independent invention became widely known.

One of Leibniz's lifelong ambitions was the development of a kind of universal language—a *lingua chacteristica* that would not only incorporate mathematical computation but also serve as a medium for communicating all scientific and mathematical knowledge.

Sadly, and in dramatic contrast to the celebrity of Newton, Leibniz was ultimately overlooked by all and died a lonely man in near total obscurity. In the words of the eighteenth-century French philosopher and scientist Fontenelle, "He was glad to observe the flowering in other people's gardens of plants whose seeds he provided."

Owen Gingerich

Isaac Newton's Principia, *published in 1687, has been called the greatest scientific book ever written. Is it true that Newton discovered many of his propositions using calculus, but then revised the proofs to disguise this?*

For many years it was commonly believed that Newton systematically recast his proofs into geometrical form, but a detailed study of his manuscripts made in the past few decades shows that what he first got is what you see. In fact, the *Principia* was written very quickly, in just over two years, and part of the reason the book is so difficult to read is that Newton was working at the very frontier, and his initial attack was often clumsy or obscure by later standards. There were probably fewer than a dozen mathematicians in Europe who could really comprehend it.

Does this mean the Principia *didn't use calculus?*

Newton didn't explicitly use differentials or integrals, but he continuously used the basic concepts of letting quantities become very small to see what happens to a ratio. The book is teeming with theory and applications of the infinitesimal calculus.

What about integration?

There is a very famous proposition in which Newton proves that the gravitational attraction from a sphere acts as if the mass is all concentrated at the center. Sometimes it is said that Newton had to invent the integral calculus to solve this particular problem. He did it by dividing the sphere into very many tiny parts and devising an ingenious method to sum over their individual attractions. In concept it is very like integration.

But Newton had invented the calculus before he wrote the Principia, *not so?*

Yes and no. He had found the binomial theorem about 20 years earlier, and he saw the relation of that infinite series to differentiation and integration. But the evidence from the vast hoard of manuscripts remaining at (the English) Cambridge shows that the real development of the notation and its application to *Principia*-type problems came in the 1690s.

For centuries there has been a controversy over who invented calculus. Was it Newton or his German rival, Gottfried Wilhelm Leibniz?

Newton put together the basic ideas first, but Leibniz published first, in 1684. The priority battle was very much fueled by Newton later on. As president of the Royal Society, he appointed an impartial committee to investigate whether Leibniz had plagiarized the invention, and the committee handed the credit to Newton. That same hoard of manuscripts shows that Newton himself wrote the committee's report, making draft after draft of it. Today Leibniz shares the credit for the invention.

In fact, Leibniz had the last word, didn't he, because today we use his notation.

Yes, Leibniz was very attentive to questions of form and elegant presentation. The use of dx and dy is his, and also the integral sign, a stylized "s" for summation. The terms *differential* and *integral calculus* also arose from his vocabulary.

What did Newton call his system? Not the calculus?

No, he called it "The Method of Fluxions' and his book of that name was finally published in 1736. His first book on it, *De analysi* ("On analysis by equations with an infinite number of terms"), was composed in 1669 and circulated in manuscript among his friends, but wasn't published until 1711.

It's often said that mathematicians usually get their most creative ideas when they're young, but if Newton was working out his "fluxions" in the 1690s, he was already about 50 years old.

Yes, it's amazing how Newton's mathematical powers were so persistent, even when he wasn't especially interested anymore. In 1696 he became Warden of the Mint, and he was very serious about overseeing the new coinage, when mill marks were introduced for the first time to the edges of the English coins. The European mathematicians including Leibniz had begun to doubt whether this strange man who had abandoned science was really such a genius after all, so the Swiss mathematician Johann Bernoulli sent out a challenge problem. Although it was addressed "to all the mathematicians of Europe," the real quarry was Newton himself. Later [it was] called the brachistochrone problem (see Chapter 9). The Secretary of the Royal Society sent the problem to Newton, who pushed it aside saying, "I have not time for trifles while on the King's business." Newton took the problem home at about four in the afternoon, and though being very tired from the work at the Mint, he didn't sleep until he had it solved. The next morning he sent the full answer back to the Royal Society, showing that the required curve was a cycloid, and his solution was then printed anonymously. When Bernoulli saw the results, he exclaimed, "Ex unge leonem!"—"By the claw, the lion is revealed." There was no question as to the author's identity.

Owen Gingerich is a senior astronomer at the Smithsonian Astrophysical Observatory and Professor of Astronomy and of the History of Science at Harvard University. He was Chair of the Department of History of Science. Professor Gingerich is a leading authority on the astronomer Johannes Kepler and the cosmologist Nicholas Copernicus. In recognition of these studies, an asteroid has been named in Professor Gingerich's honor.

CHAPTER 2

LIMITS AND CONTINUITY

Calculus is built on the two fundamental concepts of function and limit. We introduced the function concept in Chapter 1. In this chapter we take up limits of functions. We will see that the idea of limit underlies virtually all the concepts of calculus and that our understanding of the mathematics of change and motion depends on the limit.

We begin by seeing intuitively what it means for a function $f(x)$ to approach a limit L as x approaches some value a. Next, we give some of the important properties of limits of functions. We then introduce the closely related topic of continuity of a function, which is defined in terms of a limit. As will be evident later on, continuity is a key property that assures the validity of certain important calculus operations.

In our definition of limit, the numbers a and L are both understood to be (finite) real numbers. In Section 2.4 we extend the limit notion by permitting either a or L to be replaced by ∞ or $-\infty$. We refer to these extensions as "limits at infinity" and "infinite limits." As we will see, these ideas are closely related to horizontal and vertical asymptotes.

We conclude this chapter by giving precise formulations of the limit and continuity definitions. We will indicate how these formulations enable us to prove certain properties stated earlier. We defer most of the proofs, however, to Supplementary Appendix 5.

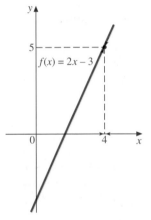

FIGURE 2.1
As $x \to 4$, $f(x) \to 5$

2.1 THE LIMIT OF A FUNCTION

Intuitively, the idea of the limit of a function is fairly simple. For example, let $f(x) = 2x - 3$, and consider what happens to $f(x)$ as x approaches 4. That is, if x is very close to 4, what can we say about $f(x)$? Clearly, the answer is that $f(x)$ is close to 5, as Figure 2.1 shows. We say that *the limit of* $2x - 3$ *as* x *approaches* 4 *is* 5 and write this statement in symbols as

$$\lim_{x \to 4}(2x - 3) = 5$$

or equivalently, we may write $(2x - 3) \to 5$ as $x \to 4$. Not all limits are this obvious, as the next example shows.

	$(x < 0)$
x	$f(x)$
-0.01	0.12501954
-0.001	0.12500195
-0.0001	0.12500020
-0.00001	0.12500002

	$(x > 0)$
x	$f(x)$
0.01	0.12498047
0.001	0.12499805
0.0001	0.12499980
0.00001	0.12499998

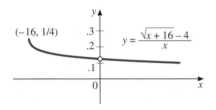

FIGURE 2.2

EXAMPLE 2.1   Use a calculator to evaluate the following function near $x = 0$. Estimate the limit as $x \to 0$.

$$f(x) = \frac{\sqrt{x + 16} - 4}{x}$$

Solution   Note first that 0 is not in the domain of the function, so we cannot substitute $x = 0$. We calculate and tabulate values of $f(x)$ for $x < 0$ and $x > 0$.

It appears that $f(x)$ approaches the limit 0.125, or 1/8, as $x \to 0$. Although our calculations give strong evidence that the limit is 1/8, we have not *proved* this result. We will soon see a way of proving it. The graph in Figure 2.2 seems to confirm our result. Notice the missing point for $x = 0$.   ■

REMARK ─────────────────────────────

■ The fact that the function f in this example is undefined at $x = 0$ has no significance in terms of the limit of $f(x)$ as x approaches 0. More generally, in finding the limit of a function as $x \to a$, we consider x values close to a, *but not equal to a*.

─────────────────────────────────────

If you continued taking smaller values of x in Example 2.1, you would expect that the function values would come still closer to 0.125. But an unexpected thing happens. If you take $x = 0.00000001$, for example, you will find that the calculator gives the value 0! The problem is that $\sqrt{x + 16}$ with this value of x is so close to 4 that the calculator gives 0 for the numerator of $f(x)$. This result shows that you cannot always rely on numerical answers given by calculators or computer software that is not storing or carrying enough digits. A Computer Algebra System (CAS) can typically carry an arbitrary number of digits and does not have the limitations of a calculator.

We are now ready for the following informal definition.

Definition 2.1
Informal Definition of Limit

The statement

$$\lim_{x \to a} f(x) = L$$

means that $f(x)$ will be as close as we please to the number L for all x values sufficiently close to a, but not equal to a.

We call this definition informal because it contains the imprecise notions "as close as we please" and "sufficiently close." We will give a precise definition in Section 2.5. For many purposes, an intuitive understanding of the limit concept is sufficient, but for proving theorems about limits, we need the precise definition.

In Definition 2.1, how close to a we should make x depends, in general, on how close to L we want $f(x)$ to be. If we first say how close to L we want the $f(x)$ values to be, then we can find how close to a (that is, we can say what "sufficiently close" is) to choose the x values. Normally, the closer to L we require $f(x)$ to be, the closer to a we will need to take x. (We are assuming here that $f(x)$ does approach a limit L.) It should be clear that if the limit exists, that limit is unique. That is, a function cannot have more than one limit as x approaches a. The $f(x)$ values cannot all be arbitrarily close to two different limits for all x values near $x = a$.

FIGURE 2.3

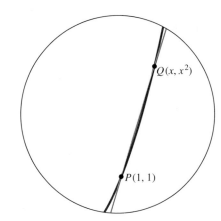

FIGURE 2.4

The Tangent Problem

As another example of the limit of a function, we consider a special case of a type of problem (called the *tangent problem*) that will motivate our definition of the derivative in the next chapter. So this problem is a preview of things to come. In Figure 2.3 we show the graph of $y = x^2$, together with the fixed point $P(1, 1)$ and the variable point $Q(x, x^2)$, with $x \neq 1$. Our objective is to define the tangent line to the graph at P. First we will find the slope of the line PQ, called a *secant line*. (More generally, a secant line to a curve is a line joining any two points on the curve.) Since P and Q are distinct points, we can compute the slope of the secant line PQ. Then we will allow Q to approach P by letting x come arbitrarily close to 1, but not equal to 1. This latter qualification deserves emphasis, for if we permitted x to be equal to 1, then Q would coincide with P, so there would not be a well-defined line joining P and Q and we would not see how to compute a slope. If x is very close to 1, then Q is close to P, but as the magnified view in Figure 2.4 shows, the two points are distinct, and so a unique secant line through them exists.

In Figure 2.5 we show several secant lines PQ for points Q closer and closer to P. Notice how these secant lines seem to be approaching some limiting position (the tangent line). For any $x \neq 1$, the slope of the secant line PQ for our example is

$$m_{\text{sec}} = \frac{\text{change in } y}{\text{change in } x} = \frac{x^2 - 1}{x - 1}$$

It seems reasonable to *define* the tangent line at P to be that line through P whose slope is the limit of m_{sec} as x approaches 1, provided this limit exists. We show this tangent line in Figure 2.5. According to this definition, the slope of the tangent line is given by

$$m_{\text{tan}} = \lim_{x \to 1} m_{\text{sec}} = \lim_{x \to 1} \frac{x^2 - 1}{x - 1}$$

The problem, then, is to find this limiting value. We examine this limit in the next example.

EXAMPLE 2.2   The slope of the tangent line to the graph of $y = x^2$ at the point $(1, 1)$ is

$$m_{\text{tan}} = \lim_{x \to 1} \frac{x^2 - 1}{x - 1}$$

Use a calculator to estimate this limit, and then find the equation of the tangent line, based on the estimate.

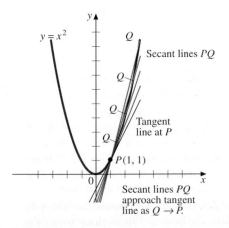

FIGURE 2.5

$(x > 1)$	
x	m_{sec}
1.1	2.1
1.01	2.01
1.001	2.001
1.0001	2.0001

$(x < 1)$	
x	m_{sec}
0.9	1.9
0.99	1.99
0.999	1.999
0.9999	1.9999

Solution   As in the preceding discussion, we know that $m_{\text{sec}} = (x^2 - 1)/(x - 1)$, and we want to find what these slopes approach as x approaches 1. We cannot simply substitute $x = 1$, since we would get 0/0, which is undefined. In the margin we show tables giving values obtained from a calculator for m_{sec}, taking x slightly greater than 1 in one case and slightly less than 1 in the other. These tables give convincing evidence that the limit is 2. In the next section we will discuss an algebraic way of showing that the limit actually is 2. (Can you see a way?)

Let us assume that $m_{\text{tan}} = 2$ based on our evidence. Then we can use the *point-slope form* for the equation of a line (Supplementary Appendix 1):

$$y - y_1 = m(x - x_1)$$

Taking (x_1, y_1) as $(1, 1)$ and m as $m_{\text{tan}} = 2$, we obtain $y - 1 = 2(x - 1)$, or

$$y = 2x - 1$$

as the equation of the tangent line.   ∎

The limit in the next example is an important one that we will have occasion to use later. Here again we rely on an estimate obtained by using a calculator. (A CAS or spreadsheet also could be used.) In Chapter 3 we will show a geometric way of verifying our result.

EXAMPLE 2.3   Use a calculator, a CAS, or a spreadsheet to conjecture whether

$$\lim_{x \to 0} \frac{\sin x}{x}$$

exists, and if so, what its value is.

x	$(\sin x)/x$
1.0	0.84147
0.5	0.95885
0.1	0.99833
0.05	0.99958
0.01	0.99998

Solution   Observe first that $(\sin x)/x$ is even, since

$$\frac{\sin(-x)}{-x} = \frac{-\sin x}{-x} = \frac{\sin x}{x}$$

So we need only consider $x > 0$, since the same value of $(\sin x)/x$ will occur for $-x$ as for x.

We show a table in the margin for selected values of x near 0. The table gives convincing evidence that

$$\lim_{x \to 0} \frac{\sin x}{x} = 1$$

The conjecture that the limit is 1 can be further supported by the graph of $f(x) = (\sin x)/x$ in Figure 2.6, generated by a computer algebra system. We have added a small circle at the point $(0, 1)$ to emphasize that $x = 0$ is not in the domain of this function.   ∎

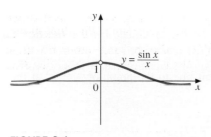

FIGURE 2.6

We emphasize that when we ask for $\lim_{x \to a} f(x)$, we want to know what value $f(x)$ *approaches* as x approaches a. We are not concerned with whether $f(x)$ actually reaches its limit, nor are we concerned with what happens when $x = a$. In fact, $f(a)$ may not even be defined, as the tangent problem illustrated. Even if $f(a)$ were defined, it may not equal the limit L that is approached by $f(x)$ as x approaches a. The next example illustrates this situation.

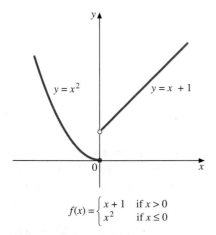

FIGURE 2.7

x	$1 + x^2$	x	$1 - x$
$(x > 0)$		$(x < 0)$	
0.5	1.25	-0.5	1.5
0.1	1.01	-0.1	1.1
0.01	1.0001	-0.01	1.01
0.001	1.000001	-0.001	1.001

EXAMPLE 2.4   Let

$$f(x) = \begin{cases} x^2 + 1 & \text{if } x > 0 \\ 0 & \text{if } x = 0 \\ 1 - x & \text{if } x < 0 \end{cases}$$

Determine whether $\lim_{x \to 0} f(x)$ exists, and if so, find its value.

Solution   We show the graph of f in Figure 2.7. When x is positive, $f(x) = x^2 + 1$, so if x is close enough to 0, we can make $f(x)$ as close to 1 as we please. Similarly, if x is negative, $f(x) = 1 - x$, which we can also make as close to 1 as we want by choosing x sufficiently small in absolute value. The tables in the margin give further evidence that $f(x)$ comes arbitrarily close to 1 as x approaches 0 through either positive or negative values. Although we have not formally proved this result, it is intuitively evident that

$$\lim_{x \to 0} f(x) = 1$$

We were given that $f(0) = 0$, but this fact has nothing to do with the limit.   ∎

One-Sided Limits

In Example 2.4 we found the limit by considering first the value approached by $f(x)$ as x approached 0 through positive values and then the value approached by $f(x)$ as x approached 0 through negative values. In each case, we found the same value, 1, and so we concluded that the limit of $f(x)$ as x approached 0 was 1. We now give names and symbols to "one-sided" limits such as those in this example. If $f(x)$ is as close to a number L_1 as we please for all $x > a$ and sufficiently close to a, then we say that L_1 is the **right-hand limit** of $f(x)$ as x approaches a, and we indicate such a limit symbolically by

$$\lim_{x \to a^+} f(x) = L_1$$

The **left-hand limit** is defined analogously. If its value is L_2, we write

$$\lim_{x \to a^-} f(x) = L_2$$

In Example 2.4 both L_1 and L_2 were equal to the same value, 1, and so we concluded that $\lim_{x \to 0} f(x) = 1$. If $L_1 \neq L_2$, as is true for the function in Figure 2.8, then there is no limit, since $f(x)$ cannot be made arbitrarily close to any *one* value for all x values close to a. We can summarize this discussion as follows.

$$f(x) = \begin{cases} x + 1 & \text{if } x > 0 \\ x^2 & \text{if } x \le 0 \end{cases}$$

FIGURE 2.8

$\lim_{x \to 0^+} f(x) = 1$ and $\lim_{x \to 0^-} f(x) = 0$, so $\lim_{x \to 0} f(x)$ does not exist.

$\lim_{x \to a} f(x) = L$ if and only if both $\lim_{x \to a^+} f(x) = L$ and $\lim_{x \to a^-} f(x) = L$

That is, the (unrestricted) limit exists if and only if the right-hand and left-hand limits both exist and are equal.

REMARK ───

■ The expression $x \to a^+$ means that x approaches a from the right, and it has nothing to do with the sign of a. For example, $x \to -2^+$ means x is approaching -2 through values to the right of -2 (such as $-1.9, -1.99, -1.999, \ldots$). Similarly, the expression $x \to a^-$ means that x approaches a from the left.

───

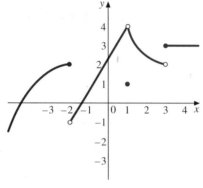

FIGURE 2.9

EXAMPLE 2.5   From the graph in Figure 2.9 determine $\lim_{x \to a^-} f(x)$ and $\lim_{x \to a^+} f(x)$, if they exist, for (a) $a = -2$, (b) $a = 1$, and (c) $a = 3$. In each case, state whether $\lim_{x \to a} f(x)$ exists.

Solution

(a) $\displaystyle\lim_{x \to -2^-} f(x) = 2$ and $\displaystyle\lim_{x \to -2^+} f(x) = -1$
Thus, $\displaystyle\lim_{x \to -2} f(x)$ does not exist.

(b) $\displaystyle\lim_{x \to 1^-} f(x) = 4$ and $\displaystyle\lim_{x \to 1^+} f(x) = 4$
Thus, $\displaystyle\lim_{x \to 1} f(x) = 4$.

(c) $\displaystyle\lim_{x \to 3^-} f(x) = 2$ and $\displaystyle\lim_{x \to 3^+} f(x) = 3$
Thus, $\displaystyle\lim_{x \to 3} f(x)$ does not exist.   ■

EXAMPLE 2.6   Let

$$f(x) = \frac{|x|}{x}$$

Find $\lim_{x \to 0^+} f(x)$ and $\lim_{x \to 0^-} f(x)$. Does $\lim_{x \to 0} f(x)$ exist?

Solution   When $x > 0$, we have $|x| = x$, so $f(x) = x/x = 1$. We can make $f(x)$ as close to 1 as we please (namely, *equal* to 1) by choosing x positive and sufficiently close to 0 (namely, *anywhere* to the right of 0). Thus,

$$\lim_{x \to 0^+} \frac{|x|}{x} = 1$$

If $x < 0$, then $|x| = -x$, so

$$f(x) = \frac{|x|}{x} = \frac{-x}{x} = -1$$

Reasoning as above, we conclude that

$$\lim_{x \to 0^-} \frac{|x|}{x} = -1$$

FIGURE 2.10

Since the one-sided limits are not equal, we see that $\lim_{x \to 0} |x|/x$ does not exist. Note that $x = 0$ is not in the domain of f, but this fact has nothing to do with our investigation of *limits* as x approaches 0. We show the graph of f in Figure 2.10.   ■

Other Functions Whose Limits Fail to Exist

In Examples 2.5 and 2.6 we saw some functions that failed to have a limit as x approached some number, because the one-sided limits were unequal. In the next example, we show some other ways that a function may fail to have a limit.

EXAMPLE 2.7   Show that the limit of each of the following functions fails to exist as $x \to 0$. Draw the graphs and explain in what way, or ways, Definition 2.1 fails.

(a) $f(x) = \dfrac{1}{x^2}$ (b) $g(x) = \sin \dfrac{1}{x}$

Solution   We show the graphs of $y = f(x)$ and $y = g(x)$ in Figure 2.11.

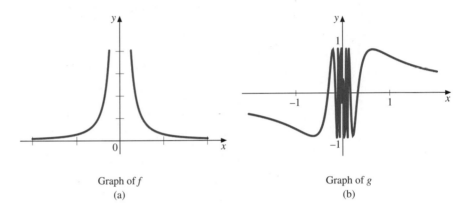

Graph of f
(a)

Graph of g
(b)

FIGURE 2.11

(a) The graph of f has a vertical asymptote at $x = 0$. That is, as $x \to 0$, $f(x) \to \infty$. But ∞ is not a real number, so $\lim_{x \to 0} f(x)$ does not exist. As we will show in Section 2.4, we often write in this case $\lim_{x \to 0} f(x) = \infty$. Saying the limit is "infinity" means that the function becomes arbitrarily large.

(b) As $x \to 0$, the function g takes on every value between -1 and 1 infinitely many times since $\sin(1/x)$ oscillates between -1 and 1 infinitely many times as $x \to 0$. In fact, $\sin(1/x) = 1$ if $1/x = \pi/2 + 2n\pi$, for any integer n, that is, if

$$x = \frac{2}{(4n + 1)\pi}$$

Similarly, $\sin(1/x) = -1$ if

$$x = \frac{2}{(4n + 3)\pi}$$

You can see that by choosing n large enough, each of these values of x can be made as close to 0 as we please. So the values of $g(x)$ cannot be made arbitrarily close to any fixed value for all x values sufficiently close to 0.   ■

LIMITS USING COMPUTER ALGEBRA SYSTEMS

 CAS 12

Compute each of the following limits using Maple, Mathematica, or DERIVE.

1. (a) $\lim_{x \to -2} (2x - 1)$ **2.** (b) $\lim_{x \to \infty} \left(\dfrac{x^3 - 3x^2 + 1000x + 25000}{3x^3 + 25000x + 10} \right)$

Maple:

limit(2*x−1,x=−2);

Output: −5, as you would expect.

Mathematica:

Limit[2*x−1,x−>−2]

Output: −5

DERIVE:

(At the □ symbol, go to the next step.)
a (author) □ 2x−1 □ c (calculus) □ l
(limit) □ [choose expression] □ x (variable) □ −2 (limit point)

Output: $\lim_{x \to -2} 2x - 1$

s (simplify) □
[choose expression]

Output: −5

Maple:

h := x−>(x^3−3*x^2+1000*x+25000)/
    (3*x^3+25000*x+10);
limit(h(x), x=infinity);

Output: $\frac{1}{3}$

Mathematica:*

g[x_]=x^3−3*x^2+1000*x+25000)/
    (3*x^3+25000*x+10)
Limit[g[x],x−>Infinity]

Output: $\frac{1}{3}$

DERIVE:

(At the □ symbol, go to the next step.)
a (author) □ x^3−3x^2+1000x+25000)/
□ (3x^3+25000x+10) □ c (calculus) □
l (limit) □ [choose expression] □ x (variable) □ inf (limit point) □ s (simplify)
□ [choose expression]
Output: $\frac{1}{3}$

 CAS 13

Determine

$$\lim_{x \to 1} \frac{1}{(x - 1)}$$

---

*We have defined h and g as functions and not just equations in Maple and Mathematica, respectively. The functions h and g can then be evaluated at any expression for x. In Maple be sure to use the minus sign followed by the greater than sign, as in $h:=x->$ (x goes to the expression on the right). In Mathematica use the underscore character following the variable, as in $g[x\_]$.

In Maple, we have:

g :=x -> 1/(x–1);
limit(g(x),x=1);

Output: undefined
This is the correct answer. Let's check the left- and right-hand limits. In Maple we have:

limit(g(x),x=1,right);

Output: infinity

limit(g(x),x=1,left);

Output: -infinity
The plot of the function makes this clear:

plot(g(x),x=−1..2);

In Mathematica, we proceed in a similar manner:

g[x_] = 1/(x–1)
Limit[g[x],x–>1]

Here we need to be cautious, since the output from our version of Mathematica is:

infinity

This is not the correct answer. If we check the left- and right-hand limits in Mathematica, we get the same answer as we did using Maple.

Limit[g[x],x–>1,
    Direction–> –1]
Limit[g[x],x–>1,Direction–>1]

The first command gives an output of infinity and the second, an output of − infinity. The −1 means from the right and 1 means from the left.

In DERIVE, we get a result similar to the one in Mathematica:

(At the □ symbol, go to the next step.)
a (author) □ 1/(x−1) □ c (calculus) □ l (limit) □ [choose expression] □ x (variable) □ 1 (limit point) □ s (simplify) □ [choose expression]

Output: $\frac{1}{0}$

Now proceed to compute the one-sided limits: c (calculus) □ l (limit) □ [choose expression] □ x (variable) □ 1 (limit point) □ left (left-hand limit) □ s (simplify) □ [choose expression]

Output: $-\infty$

c (calculus) □ l (limit) □ [choose expression] □ x (variable) □ 1 (limit point) □ right (right-hand point) □ s (simplify) □ [choose expression]

Output: ∞

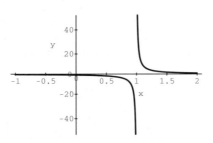

FIGURE 2.1.1
$$y = f(x) = \frac{1}{1-x}$$

Exercise Set 2.1

In Exercises 1–6, sketch the graph of the function and use it to determine the given limit.

1. $f(x) = 5 - x$; $\lim\limits_{x \to 2} f(x)$

2. $g(x) = x^2 + 2x + 1$; $\lim\limits_{x \to 1} g(x)$

3. $f(t) = \begin{cases} t^2 + 1 & \text{if } t \geq 0 \\ -2 & \text{if } t < 0 \end{cases}$; $\lim\limits_{t \to 0} f(t)$

4. $h(x) = 3$; $\lim\limits_{x \to 10} h(x)$

5. $f(x) = \begin{cases} 2x + 2 & \text{if } x \neq 1 \\ 0 & \text{if } x = 1 \end{cases}$; $\lim\limits_{x \to 1} f(x)$

6. $g(t) = \dfrac{1}{t - 5}$; $\lim\limits_{t \to 5} g(t)$

In Exercises 7–12, use the graph to determine $\lim_{x \to 2^+} f(x)$, $\lim_{x \to 2^-} f(x)$, and $\lim_{x \to 2} f(x)$, or explain why the limit fails to exist.

7.

8.

9.

10.

11.

12.

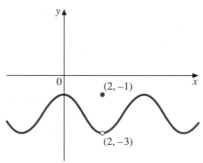

In Exercises 13–18, find $\lim_{x \to a^+} f(x)$ and $\lim_{x \to a^-} f(x)$ for the given values of a and $f(x)$. State whether $\lim_{x \to a} f(x)$ exists, and if so, give its value.

13. $a = 2;\ f(x) = \begin{cases} x^2 & \text{if } x \geq 2 \\ 6 - x & \text{if } x < 2 \end{cases}$

14. $a = 0;\ f(x) = \begin{cases} 1 & \text{if } x > 0 \\ 0 & \text{if } x = 0 \\ -1 & \text{if } x < 0 \end{cases}$

15. $a = 1;\ f(x) = \begin{cases} 2x & \text{if } x > 1 \\ 0 & \text{if } x = 1 \\ x^2 + 1 & \text{if } x < 1 \end{cases}$

16. $a = 0;\ f(x) = \dfrac{x}{\sqrt{x^2}}$

17. $a = 1;\ f(x) = \dfrac{x^2 - x}{|x - 1|}$

18. $a = 2;\ f(x) = \dfrac{|2 - x|}{x^2 - 4}$

In Exercises 19–28, use a calculator, CAS, or spreadsheet to make a table of values, and conjecture the limit based on the table.

19. $\lim_{x \to -2} \dfrac{x^3 + 8}{x + 2}$

20. $\lim_{x \to 2} \dfrac{x^2 - x - 6}{x^2 - 4}$

21. $\lim_{t \to 1} \dfrac{3 - 2t - t^2}{2t^2 - t - 1}$

22. $\lim_{x \to 4} \dfrac{\sqrt{x} - 2}{x - 4}$

23. $\lim\limits_{x \to 4} \dfrac{\sqrt{2x+1}-3}{x-4}$

24. $\lim\limits_{h \to 0} \dfrac{\sqrt{(4+h)^2+9}-5}{h}$

25. $\lim\limits_{x \to 0} \dfrac{1-\cos x}{x^2}$

26. $\lim\limits_{t \to 0} \dfrac{t^3}{t-\sin t}$

27. $\lim\limits_{x \to (\pi/2)^-} (\tan x)^{\cos x}$

28. $\lim\limits_{x \to 1} x^{1/(1-x)}$

In Exercises 29–32, use a calculator, CAS, or spreadsheet to estimate the limits. Then, after estimating that $\lim_{x \to a} f(x) = L$, find an interval centered at a such that for all x in this interval, $f(x)$ is less than 0.01 unit from L. Repeat, replacing 0.01 with 0.001.

29. $\lim\limits_{x \to 2} (3x-1)$

30. $\lim\limits_{x \to 2} \dfrac{x^2+1}{x-1}$

31. $\lim\limits_{x \to 0} x \sin \dfrac{1}{x}$

32. $\lim\limits_{x \to 0} \dfrac{\sin 5x}{x}$

2.2 LIMIT PROPERTIES AND TECHNIQUES FOR FINDING LIMITS

In Section 2.1, we made implicit use of certain properties of limits involving sums, differences, products, and quotients. We now state these properties explicitly. Each of the properties seems intuitively evident, and they can be proved formally. We will prove some of them in Section 2.5 and the others in Appendix 5.

Properties of Limits

If $\lim_{x \to a} f(x)$ and $\lim_{x \to a} g(x)$ both exist, then the following properties hold true:

1. $\lim\limits_{x \to a} cf(x) = c \lim\limits_{x \to a} f(x)$ for any constant c.
The limit of a constant times a function is the constant times the limit of the function.

2. $\lim\limits_{x \to a} [f(x) + g(x)] = \lim\limits_{x \to a} f(x) + \lim\limits_{x \to a} g(x)$
The limit of a sum is the sum of the limits.

3. $\lim\limits_{x \to a} [f(x) - g(x)] = \lim\limits_{x \to a} f(x) - \lim\limits_{x \to a} g(x)$
The limit of a difference is the difference of the limits.

4. $\lim\limits_{x \to a} [f(x) \cdot g(x)] = [\lim\limits_{x \to a} f(x)] \cdot [\lim\limits_{x \to a} g(x)]$
The limit of a product is the product of the limits.

5. $\lim\limits_{x \to a} \dfrac{f(x)}{g(x)} = \dfrac{\lim\limits_{x \to a} f(x)}{\lim\limits_{x \to a} g(x)}$, provided $\lim\limits_{x \to a} g(x) \neq 0$
The limit of a quotient is the quotient of the limits, provided the limit of the denominator is not 0.

REMARK ———————————————————————————

■ Let us emphasize that for these properties of limits to hold true, each individual limit on the right side of the equation must exist (as a finite number). To see a case where this condition fails, consider $f(x) = x$ and $g(x) = 1/x$. If we take $a = 0$ and attempt to apply Property 4, we find that $f(x) \cdot g(x) = x \cdot (1/x) = 1$

for all $x \neq 0$. So $\lim_{x \to 0}[f(x) \cdot g(x)] = 1$. But the right-hand side is

$$\left[\lim_{x \to 0} x\right] \cdot \left[\lim_{x \to 0} \frac{1}{x}\right]$$

and the second limit does not exist, so the product is meaningless. Thus, the equality in Property 4 does not hold true in this instance.

If we apply Property 4 to $[f(x)]^2 = [f(x)] \cdot [f(x)]$, we get

$$\lim_{x \to a}[f(x)]^2 = \left[\lim_{x \to a} f(x)\right] \cdot \left[\lim_{x \to a} f(x)\right] = \left[\lim_{x \to a} f(x)\right]^2$$

Similarly,

$$\lim_{x \to a}[f(x)]^3 = \left[\lim_{x \to a} f(x)\right]^2 \cdot \left[\lim_{x \to a} f(x)\right] = \left[\lim_{x \to a} f(x)\right]^3$$

and in general, for any positive integer n,

$$\lim_{x \to a}[f(x)]^n = \left[\lim_{x \to a} f(x)\right]^n$$

(The general result can be proved by the technique of mathematical induction, discussed in Supplementary Appendix 4.)

An analogous result holds for limits of roots. We list the following two results as further properties of limits.

Further Properties of Limits

If $\lim_{x \to a} f(x)$ exists and n is a positive integer, then the following properties hold true:

6. $\lim_{x \to a}[f(x)]^n = [\lim_{x \to a} f(x)]^n$

7. $\lim_{x \to a} \sqrt[n]{f(x)} = \sqrt[n]{\lim_{x \to a} f(x)}$, provided that when n is even, $\lim_{x \to a} f(x) > 0$.

In addition to these general properties, we have made use of the following special limits.

Two Special Limits

1. If c is a constant and $f(x) = c$ for all x, then $\lim_{x \to a} f(x) = c$ for every real number a. More briefly,

$$\lim_{x \to a} c = c \qquad (2.1)$$

2. If $f(x) = x$, then $\lim_{x \to a} f(x) = a$ for every real number a. More briefly,

$$\lim_{x \to a} x = a \qquad (2.2)$$

The graphs of these two special functions, shown in Figure 2.12, illustrate their limits as $x \to a$.

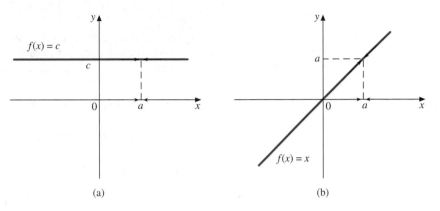

(a)   (b)

FIGURE 2.12

EXAMPLE 2.8   Show how the limit properties are used in finding each of the following limits:

(a) $\lim\limits_{x \to 2}(3x^2 - 4x + 5) = 9$ (b) $\lim\limits_{x \to -1} \dfrac{x^3 + 2x + 4}{3x^2 - 2} = 1$

Solution

(a) $\lim\limits_{x \to 2}(3x^2 - 4x + 5) = \lim\limits_{x \to 2}(3x^2) - \lim\limits_{x \to 2}(4x) + \lim\limits_{x \to 2}5$ Properties 2 and 3

$\qquad\qquad\qquad\qquad\quad = 3 \lim\limits_{x \to 2} x^2 - 4 \lim\limits_{x \to 2} x + 5$ Property 1 and Special Limit 2.1

$\qquad\qquad\qquad\qquad\quad = 3(2)^2 - 4(2) + 5$ Property 6 and Special Limit 2.2

$\qquad\qquad\qquad\qquad\quad = 12 - 8 + 5 = 9$

(b) $\lim\limits_{x \to -1} \dfrac{x^3 + 2x + 4}{3x^2 - 2} = \dfrac{\lim\limits_{x \to -1}(x^3 + 2x + 4)}{\lim\limits_{x \to -1}(3x^2 - 2)}$ Property 5

$\qquad\qquad\qquad\qquad = \dfrac{\left(\lim\limits_{x \to -1} x^3\right) + \left[\lim\limits_{x \to -1}(2x)\right] + \left(\lim\limits_{x \to -1} 4\right)}{\lim\limits_{x \to -1}(3x^2) - \lim\limits_{x \to -1}(2)}$ Properties 2 and 3

$\qquad\qquad\qquad\qquad = \dfrac{(-1)^3 + 2(-1) + 4}{3(-1)^2 - 2}$ Properties 1 and 6 and Special Limits 2.1 and 2.2

$\qquad\qquad\qquad\qquad = \dfrac{-1 - 2 + 4}{3 - 2} = \dfrac{1}{1} = 1$ ∎

EXAMPLE 2.9   Show how the limit properties are used in finding the limit

$$\lim_{x \to 0} \sqrt{\frac{1 + \sin^2 x}{\cos x}} = 1$$

Solution

$$\lim_{x \to 0} \sqrt{\frac{1 + \sin^2 x}{\cos x}} = \sqrt{\lim_{x \to 0} \left(\frac{1 + \sin^2 x}{\cos x} \right)} \qquad \text{Property 7}$$

$$= \sqrt{\frac{\lim_{x \to 0} (1 + \sin^2 x)}{\lim_{x \to 0} \cos x}} \qquad \text{Property 5}$$

$$= \sqrt{\frac{\lim_{x \to 0} 1 + \lim_{x \to 0} \sin^2 x}{1}} \qquad \text{Property 2}$$

$$= \sqrt{1 + \left(\lim_{x \to 0} \sin x \right)^2} \qquad \text{Special Limit 2.1 and Property 6}$$

$$= \sqrt{1 + (0)^2} = 1 \qquad\qquad \blacksquare$$

Property 6, when applied to the function $f(x) = x$, gives

$$\lim_{x \to a} x^n = a^n$$

and this result, together with Property 1, gives

$$\lim_{x \to a} cx^n = ca^n$$

Since polynomials consist of sums or differences of terms of the form cx^n, we can apply Properties 2 and 3 repeatedly to obtain the following result, which we state as a theorem.

THEOREM 2.1

> If p is any polynomial function, then for every real number a,
>
> $$\lim_{x \to a} p(x) = p(a)$$

For example, if $p(x) = x^2 - 5x + 7$, then

$$\lim_{x \to 2} p(x) = 2^2 - 5(2) + 7 = 1$$

Theorem 2.1 says that you can find the limit of a polynomial function as x approaches a by direct substitution of $x = a$ into the function.

Now we can use Property 5 to get another theorem.

THEOREM 2.2

If f is a rational function

$$f(x) = \frac{p(x)}{q(x)}$$

where $p(x)$ and $q(x)$ are polynomials, then

$$\lim_{x \to a} f(x) = \frac{p(a)}{q(a)}$$

for all real numbers a for which $q(a) \neq 0$.

For example, if

$$f(x) = \frac{2x^2 - 3x + 4}{x^2 - 4}$$

then

$$\lim_{x \to 1} f(x) = \frac{2 - 3 + 4}{1 - 4} = \frac{3}{-3} = -1$$

Be careful, though, to observe the restriction that $q(a) \neq 0$ before applying Theorem 2.2. If $q(a)$ does equal 0, then, as we saw in Section 2.1, the limit of the quotient *may* exist if $p(a)$ also equals 0. In the next five examples we illustrate some algebraic techniques for finding limits when both numerator and denominator are zero at the point in question.

EXAMPLE 2.10   Show that

$$\lim_{x \to 1} \frac{x^2 - 1}{x - 1} = 2$$

Solution   Recall that this is the limit in Example 2.2 giving the slope of the tangent line to the graph of $y = x^2$ at the point $(1, 1)$. In that example, we used a calculator to estimate the limit.

By factoring the numerator, we can write

$$\lim_{x \to 1} \frac{x^2 - 1}{x - 1} = \lim_{x \to 1} \frac{(x + 1)(x - 1)}{x - 1} = \lim_{x \to 1} (x + 1) = 2$$

where in the last step we used Theorem 2.1. Note carefully that we can divide out the common factor $(x - 1)$ because in taking the limit we restrict x to be different from 1. Thus, when we write "$\lim_{x \to 1}$," we know that although we are letting x come arbitrarily close to 1, it is not equal to 1.

We show the graph of $y = (x^2 - 1)/(x - 1)$ in Figure 2.13. Notice that it differs from the graph of $y = x + 1$ only in that it has no value when $x = 1$. But since we exclude $x = 1$ in the limit calculation, the limit of the given function is identical to the limit of $y = x + 1$, namely, 2.   ∎

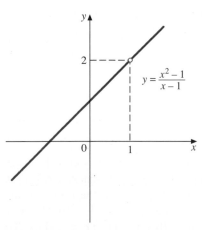

FIGURE 2.13

REMARK ————

■ Note that in the example we wrote the symbol "$\lim_{x \to 1}$" at each step until we found the limit at the last step. This symbol "$\lim_{x \to 1}$" indicates an operation to be performed and must be written until the limit is found. Similar remarks apply for other limits.

EXAMPLE 2.11   Determine whether the limit

$$\lim_{x \to -1} \frac{x^3 + 1}{x^2 - 1}$$

exists, and if it does, find its value.

x	$\dfrac{x^3 + 1}{x^2 - 1}$
-0.9	-1.43
-0.99	-1.493
-0.999	-1.4993
-1.1	-1.576
-1.01	-1.5075
-1.001	-1.50075

Sum of two cubes:

$$x^3 + a^3 = (x + a)(x^2 - ax + a^2)$$

Difference of two cubes:

$$x^3 - a^3 = (x - a)(x^2 + ax + a^2)$$

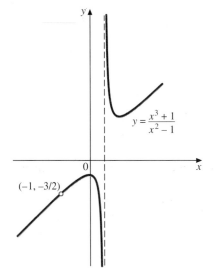

$$y = \frac{x^3 + 1}{x^2 - 1}$$

$(-1, -3/2)$

FIGURE 2.14

t	$f(t)$
± 0.5	0.533
± 0.1	0.501
± 0.01	0.50001

Solution   For this limit we see once again that we cannot simply substitute $x = -1$ into the numerator and denominator. The result of this substitution would be the undefined quantity $0/0$.

To examine the behavior of $(x^3 + 1)/(x^2 - 1)$ as x approaches -1, we can first make a table of values, such as the one in the margin. The table suggests that the limit may actually exist and equal -1.5. Note that as x approaches -1, the numerator approaches 0, causing the fraction to get smaller, while the denominator also approaches 0 but causes the fraction to get larger. Apparently these two trends cancel each other.

To see this limit algebraically, we need to simplify the expression $(x^3 + 1)/(x^2 - 1)$. By factoring the numerator and denominator, we get

$$\lim_{x \to -1} \frac{x^3 + 1}{x^2 - 1} = \lim_{x \to -1} \frac{(x + 1)(x^2 - x + 1)}{(x + 1)(x - 1)}$$

$$= \lim_{x \to -1} \frac{x^2 - x + 1}{x - 1}$$

Note carefully that the equality of the last two fractions is valid because in taking the limit as $x \to -1$, we are excluding $x = -1$, so the common factor $(x + 1)$ in the numerator and denominator is not 0. Now we can apply Theorem 2.2, since the denominator $x - 1$ is not 0 when $x = -1$. Thus,

$$\lim_{x \to -1} \frac{x^3 + 1}{x^2 - 1} = \lim_{x \to -1} \frac{x^2 - x + 1}{x - 1} = -\frac{3}{2}$$

We show the graph of $y = (x^3 + 1)/(x^2 - 1)$ in Figure 2.14. Notice the missing point when $x = -1$. The limit approached, however, is the same as that approached by the function $(x^2 - x + 1)/(x - 1)$, which is identical to the original function except that the value for $x = -1$ is defined, namely, $-3/2$. ∎

EXAMPLE 2.12   Evaluate the limit

$$\lim_{t \to 0} \frac{1 - \cos t}{\sin^2 t}$$

Solution   Since $\cos 0 = 1$ and $\sin 0 = 0$, the function assumes the form $0/0$ when $t = 0$ and so is not defined there. Using a calculator, CAS, or a spreadsheet we obtain the table in the margin, indicating that the limit appears to be 0.5. (Note that we made use of the fact that $f(t) = (1 - \cos t)/\sin^2 t$ is an even function in constructing the table. That is, $f(-t) = f(t)$.) To confirm the conclusion suggested by the table, we use the trigonometric identity $\sin^2 t + \cos^2 t = 1$ to replace $\sin^2 t$ with $1 - \cos^2 t$, getting

$$\lim_{x \to 0} \frac{1 - \cos t}{\sin^2 t} = \lim_{t \to 0} \frac{1 - \cos t}{1 - \cos^2 t} = \lim_{t \to 0} \frac{1 - \cos t}{(1 + \cos t)(1 - \cos t)}$$

$$= \lim_{t \to 0} \frac{1}{1 + \cos t} = \frac{1}{2}$$

since by the definition of $\cos t$ (Section 1.6), it is clear that $\cos t \to 1$ as $t \to 0$.

$$y = \frac{1 - \cos t}{\sin^2 t}$$

FIGURE 2.15

We show the graph of $y = (1 - \cos t)/\sin^2 t$ from $-\pi$ to π in Figure 2.15. As the graph shows, the function approaches the missing value, $1/2$, as x approaches 0. ∎

EXAMPLE 2.13   Verify the limit found in Example 2.1:

$$\lim_{x \to 0} \frac{\sqrt{x + 16} - 4}{x} = \frac{1}{8}$$

Solution   This time we change the fraction to a form in which we can use our limit properties by *rationalizing the numerator*. That is, we get rid of the radical on the numerator. To do so, we multiply both numerator and denominator by $(\sqrt{x + 16} + 4)$, using the fact that the difference of two numbers times their sum gives the difference of their squares. (Recall that we did a similar calculation in Example 1.3.)

$$\lim_{x \to 0} \frac{\sqrt{x + 16} - 4}{x} = \lim_{x \to 0} \frac{\sqrt{x + 16} - 4}{x} \cdot \frac{\sqrt{x + 16} + 4}{\sqrt{x + 16} + 4}$$

$$= \lim_{x \to 0} \frac{(x + 16) - 16}{x(\sqrt{x + 16} + 4)}$$

$$= \lim_{x \to 0} \frac{x}{x(\sqrt{x + 16} + 4)}$$

$$= \lim_{x \to 0} \frac{1}{\sqrt{x + 16} + 4} \qquad \text{Since } x \neq 0$$

Now the denominator does not approach 0 as $x \to 0$, so we can use Properties 5 and 7, together with Theorem 2.1, to get

$$\lim_{x \to 0} \frac{1}{\sqrt{x + 16} + 4} = \frac{1}{\sqrt{16} + 4} = \frac{1}{8}$$

We showed the graph of $y = (\sqrt{x + 16} - 4)/x$ in Figure 2.2. ∎

The limit in the next example can be interpreted as the slope of the tangent line to a curve, much as in Example 2.2 but with different notation. To see why, consider Figure 2.16, where we have shown the graph of $y = \sqrt{x}$. The point $P(x, \sqrt{x})$ for $x > 0$ is arbitrary, but fixed. The point $Q(x + h, \sqrt{x + h})$ with $h \neq 0$ varies as h varies. In particular, Q approaches P as h approaches 0. The slope of the secant line PQ is

$$m_{\text{sec}} = \frac{\sqrt{x + h} - \sqrt{x}}{(x + h) - x} = \frac{\sqrt{x + h} - \sqrt{x}}{h}$$

The slope of the tangent line at P, if it exists, is then the limit of m_{sec} as $h \to 0$. It is this limit that we consider in the next example.

EXAMPLE 2.14   Evaluate the limit

$$\lim_{h \to 0} \frac{\sqrt{x + h} - \sqrt{x}}{h} \qquad (x > 0)$$

Solution   Note that in finding the limit as $h \to 0$, x is considered fixed and h is the variable. Again, if we allowed h to be 0, we would have $0/0$, so we must exclude $h = 0$. Since x is not specified, we cannot make a table of values

FIGURE 2.16

to give any indication of the limit. Again, we rationalize the numerator, as in the previous example.

$$\lim_{h \to 0} \frac{\sqrt{x+h} - \sqrt{x}}{h} = \lim_{h \to 0} \frac{\sqrt{x+h} - \sqrt{x}}{h} \cdot \frac{\sqrt{x+h} + \sqrt{x}}{\sqrt{x+h} + \sqrt{x}}$$

$$= \lim_{h \to 0} \frac{(x+h) - x}{h(\sqrt{x+h} + \sqrt{x})} = \lim_{h \to 0} \frac{h}{h(\sqrt{x+h} + \sqrt{x})}$$

$$= \lim_{h \to 0} \frac{1}{\sqrt{x+h} + \sqrt{x}}$$

Now the limit of the denominator is no longer 0 as $h \to 0$, so by our limit properties we have

$$\lim_{h \to 0} \frac{1}{\sqrt{x+h} + \sqrt{x}} = \frac{1}{\sqrt{x} + \sqrt{x}} = \frac{1}{2\sqrt{x}}$$

From the discussion preceding this example, we can conclude that for any $x > 0$, the slope of the tangent line to the graph of $y = \sqrt{x}$ at $(x, \sqrt{x})$ is $1/(2\sqrt{x})$. For example, at $(1, 1)$ the slope is $1/(2\sqrt{1}) = 1/2$, and at $(4, 2)$ the slope is $1/(2\sqrt{4}) = 1/4$. ■

In each of Examples 2.10 through 2.14, we replaced the original function with a simpler one whose limit we could find. The two functions were equal except for the one point being approached by x. For instance, in Example 2.10 we replaced

$$\frac{x^2 - 1}{x - 1} \quad \text{with} \quad x + 1$$

since these two functions are equal except when $x = 1$. But in finding the limit as x approaches 1, we exclude $x = 1$ from consideration. So the limits are the same. We can state the general result as follows.

If $f(x) = g(x)$ except when $x = a$, and if $\lim_{x \to a} g(x) = L$, then $\lim_{x \to a} f(x) = L$ also.

The Squeeze Theorem

Sometimes it may be difficult to compute a limit directly, but it may be possible to compute it indirectly by "squeezing" the function between two known functions that have the same limit. For example, suppose we want to find

$$\lim_{x \to a} f(x)$$

but do not see a way of finding it directly. If we can find two functions g and h such that

$$g(x) \le f(x) \le h(x)$$

and for which $\lim_{x \to a} g(x) = \lim_{x \to a} h(x) = L$, then it is evident geometrically that $\lim_{x \to a} f(x) = L$ also (see Figure 2.17). We summarize this result in the following theorem.

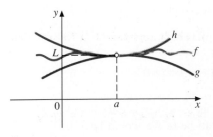

FIGURE 2.17

THEOREM 2.3

The Squeeze Theorem

If f, g, and h are functions satisfying

$$g(x) \leq f(x) \leq h(x)$$

for all x in some open interval containing a, except possibly at a itself, and if

$$\lim_{x \to a} g(x) = L \qquad \text{and} \qquad \lim_{x \to a} h(x) = L$$

then

$$\lim_{x \to a} f(x) = L$$

EXAMPLE 2.15   Prove that

$$\lim_{x \to 0} x \sin \frac{1}{x} = 0$$

We are using the following facts concerning absolute values: For all real numbers a and b,

$$|ab| = |a||b|$$

and

$$|a| \leq |b|$$

is equivalent to

$$-|b| \leq a \leq |b|$$

Solution   Since the sine of a number never exceeds 1 in absolute value and since

$$\left| x \sin \frac{1}{x} \right| = |x| \left| \sin \frac{1}{x} \right|$$

we have

$$\left| x \sin \frac{1}{x} \right| \leq |x|$$

That is,

$$-|x| \leq x \sin \frac{1}{x} \leq |x| \qquad (x \neq 0)$$

We can use limit properties and Special Limit 2.2 to show that both $|x|$ and $-|x|$ approach 0. For if $x > 0$, $|x| = x$, and we therefore have

$$\lim_{x \to 0^+} |x| = \lim_{x \to 0^+} x = 0 \qquad \text{Special Limit 2.2}$$

Whereas if $x < 0$, $|x| = -x = (-1)x$, so

$$\lim_{x \to 0^-} |x| = \lim_{x \to 0^-} (-1)x = (-1) \lim_{x \to 0^-} x = 0 \qquad \text{Property 1 and Special Limit 2.2}$$

Since these one-sided limits are equal, $\lim_{x \to 0} |x|$ must be 0. Now, using Property 1 again with $c = -1$, we get

$$\lim_{x \to 0} (-|x|) = (-1) \lim_{x \to 0} |x| = 0$$

Finally, we apply the Squeeze Theorem with $g(x) = -|x|$ and $h(x) = |x|$, both of which have limit 0 as $x \to 0$. Since

$$f(x) = x \sin \frac{1}{x}$$

is squeezed between $g(x)$ and $h(x)$, both of which approach 0, it follows that $f(x)$ also approaches 0 as $x \to 0$; that is,

$$\lim_{x \to 0} x \sin \frac{1}{x} = 0$$

The graphs of $g(x) = -|x|$, $h(x) = |x|$, and $f(x) = x \sin \frac{1}{x}$ are shown in Figure 2.18.   ■

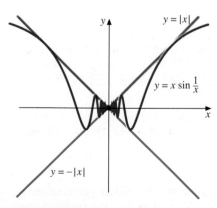

$y = |x|$

$y = x \sin \frac{1}{x}$

$y = -|x|$

FIGURE 2.18

Exercise Set 2.2

In Exercises 1–16, evaluate the limits.

1. (a) $\lim_{x \to 1}(2x^2 - 3x + 1)$

   (b) $\lim_{x \to -2} \dfrac{2}{x + 1}$

2. (a) $\lim_{x \to -3} \dfrac{2x^2 - 5x + 1}{x^2 - 2x}$ (b) $\lim_{x \to 3} \dfrac{x - 2}{\sqrt{x^2 - x - 2}}$

3. (a) $\lim_{x \to 0} \dfrac{1 - \sin x}{\cos x}$ (b) $\lim_{x \to \pi/2} \sqrt{2 \sin x - 3 \cos x}$

4. (a) $\lim_{x \to \pi} \dfrac{\sec x}{\tan x + 2}$ (b) $\lim_{x \to 1/2} \dfrac{\sin \pi x}{1 - \cos \pi x}$

5. $\lim_{x \to -2} \dfrac{x^2 - 4}{x + 2}$ **6.** $\lim_{x \to 3/2} \dfrac{4x^2 - 9}{2x - 3}$

7. $\lim_{x \to 3} \dfrac{x^2 - x - 6}{x - 3}$ **8.** $\lim_{x \to 7} \dfrac{x^2 - 5x - 14}{x - 7}$

9. $\lim_{x \to 1} \dfrac{x^2 + 3x - 5}{x^2 - 3x + 2}$ **10.** $\lim_{x \to -3} \dfrac{x^2 + 5x + 6}{x^2 + 6x + 9}$

11. $\lim_{x \to 1} \dfrac{\sqrt{x} - 1}{x - 1}$ **12.** $\lim_{x \to 2} \dfrac{\sqrt{2x - 3} - 1}{x - 2}$

13. $\lim_{x \to 5} \dfrac{\sqrt{3x + 1} - 4}{x - 5}$ **14.** $\lim_{x \to -2} \dfrac{\sqrt{2 - x} - 2}{x + 2}$

15. $\lim_{x \to \pi} \dfrac{1 + \cos x}{\sin^2 x}$ **16.** $\lim_{x \to -\pi/4} \dfrac{\sin x + \cos x}{\sin^2 x - \cos^2 x}$

In Exercises 17–24, determine the domain of f. Then write $f(x)$ in a simpler form for x in that domain and draw its graph, observing any restrictions on its domain. Then, for the given value of a, find $\lim_{x \to a^+} f(x)$, $\lim_{x \to a^-} f(x)$, and $\lim_{x \to a} f(x)$, or show they fail to exist.

17. $f(x) = \dfrac{x^2 - x - 2}{x + 1}$; $a = -1$

18. $f(x) = \dfrac{2x^2 + x - 1}{2x - 1}$; $a = \dfrac{1}{2}$

19. $f(x) = \dfrac{6x^2 + x - 2}{3x + 2}$; $a = -\dfrac{2}{3}$

20. $f(x) = \dfrac{6 + 7x - 3x^2}{x - 3}$; $a = 3$

21. $f(x) = \dfrac{x^3 + 2x^2}{x + 2}$; $a = -2$

22. $f(x) = \dfrac{x^3 - x^2 + x - 1}{x - 1}$; $a = 1$

23. $f(x) = \dfrac{x + 1}{x^2 - 1}$; $a = -1$

24. $f(x) = \dfrac{x - 2}{4 - x^2}$; $a = 2$

In Exercises 25–30, evaluate the limit

$$\lim_{h \to 0} \frac{f(a + h) - f(a)}{h}$$

for the given values of a and $f(x)$, or show that the limit does not exist. (We will see in Chapter 3 that this type of limit is the basis for defining the derivative of a function.)

25. $a = 2$; $f(x) = 2x - 1$

26. $a = 3$; $f(x) = x^2 - 1$

27. $a = 1$; $f(x) = \dfrac{1}{x}$

28. $a = -1$; $f(x) = \sqrt{3 - x}$

29. $a = 0$; $f(x) = \dfrac{x}{x + 1}$

30. $a = 2$; $f(x) = \dfrac{1}{\sqrt{x + 2}}$

31. Find functions f and g such that $\lim_{x \to a} f(x)$ and $\lim_{x \to a} g(x)$ both fail to exist but
   (a) $\lim_{x \to a}[f(x) + g(x)]$ exists.
   (b) $\lim_{x \to a}[f(x) \cdot g(x)]$ exists.

32. Do functions f and g exist such that $\lim_{x \to a} f(x)$ exists, and $\lim_{x \to a} g(x)$ does not exist, and
   (a) $\lim_{x \to a}[f(x) + g(x)]$ exists?
   (b) $\lim_{x \to a}[f(x) \cdot g(x)]$ exists?
   In each case, if your answer is yes, give examples. If your answer is no, explain why.

In Exercises 33 and 34, use the Squeeze Theorem to prove the given limit statement.

33. $\lim_{x \to 0} x^2 \cos 2x = 0$ **34.** $\lim_{x \to 1} |x - 1| \sin x = 0$

2.3 CONTINUITY

Informally stated, a function is *continuous* at a given point if its graph does not have a break there. Most functions describing physical phenomena are continuous. For example, the velocity of a moving object as a function of time is continuous. A car cannot suddenly jump from a speed of 20 mph to 40 mph. Rather, its speed varies continuously over all values between 20 and 40. Similarly, air pressure is a continuous function of height above sea level. You can probably think of many other examples of continuous functions. On the other hand, the cost of domestic first-class postage as a function of weight is *discontinuous* at certain weights. A letter weighing 1 oz or less costs 29¢ in 1994, but if it exceeds 1 oz, even slightly, but is less than 2 oz, the cost jumps to 52¢.

Defining continuity of a function at a point as meaning that the graph does not break there is not totally satisfactory. How can we tell for sure whether the graph has a break? Even CAS-generated graphs might mask a break if the scale is poorly chosen. Let us be precise about what we mean by "no break." Suppose the domain of a function f includes an open interval containing the point $x = a$. Suppose further that $\lim_{x \to a} f(x)$ exists. If this limiting value is equal to the function value $f(a)$, then we say f is continuous at $x = a$. So continuity at $x = a$ means not only that the function *approaches* a limiting value as x approaches a, but also that this limit is *reached* when $x = a$ and is exactly the value of the function at that point.

Figure 2.19 illustrates this situation. The limit approached by $f(x)$ as x approaches a is the same as the value of f at a. Functions can be continuous at some points and not at others. When f is continuous at every point of some interval, we say it is continuous on that interval. The following definition summarizes this discussion.

FIGURE 2.19

Definition 2.2 Continuity of a Function	Let f be defined in an open interval containing the point $x = a$. Then we say that f **is continuous at a**, provided that $$\lim_{x \to a} f(x) = f(a)$$ If f is continuous at each point of an open interval I, then we say that f **is continuous on I**.

Continuity of f at a implies three conditions:

1. $\lim_{x \to a} f(x)$ exists.
2. $f(a)$ exists (i.e., a is in the domain of f).
3. The values in Conditions 1 and 2 are the same.

So a function can *fail* to be continuous—that is, it can be **discontinuous**—at $x = a$ by failing to satisfy one or more of these three conditions.

When a function is defined on a *closed* interval I, we extend the definition of continuity on I by additional requirements at the endpoints where only one-sided limits are used. In particular, if $I = [a, b]$, we require that f be **continuous from the right** at a, that is, $\lim_{x \to a^+} f(x) = f(a)$, and **continuous from the left** at b, that is, $\lim_{x \to b^-} f(x) = f(b)$.

It is instructive to look at various discontinuous functions. Examples are provided in Figures 2.20 through 2.23.

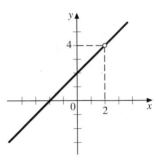

FIGURE 2.20
f is discontinuous at $x = 0$. $\lim_{x \to 0} f(x)$ does not exist and $f(0)$ does not exist.

FIGURE 2.21
f is discontinuous at $x = 1$. $f(1) = 0$, but $\lim_{x \to 1^+} f(x) \neq \lim_{x \to 1^-} f(x)$, so $\lim_{x \to 1} f(x)$ does not exist.

FIGURE 2.22
f is discontinuous at $x = 2$. $f(2) = 0$ but $\lim_{x \to 2} f(x) = 4$, so $\lim_{x \to 2} f(x) \neq f(2)$.

FIGURE 2.23
f is discontinuous at $x = 2$. $\lim_{x \to 2} f(x) = 4$, but $f(2)$ is not defined.

The type of discontinuity illustrated in Figure 2.20 is called an **infinite** discontinuity, that in Figure 2.21 is called a **simple** (or **jump**) discontinuity, and those in Figures 2.22 and 2.23 are called **removable** discontinuities (since by defining, or redefining, the function at one point only, the discontinuity can be removed). These examples illustrate the most common types of discontinuities. There are others, some of which we will see later.

The function

$$f(x) = \frac{\sin x}{x}$$

that we considered in Example 2.5 has a removable discontinuity at $x = 0$. Although $f(0)$ is not defined, $\lim_{x \to 0} f(x) = 1$. (We conjectured this limit in Example 2.5 and will give a proof later.) Thus, if we *define* $f(0)$ to be 1, in addition to $f(x) = (\sin x)/x$ for $x \neq 0$, we get a continuous function. We show again the graph of $(\sin x)/x$ given by a CAS in Figure 2.24. Although the point $(0, 1)$ is not on the graph, it is difficult to detect a missing point. So what we see appears to be the graph of the continuous function in which $f(0)$ has been defined as 1.

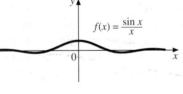

FIGURE 2.24

Based on results we have obtained, we can easily determine where polynomial and rational functions are continuous. Recall that, according to Theorem 2.1, if p is any polynomial function, then

$$\lim_{x \to a} p(x) = p(a)$$

But this is precisely the requirement for continuity at $x = a$. Thus, a polynomial function is continuous for every real number. Similarly, Theorem 2.2 states that if p and q are polynomial functions, and $f(x) = p(x)/q(x)$, then for all real numbers a for which $q(a) \neq 0$,

$$\lim_{x \to a} f(x) = f(a)$$

Thus, every rational function $p(x)/q(x)$ is continuous for all points in its domain. We state these results as a theorem.

THEOREM 2.4
> Every polynomial function is continuous on all of $\mathbb{R}$, and every rational function is continuous wherever it is defined.

EXAMPLE 2.16 Determine where each of the following functions is continuous.

(a) $f(x) = \sqrt{3}x^5 + 2.7x^3 - \pi x + 3$

(b) $g(x) = \dfrac{2x^4 + 3x^2 - 7}{x^5 - 16x}$

Solution

(a) f is a polynomial function and so is continuous for all x. Its CAS-generated graph is shown in Figure 2.25.

(b) g is a rational function whose domain includes all real numbers x except those for which $x^5 - 16x = 0$. Solving for x, we get

$$x(x^4 - 16) = 0$$
$$x(x^2 - 4)(x^2 + 4) = 0$$
$$x(x - 2)(x + 2)(x^2 + 4) = 0$$

So the three points for which g is undefined are $x = 0$, $x = 2$, and $x = -2$. Thus, g is continuous for all other values of x. As the graph (Figure 2.26) shows, g has infinite discontinuities at $x = 0$, $x = 2$, and $x = -2$. ■

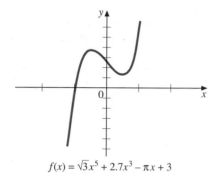

$f(x) = \sqrt{3}x^5 + 2.7x^3 - \pi x + 3$

FIGURE 2.25
f is continuous for all x

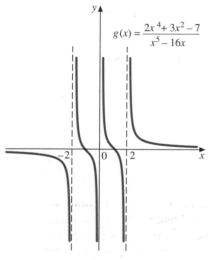

$g(x) = \dfrac{2x^4 + 3x^2 - 7}{x^5 - 16x}$

FIGURE 2.26
g has infinite discontinuities at $x = 0$, $x = 2$, and $x = -2$, and is continuous for all other x.

Combinations of Continuous Functions

Certain combinations of continuous functions result in a new continuous function. This fact enables us in many cases to recognize continuous functions without having to rely on Definition 2.2. For example, if a function is the sum of two continuous functions, it is also continuous. To see why, suppose $\lim_{x \to a} f(x) = f(a)$ and $\lim_{x \to a} g(x) = g(a)$. Then, by Property 2 of limits in Section 2.1,

$$\lim_{x \to a} [f(x) + g(x)] = \lim_{x \to a} f(x) + \lim_{x \to a} g(x) = f(a) + g(a)$$

Thus, if $h = f + g$, then $\lim_{x \to a} h(x) = h(a)$, so h is continuous at $x = a$. Similar results hold for differences, products, and quotients of continuous functions (where, as always for quotients, we must require that the denominator be nonzero at $x = a$). We therefore have the following theorem.

THEOREM 2.5
> If f and g are continuous at $x = a$, then so are $f + g$, $f - g$, $f \cdot g$, and f/g, provided that in the case of f/g, $g(a) \neq 0$.

Thus, we are able to recognize a broad class of continuous functions. This class can be extended further by the next theorem, which we will prove in Section 2.5. In essence it says that *a continuous function of a continuous function is continuous*. Recall (Section 1.3) that the domain of the composite function $f \circ g$ consists of all x in the domain of g for which $g(x)$ is in the domain of f. In the following theorem we suppose $x = a$ to be such a point.

THEOREM 2.6

> Suppose the composite function $f \circ g$ is defined in some open interval containing the point $x = a$. If g is continuous at $x = a$ and f is continuous at $g(a)$, then $f \circ g$ is continuous at $x = a$.

EXAMPLE 2.17   Show that $F(x) = \sqrt{x^2 + 1}$ is continuous on all of $\mathbb{R}$.

Solution   We can express F as the composition $f \circ g$, where $f(x) = \sqrt{x}$ and $g(x) = x^2 + 1$. Since g is a polynomial function, it is continuous on all of $\mathbb{R}$. Furthermore, $g(x) > 0$ for all x, so $(f \circ g)(x) = \sqrt{x^2 + 1}$ is defined for all x. By Limit Property 7 in Section 2.2 and by Special Limit 2.2, we can see that for $a > 0$, $\lim_{x \to a} \sqrt{x} = \sqrt{a}$. Thus, f is continuous everywhere on its domain. By Theorem 2.6 it now follows that $f \circ g$ is continuous at every real number a. That is, F is continuous on all of $\mathbb{R}$.   ∎

Continuity of the Trigonometric Functions

From the definitions of $\cos t$ and $\sin t$ as the x- and y-coordinates of a point on the unit circle, we can see geometrically that both functions are continuous for all real values of t. Since the other four trigonometric functions are defined in terms of $\sin t$ and $\cos t$, we can make use of Theorem 2.5 to conclude that they are also continuous except where their denominators equal 0. We summarize these results in the next theorem.

THEOREM 2.7

> The functions $\sin t$ and $\cos t$ are continuous for all real t. The functions $\tan t$ and $\sec t$ are continuous except where $\cos t = 0$, namely, $t = \pi/2 + n\pi$, for $n = 0, \pm 1, \pm 2, \ldots$. The functions $\cot t$ and $\csc t$ are continuous except where $\sin t = 0$, namely, $t = n\pi$, for $n = 0, \pm 1, \pm 2, \ldots$.

EXAMPLE 2.18   Determine where each of the following functions is continuous.
(a) $\sin\left(\dfrac{1 + x}{1 - x}\right)$ (b) $\cot\left(\dfrac{1}{t}\right)$

Solution

(a) Let $f(x) = \sin x$ and $g(x) = (1 + x)/(1 - x)$. Then

$$(f \circ g)(x) = f(g(x)) = \sin\left(\frac{1 + x}{1 - x}\right)$$

According to Theorem 2.6, this composite function is continuous at $x = a$ if g is continuous at a and f is continuous at $g(a)$. In this instance, g is continuous everywhere except at $x = 1$, and f is continuous everywhere. So the given composite function is continuous for all x except possibly at $x = 1$. Now we check $x = 1$. At $x = 1$, the composite function is not defined, so it is discontinuous there.

(b) Reasoning as above, the composite function $\cot(1/t)$ is continuous except where $t = 0$ (where $1/t$ is discontinuous) and where $1/t = n\pi$ (where the cotangent is discontinuous). Thus, $\cot(1/t)$ is continuous for all real t except where $t = 0$ or $t = 1/n\pi$, for $n = \pm1, \pm2, \pm3, \dots$. ∎

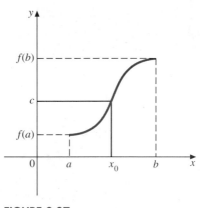

FIGURE 2.27

The Intermediate-Value Theorem

Functions that are continuous on an interval, especially on a closed and bounded interval, have many important properties. We will see some of these in later chapters. For now, we state one such property whose truth seems intuitively obvious, but which is surprisingly difficult to prove. The proof can be found in textbooks on advanced calculus.

THEOREM 2.8

> **The Intermediate-Value Theorem**
>
> Let f be continuous on the closed interval $[a, b]$ with $f(a) \neq f(b)$. If c is any number between $f(a)$ and $f(b)$, then there is at least one number x_0 in the open interval (a, b) such that $f(x_0) = c$.

Figure 2.27 illustrates the idea of this theorem. The graph crosses each horizontal line $y = c$, where c is between $f(a)$ and $f(b)$. One consequence of it is that if f is continuous on $[a, b]$ with $f(a)$ and $f(b)$ opposite in sign, then the equation $f(x) = 0$ has at least one real root. For example, every polynomial of odd degree must have at least one real zero. (You will be asked to prove this result in Exercise 45 of Exercise Set 2.3.)

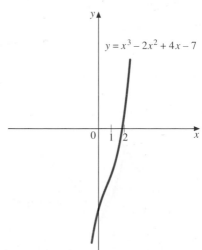

FIGURE 2.28

EXAMPLE 2.19   Show that the equation

$$x^3 - 2x^2 + 4x - 7 = 0$$

has a root between $x = 1$ and $x = 2$.

Solution   Let $f(x) = x^3 - 2x^2 + 4x - 7$. We know that f is continuous for all x since it is a polynomial function. By substituting $x = 1$ and $x = 2$, we find that $f(1) = -4$ and $f(2) = 1$. The Intermediate-Value Theorem tells us that f assumes all values between -4 and 1 as x ranges from 1 to 2. Since 0 is between -4 and 1, it follows that there is a number x_0 in the open interval $(1, 2)$ such that $f(x_0) = 0$. That is, x_0 is a root of the equation $x^3 - 2x^2 + 4x - 7 = 0$ that lies between $x = 1$ and $x = 2$. We show the graph of f in Figure 2.28. ∎

Exercise Set 2.3

In Exercises 1–10, determine from the graph where the function is continuous and where it is discontinuous. At points of discontinuity, describe in what way or ways the definition of continuity fails.

1.

2.

3.

4.

5.

6.

7.

8.

9.

10.

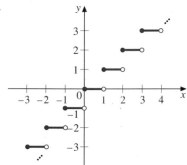

In Exercises 11–24, determine whether the function f is continuous at $x = a$. If it is discontinuous, state in what way the definition of continuity is not satisfied.

11. $f(x) = x^2 - 3x + 4$; $a = 2$

12. $f(x) = 2 - 3x + 4x^3$; $a = -1$

13. $f(x) = \dfrac{x}{x - 1}$; $a = 3$

14. $f(x) = \dfrac{2x - 3}{x^2 - 1}$; $a = 4$

15. $f(x) = \dfrac{x^2 - 2x + 1}{3x^2 - x - 3}$; $a = -1$

16. $f(x) = \dfrac{x^3 - 1}{x^2 + 2x - 8}$; $a = -4$

17. $f(x) = |x - 2|$; $a = 2$

18. $f(x) = \dfrac{|x|}{x}$; $a = 0$

19. $f(x) = \dfrac{x - 1}{x - 2}$; $a = 0$

20. $f(x) = \dfrac{x^2 - 4}{x - 2}$; $a = 2$

21. $f(x) = \dfrac{x + 3}{x^2 - 9}$; $a = -3$

22. $f(x) = \dfrac{x^2 - 2x - 3}{x^2 - 5x + 6}$; $a = 3$

23. $f(x) = \begin{cases} \sin \dfrac{1}{x} & \text{if } x \neq 0 \\ 0 & \text{if } x = 0 \end{cases}$; $a = 0$

24. $f(x) = \begin{cases} x \sin \dfrac{1}{x} & \text{if } x \neq 0 \\ 0 & \text{if } x = 0 \end{cases}$; $a = 0$

In Exercises 25–34, graph the given function and state on what intervals it is continuous. When an endpoint is included, indicate whether the function is continuous from the right or the left.

25. $f(x) = \begin{cases} x^2 + 1 & \text{if } x \geq 0 \\ 1 - x & \text{if } x < 0 \end{cases}$

26. $g(x) = \begin{cases} x + 2 & \text{if } x > 1 \\ 2x + 1 & \text{if } x \leq 1 \end{cases}$

27. $h(x) = \begin{cases} \dfrac{1}{x} & \text{if } x > 0 \\ -x & \text{if } x \leq 0 \end{cases}$

28. $F(x) = \begin{cases} \dfrac{1}{x^2} & \text{if } x \neq 0 \\ 0 & \text{if } x = 0 \end{cases}$

29. $G(x) = \begin{cases} \dfrac{|x|}{x} & \text{if } x \neq 0 \\ 1 & \text{if } x = 0 \end{cases}$

30. $H(x) = \begin{cases} x^2 - 4 & \text{if } x > 2 \\ 1 & \text{if } x = 2 \\ x - 1 & \text{if } x < 2 \end{cases}$

31. $f(x) = \begin{cases} \dfrac{x^2 - 9}{x - 3} & \text{if } x \neq 3 \\ 6 & \text{if } x = 3 \end{cases}$

32. $g(x) = \begin{cases} \dfrac{4 - x^2}{x - 2} & \text{if } x \neq 2 \\ -4 & \text{if } x = 2 \end{cases}$

33. $f(x) = \begin{cases} 2x + 1 & \text{if } x < 0 \\ 1 - x & \text{if } 0 \leq x < 2 \\ x - 2 & \text{if } x \geq 2 \end{cases}$

34. $f(x) = \begin{cases} \sqrt{x} & \text{if } x \geq 0 \\ -1 & \text{if } -2 \leq x < 0 \\ 2x + 5 & \text{if } x < -2 \end{cases}$

35. Let

$$f(x) = \frac{x^2 - 9}{x^2 - x - 6}$$

if $x \neq 3$. Define $f(3)$ so that f is continuous at $x = 3$.

36. Let

$$f(x) = \begin{cases} px & \text{if } x \leq 2 \\ qx^2 + x + 3 & \text{if } x > 2 \end{cases}$$

Determine all choices of p and q such that f is continuous at $x = 2$.

37. Let $\phi(x) = \begin{cases} 1 & \text{if } x \text{ is rational} \\ 0 & \text{if } x \text{ is irrational} \end{cases}$.
Discuss the continuity of ϕ (phi). (*Hint*: The rationals and the irrationals are *dense* in $\mathbb{R}$. That is, between any two real numbers there is at least one rational number and one irrational number.)

38. Let $\psi(x) = \begin{cases} x & \text{if } x \text{ is rational} \\ 0 & \text{if } x \text{ is irrational} \end{cases}$.
Discuss the continuity of ψ (psi). (See the hint for Exercise 37.)

39. A function is said to be **bounded** if there is some positive number M such that the graph of f lies between the horizontal lines $y = M$ and $y = -M$. Give an example of a function that is continuous on an open interval but is not bounded there.

40. Give an example of a function that has a finite value at each point of a closed interval $[a, b]$ but is not bounded. (See Exercise 39.) (*Hint*: The function will have to be discontinuous.)

41. Does there exist a function that is discontinuous at every point? Does there exist a function that is continuous at

one and only one point? (*Hint*: See Exercises 37 and 38.)

42. Suppose f is continuous on $\mathbb{R}$ and p is a polynomial function for which it is known that $f(x) = p(x)$ for all $x \in \mathbb{R}$ except the single point $x = a$. Show that $f(a) = p(a)$; that is, $f(x)$ must be identical to $p(x)$ everywhere. What conclusion can we draw if we know only that $f(x) = p(x)$ except for finitely many values of x? Explain.

43. Prove or disprove: If f and g are discontinuous at $x = a$, then (a) $f + g$ is discontinuous at $x = a$; (b) $f \cdot g$ is discontinuous at $x = a$.

44. Prove that if f is continuous on an interval I, then so is $|f|$. Is the converse of this statement also true? Prove or disprove.

45. Prove that every polynomial function p_n of *odd* degree has at least one real zero; that is, $p_n(x_0) = 0$ for some x_0.

46. Refer to the first-class postage function described at the beginning of Chapter 1 (see also Exercise 23 in Exercise Set 1.1). Where is this function continuous? Where is it discontinuous?

47. Let $f(x)$ denote the greatest integer function $[\![x]\!]$ (see Exercise 22 in Exercise Set 1.1). Determine all real x for which f is continuous and all real x for which f is discontinuous.

2.4 INFINITE LIMITS AND LIMITS AT INFINITY

Throughout our discussion of limits, we have understood that when we write $\lim_{x \to a} f(x) = L$, both a and L represent finite real numbers. It is possible, however, to assign meaning to the limit statement when either a or L is replaced (or both are replaced) by one of the symbols ∞ or $-\infty$. In a sense, we have already discussed limits of this sort when we considered vertical and horizontal asymptotes in Section 1.6, although we did not use limit terminology. Consider, for example, the function

x	$1/x^2$
± 1	1
± 0.1	100
± 0.01	10,000
± 0.001	1,000,000

$$f(x) = \frac{1}{x^2}$$

As $x \to 0$, we see that $f(x)$ becomes arbitrarily large (see the table). In Section 1.6 we used the expression

$$\text{as } x \to 0, \ f(x) \to \infty$$

to indicate this behavior. Equivalently, we can (and usually do) write

x	$1/x^2$
± 10	0.01
± 100	0.0001
± 1000	0.000001

$$\lim_{x \to 0} f(x) = \infty$$

Graphically, we know then that the line $x = 0$ is a vertical asymptote.

FIGURE 2.29
$$\lim_{x \to 0} \frac{1}{x^2} = \infty, \ \lim_{x \to \infty} \frac{1}{x^2} = 0,$$
$$\lim_{x \to -\infty} \frac{1}{x^2} = 0$$

When looking for horizontal asymptotes, we want to see if $f(x)$ approaches some finite value as x becomes arbitrarily large in absolute value. For $f(x) = 1/x^2$ the line $y = 0$ is a horizontal asymptote, since $1/x^2$ becomes arbitrarily close to 0 as $|x|$ becomes larger and larger (see the bottom table on p. 110). Previously, we indicated this behavior by saying

$$\text{as } x \to \infty, \ f(x) \to 0, \ \text{and as } x \to -\infty, \ f(x) \to 0$$

Again, we can use the alternative statements

$$\lim_{x \to \infty} f(x) = 0 \quad \text{and} \quad \lim_{x \to -\infty} f(x) = 0$$

The graph of f, given in Figure 2.29, shows the vertical and horizontal asymptotes.

We summarize this discussion in the following informal definition.

Definition 2.3
Infinite Limits and Limits at Infinity

(a) The statement

$$\lim_{x \to a} f(x) = \infty$$

means that $f(x)$ will be as large as we please for all x sufficiently close to a but not equal to a.

(b) The statement

$$\lim_{x \to \infty} f(x) = L$$

means that $f(x)$ will be as close to L as we please for all sufficiently large values of x.
Analogous definitions hold for

$$\lim_{x \to a} f(x) = -\infty \text{ and } \lim_{x \to -\infty} f(x) = L$$

REMARK ────────────────────
■ In part (a) when we say "as large as we please," we mean that $f(x)$ can be made *arbitrarily large*. That is, no matter how large a positive number M is chosen, $f(x)$ will be larger than M for all x values sufficiently close to a but different from a. In part (b) when we say "as close to L as we please," we mean that $f(x)$ comes *arbitrarily close* to L when x is large enough.

────────────────────────────

In Definition 2.3(a), by restricting x to one side of a, we get the right-hand and left-hand limits

$$\lim_{x \to a^+} f(x) = \infty \qquad \text{and} \qquad \lim_{x \to a^-} f(x) = \infty$$

and similarly when ∞ is replaced by $-\infty$.

For the function $f(x) = 1/x$, for example, the one-sided limits at 0 are (see Figure 2.30)

$$\lim_{x \to 0^+} \frac{1}{x} = \infty \qquad \text{and} \qquad \lim_{x \to 0^-} \frac{1}{x} = -\infty$$

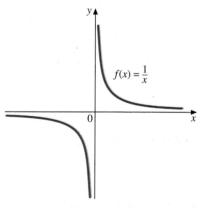

FIGURE 2.30
$$\lim_{x \to 0^+} \frac{1}{x} = \infty, \ \lim_{x \to 0^-} \frac{1}{x} = -\infty$$

REMARKS
■ We cannot write anything meaningful for $\lim_{x\to 0} 1/x$, since the left-hand and right-hand limits are not the same.
■ When we write $\lim_{x\to a} f(x) = \infty$ (or $-\infty$), we sometimes say "the limit is infinite." However, we do not mean that the limit exists but rather that it fails to exist in a particular way, namely, that the function gets large beyond all bound. When we say "the limit exists," we will always mean that it is finite.

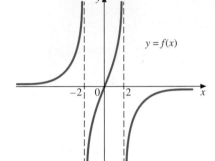

FIGURE 2.31

EXAMPLE 2.20   Refer to the function f whose graph is shown in Figure 2.31. Find each of the following:
(a) $\lim_{x\to -2^-} f(x)$ (b) $\lim_{x\to -2^+} f(x)$
(c) $\lim_{x\to 2^-} f(x)$ (d) $\lim_{x\to 2^+} f(x)$

Solution   From the graph we get the following results:
(a) $\lim_{x\to -2^-} f(x) = \infty$ (b) $\lim_{x\to -2^+} f(x) = -\infty$
(c) $\lim_{x\to 2^-} f(x) = \infty$ (d) $\lim_{x\to 2^+} f(x) = -\infty$ ■

EXAMPLE 2.21   Show that the following limits are infinite.
(a) $\lim_{x\to 1^+} \dfrac{2x+3}{\sqrt{x-1}}$ (b) $\lim_{x\to 0^-} \dfrac{1+\cos x}{\sin x}$

Solution

(a) As $x\to 1^+$, $2x+3$ comes arbitrarily close to 5, and $\sqrt{x-1}$ comes arbitrarily close to 0 through positive numbers. Thus, the quotient $(2x+3)/\sqrt{x-1}$ is positive and becomes arbitrarily large. That is,
$$\lim_{x\to 1^+} \frac{2x+3}{\sqrt{x-1}} = \infty$$
We show the graph in Figure 2.32.

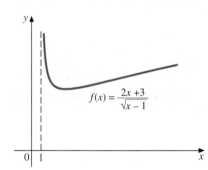

FIGURE 2.32
$\lim_{x\to 1^+} f(x) = \infty$

(b) The numerator of the fraction
$$\frac{1+\cos x}{\sin x}$$
approaches the positive number 2 as $x\to 0^-$, and the denominator approaches 0 through negative values. So we conclude that
$$\lim_{x\to 0^-} \frac{1+\cos x}{\sin x} = -\infty$$
We show the graph in Figure 2.33.   ■

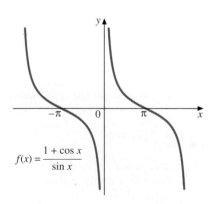

$f(x) = \dfrac{1+\cos x}{\sin x}$

FIGURE 2.33
$\lim_{x\to 0^-} f(x) = -\infty$

EXAMPLE 2.22   Find the following limits.
(a) $\lim_{x\to\infty} \dfrac{3x+5}{\sqrt{4x^2-9}}$ and $\lim_{x\to -\infty} \dfrac{3x+5}{\sqrt{4x^2-9}}$
(b) $\lim_{x\to\infty} \dfrac{\cos x}{\sqrt{x}+2}$

$$f(x) = \frac{3x+5}{\sqrt{4x^2-9}}$$

FIGURE 2.34

$\lim\limits_{x\to\infty} f(x) = \frac{3}{2}$ and $\lim\limits_{x\to-\infty} f(x) = -\frac{3}{2}$

$$f(x) = \frac{\cos x}{\sqrt{x}+2}$$

FIGURE 2.35

$\lim\limits_{x\to\infty} f(x) = 0$

Solution

(a) Using the dominant-term approach (see Section 1.3), for $|x|$ large, $3x + 5 \approx 3x$ and $4x^2 - 9 \approx 4x^2$. So

$$\frac{3x+5}{\sqrt{4x^2-9}} \approx \frac{3x}{\sqrt{4x^2}} = \frac{3x}{2|x|} \qquad \text{For } |x| \text{ large}$$

Thus,

$$\lim_{x\to\infty} \frac{3x+5}{\sqrt{4x^2-9}} = \lim_{x\to\infty} \frac{3x}{2x} = \frac{3}{2}$$

$$\lim_{x\to-\infty} \frac{3x+5}{\sqrt{4x^2-9}} = \lim_{x\to-\infty} \frac{3x}{2|x|} = \lim_{x\to-\infty} \frac{3x}{2(-x)} = -\frac{3}{2}$$

For the limit as $x \to -\infty$, we made use of the fact that $\sqrt{x^2} = |x|$ and that $|x| = -x$ when x is negative. The graph in Figure 2.34 shows the asymptote $y = 3/2$ on the right and $y = -3/2$ on the left.

(b) We can see intuitively that

$$\lim_{x\to\infty} \frac{\cos x}{\sqrt{x}+2} = 0$$

since the numerator varies between 1 and -1, while the denominator becomes arbitrarily large. So the fraction approaches 0. More formally, since $-1 \le \cos x \le 1$ for all x, we have

$$-\frac{1}{\sqrt{x}+2} \le \frac{\cos x}{\sqrt{x}+2} \le \frac{1}{\sqrt{x}+2}$$

Clearly, both $1/(\sqrt{x}+2)$ and its negative approach 0 as $x \to \infty$. The Squeeze Theorem (Theorem 2.3) continues to hold true when $x \to a$ is replaced by $x \to \infty$, so we conclude that

$$\lim_{x\to\infty} \frac{\cos x}{\sqrt{x}+2} = 0$$

(See the graph in Figure 2.35.) Note that the graph crosses the horizontal asymptote $y = 0$ infinitely many times. ∎

Vertical and Horizontal Asymptotes

Let us restate the definition of vertical and horizontal asymptotes, using the limit terminology of this section.

Definition 2.4 Vertical and Horizontal Asymptotes	(a) The line $x = a$ is a vertical asymptote to the graph of $y = f(x)$ if either of the following holds true: 　(i) $\lim_{x\to a^+} f(x) = \infty$ or $-\infty$ 　(ii) $\lim_{x\to a^-} f(x) = \infty$ or $-\infty$ (b) The line $y = b$ is a horizontal asymptote to the graph of $y = f(x)$ if either of the following holds true: 　(i) $\lim_{x\to\infty} f(x) = b$ 　(ii) $\lim_{x\to-\infty} f(x) = b$

REMARK ───

■ We can paraphrase part (a) by saying that the line $x = a$ is a vertical asymptote if $f(x)$ becomes infinite (either positively or negatively) as x approaches a from either side. Similarly, for part (b), the line $y = b$ is a horizontal asymptote if $f(x)$ comes arbitrarily close to $y = b$ as x becomes infinite (either positively or negatively).

───

In Figures 2.36 and 2.37 we show some examples of vertical and horizontal asymptotes. Notice that the graph of a function $y = f(x)$ can cross a horizontal asymptote but not a vertical one.

FIGURE 2.36
$x = -1$ and $x = 1$ are vertical asymptotes.

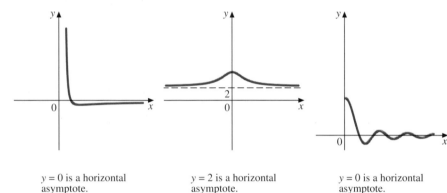

$y = 0$ is a horizontal asymptote. $y = 2$ is a horizontal asymptote. $y = 0$ is a horizontal asymptote.

(a)   (b)   (c)

FIGURE 2.37

EXAMPLE 2.23   Find all vertical and horizontal asymptotes.

(a) $f(x) = \dfrac{2x^2 + 3}{4 - x^2}$ (b) $f(x) = \dfrac{1 + \sin x}{\sqrt{x - 1}}$

Solution

(a) Note that f is even, so its graph is symmetric with respect to the y-axis. Thus, we need only consider $x \geq 0$. By factoring the denominator, we obtain

$$f(x) = \frac{2x^2 + 3}{(2 - x)(2 + x)}$$

Since

$$\lim_{x \to 2^+} \frac{2x^2 + 3}{(2 - x)(2 + x)} = -\infty \quad \text{and} \quad \lim_{x \to 2^-} \frac{2x^2 + 3}{(2 - x)(2 + x)} = \infty$$

the line $x = 2$ is a vertical asymptote. By symmetry, so is the line $x = -2$. For horizontal asymptotes, we need to calculate

$$\lim_{x \to \infty} \frac{2x^2 + 3}{4 - x^2}$$

Using dominant terms on the numerator and denominator, we see that for large x,

$$\frac{2x^2 + 3}{4 - x^2} \approx \frac{2x^2}{-x^2}$$

so

$$\lim_{x \to \infty} \frac{2x^2 + 3}{4 - x^2} = \lim_{x \to \infty} \frac{2x^2}{-x^2} = -2$$

Thus, the line $y = -2$ is a horizontal asymptote on the right, and by symmetry, also on the left. We show the graph of f in Figure 2.38.

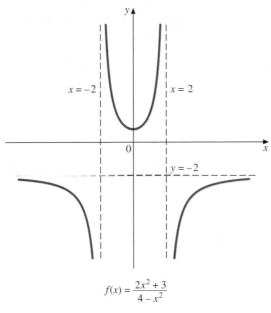

$$f(x) = \frac{2x^2 + 3}{4 - x^2}$$

FIGURE 2.38
Vertical asymptotes are $x = 2$ and $x = -2$; horizontal asymptote is $y = -2$.

(b) Note first that the domain of f is the interval $(1, \infty)$, since otherwise $\sqrt{x - 1}$ is either 0 (which is excluded since we cannot have 0 in the denominator) or not a real number. Since

$$\lim_{x \to 1^+} \frac{1 + \sin x}{\sqrt{x - 1}} = \infty$$

the line $x = 1$ is a vertical asymptote.

For horizontal asymptotes, we have

$$\lim_{x \to \infty} \frac{1 + \sin x}{\sqrt{x - 1}} = 0$$

since the numerator is always between 0 and 2 (inclusive) and the denominator becomes arbitrarily large. More explicitly, since $-1 \le \sin x \le 1$ for all x, we have

$$0 \le \frac{1 + \sin x}{\sqrt{x - 1}} \le \frac{2}{\sqrt{x - 1}}$$

Again, we can use the Squeeze Theorem to see that since

$$\lim_{x \to \infty} \frac{2}{\sqrt{x - 1}} = 0$$

$$f(x) = \frac{1 + \sin x}{\sqrt{x^2 - 1}}$$

$x = 1$

FIGURE 2.39
Vertical asymptote is $x = 1$; horizontal asymptote is $y = 0$.

the function $(1 + \sin x)/\sqrt{x - 1}$ is squeezed between 0 and something that approaches 0. So

$$\lim_{x \to \infty} \frac{1 + \sin x}{\sqrt{x - 1}} = 0$$

We show the graph of f in Figure 2.39.   ∎

In Section 1.5 we showed how to find horizontal asymptotes to graphs of rational functions. We repeat those findings here, using limit terminology.

Limits of Rational Functions at Infinity

Let $p(x)$ and $q(x)$ be polynomials. Then

(a) $\lim_{x \to \pm\infty} \dfrac{p(x)}{q(x)} = 0$ if $\deg p < \deg q$

(b) $\lim_{x \to \pm\infty} \dfrac{p(x)}{q(x)} = \dfrac{\text{leading coefficient of } p(x)}{\text{leading coefficient of } q(x)}$ if $\deg p = \deg q$

(c) $\lim_{x \to \pm\infty} \dfrac{p(x)}{q(x)}$ is infinite if $\deg p > \deg q$

REMARK

∎ When we write $\lim_{x \to \pm\infty} f(x) = L$, we mean both $\lim_{x \to \infty} f(x) = L$ and $\lim_{x \to -\infty} f(x) = L$.

EXAMPLE 2.24   Find the following limits.

(a) $\displaystyle\lim_{x \to \infty} \frac{2x^2 - 3x + 4}{5 - 4x + x^4}$ (b) $\displaystyle\lim_{x \to -\infty} \frac{5x^3 - 4x + 7}{2x^3 + 6x^2 - 3}$ (c) $\displaystyle\lim_{x \to \infty} \frac{2x^3 + 1}{x^2 - 1}$

Solution

(a) $\displaystyle\lim_{x \to \infty} \frac{2x^2 - 3x + 4}{5 - 4x + x^4} = 0$ since the degree of the numerator is less than the degree of the denominator.

(b) $\displaystyle\lim_{x \to -\infty} \frac{5x^3 - 4x + 7}{2x^3 + 6x^2 - 3} = \frac{5}{2}$ since the numerator and denominator are of the same degree.

(c) $\displaystyle\lim_{x \to \infty} \frac{2x^3 + 1}{x^2 - 1} = \infty$ since the numerator is of higher degree. Note that for large x, the fraction is approximately $2x^3/x^2 = 2x$. So as $x \to \infty$, the fraction becomes infinite.   ∎

EXAMPLE 2.25   Find all vertical and horizontal asymptotes to the graph of

$$y = \frac{x^2 + 4}{x^2 - 4}$$

Make use of these asymptotes, along with intercepts and symmetry, to sketch the graph.

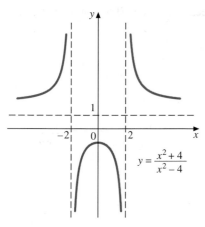

FIGURE 2.40

Solution   The vertical asymptotes occur when the denominator is 0, namely, when $x = 2$ and $x = -2$. In particular,

$$\lim_{x \to 2^+} \frac{x^2 + 4}{x^2 - 4} = \infty \quad \text{and} \quad \lim_{x \to 2^-} \frac{x^2 + 4}{x^2 - 4} = -\infty$$

The graph is symmetric with respect to the y-axis, since replacing x with $-x$ does not change the value of y. So we can limit our analysis to nonnegative values of x and then reflect the graph about the y-axis.

Since the degree of the numerator is the same as that of the denominator, we conclude that

$$\lim_{x \to \infty} \frac{x^2 + 4}{x^2 - 4} = 1$$

Thus, $y = 1$ is a horizontal asymptote.

The y-intercept is -1, found by setting $x = 0$. There is no x-intercept. We now have enough information to sketch the graph (Figure 2.40).   ∎

REMARK ────────────────────────────────

■ For a rational function

$$y = \frac{p(x)}{q(x)}$$

in *lowest terms* (that is, $p(x)$ and $q(x)$ have no nonconstant factor in common), the vertical asymptotes occur at zeros of the denominator. So if $q(a) = 0$, then the line $x = a$ is a vertical asymptote. If $p(x)$ and $q(x)$ do have a common factor, that factor should first be divided out, bearing in mind that the x value that causes the factor to be 0 is not in the domain of the original function. Recall that we illustrated cases where $p(x)$ and $q(x)$ have a common factor in Examples 2.10 and 2.11.

───

Exercise Set 2.4

In Exercises 1–6, determine whether the functions approach ∞ or $-\infty$.

1. (a) $\displaystyle\lim_{x \to 2^-} \frac{1}{x - 2}$ (b) $\displaystyle\lim_{x \to 2^+} \frac{1}{x - 2}$

2. (a) $\displaystyle\lim_{x \to 3^-} \frac{x + 1}{x - 3}$ (b) $\displaystyle\lim_{x \to 3^+} \frac{x + 1}{x - 3}$

3. (a) $\displaystyle\lim_{x \to -1^-} \frac{2x - 3}{x + 1}$ (b) $\displaystyle\lim_{x \to -1^+} \frac{2x - 3}{x + 1}$

4. (a) $\displaystyle\lim_{x \to -2^-} \frac{1 - x}{x^2 - 4}$ (b) $\displaystyle\lim_{x \to -2^+} \frac{1 - x}{x^2 - 4}$

5. (a) $\displaystyle\lim_{x \to 5^-} \frac{3 - 4x}{(x - 5)^2}$ (b) $\displaystyle\lim_{x \to 5^+} \frac{3 - 4x}{(x - 5)^2}$

6. (a) $\displaystyle\lim_{x \to 4^-} \frac{x^2 + x - 1}{x^2 - 2x - 8}$ (b) $\displaystyle\lim_{x \to 4^+} \frac{x^2 + x - 1}{x^2 - 2x - 8}$

In Exercises 7–16, find the indicated limits or show that they fail to exist.

7. $\displaystyle\lim_{x \to \infty} \frac{1}{2x + 3}$ **8.** $\displaystyle\lim_{x \to -\infty} \frac{x}{x^2 - 4}$

9. $\displaystyle\lim_{x \to -\infty} \frac{x + 1}{x - 2}$ **10.** $\displaystyle\lim_{x \to \infty} \frac{x^2 + 3}{2x^2 - 5}$

11. $\displaystyle\lim_{x \to \infty} \frac{1 - x + 2x^2}{x^2 - 4}$ **12.** $\displaystyle\lim_{x \to -\infty} \frac{3x^3 - 2x - 7}{2 + 4x^2 - 5x^3}$

13. $\displaystyle\lim_{x \to -\infty} \frac{x^2 - 1}{x + 2}$ **14.** $\displaystyle\lim_{x \to \infty} \frac{4 - x - 2x^2}{3x - 7}$

15. (a) $\displaystyle\lim_{x \to \infty} \frac{2x + 3}{\sqrt{x^2 - 1}}$ (b) $\displaystyle\lim_{x \to -\infty} \frac{2x + 3}{\sqrt{x^2 - 1}}$

16. $\displaystyle\lim_{x \to -\infty} \frac{\sqrt{2 - x}}{x + 2}$

In Exercises 17–19, determine $\lim_{x\to\infty} f(x)$ *and* $\lim_{x\to-\infty} f(x)$ *from the graph.*

17.

(a)

(b)

18.

(a)

(b)

19.

(a)   (b)

In Exercises 20–35, find all vertical and horizontal asymptotes.

20. $f(x) = \dfrac{x}{x-3}$

21. $f(x) = \dfrac{x-1}{x^2-4}$

22. $f(x) = \dfrac{x^2-1}{x^2-9}$

23. $f(x) = \dfrac{2x^2+x-1}{x^2-x-6}$

24. $f(x) = \dfrac{x-3}{6x^2+x-2}$

25. $f(x) = \dfrac{x^2-4}{x+1}$

26. $f(x) = \dfrac{x^2-9}{x+3}$

27. $f(x) = \dfrac{x+3}{x^2-9}$

28. $f(x) = \dfrac{x^2-5x+4}{3x^2-x-2}$

29. $f(x) = \dfrac{2x^2+x-10}{x^2-x-2}$

30. $f(x) = 1 - \cos\dfrac{2}{x}$

31. $f(x) = \dfrac{1}{x-1}\sin\dfrac{1}{x}$

32. $f(x) = \dfrac{\cos x}{x(1+\sin^2 x)}$

33. $f(x) = \dfrac{\sin^2 x}{\sqrt{x}+\cos^2 x}$

34. $f(x) = \dfrac{2\cos^2 x - 3}{2\sin^2 x - \sin x - 1}$ (vertical only)

35. $f(x) = \dfrac{x^2-4}{4\cos^2 x - 3}$

36. Describe the manner in which the graph of

$$f(x) = 1 + \frac{\sin x}{x}$$

approaches its horizontal asymptote. Use a graphing calculator or CAS to sketch the graph. Vary the x-scale and y-scale so that the behavior for large x is clearly displayed.

37. Use a calculator, CAS, or spreadsheet to conjecture numerically the value of the limit

$$\lim_{x\to\infty}\left(1+\frac{1}{x}\right)^x$$

Obtain the graph of this function using a graphing calculator or CAS and see if the graph is consistent with your conjecture.

38. Suppose $\lim_{x\to\infty} f(x) = 0$ and $\lim_{x\to\infty} g(x) = \infty$. Discuss the possibilities for $\lim_{x\to\infty} f(x) \cdot g(x)$. Give examples to support your answer.

39. Give examples to show that Limit Properties 2–5 in Section 2.2 need not hold true if either $\lim_{x\to a} f(x)$ or $\lim_{x\to a} g(x)$ is infinite.

40. Give examples of functions f and g such that

$$\lim_{x\to a} f(x) = \infty, \qquad \lim_{x\to a} g(x) = \infty$$

and

$$\lim_{x\to a} [f(x) - g(x)] = L$$

where L is a finite, nonzero number.

41. (a) If \$1,000 is invested at 10% interest compounded n times per year, show that the amount of money, $P(n)$, in the account at the end of one year is

$$P(n) = 1000 \left(1 + \frac{0.1}{n} \right)^n$$

(b) Complete the following table:

How compounded	Number of times (n) per year	$P(n)$ (to nearest cent)
Annually	1	\$1,100.00
Semiannually	2	\$1,102.50
Quarterly		
Monthly		
Weekly		
Daily		
Hourly		

(c) If interest is compounded *continuously*, the amount after one year is defined to be $\lim_{n\to\infty} P(n)$. Conjecture the result from the table you completed in part (b).

42. Suppose that for a certain species of fish the relationship between next year's breeding population P and the current number S that are spawning is given by

$$P(S) = \frac{aS^2}{S^2 + b}$$

where a and b are constants determined by the particular species.
(a) Graph $y = P(S)$ for several choices of a and b.
(b) What happens to $P(S)$ as the number of fish spawning increases indefinitely?

43. The number of bacteria of a certain type present in a culture at time t is given by

$$f(t) = 10,000 \left(\frac{2t^2 + 1}{t^2 + 1} \right) \qquad (t \geq 0)$$

As t increases, does the size of the bacteria colony stabilize? If so, what is the limiting number?

44. A rocket of mass m is projected upward from the earth at an initial velocity v_0. We wish to determine the smallest value of v_0 that will cause the rocket to escape from the earth's gravitational field. Denote the radius of the earth by R, and let v and r be the velocity of the rocket and its distance from the center of the earth at any time after is has been projected upward. Its kinetic energy is $\frac{1}{2}mv^2$ and its potential energy is $-GMm/r$, where M is the mass of the earth and G is the universal gravitational constant. By the law of conservation of energy, the sum of the kinetic energy and potential energy is constant. So this sum at time t is the same as when the rocket was initially projected:

$$\frac{1}{2}mv^2 - \frac{GMm}{r} = \frac{1}{2}mv_0^2 - \frac{GMm}{R} \qquad (2.3)$$

We can find the velocity v_0 for which the rocket just escapes by calculating v_0 for which $\lim_{v\to 0} r = \infty$. Carry out this calculation as follows:
(a) Solve Equation 2.3 for r.
(b) Find $\lim_{v\to 0} r$ from part (a).
(c) Find v_0 that causes the limit in part (b) to be ∞.
(d) Use $G = 6.67 \times 10^{-11}$ N $\cdot$ m^2/kg^2, $M = 5.97 \times 10^{24}$ kg, and $R = 6.37 \times 10^6$ m to find the escape velocity in m/s.

45. The radius of the moon is approximately 0.27 times the radius of the earth, and the mass is about 1/81 the mass of the earth. Find the escape velocity from the moon. (See Exercise 44.)

46. The number of stars in a spherical shell around our sun is the *number density*, N (the average number of stars per unit volume), times the volume of a shell. The volume of a shell of thickness Δr and radius r is $4\pi r^2 \Delta r$. The brightness of a star is I/r^2, where I is the intrinsic intensity.
(a) Find a formula for the brightness of a shell of radius R and find the limit as $r \to \infty$, holding Δr fixed.
(b) If we consider n shells with increasing radii and let B_n denote the sum of the brightnesses of these shells, find $\lim_{n\to\infty} B_n$.
Note: The discrepancy between the result in part (b) and the observed darkness of space is called Olbers's Paradox, after the nineteenth-century astronomer Wilhelm Olbers.

2.5 A RIGOROUS APPROACH TO LIMITS AND CONTINUITY

To motivate a precise formulation of the definition of the limit of a function, let us consider the linear function

$$f(x) = 2x - 1$$

and investigate its limit as x approaches 2. Of course, we know the answer is 3. That is,

$$\lim_{x \to 2} f(x) = 3$$

In the language of Definition 2.1, the fact that the limit is 3 means that $f(x)$ will be as close to 3 as we wish for all x sufficiently close to 2, but not equal to 2. (In this particular case the condition $x \neq 2$ is not necessary, because the function is continuous, but we have seen examples where such a restriction is necessary.) Our objective is to show precisely what we mean by "as close as we wish" and "sufficiently close."

Suppose as a first approximation we choose to make $f(x)$ within 0.5 units of 3. To accomplish this relationship, how close to 2 should we take x? Figure 2.41 helps answer this question. For $f(x)$ to be within 0.5 units of 3, it must lie between 2.5 and 3.5. That is,

$$2.5 < f(x) < 3.5$$

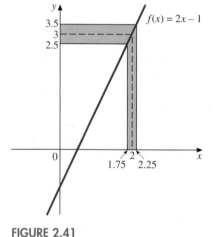

FIGURE 2.41

Graphically, we can determine corresponding x values by drawing the lines $y = 2.5$ and $y = 3.5$, determining their intersections with the graph of f, and projecting these points downward to the x-axis, as we have indicated in Figure 2.41. We should emphasize, however, that this graphical approach works in this case because f is an increasing function. It would also work with a decreasing function, but not necessarily with a function that is neither always increasing nor decreasing in the interval between the two x projections. Algebraically, we can find the x values from the inequality $2.5 < f(x) < 3.5$ by substituting the value of $f(x)$ and solving for x:

$$2.5 < 2x - 1 < 3.5$$

$$3.5 < 2x < 4.5$$

$$1.75 < x < 2.25$$

Since we are interested in knowing how far from 2 we may take x, we subtract 2 from each term of the last inequality to get

$$-0.25 < x - 2 < 0.25$$

This inequality can be written using absolute values as

$$|x - 2| < 0.25$$

Thus, if x is any number less than 0.25 units from 2, the corresponding $f(x)$ value will be less than 0.5 units from the asserted limit 3. That is,

$$|f(x) - 3| < 0.5$$

We have thus found a tolerance limit on x, given the desired tolerance for $f(x)$.

FIGURE 2.42

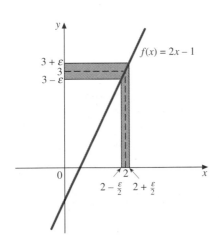

FIGURE 2.43

Suppose now that we specify a smaller tolerance for the difference between $f(x)$ and 3, say 0.1. Figure 2.42 illustrates how we can find graphically a suitable interval about $x = 2$. Algebraically, we proceed as before.

$$2.9 < f(x) < 3.1$$

$$2.9 < 2x - 1 < 3.1$$

$$3.9 < 2x < 4.1$$

$$1.95 < x < 2.05$$

If we choose any x such that

$$-0.05 < x - 2 < 0.05$$

or, equivalently,

$$|x - 2| < 0.05$$

then $f(x)$ will be within 0.1 unit of 3. That is, the inequality

$$|f(x) - 3| < 0.1$$

will be satisfied.

We could continue in this way, using smaller and smaller "degrees of closeness," or tolerance levels, but so long as we use specific values, such as 0.5 or 0.1, no matter how small we take them (greater than 0), we will not have shown that $f(x)$ can be made "as close as we please" to 3. That is, we must show that the difference between $f(x)$ and 3, in absolute value, can be made less than *any* arbitrarily small positive number by choosing x sufficiently close to 2. It is customary to use the Greek letter ε (epsilon) to designate an arbitrarily small positive number. Suppose we use ε to denote how close to 3 we want $f(x)$ to be. In our first two approximations we used $\varepsilon = 0.5$ and $\varepsilon = 0.1$, respectively. Now we are leaving ε unspecified. As before, we illustrate graphically (Figure 2.43) how to find an interval about $x = 2$ on the x-axis such that for any x in that interval, $f(x)$ will be between the lines $y = 3 - \varepsilon$ and $y = 3 + \varepsilon$. Algebraically we find this interval in the same way as before:

$$3 - \varepsilon < f(x) < 3 + \varepsilon$$

$$3 - \varepsilon < 2x - 1 < 3 + \varepsilon$$

$$4 - \varepsilon < 2x < 4 + \varepsilon$$

$$2 - \frac{\varepsilon}{2} < x < 2 + \frac{\varepsilon}{2}$$

Thus,

$$-\frac{\varepsilon}{2} < x - 2 < \frac{\varepsilon}{2}$$

or equivalently,

$$|x - 2| < \frac{\varepsilon}{2}$$

Working backward, we conclude that if x is any number less than $\varepsilon/2$ units from 2, then $f(x)$ is less than ε units from 3. That is, $|f(x) - 3| < \varepsilon$. Thus, finally and conclusively, we have shown that $f(x)$ can be made *arbitrarily* close to 3 (that is, within ε units) provided x is *sufficiently* close to 2 (that is, within $\varepsilon/2$ units). The degree of closeness of x to the number in question, in this case

2, is usually designated by the Greek letter δ (delta). So in this case we simply choose $\delta = \varepsilon/2$. Thus, whenever $|x - 2| < \delta$, we know that $|f(x) - 3| < \varepsilon$. Note that any smaller positive value of δ would also work, for if x is within a smaller interval about 2, it will also be within $\varepsilon/2$ units of 2, and so $f(x)$ will again be within ε units of 3. Note also that δ depends on ε. A smaller value of ε necessitates a smaller value of δ.

Let us use the example we just completed as a model for the precise meaning of

$$\lim_{x \to a} f(x) = L$$

For any $\varepsilon > 0$ we want to find how far from a we may allow x to be (we will call this distance δ) so that $f(x)$ will remain within ε units of L. Again, we show the situation graphically in Figure 2.44. We have shown a small circle about $x = a$, indicating that we are excluding this point from consideration, for as we have seen, there are cases in which the limit as $x \to a$ exists even though $f(a)$ does not exist, or if it exists, it may not equal L. In the figure, we show the lines $y = L - \varepsilon$ and $y = L + \varepsilon$ and their intersections with the graph of f. The x-coordinates of these points of intersection in general determine an interval about $x = a$, as shown. If f is either increasing or decreasing throughout this interval, we can determine δ graphically from this interval, as shown in Figure 2.44. The interval may not be symmetrical about $x = a$, as it was in the preceding example. If not, we choose a smaller interval that is symmetrical about a and denote it by $(a - \delta, a + \delta)$. Algebraically, we proceed as in our example, working from the inequality

$$L - \varepsilon < f(x) < L + \varepsilon$$

to solve for x in the form

$$a - \delta < x < a + \delta$$

and restricting x to be different from a. Equivalently, we can begin with the absolute value of the difference between $f(x)$ and L, $|f(x) - L|$. Then, by algebraic manipulation, we try to verify that this quantity can be made less than ε when $|x - a| < \delta$ for a suitable δ. We will illustrate this latter approach after stating the definition.

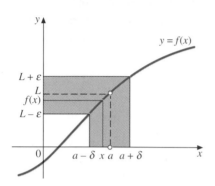

FIGURE 2.44

Definition 2.5
Formal Definition of the Limit

Let f be defined in some open interval containing the point $x = a$, except possibly at $x = a$ itself. Then the limit of $f(x)$ as x approaches a is the number L, written

$$\lim_{x \to a} f(x) = L$$

means that for each positive number ε there exists a positive number δ such that

$$\text{if } 0 < |x - a| < \delta, \text{ then } |f(x) - L| < \varepsilon$$

REMARK
■ The inequality $0 < |x - a|$ ensures that $x \neq a$.

EXAMPLE 2.26   Use Definition 2.5 to prove that

$$\lim_{x \to 2} (3x + 4) = 10$$

Solution   Let ε denote an arbitrary positive number. To find δ we will use the procedure suggested just before the statement of Definition 2.5. We let $f(x) = 3x + 4$ and begin by writing $|f(x) - 10|$ in equivalent forms, with the goal of obtaining an expression involving $|x - 2|$.

$$|f(x) - 10| = |(3x + 4) - 10| = |3x - 6| = 3|x - 2|$$

Now we can conclude that $|f(x) - 10|$ will be less than ε whenever $3|x-2| < \varepsilon$. That is, if

$$|x - 2| < \frac{\varepsilon}{3}$$

then

$$|f(x) - 10| < \varepsilon$$

So we have found a suitable value of δ, namely, $\delta = \varepsilon/3$, satisfying the requirements of the definition. Once again, in this case the restriction $x \neq 2$ is unnecessary, although imposing it does not alter the conclusion. ■

In the next example, the restriction $x \neq a$ is necessary. Also, the function involved is nonlinear.

EXAMPLE 2.27   Determine the limit

$$\lim_{x \to 1} \frac{x^3 - x^2 + x - 1}{x - 1}$$

and then prove your result using Definition 2.5.
Note: This limit would arise, for example, in finding the slope of the tangent line to the graph of $y = x^3 - x^2 + x$ at $x = 1$.

Solution   Let $f(x)$ denote the given function. Clearly $f(1)$ is not defined. In fact, both numerator and denominator equal 0 at $x = 1$. We can factor the numerator by grouping terms two by two:

$$x^3 - x^2 + x - 1 = x^2(x - 1) + (x - 1) = (x^2 + 1)(x - 1)$$

Thus, if $x \neq 1$,

$$f(x) = \frac{(x^2 + 1)(x - 1)}{x - 1} = x^2 + 1$$

We can see now that

$$\lim_{x \to 1} f(x) = \lim_{x \to 1} (x^2 + 1) = 2$$

To prove that Definition 2.5 is satisfied, let ε denote any positive number, and consider the difference $|f(x) - 2|$, for $x \neq 1$.

$$|f(x) - 2| = |(x^2 + 1) - 2| = |x^2 - 1| = |(x + 1)(x - 1)| = |x + 1||x - 1|$$

Keep in mind that our goal is to find an inequality of the form $|x-1| < \delta$ that will guarantee that $|f(x) - 2| < \varepsilon$. We have shown that $|f(x) - 2| = |x + 1||x - 1|$. This product contains $|x - 1|$, which we want, but it also contains $|x + 1|$, which we do not want. A way out of our difficulty is to require at the outset that x

be, say, between 0 and 2. This restriction is equivalent to stipulating in advance that whatever δ we eventually choose, we will not let it exceed 1. For if $\delta \leq 1$ and $|x - 1| < \delta$, then $-1 < x - 1 < 1$, or $0 < x < 2$. Placing such a restriction on δ involves no loss of generality, since whenever a value of δ is found, any smaller value will do just as well.

If we make the restriction, then, that $0 < x < 2$, it follows from the triangle inequality (see marginal note) that $|x + 1| \leq |x| + 1 < 3$. So we can write

The **triangle inequality**

$$|a + b| \leq |a| + |b|$$

holds true for all real numbers a and b (Appendix 1).

$$|f(x) - 2| = |x + 1||x - 1| < 3|x - 1|$$

Thus, if

$$|x - 1| < \frac{\varepsilon}{3} \quad (x \neq 1)$$

then

$$|f(x) - 2| < 3\left(\frac{\varepsilon}{3}\right) = \varepsilon$$

It appears, then, that we should choose $\delta = \varepsilon/3$. But remember our initial restriction that $\delta \leq 1$. We can take care of both restrictions by choosing δ to be the smaller of the numbers 1 and $\varepsilon/3$, which we write as

$$\delta = \min\left\{1, \frac{\varepsilon}{3}\right\}$$

("min" stands for "minimum"). Then, if

$$0 < |x - 1| < \delta$$

we have

$$|f(x) - 2| < \varepsilon$$

and the proof is complete. ∎

Most of the proofs of the limit properties given in Section 2.2 are relegated to Appendix 5, but we will prove the first property here, since the proof is particularly simple and it further illustrates Definition 2.5.

EXAMPLE 2.28   Prove that if $\lim_{x \to a} f(x)$ exists and c is any real number, then

$$\lim_{x \to a} cf(x) = c \lim_{x \to a} f(x)$$

Solution   Let L denote the limit of $f(x)$ as $x \to a$. Then we must show that

$$\lim_{x \to a} cf(x) = cL$$

Let ε denote any positive number. We consider two cases: $c = 0$ and $c \neq 0$. The proof for $c = 0$ is easy. As you will see, however, to prove the result for $c \neq 0$, we will need to divide by $|c|$, which is why we cannot include $c = 0$ in our argument.

Case (i): If $c = 0$, then

$$|cf(x) - cL| = |0 - 0| = 0$$

So $|cf(x) - cL| < \varepsilon$ for all $\varepsilon > 0$ whenever x is in the domain of f, which proves the result we want when $c = 0$.

Case (ii): Now suppose $c \neq 0$. Then

$$|cf(x) - cL| = |c(f(x) - L)| = |c||f(x) - L|$$

So the inequality $|cf(x) - cL| < \varepsilon$ will be satisfied if $|c||f(x) - L| < \varepsilon$, or equivalently, if

$$|f(x) - L| < \frac{\varepsilon}{|c|}$$

Now $\varepsilon/|c|$ is a positive number, and we can use it in place of ε in Definition 2.5. Because we are given that $\lim_{x \to a} f(x) = L$, we know that a number $\delta > 0$ exists such that

$$\text{if } 0 < |x - a| < \delta, \text{ then } |f(x) - L| < \frac{\varepsilon}{|c|}$$

or equivalently,

$$|cf(x) - cL| < \varepsilon$$

The proof is therefore complete. ∎

Infinite Limits

The next definition makes precise the meanings of the "infinite limits"

$$\lim_{x \to a} f(x) = \infty \quad \text{and} \quad \lim_{x \to a} f(x) = -\infty$$

Definition 2.6
Infinite Limits

Let f be defined in some open interval containing $x = a$, except possibly at $x = a$ itself.

(a) The limit statement

$$\lim_{x \to a} f(x) = \infty$$

means that for every positive number M (however large it might be) there exists a positive number δ such that

$$\text{if } 0 < |x - a| < \delta, \text{ then } f(x) > M$$

(See Figure 2.45.)

(b) The limit statement

$$\lim_{x \to a} f(x) = -\infty$$

means that for every positive number M (however large it might be) there exists a positive number δ such that

$$\text{if } 0 < |x - a| < \delta, \text{ then } f(x) < -M$$

(See Figure 2.46.)

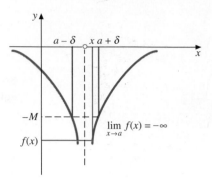

FIGURE 2.45   **FIGURE 2.46**

REMARK ───────────────────────────────

■ If $\lim_{x\to a} f(x) = \infty$, we often say "$f(x)$ approaches infinity" or "$f(x)$ becomes arbitrarily large" as x approaches a. Similarly, if $\lim_{x\to a} f(x) = -\infty$, we often say "$f(x)$ approaches negative infinity" or, colloquially, "$f(x)$ approaches minus infinity." Of course, in each case, ∞ is not a number, so the limit does not exist, but it fails to exist by becoming arbitrarily large (positively in the first case and negatively in the second).

───────────────────────────────

EXAMPLE 2.29   Find

$$\lim_{x\to -1} \frac{2}{(x+1)^2}$$

and verify the result using Definition 2.6.

Solution   Since $(x+1)^2 \to 0$ as $x \to -1$, and $(x+1)^2$ is always positive because it is a square, the quotient $2/(x+1)^2$ becomes arbitrarily large positively. That is,

$$\lim_{x\to -1} \frac{2}{(x+1)^2} = \infty$$

To show that Definition 2.6(a) is satisfied, let M denote any positive number. Then the inequality

$$\frac{2}{(x+1)^2} > M$$

will be true provided

$$(x+1)^2 < \frac{2}{M} \qquad (x \neq -1)$$

This latter inequality is satisfied if

$$|x+1| < \sqrt{\frac{2}{M}} \qquad \text{Since } \sqrt{a^2} = |a|$$

Let $\delta = \sqrt{2/M}$. We have shown that if

$$0 < |x - (-1)| < \delta$$

then

$$\frac{2}{(x+1)^2} > M$$

which, according to Definition 2.6(a), proves that

$$\lim_{x \to -1} \frac{2}{(x+1)^2} = \infty$$

Note that $x = -1$ is not in the domain of the function, so we use $0 < |x + 1|$ to restrict x from being equal to -1. ■

REMARK ──

■ Definitions 2.5 and 2.6 can easily be modified to define the right-hand and left-hand limits. For example, if we want x to approach a from the right, then in both definitions the inequality $0 < |x - a| < \delta$ is replaced with

$$a < x < a + \delta$$

For the left-hand limit we use

$$a - \delta < x < a$$

For example,

$$\lim_{x \to 0^+} \left(\frac{1}{x} \right) = \infty$$

since for any $M > 0$, $1/x > M$ provided that $0 < x < 1/M$. Similarly,

$$\lim_{x \to 0^-} \left(\frac{1}{x} \right) = -\infty$$

since for $M > 0$, $1/x < -M$ provided that $-1/M < x < 0$. In each case we are using $\delta = 1/M$.

Limits at Infinity

We have seen limits that are infinite as x approaches a finite value. Now let us define the opposite: finite limits as x becomes positively or negatively infinite, symbolized by

$$\lim_{x \to \infty} f(x) = L \qquad \text{and} \qquad \lim_{x \to -\infty} f(x) = L$$

We sometimes refer to these limits as "limits at infinity." The precise definition is as follows.

Definition 2.7
Limits at Infinity

(a) Let the domain of f contain an interval of the form (a, ∞). The limit statement

$$\lim_{x \to \infty} f(x) = L$$

means that corresponding to each positive number ε there exists a positive number N such that

$$\text{if } x > N, \text{ then } |f(x) - L| < \varepsilon$$

(See Figure 2.47.)

(b) Let the domain of f include an interval of the form $(-\infty, a)$. The limit statement

$$\lim_{x \to -\infty} f(x) = L$$

means that corresponding to each positive number ε there is a positive number N such that

$$\text{if } x < -N, \text{ then } |f(x) - L| < \varepsilon$$

(See Figure 2.48.)

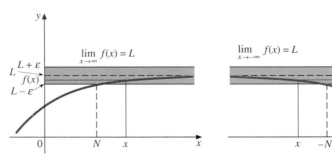

FIGURE 2.47   **FIGURE 2.48**

REMARK

■ If $\lim_{x \to \infty} f(x) = L$, we often say that $f(x)$ approaches L as x "approaches infinity," but the words "approaches infinity" mean merely that x becomes large without bound. Similar remarks apply to $\lim_{x \to -\infty} f(x) = L$.

EXAMPLE 2.30   Prove that

$$\lim_{x \to \infty} \frac{2x^2 + 1}{x^2} = 2$$

Solution   Let $f(x) = (2x^2 + 1)/x^2$. By dividing each term of the numerator by x^2, we can write $f(x)$ in the equivalent form

$$f(x) = 2 + \frac{1}{x^2} \qquad (x \neq 0)$$

Let ε denote any positive number. We can write $|f(x) - 2|$ as

$$\left| \left(2 + \frac{1}{x^2} \right) - 2 \right| = \frac{1}{x^2}$$

So we will have $|f(x) - 2| < \varepsilon$ whenever

$$\frac{1}{x^2} < \varepsilon$$

which is true provided $x > 1/\sqrt{\varepsilon}$. Thus, in Definition 2.7(a) if we take $N = 1/\sqrt{\varepsilon}$, we see that when $x > N, |f(x) - 2| < \varepsilon$, and we conclude that

$$\lim_{x \to \infty} f(x) = 2 \qquad \blacksquare$$

The $\varepsilon - \delta$ Definition of Continuity

Recall that for f to be continuous at $x = a$, the limit of $f(x)$ must exist as $x \to a$ and be equal not just to *some* number L, but specifically to the number $f(a)$. In Definition 2.5, then, we replace L by $f(a)$ and remove the restriction $0 < |x - a|$, since now we permit x to equal a.

Definition 2.8 The $\varepsilon - \delta$ Definition of Continuity	Let f be defined for all x in some interval containing the point $x = a$. Then **f is continuous at $x = a$** provided that for each positive number ε there exists a positive number δ such that $\quad$ if $|x - a| < \delta$, then $|f(x) - f(a)| < \varepsilon$ If f is continuous at every point of some open interval I, then we say that **f is continuous on I**.

For continuity at the endpoints of a closed interval $[a, b]$, we use one-sided limits. In particular, we say that **f is continuous from the left** at b if for every $\varepsilon > 0$ there is a $\delta > 0$ such that

$$\text{if } b - \delta < x \leq b, \text{ then } |f(x) - f(b)| < \varepsilon$$

Similarly, **f is continuous from the right** at a if for every $\varepsilon > 0$ there is a $\delta > 0$ such that

$$\text{if } a \leq x < a + \delta, \text{ then } |f(x) - f(a)| < \varepsilon$$

REMARK ———————————————————————————————

■ In both cases, we must have $\delta \leq b - a$ in order that x is in the interval $[a, b]$. In the future, when we say f is continuous on the closed interval $[a, b]$, we will mean f is continuous at each point between a and b and is continuous from the right at a and from the left at b.

EXAMPLE 2.31   Prove that $f(x) = \sqrt{x}$ is continuous on $[0, \infty)$.

Solution   First, we prove continuity from the right at $x = 0$. Let ε denote any positive number. Since

$$|f(x) - f(0)| = |\sqrt{x} - \sqrt{0}| = \sqrt{x}$$

the inequality $|f(x) - f(0)| < \varepsilon$ will be satisfied if $\sqrt{x} < \varepsilon$, or equivalently, if $0 \le x < \varepsilon^2$. So we let $\delta = \varepsilon^2$, and the requirement for right-hand continuity at $x = 0$ is satisfied.

Next, we prove continuity at all points in $(0, \infty)$. Let $x = a$ be in this open interval. Then $a > 0$. So we have, for $x > 0$,

$$|f(x) - f(a)| = |\sqrt{x} - \sqrt{a}| = \left| \frac{\sqrt{x} - \sqrt{a}}{1} \cdot \frac{\sqrt{x} + \sqrt{a}}{\sqrt{x} + \sqrt{a}} \right|$$

$$= \frac{|x - a|}{\sqrt{x} + \sqrt{a}} < \frac{|x - a|}{\sqrt{a}} \qquad \text{Since } \sqrt{x} > 0$$

Notice how we rationalized the expression $\sqrt{x} - \sqrt{a}$ in order to introduce the factor $|x - a|$. For any $\varepsilon > 0$ that we choose, the inequality

$$\frac{|x - a|}{\sqrt{a}} < \varepsilon$$

will be satisfied if $|x - a| < \varepsilon\sqrt{a}$. So it appears that we should take $\delta = \varepsilon\sqrt{a}$. We must also have $\delta \le a$, however, in order that x is not negative when

$$|x - a| < \delta$$

(otherwise $\sqrt{x}$ is not defined). Thus, we take $\delta = \min\{a, \varepsilon\sqrt{a}\}$. Then, whenever $|x - a| < \delta$, we have $|f(x) - f(a)| < \varepsilon$, as can be seen by reversing the preceding steps. So f is continuous at $x = a$. Since a represents *any* number in $(0, \infty)$, f is continuous for *all* points in this open interval. Finally, f is continuous from the right at 0, and we conclude that f is continuous on $[0, \infty)$.   ∎

Although the definitions we have given in this section are technical in nature, they are fundamental in proving the properties of limts and continuity on which much of calculus is based. We will see, for example, in the next chapter how the derivative is based on the limit concept, and its properties are based on those of the limit. So for a deep understanding of the derivative, the limit definitions and properties are crucial.

Continuity of Composite Functions

As indicated previously, we will defer most of the proofs of theorems concerning limits and continuity to Supplementary Appendix 5. We do choose to prove Theorem 2.6 here, however, as well as a corollary to it, because these results will be used repeatedly in subsequent chapters. We restate the theorem for completeness.

THEOREM 2.6

> Suppose the composite function $f \circ g$ is defined in some open interval containing the point $x = a$. If g is continuous at $x = a$ and f is continuous at $g(a)$, then $f \circ g$ is continuous at $x = a$.

Proof   Let ε denote any positive number. Write $y = g(x)$ and let $b = g(a)$. Since f is continuous at b, there exists a positive number η (eta) such that if $|y - b| < \eta$, then $|f(y) - f(b)| < \varepsilon$. But $|y - b| = |g(x) - g(a)|$ and since g

is continuous at a, there exists a $\delta > 0$ such that $|g(x) - g(a)| < \eta$ whenever $|x - a| < \delta$. Thus, if $|x - a| < \delta$, we have $|f(y) - f(b)| < \varepsilon$, or equivalently

$$|f(g(x)) - f(g(a))| < \varepsilon$$

So $f \circ g$ is continuous at $x = a$. ∎

COROLLARY 2.6

If $\lim_{x \to a} g(x) = b$ and f is continuous at b, then

$$\lim_{x \to a} f(g(x)) = f\left(\lim_{x \to a} g(x)\right)$$

Proof   We make g into a continuous function at $x = a$, if it is not already continuous, by defining (or redefining if necessary) $g(a)$ to equal the limiting value b. Then $f \circ g$ is continuous at $x = a$, by Theorem 2.6. Thus, by the definition of continuity,

$$\lim_{x \to a} f(g(x)) = f(g(a)) = f(b) = f\left(\lim_{x \to a} g(x)\right) \qquad ∎$$

As a consequence of this corollary and Example 2.31, we can prove Property 7 for limits, stated in Section 2.2. We can restate the property as follows: if $\lim_{x \to a} g(x)$ exists and is positive,

$$\lim_{x \to a} \sqrt{g(x)} = \sqrt{\lim_{x \to a} g(x)}$$

To see that this result is true, recall that in Example 2.31 we proved that $f(x) = \sqrt{x}$ is continuous for all $x \geq 0$. So, by Corollary 2.6,

$$\lim_{x \to a} f(g(x)) = f\left(\lim_{x \to a} g(x)\right)$$

That is,

$$\lim_{x \to a} \sqrt{g(x)} = \sqrt{\lim_{x \to a} g(x)}$$

In the result just proved, we assumed that $\lim_{x \to a} g(x) > 0$. It follows that $g(x)$ itself is positive throughout some open interval $(a - \delta, a + \delta)$ except possibly at $x = a$. (See Exercise 38 in Exercise Set 2.5.) If we permitted $\lim_{x \to a} g(x)$ to be 0, then it is possible that there would be values of x arbitrarily close to a for which $g(x)$, although close to 0, would be negative, in which case $\sqrt{g(x)}$ would not be real. However, if we also know that $g(x) \geq 0$ for all x sufficiently close to a, then Property 7 continues to hold true even when $\lim_{x \to a} g(x) = 0$. For example, since for all x,

$$g(x) = 1 + \cos x \geq 0 \qquad \text{Why?}$$

and $\lim_{x \to \pi} g(x) = 0$, we can conclude that

$$\lim_{x \to \pi} \sqrt{1 + \cos x} = \sqrt{\lim_{x \to \pi}(1 + \cos x)} = \sqrt{0} = 0$$

Note carefully that passing to the limit under the radical sign was valid in this case since $g(x) = 1 + \cos x$ is never negative. In general, the equality of $\lim_{x \to a} \sqrt{g(x)}$ and $\sqrt{\lim_{x \to a} g(x)}$ holds true only when $\lim_{x \to a} g(x)$ is greater than zero.

Exercise Set 2.5

In Exercises 1–17, prove the statements by showing that Definition 2.5 is satisfied.

1. $\lim_{x \to 1} (2x + 3) = 5$

2. $\lim_{x \to 3} (4x - 5) = 7$

3. $\lim_{x \to 2} \left(\dfrac{x + 1}{3} \right) = 1$

4. $\lim_{x \to 4} \left(\dfrac{2x - 3}{5} \right) = 1$

5. $\lim_{x \to 2} (3 - 4x) = -5$

6. $\lim_{x \to 5} \left(\dfrac{1 - x}{2} \right) = -2$

7. $\lim_{x \to -1} (2x + 5) = 3$

8. $\lim_{x \to -2} (3 - 4x) = 11$

9. $\lim_{x \to 0} \dfrac{5 - 2x}{10} = \dfrac{1}{2}$

10. $\lim_{x \to -6} \dfrac{3 - x}{12} = \dfrac{3}{4}$

11. $\lim_{x \to 2} \dfrac{x^2 - 4}{x - 2} = 4$

12. $\lim_{x \to 3} \dfrac{9 - x^2}{x - 3} = -6$

13. $\lim_{x \to -1/2} \dfrac{1 - 4x^2}{2x + 1} = 2$

14. $\lim_{x \to 4/3} \dfrac{9x^2 - 16}{3x - 4} = 8$

15. $\lim_{x \to a} x = a$

16. If $\lim_{x \to a} f(x) = 0$ and $|g(x)| \le M$, then $\lim_{x \to a} f(x) \cdot g(x) = 0$.

17. If $\lim_{x \to a} f(x) = 0$, then $\lim_{x \to a} [f(x)]^2 = 0$. (*Hint:* For x in a sufficiently small open interval containing a, excluding $x = a$ itself, show that $|f(x)| < 1$, and explain why it is true also that $[f(x)]^2 \le |f(x)|$ for x in this same open interval excluding a.)

18. Prove that if $\lim_{x \to a} f(x) = L$, then $\lim_{x \to a} |f(x)| = |L|$. Is the converse true? Prove your answer.

19. Prove that $f(x) = |x|$ is continuous on $\mathbb{R}$.

20. Use the result of Exercise 19 and Theorem 2.6 to prove that if f is continuous at $x = a$, then so is $|f|$.

In Exercises 21–24, use Definition 2.6 to prove that $\lim_{x \to a} f(x) = \infty$ or $\lim_{x \to a} f(x) = -\infty$.

21. $\lim_{x \to 0} \dfrac{3x + 2}{x^2}$

22. $\lim_{x \to 0} \left(1 - \dfrac{1}{x^2} \right)$

23. $\lim_{x \to -1} \dfrac{-2}{|x + 1|}$

24. $\lim_{x \to 1} \left(\dfrac{x + 1}{x - 1} \right)^2$ (*Hint:* Initially restrict x so that $0 \le x \le 2$.)

In Exercises 25–28, find the limit and then prove your result using Definition 2.7.

25. $\lim_{x \to \infty} \dfrac{x + 2}{2x + 1}$

26. $\lim_{x \to \infty} \dfrac{x^2}{x^2 + 3}$

27. $\lim_{x \to -\infty} \dfrac{3 - x}{x^2 + 1}$

28. $\lim_{x \to -\infty} \dfrac{x^2 - 1}{3 - x^2}$

29. Formulate a precise definition of each of the following.
(a) $\lim_{x \to \infty} f(x) = \infty$ (b) $\lim_{x \to \infty} f(x) = -\infty$
(c) $\lim_{x \to -\infty} f(x) = \infty$ (d) $\lim_{x \to -\infty} f(x) = -\infty$

In Exercises 30–35, use Definition 2.5 to prove the given limit statement.

30. $\lim_{x \to 2} x^2 = 4$ (*Hint:* First restrict x to lie in the interval $(1, 3)$.)

31. For any real number a, $\lim_{x \to a} x^2 = a^2$. (*Hint:* First restrict x to lie in the interval $(a - 1, a + 1)$.)

32. $\lim_{x \to a} \dfrac{1}{\sqrt{x}} = \dfrac{1}{\sqrt{a}}$ if $a > 0$

33. $\lim_{x \to 2} \dfrac{x}{x - 1} = 2$

34. $\lim_{x \to 0} f(x) = 1$ where $f(x) = \begin{cases} x^2 + 1 & \text{if } x > 0 \\ 3 & \text{if } x = 0 \\ 1 - 2x & \text{if } x < 0 \end{cases}$

35. $\lim_{x \to 1} g(x) = 0$ where $g(x) = \begin{cases} x^2 - 1 & \text{if } x > 1 \\ 1 & \text{if } x = 1 \\ \dfrac{x - 1}{3} & \text{if } x < 1 \end{cases}$

36. Let $\phi(x) = \begin{cases} 0 & \text{if } x \text{ is irrational} \\ 1 & \text{if } x \text{ is rational} \end{cases}$.
Prove that ϕ is discontinuous everywhere.

37. Let $f(x) = \begin{cases} 0 & \text{if } x \text{ is irrational} \\ x & \text{if } x \text{ is rational} \end{cases}$.
Prove that f is continuous at $x = 0$ but discontinuous at all other points.

38. Prove that if $\lim_{x \to a} f(x) = L$ and $L > 0$, then there is a positive number δ such that $f(x) > 0$ for all x in the interval $(a - \delta, a + \delta)$, except possibly at $x = a$.

39. Use the result of Exercise 38 to show that if f is continuous at $x = a$ and $f(a) > 0$, then there is a positive number δ such that $f(x) > 0$ for all x in $(a - \delta, a + \delta)$. State and prove an analogous result for $f(x) < 0$.

40. Prove the Squeeze Theorem (Theorem 2.3). (*Hint:* First show that $g(x) - L \le f(x) - L \le h(x) - L$. Then show how it follows that $|f(x) - L| \le |g(x) - L| + |h(x) - L|$.)

Chapter 2 Review Exercises

1. Let $f(x) = \dfrac{x^2 - x}{x^2 - 4}$. Find each of the following limits, and then use the information from parts (a) through (g) to draw the graph of f.

(a) $\lim\limits_{x \to 2^+} f(x)$ (b) $\lim\limits_{x \to 2^-} f(x)$

(c) $\lim\limits_{x \to -2^+} f(x)$ (d) $\lim\limits_{x \to -2^-} f(x)$

(e) $\lim\limits_{x \to \infty} f(x)$ (f) $\lim\limits_{x \to -\infty} f(x)$

(g) The zeros of f

In Exercises 2–10, find the indicated limits or show they do not exist.

2. (a) $\lim\limits_{x \to 1} \dfrac{1 - 2x}{(x-1)^{2/3}}$ (b) $\lim\limits_{x \to \pi/3^-} \dfrac{3 + 2\tan x}{\sec x - 2}$

3. (a) $\lim\limits_{x \to 2} \dfrac{x^2 - 4x + 4}{x^2 + 2x - 8}$ (b) $\lim\limits_{x \to 3} \dfrac{\sqrt{x+1} - 2}{x - 3}$

4. (a) $\lim\limits_{x \to 1^+} \dfrac{x^2 - 1}{\sqrt{x - 1}}$ (b) $\lim\limits_{x \to -1} \dfrac{x^{1/3} + 1}{x^{2/3} - 1}$

5. (a) $\lim\limits_{x \to 2} \dfrac{x - 2}{|x - 2|}$ (b) $\lim\limits_{x \to 0} \dfrac{x}{\sqrt{x^3 + x^2}}$

6. (a) $\lim\limits_{x \to -3^-} \dfrac{2x^2 - 1}{x^2 + 3x}$ (b) $\lim\limits_{x \to 2^-} \dfrac{\sqrt{4 - x}}{x - 2}$

7. (a) $\lim\limits_{x \to 2} \dfrac{\frac{1}{x} - \frac{1}{2}}{x - 2}$ (b) $\lim\limits_{x \to -1} \dfrac{|x - 1| - 2}{x + 1}$

8. (a) $\lim\limits_{x \to \pi/4} \dfrac{\sin x - \cos x}{\cos 2x}$ (*Hint*: Find an identity for $\cos 2x$ in Appendix 2.)

(b) $\lim\limits_{x \to \pi/2} \dfrac{\sec x - \tan x}{\sin x - 1}$ (*Hint*: Write in terms of $\sin x$ and $\cos x$.)

9. (a) $\lim\limits_{x \to -\infty} \dfrac{\sin x}{\sqrt{1 - x}}$ (b) $\lim\limits_{x \to \infty} \dfrac{x^{2/3}}{\sqrt{4x^2 - 9}}$

10. (a) $\lim\limits_{x \to \pi^-} \dfrac{1 - 2\cos x}{\sin x}$ (b) $\lim\limits_{x \to 1} \dfrac{x^2 - 1}{x^3 - 1}$

11. Let $f(x) = \sqrt{x}$, $g(x) = 2x + 3$, $F = f \circ g$, and $G = g \circ f$. Find

$$\lim\limits_{h \to 0} \dfrac{F(x + h) - F(x)}{h} \quad \text{and} \quad \lim\limits_{h \to 0} \dfrac{G(x + h) - G(x)}{h}$$

In Exercises 12–14, determine all points of discontinuity of the given function, and at each such point tell in what ways the definition of continuity fails to be satisfied.

12. (a) $f(x) = \dfrac{x + 1}{x^2 - 3x - 10}$

(b) $f(x) = \dfrac{1 - x^2}{x + 1}$

13. (a) $f(x) = \dfrac{|2 - x|}{x - 2}$ if $x \neq 2$; $f(2) = 1$

(b) $f(x) = \dfrac{|x|}{x^2 - 2}$

14. (a) $f(x) = \begin{cases} x^2 + 1 & \text{if } x > 0 \\ 0 & \text{if } x = 0 \\ 1 - x & \text{if } x \leq 0 \end{cases}$

(b) $f(x) = \begin{cases} \dfrac{2}{x} & \text{if } x > 0 \\ \dfrac{x}{2} & \text{if } x \leq 0 \end{cases}$

In Exercises 15–18, prove the statements using Definition 2.5.

15. $\lim\limits_{x \to 4} (2x - 5) = 3$ **16.** $\lim\limits_{x \to -1} \dfrac{1 - x}{2} = 1$

17. $\lim\limits_{x \to 4} |2 - x| = 2$ **18.** $\lim\limits_{x \to 0} \sqrt{2x + 1} = 1$

19. Using Definition 2.8, prove that the linear function $f(x) = mx + b$, where $m \neq 0$, is continuous everywhere.

20. Using Definition 2.6, prove that $\lim\limits_{x \to -2} \dfrac{3}{(x + 2)^2} = \infty$.

21. Using Definition 2.7, prove that $\lim\limits_{x \to \infty} \dfrac{2x^2 - 1}{x^2 + 4} = 2$.

Chapter 2 Concept Quiz

1. Define:
   (a) $\lim_{x \to a} f(x) = L$ (b) $\lim_{x \to a^+} f(x) = L$
   (c) $\lim_{x \to a^-} f(x) = L$ (d) $\lim_{x \to a} f(x) = \infty$
   (e) $\lim_{x \to a} f(x) = -\infty$ (f) $\lim_{x \to \infty} f(x) = L$
   (g) $\lim_{x \to -\infty} f(x) = L$ (h) f is continuous at a.
   (i) f is continuous on $[a, b]$.

2. Suppose $\lim_{x \to a} f(x) = L$ and $\lim_{x \to a} g(x) = M$. What can be said about each of the following?
   (a) $\lim_{x \to a} c f(x)$ (c a constant)
   (b) $\lim_{x \to a} [f(x) + g(x)]$ (c) $\lim_{x \to a} [f(x) - g(x)]$
   (d) $\lim_{x \to a} f(x) \cdot g(x)$ (e) $\lim_{x \to a} f(x)/g(x)$
   (f) $\lim_{x \to a} [f(x)]^n$ (g) $\lim_{x \to a} \sqrt[n]{f(x)}$

3. (a) State three conditions that must be satisfied in order for a function f to be continuous at a.
   (b) For each of the conditions in part (a), give a graph of a function that fails to satify the condition.

4. Fill in the blanks.
   (a) If $\lim_{x \to a} f(x) = L$ and $L \neq f(a)$ or $f(a)$ does not exist, then f has a _____ discontinuity at a.
   (b) If $\lim_{x \to a^+} f(x)$ and $\lim_{x \to a^-} f(x)$ both exist but are unequal, then f has a _____ discontinuity at a.
   (c) If f is _____ on $[a, b]$ and c is any number between $f(a)$ and $f(b)$, then there is some x_0 between a and b such that _____.
   (d) If $f(x) \leq g(x) \leq h(x)$ for all x in an open interval containing a and $\lim_{x \to a} f(x) = \lim_{x \to a} h(x)$, then _____.

5. State which of the following are true and which are false.
   (a) If f has a finite value for each x in I, then f is continuous on I.
   (b) The sum of two discontinuous functions at a point is also discontinuous at the point.
   (c) If $f(a) \cdot f(b) < 0$ and f is continuous on $[a, b]$, then the equation $f(x) = 0$ is satisfied by at least one x in (a, b).
   (d) If $\lim_{x \to a} f(x) \cdot g(x)$ exists, then so do $\lim_{x \to a} f(x)$ and $\lim_{x \to a} g(x)$.
   (e) It is impossible for a function that is continuous on an interval I to be positive at just one point in I.

APPLYING CALCULUS

1. According to Einstein's special theory of relativity, if a body of initial mass m_0 is moving at a velocity v then its mass m is

$$m = \frac{m_0}{\sqrt{1 - \dfrac{v^2}{c^2}}}$$

where c is the speed of light (3×10^8 m/s). What happens to m as v approaches c from smaller values? (Note that when v is small relative to c, as is true of most moving bodies near the earth, the mass is approximately constant.)

2. Einstein's special theory of relativity explains how an object moving close to the speed of light can appear shortened. The following formula was earlier found by the Dutch physicist H. A. Lorentz and is known as the *Lorentz contraction*:

$$L = L_0\sqrt{1 - \frac{v^2}{c^2}}$$

where L is the apparent length, L_0 the original length, v the velocity, and c the speed of light.

(a) What mathematical reason affects whether we can find $\lim_{v \to c+} L$? Could we find $\lim_{v \to c+} L$? Explain.

(b) Can we find $\lim_{v \to c-} L$? Do so, and explain your reasoning.

(c) What would be the consequence of increasing the speed to the speed of light (if it were possible to do so)?

3. In economics, the profit function P gives the profit $P(x)$ to the manufacturer when x units of a commodity are produced and sold. The **marginal profit**, when x units are produced, is defined by

$$\lim_{h \to 0} \frac{P(x+h) - P(x)}{h}$$

Suppose P is given by

$$P(x) = 10,000 + 200x - x^2$$

where $P(x)$ is in dollars. Find the marginal profit.

4. In the beginning of Chapter 1, a table showing the cost of mailing a first-class letter (in 1994 in the United States) was given. In Exercise 23 of Exercise Set 1.1, you were asked to verify that this postage function p is given by

$$p(w) = \begin{cases} 0.29 + 0.23[\![w]\!] & \text{if } w \text{ is not an integer} \\ 0.29 + 0.23(w - 1) & \text{if } w \text{ is an integer} \end{cases}$$

where w is in ounces and $p < w \le 12$. (When $w > 12$ oz, Priority Mail Rates take over.)

(a) Resketch the curve. What is the domain and range of the function?

(b) Find $\lim_{w \to 2.5} p(w)$ and $p(2.5)$. Is p continuous at $w = 2.5$?

(c) Find $\lim_{w \to 4+} p(w)$ and $\lim_{w \to 4-} p(w)$. Is p continuous at $w = 4$?

(d) For what values of w is p continuous?

5. For a given (constant) velocity v, the distance s that an object travels in a straight line in time t is given by $s(t) = vt$. For a given (constant acceleration a, the velocity v and the distance traveled s at time t are given by $v(t) = at$ and $s(t) = \frac{1}{2}at^2$. An athlete runs a 100 meter race in 10 seconds, accelerating at a constant rate a for the first 4 seconds, attaining the velocity v, and then running at that velocity for 6 seconds.

(a) Determine the distance run s, velocity v, and acceleration a as functions of time. (Hint: s, v, a will be piecewise defined; determine s first.)

(b) Graph the function s, v and a determined in part (a). Which of these is a continuous function?

(c) Find $\lim_{t \to 4} a(t)$, $\lim_{t \to 4} v(t)$, and $\lim_{t \to 4} s(t)$.

(d) What is the runner's maximum speed?

6. In the spectrum of the sun and stars, a series of colors appear dark. This series converges towards the blue, that is, toward shorter wavelengths. In the last century, Johann Balmer, a Swiss schoolteacher, found that the wavelengths λ obey the following rule: $1/\lambda = $ (constant) $\left(\dfrac{1}{2^2} - \dfrac{1}{n^2} \right)$, $n = 3,\ 4,\ 5\ldots$, "the Balmer series."

(a) Given that the wavelength of the first line ($n = 3$), known as H-alpha, is 656.3 nm, find the constant.

(b) Using a calculator, spreadsheet, or CAS, calculate $\dfrac{1}{2^2} - \dfrac{1}{n^2}$ and, to four significant figures, the wavelengths of the lines $n = 4$ (H-beta), $n = 5$ (H-gamma), and $n = 6$ (H-delta).

(c) By choosing larger and larger values of n ($n = 50$, $n = 100$, $n = 200$, $n = 500$), find the Balmer limit, the value of λ that the sequence converges.

(d) Compare this value with the limit you can find symbolically by letting $n \to \infty$ in $\dfrac{1}{2^2} - \dfrac{1}{n^2}$.

Judith V. Grabiner

We've heard that Isaac Newton and Gottfried Wilhelm Leibniz independently and almost simultaneously invented the calculus. What exactly did they invent and why was it different from what was known mathematically in that time?

By the mid-seventeenth century, methods already existed to find tangents, areas, volumes, and maxima and minima, mostly involving curves represented by polynomial functions. Newton and Leibniz saw that these separate problems could all be solved using the two general concepts that we now call derivative and integral.

Leibniz thought of the derivative of a function as the quotient of the infinitely small differences of the function and of the independent variable; he called it the *differential quotient*. His integral represents the area under a curve as the sum of rectangles of infinitely small width drawn under it. Newton thought of a function whose value changes as a "flowing quantity" or a quantity in flux; its rate of change or flow is what we call its derivative and what he called its *fluxion*. The changing function itself—that which flows—he called the *fluent*; this is the inverse of the fluxion and is what we now call the integral.

Newton and Leibniz each devised a notation for these concepts, which made calculating with them almost automatic. Because of the notation, the new methods were not only powerful, but also easy to learn and use.

Above all, both men realized—and demonstrated—that the basic processes of finding tangent and areas, that is, differentiating and integrating, are mutually inverse—a result we now call the Fundamental Theorem of Calculus. This was a revelation. Finally, their work made clear that the calculus could be used to solve a large number of problems. Mathematicians in the eighteenth century extended the calculus in many directions.

Is it true that in the eighteenth century, mathematicians didn't worry much about the underlying logic of the discipline?

Yes. Results were what counted. They could now solve the equations of motion for the whole solar system. They could analyze physical systems ranging from bridges to projectile motion to vibrating strings. Calculus is about rates of change and the results of those changes, that an area or volume is the sum of infinitely thin slices, and that the slope of the tangent is given by the ratio of an infinitely small change in a function to the corresponding change in the independent variable. They could then use these intuitions and the new notation to find distances and speeds, tangents and area, arc lengths and volumes. Calculus helped solve many important problems, so people paid less attention to its logical foundations. They certainly did not understand or care about foundations the way we do today.

Weren't eighteenth-century mathematicians making errors because they didn't have an explicitly formulated rigorous foundation for the calculus?

No. They were usually right. If one concentrates on functions of one variable, on functions that—because they arise from physical problems—are usually continuous, which take only real and not complex values, and if one considers only those infinite series that are power series with finite coefficients, the solution methods work very well without worrying about whether the derivative or integrals exist or whether the series converge. Also, mathematicians such as Euler and Laplace had a real insight into the ideas they used and were able to avoid logical pitfalls.

How did the concept of limit develop?

The term *limit* originally just meant "boundary." Think of the border of a county, or think of the circle as the "limit" of all inscribed polygons with arbitrarily many sides. Limits had been described since the time of Newton as bounds that could be approached closer and closer but not surpassed. Since average speed is defined as the ratio of the change in distance to the change in time, Newton understood an instantaneous speed to be that which is approached, closer and closer, by a ratio of finite changes in distance and time. In modern textbooks, we call this the limit of $[f(x + h) - f(x)]/h$ as h goes to zero. Thus, the instantaneous rate of change is the limit of the ratio of finite quantities. Newton called the instantaneous speed the "last value" that the ratio has right when the quantities h and $f(x + h) - f(x)$ vanish together. At first this sounds very modern but there are two differences between his concept and the modern one. First, he spoke of actually "reaching" the limit when the quantities vanish, but all that is required in proofs about limits is that the quantities "approach nearer than any given difference." Newton thought his formulation was more exact, but it later caused logical difficulties. Also, requiring that the variable "never go beyond" the limit means that it cannot oscillate about its limit. For these reasons, Newton's description of limit was not general enough to support proofs about all the concepts of the calculus. Finally, Newton and his colleagues did not translate this definition into algebraic terms and did not apply it algebraically to proofs; it was designed to make people feel more confident about the subject in general. The description is basically correct, but it needed refinement, which it got in the late eighteenth and early nineteenth centuries.

What about the derivative?

In a sense, the derivative was used before it was fully understood. In the 1630s Fermat developed a method for finding maxima and minima and a method for finding tangents. It worked, but he didn't fully understand why, or how the two problems were related. Fermat was implicitly using derivatives without realizing it. Newton and Leibniz effectively discovered the derivative by showing that this single concept unified the methods for finding maxima and minima, tangents, and velocities. But although they understood the derivative's most useful properties, even they didn't have the whole story. Mathematicians such as Taylor, Maclaurin, Euler, and Lagrange learned more about the derivative, especially through their work on series and on dif-on differential equations. Cauchy eventually gave a second algebraic definition of the concept of limit. He then used that definition to rigorously define the derivative as the limit, when it exists, of the ratio $[f(x+h)-f(x)]/h$ as h goes to zero. After Cauchy, the calculus was viewed differently. It was no longer seen just as a set of powerful methods, but a rigorous subject—like Euclidean geometry—with clear definitions and theorems whose proofs were based on those definitions. Cauchy's new rigor put the earlier results on a firm foundation, but it also provided a framework for new results, some of which could not even have been formulated before his work. However, Cauchy's original work had a subtle flaw—not distinguishing between convergence and uniform convergence—and it was left to later workers to produce a thoroughly rigorous treatment of calculus.

The historical development is the reverse of what we see in the textbooks. There, one starts with a definition, then explores some properties and results, and only then discusses applications. Historically, mathematicians began with the problems they wanted to solve and then developed the concepts they needed. When it was clear what the key properties were, they gave the subject its final, rigorous form, thus establishing the precise conditions under which the results were valid.

OK, but why did rigor suddenly become important?

Proofs have always been important to mathematicians but until the nineteenth century calculus wasn't ready to be made rigorous. Only then was there a well-developed algebra of inequalities. The key properties of calculus were better understood, and these properties could be expressed in the language of inequalities. Thus, Cauchy, Riemann, and Weierstrass gave the calculus a rigorous basis, using the already existing language of inequalities, and built a logically connected structure of theorems about the concepts of the calculus. Also, they started to think of themselves as mathematicians, not scientists, and believed that more abstract methods were appropriate to mathematics.

Another set of reasons was philosophical. In 1734 the philosopher Bishop Berkeley attacked calculus as being incorrect because it wasn't rigorous. One of his motives was to defend religion. Scientists, he said, attack religion for being unreasonable; well, let them improve their own reasoning first. He ridiculed Newton's treatment of vanishing increments, calling them "ghosts of departed quantities." He attacked the method of calculating even the simplest derivatives. Suppose, for instance, that $y = x^2$. Now take the ratio of the differences $[(x + h)^2 - x^2]/h$. When the ratio is simplified algebraically, one gets $2x + h$. To get the derivative $2x$, we have to let h "vanish." But what right did we have to get rid of the h? If h is zero, we cannot divide by it in setting up the ratio. But if it is not zero, we have no right to throw it away. A quantity is either zero or not; there is nothing in between! Berkeley described the mathematicians of his time as "rather accustomed to compute than to think." The answer to this apparent paradox requires recognizing that an equation involving limits is really a shorthand expression for a set of inequalities; the ratio of differences never really equals $2x$, but the ratio can be made as close

to $2x$ as one wishes, provided that h is taken small enough. This is a subtle and complex idea. No eighteenth-century mathematician managed to give a satisfactory response, though many tried. Maclaurin, d'Alembert, Lagrange, and Lazare Carnot, to name a few, all were concerned enough about these criticisms to think and write about the foundations of the calculus. Finally, it seemed to many late eighteenth century mathematicians that there was a limit to the results that could be obtained by their methods. The philosopher Diderot said that the mathematicians of the day had "erected the pillars of Hercules" beyond which it was impossible to go. Lagrange became interested in the foundations of calculus and influenced other mathematicians to work on the problem. Nobody solved it, though, so when Lagrange had to teach the calculus at the Ecole Polytechnique in Paris, he tried to solve it himself. And his work, in turn, decisively influenced Cauchy.

Teaching mathematics forces you to justify your concepts, and at the end of the eighteenth century there was an increased focus on teaching in Europe. It was obvious that mathematics was very useful, and therefore it was necessary to teach more of it to students of science and engineering. The need to teach calculus was the catalyst that produced the new rigor from the knowledge gained in the eighteenth century.

That's very interesting. One doesn't usually think of teaching being the impetus for research. It's usually the opposite, isn't it?

Explaining complex ideas to students makes you think deeply about what they really mean. Even in the nineteenth and twentieth centuries, important work on the foundations of analysis came from the need to teach. Weierstrass developed his ideas on foundations in his lectures at Berlin; Dedekind said he first realized the need for a deeper understanding of continuity while teaching at Zurich; Dini and Landau turned to foundations when teaching analysis. And—back to the first rigorous definition of limit and derivative—Cauchy's foundations of analysis originated as lectures at the Ecole Polytechnique.

Judith V. Grabiner teaches mathematics and the history of science at Pitzer College in Claremont, California. Among her publications are *The Origins of Cauchy's Rigorous Calculus* (MIT Press, 1981), *The Calculus as Algebra: J. L. Lagrange 1736–1813* (Garland, 1990) and "The Centrality of Mathematics in the History of Western Thought," *American Mathematical Monthly*, 1988. She is currently working on a history of the calculus of Colin Maclaurin. She received her B.S degree in Mathematics from the University of Chicago and her Ph.D. in the History of Science from Harvard University.

3

THE DERIVATIVE

In this chapter we begin the study of one of the two major branches of calculus, called **differential calculus**. It is based on the concept of the **derivative**, which is defined as a special type of limit. The second major branch of calculus, integral calculus, is also based on limits, and we will see that there is a very close connection between the two branches.

To understand the concept of the derivative we first consider two seemingly unrelated problems: (1) finding the tangent line to a given curve (a geometric problem) and (2) finding the velocity of a moving object (a physical problem). Imagine a line drawn between the points at which two consecutive wheels of a roller coaster contact the track. The slope of that line is approximately the slope of the tangent line to a curve (the track) at a given point (the position of the car at the instant the picture was taken). At that same instant the car has a specific "instantaneous" velocity. Finding the slope of the tangent line and the instantaneous velocity of an object are both problems that can be solved using the derivative. It is important to understand that although the position and speed of the roller coaster provide a useful physical model, the derivative is a general mathematical concept that is very powerful in developing applications for both the physical world and mathematics itself.

After we investigate the relationship between a function and its derivative, which is itself a function, we will develop some working rules for calculating derivatives of functions such as those introduced in Chapter 1. This development will provide the background needed for the study of applications of the derivative, which is the subject of Chapter 4.

3.1 THE TANGENT PROBLEM AND THE VELOCITY PROBLEM

We know that the tangent line to a circle at a point is perpendicular to the radius drawn to that point (see Figure 3.1), but how can we find tangent lines to other curves? More fundamentally, how is the tangent line defined? Students are often tempted to say "the line that touches the curve in only one point" or "the line that touches the curve but doesn't cross it." The curves in Figure 3.2 show, however, that neither of these descriptions suffices.

Line *l* is tangent to the curve at *P* but also crosses the curve at *Q*.

Line *l* is tangent to the curve at *P* and also crosses the curve there.

FIGURE 3.2

To see how to define the tangent line, consider the graph of $y = f(x)$. Let *P* be the point at which we want to draw the tangent line. As shown in Figure 3.3(a), let *Q* be another point on the graph, and consider the line that joins *P* and *Q*. This line is called a *secant line*. (The word *secant* comes from a Latin word meaning *to cut*.) Holding *P* fixed, we now let *Q* move along the curve toward *P*. In this way, we get a collection of secant lines through *P*, as shown in Figure 3.3(b). If these secant lines approach some limiting position as *Q* approaches *P*, then this limiting line is by definition the tangent line at *P*.

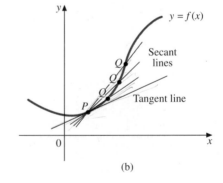

(a)

(b)

FIGURE 3.3

We can state the definition in a more useful way by considering slopes. Suppose *P* has coordinates $(a, f(a))$. The coordinates of *Q* can then be written in the form $(a + h, f(a + h))$ for some $h \neq 0$. The slope of the secant line *PQ* is then

$$m_{\text{sec}} = \frac{f(a + h) - f(a)}{h}$$

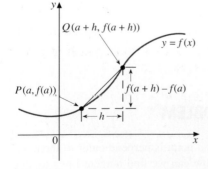

FIGURE 3.4

(See Figure 3.4.) Now we can make *Q* approach *P* simply by letting *h* approach 0. It is natural to expect that if the tangent line at *P* exists, then the slopes of these secant lines approach the slope of the tangent line at *P*, that is,

$$\lim_{h \to 0} m_{\text{sec}} = m_{\text{tan}}$$

FIGURE 3.1

Note: In Figure 3.4 we show $h > 0$. We could instead have $h < 0$, in which case Q would be on the other side of P.

This discussion leads to the following definition (made by the seventeenth-century French mathematician Pierre de Fermat).

Definition 3.1
The Tangent Line to the Graph of a Function

Let $P(a, f(a))$ be a point on the graph of the function f. If the limit

$$m_{\tan} = \lim_{h \to 0} \frac{f(a + h) - f(a)}{h} \tag{3.1}$$

exists (which implies that it is finite), the unique line through P with slope equal to $m_{\tan}$ is called the **tangent line** to the graph of f at P.

EXAMPLE 3.1   Show that a tangent line exists at $(1, 2)$ for the curve defined by $y = x^2 + 1$, and find its equation.

Solution   Let $(a, f(a)) = (1, 2)$ and investigate the limit in Equation 3.1:

$$\lim_{h \to 0} \frac{f(1 + h) - f(1)}{h} = \lim_{h \to 0} \frac{[(1 + h)^2 + 1] - 2}{h}$$

$$= \lim_{h \to 0} \frac{1 + 2h + h^2 + 1 - 2}{h}$$

$$= \lim_{h \to 0} \frac{h^2 + 2h}{h}$$

$$= \lim_{h \to 0} \frac{h(h + 2)}{h}$$

$$= \lim_{h \to 0} (h + 2) = 2 \qquad \text{Remember that } h \neq 0.$$

Since this limit exists, the tangent line exists and has slope $m_{\tan} = 2$. Thus, from the point-slope form of a straight line $y - y_1 = m(x - x_1)$, we have

$$y - 2 = 2(x - 1)$$

or, after simplification,

$$y = 2x$$

In Figure 3.5 we show the graph of $y = x^2 + 1$ and its tangent line $y = 2x$ at the point $(1, 2)$. ∎

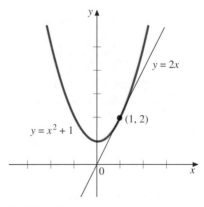

FIGURE 3.5

EXAMPLE 3.2   Find the slope of the tangent line to the graph of $f(x) = 2\sqrt{x}$ at an arbitrary point $(x, f(x))$ on the curve, with $x > 0$.

Solution   By Equation 3.1, with a replaced by x,

$$m_{\tan} = \lim_{h \to 0} \frac{f(x + h) - f(x)}{h}$$

$$= \lim_{h \to 0} \frac{2\sqrt{x + h} - 2\sqrt{x}}{h}$$

We rationalized the numerator in order to change the form of the fraction to one whose limit we could find.

$$= \lim_{h \to 0} \frac{2(\sqrt{x+h} - \sqrt{x})}{h} \cdot \frac{\sqrt{x+h} + \sqrt{x}}{\sqrt{x+h} + \sqrt{x}} \qquad \text{Rationalizing the numerator}$$

$$= \lim_{h \to 0} \frac{2(x+h-x)}{h(\sqrt{x+h} + \sqrt{x})}$$

$$= \lim_{h \to 0} \frac{2h}{h(\sqrt{x+h} + \sqrt{x})}$$

$$= \lim_{h \to 0} \frac{2}{\sqrt{x+h} + \sqrt{x}} \qquad \text{Since } h \neq 0$$

$$= \frac{2}{2\sqrt{x}} = \frac{1}{\sqrt{x}}$$

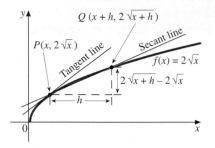

FIGURE 3.6

In Figure 3.6 we show the graph of f, along with a secant line joining $P(x, 2\sqrt{x})$ and $Q(x + h, 2\sqrt{x+h})$, as well as the tangent line at P. Note that we now have the formula for the slope of the tangent line to this curve at *any* point for which $x > 0$. For example, at the point $(1, 2)$ the slope is $1/\sqrt{1} = 1$, and at $(4, 4)$ it is $1/\sqrt{4} = 1/2$. ∎

The Velocity Problem

The problem of finding the tangent line to a curve at an arbitrary point was first studied systematically by the French mathematician Pierre de Fermat (1601–1665). His work preceded that of Newton and Leibniz and provided an important part of the background for their later formulation of the calculus. Fermat (along with René Descartes) was an originator of what is now known as analytical geometry. Fermat's work on tangent lines soon led to a way of finding the instantaneous velocity of a moving object. This instantaneous velocity can be interpreted as the slope of the tangent line to the position-versus-time curve. Thus, a purely geometric problem provided the key to solving a physical problem.

To see how the two problems are related, let us begin with an example. Suppose a skier moves downhill on a straight path for the first 3 seconds. The skier's distances $s(t)$ in meters from the starting point at half-second intervals are given in the table.

t	0	0.5	1.0	1.5	2.0	2.5	3.0	t in seconds,
$s(t)$	0	5.3	11.7	19.2	29.5	41.2	53.8	$s(t)$ in meters

Let us consider the problem of finding the skier's velocity at a particular instant, say $t = 2$. We can easily find the *average velocity* in the time interval from $t = 2.0$ to $t = 2.5$, using the familiar relationship "distance = rate × time." In this case, "rate" means average velocity. Thus, the average velocity over the time interval in question is

$$v_{\text{ave}} = \frac{\text{distance covered}}{\text{elapsed time}} = \frac{41.2 - 29.5}{2.5 - 2.0}$$

$$= \frac{11.7}{0.5} = 23.4 \text{ m/s}$$

t	$s(t)$
1.9	27.33
2.0	29.50
2.1	31.71

Since the skier's velocity is increasing with time, the velocity at $t = 2$ is somewhat less than 23.4. So v_{ave} in this case is an overestimate of the velocity at $t = 2$. Suppose now that distances are known at time intervals of 0.1 second, as shown, near $t = 2$. Then the average velocity from $t = 2.0$ to 2.1 is

$$v_{\text{ave}} = \frac{31.71 - 29.50}{2.1 - 2.0} = \frac{2.21}{0.1} = 22.1 \text{ m/s}$$

If we knew distances for still smaller time intervals, our average velocities would come closer and closer to the instantaneous velocity at $t = 2$. For example, if we know the distance is 29.720 when $t = 2.01$, then the average velocity from $t = 2.00$ to $t = 2.01$ is

$$v_{\text{ave}} = \frac{29.720 - 29.500}{2.01 - 2.00} = \frac{0.220}{0.01} = 22.0 \text{ m/s}$$

REMARK ───

■ We have used the time intervals [2.0, 2.5], [2.0, 2.1], and [2.00, 2.01] in this example. We could equally well have used intervals with 2 as the right endpoint: [1.5, 2.0], [1.9, 2.0], and [1.99, 2.00]. If we had used these intervals, we would have obtained average velocities that underestimate the instantaneous velocity at $t = 2$.

───

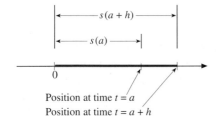

FIGURE 3.7

Let us try to generalize the procedure we have just described. Suppose an object (such as a person, a car, or a rocket) is moving along a straight path. We set up a coordinate system so that the object's position $s(t)$ on the path is known at time t. We wish to determine the velocity of the object at a particular instant, say $t = a$. As a first approximation, we calculate the average velocity over a small time interval from $t = a$ to $t = a + h$, where $h \neq 0$. Figure 3.7 shows the position at $t = a$ and at $t = a + h$. The average velocity over this time interval is

$$v_{\text{ave}} = \frac{\text{change in position}}{\text{change in time}} = \frac{s(a + h) - s(a)}{h}$$

If we now consider smaller and smaller values of h, the average velocities may approach some limiting value. If so, then this limiting value is by definition the **instantaneous velocity** at $t = a$:

$$v(a) = \lim_{h \to 0} \frac{s(a + h) - s(a)}{h} \qquad (3.2)$$

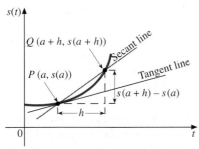

Slope of secant line $= \dfrac{s(a + h) - s(a)}{h}$

$=$ average velocity from P to Q

Slope of tangent line $= \displaystyle\lim_{h \to 0} \dfrac{s(a + h) - s(a)}{h}$

$=$ instantaneous velocity at P

FIGURE 3.8

Compare Equation 3.2 for instantaneous velocity at $t = a$, with Equation 3.1 for the slope of the tangent line to a curve. Except for the letters used to name the functions, the equations are identical. To see more clearly why these should be the same, let us consider the graph of the position function $s(t)$, shown in Figure 3.8. Let P be the point $(a, s(a))$ and Q be the point $(a + h, s(a + h))$. Then the average velocity

$$v_{\text{ave}} = \frac{s(a + h) - s(a)}{h}$$

is the slope of the secant line through P and Q. As we let $h \to 0$, we see that $Q \rightarrow P$, and the secant lines approach the tangent line at P. So the instantaneous velocity at $t = a$, denoted by $v(a)$, is exactly the same as the slope of the tangent line at P.

EXAMPLE 3.3   A particle moves in a straight line in such a way that its position from the starting point, in meters, is given by

$$s(t) = 16 \left(1 - \frac{1}{t + 1} \right)$$

where t is nonnegative and is given in seconds.

(a) Find the average velocity of the particle over the time interval from $t = 1$ to $t = 3$.

(b) Find the instantaneous velocity of the particle when $t = 3$.

Solution

(a) When $t = 1$, the particle is at a distance $s(1) = 16(1 - \frac{1}{2}) = 8$ m from the starting point. When $t = 3$, it is $s(3) = 16(1 - \frac{1}{4}) = 12$ m from the starting point. The average velocity is given by the distance traveled divided by the elapsed time:

$$v_{\text{ave}} = \frac{s(3) - s(1)}{3 - 1} = \frac{12 - 8}{2} = 2 \text{ m/s}$$

(b) In Equation 3.2 we let $a = 3$ and get

$$v(3) = \lim_{h \to 0} \frac{s(3 + h) - s(3)}{h}$$

$$= \lim_{h \to 0} \frac{16 \left(1 - \dfrac{1}{3 + h + 1} \right) - 12}{h}$$

$$= \lim_{h \to 0} \frac{16 - \dfrac{16}{4 + h} - 12}{h}$$

$$= \lim_{h \to 0} \frac{4 - \dfrac{16}{4 + h}}{h}$$

$$= \lim_{h \to 0} \frac{4(4 + h) - 16}{h(4 + h)} \qquad \text{We multiplied numerator and denominator by } 4 + h.$$

$$= \lim_{h \to 0} \frac{4h}{h(4 + h)}$$

$$= \lim_{h \to 0} \frac{4}{4 + h} = 1 \qquad \text{Dividing by } h \text{ is valid since } h \neq 0.$$

So at the instant when $t = 3$ seconds, the particle is traveling at the rate of 1 m/s.

In Figure 3.9 we have shown the graph of $s(t)$ and the solution to part (a) as the slope of the secant line joining $(1, 8)$ and $(3, 12)$. The solution to part (b) is the slope of the tangent line at $(3, 12)$. ∎

Secant line; slope = $v_{ave} = 2$ Tangent line; slope = $v(3) = 1$

$s(t) = 16\left(1 - \dfrac{1}{t+1}\right)$

$(3, 12)$

$(1, 8)$

FIGURE 3.9

From the preceding example, we note that the velocity is 1 m/s when $t = 3$, so at that instant the particle is moving in the positive direction on its path. We can see this conclusion intuitively by reasoning that if the velocity remained the same for 1 second, then the distance would be 1 meter greater after 1 second had elapsed. More generally, *when the instantaneous velocity is positive, the directed distance from the starting point is increasing*, and *when the instantaneous velocity is negative, the distance from the starting point is decreasing*. Figure 3.10 illustrates this conclusion. It should be emphasized that these are instantaneous conditions only. For example, at the instant when a particle changes direction, its velocity will be 0, but an observer would not be able to discern that it was not moving at that instant. For example, if a ball is thrown upward, gravity is pulling downward, so when the ball reaches its maximum height, $v = 0$. But v is always changing and does not remain at 0 for any finite time. It is an *instantaneous* velocity.

We will consider the motion of a particle in more detail after we have formalized the notion of a derivative.

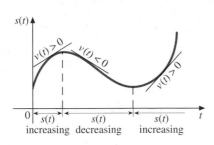

FIGURE 3.10

Exercise Set 3.1

1. Let $f(x) = x^2/2$, and let $P = (2, 2)$ and $Q = (2 + h, f(2 + h))$.
   (a) On a graph, show the curve $y = f(x)$ and the secant lines through P and Q for $h = 1$, $h = 0.5$, and $h = 0.1$, respectively.
   (b) Calculate the slopes of the secant lines in part (a).
   (c) Find the slope of the tangent line at P.
   (d) Find the equation of the tangent line at P.

2. Let $f(x) = 4/x$, and let $P = (2, 2)$ and $Q = (2 + h, f(2 + h))$. Repeat parts (a)–(d) of Exercise 1 for this function.

In Exercises 3–10, show that a tangent line to the graph of $y = f(x)$ exists at $(a, f(a))$, and find its equation. Draw the graph of the function and show its tangent line at $(a, f(a))$.

3. $f(x) = 1 - x^2;\ a = 1$ 4. $f(x) = 4 - \dfrac{2}{x};\ a = 1$

5. $f(x) = \dfrac{2}{x - 1};\ a = -1$

6. $f(x) = x^2 - x;\ a = 1$

7. $f(x) = (x - 2)^2;\ a = 3$

8. $f(x) = 3x - x^2;\ a = 2$

9. $f(x) = \dfrac{1}{2}x(x - 2);\ a = -1$

10. $f(x) = \dfrac{3}{x + 2};\ a = -4$

11. A particle moves in a straight line in such a way that its directed distance from the origin at time t is $s(t) = 2t^2 - 3$.
    (a) Find the average velocity over each of the time intervals $[1, 2]$, $[1, 1.5]$, $[1, 1.1]$, and $[1, 1.01]$.
    (b) Find the average velocity over the time interval $[1, 1 + h]$ for h arbitrary but nonzero.
    (c) Find the instantaneous velocity at time $t = 1$.
    (d) Sketch the graph of $y = s(t)$ and show the instantaneous velocity at $t = 1$.

12. A driver sees traffic stopped ahead and applies the brakes. Assume that the table below gives the distance the car travels (in feet) in t seconds after the brakes are applied.

t	0	0.5	1.0	1.5	2.0	2.5	3.0
$s(t)$	0	38	74	106	131	149	160

   (a) Find the average velocity in the time interval $[2.0, 2.5]$.
   (b) Find the average velocity in the time interval $[1.5, 2.0]$.
   (c) Estimate the instantaneous velocity at $t = 2.0$ and explain the basis for your estimate.
   (d) What is the significance of the sign of the answers in parts (a)–(c)?

13. Let $s(t)$ be the position function of a particle moving in a straight line. Consider the graph of $s(t)$ versus t, and tell what you can infer about the velocity $v(t)$ in each of the following situations, assuming that $v(t)$ exists.
(a) $s(t)$ is increasing.
(b) $s(t)$ is decreasing.
(c) The graph is concave up.
(d) The graph is concave down.
(e) The point $(t_0, s(t_0))$ is a local maximum point.

14. The accompanying graph shows the position function $s(t)$ versus time t for a particle moving in a straight line. Give the time intervals during which the particle is doing the following things:
(a) moving in the positive direction;
(b) moving in the negative direction;
(c) speeding up;
(d) slowing down.
At what times is the particle's velocity 0?

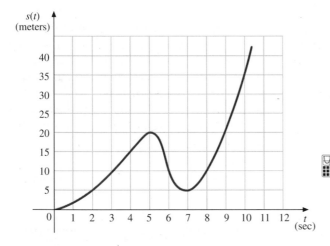

15. By estimating values from the graph in Exercise 14, obtain approximate values for the following.
(a) The average velocity during the time interval $[5, 6]$
(b) The instantaneous velocity at $t = 5$

16. A person on a bicycle leaves home, riding for several minutes on a straight, level road, gradually picking up speed. Then the road goes downhill and the cyclist picks up speed rapidly but slows down rapidly when the road goes uphill again. After reaching the top, the cyclist turns around and returns home over the same road. Sketch a graph of the position function $s(t)$ that would be consistent with the cyclist's ride.

In Exercises 17–26, use the given values of a and s(t) to find the instantaneous velocity v(a) of an object moving in a straight line in such a way that its directed distance from the origin is s(t) at time t.

17. $s(t) = 2t + 3$; $a = 2$ **18.** $s(t) = 3t - 4$; $a = 3$

19. $s(t) = \dfrac{t+1}{2}$; $a = 5$ **20.** $s(t) = t^2$; $a = 1$

21. $s(t) = t^2 - t$; $a = 3$ **22.** $s(t) = 1 - \dfrac{1}{t}$; $a = 2$

23. $s(t) = \dfrac{t+4}{t}$; $a = 3$ **24.** $s(t) = \dfrac{12}{t-1}$; $a = 4$

25. $s(t) = t^2 - \dfrac{2}{t}$; $a = 2$ **26.** $s(t) = \dfrac{8}{t+2}$; $a = 2$

In Exercises 27–30, find the slope of the tangent line to the graph of $y = f(x)$ at an arbitrary point $(x, f(x))$.

27. $f(x) = x^2 - 2x$ **28.** $f(x) = \dfrac{1}{x^2}$; $x \neq 0$

29. $f(x) = \dfrac{x}{x+1}$; $x \neq -1$

30. $f(x) = x - \dfrac{1}{x}$; $x \neq 0$

31. Find the equation of the tangent line to the graph of $y = \sqrt{2x + 1}$ at $x = 4$.

32. Find the average velocity during the time interval from $t = 0$ to $t = 3$ of a particle moving in a straight line according to the formula $s(t) = 4/\sqrt{t + 1}$. What is the instantaneous velocity at $t = 3$? In what direction is the particle moving at the instant when $t = 3$?

In Exercises 33–35, use a CAS or a graphing calculator.

33. The position function of an object moving in a straight line is given by

$$s(t) = \frac{\sin^2 t + t^2}{t^2 - 1}$$

where t is in seconds and $s(t)$ is in feet. Estimate the intervals on which the velocity is increasing and those on which it is decreasing.

34. The position function of an object moving along a line is given by

$$s(t) = \frac{\sin^2 t + t^2}{t^2 + 1}$$

where t is in seconds and $s(t)$ is in meters.
(a) Estimate the intervals on which $v(t)$ is increasing and those on which it is decreasing.
(b) Estimate the velocity at $t = 1$ using Equation 3.2 with $h = 0.1$, $h = 0.01$, and $h = 0.001$.

35. Repeat Exercise 34 with

$$s(t) = \frac{\cos^2 t + t^2}{t^2 + 1}$$

3.2 THE DERIVATIVE

We have seen that the geometric problem of finding the slope of the tangent line to the graph of a function involves the same type of limiting operation as the physical problem of finding the instantaneous velocity of a moving object. We now introduce a name and a symbol for the type of limit involved in each of these calculations. This limit is one of the fundamental concepts in calculus.

Definition 3.2 The Derivative of a Function at a Point	Let f be a function defined in some open interval containing the point $x = a$. The **derivative** of f at $x = a$, denoted by $f'(a)$, is defined by $$f'(a) = \lim_{h \to 0} \frac{f(a+h) - f(a)}{h} \qquad (3.3)$$ provided this limit exists.

If the graph of $y = f(x)$ has a tangent line at the point $(a, f(a))$, then its slope is $f'(a)$. Similarly, if $s(t)$ is the position function for a particle moving along a line, then its velocity at time $t = a$ is the derivative $s'(a)$. These are but two of the many interpretations of the derivative as the **instantaneous rate of change** of a function with respect to its independent variable. Velocity is the instantaneous rate of change of position with respect to time. The slope of the tangent line to a curve can also be thought of as a rate of change—namely, the rate at which y changes with respect to x.

Derivative as a Rate of Change

Here are some other examples that illustrate the derivative as an instantaneous rate of change.

1. Let $N(t)$ be the size of a bacteria culture at time t. Then $N'(t)$ is the rate at which the culture grows or diminishes.

2. Let $Q(t)$ be the quantity of a radioactive substance on hand at time t. Then $Q'(t)$ is the rate of radioactive decay.

3. Let $X(t)$ be the amount present at time t of a compound being formed by the chemical reaction of two other compounds. Then $X'(t)$ is the reaction rate.

4. Let $C(x)$ be the cost of manufacturing x items of a certain type. Then $C'(x)$ is the rate of change of cost with respect to the number of items produced. This rate is called **marginal cost** in economics. Similarly, there is **marginal revenue**, $R'(x)$, and **marginal profit**, $P'(x)$.

5. Let $A(t)$ be the accumulated amount of money after t years that results from a certain principal invested at continuously compounded interest. Then $A'(t)$ is the rate of growth of the account.

6. Let $p(h)$ be the atmospheric pressure at height h m above the earth's surface. Then $p'(h)$ is the rate at which pressure is changing with respect to altitude.

7. Let $L(t)$ be the current state of knowledge of a learner in an instructor/
    learner situation. Then $L'(t)$ is the rate at which the material is being
    learned.

We could give many other examples, but these should suffice to give you a
feeling for the broad applications of the derivative as an instantaneous rate of
change. In the next chapter we will return to applications. For the present, we
concentrate on how to calculate the derivative and on the relationship between
a function and its derivative.

EXAMPLE 3.4   Find $f'(3)$ if $f(x) = \sqrt{x+1}$.

Solution

$$f'(3) = \lim_{h \to 0} \frac{f(3+h) - f(3)}{h}$$

$$= \lim_{h \to 0} \frac{\sqrt{(3+h)+1} - \sqrt{3+1}}{h}$$

$$= \lim_{h \to 0} \frac{\sqrt{h+4} - 2}{h} \cdot \frac{\sqrt{h+4}+2}{\sqrt{h+4}+2} \qquad \text{Here, we are rationalizing the numerator.}$$

$$= \lim_{h \to 0} \frac{h+4-4}{h(\sqrt{h+4}+2)} = \lim_{h \to 0} \frac{h}{h(\sqrt{h+4}+2)}$$

$$= \lim_{h \to 0} \frac{1}{\sqrt{h+4}+2} = \frac{1}{4}$$

So the derivative of f at 3 is $\frac{1}{4}$. ∎

An alternative way of writing the limit in Definition 3.2 is sometimes useful.
As we show in Figure 3.11, we let $x = a + h$, so that $h = x - a$, and observe
that $h \to 0$ if and only if $x \to a$. So we can write $f'(a)$ as

$$f'(a) = \lim_{x \to a} \frac{f(x) - f(a)}{x - a} \tag{3.4}$$

We can redo Example 3.4 using this alternative form as follows. Since
$f(x) = \sqrt{x+1}$, we get $f(3) = \sqrt{4} = 2$. So, with $a = 3$, we use Equation 3.4
to obtain

$$f'(3) = \lim_{x \to 3} \frac{\sqrt{x+1} - 2}{x - 3} = \lim_{x \to 3} \frac{\sqrt{x+1} - 2}{x - 3} \cdot \frac{\sqrt{x+1}+2}{\sqrt{x+1}+2}$$

$$= \lim_{x \to 3} \frac{(x+1) - 4}{(x-3)(\sqrt{x+1}+2)} = \lim_{x \to 3} \frac{x - 3}{(x-3)(\sqrt{x+1}+2)} = \frac{1}{4}$$

The alternative form of Equation 3.4 can also be readily used to prove the
following important result.

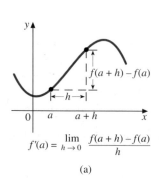

$$f'(a) = \lim_{h \to 0} \frac{f(a+h) - f(a)}{h}$$

(a)

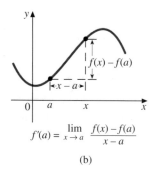

$$f'(a) = \lim_{x \to a} \frac{f(x) - f(a)}{x - a}$$

(b)

FIGURE 3.11
Two equivalent ways of writing a
formula for $f'(a)$.

THEOREM 3.1   If $f'(a)$ exists, then f is continuous at $x = a$.

Proof   According to Definition 2.9, we need to show that

$$\lim_{x \to a} f(x) = f(a)$$

or, equivalently, that $\lim_{x \to a}[f(x) - f(a)] = 0$. If $x \neq a$, we can write

$$f(x) - f(a) = [f(x) - f(a)] \cdot \left[\frac{x - a}{x - a}\right] \qquad \text{Since } \tfrac{x-a}{x-a} = 1 \text{ when } x \neq a$$

Thus, if $x \neq a$,

$$f(x) - f(a) = \frac{f(x) - f(a)}{x - a} \cdot (x - a)$$

Taking limits and using Limit Property 2 of Section 2.1, we get

$$\lim_{x \to a}[f(x) - f(a)] = \left[\lim_{x \to a} \frac{f(x) - f(a)}{x - a}\right]\left[\lim_{x \to a}(x - a)\right]$$

$$= [f'(a)] \cdot 0 = 0$$

So f is continuous at $x = a$. Notice that we made use of the fact that $f'(a)$ exists when we wrote the limit of the product as the product of the limits. ∎

The converse of Theorem 3.1 is not true. That is, if f is continuous at $x = a$, we cannot necessarily conclude that $f'(a)$ exists. To see why, consider the example $f(x) = |x|$, whose graph is shown in Figure 3.12. If $x > 0$, then $f(x) = x$ and if $x < 0$, $f(x) = -x$, both of which are continuous linear functions. We have previously shown that $\lim_{x \to 0} f(x) = 0$; since $f(0) = 0$, this function is continuous on all of $\mathbb{R}$. Now let us investigate whether $f'(0)$ exists. First, consider $h > 0$.

$$\lim_{h \to 0^+} \frac{f(0 + h) - f(0)}{h} = \lim_{h \to 0^+} \frac{f(h) - 0}{h} = \lim_{h \to 0^+} \frac{|h|}{h}$$

$$= \lim_{h \to 0^+} \frac{h}{h} = 1$$

Whereas, if $h < 0$,

$$\lim_{h \to 0^-} \frac{f(0 + h) - f(0)}{h} = \lim_{h \to 0^-} \frac{f(h) - 0}{h} = \lim_{h \to 0^-} \frac{|h|}{h}$$

$$= \lim_{h \to 0^-} \frac{-h}{h} = -1$$

Thus, the right- and left-hand limits are not the same, and so

$$\lim_{h \to 0} \frac{f(0 + h) - f(0)}{h}$$

does not exist. That is, $f'(0)$ does not exist. Geometrically, we have shown that as we approach 0 from the right, the slope of each of the secant lines is 1, and when we approach 0 from the left, the slope is -1. Thus, *at 0* there is no well-defined tangent line. At all other points, however, the tangent lines do exist, so $f'(x)$ exists everywhere except at $x = 0$ for this function.

As another example, consider the function g whose graph is shown in Figure 3.13. The curve has a vertical tangent line at $x = -3$ and at $x = 1$. Since slope is not defined for vertical lines, it follows that for this function $g'(-3)$ and $g'(1)$ do not exist. Yet from the graph we see that g is continuous for all x.

In the case of $f(x) = |x|$, $f'(x)$ fails to exist at one point only. For $g(x)$ of Figure 3.13, $g'(x)$ fails to exist at two points. You can probably think of graphs of functions that are continuous everywhere but whose derivative fails to exist at infinitely many points. (See, for example, the "sawtooth" function

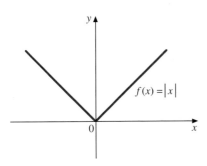

FIGURE 3.12

$f'(0)$ does not exist.

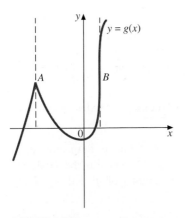

FIGURE 3.13

Tangent lines at points A and B are vertical, so slopes are undefined.

FIGURE 3.14
A "sawtooth" function that fails to have a derivative at infinitely many points.

in Figure 3.14.) It is harder to imagine a continuous function whose derivative fails to exist at *every* point. Yet such functions do exist! As you might imagine, their formulas are complicated, and they are impossible to draw, so we will not attempt to illustrate such a function. Examples can be found in advanced analysis textbooks. The first such example was given by the German mathematician Karl Weierstrass in 1861. The important point of this discussion is that the converse of Theorem 3.1 is not true. So while the existence of the derivative at a point necessarily implies continuity there, continuity at a point does not imply existence of the derivative there.

The process of calculating the derivative of a function is called **differentiation**, and to **differentiate** a function means to find its derivative. When the derivative of a function exists at a given point, we say the function is **differentiable** at that point. The quotient

$$\frac{f(a+h) - f(a)}{h} \qquad (h \neq 0)$$

that occurs in the definition of the derivative of f at a is called a **difference quotient** for f. So the derivative is a limit of a difference quotient. If a function is differentiable at all points of an open interval (a, b), we say it is **differentiable on (a, b)**.

In Definition 3.2, if we restrict h to be positive and consider the limit as $h \to 0^+$, we call the limit the **right-hand derivative at a**. Similarly, if we restrict h to be negative and consider the limit as $h \to 0^-$, we call the limit the **left-hand derivative at a**. A function is said to be differentiable on the closed interval $[a, b]$ if it is differentiable on the open interval (a, b) and the right-hand derivative at a exists and the left-hand derivative at b exists. In the future when we speak of differentiability on an interval, we will understand that if either endpoint is included, then the appropriate one-sided derivative must be used there.

The Derivative Function

So far, we have considered the derivative of a function at a fixed point, which we have designated by $x = a$. Now we want to consider the set of all points x at which the derivative of a given function f exists. These points constitute a subset of the domain of f. (The subset may even be the entire domain of f.) For each x in this subset, a value of $f'(x)$ exists. Thus, what we have is really a new function f', called the **derivative** of f. Its value at x is given by the limit that defines the derivative, namely,

$$f'(x) = \lim_{h \to 0} \frac{f(x+h) - f(x)}{h}$$

and its domain is the largest subset of the domain of f (since $f(x)$ must exist for the numerator to be defined) for which this limit exists. In the next two examples we explore the relationship between the graphs of f and f'. First, we find f', given f.

EXAMPLE 3.5   The graph of a function f is given in Figure 3.15. From it, obtain a sketch of the graph of f'. What is the domain of f'?

x	$f'(x)$
-2	2
-1	0
0	-1
1	0
2	2
3	Undefined
4	0
5	-3

Solution   At selected values of x, we estimate the values of $f'(x)$ by means of the slope of the tangent line. This method gives only a crude approximation. At the integer points on the interval $[-2, 5]$, these values appear to be approximately those in the table. Notice that at $x = 3$ the tangent line is vertical, so there is no slope. The corresponding point on the graph is an example of a **cusp**. As 3 is approached from the left, the slope gets arbitrarily large, and as 3 is approached from the right, the slope is negative but arbitrarily large in absolute value. This behavior results in a vertical asymptote for the graph of f'. We show the graph of f' in Figure 3.16. From the graph in Figure 3.15, it appears that the domain of f is all of $\mathbb{R}$. The domain of f' therefore is the set of all x in $\mathbb{R}$ except $x = 3$. ■

FIGURE 3.15

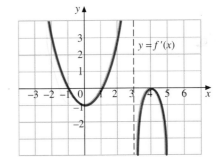

FIGURE 3.16

wherever f' crosses x axis, f has horizontal Tan. line
f is concave ↓ where f' is neg. + decreasing

Next, we reverse the process and find f, given f'.

EXAMPLE 3.6   The graph of the derivative f' of a certain function f is given in Figure 3.17. It is known that $f(0) = 2$. Sketch the graph of f.

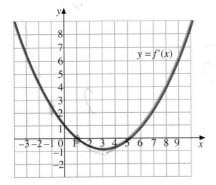

FIGURE 3.17

Solution   The y value that we read from the graph of f' corresponding to any x value represents the slope of the tangent line to the graph of f. We know that since $f(0) = 2$, the graph of f goes through the point $(0, 2)$. At $x = 0$, we read from Figure 3.17 that $f'(0) = 1$. So as the graph of f passes through $(0, 2)$, it does so with its tangent line having slope 1. As we move to the right, $f'(x)$ decreases but remains positive until $x = 1$. The graph of f therefore continues to rise on the interval $[0, 1]$ but is rising less and less rapidly. At $x = 1$, the tangent line is horizontal, since $f'(1) = 0$. From $x = 1$ to $x = 3$, the slope is more and more negative, so the graph of f goes downward more and more steeply. From $x = 3$ to $x = 5$, $f'(x)$ is still negative but approaches zero again as we move to the right, which means that f continues to decrease on this interval but is decreasing less and less rapidly. Since $f'(5) = 0$, the tangent line to the graph of f is horizontal at $x = 5$. Beyond $x = 5$, $f'(x)$ is positive and increases as we move to the right. So the graph of f rises more and more rapidly.

To the left of $x = 0$, $f'(x)$ is also positive, and as we move from left to right toward $x = 0$, the slope decreases, so the graph of f is rising on the interval $(-\infty, 0)$ but less and less rapidly.

When we put all this information together, we obtain the sketch of the graph of f in Figure 3.18. ■

FIGURE 3.18

FIGURE 3.19
At each value of x these curves have the same slope.

REMARK ————————————————————

■ In this example if we had not been given a point on the graph of f, we could have found the slope of the graph but not its location vertically. After all, "parallel" curves all have the same slopes. That is, when two functions differ only by some constant value k—so that the graph of one of them is shifted k units vertically from the other—they clearly have the same derivative at every value of x. So when we know only the graph of f' and nothing more, we cannot say anything about the location of the graph of f in the vertical direction. The graph could be one of many curves, as illustrated in Figure 3.19.

The Δ-Notation

There is nothing special about using the letter h in the difference quotient $[f(x + h) - f(x)]/h$. In fact, the symbol Δx, read "delta x," has been used historically where we have used h, and it still is frequently used. The symbol Δ (capital Greek letter delta) is often used to denote a change, so Δx is understood to convey the idea of an *increment*, or *change*, in x—that is, the change from the point x to the point $x + \Delta x$. If we write $y = f(x)$, then the y increment that results from the increment Δx in x can be denoted by Δy; that is, $\Delta y = f(x + \Delta x) - f(x)$. In this notation, then,

$$f'(x) = \lim_{\Delta x \to 0} \frac{f(x + \Delta x) - f(x)}{\Delta x} = \lim_{\Delta x \to 0} \frac{\Delta y}{\Delta x}$$

Although in most cases we shall continue to use the h notation, at times the delta notation is more convenient.

Exercise Set 3.2

In Exercises 1–10, find $f'(a)$ using Definition 3.2. Also find the equation of the tangent line to the graph of f at the point where $x = a$.

1. $f(x) = 3x + 4;\ a = 1$

2. $f(x) = \dfrac{2x - 1}{3};\ a = 6$

3. $f(x) = x^2 - 4;\ a = -2$

4. $f(x) = 4x - 3x^2;\ a = 2$

5. $f(x) = 1 - \dfrac{1}{x};\ a = 1$

6. $f(x) = \dfrac{1}{x - 1};\ a = 2$

7. $f(x) = \sqrt{x};\ a = 9$

8. $f(x) = \sqrt{x - 1};\ a = 5$

9. $f(x) = \dfrac{x + 2}{x};\ a = -1$

10. $f(x) = \dfrac{x}{x - 1};\ a = 2$

In Exercises 11–20, redo Exercises 1–10, using Equation 3.4.

In Exercises 21–26, find $f'(x)$ using Definition 3.2.

21. $f(x) = 2x + 3$

22. $f(x) = 3x^2 - 5$

23. $f(x) = \dfrac{2}{x}$

24. $f(x) = \sqrt{x + 2}$

25. $f(x) = \dfrac{x}{x + 1}$

26. $f(x) = \dfrac{1}{\sqrt{x}}$

In Exercises 27–29, sketch the graph of f' using the method suggested in Example 3.5.

27.

(a)

$y = f(x)$

(b)

$y = f(x)$

28.

(a)

$y = f(x)$

(b)

$y = f(x)$

29.

(a)

$y = f(x)$

(b)

$y = f(x)$

In Exercises 30–32, the graph of the derivative f' of a given function f is shown, and $f(a)$ is given for a certain value of a. Use the ideas of Example 3.6 to sketch a possible graph of f.

30.

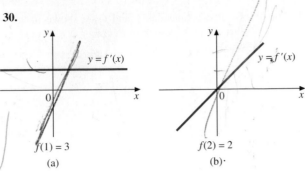

(a) $y = f'(x)$ $f(1) = 3$

(b) $y = f'(x)$ $f(2) = 2$

31.

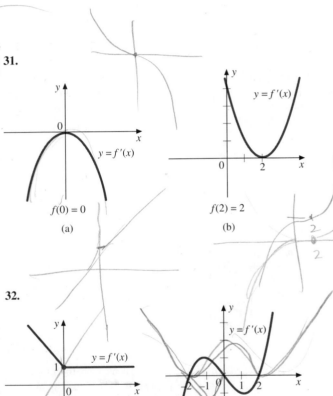

(a) $y = f'(x)$ $f(0) = 0$

(b) $y = f'(x)$ $f(2) = 2$

32.

(a) $y = f'(x)$ $f(3) = 4$

(b) $y = f'(x)$ $f(0) = -2$

In Exercises 33–36, find $f'(x)$ and give its domain.

33. $f(x) = \dfrac{1}{\sqrt{2x - 3}}$

34. $f(x) = x^3 - 2x$

35. $f(x) = x^{3/2}$

36. $f(x) = \sqrt{4 - x^2}$

37. The graphs of two functions f and g are shown in the figure. What is the derivative of $h(x) = f(x) - g(x)$?

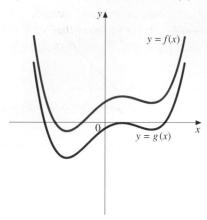

38. The graph of a function f is given. Sketch the graph of the function g that satisfies *both* of the following conditions.
(a) $g'(x) = f'(x)$ for all x
(b) $g(2) = 3$

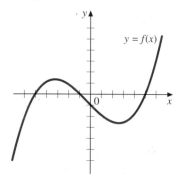

39. From the graph of $f(x) = |x|$ determine the graph of $f'(x)$. What is the domain of f'? Describe the relationship between f' and the function g defined by

$$g(x) = \frac{|x|}{x} \qquad (x \neq 0)$$

40. By changing variables in Equation 3.4, show that an alternative formula for $f'(x)$ is

$$f'(x) = \lim_{t \to x} \frac{f(t) - f(x)}{t - x}$$

and use this formula to find $f'(x)$ for $f(x) = x/(x - 1)$.

41. A particle moves in a straight line so that its position function is

$$s(t) = \frac{t}{t + 1}$$

Find its velocity function $v(t)$. [*Hint:* $v(t) = s'(t)$.]

42. The volume of water in a conical rain gauge, as the water rises, is given by

$$V(h) = \frac{1}{12}\pi h^3$$

where h is the height of the water level. Find the instantaneous rate of change of volume with respect to height when the height is 3 cm.

43. Newton's Universal Law of Gravitation states that two bodies of masses m_1 and m_2, respectively, at a distance r units apart, exert a force of attraction on each other given by the formula

$$F(r) = G\frac{m_1 m_2}{r^2}$$

where G is the universal gravitational constant. Find the instantaneous rate of change of F with respect to r. Explain the significance of the sign of your answer.

44. The revenue function $R(x)$ for selling x items of a certain type is given by

$$R(x) = 100x - 0.05x^2 \qquad (0 < x < 2000)$$

Find the marginal revenue function $R'(x)$. Show that $R'(1000) = 0$. Explain the significance of these results.

45. (a) Draw the graph of $f(x) = \sin x$ for $0 \leq x \leq 2\pi$. On the same set of axes sketch the graph of $f'(x)$, using the ideas of Example 3.5. Can you guess the formula for $f'(x)$, based on the graph you obtained?
(b) Repeat part (a), replacing $f(x)$ by $g(x) = \cos x$.

46. Let $f(x) = \ln x$. Approximate $f'(x)$ for each of the values $x = 0.5$, $x = 1$, $x = 2$, and $x = 3$, using the difference quotient

$$\frac{f(x + h) - f(x)}{h}$$

with $h = 0.001$. Obtain values from your calculator or CAS. Do your results suggest a general formula for $f'(x)$?

47. Let $g(x) = e^x$. Approximate $g'(x)$ for $x = -1$, $x = 0$, $x = 1$, and $x = 2$, using the difference quotient

$$\frac{g(x + h) - g(x)}{h}$$

for $h = 0.001$. Use your calculator or CAS. Compare your answers with $g(x)$ in each case. Make a conjecture about a general formula for $g'(x)$.

48. Use a graphing calculator or a CAS to find the approximate values of x at which the function $f(x) = |x^5 + x^4 - 2x^3 + x^2 + x - 2|$ is not differentiable.

3.3 DERIVATIVES OF SUMS, DIFFERENCES, AND POWERS

Calculating derivatives by means of the definition can become quite tedious (as you doubtless have discovered). Fortunately, in many cases the work can be greatly reduced by using results that we derive in the remainder of this chapter.

THEOREM 3.2

If $f(x) = c$, then $f'(x) = 0$ for all x; that is, *the derivative of a constant function is* 0.

FIGURE 3.20

Proof

$$f'(x) = \lim_{h \to 0} \frac{f(x + h) - f(x)}{h} = \lim_{h \to 0} \frac{c - c}{h} = \lim_{h \to 0} 0 = 0$$

Figure 3.20 shows this result geometrically. The tangent line coincides with the graph of f and has slope 0. ∎

THEOREM 3.3

If c is any constant, then the derivative of the function cf is

$$(cf)'(x) = c \cdot f'(x)$$

for all x in the domain of f'.

Proof   Since

$$\frac{(cf)(x + h) - (cf)(x)}{h} = \frac{cf(x + h) - cf(x)}{h} = c\left[\frac{f(x + h) - f(x)}{h}\right]$$

it follows that

$$(cf)'(x) = c\left[\lim_{h \to 0} \frac{f(x + h) - f(x)}{h}\right] = c \cdot f'(x) \qquad \text{By Limit Property 2} \quad ∎$$

In words, Theorem 3.3 states that *the derivative of a constant times a function is the constant times the derivative of the function.* So you can factor out constant multipliers when differentiating. Another way of looking at the result is to observe that multiplying the y-coordinates by a constant multiplies the slopes by the same constant.

THEOREM 3.4

For all x in the domain of f' and g',

$$(f + g)'(x) = f'(x) + g'(x)$$

and

$$(f - g)'(x) = f'(x) - g'(x)$$

Proof   The difference quotient for $f + g$ can be written as

$$\frac{(f + g)(x + h) - (f + g)(x)}{h} = \frac{[f(x + h) + g(x + h)] - [f(x) + g(x)]}{h}$$

$$= \frac{f(x + h) - f(x)}{h} + \frac{g(x + h) - g(x)}{h}$$

Taking limits of both sides as $h \to 0$ and using Limit Property 3, it follows that

$$(f + g)'(x) = f'(x) + g'(x)$$

A similar proof can be given for $(f - g)'$, or alternatively we may write $f - g = f + (-1)g$ and use Theorems 3.3 and 3.4.   ■

Theorem 3.4 can be read as follows: *the derivative of a sum (or difference) is the sum (or difference) of the derivatives.* This result can be extended by associativity to any finite sum or difference. For example,

$$(f + g - h)'(x) = [(f + g) - h]'(x) = (f + g)'(x) - h'(x) = f'(x) + g'(x) - h'(x)$$

For the next theorem we need to use the *Binomial Theorem*, the proof of which is given in Appendix 4. It states that for any positive integer n,

$$(a + b)^n = a^n + na^{n-1}b + \frac{n(n - 1)}{2!}a^{n-2}b^2 + \frac{n(n - 1)(n - 2)}{3!}a^{n-3}b^3 + \cdots + b^n$$

(Recall that $k!$, read "k factorial," means $1 \cdot 2 \cdot 3 \cdot \cdots \cdot k$.)

THEOREM 3.5

> **The Power Rule**
>
> Let $f(x) = x^n$, where n is any positive integer. Then for all real numbers x,
>
> $$f'(x) = nx^{n-1} \tag{3.5}$$

Proof

$$f'(x) = \lim_{h \to 0} \frac{f(x + h) - f(x)}{h} = \lim_{h \to 0} \frac{(x + h)^n - x^n}{h}$$

$$= \lim_{h \to 0} \frac{\left[x^n + nx^{n-1}h + \frac{n(n - 1)}{2!}x^{n-2}h^2 + \cdots + h^n\right] - x^n}{h}$$

$$= \lim_{h \to 0} \left[nx^{n-1} + \frac{n(n - 1)}{2!}x^{n-2}h + \cdots + h^{n-1}\right]$$

$$= nx^{n-1}$$

Note that in the next-to-last step, all terms but the first contain h as a factor and so go to 0 in the limit.   ■

EXAMPLE 3.7   Find $f'(x)$ if $f(x) = x^{10}$.

Solution   By Theorem 3.5 with $n = 10$, $f'(x) = 10x^9$. You might wish to compare this quick way with the work of finding $f'(x)$ directly from the definition.   ■

By combining Theorems 3.3 and 3.5 we see that if $f(x) = cx^n$, then

$$f'(x) = c(nx^{n-1}) = (cn)x^{n-1}$$

For example, if $f(x) = 3x^4$, then $f'(x) = 12x^3$.

We are now ready to take the derivative of any polynomial function by using Theorems 3.2 through 3.5.

EXAMPLE 3.8   If $f(x) = 3x^4 + 5x^3 - 7x^2 + 2x - 9$, find $f'(x)$.

Solution

$$\begin{aligned} f'(x) &= 3(4x^3) + 5(3x^2) - 7(2x) + 2 \\ &= 12x^3 + 15x^2 - 14x + 2 \end{aligned}$$

You should check to see how we have made use of each of the four theorems.

∎

EXAMPLE 3.9   If $f(x) = 2x^4 - 6x^2 + 7$, find $f'(2)$.

Solution   First we find $f'(x)$ in general:

$$f'(x) = 8x^3 - 12x$$

Now we substitute $x = 2$:

$$f'(2) = 8(8) - 12(2) = 64 - 24 = 40$$

∎

Do not conclude that because you have now learned the "easy way" to differentiate a polynomial function, the definition of the derivative is no longer useful. There are many important functions that are not polynomials, and each time we encounter one of these, we will have to return to Definition 3.2 to develop a differentiation formula. Further, we will see that an understanding of the definition is essential in finding new applications of the derivative. This definition also provides a key to the fundamental relationship between the derivative and the integral. So this definition is of fundamental importance and must be learned.

Exercise Set 3.3

In Exercises 1–20, find $f'(x)$, by using Theorems 3.1 through 3.5.

1. $f(x) = 3$

2. $f(x) = x$

3. $f(x) = x^4$

4. $f(x) = 3x^5$

5. $f(x) = 1 - 2x$

6. $f(x) = (3x + 4)/5$

7. $f(x) = ax + b$

8. $f(x) = 2x^2 - 3x + 4$

9. $f(x) = 4x^3 - 7x^2 + 8x - 6$

10. $f(x) = 3 - x - 7x^2 + 2x^3$

11. $f(x) = x^5 - 6x^4$

12. $f(x) = \frac{2}{3}x^6 - 5x^4 + \frac{1}{2}x^2 - 8$

13. $f(x) = ax^2 + bx + c$

14. $f(x) = (2x - 1)(x + 3)$

15. $f(x) = (2x - 1)(2 + x)$

16. $f(x) = x(1 - x)(2 + x)$

17. $f(x) = (3x + 4)^2$ **18.** $f(x) = (x + 2)^3$

19. $f(x) = 2x^3 - 3x(2 + x - x^2)$

20. $f(x) = x(2x - 3) - (3 - x)^2$

21. Find the equation of the tangent line to the graph of $y = 2x^2 - 3x + 4$ at the point $(2, 6)$.

22. Find the equation of the tangent line to the graph of $y = 3x^2 - x^4$ at the point for which $x = -2$.

23. Find k so that the slope of the tangent line to the graph of $y = 2x^2 - 3kx + 4$ is 4 at the point for which $x = 2$.

24. Find k so that the slope of the tangent line to the graph of $y = 2k^2x^2 - 2kx + 3$ is 2 at the point for which $x = 1$.

25. The line perpendicular to the tangent line to a curve at the point of tangency is called the **normal line**. Find the equations of the tangent and normal lines to $y = x^3 - 3x^2 - 7x + 18$ at the point for which $x = 3$.

26. Find all points on the graph of $y = x^4 - 4x^3 + 4x^2 + 3$ at which the tangent line is horizontal.

In Exercises 27–30, a particle moves in a straight line with position function $s(t)$. Find the instantaneous velocity of the particle at time t_1.

27. $s(t) = 3t^2 - 4t + 5; \ t_1 = 4$

28. $s(t) = t(t^2 - 3t); \ t_1 = 3$

29. $s(t) = 1 - 2t(t + 1); \ t_1 = 2$

30. $s(t) = (2t - 3)(3t + 4); \ t_1 = 5$

31. Find a and b such that the graph of $y = ax^2 + bx$ passes through the point $(2, -3)$ and such that at this point the tangent line has slope 1.

32. Find all points on the graph of $y = 2x^3 - 3x^2 - 10x + 7$ at which the slope of the tangent line is 2. Find the equations of the tangent lines at these points.

33. Two tangent lines to the curve $y = x^2$ pass through the point $(0, -4)$. Find the points of tangency and the equations of the tangent lines.

34. Find the equations of all tangent lines to the graph of $f(x) = 3x - x^2$ that pass through the point $(4, 0)$. Draw the graph of f, showing these tangent lines along with their points of tangency.

35. If $s(t) = (2t^2 - 3)^3$ is the position function for a particle moving in a straight line, find the velocity function $v(t)$. Also find the acceleration $a(t)$, defined as $v'(t)$. (*Hint:* $(a - b)^3 = a^3 - 3a^2b + 3ab^2 - b^3$.)

36. Let $f(x) = (x^2 + 2x + 1)(x^3 + x)$.
   (a) Find $f'(x)$.
   (b) Let $g(x) = x^2 + 2x + 1$ and $h(x) = x^3 + x$. Compute
$$g'(x)h(x) + g(x)h'(x)$$
   and simplify. Is there any relationship between this answer and the answer to part (a)?

3.4 THE PRODUCT AND QUOTIENT RULES

Our main objective in this section is to develop formulas for differentiating products and quotients of functions. These formulas will prove to be indispensable in our continued study of derivatives.

THEOREM 3.6

> **The Product Rule**
>
> If f and g are both differentiable at x, then so is fg, and
> $$(fg)'(x) = f'(x)g(x) + f(x)g'(x) \qquad (3.6)$$

Proof   We write the difference quotient for fg in the following equivalent ways:

$$\frac{(fg)(x + h) - (fg)(x)}{h}$$

$$= \frac{f(x+h)g(x+h) - f(x)g(x)}{h} \qquad \text{Definition of } fg$$

$$= \frac{f(x+h)g(x+h) - f(x)g(x+h) + f(x)g(x+h) - f(x)g(x)}{h}$$

We subtracted $f(x)g(x+h)$
and then added it.

$$= \left[\frac{f(x+h) - f(x)}{h} \right] g(x+h) + f(x) \left[\frac{g(x+h) - g(x)}{h} \right] \qquad \begin{array}{l}\text{Algebraic}\\\text{simplification}\end{array}$$

Since $g'(x)$ exists, we know by Theorem 3.1 that g is continuous at x. It follows that $\lim_{h \to 0} g(x+h) = g(x)$. (See Exercise 35 in Exercise Set 3.4.) Thus, on taking limits as $h \to 0$ and using Definition 3.2, as well as limit properties for sums and products, we get

$$(fg)'(x) = f'(x)g(x) + f(x)g'(x) \qquad \blacksquare$$

It is useful to learn this result in words. We call f the first function and g the second.

The derivative of a product of two functions is given by the derivative of the first times the second plus the first times the derivative of the second.

EXAMPLE 3.10   Let $F(x) = (2x^3 - x + 3)(x^4 + 3x^2 - 8)$. Find $F'(x)$.

Solution   One way to do the problem is to multiply the factors together. We leave it to you to show that the result is

$$F(x) = 2x^7 + 5x^5 + 3x^4 - 19x^3 + 9x^2 + 8x - 24$$

The derivative is therefore

$$F'(x) = 14x^6 + 25x^4 + 12x^3 - 57x^2 + 18x + 8$$

An easier way is to use the Product Rule without multiplying the factors. By Equation 3.6, we have

$$F'(x) = \underbrace{(6x^2 - 1)}_{\substack{\text{derivative}\\\text{of}\\\text{the first}}} \cdot \underbrace{(x^4 + 3x^2 - 8)}_{\text{second}} + \underbrace{(2x^3 - x + 3)}_{\text{first}} \cdot \underbrace{(4x^3 + 6x)}_{\substack{\text{derivative}\\\text{of}\\\text{the second}}}$$

The answer may be left in this form. You may wish to multiply and collect terms, however, to show that our two answers are the same.   ∎

In differentiating a product $f(x) \cdot g(x)$, it is tempting just to multiply the two derivatives and get $f'(x) \cdot g'(x)$, *but you must not yield to this temptation, because it is incorrect.* The correct result is given by Equation 3.6: $f'(x) \cdot g(x) + f(x) \cdot g'(x)$.

THEOREM 3.7

The Quotient Rule

If f and g are both differentiable at x and $g(x) \neq 0$, then f/g is differentiable at x, and

$$\left(\frac{f}{g}\right)'(x) = \frac{f'(x)g(x) - f(x)g'(x)}{[g(x)]^2} \qquad (3.7)$$

Proof   We use a similar approach to the one we used in the proof of the Product Rule. We write the difference quotient for f/g in the following equivalent ways.

$$\frac{\left(\dfrac{f}{g}\right)(x+h) - \left(\dfrac{f}{g}\right)(x)}{h}$$

$$= \frac{\dfrac{f(x+h)}{g(x+h)} - \dfrac{f(x)}{g(x)}}{h} \qquad \text{Definition of } f/g$$

$$= \frac{f(x+h)g(x) - f(x)g(x+h)}{h[g(x)g(x+h)]} \qquad \text{We simplified the complex fraction.}$$

$$= \frac{f(x+h)g(x) - f(x)g(x) + f(x)g(x) - f(x)g(x+h)}{h[g(x)g(x+h)]} \qquad \text{We subtracted } f(x)g(x) \text{ and then added it.}$$

$$= \frac{\left[\dfrac{f(x+h) - f(x)}{h}\right] \cdot g(x) - f(x)\left[\dfrac{g(x+h) - g(x)}{h}\right]}{g(x)g(x+h)} \qquad \text{Factoring}$$

Now we take the limit as $h \to 0$, using limit properties and the definitions of $f'(x)$ and $g'(x)$, as well as the continuity of g (so that $g(x+h) \to g(x)$ as $h \to 0$):

$$\left(\frac{f}{g}\right)'(x) = \frac{f'(x)g(x) - f(x)g'(x)}{[g(x)]^2} \qquad \blacksquare$$

We can state the Quotient Rule as follows:

The derivative of a quotient of two functions is given by the derivative of the numerator times the denominator minus the numerator times the derivative of the denominator, all divided by the square of the denominator.

EXAMPLE 3.11   Find $F'(x)$, where

$$F(x) = \frac{x^2 - 1}{2x^2 + 3}$$

Solution

$$F'(x) = \frac{\overbrace{(2x)}^{\substack{\text{derivative}\\\text{of}\\\text{numerator}}} \cdot \overbrace{(2x^2+3)}^{\text{denominator}} - \overbrace{(x^2-1)}^{\text{numerator}} \cdot \overbrace{(4x)}^{\substack{\text{derivative}\\\text{of}\\\text{denominator}}}}{\underbrace{(2x^2+3)^2}_{\text{square of denominator}}}$$

$$= \frac{4x^3 + 6x - 4x^3 + 4x}{(2x^2+3)^2} = \frac{10x}{(2x^2+3)^2} \qquad \blacksquare$$

The following corollary to Theorem 3.7 will enable us to extend the Power Rule (Theorem 3.5) to negative exponents.

COROLLARY 3.7

If $g'(x)$ exists and $g(x) \neq 0$, then the derivative of $1/g$ exists at x and

$$\left(\frac{1}{g}\right)'(x) = -\frac{g'(x)}{[g(x)]^2}$$

Proof   In Equation 3.7 let $f(x)$ be the constant function $f(x) = 1$. Then $f'(x) = 0$, and we get the desired result. $\blacksquare$

Consider now a function of the form $f(x) = x^n$, with n a negative integer. Write $n = -m$, where m is a positive integer $(m = |n|)$. Then

$$f(x) = x^{-m} = \frac{1}{x^m}$$

In Corollary 3.7, let $g(x) = x^m$. By Theorem 3.5,

$$g'(x) = mx^{m-1}$$

So the corollary gives

$$f'(x) = -\frac{mx^{m-1}}{(x^m)^2} = -m\frac{x^{m-1}}{x^{2m}} = (-m)x^{-m-1} = nx^{n-1}$$

Thus, Equation 3.5 continues to hold true when n is a negative integer. For example, if $f(x) = x^{-3}$, then $f'(x) = -3x^{-4}$.

EXAMPLE 3.12   If $f(x) = \dfrac{2x-4}{x^2}$, find $f'(x)$.

Solution   We could use the Quotient Rule, but in cases such as this, in which the denominator is a monomial (a polynomial with only one term), it is usually easier to divide first and then use the Power Rule. Doing it this way, we first divide, getting

$$f(x) = 2x^{-1} - 4x^{-2}$$

Then we use the Power Rule, extended to negative exponents:

$$f'(x) = -2x^{-2} + 8x^{-3} \qquad \blacksquare$$

REMARK ————————————————————————————————

■ Later in this chapter, we will see that the Power Rule holds true for fractional (rational) exponents, and in Chapter 7, after we have given meaning to irrational powers, we will prove it holds true for irrational exponents as well. So it actually is true for all real numbers as exponents.

Exercise Set 3.4

In Exercises 1–6, find $f'(x)$ in two ways: (a) by multiplying the factors first, and (b) by using the Product Rule. Show that the results are equal.

1. $f(x) = (2x - 3)(3x + 4)$

2. $f(x) = (x^2 - 2)(x^2 + 3)$

3. $f(x) = (1 - x^2)(2 + 5x^2)$

4. $f(x) = (x^3 - 3x^2)(2x^2 + 1)$

5. $f(x) = (2x^3 - 1)(x^4 + x)$

6. $f(x) = (x^2 + x - 1)(x^2 - x + 1)$

In Exercises 7–12, find $f'(x)$ by the Product Rule.

7. $f(x) = (x^2 - 2x + 3)(4 - 3x^2)$

8. $f(x) = (4x - x^3)(x^2 + 2x - 3)$

9. $f(x) = (x^4 - 2x^2 + 1)(x^2 - x + 3)$

10. $f(x) = (3x^4 - 2x^2)(2x^4 + 3x^2 + 5)$

11. $f(x) = (5x^4 + 6x^3 + 7)(2x^3 - 3x + 1)$

12. $f(x) = (4 - x - x^3)(3x + 2x^3 - x^5)$

In Exercises 13–22, find $f'(x)$ using the Quotient Rule.

13. $f(x) = \dfrac{x}{x - 2}$

14. $f(x) = \dfrac{3x + 2}{4x - 3}$

15. $f(x) = \dfrac{x^2 - 1}{x^2 + 1}$

16. $f(x) = \dfrac{2x^2 - 3x}{1 - x^2}$

17. $f(x) = \dfrac{x^2}{x^2 - 4}$

18. $f(x) = \dfrac{4 - 2x + 3x^2}{x^2 + 2}$

19. $f(x) = \dfrac{2x - 3}{x^2 + 2x}$

20. $f(x) = \dfrac{3x - x^2}{2x + 1}$

21. $f(x) = \dfrac{x^3 + 8}{x^3 - 1}$

22. $f(x) = \dfrac{4x^2 + 3x}{x^4 - 16}$

In Exercises 23–28, find $f'(x)$ using the Power Rule.

23. $f(x) = 2x^{-3}$

24. $f(x) = 3x^{-5}$

25. $f(x) = x^2 + x^{-2}$

26. $f(x) = 2 - 3x^{-1} + 4x^{-2}$

27. $f(x) = x + 3 + \dfrac{4}{x}$

28. $f(x) = 3 + 2x - \dfrac{4}{x^2}$

29. Let $g(x) = (x^2 - 4)/x^2$. Find $g'(x)$ in two ways: by the Quotient Rule and by writing $g(x)$ in the form $1 - 4x^{-2}$. Compare the two methods.

30. Let $h(x) = (3 - 4x + 5x^2)/2x$. Find $h'(x)$ without using the Quotient Rule.

31. Find the equation of the tangent line to the graph of

$$y = \frac{x + 3}{x - 1}$$

at the point $(2, 5)$.

32. Find the equation of the tangent line to the graph of

$$y = \frac{x^2}{2x - 3}$$

at the point for which $x = -3$.

33. A particle moves in a straight line so that its position function at time t is given by

$$s(t) = \frac{2t^2 + 3}{t + 1}$$

Find its velocity at the instant when $t = 3$.

34. The length of a rectangle is increasing at the rate of 5 cm/s, and the width is increasing at the rate of 3 cm/s. At what rate is the area increasing when the length is 8 cm and the width is 5 cm?

35. In the proof of the Product Rule, we used the fact that continuity of g at x implies that $\lim_{h \to 0} g(x + h) = g(x)$. Prove this result.

In Exercises 36 and 37, find $F'(x)$ by using the Quotient Rule and the Product Rule. Simplify the result.

36. $F(x) = \dfrac{(x^2 - 4)(x^2 + 1)}{x^2 - 9}$

37. $F(x) = \dfrac{2x - 1}{(3x^2 + 5)(x^2 - 2x)}$

38. Let $f(x) = (x^2 - 4)/(x^2 + 4)$. Find $f'(x)$ without using the Quotient Rule or Product Rule. (*Hint:* Divide first

and then make use of Corollary 3.7.)

39. Follow the instructions in Exercise 38 for

$$f(x) = \frac{(x^3 - x^2 + x)}{(x^2 + 1)}$$

40. Derive a formula for the derivative of the product fgh of three differentiable functions. (*Hint:* Write $(fgh)' = [(fg) \cdot h]'$ and apply the Product Rule twice.)

41. Using the result of Exercise 40, find $F'(x)$ if

$$F(x) = (2x + 1)(3x^2 - 4x)(x^2 + 4)$$

42. Using the result of Exercise 40, make a conjecture for a formula for the derivative $(f_1 \cdot f_2 \cdot \cdots \cdot f_n)'$ of the product of n differentiable functions. Prove your formula by mathematical induction. (See Supplementary Appendix 4.)

3.5 HIGHER-ORDER DERIVATIVES AND OTHER NOTATIONS

Since for a differentiable function f, the derivative f' is a function in its own right, there is no reason why it may not also have a derivative. When it does, we denote its derivative by f''. So f'' is the derivative of f'. We also call f'' the **second derivative** of f. For example, suppose $f(x) = 3x^4 - 2x^2 + 7$. Then

$$f'(x) = 12x^3 - 4x \quad \text{First derivative}$$

and

$$f''(x) = 36x^2 - 4 \quad \text{Second derivative}$$

There is no reason for stopping here, of course; we can take further derivatives, getting

$$f'''(x) = 72x \quad \text{Third derivative}$$

$$f''''(x) = 72 \quad \text{Fourth derivative}$$

Note that all derivatives higher than the fourth are identically zero in this case.

It gets rather cumbersome to write so many primes, so we use the symbol

$$f^{(k)}(x)$$

to mean the *kth-order derivative*. (Usually we go to this notation for $k > 3$, though in practice the need for such higher-order derivatives rarely arises.) For example, $f^{(5)}(x)$ means the fifth derivative. The superscript k is put in parentheses to distinguish it from an exponent.

There are also other notations for derivatives. We introduce them not to make life more difficult for you, but rather because each notation has certain advantages, or historical status, and so it is useful to know them all. Moreover, in other books where calculus is used (higher math, physics, economics, statistics, epidemiology, and so on) you might encounter these other notations.

Suppose that the rule for a function is given by an equation of the form $y = f(x)$. Then the derivative, $f'(x)$, can be symbolized using any of the following alternative forms:

$$y' \qquad \frac{dy}{dx} \qquad \frac{df(x)}{dx} \qquad D_x y \qquad D_x f$$

The second notation, dy/dx (read "dee wye dee ex"), is widely used. This notation was introduced by Leibniz. It is suggestive of the limit

$$\lim_{\Delta x \to 0} \frac{\Delta y}{\Delta x}$$

that we introduced in Section 3.2. For the present you should not think of this form as a quotient but simply as a symbol for the derivative. Later we will see that we can give meaning to dy and dx separately. Even now we can, however, give meaning to the symbol d/dx. We refer to this symbol as a *derivative operator*, and when we follow it with an expression, as in

$$\frac{d}{dx}(\quad)$$

we mean to take the derivative with respect to x of what is inside the parentheses. For example,

$$\frac{d}{dx}(3x^2 + 2x) = 6x + 2$$

In this sense, then, dy/dx can be thought of as the operator d/dx operating on y. Alternatively, we can replace y with $f(x)$ and write $df(x)/dx$, or simply df/dx. The symbol D_x can be used in the same way as d/dx. Thus,

$$D_x(3x^2 + 2x) = 6x + 2$$

Suppose we are given that $f(x) = 3x^2 + 2x$. By setting $y = f(x)$, we can use any of the following to indicate the derivative:

$$f'(x) \qquad y' \qquad \frac{dy}{dx} \qquad \frac{df(x)}{dx} \qquad \frac{d}{dx}(3x^2+2x) \qquad D_x(3x^2+2x)$$

or even $(3x^2 + 2x)'$. In each case the value is $6x + 2$. If we want to evaluate the derivative at a specific point, say $x = 2$, the $f'(x)$ notation has a clear advantage; we simply write $f'(2)$. For the other notations we resort to the following device:

$$y'\big|_{x=2} \qquad \frac{dy}{dx}\bigg|_{x=2} \qquad \frac{d}{dx}(3x^2+2x)\bigg|_{x=2}$$

and so on. We will see advantages to some of the other notations as we progress.

Indicating higher-order derivatives with these new notations, we have

$$y', \ y'', \ y''', \ y^{(4)}, \dots$$

$$\frac{dy}{dx}, \ \frac{d^2y}{dx^2}, \ \frac{d^3y}{dx^3}, \ \frac{d^4y}{dx^4}, \dots$$

$$D_x y, \ D_x^2 y, \ D_x^3 y, \ D_x^4 y, \dots$$

The notation d^2y/dx^2 can be explained by observing that the second derivative is the derivative of dy/dx and so can be obtained from it by operating on dy/dx with the derivative operator d/dx:

$$\frac{d}{dx}\left(\frac{dy}{dx}\right)$$

In the next chapter we will see some applications of second derivatives, but we can point out one application now. We know that if a particle moves in a straight line so that its position function at time t is $s(t)$, then its instantaneous velocity $v(t)$ is the derivative $s'(t)$. Since acceleration is the rate of change of velocity, the average acceleration over the time interval from t to $t+h$ is

$$\frac{v(t+h) - v(t)}{h}$$

and the limit of this quotient as $h \to 0$ is the *instantaneous acceleration* at t. So we have

$$a(t) = v'(t) = s''(t)$$

The third derivative of the position function is the rate of change of acceleration and is sometimes referred to as the "jerk."

Exercise Set 3.5

In Exercises 1–6, find y', y'', and y'''.

1. $y = 2x^4$

2. $y = 3x^{-2}$

3. $y = 4x^3 - 2x^2 + 1$

4. $y = x^5 - 2x^3 + 3$

5. $y = x^2 - x^{-2}$

6. $y = \dfrac{2}{x} - 3x^2$

In Exercises 7–12, find dy/dx and d^2y/dx^2.

7. $y = (2x + 3)(3x - 1)$

8. $y = (x^2 - 1)(2x^3 + 5)$

9. $y = x^2(x^2 - 3) + \dfrac{5}{x}$

10. $y = \dfrac{2}{x^3} - \dfrac{3}{2x^2}$

11. $y = \dfrac{x^2 - x + 3}{x}$

12. $y = \dfrac{x^2 - 1}{x^4}$

In Exercises 13–16, find the indicated derivatives.

13. $\dfrac{d}{dx}(x^3 - 3x^2 + 1)$

14. $\dfrac{d}{dx}(x^2 - 2x^{-3})$

15. $D_x\left(\dfrac{x - 2}{x + 3}\right)$

16. $D_x\left(\dfrac{3x - 4}{2x + 3}\right)$

In Exercises 17–20, find $f''(x)$.

17. $f(x) = x^3 - x^{-2}$

18. $f(x) = \dfrac{3x - 2}{x^2}$

19. $f(x) = \left(x - \dfrac{1}{2x}\right)^2$

20. $f(x) = \sqrt{x^2 + 1 + \dfrac{1}{4x^2}}$, $x > 0$ (*Hint:* Compare with Exercise 19.)

21. Let $P(x) = a_0 + a_1x + a_2x^2 + \cdots + a_nx^n$. Find $P^{(n)}(x)$. What is $P^{(k)}(x)$ for $k > n$?

22. Let $g(x) = \dfrac{1}{x}$. Find $g^{(n)}(x)$.

23. Let $f(x) = \sqrt{x}$. Assuming the Power Rule holds true for fractional exponents, find the first four derivatives of f, and devise a formula for $f^{(n)}(x)$. (The formula for the derivative of $f^{(n)}(x)$ can be proved by mathematical induction, which is discussed in Appendix 4.)

24. Let $y = f(x) \cdot g(x)$, where f and g have derivatives of order 3 or more. Find a formula for the following:

(a) $\dfrac{d^2y}{dx^2}$

(b) $\dfrac{d^3y}{dx^3}$

25. A particle moves in a straight line so that its position function at time t is given by

$$s(t) = 2t^2 - \frac{t^2 - 1}{t + 3}$$

Find formulas for its velocity $v(t)$ and its acceleration $a(t)$ at any time $t \geq 0$. Also find the velocity and acceleration after 1 second, $v(1)$ and $a(1)$, respectively.

26. Show that $y = x^3 - 2x^2 + 4$ satisfies the equation $y'' - 2y' + y = x^3 - 8x^2 + 14x$.

27. Show that $y = (2x^2 + 1)/x$ satisfies the equation $x^3 y'' + 3x^2 y' - xy = 4x^2 - 2$.

28. The accompanying figure shows portions of graphs, generated by a CAS, of a certain polynomial function, $y = p(x)$, and its first two derivatives, $y = p'(x)$ and $y = p''(x)$. Each tick mark on the x-axis represents one unit, and on the y-axis, ten units. Can this graph be an accurate description of p? That is, can p be a quadratic polynomial? Explain. If your answer is "no," what is the least degree p can be?

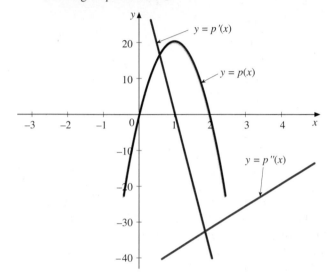

29. Compute dy/dx for each of the following by expanding and then using the Power Rule.
  (a) $y = (3x - 2)^2$
  (b) $y = (2x + 4)^2$
  (c) $y = (x^2 - 2x + 1)^2$
  (d) $y = (x^3 + 1)^2$

30. (a) Refer to Exercise 29. Notice that each equation can be written in the form $y = [f(x)]^2$. In each case find a function g so that the derivative can be written in the form

$$\frac{dy}{dx} = 2f(x)g(x)$$

  (b) From the result of part (a) conjecture a formula for

$$\frac{d}{dx}[f(x)]^2$$

3.6 DERIVATIVES OF THE TRIGONOMETRIC FUNCTIONS

Our main objective in this section is to obtain formulas for the derivatives of the sine and cosine functions. Since the other four trigonometric functions are defined in terms of these two basic ones, we can obtain formulas for their derivatives using results for derivatives of quotients and reciprocals from Section 3.4.

Before getting to the main theorem, we need a preliminary result. Such a preliminary result is called a *lemma* (a theorem used in proving another theorem). The first part of the lemma states that

$$\lim_{t \to 0} \frac{\sin t}{t} = 1$$

We have seen evidence earlier (Example 2.5) that this limit is correct. In the proof of the lemma we use a geometric argument based on the unit circle definition of $\sin t$ and $\cos t$.

LEMMA

(a) $\displaystyle \lim_{t \to 0} \frac{\sin t}{t} = 1$

(b) $\displaystyle \lim_{t \to 0} \frac{1 - \cos t}{t} = 0$

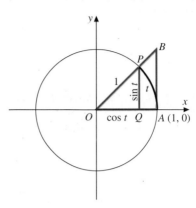

FIGURE 3.21

Proof

(a) In Figure 3.21 we show a unit circle, along with the point P that is t units along the circle from $A(1, 0)$, where $0 < t < \pi/2$. The point Q is the projection of P on the x-axis, and B is the intersection of the tangent line $x = 1$ at A and the line through O and P. So the triangles OQP and OAB are similar right triangles.

By the definitions of $\sin t$ and $\cos t$, the coordinates of P are $(\cos t, \sin t)$. Thus, the length of the line segment PQ, which we symbolize by $\overline{PQ}$, is $\sin t$, and $\overline{OQ} = \cos t$. Also, $\overline{OA} = 1$, the radius of the circle. By similarity of triangles OQP and OAB, we have the proportion

$$\frac{\overline{AB}}{\overline{PQ}} = \frac{\overline{OA}}{\overline{OQ}}$$

That is,

$$\frac{\overline{AB}}{\sin t} = \frac{1}{\cos t}$$

So

$$\overline{AB} = \frac{\sin t}{\cos t}$$

We are going to make use of the fact that the area of the circular sector OAP (the "pie-shaped" region) lies between the area of the triangle OQP and the area of the triangle OAB. We know how to find the area of a triangle ($\frac{1}{2}$base $\times$ height). The area of the sector OAP is that fractional part of the area of the entire circle given by the ratio of the arc length t to the total circumference 2π ($C = 2\pi r$ and $r = 1$). Since the area of the circle is π (area $= \pi r^2$ and $r = 1$), we obtain

$$\text{area of sector } OAB = \left(\frac{t}{2\pi}\right)\pi = \frac{t}{2}$$

The area of the triangle OQP is $\frac{1}{2}(\overline{OQ})(\overline{QP}) = \frac{1}{2}(\cos t)(\sin t)$ and the area of the triangle OAB is

$$\frac{1}{2}(\overline{OA})(\overline{AB}) = \frac{1}{2}(1)\left(\frac{\sin t}{\cos t}\right) = \frac{1}{2}\left(\frac{\sin t}{\cos t}\right)$$

So we have

$$\frac{1}{2}(\cos t)(\sin t) < \frac{t}{2} < \frac{1}{2}\left(\frac{\sin t}{\cos t}\right)$$

how did ?? flip?

Multiplying through by the positive quantity $2/\sin t$, and then inverting all members (thus reversing the sense of each inequality), we get

$$\cos t < \frac{\sin t}{t} < \frac{1}{\cos t}$$

As $t \to 0^+$, $\cos t \to 1$ and $1/\cos t \to 1$. By the Squeeze Theorem (Theorem 2.3), it follows that

$$\lim_{t \to 0^+} \frac{\sin t}{t} = 1$$

We complete the proof of part (a) by observing that $f(t) = (\sin t)/t$ is an even function. That is,

$$f(-t) = \frac{\sin(-t)}{(-t)} = \frac{-\sin t}{-t} = \frac{\sin t}{t} = f(t)$$

It follows from symmetry that

$$\lim_{t \to 0^-} \frac{\sin t}{t} = \lim_{t \to 0^+} \frac{\sin t}{t}$$

Since the limit on the right is 1, the limit on the left also is 1. Thus,

$$\lim_{t \to 0} \frac{\sin t}{t} = 1$$

(b) Since $\sin^2 t + \cos^2 t = 1$, we can write $1 - \cos^2 t = \sin^2 t$. Thus,

$$\frac{1 - \cos t}{t} = \frac{1 - \cos t}{t} \cdot \frac{1 + \cos t}{1 + \cos t} = \frac{1 - \cos^2 t}{t(1 + \cos t)} = \frac{\sin^2 t}{t(1 + \cos t)}$$

$$= \left(\frac{\sin t}{t} \right) \left(\frac{\sin t}{1 + \cos t} \right)$$

We know from part (a) that $\lim_{t \to 0}(\sin t)/t = 1$. Also,

$$\lim_{t \to 0} \frac{\sin t}{1 + \cos t} = \frac{0}{1 + 1} = \frac{0}{2} = 0$$

Finally, using the fact that the limit of a product is the product of the limits when they both exist (Limit Property 5), we have

$$\lim_{t \to 0} \frac{1 - \cos t}{t} = (1)(0) = 0 \qquad \blacksquare$$

REMARK ──

■ The limits given by the Lemma are based on the definitions of $\sin t$ and $\cos t$ as functions of the *real number t*. If t is an *angle*, then for these limits to be valid, *t must be expressed in radians*. Since the derivatives of the sine and cosine (that we will consider in the theorem that is to follow) are based on the Lemma, in calculus angles should always be expressed in radians.

──

Before going to the main theorem, we give an example to show how the Lemma can be used to obtain other related limits.

EXAMPLE 3.13   Evaluate the following limits.

(a) $\displaystyle \lim_{x \to 0} x \cot x$

(b) $\displaystyle \lim_{\theta \to 0} \frac{\sin 2\theta}{\sin 3\theta}$

Solution

(a) $x \cot x = x \left(\dfrac{\cos x}{\sin x} \right) = \left(\dfrac{x}{\sin x} \right) \cos x = \dfrac{1}{\left(\dfrac{\sin x}{x} \right)} \cdot \cos x \qquad (x \neq 0)$

So

$$\lim_{x \to 0} x \cot x = \lim_{x \to 0} \left(\frac{1}{\dfrac{\sin x}{x}} \right) \cdot \lim_{x \to 0} \cos x$$

$$= \frac{1}{\displaystyle \lim_{x \to 0} \left(\dfrac{\sin x}{x} \right)} \cdot 1$$

$$= \frac{1}{1} \qquad \text{Lemma, part (a)}$$

$$= 1$$

(b) We write $(\sin 2\theta)/(\sin 3\theta)$ in the form

$$\frac{\sin 2\theta}{\sin 3\theta} = \frac{\sin 2\theta}{2\theta} \cdot \frac{3\theta}{\sin 3\theta} \cdot \frac{2}{3} \qquad (\theta \neq 0 \text{ and } |\theta| < \pi/3)$$

By part (a) of the Lemma, replacing t by 2θ in one case and by 3θ in the other, we know that

$$\lim_{\theta \to 0} \frac{\sin 2\theta}{2\theta} = 1 \qquad \text{and} \qquad \lim_{\theta \to 0} \frac{\sin 3\theta}{3\theta} = 1$$

So

$$\lim_{\theta \to 0} \frac{\sin 2\theta}{\sin 3\theta} = \left(\lim_{\theta \to 0} \frac{\sin 2\theta}{2\theta} \right) \left(\lim_{\theta \to 0} \frac{3\theta}{\sin 3\theta} \right) \cdot \frac{2}{3}$$

$$= (1) \left(\frac{1}{1} \right) \cdot \frac{2}{3} = \frac{2}{3} \qquad \blacksquare$$

In the proof of the main theorem we also need the following two trigonometric identities, called the *addition formulas* for the sine and cosine. They are proved in Supplementary Appendix 2.

$$\sin(\alpha + \beta) = \sin \alpha \cos \beta + \cos \alpha \sin \beta \qquad (3.8)$$

$$\cos(\alpha + \beta) = \cos \alpha \cos \beta - \sin \alpha \sin \beta \qquad (3.9)$$

Both formulas are valid for all real numbers α and β.

Before stating the theorem, let us try to anticipate the result from the graphs of $y = \sin x$ and $y = \cos x$. In the top graph of Figure 3.22 we have shown the graph of $f(x) = \sin x$. Using slopes of tangent lines drawn at selected points on this curve, we have obtained points on the graph of $y = f'(x)$, shown below the sine curve. The graph of f' looks suspiciously like the graph of $y = \cos x$. So a good conjecture is that the derivative of $\sin x$ is $\cos x$. In the theorem we prove that this conjecture is correct. Refer again to Figure 3.22, and assume the lower curve is the graph of $y = \cos x$. By considering slopes of tangent lines to this curve, can you conjecture what the derivative of the cosine is? Try it before going on to the theorem.

THEOREM 3.8

(a) $\dfrac{d}{dx} \sin x = \cos x$

(b) $\dfrac{d}{dx} \cos x = -\sin x$

Proof

(a) Let $f(x) = \sin x$. Then, by Definition 3.2,

$$f'(x) = \lim_{h \to 0} \frac{f(x + h) - f(x)}{h} = \lim_{h \to 0} \frac{\sin(x + h) - \sin x}{h}$$

FIGURE 3.22

We use Identity 3.8 with $\alpha = x$ and $\beta = h$ to write the difference quotient on the right in the following equivalent ways.

$$\frac{\sin(x+h) - \sin x}{h} = \frac{\sin x \cos h + \cos x \sin h - \sin x}{h}$$

$$= \frac{\sin x(\cos h - 1) + \cos x \sin h}{h}$$

$$= -\sin x \cdot \left(\frac{1 - \cos h}{h}\right) + \cos x \left(\frac{\sin h}{h}\right)$$

Applying the Lemma we have, on taking the limit as $h \to 0$,

$$f'(x) = (-\sin x)(0) + (\cos x)(1) = \cos x$$

That is,

$$\frac{d}{dx} \sin x = \cos x$$

(b) Let $g(x) = \cos x$. Then

$$g'(x) = \lim_{h \to 0} \frac{g(x+h) - g(x)}{h} = \lim_{h \to 0} \frac{\cos(x+h) - \cos x}{h}$$

Using Identity 3.9, we write

$$\frac{\cos(x+h) - \cos x}{h} = \frac{\cos x \cos h - \sin x \sin h - \cos x}{h}$$

$$= \frac{\cos x(\cos h - 1) - \sin x \sin h}{h}$$

$$= -\cos x \left(\frac{1 - \cos h}{h}\right) - \sin x \left(\frac{\sin h}{h}\right)$$

Letting $h \to 0$ and applying the Lemma we get

$$g'(x) = (-\cos x)(0) - (\sin x)(1) = -\sin x$$

That is,

$$\frac{d}{dx} \cos x = -\sin x$$ ∎

EXAMPLE 3.14   Let $f(x) = x \sin x + \cos x$. Find $f'(x)$ and $f''(x)$.

Solution

$$f'(x) = \underbrace{(\sin x + x \cos x)}_{\substack{\text{derivative of} \\ x \sin x}} + (-\sin x) = x \cos x$$

$$f''(x) = \cos x + x(-\sin x) = \cos x - x \sin x$$ ∎

EXAMPLE 3.15   Find

$$\frac{d}{dt} \left(\frac{1 - \cos t}{\sin t} \right)$$

Solution   By the Quotient Rule, we have

$$\frac{d}{dt} \left(\frac{1 - \cos t}{\sin t} \right) = \frac{(\sin t)(\sin t) - (1 - \cos t) \cos t}{\sin^2 t}$$

$$= \frac{\sin^2 t - \cos t + \cos^2 t}{\sin^2 t}$$

$$= \frac{1 - \cos t}{\sin^2 t} \qquad \text{Since } \sin^2 t + \cos^2 t = 1$$

$$= \frac{1 - \cos t}{1 - \cos^2 t} \qquad \text{Since } \sin^2 t = 1 - \cos^2 t$$

$$= \frac{1 - \cos t}{(1 - \cos t)(1 + \cos t)}$$

$$= \frac{1}{1 + \cos t}$$ ∎

REMARK ─────────────────────────────────────

∎ In problems such as the one in Example 3.15 we will always assume any restrictions that are necessary on the independent variable so that denominators are not 0.

──

Derivatives of tan x, cot x, sec x, and csc x

Using the definitions

$$\tan x = \frac{\sin x}{\cos x} \qquad \cot x = \frac{\cos x}{\sin x} \qquad \sec x = \frac{1}{\cos x} \qquad \csc x = \frac{1}{\sin x}$$

we can get the derivatives of these other four trigonometric functions from those for the sine and cosine simply by using our formulas for differentiating quotients and reciprocals. We illustrate the calculation for the tangent and leave the others as exercises:

$$\frac{d}{dx}(\tan x) = \frac{d}{dx}\left(\frac{\sin x}{\cos x}\right)$$

$$= \frac{(\cos x)(\cos x) - (\sin x)(-\sin x)}{\cos^2 x} \qquad \text{Quotient Rule}$$

$$= \frac{\cos^2 x + \sin^2 x}{\cos^2 x}$$

$$= \frac{1}{\cos^2 x}$$

$$= \sec^2 x$$

We summarize here the derivatives of all six trigonometric functions.

Derivatives of the Trigonometric Functions

$$\frac{d}{dx}(\sin x) = \cos x \qquad\qquad \frac{d}{dx}(\cot x) = -\csc^2 x$$

$$\frac{d}{dx}(\cos x) = -\sin x \qquad\qquad \frac{d}{dx}(\sec x) = \sec x \tan x$$

$$\frac{d}{dx}(\tan x) = \sec^2 x \qquad\qquad \frac{d}{dx}(\csc x) = -\csc x \cot x$$

The formulas for derivatives of the sine and cosine are valid for all real x, those for the tangent and secant for all $x \neq \frac{\pi}{2} + n\pi$, and those for the cotangent and cosecant for all $x \neq n\pi$, with $n = 0, \pm 1, \pm 2, \ldots$.

EXAMPLE 3.16   Find $\dfrac{dy}{dx}$.

(a) $y = \sec x(1 + \tan x)$ (b) $y = \dfrac{\cot x}{1 - \csc x}$

Solution

(a) Using the Product Rule, we get

$$\frac{dy}{dx} = (\sec x \tan x)(1 + \tan x) + \sec x(\sec^2 x)$$

$$= \sec x \tan x + \sec x \tan^2 x + \sec^3 x$$

(b) By the Quotient Rule,

$$\frac{dy}{dx} = \frac{(-\csc^2 x)(1 - \csc x) - \cot x(\csc x \cot x)}{(1 - \csc x)^2}$$

$$= \frac{-\csc^2 x + \csc^3 x - \csc x \cot^2 x}{(1 - \csc x)^2}$$

We can simplify the answer further by making use of the identity

$$\cot^2 x = \csc^2 x - 1$$

given in Supplementary Appendix 2. (You can get it from $\sin^2 x + \cos^2 x = 1$ by dividing both sides by $\sin^2 x$.) Replacing $\cot^2 x$ by $\csc^2 x - 1$ gives

$$\frac{dy}{dx} = \frac{-\csc^2 x + \csc^3 x - \csc x(\csc^2 x - 1)}{(1 - \csc x)^2}$$

$$= \frac{-\csc^2 x + \csc x}{(1 - \csc x)^2}$$

$$= \frac{\csc x(1 - \csc x)}{(1 - \csc x)^2} = \frac{\csc x}{1 - \csc x} \qquad \blacksquare$$

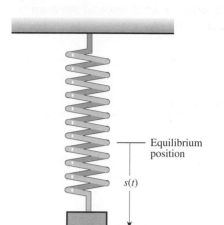

FIGURE 3.23

Equilibrium position

$s(t)$

W

EXAMPLE 3.17   In Figure 3.23 a spring is attached to a vertical support, and a weight W is attached to the spring. After reaching the equilibrium position, the spring is set in motion when the weight is pulled down and released. If we assume no retarding force (such as air resistance), the spring and weight will oscillate up and down with what is known as **simple harmonic motion**. Suppose that the position function $s(t)$ of the weight, measured positively downward from the equilibrium position, is given by

$$s(t) = 8\cos t$$

where t is in seconds and $s(t)$ is in centimeters. Discuss the motion.

Solution   We know that the first derivative of s gives velocity and the second derivative gives acceleration. So we have

$$v(t) = -8\sin t$$

$$a(t) = -8\cos t$$

At time $t = 0$, the position of the weight is $s(0) = 8$ cm below the equilibrium position, and the velocity $v(0) = 0$. So the weight was initially pulled down 8 cm from the equilibrium position and released from rest. The amplitude of the motion is 8, and the period is 2π. So the weight will oscillate from the lowest point 8 cm below equilibrium to its highest point 8 cm above equilibrium, and it will go through a complete cycle every 2π seconds.

From $t = 0$ to $t = \pi$, the velocity is negative, indicating the weight is moving upward (remember that downward is the positive direction). At $t = \pi$ it reaches its highest position. Then, from $t = \pi$ to $t = 2\pi$, the velocity is positive, and the weight is moving downward. It passes through the equilibrium position at times $t = \pi/2$, $3\pi/2$, $5\pi/2$, ... (See Figure 3.24.)

The acceleration is opposite to the position; that is, $a(t) = -s(t)$. Thus, when the weight is below equilibrium, the acceleration is positive, and when it is above equilibrium, the acceleration is negative. From Newton's Second Law of Motion, we know that force = mass × acceleration. So when the weight is below equilibrium, the force exerted by the spring is negative (upward), and when it is above equilibrium, the force is positive (downward). We summarize the nature of the motion in Figure 3.25. $\qquad \blacksquare$

$v(t)$

$v(t) = -8\sin t$

Velocity negative, moving upward

Velocity positive, moving downward

FIGURE 3.24

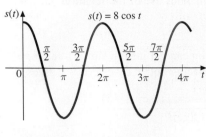

— Moving up, speeding up
— Moving up, slowing down
— Moving down, speeding up
— Moving down, slowing down

FIGURE 3.25

REMARK ────────────────────────────

■ All simple harmonic motion can be modeled by an equation of the form

$$s(t) = a\cos(\omega t) + b\sin(\omega t)$$

which can also be written in the form

$$s(t) = A\sin(\omega t + \alpha)$$

using Identity 3.8 (see Exercise 36 in this section), where a, b, ω, as well as A and α are constants that depend on the initial position and velocity, the mass of the object in motion, and a property of the spring known as the spring constant. In practice, pure simple harmonic motion is never achieved because of resisting forces, but the idealized model is nevertheless useful in approximating the motion.

Exercise Set 3.6

In Exercises 1–10, make use of the Lemma (on page 166) to find the limits.

1. $\displaystyle\lim_{t\to 0}\frac{\sin 2t}{t}$

2. $\displaystyle\lim_{x\to 0}\frac{\sin^2 x}{x}$

3. $\displaystyle\lim_{\theta\to 0}\frac{\tan\theta}{\theta}$

4. $\displaystyle\lim_{t\to 0}\frac{3t - \sin t}{t}$

5. $\displaystyle\lim_{x\to 0}\frac{\sec x - 1}{x}$

6. $\displaystyle\lim_{\theta\to 0}\frac{\sin\theta}{\theta + \sin\theta}$

7. $\displaystyle\lim_{t\to 0}\frac{\sin 3t}{\tan 2t}$

8. $\displaystyle\lim_{x\to 0}\frac{1 - \cos x}{\sin x}$

9. $\displaystyle\lim_{h\to 0}(\sin 2h)(\cot 5h)$

10. $\displaystyle\lim_{x\to\infty} x\sin\frac{1}{x}$ (*Hint:* Let $t = 1/x$.)

In Exercises 11–20, find dy/dx.

11. $y = x^2\sin x$

12. $y = 3\sin x - 2\cos x$

13. $y = \dfrac{1 - \sin x}{1 + \sin x}$

14. $y = \dfrac{\sec x}{1 + \tan x}$

15. $y = \dfrac{\sin x}{x}$

16. $y = x\cos x - \sin x$

17. $y = \dfrac{2\cot x + 1}{\csc x}$

18. $y = \sec x\tan x$

19. $y = \dfrac{\tan x}{1 - \tan x}$

20. $y = \dfrac{x\sin x}{1 - \cos x}$

21. (a) By writing $\sin^2 x = (\sin x)(\sin x)$, find a formula for $\dfrac{d}{dx}(\sin^2 x)$.
   (b) Use the idea of part (a) to find a formula for $\dfrac{d}{dx}(\cos^2 x)$.

22. Use the results of Exercise 21 to find $f'(x)$ if $f(x) = \cos^2 x - \sin^2 x$.

23. Find the equation of the tangent line to the graph of $y = x\sin x + \cos x$ at the point for which $x = \pi/2$.

24. Find all points on the graph of $y = \sqrt{3}\sin x + \cos x$ between $x = 0$ and $x = 2\pi$ where the tangent line is horizontal.

In Exercises 25–28, find y''.

25. $y = x\cos x - \sin x$

26. $y = \dfrac{\sin x}{1 - \cos x}$

27. $y = \dfrac{x}{\cos x}$

28. $y = \sin^2 x + x$

29. The position function $s(t)$ for a particle moving on a line is given by

$$s(t) = (t + 1)\sin t$$

Find its velocity $v(t)$ and its acceleration $a(t)$ at time t.

30. Show that for all constants C_1 and C_2, the function $y = C_1\cos x + C_2\sin x$ satisfies the equation $y'' + y = 0$.

31. Show that $y = x(2\cos x + \sin x)$ satisfies the equation $y'' + y = 2\cos x - 4\sin x$.

In Exercises 32–34, prove the given derivative formula.

32. $\dfrac{d}{dx}\cot x = -\csc^2 x$

33. $\dfrac{d}{dx}\sec x = \sec x \tan x$

34. $\dfrac{d}{dx}\csc x = -\csc x \cot x$

35. A spring with an attached weight, as in Example 3.17, moves in simple harmonic motion according to the formula

$$s(t) = 3\cos t + 2\sin t$$

(a) Where is the weight at time $t = 0$, and what is its velocity?
(b) At time $t = 2\pi/3$ give the location of the weight, its velocity, and its acceleration. In what direction is it moving? Is it speeding up or slowing down?
(c) When does the weight go through the equilibrium position for the first time?

36. (a) Show that the equation

$$s(t) = a\cos(\omega t) + b\sin(\omega t)$$

can be written in the equivalent form

$$s(t) = A\sin(\omega t + \alpha)$$

where $A = \sqrt{a^2 + b^2}$, $\sin\alpha = a/\sqrt{a^2 + b^2}$, and $\cos\alpha = b/\sqrt{a^2 + b^2}$. (*Hint:* Multiply and divide the right-hand side of the original equation by $\sqrt{a^2 + b^2}$. Draw a figure for the angle α. Then use Identity 3.8.)
(b) Rewrite the equation from Exercise 35 in the form $s(t) = A\sin(\omega t + \alpha)$. Sketch its graph, showing the amplitude and period.

37. If x is the *degree* measure of an angle, is the following statement still true?

$$\lim_{x \to 0} \frac{\sin x}{x} = 1$$

Prove your answer. If the limit is not 1, see if you can find what it is, and prove it.

38. Prove the following statement.

$$\frac{d}{dx}(\sin 2x) = 2\cos 2x$$

(*Hint:* $\sin 2x = 2\sin x \cos x$.)

39. Prove the following statement.

$$\frac{d}{dx}(\cos 2x) = -2\sin 2x$$

(*Hint:* $\cos 2x = 2\cos^2 x - 1$.)

40. A hot-air balloon lifts off from a point 2500 ft from an observer and rises vertically. At what rate (in feet per degree) is the distance from the observer to the balloon changing when the angle of elevation θ is $30°$? (See the figure.)

41. Let $f(x) = (x + \sin x)/\cos x$, $-\pi/2 < x < \pi/2$.
(a) Is f an even function, odd function, or neither? Prove your answer.
(b) Determine $f'(x)$.
(c) Find the equation of the tangent line to the curve $y = f(x)$ at the point for which $x = \pi/3$.

3.7 THE CHAIN RULE

In this section we will develop a formula, known as the **Chain Rule**, for the derivative of the composition of two functions. This is a powerful result that enables us to differentiate many otherwise intractable functions.

Consider the system of five gears shown in the accompanying photograph, where gear A is topmost and gears B, C, D, and E are clockwise from A. Gear A (12 teeth) drives B (14 teeth), B drives C (22 teeth), C drives D (36 teeth), and D drives E (17 teeth). Let's use the number of teeth on the gears as a measure of their relative circumference, assign a rate of rotation to each gear, and find the rate of rotation of gear E with respect to gear A.

Since the teeth on the gears are equally spaced, A is 12/14 the circumference of B, and therefore B turns 12/14 times every time A rotates once. Similarly, C rotates 14/22 times for every rotation of B and D rotates 22/36 times for every rotation of C. Gear E's rotation is 36/17 that of D. Visualize the gears turning. Do you see that the rate of rotation of gear E with respect to the drive gear, A, is the product of the rates of rotation?

$$\frac{\text{rate of rotation of E}}{\text{rate of rotation of A}} = \left(\frac{36}{17}\right)\left(\frac{11}{18}\right)\left(\frac{7}{11}\right)\left(\frac{6}{7}\right)$$

$$= \frac{12}{17} \approx .706$$

That is, gear E turns 12/17 times every time gear A turns once. We will see in this section that this example is a special case of the Chain Rule.

Before stating the rule, let us consider another example. Suppose we are given the function F defined by

$$F(x) = \sin(x^2 + 2)$$

and wish to find $F'(x)$. If we let $f(x) = \sin x$ and $g(x) = x^2 + 2$, then

$$F(x) = f(g(x))$$

That is, F is the composite function $f \circ g$. We already know how to differentiate f and g individually. The question is how to differentiate their composition. The Chain Rule provides the answer; namely, we *multiply* the individual derivatives of f and g, evaluated at appropriate points. We will return to this example after stating the Chain Rule.

The Chain Rule is one of the most important rules for differentiation. It is essential that you learn it and understand how to apply it.

THEOREM 3.9

The Chain Rule

Let g be differentiable at x and f be differentiable at $g(x)$. Then the composition $f \circ g$ is differentiable at x, and its derivative there is

$$(f \circ g)'(x) = f'(g(x)) \cdot g'(x) \qquad (3.10)$$

REMARK ─────────────────

■ Since $(f \circ g)(x) = f(g(x))$, it is useful to refer to f as the *outer* function and g as the *inner* function. By the Chain Rule we first differentiate the outer function and evaluate it at the inner, getting $f'(g(x))$. Then we multiply this result by the derivative of the inner function at x, $g'(x)$.

Before we consider the proof of the Chain Rule, we give some more examples to show how it works.

EXAMPLE 3.18   Find $F'(x)$ if $F(x) = \sin(x^2 + 2)$.

Solution   As we showed in our preliminary discussion, $F = f \circ g$, where $f(x) = \sin x$ (the outer function) and $g(x) = x^2 + 2$ (the inner function). Thus, by the Chain Rule,

$$F'(x) = \overbrace{[\cos(x^2 + 2)]}^{f'(g(x))} \cdot \overbrace{(2x)}^{g'(x)} = 2x \cos(x^2 + 2) \qquad \blacksquare$$

EXAMPLE 3.19   Use the Chain Rule to find dy/dx if $y = (x^2 + 4)^{10}$.

Solution   We can write y in the form $y = (f \circ g)(x)$, where $f(x) = x^{10}$ and $g(x) = x^2 + 4$. Since $f'(x) = 10x^9$, we see that $f'(g(x)) = 10(x^2 + 4)^9$. Thus, by the Chain Rule,

$$\frac{dy}{dx} = f'(g(x)) \cdot g'(x) = 10(x^2 + 4)^9 \cdot 2x = 20x(x^2 + 4)^9 \qquad \blacksquare$$

EXAMPLE 3.20   Find y' if $y = \cos^3(2x)$.

Solution   Since $\cos^3(2x)$ means $[\cos(2x)]^3$, we can write y as the composition of the three functions $f(x) = x^3$, $g(x) = \cos x$, and $h(x) = 2x$. We can do this composition in stages, two at a time. For example, we can write y as

$$y = [f \circ (g \circ h)](x)$$

and first apply the Chain Rule separately to the two functions f and $g \circ h$. Then we apply the Chain Rule again to $g \circ h$. Thus,

$$
\begin{aligned}
y' &= f'[(g \circ h)(x)] \cdot (g \circ h)'(x) && \text{First application} \\
&= f'[(g \circ h)(x)] \cdot g'(h(x)) \cdot h'(x) && \text{Second application} \\
&= 3[\cos(2x)]^2 \cdot [-\sin(2x)] \cdot 2 \\
&= -6\cos^2(2x) \sin(2x) && \blacksquare
\end{aligned}
$$

REMARK ————————————————————————————

■ Although the analysis we went through in the preceding example is correct, in actual practice one would likely shortcut the procedure, as follows:

$$
\begin{aligned}
y' &= 3[\cos(2x)]^2 \cdot \frac{d}{dx}[\cos(2x)] \\
&= 3[\cos(2x)]^2 [-\sin(2x)] \cdot \frac{d}{dx}(2x) \\
&= 3[\cos(2x)]^2 [-\sin(2x)]2 \\
&= -6\cos^2(2x) \sin(2x)
\end{aligned}
$$

Note that we work from the outside in, continuing to take derivatives until we have differentiated the innermost function. With practice, you can shortcut the procedure even more, going from the original function to the final answer in one or two steps.

Partial Proof of the Chain Rule

Proof   Let $F(x) = (f \circ g)(x)$. That is, $F(x) = f(g(x))$. In applying the definition of the derivative (Definition 3.2), it is convenient here to use the delta notation, introduced in Section 3.2, writing Δx in place of h. Thus,

$$F'(x) = \lim_{\Delta x \to 0} \frac{F(x + \Delta x) - F(x)}{\Delta x} = \lim_{\Delta x \to 0} \frac{f(g(x + \Delta x)) - f(g(x))}{\Delta x}$$

Let $y = f(g(x))$ and $\Delta y = f(g(x + \Delta x)) - f(g(x))$. Then we have

$$F'(x) = \lim_{\Delta x \to 0} \frac{\Delta y}{\Delta x}$$

If we also write $u = g(x)$ and $\Delta u = g(x + \Delta x) - g(x)$, we obtain

$$f'(u) = \lim_{\Delta u \to 0} \frac{f(u + \Delta u) - f(u)}{\Delta u} = \lim_{\Delta u \to 0} \frac{\Delta y}{\Delta u}$$

and

$$g'(x) = \lim_{\Delta x \to 0} \frac{g(x + \Delta x) - g(x)}{\Delta x} = \lim_{\Delta x \to 0} \frac{\Delta u}{\Delta x}$$

By hypothesis, $g'(x)$ exists, so we know by Theorem 3.1 that g is continuous at x. Thus, $g(x + \Delta x) \to g(x)$ as $\Delta x \to 0$, and since $\Delta u = g(x + \Delta x) - g(x)$, it follows that $\Delta u \to 0$ as $\Delta x \to 0$. If $\Delta u \neq 0$ for all sufficiently small (nonzero) values of $|\Delta x|$, then we can write, for such values,

$$\frac{\Delta y}{\Delta x} = \frac{\Delta y}{\Delta u} \cdot \frac{\Delta u}{\Delta x}$$

and thus,

$$\begin{aligned}
F'(x) = \lim_{\Delta x \to 0} \frac{\Delta y}{\Delta x} &= \lim_{\Delta x \to 0} \left(\frac{\Delta y}{\Delta u} \right) \left(\frac{\Delta u}{\Delta x} \right) \\
&= \left(\lim_{\Delta u \to 0} \frac{\Delta y}{\Delta u} \right) \left(\lim_{\Delta x \to 0} \frac{\Delta u}{\Delta x} \right) \qquad \text{Since } \Delta u \to 0 \text{ as } \Delta x \to 0 \\
&= f'(u)g'(x) \\
&= f'(g(x))g'(x)
\end{aligned}$$

The theorem is therefore proved under the assumption $\Delta u \neq 0$ stated above.   ∎

Unfortunately, there are functions $u = g(x)$ for which $\Delta u = 0$ for some value of Δx in every interval about $\Delta x = 0$. In these cases, the proof we have given is not valid. A proof that works in all cases is given in Supplementary Appendix 5.

In the proof just given we introduced the notation $y = f(u)$ and $u = g(x)$. The result (Equation 3.10) of the Chain Rule can thus be written in the alternative form

---

The Chain Rule (Alternative Notation)

$$\frac{dy}{dx} = \frac{dy}{du} \cdot \frac{du}{dx} \qquad\qquad (3.11)$$

For example, consider again the function $y = (x^2 + 4)^{10}$ of Example 3.19. We can let $u = x^2 + 4$. Then $y = u^{10}$. So, by Equation 3.11,

$$\frac{dy}{dx} = \underbrace{(10u^9)}_{\frac{dy}{du}} \underbrace{(2x)}_{\frac{du}{dx}} = 20x(x^2 + 4)^9$$

Powers of Differentiable Functions

Derivatives of functions of the form $[g(x)]^n$ occur with sufficient frequency to warrant special consideration. A function of this form can be regarded as the composition $(f \circ g)(x)$, where $f(x) = x^n$. Assuming g is differentiable and that $g(x) \neq 0$ if $n < 1$, we get from the Chain Rule

$$\frac{d}{dx}[g(x)]^n = n[g(x)]^{n-1} \cdot g'(x)$$

This special case of the Chain Rule is sometimes called the **Generalized Power Rule**. Our previous results show it to be valid for all integers n, and we will see later that it is true if n is any real number. By writing $u = g(x)$, the formula assumes the form given in Equation 3.12.

Generalized Power Rule

If $u = g(x)$ is differentiable at x, then for all integers n,

$$\frac{d}{dx}(u^n) = nu^{n-1}\frac{du}{dx} \qquad\qquad (3.12)$$

provided that when $n < 1$, $u \neq 0$.

Using this result on the function $y = (x^2 + 4)^{10}$ of Example 3.19, we get the answer immediately:

$$\frac{dy}{dx} = 10(x^2 + 4)^9 \cdot 2x = 20x(x^2 + 4)^9$$

The next two examples further illustrate the Generalized Power Rule.

EXAMPLE 3.21   Find dy/dx if

$$y = \frac{2}{(3x - x^2)^4}$$

Solution   First write y in the form $y = 2(3x - x^2)^{-4}$. Then, by Equation 3.12,

$$\frac{dy}{dx} = -8(3x - x^2)^{-5}(3 - 2x)$$ ∎

EXAMPLE 3.22   Find dy/dx if $y = \sin^3 x \cos^2 x$.

Solution   Here y is a product of two functions, each of which is of the form u^n. By using the Product Rule together with Equation 3.12, we get

$$\frac{dy}{dx} = \left[3\sin^2 x (\cos x)\right]\cos^2 x + \sin^3 x \left[2\cos x(-\sin x)\right]$$

$$= 3\sin^2 x \cos^3 x - 2\sin^4 x \cos x$$ ∎

One of the most frequent mistakes students make in differentiation is failure to apply the last step of the Chain Rule. For example, consider $\frac{d}{dx}(\sin 2x)$. All too frequently students give the answer as $\cos 2x$ and forget the final du/dx of Equation 3.11, which in this case is 2. The correct answer is $(\cos 2x) \cdot 2 = 2\cos 2x$. So be alert to this danger. *Do not forget the final du/dx.*

The next example mixes algebraic and trigonometric functions.

EXAMPLE 3.23   Find $f'(x)$ for each of the following.

(a) $f(x) = x^2 \sin 2x$
(b) $f(x) = (\tan x - x)^2$

Solution

(a) The function is a product, so we have

We used the
Chain Rule here.

$$f'(x) = (2x)(\sin 2x) + x^2 \overbrace{(2\cos 2x)}$$

$$= 2x(\sin 2x + x \cos 2x)$$

(b)

$$f'(x) = 2(\tan x - x)(\sec^2 x - 1)$$

$$= 2(\tan x - x)(\tan^2 x) \qquad \text{By the identity } 1 + \tan^2 x = \sec^2 x$$

$$= 2\tan^2 x(\tan x - x)$$ ∎

Exercise Set 3.7

In Exercises 1–40, find dy/dx.

1. $y = (x^2 + 2)^4$

2. $y = (2x^3 - 3)^5$

3. $y = (1 - 2x^2)^5$

4. $y = (x^3 - x)^{-1}$

5. $y = (3x^4 - 2)^{10}$

6. $y = \left(\dfrac{x-1}{\sqrt{x}}\right)^{12}$

7. $y = (x^2 - 3x + 4)^3$

8. $y = 4(x^3 - 3x^2 + 5)^2$

9. $y = (3x^4 - 2x^2 + 1)^{-2}$

10. $y = (x^2 - 4x - 7)^{-3}$

11. $y = (2x + 1)^2(x^2 + 2)^3$

12. $y = x^5(x^3 - 2x^2 + 3)^4$

13. $y = \dfrac{2}{(x^2 + 1)^3}$

14. $y = \dfrac{3}{(2x^2 - 5)^4}$

15. $y = \dfrac{(3x - 4)^3}{x + 1}$

16. $y = \dfrac{(x^2 + 2)^2}{2x - 1}$

17. $y = \dfrac{x^2 + 1}{(x^2 - 1)^2}$

18. $y = \dfrac{2x^2 - x}{(3x - 4)^3}$

19. $y = \left(\dfrac{x-1}{x+1}\right)^3$

20. $y = \left(\dfrac{x^2}{x^2 - 1}\right)^{-2}$

21. $y = \sin 3x$

22. $y = \sin(2x + 3)$

23. $y = \cos(1 - 2x)$

24. $y = 2\cos(x^2 + 1)$

25. $y = \tan(3x^2)$

26. $y = 3\cot(1 - x)$

27. $y = 2\csc 3x$

28. $y = \sec\left(\dfrac{1}{x}\right)$

29. $y = 5\sin^2 x$

30. $y = 1 - 2\cos^2 x$

31. $y = \tan^2(2x + 3)$

32. $y = \csc 2x - \cot 2x$

33. $y = \tan^3 x^2$

34. $y = (1 + \sin 2x)^2$

35. $y = (\cos x - \sin x)^2$

36. $y = x^2 \cos 2x$

37. $y = (3x + 2)\csc 3x$

38. $y = (x + \tan x)^3$

39. $y = \sin(\cos x)$

40. $y = \tan(\sec^2 x)$

In Exercises 41–46, find $(f \circ g)'(x)$ for the given functions f and g.

41. $f(x) = x^3$; $g(x) = 2x + 3$

42. $f(x) = 2x^4$; $g(x) = 1 - 3x$

43. $f(x) = x^{-2}$; $g(x) = x^2 + \sin x$

44. $f(x) = \dfrac{1}{x}$; $g(x) = \cos x - 1$

45. $f(x) = 3x^2$; $g(x) = x^3 - 8$

46. $f(x) = x^{10}$; $g(x) = \tan x$

In Exercises 47–52, use Equation 3.11 to find dy/dx. Express the answer in terms of x.

47. $y = 2u^3$; $u = x^2 - 3$

48. $y = u^2 - 1$; $u = \cos 2x$

49. $y = 3/u^2$; $u = x^3 + 4$

50. $y = 4u^2 - 3$; $u = \sin 3x$

51. $y = u^2 - 2u$; $u = \cos x - \sin x$

52. $y = 1 - u^5$; $u = 1 - x^5$

In Exercises 53–60, find $f''(x)$.

53. $f(x) = (x^2 + 1)^4$

54. $f(x) = (2x^2 - 7)^5$

55. $f(x) = x/(1 - x)$

56. $f(x) = 1/(x^2 - 1)^2$

57. $f(x) = x^2 \cos 2x$

58. $f(x) = \tan x^2$

59. $f(x) = \sec^3 2x$

60. $f(x) = \cos x^3$

61. Let $h(t) = (t + 1)/(t - 1)$ and $g(t) = t^2$. Use the Chain Rule to find $(h \circ g)'(t)$ and $(g \circ h)'(t)$.

62. Let $\phi(u) = u^2 - 2u + 3$ and $\psi(u) = (2u - 1)/(3u + 1)$. Use the Chain Rule to find $(\phi \circ \psi)'(u)$ and $(\psi \circ \phi)'(u)$.

63. Let $z = t^3 - 3t^2 + 4$ and $t = y^2 + 3y - 1$. Use the Chain Rule to find dz/dy.

64. Find the equation of the tangent line and normal line (see Exercise 25 in Exercise Set 3.3) to the graph of

$$y = \dfrac{(3x - 4)^2}{2(x^2 - 2)^2}$$

at the point for which $x = 2$.

65. Find $f''(x)$ if $f(x) = (2x - 1)/(x^2 + 1)^2$.

66. Derive a formula for $(f \circ g \circ h)'(x)$, assuming appropriate differentiability properties by writing $f \circ g \circ h = (f \circ g) \circ h$.

67. Use the result of Exercise 66 to find y' if $y = [(2x^3 - 3)^2 + 4]^3$. What are f, g, and h in this case?

68. Show that $y = \cos 2x + \frac{1}{2}(1 - x \sin 2x)$ satisfies the equation $y'' + 4y = 4 \sin^2 x$.

69. A spring with a weight attached is set in motion, and its position $x(t)$ (in centimeters) from the equilibrium position at time t (in seconds) is given by

$$x(t) = \frac{1}{4}(1 - 8t) \cos 16t + \frac{1}{8} \sin 16t$$

Find its velocity and acceleration when $t = \dfrac{5\pi}{3}$.

70. Suppose f is differentiable on $\mathbb{R}$. Determine expressions for $\dfrac{d}{dx} f(\sin x)$ and $\dfrac{d}{dx} \sin(f(x))$.

71. Suppose f is differentiable on $\mathbb{R}$, and let $y = f(x)$. Determine an expression for $\dfrac{d}{dx} \cos(y^2 + 2xy)$.

72. Suppose f is differentiable on $\mathbb{R}$. If f is an even function, what can you say about f'? If f is an odd function, what can you say about f'? Prove your answers.

73. Let f be a differentiable function, and let

$$f_1(x) = f(x)$$
$$f_2(x) = (f \circ f)(x)$$
$$f_3(x) = (f \circ f \circ f)(x)$$
$$\vdots$$
$$f_n(x) = \underbrace{(f \circ f \circ \cdots \circ f)}_{n \text{ times}}(x)$$

Determine a formula for $f_n'(x)$ for $n \geq 2$.

74.(a) Use the fact that

$$\frac{d}{dx}|x| = \begin{cases} 1 & \text{if } x > 0 \\ -1 & \text{if } x < 0 \end{cases}$$

(see Exercise 39 in Exercise Set 3.2) and the Chain Rule to find $\dfrac{d}{dx}|\sin x|$. What is the domain of this derivative?

(b) On the same axes sketch the graphs of $y = |\sin x|$ and $y = \dfrac{d}{dx}|\sin x|$.

75. Let

$$f(x) = \begin{cases} x \sin \dfrac{1}{x} & \text{if } x \neq 0 \\ 0 & \text{if } x = 0 \end{cases}$$

(a) Find $f'(x)$ for $x \neq 0$.
(b) Use Definition 3.2 to prove that $f'(0)$ does not exist.
(c) Prove that f is continuous on all of $\mathbb{R}$.

76. Let

$$g(x) = \begin{cases} x^2 \sin \dfrac{1}{x} & \text{if } x \neq 0 \\ 0 & \text{if } x = 0 \end{cases}$$

(a) Find $g'(x)$ for $x \neq 0$.
(b) Use Definition 3.2 to prove that $g'(0)$ exists, and find its value.
(c) Prove that g' is discontinuous at $x = 0$.

3.8 IMPLICIT DIFFERENTIATION

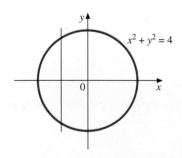

FIGURE 3.26

Certain equations in two variables x and y define y *implicitly* as a function of x. Consider, for example, the equation

$$2x^2 - 3y = 5$$

Although this equation is not in the form $y = f(x)$, we can easily solve for y to put it in this form:

$$y = \frac{2x^2 - 5}{3}$$

Now y is expressed *explicitly* as a function of x.

Consider now the equation

$$x^2 + y^2 = 4$$

of a circle of radius 2, centered at the origin, as shown in Figure 3.26. Since the graph fails the vertical line test, the equation does not define y as a function

of x. If we solve for y, we get

$$y = \pm\sqrt{4 - x^2}$$

Suppose we let $f_1(x) = \sqrt{4 - x^2}$ and $f_2(x) = -\sqrt{4 - x^2}$. Then, the graph of $y = f_1(x)$ is the upper semicircle in Figure 3.27(a) and the graph of $y = f_2(x)$ is the lower semicircle in Figure 3.27(b). Thus, we can say that the equation $x^2 + y^2 = 4$ defines y implicitly as two functions of x, namely, f_1 and f_2.

(a)

FIGURE 3.27

(b)

FIGURE 3.28

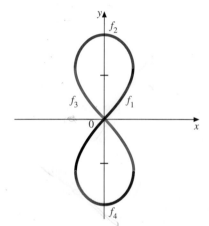

FIGURE 3.29

As another example of implicitly defined functions, consider the equation

$$(x^2 + y^2)^2 = 4(y^2 - x^2)$$

whose graph is shown in Figure 3.28. This curve is known as a **lemniscate**. (It is one of the curves we will consider in Chapter 9.) It would be difficult, to say the least, to solve the equation for y in terms of x. However, as we indicate in Figure 3.29 with the different colors, we can divide the graph into four parts, each of which passes the vertical line test and thus represents a function.

You might be thinking at this point that every equation in x and y defines y implicitly as one or more functions of x. This conclusion is not correct, as the next two examples illustrate.

EXAMPLE 3.24   Show that the equation $x^2 + y^2 + 1 = 0$ does not define y as a function of x.

Solution   Whatever real values x and y may be, the left-hand side of this equation is always positive. So it cannot equal 0. The equation is therefore not satisfied by any real values of x and y and hence cannot define y as a (real-valued) function of x. ■

EXAMPLE 3.25   Show that the equation $(x + y)^2 = y^2 + 2xy + 4$ does not define y as a function of x.

Solution   If the left-hand side is expanded and terms are collected, the result is (verify it)

$$x^2 - 4 = 0$$

or equivalently,

$$x = 2 \quad \text{or} \quad x = -2$$

The graph therefore consists of the two vertical lines shown in Figure 3.30. This equation therefore does not define y as a function of x. ■

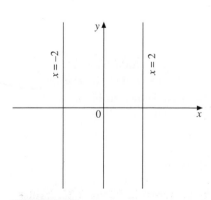

FIGURE 3.30

Let us return now to the example of the circle $x^2 + y^2 = 4$, which defines y implicitly as $y = f_1(x)$ and $y = f_2(x)$, where $f_1(x) = \sqrt{4 - x^2}$ and $f_2(x) = -\sqrt{4 - x^2}$. We can find $f_1'(x)$ and $f_2'(x)$ using the Generalized Power Rule (it is true for nonintegral n, as remarked earlier). That is,

$$f_1'(x) = \frac{1}{2}(4 - x^2)^{-1/2}(-2x) = \frac{-x}{\sqrt{4 - x^2}}$$

Similarly,

$$f_2'(x) = \frac{x}{\sqrt{4 - x^2}}$$

We now illustrate an alternative method, called **implicit differentiation**, whereby we can find both derivatives at once, using the original equation.

We reason as follows. Let $f(x)$ be either of the functions defined implicitly by the equation $x^2 + y^2 = 4$. That is, $f(x)$ can be either $f_1(x)$ or $f_2(x)$. If we substitute $f(x)$ for y in the equation, we get

$$x^2 + [f(x)]^2 = 4$$

and this equation is an *identity* in x. That is, the equation is true for all values of x in the domain of f, namely, $[-2, 2]$. The left-hand side is therefore the constant function 4 and so its derivative is 0, wherever that derivative exists. Thus, by the Chain Rule,

$$2x + 2[f(x)]f'(x) = 0$$

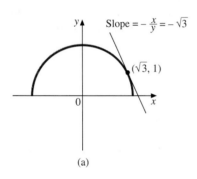

Slope $= -\dfrac{x}{y} = -\sqrt{3}$

$(\sqrt{3}, 1)$

(a)

Solving for $f'(x)$, we obtain the result

$$f'(x) = -\frac{x}{f(x)} = -\frac{x}{y}$$

which is valid except where $y = 0$. (Note that when $y = 0$, the tangent lines are vertical.) Since $f(x)$ represents either $f_1(x)$ or $f_2(x)$, we have found a formula for the derivative of each of these functions. For example, consider the point $(\sqrt{3}, 1)$ on the circle $x^2 + y^2 = 4$. We can find the slope of the tangent line to the circle at this point simply by substituting the x- and y-coordinates of the point into the formula just derived. That is,

$$m_{\tan} = -\frac{x}{y} = -\frac{\sqrt{3}}{1} = -\sqrt{3}$$

Slope $= -\dfrac{x}{y} = \sqrt{3}$

$(\sqrt{3}, -1)$

(b)

FIGURE 3.31

Since we know the point $(\sqrt{3}, 1)$ is on the upper semicircle, we could have computed the slope of the tangent line using the formula for $f_1'(x)$. That is,

$$m_{\tan} = f_1'(\sqrt{3}) = \frac{-\sqrt{3}}{\sqrt{4 - (\sqrt{3})^2}} = -\sqrt{3}$$

As suspected, the answers are the same. Similar computations can be done with the point $(\sqrt{3}, -1)$ on the lower semicircle, replacing f_1' by f_2'. The result in this case for the slope is $\sqrt{3}$. Both tangent lines are shown in Figure 3.31.

In practice, we usually shortcut the above procedure as follows.

$$x^2 + y^2 = 4$$

$$2x + 2yy' = 0 \qquad \text{We differentiated each term with respect to } x.$$

$$y' = -\frac{x}{y} \qquad \text{Solve for } y'.$$

Here, it is important to understand that y is a function of x, namely, one of the functions f_1 or f_2 determined implicitly by the original equation. Thus, when we differentiate y^2, we must use the Chain Rule

$$\frac{d}{dx}(y)^2 = 2(y)\frac{dy}{dx}$$

That is,

$$\frac{d}{dx}(y^2) = 2yy'$$

As we have seen, not all equations in x and y define y implicitly as one or more functions of x. Conditions that ensure the equation does define y as such a function, or functions, are usually given in advanced calculus courses. In our examples and exercises, we will *assume* that all the equations we use do define y as one or more functions that are differentiable on some suitable domain. The importance of implicit differentiation lies in the fact that we can find formulas for the derivatives when it is difficult, or even impossible, to solve for the explicit functions. In the case of $x^2 + y^2 = 4$, we could have found $f_1'(x)$ and $f_2'(x)$ directly, although doing so with implicit differentiation is easier. In the case of the equation $(x^2 + y^2)^2 = 4(y^2 - x^2)$ of the lemniscate we illustrated in Figures 3.28 and 3.29, solving for f_1, f_2, f_3, and f_4 is very difficult (try it). Yet we can still find y' implicitly. (See Exercise 17 in this section.) In the case of the equation $x^3 - 2x^2y + y^4 - 3y^5 = 0$, it is *impossible* to solve for y, but finding y' implicitly is straightforward. (See Exercise 18 in this section.)

EXAMPLE 3.26   Find y'.
(a) $x^3 - 2xy + y^3 = 4$ (b) $y = x \sin y$

Solution

(a) Proceeding as above, we have

$$3x^2 - 2(y + xy') + 3y^2 \cdot y' = 0 \qquad \text{The Product Rule is used on the } xy \text{ term.}$$

Now we rearrange terms and solve for y':

$$y'(3y^2 - 2x) = 2y - 3x^2$$

$$y' = \frac{2y - 3x^2}{3y^2 - 2x}$$

Of course, any points (x, y) for which the denominator is 0 must be excluded.

(b) At first, this looks like an explicit function, since it is of the form $y = $ something. But the "something" involves y, so it must be treated implicitly. Differentiating both sides, we get

$$y' = \sin y + x(\cos y)y'$$

$$y'(1 - x \cos y) = \sin y$$

$$y' = \frac{\sin y}{1 - x \cos y} \qquad (x \cos y \neq 1)$$

Note that the derivative of the term $x \sin y$ involved both the Product Rule and the Chain Rule.   ∎

EXAMPLE 3.27   Find y' and y'' if $x^3 - y^3 = 8$.

Solution   We differentiate both sides with respect to x:

$$3x^2 - 3y^2 \cdot y' = 0$$

$$y' = \frac{x^2}{y^2} \qquad (y \neq 0)$$

Now we differentiate again, using the Quotient Rule on the right:

$$y'' = \frac{(2x)y^2 - x^2(2y \cdot y')}{y^4}$$

We next divide out the common factor y on the numerator and denominator, replace y' by its value we have just found, and simplify:

$$y'' = \frac{2xy - 2x^2\left(\dfrac{x^2}{y^2}\right)}{y^3} = \frac{2x(y^3 - x^3)}{y^5}$$

Finally, we observe that since the original equation was $x^3 - y^3 = 8$, it follows that $y^3 - x^3 = -8$, and the final answer can be written in the form

$$y'' = \frac{-16x}{y^5} \qquad (y \neq 0)$$

You should be alert to the possibility of simplifying final answers in this way. It frequently happens in finding second derivatives. ∎

Derivatives of Rational Powers

Implicit differentiation allows us to extend the Power Rule for differentiation to rational exponents, as the next theorem shows.

THEOREM 3.10

> **Power Rule for Rational Exponents**
>
> For any rational number r, if $y = x^r$, then for all x for which y is differentiable
>
> $$\frac{dy}{dx} = rx^{r-1}$$

Proof   Since r is rational, it can be written in the form $r = m/n$, where m and n are integers. So $y = x^{m/n}$. We raise both sides to the nth power and differentiate implicitly:

$$y^n = x^m$$

$$ny^{n-1} \cdot y' = mx^{m-1} \qquad \text{Since we know the Power Rule holds for integers}$$

$$y' = \frac{m}{n}\frac{x^{m-1}}{y^{n-1}} \qquad \text{If } y^{n-1} \neq 0$$

Replacing y with $x^{m/n}$ and simplifying, we get

$$y' = \frac{m}{n} \frac{x^{m-1}}{x^{m(n-1)/n}} = \frac{m}{n} \frac{x^{m-1}}{x^{m-(m/n)}} = \frac{m}{n} x^{(m-1)-m+(m/n)}$$

$$= \frac{m}{n} x^{(m/n)-1}$$

which is the desired result, since $r = \frac{m}{n}$.   ■

REMARK ───
■ In Chapter 7 we will show that $y = x^r$ is differentiable for all x, except that when $x = 0$, we must have $r \geq 1$.
───

By combining Theorem 3.10 with the Chain Rule, we can extend the Generalized Power Rule (Equation 3.12) for rational exponents. That is, if $u = g(x)$ is differentiable at x and r is any rational number,

$$\frac{d}{dx} u^r = r u^{r-1} \frac{du}{dx}$$

provided $u \neq 0$ if $r < 1$.

EXAMPLE 3.28   Find $f'(x)$:

(a) $f(x) = \dfrac{\sqrt{x^2 + 1}}{x}$ (b) $f(x) = \sqrt{\cos x}$

Solution

(a) We write $f(x)$ in the form

$$f(x) = \frac{(x^2 + 1)^{1/2}}{x}$$

and use the Quotient Rule, along with Theorem 3.10 and the Chain Rule, to get

$$f'(x) = \frac{\left[\frac{1}{2}(x^2 + 1)^{-1/2} \cdot 2x\right] x - (x^2 + 1)^{1/2}}{x^2}$$

This result can be simplified considerably by multiplying numerator and denominator by the factor $(x^2 + 1)^{1/2}$, since doing so eliminates the negative exponent and also brings the other fractional exponent up to 1. You should verify that the result is

$$f'(x) = \frac{x^2 - (x^2 + 1)}{x^2(x^2 + 1)^{1/2}} = \frac{-1}{x^2\sqrt{x^2 + 1}}$$

(b)

$$f(x) = \sqrt{\cos x} = (\cos x)^{1/2}$$

$$f'(x) = \frac{1}{2}(\cos x)^{-1/2}(-\sin x) = -\frac{\sin x}{2\sqrt{\cos x}}$$ ■

EXAMPLE 3.29   Find y' and y'' if

$$x^{2/3} + y^{2/3} = a^{2/3}$$

Solution   Note first of all that we *cannot* eliminate the fractional exponents by cubing each term (however tempting this might be). Differentiating implicitly, we obtain

$$\frac{2}{3}x^{-1/3} + \frac{2}{3}y^{-1/3} \cdot y' = 0 \quad \text{Since } a \text{ is a constant}$$

$$y' = -\frac{y^{1/3}}{x^{1/3}} \quad (x \neq 0)$$

We find y'' as in Example 3.27:

$$y'' = -\frac{\left(\frac{1}{3}y^{-2/3} \cdot y'\right) \cdot x^{1/3} - y^{1/3} \cdot \frac{1}{3}x^{-2/3}}{x^{2/3}}$$

$$= -\frac{x^{1/3}y^{-2/3}\left(\dfrac{-y^{1/3}}{x^{1/3}}\right) - y^{1/3}x^{-2/3}}{3x^{2/3}} \quad \text{Substituting for } y'$$

$$= -\frac{-y^{-1/3} - y^{1/3} \cdot x^{-2/3}}{3x^{2/3}} \cdot \frac{x^{2/3}y^{1/3}}{x^{2/3}y^{1/3}} \quad \begin{array}{l}\text{This multiplication eliminates}\\\text{negative exponents.}\end{array}$$

$$= \frac{x^{2/3} + y^{2/3}}{3x^{4/3}y^{1/3}}$$

$$= \frac{a^{2/3}}{3x^{4/3}y^{1/3}} \quad \text{Since } x^{2/3} + y^{2/3} - a^{2/3}$$

Observe that y'' does not exist if $x = 0$ or $y = 0$.   ∎

Exercise Set 3.8

Use implicit differentiation to find y' in Exercises 1–18.

1. $x^2 - y^2 = 1$

2. $x^2 + 2xy - 3y^2 = 0$

3. $x^2y^3 - 2x + y = 3$

4. $y^2 - 3xy - x^3 = 8$

5. $y = x^3 - y^3$

6. $x(y^2 - 3x) = 5$

7. $y(x^2 - 2y) = 3$

8. $\sqrt{x} + \sqrt{y} = 1$

9. $y = \sqrt{x + y}$

10. $x + 2\sqrt{xy} + y = 3$

11. $x \sin y + y \sin x = 1$

12. $y = \tan(x + y)$

13. $\sec(xy) = 3x$

14. $x = \cos\sqrt{xy}$

15. $\sin x \cos y = x - y$

16. $\sin^2 x + \cos^2 y = x^2 + y^2$

17. $(x^2 + y^2)^2 = 4(x^2 - y^2)$

18. $x^3 - 2x^2y + y^4 - 3y^5 = 0$

In Exercises 19–36, find dy/dx.

19. $y = \sqrt{x^2 + 1}$

20. $y = x\sqrt{1 - x}$

21. $y - \dfrac{1}{\sqrt{4 - x^2}}$

22. $y = \dfrac{x}{\sqrt{x + 4}}$

23. $y = \sqrt[3]{(2x^3 + 5)^2}$

24. $y = 1 + (1 - 3x^3)^{2/3}$

25. $y = \dfrac{x}{\sqrt{1 - x^2}}$

26. $y = \sin\sqrt{x + 1}$

27. $y = \sqrt{\sin x}$

28. $y = (\sqrt{\cos x})^3$

29. $y = \tan\left(\dfrac{1}{\sqrt{x}}\right)$

30. $y = \sin^2 x^2$

31. $y = \dfrac{\sec(2x+3)}{\tan(2x+3)}$

32. $y = (1 - \sin^2 x)^{3/2}$

33. $y = \sqrt{2(1 - \cos 2x)}$ $(0 \le x \le \pi)$

34. $y = \sin^4 x + \cos^4 x$

35. $y = \dfrac{\sqrt{4 - x^2}}{x}$

36. $y = \sqrt{\dfrac{1-x}{1+x}}$

In Exercises 37–44, find y''.

37. $x^2 + y^2 = 1$

38. $x^3 + y^3 = 8$

39. $x^{3/2} - y^{3/2} = 8$

40. $\sqrt{x} - \sqrt{y} = 1$

41. $y - \sqrt{1 - x^2}$

42. $y = 1/\sqrt{x^2 + 1}$

43. $y = \sin \sqrt{x}$

44. $y = \sqrt{\cos x}$

45. Find the equation of the tangent line to the ellipse defined by
$$\frac{x^2}{a^2} + \frac{y^2}{b^2} = 1$$
at an arbitrary point (x_0, y_0) on the ellipse for which $y_0 \ne 0$.

46. Find dy/dx and d^2y/dx^2 if $x^{4/3} - 2y^{4/3} = a^{4/3}$, where a is a constant.

47. Find $f'(x)$ and $f''(x)$ if $y = f(x)$ is a twice-differentiable solution of $x = \tan(x + y)$.

48. Assuming g is a differentiable function for which $x = g(t)$ is a solution of
$$x^2 t^3 - 4x^3 t^2 + 3x^2 - 2t^5 - 7 = 0$$
find $g'(t)$.

49. (a) Find two functions f_1 and f_2 that are differentiable on $(-2, 2)$ and such that $y = f_1(x)$ and $y = f_2(x)$ satisfy the equation $y^2 - 2xy + 2x^2 - 4 = 0$. (*Hint:* Use the Quadratic Formula.)
(b) Find $f_1'(x)$ and $f_2'(x)$.
(c) Using the equation from part (a), differentiate implicitly to find y'.
(d) Show that the answers in parts (b) and (c) are the same.

50. (a) Show that $y = x$ satisfies the equation $y^3 - xy^2 - x^4 y + x^5 = 0$, and find two other solutions. Differentiate each solution.
(b) Use implicit differentiation to find y' for the equation in part (a), and show that this answer agrees with the derivatives found in part (a).

51. A particle moves in a straight line so that its position function at time t is $s(t) = (t^2 - 1)^{2/3}$. Find its velocity $v(t)$ and its acceleration $a(t) = v'(t)$ at time t. What are $v(3)$ and $a(3)$?

52. Two curves are said to be **orthogonal** at a point of intersection if their tangent lines at that point are perpendicular to each other. Prove that the graphs of $x^2 = 4y$ and $2y^2 + x^2 = 6$ are orthogonal at each of their points of intersection. Draw both curves, showing their tangent lines at their points of intersection.

53. Prove that the *family* of curves $y^2 = c_1 x^3$ is orthogonal to the family $2x^2 + 3y^2 = c_2$ in the sense that for all values of the constants c_1 and c_2, each member of the first family is orthogonal to each member of the second at each point of intersection (see Exercise 52). Use a CAS or graphing calculator to draw several members of each family on the same set of axes. (*Hint:* Before differentiating, write the equation of the first family in the form $y^2/x^3 = c_1$.)

3.9 THE DIFFERENTIAL AND LINEAR APPROXIMATION

We have used the Leibniz notation dy/dx as one of several ways to designate the derivative $f'(x)$, where $y = f(x)$, but until now we have never given meaning to the symbols dx and dy individually. It turns out, though, that we can give them meanings that are consistent with the derivative notation, but which allow us to treat dx and dy as separate entities. These separate entities are called *differentials*.

Suppose f is differentiable at x. Let dx denote an independent variable that can take on any real value. Then the **differential** of y, designated dy, is defined as follows.

The Differential of $y = f(x)$

$$dy = f'(x)\,dx \qquad\qquad (3.13)$$

Note that dy is a dependent variable, dependent on dx for each fixed value of x. We call dx the differential of x.

EXAMPLE 3.30   Find dy if $y = \sin^2 x$.

Solution   Let $f(x) = \sin^2 x$. Then $f'(x) = 2\sin x \cos x = \sin 2x$. So by Equation 3.13,

$$dy = (\sin 2x)dx \qquad\qquad \blacksquare$$

If $dx \neq 0$ and we divide both sides of Equation 3.13 by dx, we get

$$\frac{dy}{dx} = f'(x)$$

Now this equation appears to be just a restatement of a familiar fact. But there really is something new here, since now the left-hand side is a quotient of two differentials rather than simply a symbol for the derivative, as we have previously used it. The two interpretations give the same result, however.

The notations dy/dx and $f'(x)$ for the derivative were devised by Leibniz and Newton, respectively, who took different approaches to the development of the ideas of calculus. It is interesting that the formal verification of the *correctness* of Leibniz's approach was not understood fully until the 1960s when Abraham Robinson introduced what is called *nonstandard analysis*.

All formulas for derivatives can now be phrased in terms of differentials as well. Suppose, for example, that $u = f(x)$ and $v = g(x)$, where f and g are differentiable in some common domain. Then, since $du = f'(x)dx$ and $dv = g'(x)dx$, we have the following formulas for differentials of sums, differences, products, and quotients:

(i) $d(u \pm v) = du \pm dv$

(ii) $d(uv) = u\,dv + v\,du$

(iii) $d\left(\dfrac{u}{v}\right) = \dfrac{v\,du - u\,dv}{v^2} \qquad (v \neq 0)$

To illustrate how to prove these differential forms, consider equation (ii). Since $uv = (fg)(x)$, by the definition of the differential,

$$d(uv) = \left[f'(x)g(x) + f(x)g'(x)\right]dx$$
$$= \left[f(x)g'(x) + g(x)f'(x)\right]dx$$
$$= \underbrace{f(x)}_{u}\,\underbrace{g'(x)dx}_{dv} + \underbrace{g(x)}_{v}\,\underbrace{f'(x)dx}_{du}$$

These results provide an alternative means of finding derivatives implicitly, as the next example shows.

EXAMPLE 3.31   Use differentials to find $\dfrac{dy}{dx}$ if

$$x^2 + xy - 3y^2 = 5$$

Solution   We take the differential of each term, getting

$$2x\,dx + (x\,dy + y\,dx) - 6y\,dy = 0$$

We collect the terms with dx on one side and those with dy on the other and then solve for the quotient dy/dx:

$$(2x + y)dx = (6y - x)dy$$

$$\frac{dy}{dx} = \frac{2x + y}{6y - x} \qquad (x \neq 6y) \qquad ∎$$

Suppose now that $y = f(x)$ and $x = g(t)$, so that $y = f(g(t)) = (f \circ g)(t)$. Then, by Equation 3.13 and the Chain Rule, we have

$$dy = (f \circ g)'(t)dt$$

$$= f'(g(t)) \cdot g'(t)dt$$

But again by Equation 3.13 applied to $x = g(t)$, we see that $dx = g'(t)dt$. So

$$dy = \underbrace{f'(g(t))}_{x} \cdot \underbrace{g'(t)dt}_{dx} = f'(x)dx$$

This calculation shows that when $y = f(x)$, the differential of y is given by Equation 3.13 even when x is dependent on some other variable (assuming appropriate differentiability conditions).

Geometric Interpretation

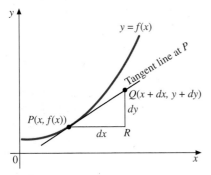

FIGURE 3.32

The differentials dx and dy have a useful geometric interpretation, as shown in Figure 3.32. Suppose $y = f(x)$. Let $P(x, f(x))$ be a point on the graph at which f' exists, and let l be the tangent line at P. Since l has slope dy/dx, it follows that the point $Q(x + dx, y + dy)$ always lies on l for all nonzero values of dx. Thus, for any horizontal displacement dx from P, dy is the vertical displacement RQ to the tangent line l. Recall that we have previously used the symbol Δx to indicate an increment, or change, in x, and correspondingly we have used Δy to be the change in the function:

$$\Delta y = f(x + \Delta x) - f(x)$$

FIGURE 3.33

Suppose such an increment Δx is given. Since dx may be chosen arbitrarily, let us choose $dx = \Delta x$. Now we want to examine the relationship between dy and Δy. As we have seen, dy gives the vertical displacement from $P(x, f(x))$ *to the tangent line* drawn at P. The increment Δy, on the other hand, gives the vertical displacement from P *to the curve itself*. The relationship between dy and Δy is shown in Figure 3.33. Observe carefully that, although we are free to choose the differential dx of the *independent* variable equal to the increment Δx, the differential dy of the *dependent* variable will not, in general, equal the increment Δy. However, if $\Delta x \ (= dx)$ is sufficiently small, it seems reasonable to suppose that $\Delta y \approx dy$. To illustrate this approximation, consider the following two examples.

EXAMPLE 3.32   Compare Δy and dy for $y = 2x^2 + 3x - 1$ when $x = 3$ and $\Delta x = 0.1$.

Solution   Denote the given function by f, and let $dx = \Delta x = 0.1$. Then

$$\Delta y = f(x + \Delta x) - f(x) = f(3.1) - f(3)$$
$$= [2(3.1)^2 + 3(3.1) - 1] - [2(3)^2 + 3(3) - 1]$$
$$= 27.52 - 26 = 1.52$$

Since $f'(x) = 4x + 3$, we get

$$dy = f'(3)dx = (15)(0.1) = 1.5$$

So in this case, the error in using dy to approximate Δy is only 0.02.   ■

EXAMPLE 3.33   A spherical natural gas storage tank is designed with a radius of 24 m. The allowable error in the actual radius is ± 10 cm. Use differentials to find the approximate excess capacity the tank could have over its designed volume. Compare this result with the volume of the maximum excess capacity as calculated from the given dimensions.

Solution   The volume $f(r)$ of a sphere is given by $V = \frac{4}{3}\pi r^3$. We take $r = 24$ and $\Delta r = dr = 0.10$, since 10 cm $= 0.10$ m. The approximate change in volume if the radius is 24.1 instead of 24 is

$$dV = f'(r)dr = 4\pi r^2 dr = 4\pi(24)^2(0.10)$$
$$\approx 723.8 \text{ m}^3$$

The change calculated from the given measurements is

$$\Delta V = f(r + \Delta r) - f(r) = f(24.1) - f(24)$$
$$= \frac{4}{3}\pi(24.1)^3 - \frac{4}{3}\pi(24)^3$$
$$\approx 726.8 \text{ m}^3$$

So the error in using dV to approximate ΔV is about 3 m^3.   ■

Relative Error and Percentage Error

If $y = f(x)$ and x is in error by an amount Δx, then the corresponding error in y is $\Delta y = f(x + \Delta x) - f(x)$. As we have seen, Δy can be approximated by $dy = f'(x)\,\Delta x$ when Δx is small. Often it is more useful to know how large Δy, or its approximation dy, is in relation to y itself. We define the ratio of Δy to y as the **relative error** in y. The **percentage error** is the relative error multiplied by 100. So we have the following:

Absolute error	$\Delta y \approx dy$
Relative error	$\dfrac{\Delta y}{y} \approx \dfrac{dy}{y}$
Percentage error	$100\dfrac{\Delta y}{y} \approx 100\dfrac{dy}{y}$

EXAMPLE 3.34   Find the approximate relative errors and percentage errors in Examples 3.32 and 3.33.

Solution   In Example 3.32 we were given that $y = 2x^2 + 3x - 1$, $x = 3$, and $\Delta x = 0.1$. We found $dy = 1.5$. To find y, we substitute $x = 3$ in the given equation:

$$y = 2(3)^2 + 3(3) - 1 = 26$$

So the relative error and percentage error are approximately

$$\text{relative error} \approx \frac{dy}{y} = \frac{1.5}{26} \approx 0.0577$$

$$\text{percentage error} \approx 5.77\%$$

In Example 3.33, the volume when $r = 24$ is $\frac{4}{3}\pi(24)^3 \approx 57{,}905.84$. We found $dV \approx 723.8$. So we have

$$\text{relative error} \approx \frac{dV}{V} \approx \frac{723.8}{57{,}906} \approx 0.0125$$

$$\text{percentage error} \approx 1.25\%$$ ∎

Linear Approximation

If we know the value of a differentiable function f at a given point $x = a$, then we can get its value at a nearby point $a + \Delta x$ by adding Δy to the original value; that is,

$$f(a + \Delta x) = f(a) + \Delta y$$

Furthermore, since $f'(a)$ exists, we can approximate Δy by dy if we take $dx = \Delta x$. So we have

$$f(a + \Delta x) \approx f(a) + dy$$

or equivalently,

$$f(a + \Delta x) \approx f(a) + f'(a)\Delta x \qquad (3.14)$$

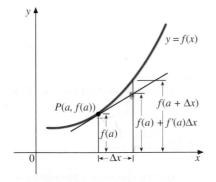

FIGURE 3.34
The distance from the x-axis to the tangent line at P approximates the distance to the curve if Δx is small.

This result is known as **linear approximation,** since we are, in effect, using values along the *tangent line* to approximate values on the *curve*. (See Figure 3.34.)

As we saw in Examples 3.33 and 3.34, it is frequently easier to compute dy than Δy, and so approximating $f(x)$ near a point where its value is known by the linear approximation formula (Equation 3.14) is easier than calculating its actual value.

EXAMPLE 3.35   Without using a calculator, find an approximation to $\sqrt[3]{65}$. Then check your result with a calculator.

Solution   Let $f(x) = \sqrt[3]{x} = x^{1/3}$. The point nearest 65 at which we know the exact value of f is $x = 64$, since $\sqrt[3]{64} = 4$. So in Equation 3.14 we take $x = 64$ and $\Delta x = 1$. Since $f'(x) = \frac{1}{3}x^{-2/3}$, we get

$$f(65) \approx f(64) + f'(64)\Delta x = 4 + \frac{1}{3}(64)^{-2/3}(1) = 4 + \frac{1}{3}\left(\frac{1}{16}\right) \approx 4.0208$$

With a calculator, we find $\sqrt[3]{65} = (65)^{1/3} = 4.02073$ (to five decimal places). So our linear approximation is quite good in this case.   ∎

Refer again to Figure 3.34. Since the tangent line at P has slope $f'(a)$, its equation is

$$y = f(a) + f'(a)(x - a)$$

(you can verify this result). Let us designate the function defined by this equation by L (for *linear*), so that

$$L(x) = f(a) + f'(a)(x - a) \tag{3.15}$$

If in Equation 3.14 we replace Δx by $x - a$, we see that the equation becomes

$$f(x) \approx f(a) + f'(a)(x - a)$$

That is,

$$f(x) \approx L(x)$$

The closer x is to a (or equivalently, the smaller Δx is), the better the approximation. We call $L(x)$ the **linearization** of $f(x)$ at $x = a$.

In the next example we find the linearization of $f(x) = \sqrt{1 + x}$ at $x = 0$ and at $x = 8$, and use the results to approximate values of f near these two points.

EXAMPLE 3.36   Let $f(x) = \sqrt{1 + x}$.

(a) Find the linearization of $f(x)$ at $x = 0$ and use the result to approximate $\sqrt{1.1}$, $\sqrt{1.01}$, and $\sqrt{1.001}$.

(b) Find the linearization of $f(x)$ at $x = 8$ and use the result to approximate $\sqrt{8.9}$, $\sqrt{9.1}$, and $\sqrt{10}$.

Solution   First, we calculate $f'(x)$, since it will be used in both part (a) and part (b).

$$f'(x) = \frac{1}{2}(1 + x)^{-1/2}(1) = \frac{1}{2\sqrt{1 + x}}$$

(a) Taking $a = 0$, $f'(a) = \dfrac{1}{2\sqrt{1 + 0}} = \dfrac{1}{2}$. By Equation 3.15, the linearization at 0 is

$$L(x) = \sqrt{1 + 0} + \frac{1}{2}(x - 0) = 1 + \frac{x}{2}$$

Thus, for x near 0,

$$f(x) \approx 1 + \frac{x}{2}$$

Since $\sqrt{1.1} = \sqrt{1 + 0.1} = f(0.1)$, we have

$$\sqrt{1.1} \approx 1 + \frac{0.1}{2} = 1.05$$

Similarly,

$$\sqrt{1.01} \approx 1 + \frac{0.01}{2} = 1.005$$

and

$$\sqrt{1.001} \approx 1 + \frac{0.001}{2} = 1.0005$$

By calculator, $\sqrt{1.1} \approx 1.0488$, $\sqrt{1.01} \approx 1.004988$, and $\sqrt{1.001} \approx 1.0004999$. Our approximations are therefore correct to two, three, and four decimal places, respectively. The closer x is to 0, the more accurate is the approximation.

(b) Since

$$f'(8) = \frac{1}{2\sqrt{1+8}} = \frac{1}{6}$$

and $f(8) = \sqrt{1+8} = 3$, the linearization at $x = 8$ is

$$L(x) = 3 + \frac{1}{6}(x - 8)$$

So for x near 8,

$$\sqrt{1+x} \approx 3 + \frac{1}{6}(x - 8)$$

In particular, with $x = 7.9$,

$$\sqrt{8.9} \approx 3 + \frac{1}{6}(-0.1) \approx 2.98333 \quad \text{By calculator, 2.98329}$$

Similarly, with $x = 8.1$,

$$\sqrt{9.1} \approx 3 + \frac{1}{6}(0.1) \approx 3.01666 \quad \text{By calculator, 3.01662}$$

Finally, with $x = 9$,

$$\sqrt{10} \approx 3 + \frac{1}{6}(1) = 3.16666 \quad \text{By calculator, 3.1623}$$

Notice that the first two results are correct to at least four places, but the last is correct only to one place. The error is greater in $\sqrt{10}$ because $x = 9$ is farther from 8 than either $x = 7.9$ or 8.1.

Figure 3.35 shows the graph of f along with the tangent lines $y = L(x)$ at the points $(0, 1)$ and $(8, 3)$. Notice that because the curve is flatter near $(8, 3)$ than it is near $(0, 1)$, the tangent line at $(8, 3)$ provides a better approximation to $f(x)$ than it does at $(0, 1)$. The figure clearly shows that it would be inappropriate to use the linearization at $(0, 1)$ to approximate values of $f(x)$ in part (b) of this example, since the tangent line at $(0, 1)$ is nowhere near the actual curve in the vicinity of $x = 8$. ■

FIGURE 3.35

REMARK

■ You should bear in mind that values given by the calculator (or numerical software) are also approximations in most cases.

The Error in the Approximation of Δy by dy

We conclude this section by examining the size of the error in using the linear approximation given by Equation 3.14—that is, the error in using dy to approximate Δy. Assume $f'(x)$ exists and let $dx = \Delta x$. Then, $dy/dx = f'(x)$. Also,

$$\lim_{\Delta x \to 0} \frac{\Delta y}{\Delta x} = \lim_{\Delta x \to 0} \frac{f(x + \Delta x) - f(x)}{\Delta x} = f'(x)$$

and

$$\lim_{\Delta x \to 0} \frac{\Delta y - dy}{\Delta x} = \lim_{\Delta x \to 0} \left(\frac{\Delta y}{\Delta x} - \frac{dy}{\Delta x} \right)$$

Since $dx = \Delta x$, this limit is the same as

$$\lim_{\Delta x \to 0} \left(\frac{\Delta y}{\Delta x} - \frac{dy}{dx} \right)$$

which equals

$$\left(\lim_{\Delta x \to 0} \frac{\Delta y}{\Delta x} \right) - f'(x) = f'(x) - f'(x) = 0$$

Thus,

$$\lim_{\Delta x \to 0} \frac{\Delta y - dy}{\Delta x} = 0$$

This result tells us not only that the difference $\Delta y - dy$ approaches 0 as $\Delta x \to 0$, but also that it does so more rapidly than Δx itself. So the error $|\Delta y - dy|$ is small in comparison with the size of $|\Delta x|$. We now state a theorem that gives a more explicit estimate of the error. Its proof depends on results we will obtain in Chapter 10. The theorem refers to a *bounded second derivative* in an interval and an *upper bound* on $|f''(t)|$. To say that f has a bounded second derivative means that there is some positive constant M such that $|f''(t)| \le M$ for all t in the interval in question. Such a constant M is then called an upper bound for $|f''(t)|$.

THEOREM 3.11

If f has a bounded second derivative in some open interval $(x - r, x + r)$ centered on x, then for all Δx such that $|\Delta x| < r$,

$$|\Delta y - dy| \le \frac{M}{2}(\Delta x)^2 \tag{3.16}$$

where M is an upper bound on $|f''(t)|$ for t in $[x, x + \Delta x]$ when $\Delta x > 0$ and in $[x - |\Delta x|, x]$ when $\Delta x < 0$.

To illustrate this inequality, recall Example 3.35. There we had $f(x) = x^{1/3}$ and $f'(x) = \frac{1}{3}x^{-2/3}$, and so

$$f''(x) = -\frac{2}{9}x^{-5/3} = -\frac{2}{9x^{5/3}}$$

In that example we set $x = 64$ and $\Delta x = 1$, so that we want to know a bound on $|f''(t)|$ for t in the interval $[64, 65]$. Since

$$|f''(t)| = \frac{2}{9t^{5/3}}$$

we obtain the largest value of $|f''(t)|$ by substituting the smallest value of t—namely, 64—since t is in the denominator. Thus, for t in the interval $[64, 65]$,

$$|f''(t)| \leq \frac{2}{9(64)^{5/3}} = \frac{2}{9(4)^5} = \frac{1}{4608} \approx 0.000217$$

So, by Inequality 3.16,

$$|\Delta y - dy| \leq \frac{0.000217}{2}(1)^2 \approx 0.0001$$

This result confirms that using dy to approximate Δy gives accuracy to at least three decimal places.

SYMBOLIC DIFFERENTIATION USING COMPUTER ALGEBRA SYSTEMS

Computer algebra systems have the capability for performing symbolic operations including differentiation. We have seen in this chapter that to differentiate a variety of functions, we need a collection of formulas. Even with the formulas available, many functions are difficult to differentiate because of the algebraic manipulation that is necessary. Computer algebra systems can be used to do the tedious work of finding derivatives, allowing us to investigate a greater number of problems.

CAS 14

Find the derivative of

$$f(x) = x^2 \cos\left(\sqrt{x^3 - x^{1/2}}\right)$$

This problem is easy to solve using Maple, Mathematica, or DERIVE.

Maple:

diff(x^2*cos(sqrt(x^3 − x^(1/2))),x);

where "diff" means "differentiate." The expression for the derivative given in Maple is:

$$2x \cos\left(\sqrt{x^3 - \sqrt{x}}\right) - \frac{1}{2} \frac{x^2 \sin\left(\sqrt{x^3 - \sqrt{x}}\right)\left(3x^2 - \frac{1}{2}\frac{1}{\sqrt{x}}\right)}{\sqrt{x^3 - \sqrt{x}}}$$

Mathematica:*

f[x_]=x^2*Cos[Sqrt[x^3 − x^(1/2)]]

f'[x]

DERIVE:

(At the □ symbol, go to the next step.)

a (author) □ x^2cos(sqrt(x^3 − x^(1/2)))

□ c (calculus) □ d (differentiate) □ [choose expression] □ x (variable) □ 1 (order)

□ s (simplify) □ [choose expression]

---

*In Mathematica the derivative of f can then be computed by f'[x], as we did, or by using D[f[x],x].

CAS 15

Plot the function $f(x) = \dfrac{x^4}{4} + \dfrac{x^3}{3} - x^2 + 1$ and the tangent line to the curve where $x = 2$. Find all points where the tangent line is horizontal.

Maple:[†]

f := x->x^4/4+x^3/3–x^2+1;
m := x->D(f)(x);

Output: $m := x \rightarrow x^3 + x^2 - 2x$

L := m(2)*(x–2)+f(2);

Output:

$$L := 8x - \frac{37}{3}$$

plot({f(x),L},x=–3..3,y=–5..5);

The third command defines the line from the point-slope formula $y = \text{slope} \cdot (x - x_0) + y_0$; we used m to stand for the slope.

Mathematica:

f[x_]=x^4/4+x^3/3–x^2+1
L[x_]=f'[2]*(x–2)+f[2]
Plot[{L[x],f{x]},{x,–3,3}]

DERIVE:

(At the □ symbol, go to the next step.)
a (author) □ f(x) := x^4/4+x^3/3–x^2+1
□ c (calculus) □ d (differentiate) □
[choose expression for f] □ x (variable) □ 1 (order) □ s (simplify) □
[choose expression] □ a (author) □ m(x)
:= [previous expression] □ a (author) □
m(2)(x–2)+f(2) □ s (simplify) □ [choose expression] □ p (plot window) □ p (plot)
□ a (algebra window) □ [highlight f]
□ p (plot window) □ p (plot)

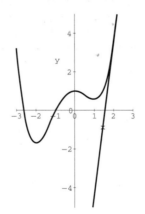

FIGURE 3.9.1.
$y = f(x) = \dfrac{x^4}{4} + \dfrac{x^3}{3} - x^2 + 1,$
$y = 8x - \dfrac{37}{3}$

To find the points where the tangent line is horizontal, we need to find the x and y coordinates of the points where the tangent line has slope 0.

[†] In Maple the operator **D** computes derivatives of operators, whereas **diff** (see CAS 14) computes derivatives of expressions. That is, the argument and result of **D** are functions, whereas the argument and result of **diff** are expressions.

Maple:

solve(m(x)=0,x);

The output of the x-coordinates where the tangent has slope 0 is

0, 1, –2

Now find the y-coordinates of the three points:

f(0);
f(1);
f(–2);

The three points are given by:

(0, 1), (1, 7/12), and, (–2, –5/3)

Mathematica:

Solve[f'[x] == 0,x]

Output:

{{x –> 0}, {x –> –2}, {x –> 1}}

To find the y-coordinates:

f[0]
f[1]
f[–2]

DERIVE:

(At the □ symbol, go to the next step.)
1 (solve) □ [choose expression for $m(x)$]
Output:

$$x = 0$$

$$x = 1$$

$$x = -2$$

To find the y-coordinates: a (author) □ f(0) □ s (simplify) □ [choose expression] □ a (author) □ f(1) □ s (simplify) □ [choose expression] □ a (author) □ f(–2) □ s (simplify) □ [choose expression]

CAS 16

Define the function f by

$$f(x) = \begin{cases} x^2 \sin(1/x) & \text{if } x \neq 0 \\ 0 & \text{if } x = 0 \end{cases}$$

Investigate the derivative of f at 0.

In this example the derivative at 0 cannot be computed by substituting 0 into the expression for the derivative. We need to use the definition of the derivative at a point as the limit of the difference quotient. Notice that since $f(0) = 0$, the numerator of the difference quotient is simply $f(x + h)$.

Maple:

f := x–>x^2*sin(1/x);
df := x–>f(x+h)/h;
limit(df(0),h=0);

The value of this last limit is 0, so the derivative is equal to 0 when $x = 0$.

Mathematica:

f[x_] = x^2*Sin[1/x]
f'[x]
df = (f[0+h])/h
Limit[df,h–>0]

DERIVE:

(At the □ symbol, go to the next step.)
a (author) □ x^2sin(1/x) □ m (manage) □ s (substitute) □ [choose expression] □ 0+h (substitute for x) □ c (calculus) □ l (limit) □ [choose expression] □ h (limit with respect to h) □ 0 (limit point) □ s (simplify) □ [choose expression]

Exercise Set 3.9

In Exercises 1–16, find dy.

1. $y = x^2 + 2x - 3$

2. $y = 2x - x^3$

3. $y = \sqrt{4 - x^2}$

4. $y = 1/\sqrt{3x + 4}$

5. $y = x/\sqrt{1 - x}$

6. $y = x/\sqrt{1 - x^2}$

7. $y = x \sin x$

8. $y = \cot \sqrt{x}$

9. $y = \sec x + \tan x$

10. $y = \cos^2 3x$

11. $y = (x^6 + 1)^{2/3}$

12. $y = (x^{2/3} + a^{2/3})^{3/2}$

13. $y = (\tan x)/x$

14. $y = 2 \csc^3 2x$

15. $y = \sin^2 x^2$

16. $y = (\sin 2x)/(1 - \cos 2x)$

In Exercises 17–24, find dy/dx implicitly using differentials.

17. $x^3 - 2xy^2 = 6$

18. $y^2 - x \sin y = \cos x$

19. $2x^2 - 3xy + 4y^2 - 10 = 0$

20. $\sqrt{x^2 + y^2} - 3 = 2xy$

21. $x^{2/3} + y^{2/3} = a^{2/3}$

22. $x^4 - 3x^2 y^3 + y^5 = 10$

23. $x \sin y - y \sin x = 1$

24. $x^{4/3} - y^{4/3} = a^{4/3}$

In Exercises 25–27, use differentials for the approximations.

25. The side of a square is measured to be 18 cm. If there is an error in the measurement of 0.05 cm, approximate the resulting error in (a) the area and (b) the perimeter.

26. Find the approximate error in the calculated area of a circle with a diameter measured as 20 in. when in fact it is 20.01 in. What is the error in the circumference?

27. Find the approximate relative error and percentage error in each part of Exercise 26.

In Exercises 28–31, estimate the value using linear approximation, and compare your result with the answer obtained by using a calculator. Find the percentage error.

28. $\sqrt{10}$

29. $\sqrt[3]{7.5}$

30. $\sqrt[3]{-25}$

31. $1/\sqrt{102}$

In Exercises 32–37, find the linearization $L(x)$ of f at $x = a$. Then graph f and L in the vicinity of $x = a$. Confirm your graph using a CAS or a graphing calculator.

32. $f(x) = \sin x$; $a = 0$

33. $f(x) = \tan x$; $a = \dfrac{\pi}{4}$

34. $f(x) = \sqrt{x}$; $a = 1$

35. $f(x) = x^{3/2}$; $a = 4$

36. $f(x) = 2 \cos x - 1$; $a = \dfrac{\pi}{3}$

37. $f(x) = \dfrac{x^3}{3} + \dfrac{x^2}{2} - 2x - 1$; $a = 2$

38. Show that for x near 0, $(1 + x)^k \approx 1 + kx$ for all rational numbers k.

In Exercises 39–44, use the result of Exercise 38 to find a linear approximation of the given function near zero.

39. $(1 + x)^3$

40. $\sqrt[3]{1 + x}$

41. $\dfrac{1}{(1 + x)^2}$

42. $\dfrac{1}{\sqrt{1 + x}}$

43. $(1 - 2x)^{3/2}$ (*Hint:* In the formula in Exercise 38, replace x by $(-2x)$.)

44. $\dfrac{1}{\sqrt{4 - x}}$ (*Hint:* Write as $\dfrac{1}{2}\left(1 - \dfrac{x}{4}\right)^{-1/2}$ and use Exercise 38 with x replaced by $-x/4$.)

45. A cardboard box has a square base, and its height is double the length of a side of the base. The box is supposed to be 2 ft × 2 ft × 4 ft, but it is subject to an error of $\pm\frac{1}{8}$ inch in each dimension. Find the approximate maximum and minimum values for the actual volume. Then use a calculator to find the exact maximum and minimum values and compare the two sets of results.

46. The radius of the orbit (assumed circular) of communications satellite B is 1 mi larger than that of satellite A. Approximately how much farther does satellite B travel in completing one orbit than satellite A? Show that linear approximation gives the exact value in this case. Why doesn't the answer depend on whether the satellite is in a low earth orbit or in a synchronous orbit 22,600 mi high?

47. The cost $C(x)$ in dollars of manufacturing x television sets at a certain plant is given by

$$C(x) = 4000 - 15x + 0.015x^2$$

If 600 sets are typically manufactured each day, find the approximate cost of manufacturing one more set. (*Note:* Although x actually takes on only integer values, for purposes of calculation we can treat it as though it were a continuous variable that takes on all positive real values.)

48. The profit function $P(x)$ for manufacturing and selling x items of a certain type is given by

$$P(x) = 100\sqrt{x} - 0.75x - 600$$

Approximately what additional profit would result from manufacturing and selling 405 items instead of 400 items?

49. Suppose $y = x\sqrt{4 - x^2}$ and $x = 2\sin t$. Calculate dy in the following two ways and show that they are the same: (a) by replacing x in the first equation by its value from the second equation, and (b) by applying Equation 3.13.

50. Verify the differential formula

$$d\left(\frac{u}{v}\right) = \frac{v\,du - u\,dv}{v^2}$$

where u and v are differentiable functions of x, with $v \neq 0$.

51. Find du if

$$u = \frac{z \tan z}{\sqrt{1 + z^2}}$$

52. The inner radius of a section of steel pipe for an oil pipeline is 2 ft, and the thickness of the metal is $\frac{1}{4}$ in. If the section is 32 ft long, use differentials to approximate the number of cubic feet of metal in the pipe.

53. The height of Pike's Peak is calculated by measuring the angles of elevation α and β at points A and B, d meters apart, then using trigonometry. (See the figure.) Show that

$$h = \frac{d}{\cot \alpha - \cot \beta}$$

Suppose B is chosen so that $\beta = 2\alpha$. If $d = 2000$ (assumed exact for purposes of this problem) and $\alpha = 20°$, with a possible error of $\pm 0.5°$, find the approximate possible error in the calculation of h.

54. Use linear approximation to estimate each of the following, and find an upper bound on the error using Theorem 3.11:

(a) $\sqrt{4.2}$

(b) $(30)^{2/3}$

Chapter 3 Review Exercises

In Exercises 1–4, find $f'(a)$ using Definition 3.2.

1. $f(x) = 2x^2 - 3x$; $a = 3$

2. $f(x) = 2/\sqrt{x}$; $a = 1$

3. $f(x) = x/(x + 1)$; $a = 2$

4. $f(x) = \sqrt{2x + 3}$; $a = -1$

In Exercises 5–14, find $f'(x)$.

5. $f(x) = (3x^2 - 2)^{10}$

6. $f(x) = \dfrac{2x^2 - 3x}{x^2 + 1}$

7. $f(x) = \dfrac{x}{\sqrt{1 - 2x}}$

8. $f(x) = x^2 \sin x^3$

9. $f(x) = \dfrac{\sin x}{1 + \cos x}$

10. $f(x) = \sqrt{\dfrac{1 + x}{1 - x}}$

11. $f(x) = (a^{2/3} - x^{2/3})^{3/2}$

12. $f(x) = \sec^3 x + \tan^3 x$

13. $f(x) = \dfrac{\sqrt{1 - \cos 2x}}{\cos x}$, $0 \leq x < \dfrac{\pi}{2}$

14. $f(x) = \dfrac{x\sqrt{1 + x^2}}{1 - x^2}$

In Exercises 15 and 16, find the equation of the tangent line and normal line (line perpendicular to the tangent line) to the graph of f at the point $(a, f(a))$.

15. $f(x) = x^2/(x - 4)$; $a = 2$

16. $f(x) = (\sqrt{x} - 2)^3$; $a = 1$

In Exercises 17 and 18, the position function $s(t)$ of a particle moving on a straight line is given. Find its instantaneous velocity and acceleration when $t = a$.

17. $s(t) = 100\left(\dfrac{t^2 - 1}{t^2 + 1}\right)$; $a = 3$

18. $s(t) = 16\sqrt[3]{t^3 - 3t + 6}$; $a = 2$

In Exercises 19–22, find y' using implicit differentiation.

19. $2xy^2 + x^4 = y^3$

20. $y^3 + (x - y)^2 = 3$

21. $y \tan x - x \sec y = 1$

22. $\sqrt{x^2 + y^2}/(x + y) = 4$

23. Find y' and y'' if $x^{1/2} + y^{1/2} = a^{1/2}$, where $x > 0$, $y > 0$, and a is a positive constant.

24. Find the equations of the tangent line and normal line (line perpendicular to the tangent line) to the graph of $x^3 + x^2 y^2 - y^3 = 1$ at the point $(-2, 3)$.

25. The height of a model of an ice cream cone is three times the radius of the top. The top is supposed to have a radius of 6 cm, but this measurement may be in error by as much as 1 mm. Use differentials to approximate the possible error in (a) the volume and (b) the lateral surface area. Also find the percentage error in each case. Assuming that the cone would have 50 calories if the radius measurement had no error, how uncertain is the calorie count?

26. Use differentials to approximate the area of the shaded circular segment in the figure. Leave your answer in exact form.

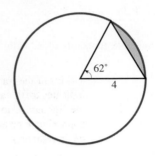

27. Verify that $y = x^{-1/2}(1 + \sin x)$ is a solution of the equation

$$x^2 y'' + xy' + \left(x^2 - \frac{1}{4} \right) y = x^{3/2}$$

Chapter 3 Concept Quiz

1. Let $y = f(x)$. Define each of the following:
  (a) $f'(a)$
  (b) the derivative function f'
  (c) dy
  (d) the linearization $L(x)$ of f at a

2. State each of the following:
  (a) the Product Rule
  (b) the Quotient Rule
  (c) the Chain Rule
  (d) the Power Rule
  (e) the relationship between differentiability and continuity

3. Write the formulas for the derivatives of the six trigonometric functions.

4. (a) Write the equation of the tangent line to the graph of a differentiable function f at $x = a$.
  (b) Show graphically the relationship between Δy and dy for $y = f(x)$, where f is differentiable at $x = a$.

5. Fill in the blanks.
  (a) If f is differentiable at $x = a$, then $f'(a)$ is the _____ of the _____ _____ to the graph of f at $(a, f(a))$.
  (b) If $s''(t)$ exists and s is the position function

of a particle moving along a line, then $s'(t)$ is the _____ of the particle and $s''(t)$ is its _____.

  (c) In words, the Product Rule states that the derivative of a product is the _____ of the _____ times the _____, plus the _____ times the _____ of the _____.
  (d) In words, the Quotient Rule states that the derivative of a quotient is the _____ of the _____ times the _____, minus the _____ times the _____ of the _____, divided by the _____ of the _____.
  (e) Linear approximation means that if f is differentiable at $x = a$, then $f(x)$ is approximated near a by _____.

6. State which of the following are true and which are false.
  (a) If $y = f(x)g(x)$, then $y' = f'(x)g'(x)$.
  (b) If f is not continuous at a, then f is not differentiable at a.
  (c) If $y = f(x)/g(x)$, then $y' = f'g - fg'$.
  (d) If $f(x) = \sin x$ and $g(x) = \cos x$, then $(f + g)'(x) = g(x) - f(x)$.
  (e) If $f''(a)$ exists, then f' is continuous at a.

APPLYING CALCULUS

75 ft

θ

Buoy

1. From the top of a 75-feet-high lighthouse, the angle of depression θ of a buoy that indicates the location of a sunken ship is measured to be $23°15'$, and the distance x from the lighthouse to the buoy is calculated using trigonometry (see the figure). Find the approximate error in x if the measurement of θ is in error by $5'$. (*Note:* Express θ and $d\theta$ in radians.)

2. A rocket fired at an angle θ from the horizontal with an initial velocity of v_0 meters per second will rise to a maximum height of

$$y = \frac{v_0^2 \sin \theta}{2g}$$

where $g = 9.80$ m/s^2 is the acceleration caused by gravity. Approximately how much higher will a rocket fired at an angle of $31°$ go than one that is fired at an angle of $30°$, if the initial velocity in each case is 80 m/s?

3. A new computer workstation costs $25,000. Its useful lifetime will be 5 years, at which time it will be worth an estimated $5,000. The company calculates its depreciation using the linear decline that is an option in the tax laws.

   (a) Find the linear equation that expresses the value V of the equipment as a function of time t, $0 \leq t \leq 5$.

   (b) How much will the equipment be worth after 2.5 years?

   (c) What is the average rate of change in the value of the equipment from 1 to 3 years?

   (d) What is the instantaneous rate of change in the value of the equipment at any time $t = a$?

4. According to Boyle's Law, if the temperature of a confined gas is held fixed, then the product of the pressure P (in lb./in.^{2}) and the volume V (in m^3) is a constant. Suppose, for a certain gas, $PV = 8000$.

   (a) Determine the average rate of change of P as V increases from 200 cu. in. to 250 cu. in.

   (b) Express V as a function of P and show that the instantaneous rate of change of V with respect to P is proportional to the inverse square of P.

5. Consider the Earth as a sphere of radius $R = 6378$ km. Given the geometric formulas for the surface area and volume of a sphere and the formula for the circumference of a circle, use differentials to estimate:

   (a) The increase in the Earth's surface area if its radius were increased by 1 km.

   (b) The decrease in the Earth's volume if its radius were decreased by 1 km.

   (c) The difference in the circumference of the orbital path of the Hubble Space Telescope and that of the Earth, given that the height of the telescope's orbit is approximately 300 km above the Earth's surface. Express the answer both as a formula and in meters.

The Hubble Space Telescope being placed into an orbit 300 km above the Earth's surface by a space shuttle.

6. Let $P : (x_0, y_0)$ be a point on the ellipse $\frac{x^2}{a^2} + \frac{y^2}{b^2} = 1$.

   (a) Determine an equation for the tangent line to the ellipse at the point P.

   (b) Assume $a^2 > b^2$. Let $c = \sqrt{a^2 - b^2}$ and let F_1 and F_2 be the points $(c, 0)$ and $(-c, 0)$, respectively. Show that the angle between the line $F_1 P$ and the tangent line equals the angle between the line $F_2 P$ and the tangent line.

   Note: The points $(c, 0)$ and $(-c, 0)$ are the foci of the ellipse. The property suggested in (b) is the "reflection" property of ellipses—a line emanating from one focus is "reflected" through the other focus. Elliptically shaped rooms are known as "whispering galleries."

7. The rate of change of electric charge with respect to time is called current. Thus, if $q = q(t)$ denotes the charge (in coulombs) at time t, then the current i is given by $i = dq/dt$.

(a) Suppose $q = 2t^2 - 6t + 3$ coulombs of charge flow through a conducting wire in t seconds.

(i) Determine the current i (measured in amperes = coulombs per sec) after t seconds. What is the current after 4 seconds?

(ii) When does the current reverse itself, i.e., when does i change sign?

(iii) Suppose there is a 15-amp fuse in the line. How long will it last?

(b) Suppose $q = (1/3)t^3 - t^2 + t + 4$.

(i) Determine the current i after t seconds.

(ii) Does the current ever reverse itself?

(iii) If there is a 25 amp fuse in the line, how long will it last?

8. Telescopes and solar collectors are often made in paraboloidal shapes (the volume generated by rotating a parabola around its axis), since then on-axis parallel rays are reflected to the same point, called the focus. Consider the parabola $y^2 = 4px$.

(a) Show that the slope of the tangent to the curve is given by $dy/dx = \sqrt{p/x}$.

(b) For a ray of light approaching the parabola at angle β, with α indicating the angle to the tangent and θ indicating the angle from the axis to the point where the incoming ray is incident, show that $\alpha + \beta = \theta$.

(c) Show that the angle of incidence = the angle of reflection, using the trigonometric formula

$$\tan(a - b) = \frac{\tan a - \tan b}{1 + \tan a \tan b}$$

(d) By finding the equation of the reflected line (the incoming line will be horizontal in the figure), show that the incoming signal is reflected to the focus F.

9. National health care spending, in billions of dollars, has taken the shape of a parabola over the last few decades, increasing at an alarming rate. (See the table.)

Year	Dollars (in billions)
1965	30
1970	80
1975	120
1980	250
1985	400
1990	690
1992	820 (estimated)

(a) Using the 1965, 1980, and 1990 data only, fit a parabola of the form $y = ax^2 + bx + c$.

(b) Calculate the values for the other years given, and assess the error as a percentage of the value.

(c) Differentiating, calculate the slope of the parabola, and substitute values to get the numerical value of the derivative at each of the years listed.

(d) What percentage increase does the parabolic increase predict for 1993?

(e) A newspaper article reports that employers have budgeted for increases of 15% to 18% for 1993, whereas preliminary indications are that the costs have increased by only 8% to 12%. Compare these values and the slopes they imply with the predictions of part (d).

10. It has recently been discovered that "when fielders run backwards or forwards to catch a ball they run at a speed which keeps $d^2(\tan \alpha)/dt^2$ zero, where α is the angle of elevation of gaze from catcher to ball. This ensures that they intercept the ball before it reaches the ground (provided they can run fast enough to keep $d^2(\tan \alpha)/dt^2$ zero) whatever the effect of aerodynamic drag on the ball's trajectory."[†] Find the derivative of $\tan \alpha$ with respect to t, and make use of the given information about the second derivative to solve for $d\alpha/dt$. (Your answer will involve a constant.)

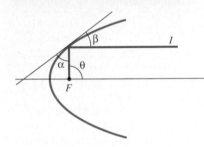

[†] (Reference: Peter McLeod and Zoltan Dienes, *Nature*, vol. 362, p. 23, 1993.)

A PERSONAL VIEW

Mark J. T. Smith

What do you remember of your first calculus class?

I took precalculus in high school and calculus when I attended the Massachusetts Institute of Technology. I remember finding it more abstract than the mathematics courses I'd taken previously. There were aspects of the course that were interesting but overall I wouldn't call it exciting. I remember thinking that I would probably forget all the tricks for integration and was glad to learn that there were integral tables available. I had no real understanding of applications then. Calculus seemed to me to be about solving contrived problems, many of which were fun, but they still seemed contrived. I basically accepted the fact that calculus was an important tool and that I would use it later on.

Did you already intend to be an engineer at that point in your life?

I decided in junior high that I wanted to be either an engineer or a physicist. For a short period of time as a freshman, I thought I wanted to be a biomedical engineer, but by the end of my freshman year, I had chosen electrical engineering. I was drawn to it because I had a fascination with electronics and building electrical things. I started in electrical engineering because I had this naive idea that it was all about electronics. As I learned more I became more interested in aspects of electrical engineering that have nothing to do with electronics.

Are there particular mathematics courses that a freshman should pay special attention to if they intend to major in engineering?

That depends dramatically on what type of engineering you are going to be doing. The mathematics courses that I remember taking for which I needed calculus at the undergraduate level were differential equations and probability theory and at the graduate level, random processes and real analysis. I also took a course in special functions that used calculus a lot. Many electrical engineering students have trouble in electromagnetics. At M.I.T. it had a reputation of being a killer course, but I found all this to be exaggerated. It was actually quite an enjoyable course. Electromagnetics makes extensive use of differentiation and integration. It wasn't until I took this course that I felt I was really proficient in calculus.

Having said that, the use of calculus in electrical engineering is not evenly distributed throughout the discipline. Electromagnetics is very calculus intensive, and it's a required course, but digital logic design, for instance, requires other types of mathematics. My own area is digital signal processing. I rely on discrete mathematics, time series, convergence theorems, and numerical methods. The concept of approximating a continuous time function is one that seems essential in virtually any of the problems I do. In this context differentiation and integration are used heavily.

So you still use calculus in your current research?

My research area is digital signal processing, image processing, data compression and speech processing. We are always using mathematics to optimize our systems. This is typically done by computing or estimating derivatives.

Could you give us an example?

Engineers often first model systems with mathematical equations that they intend to build later on. Consider an automatic volume control device for a car radio as a hypothetical example. The purpose of the device is to keep the volume of the radio at a pleasing level, regardless of the background noise. It should gently raise the radio volume when you drive in noisy traffic and should lower the volume when you come to a stop light in a quiet hospital zone. This system has as its input a sensor input $x(t)$ that measures all the noise in the car for each instant in time t'. The output of the system is the radio signal you hear. The control parameter c is the volume control. Assume that the radio signal output $y(t)$ is represented by an equation where one of the variables is the volume control c. So $y(t)$ is a function in the variables c and t. The idea is to have the system automatically determine c for each instant in time t so that the radio level is pleasing.

This optimal control value c can be found by minimizing an error function E, which measures performance. In particular, the error function value is large if the radio volume is too loud or too soft and is zero (or relatively small) if the volume is just right.

We can find the value of c by taking the derivative of the error function with respect to c and setting the result equal to zero. Solving this equation for c gives us the optimal value of c resulting in the perfect volume for the radio. In practice, the system would update the value of c frequently (say every second) thereby keeping the radio volume at a pleasing level for the duration of the car trip. Often, we have to introduce constraints in our problems. Let's assume now that our suggested system has problems. The value computed for c is sometimes negative. Clearly having a negative value for the volume does not make sense. This problem can be solved by introducing a constraint into the minimization equation. We can modify the formulation so that we minimize the error subject to the constraint that c is always positive. This is called *constrained optimization* and is used extensively in engineering.

To what extent is it important to understand the rigorous foundations of calculus if you intend to be an engineer?

Well, it's not as important as it would be to someone in a pure mathematics concentration. I find that proofs, per se, in calculus are not as important as having a functional understanding of the mathematics and how to apply it. However, engineers rely on proving theorems and understanding their properties all the time. This is because engineers are very often seeking optimal solutions to problems. We would like to first determine the best

2

2

situation that is possible in theory. Even though we may not be able to implement this ideal level of performance in practice, knowing the optimal situation helps us set realistic goals and provides a measure we can use for comparison. Such optimal solutions are often derived using calculus.

To illustrate the importance of determining optimal upper bounds on performance, consider the design of a solar energy system for the home. We can model the atmosphere, earth position and rotation, and properties of the materials, and determine reasonable upper bounds on the amount of energy that can be extracted from such a solar collection system. This information can be extremely useful. For example, it may tell you immediately that it is impossible for this technology to provide enough supplementary energy to justify the cost of the system or that the design of the system cannot meet some target energy requirements. This kind of analysis is critical in effective engineering design.

You mentioned differentiation and integration, but what about sequences and series? How significant are they in understanding engineering principles?

The study of sequences and series is the foundation of digital signal processing and discrete systems. There is a rapidly growing trend toward digital implementations. Devices from music systems to hand-held personal communications systems are all starting to employ digital representations of signals. Sequences and series are indispensable in carrying out numerical computations because they can be used to approximate numbers like π and e or evalute functions like $\ln x$ or $\cos x$ to any desired degree of accuracy. That is, you can use an arbitrary number of terms to get as close as you like to the number or function's true value. Approximations to these constants and functions are repeatedly used in electrical engineering models. In other words, signals are digitally represented as sequences of numbers. These numbers are processed by computers to, perhaps, enhance the signal quality by filtering out background noise, or compress the data for efficient transmission. Digital signal processing is one of the most important areas of technology for the twenty-first century.

Some relatively new mathematical techniques such as wavelets are being applied to communications and coding problems. Is calculus necessary in understanding these new mathematical techniques?

In the areas of systems, telecommunications, controls, and signal processing, we often rely on transformations or transforms as they are called. These transforms convert a signal from its original form to another form that can be interpreted more easily. For example, the Fourier transform converts a signal into a frequency domain representation. In the Fourier domain we represent a signal in terms of its frequency components. Fourier analysis is a fundamental part of engineering. The Fourier transform as well as many other transforms are based on taking integrals. [See Section 10.10.]

Engineers always try to make the best use of new advances in mathematics, science, and technology. For example, the introduction of fractals and wavelets have created a tremendous amount of interest. Wavelets, for instance, are extremely important in data compression so that data can be transmitted far more efficiently than before. There are many engineers and scientists who believe that these techniques have the potential to vastly improve performance over conventional approaches.

Do you think it's a good idea for students to learn to use mathematical software such as computer algebra systems if they intend to go on in engineering?

I strongly recommend that students start to learn to use these tools when they are undergraduates. They will certainly use them later on and if they can help learn calculus, so much the better. Our engineering undergraduates and graduate students at Georgia Tech use Maple, Mathematica, and Matlab very frequently. In the future, I see mathematical software as being as important to calculus and engineering as the calculator is to arithmetic. It is a tool that makes our work easier, faster, and more reliable. It allows us to handle very large problems that could not be done easily by hand. These tools are by no means a substitute for learning the basics. Just because you know how to use a calculator does not mean that you should not know how to add or multiply. This is even more critical with calculus. Understanding the fundamentals of calculus provides insight into formulating the problems and their solutions. This is something that a computer program can't do for you.

I recall learning how to do matrix inversion when I was an undergraduate. It's a tedious process, particularly for matrix sizes 3×3 and larger. Computer tools now do matrix inversion for you instantaneously. By working with software tools like Matlab, Maple, and Mathematica, many of the more mundane aspects of the computation are removed. It's more than just a convenience. It gives you more time to concentrate on what the results of the computation mean and how they are applicable to your original problem.

Mark J. T. Smith is an Associate Professor in the School of Electrical Engineering at the Georgia Institute of Technology and a member of the faculty at Georgia Institute of Technology in Lorraine, France. He has received several awards for teaching and research and is the co-author of two introductory books in digital signal processing. Professor Smith has authored more than 80 papers in this area and is presently the Chairman of the IEEE Digital Signal Processing Technical Committee. Professor Smith's research interests are in the areas of speech, image, and video computer processing. He received the S.B. degree from MIT in 1978, and the M.S. and Ph.D. degrees from the Georgia Institute of Technology in 1979 and 1984.

4

APPLICATIONS OF THE DERIVATIVE

Among the many different applications of derivatives are projectile problems. For instance, the derivative can determine the optimal angle at which a football should be kicked to give maximum distance.

We have already seen two applications of the derivative. First, we found that it gave the slope of the tangent line to a curve. Second, we found that it gave the velocity of a body moving along a line. More generally, we have seen that the derivative of a function gives its instantaneous rate of change with respect to its independent variable, whatever the independent variable and the function may represent. In this chapter we explore further this interpretation of the derivative as a "rate of change."

We will show that if we know only where $f'(x)$ and $f''(x)$ are positive and where they are negative, we will then know where the function f is increasing and where it is decreasing, where its graph is concave up and concave down, and where f attains maximum and minimum values. We will make use of this knowledge of the behavior of functions to solve a wide variety of practical problems. We will emphasize those that involve maximizing or minimizing a function. For example, if you owned a pizza parlor, you would want to know what price per pizza to charge to maximize your profit.

Rates of change with respect to time (of which velocity is but one example) are especially important. We will show how knowing such a rate of change for one or more variables can often enable us to find the rate of change of another. As a further application, we will show how the derivative of a function can be used to approximate the zeros of the function. This process provides a way of approximating solutions to equations for which it is difficult or impossible to find exact solutions.

As another application of the derivative, we show a procedure, known as L'Hôpital's Rule, for finding limits (if they exist) of quotients in which the numerator and denominator both approach 0 or both become infinite. Limits of these types are examples of *indeterminate forms*. We will encounter other types of indeterminate forms in Chapter 7 and will see how L'Hôpital's Rule can again be used.

We conclude the chapter by showing how we can work backward from the derivative of a function to the function itself, provided we know the value of the function at one point. This process leads naturally to the subject of differential equations, which we introduce here and study in more detail later.

At the outset we proceed intuitively, relying mainly on geometrical reasoning. Toward the end of the chapter, however, we provide the theoretical basis for our conclusions in the form of a theorem with far-reaching consequences, called the Mean-Value Theorem. We will encounter the need for this theorem in other chapters as well, especially in Chapter 5 on the integral.

4.1 EXTREME VALUES

Rising curves; slopes are positive

Falling curves; slopes are negative

FIGURE 4.1

One of the most important applications of the derivative is in finding the largest or smallest value of a function. For example, we might wish to know the dimensions of a container that can hold a given volume yet requires the least amount of material. Or we might wish to know what price a theater owner should charge to obtain maximum revenue. As another example, the owner of an orchard may be interested in knowing how many apple trees to plant per acre to maximize the yield of apples. These problems and many others, when properly formulated, can be solved by making use of differentiation. Because we can find out a lot about a function by looking at its graph, we begin by considering how the derivative of a function can be used to determine certain information about its graph.

We know already that if $y = f(x)$, where f is a differentiable function, then $f'(x)$ gives the slope of the tangent line to the graph of the function at the point $(x, f(x))$. When this slope is positive at each point of some interval, we would expect the curve to be rising on that interval as we move from left to right. When the slope is negative, we expect the curve to be falling. We illustrate this relationship between the sign of the slope and direction of the curve in Figure 4.1. We have previously used the terms *increasing* and *decreasing* to describe a function whose graph is rising or falling, respectively. We now give a precise definition of these terms.

Definition 4.1 Increasing and Decreasing Functions; Monotone Functions	Let a function f be defined on an interval I. Then f is said to be **increasing** on I if $f(x_1) < f(x_2)$ for any two numbers x_1 and x_2 in I with $x_1 < x_2$. Similarly, f is said to be **decreasing** on I if $f(x_1) > f(x_2)$ for any two numbers x_1 and x_2 in I with $x_1 < x_2$. A function that is either increasing on I or decreasing on I is said to be **monotone** on I.

For example, if $f(x) = x^3$, then f is an increasing function on all of $\mathbb{R}$, since if $x_1 < x_2$, then $x_1^3 < x_2^3$. The function g for which $g(x) = x^2$ is decreasing on $(-\infty, 0)$ and increasing on $(0, \infty)$. The graphs of these two functions are shown in Figure 4.2.

We formalize our observation about the relationship between derivatives and monotone functions in the following theorem.

THEOREM 4.1	If $f'(x) > 0$ for all x in an open interval I, then f is increasing on I, and if $f'(x) < 0$ for all x in I, then f is decreasing on I.

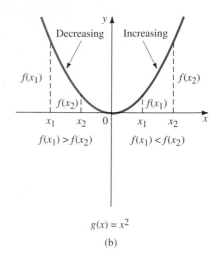

$f(x) = x^3$
$f(x_1) < f(x_2)$ when $x_1 < x_2$

(a)

$g(x) = x^2$

(b)

FIGURE 4.2

As we have seen, Theorem 4.1 is a plausible result, since $f'(x) > 0$ in I means that the tangent line is always sloping upward from left to right. This upward slope implies that the function is increasing. Similar remarks apply when $f'(x) < 0$. We will prove these statements in Section 4.7 after we have developed more background. For now, we will proceed intuitively. Let us emphasize that knowing where a function is increasing and where it is decreasing is not sufficient to describe the shape of the curve. For example, although the function $f(x) = x^3$ in Figure 4.2(a) is always increasing, its shape for $x < 0$ is quite different from its shape for $x > 0$. We will discuss the additional information we need in Section 4.2 when we take up the notion of concavity.

EXAMPLE 4.1   Determine the intervals on which the function f is increasing and the intervals on which it is decreasing:

$$f(x) = 2x^3 - 3x^2 - 12x + 4$$

Solution   First calculate $f'(x)$:

$$f'(x) = 6x^2 - 6x - 12 = 6(x + 1)(x - 2)$$

From the sign graph for f' (Figure 4.3), we see that $f'(x) > 0$ on $(-\infty, -1) \cup (2, \infty)$ and $f'(x) < 0$ on $(-1, 2)$. Thus,

$$f \text{ is increasing on } (-\infty, -1) \text{ and on } (2, \infty)$$

$$f \text{ is decreasing on } (-1, 2)$$

Figure 4.4 illustrates these results. ■

High points and low points on the graph of a continuous function occur when the function changes from increasing to decreasing, or vice versa. They might also occur at endpoints of intervals. Figure 4.5 illustrates some possibilities. The points P_2, P_4, and P_7 are all high points in the sense that they are higher than other points in their immediate vicinity. Similarly, P_1, P_3, P_5, and P_8 are low points. If we ask for the *highest* point, it is P_7, and the *lowest* point is P_8. Note that except for P_1 and P_8, which are at the endpoints of the interval, the tangent lines at these high and low points appear to be either horizontal or

FIGURE 4.3
Sign graph for $f'(x) = 6(x + 1)(x - 2)$

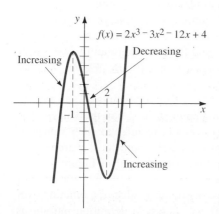

$f(x) = 2x^3 - 3x^2 - 12x + 4$

FIGURE 4.4

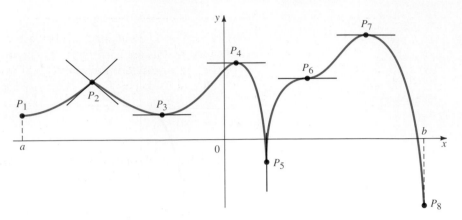

FIGURE 4.5

vertical, or else there is no tangent line. In particular, at P_2 there is no tangent line, since from the left the slopes approach one value and from the right they approach another. At P_3, P_4, and P_7 the tangent line is horizontal, and at P_5 it is vertical. Note also, however, that at P_6 there is a horizontal tangent line, even though P_6 is neither a high point nor a low point. The next definition makes these ideas more precise.

Definition 4.2
Local Maximum and Local Minimum

A function f is said to have a **local maximum** at the point $x = a$ if there is an open interval I containing a such that $f(a) \geq f(x)$ for all x in I. Similarly, f has a **local minimum** at $x = a$ if there is an open interval I containing a such that $f(a) \leq f(x)$ for all x in I.

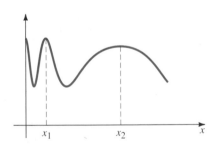

FIGURE 4.6
f has local maximum points when $x = x_1$ and when $x = x_2$. For x_1, the interval throughout which $f(x_1) \geq f(x)$ is small, whereas for x_2, the interval throughout which $f(x_2) \geq f(x)$ is large.

Note: There is no requirement on the size of the interval I in this definition. In some cases it can be very large, and in others it may be very small. (See Figure 4.6.)

If f has a local maximum at a, then we say the point $(a, f(a))$ is a **local maximum point** on the graph of f, and $f(a)$ is called a **local maximum value** of f. Similar terminology is used for the local minimum. The terms **maxima** and **minima** are the plural forms of maximum and minimum, respectively. When we wish to speak of maxima and minima collectively, we say either **extreme values** or **extrema**. Thus, if we ask for all local extrema, we mean the collection of all local maximum values and all local minimum values. (For example, the local extrema in Figure 4.5 occur at P_1, P_2, P_3, P_4, P_5, P_7, and P_8, but not at P_6.)

When the domain of f includes one or both of the endpoints of an interval, then the value of f at such a point may be greater or less than the values of f at all nearby points in the domain. It is convenient to refer to such values as **endpoint extrema** and to investigate them separately from local extrema that occur at interior points. For example, the function pictured in Figure 4.5 has endpoint minima at both $x = a$ and $x = b$.

Frequently we are interested in the largest or smallest value of a function over its entire domain, as opposed to local extrema. The next definition introduces terminology for such global extrema.

Definition 4.3
Absolute Maximum and
Absolute Minimum

A function f is said to have an **absolute maximum** (or **global maximum**) at the point a of its domain if $f(a) \geq f(x)$ for all x in its domain, and it is said to have an **absolute minimum** (or **global minimum**) at a if $f(a) \leq f(x)$ for all x in its domain.

If f has an absolute maximum at a, then $f(a)$ is called the **absolute maximum value of f** (or simply the maximum value of f), and the point $(a, f(a))$ is called the absolute maximum point on the graph of f. We use analogous terminology for the absolute minimum.

If a function has an absolute maximum value, this value is the largest among all local maximum values and endpoint values. But as the graphs in Figures 4.7 through 4.9 show, not all functions have absolute maximum and minimum values. There may not even be local maxima or minima.

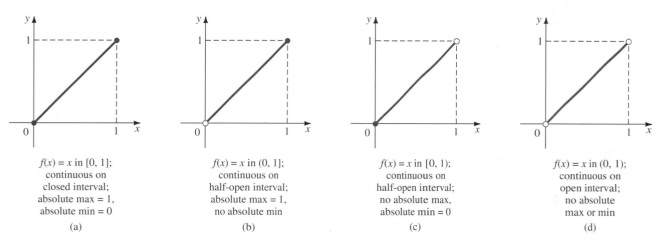

$f(x) = x$ in [0, 1];	$f(x) = x$ in (0, 1];	$f(x) = x$ in [0, 1);	$f(x) = x$ in (0, 1);
continuous on	continuous on	continuous on	continuous on
closed interval;	half-open interval;	half-open interval;	open interval;
absolute max = 1,	absolute max = 1,	no absolute max,	no absolute
absolute min = 0	no absolute min	absolute min = 0	max or min
(a)	(b)	(c)	(d)

FIGURE 4.7

In Figure 4.7 we show the function $f(x) = x$ on intervals from $x = 0$ to $x = 1$, including both endpoints in (a), one endpoint in (b) and (c), and neither endpoint in (d). Notice that only in (a) where f is defined on the closed interval [0, 1] does f attain both an absolute maximum and an absolute minimum value. In (b) and (d) the function comes arbitrarily close to 0 but does not get there, so there is no local or absolute minimum. Similarly, in (c) and (d) there is no local or absolute maximum. In Figure 4.8 both f and g are defined on the closed interval [0, 2] but f fails to have an absolute maximum and g fails to have an absolute minimum. Each of the functions in Figures 4.7 and 4.8 is bounded, that is, their graphs are contained between two horizontal lines. In Figure 4.9 the function h is unbounded both above and below and hence has no absolute maximum or minimum, although it is continuous on the domain $(-3, 3)$. There is a local minimum at P_1 and a local maximum at P_2.

The following theorem gives conditions that guarantee that a function will have an absolute maximum and minimum. The theorem is thus of paramount importance in the study of extreme values. Its proof requires concepts we have not studied, but it can be found in texts on advanced calculus.

$$f(x) = \begin{cases} x \text{ if } 0 \leq x < 1 \\ 0 \text{ if } 1 \leq x \leq 2 \end{cases};$$
discontinuous at $x = 1$;
no absolute max,
absolute min = 0
(a)

$$g(x) = \begin{cases} 1 - x \text{ if } 0 \leq x < 1 \\ \frac{1}{2} \text{ if } 1 \leq x \leq 2 \end{cases};$$
discontinuous at $x = 1$;
absolute max = 1,
no absolute min
(b)

FIGURE 4.8

$$h(x) = \frac{x(x^2 - 4)}{x^2 - 9} \text{ on } (-3, 3)$$

FIGURE 4.9

THEOREM 4.2	**Extreme-Value Theorem**
	If f is continuous on the closed interval I, then f attains both an absolute maximum value and an absolute minimum value on I.

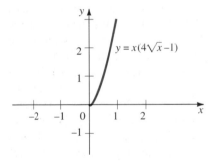

FIGURE 4.10

REMARK

■ Theorem 4.2 is an example of an *existence theorem*. When the conditions of the theorem are satisfied, we can conclude that the function does have an absolute maximum and an absolute minimum, but the theorem does not tell us what the extreme values are nor where they occur. The next theorem will provide a way for finding these values.

Re-examining the functions shown in Figures 4.7 through 4.9 reveals in what ways one of the conditions of this theorem fails to hold true in each case. The function f in Figure 4.7 is continuous on an interval, but the interval is not closed in parts (b), (c), and (d). The functions f and g in Figure 4.8 are defined on a closed interval but are not continuous. The function h in Figure 4.9 is continuous, but the interval is open.

Suppose now that we are given a function and we want to find its local maximum and minimum points, if any exist. One way of finding them is to obtain an accurate graph by using a graphing calculator or a CAS, then zooming in on the points that appear to be local maximum or minimum points and finding their coordinates to any desired degree of accuracy. For many functions this graphical approach is quite satisfactory. There are some possible pitfalls, however. Depending on the scale you use (determining the x range and y range), you may or may not correctly identify all the local extrema. Consider, for example, the graph of $f(x) = x(4\sqrt{x} - 1)$ shown in Figure 4.10. This graph was generated with a CAS using a normal scale in which each tick mark is 1 unit on each axis. Unless you look very carefully, you would probably conclude that the function reaches its lowest point at the origin. This conclusion is not correct, however, as the graph in Figure 4.11 shows. Here we used a scale of 0.1

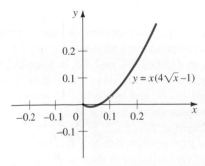

FIGURE 4.11

for x and 0.1 for y. Clearly, there is a local (also absolute) minimum point for some value of x between 0 and 0.1. In fact, by methods we will soon discuss, we can find the exact minimum point. Its coordinates are $(1/36, -1/108)$.

The point of the preceding example is that care must be taken when drawing conclusions based solely on graphs generated by a computer or calculator. The analysis we provide here is one way to be certain of the nature of the function under consideration.

An important key to locating local extrema is given in the next theorem.

THEOREM 4.3

Let f be defined on an interval I. If f has a local maximum or minimum at an interior point a of I, and if $f'(a)$ exists, then $f'(a) = 0$.

Note carefully that the theorem does *not* say that if $f'(a) = 0$, then f has a local maximum or minimum at a. Figure 4.12 illustrates the various possibilities that can occur when $f'(a) = 0$. In each case $f'(0) = 0$, so the tangent line at $(0, 0)$ is horizontal. In parts (c) and (d), however, the point $(0, 0)$ is neither a local maximum nor a local minimum (it is called a *point of inflection*, as we will learn in the next section). The theorem also does not say anything about the nature of points where the derivative fails to exist. In Figure 4.13(a) (on the following page) we show the graph of $f(x) = |x|$, which we have previously shown to be nondifferentiable at $x = 0$. Still, the function does have a minimum value there. On the other hand, in Figure 4.13(b) we show the graph of $g(x) = x^{1/3}$. It also fails to have a derivative at $x = 0$ (the slope is "infinite") and there is no maximum or minimum at $x = 0$. Thus, when $f'(a)$ fails to exist, there may or may not be a local extreme point when $x = a$. Finally, if we omit the hypothesis that $x = a$ be an interior point of an interval, the function can attain a maximum or minimum value even though the derivative is not zero. The function $f(x) = x$ on the closed interval $[0, 1]$ in Figure 4.14 clearly has an endpoint (and absolute) maximum at $x = 1$ and an endpoint (and absolute) minimum at $x = 0$, yet at both points the (one-sided) derivative equals 1. So the conclusion of the theorem may not be true if $x = a$ is an endpoint of the interval.

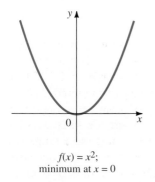

$f(x) = x^2$;
minimum at $x = 0$

(a)

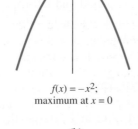

$f(x) = -x^2$;
maximum at $x = 0$

(b)

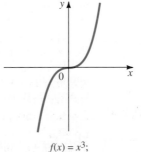

$f(x) = x^3$;
neither maximum
nor minimum at $x = 0$

(c)

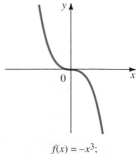

$f(x) = -x^3$;
neither maximum
nor minimum at $x = 0$

(d)

FIGURE 4.12

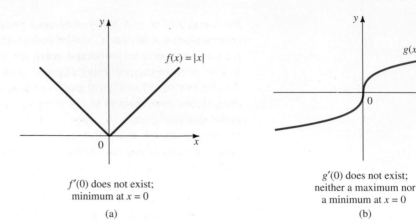

$f'(0)$ does not exist;
minimum at $x = 0$

(a)

$g'(0)$ does not exist;
neither a maximum nor
a minimum at $x = 0$

(b)

FIGURE 4.13

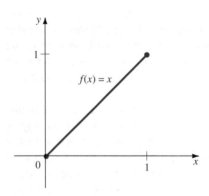

FIGURE 4.14
f has a maximum at $x = 1$ and a minimum at $x = 0$, but $f'(0) = 1$ and $f'(1) = 1$.

Proof of Theorem 4.3   We prove the theorem for a local maximum only. The proof for a local minimum is similar. We suppose that the derivative at a exists. Using the form for $f'(a)$ from Equation 3.4, we have

$$f'(a) = \lim_{x \to a} \frac{f(x) - f(a)}{x - a}$$

So the limit from the right and the limit from the left both exist and are the same. We consider these one-sided limits separately. First, consider $x > a$. By hypothesis, the point $(a, f(a))$ is a local maximum point, so if x is near a, $f(a) \geq f(x)$. Thus, $f(x) - f(a) \leq 0$. Since we are taking $x > a$, it follows that $x - a > 0$, and

$$\frac{f(x) - f(a)}{x - a} \leq 0 \qquad \text{Numerator negative, denominator positive}$$

So

$$\lim_{x \to a^+} \frac{f(x) - f(a)}{x - a} \leq 0 \qquad \text{See Exercise 69 in Exercise Set 4.1.}$$

Now consider $x < a$. Then $x - a < 0$, and

$$\frac{f(x) - f(a)}{x - a} \geq 0 \qquad \text{Numerator negative, denominator negative}$$

So

$$\lim_{x \to a^-} \frac{f(x) - f(a)}{x - a} \geq 0$$

Since $f'(a)$ exists, the right-hand and left-hand limits are equal. But the only number that is both greater than or equal to zero and less than or equal to zero is zero itself. Thus, $f'(a) = 0$.   ∎

The First-Derivative Test

As we saw in Figures 4.12(c) and 4.12(d), the fact that $f'(a) = 0$ does not necessarily imply that f has a maximum or minimum value at $x = a$. We can, however, treat such a point as a *candidate* for an x-coordinate at which f has an extreme value. The theorem does tell us that if a is an interior point of the domain of f and $f'(a) \neq 0$, then $f(a)$ is definitely not a local maximum or minimum value. So as a first step in locating points at which f has local extrema, we set $f'(x) = 0$ and solve for x. Then we test each such value to determine the nature of f at that point. Another possibility, as we saw in Figure 4.13(a), is that f has a local extreme value at a point where $f'(x)$ fails to exist. So a second candidate to test is any point at which $f'(x)$ fails to exist. Finally, as we saw in Figure 4.14, when the domain of f includes one or both endpoints of an interval, we must also test these points. In summary, we conclude that maximum and minimum values for a function defined on an interval can occur only at the following types of points in the interval:

1. An interior point where $f'(x) = 0$
2. An interior point where $f'(x)$ does not exist
3. An endpoint of the interval

Types 1 and 2 are called **critical points** of f. Critical points are not themselves maximum or minimum points but are x-coordinates of points on the graph that *may* be maximum or minimum points. So if a is a critical point, we substitute into the given function, f, to get $f(a)$, which is then a candidate for a maximum or minimum value. The endpoint values, if they are in the domain, must also be checked.

We saw in Figure 4.12 the four possibilities that can exist at critical points of type 1. A critical point of type 2, for which f' does not exist, also may or may not be the x-coordinate of a maximum or minimum point. We saw such a situation in Figure 4.13. Some of the many other possibilities when f' does not exist are illustrated in Figure 4.15. In parts (a) and (c) the tangent line is vertical and so the slope is undefined. In (b) the slopes from the left and from the right are unequal. In (d), (e), (f), and (g) the functions are discontinuous and so the derivatives cannot exist. The function in part (h) is given by $f(x) = x \sin \frac{1}{x}$ for $x \neq 0$, with $f(0)$ defined as 0. You will be asked in the exercises to show that $f'(0)$ does not exist in this latter case.

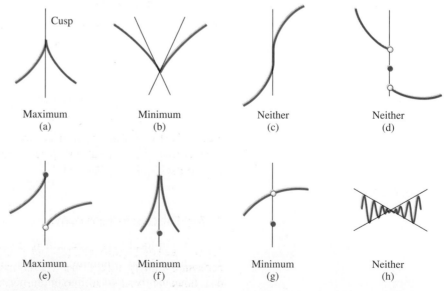

Cusp

| Maximum | Minimum | Neither | Neither |
| (a) | (b) | (c) | (d) |

| Maximum | Minimum | Minimum | Neither |
| (e) | (f) | (g) | (h) |

FIGURE 4.15

After finding the critical points, we must test in some way to determine the nature of the corresponding points on the graph. For a continuous function f there are two ways of doing this, one using the first derivative and one using the second derivative. We will discuss the latter in the next section. When f is discontinuous at a critical value, the possibilities are numerous, as Figures 4.15(d)–(g) show. The only feasible test is to apply Definition 4.2 directly, comparing $f(a)$ with $f(x)$ for x near a.

The basis for the First-Derivative Test is that for a continuous function, a maximum occurs when the function is increasing to the left of the critical value and decreasing to the right. The reverse is true at a minimum. Thus, by Theorem 4.1 we have the following test.

THEOREM 4.4

First-Derivative Test for Local Maxima and Minima

Let $x = a$ be a critical point of the function f, and suppose that f is continuous at $x = a$. Suppose further that $f'(x)$ exists for all x in some open interval I containing a, except possibly at $x = a$.

1. If for x in I, $f'(x) > 0$ when $x < a$ and $f'(x) < 0$ when $x > a$, then f has a local maximum at a.
2. If for x in I, $f'(x) < 0$ when $x < a$ and $f'(x) > 0$ when $x > a$, then f has a local minimum at a.
3. If for x in I, $f'(x)$ is of the same sign for $x < a$ as for $x > a$, then f has neither a local maximum nor a local minimum at a.

Figure 4.16 illustrates the situation when $x = a$ is a critical point for which $f'(a) = 0$ (type 1).

Maximum
(a)

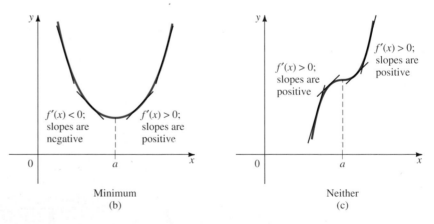

Minimum
(b)

Neither
(c)

FIGURE 4.16

We can state the First-Derivative Test more briefly as follows:

1. f has a local maximum at a if the sign of f' goes from $+$ to $-$.
2. f has a local minimum at a if the sign of f' goes from $-$ to $+$.
3. f has neither a local maximum nor a local minimum if the sign of f' goes from $-$ to $-$ or from $+$ to $+$.

In applying this briefer version, we must understand that f is continuous at a and the signs indicated occur as we move from left to right through a.

We illustrate this test in the next four examples.

EXAMPLE 4.2   Find all local maxima and minima for the function f for which

$$f(x) = 2x^3 - 3x^2 - 12x + 4 \qquad (x \in \mathbb{R})$$

Solution   To find the critical values we first calculate $f'(x)$:

$$f'(x) = 6x^2 - 6x - 12 = 6(x+1)(x-2)$$

FIGURE 4.17
Sign graph for $f'(x) = 6(x+1)(x-2)$

Since $f'(x) = 0$ when $x = -1$ or $x = 2$, and $f'(x)$ exists everywhere, it follows that -1 and 2 are the only critical points. To test whether these are extrema, we use the First-Derivative Test. A useful device for doing this test is the sign graph for f' in Figure 4.17. Since f' changes from $+$ to $-$ as we pass through -1 and from $-$ to $+$ as we pass through 2, we conclude that f has a local maximum at -1 and a local minimum at 2. To determine these maximum and minimum values, we substitute into the original function:

$$f(-1) = 11 \text{ is a local maximum value.}$$

$$f(2) = -16 \text{ is a local minimum value.}$$

Alternatively, we can say $(-1, 11)$ is a local maximum point on the graph of f and $(2, -16)$ is a local minimum point on the graph.

If we take $|x|$ large enough, $f(x) \approx 2x^3$, so $f(x)$ can be made arbitrarily large for large positive values of x. By taking negative x values large in absolute value, $f(x)$ can be made arbitrarily large in absolute value and negative. So since the domain of f is unlimited, it follows that there is no absolute maximum or minimum value of f. We showed the graph of f in Example 4.1 (Figure 4.4). ∎

EXAMPLE 4.3   Find all local maxima and minima for the function

$$f(x) = \frac{x+1}{x^2 - 3x} \qquad (x \neq 0, 3)$$

Solution   The domain of f consists of the union of the three intervals $(-\infty, 0)$, $(0, 3)$, and $(3, \infty)$. So we can apply our theory to each of these intervals. First we find the critical points:

$$f'(x) = \frac{1(x^2 - 3x) - (x+1)(2x-3)}{(x^2 - 3x)^2} = -\frac{x^2 + 2x - 3}{(x^2 - 3x)^2}$$

$$= -\frac{(x+3)(x-1)}{(x^2 - 3x)^2} = -\frac{(x+3)(x-1)}{[x(x-3)]^2}$$

FIGURE 4.18
Sign graph for $f'(x) = -\dfrac{(x+3)(x-1)}{[x(x-3)]^2}$

Since $f'(x) = 0$ when $x = -3$ or when $x = 1$, these are critical points. There are no values of x in the domain of f for which $f'(x)$ is undefined, so there are no other critical points. (We have already omitted from the domain of f the points $x = 0$ and $x = 3$, where the denominator is zero.)

To test $x = -3$ and $x = 1$, we look at a sign graph for f' in Figure 4.18. Notice that we show the points 0 and 3 on the line, since they are endpoints of intervals on which f is defined.

$$f(x) = \frac{x+1}{x^2 - 3x}$$

$(1, -1)$

$(-3, -\frac{1}{9})$

FIGURE 4.19

Since f' changes from $-$ to $+$ as we pass through -3 and from $+$ to $-$ as we pass through 1, we conclude that f has a local minimum at -3 and a local maximum at 1. The corresponding points on the graph are

$$\left(-3, -\frac{1}{9}\right), \quad \text{local minimum point}$$

$$(1, -1), \quad \text{local maximum point}$$

The graph of f is shown in Figure 4.19.   ■

EXAMPLE 4.4   Find all local and absolute extrema of the function f defined on the interval $[0, 2\pi]$ by $f(x) = \sin 2x + 2\sin x$.

Solution $f'(x) = 2\cos 2x + 2\cos x$

To write $f'(x)$ in an equivalent form in which each trigonometric function is evaluated at x only, rather than at both $2x$ and x, we can use the identity $\cos 2x = 2\cos^2 x - 1$ (see Supplementary Appendix 2) to get

$$f'(x) = 2(2\cos^2 x - 1) + 2\cos x$$
$$= 2(2\cos^2 x + \cos x - 1)$$
$$= 2(\cos x + 1)(2\cos x - 1)$$

FIGURE 4.20
Sign graph for
$f'(x) = 2(\cos x + 1)(2\cos x - 1)$

The critical points are those in the closed interval $[0, 2\pi]$ for which $\cos x = -1$ or $\cos x = \frac{1}{2}$, namely, $x = \pi$, $\pi/3$, and $5\pi/3$. The sign graph for f' (Figure 4.20) shows that f has a local maximum at $\pi/3$ and a local minimum at $5\pi/3$. At $x = \pi$ there is neither a maximum nor a minimum. Substituting into the formula for $f(x)$, we find

$$f\left(\frac{\pi}{3}\right) = \frac{3\sqrt{3}}{2} \text{ is a local maximum value.}$$

$$f\left(\frac{5\pi}{3}\right) = -\frac{3\sqrt{3}}{2} \text{ is a local minimum value.}$$

The endpoint values are $f(0) = 0$ and $f(2\pi) = 0$. Thus, the absolute maximum value of f is $3\sqrt{3}/2$ and the absolute minimum value is $-3\sqrt{3}/2$. We show the graph of f in Figure 4.21.   ■

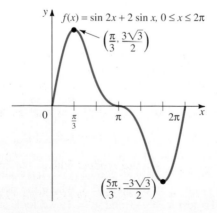

$f(x) = \sin 2x + 2\sin x,\ 0 \le x \le 2\pi$

$\left(\frac{\pi}{3}, \frac{3\sqrt{3}}{2}\right)$

$\left(\frac{5\pi}{3}, \frac{-3\sqrt{3}}{2}\right)$

FIGURE 4.21

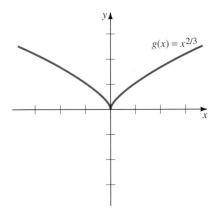

FIGURE 4.22

FIGURE 4.23

EXAMPLE 4.5   **Critical points of type 2.** Find and test the critical points of the following functions:

(a) $f(x) = x^{1/3}$

(b) $g(x) = x^{2/3}$

Solution

(a) $f'(x) = \dfrac{1}{3}x^{-2/3} = \dfrac{1}{3x^{2/3}}$

Since $f'(0)$ does not exist, $x = 0$ is a critical point of type 2. There are no other critical points. Since $x^{2/3}$ is positive for all $x \neq 0$, $f'(x)$ does not change sign at 0. So f does not have a local maximum nor a local minimum there. The graph of f is shown in Figure 4.22.

(b) $g'(x) = \dfrac{2}{3}x^{-1/3} = \dfrac{2}{3x^{1/3}}$

Again, $x = 0$ is a critical point of type 2. If $x < 0$, $g'(x) < 0$ and if $x > 0$, $g'(x) > 0$. By the First-Derivative Test, we conclude that g has a local minimum at $x = 0$. Since $g(0) = 0$, and for $x \neq 0$, $g(x) > 0$, we see that $(0, 0)$ is also the absolute minimum point on the graph. The graph is shown in Figure 4.23. ∎

If a function f is continuous on a closed interval $[a, b]$, we know by Theorem 4.2 that f has an absolute maximum and an absolute minimum on $[a, b]$. The absolute maximum is the largest of the local and endpoint maxima, and similarly for the absolute minimum. Thus, if the only objective is to find the absolute extrema, the following procedure is sufficient:

Procedure for Finding Absolute Extrema

If f is continuous on $[a, b]$:

1. Find all critical points.

2. Evaluate f at each critical point and at each endpoint.

3. Select the largest and smallest values in step 2.

Note that it is not necessary in this case to test the critical points to see if they give local maxima or minima.

We illustrate this procedure in the next example.

EXAMPLE 4.6   Find the absolute extrema of f, where

$$f(x) = (2x - 5)(x - 5)^{2/3} \quad \text{on } [3, 6]$$

Solution   By the Product Rule,

$$f'(x) = 2(x - 5)^{2/3} + (2x - 5) \cdot \frac{2}{3}(x - 5)^{-1/3}$$

$$= \frac{2}{3}(x - 5)^{-1/3}[(3(x - 5) + (2x - 5)]$$

$$= \frac{2}{3}(x-5)^{-1/3}(3x - 15 + 2x - 5)$$

$$= \frac{10(x-4)}{3(x-5)^{1/3}}$$

The critical points are $x = 4$ and $x = 5$.

$$f(4) = (3)(-1)^{2/3} = 3$$

$$f(5) = (5)(0) = 0$$

$$f(3) = (1)(-2)^{2/3} = \sqrt[3]{4} \quad \text{Left endpoint}$$

$$f(6) = (7)(1)^{2/3} = 7 \quad \text{Right endpoint}$$

Conclusions:

$$\text{Absolute maximum value} = 7 \text{ when } x = 6.$$

$$\text{Absolute minimum value } = 0 \text{ when } x = 5. \quad \blacksquare$$

EXAMPLE 4.7   Find the absolute extreme values of $f(x) = x^3 - 3x + 2$ on the interval $[0, 2]$.

Solution $f'(x) = 3x^2 - 3 = 3(x^2 - 1) = 3(x + 1)(x - 1)$
The critical points are $x = 1$ and $x = -1$, but $x = -1$ is not in the interval $[0, 2]$, so we ignore it. We calculate $f(x)$ at $x = 1$ and at the endpoints $x = 0$ and $x = 2$:

$$f(1) = 0$$

$$f(0) = 2$$

$$f(2) = 4$$

Thus, the absolute minimum value is 0, attained when $x = 1$. The absolute maximum value is 4, attained when $x = 2$. We show the graph of f in Figure 4.24. $\quad \blacksquare$

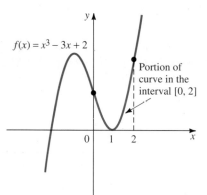

$f(x) = x^3 - 3x + 2$

Portion of curve in the interval $[0, 2]$

FIGURE 4.24

Exercise Set 4.1

In Exercises 1–12, find the intervals on which the function f is increasing and those on which it is decreasing.

1. $f(x) - 3 - 2x - x^2$

2. $f(x) - 2x^2 - x - 3$

3. $f(x) = x^3 - x^2 - 8x$

4. $f(x) = 2x^3 + 3x^2 - 36x + 5$

5. $f(x) = x^3 - 2x^2 - 4x + 7$

6. $f(x) = 12 + 40x + 7x^2 - 4x^3$

7. $f(x) = 3x^4 - 4x^3$

8. $f(x) = x^4 - 2x^3 - 5x^2$

9. $f(x) = \dfrac{1}{(x-2)^2}$

10. $f(x) = \dfrac{x}{x+3}$

11. $f(x) = \dfrac{1-x}{3+x^2}$

12. $f(x) = \dfrac{x+3}{(x-1)^2}$

In Exercises 13–24, find all local maximum and minimum points.

13. $f(x) = x^2 - 8x + 7$

14. $f(x) = 3 - x - 2x^2$

15. $f(x) = x^3 - 3x + 6$

16. $f(x) = 2x^3 + 3x^2 - 12x$

17. $f(x) = x^3 - 3x^2 - 24x + 4$

18. $f(x) = x^3 + 5x^2 - 8x + 3$

19. $f(x) = x^4 - 8x^2 + 12$

20. $f(x) = x^4 - 4x^3 - 8x^2 + 12$

21. $f(x) = x^4 - 4x^3 + 8$

22. $f(x) = 2x^5 + 5x^4 + 10$

23. $f(x) = 1 + 2(x - 3)^{2/3}$

24. $f(x) = (x^2 - 1)^{2/3}$

In Exercises 25–36, find local as well as absolute maximum and minimum values on the given interval.

25. $f(x) = x^2 - 4x + 3$ on $[0, 5]$

26. $f(x) = 5 - 3x - 2x^2$ on $[-6, 0]$

27. $f(x) = x^3 - 2x^2 - 4x + 8$ on $[-2, 3]$

28. $f(x) = 12x - x^3 - 9$ on $[-1, 3]$

29. $f(x) = x^4 - 2x^3 - 9x^2 + 27$ on $[-2, 4]$

30. $f(x) = 3x^4 - 16x^3 + 24x^2 - 21$ on $[-1, 3]$

31. $f(x) = \sin x + \cos x$ on $[0, \pi]$

32. $f(x) = \sin 2x - x$ on $[-\pi, \pi]$

33. $f(x) = \tan x - 2x + 1$ on $[-\frac{\pi}{3}, \frac{\pi}{3}]$

34. $f(x) = 2x + \sin 2x - 4\sin x$ on $[-\pi, \pi]$

35. $f(x) = \sin 2x + 4\cos x + x - 1$ on $[0, 2\pi]$

36. $f(x) = x\sin x + \cos x - \pi\sin x$ on $[0, 2\pi]$

Exercises 37–46 refer to earlier problems in this exercise set. In each case find the absolute maximum and absolute minimum values of f on the indicated interval. Make use of previously obtained results.

37. Exercise 15, on $[-3, 3]$

38. Exercise 16, on $[-3, 2]$

39. Exercise 17, on $[-3, 6]$

40. Exercise 18, on $[-6, 1]$

41. Exercise 19, on $[-3, 3]$

42. Exercise 20, on $[-2, 3]$

43. Exercise 21, on $[-4, 4]$

44. Exercise 22, on $[-3, 1]$

45. Exercise 23, on $[-5, 4]$

46. Exercise 24, on $[-3, 3]$

47. Show that $f(x) = x^7$ has a critical point at $x = 0$ but that f does not have a local extremum there. For which positive integers n does $f(x) = x^n$ satisfy this same property? If, in addition, a and b are real numbers, what can be said about $f(x) = (x - a)^n + b$ at the point $x = a$?

48. Find all critical points of the greatest integer function $f(x) = [\![x]\!]$.

49. If f has a maximum value at $x = a$, show that $g(x) = -f(x)$ has a minimum value at $x = a$.

50. The graph of the derivative, $y = f'(x)$, of a certain function f is shown in the figure. How many solutions can the equation $f(x) = 0$ have? Explain.

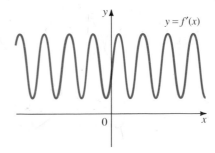

51. Repeat Exercise 50 for the graph of $y = f'(x)$ shown here.

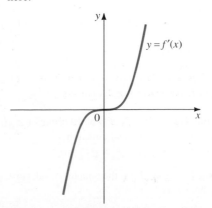

52. The accompanying figure shows the graphs of two functions f and g. If $x = c$ is the point where the vertical distance between the curves is greatest for x in the interval $[a, b]$, what can you say about the tangent lines to the curves $y = f(x)$ and $y = g(x)$ at $x = c$? Prove your answer. Assume that $f'(x)$ and $g'(x)$ exist for all x. (*Hint:* Consider $h(x) = f(x) - g(x)$.)

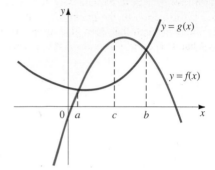

53. Let f be defined on $(0, 1)$, and suppose that $f'(x)$ exists for all x in $(0, 1)$. Define a function g on $(0, 1)$ by $g(x) = f(x^3)$. If f is increasing on $(0, 1)$, what can be said about g? Explain.

54. The graph of the derivative, $y = f'(x)$, of a function f is shown. From it, determine where f is increasing and where it is decreasing. Find the x-coordinates of all maximum and minimum points (indicate which) on the graph of f. If $f(0) = 0$, sketch a possible graph of f.

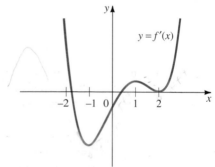

55. Sketch the graph of a function that satisfies all of the following conditions:
(a) $f'(-2) = f'(4) = 0$
(b) $f'(x) > 0$ for $x < -2$ or $x > 4$
(c) $f'(x) < 0$ for $-2 < x < 4$
(d) $\lim_{x \to \infty} f(x) = 5$ and $\lim_{x \to -\infty} f(x) = -\infty$

56. **Cubic Polynomials.** The problem of describing the graph of the general cubic polynomial function

$$f(x) = ax^3 + bx^2 + cx + d$$

can be reduced to describing the graphs of all curves $y = g(x)$, where g is given by

$$g(x) = x^3 + kx \qquad k \in \mathbb{R}$$

(See Exercise 27 in Exercise Set 1.4.)
(a) Show that $g(x) = x^3 + kx$ can have no critical points, one critical point, or two critical points. Sketch graphs to illustrate each possibility.
(b) How many local extreme values can the general cubic function f have?
(c) Determine where the function g is increasing and where it is decreasing for each of the cases $k \geq 0$ and $k < 0$.
(d) How do the slopes of the tangent lines to the curves $y = x^3$ and $y = x^3 + kx$, for $k > 0$, differ at $x = 0$?

In Exercises 57–68, find the local and absolute extrema and state where they occur.

57. $f(x) = x\sqrt{x^2 - 1}$ on $[1, \infty)$

58. $f(x) = (x - 1)^{2/3}(6 - x)$ on $[0, 6]$

59. $f(x) = \dfrac{x}{(x - 1)^2}$ on $(-\infty, \frac{1}{2}]$. Also, discuss absolute extrema on $(-\infty, 1)$ and on $(1, \infty)$.

60. $f(x) = x\sqrt[3]{x - 4}$ on $[0, 5]$

61. $f(x) = \dfrac{(x - 4)^{2/3}}{x}$ on $[3, 12]$

62. $f(x) = \sin^2 x - 2\cos x + \sin x + x$ on $\left[\dfrac{\pi}{2}, \dfrac{3\pi}{2}\right]$

63. $f(x) = x^4 - 4x^3 - 18x^2 + 108x + 120$ on $[-5, 4]$

64. $f(x) = 2x^4 + 4x^3 - 9x^2 - 27x + 5$ on $[-2, 2]$

65. $f(x) = x^4 - 4x^3 + 16x - 8$ on $[-2, 3]$

66. $f(x) = x^4 - 2x^3 - 23x^2 + 24x + 64$ on $[-4, 5]$

67. $f(x) = 2.13x^3 + 3.02x^2 - 11.9x + 32.4$ on $[-3, 2]$

68. $f(x) = x^3 - 3.256x^2 + 1.432x + 5.875$ on $[-1.256, 2.435]$

69. Suppose $f(x) \geq 0$ for all x in some open interval containing a, except possibly at $x = a$ itself, and suppose $\lim_{x \to a} f(x) = L$ exists. Prove that $L \geq 0$. Show by an example that L may be 0 even if $f(x) > 0$ in the interval. (*Hint:* For the first part, assume $L < 0$ and use Definition 2.5 to arrive at a contradiction.)

70. Complete the proof of Theorem 4.3 by showing that if f' exists at an interior point a at which f attains a local minimum, then $f'(a) = 0$.

71. Use the definition of the derivative to show that $f'(0)$ does not exist for the function defined by

$$f(x) = \begin{cases} x \sin \dfrac{1}{x} & \text{if } x \neq 0 \\ 0 & \text{if } x = 0 \end{cases}$$

72. Use the definition of the derivative to show that for the function defined by

$$f(x) = \begin{cases} x^2 \sin \dfrac{1}{x} & \text{if } x \neq 0 \\ 0 & \text{if } x = 0 \end{cases}$$

$f'(0)$ exists, and find its value.

73. A function is **bounded** on its domain D if there exists a positive real number M such that $|f(x)| \leq M$ for all $x \in D$. Give an example of a function, either by a graph or by a formula, that is bounded on a closed interval I but does not attain a maximum value on I.

74. Is it possible for a continuous function on a closed interval to be unbounded? Explain. (See Exercise 73.)

75. Suppose both f and g are defined on an interval I and each has an absolute maximum there. Let $\max_{(I)} f$ and $\max_{(I)} g$ denote these maximum values. State everything you can about $\max_{(I)}(f + g)$ and $\max_{(I)}(fg)$.

76. The accompanying figure shows a portion of the graph of

$$f(x) = \frac{1}{4}x^4 - 10x^3 + 100x^2 + 100$$

Use information from $f'(x)$ to explain why this is or is not an accurate depiction of the graph of f. If it is not accurate, obtain a graph that does give an accurate description. What scales did you use on the x- and y-axes?

$y = \frac{1}{4}x^4 - 10x^3 + 100x^2 + 100$

4.2 CONCAVITY

FIGURE 4.25

In Chapter 1 we used the terms *concave up* and *concave down* to describe curves that bend upward or downward, respectively. In this section we give a more precise definition of concavity and show how it is related to the second derivative of a function.

In Figure 4.25(a) we show a portion of the graph of a function that is concave up and in part (b) a portion of a graph that is concave down. Observe the slopes of the tangent lines as we move from left to right in each case. In the case of upward concavity these slopes increase, and in the case of downward concavity the slopes decrease. Since the slopes of the tangent lines are given by the derivative of the function, we can say that upward concavity corresponds to an increasing derivative and downward concavity to a decreasing derivative. In fact, we take this characterization as the definition of upward and downward concavity.

Definition 4.4 Concavity	Let f be differentiable on an open interval I. Then the graph of f is said to be **concave up** on I if f' is an increasing function on I, and the graph of f is said to be **concave down** on I if f' is a decreasing function on I.

Consider again the graphs in Figure 4.25. Notice that in part (a) the graph is always above the tangent line at any point, whereas in part (b) the graph is below the tangent line. These relationships hold true in general, although we will not give a proof.

Note also that concavity is independent of whether the function itself is increasing, decreasing, or neither. In Figure 4.26 each of the graphs is concave up, for example.

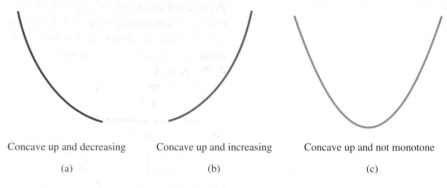

Concave up and decreasing   Concave up and increasing   Concave up and not monotone

(a)   (b)   (c)

FIGURE 4.26

There is a simple test for concavity when $f''(x)$ exists. According to Theorem 3.1 (with f' replacing f), f' is an increasing function if $f''(x) > 0$ and a decreasing function if $f''(x) < 0$. So we have the following theorem.

THEOREM 4.5

Suppose $f''(x)$ exists for all x on an interval I.

1. If $f''(x) > 0$ for all x in I, then the graph of f is concave up on I.
2. If $f''(x) < 0$ for all x in I, then the graph of f is concave down on I.

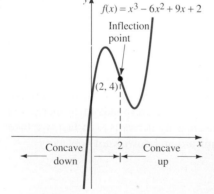

FIGURE 4.27

EXAMPLE 4.8   Determine where the graph of the function f defined by $f(x) = x^3 - 6x^2 + 9x + 2$ is concave up and where it is concave down.

Solution   We need to determine where $f''(x) > 0$ and where $f''(x) < 0$. So we first calculate $f''(x)$:

$$f'(x) = 3x^2 - 12x + 9$$
$$f''(x) = 6x - 12 = 6(x - 2)$$

Thus, $f''(x) > 0$ for all $x > 2$ and $f''(x) < 0$ for all $x < 2$. By Theorem 4.5 we conclude that the graph of f is concave up on $(2, \infty)$ and concave down on $(-\infty, 2)$. The graph in Figure 4.27 shows these results.   ∎

Knowing where the graph of a function is concave up and where it is concave down is very useful in drawing the graph. Consider the graph of $f(x) = x^3$ in Figure 4.28, which we have encountered earlier. Up to now, we have simply

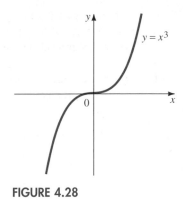

FIGURE 4.28

assumed that the curve bends upward to the right of $x = 0$ and downward to the left. Knowing only that it increases everywhere is not enough. From the second derivative $f''(x) = 6x$, though, we see that the graph is concave up for all $x > 0$ and concave down for all $x < 0$, confirming the shape shown in Figure 4.28.

Inflection Points

A point on the graph of a continuous function at which the concavity changes is called an **inflection point**. In Example 4.8 the point $(2, 4)$ is such a point. To the left of $x = 2$, the graph is concave down, and to the right it is concave up. So the point for which $x = 2$ is an inflection point. Since $f(2) = 2^3 - 6(2^2) + 9(2) + 2 = 4$, the point is $(2, 4)$.

If the graph of a twice-differentiable function has an inflection point at $x = a$, we know that the sign of $f''(x)$ for $x < a$ is opposite to that for $x > a$, since the concavity changes at $x = a$. There is an Intermediate-Value Theorem for Derivatives (see Exercise 34 of Section 4.7) that guarantees $f''(a) = 0$ in this case. So when $f''(x)$ exists for all x, we can find all possible x-coordinates of points of inflection simply by setting $f''(x) = 0$ and solving for x. After finding these values, we must test to see if the concavity changes.

Figure 4.29 illustrates a graph with several inflection points. Observe the change in concavity at each of the points P_1 through P_5; each of these is an inflection point. The points P_1 and P_4 are the most common types. At each of these both f' and f'' exist, with $f' \neq 0$ and $f'' = 0$. The points P_2, P_3, and P_5 all occur at critical points of f. At P_2 and P_5, f' is not defined, and at P_3, $f' = 0$. The tangent line is vertical at P_2 (and so its slope is not defined), and no tangent line exists at P_5. At P_3 both f' and f'' equal 0. This type of inflection point always has an appearance similar to one of those in Figure 4.30.

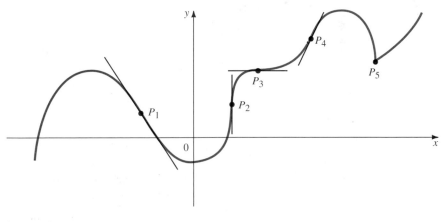

FIGURE 4.29

EXAMPLE 4.9

(a) Show that the graph of $y = x^4 - 4x^3 + 12x + 3$ has two inflection points where $y'' = 0$.

(b) Show that the graph of $y = (x - 2)^{5/3}$ has an inflection point but $y'' \neq 0$.

(c) Show that the graph of $y = x^4 - 2x + 3$ has no inflection points even though y'' can be 0.

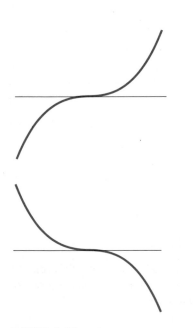

FIGURE 4.30
Inflection points at which both $f' = 0$ and $f'' = 0$

FIGURE 4.31
Sign graph for $y'' = 12x(x - 2)$

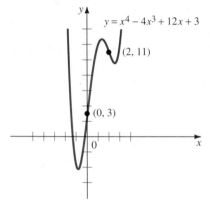

FIGURE 4.32
Inflection points $(0, 2)$ and $(2, 11)$

Solution

(a) Since y'' exists everywhere in this case, all inflection points can be found by setting $y'' = 0$. So we first calculate y''.

$$y' = 4x^3 - 12x^2 + 12$$

$$y'' = 12x^2 - 24x = 12x(x - 2)$$

Setting $y'' = 0$ gives $x = 0$ or $x = 2$. These two values are x-coordinates of *possible* points of inflection. We must test to see if they correspond to actual points of inflection. As the sign graph for y'' in Figure 4.31 shows, y'' changes sign at $x = 0$ and at $x = 2$. Thus, there is a change in concavity both at $x = 0$ and at $x = 2$. We calculate the y values in each case from the original equation by substituting the x value and solving for y. We find that the points of inflection are $(0, 3)$ and $(2, 11)$. We show the graph in Figure 4.32.

(b) First, we compute y''.

$$y' = \frac{5}{3}(x - 2)^{2/3}$$

$$y'' = \frac{10}{9}(x - 2)^{-1/3}$$

In this case y'' is never zero. It is undefined at $x = 2$, but the original function $y = (x - 2)^{5/3}$ is defined there and equals 0. Since $y'' > 0$ for $x > 2$ and $y'' < 0$ for $x < 2$, we conclude that $(2, 0)$ is a point of inflection. This example shows that a point of inflection *may* occur at $x = a$ even if $f''(a) \neq 0$, in particular, where $f''(a)$ is not defined. We show the graph in Figure 4.33.

(c) Again, we find y''.

$$y' = 4x^3 - 2$$

$$y'' = 12x^2$$

Thus, $y'' = 0$ when $x = 0$. But for $x > 0$, $y'' > 0$ and for $x < 0$, $y'' > 0$. So there is no change in concavity, and hence no inflection point, at $x = 0$. This example shows that having $f''(a) = 0$ does not guarantee that an inflection point occurs at $x = a$. We show the graph in Figure 4.34. ∎

FIGURE 4.33
Inflection point $(2, 0)$

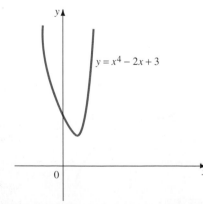

FIGURE 4.34
No inflection points

Be careful to note that *an inflection point must be on the graph of the function.* For example, consider $f(x) = 1/x$. Since $f'(x) = -1/x^2$, we see that $f''(x) = 2/x^3$. Thus, for $x > 0$, the graph is concave up, and for $x < 0$, the graph is concave down. Yet there is no inflection point at $x = 0$, because $f(0)$ is not defined. (See Figure 4.35.)

The Second-Derivative Test

Suppose that $x = a$ is a critical point for which $f'(a) = 0$ (type 1). In addition, suppose that $f''(a) > 0$. We can then show that the graph of f is concave up in some open interval containing a. So we know two things: since $f'(a) = 0$, the graph has a horizontal tangent line at the point $(a, f(a))$, and since $f''(a) > 0$, the graph is concave up in an interval containing a. Figure 4.36(a) illustrates this situation. Clearly, the point $(a, f(a))$ is a local minimum point on the graph. If $f''(a) < 0$, then the graph is concave down near a, and the situation is as pictured in Figure 4.36(b). The point $(a, f(a))$ is a local maximum point in this case. We therefore have an alternative to the First-Derivative Test for finding local extrema in the case of critical points of type 1, provided the second derivative exists at $(a, f(a))$.

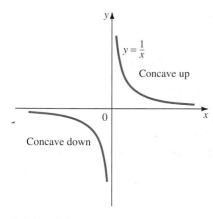

FIGURE 4.35
No inflection point since $x = 0$ is not in the domain

FIGURE 4.36

THEOREM 4.6

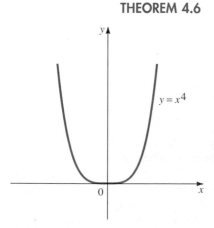

FIGURE 4.37

> **Second-Derivative Test for Maxima and Minima**
>
> Let $x = a$ be a critical point of f for which $f'(a) = 0$ and $f''(a)$ exists.
>
> **1.** If $f''(a) > 0$, the point $(a, f(a))$ is a local minimum point.
> **2.** If $f''(a) < 0$, the point $(a, f(a))$ is a local maximum point.

REMARK ——————————————————————————————————
■ If $f''(a) = 0$, this test is inconclusive. The point $(a, f(a))$ may be an inflection point of the type shown in Figure 4.30, or it may be a local maximum or minimum. For example, if $f(x) = x^3$, then $f''(0) = 0$, and as we have seen, the point $(0, 0)$ is an inflection point (see Figure 4.28). If, however, $f(x) = x^4$, again $f''(0) = 0$, but since $f'(x) = 4x^3$, the slope changes from $-$ to $+$ as we move through 0, so the point $(0, 0)$ is a local (and absolute) minimum point. (See Figure 4.37.)

EXAMPLE 4.10   Use the Second-Derivative Test to find local maxima and minima for the function given by

$$f(x) = x^3 + 3x^2 - 24x - 20$$

Solution   First, we find the critical points.

$$
\begin{aligned}
f'(x) &= 3x^2 + 6x - 24 \\
&= 3(x^2 + 2x - 8) \\
&= 3(x + 4)(x - 2)
\end{aligned}
$$

The critical points are $x = -4$ and $x = 2$.

Next, we find $f''(x)$ and substitute the critical points for x.

$$f''(x) = 6x + 6 = 6(x + 1)$$

$$f''(-4) = -18 < 0$$

$$f''(2) = 18 > 0$$

By Theorem 4.6 we conclude that f has a local maximum at $x = -4$ and a local minimum at $x = 2$.

Finally, we substitute -4 and 2, respectively, into the original function to find the points on the graph:

$$(-4, 60) \quad \text{is a local maximum point.}$$

$$(2, -48) \quad \text{is a local minimum point.} \qquad \blacksquare$$

Choice of Tests

Whether to use the First-Derivative Test or the Second-Derivative Test depends on several factors. First, we have no choice if a is a critical point for which either $f''(a) = 0$ or $f'(a)$ is not defined; the Second-Derivative Test will not work. When the function is a polynomial, the Second-Derivative Test often is simpler to apply. For other functions, it depends on the difficulty of calculating f''. If you need f'' anyway (for example, to determine concavity and inflection points), then use the Second-Derivative Test. If the only objective is to determine maxima and minima, and if it appears that the calculation of f'' would be quite laborious, then it is probably easier to use the First-Derivative Test.

A Procedure for Sketching Graphs

From the first two derivatives of a function, we now know how to determine where the function is increasing and where it is decreasing, the local maximum and minimum points on its graph, the intervals where it is concave up and where it is concave down, and the points of inflection. When we combine this information with the aids to graphing we studied in Chapter 1, such as intercepts, symmetry, and asymptotes, we have a wealth of information about the nature of the function and its graph. In the next three examples we illustrate how to use all this information to draw the graph, showing all of its important features. In most cases we will need to plot few, if any, additional points.

EXAMPLE 4.11   For the function f, given by

$$f(x) = 2x^3 - 3x^2 - 12x + 8 \qquad (x \in \mathbb{R})$$

find the intervals on which f is increasing and on which f is decreasing, the local maximum and minimum points, the intervals where the graph of f is concave up and where concave down, and the points of inflection. Use the information obtained, along with intercepts, symmetry, and asymptotes, to draw the graph of f.

Solution   Since $f(x)$ is a cubic polynomial, the graph of f has no asymptotes. To find the x-intercepts, we would need to solve a cubic equation. We will see a way of approximating the roots in Section 4.5, but for now we will not attempt to find them. As we get other information about the graph, the approximate location of the places where it crosses the x-axis will become evident. The y-intercept is 8, found by setting $x = 0$.

To locate maximum and minimum points, to find where f is increasing and where it is decreasing, and to see where the graph is concave up and where it is concave down, we calculate $f'(x)$ and $f''(x)$.

$$f(x) = 2x^3 - 3x^2 - 12x + 8$$
$$f'(x) = 6x^2 - 6x - 12 = 6(x^2 - x - 2) = 6(x - 2)(x + 1)$$
$$f''(x) = 12x - 6 = 6(2x - 1)$$

FIGURE 4.38
Sign graph for $f'(x) = 6(x - 2)(x + 1)$

The critical points are $x = 2$ and $x = -1$, where $f'(x) = 0$. From the sign graph for f' (Figure 4.38), we see that f is increasing on $(-\infty, -1)$ and on $(2, \infty)$, and f is decreasing on $(-1, 2)$.

Since $f''(-1) = -18$ and $f''(2) = 18$, we see by the Second-Derivative Test that f has a local maximum at -1 and a local minimum at 2. We calculate $f(-1) = 15$ and $f(2) = -12$. Thus, $(-1, 15)$ is a local maximum point and $(2, -12)$ is a local minimum point.

For concavity, we consider $f''(x)$. When $x > 1/2$, $f''(x) > 0$, and when $x < 1/2$, $f''(x) < 0$. So the graph is concave up on $(1/2, \infty)$ and concave down on $(-\infty, 1/2)$. Furthermore, the point $(1/2, f(1/2)) = (1/2, 3/2)$ is a point of inflection.

This information enables us to draw the graph of f with reasonable accuracy (see Figure 4.39 on the following page). Plotting the additional points $(3, -1)$ and $(-2, 4)$ increases the accuracy. ∎

EXAMPLE 4.12   Draw the graph of the function g given by $g(x) = x^4 - 4x^3 + 16$, making use of the information obtained from g' and g''.

Solution   Since $g(0) = 16$, the y-intercept is 16. Again, we will not attempt to find the x-intercepts (although you might observe that $g(2) = 0$). There are no asymptotes and no symmetry. We calculate $g'(x)$ and $g''(x)$.

$$g(x) = x^4 - 4x^3 + 16$$
$$g'(x) = 4x^3 - 12x^2 = 4x^2(x - 3)$$
$$g''(x) = 12x^2 - 24x = 12x(x - 2)$$

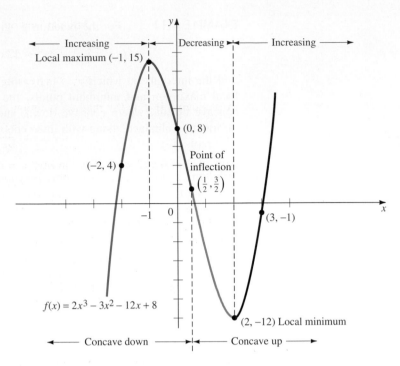

$f(x) = 2x^3 - 3x^2 - 12x + 8$

FIGURE 4.39

FIGURE 4.40
Sign graph for $g'(x) = 4x^2(x - 3)$

FIGURE 4.41
Sign graph for $g''(x) = 12x(x - 2)$

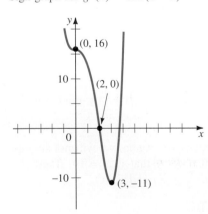

FIGURE 4.42

The critical points are $x = 0$ and $x = 3$. Since $g''(0) = 0$ and $g''(3) > 0$, we see that g has a local minimum at $x = 3$. We cannot test $x = 0$ by the Second-Derivative Test since $g''(0) = 0$, so we use the First-Derivative Test. The sign graph for g' (Figure 4.40) shows that the slope does not change at $x = 0$, so g does not have a local maximum or a local minimum there. It is, rather, decreasing both to the left and to the right of $x = 0$.

Next, look at the sign graph for g'' (Figure 4.41). We see that the concavity changes at $x = 0$ and $x = 2$, so we conclude that g has points of inflection at $x = 0$ and $x = 2$.

We calculate $g(x)$ at the points in question and summarize our findings:

g is increasing on $(3, \infty)$.

g is decreasing on $(-\infty, 0)$ and on $(0, 3)$.

g is concave up on $(-\infty, 0)$ and on $(2, \infty)$.

g is concave down on $(0, 2)$.

$(3, -11)$ is a local minimum point.

$(0, 16)$ is a point of inflection where the tangent line is horizontal.

$(2, 0)$ is a point of inflection.

The graph is shown in Figure 4.42. ∎

Asymptotes

Before we go to the next example, let us review briefly horizontal and vertical asymptotes. If any one of the following holds true, then the line $x = a$ is a vertical asymptote:

$$\lim_{x \to a} f(x) = \infty \text{ or } -\infty,$$

$$\lim_{x \to a^+} f(x) = \infty \text{ or } -\infty,$$

$$\lim_{x \to a^-} f(x) = \infty \text{ or } -\infty.$$

In particular, if f is the rational function

$$f(x) = \frac{p(x)}{q(x)}$$

with $q(a) = 0$ and $p(a) \neq 0$, the line $x = a$ is an asymptote. For example, if

$$f(x) = \frac{x^2 - 4}{x^2 - 9}$$

then both $x = 3$ and $x = -3$ are vertical asymptotes.

A line $y = b$ is a horizontal asymptote to the graph of f if

$$\lim_{x \to \infty} f(x) = b \qquad \text{or} \qquad \lim_{x \to -\infty} f(x) = b$$

For a rational function f, such that

$$f(x) = \frac{p(x)}{q(x)}$$

we saw in Chapter 1 that by considering dominant terms on numerator and denominator, we can draw the following conclusions.

(i) If $\deg p < \deg q$, then $y = 0$ is a horizontal asymptote (we are using "deg" to mean "degree of").

(ii) If $\deg p = \deg q$, then $y = b$ is a horizontal asymptote, where b is the leading coefficient of p divided by the leading coefficient of q.

(iii) If $\deg p > \deg q$, there is no horizontal asymptote.

We also saw that if $\deg p = \deg q + 1$, there is an inclined asymptote, $y = mx + b$, where $mx + b$ is the quotient obtained by dividing $p(x)$ by $q(x)$. To illustrate, the graph of

$$f(x) = \frac{x^2 - 4}{x^2 - 9}$$

has the horizontal asymptote $y = 1$ [case (ii) above]. The graph of

$$g(x) = \frac{x^2 - 4}{x + 1} = x - 1 - \frac{3}{x + 1} \qquad \text{See the long division in the margin.}$$

has no horizontal asymptote [case (iii)], but since

$$\lim_{x \to \infty} [g(x) - (x - 1)] = \lim_{x \to \infty} \left(-\frac{3}{x + 1} \right) = 0$$

it has the inclined asymptote

$$y = x - 1$$

EXAMPLE 4.13   Discuss and sketch the graph of

$$f(x) = \frac{x^2 - 1}{x^2 - 4}$$

Solution   Let $y = f(x)$.

(a) *Intercepts.* Setting $y = 0$, we get $x = \pm 1$. Setting $x = 0$, we get $y = 1/4$.

(b) *Symmetry.* Since $f(x) = f(-x)$, the graph is symmetric with respect to the y-axis.

(c) *Asymptotes.* The vertical asymptotes are $x = 2$ and $x = -2$, and the horizontal asymptote is $y = 1$.

(d) *Intervals of Increase and Decrease.* The derivative is found by the Quotient Rule to be

$$f'(x) = \frac{-6x}{(x^2 - 4)^2}$$

so $x = 0$ is the only critical point. Note that although $f'(x)$ is not defined at $x = \pm 2$, these points are not critical points of the second kind. In fact, they are not in the domain of the function, since $f(x)$ also is not defined for $x = \pm 2$. If $x \neq 2$, and $x > 0$, $f'(x) < 0$, so f is decreasing for all positive x values in the domain of f, namely, $(0, 2) \cup (2, \infty)$. Similarly, when $x \neq -2$ and $x < 0$, $f'(x) > 0$, so f is decreasing on $(-\infty, -2) \cup (-2, 0)$.

(e) *Maxima and Minima.* We have seen that $x = 0$ is the only critical point. Since $f'(x)$ changes from $+$ to $-$ as x goes from left to right through 0, we conclude from the First-Derivative Test that f has a local maximum value at $x = 0$. Since $f(0) = 1/4$, the point $(0, 1/4)$ is a local maximum point.

(f) *Concavity and Inflection Points.* Calculating $f''(x)$, we get, after simplifying,

$$f''(x) = \frac{6(3x^2 + 4)}{(x^2 - 4)^3} = \frac{6(3x^2 + 4)}{(x - 2)^3(x + 2)^3}$$

We show a sign graph for f'' in Figure 4.43. Note that the numerator for $f''(x)$ is always positive, so it does not need to be considered. We see from the sign graph that when $x > 2$ or $x < -2$, $f''(x) > 0$, so the graph is concave up. For $-2 < x < 2$, $f''(x) < 0$, and the graph is therefore concave down. Since $x = 2$ and $x = -2$ are not in the domain of f, there are no points on the graph where the concavity changes, so there are no inflection points.

By plotting a few additional points, we can now draw the graph, as shown in Figure 4.44.

FIGURE 4.43

Sign graph for $f''(x) = \dfrac{6(3x^2 + 4)}{(x - 2)^3(x + 2)^3}$

FIGURE 4.44

Infinite limits for f' can also provide useful information about the graph. Suppose that f is continuous at a and $\lim_{x \to a} f'(x) = \infty$. Since $f'(x)$ gives the slope of the tangent line, we know that the slope is positive near a, and the tangent line becomes steeper and steeper as x approaches a. At a itself, the tangent line is vertical. So the graph must be similar to Figure 4.45(a) in the vicinity of a. The other possibilities are shown in parts (b), (c), and (d).

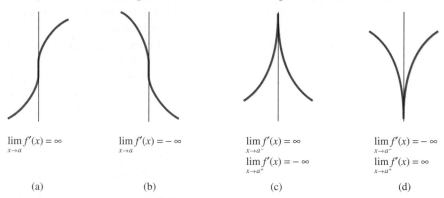

$\lim_{x \to a} f'(x) = \infty$	$\lim_{x \to a} f'(x) = -\infty$	$\lim_{x \to a^-} f'(x) = \infty$	$\lim_{x \to a^-} f'(x) = -\infty$
		$\lim_{x \to a^+} f'(x) = -\infty$	$\lim_{x \to a^+} f'(x) = \infty$
(a)	(b)	(c)	(d)

FIGURE 4.45

In parts (a) and (b) of Figure 4.45, an inflection point occurs at $(a, f(a))$. The maxima and minima in parts (c) and (d) occur at cusps. The four possibilities shown assume that $\lim_{x \to a^+} f'(x)$ and $\lim_{x \to a^-} f'(x)$ are both infinite. If only one of these one-sided limits is infinite, or if f is discontinuous, various possibilities exist, some of which are shown in Figure 4.46.

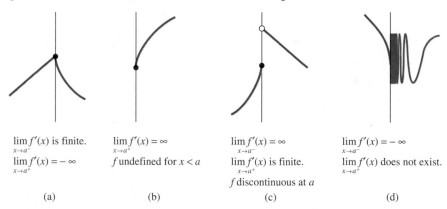

$\lim_{x \to a^-} f'(x)$ is finite.	$\lim_{x \to a^+} f'(x) = \infty$	$\lim_{x \to a^-} f'(x) = \infty$	$\lim_{x \to a^-} f'(x) = -\infty$
$\lim_{x \to a^+} f'(x) = -\infty$	f undefined for $x < a$	$\lim_{x \to a^+} f'(x)$ is finite.	$\lim_{x \to a^+} f'(x)$ does not exist.
		f discontinuous at a	
(a)	(b)	(c)	(d)

FIGURE 4.46

EXAMPLE 4.14   Discuss and sketch the graph of

$$f(x) = x + 3(1 - x)^{1/3}$$

Solution

(a) *Intercepts.* $f(0) = 3$, so the y-intercept is 3. The x-intercepts are not easily obtained, but with the aid of a calculator it can be shown that one x-intercept is between -5 and -6 and another is between 4 and 5.

(b) *Symmetry.* There is no symmetry, since $f(-x) = -x + 3(1 + x)^{1/3}$, so $f(-x)$ is not equal to $f(x)$ or $-f(x)$.

(c) *Asymptotes.* There are no asymptotes.

(d) *Intervals of Increase and Decrease.* We calculate $f'(x)$ to determine the critical points.

$$f'(x) = 1 - (1 - x)^{-2/3}$$

The critical points are x values in the domain of f for which $f'(x) = 0$ or for which $f'(x)$ does not exist. Clearly, $f'(1)$ does not exist, and since $f(1)$ does exist (that is, $x = 1$ is in the domain of f), we conclude that $x = 1$ is a critical point. To see if there are critical points for which $f'(x) = 0$, we solve

$$1 - (1 - x)^{-2/3} = 0$$

or, equivalently,

$$1 = \frac{1}{(1-x)^{2/3}}$$

Inverting, we get

$$(1 - x)^{2/3} = 1 \quad \text{Cube both sides.}$$

$$(1 - x)^2 = 1 \quad \text{Take square roots.}$$

$$1 - x = \pm 1$$

$$x = 0 \text{ or } 2$$

The critical points are therefore 0, 1, and 2. To determine the sign graph for f', we can test a value in each of the four regions into which the critical points divide the number line. For example, we can use $x = 3$ in the interval $(2, \infty)$, $x = 1.5$ in the interval $(1, 2)$, $x = 0.5$ in the interval $(0, 1)$, and $x = -1$ in the interval $(-\infty, 0)$. With the aid of a calculator, we obtain the signs as indicated in Figure 4.47. We conclude that

$$f \text{ is increasing in } (2, \infty) \text{ and in } (-\infty, 0)$$

$$f \text{ is decreasing in } (0, 1) \text{ and in } (1, 2)$$

(e) *Maxima and Minima.* From the sign graph in Figure 4.47, we conclude that f has a local maximum at $x = 0$ and a local minimum at $x = 2$. There is no local maximum or minimum at $x = 1$. Substituting $x = 0$ and $x = 2$ into the original formula for $f(x)$, we obtain the following points:

$$(0, 3) \text{ is a local maximum point.}$$

$$(2, -1) \text{ is a local minimum point.}$$

We can gain further insight into the nature of the graph near the critical point $x = 1$ by examining the right-hand and left-hand limits of $f'(x)$ as x approaches 1:

$$\lim_{x \to 1^+} f'(x) = \lim_{x \to 1^+} \left[1 - \frac{1}{(1-x)^{2/3}} \right] = -\infty$$

$$\lim_{x \to 1^-} f'(x) = \lim_{x \to 1^-} \left[1 - \frac{1}{(1-x)^{2/3}} \right] = -\infty$$

Thus, the tangent line is vertical at $x = 1$.

FIGURE 4.47

Sign graph for $f'(x) = 1 - \dfrac{1}{(1-x)^{2/3}}$

(f) *Concavity and Inflection Points.* We first calculate $f''(x)$.

$$f''(x) = -\frac{2}{3}(1-x)^{-5/3} = -\frac{2}{3(1-x)^{5/3}}$$

Don't forget to use the Chain Rule.

For $x > 1$, $f''(x) > 0$, so the graph is concave up. For $x < 1$, $f''(x) < 0$, so the graph is concave down. Since $x = 1$ is in the domain of f, we conclude that there is an inflection point at $x = 1$. Since $f(1) = 1$, the coordinates of the inflection point are $(1, 1)$.

In summary, the point $(0, 3)$ is a local maximum point, $(2, -1)$ is a local minimum point, and $(1, 1)$ is an inflection point at which the tangent line is vertical. Furthermore, the graph is concave down to the left of 1 and concave up to the right of 1. We get a few additional points and draw the graph as shown in Figure 4.48. ∎

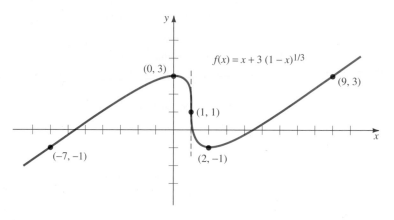

FIGURE 4.48

ANALYSIS OF FUNCTIONS USING A COMPUTER ALGEBRA SYSTEM

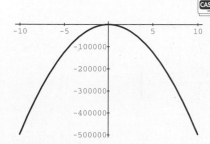

FIGURE 4.2.1

$y = f(x) = \dfrac{x^4}{4} - 5000x^2 + 5,$

$-10 \le x \le 10$

CAS 17

Let $f(x) = \frac{x^4}{4} - 5000x^2 + 5$. Give a complete analysis of the graph of f.

We can begin by simply graphing the curve, specifying an initial range for x (Figure 4.2.1). In Maple we could enter

```
f:=x->x^4/4–5000*x^2+5;
plot(f(x), x=–10..10);
```

The plot suggests the curve is a parabola. We should be wary of simply accepting this initial scale—all the important information may not be displayed. Rather than trying to guess a more appropriate scale, we can use the tools of calculus to sketch the curve more accurately. We start with finding the critical values of the function.

In Maple enter:

fprime := diff(f(x),x);   fprime is a name we give to the derivative of f with respect to x.
solve(fprime=0);

Maple gives the output 0, 100, −100 for the three critical values for f. We could see this set of critical values immediately if we could factor the derivative. Enter

factor(fprime);

and the output is

x(x − 100) (x + 100)

The location of the critical values suggest that we should sketch the curve over an interval such as $[-200, 200]$ (Figure 4.2.2). Enter

plot(f(x),x=−200..200);

The curve now shows more complicated behavior. The complete analysis of the curve can be done as easily using Maple, Mathematica, or DERIVE. We complete the analysis using Maple.

Increasing and Decreasing Intervals:

solve(fprime > 0,x);

Output: $\{x < 0, -100 < x\}, \{100 < x\}$

solve(fprime < 0,x);

Output: $\{x < -100\}, \{x < 100, 0 < x\}$

Local Extremes: They can be read off the graph or the First-Derivative Test can be invoked.

$$\text{Local maxima:} \quad (0, f(0)) = (0, 0)$$

$$\text{Local minima:} \quad (-100, f(-100)) \text{ and } (100, f(100))$$

Use Maple to determine the y-coordinates.

f(−100); −24999995
f(100);   −24999995

Concavity: First, compute the second derivative.

f2prime := diff(fprime,x);

Concave up:

solve(f2prime > 0,x);

Output:

$$\left\{ x < -\frac{100}{3}\sqrt{3} \right\}, \left\{ \frac{100}{3}\sqrt{3} < x \right\}$$

Concave down:

solve(f2prime < 0,x);

Output:

$$\left\{ -\frac{100}{3}\sqrt{3} < x, x < \frac{100}{3}\sqrt{3} \right\}$$

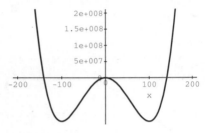

FIGURE 4.2.2

$y = f(x) = \dfrac{x^4}{4} - 5000x^2 + 5,$
$-200 \le x \le 200$

Inflection Points: They can now be read off as

$$\left(\frac{100}{3}\sqrt{3}, f\left(\frac{100}{3}\sqrt{3}\right)\right) \quad \text{and} \quad \left(-\frac{100}{3}\sqrt{3}, f\left(-\frac{100}{3}\sqrt{3}\right)\right)$$

Use the CAS to determine the y-coordinates:

$$f(100/3*\text{sqrt}(3)); \ -\frac{124999955}{9}$$

$$f(-100/3*\text{sqrt}(3)); \ -\frac{124999955}{9}$$

Asymptotes: There are none.

→Exercise Set 4.2

In Exercises 1–10, determine the intervals on which the graph of the given function is concave up and the intervals on which it is concave down.

1. $f(x) = 2x^2 - 3x + 4$ **2.** $g(x) = 5 - 6x - x^2$

3. $f(x) = ax^2 + bx + c$ **4.** $f(x) = x^3 - 3x^2 + 4x$

5. $h(x) = x^3 + 6x^2 - 5x + 1$

6. $F(x) = 2x^3 - 3x^2 - x + 5$

7. $G(x) = x^4 - 2x^3$

8. $g(x) = x^4 + 2x^3 - 12x^2 + 10$

9. $f(x) = x^4 - 6x^2 + 8$

10. $F(x) = x^4 - 3x^3 - 15x^2 - 12x + 9$

11. Refer to the points A, B, C, and D on the accompanying graph of the function f.
  (a) At which points is $f'(x) > 0$?
  (b) At which points is $f''(x) > 0$?

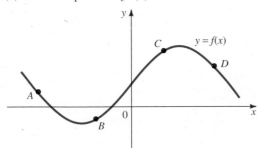

In Exercises 12–21, find all local maximum and minimum points, using the Second-Derivative Test when applicable and the First-Derivative Test otherwise.

12. $f(x) = 2x^3 - 6x + 4$

13. $g(x) = x^3 + 2x^2 - 4x + 5$

14. $h(x) = 2x^3 - 3x^2 + 7$

15. $\phi(x) = 2 - 3x + 5x^2 - x^3$

16. $F(x) = x^4 + 4x^3 - 8x^2 + 4$

17. $G(x) = 3x^5 - 25x^3 + 60x$

18. $f(x) = x + \dfrac{1}{x}$ **19.** $f(x) = \dfrac{1}{x^2} + \dfrac{2}{x} + 3$

20. $g(x) = 3x^4 - 8x^3 + 6x^2$

21. $h(x) = 3x^5 - 20x^3 + 7$

In Exercises 22–41, determine the intervals where the function is increasing and where decreasing, local maxima and minima, concavity, and all points of inflection. Use the information obtained to draw the graph of $y = f(x)$.

22. $f(x) = x^3 - 3x^2$ **23.** $f(x) = x^3 - 12x$

24. $f(x) = x^3 + 3x^2 - 9x - 15$

25. $f(x) = x^3 - x^2 - x + 1$

26. $f(x) = 4 + 3x - x^3$

27. $f(x) = 2x^3 + 4x^2 - 8x + 2$

28. $f(x) = x^4 - 4x^3 + 5$

29. $f(x) = x^3 - 6x^2 + 12x - 3$

30. $f(x) = x^4 - 6x^2 + 10$

31. $f(x) = 3 + 4x - x^4$

32. $f(x) = \dfrac{x^5}{5} - \dfrac{10x^3}{3} + 9x$

33. $f(x) = (x^2 - 4)(x - 2)$

34. $f(x) = (x - 3)^2(x + 3)$

35. $f(x) = (x - 2)^2(2x + 5)$

36. $f(x) = 2\sin\dfrac{x}{2} + 1$ on $[0, 4\pi]$

37. $f(x) = \sin x + \cos x$ on $[0, 2\pi]$

38. $f(x) = x + \sin x$ on $[0, 4\pi]$

39. $f(x) = \cos x - \sqrt{3}\sin x$ on $\left[-\dfrac{\pi}{3}, \dfrac{5\pi}{3}\right]$

40. $f(x) = \cos 2x - 4\cos x$ on $[-\pi, \pi]$

41. $f(x) = \sin 2x - 2\sin^2 x$ on $[-\pi, \pi]$

In Exercises 42–51, discuss intercepts, symmetry, vertical and horizontal asymptotes, intervals where f is increasing and where decreasing, local maxima and minima, concavity, and points of inflection; then sketch the graph.

42. $f(x) = \dfrac{x - 1}{x + 2}$ **43.** $f(x) = \dfrac{x}{x^2 - 1}$

44. $f(x) = \dfrac{2}{x^2 + 1}$ **45.** $f(x) = \dfrac{x}{x^2 + 1}$

46. $f(x) = \dfrac{x^2}{x^2 - 4}$ **47.** $f(x) = \dfrac{x^2 - 4}{x^2 - 1}$

48. $f(x) = \dfrac{x^2}{x^2 + 4}$ **49.** $f(x) = \dfrac{1 - x^2}{x^2 - 4}$

50. $f(x) = \dfrac{1}{x - 1} + \dfrac{1}{x + 1}$

51. $f(x) = \dfrac{1}{x} - \dfrac{1}{x + 1}$

52. Sketch a possible graph of a function f satisfying all of the following properties.
   (a) $f''(x) > 0$ for $-2 < x < 2$
   (b) $f'(0) = 0$, $f(0) = 1$, $f'(3) = 0$, $f'(5) = 0$
   (c) $f'(x) > 0$ for $2 < x < 3$ or $x > 5$
   (d) $f'(x) < 0$ for $3 < x < 5$
   (e) $\lim_{x \to -2} f(x) = \infty$
   (f) $\lim_{x \to 2^-} f(x) = \infty$ and $\lim_{x \to 2^+} f(x) = -\infty$
   (g) $\lim_{x \to -\infty} f(x) = 0$

In Exercises 53–60, show that the function has a vertical tangent line, and determine the nature of the graph at that point.

53. $f(x) = x^{2/3}$ **54.** $f(x) = 1 - x^{1/3}$

55. $f(x) = \sqrt[3]{1 - x}$ **56.** $f(x) = x^{3/5} - 2x^{2/5}$

57. $f(x) = \dfrac{x^{2/3}}{x - 1}$

58. $f(x) = \sqrt{1 - x^2}$, $0 \le x \le 1$

59. $f(x) = 1 - x^{1/2}$, $x \ge 0$

60. $f(x) = \sqrt{\dfrac{x - 1}{x + 1}}$, $x \ge 1$

In Exercises 61–66, find all local extrema, using the Second-Derivative Test when applicable and the First-Derivative Test otherwise.

61. $f(x) = 3 + 4x - 2x^2 - \dfrac{x^3}{3} + \dfrac{x^4}{4}$

62. $f(x) = \dfrac{x^2 - x + 2}{x + 1}$

63. $f(x) = \dfrac{x^2}{2} - \dfrac{1}{x} + 2$

64. $f(x) = x^4 - 6x^2 + 8x - 3$

65. $f(x) = 3x + 2(4 - x)^{3/2}$

66. $f(x) = x^{2/3}(2x - 5)$

In Exercises 67–82, discuss intercepts, symmetry, asymptotes, intervals where f is increasing and where decreasing, local maxima and minima, concavity, and points of inflection; then draw the graph.

67. $f(x) = \frac{1}{16}(3x^4 - 8x^3 - 24x^2 + 96x)$

68. $f(x) = \dfrac{x^4}{2} - \dfrac{x^3}{3} - 4x^2 + 4x + 7$

69. $f(x) = x^5 - 5x^3 + 10x - 3$

70. $f(x) = 2 - 8x + 6x^2 - x^4$

71. $f(x) = x^4 - 4x^3 + 16x - 4$

72. $f(x) = x^4 - 2x^3 + 2x + 3$

73. $f(x) = \dfrac{3\cos x}{2 - \sin x}$ on $[-\pi, \pi]$

74. $f(x) = \dfrac{\tan x}{2\sec x - 1}$ on $[0, 2\pi]$

75. $f(x) = \cos^4 x - 4\cos x + 3$ on $\left[-\dfrac{3\pi}{2}, \dfrac{3\pi}{2}\right]$

76. $f(x) = 2x + 3(1 - x)^{2/3}$

77. $f(x) = \dfrac{3 - x^2}{x^3}$

78. $f(x) = \dfrac{x^2 - 1}{x^3 - 3x}$

79. $f(x) = \dfrac{3x^{1/3}}{2 - x}$

80. $f(x) = \dfrac{1 - \sin x}{\cos x - x}$

81. $f(x) = \dfrac{x^3 + 1}{x^2}$

82. $f(x) = \dfrac{x^2 - 8}{x - 3}$

83. Cubic Polynomials. Use information from the first and second derivatives to give a complete description of all possible shapes of the graphs of the family of cubic functions

$$f(x) = x^3 + kx, \qquad k \in \mathbb{R}$$

Consider the cases $k = 0$, $k > 0$, and $k < 0$ separately. Use your results to discuss possible shapes of the general cubic function

$$g(x) = ax^3 + bx^2 + cx + d \qquad a, b, c, d \in \mathbb{R}, \ a \neq 0$$

(See Exercise 27 in Section 1.4 and Exercise 56 in Section 4.1.)

84. Let $f(x) = -x^3 - 2x^2 + x + 2$.
 (a) Obtain the graphs of f, f', and f''.
 (b) Using the graphs in part (a), approximate where $f'(x)$ has its maximum value.
 (c) Find the exact value of x where $f'(x)$ has its maximum value. Justify your answer, and compare it with the approximation in part (b).

85. Investigate the family of curves given by

$$f(x) = \frac{1}{1 + x^2} + cx, \qquad c \in \mathbb{R}$$

 (a) Conjecture for which values of c the function f is increasing for all $x \in \mathbb{R}$ and for which values of c the function f is decreasing for all $x \in \mathbb{R}$.
 (b) Determine the inflection points for the graph of f.
 (c) Is there any relationship between the location of the inflection points and the answer you found in part (a)?

86. Let

$$f(x) = \frac{ax + b}{cx + d}$$

where a, b, c, and d are positive constants. Use a graphing calculator or a computer algebra system to obtain several graphs by varying a, b, c, and d. Do the graphs have any common features? If so, what are they? What can be said about the graph of the given function f in general?

4.3 APPLIED MAXIMUM AND MINIMUM PROBLEMS

As we stated at the beginning of Section 4.1, one of the most important applications of the derivative is in finding maximum or minimum values of functions. Usually certain restrictions, known as "constraints," are imposed on the variable (or variables) involved. Such problems are referred to collectively as **optimization problems**, since we are trying to find the best (optimal) result under the given constraints. Some optimization problems can be formulated in terms of one *continuous* variable (one that can assume any real value on an interval), and some require more than one such variable. We will concentrate here on the first type and study the second type later. It should also be mentioned that certain problems involve *discrete* variables (for example, variables that can assume only integer values, such as the number of seats in a theater). Such problems often can be solved by supposing the variable to be continuous and interpreting the final result appropriately. Other problems require totally different approaches, as are studied in discrete mathematics courses.

Our previous studies have given us the capability of finding extreme values, *provided the quantity to be maximized or minimized is expressed as a function of one continuous variable*. This restriction is what we will concentrate on in this section. The main difficulty is that in real-life situations, problems are usually presented in words, not as neat "textbook-type" mathematical formulas. So our first step is to translate the word problem into an appropriate mathematical form. Then we solve the mathematical problem, and finally, we interpret the answer in terms of the original problem. This process is known as **mathematical modeling**, and it is one of the most important aspects of the work of applied mathematicians and others who use mathematics in their work. Although there are no guaranteed paths to success in setting up a model of an optimization problem for which calculus is appropriate, the following steps can serve as a general guide.

1. Read the problem through (more than once if necessary) and identify the function to be maximized or minimized. We refer to this function as the **objective function**. Give it a name. (Using initial letters such as C for cost, P for profit, V for volume, and so on, is a good idea.)

2. Determine a variable on which the objective function depends and give it a name, such as x or t. This variable is the *independent variable*. Note any limitations on its size. Sometimes there may be two (or more) such variables, but in the present chapter we will always give enough information to express any additional variables in terms of the first.

3. Express the objective function in terms of the independent variable. This is a key step and will be made easier if steps 1 and 2 have been carried out carefully. Frequently it is helpful to draw a figure. Sometimes well-known relationships such as "distance = rate × time" or "profit = revenue − cost" are used. Also, mensuration formulas, such as those for area or volume, may be useful.

4. Find the derivative of the objective function with respect to the independent variable identified in step 2, and determine the critical points.

5. Test critical points that lie in the allowable domain to see if they yield maximum points, minimum points, or neither. Often there is only one critical point in the allowable domain. The nature of the problem might ensure that there is an extreme value interior to the interval in question, so that this extreme value must occur at the critical point. In such cases it is not necessary to test the critical point, but it is still a good idea to test it as a check.

6. Find the value of the objective function at each critical point identified in step 5. If the endpoint values also are feasible solutions (that is, if they are in the domain of the objective function), find the value of the objective function at these points as well. The value that gives the desired extreme can now be determined by direct computation and comparison of the functional values at the critical points and endpoints (if any).

7. Mentally check to see that the answer you found is reasonable. Sometimes it is possible to see that you made a mistake because the answer you found is clearly incorrect. (If you were to find that the price of a one-family house comes out as 2¢ or 2 billion dollars, for example, or if you find that the length of a rectangle is −10 ft, you would clearly know you had made a mistake.)

The following examples illustrate this procedure.

EXAMPLE 4.15   An open-top box is to have a square base and a volume of 10 ft³. If the bottom costs 15¢ per square foot and the sides 6¢ per square foot, find the dimensions of the most economical box, and find the minimum cost.

Solution   The most economical box is the one with the minimum cost. So the objective function is the cost C, and we want to find its minimum value. We introduce as the independent variable the side x of the square base (Figure 4.49). The height y is also a relevant variable, but it can be written in terms of x as follows. We know the volume of the box is 10 ft³, and since the volume of a box is the area of the base times the height, we have

$$x^2 y = 10$$

So

$$y = \frac{10}{x^2}$$

FIGURE 4.49

Clearly we must have $x > 0$, because we are discussing a real box.

Now we can find $C(x)$ from the given information. The base of the box has area x^2 ft², and each square foot costs 15¢. There are four vertical sides, each having area $xy = x(10/x^2) = 10/x$, each costing 6¢ per square foot. Thus, the value $C(x)$ of the cost function, in cents, is

$$C(x) = 15x^2 + 4(6)\left(\frac{10}{x}\right) = 15x^2 + \frac{240}{x}$$

The computer-generated graph of $y = C(x)$, for $x > 0$ (Figure 4.50), clearly shows the existence of a minimum value. To determine exactly where this minimum occurs, we find the critical points by taking the derivative of $C(x)$.

$$C'(x) = 30x - \frac{240}{x^2}$$

Set $C'(x) = 0$:

$$30x - \frac{240}{x^2} = 0 \qquad \text{Multiply by } x^2 \text{ and divide by 30.}$$

$$x^3 - 8 = 0$$

$$x^3 = 8$$

$$x = 2$$

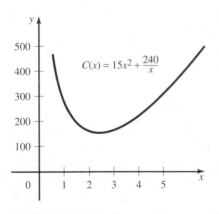

$$C(x) = 15x^2 + \frac{240}{x}$$

FIGURE 4.50

It seems reasonable that this value of x gives a minimum, since mathematically we can see from $C(x)$ that for large x or x close to 0, the cost is high. The graph of $C(x)$ in Figure 4.50 confirms this observation. We should check to see that the value $x = 2$ gives a minimum by substituting in $C''(x)$:

$$C''(x) = 30 + \frac{480}{x^3}$$

$$C''(2) = 30 + \frac{480}{8} = 90$$

Thus, $C''(2) > 0$, so the minimum value of C indeed occurs when $x = 2$. At this value of x we have

$$y = \frac{10}{2^2} = \frac{5}{2}$$

and

$$C(2) = 15(2^2) + \frac{240}{2} = 60 + 120 = 180$$

Thus, the most economical box has dimensions (in feet) of $2 \times 2 \times 2\frac{1}{2}$, and the minimum cost is 180¢, or \$1.80. ∎

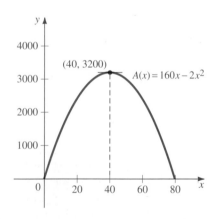

Building

3   3

x

$x+3$ Flower bed

3

y

$y+6$

FIGURE 4.51

EXAMPLE 4.16   A rectangular flower bed is to be designed with one side adjacent to a building and the other three sides bordered by a walkway 3 ft wide. If the outer perimeter of the walkway is to be 172 ft (we have 172 linear feet of decorative edging available), what should be the dimensions of the flower bed if it is to have maximum area? What is this maximum area?

Solution   The objective function is the area A of the flower bed. As in Figure 4.51, let x be the width of the bed and y its length. Since the outer perimeter of the walkway is $2(x + 3) + (y + 6)$ (verify this), we must have

$$2x + 6 + y + 6 = 172$$
$$y = 160 - 2x$$

Since x and y are both positive, we must have $0 < x < 80$. The area of the bed is $xy = x(160 - 2x)$, so

$$A(x) = 160x - 2x^2$$
$$A'(x) = 160 - 4x$$

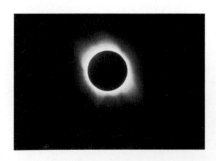

y

4000

(40, 3200)

3000 $A(x) = 160x - 2x^2$

2000

1000

0   20   40   60   80 x

FIGURE 4.52

To find the critical points, we set $A'(x) = 0$, getting $x = 40$. To test whether this value gives a maximum or a minimum area, we calculate the second derivative and get $A''(x) = -4$. The only critical point is $x = 40$, and since $A''(40) < 0$, this value produces a maximum value of the area. Thus, the dimensions that produce the maximum area are

$$x = 40 \quad \text{and} \quad y = 160 - 2(40) = 80$$

and the maximum area is

$$A(40) = (40)(80) = 3200 \text{ ft}^2$$

Figure 4.52 illustrates our results. ∎

EXAMPLE 4.17   An airplane is chartered for a flight to Hawaii to view a total solar eclipse for 100 passengers at \$1,000 per passenger. The price per ticket will be reduced by \$5 per passenger for each person in excess of 100 who goes on the trip. Find the number of passengers that will provide maximum revenue, and find the maximum revenue. The plane will accommodate up to 200 passengers.

Solution   Let R denote the revenue, the function we want to maximize. The revenue is calculated by multiplying the number of passengers by the price of each ticket. Each of these quantities depends on the number of passengers *in excess* of 100. Call this number x. Then we have

$$100 + x = \text{total number of passengers}$$
$$1,000 - 5x = \text{price of each ticket, in dollars}$$

Because of the capacity of the airplane, we must have $0 \leq x \leq 100$. Thus, we have

$$R(x) = (100 + x)(1,000 - 5x) = 100,000 + 500x - 5x^2$$

$$R'(x) = 500 - 10x$$

Setting $R'(x) = 0$, we find the critical value $x = 50$. Since $R''(x) = -10$, we conclude from the Second-Derivative Test that $x = 50$ gives a local maximum value. The revenue when $x = 50$ is

$$R(50) = (150)(1,000 - 250) = (150)(750) = 112,500$$

We need to check the endpoint values also:

$$R(0) = (100)(1,000) = 100,000 \qquad R(100) = (200)(500) = 100,000$$

We conclude that a maximum revenue of \$112,500 is obtained when 50 passengers in excess of 100 go—that is, when 150 passengers are on the flight.

■

REMARK ———————————————————————————

■ We could have used x for the total number of passengers rather than the number in excess of 100. The calculations would have been different, but the final result would have been the same. You might try this alternative and compare the difficulty with the method we used.

_____

EXAMPLE 4.18   Towns A and B are 5 km apart and are located on the same side of a straight river. Town A is 1 km from the river and town B is 4 km from the river. A pumping station is to be built at the river's edge and pipes are to go to the two towns. Where should the pumping station be located to minimize the amount of pipe used?

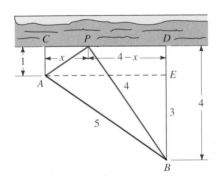

FIGURE 4.53

Solution   In Figure 4.53, C is the point on the shore nearest to A and D is the point on the shore nearest to B. Then the distance CD is the same as AE, which by the Pythagorean Theorem is 4. Let P be the point at which the pumping station is to be located, and use x to denote its distance from C. Then $4 - x$ is its distance from D. Clearly, $0 \leq x \leq 4$. If $L(x)$ denotes the total length of pipe, then $L(x) = \overline{PA} + \overline{PB}$. Using the Pythagorean Theorem on triangles ACP and BDP, we get

$$L(x) = \sqrt{x^2 + 1} + \sqrt{(4 - x)^2 + 16}$$

Differentiating,

$$L'(x) = \frac{x}{\sqrt{x^2 + 1}} - \frac{4 - x}{\sqrt{(4 - x)^2 + 16}}$$

Setting $L'(x) = 0$, we get

$$x\sqrt{(4 - x)^2 + 16} = (4 - x)\sqrt{x^2 + 1}$$

and on squaring and simplifying, this gives (check the details)

$$15x^2 + 8x - 16 = 0$$

$$(5x - 4)(3x + 4) = 0$$

$$x = \frac{4}{5} \text{ or } x = -\frac{4}{3}$$

The answer $x = -4/3$ is ruled out on two counts. First, it is outside the admissible domain. Second, it fails to satisfy the equation we obtained by setting $L'(x) = 0$, since it would cause one side to be positive and the other side to be negative. The solution $x = 4/5$ does satisfy this equation (check this), so it is a critical point. It can be shown that $L''(4/5) > 0$, so a minimum value of L occurs at $x = 4/5$. However, calculating L'' is a little messy, and we can avoid finding it (and also avoid using the First-Derivative Test) by calculating $L(4/5)$ and comparing it with the endpoint values $L(0)$ and $L(4)$. After a bit of arithmetic, we get

$$L\left(\tfrac{4}{5}\right) = \sqrt{41} \approx 6.40 \qquad L(0) = 1 + 4\sqrt{2} \approx 6.66 \qquad L(4) = 4 + \sqrt{17} \approx 8.12$$

Since L is continuous on $[0, 4]$, it reaches both a maximum and a minimum value there, and the only candidates for these extreme values are the three given here. So $L(4/5)$ is the minimum value. Thus, to use the least amount of pipe, the pumping station should be located $4/5$ km from C and $16/5$ km from D. ∎

REMARK ────────────────────────────────

■ It is not unusual for algebraic solutions of the equation obtained by setting the derivative of the objective function equal to 0 (to get critical points) to be invalid for the applied problem. For example, in the pumping station problem, we had to rule out $x = -4/3$. You should be alert to such limitations.

────────────────────────────────

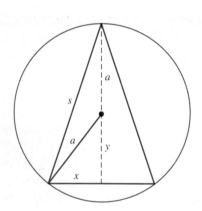

FIGURE 4.54

EXAMPLE 4.19   Prove that the isosceles triangle of maximum area that can be inscribed in a circle of fixed radius a is equilateral.

Solution   Referring to Figure 4.54, we see by the Pythagorean Theorem that $y = \sqrt{a^2 - x^2}$, so the area $A(x)$ of the isosceles triangle is

$$A(x) = x(\sqrt{a^2 - x^2} + a), \qquad 0 \le x \le a$$

We calculate $A'(x)$ and set it equal to 0. (We omit some of the algebra.)

$$A'(x) = \frac{-x^2}{\sqrt{a^2 - x^2}} + \sqrt{a^2 - x^2} + a = \frac{a^2 - 2x^2 + a\sqrt{a^2 - x^2}}{\sqrt{a^2 - x^2}}$$

$$A'(x) = 0 \text{ if } 2x^2 - a^2 = a\sqrt{a^2 - x^2}$$

After squaring both sides and collecting terms, we get

$$x^2(4x^2 - 3a^2) = 0$$

so that $x = 0$ or $x = \pm a\sqrt{3}/2$. Since $x = 0$ gives an area of 0 and since x cannot be negative, the only critical point of interest is $a\sqrt{3}/2$. We calculate

$$A\left(\frac{a\sqrt{3}}{2}\right) = \frac{3a^2\sqrt{3}}{4}$$

which is greater than the endpoint value $A(a) = a^2$ because $3\sqrt{3}/4 > 1$, so $x = a\sqrt{3}/2$ does produce the maximum area.

To show that the triangle with this value of x is equilateral, we calculate the height,

$$y + a = \sqrt{a^2 - x^2} + a = \sqrt{a^2 - \frac{3a^2}{4}} + a = \sqrt{\frac{a^2}{4}} + a = \frac{a}{2} + a = \frac{3a}{2}$$

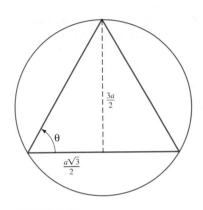

FIGURE 4.55

The distance formula:

$$d = \sqrt{(x_2 - x_1)^2 + (y_2 - y_1)^2}$$

FIGURE 4.56

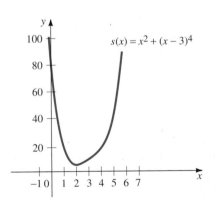

FIGURE 4.57

The tangent of the base angle θ (Figure 4.55) is

$$\tan \theta = \frac{\dfrac{3a}{2}}{\dfrac{a\sqrt{3}}{2}} = \sqrt{3}$$

so $\theta = 60°$. The base angles are equal, and since the three angles add to 180°, all three angles are equal to 60°. So the triangle is equiangular and hence equilateral. ∎

EXAMPLE 4.20   Find the point on the parabola $y = (x - 3)^2$ that is nearest the origin.

Solution   An arbitrary point P on the parabola has coordinates $(x, (x - 3)^2)$, and its distance $d(x)$ from the origin is

$$d(x) = \sqrt{x^2 + (x - 3)^4}$$

(Figure 4.56). This is the function we want to minimize. We can make the problem simpler, however, by observing that the distance will be minimum if and only if the square of the distance is a minimum. We denote the squared distance by $s(x)$:

$$s(x) = x^2 + (x - 3)^4$$

The computer-generated graph of $s(x)$ in Figure 4.57 shows clearly the existence of a minimum. To find where it occurs, we first calculate $s'(x)$:

$$s'(x) = 2x + 4(x - 3)^3 = 4x^3 - 36x^2 + 110x - 108$$

We set $s'(x) = 0$ and divide through by 2 to get

$$2x^3 - 18x^2 + 55x - 54 = 0$$

Various techniques exist for solving equations of degree 3 or higher, some of which you may have studied in your precalculus course. If there are any rational roots, we can find them by trial and error, with the trials limited as follows. If $x = p/q$ is a root, then p must be a factor of the constant term, in this case 54, and q must be a factor of the leading coefficient, in this case 2. Starting with integer possibilities, we eliminate $x = 1$ but find that $x = 2$ works. (A calculator is useful here.) After dividing out the factor $x - 2$, the other factor is found to be $2x^2 - 14x + 27$, which has no real zeros (you can verify that the zeros are complex by using the quadratic formula). So $x = 2$ is the only critical point. Since the geometry of the problem assures us that a minimum distance exists, we can conclude that $x = 2$ produces this minimum. (We could also check this conclusion by observing that $s''(2) > 0$.) The point on the graph nearest the origin is therefore $(2, 1)$, and its distance from the origin is $d(2) = \sqrt{5}$. ∎

Concepts from Economics

Economists often use calculus, so before giving the last two examples of this section, we discuss briefly some basic concepts of economics. We will be concerned here with maximizing the profit from the manufacture and sale of some commodity. If we let $C(x)$ denote the cost of producing x items and $R(x)$ denote the gross revenue obtained from selling them, then the profit, $P(x)$, is

$$P(x) = R(x) - C(x)$$

Although x in this situation can actually take on only positive integer values, we can treat it as if it could take on any positive real value. By doing so, we can apply the techniques of calculus to the functions C, P, and R. Then when we obtain the critical point x that maximizes the profit, if x is not an integer, we test the two closest integers to see which gives a maximum.

The cost function C usually consists of a **fixed cost** and an additional **variable cost**. The fixed cost is a constant, independent of the number of items produced. Its value is determined by such things as rent, mortgage payments, taxes, maintenance, basic utility bills, and any salaries that are not based on the number of items produced. The variable cost is dependent on the number of items produced. For example, the cost of plastic for making pitchers is a variable cost, since it depends on how many pitchers you make. On the other hand, the salary of the company president is a fixed cost. The formula for the variable cost may be obtained by analyzing the factors that affect the cost, or it may be obtained empirically. In the latter case, records may be kept over a period of time and a graph plotted showing the cost versus the number of items produced. Then by a process known as "curve fitting," a function whose graph matches as closely as possible the one obtained experimentally is determined. The resulting function can take a variety of forms, but it is often a polynomial of first, second, or third degree.

To see how the variable cost might be calculated in one situation, consider the following example.

EXAMPLE 4.21   The admissions office of a college expects to need between 20,000 and 40,000 catalogs, and it receives the following bid from a printer: For the first 20,000 copies the cost will be \$3.00 per catalog. For additional catalogs ordered, the cost per copy for these additional catalogs will be reduced by 5¢ per 1,000 catalogs ordered. For example, if the college orders 26,000 catalogs, the cost will be \$3.00 per copy for the first 20,000; then for the next 6,000 the cost will be \$2.70 per copy, a reduction of 30¢ (six times 5¢) per copy. Suppose x thousand catalogs are ordered, where $20 \le x \le 40$. Find the variable cost function $C_v(x)$.

Solution   The cost in thousands of dollars is

$$C_v(x) = x[\overbrace{3 - 0.05 \underbrace{(x - 20)}_{\substack{\text{number of} \\ \text{thousands in} \\ \text{excess of } 20}}}^{\substack{\text{cost per catalog} \\ \text{of } x \text{ thousand}}}] = -0.05x^2 + 4x$$

In this case, then, the variable cost function is quadratic over the interval in question. We show its graph in Figure 4.58.   ■

Quadratic cost functions are common, as are linear ones. Note that in Figure 4.58, although the curve is rising, it is concave down, reflecting the fact that the unit cost decreases as more units are produced. In some similar situations it may happen that if x increases still further, the cost curve turns upward again,

$C_v(x) = -0.05x^2 + 4x,$
$20 \le x \le 40$

(thousands of dollars)

80
60
40
20

10 20 30 40

(thousands of copies)

FIGURE 4.58

since more machines, more employees, more space, and so on, may be required. These considerations would probably lead to a cubic cost function.

The **revenue function** R is obtained by multiplying the number of items sold by the price p at which they are sold, so that

$$R(x) = xp(x)$$

We have indicated p as a function of x, since there is a definite functional relationship between price and number of items sold. The function $p(x)$ is sometimes called the **demand function**, since it gives the price per unit when there is demand for x units. Typically, when the price goes down, the number of items that can be sold at that price goes up.

The next example illustrates these ideas.

EXAMPLE 4.22   The Sun-Ray Company produces cells for solar collectors. It finds through experience that on the average it can sell x cells per day when the price $p(x) = 100 - 0.05x$, where $250 \leq x \leq 800$. The fixed cost is $4000 per day and the variable cost *per unit* is $60 - 0.01x$. Find the number of cells that Sun-Ray should produce each day to maximize profit, and find the maximum profit.

Solution   The revenue is the number of units times the price per unit:

$$R(x) = xp(x) = x(100 - 0.05x) = 100x - 0.05x^2$$

The variable cost is the number of units times the cost per unit. So the total cost is

$$C(x) = 4000 + x(60 - 0.01x) = 4000 + 60x - 0.01x^2$$

Thus,

$$
\begin{aligned}
P(x) &= R(x) - C(x) \\
&= 100x - 0.05x^2 - (4000 + 60x - 0.01x^2) \\
&= -0.04x^2 + 40x - 4000 \\
P'(x) &= -0.08x + 40 \\
P'(x) &= 0 \text{ if } x = 500
\end{aligned}
$$

Since $P''(x) < 0$, the critical point $x = 500$ gives a maximum. We calculate $P(500) = 6000$ and the endpoint values $P(250) = 3500$ and $P(800) = 2400$. So a maximum profit of $6000 per day is obtained when 500 cells are produced. ∎

Marginal Revenue and Marginal Cost

In general, as in the preceding example, the maximum profit will occur at a critical value where $P'(x) = 0$. Let x_0 be such a point. Then we have

$$P'(x_0) = R'(x_0) - C'(x_0) = 0$$

so that

$$R'(x_0) = C'(x_0)$$

FIGURE 4.59

That is, maximum profit occurs where R' and C' are equal. We illustrate this result in Figure 4.59. As we mentioned in Section 3.2, these quantities, $R'(x)$ and $C'(x)$, are called, respectively, **marginal revenue** and **marginal cost**. If we let x change by a small amount, say dx, then the corresponding changes in $R(x)$ and $C(x)$ are approximated by their differentials:

$$\Delta R \approx dR = R'(x)\, dx \qquad \text{and} \qquad \Delta C \approx dC = C'(x)\, dx$$

In particular, if $dx = 1$, we have $\Delta R \approx R'(x)$ and $\Delta C \approx C'(x)$. Thus, the marginal revenue is approximately the change in revenue caused by selling one additional item, and the marginal cost is approximately the change in cost caused by producing one additional item. Also, we have

$$\Delta P \approx dP = [R'(x) - C'(x)]dx$$

As long as the marginal revenue $R'(x)$ exceeds the marginal cost $C'(x)$, there is an increase in profit by increasing the number of items. The value x_0 at which $R'(x_0) = C'(x_0)$ is the value where no further increase in profit would occur by increasing x, and so the profit is then a maximum.

Exercise Set 4.3

1. A rectangular plot of ground is to be fenced on three sides to form a garden, with the fourth side bounded by a retaining wall. If there are 50 ft of fencing available, what should be the dimensions of the garden to produce the maximum area?

2. A rectangular lot adjacent to a road is to be enclosed by a fence. The side along the road requires reinforced fencing, which costs $7 per foot. Fencing that costs $5 per foot can be used for the other three sides. What is the largest area that can be enclosed for a cost of $3600, and what are the dimensions that give this maximum area?

3. A farmer wishes to fence a rectangular pasture having a total area of 12,000 m², and wants to divide it into two parts with a fence across the middle. Fencing around the outside costs $7.50 per meter, but the farmer can use less expensive fencing at $3 per meter as the divider. What dimensions will result in the least cost?

4. A water trough is to be formed from a sheet of metal 2 ft wide and 10 ft long by bending up at a right angle equal amounts from each long side. (Pieces will then be welded onto the ends.) How many inches on each side should be bent up to give the maximum volume?

5. A cardboard box is to be constructed having a volume of 9 ft³ and such that the bottom is a rectangle twice as long as it is wide. What should be the dimensions so that the least amount of cardboard is used?

6. An open-top box is to be made from a piece of cardboard 4 ft long and $2\frac{1}{2}$ ft wide by cutting out squares of equal size at each corner and bending up the flaps. What size squares should be removed in order to produce the maximum volume?

7. Find the dimensions of the rectangle of maximum area that can be inscribed in an isosceles triangle of height 6 and base 4 (see the figure). (*Hint:* Use similar triangles to find y in terms of x.)

8. Find the dimensions of the rectangle of maximum area that can be inscribed as shown in a right triangle with legs of lengths 6 and 8, respectively.

9. Show that the rectangle of maximum area that can be inscribed in a circle with fixed radius a is a square.

10. Find the dimensions of the rectangle of maximum area that can be inscribed in a semicircle of radius a as shown in the figure.

11. Find the dimensions of the right circular cylinder of maximum volume that can be inscribed in a sphere of fixed radius a. (*Hint:* Consider a cross section as shown in the figure to find the altitude h of the cylinder in terms of its base radius r.)

(a)

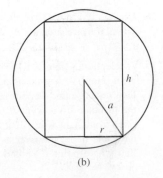

(b)

12. In Exercise 11, maximize the lateral surface area instead of the volume.

13. The unit price of a microwave oven is constant at \$250. The fixed cost is \$80, and the variable cost per unit is found by experience to be approximately $x + 90$, when the number x of ovens is between 10 and 100. Assuming x is in this range and that all ovens manufactured are sold, find the number that should be manufactured to produce the maximum profit.

14. The total cost function for producing x tables in a certain furniture factory is determined to be $C(x) = 0.02x^2 + 20x + 100$ for $10 \le x \le 100$, and the unit price function $p(x) = 25 + \frac{120}{x} - 0.03x$ for x in this range. Find the number of tables that should be produced to maximize profit. What is the maximum profit?

15. The owner of an orchard has a 10-acre plot planted with 200 apple trees, and the average yield is 500 apples per tree. The orchard owner plans to put apple trees of the same type on another 6-acre plot that has the same soil conditions. If the trees are planted closer together on the second plot than on the first plot, the yield per tree will be less, but it is possible that with more trees the total yield per acre will be greater. For each additional tree per acre planted, the average yield per tree will be reduced by 10 apples. How many trees should be planted on the 6-acre plot to give the maximum total yield?

16. An excursion train is to be run to the Super Bowl. Amtrak sets the fare at \$100 per ticket if 200 people go but agrees to lower the cost of all tickets by 20¢ each for every passenger in excess of 200. How many passengers will produce the greatest revenue, and what is the maximum revenue? (The train has a capacity of 450 passengers.)

17. A yacht is rented for an excursion for 60 passengers at \$20 per passenger. For each person in excess of 60 who goes, up to an additional 40 passengers, the fare of each passenger will be reduced by 25¢. Find the number of passengers that will produce the maximum revenue.

18. A restaurant has a fixed price of \$24 for a complete dinner. The average number of customers per evening is 160. The owner estimates on the basis of experience that for each \$1 increase in the cost of a dinner, there will be an average of 5 fewer customers per evening. What price should be charged to produce the maximum revenue, and what is the maximum revenue?

19. At a certain movie theater the price of admission is $5, and the average daily attendance is 200. As an experiment, the manager reduces the price by 10¢ and finds that the average attendance increases by 5 people per day. Assuming that for each additional 10¢ reduction the average attendance rises by 5, find the number of 10¢ reductions that would result in the maximum revenue.

20. A manufacturer makes aluminum cups in the form of right circular cylinders open at the top (no handle), having a volume of 16π in.^{3}. If the cost of the material for the bottom is twice that for the sides, find the dimensions that will give the lowest cost.

21. The strength of a rectangular beam of fixed length is jointly proportional to the width and the square of the depth. Find the dimensions of the strongest beam that can be cut from a log 3 ft in diameter.

22. Find the point on the graph of $y = \sqrt{2x-3}$ that is nearest the point $(3, 0)$.

23. A picture with an area of 120 in.^{2} is rectangular and is framed so that there is a $2\frac{1}{2}$-in. matting on each side and a 3-in. matting at the top and bottom. Find the dimensions of the picture if the total area of the framed picture is to be as small as possible. (Test to see that you have found the minimum.)

24. Find the point on the upper half of the ellipse $4x^2 + 9y^2 = 36$ that is nearest the point $(1, 0)$.

25. The printed matter on a page is to occupy an area of 80 in.^{2}, and there are to be margins of 1 in. at the top, bottom, and right side, and $1\frac{1}{2}$ in. at the left. Find the overall dimensions of the page that requires the least amount of paper.

26. A rain gutter is to be formed as shown in the figure by bending up at equal angles sides of 10 cm from a long metal sheet 30 cm wide. Find the angle θ that will enable the gutter to handle the greatest flow of water without overflowing.

27. Find the dimensions of the right circular cone of maximum volume that can be inscribed in a sphere of fixed radius a.

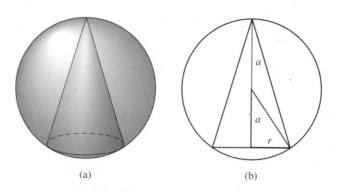

(a)   (b)

28. Change Exercise 27 so that the cone is to have maximum lateral surface area.

29. A telephone cable is to be laid under a river from point A to point C (see the figure at the top of page 251) and then underground to point B. The river is 20 m wide, and the distance from point B to the point directly across the river from A is 40 m. If it costs $1\frac{1}{2}$ times as much to lay the cable under the river as it does to put it underground, find where point C should be located to minimize the cost.

30. A window is to be designed in the shape of a rectangle surmounted by an equilateral triangle. The overall outside perimeter is to be 12 ft. Find the dimensions of the window if the area is to be a maximum.

31. Repeat Exercise 30 with the triangle replaced by a semicircle.

32. A silo is constructed in the form of a right circular cylinder surmounted by a hemisphere of the same radius as the cylinder. Find the dimensions that give the minimum cost if it is to have a volume of 600 m³, and the unit cost of the material for the hemisphere is twice that for the cylinder.

33. A car leaves town A traveling at 80 km/hr due east on a straight road toward town B, 50 km away. At the same time, a second car leaves town B traveling 60 km/hr due south on a straight road (see the figure). How much time must the cars travel until the distance between them is a minimum? At that instant where are the cars and how far apart are they?

34. The cost of operating a certain truck is determined from experience to be $20 + x/12$ cents per kilometer when the truck is driven at a speed of x km/hr. The truck driver earns $9 per hour of actual driving time. What is the most economical speed at which to operate the truck on a 600-km trip?

35. Find the length of the longest piece of straight pipe that can be carried horizontally around the corner from a 5-ft-wide hallway into a 3-ft-wide hallway. Assume the pipe has negligible thickness. (*Hint:* Choose as independent variable the angle θ as shown on the following page.)

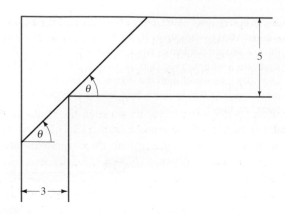

37. A long sheet of paper 6 in. wide is to be folded up as shown in the figure. Find where the point P of the fold should be so that (a) the area of the folded part is a minimum and (b) the length PQ of the fold is a minimum.

36. The illumination from a source of light varies directly as the intensity of the light source and inversely as the square of the distance from the source. One light is 1 m directly above one end of a table 3 m in length, and another light of twice the candlepower is directly opposite the first at the other end of the table and 2 m above it (see the figure). At what point on the table directly between the lights will the sum of the intensities of lights be the least?

38. Reconsider the situation in Exercise 36 with two identical light sources s units apart, positioned at the same distance above the table. Show that the point at which the intensity of illumination on the table is a minimum is *not* $s/2$, but rather depends on the distance, d, of the light sources from the table. This problem is more typical of those encountered in science and engineering in that it involves an additional variable, the parameter d, and branching solutions.

4.4 RELATED RATES

In this section we will apply the derivative to functions that have *time* as the independent variable. We already know that if a function represents the position of a particle moving in a straight line, then the derivative with respect to time is the velocity of the particle, and the derivative of the velocity is the acceleration. But many other time-dependent functions exist. The type of problem we wish to consider is one in which we know one or more rates of change and want to find another rate of change that is dependent on those we know. Such problems are called **related-rate problems**. The next two examples illustrate how one rate of change is influenced by another.

FIGURE 4.60

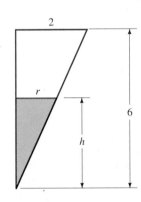

FIGURE 4.61

EXAMPLE 4.23   Suppose a cone-shaped container is 6 m tall and the radius of the top is 2 m. (See Figure 4.60.) Water is being pumped into the container at the constant rate of 500 L/min (L is the symbol for liters, and $1 \text{ L} = 10^{-3} \text{ m}^3$). Find the rate at which the water level is rising (a) when the water level is 3 m and (b) when the container is half full.

Solution   We denote by V, h, and r the volume of water in the container, the height of the water, and the radius of the circular surface of the water, respectively, all at time t. It is important to note that all three variables are functions of time and that they are not constants. It would be incorrect at this stage, for example, in solving part (a) to put $h = 3$, because h is not always 3, and we must first find the relationship between the relevant rates of change before we consider what they are at a particular instant. *In short, before substituting a particular value for a variable quantity, we must differentiate.*

By the formula for the volume of a cone, the volume of water at any instant is given by

$$V = \frac{1}{3}\pi r^2 h \qquad (4.1)$$

In both parts (a) and (b) we want to know dh/dt at a particular instant, so we will need to differentiate. First, however, it will simplify matters to replace r by its equivalent in terms of h, since h is the variable of interest in this problem. We can establish a relation between r and h using similar triangles. Figure 4.61 is the right-hand half of a vertical slice along the axis of the cone. From it, we get

$$\frac{r}{2} = \frac{h}{6}$$

so $r = \frac{1}{3}h$. Substituting this equivalent value for r in Equation 4.1 and simplifying, we get

$$V = \frac{1}{27}\pi h^3 \qquad (4.2)$$

Now we can differentiate both sides with respect to time t:

$$\frac{dV}{dt} = \frac{1}{9}\pi h^2 \frac{dh}{dt}$$

(Note the use of the Chain Rule.) Since dV/dt is the rate of change of the volume of water, and since we are given that water is entering the container at the rate of 500 L/min, or $\frac{1}{2}$ m³/min, we have

$$\frac{dV}{dt} = \frac{1}{2}$$

So

$$\frac{1}{2} = \frac{1}{9}\pi h^2 \frac{dh}{dt}$$

or

$$\frac{dh}{dt} = \frac{9}{2\pi h^2} \qquad (4.3)$$

Now we can find dh/dt for any value of h. For part (a) we have

$$\left.\frac{dh}{dt}\right|_{h=3} = \frac{9}{2\pi(3^2)} = \frac{1}{2\pi} \approx 0.159 \text{ m/min}$$

For part (b) we must find h when the container is half full. The volume of the entire container is $\frac{1}{3}\pi(2)^2 \cdot 6 = 8\pi$, so when it is half full, the volume is 4π. Substituting in Equation 4.2, we get

$$4\pi = \frac{1}{27}\pi h^3$$

$$h^3 = 4(27)$$

$$h = 3\sqrt[3]{4}$$

Now we use Equation 4.3 again:

$$\left.\frac{dh}{dt}\right|_{h=3\sqrt[3]{4}} = \frac{9}{2\pi(3\sqrt[3]{4})^2} = \frac{1}{4\pi\sqrt[3]{2}} \approx 0.0632 \text{ m/min} \qquad \blacksquare$$

EXAMPLE 4.24   A car leaves town A and travels due north at an average speed of 40 km/hr. At the same instant, another car leaves town B, 50 km due east of town A, and travels due south at an average speed of 80 km/hr. Find how fast the distance between the cars is increasing 1 hr later.

Solution   As in Figure 4.62 let x denote the distance traveled by the first car, y the distance traveled by the second car, and z the distance between them, all after t hr. Note that x, y, and z are all variable quantities. The only constant is the distance 50 between A and B. From the given information, we know that

$$\frac{dx}{dt} = 40 \qquad \text{and} \qquad \frac{dy}{dt} = 80$$

and we want to find dz/dt at a particular time. The first thing we must do is to find an equation relating the three variables. Then we will differentiate with respect to t.

By means of the dashed lines in Figure 4.62 we have constructed a right triangle with legs of length 50 and $x+y$, and hypotenuse z. By the Pythagorean Theorem we have

$$z^2 = (50)^2 + (x + y)^2 \qquad (4.4)$$

We differentiate both sides with respect to time, t:

$$2z\frac{dz}{dt} = 2(x + y)\left(\frac{dx}{dt} + \frac{dy}{dt}\right)$$

Substituting the known values of dx/dt and dy/dt and solving for dz/dt gives

$$\frac{dz}{dt} = 120\left(\frac{x + y}{z}\right) \qquad (4.5)$$

which is the rate of change of the distance between the cars at any time. Now we wish to let $t = 1$. At that instant $x = 40$ and $y = 80$, and so from Equation 4.4,

$$z = \sqrt{(50)^2 + (120)^2} = \sqrt{16,900} = 130$$

Thus, by Equation 4.5,

$$\left.\frac{dz}{dt}\right|_{t=1} = (120)\left(\frac{40 + 80}{130}\right) = \frac{1440}{13} \approx 111 \text{ km/hr} \qquad \blacksquare$$

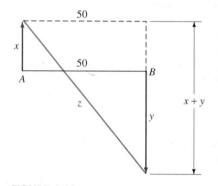

FIGURE 4.62

The General Procedure

Examples 4.23 and 4.24 are typical of "related-rate" problems. Let us examine what they have in common. In both cases we were given one or more rates of change, and we wished to find another at a given instant. We were able to find a relationship between the variables whose rates of change we knew and the variable whose rate of change we were seeking. In Example 4.23 we found a relationship by using the formula for the volume of a cone and then using similar triangles to get r in terms of h. In Example 4.24 we obtained a relationship by using the Pythagorean Theorem. In both cases drawing a figure was an important aid in finding the relationships. Next, we differentiated with respect to time and substituted the known rates of change. Finally, we substituted the values of all variables at the particular instant in question and solved for the unknown rate of change.

We summarize these steps here.

Procedure for Solving Related-Rate Problems

1. If it is appropriate, draw a figure. On it label all variable parts with letters and all fixed parts with numbers.

2. Find an equation relating the variables whose rates of change are known and the variable whose rate of change is desired. Look for such things as areas or volumes of geometric figures, similar triangles, the Pythagorean Theorem, or trigonometric relationships.

3. Differentiate both sides of the equation found in step 2 with respect to time, and substitute the known rates of change. Don't forget to apply the Chain Rule.

4. Calculate the values of all variables involved at the particular instant in question, and substitute them in the result of step 3.

5. Solve for the unknown rate of change.

We cannot overemphasize the importance of waiting until step 4 to substitute instantaneous values of variable quantities. This substitution is to be done only *after* differentiation. Substituting such values too soon is the most frequent mistake students make.

We will give one further example of a related-rate problem.

EXAMPLE 4.25   A television camera is 30 m directly opposite the finish line at an automobile racetrack and is trained on the lead car as it approaches the finish. If the car is traveling at the rate of 150 km/hr, find the rate in radians per second at which the camera is turning when the car is 40 m from the finish line. What is the rate in degrees per second?

Solution   In Figure 4.63 we have labeled the relevant variables x and θ. We will measure x in meters and θ in radians. Note carefully that the distance from the car to the finish line is not shown as 40 because this distance is a *variable* quantity. We will put $x = 40$ only *after* we differentiate.

FIGURE 4.63

We are given that the car is approaching the finish line at 150 km/hr. Since other units are in meters and we want the answer to be in radians per second, we convert 150 km/hr to meters per second:

$$150 \text{ km/hr} = \frac{150 \text{ km} \times 1000 \text{ m/km}}{1 \text{ hr} \times 3600 \text{ s/hr}}$$

$$= \frac{125}{3} \text{ m/s}$$

Since x decreases with time, dx/dt must be negative. So

$$\frac{dx}{dt} = -\frac{125}{3}$$

An equation relating x and θ is

$$\tan \theta = \frac{x}{30}$$

We differentiate both sides with respect to time, recalling that $\dfrac{d}{d\theta} \tan \theta = \sec^2 \theta$:

$$\sec^2 \theta \frac{d\theta}{dt} = \frac{1}{30} \frac{dx}{dt} = \frac{1}{30}\left(-\frac{125}{3}\right) = -\frac{25}{18}$$

So

$$\frac{d\theta}{dt} = \frac{1}{\sec^2 \theta}\left(-\frac{25}{18}\right) = -\frac{25}{18}\cos^2 \theta$$

Now we are ready to consider the situation when $x = 40$. At that instant, the right triangle is as shown in Figure 4.64, so the hypotenuse is 50. Thus, $\cos \theta = 3/5$. Therefore, we have

$$\left.\frac{d\theta}{dt}\right|_{x=40} = -\frac{25}{18}\left(\frac{3}{5}\right)^2 = -\frac{25}{18}\cdot\frac{9}{25} = -\frac{1}{2} \text{ radian/s}$$

The significance of the negative sign is that θ decreases with time. Since 1 radian $= (180/\pi)°$, or approximately $57.3°$, the camera is turning at about $-28.6°/\text{s}$, which would lead to one revolution in about $12\frac{1}{2}$ seconds, and this rate seems reasonable. (If the rate of turning had been several revolutions per second, we would have rejected it as being unreasonable and rechecked.) ■

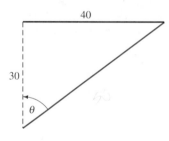

FIGURE 4.64

Exercise Set 4.4

1. Two joggers leave simultaneously from the same point, one going east at 3 m/s and the other going south at 4 m/s. Find how fast the distance between them is increasing 30 min later.

2. The commander of a Coast Guard cutter is notified from an aircraft that a boat suspected of carrying illegal cargo is headed north in the Gulf of Mexico at the rate of 15 knots (1 knot is 1 nautical mile per hour). By heading east and traveling at 36 knots, the commander estimates that the cutter can intercept the boat in 50 min. Find how fast the two boats are approaching each other after 30 min.

3. A child 5 ft tall is walking away from a lamppost at the rate of 3 ft/s. If the height of the lamppost is 20 ft, find the rate at which the child's shadow is lengthening.

4. In Exercise 3, find the rate at which the tip of the child's shadow is moving.

5. A child throws a rock into a pond, causing a circular ripple. If the radius of the circle increases at the constant rate of 1 m/s, find how fast the area is increasing when the radius is 5 m.

6. A V-shaped trough has a cross section in the form of an inverted isosceles triangle with base 40 cm and altitude 50 cm. If the trough is 3 m long and is being filled with water at the rate of 200 L/min ($1 \text{ L} = 10^{-3} \text{ m}^3$), find how fast the water level is rising when the water level is 30 cm.

7. A lighthouse is 500 m from a straight shore. If its light is revolving at the rate of 10 rev/min, find how fast the beam of light is moving along the shoreline when the angle between the light beam and the shoreline is 30°.

8. Helium is being pumped into a spherical balloon at the rate of 600 cm³/s. At what rate is the radius increasing when the radius is 20 cm? At what rate is the surface area of the balloon increasing at this same instant?

9. Two runners are 7 m apart at the starting line, and they run on parallel paths, one averaging 8 m/s and the other

7.6 m/s. At what rate is the distance between the runners increasing after 1 min?

10. Coal is being transported on a conveyor to conical piles each of whose height is one-fourth the diameter of the base. If the coal is being dropped at the rate of 3m³/min, find how fast the height of a pile is increasing when it is 10 m high. How fast is the circumference of the base increasing at this same instant?

11. A watering trough 10 ft long has a cross section in the form of an isosceles trapezoid with dimensions as shown in the figure. If water is being pumped into the tank at the rate of 8 ft³/min, find how fast the water level is rising when it is 1 ft deep.

12. In Exercise 11, find how fast the water level is rising when the trough is half full.

13. A boat is being pulled in by means of a rope attached to a windlass on the dock. The windlass is 4.5 m above the level at which the rope is attached to the boat, and the rope is being pulled in at the rate of 2 m/s. Find how fast the boat is approaching the base of the dock when it is 6 m from the base.

14. A helicopter is 500 m directly above an observer on the ground at a given instant and is flying horizontally at the rate of 50 m/s. Find how fast the angle at the observer from the vertical to the line of sight of the helicopter is changing after 10 seconds (see the figure on the following page).

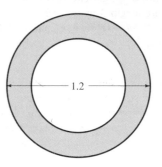

Through how many radians per second is the searchlight turning when the car is 20 m past the point on the road nearest the searchlight?

21. A child flying a kite is letting string out at the rate of 2 m/s, and the kite is 30 m high (from hand level) moving horizontally. Find the speed of the kite at the instant when 34 m of string are out.

22. The buildup of mineral deposits in a water pipe in an area of hard water gradually restricts the rate of flow. Assume that the buildup forms a uniform circular ring around the inside of a pipe (see the figure), and suppose that in a pipe of inside diameter 1.2 cm the area of the annular ring increases at the rate of 10π mm^2 per year. At what rate is the effective diameter of the pipe decreasing when the thickness of the deposit is 2 mm?

15. When a gas under pressure expands or contracts with no gain or loss of heat, the relationship between pressure p and volume v is given by $pv^k = C$, where k and C are constants. This relation is known as the *adiabatic law*. Suppose for a certain gas $k = 1.4$ and at a given instant the pressure is 2 N/m^2 (newtons per square meter), the volume is 0.064 m^3, and the pressure is increasing at the rate of 0.5 N/m^2 each second. Find the rate of change of the volume at this instant.

16. An oil spill from a drilling platform in the Gulf of Mexico is approximately circular, and the diameter is observed to be increasing at the constant rate of 100 m per day. How fast is the area of the spill increasing when the diameter is 500 m?

17. Newton's Universal Law of Gravitation states that two bodies of masses m_1 and m_2, at a distance r units apart, exert a force of attraction on each other given by the formula

$$F = G\frac{m_1 m_2}{r^2}$$

where G is the universal gravitational constant. Suppose that for two bodies of fixed masses m_1 and m_2, respectively, the force of attraction is 20 N (newtons) when they are 10 m apart, and at that instant they are separating at the rate of 5 m/s. How fast is the force of attraction decreasing?

18. A small airplane is flying 120 mi/hr at an altitude of 5000 ft. An observer on the ground is keeping a transit (a surveyor's instrument that measures angles) trained on the plane. Through how many radians per second must the observer rotate the scope of the transit when the plane is flying away from the observer and the angle of elevation of the plane is 60°? Neglect the height of the transit. (Be careful with units.)

19. Some children are rolling a ball of snow to make a snowman. If the radius of the ball is increasing at the rate of 20 cm/min, how fast is the volume of the ball increasing when it is 60 cm in diameter?

20. A searchlight located 30 m from a straight road is kept trained on a car traveling on the road at 50 km/hr.

23. The larvae of a certain insect have the shape of a right circular cylinder with hemispheres at both ends and with length five times the diameter (see the figure). During the first stages of growth, the larvae increase in diameter at the rate of 0.2 mm per day. At what rate does the volume increase when the diameter is 3 mm?

24. A water tank in the form of an inverted cone of height 6 m and base radius 3 m is being drained at the rate of 2 m^3/min. How fast is the water level falling when there are $(2\pi)/3$ m^3 of water in the tank?

25. In a certain mechanical system, parts A and B are connected by a cable 84 cm long that passes over a pulley at C 30 cm above the point D on a horizontal track on which parts A and B are constrained to move (see the figure). The parts move back and forth in such a way that the cable is always taut. At a certain instant, part B

is 16 cm from D and is moving to the right at 17 cm/s. How fast is part A moving at that instant?

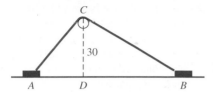

26. One end of an 85-ft rope is attached to a weight on the ground. The rope goes over a pulley 45 ft above the ground, with the other end held by a worker standing beside the weight. The end of the rope held by the worker is 5 ft above the ground, and continuing to hold it at this height, the worker begins to walk away at the constant rate of 5 ft/s, thereby raising the weight. How fast is the weight rising when the worker has walked 30 ft?

27. An oil tank is in the form of a right circular cylinder with a horizontal axis, having length 3 m and radius 1 m. If the tank is being filled at the rate of 60 L/min, find the rate at which the oil depth is increasing at the instant when the depth is 20 cm.

28. Cars A and B leave from the same point, traveling on roads that are $120°$ apart. Car A travels at an average speed of 60 km/hr. Car B leaves the intersection one-half hour after car A and travels at an average speed of 45 km/hr. How fast are the cars separating when car A has traveled 150 km?

29. A water tank is in the form of a frustum of a cone (portion of a cone between two parallel planes; see the figure) with upper radius 3 m, lower radius 1 m, and height 4 m. Water is being pumped into the tank at the rate of 5 m³/min and is being taken from the

tank at the rate of 2 m³/min. Find the rate at which the water level is rising when the tank contains 30 m³ of water.

30. A train passes over the middle of a bridge at the same time a car goes under the middle on a path at right angles to the bridge. The bridge is 24 m above the road. If the train's speed is 10 m/s and the car's speed is 15 m/s, find the rate at which the distance between them is increasing 4 s later.

4.5 APPROXIMATING ROOTS OF EQUATIONS

In this section we show how to use derivatives to find approximate roots of equations. If an equation is of the form $f(x) = 0$, then $x = r$ is a **root** of the equation provided $f(r) = 0$. So a root of the equation is a zero of the function f, or equivalently, an x-intercept of its graph. One approach to solving the equation $f(x) = 0$, then, is to draw the graph of the function f and estimate the values where it crosses the x-axis. Obtaining the graph with the aid of a CAS or graphing calculator and zooming in on the x-intercepts repeatedly is one way of approximating roots of equations. The method we will describe is often more efficient and will yield answers to a high degree of accuracy in just a few steps in most cases.

Of course, there are equations for which exact solutions can be found. For example, if f is a polynomial function of degree 1 or 2, we can obtain exact roots. For the first-degree equation $ax + b = 0$, the solution is simply $x = -b/a$, and for the second-degree equation $ax^2 + bx + c = 0$, the solutions are given by the quadratic formula

$$x = \frac{-b \pm \sqrt{b^2 - 4ac}}{2a}$$

Just as the quadratic formula works on every second-degree polynomial equation, formulas exist for solving third- and fourth-degree equations. These formulas are seldom used, however, as they are rather complicated. Surprisingly, though, no such formulas exist for polynomial equations of degree 5 or higher. It is not just that no one has been clever enough to find such formulas; they do not exist! This amazing result was proved in 1824 by the brilliant young Norwegian mathematician Niels Henrik Abel (1802–1829). So, for example, if you wanted to solve the equation

$$x^5 - 3x^2 + 4x - 7 = 0$$

you would have to use some approximating technique. (It should be noted, however, that some fifth-degree, or higher-degree, equations have rational roots, which can be obtained by methods you may have studied. For example, you can verify that the roots of $x^5 + 3x^4 - 5x^3 - 15x^2 + 4x + 12 = 0$ are $x = \pm 1$, ± 2, and -3.) For nonpolynomial functions, the situation is likely to be even worse, as only very special equations can be solved in an exact way. Even the simple equation $\tan x - x = 0$ cannot be solved in an exact form for any root other than the obvious one, $x = 0$.

The basic idea of the method we describe originated with Newton, and it is generally referred to as **Newton's Method**. However, the form in which it is presently used was given by Joseph Raphson (1648–1715), so it is also known as the **Newton–Raphson Method**. The method is easy to describe geometrically. Suppose that by graphical means or otherwise, we find that the equation $f(x) = 0$ has at least one real root. Let r be the true value of one such root. We assume that f is differentiable on an interval I containing r. We make an initial guess at the value of r and denote this first guess by x_0. Sketching the graph of f or making a table of values is usually sufficient to provide a reasonable guess for x_0. Unless we are unusually lucky, x_0 will not equal r. So we locate the point $(x_0, f(x_0))$ and draw the tangent line to the graph of f there. As in Figure 4.65 we let x_1 be the x-intercept of this tangent line. This value, x_1, is our next approximation to r. In general, if x_0 is close enough to r, x_1 will be closer to r than x_0. (If this is not so, try another x_0 as the starting point.) Now we repeat the process. We draw the tangent line to the graph of f at $(x_1, f(x_1))$ and obtain the next approximation to r from its x-intercept, which we call x_2. We continue this process until we obtain the desired degree of accuracy.

To develop a formula for x_{n+1}, once we know x_n, we need to find the equation of the tangent line at $(x_n, f(x_n))$. Its slope is $f'(x_n)$. So by the point-slope form of the equation of a line, its equation is

$$y - f(x_n) = f'(x_n)(x - x_n)$$

To get the x-intercept, we set $y = 0$:

$$-f(x_n) = f'(x_n)(x - x_n)$$

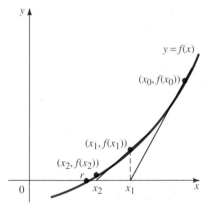

FIGURE 4.65

We solve for x and name the result x_{n+1}:

$$x_{n+1} = x_n - \frac{f(x_n)}{f'(x_n)} \qquad (4.6)$$

Of course, $f'(x_n)$ must be different from 0. The successive approximations $x_0, x_1, x_2, \ldots$ form what is called a **sequence**, and the approximations are called **terms** of the sequence. Equation 4.6, giving the term x_{n+1} when x_n is known, is called a **recursive formula** for the sequence.

Newton's Method (The Newton–Raphson Method)

1. Make an initial guess x_0 for a root r of the equation $f(x) = 0$.
2. Find $x_1, x_2, x_3, \ldots$ using

$$x_{n+1} = x_n - \frac{f(x_n)}{f'(x_n)} \qquad (f'(x_n) \neq 0)$$

by taking $n = 0, 1, 2, 3, \ldots$

If the terms of the sequence of approximations come arbitrarily close to the root r as n increases, we say the sequence **converges** to r. In general, the sequence of approximations will converge to the root r you are seeking if your initial choice x_0 is close to r. There are various tests to guarantee convergence, and we will mention one of these in a moment, but in practice it is usually sufficient to try a starting point x_0 and then to observe whether subsequent approximations seem to be approaching a constant value. We illustrate this approach in the next two examples.

EXAMPLE 4.26   Approximate the positive root of $x^2 - 2x - 2 = 0$ using Newton's Method. Continue the process until two consecutive approximations agree to the first five decimal places, and round the answer to four places. Compare with the exact answer.

Solution   We show the graph of $f(x) = x^2 - 2x - 2$ in Figure 4.66. Since $f(2) = -2$ and $f(3) = 1$, the root we are seeking is between 2 and 3, and from the graph it appears to be closer to 3. A good first approximation might therefore be $x_0 = 2.6$. To apply Newton's Method we need f':

$$f'(x) = 2x - 2 = 2(x - 1)$$

Now we apply the recursion formula (4.6), using a calculator:

$$n = 0: \quad x_1 = x_0 - \frac{f(x_0)}{f'(x_0)} = 2.6 - \frac{f(2.6)}{f'(2.6)} = 2.7375$$

$$n = 1: \quad x_2 = x_1 - \frac{f(x_1)}{f'(x_1)} = 2.72918$$

$$n = 2: \quad x_3 = x_2 - \frac{f(x_2)}{f'(x_2)} = 2.73205$$

$$n = 3: \quad x_4 = x_3 - \frac{f(x_3)}{f'(x_3)} = 2.7320508$$

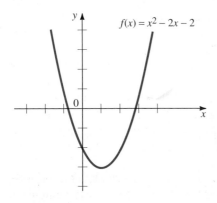

$f(x) = x^2 - 2x - 2$

FIGURE 4.66

Since x_3 and x_4 agree to the first five decimal places, we stop here and say that $r \approx 2.7321$. The fact that x_3 and x_4 agree to five places does not guarantee that the answer is correct to four places, but it strongly suggests that is true.

To compare with the exact answer, we solve for x by the quadratic formula, getting

$$x = \frac{2 \pm \sqrt{4 + 8}}{2} = 1 \pm \sqrt{3}$$

The positive root is $1 + \sqrt{3}$, which to seven places is 2.7320508. Note that x_4 agrees with this answer to all seven places, so four iterations of Newton's Method yielded very good results. ∎

EXAMPLE 4.27   Use Newton's Method to approximate the smallest positive root of the equation $\tan x - x = 0$.

Solution   Let $f(x) = \tan x - x$. We will need f':

$$f'(x) = \sec^2 x - 1 = \tan^2 x$$

Rather than graphing f, it is easier to graph $y_1 = \tan x$ and $y_2 = x$ and estimate where $y_1 = y_2$, as we have done in Figure 4.67. It appears that the root r we are seeking is close to $3\pi/2 \approx 4.71$. We try $x_0 = 4.5$ as a first approximation of r, and proceed with Newton's Method. The recursion formula is

$$x_{n+1} = x_n - \frac{f(x_n)}{f'(x_n)} = x_n - \frac{\tan x_n - x_n}{\tan^2 x_n}$$

This time we use a table:

n	x_n	x_{n+1}
0	4.5	4.4936
1	4.4936	4.49341
2	4.49341	4.49341

So in just three iterations it appears that we have five-place accuracy and can say that $r \approx 4.49341$. ∎

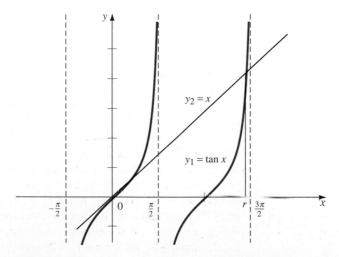

FIGURE 4.67

REMARK ──

■ It was essential in the preceding example that we have the calculator set in the *radian* mode.

──

Pitfalls of Newton's Method

We turn our attention now to ways in which Newton's Method can fail. We know already that if $f'(x_n) = 0$ for some n, then the method fails. In such a case we try another starting point. As shown in Figure 4.68, if for some n, x_n is simply near a point where the derivative is 0, it can be sufficient to cause all subsequent approximations to be farther from r than x_n because of the flatness of the slope of the tangent line at $(x_n, f(x_n))$. Some other ways in which the process may go awry are shown in Figure 4.69. In part (a) the approximations again get farther and farther from r. In part (b) the process is in an endless loop, and in part (c) Newton's Method does converge, but to a different root from the one we expect.

FIGURE 4.68

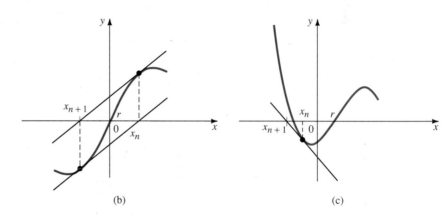

(a)   (b)   (c)

FIGURE 4.69

Criteria for Convergence

If there is an open interval containing the root r that we are trying to approximate throughout which f is monotone and the concavity is always in the same direction, then r is the only root in the interval. Furthermore, if we choose x_0 such that $f(x_0)$ is positive when the graph is concave up and negative when the graph is concave down, then it can be proved that the successive approximations in Newton's Method always converge to r. Figure 4.70 illustrates such a choice of x_0 for each type of concavity for both increasing and decreasing functions. Notice that in each case as you proceed along the curve from the point $(x_0, f(x_0))$ toward $(r, 0)$, the curve bends toward the x-axis. We emphasize that the conditions stated here for convergence are *sufficient* only. So even if these conditions are not met, the approximations may (and usually will) converge. That is, if the conditions are satisfied, then approximations produced by Newton's Method definitely will converge to the root, but it is possible they will converge to the root in the absence of one or more of the conditions stated.

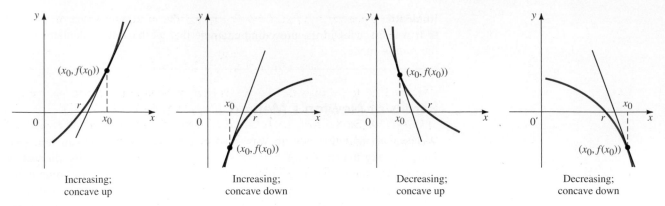

Increasing;
concave up

Increasing;
concave down

Decreasing;
concave up

Decreasing;
concave down

FIGURE 4.70

Suppose that the conditions stated in the preceding paragraph are met. If there are positive constants m and M such that $|f'(x)| > m$ in the interval (the slope is not too small) and $|f''(x)| \leq M$ in the interval (the concavity is not too great), then the error in x_{n+1} is related to the error in x_n by the inequality

$$E_{n+1} \leq \frac{M}{2m} E_n^2 \tag{4.7}$$

where $E_n = |x_n - r|$ and $E_{n+1} = |x_{n+1} - r|$. We will see how to prove this result in Chapter 10. One consequence of Inequality 4.7 is that when $M/2m < 2$, if x_n approximates r correctly to k decimal places, then x_{n+1} approximates r correctly to at least $2k$ decimal places. (See Exercise 29 in Exercise Set 4.5.) That is, the precision is doubled with each successive approximation, which explains the rapidity of convergence of Newton's Method.

EXAMPLE 4.28   Find a starting point x_0 for Newton's Method that will guarantee convergence to the largest positive root of $x^3 - 3x^2 - 24x + 48 = 0$. Find an upper bound on the error in using x_3 to approximate the root.

Solution   Let $f(x) = x^3 - 3x^2 - 24x + 48$. Then

$$f'(x) = 3x^2 - 6x - 24 = 3(x + 2)(x - 4)$$

$$f''(x) = 6x - 6 = 6(x - 1)$$

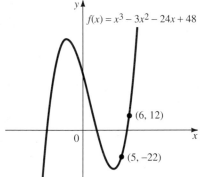

$f(x) = x^3 - 3x^2 - 24x + 48$

(6, 12)

(5, −22)

FIGURE 4.71

We analyze the signs of $f'(x)$ and determine that f is increasing on $(4, \infty)$ and on $(-\infty, -2)$, and f is decreasing on $(-2, 4)$. From the sign of $f''(x)$, we see that the graph is concave up for $x > 1$ and concave down for $x < 1$. A sketch of the graph of f is given in Figure 4.71. By calculator, we find that $f(5) = -22$ and $f(6) = 12$. So a root lies in the interval $I = (5, 6)$. No root can be to the right of this one since f is increasing and concave up everywhere to the right of $x = 5$. According to the convergence criteria we have stated, we should choose x_0 so that $f(x_0) > 0$, since the concavity is upward in the interval $(5, 6)$ containing the root. Checking $f(5.8)$, we find its value to be 2.992. So $x_0 = 5.8$ is a suitable starting point. The table shows the calculated values of x_1, x_2, and x_3.

n	x_n	x_{n+1}
0	5.8	5.729
1	5.729	5.72716
2	5.72716	5.72716014

Now we wish to estimate the error in the approximation to r given by x_3, which is the last entry in the table. To use Inequality 4.7, we need positive constants m and M such that for x in the original interval $I = (5, 6)$, $|f'(x)| \geq m$ and $|f''(x)| \leq M$. Both f' and f'' are increasing and nonnegative in I. So the minimum value of $f'(x)$ is $f'(5) = 21$, and the maximum value of $f''(x)$ is $f''(6) = 30$. Thus, Inequality 4.7 becomes

$$E_{n+1} \leq \frac{30}{2(21)} E_n^2 = \frac{5}{7} E_n^2$$

We need E_0 to begin with. To get as small a bound on the error as we can, we should make E_0 small. We know that $f(5.8) > 0$. Checking 5.7, we find $f(5.7) < 0$. So the root lies between 5.7 and 5.8, which means that $x_0 = 5.8$ is less than 0.1 unit from the root. That is, $E_0 \leq 0.1$. Thus,

$$E_1 \leq \frac{5}{7}(0.1)^2 \approx 0.00714$$

$$E_2 \leq \frac{5}{7}(0.00714)^2 \approx 0.0000364$$

$$E_3 \leq \frac{5}{7}(0.0000364)^2 \approx 0.000000000946$$

Clearly, the answer we have found for x_3 is accurate to at least eight decimal places. ■

We will mention briefly two other methods for approximating roots of equations. Although these methods do not involve the derivative, they are easily understood and simple to apply. Furthermore, the first method provides an alternative way of estimating the error in Newton's Method.

The Method of Secants

Suppose the interval I contains only one root r of the equation $f(x) = 0$. For example, this condition will be met, as we have previously seen, when f is monotone in I and the direction of its concavity does not change there. Choose as a starting point, x_0, any point in I such that $f(x_0) > 0$ when the concavity is upward and $f(x_0) < 0$ when the concavity is downward. In Figure 4.72, we illustrate such a choice for a graph that is concave up. Now choose any other point z_0 in I such that $f(z_0)$ is opposite in sign to $f(x_0)$. Then, by the Intermediate-Value Theorem we know that the function takes on the value 0 somewhere between z_0 and x_0; that is, the root r lies between z_0 and x_0. As shown in Figure 4.72 we now draw the *secant line* joining the points $(x_0, f(x_0))$ and $(z_0, f(z_0))$. Its x-intercept is the next approximation to the root. Call it z_1. Then, let z_2 be the x-intercept of the secant line joining $(x_0, f(x_0))$ and $(z_1, f(z_1))$. Continue this process in such a way that z_{n+1} is the x intercept of the secant line joining $(x_0, f(x_0))$ and $(z_n, f(z_n))$ for $n = 0, 1, 2, 3, \ldots$ In Exercise 30 of this section you will be asked to derive the formula

$$z_{n+1} = z_n - \frac{f(z_n)}{m_n} \tag{4.8}$$

where m_n is the slope of the secant line joining $(x_0, f(x_0))$ and $(z_n, f(z_n))$. The method we have described is called the **Method of Secants**, for obvious reasons.

It is evident geometrically (see Figure 4.73) that at the nth stage of the Method of Secants, the point z_n and the point x_n at the nth stage of Newton's Method are on opposite sides of the root r. It follows, then, that the error $E_n = |x_n - r|$ at the nth stage of Newton's Method does not exceed the distance between x_n and z_n: $E_n \leq |x_n - z_n|$.

FIGURE 4.72

FIGURE 4.73

FIGURE 4.74

FIGURE 4.75

The Bisection Method

The other method we describe is called the **Bisection Method**. Assume again that f has only one root r in an interval I throughout which f is continuous. Let a_0 and b_0 (with $a_0 < b_0$) be two points in I such that $f(a_0)$ and $f(b_0)$ are opposite in sign so that the root r lies in the interval $[a_0, b_0]$. What we are going to do is bisect the interval $[a_0, b_0]$ and choose the half containing the root r. Then we will bisect that half and again choose the half containing r. By continuing in this manner, we get a sequence (called a *nest*) of intervals, each half as long as the preceding one, and each containing the root r. Since each interval is half as long as the previous one, the lengths become arbitrarily small as we continue the process indefinitely. Thus we can obtain an approximation to the root to whatever degree of accuracy we choose. In Figure 4.74 we show a possible sequence of intervals such as we have described. We designate the half of $[a_0, b_0]$ containing r by $[a_1, b_1]$, and we show it below $[a_0, b_0]$, although it actually coincides with the right half of $[a_0, b_0]$. Similarly, $[a_2, b_2]$ is the half of $[a_1, b_1]$ containing r, and so on. If we stop with the interval $[a_n, b_n]$, then we use its midpoint

$$M_n = \frac{a_n + b_n}{2}$$

as the approximation to r.

Since the width of each new interval in the Bisection Method is half of the preceding one, the width of the nth interval is $(b_0 - a_0)/2^n$. The root r lies in the nth interval. So the distance from the midpoint M_n to r is less than half the width of the interval (see Figure 4.75). Thus, the error, E_n, in stopping at the nth stage satisfies

$$E_n \le \frac{b_0 - a_0}{2^{n+1}} \tag{4.9}$$

REMARK ──────────────────────────────

■ It may happen at some stage that the midpoint of some interval is the exact root r. If so, we stop the process at that stage.

──

The Bisection Method is easy to understand and is easy to program on a computer. The convergence is quite slow, however, compared with Newton's Method or the Method of Secants.

EXAMPLE 4.29   Approximate the largest positive root of $f(x) = 0$, where

$$f(x) = x^3 - 3x^2 - 24x + 48$$

Use (a) the Method of Secants and (b) the Bisection Method. Compare the results with the answer obtained by Newton's Method in Example 4.28.

Solution   In Example 4.28 we found that the largest positive root was in the interval $(5, 6)$ and that $f(5) = -22$ and $f(6) = 12$.

(a) By the Method of Secants we can take $x_0 = 6$ and $z_0 = 5$. The slope of the secant line joining the two points $(5, -22)$ and $(6, 12)$ is

$$m_0 = \frac{12 - (-22)}{6 - 5} = 34$$

So by the recursion formula (4.8)

$$z_1 = 5 - \frac{-22}{34} = 5.647$$

We calculate $f(z_1)$ to be

$$f(5.647) = -3.119$$

The slope of the secant line joining $(z_1, f(z_1))$ and $(x_0, f(x_0))$ is

$$m_1 = \frac{12 + 3.119}{6 - 5.647} = 43.376$$

By the recursion formula (4.8), we find that

$$z_2 = 5.647 - \frac{-3.119}{43.376} = 5.7198$$

Continuing in this way, we find (omitting the details) that

$$z_3 = 5.7264 \qquad z_4 = 5.727095 \qquad z_5 = 5.727154$$

(b) By the Bisection Method, we take the initial interval $[a_0, b_0]$ to be $[5, 6]$. Its midpoint is 5.5. We know already that $f(6)$ is positive, and we find that $f(5.5)$ is negative. So we take $[a_1, b_1]$ to be $[5.5, 6]$. The value of f at the midpoint, 5.75, is found to be positive, so the root lies in $[5.5, 5.75]$, which we call $[a_2, b_2]$. The midpoint of $[a_2, b_2]$ is 5.625, and $f(5.625)$ is negative, so we take $[a_3, b_3]$ to be $[5.625, 5.75]$. Continuing in this way, we find after 14 bisections that the root lies in the interval $[5.727155, 5.727185]$. If we stop here, we take the midpoint M_{14} as the approximation to r:

$$M_{14} = 5.72717$$

The error, according to the estimate (4.9), is

$$E_{14} \leq \frac{6-5}{2^{15}} = 0.0000305$$

Thus, M_{14} is accurate to at least four decimal places. We round off this estimate to 5.7272, to indicate this level of accuracy. ∎

In Example 4.28, we found that after only three iterations by Newton's Method we had accuracy to at least eight decimal places. In fact, the accuracy after only two iterations was to four places. By contrast, we had to carry the Method of Secants through five iterations to get four-place accuracy, and we had to go through 14 bisections in the Bisection Method to obtain the same accuracy. Newton's Method would therefore seem to be the one to use. Aside from the pitfalls of this method that we already discussed, however, it requires that the function be differentiable, whereas the other two do not. So there are times when Newton's Method simply will not work. The Bisection Method is the least restrictive of the three, so despite the relative slowness of its convergence, it may be the best of the three choices in certain cases.

REMARK

■ We should note that there are still other methods for approximating roots of equations, some of which are studied in numerical analysis courses.

APPROXIMATING ROOTS USING A COMPUTER ALGEBRA SYSTEM—NEWTON'S METHOD

CAS 18

Use Newton's Method to approximate a root of the function $f(x) = x^3 - x - 1$.

If we first plot the function we can determine where a root lies and then choose an appropriate starting value for Newton's Method.

Maple:

f: = x–>x^3–2*x–1;
plot(f(x),x=–3..3,y=–5..5);

The plot suggests there is a root in the interval $[1, 2]$. We will start Newton's Method with a value of $x = 2$. First we define a function we call newton.

newton: = x–>x–f(x)/D(f)(x);

Now start the algorithm.

evalf(newton(2));

The output is 1.7. Now iterate by placing the previous value into newton, using the ditto mark.

evalf(newton("));	1.623088456
evalf(newton("));	1.618055040
evalf(newton("));	1.618033989
evalf(newton("));	1.618033989

When the values stabilize the computation is complete. Calculating these 6 terms could have been done quicker by entering:

x:=2;
for i from 1 by 1 to 6 do
evalf(newton(x));
x:=";
od;

The end of a loop that begins with "do" is marked by the reverse letters, "od."

Mathematica:

f[x_] = x^3–2*x–1
Plot [f[x],{x,–3, 3}]
newton[x_] = x – f[x]/f'[x]
newton[2]

Output:

$$\frac{17}{10}$$

For decimal output:

newton[N[2,20]]

Now iterate:

newton[%]
newton[%]
newton[%]
newton[%]
newton[%]

To calculate all six terms:

NestList[newton,N[2,20],7]

DERIVE:*

(At the □ symbol, go to the next step.)
a (author) □ f(x): =x^3–2x–1 □ p (plot window) □ p (plot) □ a (author) □ newton(x) := x–f(x)/dif(f(x),x) □ a (author) □ iterates(newton(x),x,2,6) □ x (approximate) □ [choose expression]

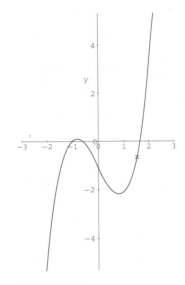

FIGURE 4.5.1
$y = f(x) = x^3 - 2x - 1$

---

*Note that in DERIVE, dif(f(x),x) can be used for the derivative of the function f with respect to x.

Exercise Set 4.5

 In Exercises 1–8, approximate each root using Newton's Method, continuing until two successive approximations agree to the first five decimal places. Round the answer to four places.

1. The positive root of $2x^2 - 4x - 1 = 0$
Compare with the exact answer.

2. The real root of $x^3 - 4x^2 + 7x - 5 = 0$

3. The negative root of $x^4 - 2x^3 - 2 = 0$

4. The positive root of $2 \sin x - x = 0$

5. The real root of $e^x + x = 2$

6. The smallest positive root of $\ln x - x + 2 = 0$

7. Both real roots of $2x^4 - 3x^3 - 2x + 2 = 0$

8. The positive root of $x^2 - \cos x = 0$

 In Exercises 9–12, approximate the root to four decimal places of accuracy by writing an equation that has the given root as a solution, and using Newton's Method.

9. $\sqrt{2}$ **10.** $\sqrt[3]{4}$

11. $\sqrt[3]{-9}$ **12.** $\sqrt[4]{6}$

In Exercises 13–16, without the aid of a calculator or CAS, find an interval I containing the indicated root r of $f(x) = 0$ such that f is monotone in I and the direction of the concavity of the graph of f does not change in I. Then find a starting point x_0 in I such that $f(x_0) > 0$ if the concavity is upward and $f(x_0) < 0$ if the concavity is downward. Sketch the graph of f for x in I.

13. $f(x) = x^3 - 5x^2 - 8x - 3;\ r > 0$

14. $f(x) = x^4 - 8x^2 - 3;\ r > 0$

15. $f(x) = 3(x-1)^{2/3} - x;\ 1 < r < 9$

16. $f(x) = 2 \cos x + x - 2;\ 0 < r < \pi$

 In Exercises 17–20, show that the interval I contains a root of $f(x) = 0$ and that I and the starting point x_0 satisfy the conditions stated in the instructions for Exercises 13–16. Find the approximation x_3 in Newton's Method, and use Inequality 4.7 to obtain a bound on the error.

17. $f(x) = 2x^3 - 3x^2 + 4x - 8;\ I = (1, 2),\ x_0 = 1.8$

18. $f(x) = x^4 - 2x^3 - 3x - 8;\ I = (-2, -1),\ x_0 = -1.2$

19. $f(x) = 2 \sin x - \dfrac{x^2}{2};\ I = (\pi/2, \pi),\ x_0 = 2.1$

20. $f(x) = x - \cos x;\ I = (0, \pi/2),\ x_0 = 0.9$

 In Exercises 21–24, approximate the root in each of Exercises 17–20 by the Method of Secants with $n = 3$. Begin by choosing z_0 in I such that $f(z_0)$ and $f(x_0)$ are opposite in sign. Compare your answer with x_3 as obtained by Newton's Method. Which answer is more accurate?

 In Exercises 25 and 26, use the Method of Bisection to approximate the real root correct to two decimal places.

25. $x^3 - 3x^2 + 3x - 4 = 0$

26. $xe^{-x} + 1 = 0$ (Use your calculator to evaluate e^{-x}.)

 27. Find all local maximum and minimum points correct to four decimal places, by Newton's Method, on the graph of the function f defined by

$$f(x) = 3x^4 - 8x^3 - 5x^2 + 4x - 20$$

28. How many real zeros does the function f of Exercise 27 have? Find them, using Newton's Method, correct to four decimal places. Draw the graph of f.

29. Prove that if f satisfies the conditions for Inequality 4.7 and $M < 4m$, then in Newton's Method whenever x_n is correct to k decimal places, x_{n+1} is correct to $2k$ decimal places. (*Hint:* The approximation x_n is correct to k places if $E_n \le 5 \times 10^{-(k+1)}$.)

30. For the Method of Secants, derive the recursion relation

$$z_{n+1} = z_n - \frac{f(z_n)}{m_n}$$

where

$$m_n = \frac{f(z_n) - f(x_0)}{z_n - x_0}$$

4.6 L'HÔPITAL'S RULE

Sometimes the limit properties we discussed in Chapter 1 are not applicable. For example, in a quotient both numerator and denominator might approach 0, or they might both become infinite. In this section we show how derivatives can often be used to find limits of such functions. Actually, we have already encountered some limits of this type. For example,

$$\lim_{x \to 1} \frac{x^2 - 1}{x - 1} \qquad \text{and} \qquad \lim_{x \to 0} \frac{\sin x}{x}$$

cannot be evaluated by taking limits of the numerator and denominator individually and then dividing. If we did so, in both cases the division would give us the meaningless result 0/0. Such limits are called "indeterminate" as they stand, since the limit may or may not exist. Some other procedure is needed. In the two examples just cited we were able to determine the limits. In the first case (Chapter 1) we factored the numerator to get

$$\lim_{x \to 1} \frac{x^2 - 1}{x - 1} = \lim_{x \to 1} \frac{(x + 1)(x - 1)}{(x - 1)} = \lim_{x \to 1}(x + 1) = 2$$

(remember that in the limiting process we do not let $x = 1$, so $x - 1 \neq 0$). In the second case we used a geometric argument (the Lemma on p. 166, part (a)) and found that

$$\lim_{x \to 0} \frac{\sin x}{x} = 1$$

These techniques will not work on many other indeterminate limits. For example, the limit

$$\lim_{x \to \pi/4} \frac{\tan x - 1}{4x - \pi}$$

also would assume the form 0/0 if we tried to take limits of the numerator and denominator individually. Further, it is not clear that any geometric argument or factoring technique would enable us to find the limit or show that it fails to exist. The method we will introduce shortly will enable us to evaluate this limit, as well as many others.

Closely related to quotients in which both numerator and denominator approach 0 is the case where they both become infinite. Again, we have encountered some limits of this type in finding horizontal asymptotes to graphs of rational functions. For example, to find the horizontal asymptote to the graph of

$$f(x) = \frac{x^3 - 2x}{3x^3 + 10x^2 - 5}$$

we need to determine the limit

$$\lim_{x \to \infty} \frac{x^3 - 2x}{3x^3 + 10x^2 - 5}$$

As it stands, this limit is indeterminate because both numerator and denominator become infinite. Recall that we found the limit in problems of this type by using dominant terms in numerator and denominator, getting

$$\lim_{x \to \infty} \frac{x^3 - 2x}{3x^3 + 10x^2 - 5} = \lim_{x \to \infty} \frac{x^3}{3x^3} = \frac{1}{3}$$

There are many other cases, however, where this dominant-term approach will not work. An example is

$$\lim_{x \to (\pi/2)^-} \frac{\sec x}{1 + \tan x}$$

Here again, both numerator and denominator become infinite as $x \to (\pi/2)^-$. We need some new procedure, then, for evaluating the limit.

Indeterminate Forms of Types 0/0 and ∞/∞

In the following definition we make the notion of the two types of indeterminate limits we have discussed more precise.

Definition 4.5
Indeterminate Forms of
Type 0/0 and ∞/∞

The limit

$$\lim_{x \to a} \frac{f(x)}{g(x)}$$

is said to be an **indeterminate form of type 0/0** if

$$\lim_{x \to a} f(x) = 0 \text{ and } \lim_{x \to a} g(x) = 0$$

It is **indeterminate of type ∞/∞** if

$$\lim_{x \to a} f(x) = \pm\infty \quad \text{and} \quad \lim_{x \to a} g(x) = \pm\infty$$

In each case, $x \to a$ can be replaced by any of the following: $x \to a^+$, $x \to a^-$, $x \to \infty$, or $x \to -\infty$.

We are now ready to state the main result of this section. It appeared in the first calculus book ever published, written by the French mathematician G. F. L'Hôpital (the Marquis de L'Hôpital, 1661–1704). The result has been known ever since as **L'Hôpital's Rule**. (L'Hôpital is pronounced "lo-pee-tal'"). The result had actually been discovered earlier by his teacher, Johann Bernoulli (1667–1748), who communicated it to L'Hôpital in 1694.

The idea of the rule is very simple. When we have an indeterminate limit of the form 0/0 or ∞/∞ we simply replace the given limit with the limit of the *derivative* of the numerator over the *derivative* of the denominator. For example, we write

$$\lim_{x \to 0} \frac{\sin x}{x} = \lim_{x \to 0} \frac{\cos x}{1}$$

Now the second limit is no longer indeterminate. Since $\cos x \to 1$ as $x \to 0$, the answer is 1. It is not at all obvious why the rule works. We will prove it for a special case (which is a commonly occurring case) shortly, but first we state the rule formally and give some more examples.

THEOREM 4.7

L'Hôpital's Rule

Let

$$\lim_{x \to a} \frac{f(x)}{g(x)}$$

be an indeterminate form of type $\frac{0}{0}$ or $\frac{\infty}{\infty}$. If $f'(x)$ and $g'(x)$ exist and $g'(x) \neq 0$ at all points in some open interval I containing a, except possibly at a itself, then

$$\lim_{x \to a} \frac{f(x)}{g(x)} = \lim_{x \to a} \frac{f'(x)}{g'(x)} \qquad (4.10)$$

provided the limit on the right exists or is ∞ or $-\infty$.

L'Hôpital's Rule is also true when a is an endpoint of the interval I, in which case $x \to a$ is to be replaced by $x \to a^-$ or $x \to a^+$, as appropriate. The rule is also true when a is replaced by ∞ or $-\infty$, provided I is infinite.

Equation 4.10 does *not* say that $f(x)/g(x) = f'(x)/g'(x)$, but rather that their *limits* are the same, when the hypotheses are satisfied. Note also that $f'(x)/g'(x)$ is *not* the derivative of the quotient $f(x)/g(x)$. In other words, do not use the Quotient Rule for differentiation here.

We illustrate the rule with two of the indeterminate forms mentioned earlier.

EXAMPLE 4.30   Use L'Hôpital's Rule to evaluate the following limits.

(a) $\lim\limits_{x \to \pi/4} \dfrac{\tan x - 1}{4x - \pi}$

(b) $\lim\limits_{x \to (\pi/2)^-} \dfrac{\sec x}{1 + \tan x}$

Solution

(a) This limit is indeterminate of the form $0/0$, and the hypotheses of L'Hôpital's Rule are satisfied. So we have

$$\lim_{x \to \pi/4} \frac{\tan x - 1}{4x - \pi} = \lim_{x \to \pi/4} \frac{\sec^2 x}{4} = \frac{2}{4} = \frac{1}{2}$$

(b) The limit is indeterminate of the form ∞/∞, and again, the hypotheses of L'Hôpital's Rule are satisfied. Applying the rule, we get

$$\lim_{x \to (\pi/2)^-} \frac{\sec x}{1 + \tan x} = \lim_{x \to (\pi/2)^-} \frac{\sec x \tan x}{\sec^2 x} = \lim_{x \to (\pi/2)^-} \frac{\tan x}{\sec x}$$

$$= \lim_{x \to (\pi/2)^-} \frac{\dfrac{\sin x}{\cos x}}{\dfrac{1}{\cos x}} = \lim_{x \to (\pi/2)^-} (\sin x) = 1$$

Note that we simplified the fraction $\tan x / \sec x$ by changing to sines and cosines before taking the limit. In applying L'Hôpital's Rule you should simplify fractions as much as possible at each stage. ∎

Proof of L'Hôpital's Rule for a Special Case

We give here the proof of a special case of L'Hôpital's Rule for a 0/0 indeterminate form. A proof for the general case is given in Appendix 5. The proof for the form ∞/∞ is more difficult, and we omit it. For the special case, we assume $f(a) = g(a) = 0$ (so that we are dealing with a 0/0 form), and we assume further that f'/g' is a continuous function at $x = a$. This continuity hypothesis implies that $f'(x)$ and $g'(x)$ both exist in some open interval I containing a (including a itself) and that $g'(x) \neq 0$ in I. By the definition of continuity, it also implies that

$$\lim_{x \to a} \frac{f'(x)}{g'(x)} = \frac{f'(a)}{g'(a)}$$

Although the continuity hypothesis on f'/g' is more restrictive than the hypotheses stated in the rule, in actual application of the rule the continuity hypothesis often is satisfied. We proceed now to the proof.

$$\lim_{x \to a} \frac{f(x)}{g(x)} = \lim_{x \to a} \frac{f(x) - f(a)}{g(x) - g(a)} \qquad \text{Since } f(a) = g(a) = 0.$$

$$= \lim_{x \to a} \frac{\dfrac{f(x) - f(a)}{x - a}}{\dfrac{g(x) - g(a)}{x - a}} \qquad \begin{array}{l}\text{Since } x \neq a, \text{ we can divide} \\ \text{numerator and denominator by } x - a.\end{array}$$

$$= \frac{\displaystyle\lim_{x \to a} \dfrac{f(x) - f(a)}{x - a}}{\displaystyle\lim_{x \to a} \dfrac{g(x) - g(a)}{x - a}} \qquad \begin{array}{l}\text{Since the limit of the denominator is } g'(a) \neq 0, \\ \text{we can apply the limit property for quotients.}\end{array}$$

$$= \frac{f'(a)}{g'(a)} \qquad \text{By definition of the derivative}$$

$$= \lim_{x \to a} \frac{f'(x)}{g'(x)} \qquad \text{By continuity of } f'/g' \qquad \blacksquare$$

As the next example shows, we sometimes have to apply L'Hôpital's Rule more than once.

EXAMPLE 4.31   Find the limit

$$\lim_{x \to 0^-} \frac{\cos x - 1}{x^3}$$

or show it does not exist.

Solution   The given limit is indeterminate of the type 0/0, and the hypotheses of L'Hôpital's Rule are satisfied. So we have

$$\lim_{x \to 0^-} \frac{\cos x - 1}{x^3} = \lim_{x \to 0^-} \frac{-\sin x}{3x^2}$$

Now we have another 0/0 indeterminate form. So we apply the rule again.

$$\lim_{x \to 0^-} \frac{-\sin x}{3x^2} = \lim_{x \to 0^-} \frac{-\cos x}{6x}$$

The last limit is not indeterminate, since the numerator approaches 1 and the denominator approaches 0, both through negative values. Thus,

$$\lim_{x \to 0^-} \frac{\cos x - 1}{x^3} = \infty \qquad \blacksquare$$

 If we had carelessly applied L'Hôpital's Rule a third time in the preceding example, we would have gotten

$$\lim_{x \to 0^-} \frac{\cos x}{6x} = \lim_{x \to 0^-} \frac{\sin x}{6} = 0$$

But this result is *incorrect*. The problem lies in the fact that the limit on the left is not of the form $0/0$, since $\cos x \to 1$ as $x \to 0^-$. Thus, L'Hôpital's Rule does not apply. Remember that L'Hôpital's Rule works only on $0/0$ or ∞/∞ indeterminate forms.

The Indeterminate Forms $0 \cdot \infty$ and $\infty - \infty$

Many times an indeterminate limit that does not fit L'Hôpital's Rule can be transformed to one that does. For example,

$$\lim_{x \to \infty} x^3 \sin\left(\frac{1}{x}\right)$$

involves the product of two factors, one of which goes to ∞ (the x^3 factor) and the other to 0 (the $\sin(1/x)$ factor, since $1/x$ goes to 0 and $\sin 0 = 0$). Because one factor becomes arbitrarily large and the other arbitrarily small, it is not clear what, if anything, the product approaches. The limit is thus indeterminate. By rewriting the limit as

$$\lim_{x \to \infty} \frac{\sin(1/x)}{1/x^3}$$

we change it to the form $0/0$, and L'Hôpital's Rule can be applied.

More generally, if $\lim_{x \to a} f(x) = 0$ and $\lim_{x \to a} g(x) = \pm\infty$, we say that $\lim_{x \to a} f(x) \cdot g(x)$ is an **indeterminate form of the type** $0 \cdot \infty$. As we just illustrated, to handle this type of indeterminate form, we rewrite the product as a quotient, either

$$f(x) \cdot g(x) = \frac{f(x)}{1/g(x)} \qquad \text{or} \qquad f(x) \cdot g(x) = \frac{g(x)}{1/f(x)}$$

and then see if L'Hôpital's Rule can be applied.

EXAMPLE 4.32   Evaluate the following limits or show that they fail to exist.

(a) $\lim_{x \to 0} x \cot x$

(b) $\lim_{x \to \infty} x^3 \sin\left(\frac{1}{x}\right)$

Solution

(a) As it stands, the limit is indeterminate of the form $0 \cdot \infty$. Since $\cot x = 1/\tan x$, we can write

$$\lim_{x \to 0} x \cot x = \lim_{x \to 0} \frac{x}{\tan x}$$

The last limit is of the form 0/0, so we can apply L'Hôpital's Rule to get

$$\lim_{x \to 0} x \cot x = \lim_{x \to 0} \frac{x}{\tan x} = \lim_{x \to 0} \frac{1}{\sec^2 x} = 1$$

(b) This limit is the one we discussed earlier. We rewrite it as a quotient and apply L'Hôpital's Rule.

$$\lim_{x \to \infty} x^3 \sin\left(\frac{1}{x}\right) = \lim_{x \to \infty} \frac{\sin(1/x)}{1/x^3} = \lim_{x \to \infty} \frac{\cos(1/x) \cdot (-1/x^2)}{-3/x^4}$$

$$= \lim_{x \to \infty} \frac{1}{3} x^2 \cos\left(\frac{1}{x}\right)$$

As $x \to \infty$, $1/x \to 0$, so $\cos 1/x \to \cos 0 = 1$. Thus, the limit is ∞. That is,

$$\lim_{x \to \infty} x^3 \sin\left(\frac{1}{x}\right) = \infty \qquad \blacksquare$$

Another type of indeterminate form occurs when we have the limit of the difference of two functions, both of which become infinite. In particular, if

$$\lim_{x \to a} f(x) = \infty \qquad \text{and} \qquad \lim_{x \to a} g(x) = \infty$$

then we say that $\lim_{x \to a}[f(x) - g(x)]$ is an **indeterminate form of the type** $\infty - \infty$. We use the same designation when $\lim_{x \to a} f(x) = -\infty$ and $\lim_{x \to a} g(x) = -\infty$.

REMARK

■ If $f(x)$ and $g(x)$ become infinite with opposite signs, then $\lim_{x \to a}[f(x) - g(x)]$ is not indeterminate (but $\lim_{x \to a}[f(x) + g(x)]$ is). For example, symbolically, $\infty - (-\infty) = \infty$ and $-\infty - (\infty) = -\infty$.

You might be tempted to say that $\infty - \infty = 0$, but remember that ∞ is not a number, so ordinary rules of arithmetic do not apply. We can often handle $\infty - \infty$ forms by expressing $f(x) - g(x)$ as a single fraction and then using L'Hôpital's Rule. We illustrate this approach in the next example.

EXAMPLE 4.33 Evaluate the limits

(a) $\displaystyle\lim_{x \to 1} \left(\frac{2x}{x^2 - 1} - \frac{1}{x - 1}\right)$

(b) $\displaystyle\lim_{x \to 0} \left(\frac{1}{x} - \csc x\right)$

Solution

(a) As it stands, the limit is of the form $\infty - \infty$, since both fractions approach ∞ as $x \to 1^+$ and both approach $-\infty$ as $x \to 1^-$. By getting a common denominator, we can write

$$\lim_{x \to 1} \left(\frac{2x}{x^2 - 1} - \frac{1}{x - 1} \right) = \lim_{x \to 1} \frac{2x - (x + 1)}{x^2 - 1} = \lim_{x \to 1} \frac{x - 1}{x^2 - 1}$$

We then find the last limit, which is of the form 0/0, either by factoring the denominator or by using L'Hôpital's Rule. By the latter method, we have

$$\lim_{x \to 1} \frac{x - 1}{x^2 - 1} = \lim_{x \to 1} \frac{1}{2x} = \frac{1}{2}$$

(b) Again, we have an $\infty - \infty$ form. To express the difference as a fraction, we write $\csc x = 1 / \sin x$ and then get a common denominator.

$$\lim_{x \to 0} \left(\frac{1}{x} - \frac{1}{\sin x} \right) = \lim_{x \to 0} \left(\frac{\sin x - x}{x \sin x} \right) \qquad \text{0/0 form}$$

$$= \lim_{x \to 0} \left(\frac{\cos x - 1}{\sin x + x \cos x} \right) \qquad \text{Still 0/0}$$

$$= \lim_{x \to 0} \frac{-\sin x}{\cos x - x \sin x + \cos x} = 0 \qquad \blacksquare$$

In Chapter 7 we will discuss still other types of indeterminate forms. There, too, we will have to transform the limit to one of the forms 0/0 or ∞/∞ before we can apply L'Hôpital's Rule.

Exercise Set 4.6

In Exercises 1–25, use L'Hôpital's Rule to evaluate the limits or to show that they do not exist.

1. $\lim_{x \to 3} \dfrac{x^2 - 9}{3 - x}$

2. $\lim_{x \to 2} \dfrac{x^2 + x - 6}{x^2 - 4}$

3. $\lim_{x \to 1} \dfrac{\sqrt{x} - 1}{x - 1}$

4. $\lim_{x \to \infty} \dfrac{2x^4 - 3x}{1 - x - x^4}$

5. $\lim_{x \to 0} \dfrac{\tan x}{x}$

6. $\lim_{x \to 0} \dfrac{1 - \cos x}{x^2}$

7. $\lim_{x \to 1} \dfrac{x - 5x^4 + 4x^6}{(x - 1)^2}$

8. $\lim_{x \to (\pi/2)^-} \dfrac{\cot x}{\sin 2x}$

9. $\lim_{x \to 0} \dfrac{x^3}{x - \sin x}$

10. $\lim_{x \to 1} \dfrac{x^2 + x - 2}{\sin \pi x}$

11. $\lim_{x \to 2} \dfrac{x - 2}{\sqrt{x + 2} - 2}$

12. $\lim_{x \to 0} \dfrac{x - \sin x}{\cos 2x - 1}$

13. $\lim_{x \to 0} \dfrac{\cos x - 1}{\cos 2x - 1}$

14. $\lim_{x \to 0} \dfrac{\cot x + x}{\csc x}$

15. $\lim_{x \to 0} \dfrac{1 - \cos x}{\sin x^2}$

16. $\lim_{x \to 0} \dfrac{x \sin x}{\sec x - 1}$

17. $\lim_{x \to 0} x \csc x$

18. $\lim_{x \to 1/2} (2x - 1) \tan \pi x$

19. $\lim_{x \to 0} (\sin 2x)(\cot x)$

20. $\lim_{x \to \pi} (\cos x + 1)(\csc x)$

21. $\lim_{x \to 0} \left(\cot x - \dfrac{1}{x} \right)$

22. $\lim_{x \to 0} (\csc x - \cot x)$

23. $\lim_{x \to 1} \left(\dfrac{3}{x^2 - 1} + \dfrac{\sqrt{4x + 5}}{1 - x^2} \right)$

24. $\lim_{x \to 0^+} \left(\dfrac{x + 1}{x} - \dfrac{2x + 3}{x^2} \right)$

25. $\lim_{x \to \infty} \left(\sqrt{x^2 + x} - x \right)$ (*Hint:* Let $x = 1/t$.)

26. In each of the following, find the limit without using L'Hôpital's Rule. Then see what happens when you try to apply L'Hôpital's Rule. Explain why you do not get the correct result by L'Hôpital's Rule.

(a) $\lim\limits_{x\to 0} \dfrac{x^2 \sin(1/x)}{\sin x}$ (*Hint:* Write as $\left(\frac{x}{\sin x}\right)\left(x \sin \frac{1}{x}\right)$.)

(b) $\lim\limits_{x\to\infty} \dfrac{x + \cos x}{x}$

27. Use a calculator, CAS, or spreadsheet with $x = 0.1, 0.01, 0.001,$ and 0.0001 in part (a) and $x = 10, 100, 1,000,$ and $10,000$ in part (b) to give further evidence that the limits you found in Exercise 26 are correct.

28. Let
$$f(x) = \frac{\cos x - 1}{x^3}$$

(a) Determine all horizontal and vertical asymptotes to the graph of f. (See Example 4.31.)

(b) Use a CAS or graphing calculator to obtain the graph of f. Pay particular attention to values of x near 0 and to the behavior of f for large values of x.

29. Let
$$f(x) = \frac{2x^2 - 2x + 1}{x^3 + 2x - 1}$$

(a) Use L'Hôpital's Rule to determine the horizontal asymptote to the graph of f.

(b) Use Newton's Method to approximate any vertical asymptotes to the graph.

(c) Use a CAS or graphing calculator to obtain the graph of f.

(d) Approximate all local extrema of the graph.

4.7 THE MEAN-VALUE THEOREM

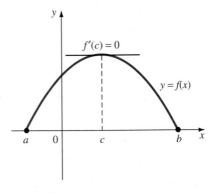

FIGURE 4.76

In this section we prove a theorem that has far-reaching consequences, known as the **Mean-Value Theorem** for differential calculus, or the **Law of the Mean**. As one application of this theorem, we will use it to prove some of the results of this chapter that we accepted earlier on intuitive grounds. We precede the main theorem with another result, which is used in the proof and so could be referred to as a lemma. This preliminary result is important in its own right, however, and it is known as **Rolle's Theorem**, after the seventeenth-century French mathematician Michel Rolle.

Before we state Rolle's Theorem, we will describe what it says informally. Consider the graph in Figure 4.76. There we have pictured a continuous function on the closed interval $[a, b]$ that equals 0 at each of the endpoints $x = a$ and $x = b$. We assume also that f has a derivative at all points between a and b. That is, the graph has a tangent line (nonvertical) at every point, except possibly at the endpoints a and b (where the tangent line could be vertical). Then, as we indicate, there must be some point on the curve (we have shown this point as having x-coordinate c) where the tangent line is horizontal. So when $x = c$, the slope is 0. This seemingly obvious result has some important consequences. As we have noted, one application of it is in proving the Mean-Value Theorem. There are many others.

THEOREM 4.8

Rolle's Theorem

Suppose the function f is continuous at every point of $[a, b]$ and differentiable at every point of (a, b), with $f(a) = 0$ and $f(b) = 0$. Then there exists a number c in the open interval (a, b) such that $f'(c) = 0$.

Proof   Since f is continuous on $[a, b]$, we know by Theorem 4.2 that it attains both an absolute maximum and an absolute minimum there. If $f(x) = 0$ for all x on $[a, b]$, then any number c between a and b satisfies the conclusion. If f is not constant, then its graph goes above the x-axis or below it, or both. Thus, either its maximum is greater than 0 or its minimum is less than 0, or both, and hence at least one of these extremes must occur at a point c between a and b. But since $f'(c)$ exists, it must be 0 by Theorem 4.3. The proof is therefore complete.   ■

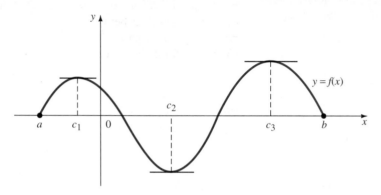

FIGURE 4.77

Rolle's Theorem guarantees the existence of *at least one* number c such that $f(c) = 0$, but there may be more than one. In Figure 4.77 we show a graph having three points satisfying this property.

If one or more of the hypotheses of Rolle's Theorem fail to hold true, then the conclusion may or may not be true. In Figure 4.78 we show three functions for which there is no number c in the interval in question such that $f'(c) = 0$. In part (a) the function is continuous in $[0, 2]$ and the derivative exists in

$f(0) \neq 0$, so $f(a)$ and $f(b)$ are not both 0, where $a = 0$ and $b = 2$.

(a)

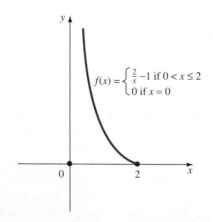

f is not continuous at $x = 0$, so f is not continuous on $[a, b]$, where $a = 0$ and $b = 2$.

(b)

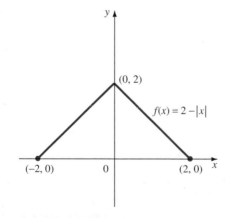

$f'(0)$ does not exist, so $f'(x)$ does not exist at all points of (a, b), where $a = -2$ and $b = 2$.

(c)

FIGURE 4.78

(0, 2), but $f(0) \neq 0$. In part (b) the function is 0 at both endpoints and its derivative exists in (0, 2), but the function is not continuous at the left endpoint, $x = 0$. In part (c) the function is continuous on $[-2, 2]$ and it is 0 at both endpoints, but the derivative does not exist at $x = 0$. These three examples show that all the hypotheses of Rolle's Theorem are needed to guarantee the conclusion.

Let us now consider what conclusion we can draw if we drop the hypothesis in Rolle's Theorem that $f(a)$ and $f(b)$ both equal 0. We have pictured such a function in Figure 4.79. The function is continuous on the closed interval $[a, b]$ and differentiable on the open interval (a, b), but no requirement is placed on the value of $f(a)$ or of $f(b)$. As we indicate in the figure, there appears to be some number c between a and b such that the tangent line drawn at the point $(c, f(c))$ is parallel to the secant line joining the endpoints $(a, f(a))$ and $(b, f(b))$. So their slopes are the same:

$$f'(c) = \frac{f(b) - f(a)}{b - a}$$

What we have observed is exactly the content of the Mean-Value Theorem, which we now state formally.

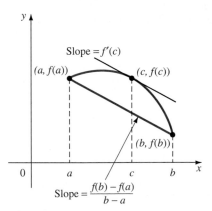

FIGURE 4.79

THEOREM 4.9

Mean-Value Theorem

Let f be continuous on the closed interval $[a, b]$ and differentiable on the open interval (a, b). Then there exists a point c in the open interval (a, b) such that

$$f'(c) = \frac{f(b) - f(a)}{b - a}$$

Proof   For convenience, let

$$m = \frac{f(b) - f(a)}{b - a}$$

Then the equation of the secant line that joins the endpoints is

$$y = f(a) + m(x - a) \qquad \text{Verify.}$$

We define a new function F on $[a, b]$ as follows:

$$F(x) = f(x) - [f(a) + m(x - a)] \tag{4.11}$$

Note that $F(x)$ is the vertical (directed) distance from the secant line to the curve, which is positive when the curve is above the secant line and negative when it is below. Now F is the difference of two functions, each of which is continuous on $[a, b]$ and differentiable on (a, b), and so F also has these properties. Furthermore, $F(a) = 0$ and $F(b) = 0$, as can be seen by direct substitution. Thus, F satisfies all the hypotheses of Rolle's Theorem and hence its conclusion; that is, there exists a number c in (a, b) such that $F'(c) = 0$. But from Equation 4.11,

$$F'(x) = f'(x) - m$$

So $F'(c) = 0$ implies that $f'(c) - m = 0$, or equivalently,

$$f'(c) = m$$

That is,

$$f'(c) = \frac{f(b) - f(a)}{b - a}$$

∎

EXAMPLE 4.34   Find a number c in the interval $(-1, 3)$ that satisfies the conclusion of the Mean-Value Theorem for the function

$$f(x) = x^3 - x^2 + 3$$

Solution   Note first that the hypotheses of Theorem 4.9 are satisfied, since f is continuous on $[-1, 3]$ and differentiable on $(-1, 3)$. In fact, f is differentiable and hence continuous everywhere, since it is a polynomial. We want to show the existence of a number c in $(-1, 3)$ for which

$$f'(c) = \frac{f(3) - f(-1)}{3 - (-1)} = \frac{21 - 1}{4} = 5$$

Since $f'(x) = 3x^2 - 2x$, we have $f'(c) = 3c^2 - 2c$, so we solve the equation

$$3c^2 - 2c = 5$$

$$3c^2 - 2c - 5 = 0$$

$$(3c - 5)(c + 1) = 0$$

$$c = \frac{5}{3} \quad \text{or} \quad c = -1$$

But the requirement is for c to be in the *open* interval $(-1, 3)$, so we must have $-1 < c < 3$. The solution $c = 5/3$ is therefore the only one that satisfies the theorem.   ∎

REMARK ───

■ If f fails to satisfy the hypotheses of the Mean-Value Theorem, then the conclusion may or may not be true. Consider, for example,

$$f(x) = 2 + (1 - x)^{2/3} \quad \text{on } [0, 2]$$

Here, $f(0) = 3$ and $f(2) = 3$. So $[f(2) - f(0)]/2 = 0$. Since

$$f'(x) = -\frac{2}{2(1 - x)^{1/3}}$$

can never be 0, the conclusion of the Mean-Value Theorem cannot be satisfied. This failure does not contradict the theorem, however, because $f'(x)$ does not exist at $x = 1$, which is in the open interval $(0, 2)$.

Average versus Instantaneous Velocity

An interesting physical interpretation of the Mean-Value Theorem relates the instantaneous velocity of a moving object and its average velocity over some time interval. Suppose that $s(t)$ represents the position of the object along a line at time t, and that we are interested in the time interval from $t = a$ to $t = b$. Assume that s satisfies the hypotheses of the Mean-Value Theorem (replacing

$f(x)$ by $s(t)$). Then the conclusion of that theorem states that there is some value $t = c$ between $t = a$ and $t = b$ such that

$$s'(c) = \frac{s(b) - s(a)}{b - a}$$

But $s'(c)$ is the instantaneous velocity of the object at time $t = c$, and the right-hand side of the equation is the average velocity over the time interval from $t = a$ to $t = b$ (change in distance divided by change in time). Thus, *at some value of t between a and b, the instantaneous velocity equals the average velocity over the time interval from $t = a$ to $t = b$.* We illustrate this result in the next example.

EXAMPLE 4.35   A car starts from point A and accelerates, reaching point B 10 seconds later. Its position, measured from A, is given by

$$s(t) = 0.8t^2 \qquad (0 \le t \le 10)$$

where t is in seconds and $s(t)$ is in meters. Find (a) the car's average velocity in going from A to B, and (b) the time and location of the car when its instantaneous velocity equals its average velocity.

Solution   Note first that $s(t)$ is differentiable, and hence continuous, everywhere, so the hypotheses of the Mean-Value Theorem are satisfied.

(a) $\quad v_{\text{ave}} = \dfrac{s(10) - s(0)}{10 - 0} = \dfrac{(0.8)(100)}{10} = 8$ m/s

(b) The instantaneous velocity $v(t)$ is given by

$$v(t) = s'(t) = 1.6t$$

To find where $v(t)$ equals v_{ave}, we set $1.6t$ equal to 8 and solve for t:

$$1.6t = 8$$

$$t = 5$$

So the instantaneous velocity after 5 seconds equals the average velocity. At that time the car is $(0.8)(5)^2 = 20$ m from A.   ∎

Using the Mean-Value Theorem

We will now use the Mean-Value Theorem to prove Theorem 4.1, which we accepted on intuitive grounds earlier. We will also establish two important corollaries to the Mean-Value Theorem. First, we restate Theorem 4.1.

THEOREM 4.1

> **(restated)**
>
> If $f'(x) > 0$ for all x in an open interval I, then f is increasing on I, and if $f'(x) < 0$ for all x in I, then f is decreasing on I.

Proof   Suppose first that $f'(x) > 0$ on I. Choose x_1 and x_2 in I, with $x_1 < x_2$. We want to show that $f(x_2) > f(x_1)$. Since $f'(x)$ exists on the interval I, f is differentiable and therefore continuous on the interval $[x_1, x_2]$, so the hypotheses of the Mean-Value Theorem are definitely satisfied. In that theorem we replace a by x_1 and b by x_2 and conclude that a number c exists in the open interval (x_1, x_2) such that

$$f'(c) = \frac{f(x_2) - f(x_1)}{x_2 - x_1}$$

Since $f'(c) > 0$ by hypothesis, and since $x_2 > x_1$ by choice, it follows that the numerator on the right must be positive; that is, $f(x_2) > f(x_1)$. In summary, we have shown that if $x_1 < x_2$, then $f(x_1) < f(x_2)$. Thus, by definition, f is an increasing function. The proof of the first part of the theorem is therefore complete. We leave the second part as an exercise.   ■

We conclude this section with two important corollaries of the Mean-Value Theorem.

COROLLARY 4.9(a)

If $f'(x) = 0$ for all x in an interval I, then f is a constant function on I.

Proof   Let a denote any fixed point in I, and let x denote any other point in I. We will show that $f(x) = f(a)$, and since $f(a)$ is a constant, this will tell us that all functional values are equal to this constant. We prove the result for $x > a$ only. The proof for $x < a$ is similar. We can apply the Mean-Value Theorem to the interval $[a, x]$. The existence of f' throughout I guarantees that the hypotheses of the Mean-Value Theorem are satisfied. (Why?) Thus, there is a number c in (a, x) such that

$$f'(c) = \frac{f(x) - f(a)}{x - a}$$

But since c is in I, $f'(c) = 0$. Thus, since $x - a \neq 0$ it follows that $f(x) - f(a) = 0$, and $f(x) = f(a)$. Thus, $f(x)$ is constant for all x in I.   ■

REMARK ————————————————————————————————
■ We already know that if $f(x)$ is a constant, then $f'(x) = 0$. Thus, we now know that $f(x)$ is a constant if and only if $f'(x) = 0$.

COROLLARY 4.9(b)

If $f'(x) = g'(x)$ for all x in an interval I, then for all x in I, $f(x) = g(x) + C$ for some constant C.

Proof   Define $h(x) = f(x) - g(x)$. Then

$$h'(x) = f'(x) - g'(x) = 0$$

for all x in I. So by Corollary 4.9(a), $h(x)$ is a constant, that is, $h(x) = C$. Thus,

$$f(x) - g(x) = C \qquad (x \in I)$$

which is equivalent to the desired conclusion.   ■

This corollary says that when two functions have the same slopes at all points, the curves are separated by a constant. In a certain sense we can think of the curves as being parallel. Figure 4.80 illustrates this relationship.

FIGURE 4.80

Exercise Set 4.7

In Exercises 1–8, show that the conditions of Rolle's Theorem are satisfied by the function on the given interval, and find a number c that satisfies the conclusion of the theorem.

1. $f(x) = x^2 - 3x - 4$ on $[-1, 4]$

2. $f(x) = x^2 - 4x + 3$ on $[1, 3]$

3. $f(x) = 2x^2 - 3x - 2$ on $[-\frac{1}{2}, 2]$

4. $f(x) = x^3 + x^2 - x - 1$ on $[-1, 1]$

5. $f(x) = x^3 - 9x$ on $[0, 3]$

6. $f(x) = 2x^3 - x^2 - 8x + 4$ on $[-2, 2]$

7. $f(x) = (x^2 - 9)\sqrt{x + 1}$ on $[-1, 3]$

8. $f(x) = (x - 1)\sqrt{3 - x^2}$ on $[1, \sqrt{3}]$

In Exercises 9–16, show that the conditions of the Mean-Value Theorem are satisfied by the function on the given interval, and find a number c that satisfies the conclusion of the theorem.

9. $f(x) = x^2 - 3x + 1$ on $[-1, 2]$

10. $f(x) = 4 - x - 3x^2$ on $[0, 3]$

11. $f(x) = x^3 - 2x^2 + 3$ on $[-1, 4]$

12. $f(x) = x^3 - 2x^2 + 5x + 8$ on $[-2, 2]$

13. $f(x) = x^3 + 5x^2 - 2x - 5$ on $[-1, 2]$

14. $f(x) = x - \frac{1}{x}$ on $[\frac{1}{2}, 2]$

15. $f(x) = \sqrt{2x + 1}$ on $[0, 4]$

16. $f(x) = \frac{x - 1}{x + 1}$ on $[0, 5]$

In Exercises 17–20, show that the conclusion of Rolle's Theorem is not satisfied by the function on the given interval, and explain why there is no contradiction with the theorem.

17. $f(x) = 1 - x^{2/3}$ on $[-1, 1]$

18. $f(x) = \frac{x^2 - 4}{x + 1}$ on $[-2, 2]$

19. $f(x) = \frac{2\sin x - 1}{\cos x}$ on $\left[0, \frac{5\pi}{6}\right]$

20. $f(x) = \frac{1 - 2\cos x}{\sin x}$ on $\left[\frac{-\pi}{3}, \frac{\pi}{3}\right]$

In Exercises 21–24, show that the conclusion of the Mean-Value Theorem is not satisfied by the function on the given interval, and explain why there is no contradiction with the theorem.

21. $f(x) = \frac{x^2 - 1}{x}$ on $[-1, 2]$

22. $f(x) = x - \frac{2}{x} + 3$ on $[-2, 2]$

23. $f(x) = (x - 1)^{2/3}$ on $[0, 2]$

24. $f(x) = x - 3x^{2/3}$ on $[-8, 8]$

25. Show that $f(x) = 1 - x + (3x - 1)^{1/3}$ fails to satisfy one or more of the hypotheses of Rolle's Theorem on $[0, 3]$, yet the conclusion of the theorem is true. Does this represent a contradiction? Explain.

26. Show that $f(x) = x + 3(1 - x)^{1/3}$ fails to satisfy one or more of the hypotheses of the Mean-Value Theorem on $[0, 2]$, yet the conclusion of the theorem is true. Does this represent a contradiction? Explain.

27. Find two points in the interval $(-1, 3)$ for which the tangent line to the graph of $f(x) = 2x^3 - 5x^2 - 3x + 1$ is parallel to the secant line that joins $(-1, f(-1))$ and $(3, f(3))$. Draw a figure showing the graph of f, the two tangent lines, and the secant line.

28. Let f satisfy the conditions of the Mean-Value Theorem on $[a, b]$, and let x be any number in $[a, b]$. Show that for some number c in (a, x),

$$f(x) = f(a) + f'(c)(x - a)$$

29. Let f satisfy the conditions of the Mean-Value Theorem on $[a, b]$, and let h be any number that satisfies $0 < h \leq b - a$. Show that $f(a+h) = f(a) + h \cdot f'(a+\theta h)$ for some number θ such that $0 < \theta < 1$.

30. Use the Mean-Value Theorem to show that for any two numbers x_1 and x_2,

$$|\sin x_1 - \sin x_2| \leq |x_1 - x_2|$$

31. Use the Mean-Value Theorem to show that $\tan x > x$ for all x that satisfy $0 < x < \frac{\pi}{2}$. (*Hint:* First pick an x in the given interval and then apply the Mean-Value Theorem to $f(x) = \tan x$ on $[0, x]$.)

32. Complete the proof of Theorem 4.1 by showing that if $f'(x) < 0$ for all x on an open interval I, then f is a decreasing function.

33. Prove that if f is differentiable on $[a, b]$ with $f'(a)$ and $f'(b)$ of opposite sign, then there is a number c between

a and b such that $f'(c) = 0$. [*Hint:* Show that f must attain its maximum or its minimum (or both) somewhere between a and b. Then use Theorem 4.2.]

34. Prove the following **Intermediate-Value Theorem for Derivatives**: If f is differentiable on $[a, b]$ and k is any number between $f'(a)$ and $f'(b)$, then there is a number c between a and b such that $f'(c) = k$. (Note that f' need not be a continuous function.) (*Hint:* Show that the result of Exercise 33 can be applied to $g(x) = f(x) - kx$.)

4.8 ANTIDERIVATIVES AND DIFFERENTIAL EQUATIONS

It is very important to be able to do the opposite of differentiating—that is, to find a function whose derivative we know. This process is called **antidifferentiation**, and a function arrived at in this way is called an **antiderivative** of the given function. More formally we have the following definition.

Definition 4.6
Antiderivative of a Function

A function F is called an **antiderivative** of f if $F'(x) = f(x)$ for all x in the domain of f.

For example, suppose we are given the function $3x^2$, and we are asked to find a function for which $3x^2$ is the derivative. We can do this calculation mentally and arrive at x^3 as an answer. So we can say that x^3 is an antiderivative of $3x^2$. We are being very careful in using the article *an* here rather than *the* because there will always be many (infinitely many, in fact) antiderivatives of a function if there are any at all. In the present case, for example, we can see that each of the following is an antiderivative of $3x^2$: $x^3 + 2$, $x^3 - 4$, and $x^3 + 100$. In fact, $x^3 + C$, where C is *any* real number, is an antiderivative of $3x^2$. Whether there can be any that are not of this form is answered by the next theorem, whose proof further illustrates the importance of the Mean-Value Theorem and its corollaries.

THEOREM 4.10

Let $F(x)$ be an antiderivative of $f(x)$ on an interval I. Then every other antiderivative of $f(x)$ on I is of the form $F(x) + C$ for some constant C.

Proof   We are given that $F'(x) = f(x)$ for all $x \in I$. Suppose G is any other antiderivative of $f(x)$ on I, so that $G'(x) = f(x)$ on I. Then for all x in I, $G'(x) = F'(x)$. Therefore, by Corollary 4.9(b),

$$G(x) = F(x) + C$$

for some constant C.   ∎

If $F(x)$ is an antiderivative of $f(x)$, then we sometimes call $F(x) + C$ the **general antiderivative** of $f(x)$, so called because we leave the constant C unspecified. In the sense of Theorem 4.10, $F(x) + C$ represents the *family of all* antiderivatives of $f(x)$. We will soon learn certain techniques for finding antiderivatives that are less obvious than the one we have considered so far. For the present, we will concentrate on certain basic types.

REMARK ——————————————————————————————————

■ Strictly speaking, we should write the family of all antiderivatives of f as the set

$$\{F(x) + C : C \in \mathbb{R} \text{ and } F'(x) = f(x)\}$$

For brevity, though, and in accordance with customary usage, we will write simply $F(x) + C$.

——————————————————————————————————

We know that to differentiate a rational power of x, say x^n, we multiply by the exponent and then subtract 1 from the exponent, getting nx^{n-1}. To find an antiderivative of x^n we do the reverse; that is, we *add* 1 to the exponent and then *divide* by the resulting exponent, getting

$$\frac{x^{n+1}}{n+1}$$

For this quotient to have meaning we must have $n \neq -1$. With this restriction, we can check the result by differentiating:

$$\frac{d}{dx}\left(\frac{x^{n+1}}{n+1}\right) = (n+1) \cdot \frac{x^n}{n+1} = x^n$$

For example, an antiderivative of x^3 is $x^4/4$, and an antiderivative of $x^{-1/2}$ is $2x^{1/2}$.

Because the derivative of a constant times a function is the constant times the derivative of the function, the same is true of antiderivatives. For example, an antiderivative of $5x^3$ is $5x^4/4$. Also, since the derivative of the sum or difference of functions is the sum or difference of the derivatives, the same is true for antiderivatives. These facts enable us to find antiderivatives of polynomials. For example, an antiderivative of $2x^2 - 3x + 4$ is

$$\frac{2x^3}{3} - \frac{3x^2}{2} + 4x$$

The following table summarizes some of the basic antiderivatives we already know.

Function	Antiderivative
x^n	$\dfrac{x^{n+1}}{n+1} + C \qquad (n \neq -1)$
$\sin x$	$-\cos x + C$
$\cos x$	$\sin x + C$
$\sec^2 x$	$\tan x + C$
$\csc^2 x$	$-\cot x + C$
$\sec x \tan x$	$\sec x + C$
$\csc x \cot x$	$-\csc x + C$

All these antiderivatives can be verified by differentiating the function in the right-hand column to get the function on the left. It should be clear that, whenever we find a formula for the derivative of a function, we also have found a formula for an antiderivative.

If we denote by F any antiderivative of f, then we also have the following:

Function	Antiderivative
$cf(x)$	$cF(x)$
$f_1(x) \pm f_2(x)$	$F_1(x) \pm F_2(x)$

FIGURE 4.81
The family $y = x^3 - x^2 - 2x + C$ of antiderivatives of $f(x) = 3x^2 - 2x - 2$

EXAMPLE 4.36   Find the general antiderivative of the function

$$f(x) = 3x^2 - 2x - 2$$

and draw graphs for several different values of the constant C.

Solution   The general antiderivative of f is

$$x^3 - x^2 - 2x + C$$

We show the graphs for $C = -1, 0, 1, 2,$ and 3 in Figure 4.81.   ∎

The preceding example further illustrates the nature of the general antiderivative of a function as being not just one function but a *family* of functions. The members of the family differ from one another only by constant values. The graphs in Figure 4.81 clearly show this property. At each value of x, the slopes of all members of the family are the same.

Often, we will want to single out one particular member of the family of antiderivatives of a function. For example, we may want the member of the family whose graph goes through a particular point. The next example shows how to find such a particular antiderivative.

EXAMPLE 4.37   Find the particular antiderivative of the function

$$f(x) = 3x^2 - 2x - 2$$

whose graph passes through the point $(2, -1)$.

Solution   In Example 4.36 we found the general antiderivative of f to be

$$x^3 - x^2 - 2x + C$$

Let us designate by $F(x)$ the particular member we are seeking. Thus,

$$F(x) = x^3 - x^2 - 2x + C$$

where the value of C is yet to be determined. Since the point $(2, -1)$ is on the graph of F, we must have $F(2) = -1$. So to evaluate the constant C we substitute $x = 2$ and $F(x) = -1$:

$$-1 = (2)^3 - (2)^2 - 2(2) + C$$

$$-1 = 8 - 4 - 4 + C$$

$$C = -1$$

The particular antiderivative we are seeking is therefore

$$F(x) = x^3 - x^2 - 2x - 1$$ ∎

Differential Equations

An equivalent way of asking for the general antiderivative of a function f is to ask for all functions F such that $F'(x) = f(x)$. Thus, we could have rephrased the instructions in Example 4.36 as follows: Find all functions F such that

$$F'(x) = 3x^2 - 2x - 2 \qquad (4.12)$$

The answer is what we found in the example, namely, $F(x) + C$, where

$$F(x) = x^3 - x^2 - 2x$$

Equation 4.12 is called a **differential equation**. In general, differential equations involve one or more derivatives of some unknown function. To *solve* a differential equation means to find the unknown function or family of functions.

REMARK ───────────────────────────────────────
■ Leibniz used the term *differential equation* to mean an equation relating differentials dx and dy. This terminology has stayed with us. A better name would be *derivative* equation.

───

Equation 4.12 is the simplest type of differential equation and is the only type we will consider in this section. Here are some other examples of differential equations:

$$x\frac{dy}{dx} - 3y = 4$$

$$(y + 1)dy = (x - 2)dx$$

$$y'' - 2y' + y = \sin x$$

(We will show how to solve the first two of these in Chapter 7.)

Differential equations are of fundamental importance in virtually all the sciences. They usually arise as mathematical models involving rates of change. The subject of differential equations is very broad, and we can only touch on it here. Whole courses at both the undergraduate and graduate levels are devoted to the subject.

If we use the Leibniz notation dy/dx for the derivative, simple differential equations such as Equation 4.12 can be written in the form

$$\frac{dy}{dx} = f(x)$$

The solution is the general antiderivative

$$y = F(x) + C$$

where $F'(x) = f(x)$, and it is called the **general solution** of the differential equation. For example, Equation 4.12 can be written as

$$\frac{dy}{dx} = 3x^2 - 2x - 2$$

and its general solution is

$$y = x^3 - x^2 - 2x + C$$

We showed the graphs of several members of this family in Figure 4.81.

If we know a particular value of y for some x value, the differential equation, together with the known pair (x, y), is called an **initial-value problem**. In the next example we rephrase Example 4.37 as such an initial-value problem.

EXAMPLE 4.38   Solve the initial-value problem

$$\frac{dy}{dx} = 3x^2 - 2x - 2; \quad y = -1 \text{ when } x = 2$$

Solution   The general solution of the differential equation is

$$y = x^3 - x^2 - 2x + C$$

Now we substitute $x = 2$ and $y = -1$ to find C:

$$-1 = (2)^3 - (2)^2 - 2(2) + C$$

$$C = -1$$

Thus, the particular solution we are seeking is

$$y = x^3 - x^2 - 2x - 1 \qquad \blacksquare$$

Often differential equations involve derivatives with respect to *time*. For example, equations relating to an object moving along a line might involve its velocity or its acceleration. If $s(t)$ represents the position of the object from a designated origin at time t, then we know that $v(t) = s'(t)$ and $a(t) = v'(t)$, or equivalently $a(t) = s''(t)$. The next two examples illustrate differential equations of this type.

EXAMPLE 4.39   From just off the top edge of a 10-story building 96 ft high, a ball is thrown vertically upward with an initial velocity of 40 ft/s. Find the position $s(t)$ of the ball above the ground t seconds later. What is the velocity after 2 seconds? Find the maximum height of the ball and also the ball's velocity when it hits the ground. Assume that acceleration caused by gravity is 32 ft/s², acting downward, and that air resistance is negligible.

Solution   We take the distance as 0 at the ground level and choose the positive direction upward, as indicated in Figure 4.82. Since the only force on the ball is assumed to be gravity, and it acts downward, the acceleration $a(t)$ of the ball is $a(t) = -32$. Since $a(t) = v'(t)$, we can find $v(t)$ by solving the initial-value problem

$$v'(t) = -32; \qquad v(0) = 40$$

The antiderivative is

$$v(t) = -32t + C$$

and we find C by the initial condition $v(0) = 40$. That is, we let $t = 0$ and $v(t) = 40$:

$$40 = -32(0) + C$$

$$C = 40$$

Thus,

$$v(t) = -32t + 40 \qquad (4.13)$$

96 ft

$s(t)$

FIGURE 4.82

Also, we know that $v(t) = s'(t)$, so we have a second initial-value problem to solve:

$$s'(t) = -32t + 40; \qquad s(0) = 96$$

(since the initial height is 96 ft). Again taking an antiderivative, we get

$$s(t) = -16t^2 + 40t + C_1$$

We denote the constant by C_1, since it is not necessarily the same constant we had in the velocity problem. The initial condition $s(0) = 96$ gives

$$C_1 = 96$$

So the position of the ball at time t is

$$s(t) = -16t^2 + 40t + 96 \qquad (4.14)$$

To get the velocity after 2 seconds, we use Equation 4.13 with $t = 2$:

$$v(2) = -32(2) + 40 = -24$$

The negative sign indicates that $s(t)$ is decreasing when $t = 2$. So at that instant the ball is falling at the rate of 24 ft/s, which is one of the results requested.

Now let us find the maximum height the ball reaches. The distance $s(t)$ is maximum when its derivative, $v(t)$, is 0. So we set $v(t) = 0$ and solve for t:

$$-32t + 40 = 0$$

$$t = \frac{5}{4}$$

The velocity is positive for $t < 5/4$ and negative for $t > 5/4$, confirming that the critical value 5/4 produces a maximum value of the position function. From Equation 4.14, we get

$$s_{\max} = s\left(\frac{5}{4}\right) = -16\left(\frac{25}{16}\right) + 40\left(\frac{5}{4}\right) + 96 = 121$$

So the ball reaches a maximum height above the ground of 121 ft (25 ft above the building). In Figure 4.83 we show the graph of the position function $s(t)$ versus time t. The graph clearly shows the maximum value of $s(t)$ when $t = 5/4$.

Finally, we want to know the velocity with which the ball hits the ground. First we find the time when the ball hits the ground by setting $s(t) = 0$ and solving for t:

$$-16t^2 + 40t + 96 = 0$$

$$2t^2 - 5t - 12 = 0$$

$$(2t + 3)(t - 4) = 0$$

$$t = -\frac{3}{2} \quad \text{or} \quad t = 4$$

We reject $t = -\frac{3}{2}$ as being meaningless in the context of the problem. So the ball hits the ground when $t = 4$ seconds. From Equation 4.13 we therefore have

$$v(4) = -32(4) + 40 = -88$$

The ball is traveling (downward) at 88 ft/s at the instant it strikes the ground. (Note that 88 ft/s is 60 mph, so you wouldn't want to let the ball hit you on the head!) ∎

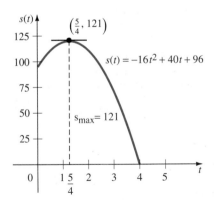

FIGURE 4.83

It is useful to note that the constant C in the solution for $v(t)$ is equal to the initial velocity $v(0)$. Similarly, the constant C_1 in the solution for $s(t)$ is equal to the initial position $s(0)$. These values for the constants are typical for problems involving motion along a line.

REMARK ——————————————————————————————————————

■ An equivalent way of setting up the initial-value problem in Example 4.39 is

$$s''(t) = -32; \qquad s'(0) = 40, \; s(0) = 96$$

The differential equation now is of *second order* (it involves a second derivative), and two initial conditions are required to find the constants that arise in the solution. The solution in this case would proceed just as in the example, with $s'(t)$ replacing $v(t)$.

——

EXAMPLE 4.40   A particle moves on a line so that its acceleration at time $t \geq 0$ is given by $a(t) = 6(1 - t)$. The initial conditions are $v(0) = 9$ and $s(0) = 0$.

(a) Find $v(t)$ and $s(t)$.

(b) Find when and where the particle reaches its greatest positive distance from the origin.

(c) Determine the time intervals and corresponding distance intervals in which the particle is moving to the right (positive direction) and those in which it is moving to the left.

(d) Find the time and distance intervals when it is slowing down and those in which it is speeding up.

Solution

(a) Solving the initial-value problem

$$v'(t) = 6(1 - t); \qquad v(0) = 9$$

we obtain

$$v(t) = 6t - 3t^2 + C$$

By the comment above concerning constants in motion problems such as this one, we have $C = v(0)$. So, since we are given that $v(0) = 9$,

$$v(t) = -3t^2 + 6t + 9$$

Thus, we have the second initial-value problem

$$s'(t) = -3t^2 + 6t + 9; \qquad s(0) = 0$$

with the following solution:

$$s(t) = -t^3 + 3t^2 + 9t + s(0) \qquad \text{The constant } C_1 = s(0).$$

The answer to part (a) is therefore

$$v(t) = -3t^2 + 6t + 9$$
$$s(t) = -t^3 + 3t^2 + 9t$$

(b) To find the greatest positive distance from the origin, we look for the absolute maximum value of $s(t)$. First, we find critical points by setting $v(t) = s'(t) = 0$:

$$-3t^2 + 6t + 9 = 0$$

$$t^2 - 2t - 3 = 0$$

$$(t - 3)(t + 1) = 0$$

$$t = 3 \text{ or } t = -1$$

We reject $t = -1$, since we are given that $t \geq 0$. Since $s(3) = -27 + 3(9) + 9(3) = 27$, and $s''(3) = a(3) < 0$, it follows that 27 is a local maximum value of $s(t)$. The only endpoint value is $s(0) = 0$, so 27 is the absolute maximum value of $s(t)$.

(c) Since $v(t) = -3(t+1)(t-3)$, we see that for $t \geq 0$, the velocity is positive for $t < 3$ and negative for $t > 3$. So the particle moves to the right from $s(0) = 0$ to $s(3) = 27$. Then it turns around and moves to the left for all $t > 3$.

(d) Since acceleration measures the rate of change of velocity, we know that if $a(t) > 0$, the velocity is increasing. Thus, if both $v(t) > 0$ and $a(t) > 0$, the particle is moving to the right and is speeding up. On the other hand, if $v(t) < 0$ and $a(t) > 0$, the particle is moving to the left but slowing down, since the velocity is negative but is tending toward becoming positive. Suppose now that $a(t) < 0$. Then, if $v(t)$ also is negative, the particle is moving to the left, and its velocity is becoming more negative; that is, the particle is speeding up. On the other hand, if $v(t) > 0$ but $a(t) < 0$, the particle is moving to the right but with decreasing velocity. We can summarize this discussion as follows:

1. If $v(t)$ and $a(t)$ are *like* in sign, the particle is speeding up.
2. If $v(t)$ and $a(t)$ are *opposite* in sign, the particle is slowing down.

In either case the direction of motion is determined by $v(t)$: to the right if $v(t) > 0$ and to the left if $v(t) < 0$. A comparison of sign graphs (Figure 4.84) for v and a makes clear when each situation occurs. We see that from $t = 0$ to $t = 1$, the particle is speeding up and moving to the right ($v > 0$, $a > 0$); from $t = 1$ to $t = 3$ it continues to the right but is slowing down ($v > 0$, $a < 0$), and for $t > 3$ it moves to the left faster and faster ($v < 0$, $a < 0$). Figure 4.85 summarizes our findings. The motion actually takes place on a line, but in this diagram we show it on two levels to see more clearly the nature of the motion in each direction. ∎

For $0 < t < 1$, $v(t) > 0$ and $a(t) > 0$;
for $1 < t < 3$, $v(t) < 0$ and $a(t) > 0$;
for $t > 3$, $v(t) < 0$ and $a(t) < 0$

FIGURE 4.84

FIGURE 4.85

Exercise Set 4.8

In Exercises 1–20, find the general antiderivative of the given function.

1. $f(x) = x^5$

2. $f(x) = 3x^4$

3. $f(x) = 2x^2 - 4x + 5$

4. $f(x) = 6x^2 - 2x + 3$

5. $f(x) = 3x^3 - 5x^2 + 2x - 3$

6. $f(x) = 2 - x + 4x^2 - 6x^3$

7. $f(x) = 1/(2x^2) - x^2/2$

8. $f(x) = 2x^{-3} - 4x^{-2} + 1$

9. $f(x) = 2x^{1/2} - 3x^{-1/2}$

10. $f(x) = 1/\sqrt{x} - 2/x^2 + 1$

11. $f(x) = 3x - 4\sqrt{x} + 7$

12. $f(x) = x^{3/2} + 2x^{1/2}$

13. $f(x) = x(\sqrt{x} + 3)$

14. $f(x) = (x - 1)(1/\sqrt{x} + 1)$

15. $f(x) = 2\sin x - 3\cos x$

16. $f(x) = \sec x(\sec x + \tan x)$

17. $f(x) = x - 1/\sin^2 x$

18. $f(x) = \cos x + 1/\cos^2 x$

19. $f(x) = 1 - \cos x/\sin^2 x$

20. $f(x) = \tan x \cos x - \sin x \cot x$

In Exercises 21–28, find the general solution of the differential equation.

21. $f'(t) = \sqrt{t}(t - 1)$

22. $g'(x) = \cos x - x$

23. $\dfrac{dy}{dx} = \dfrac{1 - x}{x^3}$

24. $dy = (x^2 + 2)dx$

25. $(\cos^2 x)dy = (\sin x + 1)dx$

26. $\dfrac{dx}{dt} = 1 - t^{-0.2}$

27. $v'(t) = (t - 1)(t - 2)$

28. $s'(t) = t^{3/2} - 2t^{1/2} + 3$

In Exercises 29–38, solve the given initial-value problem.

29. $f'(x) = 2x - 3; \; f(0) = 5$

30. $f'(x) = 3x^2 - 2x + 1; \; f(-2) = 3$

31. $g'(x) = 1 - x - 4x^2; \; g(3) = 2$

32. $g'(x) = \dfrac{3}{x^2} - 2x + 3; \; g(1) = 5$

33. $y' = 2x - x^2; \;$ when $x = 1, \; y = -2$

34. $dy = (\sqrt{x} - 1)dx; \;$ when $x = 1, \; y = 3$

35. $f''(x) = 3x; \; f'(0) = 0, \; f(0) = 2$

36. $f''(x) = 1 - x; \; f'(1) = 2, \; f(1) = -1$

37. $f''(x) = x^2 - 1; \; f'(-1) = 0; \; f(0) = 2$

38. $f'''(x) = 2; \; f''(0) = 2; \; f'(0) = -1; \; f(0) = 4$

39. Find y as a function of x if $dy/dx = x + \sin x$ and the graph of the function goes through the point $(0, 3)$.

40. The graph of $y = f(x)$ goes through the point $(1, -2)$ and has a slope of $\frac{1}{2}$ at this point. If $y'' = 1 - x$, find $f(x)$.

In Exercises 41 and 42, assume that the acceleration due to gravity is 32 ft/s² downward, and assume there is no air resistance.

41. A ball is thrown vertically upward from the ground with an initial velocity of 64 ft/s. Find expressions for its height $s(t)$ and its velocity $v(t)$ after t seconds. How high will the ball rise? When will it strike the ground, and what will its velocity be at that instant?

42. From a balloon 1600 ft above the ground, a projectile is fired vertically downward at an initial velocity of 240 ft/s. How long will it take to reach the ground, and what will be its terminal velocity?

43. From just over the top edge of a sheer cliff 58.8 m above the canyon floor, a rock is thrown vertically upward at an initial velocity of 19.6 m/s. How high will the rock rise, and when will it strike the canyon floor? (Take acceleration caused by gravity to be 9.8 m/s^2.)

44. The accompanying figure shows the graphs of three functions f, g, and h. The function f is an antiderivative of one of the functions g or h. Which is it?

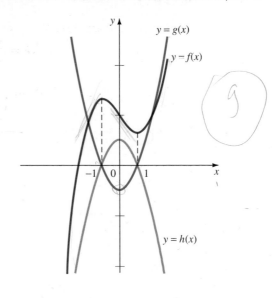

Chapter 4 Review Exercises

In Exercises 1 and 2, find the intervals on which f is increasing and those on which it is decreasing.

1. (a) $f(x) = x^4 - 4x^3 - 2x^2 + 12x$
   (b) $f(x) = (x + 2)^3(2x - 1)^2$

2. (a) $f(x) = (3x^2 - 8x)/(x^2 + 4)$
   (b) $f(x) = (x - 1)^{2/3}(x + 7)^2$

In Exercises 3–10, find all local maxima and minima. Also find absolute maximum and minimum values if they exist.

3. $f(x) = 5 + 4x + 2x^2 - x^3$

4. $f(x) = 3x^5 - 25x^3 + 60x$

5. $f(x) = x^2/(x - 1)$

6. $f(x) = (x^2 + 3)/(x^3 - 5x)$ on $[-2, 2]$

7. $f(x) = x^4 - 6x^2 - 8x + 5$ on $[-2, 3]$

8. $f(x) = \sqrt{x - 1}\,(x - 6)^2$ on $[1, 10]$

9. $f(x) = \cos 2x + x$ on $\left[-\dfrac{\pi}{2}, \dfrac{\pi}{2}\right]$

10. $f(x) = 2\cos x + \sin 2x$ on $\left[-\pi, \dfrac{\pi}{2}\right]$

In Exercises 11 and 12, determine intervals on which the graph of f is concave up and those on which it is concave down.

11. (a) $f(x) = x^4 - x^3 - 3x^2 - 5x + 7$
   (b) $f(x) = 3x^5 - 15x^4 - 10x^3 + 90x^2 - 100$

12. (a) $f(x) = (x - 2)/(x + 1)^2$
   (b) $f(x) = (x - 2)^4(x + 1)^2$

In Exercises 13–30, discuss intercepts, symmetry, asymptotes, local maxima and minima, and points of inflection. Use this information to draw the graph of f.

13. $f(x) = (x^2 - 1)(x + 1)$

14. $f(x) = 2x^3 - 3x^2 - 12x + 4$

15. $f(x) = x^4 - 18x^2 + 60$

16. $f(x) = \dfrac{x^2}{x^2 - 4}$ **17.** $f(x) = \dfrac{8x}{x + 4}$

18. $f(x) = 3x^5 - 5x^3$ **19.** $f(x) = x^4 - 4x^3 + 15$

20. $f(x) = \frac{1}{3}(x - 1)^3(x + 3)$

21. $f(x) = x - \dfrac{2}{x-1}$ **22.** $f(x) = x^{1/3}(4+x)$

23. $f(x) = x - (1-3x)^{2/3}$

24. $f(x) = 3x^4 - 8x^3 + 6x^2 - 1$

25. $f(x) = \dfrac{4x}{(x-1)^2}$ **26.** $f(x) = \dfrac{4(1-x)}{x^2}$

27. $f(x) = 3x^{-1} - x^{-3}$ **28.** $f(x) = \dfrac{\sin x}{\cos x - 2}$

29. $f(x) = \dfrac{2\sin^2 x + 1}{2\sin^2 x - 1}$

30. $f(x) = \frac{1}{12}(3x^4 + 4x^3 - 30x^2 + 36x)$

31. A closed box with a square base is to be constructed so that its volume is 324 ft³. The material for the top and bottom costs $3 per square foot, and that for the sides costs $2 per square foot. Find the dimensions so that the cost will be a minimum.

32. Find the dimensions of the right circular cylinder of maximum volume that can be inscribed in a cone of base radius r and altitude h.

33. The largest box the United Parcel Service will accept is one for which the sum of the length and the girth (distance around) is 108 in. Find the dimensions of the box with square cross section having the greatest volume that can be sent via the United Parcel Service.

34. A car dealer sells an average of 100 cars of a certain type per month when the selling price is $15,000. The cost to the dealership is $11,000 per car. The dealer estimates that for every $200 in rebate, they will sell 10 more cars per month. What total rebate should the dealer offer to obtain the maximum profit?

35. An open-top cylindrical boiler pan is to have a volume of 250π in.³. The lateral surface is to be made of stainless steel and the bottom of copper. If copper costs twice as much as stainless steel, what should be the dimensions of the pan for minimum cost?

36. An oil drilling platform in the Gulf of Mexico is 9 km from point A, the nearest point on shore. A second oil drilling platform is 3 km from B, the nearest point on shore to it. The distance from A to B is 5 km. A supply depot is located at a point C on the shore between A and B in such a way that the sum of the distances from C to the two platforms is a minimum. What is the distance from A to C? (See the figure.)

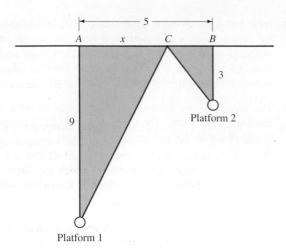

37. According to one model of population growth, the rate dx/dt at which a population grows is jointly proportional to the current size x of the population and the remaining capacity to grow, $m - x$, where m is the theoretical maximum size of the population. Find how large the population is when it is growing most rapidly.

38. To produce a certain manufactured item there is for each unit a fixed cost of $30 and a variable cost of $20 - 0.1x$, where the number x of items does not exceed 100. The unit price function is

$$p(x) = 700 + \frac{100}{x} - 0.3x$$

Find how many units should be produced and sold to maximize profit.

39. An 8-ft-high fence stands 1 ft from a building. A ladder is to be placed against the building and over the fence (see the figure). What is the length of the shortest ladder that can be used?

40. Car A and Car B are approaching a town on roads that are at right angles to each other. Car A is traveling at 80 km/hr and Car B at 55 km/hr. At the instant when Car A is 5 km from the town, Car B is 12 km from the town. How fast is the distance between the cars decreasing at that instant?

41. At the instant a hot-air balloon begins rising, a runner leaves from the takeoff point, running in a straight line. If the balloon rises at the constant rate of 22 ft/s and the runner goes 8 mph, find the rate at which the distance between the runner and the balloon is increasing after 10 min.

42. A yacht is cruising at 12 mph. A person standing in the stern 15 ft above the water is pulling in a rope (assumed taut) attached to a rowboat at the rate of 50 ft/min. How fast is the rowboat moving through the water when there are 39 ft of rope out?

43. A cone-shaped container is being filled with water at the rate of 4 ft^3/min, at the same time that water is being drained out through the bottom at the rate of 2 ft^3/min. The top radius of the container is 2 ft and its height is 8 ft. Find how fast the water level is rising when (a) the water level is 4 ft and (b) the container is half full.

44. At a track meet a TV camera operator is focusing the camera on the lead runner in a 100 m dash. The camera is positioned 9 m from the track, directly opposite the finish line. If the runner is going at the constant rate of 10 m/s, how many radians is the camera turning through per second at the instant the runner is 12 m from the finish line?

In Exercises 45–48, approximate the specified roots of the equation using Newton's Method, continuing until two successive approximations agree to the first five decimal places. Round the answer to four places.

45. The real root of $x^3 + 2x^2 + 2x - 3 = 0$

46. All real roots of $x^4 - 2x^2 + 5x - 8 = 0$

47. All real roots of $x^2 - \frac{1}{x} = 4$

48. All real roots of $x^{-2} = \sec x$

49. Use Newton's Method to find the maximum value of $f(x) = x \cos x$ on $(0, \pi/2)$ correct to six decimal places.

50. Show that if $a \neq 0$, $\sqrt[3]{a}$ can be approximated by the sequence of iterations
$$x_{n+1} = \frac{2x_n^3 + a}{3x_n^2}$$
and this will converge as long as x_0 has the same sign as a.

51. What value of n will ensure that in using Newton's Method, x_n will approximate the root of $x^3 - 3x^2 + 2x - 4 = 0$ that lies in $[2, 3]$ to an accuracy of ten decimal places, if x_0 is taken as 2.8? (*Hint:* First show that the root lies between 2.7 and 2.8.)

In Exercises 52 and 53, show that a root r of $f(x) = 0$ lies in the given interval and that f is monotone and its graph does not change direction of concavity there. Then find a suitable starting point x_0 that will ensure convergence of Newton's Method to the root r.

52. $f(x) = \dfrac{3}{x-1} - x$ on $(2, 3)$

53. $f(x) = \cos x - \sqrt{x}$ on $\left(\dfrac{\pi}{6}, \dfrac{\pi}{3}\right)$

54. Use Inequality 4.7 to find n such that the nth approximation by Newton's Method to the root of $\tan x - \sqrt{2x} = 0$ that lies in $(0, \pi/2)$ is correct to eight decimal places. What is x_n?

In Exercises 55–58, use L'Hôpital's Rule to find each limit.

55. (a) $\lim\limits_{x \to 1} \dfrac{x^3 - 3x + 2}{2x^3 - 3x^2 + 1}$ (b) $\lim\limits_{x \to 0} \dfrac{\sin x - \tan x}{x}$

56. (a) $\lim\limits_{x \to \infty} \dfrac{1 - 3x + 2x^3}{3x^3 + 2x^2 + 5}$ (b) $\lim\limits_{x \to 4} \dfrac{\sqrt[3]{2x} - 2}{\sqrt{x} - 2}$

57. (a) $\lim\limits_{x \to 0} x^2 \cot 2x$ (b) $\lim\limits_{x \to 0} (\sin 3x)(\csc 2x)$

58. (a) $\lim\limits_{x \to -1} \left(\dfrac{x}{x+1} - \dfrac{4 - 3x^2}{x^2 + x}\right)$
 (b) $\lim\limits_{x \to \pi/4} (\sec 2x - \tan 2x)$

In Exercises 59 and 60, show that Rolle's Theorem is satisfied by f on the given interval, and find a number c that satisfies the conclusion of that theorem.

59. (a) $f(x) = x^3 - 3x - 2$ on $[-1, 2]$
 (b) $f(x) = (x^2 - 1)/(x^2 + 2)$ on $[-1, 1]$

60. (a) $f(x) = x^{2/3}(1 - x)$ on $[0, 1]$
 (b) $f(x) = \sqrt{1 - x}(x + 3)$ on $[-3, 1]$

In Exercises 61 and 62, show that the Mean-Value Theorem is satisfied by f on the given interval, and find a number c that satisfies the conclusion of that theorem.

61. (a) $f(x) = 2x^2 - 4x + 3$ on $[-1, 2]$

(b) $f(x) = x^3 - 2x^2 + 4x - 4$ on $[1, 3]$

62. (a) $f(x) = (x + 3)/(x - 2)$ on $[0, 1]$

(b) $f(x) = (3x - 4)^{4/3}$ on $[1, 4]$

63. (a) Suppose $f'(x) = 0$ for all x, and $f(2) = 7$. What is $f(5)$, and why?

(b) Suppose $f'(x) = g'(x)$ for all x, and $f(x) = x^2 + 4$. If $g(1) = 2$, what is $g(x)$, and why?

In Exercises 64–66, find the general antiderivative of f.

64. (a) $f(x) = 3x^2 - 2x + \dfrac{4}{x^2}$

(b) $f(x) = \dfrac{x^2 - 1}{\sqrt{x}}$

65. (a) $f(x) = 4\sin x - 3\cos x$

(b) $f(x) = \dfrac{\sec x + 2\tan x}{\cos x}$

66. (a) $f(x) = \tan^2 x + \sec^2 x$

(b) $f(x) = \sin^2 x - \cos^2 x$

In Exercises 67–69, find the solution to the initial-value problem.

67. (a) $f'(x) = 3x^2 - 2/x^2$; $f(1) = 4$

(b) $\dfrac{dy}{dx} = 2\sin x - x$; when $x = 0$, $y = 5$

68. (a) $f''(x) = 2x$; $f(1) = -3$, $f'(1) = 2$

(b) $y'' = 4$; when $x = 0$, $y = 7$, and $y' = 0$

69. (a) $f'''(x) = 1/x^4$; $f(-1) = 2$, $f'(-1) = 1$, $f''(-1) = 0$

(b) $\dfrac{d^4y}{dx^4} = \cos x$; when $x = 0$, $y = 2$, $y' = -1$, $y'' = 1$, and $y''' = 0$

70. Find the function f such that $y = f(x)$ satisfies the differential equation

$$\frac{d^2y}{dx^2} + x = 2$$

so that the graph of f passes through the point $(-1, 3)$ and has slope $\frac{1}{2}$ at that point.

71. A particle moves on a line directed positively to the right so that its acceleration at time t is $a(t) = 6t$. At time $t = 0$ it is at the origin and has initial velocity $v(0) = -12$. Consider $t \geq 0$.

(a) Find its velocity and position at time t.

(b) Find when and where it reaches its leftmost position.

(c) Find the intervals in which it is speeding up and those in which it is slowing down.

(d) Indicate its motion by means of a sketch.

72. From the edge of a building 128 ft high, a rock is thrown vertically upward at a speed of 32 ft/s. Find (a) how high the rock rises and (b) the speed with which it strikes the ground.

Chapter 4 Concept Quiz

1. Define each of the following:

   (a) an increasing function on I
   (b) a decreasing function on I
   (c) a monotone function on I
   (d) a local maximum value of f
   (e) a local minimum value of f
   (f) an absolute maximum value of f on I
   (g) an absolute minimum value of f on I
   (h) a critical point
   (i) concave up
   (j) concave down
   (k) an inflection point
   (l) an antiderivative of f

2. State each of the following:

   (a) The Extreme-Value Theorem
   (b) The First-Derivative Test for Extrema
   (c) The Second-Derivative Test for Extrema
   (d) L'Hôpital's Rule
   (e) Rolle's Theorem
   (f) The Mean-Value Theorem
   (g) Newton's Method
   (h) Two corollaries to the Mean-Value Theorem

3. Describe the graph of f in each case.

   (a) $f'(x) > 0$, $f''(x) > 0$ on I
   (b) $f'(x) > 0$, $f''(x) < 0$ on I
   (c) $f'(x) < 0$, $f''(x) > 0$ on I
   (d) $f'(x) < 0$, $f''(x) < 0$ on I
   (e) $f'(x_0) = 0$, $f''(x_0) < 0$
   (f) $f'(x_0) = 0$, $f''(x_0) > 0$
   (g) $f(x_0)$ finite but $\lim_{x \to x_0} f'(x) = \infty$
   (h) $f'(x_0) = 0$, $f'(x) < 0$ for $x \neq x_0$

4. Fill in the blanks.

   (a) If f is continuous at x_0 and _____ for $x < x_0$ and _____ for $x > x_0$, then f has a minimum value at x_0.
   (b) The limit $\lim_{x \to a}[f(x)/g(x)]$ is an indeterminate form of type $0/0$ if _____ and _____.
   (c) In Newton's Method, a root of $f(x) = 0$ is approximated using x-intercepts of the _____ line to the graph of f.
   (d) If $f(x_0)$ exists but $\lim_{x \to x_0^+} f'(x) = \infty$ and $\lim_{x \to x_0^-} f'(x) = -\infty$, then the tangent line to the graph of f at x_0 is _____ and the point $(x_0, f(x_0))$ is a _____ point on the graph.
   (e) According to the Mean-Value Theorem, if f is continuous on $[a, b]$ and differentiable on (a, b), there is a point c between a and b such that the _____ line at $x = c$ is parallel to the _____ line joining the points _____ and _____.

5. Which of the following are true and which are false?

   (a) If $f'(x) = g'(x)$ for all x, then $f(x) = g(x)$.
   (b) L'Hôpital's Rule states that when $\lim_{x \to a} f(x) = 0$ and $\lim_{x \to a} g(x) = 0$, then

   $$\lim_{x \to a} \frac{f(x)}{g(x)} = \lim_{x \to a} \left(\frac{f}{g}\right)'(x)$$

   (c) If $f''(x_0) = 0$, then $(x_0, f(x_0))$ is an inflection point.
   (d) If $f'(x) \cdot f''(x) < 0$ for all x, then f is either increasing and concave down or decreasing and concave up.
   (e) If f is continuous and has a maximum at $x = x_0$, then $f'(x_0) = 0$.

APPLYING CALCULUS

1. A projectile is fired from the ground and follows the path $y = 0.5x - 0.006x^2$, where x is in kilometers, and the origin is taken as the point of firing, as shown in the figure. If the horizontal speed of the projectile is constant at $\frac{80}{3}$ km/min, find the rate at which the angle θ is changing when $x = 50$ km.

2. The formula

$$\frac{1}{f} = \frac{1}{u} + \frac{1}{v}$$

expresses the relationship between the focal length f of a concave mirror or lens, the distance u of an object from the lens, and the distance v of its image from the lens. Suppose the focal length of a certain lens is 30 cm and an object is moving toward the lens at the rate of 3 cm/s. Find the rate at which the image is receding from the lens when the object is 90 cm away.

	Progresso	Campbell
Outside sizes		
diameter	3 3/8″	3 1/8″
height	4 5/16″	5″
calc. volume	38.6	38.4
On label		
net weight	19 oz;	19 oz;
	538 g	539 g

3. Campbell's and Progresso both sell soup in cans designed to hold an equivalent weight, approximately 19 ounces or 538 grams. The cans have the dimensions shown in the table. Assume that the soups are of equal density (weight per volume) and that both companies manufacture an equivalent amount of soup at equivalent costs. Thus, differences in production costs should be a result of differences in the costs of the cans, which include the tops, bottoms, and sides.

(a) Which company should have the lower production costs?

(b) Suppose the materials used to make the cans are such that the tops and bottoms cost $0.14/ft^2$ and the sides cost $0.08/ft^2$. Which company has the lowest costs under these conditions?

(c) Suppose both companies require that the cans have a volume of 33.4 cu. in. Assuming that the same material is used for the tops, bottoms, and sides, what dimensions should the cans have to minimize the cost?

4. An engineer is designing a drainage canal that has a trapezoidal cross section. The bottom and sides of the channel are each L feet long, and the sides make an angle θ with the horizontal.

(a) Find an expression for the cross-sectional area of the channel.

(b) Determine the angle θ that will maximize the cross-sectional area.

5. A manufacturer has outlets at 3 locations A, B, and C (see the figure). The owner wishes to locate the plant at a point X so that total shipping costs to all outlets are as small as possible. Assuming that the number of shipments to all 3 outlets are equal, and that the cost of transportation is proportional to the distance, shipping costs will be minimized if x is chosen so that the sum $S(x)$ of the distances from X to A, B, and C is as small as possible. Using the simplifying assumption that the points A and B are symmetric with respect to C (i.e., the line connecting C to the midpoint M of segment AB is perpendicular to that segment as in the figure), you are asked to locate the point X by following the steps below.

(a) Show that the minimizing point X must lie on the segment CM.

(b) Introduce rectangular coordinates so that C is the origin and the ray CM is the positive x-axis. If the coordinates of A are (a, b), what are the coordinates of B?

(c) If we denote the coordinates of X by $(x, 0)$, what inequality(ies) must x satisfy? Show that S, which depends on x, is given by the formula

$$S(x) = x + 2\sqrt{(a - x)^2 + b^2}$$

(d) Finally, show that if $a \leq b/\sqrt{3}$, $S(x)$ achieves its minimum at $x = 0$, and if $a > b/\sqrt{3}$, $S(x)$ achieves its minimum at $x = a - b/\sqrt{3}$.

6. The Massachusetts Turnpike is 123 miles long. When entering the turnpike the driver of a passenger car is given a card indicating the date and time and listed are the toll charges at each exit off the turnpike. Suppose you enter the turnpike on the east end at 1:00 P.M. and travel the entire 123 miles of the turnpike. You arrive at the toll gate on the west end of the turnpike at 2:30 P.M. When the toll gate attendant examines your toll card he informs you that not only must you pay the toll but an additional $80 for speeding. He claims you have driven at least 75 mph at one point in your journey and the maximum turnpike speed limit is only 65 mph. When you protest, he insists your arguments are useless since he has calculus on his side. Why is the toll booth attendant correct in his claim that you were speeding?*

7. A public-works department is to construct a new road between towns A and B. Town A lies on an abandoned road running east-west. Town B is 20 miles north of this road and 40 miles east of town A. An engineer proposes that the road be constructed by restoring a section of the old road leaving town A and joining it to an entirely new section of the old road at a point to be determined, connecting that point with town B (see the figure).

(a) If the cost of restoring the old road is $200,000 per mile and the cost of the new road is $400,000 per mile, how much of the old road should be restored in order to minimize the department's cost?

(b) If the cost of restoring the old road is $200,000 per mile, what would the maximum cost per mile of the new road have to be in order for the department to construct a direct route from A to B? [Hint: Treat the cost of the new road as an unknown parameter.]

(c) Under what conditions would it make sense to construct an L-shaped route connecting A and B?

(d) Extending the problem: Treat both the cost of restoring the old road and the construction of the new road as unknown parameters. Determine a relationship between the two costs that makes a direct link between A and B the most economical.

8. A trucking company would like to determine the highway speed that it should require of its drivers. The decision is to be made purely on economical grounds, and the two primary factors to be considered are driver wages and fuel consumption. Wage information is easily obtained: drivers earn from $11.00 to $15.00 an hour, depending on experience. Incorporating the fuel consumption question is much more difficult and the company has hired you as consultants in order to solve the problem for them. Correctly assuming that fuel consumption is closely related to fuel economy at various highway speeds, they have provided you with the following statistics taken from a U.S. Department of Transportation study:†

---

*Adapted from J. Chover, *The Green Book of Calculus*.

†U.S. Department of Transportation. *The Effect of Speed on Truck Fuel Consumption Rates*, by E. M. Cope. [Washington]: U. S. Dept. of Transportation, Federal Highway Administration, Office of Highway Planning, Highway Statistics Division. 1974. (TD2.2:Sp3)

Miles per Gallon at Selected Speeds

Vehicle	50 mph	55 mph	60 mph	65 mph
Truck #1	5.12	5.06	4.71	*
Truck #2	5.41	5.02	4.59	4.08
Truck #3	5.45	4.97	4.52	*
Truck #4	5.21	4.90	4.88	4.47
Truck #5	4.49	4.40	4.14	3.72
Truck #6	4.97	4.51	4.42	*

*The truck could not go this fast.

Driver wage	Ideal speed
$11.00	
$12.00	
$13.00	
$14.00	
$15.00	

Compute the average mpg at each speed and fit a cubic polynomial through the points. This gives a function relation f between miles per gallon and miles per hour. Then let D represent distance traveled, F the cost of fuel per gallon, and W, the driver's hourly wage. Write the total cost in terms of W, D, $F(x)$, and x, where x represents speed. Apply extrema theory to the cost function obtained by letting $F = 1$ and $W = 11, 12, 13, 14$, and then 15. Complete the table in the margin.

9. The Babylonians discovered a rule for approximating the square root of a positive number k. Begin by making a reasonable estimate a of $\sqrt{k}$. Then

$$b = \tfrac{1}{2}\left(a + \frac{k}{a}\right)$$

will be a better estimate. Continue the process.

(a) Use this method to approximate $\sqrt{13}$ to five decimal places.

(b) Show that this method of approximating square roots is a special case of Newton's method.

10. Given a function F defined on an interval I a point $x \in I$ is a *fixed point* of F if $F(x) = x$. Many important applications of mathematics involve finding fixed points of functions. It can be shown using the Intermediate-Value Theorem that if F is continuous on the interval $[a, b]$ and satisfies

$$a \leq F(x) \leq b$$

for all $x \in [a, b]$, then F has at least one fixed point on the interval. Moreover, it can be shown that if F is differentiable and satisfies $|F'(x)| < r$, for $a < x < b$, where $0 < r < 1$, then F has exactly one fixed point on $[a, b]$.

(a) One method for finding a fixed point of a given function F is by *iteration*. If F satisfies the second condition above, the sequence (x_n) defined by x_1, $x_2 = F(x_1), x_3 = F(x_2), \ldots, x_{n+1} = F(x_n), \ldots$, where x_1 is an initial "guess," will converge to the unique fixed point. Show that $F(x) = \cos x$, $-3 \leq x \leq 3$, satisfies the two conditions stated above and find an approximate solution of $F(x) = x$, accurate to 4 decimal places.

(b) Since a fixed point of a given function F is a root of the equation $G(x) = 0$, where $G(x) = F(x) - x$, fixed points can be found by Newton's method. Indeed, in most instances, Newton's method is faster than iteration. Determine the recursion formula corresponding to Newton's method for finding the roots of $G(x) = F(x) - x = 0$. Use this method to approximate the fixed point of $F(x) = \cos x$ on $[0, 3]$.

Anna J. Schwartz

Is mathematics necessary or merely useful in economics? For instance, is calculus an essential course to take if you plan to become an economist?

It's definitely more than just useful. Mathematics helps explain economic phenomena. We express the problem of how one variable changes in relation to another by means of a functional relationship between variables. The derivative measures the instantaneous rate of growth of one quantity with respect to another and approximates the actual average rate of growth when the variation considered is small. In economics, derivatives are of special importance, especially in dealing with cost and profit functions. In econometrics, which is the main empirical [experimental] way in which economics is done, the basic mathematical tool is calculus. If you plan to major in economics, it makes sense to take both calculus and linear algebra, especially if you want to go on to graduate school.

What exactly is econometrics?

Econometrics is a combination of economic theory, mathematics, and statistics. All three disciplines are needed to understand the quantitative relationships that are inherent in modern economic life. Economic theory, which is derived from statistical data and other facts, suggests hypotheses to be tested by empirical analysis. Theory identifies the factors thought to be involved in the behavior of individual economic units and the markets in which they interact, as well as economic aggregates, such as aggregate consumption and aggregate investment. Econometric methods apply mathematics, such as calculus-based probability, to the analysis of economic data. The aim is to test the validity of the theory under consideration with appropriate methods and data. Empirical economic research deals with functions with variables expressed as levels or as rates of change. For this purpose, one needs familiarity with calculus.

Calculus deals in part with rates of growth. In what sense is the economy a result of many competing rates of growth?

The question is not so much rates of growth, per se, but the functional relationship among variables with different rates of growth. In the United States, until the end of World War II, the rate of growth of (per-capita) money holdings was greater than the rate of growth of (per-capita) income. Since then, on average, the rate of growth of money holdings has been lower than the rate of growth of income. It is an interesting economic question to explain this difference in behavior.

Do you remember a particular problem?

One such problem is hyperinflation, the radical devaluing of currencies, in the countries of post–World War II Europe. In the problem of hyperinflation you want to know how acceleration (of growth) in the money supply interacts with the demand for money. The empirical problem is to identify the variables that determine the amount of money people want to hold, rather than

spend. Tests of various hyperinflations suggest that people have a well-defined demand for money in terms of the cost of holding it. Two expected costs of holding money are the expected yield on some alternative asset, and the rate at which the value of money is expected to decline due to rising prices. To estimate the second component of costs, we assume that individuals modify their expectations about future prices on the basis of the error in the period before. Therefore, previous price changes determine expectations of current prices. In Europe, hyperinflation was a really daunting problem that was eventually solved by replacing the money that had lost value, because of oversupply and the public's psychological rejection of that money, with substitute currency that had real purchasing power.

Could you give us some examples of the use of the derivative in economics?

In economics, two central concepts are *average* and *marginal* value. The average variation of total cost (C_t), with respect to total output (O_t), is the total cost to the firm (at any level of output) divided by the total output; and the marginal variation is the variation of C_t for small values of O_t. Marginal cost, however, is the change in C_t at a certain level of output brought about by a very small increase in the quantity of output from that level. The marginal value concept is basic to many economic ideas about supply and demand, costs and prices, production rates, and revenues. Of course it is the derivative of the function that relates C_t and O_t. The numerical value of the derivative measures how rapidly the function C_t increases or decreases with respect to O_t. The results indicate the nature of the function over a whole range of values of the variable O_t. Determination of maximum or minimum positions is important in economic theory. For example, a firm chooses output or price that will produce the largest net revenue; similarly, the consumer seeks the highest position on a scale of consumption possibilities. The variation of a function with more than one independent variable involves differentiation to obtain the differential of a function. For example, production of a certain item requires inputs of capital and labor. Via differentiation we can calculate the marginal rate of substitution of capital for labor in the production of the item. Similar calculations enable us to determine the marginal rate of substitution by the consumer of one item for another.

What about integration? Is it useful as well?

Very. Economists use integration to measure the area under a curve, just as other scientists do. We commonly plot graphs of functions using known or assumed values of variables. The market demand for an item can be represented by three functions and curves showing total, average, and marginal revenue, respectively, for various outputs of the item. Total revenue from any output is then the area (the integral) under either the average or marginal revenue curves. A similar result holds for any pair of average and marginal curves.

Integration is an important tool in both microeconomics, the study of the economic behavior of individuals and of firms, and macroeconomics, the study of the economy as a whole. In microeconomics you might want to study an integral that maximizes profit. In macroeconomics you might want to use integration to evaluate aggregate supply and demand. For example, supply and demand of a commodity may depend not only on present prices but also on prices in the recent past. Farmers might plant more wheat this year if last year's price was high and plant less this year if last year's price was low. This relationship can be expressed and computed in integral form, so calculus makes this kind of application possible. This kind of decision making is very important to industry and the economy as a whole.

Calculus is also useful in problems involving capital and interest. Logarithmic functions are particularly important for this purpose. Another application of the integral is the capital value at the present time of an income stream from some asset. The capital value depends on the size of the income stream, the number of years of the income flow, and the rate of interest. Yet another example is consumer surplus, which is found by the area under a demand curve that measures the consumer value of an item in excess of its price.

Is calculus used in developing theoretical economic models?

Yes. Calculus is useful both in an empirical way and in setting up theoretical structures or models. For instance, Arthur Laffer, the originator of supply-side economics, drew a curve that showed the revenue that a government obtained with various marginal rates of income tax. He used this curve to assert that governments could obtain more revenue within a certain range of marginal rates of taxation and that the revenue would diminish as the rate of marginal taxation rose. This "Laffer curve" was used as a theoretical basis and computed by measuring different points at different marginal rates of taxation using derivatives. These models may also involve the use of integration. For instance, you could set up a model that assumes people will try to maximize what turns out to be the integral of three elements that determine the behavior of economic time series: (1) maximizing quantities and prices subject to budget constraints as a function of time, (2) minimizing the effect of erratic shocks, and (3) minimizing the effect of excessive inventories. These abstract concepts have become part and parcel of the way economists think, and work.

When did you realize how important the tools of calculus were to your profession?

Not soon enough. When I was a graduate student, it just wasn't part of prescribed graduate training. Now it's taken for granted. Every graduate student has to take a course in mathematics for economists, which includes at least some calculus. I learned by doing and by working with people who had better mathematical training. They introduced me to the kinds of exercises and applications that one would ordinarily encounter in class. My exposure was on the job where there were real problems to solve.

I wouldn't suggest that the use of calculus is the end of the story on how economists use mathematics in their work. There are economists who are working with chaos theory or fractals, for instance, which is a different mathematical subdiscipline. These methods don't seem to have the generality of calculus, however.

What if a student decided not to go to graduate school? Could this person still become a working economist?

Well, the calculus course they're taking now would give them the basic concepts—I don't want to alarm anyone about needing to have exceptional mathematical aptitude, but I think a graduate degree is very important in this day and age. You might get an entry-level position without a graduate degree but remain in that position throughout your career, while those with Ph.Ds were assigned the more interesting work. Your salary wouldn't grow at the same pace as someone with a graduate degree either.... There's an economic fact that you don't need calculus to understand.

Do you see these economic issues as being more understandable if you understand calculus? Would Isaac Newton, for instance, have understood the application of calculus to today's economic issues?

Well, I'd be speculating of course, but Newton was a genius and in his later years was Britain's Master of the Mint. As such he had a major influence over the money supply. He took his job very seriously, personally supervising the hanging of guilty counterfeiters. At one point he established a "Mint price" for guineas, which overvalued them and therefore inadvertently displaced the then-current silver standard by the gold standard. If the Mint price for guineas had been lower or the price of silver higher, no change in the standard would have occurred.

Anna J. Schwartz is a native New Yorker and a member of the research staff at the National Bureau of Economic Research. Recently named a Distinguished Fellow of the American Economic Association, she has collaborated with Milton Friedman on studies of monetary economics, including *A Monetary History of the United States, 1867–1960* (1963). She currently serves on the editorial boards of three economics journals. Dr. Schwartz holds a B.A. from Barnard and M.A. and Ph.D. degrees in Economics from Columbia University.

5

THE INTEGRAL

In Chapters 3 and 4 we explored **differential calculus**, based on the concept of the derivative. In this chapter we begin the study of the other major branch of calculus, **integral calculus**, based on the concept of the *definite integral*, which we will define in Section 5.2.

In differential calculus, we used the problem of defining tangent lines to curves (a geometric problem) and the problem of computing the velocity of a moving body from its change in position (a physical problem) to develop the definition of the derivative in terms of a limit. In a similar manner we will use the problem of determining the area under a curve (a geometric problem) and the problem of determining the total distance traveled by an object with a given velocity (a physical problem) to develop the definition of the integral of a function, also in terms of a limiting process. Table 5.1 shows the relationship between these fundamental problems and the two branches of calculus. It is critically important to understand the connection between the two branches of calculus, which we will make precise in Section 5.3.

TABLE 5.1

	Differential calculus	Integral calculus
Geometric problem	Tangent lines to curves	Area under a curve
Physical problem	Velocity of a moving body (from its distance traveled over time)	Total distance traveled by a moving body with a given velocity

The study of the area problem—computing the area under a curve—and the theory of integration that leads from it have intrinsic mathematical value. That is, they are worth studying in their own right. They become even more important when we realize we can use integrals to solve an incredible range of diverse problems. For example, we can find volumes of solids, lengths of curves, forces on dams, the work done by variable forces, and probabilities of certain events. Still more applications of the integral occur in such fields as economics, biology, and chemistry. It is testimony to the power of mathematics that a single abstract concept, such as the integral, applies to a huge variety of real-world problems.

5.1 THE AREA PROBLEM AND THE DISTANCE PROBLEM

Before beginning study of the integral, let us consider two geometric problems involving area and distance. Imagine that we want to estimate the *area* of the "two-point zone" at one end of the basketball court in Boston Garden, as shown in the photograph. The difficulty in estimating this area results from its shape, which appears to be a rectangle surmounted by the arc of a circle (of unknown diameter).

Instead of trying to estimate the area directly, we establish an upper and a lower bound for the area in question by counting the wooden floor tiles that enclose the area (A_{max}) and those that are entirely inside the area (A_{min}), as shown in Figure 5.1(a). Since the court measures 94 ft by 50 ft, and ten tiles span the width of the court, each tile is 5 ft square and has area 25 ft^2. There are 34 whole tiles inside the two-point zone; therefore, A_{min} is 34×25 ft$^2 = 850$ ft^2. Similarly, there are 48 whole tiles that include all or part of the two-point zone; therefore, A_{max} is 48×25 ft$^2 = 1200$ ft^2. From our crude visual estimate we know only that $850 < A < 1200$. How can we improve our estimate? What if we had smaller tiles? If we subdivide each tile to create 25 tiles of 1 ft^2 each, can we use the smaller tiles to approximate the area of the two-point zone more accurately? Why? What would happen if we made the tiles very small so that their areas approached zero? Could the limit of this process somehow be used to find the area we are trying to estimate? Before we address these questions in this section, let's look at a situation involving *distance*.

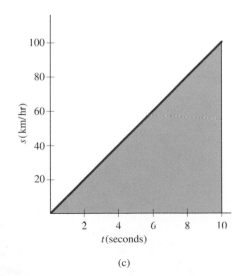

(a)   (b)   (c)

FIGURE 5.1

The graphs in Figures 5.1(b) and 5.1(c) represent the speed of a car in km/hr over the given period of time. In Figure 5.1(b) we see the average speed of the automobile over a several *hour* period and in 5.1(c) we see the automobile's speed over a several *second* period. By inspecting Figure 5.1(b), we know that the car was traveling at 100 km/hr five hours into the trip (we ignore the time spent accelerating to the average speed). What total distance did the car travel during the nine-hour trip? We can see from the graph that the area under the speed "curve" is the product of the speed of the car and the time elapsed (kilometers/hour × hours = kilometers). If we sum the individual areas of 10 km/hr increments ×1 hr periods, we get the total distance traveled, in this case 59 km. Note that the area under the speed curve represents distance, not distance squared. Similarly, from Figure 5.1(c), we see that the car accelerates from 0 to 100 km/hr in 10 seconds (1/360 hr). What distance did the car travel during this 10-second period? From the right-triangular area under the curve, $A = (1/2)(\text{base})(\text{height})$; that is, $(1/2)(1/360 \text{ hr}) \times (100 \text{ km/hr}) = 5/36$, or about 0.139 km. This distance is not as simple to conceive as that for the average speed situation in Figure 5.1(b), but the calculation is the same in that we find distance traveled by computing the area under a curve.

Both the area problem and the distance problem are solved using the general mathematical notion of the integral (although an integral does not have to be an area or a distance). In this chapter we will make these ideas more precise. Let's look at each type of problem in more detail.

The Area Problem—A Geometric Problem

We continue our discussion of the area problem. Specifically, we want to find the area, as shown in Figure 5.2, bounded above by the graph of a continuous nonnegative function f, below by the x-axis, and on the left and right by the vertical lines $x = a$ and $x = b$. We will refer to the area of this shaded region as *the area under the graph of f from x = a to x = b*. When there is no ambiguity about the curve and interval being discussed, we can simply say *the area under the curve*. The real question is: How do we *define* what we mean by the area of such a region?

Area formulas for many of the basic geometric figures date back to the earliest mathematical writings thousands of years ago. The area of a rectangle is "base times height," a triangle has area "one-half base times height," and the area of a polygon can be found by dividing it into triangles (Figure 5.3).

FIGURE 5.2

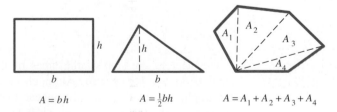

$A = bh$ $A = \frac{1}{2}bh$ $A = A_1 + A_2 + A_3 + A_4$

FIGURE 5.3
A parabolic region

FIGURE 5.4

FIGURE 5.5
Area approximated by triangles

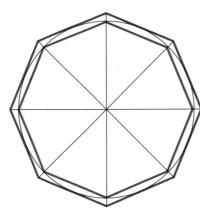

FIGURE 5.6

The problem of finding areas of regions with curved boundaries was of great interest to the Greeks, especially Archimedes (287–211 B.C.), who devised the "method of exhaustion" to solve special cases of this problem. For example, Archimedes was able to find the areas of certain parabolic regions (Figure 5.4) by inscribing an increasing number of triangles in the region (Figure 5.5), getting at each stage a better approximation to the actual area. In this manner the region is eventually exhausted, and the area is the limit of the approximations.

Archimedes also gave arguments for the areas of circles and spirals. In the case of circles, he extended his "method of exhaustion" to what has been called his "method of compression." Here the circle is inscribed (red lines) and circumscribed (blue lines) with polygons constructed from isosceles triangles (Figure 5.6). The area of the circle is then compressed between the larger and smaller approximations, and once again the area is the limit of these approximations.

The first solutions to the general area problem took almost another 2000 years when, independently, solutions were given in 1669 by Sir Isaac Newton, the great British mathematician and physicist, and in 1673 by the German mathematician Gottfried Wilhelm Leibniz. The technique we will describe and use was developed during the seventeenth and eighteenth centuries, culminating with the work of another German mathematician, Georg Riemann (1826–1866).

There are several reasonable ways of approximating the area of a region such as the one in Figure 5.2, but we concentrate on only two. The approach we will take is similar to Archimedes' "method of compression." The first method is called the method of **inscribed rectangles**. In this method we divide the interval $[a, b]$ into a number of subintervals. For simplicity, we take them to be of equal width, though this specification is not required. Next, we construct rectangles that have these subintervals as bases and such that the height of each rectangle is the minimum value of $f(x)$ for all x in the subinterval forming the base. We know by the theorem on extreme values for continuous functions (Theorem 4.2) that f does assume a minimum value (as well as a maximum value) on each of the closed subintervals. In Figure 5.7 we have shown two sets of inscribed rectangles. In part (b) we have used a finer subdivision than in part (a), causing the rectangles to be thinner and more numerous.

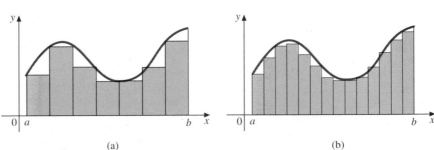

(a)   (b)

FIGURE 5.7

We refer to the sum of the areas of the inscribed rectangles for a given subdivision as a **lower sum**. Although we have not yet defined the area under the graph, intuitively it would appear that a given lower sum approximates that area from below. That is, each lower sum is an underestimate of the area. Furthermore, as we increase the number of rectangles, thereby making the widths smaller and smaller, the rectangles exhaust more and more of the area, so the approximations become increasingly better.

A second method is to use the maximum value of f on each subinterval, instead of the minimum value. Then we arrive at what we call **circumscribed rectangles**, as in Figure 5.8, and the sum of their areas, called an **upper sum**, approximates the area from above.

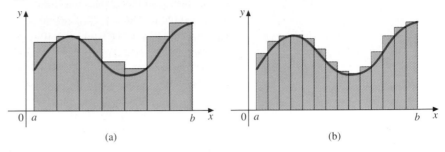

(a)   (b)

FIGURE 5.8

For any particular subdivision of $[a, b]$, we always have

$$\text{lower sum} \le \text{upper sum}$$

It appears that as we allow the number n of subintervals to increase indefinitely (that is, let $n \to \infty$), the lower and upper sums approach each other, and their common limit is the area we are seeking.

Before we develop more general methods of approximating the area under a curve, we will illustrate the procedure we have described with an example of a region whose area we already know, to show that our method does give the correct answer in this case.

EXAMPLE 5.1   Approximate the area under the graph of the function $f(x) = x - 1$ from $x = 1$ to $x = 3$, using both inscribed and circumscribed rectangles.

Solution   The graph is shown in Figure 5.9. Notice that the region is triangular, so we know in advance that the area is

$$A = \frac{1}{2}bh = \frac{1}{2}(2)(2) = 2$$

Let us introduce the rectangles by partitioning the interval $[1, 3]$ into subintervals of equal lengths. Figure 5.10 shows the interval $[1, 3]$ divided into two such subdivisions, figure (a) with 8 equal parts and figure (b) with 16 equal parts; the resulting inscribed and circumscribed rectangles are displayed. We

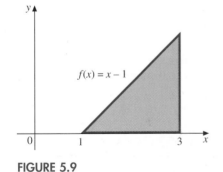

$f(x) = x - 1$

FIGURE 5.9

(a)

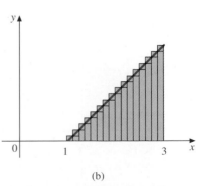

(b)

FIGURE 5.10

can now compute the lower sum by multiplying the base of each rectangle by its height, and then summing the areas of the rectangles. To find the base, we divide the width of the total interval from 1 to 3 by the number of rectangles. So for 8 rectangles, each base is $(3-1)/8 = 2/8 = 1/4$. The height of each lower rectangle is computed by substituting the left endpoint of each subinterval for x in $f(x) = x - 1$. For the upper sum, we use the right endpoints of the subintervals instead. Choosing left endpoints to get the heights of lower rectangles and right endpoints to get the heights of upper rectangles is typical for all *increasing* functions, such as the one in this example. In the case of *decreasing* functions these choices would be reversed. For our function, using 8 rectangles, the lower sum (L_8) and the upper sum (U_8) are then, respectively,

$$L_8 = \tfrac{1}{4}(1-1) + \tfrac{1}{4}(\tfrac{5}{4}-1) + \tfrac{1}{4}(\tfrac{6}{4}-1) + \tfrac{1}{4}(\tfrac{7}{4}-1) + \tfrac{1}{4}(\tfrac{8}{4}-1)$$
$$+ \tfrac{1}{4}(\tfrac{9}{4}-1) + \tfrac{1}{4}(\tfrac{10}{4}-1) + \tfrac{1}{4}(\tfrac{11}{4}-1)$$
$$= \tfrac{1}{4}(0 + \tfrac{1}{4} + \tfrac{2}{4} + \tfrac{3}{4} + \tfrac{4}{4} + \tfrac{5}{4} + \tfrac{6}{4} + \tfrac{7}{4}) = 1.75$$

and

$$U_8 = \tfrac{1}{4}(\tfrac{5}{4}-1) + \tfrac{1}{4}(\tfrac{6}{4}-1) + \tfrac{1}{4}(\tfrac{7}{4}-1) + \tfrac{1}{4}(\tfrac{8}{4}-1) + \tfrac{1}{4}(\tfrac{9}{4}-1)$$
$$+ \tfrac{1}{4}(\tfrac{10}{4}-1) + \tfrac{1}{4}(\tfrac{11}{4}-1) + \tfrac{1}{4}(3-1)$$
$$= \tfrac{1}{4}(\tfrac{1}{4} + \tfrac{2}{4} + \tfrac{3}{4} + \tfrac{4}{4} + \tfrac{5}{4} + \tfrac{6}{4} + \tfrac{7}{4} + \tfrac{8}{4}) = 2.25$$

The error involved is less than or equal to $U_8 - L_8 = 0.50$.

Table 5.2 shows the computation using the number of rectangles n doubling in sequence from 4 to 512. It is easy to use a spreadsheet or a CAS to compute the values in the table. Notice that as the number of subintervals increases, the upper sums decrease toward 2, the lower sums increase toward 2, and the differences between them decrease. The data suggest that the limit of these approximations is the actual area. ∎

TABLE 5.2

n	U_n	L_n	$U_n - L_n$
4	2.5	1.5	1
8	2.25	1.75	0.5
16	2.125	1.875	0.25
32	2.063	1.938	0.125
64	2.031	1.969	0.0625
128	2.016	1.984	0.0312
256	2.008	1.992	0.0156
512	2.004	1.996	0.0078

Formulas for Upper and Lower Sums

To arrive at some general computational formulas for upper and lower sums, we introduce the following notation. Let A denote the area under the graph of the nonnegative, continuous function f from $x = a$ to $x = b$. Let the points of the subdivision of $[a, b]$ into n subintervals be denoted by $x_0, x_1, x_2, \ldots, x_n$, with endpoints $a = x_0$ and $b = x_n$, such that $x_0 < x_1 < x_2 < \cdots < x_n$. Such a set of ordered points is said to form a **partition** of $[a, b]$. Denote the common length of the subintervals by Δx. Since the distance on the x-axis from a to b is $b - a$, and there are n subintervals, it follows that the length of each subinterval is the total length divided by the number of subintervals, which we can write as

$$\Delta x = \frac{b - a}{n}$$

Let m_k denote the minimum value of $f(x)$ on the kth subinterval, $[x_{k-1}, x_k]$, and let M_k denote the maximum value of $f(x)$ on this subinterval. So m_1 is the minimum height from the x-axis to the graph of f on the interval $[x_0, x_1]$, m_2 the minimum height on the interval $[x_1, x_2]$, and so on. Similarly, $M_1, M_2, \ldots, M_n$ are the maximum heights. These meanings are indicated in Figure 5.11, where we have shown a typical subinterval.

FIGURE 5.11

Denote the lower sum for this subdivision by L_n and the upper sum by U_n, where the subscript n indicates the number of subintervals. Then

$$L_n = m_1 \Delta x + m_2 \Delta x + m_3 \Delta x + \cdots + m_n \Delta x$$
$$= (m_1 + m_2 + m_3 + \cdots + m_n)\Delta x \qquad (5.1)$$

Similarly,

$$U_n = M_1 \Delta x + M_2 \Delta x + M_3 \Delta x + \cdots + M_n \Delta x$$
$$= (M_1 + M_2 + M_3 + \cdots + M_n)\Delta x \qquad (5.2)$$

EXAMPLE 5.2   Find L_n and U_n for the area under $y = f(x) = x^2 + 1$ from $x = 0$ to $x = 2$, for $n = 4$, 10, 20, 100, and 500.

Solution   Figure 5.12 shows the inscribed and circumscribed rectangles for $n = 4$. In this case, $\Delta x = \frac{b-a}{4} = \frac{2}{4} = \frac{1}{2}$. Since the graph is increasing on the given interval, the minimum value of the function on each subinterval occurs at the left endpoint and the maximum value at the right endpoint. So we have

$$m_1 = f(0) = 1 \qquad\qquad M_1 = f(0.5) = 1.25$$
$$m_2 = f(0.5) = 1 + (0.5)^2 = 1.25 \qquad M_2 = f(1) = 2$$
$$m_3 = f(1) = 1 + (1)^2 = 2 \qquad M_3 = f(1.5) = 3.25$$
$$m_4 = f(1.5) = 1 + (1.5)^2 = 3.25 \qquad M_4 = f(2) = 5$$

Thus, by Equation 5.1,

$$L_4 = (1 + 1.25 + 2 + 3.25)(0.5) = 3.75$$

and by Equation 5.2,

$$U_4 = (1.25 + 2 + 3.25 + 5)(0.5) = 5.75$$

The calculations for the other values of n are carried out in a similar way. The results are shown in Table 5.3. (We used a spreadsheet to compute these values, which could also be easily computed with a CAS or a programmable calculator.) We conclude that the area under the curve is between 4.6587 and 4.6747. ∎

TABLE 5.3

n	L_n	U_n
4	3.75	5.75
10	4.28	5.08
20	4.47	4.87
100	4.6268	4.7068
500	4.6587	4.6747

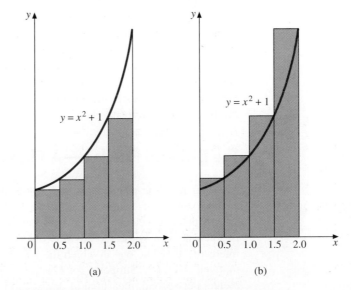

(a)   (b)

FIGURE 5.12

We have been discussing the area under the graph of a continuous, non-negative function without having *defined* the area, proceeding on an intuitive basis. Now we are ready to give a definition. The upper sums $U_1, U_2, U_3, \ldots$ form a *sequence*, which we indicate briefly by $\{U_n\}$. Similarly, we denote the sequence $L_1, L_2, L_3, \ldots$ of lower sums by $\{L_n\}$. When the terms of a sequence come arbitrarily close to some fixed value, say A, as $n \to \infty$, we say that the sequence *converges* to A. (Recall that we mentioned sequences in Chapter 4 in connection with Newton's Method. We will study sequences in more detail in Chapter 10.) As our examples illustrate, and as our intuition tells us, the two sequences $\{U_n\}$ and $\{L_n\}$, of upper sums and lower sums, respectively, both converge to a common limit. It is this common limit that we take as the definition of the area. The formal proof that $\{U_n\}$ and $\{L_n\}$ converge to a common limit for continuous functions involves a concept called *uniform continuity* that is studied in advanced calculus courses.

Definition 5.1 Area Under a Curve	Let f be a continuous, nonnegative function on the closed interval $[a, b]$. Then the area A of the region under the graph of f and above the x-axis from $x = a$ to $x = b$ is the common limit of the sequences $\{U_n\}$ and $\{L_n\}$ of upper and lower sums, as $n \to \infty$. That is, $$A = \lim_{n \to \infty} U_n, \text{ or equivalently, } A = \lim_{n \to \infty} L_n$$

REMARK ────────────────────────────────

■ Since the area A lies between U_n and L_n for every n, it follows that the **error** in using either U_n or L_n to approximate A cannot exceed $U_n - L_n$.

──

Except in the case of relatively simple functions, such as polynomials of low degree, finding a general expression for L_n or U_n and determining the limit as $n \to \infty$ is prohibitively difficult. (See Example 5.5 and Exercises 10, 12, and 14 in Exercise Set 5.1 for some cases in which this *can* be done.) As in Example 5.1, we can approximate the limit (and hence the area) to any desired number of decimal places by taking n large enough. In Section 5.3, we will prove a remarkable result that enables us to find exact values of many such limits, using antiderivatives.

The Distance Problem—A Physical Problem

Consider next the problem of finding the distance covered by an object (such as a person, a car, a train, or a falling body) moving in a straight line, with a known positive velocity, for a fixed period of time. Of course, if the velocity is constant, you know the answer (distance = rate × time), so the problem of interest to us is the one in which the velocity (the rate) changes with time. Call this velocity $v(t)$. Suppose, for example, that the velocity of a relay runner in meters per second, measured at one-second intervals for a period of 5 seconds, is as shown in the table. We want to estimate the distance the runner covers.

t	0	1	2	3	4	5
$v(t)$	3.3	3.9	4.5	5.0	5.5	5.8

We assume that throughout the time interval, the velocity is increasing. Thus, during the first one-second interval, the runner went at least 3.3 m/s and at most 3.9 m/s. So the runner covered a distance of no less than $(3.3)(1) = 3.3$ m (using $d = rt$) and no more than $(3.9)(1) = 3.9$ m. Similarly, during the next one-second interval the runner went at least 3.9 m and at most 4.5 m. Continuing in this way, we see that the total distance is between

$$3.3 + 3.9 + 4.5 + 5.0 + 5.5 = 22.2 \text{ m}$$

and

$$3.9 + 4.5 + 5.0 + 5.5 + 5.8 = 24.7 \text{ m}$$

Suppose now that the runner's velocities are known at half-second intervals, as shown in the table below. Reasoning as above, in the first half-second interval the runner's speed was at least 3.3 m/s, so the distance covered was at least $(3.3)(0.5) = 1.65$ m. Similarly, the runner's speed during this time interval was at most 3.59 m/s, so the distance covered was at most $(3.59)(0.5) = 1.795$ m.

t	0	0.5	1	1.5	2	2.5	3	3.5	4	4.5	5
$v(t)$	3.3	3.59	3.9	4.26	4.5	4.90	5.0	5.26	5.5	5.66	5.8

Continuing in this way, using the velocity at the beginning of each half-second interval to get the underestimate of distance and the velocity at the end to get the overestimate, we obtain

$$22.93 \text{ m} \leq \text{distance} \leq 24.19 \text{ m}$$

If we knew velocities at even smaller time intervals, we could come closer with our low and high estimates.

If the underestimates for the distance in this example seem to resemble a lower sum for an area under a curve and the overestimates an upper sum, it is no accident. In fact, when we draw the graph of velocity versus time, as we have done in Figure 5.13, the reason becomes clear. We have shown inscribed rectangles for subintervals of width 1 (representing a time of 1 second). The area of the first rectangle is $(3.3)(1) = 3.3$, but this value is also the distance that we obtained previously when we were underestimating the distance covered in the first second. In a similar way, the area of each rectangle can be interpreted as a distance (its height represents velocity and its base, time, so the product

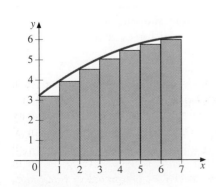

FIGURE 5.13

represents distance). So the lower sum L_5 for the area is precisely the underestimate, 22.2, that we obtained earlier. In a similar way, by using circumscribed rectangles, we find that the upper sum U_5 is the overestimate of distance, 24.7.

You should be convinced by now that the problem of finding area under a curve and the problem of finding distance when velocities are known are basically the same. In the next section we will look at the common aspects of these problems independently of their geometrical or physical interpretations.

In our examples for both area and distance, we were dealing with increasing functions. The calculations are basically the same for decreasing functions. However, if a function is not monotone, finding the minimum and maximum values on subintervals may be difficult. In the next section we will show that it is not essential to use inscribed or circumscribed rectangles. Instead, we can use any rectangles lying between these two extremes. Our examples thus far also consider only nonnegative functions. In the next section we will show that this requirement may also be relaxed. An example may help motivate how this is done.

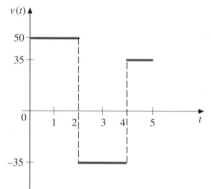

FIGURE 5.14

EXAMPLE 5.3   Suppose the velocity of an automobile on a straight road is given by Figure 5.14. Draw a graph of the distance, $s(t)$, of the car from its starting point as a function of time.

Solution   The graph in Figure 5.14 tells us that the driver travels at 50 mph for 2 hr, then goes 35 mph in the opposite direction for 2 hr, and then travels 35 mph for another hour in the starting direction. We cannot simply compute the area between the velocity curve and the t-axis as we did in the case of the runner, since this area would be the total distance traveled by the car, that is, the distance shown on the odometer, namely,

$$(50)(2) + (35)(2) + (35)(1) = 205 \text{ mi}$$

We know this result cannot be the correct answer for the distance of the car from its starting position, though, because we would be calculating the distance as though the car went only in one direction. However, the distance from the starting point can be computed as "net area" (that is, the area bounded by the t-axis and the portion of the curve above the t-axis, minus the area bounded by the t-axis and the portion of the curve below the t-axis):

$$\text{net area} = (50)(2) + (-35)(2) + (35)(1) = 65$$

Figure 5.15 is a graph of the distance from the starting point versus time.   ■

FIGURE 5.15

We shall see in the next section that the concept of area under a curve can be extended to "net area" under a curve. Before showing this fact, we will introduce a more compact notation that can be used for L_n and U_n as well as for sums lying between them.

A Useful Notation—Sigma Notation

The Greek capital letter sigma, $\sum$, is used to denote *sum*. To illustrate how sigma notation works, consider the sum of the cubes of the first n positive integers.

$$1^3 + 2^3 + 3^3 + \cdots + n^3$$

For example, if $n = 3$, the sum is $1^3 + 2^3 + 3^3$ and if $n = 6$, the sum is $1^3 + 2^3 + 3^3 + 4^3 + 5^3 + 6^3$. In general, we add up all terms of the form k^3, where k assumes all positive integer values from 1 to n. In sigma notation this sum becomes

$$\sum_{k=1}^{n} k^3$$

and is read "the summation of k^3, where k goes from 1 to n."

If $x_1, x_2, \ldots, x_n$ represents n numbers, then

$$x_1 + x_2 + \cdots + x_n$$

is written as

$$\sum_{k=1}^{n} x_k$$

and is read "the summation of x sub k as k goes from 1 to n." The letter k is called the *index of summation*. Other letters can be used as the index; it is most common to use i, j, k, m, n.

EXAMPLE 5.4   Evaluate each of the following expressions.

(a) $\sum_{i=1}^{5} i$ (b) $\sum_{k=1}^{4} (2k^2 - 1)$ (c) $\sum_{m=1}^{10} 5$

Solution

(a) $\displaystyle\sum_{i=1}^{5} i = 1 + 2 + 3 + 4 + 5 = 15$

(b) $\displaystyle\sum_{k=1}^{4} (2k^2 - 1) = (2(1) - 1) + (2(4) - 1) + (2(9) - 1) + (2(16) - 1) = 56$

(c) $\displaystyle\sum_{m=1}^{10} 5 = 5 + 5 + 5 + 5 + 5 + 5 + 5 + 5 + 5 + 5 = 5(10) = 50$ ∎

Calculations involving summation symbols often can be simplified by using the following properties.

Properties of Sigma Notation

(i) $\displaystyle\sum_{k=1}^{n} cx_k = c \sum_{k=1}^{n} x_k$, where c is any constant

(ii) $\displaystyle\sum_{k=1}^{n} (x_k + y_k) = \sum_{k=1}^{n} x_k + \sum_{k=1}^{n} y_k$

(iii) $\displaystyle\sum_{k=1}^{n} (x_k - y_k) = \sum_{k=1}^{n} x_k - \sum_{k=1}^{n} y_k$

These properties are simple extensions of the distributive and associative properties of real numbers.

Carl Friedrich Gauss (1777–1855) came from a poor and uneducated family. Despite this background, he entered advanced study at the age of 11, and by 15 he was receiving a stipend (allowance) from a Duke.

Gauss's talent with calculations enabled him to make interesting conjectures. At college, he discovered that some of his independent inventions had been made previously by others. At 19, he found how to construct regular polygons using only ruler and compass, a problem dating from the ancient Greeks. His discovery that one could construct a regular polygon of 17 sides in this way was the first advance in this area since Euclid. Gauss proved the Fundamental Theorem of Algebra at age 22.

Gauss was unbelievably prolific. Some of his work systematized whole fields of mathematics. His discoveries of non-Euclidean geometry and of noncommutative algebra remain especially noteworthy.

In 1801, an object was discovered that was thought to be a new planet, but not enough observations had been made and the object could no longer be found in the sky. Gauss worked out new ways of calculating orbits, and within a few months, the object was located again, right where Gauss had predicted. We know it now as the first asteroid, Ceres, a so-called "minor planet." Because such a remarkable prediction had been made on the basis of such a small amount of information, Gauss's reputation was even more enhanced.

Gauss became an astronomer but continued his work on mathematics and on other subjects simultaneously. In middle age, he became a geologist, measuring the shape of the Earth, and later provided basic discoveries that are still at the heart of physics. Indeed, a unit of magnetic field—the gauss—is named for him. The average magnetic field of the Earth is about 1 gauss.

It is possible to find the sums of certain expressions in what is called *closed form*; that is, we can find a formula for the given sum. One of the simplest examples is the sum of the first n natural numbers, $\sum_{k=1}^{n} k$. It can be shown (see Exercise 9 in Exercise Set 5.1) that the sum is $\frac{1}{2}n(n+1)$. That is,

$$1 + 2 + 3 + \cdots + n = \frac{n(n+1)}{2} \tag{5.3}$$

The right-hand side of Equation 5.3 is a closed form for the sum on the left.

The finite sequence

$$1, 2, 3, \ldots, n$$

whose terms are added on the left-hand side of Equation 5.3 is an example of an *arithmetic progression*, in which the difference between any two consecutive terms remains constant (in this case, the constant difference is 1).

An interesting story about Carl Friedrich Gauss involves an arithmetic progression. In 1787, when Gauss was 10 years old, his teacher gave the class a list of 100 large numbers to add up, thinking it would keep them occupied for the entire period. The numbers formed an arithmetic progression, with a constant difference between consecutive terms. The teacher knew what the answer was, but the pupils had been told nothing of arithmetic progressions. The teacher's instructions were for the first pupil to get the answer to put his slate down on his desk. Then, as others finished, they were to place their slates, in order, on the stack. Imagine the teacher's surprise when Gauss placed his slate down only seconds after the problem was given. His surprise was even greater when he saw but one number written on Gauss's slate—the correct answer for the sum!

Gauss reasoned as follows: If I were to write the sum a second time, with the order reversed, then add, term-by-term, the second sum to the first sum, I would get 100 terms, each having the same value. So all I need to do is add the first term of the original sum to the last and multiply by 100. This will give twice what I want; so I will divide the result by 2.

Let us apply Gauss's reasoning to the sum on the left-hand side of Equation 5.3, with $n = 100$ (assuming we do not know the closed form on the right-hand side):

$$1 + 2 + 3 + \cdots + 98 + 99 + 100$$

In reverse we get

$$100 + 99 + 98 + \cdots + 3 + 2 + 1$$

Adding, term-by-term, we get

$$101 + 101 + 101 + \cdots + 101$$

where there are 100 terms, each equal to 101. So the sum is 10100. This is double what we want. The answer is thus 5050.

Gauss's teacher recognized his extraordinary ability and encouraged him to move on to higher mathematics. Gauss is generally recognized as one of the greatest mathematical geniuses who have ever lived.

We show the result (Equation 5.3) again here, along with two others that you will be asked to verify in the exercises (see Exercises 11 and 13 in Exercise Set 5.1).

Some Useful Summation Formulas

$$\sum_{k=1}^{n} k = \frac{n(n+1)}{2} \tag{5.4}$$

$$\sum_{k=1}^{n} k^2 = \frac{n(n+1)(2n+1)}{6} \tag{5.5}$$

$$\sum_{k=1}^{n} k^3 = \left[\frac{n(n+1)}{2}\right]^2 \tag{5.6}$$

Upper and Lower Sums Using Sigma Notation

We can rewrite Equations 5.1 and 5.2 using sigma notation. For Equation 5.1 we have

$$L_n = m_1 \Delta x + m_2 \Delta x + m_3 \Delta x + \cdots + m_n \Delta x$$

$$= \Delta x (m_1 + m_2 + m_3 + \cdots + m_n)$$

$$= \Delta x \sum_{k=1}^{n} m_k$$

and for Equation 5.2,

$$U_n = M_1 \Delta x + M_2 \Delta x + M_3 \Delta x + \cdots + M_n \Delta x$$

$$= \Delta x (M_1 + M_2 + M_3 + \cdots + M_n)$$

$$= \Delta x \sum_{k=1}^{n} M_n$$

If an interval $[a, b]$ is partitioned into n subintervals of equal length $\Delta x = (b-a)/n$, then the partition points $x_1, x_2, x_3, \ldots, x_n$ are given by

$$x_0 = a$$

$$x_1 = a + \Delta x$$

$$x_2 = x_1 + \Delta x = a + 2\Delta x$$

$$x_3 = x_2 + \Delta x = a + 3\Delta x$$

$$\vdots$$

$$x_{n-1} = x_{n-2} + \Delta x = a + (n-1)\Delta x$$

$$x_n = x_{n-1} + \Delta x = a + n\Delta x = b$$

For an increasing function, as in Example 5.2, the minimum height of the function occurs at the left endpoint of each subinterval, and the maximum height occurs at the right endpoint. For a typical subinterval $[x_{k-1}, x_k]$, therefore,

$$m_k = f(x_{k-1}) \qquad \text{and} \qquad M_k = f(x_k)$$

Thus, for such an increasing function the lower and upper sums are

$$L_n = \sum_{k=1}^{n} f(x_{k-1})\Delta x = f(x_0)\Delta x + f(x_1)\Delta x + f(x_2)\Delta x + \cdots + f(x_{n-1})\Delta x$$

$$U_n = \sum_{k=1}^{n} f(x_k)\Delta x = f(x_1)\Delta x + f(x_2)\Delta x + f(x_3)\Delta x + \cdots + f(x_n)\Delta x$$

If f is a decreasing function, then the sums for L_n and U_n are reversed.

We have seen that the error in approximating the area using either L_n or U_n is at most $U_n - L_n$. For an increasing function we can write $U_n - L_n$ in the form

$$U_n - L_n = \sum_{k=1}^{n} f(x_k)\Delta x - \sum_{k=1}^{n} f(x_{k-1})\Delta x$$

$$= \sum_{k=1}^{n} \left[f(x_k) - f(x_{k-1}) \right] \Delta x$$

In Exercise 15 (in Exercise Set 5.1) you will be asked to show that, for a decreasing function, the result is the negative of the sum on the right, and that the result in both cases (increasing or decreasing) can be written as follows:

$$\text{error} \le |f(b) - f(a)|\Delta x \qquad \text{If } f \text{ is monotone}$$

The next example illustrates how with the aid of Equation 5.5 we can find the exact area under the graph of a portion of the parabola $y = x^2$ by calculating U_n and letting $n \to \infty$. (We could use L_n instead.)

EXAMPLE 5.5   Find the upper sum U_n for the area under the graph of the function $f(x) = x^2$ from $x = 0$ to $x = 2$. Then obtain the exact area by letting $n \to \infty$.

Solution   We show the graph, together with some circumscribed rectangles, in Figure 5.16. As we have shown, for a monotone increasing function such as this one,

$$U_n = \sum_{k=1}^{n} f(x_k)\,\Delta x$$

where $x_k = a + k\,\Delta x$ and $\Delta x = (b - a)/n$. In our case, $a = 0$ and $b = 2$. So $\Delta x = 2/n$ and $x_k = 0 + k \cdot 2/n = 2k/n$. Thus,

$$U_n = \sum_{k=1}^{n} f\left(\frac{2k}{n}\right) \cdot \frac{2}{n} = \sum_{k=1}^{n} \left(\frac{2k}{n}\right)^2 \cdot \frac{2}{n} = \sum_{k=1}^{n} \frac{8k^2}{n^3}$$

We can factor out $8/n^3$ to get

$$U_n = \frac{8}{n^3} \sum_{k=1}^{n} k^2$$

By Equation 5.6,

$$\sum_{k=1}^{n} k^2 = \frac{n(n+1)(2n+1)}{6}$$

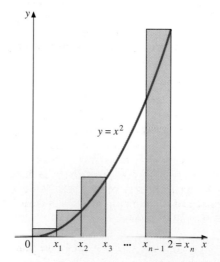

FIGURE 5.16

Thus,

$$U_n = \frac{8}{n^3}\left[\frac{n(n+1)(2n+1)}{6}\right] = \frac{4(n+1)(2n+1)}{3n^2} = \frac{4(2n^2+3n+1)}{3n^2}$$

To find the limit as $n \to \infty$, we can use the dominant-term approach or we can divide numerator and denominator by n^2 to get

$$U_n = \frac{4(2 + \frac{3}{n} + \frac{1}{n^2})}{3}$$

Thus,

$$\lim_{n\to\infty} U_n = \frac{8}{3}$$

since both $3/n$ and $1/n^2$ approach 0. The exact area is therefore 8/3. ■

Exercise Set 5.1

In Exercises 1–5, use a calculator, CAS, or spreadsheet to find L_n and U_n for the area under the graph of f on the specified interval for n = 5, 10, 20, and 100.

1. $f(x) = \dfrac{x+3}{2}$ on [0, 5]

2. $f(x) - \dfrac{x^2}{4}$ on [0, 4]

3. $f(x) = 4 - x^2$ on [0, 2]

4. $f(x) = 2x - 1$ on [1, 6]

5. $f(x) = x^2 + 1$ on [−1, 2]

6. By reading values from the graph shown, find upper and lower bounds, U_n and L_n, for the area under the curve from $x = -2$ to $x = 5$, using $n = 16$.

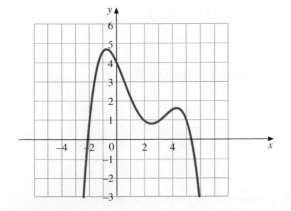

7. A driver traveling on an interstate highway sees traffic coming to a stop ahead and applies the brakes. The car's speed, in feet per second (60 mph = 88 ft/s), is given at half-second intervals in the accompanying table. Use upper sums and lower sums to give an upper bound and a lower bound on the distance covered.

t	0	0.5	1	1.5	2	2.5	3	3.5	4
$v(t)$	80	72	63	53	42	29	17	7	0

8. A container of food is dropped from a helicopter high above a refugee camp. When air resistance is taken into consideration, the container's velocity, $v(t)$, t seconds after it is dropped, is approximately given by the formula

$$v(t) = 320(1 - e^{-0.1t})$$

(See Section 1.3, where e^x is introduced.) Use a calculator, a CAS, or a spreadsheet to make a table of values showing t and $v(t)$ at half-second intervals for $t = 0$ to $t = 5$. Find an underestimate and an overestimate of the distance the container falls in the first 5 seconds. Sketch a graph of $v(t)$ versus t and show these underestimates and overestimates as sums of areas of rectangles.

9. Using Gauss's reasoning, derive Equation 5.3 for the sum of the first n natural numbers:

$$1 + 2 + 3 + \cdots + n = \frac{n(n+1)}{2}$$

10. Make use of the formula derived in Exercise 9 to find the upper sum U_n for an unspecified n for the area under the graph of $f(x) = 2x + 3$ from $x = 0$ to $x = 4$. Then find the exact area by letting $n \to \infty$. Check your answer, using what you know about the area of a trapezoid. (*Hint:* Draw the figure.)

11. Prove Equation 5.5,

$$1^2 + 2^2 + 3^2 + \cdots + n^2 = \frac{n(n+1)(2n+1)}{6}$$

using mathematical induction. (See Appendix 4.)

12. Use the result of Exercise 11 to find the lower sum L_n for the area under the graph of $f(x) = 4 - x^2$ from $x = 0$ to $x = 2$. Then find the exact area by letting $n \to \infty$.

13. Prove Equation 5.6,

$$1^3 + 2^3 + 3^3 + \cdots + n^3 = \left[\frac{n(n+1)}{2}\right]^2$$

using mathematical induction. (See Appendix 4.)

14. Use the result of Exercise 13 to find the upper sum U_n for the area under the graph of $f(x) = x^3$ from $x = 0$ to $x = 2$. Then let $n \to \infty$ to get the exact area.

15. Let f be a decreasing, nonnegative function on $[a, b]$, and let $x_0, x_1, x_2, \ldots, x_n$ be a partition of $[a, b]$ with all subintervals of equal width $\Delta x = (b - a)/n$. Express the minimum values $m_1, m_2, \ldots, m_n$ and maximum values $M_1, M_2, \ldots, M_n$ of the function on the subintervals in terms of $f(x_k)$ for appropriate values of k. Then substitute these in Equations 5.1 and 5.2 for L_n and U_n, and calculate $U_n - L_n$. Using your result, together with the result of $U_n - L_n$ for an increasing function found in this section, show that the error in using L_n or U_n to approximate the area under a monotone function satisfies

$$\text{error} \le |f(b) - f(a)|\Delta x$$

16. Use the result of Exercise 15 to determine a value of n that will guarantee an error that is less than 0.05 when approximating the area under the graph of f on $[a, b]$ by L_n or U_n, and find the corresponding area, where
    (a) $f(x) = x^2$ on $[0, 4]$
    (b) $f(x) = \sqrt{x}$ on $[1, 9]$

17. Repeat Exercise 16 for the functions given. Recall that we introduced e^x and $\ln x$ in Section 1.3. Their values can be obtained from a calculator. (We will study these functions in Chapter 7.)
    (a) $f(x) = \ln x$ on $[1, 7]$
    (b) $f(x) = 6e^{-0.24x}$ on $[2, 10]$

Exercises 18 and 19 illustrate Archimedes' "method of compression" for calculating the area of a circle.

18. The accompanying figure shows an n-sided regular polygon (that is, all sides are of equal length) inscribed in a circle of radius r. The polygon has been divided into n congruent triangles, each with central angle $2\pi/n$ radians.
    (a) Show that the area of each triangle is

$$\frac{1}{2}r^2 \sin\left(\frac{2\pi}{n}\right)$$

   (*Hint:* Use the identity $\sin 2t = 2 \sin t \cos t$.)

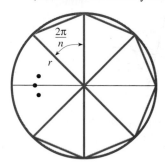

   (b) Show that as $n \to \infty$ the area of the polygon approaches πr^2, which is the area of the circle. That is, show that

$$\lim_{n \to \infty} \frac{1}{2}nr^2 \sin\left(\frac{2\pi}{n}\right) = \pi r^2$$

   (*Hint:* Use the fact that $\lim_{t \to 0}(\sin t/t) = 1$.)

19. Rather than using inscribed polygons as in Exercise 18, we can use circumscribed polygons, as in the accompanying figure, where we have again divided the polygon into n congruent triangles.

   (a) Show that the area of each triangle is

$$r^2 \tan\left(\frac{\pi}{n}\right)$$

   (b) Show that the areas of the circumscribed polygons approach πr^2 as $n \to \infty$. That is, show that

$$\lim_{n \to \infty} nr^2 \tan\left(\frac{\pi}{n}\right) = \pi r^2$$

   (See the hint for Exercise 18(b).)

20. The Keck Telescope in Hawaii is made of 36 hexagons (see the figure), each 1.8 m in its largest diagonal. By dividing a typical hexagon into triangles, calculate its area. Then give the area of the total mirror. Find the diameter of a circular mirror of the same total area. Compare the result with that of the venerable Hale Telescope on Palomar Mountain in California, whose mirror is 5 m (200 in.) in diameter.

A model of the Keck Telescope, the largest telescope in the world

 21. Use Equations 5.4, 5.5, and 5.6, along with a calculator (or CAS or spreadsheet), to find the smallest positive integer n such that

(a) $\displaystyle\sum_{k=1}^{n} k \geq 1{,}000$

(b) $\displaystyle\sum_{k=1}^{n} k^2 \geq 10{,}000$

(c) $\displaystyle\sum_{k=1}^{n} k^3 \geq 100{,}000$

5.2 THE DEFINITE INTEGRAL

Our objective in this section is to define the *definite integral* and to investigate its properties. This concept is the unifying idea that relates many diverse problems. In particular, we will see that the area problem we considered in Section 5.1 can be solved by a definite integral. In Section 5.4 we will define an *indefinite integral* and show how the two types of integrals are related.

As a first step in arriving at a definition of the definite integral, let us return to the problem of finding the area under the graph of a nonnegative, continuous function f from $x = a$ to $x = b$. We have seen that the area A is given by either

$$\lim_{n \to \infty} U_n \qquad \text{or} \qquad \lim_{n \to \infty} L_n$$

where

$$U_n = \sum_{k=1}^{n} M_k \,\Delta x \qquad \text{and} \qquad L_n = \sum_{k=1}^{n} m_k \,\Delta x$$

For monotone functions, finding the maximum values M_k and the minimum values m_k of $f(x)$ on the subintervals is straightforward, since these extreme values occur at the endpoints of the subintervals. For functions that are not monotone, it may be very difficult to find m_k and M_k, other than by estimating the values from the graph. (See Figure 5.17.) As the following argument shows, we do not need to use either m_k or M_k in approximating the area. Instead, we can use any value between these two extremes.

FIGURE 5.17

Maximum and minimum values of f do not necessarily occur at endpoints of subintervals.

FIGURE 5.18

Select arbitrary points $c_1, c_2, \ldots, c_n$ such that for each k, c_k lies in the kth subinterval. Thus,

$$x_{k-1} \le c_k \le x_k$$

for $k = 1, 2, \ldots, n$. Such a choice is indicated for the kth subinterval in Figure 5.18. Since M_k is the maximum value of $f(x)$ and m_k the minimum value of $f(x)$ on the kth subinterval, $f(c_k)$ must be between these extremes. That is,

$$m_k \le f(c_k) \le M_k$$

Thus,

$$m_k \Delta x \le f(c_k) \Delta x \le M_k \Delta x$$

Geometrically, we are saying that the area of the rectangle with height $f(c_k)$ and base Δx lies between the area of the rectangle with height m_k and the one with height M_k (see Figure 5.18), each with base Δx. When we add up all the areas, we get

$$\sum_{k=1}^{n} m_k \Delta x \le \sum_{k=1}^{n} f(c_k) \Delta x \le \sum_{k=1}^{n} M_k \Delta x$$

and so

$$L_n \le \sum_{k=1}^{n} f(c_k) \Delta x \le U_n \tag{5.7}$$

We know that L_n and U_n approach the area under the curve A as $n \to \infty$. Thus, by Equation 5.7, the sum in the middle also approaches A, since it is squeezed between L_n and U_n (the Squeeze Theorem (2.3) is true also for sequences). That is,

$$\lim_{n \to \infty} \sum_{k=1}^{n} f(c_k) \Delta x = A \tag{5.8}$$

The importance of this result is that we do not have to know the maximum or minimum values of the function on each of the subintervals but can use any value in between. For example, we could take c_k as the left endpoint, the right endpoint, the midpoint, or any other point of the kth subinterval. The limiting value in Equation 5.8 is exactly the same as the limit of L_n or U_n, namely, the area A.

EXAMPLE 5.6   Find an expression in the form of Equation 5.8 for the area under the curve $y = f(x) = \sin(\pi x)/2 + 2$ from $x = 0$ to $x = 4$, using c_k as the right endpoint of the kth subinterval.

Solution   Figure 5.19 shows the graph of f, along with a typical subdivision of the interval $[0, 4]$. Note that since the curve is neither increasing nor decreasing throughout the whole interval, building rectangles with heights equal to values of the function at the right endpoint does not give either the lower sum L_n or the upper sum U_n, but something in between.

For an arbitrary choice of n, we have

$$\Delta x = \frac{4 - 0}{n} = \frac{4}{n}$$

$$y = \sin \frac{\pi x}{2} + 2$$

FIGURE 5.19

Thus,

$$x_1 = 0 + \Delta x = \frac{4}{n}$$

$$x_2 = 0 + 2\Delta x = \frac{8}{n}$$

$$x_3 = 0 + 3\Delta x = \frac{12}{n}$$

and so on. The right endpoint of the kth subinterval $[x_{k-1}, x_k]$ is x_k, so

$$x_k = 0 + k\Delta x = \frac{4k}{n}$$

This term is the c_k of Equation 5.8. That is, $c_1 = x_1$, $c_2 = x_2, \ldots, c_n = x_n$. Thus,

$$f(c_k) = f(x_k) = 2 + \sin\frac{\pi x_k}{2} = 2 + \sin\frac{2k\pi}{n}$$

and so by Equation 5.8,

$$A = \lim_{n\to\infty} \sum_{k=1}^{n} \underbrace{\left(2 + \sin\frac{2k\pi}{n}\right)}_{\text{This is } f(c_k)} \underbrace{\left(\frac{4}{n}\right)}_{\text{This is } \Delta x}$$

If we wanted to approximate the area, we could take some large value of n, say $n = 100$, and calculate the sum on the right using a CAS or a spreadsheet. At this stage we have no way of finding the exact value of the limit as $n \to \infty$. We will soon see a way, however, of getting this exact answer. ■

Up to this point we have considered only continuous functions that are nonnegative. Now we want to relax this latter requirement. If the continuous function f is sometimes positive and sometimes negative, what geometric interpretation can be given to the sum $\sum_{k=1}^{n} f(c_k)\Delta x$? When $f(c_k)$ is negative, we can think of the rectangle with height $f(c_k)$ and base Δx as contributing "negative area," since it extends below the x-axis (Figure 5.20). So the limit in Equation 5.8 would give what we have called "net area," that is, the algebraic sum of positive and negative areas. In Figure 5.21, we show a typical situation. Regions I and III would be counted as positive area and region II as negative area. If we wanted the actual total area between the curve and the x-axis (to compute, for example, the weight of a sheet of metal cut to that shape), we would have to take the absolute value of the negative area of II and add it to the areas of I and III (area = area I + |area II| + area III).

To interpret the situation in Figure 5.21 in terms of distance, suppose x represents time and $f(x)$ represents velocity. Then, since the velocity is sometimes positive and sometimes negative, the object is sometimes moving in the direction we have designated as positive and sometimes moving in the opposite direction. The areas of I and III give the total distance covered in the positive direction, and the area of II gives the distance covered in the negative direction. So Equation 5.8 could be interpreted as the "net distance" covered—that is, the distance from the starting point of the object to its final position (see Example 5.3 in Section 5.1). For example, if you drive forward and then back to your starting point, the distance from the starting point (the "net distance") is 0. The area above the x-axis under the distance function and the area below would be equal. If you wanted to find the actual total distance covered (say, to calculate your gas mileage), you would need to add the absolute values of the areas.

FIGURE 5.20

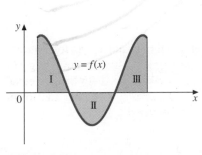

FIGURE 5.21

We indicated earlier that it is not necessary to use subintervals of equal width. We did so only for convenience. Choosing division points in this way gave the common width $\Delta x = (b - a)/n$ for all subintervals, and by letting $n \to \infty$, we were assured that the widths of all the rectangles became arbitrarily small. We can now generalize our process by allowing the points of division $x_0, x_1, x_2, \ldots, x_n$ to be spaced in any way whatsoever in $[a, b]$, subject only to the requirements that $x_0 = a$, $x_n = b$, and $x_0 < x_1 < x_2 < \cdots < x_n$. Denote the width of the kth subinterval by Δx_k, so that

$$\Delta x_k = x_k - x_{k-1} \qquad (k = 1, 2, \ldots, n)$$

We may no longer be guaranteed that as $n \to \infty$, the rectangles exhaust the area under the curve. For example, we can define points that divide only a portion of the interval into more and more subintervals and leave the remainder of the interval unchanged. As an example, half the interval could always be considered as one subinterval in each successive finer subdivision (Figure 5.22). In this case, the number of rectangles increases to infinity, but the sum of the areas does not approach the area under the curve. To remedy this problem, we also require that as $n \to \infty$, the largest of the widths, Δx_k, approaches 0. We indicate the largest Δx_k by max Δx_k. With this notation, the area is given by

$$A = \lim_{\max \Delta x_k \to 0} \sum_{k=1}^{n} f(c_k) \Delta x_k \qquad (5.9)$$

where, again, A is to be interpreted as "net area." A sum of the type appearing on the right,

$$\sum_{k=1}^{n} f(c_k) \Delta x_k$$

is called a **Riemann sum** for f in honor of the German mathematician G. F. B. Riemann, who made major contributions to the theory of the integral.

The limit $\lim_{n \to \infty} \sum_{k=1}^{n} f(c_k) \Delta x$ in Equation 5.8 is a special case of the limit in Equation 5.9, in which $\Delta x_k = \Delta x$ for all k. Since $\Delta x = (b - a)/n$, we see that $\Delta x \to 0$ as $n \to \infty$, so it follows that max $\Delta x_k \to 0$ as $n \to \infty$. (In fact, there is no need to speak of the maximum Δx_k in this case, since all subintervals are of the same width.)

EXAMPLE 5.7   Form the Riemann sum for the function $f(x) = x^2 - 2x$ on the interval $[-1, 3]$, using the points of division

$$x_0 = -1, \; x_1 = -0.2, \; x_2 = 0.6, \; x_3 = 1.8, \; x_4 = 2.4, \; x_5 = 3$$

Choose c_k as the midpoint of the kth subinterval.

Solution   The points c_k are found by averaging x_{k-1} and x_k, as follows:

$$c_1 = \frac{-1 - 0.2}{2} = -0.6 \qquad c_2 = \frac{-0.2 + 0.6}{2} = 0.2$$

$$c_3 = \frac{0.6 + 1.8}{2} = 1.2 \qquad c_4 = \frac{1.8 + 2.4}{2} = 2.1$$

$$c_5 = \frac{2.4 + 3}{2} = 2.7$$

FIGURE 5.22

In Figure 5.23 we show the graph of f, the division points, the midpoints c_k, and the rectangles whose (signed) areas make up the Riemann sum. The Δx_k values are as follows:

$$\Delta x_1 = x_1 - x_0 = -0.2 - (-1) = 0.8$$

$$\Delta x_2 = x_2 - x_1 = 0.6 - (-0.2) = 0.8$$

$$\Delta x_3 = x_3 - x_2 = 1.8 - 0.6 = 1.2$$

$$\Delta x_4 = x_4 - x_3 = 2.4 - 1.8 = 0.6$$

$$\Delta x_5 = x_5 - x_4 = 3 - 2.4 = 0.6$$

In this case max $\Delta x_k = 1.2$, and

$$\sum_{k=1}^{5} f(c_k)\Delta x_k = [f(-0.6)](0.8) + [f(0.2)](0.8) + [f(1.2)](1.2)$$

$$+ [f(2.1)](0.6) + [f(2.7)](0.6)$$

$$= (1.56)(0.8) + (-0.36)(0.8) + (-0.96)(1.2) + (0.21)(0.6)$$

$$+ (1.89)(0.6)$$

$$= 1.068$$

This result is an approximation of the net area. Since our division points were not very close together, we would not expect this to be a good approximation. The exact value, as we will show in the next section, is 4/3. ∎

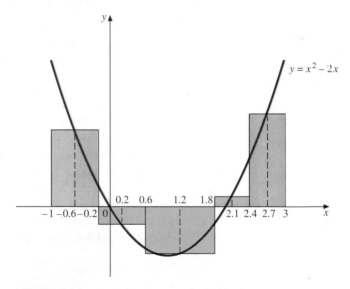

FIGURE 5.23

Definition of the Integral

To define the integral, we drop the assumption that f be continuous and assume only that f is *defined* and *bounded* on $[a, b]$. To say that f is "defined" on $[a, b]$ means that $f(x)$ has a finite value for every x on the closed interval $[a, b]$. That is, the closed interval $[a, b]$ is included in the domain of f. To say that f is "bounded" means that $f(x)$ does not get arbitrarily large either positively or negatively. More precisely, f is bounded on $[a, b]$ provided there is some positive constant M such that $|f(x)| \leq M$ for all x in the closed interval $[a, b]$. Geometrically, the graph of f lies between the horizontal lines $y = M$ and $y = -M$. Although we do not require that f be continuous, most of the functions we deal with will be continuous. We set aside for the moment the idea of area (or distance) and simply look at Riemann sums for f. For every subdivision of $[a, b]$ and for every choice of the intermediate points c_k, such a Riemann sum exists. The question is, as we take finer and finer subdivisions (that is, points closer and closer together), do these Riemann sums approach some limit? If the answer is yes, then this limit is called the *integral* of f over $[a, b]$. It is also called the *definite integral*, or the *Riemann integral*. We state this definition formally, with the symbols and terminology understood to be as we have previously defined them.

Definition 5.2 The Definite Integral	Let f be defined and bounded on $[a, b]$. If $$\lim_{\max \Delta x_k \to 0} \sum_{k=1}^{n} f(c_k)\Delta x_k$$ exists, independently of the choices of the division points and the choices of the intermediate points c_k, then this limit is called the **definite integral of f over $[a, b]$**, and is denoted by $$\int_{a}^{b} f(x)dx$$ It is read "the integral of f of x with respect to x from a to b."

The symbol $\int$ is called an **integral sign**. It has the appearance of an elongated S, the first letter of the Latin word *summa*, meaning sum. The letters a and b at the bottom and top of the integral sign $\int_{a}^{b} f(x)dx$ are called **limits of integration**, and $f(x)$ is called the **integrand**.

The symbol $\int_{a}^{b} f(x)dx$ looks something like a Riemann sum $\sum_{k=1}^{n} f(c_k)\Delta x_k$, with the integral sign replacing the summation symbol, $f(x)$ replacing $f(c_k)$, and dx replacing Δx_k. We can think of the integral as a sort of idealized sum.

The precise meaning of the limit in Definition 5.2 is as follows:

$$\lim_{\max \Delta x_k \to 0} \sum_{k=1}^{n} f(c_k)\Delta x_k = L$$

provided that corresponding to each positive number ε (however small it may be), there exists a (sufficiently small) positive number δ such that for every partition of $[a, b]$ for which $\max \Delta x_k < \delta$ (so that all subintervals are of length less than δ), then

$$\left| \sum_{k=1}^{n} f(c_k)\Delta x_k - L \right| < \varepsilon$$

regardless of how c_k is chosen in the interval $[x_{k-1}, x_k]$ for $k = 1, 2, 3, \ldots, n$. In simpler (but less precise) language, all Riemann sums for f on $[a, b]$ can be made as close as we please to L, provided only that all of the widths Δx_k are sufficiently small. This type of limit is more complex than those we studied in Chapter 2.

Because the definite integral is defined by a limit, the integral may or may not exist, depending on the behavior of the function f. When the integral exists on $[a, b]$, the function f is said to be **integrable** on $[a, b]$. Determining which functions are integrable is generally a very difficult problem. However, we state without proof that if f satisfies any of the following conditions, it is integrable on $[a, b]$:

1. f is continuous on $[a, b]$.
2. f is either only increasing or only decreasing (i.e., f is monotone) on $[a, b]$.
3. f is bounded on $[a, b]$ and has only finitely many points of discontinuity.

In the next two examples we show how the limit of a Riemann sum can be recognized as an integral.

EXAMPLE 5.8   Write the following limit as an integral.

$$\lim_{\max \Delta x_k \to 0} \sum_{k=1}^{n} (c_k^2 + 3c_k + 1)\Delta x_k, \text{ for the interval } [-1, 3]$$

Solution   This limit conforms exactly to the definition of the integral of $f(x) = x^2 + 3x + 1$ on $[-1, 3]$. So the limit is

$$\int_{-1}^{3} (x^2 + 3x + 1)dx \qquad \blacksquare$$

EXAMPLE 5.9   Write the following limit as an integral.

$$\lim_{n \to \infty} \sum_{k=1}^{n} \left[\cos\left(1 + \frac{3k}{n}\right) \right] \left(\frac{3}{n}\right)$$

Solution   The sum appears to be a Riemann sum in which the division points are equally spaced, with $\Delta x_k = 3/n$ for all k. Also, $c_k = 1 + 3k/n$. So

$$c_1 = 1 + \frac{3}{n}, \ c_2 = 1 + 2\left(\frac{3}{n}\right), \dots, \ c_n = 1 + n\left(\frac{3}{n}\right) = 4$$

The c_k's are the right endpoints of an equally spaced subdivision of $[1, 4]$. Since $\cos x$ is a continuous function everywhere, its integral from 1 to 4 exists, and according to the definition, no restriction is placed on subdivisions except that the maximum widths of the subintervals approach 0. In our case this maximum width is $3/n$, which does approach 0 as $n \to \infty$. We conclude that the limit is the integral

$$\int_{1}^{4} \cos x \, dx \qquad \blacksquare$$

In the next section we will see how to find the exact values of the integrals in Examples 5.8 and 5.9.

If $f(x) \geq 0$, we know that one way of interpreting the integral $\int_{a}^{b} f(x)dx$ is as the area under the graph of f from $x = a$ to $x = b$. In certain cases, we can calculate this area by elementary geometry, and thus the value of the integral

FIGURE 5.24

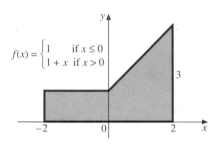

FIGURE 5.25

Recall that the area of a trapezoid is the average of the lengths of the two parallel sides times the distance between them.

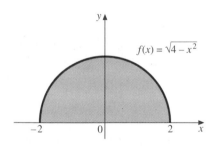

FIGURE 5.26

will have been found. In the next example we illustrate this "backdoor" method of evaluating certain integrals.

EXAMPLE 5.10   Evaluate the integral

$$\int_{-2}^{2} f(x)\,dx$$

for each of the following functions.

(a) $f(x) = 2 - |x|$

(b) $f(x) = \begin{cases} 1 & \text{if } x \le 0 \\ 1+x & \text{if } x > 0 \end{cases}$

(c) $f(x) = \sqrt{4 - x^2}$

Solution

(a) The graph of f is shown in Figure 5.24. The region below it is a triangle with base 4 and altitude 2. The area is therefore $\frac{1}{2}(4)(2) = 4$. Thus,

$$\int_{-2}^{2} (2 - |x|)\,dx = 4$$

(b) The graph is shown in Figure 5.25. The region under it consists of a rectangle and a trapezoid. The combined area is

$$(2)(1) + \left(\frac{1+3}{2}\right)(2) = 2 + 4 = 6$$

So,

$$\int_{-2}^{2} f(x)\,dx = 6$$

(c) The graph of $y = \sqrt{4 - x^2}$ is the upper half of $y^2 = 4 - x^2$, or equivalently of $x^2 + y^2 = 4$, which is a circle of radius 2. We know that the area of the entire circle is $\pi r^2 = \pi(2)^2 = 4\pi$. So the area we want (shown in Figure 5.26) is 2π. Thus,

$$\int_{-2}^{2} \sqrt{4 - x^2}\,dx = 2\pi \qquad \blacksquare$$

In part (c) of the preceding example, we assumed that you knew the formula $A = \pi r^2$ for the area enclosed by a circle of radius r. If you completed Exercise 18 or 19 in Exercise Set 5.1, you have seen a *proof* of this area formula. We will present another proof later, but for now we assume that the result is understood and we will make use of it from time to time.

By proceeding as in part (c) above, we can give the value of the more general integral of the form

$$\int_{-a}^{a} \sqrt{a^2 - x^2}\,dx$$

where a is an arbitrary positive constant. The graph of $y = \sqrt{a^2 - x^2}$ is the upper half of the circle $x^2 + y^2 = a^2$, whose total area is πa^2. Thus we have the following result.

Area Under the Semicircle with Radius a

$$\int_{-a}^{a} \sqrt{a^2 - x^2}\, dx = \frac{\pi a^2}{2} \qquad (5.10)$$

We will have several occasions later to use this result.

Unfortunately, most integrals cannot be evaluated by the method of Example 5.10, since the area under the curve cannot usually be found by elementary geometry. With a few exceptions (see Example 5.5 and Exercises 10, 12, and 14 in Exercise Set 5.1), the best we can do at this point is to approximate the integral using a Riemann sum made up of many small subintervals. For this purpose, when we know the integral exists, we might as well take subintervals of equal width, as we did in Section 5.1. For some functions it may be necessary to take a large number of subintervals to get much accuracy in approximating an integral. Of course, with a computer or programmable calculator, this requirement does not present much of a problem. Fortunately, we can often avoid any approximating procedure and get the exact value of the integral using one of the main results we will prove in the next section. There will still be some functions, however, whose integrals can only be approximated. In Section 5.5 we will give approximating techniques that are more efficient than using Riemann sums.

Using Different Variables of Integration

The letter x in the integral

$$\int_{a}^{b} f(x)dx$$

is called the **variable of integration**. The value of the integral is unchanged if we use a different letter. For example, for any integrable function f on $[a, b]$,

$$\int_{a}^{b} f(x)dx = \int_{a}^{b} f(t)dt$$

The variable of integration is called a "dummy variable," which means that changing to a different variable does not alter the value of the integral. You can see why the letter makes no difference when you think of the integral as representing area. The area is the same no matter which letter is used to name the axis of the independent variable.

In the definition of the integral $\int_{a}^{b} f(x)dx$, we have assumed that $a < b$. To take care of the other possibilities we adopt the following conventions:

$$\int_{a}^{a} f(x)dx = 0 \qquad \text{Provided } f(a) \text{ exists} \qquad (5.11)$$

(That is, the area of a vertical line segment is zero.)

If $a > b$ and f is integrable on $[b, a]$, we define

$$\int_{a}^{b} f(x)dx = -\int_{b}^{a} f(x)dx \qquad (5.12)$$

(That is, integrating from right to left is the negative of integrating from left to right.) For example, if f is integrable on $[0, 3]$,

$$\int_3^3 f(x)dx = 0 \qquad \text{and} \qquad \int_3^0 f(x)dx = -\int_0^3 f(x)dx$$

Some Properties of Integrals

We conclude this section by listing some of the properties of integrals. They can all be justified on the basis of the definition as a limit of Riemann sums, and in the exercises you will be asked to carry out the arguments for some of the simpler cases. We show by means of the figures the plausibility of most of the properties based on the area interpretation when the functions are nonnegative. We assume that all the functions are integrable on $[a, b]$.

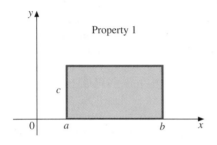

Property 1

Area $= c(b - a)$

FIGURE 5.27

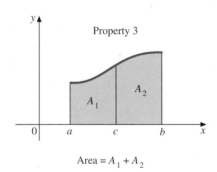

Property 3

Area $= A_1 + A_2$

FIGURE 5.28

Properties of the Integral

1. For any constant c, $\int_a^b c\, dx = c(b - a)$. (See Figure 5.27.)

2. For any constant c, $\int_a^b c\, f(x)dx = c \int_a^b f(x)dx$.

3. If $a < c < b$, $\int_a^b f(x)dx = \int_a^c f(x)dx + \int_c^b f(x)dx$. (See Figure 5.28.)

4. $\int_a^b [f(x) + g(x)]dx = \int_a^b f(x)dx + \int_a^b g(x)dx$

5. If $f(x) \geq 0$ on $[a, b]$, then $\int_a^b f(x)dx \geq 0$. (See Figure 5.29.)

6. If $f(x) \leq g(x)$ on $[a, b]$, then $\int_a^b f(x)dx \leq \int_a^b g(x)dx$. (See Figure 5.30.)

7. $\left| \int_a^b f(x)dx \right| \leq \int_a^b |f(x)|dx$

8. If $m \leq f(x) \leq M$ on $[a, b]$, then $m(b-a) \leq \int_a^b f(x)dx \leq M(b-a)$. (See Figure 5.31.)

Property 5

Area ≥ 0 if $f(x) \geq$

FIGURE 5.29

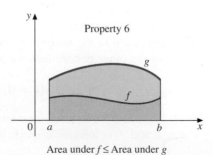

Property 6

Area under $f \leq$ Area under g

FIGURE 5.30

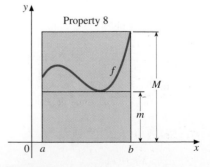

Property 8

FIGURE 5.31

Exercise Set 5.2

InExercises 1–4, find the Riemann sum for the given function on the interval specified, corresponding to the given division points and choices of c_k. Simplify the result.

1. $f(x) = 2x - 3$ on $[-3, 2]$; $x_0 = -3$, $x_1 = -2$, $x_2 = -1$, $x_3 = 0$, $x_4 = 1$, $x_5 = 2$; $c_1 = -\frac{5}{2}$, $c_2 = -1$, $c_3 = -\frac{1}{4}$, $c_4 = 0$, $c_5 = \frac{7}{4}$

2. $f(x) = 5 - 2x$ on $[0, 4]$; $x_0 = 0$, $x_1 = 1$, $x_2 = 3$, $x_3 = 4$; $c_1 = \frac{1}{2}$, $c_2 = \frac{5}{2}$, $c_3 = 4$

3. $f(x) = 10(x - 2)$ on $[1, 2]$; $x_0 = 1$, $x_1 = 1.2$, $x_2 = 1.3$, $x_3 = 1.6$, $x_4 = 1.9$, $x_5 = 2$; $c_1 = 1.1$, $c_2 = 1.225$, $c_3 = 1.4$, $c_4 = 1.8$, $c_5 = 1.95$

4. $f(x) = x(3 - x)$ on $[-1, 3]$; $x_0 = -1$, $x_1 = 0$, $x_2 = \frac{3}{2}$, $x_3 = 3$; $c_1 = -\frac{1}{2}$, $c_2 = \frac{1}{2}$, $c_3 = 2$

In Exercises 5–8, write the Riemann sum for f on $[a, b]$ using n subintervals of equal width. Take c_k as the right endpoint of the kth subinterval.

5. $f(x) = \frac{1}{x}$ on $[1, 2]$

6. $f(x) = 1 + \cos x$ on $[0, \pi]$

7. $f(x) = \sqrt[3]{3x - 4}$ on $[1, 4]$

8. $f(x) = \sin(\pi x)$ on $[0, 2]$

In Exercises 9–12, repeat Exercises 5–8, using as c_k the left endpoint of the kth subinterval.

In Exercises 13–18, write the given limit as an integral.

13. $\displaystyle\lim_{\max \Delta x_k \to 0} \sum_{k=1}^{n} \sqrt{1 + c_k^3}\, \Delta x_k$, for the interval $[0, 2]$

14. $\displaystyle\lim_{\max \Delta x_k \to 0} \sum_{k=1}^{n} \left[\frac{\ln c_k}{c_k}\right] \Delta x_k$, for the interval $[1, 3]$

15. $\displaystyle\lim_{n \to \infty} \sum_{k=1}^{n} \left[\frac{2k}{n}\right]^2 \left[\frac{2}{n}\right]$

16. $\displaystyle\lim_{n \to \infty} \sum_{k=1}^{n} \left[\frac{1}{1 + \frac{4(k-1)}{n}}\right]\left[\frac{4}{n}\right]$

17. $\displaystyle\lim_{n \to \infty} \sum_{k=1}^{n} (\sin c_k)\left[\frac{5}{n}\right]$, where $-2 + \dfrac{5(k-1)}{n} \le c_k \le -2 + \dfrac{5k}{n}$

18. $\displaystyle\lim_{n \to \infty} \sum_{k=1}^{n} \left[\frac{c_k}{1 + c_k}\right]\left[\frac{1}{5}\right]$, where $1 + \dfrac{k-1}{n} \le c_k \le 1 + \dfrac{k}{n}$

In Exercises 19–21 express the area of the shaded region as a definite integral.

19.

20.

Semicircle

21.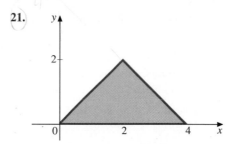

In Exercises 22–30, use geometric formulas for area to evaluate the integral.

22. $\int_{-1}^{3} (2x + 1)\,dx$

23. $\int_{-3}^{3} |x|\,dx$

24. $\int_{0}^{3} |x - 2|\,dx$

25. $\int_{0}^{\sqrt{8}} \sqrt{8 - x^2}\,dx$

26. $\int_{-2}^{4} (|2x - 3| - 3)\,dx$

27. $\int_{-2}^{0} (3 + \sqrt{4 - x^2})\,dx$

28. $\int_{-2}^{4} f(x)dx$, where $f(x) = \begin{cases} x+2 & \text{if } -2 \leq x < 0 \\ 2 & \text{if } 0 \leq x \leq 2 \\ 4-x & \text{if } 2 < x \leq 4 \end{cases}$

29. $\int_{-3}^{6} f(x)dx$, where $f(x) = \begin{cases} \sqrt{9-x^2} & \text{if } -3 \leq x < 0 \\ \frac{1}{2}(6-x) & \text{if } 0 \leq x \leq 6 \end{cases}$

30. $\int_{0}^{7} g(t)dt$, where $g(t) = \begin{cases} 2(1+t) & \text{if } 0 \leq x < 3 \\ 4(5-t) & \text{if } 3 \leq x \leq 7 \end{cases}$

In Exercises 31 and 32, the graph of f is given. Make use of areas to evaluate $\int_{a}^{b} f(x)dx$.

31.

32.

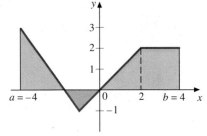

33. Evaluate $\int_{a}^{b} |f(x)|dx$ for each of the functions given in Exercises 31 and 32.

In Exercises 34–36, use the graph to estimate, by upper and lower sums, with $\Delta x = 1$, the integrals $\int_{1}^{4} f(x)dx$ and $\int_{-3}^{2} f(x)dx$. Each grid square has side 1.

34.

35.

36.

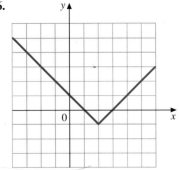

In Exercises 37–40, use a CAS, a spreadsheet, or a calculator to construct a table of left-hand and right-hand Riemann sums with $n = 2, 8, 16, 32, 100,$ and 200, respectively. Estimate the value of the integral.

37. $\int_{0}^{1} x^2 dx$

38. $\int_{0}^{\pi/2} \cos x \, dx$

39. $\int_{0.2}^{3} \sin\left(\frac{1}{x}\right) dx$

40. $\int_{0}^{3} \sqrt{x} \, dx$

In Exercises 41 and 42, use a CAS, a spreadsheet, or a calculator to compute the definite integral to one decimal place of accuracy. In each case determine the number of subdivisions that will guarantee this accuracy. Show that each function is increasing on the interval of integration.

41. $\int_{0}^{2} x^2 dx$

42. $\int_{0}^{1/2} (x - x^2)dx$

43. Use the figure (shown at top of page 331) to give an approximate value of $\int_{0}^{2.2} \cos(x^2)dx$.

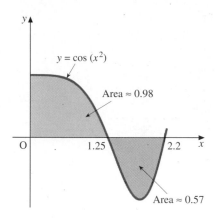

$y = \cos(x^2)$

Area ≈ 0.98

O   1.25   2.2   x

Area ≈ 0.57

44. From the areas shown in the figure, give the approximate value of each of the following.

(a) $\int_a^b f(x)dx$

(b) $\int_b^c f(x)dx$

(c) $\int_a^c f(x)dx$

(d) $\int_a^d f(x)dx$

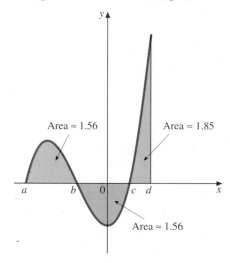

Area ≈ 1.56   Area ≈ 1.85

a   b   0   c d   x

Area ≈ 1.56

45. For the function f whose graph is given in Exercise 44, evaluate $\int_a^d |f(x)|dx$.

46. Obtain an approximation to π as follows. Use a CAS, a spreadsheet, or a calculator to compute $\int_{-1}^{1} \sqrt{1-x^2}dx$ using left-hand sums with 100 subdivisions. Explain why π is equal to twice the value of this integral. (*Hint:* Draw the graph of $y = \sqrt{1-x^2}$ on $[-1, 1]$.)

47. (a) Use a CAS or a graphing calculator to obtain graphs of

$$f(x) = \cos^5 x \quad \text{and} \quad g(x) = \sin^5 x$$

(b) Using the graphs obtained in part (a), what can you say about the value of the following integrals?

$$\int_0^{2\pi} \cos^5 x \, dx \quad \text{and} \quad \int_0^{2\pi} \sin^5 x \, dx$$

48. Use geometric reasoning to explain why

$$\int_0^1 x^n dx + \int_0^1 x^{1/n} = 1$$

for $n \geq 1$. (*Hint:* $y = x^n$ and $y = x^{1/n}$ are inverse functions.)

Exercises 49–56 refer to properties of the integral on page 328. In Exercises 49–52, use Riemann sums to prove the indicated property.

49. Property 1

50. Property 2

51. Property 4

52. Property 8

53. Assuming Property 5, prove Property 6. [*Hint:* Consider $f(x) - g(x)$.]

54. Prove Property 7 as follows. Explain why $f(x) \leq |f(x)|$, and then use Property 6 to conclude that $\int_a^b f(x)dx \leq \int_a^b |f(x)|dx$. Next, explain why $-f(x) \leq |f(x)|$, and then use Property 6 and Property 2 to conclude that $-\int_a^b f(x)dx \leq \int_a^b |f(x)|dx$. How does Property 7 follow?

55. Use mathematical induction to show that Property 4 can be extended to any finite number of functions. That is, show that if $f_1, f_2, \ldots, f_n$ are integrable on $[a, b]$,

$$\int_a^b \left(\sum_{k=1}^n f_k(x) \right) dx = \sum_{k=1}^n \left(\int_a^b f_k(x)dx \right)$$

56. Use Property 6 to show that $\int_0^1 \sin x \, dx \leq \frac{1}{2}$.

In Exercises 57–59, verify the result by making use of Equations 5.4, 5.5, or 5.6, as appropriate, together with the definition of the integral. In the Riemann sums use c_k as the right endpoint of the kth subinterval. (Hint: $\Delta t = x/n$.)

57. $\int_0^x t \, dt = \dfrac{x^2}{2}$

58. $\int_0^x t^2 dt = \dfrac{x^3}{3}$

59. $\int_0^x t^3 dt = \dfrac{x^4}{4}$

60. Let f be the function whose graph is shown. Define a new function A by

$$A(x) = \int_0^x f(t)dt \qquad (0 \le x \le 10)$$

(a) Where is the function A increasing?

(b) Where is the function A decreasing?

(c) Is A constant on any interval? If so, where?

(d) Where do local maximum values of A occur?

5.3 THE FUNDAMENTAL THEOREMS OF CALCULUS

The concepts of the derivative, derived from the tangent line problem, and the integral, derived from the area problem, appear to be completely different from each other. There is, however, a very close relationship between the two concepts. Isaac Newton and Gottfried Leibniz are credited with its discovery—one of the most profound insights in the history of mathematics. We will show this relationship in two different ways. The results are two theorems, called the First and Second Fundamental Theorems of Calculus. These two theorems provide the vital links between differential and integral calculus, and we cannot overemphasize their importance. We will see a much simpler way of evaluating certain integrals than by using the definition as a limit of Riemann sums. Before we get to the main theorems, we need a preliminary result called the Mean-Value Theorem of Integral Calculus. (It states that the value of a definite integral is equal to some mean, or intermediate, value of the integrand times the length of the interval of integration.)

THEOREM 5.1

Mean-Value Theorem of Integral Calculus

If f is continuous on $[a, b]$, there exists a point c in $[a, b]$ such that

$$\int_a^b f(x)dx = f(c)(b - a)$$

Proof   Since f is continuous on the interval $[a, b]$, we know by the Extreme-Value Theorem (Theorem 4.2) that f attains somewhere an absolute maximum value M and somewhere an absolute minimum value m on $[a, b]$.

By Property 8 of integrals (Section 5.2), since $m \le f(x) \le M$, we have

$$m(b - a) \le \int_a^b f(x)dx \le M(b - a)$$

As long as $b - a > 0$, we can divide by $b - a$ to get

$$m \le \frac{\int_a^b f(x)dx}{b - a} \le M$$

Thus, the number

$$\frac{\int_a^b f(x)dx}{b-a}$$

is between the smallest and largest values of f on the interval. We can therefore apply the Intermediate-Value Theorem (Theorem 2.8) and conclude that there exists a number c in $[a, b]$ such that

$$f(c) = \frac{\int_a^b f(x)dx}{b-a}$$

Multiplying both sides by $b-a$ yields the desired result.   ∎

This theorem has a simple geometric interpretation. Suppose f is continuous and $f(x) \geq 0$ on $[a, b]$. We know, then, that $\int_a^b f(x)dx$ gives the area under the graph of f. The Mean-Value Theorem for integrals says this area is equal to $f(c)(b-a)$ for some number c in $[a, b]$. But $f(c)(b-a)$ is just the area of the rectangle with height $f(c)$ and base equal to the interval $[a, b]$, as shown in Figure 5.32. The height $f(c)$ is therefore a sort of average (mean) value of the ordinates $f(x)$ of the graph; that is, if we replace the curve by the horizontal line $f(c)$ units high, the area under the line is the same as the area under the curve. Although we have assumed f is nonnegative and continuous on $[a, b]$, we extend this notion of the average value of f to any integrable function by the following definition.

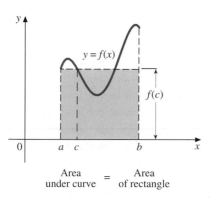

Area
under curve = Area
of rectangle

FIGURE 5.32

Definition 5.3 Average Value of a Function	Let f be integrable on $[a, b]$. The **average value of f** on $[a, b]$ is defined as $$f_{\text{ave}} = \frac{1}{b-a} \int_a^b f(x)dx$$

EXAMPLE 5.11   Find the average value of the function

$$f(x) = \sqrt{4 - x^2}$$

on the interval $[-2, 2]$.

Solution   We found in Example 5.10(c) that

$$\int_{-2}^2 \sqrt{4 - x^2}\, dx = 2\pi$$

Thus, by Definition 5.3,

$$f_{\text{ave}} = \frac{1}{2 - (-2)} \int_{-2}^2 \sqrt{4 - x^2}\, dx$$

$$= \frac{1}{4}(2\pi)$$

$$= \frac{\pi}{2}$$ ∎

The First Fundamental Theorem

Suppose now that f is continuous on the closed interval $[a, b]$ and that x is any point in this interval. We consider the integral from a to x, which we will call $A(x)$. So

$$A(x) = \int_a^x f(t)dt \tag{5.13}$$

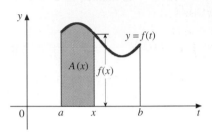

FIGURE 5.33

We are using t as a dummy variable of integration to avoid confusion with the independent variable x, which appears in the upper limit of integration. (We saw an integral of this type in Exercise 60 of Exercise Set 5.2.) If f is nonnegative on $[a, b]$, then we know that $A(x)$ is the area under the graph of f from a to x. (See Figure 5.33.) For this reason, we call $A(x)$ the *area function* for f on $[a, b]$. Since f is assumed to be continuous, we know that the integral exists on $[a, b]$ and is a unique number for each $a \leq x \leq b$. Consequently, we can see that $A(x)$ is a function of x. Even if f is of variable sign, we will still call $A(x)$ the area function, although we will then have to understand "area" to mean what we have called "net area." Before we state the first theorem, let us give an indication of what it will tell us. For instance, if we take $f(t) = t^2$ and $a = 0$, then by Exercise 58 in Exercise Set 5.2, we have

$$A(x) = \int_0^x t^2 dt = \frac{x^3}{3}$$

Now notice that $A'(x) = f(x)$ and hence we see $A' = f$, or equivalently, A is an antiderivative of f.

Now we can state the First Fundamental Theorem of Calculus. It says that for continuous functions, the derivative of this area function—that is, the derivative of the integral of f from a to x—is always $f(x)$. Or, in symbols,

$$A'(x) = f(x)$$

Referring to Figure 5.33, this theorem states that the area function is differentiable and the rate at which the area $A(x)$ is changing is exactly equal to the ordinate $f(x)$. In the statement of the theorem we replace $A(x)$ by the integral defining it in Equation 5.13.

THEOREM 5.2

> **The First Fundamental Theorem of Calculus**
>
> Let f be continuous on $[a, b]$. Then for x in $[a, b]$, the derivative of the integral $\int_a^x f(t)dt$ exists, and
>
> $$\frac{d}{dx} \int_a^x f(t)dt = f(x) \tag{5.14}$$

This is a remarkable result and is worth studying carefully. It provides a crucial link between integration and differentiation. It tells us that every function f continuous on an interval $[a, b]$ has an antiderivative there, namely, $\int_a^x f(t)dt$.

Proof   We show that the derivative of $A(x) = \int_a^x f(t)dt$ exists by showing the existence of the limit

$$\lim_{h \to 0} \frac{A(x+h) - A(x)}{h}$$

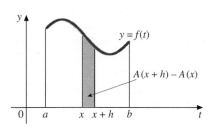

FIGURE 5.34

If $h > 0$, the numerator is the area under the graph of f from a to $x + h$ minus the area from a to x, and therefore is the shaded area from x to $x + h$, as indicated in Figure 5.34. As we have pictured it, f is nonnegative, but our result is valid without this restriction. That is,

$$A(x+h) - A(x) = \int_x^{x+h} f(t)dt$$

(This equality is justified without reference to the graph by Property 3 of integrals.) By the Mean-Value Theorem for integrals (Theorem 5.1), there is a number c between x and $x + h$ such that

$$\int_x^{x+h} f(t)dt = f(c) \cdot h$$

Thus, we obtain, on substituting $f(c) \cdot h$ for $A(x + h) - A(x)$,

$$\lim_{h \to 0} \frac{A(x+h) - A(x)}{h} = \lim_{h \to 0} \frac{f(c) \cdot h}{h} = \lim_{h \to 0} f(c)$$

Since c lies between x and $x+h$, it follows from the Squeeze Theorem (Theorem 2.3) that as $h \to 0$, $c \to x$, and so by continuity (Definition 2.2) $f(c) \to f(x)$. Thus,

$$\lim_{h \to 0^+} \frac{A(x+h) - A(x)}{h} = f(x)$$

The argument is similar for $h < 0$, and you will be asked to supply it in Exercise 69 of Exercise Set 5.3. We conclude from the definition of the derivative that $A'(x)$ exists and that $A'(x) = f(x)$. ■

EXAMPLE 5.12   Find the derivative of $\int_2^x \sin t \, dt$.

Solution   Since $\sin x$ is continuous everywhere, by Equation 5.14,

$$\frac{d}{dx} \int_2^x \sin t \, dt = \sin x$$ ■

EXAMPLE 5.13   Find the derivative of $\int_0^x \frac{du}{\sqrt{1+u^3}}$, $x > 0$.

Solution   The function $1/\sqrt{1+u^3}$ is continuous on $[0, \infty)$, so by Equation 5.14,

$$\frac{d}{dx} \int_0^x \frac{du}{\sqrt{1+u^3}} = \frac{1}{\sqrt{1+x^3}}$$

Note that u is merely a dummy variable of integration. The result would be the same no matter which letter we used, as long as we used the same letter for the function $1/\sqrt{1+u^3}$ and the differential du. The integral is a function of x, not of u. ■

The Second Fundamental Theorem

In a certain sense, Theorem 5.2 expresses an inverse relationship between integration and differentiation. If we begin with a continuous function f on $[a, b]$, integrate it from a to x, then differentiate the result, we come back to the original function, evaluated at x. The next theorem turns this procedure around and shows that under certain hypotheses, if we differentiate first and then integrate the result, we also return to the original function, evaluated at the endpoints of the interval.

THEOREM 5.3

The Second Fundamental Theorem of Calculus

Let f be continuous on $[a, b]$ and let F be any antiderivative of f on $[a, b]$; that is, $F'(x) = f(x)$. Then

$$\int_a^b f(x)dx = F(b) - F(a) \tag{5.15}$$

Proof   By Theorem 5.2 we know that $\int_a^x f(t)dt$ is one antiderivative of f on $[a, b]$, and we are given that F is also an antiderivative of f there. So by Corollary 4.9b, these two antiderivatives differ by a constant; that is,

$$\int_a^x f(t)dt = F(x) + C$$

for some constant C. In fact, we can determine C. If we put $x = a$, we get

$$\int_a^a f(t)dt = F(a) + C$$

But the integral on the left is 0, so $C = -F(a)$. Thus,

$$\int_a^x f(t)dt = F(x) - F(a) \tag{5.16}$$

Finally, we put $x = b$ in Equation 5.16 to get

$$\int_a^b f(t)dt = F(b) - F(a)$$

which is what we wanted to prove.   ∎

To give an idea of the significance of this theorem, recall the work involved in evaluating an integral such as $\int_0^2 x^2 dx$ by making use of the limit of Riemann sums, as we did in Example 5.5. Using the Second Fundamental Theorem, we can get the value of $\int_0^2 x^2 dx$ quickly and easily by first finding *any* antiderivative of x^2—for example, $F(x) = x^3/3$—and then calculating $F(2) - F(0)$:

$$\int_0^2 x^2 dx = F(2) - F(0) = \frac{8}{3} - 0 = \frac{8}{3}$$

It is convenient to introduce the notation $F(x)]_a^b$ to mean $F(b) - F(a)$. Then we can write

$$\int_a^b f(x)dx = F(x)]_a^b$$

for the result of the Second Fundamental Theorem. For example,

$$\int_0^2 x^2 dx = \left.\frac{x^3}{3}\right]_0^2 = \frac{8}{3} - 0 = \frac{8}{3}$$

EXAMPLE 5.14   Evaluate each of the following using the Second Fundamental Theorem:

(a) $\displaystyle\int_{-1}^2 (3x^2 - 2x + 4)\,dx$ (b) $\displaystyle\int_0^{\pi/2} \cos x\,dx$ (c) $\displaystyle\int_{-\pi/4}^{\pi/4} \sec^2 x\,dx$

Solution   Note first that all the integrands are continuous on the intervals involved, so the Second Fundamental Theorem applies.

(a) $\displaystyle\int_{-1}^2 (3x^2 - 2x + 4)\,dx = x^3 - x^2 + 4x\big]_{-1}^2 = (8 - 4 + 8) - (-1 - 1 - 4)$

$$= 12 + 6 = 18$$

(b) $\displaystyle\int_0^{\pi/2} \cos x\,dx = \sin x]_0^{\pi/2} = \sin\frac{\pi}{2} - \sin 0 = 1$

(c) $\displaystyle\int_{-\pi/4}^{\pi/4} \sec^2 x\,dx = \tan x]_{-\pi/4}^{\pi/4} = \tan\frac{\pi}{4} - \tan\left(\frac{-\pi}{4}\right) = 1 - (-1) = 2$ ■

REMARK ───

■ Since in the Second Fundamental Theorem we are allowed to use *any* antiderivative of the integrand, we might as well choose the simplest one. Any other would give the same result. For example,

$$\int_{-1}^2 x^2 dx = \left.\frac{x^3}{3}\right]_{-1}^2 = \frac{8}{3} - \frac{(-1)}{3} = 3$$

If we had used another antiderivative, instead of the simplest, it would necessarily have been of the form $x^3/3 + C$, so

$$\int_{-1}^2 x^2 dx = \left.\frac{x^3}{3} + C\right]_{-1}^2 = \left[\frac{8}{3} + C\right] - \left[\frac{-1}{3} + C\right]$$

$$= \frac{8}{3} + \frac{1}{3} = 3$$

In other words, the constant C will always cancel out.

───

Do not infer from this result that it is no longer necessary to add the constant when forming an antiderivative. We are speaking here of one particular use of the antiderivative where the constant can be deleted, namely, in evaluating definite integrals by the Second Fundamental Theorem. In most cases the constant is essential.

The Integral of the Derivative as Net Change in the Function

In the Second Fundamental Theorem, F is an antiderivative of f; that is, $F'(x) = f(x)$. Therefore, we could write the result in the alternative form

$$\int_a^b F'(x)dx = F(b) - F(a) \tag{5.17}$$

as long as F' is continuous on $[a, b]$. It can be proved, in fact, that Equation 5.17 holds true even if F' is not continuous as long as it is integrable on $[a, b]$. (See Exercise 70 in Exercise Set 5.3.) Going back to Equation 5.16 in the proof of the theorem, if we also replace f by F', we get

$$\int_a^x F'(t)dt = F(x) - F(a)$$

Since the only function involved here is F, we can also write the result with another letter to designate the function. In particular, we can use our usual letter f and get

$$\int_a^x f'(t)dt = f(x) - f(a) \tag{5.18}$$

which is true for all $x \in [a, b]$ provided f' is integrable there. Let us compare this result with the result of the First Fundamental Theorem, Equation 5.14, which states that if f is continuous on $[a, b]$, then

$$\frac{d}{dx}\int_a^x f(t)dt = f(x), \qquad x \in [a, b]$$

Here we differentiate the integral of f from a to x and get $f(x)$ as the result. In Equation 5.18, we integrate from a to x the derivative of f and get $f(x)$ minus a constant, $f(a)$. It is in this sense that integration and differentiation can be considered inverse processes.

Equation 5.17 provides a useful way of interpreting the definite integral. Since $F(b) - F(a)$ is the change in the function values $F(x)$ from $x = a$ to $x = b$, and $F'(x)$ is the rate of change of the function, we reach the following conclusion.

The integral from a to b of the *rate of change* of a function is the *net change* in the function from a to b.

We have seen this interpretation already in Example 5.3. The velocity function $v(t)$ is the rate of change of the position function $s(t)$ (that is, $v(t) = s'(t)$). So

$$\int_a^b v(t)dt = s(b) - s(a) = \text{net change in distance}$$

The next two examples further illustrate this interpretation of the integral of rate of change as the net change in the function.

EXAMPLE 5.15   A pump is pumping water into a tank at the rate of $2t$ liters per minute (L/min), where t is in minutes. Find the amount of water pumped into the tank during the first half-hour the pump is in operation.

Solution   Let $w(t)$ denote the number of liters of water pumped into the tank during the first t min of pumping. Then we can conclude from the given information that

$$w'(t) = 2t$$

An antiderivative of $2t$ is t^2. So the total amount of water pumped into the tank in the first half-hour (30 min) is

$$\int_0^{30} 2t\,dt = t^2\Big]_0^{30} = 900 \text{ L} \qquad \blacksquare$$

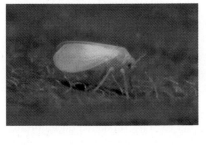

EXAMPLE 5.16   In the fall of 1991, the poinsettia white fly began devastating vegetable crops in the Imperial Valley of California. Assume that the graph in Figure 5.35 gives the rate at which fields of broccoli were being destroyed, in acres per day, for the month of November. Approximate the total number of acres of broccoli destroyed during the first 10 days of the month and during the last 10 days.

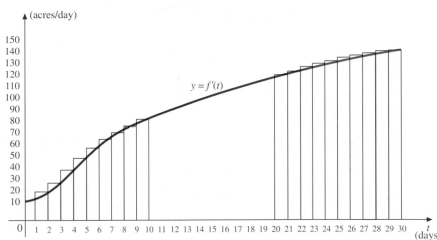

$f'(t)$ = rate of destruction of broccoli fields by
the poinsettia white fly in the Imperial Valley, in
acres per day, during November 1991

FIGURE 5.35

Solution   Let $f(t)$ denote the number of acres of broccoli destroyed in the first t days. Then the graph gives $f'(t)$. In the first 10 days, the total number of acres destroyed is given by

$$\int_0^{10} f'(t)dt$$

If we knew a formula for f, we could evaluate this integral using the Second Fundamental Theorem. Since we do not know f, we can approximate the integral with a Riemann sum

$$\sum_{k=1}^n f'(c_k)\Delta t_k$$

We might choose to use subintervals of equal length, with $\Delta t = 1$. A reasonable choice for c_k is the midpoint of the kth subinterval. When we do this, we get (reading approximate values of $f'(c_k)$ from the graph)

$$\int_0^{10} f'(t)dt \approx (11)(1) + (19)(1) + (27)(1) + (36)(1) + (48)(1) + (58)(1)$$
$$+ (65)(1) + (70)(1) + (75)(1) + (81)(1)$$
$$= 490$$

So approximately 490 acres were destroyed in the first 10 days of November.
For the last 10 days we approximate the value in a similar way, getting

$$\int_{20}^{30} f'(t)dt \approx (121)(1) + (122)(1) + (128)(1) + (130)(1) + (133)(1)$$
$$+ (135)(1) + (138)(1) + (140)(1) + (144)(1) + (142)(1)$$
$$= 1{,}333$$

Thus, approximately 1,333 acres of broccoli were destroyed during the last 10 days of November.   ■

Let us conclude this section by summarizing the two fundamental theorems. It is important that you learn these, as we will use them many times throughout the remainder of this textbook. We use the abbreviations FTC1 and FTC2 for the First and Second Fundamental Theorems of Calculus. Below each statement we paraphrase what it says in simple language.

The Two Fundamental Theorems

FTC1   If f is continuous on $[a, b]$, then

$$\frac{d}{dx} \int_a^x f(t)dt = f(x)$$

for each x in $[a, b]$.

That is, when you differentiate an integral (with respect to a variable upper limit), you get the integrand (evaluated at the upper limit).

FTC2   If f is continuous on $[a, b]$ and F is any antiderivative of f, then

$$\int_a^b f(x)dx = F(b) - F(a)$$

That is, to evaluate a definite integral, find an antiderivative of the integrand and subtract its value at the lower limit from its value at the upper limit.

Since for an antiderivative F of f, we have $F' = f$, an alternative form of FTC2 is

$$\int_a^b F'(x)dx = F(b) - F(a) \qquad \text{If } F' \text{ is continuous on } [a, b]$$

In this form we can say that to integrate the derivative of a function, you subtract its value at the lower limit from its value at the upper limit.

Exercise Set 5.3

Use the Second Fundamental Theorem to evaluate the integrals in Exercises 1–30.

1. $\int_1^3 (x^2 + 2x - 1)dx$

2. $\int_{-1}^1 (3x - x^2)dx$

3. $\int_{-2}^1 (x - 1)^2 dx$

4. $\int_3^5 x(2x - 3)dx$

5. $\int_0^4 (x^3 - 3x^2 + 4)dx$

6. $\int_{-2}^0 (x + 2)^3 dx$

7. $\int_0^1 (x - \sqrt{x} + 3)dx$

8. $\int_{-1}^1 (x^5 - 2x^3 + 4x)dx$

9. $\int_{-2}^2 (x^2 - 3)^2 dx$

10. $\int_4^1 (x - 1)(x + 2)dx$

11. $\int_{-2}^{-1} \dfrac{x^3 - 1}{x^2} dx$

12. $\int_1^4 \dfrac{x + 1}{\sqrt{x}} dx$

13. $\int_0^\pi 2 \sin x \, dx$

14. $\int_{-\pi/2}^{\pi/2} 3 \cos x \, dx$

15. $\int_0^{\pi/3} \sec^2 x \, dx$

16. $\int_{-\pi/3}^{\pi/4} \sec x \tan x \, dx$

17. $\int_{\pi/6}^{\pi/2} 2 \csc x \cot x \, dx$

18. $\int_{\pi/4}^{3\pi/4} 4 \csc^2 x \, dx$

19. $\int_4^9 (t^{3/2} - 2t^{1/2} + 1)dt$

20. $\int_1^8 (t^{2/3} + 2t^{-1/3})dt$

21. $\int_1^8 \dfrac{u - 2}{\sqrt[3]{u}} du$

22. $\int_1^4 \dfrac{(v - 1)^3}{\sqrt{v}} dv$

23. $\int_{\pi/6}^{5\pi/6} (x + \sin x)dx$

24. $\int_{\pi/4}^{\pi/2} \dfrac{1 - \cos \theta}{\sin^2 \theta} d\theta$

25. $\int_0^{\pi/4} (1 + \sin^2 \phi \sec^2 \phi)d\phi$

26. $\int_{\pi/6}^{\pi/3} \tan^2 x \, dx$

27. $\int_{2\pi/3}^{5\pi/6} \dfrac{\sec t - \csc t}{\sec t \csc t} dt$

28. $\int_{\pi/4}^{\pi/2} \cot^2 v \, dv$

29. $\int_1^3 \left(x - \dfrac{1}{x} \right) \left(2x - \dfrac{3}{x} \right) dx$

30. $\int_1^9 \dfrac{(\sqrt{t} - 2)(\sqrt{t} + 3)}{2\sqrt{t}} dt$

In Exercises 31–34, find the average value of the function on the given interval.

31. $f(x) = x^2 - 2x - 3$ on $[-2, 4]$

32. $f(x) = 1 + x - 2x^2$ on $[-1, 2]$

33. $f(x) = x^3 - 3x^2 - 4x + 2$ on $[-2, 3]$

34. $f(x) = 2x^3 - 3x - 4$ on $[-2, 2]$

35. Suppose an object moving along a line has distance function $s(t)$.
  (a) Give its average velocity from time $t = a$ to time $t = b$.
  (b) Find the average value of the velocity function $v(t) = s'(t)$ by using Definition 5.3 together with the Second Fundamental Theorem. Show that the result is the same as in part (a).

In Exercises 36–39, find a number c in [a, b] that satisfies $\int_a^b f(x)dx = f(c) \cdot (b - a)$.

36. $f(x) = \sqrt{x}$ on $[0, 9]$

37. $f(x) = 4/x^2$ on $[1, 4]$

38. $f(x) = x^2 - 1$ on $[0, 3]$

39. $f(x) = x^2 - 2x$ on $[-2, 1]$

In Exercises 40–43, use the First Fundamental Theorem to find the derivative. State on what intervals the result is valid.

40. $\dfrac{d}{dx} \int_1^x \dfrac{2}{\sqrt{1 + t^2}} dt$

41. $\dfrac{d}{dx} \int_2^x \dfrac{t}{t - 1} dt$

42. $F'(x)$ if $F(x) = \int_0^x \sqrt{t^3 + 8} \, dt$

43. $F'(x)$ if $F(x) = \int_{-3}^x \dfrac{\sin t}{2 + \cos t} dt$

44. Find $\dfrac{d}{dx} \int_{-1}^x (t^3 - 3t^2 + 1)dt$ by two methods: (a) the First Fundamental Theorem and (b) first performing the integration using the Second Fundamental Theorem and then differentiating.

45. The accompanying figure shows the graph of a function f on the interval $[0, 3]$. Let $A(x) = \int_0^x f(t)dt$. Sketch, as well as you can, the graph of $y = A(x)$ for $0 \le x \le 3$.

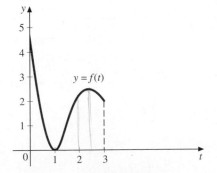

46. If a population of elephants is increasing at the rate of $500 + 2t$, where t is measured in years, find the amount by which the elephant population increases between the third and seventh years.

47. In recent years the Black Sea has become one of the most polluted bodies of water in the world. Assume that since 1980 the rate at which pollutants have been entering the sea is approximately $6300t^{5/2}$ kg/yr, where t is measured in years since 1980. Find approximately how many kilograms of pollutants were added from 1980 to 1990. If this rate continues, how many kilograms would be added from 1990 to 2000?

48. The following graph shows the inflation rate, $r(t)$, of a certain country in percent per year for the past 15 years. Estimate the total percentage increase in the cost of living during the same time period.

49. The accompanying graph shows the power consumption by a certain city on a given day in millions of kilowatts. Approximate the energy used on that day in kilowatt-hours. (The relationship between power, P, and energy, E, is $P = dE/dt$.)

50. If $f(x) = x/(x + 1)$, find $\int_0^2 f'(x)dx$.

51. Find $\displaystyle\int_0^2 \frac{d}{dt}\left(1/\sqrt{1+t^3}\right) dt$.

52. Derive a formula for

$$\frac{d}{dx}\int_x^b f(t)dt$$

where f is continuous on $[a, b]$ and $x \in [a, b]$.

53. Use the result of Exercise 52 to find $g'(x)$ if

$$g(x) = \int_x^4 \frac{\sin t}{t}dt$$

where $x > 0$.

54. Derive a formula for

$$\frac{d}{dx}\int_a^{g(x)} f(t)dt$$

where f is continuous and g is differentiable. [*Hint:* Let $u = g(x)$, and use the chain rule.]

55. Use the result of Exercise 54 to evaluate

$$\frac{d}{dx}\int_1^{x^2-2x} \sin^3 t\, dt$$

56. Find the number a for which the line $x = a$ divides the area under the graph of $y = (x - 3)^2$ from $x = -1$ to $x = 3$ in a $2 : 1$ ratio.

57. Evaluate the integral

$$\int_0^{3\pi/4} |\cos x|dx$$

58. Evaluate $\int_{-1}^2 |x^3 - 4x|dx$. (*Hint:* Draw the graph of $y = x^3 - 4x$.)

59. Find the area bounded by the curve $y = x^2 - 4x + 3$ and the x-axis.

60. In the accompanying figure, what is the ratio of the area of the shaded region to the area of the rectangle $ABCD$?

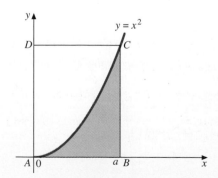

61. Find the area of the parabolic segment between the graph of $y = x^2$ and the line AB in the figure by subtracting the area under the graph of $y = x^2$ from the area of the triangle ABC.

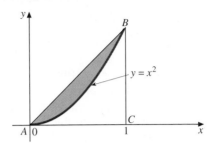

62. On the surface of the earth acceleration caused by gravity is approximately -10 m/s^2, whereas on the surface of the moon it is approximately -1.6 m/s^2.

(a) If an astronaut on the moon throws a baseball directly upward with an initial velocity of 15 m/s, what will be its velocity and height 2.5 seconds later? When will it reach its maximum height? What is the maximum height?

(b) If, on returning to the earth, the astronaut throws the ball directly upward with the same initial velocity, answer the questions of part (a).

In Exercises 63–68, simplify the integrands and then use the Second Fundamental Theorem to evaluate the integrals.

63. $\displaystyle\int_2^3 \sqrt{\frac{1}{x^4} - 1 + \frac{x^4}{4}}\; dx$ **64.** $\displaystyle\int_{-1}^2 \frac{2t^3 + 3t^2 + 4}{t + 2}\; dt$

65. $\displaystyle\int_{-\pi/6}^{\pi/3} \left(\frac{1}{1 + \sin\theta} + \frac{1}{1 - \sin\theta} \right) d\theta$

66. $\displaystyle\int_{\pi/2}^{5\pi/6} \frac{2}{1 - \cos 2x}\; dx$

67. $\displaystyle\int_{-\pi/4}^{\pi/3} \frac{1 - \sin\theta}{1 + \sin\theta}\; d\theta$ (*Hint:* Multiply numerator and denominator by $1 - \sin\theta$.)

68. $\displaystyle\int_{5\pi/6}^{\pi} \frac{1 - \sin 2x}{\sin x - \cos x}\; dx$ (*Hint:* Replace 1 by $\sin^2 x + \cos^2 x$ and use an identity for $\sin 2x$.)

69. Complete the proof of Theorem 5.2 by considering $h < 0$.

70. Prove that Equation 5.17 holds true if we assume only that F' is integrable on $[a, b]$. (*Hint:* Use the Mean-

Value Theorem of Differential Calculus to show that for any partition of $[a, b]$, there is a point $c_k \in (x_{k-1}, x_k)$, $k = 1, 2, \ldots, n$, such that

$$\sum_{k=1}^{n} [F(x_k) - F(x_{k-1})] = \sum_{k=1}^{n} F'(c_k)\Delta x_k$$

Show that the left-hand side reduces to $F(b) - F(a)$. Then let $\max \Delta x_k \to 0$.)

71. Let

$$f(x) = \begin{cases} x^2 \sin \dfrac{1}{x^2} & \text{if } x \neq 0 \\[2mm] 0 & \text{if } x = 0 \end{cases}$$

Show that $f'(x)$ exists on $[0, 1]$ but that

$$\int_0^1 f'(x)\, dx \neq f(1) - f(0)$$

Why does this not contradict Exercise 70? (*Hint:* Show that f' is unbounded on $[0, 1]$.)

72. Prove that if f is integrable on $[a, b]$, the function $F(x) = \int_a^x f(t)\, dt$ is continuous on $[a, b]$. [*Hint:* Let M be such that $|f(x)| \leq M$ on $[a, b]$. (Justify.) Then use the $\varepsilon - \delta$ definition of continuity (Definition 2.9) of F at x_0, making use of the properties of the integral.]

73. Show that the Mean-Value Theorems of Integral Calculus and of Differential Calculus are essentially different formulations of the same result. [*Hint:* In $\int_a^b f(x)\, dx = f(c)(b - a)$, let F be an antiderivative of f so that $f = F'$.]

74. Prove that if f and g are defined on $[a, b]$, with $f(x) = g(x)$ on $[a, b)$ and g continuous on $[a, b]$, then $\int_a^b f(x)\, dx$ exists and equals $\int_a^b g(x)\, dx$ regardless of how $f(b)$ is defined. (*Hint:* Use the result of Exercise 72.)

75. Prove that if f is integrable on $[a, b]$, then $F(x) = \int_x^b f(t)\, dt$ is continuous on $[a, b]$. (*Hint:* Use the hint for Exercise 72.)

76. Prove that if f and g are defined on $[a, b]$, with $f(x) = g(x)$ on $(a, b]$, and g is continuous on $[a, b]$, then $\int_a^b f(x)\, dx$ exists and equals $\int_a^b g(x)\, dx$ regardless of how $f(a)$ is defined. (*Hint:* Use the result of Exercise 75.)

77. Suppose f is defined on $[a, b]$ and is continuous on (a, b), and $\lim_{x \to a^+} f(x)$ and $\lim_{x \to b^-} f(x)$ both exist. Prove that $\int_a^b f(x)\, dx$ exists, regardless of how $f(a)$ and $f(b)$ are defined. (*Hint:* Use the results of Exercises 74 and 76.)

5.4 THE INDEFINITE INTEGRAL; THE TECHNIQUE OF SUBSTITUTION

According to the Second Fundamental Theorem of Calculus, we can evaluate the integral of a continuous function f over $[a, b]$ quickly and efficiently by the formula

$$\int_a^b f(x)dx = F(b) - F(a)$$

provided we can find F, where F is an antiderivative of f (that is, F is a function such that $F' = f$). To appreciate the importance of this result, recall the work involved in evaluating the integral as a limit of Riemann sums. There is only one hitch. To apply the theorem, you must be able to find a function F for which $F' = f$. Now we know by the First Fundamental Theorem that every continuous function on a closed interval has an antiderivative there, but finding such an antiderivative in the form of a recognizable elementary function is another matter. In some cases we can recognize the integrand as the derivative of a known function. In Section 4.7 we summarized a number of such functions along with their general antiderivatives, based on known derivatives of power functions and trigonometric functions. In this section we will learn a method that will enable us to greatly expand the list, and in Chapter 8 we will learn additional ways of finding antiderivatives. Unfortunately, no matter how many techniques we learn, there will be some continuous functions for which no one can find antiderivatives in the form of elementary functions. The Second Fundamental Theorem is no help in these cases, and we can only approximate the integral. In the next section we consider approximation techniques. Many of the functions used in describing scientific concepts fall into this category.

We need a symbol for the antiderivative:

Definition 5.4
The Indefinite Integral

The symbol $\int f(x)dx$ is called the **indefinite integral** of f and means the general antiderivative of f. That is,

$$\int f(x)dx = F(x) + C$$

where F is any function whose derivative is f. The arbitrary constant C is called the **constant of integration**.

Although definite and indefinite integrals differ only slightly in appearance, their meanings differ significantly. Consider the following:

Definite integral: $\quad \int_{-1}^{2} x^2 dx = \left. \frac{x^3}{3} \right]_{-1}^{2} = \frac{8}{3} - \left(-\frac{1}{3} \right) = 3$

Indefinite integral: $\quad \int x^2 dx = \frac{x^3}{3} + C$

In the first case the answer is a *number*, whereas in the second it is a *family of functions*. The definite and indefinite integrals are related by the Second Fundamental Theorem. In fact, we could write the conclusion of that theorem in the following form.

Restatement of FTC2

$$\int_a^b f(x)dx = \int f(x)dx \Bigg]_a^b$$

Just as with a definite integral, we refer to $f(x)$ in $\int f(x)dx$ as the *integrand*, and the process by which we obtain the answer is called *integration*. To *evaluate* an indefinite integral means to find the general antiderivative.

The following properties of indefinite integrals hold true on any interval I where the functions involved have antiderivatives.

Properties of Indefinite Integrals

1. $\dfrac{d}{dx} \int f(x)dx = f(x)$
2. $\int f'(x)dx = f(x) + C$, if f is differentiable on I
3. $\int cf(x)dx = c \int f(x)dx$, where c is any constant
4. $\int [f(x) \pm g(x)]dx = \int f(x)dx \pm \int g(x)dx$

The proofs of these properties follow from Definition 5.4 and are left as exercises.

All the antidifferentiation formulas summarized in Section 4.7 can now be restated using indefinite integral notation. In particular, we have

$$\int x^n dx = \frac{x^{n+1}}{n+1} + C, \text{ if } n \neq -1 \quad \text{Power Rule}$$

$$\int \sin x \, dx = -\cos x + C$$

$$\int \cos x \, dx = \sin x + C$$

The Guess-and-Check Method

We now turn to the question of how to evaluate an indefinite integral when the integrand is not among those in our list in Section 4.7. We want to find a function F for which

$$\int f(x)dx = F(x) + C$$

We know that $F'(x)$ must equal $f(x)$. Sometimes we can recognize the integrand as being the derivative of a composite function. Consider the following examples.

EXAMPLE 5.17   Evaluate the indefinite integral $\int 2x \cos x^2 dx$.

Solution   We know that an antiderivative of $\cos x$ is $\sin x$, so we might guess that the antiderivative we are looking for involves the sine function. Also, since we know that $d(x^2)/dx = 2x$, we might further guess that $2x \cos(x^2)$ is the derivative of the composite function $\sin x^2$, by recalling how the Chain Rule works. Checking, we see that

$$\frac{d}{dx} \sin x^2 = (\cos x^2)(2x) = 2x \cos x^2$$

So our guess is correct, and we can write the answer as

$$\int 2x \cos x^2 dx = \sin x^2 + C$$ ∎

EXAMPLE 5.18   Evaluate the indefinite integral $\int (3x^2 + 1)(x^3 + x + 1)^5 dx$.

Solution   Because $3x^2 + 1$ is the derivative of $x^3 + x + 1$, it looks as though the integrand might be the derivative of the composite function $(x^3 + x + 1)^6$, since by the Generalized Power Rule the derivative of a function to the nth power equals n times the function to the $(n - 1)$st power multiplied by the derivative of the function. So we check and find that

$$\frac{d}{dx}(x^3 + x + 1)^6 = 6(x^3 + x + 1)^5(3x^2 + 1)$$

$$= 6(3x^2 + 1)(x^3 + x + 1)^5$$

We were almost right in our guess, but it is off because of the factor 6. We can easily remedy the situation by dividing our original guess by 6. That is, we use

$$\frac{1}{6}(x^3 + x + 1)^6$$

So the correct result is

$$\int (3x^2 + 1)(x^3 + x + 1)^5 dx = \frac{1}{6}(x^3 + x + 1)^6 + C$$ ∎

We refer to the reasoning used in Examples 5.17 and 5.18 as the *guess-and-check method*. When the integrand looks as though it might be the result of applying the Chain Rule to some composite function, we guess what the function is and check by differentiating it to see if we get the integrand. If so, we add an arbitrary constant C, and we have the answer. If the derivative of our guess is off only because it is k times what we want, where k is a constant (obviously nonzero), then the correct antiderivative is our original guess divided by k.

The Method of Substitution

The guess-and-check method relies on the fact that if $F' = f$, then by the Chain Rule,

$$\frac{d}{dx}F(g(x)) = F'(g(x))g'(x) = f(g(x))g'(x)$$

so that

$$\int f(g(x))g'(x)dx = F(g(x)) + C \qquad (5.19)$$

An alternative to guessing is to substitute a new variable such as u for $g(x)$ in Equation 5.19. Then we have $u = g(x)$, $du = g'(x)dx$, so that Equation 5.19 becomes

$$\int f(u)du = F(u) + C \quad \text{(where } F' = f\text{)}$$

Then, to get the final answer we replace u by $g(x)$ in the right-hand side of this equation. As you might expect, we call this technique the *method of substitution*.

REMARK ——————————————————————————

■ The substitution method shows that the dx appearing in an indefinite integral can be treated as a differential, since $du = g'(x)dx$. The same is true in a definite integral, as we will see shortly.

In the substitution method the goal is to choose a substitution u that will make $f(u)$ the derivative of a known function $F(u)$. That is, we want $\int f(u)du$ to be among our list of known antiderivatives.

Let us redo the problems in Examples 5.17 and 5.18 using the method of substitution. In Example 5.17 we let $u = x^2$. Then $du = 2x\,dx$. Thus,

$$\int 2x\cos(x^2)dx = \int \cos u\,du = \sin u + C$$

Now all that remains is to replace u in the final answer with its value in terms of x, namely, $u = x^2$. So we get

$$\int 2x\cos(x^2)dx = \sin x^2 + C$$

In Example 5.18 we let $u = x^2 + x + 1$, so that $du = (3x^2 + 1)dx$. Then we have

$$\int (3x^2 + 1)(x^3 + x + 1)^5 dx = \int u^5\,du = \frac{u^6}{6} + C = \frac{1}{6}\left(x^3 + x + 1\right)^6 + C$$

Here are some more examples.

EXAMPLE 5.19   Use the method of substitution to evaluate $\int (x^2 - 1)^3 \cdot 2x\,dx$.

Solution   When the integrand involves a function of x raised to a power, such as $(x^2 - 1)^3$, a general rule of thumb is to substitute u for the function on the inside. So in this case we let $u = x^2 - 1$. Then $du = 2x\,dx$. So we get

$$\int \underbrace{(x^2 - 1)^3}_{u^3} \cdot \underbrace{2x\,dx}_{du} = \int u^3 du = \frac{u^4}{4} + C$$

Now we replace u by $x^2 - 1$ to get the final answer

$$\int (x^2 - 1)^3 \cdot 2x\,dx = \frac{(x^2 - 1)^4}{4} + C \qquad ■$$

EXAMPLE 5.20   Use the method of substitution to evaluate $\int x\sqrt{1-x^2}\,dx$.

Solution   First, write the integral in the form

$$\int x(1-x^2)^{1/2}dx$$

Again, the integrand involves a function raised to a power, namely, $(1-x^2)^{1/2}$. So we try $u = 1 - x^2$. Then $du = -2x\,dx$. Note that the x and dx are present in the integral. The only thing missing is the factor (-2). We can therefore get the correct differential du if we multiply by (-2) inside the integral. To compensate, we multiply by its reciprocal, $(-\frac{1}{2})$, outside the integral. Then we have

$$-\frac{1}{2}\int \underbrace{(1-x^2)^{1/2}}_{u^{1/2}}\,\underbrace{(-2x\,dx)}_{du} = -\frac{1}{2}\int u^{1/2}du$$

$$= -\frac{1}{2}\frac{u^{3/2}}{\frac{3}{2}} + C$$

$$= -\frac{1}{2}\left(\frac{2}{3}\right)u^{3/2} + C$$

$$= -\frac{1}{3}u^{3/2} + C$$

$$= -\frac{1}{3}(1-x^2)^{3/2} + C \qquad\blacksquare$$

REMARK ─────────────────────────────────

■ By Property 2 of integrals in Section 5.2, for any constant c,

$$c\int_a^b f(x)dx = \int_a^b cf(x)dx$$

It is this property that we used in Example 5.20. We can show the steps in more detail by noting that since $(-\frac{1}{2})(-2) = 1$,

$$\int (1-x^2)^{1/2}x\,dx = \int (1-x^2)^{1/2}\left(-\frac{1}{2}\right)(-2x\,dx)$$

$$= -\frac{1}{2}\int (1-x^2)^{1/2}(-2x\,dx)$$

by Property 2. Thus, multiplying by (-2) inside the integral and by its reciprocal $(-1/2)$ outside the integral leaves the integral unchanged.

EXAMPLE 5.21   Evaluate the integral

$$\int \frac{\sin\sqrt{x}}{\sqrt{x}}\,dx$$

Solution   If we could make a substitution that would result in an integral of the form $\int \sin u \, du$, we would know the antiderivative. So we try the substitution $u = \sqrt{x}$. Then

$$du = \frac{1}{2\sqrt{x}} \, dx$$

All that is missing in the original integral is the factor $\frac{1}{2}$. So we multiply by $\frac{1}{2}$ on the inside of the integral and by 2 on the outside.

$$\int \frac{\sin \sqrt{x}}{x} \, dx = 2 \int \sin \sqrt{x} \left(\frac{1}{2\sqrt{x}} \, dx \right)$$

$$= 2 \int \sin u \, du$$

$$= -2 \cos u + C$$

$$= -2 \cos \sqrt{x} + C \qquad \blacksquare$$

EXAMPLE 5.22   Evaluate the integral $\int \cos^3(2x) \sin(2x) \, dx$.

Solution   Since $\cos^3(2x)$ means $(\cos 2x)^3$, we have a function raised to a power. So, as in Examples 5.19 and 5.20, we substitute u for the function. That is, we let $u = \cos 2x$. Then $du = -2 \sin 2x \, dx$. Multiplying and dividing by -2 gives

$$\int \cos^3(2x) \sin(2x) \, dx = -\frac{1}{2} \int (\cos 2x)^3 (-2 \sin 2x \, dx)$$

$$= -\frac{1}{2} \int u^3 \, du$$

$$= -\frac{1}{2} \frac{u^4}{4} + C$$

$$= -\frac{1}{8} \cos^4(2x) + C \qquad \blacksquare$$

It is important that the substitution be chosen in such a way that everything inside the integral sign is expressed in terms of the new variable u (or whatever letter is used), including the differential. If the only thing missing is a *constant* factor, then you can multiply by that factor on the inside and by its reciprocal on the outside. We will refer to this process as "adjusting the differential."

Only constants can be factored out of an integral. So you *cannot* adjust the differential if it requires multiplying by an expression containing the variable of integration.

There is nothing special about using the letter u for a substitution, any more than there is in using x for the original variable, although both are traditional. In the following example we use different letters.

EXAMPLE 5.23   Evaluate the integral $\int \cos(3+4t)dt$.

Solution   Let $v = 3 + 4t$. Then $dv = 4\,dt$. So we adjust the differential to get

$$\int \cos(3+4t)dt = \frac{1}{4}\int \cos(3+4t)(4\,dt)$$

$$= \frac{1}{4}\int \cos v\,dv$$

$$= \frac{1}{4}\sin v + C$$

$$= \frac{1}{4}\sin(3+4t) + C \qquad\blacksquare$$

With a little practice, you will probably be able to do some substitutions mentally, without actually writing the substitution. Consider, for example, the following:

$$\int \frac{x}{\sqrt{1-x^2}}dx = -\frac{1}{2}\int (1-x^2)^{-1/2}(-2x\,dx)$$

$$= -\frac{1}{2}\frac{(1-x^2)^{1/2}}{1/2} + C$$

$$= -\sqrt{1-x^2} + C$$

Here we *mentally* let $u = 1 - x^2$ and on observing that its differential is $-2x\,dx$, we saw that we could adjust the differential by multiplying and dividing by -2. Finally, we saw that since our integral was in the form $-\frac{1}{2}\int u^{-1/2}du$, we could use the Power Rule to get the answer, with u being $1 - x^2$. After you have done several problems by writing the substitutions, you should try others mentally.

Evaluating Definite Integrals by Substitution

When evaluating an indefinite integral by the method of substitution, always express the final answer in terms of the original variable. Substituting a new variable is a means to an end only, and the new variable should not appear in the final answer. In the case of a *definite* integral, however, the answer can be obtained from the Second Fundamental Theorem using the antiderivative in terms of the new variable, *provided* the limits of integration are expressed in terms of the new variable. In the next example we illustrate this method of handling a substitution in a definite integral, as well as an alternative procedure.

EXAMPLE 5.24   Evaluate the integral

$$\int_0^2 x^2\sqrt{1+x^3}dx$$

Solution   *Method (i):* We begin, as with an indefinite integral, by letting $u = 1 + x^3$. Then $du = 3x^2 dx$. We can adjust the differential to get the proper form of du. The new thing to consider is that there are limits of integration. The lower and upper limits 0 and 2 are x values. We now find the corresponding u

values. Since $u = 1 + x^3$, we see that when $x = 0$, $u = 1$, and when $x = 2$, $u = 9$. Thus, we have

$$\int_0^2 x^2 \sqrt{1 + x^3} \, dx = \frac{1}{3} \int_0^2 (1 + x^3)^{1/2} (3x^2 \, dx)$$

$$= \frac{1}{3} \int_1^9 u^{1/2} \, du \qquad \text{Note the new limits.}$$

$$= \frac{1}{3} \cdot \frac{2}{3} u^{3/2} \Big]_1^9 = \frac{2}{9}(9^{3/2} - 1^{3/2}) = \frac{2}{9}(27 - 1) = \frac{52}{9}$$

The answer was obtained using the antiderivative in terms of u, because the limits of integration were changed to u limits. We did not return to the original x variable.

Method (ii): First find the indefinite integral, just as in Examples 5.19 through 5.23. That is, find $\int x^2 \sqrt{1 + x^3} \, dx$. Use $u = 1 + x^3$ as before, with $du = 3x^2 dx$. Adjusting the differential, we get

$$\int x^2 \sqrt{1 + x^3} \, dx = \frac{1}{3} \int (1 + x^3)^{1/2} (3x^2 \, dx) = \frac{1}{3} \int u^{1/2} \, du$$

$$= \frac{1}{3} \cdot \frac{2}{3} u^{3/2} + C = \frac{2}{9}(1 + x^3)^{3/2} + C$$

Note that here, as with all substitutions in indefinite integrals, we wrote the answer in terms of the original variable x.

Now to evaluate the given definite integral, we need an antiderivative of the integrand. But we have just found its general antiderivative. Remember that in applying the Second Fundamental Theorem, any antiderivative of the integrand will do. So we use the one with the constant $C = 0$ to simplify the calculation. (It would "drop out" anyway.) The result is

$$\int_0^2 x^2 \sqrt{1 + x^3} \, dx = \frac{2}{9}(1 + x^3)^{3/2} \Big]_0^2 = \frac{2}{9}\left[(1 + 8)^{3/2} - 1^{3/2}\right]$$

$$= \frac{2}{9}(9^{3/2} - 1) = \frac{52}{9} \qquad \blacksquare$$

Of the two methods illustrated in the preceding example, the first is usually easier. However, in "mental" substitution the second method is in effect being used.

When you use Method (i) for evaluating a definite integral in which you have made a substitution, you *must* change the limits of integration to those for the new variable. Failure to do so is a mistake students often make. For example, after making the substitution $u = 1 + x^3$ in the integral of Example 5.24, note carefully that

$$\int_0^2 x^2 \sqrt{1 + x^3} \, dx \neq \frac{1}{3} \int_0^2 u^{1/2} \, du$$

since the integrand on the right is in terms of u and the upper and lower limits on the integral are x limits.

Method (i) used in Example 5.24 is summarized in the formula

$$\int_a^b f(g(x))g'(x)dx = \int_{g(a)}^{g(b)} f(u)du = F(g(b)) - F(g(a)) \qquad (5.20)$$

where $u = g(x)$ and $F' = f$. We assume that both $f(g(x))$ and $g'(x)$ are continuous on $[a, b]$.

In the next example we show how to find the area inside an ellipse using a substitution that changes the problem to that of the area inside a circle, multiplied by an appropriate constant.

EXAMPLE 5.25   Find the area enclosed by the ellipse

$$\frac{x^2}{a^2} + \frac{y^2}{b^2} = 1$$

Solution   The graph of a typical ellipse is shown in Figure 5.36. By symmetry we can find the area above the x-axis and double it. Note that the entire ellipse is not the graph of a function, but the upper half (as well as the lower half) is the graph of a function. The equation of the upper semi-ellipse is found by solving for y and using the positive square root, giving

$$y = b\sqrt{1 - \frac{x^2}{a^2}}$$

The area under this curve from $-a$ to a is

$$b \int_{-a}^a \sqrt{1 - \frac{x^2}{a^2}}\, dx$$

We make the substitution $u = x/a$. Then $du = (1/a)dx$, so that $dx = a\,du$. When $x = -a$, $u = -1$, and when $x = a$, $u = 1$. So the integral becomes

$$b \int_{-1}^1 \sqrt{1 - u^2} \cdot a\,du = ab \int_{-1}^1 \sqrt{1 - u^2}\, du$$

Now the integral $\int_{-1}^1 \sqrt{1 - u^2}\, du$ gives the area under the semicircle $y = \sqrt{1 - u^2}$ of radius 1. (See Figure 5.37.) But we know this area is $\pi/2$ (half of πr^2, with $r = 1$). Note that the integral is of the form expressed by Equation 5.10 in Section 5.2, with $a = 1$. Thus, the area under the semi-ellipse is $\pi ab/2$, and the area enclosed by the entire ellipse is πab.   ∎

FIGURE 5.36

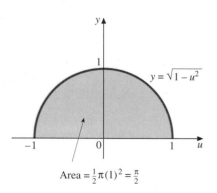

Area $= \frac{1}{2}\pi(1)^2 = \frac{\pi}{2}$

FIGURE 5.37

Integrals of Even and Odd Functions over Symmetric Intervals

We can use the method of substitution to prove an interesting and useful result concerning integrals of even and odd functions over symmetric intervals. Recall that f is an even function provided $f(-x) = f(x)$, and f is odd provided $f(-x) = -f(x)$. Even functions are symmetric about the y-axis, and odd functions are symmetric about the origin. We want to consider integrals of these two types of functions on an interval of the type $[-a, a]$. Figure 5.38 illustrates a typical situation.

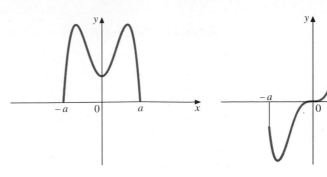

(a) Even function on symmetric
    interval $[-a, a]$

$$\int_{-a}^{a} f(x)\,dx = 2\int_{0}^{a} f(x)\,dx$$

(b) Odd function on symmetric
    interval $[-a, a]$

$$\int_{-a}^{a} f(x)\,dx = 0$$

FIGURE 5.38

It is evident geometrically, using the area interpretation of the integral, that when f is even,

$$\int_{-a}^{a} f(x)\,dx = 2\int_{0}^{a} f(x)\,dx$$

and when f is odd,

$$\int_{-a}^{a} f(x)\,dx = 0$$

The fact that the integral is 0 when f is odd follows because the area below the x-axis equals the area above, making the "net area" 0.

We can prove these results without reference to geometry as follows. By Property 3 in Section 5.2 we can write

$$\int_{-a}^{a} f(x)\,dx = \int_{-a}^{0} f(x)\,dx + \int_{0}^{a} f(x)\,dx \tag{5.21}$$

In the first integral on the right, make the substitution $u = -x$. Then $du = -dx$, or equivalently, $dx = -du$. Also, when $x = -a$, $u = a$, and when $x = 0$, $u = 0$. So we have

$$\int_{-a}^{0} f(x)\,dx = \int_{a}^{0} f(-u)(-du) = -\int_{a}^{0} f(-u)\,du$$

$$= \int_{0}^{a} f(-u)\,du$$

(Remember that interchanging upper and lower limits changes the sign of the integral.) If f is even, $f(-u) = f(u)$, so the last integral becomes

$$\int_{0}^{a} f(u)\,du$$

whereas, if f is odd, $f(-u) = -f(u)$, and the integral becomes

$$-\int_{0}^{a} f(u)\,du$$

A definite integral is unchanged if some other letter is used for the dummy variable. In particular, we can replace u by x (ignoring at this stage our earlier

substitution). Thus, for f even,

$$\int_{-a}^{0} f(x)\,dx = \int_{0}^{a} f(x)\,dx$$

and for f odd,

$$\int_{-a}^{0} f(x)\,dx = -\int_{0}^{a} f(x)\,dx$$

Returning now to Equation 5.21, we have the result that we summarize below.

If f is integrable on the symmetric interval $[-a, a]$, then

$$\int_{-a}^{a} f(x)\,dx = 2\int_{0}^{a} f(x)\,dx \qquad \text{if } f \text{ is even} \qquad (5.22)$$

$$\int_{-a}^{a} f(x)\,dx = 0 \qquad \text{if } f \text{ is odd} \qquad (5.23)$$

 When using a CAS it is critical to remember these results. Most CAS occasionally fail to evaluate integrals correctly, especially when the integrand involves a square root.

EXAMPLE 5.26   Evaluate the following integrals.

(a) $\int_{-2}^{2} (x^2 + 1)\,dx$

(b) $\int_{-1}^{1} x^2 \sin x\,dx$

Solution

(a) Since $f(x) = x^2 + 1$ is even, we can write

$$\int_{-2}^{2} (x^2 + 1)\,dx = 2\int_{0}^{2} (x^2 + 1)\,dx = 2\left[\frac{x^3}{3} + x\right]_0^2 = 2\left(\frac{8}{3} + 2\right) = \frac{28}{3}$$

One advantage of evaluating the integral from -2 to 2 by doubling its value from 0 to 2 is that it is easier to substitute 0 in the antiderivative than -2.

(b) Since $\sin x$ is odd, and x^2 is even, the product $x^2 \sin x$ is odd. So

$$\int_{-1}^{1} x^2 \sin x\,dx = 0 \qquad\qquad \blacksquare$$

 If the interval of integration is not symmetric about 0, then integrals of even and odd functions cannot be simplified as we have shown in Equations 5.22 and 5.23. For example,

$$\int_{-2}^{3} x^2\,dx \neq 2\int_{0}^{3} x^2\,dx$$

even though the integrand is even. Similarly,

$$\int_{-2}^{3} x^3\,dx \neq 0$$

even though the integrand is odd. Each of these problems would be solved by a direct application of the Second Fundamental Theorem. To apply Equations 5.22 and 5.23, *the interval of integration must be symmetric and of the form* −*a to a*.

Differential Equations Revisited

In Section 4.7 we saw how antiderivatives arise in the solutions of certain differential equations. Since an indefinite integral is a general antiderivative, we can use indefinite integrals in solving differential equations. In the next two examples we illustrate how this is done.

EXAMPLE 5.27   The slope of the tangent line to the graph of a certain function at an arbitrary point (x, y) is $2x^{-3}$, and the graph passes through the point $(1, 3)$. Find the function.

Solution   Let $y = f(x)$ denote the unknown function. Then we know that $f'(x) = 2x^{-3}$ and that $f(1) = 3$. So we have an initial-value problem to solve. The general solution of the differential equation is

$$y = \int f'(x)dx = \int 2x^{-3}dx = -x^{-2} + C = -\frac{1}{x^2} + C$$

To find C, we use the initial value $f(1) = 3$. That is, when $x = 1$, $y = 3$:

$$3 = -1 + C$$

$$C = 4$$

The particular solution we want is therefore

$$y = -\frac{1}{x^2} + 4$$

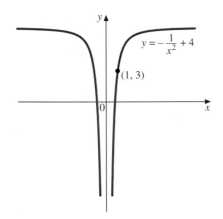

We show its graph in Figure 5.39.   ∎

FIGURE 5.39

EXAMPLE 5.28   A projectile is fired vertically upward from the ground at an initial velocity of 196 m/s. Ignoring air resistance, find formulas for (a) the velocity, and (b) the distance above the ground at any time t before it strikes the ground. How high does the projectile rise?

Solution   The acceleration of the projectile at any time t is the acceleration of gravity, which is approximately 9.8 m/s^2 acting downward. Thus, since acceleration is the derivative of the velocity, $v(t)$, we have

$$v'(t) = -9.8$$

Also, we have the initial condition $v(0) = 196$. So

$$v(t) = \int (-9.8)dt = -9.8t + C_1$$

Using the initial condition, we get

$$196 = (-9.8)(0) + C_1$$

$$C_1 = 196$$

The velocity is therefore

$$v(t) = -9.8t + 196$$

which is the answer to part (a).

Let $s(t)$ denote the distance above the ground at time t. Then $s'(t) = v(t)$, and we have the differential equation

$$s'(t) = -9.8t + 196$$

together with the initial condition $s(0) = 0$. Solving, we get

$$s(t) = \int (-9.8t + 196)dt = -4.9t^2 + 196t + C_2$$

The initial condition gives $C_2 = 0$. Thus,

$$s(t) = -4.9t^2 + 196t$$

which is the answer for part (b).

At the instant the projectile reaches its maximum height, the velocity is 0. So we set $v(t) = 0$ and solve for t to see when the maximum height is attained:

$$-9.8t + 196 = 0$$

$$t = 20$$

Now we substitute $t = 20$ into the distance formula $s(t)$:

$$s(20) = (-4.9)(400) + (196)(20) = 1,960$$

The projectile therefore rises to a maximum height of 1,960 m.   ■

Exercise Set 5.4

In Exercises 1–28, evaluate each indefinite integral by making an appropriate substitution.

1. $\int x(1 + x^2)^3 dx$

2. $\int x\sqrt{x^2 - 4}\, dx$

3. $\int x^2(x^3 - 1)^5 dx$

4. $\int \frac{x}{\sqrt{9 - x^2}}\, dx$

5. $\int x\sqrt{3x^2 + 4}\, dx$

6. $\int \frac{(\sqrt{x} + 1)^5}{\sqrt{x}}\, dx$

7. $\int \sqrt{1 + 3x}\, dx$

8. $\int (4x - 3)^{2/3}\, dx$

9. $\int (x - 1)\sqrt{x^2 - 2x + 3}\, dx$

10. $\int \frac{2x + 1}{\sqrt{x^2 + x - 4}}\, dx$

11. $\int \sin 2x\, dx$

12. $\int \cos 3x\, dx$

13. $\int \sin kx\, dx$, k a nonzero real constant

14. $\int \cos kx\, dx$, k a nonzero real constant

15. $\int 2 \sin x \cos x\, dx$

16. $\int \sin x \cos^2 x\, dx$

17. $\int x \sec^2 x^2 dx$

18. $\int \sec 2x \tan 2x\, dx$

19. $\int \sqrt{\tan x}\, \sec^2 x\, dx$

20. $\int \sqrt{1 + \sin x}\, \cos x\, dx$

21. $\int \frac{\sin x}{(1 - \cos x)^2}\, dx$

22. $\int \frac{x + \cos 2x}{\sqrt{x^2 + \sin 2x}}\, dx$

23. $\displaystyle\int \frac{\csc^2 \sqrt{x}}{\sqrt{x}}\, dx$

24. $\displaystyle\int \frac{1}{x^2} \cos \frac{1}{x}\, dx$

25. $\displaystyle\int \left(\frac{x-1}{x}\right)^5 \cdot \frac{1}{x^2}\, dx$

26. $\displaystyle\int (x^2 - x)(2x^3 - 3x^2 + 1)^{3/2}\, dx$

27. $\displaystyle\int \sqrt{x}(x^{3/2} - 1)^{2/3}\, dx$

28. $\displaystyle\int \frac{(x^{2/3} + a^{2/3})^{3/2}}{\sqrt[3]{x}}\, dx$, a a nonzero real constant

In Exercises 29–44, evaluate each definite integral by making an appropriate substitution.

29. $\displaystyle\int_0^2 (1 + 4x)^{3/2}\, dx$

30. $\displaystyle\int_0^4 x\sqrt{x^2 + 9}\, dx$

31. $\displaystyle\int_{-1}^2 x(x^2 - 2)^4\, dx$

32. $\displaystyle\int_3^4 \frac{x}{\sqrt{25 - x^2}}\, dx$

33. $\displaystyle\int_4^9 \frac{(\sqrt{x} - 4)^6}{\sqrt{x}}\, dx$

34. $\displaystyle\int_0^2 x(3x^2 + 4)^{3/2}\, dx$

35. $\displaystyle\int_1^5 (2x + 1)(x^2 + x - 3)^{1/3}\, dx$

36. $\displaystyle\int_1^2 \left(\frac{5x^2 - 8}{3x^2}\right)^{2/3} \frac{dx}{x^3}$

37. $\displaystyle\int_0^{\pi/6} \cos 3x\, dx$

38. $\displaystyle\int_0^{\pi/2} \sin \frac{x}{2}\, dx$

39. $\displaystyle\int_{\pi/6}^{2\pi/3} \sin^3 x \cos x\, dx$

40. $\displaystyle\int_0^{\pi/2} \sqrt{\cos x} \sin x\, dx$

41. $\displaystyle\int_{-\pi/6}^{\pi/8} \tan^3 2x \sec^2 2x\, dx$

42. $\displaystyle\int_{-\pi/3}^{\pi/2} \frac{\sin x}{(1 + \cos x)^2}\, dx$

43. $\displaystyle\int_0^{\pi/3} \sec^3 x \tan x\, dx$

44. $\displaystyle\int_{\pi^2/4}^{\pi^2} \frac{\sin \sqrt{x}}{\sqrt{x}}\, dx$

45. Find the area under the graph of
$$y = \frac{1}{(2x - 3)^3}$$
over the interval [2, 3].

46. Find $\int 9x^2(x^3 + 1)\, dx$ by the following two methods:
(a) Multiply the factors in the integrand and integrate term by term.
(b) Make the substitution $u = x^3 + 1$.

Explain the difference in the answers in parts (a) and (b). Are they both correct?

47. Find $\int \sec^2 x \tan x\, dx$ in the following two ways:
(a) Make the substitution $u = \tan x$.
(b) Make the substitution $u = \sec x$.
Explain the difference in the answers in parts (a) and (b). Are they both correct?

48. Find $f(x)$ if $\int f(x)\, dx = 6x^4 - 2x^2 + x + C$.

49. Find a function g such that $g'(x) = \sqrt{4x + 2}$ and $g(1/2) = 1$.

50. The marginal profit for a company is given by
$$P'(x) = \frac{x}{(100 + x^2)^{2/3}}$$
where x is the number of units sold and $P'(x)$ is in hundreds of dollars. If the company breaks even by making 30 units, find the profit function $P(x)$.

51. The ABC Company has determined that its marginal cost, in hundreds of dollars, for producing x dishwashers per month is given by
$$C'(x) = \frac{x}{\sqrt{10,000 + x^2}}$$
If the fixed cost of production per month is \$50,000, find the cost of producing 500 dishwashers per month.

52. The cost, in hundreds of dollars, of producing x units of a certain commodity is increasing at the rate of $x^2/(1 + x^2)$. Use a CAS, a spreadsheet, or a calculator to determine an upper and lower estimate, accurate to one decimal place, of the cost of increasing production from 10 to 50 units.

In Exercises 53–58, find the general solution of the differential equation.

53. $\displaystyle\frac{dy}{dx} = \frac{x}{\sqrt{x^2 + 1}}$

54. $\displaystyle\frac{dy}{dt} = (2t - 1)(t^2 - t + 4)^{2/3}$

55. $y' - \cos 2x = x$

56. $\sqrt{x}\, y' = (\sqrt{x} - 2)^4$

57. $\displaystyle f'(t) = \frac{(1 + \tan t)^{3/2}}{\cos^2 t}$

58. $g'(u) = \sqrt{u^4 - 9u^2}$ $(u > 0)$

In Exercises 59–62, solve the initial-value problem.

59. $\displaystyle\frac{dy}{dx} = \frac{x^2}{\sqrt{x^3 + 1}}$; when $x = 2$, $y = 5$

60. $dx/dt = \sin^3 t \cos t$; when $t = \pi/2$, $x = 1$

61. $y' = \sin 2x\sqrt{1 + \sin^2 x}$; when $x = 0$, $y = 2$

62. $f'(t) = t \cos t^2 \sin t^2$; $f(0) = -1$

63. The graph of a certain function passes through the point $(0, -1)$. Its slope at an arbitrary point (x, y) is $x/(x^2 + 1)^2$. Find the function.

64. A certain function has the following properties: $f''(x) = -2(3x + 2)^{-4/3}$, $f'(2) = 0$, $f(2) = 5$. Find $f(x)$.

65. (a) When the brakes of a car, initially traveling at 65 mph, are applied, they produce a constant deceleration of 11 ft/s². What is the shortest distance in which the car can be braked to a complete stop?
  (b) Redo part (a) if the car is initially traveling at 100 km/h and the deceleration is 3 m/s².

66. If a car starts from a dead stop, what constant acceleration will allow the car to travel 300 m in 12.8 seconds?

67. From a balloon 1,176 m above the ground an object is thrown downward with a velocity of 39.2 m/s. Assuming no wind and no air resistance, find (a) the velocity, and (b) the distance of the object from the balloon at any time t before it strikes the ground. How long does it take for the object to reach the ground? (Take $g = 9.8$ m/s².)

In Exercises 68–73, evaluate the indefinite integral.

68. $\displaystyle\int x\sqrt{(x^4 + 12x^2 + 36)^3}\, dx$

69. $\displaystyle\int \frac{\sec^2 x}{\sec^2 x + 2\tan x}\, dx$

70. $\displaystyle\int \cos\frac{x}{2} \sin x\, dx$

71. $\displaystyle\int \frac{dx}{\sqrt{x - x\sqrt{x}}}$ **72.** $\displaystyle\int (\sin 2x - \cos 2x)^2\, dx$

73. $\displaystyle\int \cos 2x \cos x\, dx$

Verify the equations in Exercises 74 and 75.

74. $\displaystyle\int x \sin 2x\, dx = -\frac{x \cos 2x}{2} + \frac{\sin 2x}{4} + C$

75. $\displaystyle\int \frac{1}{x^2} \cos\frac{1}{\sqrt{x}}\, dx = -2\left[\frac{1}{\sqrt{x}} \sin\frac{1}{\sqrt{x}} + \cos\frac{1}{\sqrt{x}}\right] + C$

76. Let $f(x) = \begin{cases} 1 & \text{if } x \geq 0 \\ -1 & \text{if } x < 0 \end{cases}$. Show that $\int_{-1}^{1} f(x)\,dx$ exists, but $\int f(x)\,dx$ does not exist on $[-1, 1]$. (*Hint:* Consider the result of Exercise 34 in Exercise Set 4.7.)

77. Prove Properties 1–4 of the indefinite integral.

5.5   APPROXIMATING DEFINITE INTEGRALS

We have already noted that some continuous elementary functions have no antiderivative in the form of an elementary function. If we want to evaluate a definite integral of such a function, the Second Fundamental Theorem is of no help. The best we can do is to approximate its value. Here are three integrals of this type:

$$\int_{1}^{2} \frac{\sin x}{x}\, dx \qquad \int_{0}^{1} \frac{dx}{\sqrt{1 + x^3}} \qquad \int_{0}^{4} \cos \sqrt{x}\, dx$$

In each of these cases the integrand is continuous on the interval in question, so each integral exists, but we cannot find its exact value. These are but three examples; there are many others for which approximation is also necessary.

Functions arising in applications frequently are known only graphically or by tables of values, rather than by formulas. If we need to find the integral of such a function, our only recourse is to approximate its value by some numerical procedure. Even when the exact value of an integral *can* be found, the work involved in obtaining it may not be justified, especially if accuracy to one or two decimal places is all that is needed. In such a case it may be more efficient, and equally useful, to approximate the integral numerically rather than to obtain the exact answer.

Left- and Right-Endpoint Rules

We have already seen that we can approximate a definite integral by a Riemann sum. Since, by definition, the integral is the limit of Riemann sums as the widths of the subintervals shrink to zero, any particular Riemann sum can be used to approximate the integral. That is,

$$\int_a^b f(x)dx \approx \sum_{k=1}^n f(c_k)\Delta x_k$$

where the Riemann sum on the right has the meaning we assigned in Section 5.2. Recall that the points $c_1, c_2, \ldots, c_n$ can be arbitrarily chosen in the n consecutive subintervals. For convenience, throughout this section we will consider equally spaced partition points only, so that all subintervals have the same width, $\Delta x = (b-a)/n$.

If for each integer k from 1 to n we take c_k to be the left endpoint of the kth subinterval $[x_{k-1}, x_k]$, the approximation to the integral is called the **Left-Endpoint Rule.**

$$\int_a^b f(x)dx \approx \Delta x\big[f(x_0) + f(x_1) + \cdots + f(x_{n-1})\big] \qquad (5.24)$$

Similarly, if for each integer k from 1 to n we take c_k to be the right endpoint of the kth subinterval, the approximation is called the **Right-Endpoint Rule.**

$$\int_a^b f(x)dx \approx \Delta x\big[f(x_1) + f(x_2) + \cdots + f(x_n)\big] \qquad (5.25)$$

Midpoint Rule

Instead of using the right or left endpoints of the subintervals, we might choose to use the midpoints. That is, for each k from 1 to n, we can take c_k to be $\frac{1}{2}(x_{k-1} + x_k)$. Let us denote this midpoint by $\bar{x}_k$. Approximating the integral by a Riemann sum with $c_k = \bar{x}_k$ results in what is frequently called the **Midpoint Rule** and is stated as follows.

The Midpoint Rule

$$\int_a^b f(x)dx \approx \Delta x\big[f(\bar{x}_1) + f(\bar{x}_2) + \cdots + f(\bar{x}_n)\big]$$

where $\Delta x = \dfrac{b-a}{n}$ and $\bar{x}_k = \dfrac{x_{k-1} + x_k}{2}$ for $k = 1, 2, \ldots, n$.

In Figure 5.40 we show a nonnegative continuous function f on an interval $[a, b]$ with its integral (which equals the area under the graph) approximated by left-endpoint, right-endpoint, and midpoint Riemann sums. Although it may not be obvious, the Midpoint Rule generally is more accurate than the right- and left-endpoint approximations.

Left-Endpoint Approximation   Right-Endpoint Approximation   Midpoint Approximation

FIGURE 5.40

Trapezoidal Rule

Yet another method, which might have occurred to you already, is to average the approximations given by the Left-Endpoint and Right-Endpoint Rules (5.24 and 5.25). Doing so, we get

$$\int_a^b f(x)dx \approx \frac{1}{2}\left[\sum_{k=1}^{n} f(x_{k-1})\Delta x + \sum_{k=1}^{n} f(x_k)\Delta x\right]$$

$$= \frac{\Delta x}{2}\sum_{k=1}^{n}\left[f(x_{k-1}) + f(x_k)\right]$$

$$= \frac{\Delta x}{2}\left[\left(f(x_0) + f(x_1)\right) + \left(f(x_1) + f(x_2)\right) + \cdots\right.$$

$$\left. + \left(f(x_{k-1}) + f(x_k)\right)\right]$$

$$= \frac{\Delta x}{2}\left[f(x_0) + 2f(x_1) + 2f(x_2) + \cdots + 2f(x_{k-1}) + f(x_k)\right]$$

This result is known as the **Trapezoidal Rule**, for reasons we explain below. First we restate the rule for emphasis.

The Trapezoidal Rule

$$\int_a^b f(x)dx \approx \frac{\Delta x}{2}\left[f(x_0) + 2f(x_1) + 2f(x_2) + \cdots + 2f(x_{n-1}) + f(x_n)\right]$$

where $\Delta x = \dfrac{b-a}{n}$.

REMARK

■ The Left-Endpoint, Right-Endpoint, and Midpoint Rules are all special cases of Riemann sums, but the Trapezoidal Rule is not.

To see why we use the name "Trapezoidal" Rule, consider a nonnegative continuous function f over an interval $[a, b]$, as shown in Figure 5.41. Partition the interval $[a, b]$ in the usual way with the equally spaced points $x_0 = a$, $x_1, x_2, \ldots, x_n = b$. Instead of approximating the integral $\int_a^b f(x)dx$ (that is,

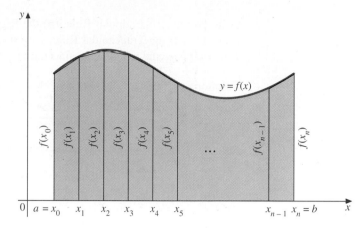

FIGURE 5.41

the area under the graph) with rectangular regions, as we do in Riemann sums, we do so with areas of trapezoids (quadrilaterals with two parallel sides), as shown. Now the area inside a trapezoid is the average of the lengths of the two parallel sides times the distance between them. For example, the leftmost trapezoid in Figure 5.41 has area

$$\frac{1}{2}\big[f(x_0) + f(x_1)\big]\Delta x$$

When we add the areas of all the trapezoids we get the approximate value of the integral:

$$\int_a^b f(x)\,dx \approx \frac{1}{2}\big[f(x_0) + f(x_1)\big]\Delta x + \frac{1}{2}\big[f(x_1) + f(x_2)\big]\Delta x + \cdots$$

$$+ \frac{1}{2}\big[f(x_{n-1}) + f(x_n)\big]\Delta x$$

$$= \frac{\Delta x}{2}\big[f(x_0) + 2f(x_1) + 2f(x_2) + \cdots + 2f(x_{n-1}) + f(x_n)\big]$$

which is exactly the value given by the Trapezoidal Rule. We have shown, then, that averaging the approximations given by the Left-Endpoint Rule and the Right-Endpoint Rule gives the same result as finding the sum of the areas of the trapezoids we have described.

REMARK ───────────────────────────────

■ Although we used a nonnegative continuous function to explain the origin of the name "Trapezoidal" Rule, the rule is applicable as a means of approximating the definite integral of any integrable function. (See Figure 5.42.)

───

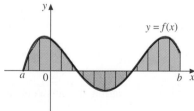

FIGURE 5.42
The Trapezoidal Rule approximates "net area" when f takes on both positive and negative values.

In the next example we use a function with a known antiderivative so that we can compare our approximations with the exact value.

EXAMPLE 5.29   Approximate the integral

$$\int_1^3 \cos x \, dx$$

using (a) the Left-Endpoint Rule, (b) the Right-Endpoint Rule, (c) the Midpoint Rule, and (d) the Trapezoidal Rule, with $n = 10$. Repeat, with $n = 20$, 50, and 100. Compare the results, and find the error in each case.

Solution   With $n = 10$, we find that

$$\Delta x = \frac{3 - 1}{10} = \frac{2}{10} = 0.2$$

So the points of division are (see Figure 5.43)

$$1.0, \ 1.2, \ 1.4, \ 1.6, \ 1.8, \ 2.0, \ 2.2, \ 2.4, \ 2.6, \ 2.8, \ 3.0$$

(a) *Left-Endpoint Rule.* By Formula 5.24,

$$\int_1^3 \cos x \, dx \approx 0.2\big[\cos(1.0) + \cos(1.2) + \cos(1.4) + \cdots + \cos(2.8)\big]$$

$$\approx -0.5450$$

(b) *Right-Endpoint Rule.* By Formula 5.25,

$$\int_1^3 \cos x \, dx \approx 0.2\big[\cos(1.2) + \cos(1.4) + \cos(1.6) + \cdots + \cos(3.0)\big]$$

$$\approx -0.8510$$

(c) *Midpoint Rule.* The midpoints of the intervals are

$$m_1 = 1.1, \ m_2 = 1.3, \ m_3 = 1.5, \ m_4 = 1.7, \ m_5 = 1.9,$$

$$m_6 = 2.1, \ m_7 = 2.3, \ m_8 = 2.5, \ m_9 = 2.7, \ m_{10} = 2.9$$

So the Midpoint Rule gives

$$\int_1^3 \cos x \, dx \approx 0.2\big[\cos(1.1) + \cos(1.3) + \cos(1.5) + \cdots + \cos(2.9)\big]$$

$$\approx -0.7015$$

(d) *Trapezoidal Rule*

$$\int_1^3 \cos x \, dx \approx \frac{0.2}{2}\big[\cos(1.0) + 2\cos(1.2) + 2\cos(1.4) + \cdots + 2\cos(2.8)$$

$$+ \cos(3.0)\big]$$

$$\approx -0.6980$$

Note that we could have obtained this result by averaging the answers in parts (a) and (b).

We use the Second Fundamental Theorem to get the exact value:

$$\int_1^3 \cos x \, dx = \sin 3 - \sin 1$$

Although $\sin 3 - \sin 1$ is the exact value, it is not in a useful form to compare with our approximations. We can estimate the value, however, using a calculator. Of course, the calculator also gives only an approximate value, but it is accurate to more decimal places than any of our approximations and so gives

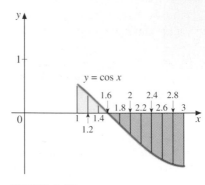

FIGURE 5.43

a suitable answer for comparison purposes. We find that, to six decimal places of accuracy,

$$\int_1^3 \cos x \, dx \approx -0.700351$$

To get the error in each of our approximations, we subtract this six-place true answer from each of the four approximations we found in parts (a)–(d). We show the results in the table in the margin, where we have used the abbreviations LEFT, RIGHT, MID, and TRAP for the approximations in parts (a), (b), (c), and (d), respectively.

Now we repeat the process with $n = 20$, 50, and 100. A computer algebra system, spreadsheet, or programmable calculator is almost a necessity for working with these larger values of n. We summarize all the results in the following two tables.

Method	Error, $n = 10$
LEFT	0.1554
RIGHT	−0.1506
MID	−0.00115
TRAP	0.00235

Method	Approximations			
	$n = 10$	$n = 20$	$n = 50$	$n = 100$
LEFT	−0.5450	−0.6232	−0.6697	−0.6850
RIGHT	−0.8510	−0.7763	−0.7309	−0.7156
MID	−0.7015	−0.7006	−0.700397	−0.700362
TRAP	−0.6980	−0.6998	−0.700257	−0.700327

Method	Errors			
	$n = 10$	$n = 20$	$n = 50$	$n = 100$
LEFT	0.1554	0.0772	0.03065	0.15535
RIGHT	−0.1506	−0.0760	−0.03055	−0.15249
MID	−0.00115	−0.000249	−0.000046	−0.000011
TRAP	0.00235	0.000551	0.000094	0.000024

Several conclusions about the methods are immediately evident for this integral:

1. LEFT and RIGHT are much less accurate than MID or TRAP.
2. The magnitudes of the errors for LEFT and RIGHT are about the same, but the errors are opposite in sign.
3. The errors for MID and TRAP are opposite in sign.
4. The magnitude of the error for MID is about half that for TRAP. ∎

The Midpoint Rule uses areas of rectangles to approximate the area under a curve. These rectangles, however, can be replaced by certain trapezoids, as we indicate in Figure 5.44. There we show a typical subinterval and the rectangle used in the Midpoint Rule to approximate the area under the curve. By rotating the top edge of the rectangle about the point $(\bar{x}_k, f(\bar{x}_k))$ until it coincides with the tangent line to the curve at that point, we form a trapezoid whose area is the same as the area of the rectangle, since the area added by the rotation equals that subtracted.

FIGURE 5.44

Error in MID

Error in TRAP

x_{k-1} x_k

FIGURE 5.45

Now consider Figure 5.45. There again we show a typical subinterval and two trapezoids, the one described above, equivalent to the rectangle in the Midpoint Rule, and the one used in the Trapezoidal Rule. You can see now that the error in the Midpoint Rule is less than that in the Trapezoidal Rule, appearing from the figure to be about half as much. In the figure we show the curve as concave down, and the Midpoint Rule overestimates, whereas the Trapezoidal Rule underestimates the actual area. If the graph were concave up, the opposite would be true.

Of the four approximation methods we have considered so far, the Midpoint Rule is clearly the most accurate in most cases. (There are certain cases that we will discuss later when two or more of the methods give the exact answer.) So why do we bother with the others? One reason is that when the function is known only by means of a table, we might not know the function values at the midpoints of the subintervals. In this case we would choose the Trapezoidal Rule.

The methods we have considered so far are called *linear approximation methods*, because in each case the actual curve is replaced by a straight line on each of the subintervals. The next method is a *quadratic approximation method*. Here we replace segments of the curve by second-degree polynomials (parabolas).

Simpson's Rule

The quadratic approximation method we consider here is called **Simpson's Rule** (after the eighteenth-century English mathematician Thomas Simpson). In it, we use a series of parabolic arcs (rather than the straight line segments of our other methods) to approximate the curve. As with the other methods, we divide the interval $[a, b]$ into n subintervals with common width $\Delta x = (b - a)/n$. For Simpson's Rule, we require that n be an *even* number, since we are considering subintervals by pairs.

Let $y_k = f(x_k)$ for $k = 0, 1, 2, \ldots, n$. We are going to fit a second-degree polynomial (whose graph is a parabola) to the three points (x_0, y_0), (x_1, y_1), and (x_2, y_2). Then we will fit another second-degree polynomial to the next group of three, (x_2, y_2), (x_3, y_3), and (x_4, y_4), and so on. In Figure 5.46 we show the parabolic arcs for a nonnegative function f, where $P_0, P_1, \ldots, P_n$ designate the points $(x_0, y_0), (x_1, y_1), \ldots, (x_n, y_n)$.

FIGURE 5.46

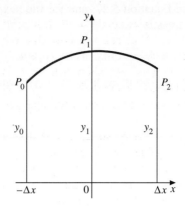

FIGURE 5.47

Consider first the problem of fitting a second-degree polynomial to the points P_0, P_1, and P_2. What we really want is the integral, which is equal to the area under the graph, and this area is unchanged if we translate the origin of our coordinate system to coincide with the point $(x_1, 0)$, as in Figure 5.47. With respect to the new axes, the coordinates of P_0, P_1, and P_2 are $(-\Delta x, y_0)$, $(0, y_1)$, and $(\Delta x, y_2)$. Let

$$y = ax^2 + bx + c$$

be the equation of the parabola in the new coordinate system that passes through P_0, P_1, and P_2. The coordinates of P_0, P_1, and P_2 each satisfy this equation, so we have

$$\begin{cases} y_0 = a(-\Delta x)^2 + b(-\Delta x) + c = a(\Delta x)^2 - b(\Delta x) + c \\ y_1 = a(0)^2 + b(0) + c = c \\ y_2 = a(\Delta x)^2 + b(\Delta x) + c \end{cases} \tag{5.26}$$

Since it is the area we really want, we calculate

$$\int_{-\Delta x}^{\Delta x} (ax^2 + bx + c)\,dx$$

Observe that ax^2 is even, bx is odd, and c is even. Since the integration is over a symmetric interval, we know by Equations 5.22 and 5.23 that the integral can be written as

$$\int_{-\Delta x}^{\Delta x} ax^2\,dx + \int_{-\Delta x}^{\Delta x} bx\,dx + \int_{-\Delta x}^{\Delta x} c\,dx = 2\int_0^{\Delta x} ax^2\,dx + 0 + 2\int_0^{\Delta x} c\,dx$$

$$= 2\frac{ax^3}{3} + 2cx\Big]_0^{\Delta x}$$

$$= 2a\frac{(\Delta x)^3}{3} + 2c\Delta x$$

$$= \frac{\Delta x}{3}\left[2a(\Delta x)^2 + 6c\right]$$

From Equations 5.26 we see that $c = y_1$, and by adding the first and third equations, we get

$$y_0 + y_2 = 2a(\Delta x)^2 + 2c$$

or

$$2a(\Delta x)^2 = y_0 + y_2 - 2y_1$$

Thus,

$$\int_{-\Delta x}^{\Delta x} (ax^2 + bx + c)\,dx = \frac{\Delta x}{3}[(y_0 + y_2 - 2y_1) + 6y_1]$$

$$= \frac{\Delta x}{3}[(y_0 + 4y_1 + y_2)] \tag{5.27}$$

This result is the area under the parabolic arc from P_0 to P_2, passing through P_1. Notice that it depends only on the width Δx and the ordinates of the three points.

Exactly the same reasoning could be used to get the area under the parabolic arc that passes through the next group of three points, P_2, P_3, and P_4. But there is no need to go through the derivation again because we already know the result. We simply change the subscripts in Equation 5.27, and for the next group of three, and so on. The sum of these areas is

$$\frac{\Delta x}{3}(y_0 + 4y_1 + y_2) + \frac{\Delta x}{3}(y_2 + 4y_3 + y_4) + \cdots + \frac{\Delta x}{3}(y_{n-2} + 4y_{n-1} + y_n)$$

When we factor out $\Delta x/3$ and combine terms, replacing each y_k by $f(x_k)$, we have the result known as Simpson's Rule.

Simpson's Rule

$$\int_a^b f(x)\,dx \approx \frac{\Delta x}{3}\big[f(x_0) + 4f(x_1) + 2f(x_2) + 4f(x_3) + \cdots$$
$$+ 2f(x_{n-2}) + 4f(x_{n-1}) + f(x_n)\big]$$

where $\Delta x = (b-a)/n$ and n is even.

REMARK ────────────────────────────────

■ Note the pattern of the coefficients here:

$$1, 4, 2, 4, 2, \ldots, 2, 4, 1$$

Contrast this pattern with the one for the Trapezoidal Rule:

$$1, 2, 2, 2, \ldots, 2, 1$$

Also, note that for Simpson's Rule the initial factor is $\Delta x/3$, whereas for the Trapezoidal Rule it is $\Delta x/2$.

───

Although our derivation of Simpson's Rule was based on area under a non-negative function, it is valid for any integrable function, as is also true for the other methods we have studied.

REMARK ────────────────────────────────

■ Simpson's Rule can be shown to be a weighted average of the Midpoint and Trapezoidal Rules. In fact, if we denote by M_n the value given by the Midpoint Rule with n subintervals, T_n the value given by the Trapezoidal Rule, and S_{2n} the value given by Simpson's Rule for $2n$ subintervals, we can show (see Exercise 22 in this section) that

$$S_{2n} = \frac{2}{3}M_n + \frac{1}{3}T_n$$

───

For comparison with the other methods, in the next example we apply Simpson's Rule to the same integral we used in Example 5.29.

EXAMPLE 5.30   Use Simpson's Rule with $n = 10$ to approximate the integral

$$\int_1^3 \cos x \, dx$$

Solution   Since $\Delta x = (3-1)/10 = 0.2$, we have

$$\int_1^3 \cos x \, dx \approx \frac{0.2}{3}\Big[\cos(1.0) + 4\cos(1.2) + 2\cos(1.4) + 4\cos(1.6)$$

$$+ \cdots + 2\cos(2.6) + 4\cos(2.8) + \cos(3.0)\Big]$$

$$\approx -0.700357$$

In Example 5.29 we found that the value of the integral, to six decimal places, is -0.700351. So with the relatively small value of $n = 10$, we have an error of only -0.000006, which is considerably more accurate than any of the other methods, even with $n = 100$.   ■

Error Analysis of the Approximation Methods

In Examples 5.29 and 5.30 we applied our methods to an integral we could evaluate in an exact form. Of course, the reason for studying approximation methods is primarily (but not solely) for integrals that cannot be evaluated exactly. For such integrals we need some way of determining the accuracy of our approximations. We state below, without proof, formulas for determining bounds on the errors for the Midpoint Rule, the Trapezoidal Rule, and Simpson's Rule. In each case E_n denotes the absolute value of the difference between the value given by the method and the exact value of the integral, for a fixed n. We again use MID and TRAP for Midpoint and Trapezoidal Rules, and we use SIMP for Simpson's Rule. By means of these formulas, we will also be able to determine how large an n will guarantee a given degree of accuracy.

Error Bounds for MID, TRAP, and SIMP

MID	TRAP
$E_n \le \dfrac{M_2(b-a)(\Delta x)^2}{24}$	$E_n \le \dfrac{M_2(b-a)(\Delta x)^2}{12}$

SIMP

$$E_n \le \frac{M_4(b-a)(\Delta x)^4}{180}$$

where M_2 and M_4 are nonnegative constants satisfying

$$|f''(x)| \le M_2 \qquad \text{and} \qquad |f^{(4)}(x)| \le M_4$$

for all x in $[a, b]$, and $\Delta x = (b-a)/n$.

We illustrate how to find these error bounds in the next example.

EXAMPLE 5.31   Determine n so that the approximation of the integral

$$\int_1^2 \frac{1}{x} \, dx$$

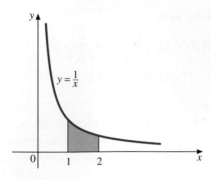

$$y = \frac{1}{x}$$

FIGURE 5.48

When the error is less than or equal to 0.00005, the answer is correct as rounded to the fourth decimal place. In general, to get accuracy to n decimal places, we want

$$E_n \le 0.\underbrace{00 \cdots 05}_{n \text{ zeros}}$$

given by (a) the Midpoint Rule, (b) the Trapezoidal Rule, and (c) Simpson's Rule is accurate to four decimal places.

Solution   The integral represents the area under the graph of $f(x) = 1/x$ from $x = 1$ to $x = 2$. (See Figure 5.48.) For accuracy to four places, we want the error E_n in each approximation not to exceed 0.00005. To find numbers M_2 and M_4, we need the first four derivatives of $f(x) = 1/x$.

$$f'(x) = -\frac{1}{x^2}$$

$$f''(x) = \frac{2}{x^3}$$

$$f'''(x) = -\frac{6}{x^4}$$

$$f^{(4)}(x) = \frac{24}{x^5}$$

For x in the interval $[1, 2]$, each derivative takes on its largest absolute value when the denominator is smallest, that is, when $x = 1$ (since dividing by a smaller value yields a larger quotient). So we have,

$$|f''(x)| \le 2 \qquad \text{and} \qquad |f^{(4)}(x)| \le 24$$

for x in $[1, 2]$. Therefore, we can take $M_2 = 2$ and $M_4 = 24$.

(a)   Midpoint Rule   Since $\Delta x = (b - a)/n$, we have

$$E_n \le \frac{M_2(b - a)(\Delta x)^2}{24} = \frac{M_2(b - a)^3}{24n^2} = \frac{2(1)^3}{24n^2} = \frac{1}{12n^2}$$

We want to find the smallest integer n for which

$$\frac{1}{12n^2} \le 0.00005$$

or equivalently,

$$n^2 \ge \frac{1}{12(0.00005)}$$

$$n \ge 40.8$$

Since n must be an integer, we take $n = 41$.

(b)   Trapezoidal Rule   In this case, again replacing Δx by $(b - a)/n$, we have

$$E_n \le \frac{M_2(b - a)(\Delta x)^2}{12} = \frac{M_2(b - a)^3}{12n^2} = \frac{2(1)^3}{12n^2} = \frac{1}{6n^2}$$

So we want to find n such that

$$\frac{1}{6n^2} \le 0.00005$$

or,

$$n^2 \ge \frac{1}{6(0.00005)}$$

$$n \ge 57.7$$

Thus, we take $n = 58$.

(c)   Simpson's Rule   The error bound in this case is

$$E_n \leq \frac{M_4(b-a)(\Delta x)^4}{180} = \frac{M_4(b-a)^5}{180n^4} = \frac{24(1)^5}{180n^4} = \frac{2}{15n^4}$$

For the desired accuracy, we want

$$\frac{2}{15n^4} \leq 0.00005$$

$$n^4 \geq \frac{2}{15(0.00005)}$$

$$n \geq 7.186$$

We take $n = 8$, which is even, as required. In general, for Simpson's Rule, we select n as the smallest *even* integer greater than or equal to the value calculated from the error bound.

In summary, we can be guaranteed of accuracy to four places by the Trapezoidal Rule with $n = 58$, by the Midpoint Rule with $n = 41$, and by Simpson's Rule with $n = 8$. Clearly, Simpson's Rule is much more efficient than either of the other methods in this case. ■

EXAMPLE 5.32   Approximate the integral

$$\int_1^2 \frac{1}{x}\, dx$$

correct to four decimal places by Simpson's Rule.

Solution   In part (c) of the preceding example we found that $n = 8$ will guarantee accuracy to at least four places for this integral. Since the interval of integration is from $x = 1$ to $x = 2$, we have

$$\Delta x = \frac{1}{8} = 0.125$$

So the points of division are 1, 1.125, 1.250, 1.375, 1.500, 1.625, 1.750, 1.875, 2. Thus, by Simpson's Rule,

$$\int_1^2 \frac{1}{x}\, dx \approx \frac{0.125}{3}\left[\frac{1}{1} + 4\left(\frac{1}{1.125}\right) + 2\left(\frac{1}{1.250}\right) + 4\left(\frac{1}{1.375}\right) + 2\left(\frac{1}{1.500}\right)\right.$$

$$\left. + 4\left(\frac{1}{1.625}\right) + 2\left(\frac{1}{1.750}\right) + 4\left(\frac{1}{1.875}\right) + \frac{1}{2}\right]$$

$$\approx 0.69315$$

Rounding to four places, we get 0.6932.

In Chapter 7 we will find that the exact value of this integral is the natural logarithm of 2, denoted $\ln 2$. Using a calculator, we find $\ln 2 \approx 0.693147$. So our approximation by Simpson's Rule with $n = 8$ is actually correct to five decimal places. ■

The constants M_2 and M_4 in the error bounds for our methods are determined from the second and fourth derivatives, respectively, of $f(x)$ on the interval $[a, b]$ in question. In Example 5.31, calculating these derivatives was straightforward and relatively simple. Also, it was easy to determine their largest absolute values on the interval in question. Unfortunately, for some functions calculation of the derivatives, especially the fourth, can be quite tedious. Then, determining upper bounds on their absolute values can again be difficult. One drawback of Simpson's Rule, in fact, is this difficulty in finding the number M_4 that is a bound for $|f^{(4)}(x)|$. The next example illustrates what we are talking about.

EXAMPLE 5.33   Find numbers M_2 and M_4 for the function

$$f(x) = \frac{1}{\sqrt{1 + x^3}}$$

on $[0, 1]$ satisfying

$$|f''(x)| \le M_2 \quad \text{and} \quad |f^{(4)}(x)| \le M_4$$

Solution   Write $f(x) = (1 + x^3)^{-1/2}$. Using the Generalized Power Rule, we find

$$f'(x) = -\frac{1}{2}(1 + x^3)^{-3/2}(3x^2)$$

$$= \frac{-3x^2}{2(1 + x^3)^{3/2}}$$

Then applying the Quotient Rule, we get (omitting some details)

$$f''(x) = \frac{15x^4 - 12x}{4(1 + x^3)^{5/2}}$$

As you can see, the details of calculating $f'''(x)$ and then $f^{(4)}(x)$ are tedious. The final result is

$$f^{(4)}(x) = \frac{135x^2(7x^6 - 40x^3 + 16)}{16(x^3 + 1)^{9/2}}$$

We would like to know the absolute maximum values of $|f''(x)|$ and $|f^{(4)}(x)|$ on $[0, 1]$, but a good bit of work is involved in these computations. We might settle for finding upper bounds on these derivatives, that is, numbers at least as large as their maximum values. One way is to *maximize the numerator* and *minimize the denominator*. In each case the numerator cannot exceed the sum of the absolute values of the terms. That is, for $f''(x)$, $|15x^4 - 12x| \le 15x^4 + 12|x|$. As x varies from 0 to 1, $15x^4 + 12|x|$ is at most $15 + 12 = 27$, since the maximum value occurs at $x = 1$. Similarly, the numerator of $f^{(4)}(x)$ does not exceed $135(7 + 40 + 16) = 8,505$ in absolute value. In each case the minimum value of the denominator occurs when $x = 0$. Thus, for all x in $[0, 1]$,

$$|f''(x)| \le \frac{27}{4} = 6.75 \quad \text{and} \quad |f^{(4)}(x)| \le \frac{8505}{16} \approx 531.6$$

so to three significant figures we can take $M_2 = 6.75$ and $M_4 = 532$.

An alternative procedure is to use a CAS or graphing calculator to obtain graphs of $|f''(x)|$ and $|f^{(4)}(x)|$ and to estimate from the graphs the maximum values for x in $[0, 1]$. We show these graphs in Figures 5.49 and 5.50. From the graphs we read the maximum values to be approximately 0.97 and 14.4, respectively. So we can take $M_2 = 1$ and $M_4 = 15$. These are much better values to use than the ones we previously obtained, since they give smaller bounds on the errors.   ∎

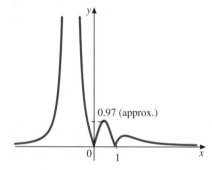

FIGURE 5.49

Graph of $|f''(x)|$ for $f(x) = \dfrac{1}{\sqrt{1 + x^3}}$

FIGURE 5.50

Graph of $|f^{(4)}_{(x)}|$ for $f(x) = \dfrac{1}{\sqrt{1 + x^3}}$

REMARK ───────────────────────────────

■ With a computer algebra system (CAS) such as DERIVE, Maple, or Mathematica, derivatives such as $f^{(4)}(x)$ above can be calculated with ease. In fact, any of these systems could be used to approximate the value of the integral.

───────────────────────────────

Cases Where Approximations Are Exact

If f is a polynomial function of degree 0 or 1, its second derivative is 0 for all x. For example, if $f(x) = 2x + 3$, then $f'(x) = 2$, and $f''(x) = 0$. Thus, in the error bounds for the Midpoint and Trapezoidal Rules, we may take $M_2 = 0$ (since $|f''(x)| = 0$ for all x). This choice of M_2 implies that $E_n = 0$ for every value of n. That is, both the Midpoint Rule and the Trapezoidal Rule give the *exact* value of the integral $\int_a^b f(x)dx$ when $f(x)$ is a polynomial of degree 0 or 1.

Similarly, if f is a polynomial function of degree 3 or less, its fourth derivative is identically 0, so in the formula for the error bound for Simpson's Rule we may take $M_4 = 0$, which implies that $E_n = 0$ for every choice of n. Thus, for such polynomial functions, Simpson's Rule gives the exact value of the integral. This result is not surprising if f is of second degree, since in this case, the graph of f is a parabola. After all, in Simpson's Rule we use parabolas to approximate the actual curve. So the approximation and the actual curve are the same. But it is not intuitively evident why fitting parabolas to cubic curves should result in the true value of the integral.

REMARK ───────────────────────────────

■ The fact that the approximation rules give exact answers for certain polynomial functions—those of degree 0 or 1 for the Midpoint and Trapezoidal Rules, and those of degree 0, 1, 2, or 3 for Simpson's Rule—is of more theoretical than practical interest. After all, if f is *any* polynomial, we can easily find an antiderivative of it and evaluate its integral by means of the Second Fundamental Theorem.

───────────────────────────────

There are some cases in which the approximation methods yield exact answers for integrals even when the integrand is not a polynomial. The next example shows one such integral. The error analysis in this case is very misleading.

EXAMPLE 5.34   Consider the integral

$$\int_0^\pi \sin^2 x \, dx$$

(a) Determine values of n for both the Trapezoidal Rule and Simpson's Rule that will guarantee $E_n \leq 10^{-7}$ for the approximations of the integral by these rules.

(b) Show that both the Trapezoidal Rule and Simpson's Rule actually give the exact answer when $n = 6$.

Solution

(a) First, we find $f''(x)$ and $f^{(4)}(x)$ for the integrand $f(x) = \sin^2 x = (\sin x)^2$

$$f'(x) = 2(\sin x)\cos x = \sin 2x \qquad \text{By the trigonometric identity}$$
$$\text{sin } 2x = 2\sin x \cos x$$
$$f''(x) = 2\cos 2x$$

$$f'''(x) = -4\sin 2x$$

$$f^{(4)}(x) = -8\cos 2x$$

Since $|\cos 2x| \le 1$ for all x, we have $|f''(x)| \le 2$ and $|f^{(4)}(x)| \le 8$. Thus, in the formulas for E_n we can take $M_2 = 2$ and $M_4 = 8$. Our interval of integration is from $x = 0$ to $x = \pi$, so in the formulas, $b - a = \pi$. We show the calculations below for $E_n \le 10^{-7} = 1/10^7$. Since $\Delta x = \pi/n$, we have

TRAP	SIMP

$$E_n \le \frac{M_2(b-a)(\Delta x)^2}{12} \qquad E_n \le \frac{M_4(b-a)(\Delta x)^4}{180}$$

$$\frac{2\pi^3}{12n^2} \le \frac{1}{10^7} \qquad \frac{8\pi^5}{180n^4} \le \frac{1}{10^7}$$

$$n^2 \ge \frac{10^7\pi^3}{6} \qquad n^4 \ge \frac{2(10^7)\pi^5}{45}$$

$$n \ge 7,189 \qquad n \ge 108$$

(b) To find the exact value of the integral, we use the trigonometric identity (see Appendix 2)

$$\sin^2 x = \frac{1}{2}(1 - \cos 2x)$$

With this identity we can write

$$\int_0^\pi \sin^2 x\, dx = \frac{1}{2}\int_0^\pi (1 - \cos 2x)\, dx = \frac{1}{2}\left[x - \frac{\sin 2x}{2}\right]_0^\pi = \frac{\pi}{2}$$

Using $n = 6$ in the Trapezoidal Rule and in Simpson's Rule, we have $\Delta x = \pi/6$ in each case. So the division points are

$$x_0 = 0, \ x_1 = \pi/6, \ x_2 = \pi/3, \ x_3 = \pi/2, \ x_4 = 2\pi/3, \ x_5 = 5\pi/6, \ x_6 = \pi$$

The approximations of the integral by the two methods are as follows.
TRAP:

$$\frac{\pi}{12}\left[\sin^2 0 + 2\sin^2\frac{\pi}{6} + 2\sin^2\frac{\pi}{3} + 2\sin^2\frac{\pi}{2} + 2\sin^2\frac{2\pi}{3}\right.$$

$$\left. + 2\sin^2\frac{5\pi}{6} + \sin^2\pi\right]$$

$$= \frac{\pi}{12}\left[0 + 2\left(\frac{1}{4}\right) + 2\left(\frac{3}{4}\right) + 2(1) + 2\left(\frac{3}{4}\right) + 2\left(\frac{1}{4}\right) + 0\right]$$

$$= \frac{\pi}{12}(6) = \frac{\pi}{2}$$

SIMP:

$$\frac{\pi}{18}\left[\sin^2 0 + 4\sin^2\frac{\pi}{6} + 2\sin^2\frac{\pi}{3} + 4\sin^2\frac{\pi}{2} + 2\sin^2\frac{2\pi}{3}\right.$$
$$\left. + 4\sin^2\frac{5\pi}{6} + \sin^2\pi\right]$$
$$= \frac{\pi}{18}\left[0 + 4\left(\frac{1}{4}\right) + 2\left(\frac{3}{4}\right) + 4(1) + 2\left(\frac{3}{4}\right) + 4\left(\frac{1}{4}\right) + 0\right]$$
$$= \frac{\pi}{18}(9) = \frac{\pi}{2}$$

Both approximations give the exact value of the integral! That is, the actual error with $n = 6$ is 0. Yet in part (a) we found by our formulas that to be assured of an error of less than 10^{-7}, we should take $n = 7,189$ for the Trapezoidal Rule and $n = 108$ for Simpson's Rule. The lesson to be learned from this example is that the value of n obtained from the formula for the bound on the error, E_n, for each method will indeed guarantee that the error does not exceed the specified amount, but it is possible that a smaller value of n will also work (and might even give the exact value, as in this example).

We should note that the value $n = 6$ in this example is special. A larger value of n need not yield the exact answer. You might find it interesting to try some larger values. In fact, even with the values of n obtained in part (a), the answers would not have been exact (but the errors would have been less than 10^{-7}).   ■

Approximating an Integral of a Function with No Known Formula

We conclude this section with an example to illustrate that Simpson's Rule, as well as the Trapezoidal Rule, can be used even if we do not know the function to be integrated, provided we know the function's values at equally spaced intervals. For the Midpoint Rule, we would also need the function's values at the midpoints of the intervals.

EXAMPLE 5.35   A flat iceberg (a sheet of ice) is measured across its width at 3 m intervals, as shown in Figure 5.51. A bore hole through the iceberg shows

FIGURE 5.51

the thickness of the ice sheet to be 6 m. Use Simpson's Rule to approximate the volume of the ice sheet. How many ice cubes, 2 cm on an edge, would the iceberg make?

Solution   The volume is the area of the surface multiplied by the thickness. Let $w(x)$ denote the width of the ice sheet x m from the left end in Figure 5.51. The area of the ice sheet is then given by the integral $\int_0^{30} w(x)dx$. (You can see why, by thinking of Riemann sums composed of areas of narrower and narrower rectangles, so that in the limit you get the integral.) Although we do not know the formula for $w(x)$, we do know its values at the endpoints of 10 equally spaced intervals, 3 m wide. So we can use Simpson's Rule with $\Delta x = 3$.

$$\int_0^{30} w(x)dx \approx \frac{3}{3}\big[3.2 + 4(12.2) + 2(15.7) + 4(18.3) + 2(21.9)$$

$$+ 4(25.6) + 2(30.7) + 4(32.3) + 2(26.2) + 4(16.1) + 7.2\big]$$

$$= 617.4$$

The thickness of the ice is 6 m, so the volume is approximately

$$(617.4)(6) \approx 3700 \text{ m}^3$$

Since we do not know the formula for $w(x)$, there is no way to check the error in our approximation.

The volume of an ice cube 2 cm on an edge is 8 cm^3. The volume of the ice sheet in cubic centimeters is approximately $(3700)(100 \text{ cm/m})^3$. So the number of ice cubes is

$$\frac{3,700,000,000}{8} \approx 460,000,000$$

which is enough for more than 10 million soft drinks!   ∎

Approximating Integrals by Computer

With a CAS such as DERIVE, Maple, or Mathematica, many definite integrals can be approximated rapidly and with a high degree of accuracy. So one might well question the need for the approximation techniques we have studied in this section. For one thing, it is worthwhile knowing how the computers do it. In fact, most of these programs use Simpson's Rule. Even if you do use a CAS, it is still important to understand the underlying theory that we have presented. It will serve you well when the software eventually makes a mistake.

INTEGRATION USING A COMPUTER ALGEBRA SYSTEM

Computer algebra systems can be used to perform both exact symbolic integration and approximations to integrals. The fact that we know that an antiderivative of a function exists does not mean it is always possible to determine the antiderivative explicitly. In some cases, the best that can be done is to approximate the integral. Even when it is possible to express the antiderivative in terms of simple functions, the algebra required may be very difficult. In both cases a computer algebra system can prove very useful.

FIGURE 5.5.1
$y = f(x) = \sin x, -2\pi \leq x \leq 2\pi$

CAS 19

Approximate $\int_0^\pi \sin x \, dx$.

We start by using Maple, which allows each of the types of approximation discussed in Section 5.5. We need to load the package called "student" in order to access some additional Maple commands.

```
with (student):
f:=x–>sin(x);
plot (f(x),x=–2*Pi..2*Pi);
leftsum (f(x),x=0..Pi,10);
```

The output from this command is the expression for the left sum, where the interval $[0, \pi]$ is partitioned into 10 subintervals of equal width.

$$\frac{1}{10}\pi \left(\sum_{i=0}^{9} \sin \left(\frac{1}{10} i\pi \right) \right)$$

To get a numerical value we need to evaluate this sum.

```
evalf(value(leftsum(f(x),x=0..Pi,10)));
```

Output: 1.983523538

If we had not used the evalf command, the expression output would be:

$$\frac{1}{10}\pi \left(\sqrt{5} + 1 + \frac{1}{2}\sqrt{2}\sqrt{5 - \sqrt{5}} + \frac{1}{2}\sqrt{2}\sqrt{5 + \sqrt{5}} \right)$$

To increase the number of subintervals over which the sum is taken, we change the last value specified in the command.

```
evalf(value(leftsum(f(x),x=0..Pi,100)));   1.999835643
```

Maple also allows graphic visualization of the process (see Figure 5.5.2).

```
leftbox (f(x),x=0..Pi,50);
```

To show the Right-Endpoint Rule, use:

```
rightbox(f(x),x = 0..Pi,50);
```

The next output shows the approximations for $n = 100$ and 500 using the Left- and Right-Endpoint Rules, the Midpoint Rule, the Trapezoid Rule, and Simpson's Rule.

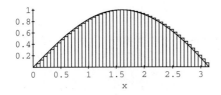

FIGURE 5.5.2
Left Riemann sum for $f(x) = \sin x$ on $[0, \pi]$ with $n = 50$

```
evalf(value(leftsum(f(x),x = 0..Pi,200)));      1.999959174
evalf(value(rightsum(f(x),x = 0..Pi,200)));     1.999959174
evalf(value(middlesum(f(x),x = 0..Pi,200)));    2.000020860
evalf(value(trapezoid(f(x),x = 0..Pi,200)));    1.999959174
evalf(value(simpson(f(x),x = 0..Pi,200)));      2.000000092
evalf(value(leftsum(f(x),x = 0..Pi,500)));      1.999997865
evalf(value(rightsum(f(x),x = 0..Pi,500)));     1.999997865
evalf(value(middlesum(f(x),x = 0..Pi,500)));    2.000003290
evalf(value(trapezoid(f(x),x = 0..Pi,500)));    1.999997865
evalf(value(simpson(f(x),x = 0..Pi,500)));      1.999999905
```

We can take this one step further and let Maple compute the integral, in this case the area under the sine curve on $[0,\pi]$ as the limit of Reimann sums.

Enter

leftsum(f(x),x=0..Pi,n);

Output:

$$\frac{\pi \left(\sum_{i=0}^{n-1} \sin \left(\frac{i\pi}{n} \right) \right)}{n}$$

limit(",n=infinity);

Output:

$$\lim_{n \to \infty} \frac{\pi \left(\sum_{i=0}^{n-1} \sin \left(\frac{i\pi}{n} \right) \right)}{n}$$

To compute the actual limit in Maple enter:

leftsum(f(x),x=0..Pi,n);   (same output as above)
value (");

Output:

$$-\frac{\pi \sin \left(\frac{\pi}{n} \right)}{n \left(\cos \left(\frac{\pi}{n} \right) - 1 \right)}$$

limit(", n = infinity);

Output: 2

In Mathematica and DERIVE, similar analysis can be carried out by defining the methods to be used. For example, to use the Left-Endpoint Rule:

Mathematica:

f[x_] = Sin[x]
a = 0
b = Pi
n = 50
dx = (b–a)/n
x[k_] = a+k*dx
leftsum = N[dx*Sum[f[x[k–1]],
{k,1,n}],10]

Output: 1.999341983

DERIVE:

(At the □ symbol, go to the next step.)
a (author) □ f(x) := sin(x) □ a (author) □ dx := (b−a)/n □ a (author) □ dxf(a+(k−1)dx) □ a (author) □ a := 0 □ a (author) □ b := Pi □ a (author) □ n := 50 □ c (calculus) □ s (sum) □ [choose expression] □ k (variable) □ 1 (lower limit of sum) □ n (upper limit of sum] □ s (simplify) □ [choose expression] □ x (approximate) □ [choose expression]
Output: 1.98352
c (calculus) □ 1 (limit) □ [choose expression for Reimann sum] □ n (variable) □ inf (limit point ∞) □ s (simplify) □ [choose expression]

CAS 20

Use the Fundamental Theorem of Calculus to find

$$\int_0^\pi \sin x \, dx \quad \text{and} \quad \int_0^4 \sin x \, dx$$

Computer algebra systems can perform symbolic integration as well as integral approximation. Both antiderivatives (indefinite integrals) and definite integrals can be computed in most cases.

To find the indefinite integral $\int \sin x \, dx$:

Maple:

f: = x–> sin(x);
int(f(x),x):

Output: $-\cos(x)$

Mathematica:

f[x_] = Sin[x]
Integrate[f[x],x]

DERIVE:

(At the □ symbol, go to the next step.)
a (author) □ sin(x) □ c (calculus) □
i (integrate) □ [choose expression] □
x (variable) □ [enter] (no limits) □ s
(simplify) □ [choose expression]

To find the definite integral $\int_0^{\pi} \sin x \, dx$:

Maple:

int(f(x),x = 0..Pi);

The value returned for this definite integral is 2. Compare this result with the results that were given in CAS 19.

Mathematica:

Integrate [f[x],{x,0,Pi}]

DERIVE:

(At the □ symbol, go to the next step.)
a (author) □ sin(x) □ c (calculus) □
i (integrate) □ [choose expression] □ x
(variable) □ 0 (lower limit) □ Pi (upper
limit) □ s (simplify) □ [choose expression]

To find the definite integral $\int_0^4 \sin x \, dx$:

Maple:

evalf (int(f(x),x = 0..4));

Mathematica:

N[Integrate[f[x],{x,0,4}],10]

DERIVE:

(At the □ symbol, go to the next step.)
a (author) □ sin(x) □ c (calculus) □ i
(integrate) □ [choose expression] □ x
(variable) □ 0 (lower limit) □ Pi (upper limit) □ s (simplify) □ [choose expression] □ x (approximate) □ [choose expression]

The output from Maple and Mathematica is 1.653643621 whereas DERIVE gives the value 1.645364. Notice that the value of the integral is less than 2, because the graph of sin x between π and 4 lies below the x-axis and the contribution to the integral will be negative. We can see this clearly by executing the Maple command:

leftbox(f(x),x=0..4,50);

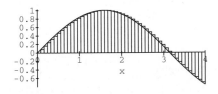

FIGURE 5.5.3
Left Riemann sum for $f(x) = \sin x$ on [0,4] with $n = 50$

Exercise Set 5.5

In Exercises 1–6, approximate the integral using (a) the Trapezoidal Rule, (b) the Midpoint Rule, and (c) Simpson's Rule, for the given value of n. Find the exact answer using the Second Fundamental Theorem, and determine the error in each approximation.

1. $\int_0^1 \dfrac{dx}{(x+1)^2}$; $n = 4$

2. $\int_2^4 \dfrac{dx}{\sqrt{2x+1}}$; $n = 10$

3. $\int_1^3 x \sin x^2 dx$; $n = 8$

4. $\int_{-1}^3 x\sqrt{x^2+1} \, dx$; $n = 8$

5. $\int_{-2}^{1}(1+x^3)dx; \ n=6$

6. $\int_{0}^{1}\sin\pi x\sqrt{1-\cos\pi x}\,dx; \ n=5$

In Exercises 7–12, determine a value of n for (a) the Trapezoidal Rule, (b) the Midpoint Rule, and (c) Simpson's Rule that will ensure that the error in approximating the integral does not exceed 10^{-4}.

7. $\int_{1}^{3}\dfrac{dx}{\sqrt{x+1}}$

8. $\int_{-1/2}^{2}\sqrt[3]{x+1}\,dx$

9. $\int_{3}^{5}\cos^2 x\,dx$

10. $\int_{1}^{3}\dfrac{1}{(4-x)^2}\,dx$

11. $\int_{0}^{2}\sin^3 x\,dx$

12. $\int_{0}^{2}(2x+1)^{-3/2}dx$

13. From the table given, approximate $\int_{0}^{5}f(x)dx$ using (a) the Trapezoidal Rule, and (b) Simpson's Rule.

x	1.0	1.4	1.8	2.2	2.6	3.0
$f(x)$	2.15	1.82	0.43	−0.15	−0.92	−1.27

x	3.4	3.8	4.2	4.6	5.0
$f(x)$	−0.84	0.16	1.18	2.35	3.76

14. By estimating values from the accompanying graph, approximate the area under the curve from $x=2$ to $x=6$ using Simpson's Rule with $n=8$.

15. The average depth of a pond is known to be approximately 2 m. The length of the pond is 16 m, and the width of the pond at 2-m increments is measured and found to be as shown in the figure (in meters). Use Simpson's Rule to estimate the volume of water in the pond.

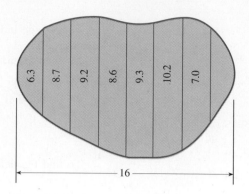

In Exercises 16 and 17, approximate the integral using (a) the Trapezoidal Rule, (b) the Midpoint Rule, and (c) Simpson's Rule, with $n=10$. Give an upper bound on the error in each case.

16. $\int_{-1}^{1}\sin x^2 dx$

17. $\int_{1}^{2}\dfrac{\sin x}{x}\,dx$

18. Using the values of M_2 and M_4 found from the graphs of $|f''(x)|$ and $|f^{(4)}(x)|$ in Example 5.33, find a value of n for (a) the Midpoint Rule and (b) Simpson's Rule that will ensure an error of less than 0.00005 in approximating the integral

$$\int_{0}^{1}\frac{dx}{\sqrt{1+x^3}}$$

In each case use the value of n that you find to obtain the approximation to the integral.

19. Use the Midpoint Rule to approximate the integral

$$\int_{0}^{2}\sqrt{\sin^2 x+3\cos^2 x}\,dx$$

with $n=10$. Use a graphing calculator or a CAS to graph $|f''(x)|$, where $f(x)$ is the integrand in this integral. Estimate the smallest number M_2 for which $|f''(x)|\le M_2$ for x in [0, 2], and use the value obtained to find a bound on the error in approximating the integral.

20. Consider the region bounded above by the graph of $y=\sin(x^2)$, below by the x-axis, and on each side by the vertical lines $x=-1$ and $x=1$. (See the figure.)
(a) Use Simpson's Rule to approximate the area of the region using $n=10$ and $n=20$. In each case find an upper bound on the error.

(b) Find a value of n so that the approximation to the error using Simpson's Rule is correct to seven decimal places.

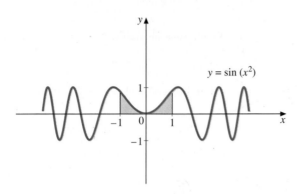

$y = \sin(x^2)$

21. Use Simpson's Rule to approximate

$$\int_0^1 \frac{dx}{1+x^2}$$

with $n = 10$, 20, 100, and 1000. Conjecture the exact value of the integral. (*Hint:* Multiply your approximations by 4 and see if you recognize the value being approached.)

22. Prove that

$$S_{2n} = \frac{2}{3} M_n + \frac{1}{3} T_n$$

where S_{2n} is the value given by Simpson's Rule with $2n$ subdivisions, and M_n and T_n are the values given by the Midpoint and Trapezoidal Rules, respectively, for n subintervals.

Chapter 5 Review Exercises

1. Let $f(x) = x^2 + 1$ on $[0, 2]$. Find the upper sum U_n and the lower sum L_n for n arbitrary, and show that they have the same limit as $n \to \infty$. Interpret the result geometrically.

2. Let $f(x) = x(1+x)(3-x)$ on $[-1, 4]$. Find the Riemann sum for f using the partition points -1, 0, 2, 3, 4, with $c_1 = -0.5$, $c_2 = 0.6$, $c_3 = 2.3$, and $c_4 = 3.4$. Compare the result with the exact value of $\int_{-1}^{4} f(x)\,dx$.

3. Find the exact value of the integral $\int_{-1}^{3}(x^2 - 2x - 1)\,dx$ in two ways: (a) by taking the limit of the Riemann sum with equally spaced partition points, where $c_k = x_k$ for $k = 1, 2, \ldots, n$, and (b) by using the Second Fundamental Theorem of Calculus.

4. Let

$$f(x) = \begin{cases} 2x + 1 & \text{if } -1 \le x < 1 \\ 3 & \text{if } 1 \le x < 4 \\ 3(5 - x) & \text{if } 4 \le x \le 5 \end{cases}$$

Find $\int_{-1}^{5} f(x)\,dx$ making use of areas of known geometric figures.

5. A tank originally full of water was observed after one minute to be leaking at the rate of 12.6 L/min. Continued efforts to stop the leak were only partially successful, as the table shows. The table gives the rate $R(t)$ of leakage at half-minute intervals. Give an overestimate and an underestimate of the amount of water that leaked out from $t = 1$ to $t = 4$.

t	1.0	1.5	2.0	2.5	3.0	3.5	4.0
$R(t)$	12.6	8.4	6.3	5.2	4.2	3.6	3.2

6. Find the exact value of the limit

$$\lim_{n \to \infty} \sum_{k=1}^{n} x_k^3 \Delta x$$

where $\{x_0, x_1, x_2, \ldots, x_n\}$ is a partition of the interval $[0, 4]$ with equally spaced partition points, so that $\Delta x = 4/n$. (*Hint:* Write the limit as an integral.)

7. Use the values of the areas shown in the figure for the graph of f to find the following:

(a) $\int_{-2}^{3} f(x)\,dx$

(b) $\int_{-2}^{3} |f(x)|\,dx$

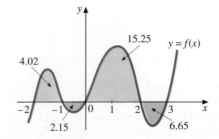

15.25 $y = f(x)$

4.02

2.15   6.65

In Exercises 8–19, use the Second Fundamental Theorem to evaluate the integral.

8. $\displaystyle\int_2^4 x(1-2x)\,dx$

9. $\displaystyle\int_{-3}^0 \left(x-\sqrt{1-x}\right)dx$

10. $\displaystyle\int_1^4 \left(x-\frac{1}{\sqrt{x}}\right)^2 dx$

11. $\displaystyle\int_0^{\pi/3} \frac{1-\sin x}{\cos^2 x}\,dx$

12. $\displaystyle\int_{-8}^{-1} (u^{1/3}-u^{-2/3})\,du$

13. $\displaystyle\int_0^{\pi/4} \sin^3 t\cos t\,dt$

14. $\displaystyle\int_{-\pi/3}^{\pi/3} (\theta+\cos\theta)\,d\theta$

15. $\displaystyle\int_{-1}^2 (2x-1)(x+4)\,dx$

16. $\displaystyle\int_0^2 \frac{x^2}{\sqrt{1+x^3}}\,dx$

17. $\displaystyle\int_9^{16} \frac{(\sqrt{x}-2)^5}{\sqrt{x}}\,dx$

18. $\displaystyle\int_0^{\pi/4} \frac{\sin 2x}{(1+\cos 2x)^3}\,dx$

19. $\displaystyle\int_{27}^{125} t^{-1/3}(t^{2/3}-9)^{3/2}\,dt$

In Exercises 20–27, find the indicated antiderivatives.

20. $\displaystyle\int \frac{1}{x^2}\sec^2\left(\frac{2}{x}\right)dx$

21. $\displaystyle\int \frac{\tan^3\theta-\sin\theta}{\cos^2\theta}\,d\theta$

22. $\displaystyle\int \frac{x-1}{\sqrt{x^2-2x+4}}\,dx$

23. $\displaystyle\int x(x^4+4x^2+4)^{5/2}dx$ (*Hint:* x^4+4x^2+4 is a perfect square.)

24. $\displaystyle\int \frac{\sqrt{1+\sqrt{x}}}{\sqrt{x}}\,dx$

25. $\displaystyle\int \frac{x(1+\cos^2 x^2)}{\sin^2 x^2}\,dx$

26. $\displaystyle\int \left[\frac{1}{1+\cos\theta}-\frac{1}{1-\cos\theta}\right]d\theta$ (*Hint:* Combine fractions.)

27. $\displaystyle\int \sin 2t(1+\cos^2 t)^3\,dt$ (*Hint:* Use an identity.)

28. Verify that
$$\int x^2\cos 3x\,dx = \frac{1}{27}\left[(9x^2-2)\sin 3x + 6x\cos 3x\right]+C$$

29. Evaluate the integral $\int_0^2 f'(t)\,dt$ if it is known that $f(x)=(3x^2-4)\sqrt{x^3+1}$.

30. Find $f'(t)$ if $\displaystyle f(t)=\int_0^t \frac{x}{\sin^2 x+1}\,dx$.

31. (a) Find $F'(\pi)$ if $\displaystyle F(x)=\int_{\pi/6}^x \frac{\sin t}{t}\,dt$.

   (b) Find the value of $\displaystyle\int_{\pi/6}^\pi \frac{d}{dx}\left(\frac{\sin x}{x}\right)dx$.

32. Find the average value of $f(x)=2x^3-4x^2-5x+2$ on $[-2,2]$.

33. Find the average value of $g(x)=x/\sqrt{x^2+9}$ on $[0,4]$.

34. Find c such that $\int_{-1}^4 \sqrt{3x+4}\,dx = 5\sqrt{3c+4}$.

In Exercises 35–38, find the general solution of the differential equation.

35. $\dfrac{dy}{dx}=2x-\sin 2x$

36. $(\cos^2 x)y'=1-\sin x$

37. $f'(t)=\sqrt{t^4+4t^2}$, $(t>0)$

38. $(x^2+1)^2 dy = x\,dx$

In Exercises 39–41, solve the initial-value problem.

39. $g'(t)=t(t^2+4)^{3/2}$; $g(0)=2$

40. $\dfrac{dy}{dx}=\sin x\cos^2 x$; $y=3$ when $x=\dfrac{\pi}{2}$

41. $\sqrt{x}\,y'=(\sqrt{x}+2)^3$; $y=5$ when $x=1$

In Exercises 42–45, approximate the integral using (a) the Trapezoidal Rule, (b) the Midpoint Rule, and (c) Simpson's Rule, for the given value of n. Find an upper bound on the error in each case. Compare with the exact value.

42. $\displaystyle\int_0^2 \sin^2\frac{\pi x}{2}\,dx$; $n=8$

43. $\displaystyle\int_0^1 \frac{x}{\sqrt{x^2+1}}\,dx$; $n=10$

44. $\displaystyle\int_0^1 \tan^2 x\,dx$; $n=10$

45. $\displaystyle\int_1^8 \frac{x^{2/3}+1}{\sqrt[3]{x}}\,dx$; $n=8$

In Exercises 46 and 47, find n such that the error in using (a) the *Trapezoidal Rule, (b) the Midpoint Rule, and (c) Simpson's Rule to approximate the given integral does not exceed 10^{-4}.*

46. $\displaystyle\int_{1}^{4} \frac{dx}{\sqrt{5-x}}$ **47.** $\displaystyle\int_{0}^{3} \sin^3 x \, dx$

 48. Use the Trapezoidal Rule with $n = 20$ to approximate

$$\int_{-1}^{1} \tan x \, dx$$

Give an upper bound on the error.

 49. Repeat Exercise 48 using the Midpoint Rule.

 50. Repeat Exercise 48 using Simpson's Rule.

51. The accompanying table gives velocities of an object moving along a line, at half-second intervals. Use Simpson's Rule to approximate the total distance covered by the object. Assume t is in seconds and $v(t)$ is in meters per second.

t	0	0.5	1.0	1.5	2.0	2.5	3.0
$v(t)$	11.9	10.4	8.5	11.2	15.6	12.3	9.2

52. Approximating Finite Sums with Integrals. Approximate the sum of the cube roots of the first n positive integers,

$$S_n = \sqrt[3]{1} + \sqrt[3]{2} + \sqrt[3]{3} + \cdots + \sqrt[3]{n}$$

by completing the following steps:

(a) Let $f(x) = \sqrt[3]{x}$. Partition the interval $[0, 1]$ into n subintervals of equal width. (See the figure.) Compute the upper sum U_n.

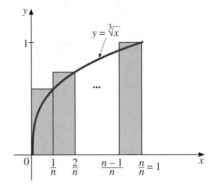

(b) Show that $S_n = n^{4/3} U_n$. Explain how it follows that for large n,

$$S_n \approx \frac{3}{4} n^{4/3}$$

53. Use a CAS to approximate the following:

(a) $\displaystyle\int_{3}^{4} \frac{\sin^3 x}{x^2} \, dx$

(b) $\displaystyle\int_{-1}^{1} \cos^4 x \sin^2 x \tan^5 x \, dx$

(c) $\displaystyle\int_{2}^{5} \frac{t^4 + t^3}{\sqrt{t^4 + 17}} \, dt$

Chapter 5 Concept Quiz

1. Define the following for a continuous function f on $[a, b]$:

(a) The upper sum U_n

(b) The lower sum L_n

(c) The area under the graph of f from $x = a$ to $x = b$, if f is nonnegative

(d) A Riemann sum for f

(e) The definite integral $\int_a^b f(x) \, dx$

(f) The average value of f on $[a, b]$

(g) The indefinite integral $\int f(x) \, dx$

2. State the following:

(a) The First Fundamental Theorem of Calculus

(b) The Second Fundamental Theorem of Calculus

(c) The Mean-Value Theorem of Integral Calculus

(d) The Midpoint Rule

(e) The Trapezoidal Rule

(f) Simpson's Rule

3. Assuming appropriate continuity conditions, give the value of each of the following in a form that does not involve an integral.

(a) $\int_{t_1}^{t_2} r(t) \, dt$, where R is a function for which $R'(t) = r(t)$

(b) $\int_a^b f'(x) \, dx$

(c) $\dfrac{d}{dt} \displaystyle\int_a^t h(x) \, dx$

(d) $\int_a^b f(g(x))g'(x) \, dx$, given that $F'(x) = f(x)$

4. Discuss in what sense integration and differentiation can be considered inverse processes.

5. Fill in the blanks. (More than one word may be needed.)

(a) If f is nonnegative and continuous on $[a, b]$, the area under the graph of f from $x = a$ to $x = b$ is given by _____ .

(b) If $A(x)$ is the area under the graph of the continuous, nonnegative function f from a to x, then $A'(x)$ exists and equals _____ .

(c) The integral $\int_a^b f(x)\, dx$ is the limit of _____ , provided this limit exists.

(d) The _____ Rule uses parabolic arcs to appoximate the integral.

(e) The _____ Rule is the average of the values given by the Left-Endpoint Rule and the Right-Endpoint Rule.

(f) The integral of the _____ of a function gives the net change in the function.

(g) In general, the Midpoint Rule is more accurate than _____ but less accurate than _____ .

6. Indicate which of the following statements are true and which are false.

(a) If $\int_a^b f(x)\, dx = \int_a^b g(x)\, dx$, then $f(x) = g(x)$.

(b) If $\int f(x)\, dx = \int g(x)\, dx$, then $f(x) = g(x)$.

(c) If $g(t) = \int_0^t \sin u^2\, du$, then $g'(t) = \sin t^2$.

(d) If $\int_a^b f(x)\, dx = 0$, then $f(x) = 0$.

(e) If $v(t)$ is the velocity of a particle moving along a line, and its acceleration $a(t)$ is continuous on $[a, b]$, then

$$\int_a^b a(t)\, dt = v(b) - v(a)$$

APPLYING CALCULUS

1. The Fish and Game Warden of a rural county is planning to stock a small pond with fish prior to the opening of the fishing season. However, before stocking the pond the warden has been advised to allow 1000 cubic feet of water for each fish and therefore needs to know the volume of water in the pond. The warden makes the measurements indicated in the figure and determines that the average depth of the pond is 20 feet.

(a) Use the given information to estimate the volume of water in the pond.

(b) How many fish will be needed to stock the pond?

(c) If the warden wants to have at least 20% of the original fish population left in the pond at the end of the season, what is the maximum number of fishing licenses that can be sold, assuming that the average seasonal catch per license is 20 fish?

2. The town council wants to build a new parking lot using the area with the measurements shown in the figure, but only $11,000 is allocated in the budget for new parking. The cost to clear the land is $.15 per square foot and the cost to pave is $2.00 per square foot. It's your job as town engineer to report on whether it can be done within the budget. What would you say and why?

3. The rate at which a person's heart pumps blood is called the *cardiac output*. For a person at rest, this rate is normally 5 liters per minute. Strenuous exercise significantly raises this rate. Also, certain diseases of the heart, circulatory system or nervous system have a significant effect on the cardiac output of a person's heart. There is a technique for measuring cardiac output called *dye dilution* which can be described as follows: At time $t = 0$, a known amount D of dye is injected into a vein near the heart. The dyed blood circulates through the right side of the heart, to the lungs, then through the left side of the heart and into the arterial system. The concentration of dye is measured at fixed time intervals at some convenient point in the arterial system, e.g., the aorta. If the heart pumps blood at the constant rate of R liters per second and the concentration of dye at time t is $C(t)$ milligrams per liter, then the amount of dye flowing through the aorta is $RC(t)$ milligrams per second. It takes about 20 seconds for blood to circulate through the body, so

$$D = \int_0^{20} RC(t)dt = R \int_0^{20} C(t)dt$$

is the total amount of dye injected. Since D is known, this equation can be solved for R to obtain

$$R = \frac{D}{\int_0^{20} C(t)dt}$$

(a) Suppose that at time t (in seconds), the dye concentration is given by

$$C(t) = -bt(t - 20)$$

where b is a positive constant. Graph C. Determine R in terms of b and D, the total amount of dye injected.

(b) Suppose $D = 5$ milligrams of dye is injected and $C = C(t)$ is measured at 1 second intervals to obtain the following data:

t	1	2	3	4	5	6	7	8	9	10	11	12	13	14	15	16	17	18	19	20
$C(t)$	0	0	.1	.6	.9	1.4	1.9	2.7	3.0	3.7	4.0	4.1	4.0	3.8	3.7	2.9	2.2	1.5	1.1	.9

Approximate the integral $\int_0^{20} C(t)dt$ using a Riemann sum with right endpoints. Estimate the cardiac output R.

An erupting volcano

4. Geologists studying an eruption of a volcano want to measure the height of the curtain of fire without getting too close. They do so by using a stopwatch to time how long a bit of molten lava takes to fall at a constant acceleration of gravity 10 m/s². In the photograph lava at the top takes 5 seconds to fall. How high is the curtain of fire?

5. (CAS) A pendulum consists of a mass at the end of a rod or string of length L. The angle, θ, that the string makes with the vertical changes as the pendulum swings. Let β be the maximum angle that the pendulum makes with the vertical (called the amplitude). Then the time T (called the period) it takes for one complete swing of the pendulum is given by

$$T = 4\sqrt{\frac{L}{g}} \int_0^{\pi/2} \frac{d\phi}{\sqrt{1 - k^2 \sin^2 \phi}}$$

where g is the acceleration due to gravity, $g = 32$ ft/s², and $k = \sin(\beta/2)$. This integral is an elliptic integral of the first kind whose antiderivative cannot be computed as a combination of simple functions.

(a) Show that $\lim_{\beta \to 0} T = 4\sqrt{\frac{L}{g}}$. Note that T is a function of β. (Hint: Use the fact that T is continuous at $\beta = 0$, or approximate $\sin \phi$ by ϕ.)

(b) Approximate the period T corresponding to $\beta = 0.25$.

(c) Approximate the amplitude, β, if the pendulum is to have a period of 2.5 s. Find a function to fit T by plotting 10 points $(\beta, T(\beta))$ for $0 \leq \beta \leq \pi/2$ and having the computer algebra system connect the points by a curve.

6. The Universe is described by a measure of four-dimensional space-time known as the Robertson-Walker metric:

$$ds^2 = c^2 ds^2 - a^2(t)[(d\Theta)^2 + r^2 d\Omega^2]$$

where s is a measure of distance known as the "proper distance," c is the speed of light, t is the time, and $a(t)$ is a scale factor that describes the expansion of the universe. We can derive that the equation of motion that relates the radial coordinate Θ (capital theta) to the time is:

$$d\Theta = \pm \frac{cdt}{a(t)}$$

For a set of galaxies expanding with the Universe, light emitted by a galaxy at position θ_g at time t_e will be observed at $\Theta = 0$ at a later time t_0 where Θ_g is the integral from t_e to t_0 of $d\Theta$.

(a) Write the integral and use one of the Fundamental Theorems of Calculus (specify which one) to differentiate it with respect to t_0.

(b) Since the galaxy's position is constant in its moving frame of reference, solve the equation of the previous part to find the relation between the time intervals dt_0 and dt_e in terms of the scale factors at the two times.

(c) Since the frequencies at which we observe the light depends on the time interval, derive the relation between the scale factors and the wavelengths, which are the inverses of the frequencies. Do the wavelengths become larger or smaller over time?

7. A beam of energy in a star, as it flows through a cubic meter of gas, loses a portion of its energy to absorption. This portion is proportional to the incoming intensity I, the density of the gas ρ, and a property of the gas known as absorptivity that is measured by the absorption coefficient κ. Also, the same volume of gas emits additional radiation in an amount proportional to both the density and to an emission coefficient ε.

(a) Express the differential of the optical depth τ in terms of its proportionality to κ, ρ, and the differential of radial distance x from the center of the star. Use a minus sign to show that the optical depth measured by us from Earth increases as the distance from the center of the star toward us decreases. Then integrate as x goes from a to b to find τ in terms of κ, ρ, a, and b.

(b) Derive the Equation of Transfer, which shows the change in intensity as a function of the absorption and emission coefficients and of the incoming intensity. To do so, express the differential of intensity through a small volume in terms of the radiation added to and subtracted from the radiation hitting that volume. Then express the differential equation in terms of optical depth to get the Equation of Transfer.

The Equation of Transfer has no simple solution, and is either solved with simplifying assumptions or using numerical methods.

Phillip A. Griffiths

Mathematics is still a "small science," an individual activity done with a lively imagination. In this sense it is highly interactive. The most productive environment for doing mathematics is one in which an individual can share ideas with others when appropriate, then retire into privacy for quiet reflection. This approach allows mathematicians to span many topics, remain flexible, and respond effectively and quickly to the needs of the rest of science.

But mathematics is fundamentally different from the sciences and the relationship between mathematics and other disciplines has changed substantially in recent years. Mathematics has been called both the "Queen of the Sciences" and the "Servant of the Sciences," but neither description seems adequate today. Mathematics is not so much above and below other disciplines, but within and around them. It has become a full and interactive partner. This is healthy not only for mathematics and associated fields like physics, but also for fields like business, psychology, and health policy analysis.

Mathematics is also a language—one on which the sciences depend when they need to quantify what they are doing. But it is more than that. Richard Feynman, a Nobel laureate in physics, said that the universe seems indescribable except in the language of mathematics. Sir Isaac Newton wanted a theoretical framework to express the motion of objects under the influence of gravity. This desire led to his law of universal gravitation as well as the calculus, one of the great achievements in the history of science. Albert Einstein spent years trying to formulate his insight that gravitation is really a reflection of the curvature of space–time, but couldn't express it mathematically until Marcell Grossman told him about Bernhard Riemann's work with curved space, following earlier work by Gauss, Bolyai, and Lobachevsky. This foundation enabled Einstein to develop the theory of relativity. Another Nobel laureate in physics, Steven Weinberg, has spoken of "spooky" coincidence, noting that physicists often find that mathematicians have been there before them. Group theory, for instance, was invented in the early nineteenth century to find the roots of polynomial equations. Evariste Galois made one of the most original leaps in the history of mathematics in introducing the concept of a group to describe symmetry. When physicists eventually discovered group theory in the twentieth century, they found it ideal for describing the conservation laws of energy, spin, momentum, and so on. The point is that even though Galois' motivation was purely internal to mathematics, his ideas eventually found application in the "real world."

Mathematicians have long been accused of losing themselves in the abstract beauty of their own conjectures. We all know that in geometry, for instance, we are studying infinitely small points, infinitely narrow lines, and perfectly round circles, all ideal objects. The concept of the ideal is as old as Plato, but in popular opinion is of little relevance to the real world. But abstraction is not a bad thing. Mathematicians concern themselves with internal consistency—that is, they are absolutely faithful to their own rules. They model the natural world on an idealized mathematical basis. They are criticized because they look at the world not as it is, but as it might be. This seems suspect until you realize that the world is not what it appears to be anyway. Who would imagine, for instance, by looking out a window, that the mass of an object becomes infinite when it is accelerated to near the speed of light? A mathematician might care about this phenomenon years before anyone else saw the need to care. Riemannian Geometry, for instance, was worked out 60 years before Einstein needed it to describe relativity and the algebra of Lie groups predated their application to particle physics by at least 30 years.

Another example of mathematics turning out to be "realer than real" was the discovery of the positron by the mathematical physicist Paul Dirac. While working on an equation describing the motion of the electron, Dirac realized his equation predicted the existence of a particle identical to the electron in every respect except charge. Experimental physicists then set out to look for this never-before-dreamed-of particle. So the discovery of the positron was a triumph of physics, but equally a triumph of mathematics.

The ulimate relevance of mathematical research is unpredictable. Armand Borel has likened the "shape" of mathematics to an iceberg. The portion above the surface is like applied mathematics, visible and apparent, but the bulk is below the surface, in the realm of pure mathematical research.

Around 1950, a mathematician named Herbert Hauptman became interested in the structure of crystals. Chemists knew about atoms in a crystal scattering X rays but couldn't pinpoint their position in the crystal itself. Hauptman was able to show that a special function called the *electron density function* could yield information about the phase of the X rays traveling through the crystal. From the phase information, one could then determine the atomic arrangement of the crystal. Even though he had taken only one chemistry course in his life, Hauptman won the 1985 Nobel Prize in Chemistry for this application of "classical" mathematics that had been available for over a century.

In another situation, a physicist, Allan Cormack, was trying to pinpoint the location and density of an object in the human body without surgery. At the time, only X rays were available and they could only give two-dimensional pictures. As it turned out, the mathematical solution of this problem had been around for many years, the work of a mathematician named Johann Radon. Using Radon's solutions, Cormack saw that by analyzing X rays from many different angles you can construct a three-dimensional image of an object in the body. This work led first to CAT scans (computer-assisted tomography) and later to MRI (magnetic resonance imaging), and then to the PET scan (positron emission tomography), an even more accurate technique.

Dirac's positron particle, itself predicted from mathematics, has become a tool used to measure both the metabolism and the anatomy of a living organism. Even more surprising, the analysis of this new emission tomography uses mathematics originally developed in the analysis of Soviet communication codes. The same techniques have now moved from medicine to paleoanthropology, where they have been used to determine that human ancestors walked around in an upright position. Radon's technique has also been applied to oceanography, where it is used to find ocean temperature in places where direct measurement is inconvenient or impossible, and has also been used to determine the distribution of stars near the sun, and the brightness of the moon. In 1979 Cormack won his own Nobel Prize—in Medicine.

Another instance of classical mathematics finding application in new technology comes from the mature subject of harmonic analysis, which has recently proved very useful in compressing data. Researchers at Yale have found that they can compress and restore almost any kind of image or sound by using mathematically generated shapes that resemble tiny waves—wavelets. These wavelets are so efficient at encapsulating information that it is estimated that the FBI's collection of 300 million fingerprints could be reduced by a factor of 20, and the time required to send the fingerprint information over the phone lines can be reduced from 20 minutes to one minute. This could save over $25 million that would otherwise be spent on optical storage disks alone. Scotland Yard in Britain plans to use wavelets for the same purpose.

Mathematical research is healthy today because internal barriers to its use are being broken down. After World War II, mathematics became more and more specialized to the point where a mathematician in one field often cannot communicate effectively with a mathematician in another field. This trend is partly offset by interesting problems that attract mathematicians in many different subfields. For example, Yang-Mills equations are differential equations that extend the famous equations of James Clerk Maxwell, who unified the phenomena of electricity and magnetism under the single theory of electromagnetism. Maxwell's equations and the Yang-Mills equations are so useful that they have been studied by mathematicians in many diverse fields, and this common interest has led to a rich exchange of mathematical ideas.

Mathematics is making numerous contributions to a whole range of other disciplines. These disciplines are challenging mathematicians with interesting problems that in turn lead to new mathematical applications. The more fundamental the mathematics involved, the wider the range of applications. A good example is the field of fluid dynamics. Underlying much of this field are the so-called Navier-Stokes equations. The "fluids" that they describe can be air, liquids, or even some solids. Mathematicians use these equations to study an incredible range of phenomena: blood flow in the heart; oil moving through porous ground; fuel mixing in carburetors; aircraft flying; crystals forming from liquids; plasma in fusion reactors; the motion of galaxies, clouds, winds, currents, hurricanes; and on and on. Even though a full description of the movement of fluids is too complex for complete theoretical understanding or even computer simulation, researchers make progress by combining theoretical

modeling, computer simulations, and experiments. These techniques are possible because of advances in computer hardware and software.

Computers are now far more than fast number crunchers. They allow mathematicians and scientists to do modeling in ways that were inconceivable a generation ago. Costly physical experiments can now be simulated more quickly, more cheaply, and with more confidence on a computer than by building a real model. In aircraft designs, for instance, wind tunnels have given way to shape optimization using software alone. These kinds of models have opened doors to techniques too complex to test in theory, such as the ways in which proteins fold and unfold, or how oil flows through deep rock. In recent years the hull design of yachts has been done by computer models. One specifies the characteristic to be optimized, like speed, and the computer generates the best design from that model. Other characteristics, like seaworthiness, are still best left to human designers. The point is that both techniques can now be combined to get optimum results.

New scientific problems are inspiring new mathematical work, and new discoveries in mathematics are finding application in science, and not just physical science. Mathematics is finally able to deal with some of the complexities of biological organisms. For instance, the unique capability of mathematics to discern patterns and organize information is starting to penetrate systems such as neural networks, the basic communication pathways of the brain. In yet another partnership, mathematicians and biologists are exploring the mechanisms of DNA replication. The mathematical subfields of knot theory, probability theory, and combinatorics, are all helping us understand the complex three-dimensional mechanics of DNA replication. Another important application for mathematics is epidemiology, which is taking advantage of high-speed computers to model epidemics. It's been found that AIDS, for instance, does not follow the usual pattern of infection found in other epidemics. The complexity of the models here is so great that even the fastest supercomputers are struggling, so mathematicians are engaged in using their skills to simplify the problems and models used so that the computers can handle the computations.

Computer modeling in industry has become so critical to economic progress that industries that do not use it will fall further and further behind. It is made possible by rapid advances in mathematical modeling as well as improvements in computer hardware and mathematical algorithms—that is, improved software. The design of microprocessor chips themselves is a highly mathematical process involving testing circuitry at hundreds or thousands of points. New algorithms can give quick solutions to problems involving thousands of points to within one percent of perfection. Minimizing the time involved in testing is crucial and advances in graph theory and computability have yielded better ways to do this.

One of the most interesting areas in which mathematics has formed a partnership with technology is the subject of *pattern theory*. The term was first introduced in the 1970s as a subfield of mathematics associated with computer vision, speech recognition, and artificial intelligence. Workers in this field are beginning to believe that this subject might encompass the basis of a universal theory of thought itself. From a practical point of view, pattern theorists are faced with the challenge of designing a machine that will translate speech into printed text. If everyone pronounced words in exactly the same way this would be an easy task, but we don't and it isn't.

In the past, theorists have tried to model the human brain as a kind of very complex computer, but this approach has failed in practice. A computer is designed to respond to logical input and computers are superb at this task. Some researchers now believe that the operation of the brain is fundamentally different from the design of a computer. What our brains perceive is not the raw sensory signal, which is usually fuzzy and ambiguous, but a remarkable reconstruction of that signal. The reconstruction makes use of memory, expectation, and logic. In this analysis, we do our thinking by use of pattern recognition and even though this is far from a full theory of cognition, it is already more successful than any competing theory. So we now find mathematics in the company of biology and behavioral psychology, propelling a new, broad, and intriguing theory that is only in its infancy.

I hope I have conveyed reasons for my belief that mathematics is extremely useful to our society. If this is true, one would assume that as a society we would vigorously support the research that leads to new uses of mathematics and that student interest would be at an all time high. Unfortunately this is not the case. The mathematics community has not effectively communicated to the public and their elected representatives that our subject is different, but no less valid, than the sciences. We do not design widgets or cure diseases, but our impact on science and engineering is tremendous and fundamental. It seems that mathematicians have dwelled so long in isolation that the public doesn't understand what it is we do.

In many cases mathematicians are disadvantaged by their training—it is too compressed to allow the development of both deep interests and broad knowledge, both of which are necessary for the kind of work I've described. Instead, there is much pressure for mathematicians to specialize early in careers in an area that has not been adequately worked, to apply for support to develop that research, and to get professional recognition for doing something of interest only to mathematics. This is true even though the intellectual trends within mathematics and related fields are very positive. There is an emerging balance between looking outward and looking inward that has not always been present, or even possible. Mathematicians are not accustomed to explaining to others what it is we do, much less promoting our skills as being useful to others.

One step toward better support from society is in producing better mathematics teachers. I can honestly say that the most important influence on my deciding to become a mathematician was my high school teacher, Lottie Wilson. Mrs. Wilson had the two essential qualities for getting her message across—she understood the majesty and the mystery of mathematics, and she knew there is no substitute for getting the right answer.

These are principles that came back to me when I found myself teaching calculus years later. Students expected "partial credit" on exams just because they got started on the solution or arrived at some "reasonable" but incorrect answer. I asked my class one day, "Suppose you end up being a doctor, are your patients going to be satisfied with a diagnosis that is partially correct?" This question was not well received, but my point was a serious one. Mathematics is one subject where there are definite answers and students can take real satisfaction in "getting it right." However, teachers have a responsibility in communicating the beauty as well as the utility of the subject.

It is a well-kept secret that mathematics is fun—at least to those who know some mathematics. I am still amazed at how often I hear the term "beautiful" to describe work satisfying to a mathematician. I am reminded of a story told by a mathematician named Jacques Tits. He was discussing the origin of human experiments with fire with a group of anthropologists. One said that he thought that humans were motived to understand fire by a need for cooking food. Another said that they were probably after a dependable source of heat. But Jacques said that he believed fire came under human control because of a fascination with the flame. I agree that the best mathematicians are fascinated by the flame. Fortunately for society, their fascination has in the end provided both the good cooking and the reliable heat that we all need.

Phillip A. Griffiths is a native of Raleigh, North Carolina, and is director of the Institute for Advanced Study in Princeton, New Jersey. Prior to joining the Institute he was professor of mathematics at both Harvard and Duke Universities. Dr. Griffiths is Chairman of the Committee on Science, Engineering, and Public Policy of the National Academy of Sciences, National Academy of Engineering, and Institute of Medicine. He is a member of the National Science Board and the American Philosophical Society. He received his doctorate in mathematics from Princeton University.

6

APPLICATIONS OF THE INTEGRAL

In the last chapter we saw a geometric interpretation of the definite integral: for a nonnegative continuous function over a specified finite interval, the integral gives the area under the graph of the function. Then we extended this geometric interpretation to what we referred to as "net area," for functions that took on both positive and negative values. We also gave a physical interpretation: when the integrand represents the velocity function of an object moving along a line, the integral represents the net distance covered by the object. More generally, we saw that the net change in any function is given by the integral of its rate of change with respect to its independent variable (that is, the integral of the derivative of the function).

In this chapter we give two further extensions of the area interpretation of the integral. We will also show how certain types of **volumes** can be calculated by integration. As some further geometric applications, we show how the integral can be used to find the **length of a curve**, and how we can use integration to find the areas of certain types of **surfaces**. We also discuss interpretations of the integral as the **work** done by a variable force acting along a line, the total **force** on an object submerged in a liquid (such as the force on a dam), and the **center of mass** of a continuously distributed one-dimensional mass (such as a wire) or two-dimensional mass (such as a thin plate). Finally, we show some applications of integration in **probability theory** and in **economics**. These applications, although by no means exhaustive, illustrate some of the many areas where integration is useful.

In all the applications mentioned and many more, the integrals involved arise in a natural way as the limit of Riemann sums of continuous functions on closed intervals. Once you master the techniques for arriving at the integrals in our applications, you should be able to apply them to many other types of problems. Scientists and engineers routinely use integrals in solving problems related to their work or areas of research.

6.1 MORE ON AREAS

$$\text{Area} = \int_a^b f(x)\,dx$$

FIGURE 6.1

We know already that if f is a nonnegative continuous function on the closed interval $[a, b]$, then the area of the region bounded by the graph of f, the x-axis, and the lines $x = a$ and $x = b$ (the shaded region in Figure 6.1) is given by the definite integral

$$A = \int_a^b f(x)\,dx$$

In this section we give two extensions of this result.

Regions Bounded by Piecewise-Continuous Functions

FIGURE 6.2
A piecewise-continuous function

First, suppose f is not continuous but is still nonnegative and bounded on $[a, b]$. Suppose further that it has a finite number of discontinuities, and that for each point of discontinuity c in the open interval (a, b), $\lim_{x \to c^+} f(x)$ and $\lim_{x \to c^-} f(x)$ both exist. We also assume that at the endpoints the one-sided limits, $\lim_{x \to a^+} f(x)$ and $\lim_{x \to b^-} f(x)$, exist. Such a function is said to be **piecewise-continuous**. An example is shown in Figure 6.2.

Suppose f is such a piecewise-continuous function. Let us designate a by x_0, b by x_n, and the consecutive points of discontinuity between a and b by $x_1, x_2, \ldots, x_{n-1}$. Then we define the area of the region under the graph of f between a and b to be

$$A = \sum_{k=1}^{n} \int_{x_{k-1}}^{x_k} f(x)\,dx$$

The fact that each of the integrals $\int_{x_{k-1}}^{x_k} f(x)\,dx$ exists follows from the continuity of f on (x_{k-1}, x_k) and the existence of the right-hand and left-hand limits, $\lim_{x \to x_{k-1}^+} f(x)$ and $\lim_{x \to x_k^-} f(x)$. (See Exercise 77 in Exercise Set 5.3.)

EXAMPLE 6.1   Find the area under the graph of

$$f(x) = \begin{cases} 1 - x^2 & \text{if } -1 \le x < 0 \\ 2 - x & \text{if } 0 \le x < 1 \\ 0 & \text{if } x = 1 \\ x^2 - 2x + 3 & \text{if } 1 < x \le 2 \end{cases}$$

from $x = -1$ to $x = 2$.

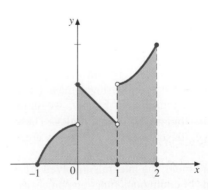

FIGURE 6.3

Solution   The graph of f is shown in Figure 6.3. Since f is nonnegative and piecewise-continuous, the area is given by

$$A = \int_{-1}^{0} f(x)\,dx + \int_{0}^{1} f(x)\,dx + \int_{1}^{2} f(x)\,dx$$

$$= \int_{-1}^{0} (1 - x^2)\,dx + \int_{0}^{1} (2 - x)\,dx + \int_{1}^{2} (x^2 - 2x + 3)\,dx$$

$$= \left[x - \frac{x^3}{3} \right]_{-1}^{0} + \left[2x - \frac{x^2}{2} \right]_{0}^{1} + \left[\frac{x^3}{3} - x^2 + 3x \right]_{1}^{2}$$

$$= \left[0 - \left(-1 + \frac{1}{3} \right) \right] + \left[\left(2 - \frac{1}{2} \right) - 0 \right] + \left[\left(\frac{8}{3} - 4 + 6 \right) - \left(\frac{1}{3} - 1 + 3 \right) \right]$$

$$= \frac{9}{2} \qquad\blacksquare$$

Areas of Regions Between Two Curves

FIGURE 6.4

FIGURE 6.5

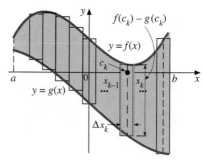

FIGURE 6.6

As a second extension of the use of integration to find areas, we consider the area of the region between the graphs of two continuous functions f and g from $x = a$ to $x = b$, where $f(x) \geq g(x)$ for all $x \in [a, b]$. Figure 6.4 illustrates a typical situation. Note that neither function needs to be nonnegative. To find the area, we will raise both curves by a constant amount k so that the entire region between them lies above the x-axis (Figure 6.5). The region R' between the graphs of $f + k$ and $g + k$ has the same area as the region R between the graphs of f and g.

To get the area of R', we subtract the area under $g + k$ from the area under $f + k$ to get

$$A = \int_a^b [f(x) + k]dx - \int_a^b [g(x) + k]dx$$

$$= \int_a^b [f(x) + k - g(x) - k]dx$$

or

$$A = \int_a^b [\overbrace{f(x)}^{\substack{\text{upper}\\\text{curve}}} - \overbrace{g(x)}^{\substack{\text{lower}\\\text{curve}}}]dx \qquad (6.1)$$

Observe that the constant k has disappeared. We could have obtained the integrand simply by subtracting the ordinate (the y value) of the original lower curve from the ordinate of the original upper curve. This observation means that it is no longer necessary to raise the curves. We did so to arrive at Equation 6.1, but now that we have the result, we can forget about the shift upward.

It is instructive to see how we can also arrive at the result in Equation 6.1 directly, as a limit of Riemann sums, by partitioning the closed interval $[a, b]$ and forming rectangles as shown in Figure 6.6. The kth rectangle has width $\Delta x_k = x_k - x_{k-1}$ and height $f(c_k) - g(c_k)$, where c_k is any point in the interval $[x_{k-1}, x_k]$, and so its area is $[f(c_k) - g(c_k)]\Delta x_k$. The sum of all such areas,

$$\sum_{k=1}^{n} [f(c_k) - g(c_k)]\Delta x_k$$

is a Riemann sum for the continuous function $f - g$, and it approximates the area we want. Its limit, as the width of the largest subinterval approaches 0, is exactly this area; that is,

$$A = \lim_{\max \Delta x_k \to 0} \sum_{k=1}^{n} [f(c_k) - g(c_k)]\Delta x_k = \int_a^b [f(x) - g(x)]dx$$

We summarize our result as follows.

Area Between Two Curves

Let f and g be continuous, with $f(x) \geq g(x)$, for $a \leq x \leq b$. Then the area A of the region between the graphs of f and g from $x = a$ to $x = b$ is given by

$$A = \int_a^b [f(x) - g(x)] \, dx$$

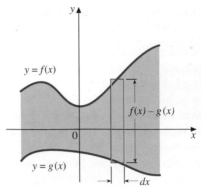

FIGURE 6.7

A rectangle such as the one shown in blue in Figure 6.6 is called a *typical element of area*. The sum of such typical elements is a Riemann sum, and the limit of this sum as max $\Delta x_k \to 0$ is the integral. A useful informal device for setting up the integral is shown in Figure 6.7. There we indicate a typical element of area on the figure but label its width as the differential dx, rather than Δx_k, and its height as $f(x) - g(x)$, rather than $f(c_k) - g(c_k)$. The area of the typical element is then $[f(x) - g(x)]\,dx$. To get the total area, all we need to do is affix the integral sign $\int_a^b$. Since the integral is a limit of Riemann sums, we can think of the integral sign as a sort of idealized summation symbol. So we are thinking of adding up (in the limiting sense) all typical elements of area. We can summarize the procedure as follows.

Shorthand Technique for Finding the Area Between Two Curves

To find the area between $y = f(x)$ and $y = g(x)$ from $x = a$ to $x = b$, where $f(x) \geq g(x)$:

Step 1. Draw a figure.

Step 2. Find a and b if they are not given.

Step 3. Show a typical element of area with its width labeled dx and its height $f(x) - g(x)$.

Step 4. Write the area of the typical element as $[f(x) - g(x)]dx$.

Step 5. Affix the integral sign $\int_a^b$ to the area in step 3 and perform the integration.

REMARK
■ The procedure we have outlined is a useful memory device for setting up the correct integral for the area. Let us emphasize, however, that the mathematically precise way of arriving at this result is by Riemann sums, as we previously indicated.

EXAMPLE 6.2   Find the area of the region bounded by the curves $y = 4 - x^2$ and $y = x^3 - x^2 - 3x$ between $x = -1$ and $x = 2$.

Solution   Let $f(x) = 4 - x^2$ and $g(x) = x^3 - x^2 - 3x$. We show the graphs of f and g in Figure 6.8, along with a typical element of area. We can either use Equation 6.1 directly or follow the five-step shorthand procedure just given. Let us follow the latter procedure. We have already completed the first three steps. The typical element of area is

$$[f(x) - g(x)]dx = [(4 - x^2) - (x^3 - x^2 - 3x)]dx$$

$$= (-x^3 + 3x + 4)dx$$

So the area is

$$\int_{-1}^{2}(-x^3 + 3x + 4)dx = -\frac{x^4}{4} \;\bigg|\; \frac{3x^2}{2} \;\bigg|\; 4x \bigg]_{-1}^{2}$$

$$= \left(-\frac{16}{4} + \frac{12}{2} + 8\right) - \left(-\frac{1}{4} + \frac{3}{2} - 4\right)$$

$$= \frac{51}{4}$$

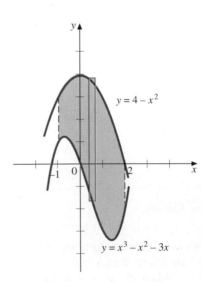

FIGURE 6.8

Sometimes the upper or lower boundaries of the region may consist of more than one curve. In such a case it may be necessary to divide the region into two or more subregions. The next example illustrates this situation.

EXAMPLE 6.3   Find the area of the region bounded by the curves $y = \sqrt{2x}$, $x + y = 4$, and $x = 3y$ that lies to the left of the line $x + y = 4$.

Solution   We find the points of intersection of the bounding curves to be $(0, 0)$, $(2, 2)$, and $(3, 1)$. As we see in Figure 6.9, the upper boundary of the region is $y = \sqrt{2x}$ if $0 \le x \le 2$, and it is $y = 4 - x$ if $2 \le x \le 3$. The lower boundary is $y = x/3$ ($x = 3y$, solved for y) for the entire region. We designate the two parts of the region by R_1 and R_2, as shown, and we find the areas of R_1 and R_2 and add the two results. This time, let us use Equation 6.1 directly.

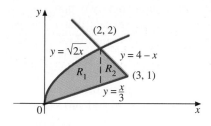

$y = \sqrt{2x}$

$(2, 2)$

$y = 4 - x$

R_1 R_2 $(3, 1)$

$y = \dfrac{x}{3}$

FIGURE 6.9

$$\text{area of } R_1 = \int_0^2 \left(\overbrace{\sqrt{2x}}^{\text{upper}} - \overbrace{\frac{x}{3}}^{\text{lower}} \right) dx = \int_0^2 \left(\sqrt{2}\,x^{1/2} - \frac{x}{3} \right) dx$$

$$= \sqrt{2}\,x^{3/2} \cdot \frac{2}{3} - \frac{x^2}{6} \Big]_0^2 = \frac{2\sqrt{2}}{3}(2^{3/2}) - \frac{4}{6} = \frac{8}{3} - \frac{2}{3} = 2$$

$$\text{area of } R_2 = \int_2^3 \left(\overbrace{4 - x}^{\text{upper}} - \overbrace{\frac{x}{3}}^{\text{lower}} \right) dx = \int_2^3 \left(4 - \frac{4x}{3} \right) dx$$

$$= 4x - \frac{2x^2}{3} \Big]_2^3 = (12 - 6) - \left(8 - \frac{8}{3} \right) = \frac{2}{3}$$

The total area is therefore $2 + \frac{2}{3} = \frac{8}{3}$.   ∎

Horizontal Elements of Area

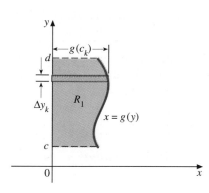

$g(c_k)$

d

Δy_k R_1

$x = g(y)$

c

FIGURE 6.10

Sometimes when we want to find the area of a region, the roles of x and y are reversed; that is, the region whose area we want is between the y-axis and a curve whose equation is of the form $x = g(y)$ from $y = c$ to $y = d$, where $g(y) \ge 0$, as in Figure 6.10. Or it may lie between two curves, $x = g(y)$ and $x = h(y)$, with $g(y) \ge h(y)$, as in Figure 6.11. All the theory we have had so far on areas can be modified in obvious ways to get the areas

$$A = \int_c^d g(y)\,dy \tag{6.2}$$

in the first case, and

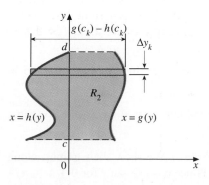

$g(c_k) - h(c_k)$

d

Δy_k

R_2

$x = h(y)$ $x = g(y)$

c

FIGURE 6.11

$$A = \int_c^d [\overbrace{g(y)}^{\substack{\text{rightmost}\\\text{curve}}} - \overbrace{h(y)}^{\substack{\text{leftmost}\\\text{curve}}}]\,dy \tag{6.3}$$

in the second.

The shorthand technique using typical elements of area, appropriately labeled, can be modified in obvious ways for the situations shown in Figures 6.10 and 6.11, interchanging the roles of x and y. When the typical rectangular element is vertical, then the integration is with respect to x (indicated by dx). When it is horizontal, the integration is with respect to y (indicated by dy).

EXAMPLE 6.4   Find the area of the region bounded by the curves $y = x^2$, $y = 6 - x$, and $y = 0$.

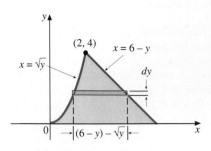

FIGURE 6.12

Solution   We show the region in Figure 6.12. We first find the point $(2, 4)$ of intersection of the curves $y = x^2$ and $y = 6 - x$ by solving the two equations simultaneously. One way of solving the problem is to use vertical elements (integration with respect to x) and divide the region into two parts, as we did in Example 6.3. Looking at the graph, however, reveals that if we use horizontal elements (integration with respect to y), the region does not have to be divided. Its right-hand boundary is always $y = 6 - x$ and its left-hand boundary is always $y = x^2$. These boundaries must be expressed as functions of y, namely, $x = 6 - y$ and $x = \sqrt{y}$, respectively. Using either Equation 6.3 with $g(y) = 6 - y$ and $h(y) = \sqrt{y}$ or the five-step shorthand procedure, we arrive at the result

$$A = \int_0^4 (6 - y - \sqrt{y})dy = 6y - \frac{y^2}{2} - \frac{2}{3}y^{3/2} \Big]_0^4$$

$$= 24 - 8 - \frac{16}{3} = \frac{32}{3}$$ ∎

Areas Between Curves That Cross Each Other

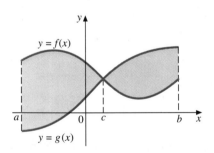

FIGURE 6.13

Finally, we note that if two curves cross each other at some point, then finding the area between them requires that we split the interval of integration at the crossing point, since the upper and lower (or rightmost and leftmost) functions are interchanged at this point. Only in this way is each area positive. For example, consider the region pictured in Figure 6.13. To get the area between f and g from $x = a$ to $x = b$, we use

$$A = \int_a^c \overbrace{[f(x)}^{\text{upper}} - \overbrace{g(x)]}^{\text{lower}}dx + \int_c^b \overbrace{[g(x)}^{\text{upper}} - \overbrace{f(x)]}^{\text{lower}}dx \qquad (6.4)$$

Equation 6.4 can be written in the compact form

$$A = \int_a^b |f(x) - g(x)|dx \qquad (6.5)$$

since

$$|f(x) - g(x)| = \begin{cases} f(x) - g(x) & \text{if } a \le x \le c \\ g(x) - f(x) & \text{if } c \le x \le b \end{cases}$$

To apply Equation 6.5, we must split the integral at the crossing point, so in effect we are applying Equation 6.4.

EXAMPLE 6.5   Find the area between the curves $y = \cos x$ and $y = \sin x$ for $0 \le x \le \pi$.

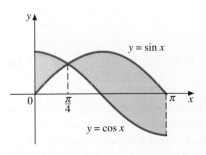

FIGURE 6.14

Solution   Let $f(x) = \cos x$ and $g(x) = \sin x$. We show their graphs, along with the region in question, in Figure 6.14. Since $\cos x \ge \sin x$ for $0 \le x \le \pi/4$ and $\sin x \ge \cos x$ for $\pi/4 \le x \le \pi$, we have

$$|f(x) - g(x)| = \begin{cases} \cos x - \sin x & \text{if } 0 \le x \le \pi/4 \\ \sin x - \cos x & \text{if } \pi/4 \le x \le \pi \end{cases}$$

So by Equation 6.5, since $\sin\frac{\pi}{4} = \cos\frac{\pi}{4} = \frac{\sqrt{2}}{2}$, $\sin 0 = \sin\pi = 0$, $\cos 0 = 1$, and $\cos\pi = -1$,

$$A = \int_0^\pi |\cos x - \sin x| dx = \int_0^{\pi/4}(\cos x - \sin x)dx + \int_{\pi/4}^\pi (\sin x - \cos x)dx$$

$$= [\sin x + \cos x]_0^{\pi/4} + [-\cos x - \sin x]_{\pi/4}^\pi$$

$$= \left(\frac{\sqrt{2}}{2} + \frac{\sqrt{2}}{2} - 0 - 1\right) + \left(1 - 0 + \frac{\sqrt{2}}{2} + \frac{\sqrt{2}}{2}\right)$$

$$= 2\sqrt{2}$$ ∎

FINDING THE AREA BETWEEN CURVES USING COMPUTER ALGEBRA SYSTEMS

A computer algebra system can be used as an aid in exploring virtually all of the applications presented in this chapter. As an example, we show how a computer algebra system can be useful when finding the area bounded by several curves.

CAS 21

Estimate the area bounded by the curves $y = \sin x$ and $y = x^2$ on the interval $[0, 2]$.

A sketch of the two curves would be a good place to start.

Maple:

```
f:=x–>sin(x);
g:=x–>x^2;
plot({f(x),g(x)},x=–2..2);
```

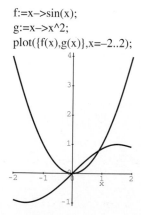

Mathematica:

```
f[x_] = Sin[x]
g[x_] = x^2
Plot[{f[x],g[x]},{x,–2,2}]
```

DERIVE:

(At the □ symbol, go to the next step.)
a (author) □ sin(x) □ p (plot window)
□ p (plot) □ a (algebra) □ a (author) □
x^2 □ p (plot window) □ p (plot)

FIGURE 6.1.1
$y = f(x) = \sin x$, $y = g(x) = x^2$

To estimate the area between the two curves, we see that the point of intersection near $x = 1$ must be found or estimated. Then the interval $[0, 2]$ can be split into two intervals, in which we integrate $f - g$ on the first part and $g - f$ on the second.

Maple:

P:=fsolve(f(x)=g(x),x);

Output: P := .8767262154

A:=int(f(x)–g(x),x=0..P)
  +int(g(x)–f(x),x=P..2);
evalf(A);

Output: 1.521914845

Mathematica:

P = FindRoot[f[x] == g[x],
  {x,1}]

Output: {x—> 0.876726}

P = 0.876726
A = Integrate[f[x]–g[x],
  {x,0,P}]+
  Integrate[g[x]–f[x],
  {x,P,2}]
N[%,10]

Output: 1.521914845

DERIVE:*

(At the □ symbol, go to the next step.)
a (author) □ x^2–sin(x) □ 1 (solve) □
s (simplify) □ [choose expression]
DERIVE does not seem to be able to
solve this. Let's try approximation.
x (approximate) □ [choose expression]
This also fails to work. Let's use Newton's Method as in CAS 18.
a (author) □ newton(x):=x–((x^2–
sin(x)))/dif(x^2–sin(x),x) □ iterates
(newton(x),x, 1,6) □ x (approximate)
□ [choose expression]

Output: 0.876726

a (author) □ P:=0.876726 □ a (author)
□ int(sin(x)–x^2,x,0,P)+int(x^2–sin(x),
x,P,2) □ x (approximate) □ [choose
expression]

Output: 1.52191

---

*Notice that in DERIVE an integral can be computed by selecting the "integrate" option from the calculus
menu or it can be authored using **int** as we did above.

$\dfrac{3}{2}$

Exercise Set 6.1

In Exercises 1–10, find the area of the region between the x-
axis and the graph of f from x = a to x = b, where [a, b] is
the interval given. Sketch the graph in each case.

1. $f(x) = \begin{cases} x^2 & \text{if } 0 \le x < 2 \\ 4 - x & \text{if } 2 \le x \le 4 \end{cases}$ on $[0, 4]$

2. $f(x) = \begin{cases} -x & \text{if } -1 \le x < 0 \\ 1 & \text{if } x = 0 \\ 2 - x & \text{if } 0 < x \le 1 \end{cases}$ on $[-1, 1]$

3. $f(x) = \begin{cases} 4 - x^2 & \text{if } 0 \le x < 2 \\ x - 2 & \text{if } 2 \le x \le 4 \end{cases}$ on $[0, 4]$

4. $f(x) = \begin{cases} x^2 & \text{if } 0 \le x < 1 \\ 2x - x^2 & \text{if } 1 \le x \le 2 \end{cases}$ on $[0, 2]$

5. $f(x) = \begin{cases} \sqrt{x + 4} & \text{if } -4 \le x < 0 \\ 1 & \text{if } x = 0 \\ \dfrac{x^2}{2} & \text{if } 0 < x \le 2 \end{cases}$ on $[-4, 2]$

6. $f(x) = \begin{cases} x^3 + 1 & \text{if } -1 \le x < 1 \\ x^2 - 4x + 5 & \text{if } 1 \le x \le 3 \end{cases}$ on $[-1, 3]$

7. $f(x) = \begin{cases} \cos x & \text{if } -\dfrac{\pi}{2} \le x < 0 \\ \sin x & \text{if } 0 \le x \le \pi \end{cases}$ on $\left[-\dfrac{\pi}{2}, \pi\right]$

8. $f(x) = \begin{cases} \cos \dfrac{x}{2} & \text{if } -\pi \le x < \dfrac{\pi}{2} \\ \sin \dfrac{x}{2} & \text{if } \dfrac{\pi}{2} \le x \le 2\pi \end{cases}$ on $[-\pi, 2\pi]$

9. $f(x) = \begin{cases} 2 \sin \dfrac{3x}{2} & \text{if } 0 \le x < \dfrac{2\pi}{3} \\ 4 + 2 \sin \dfrac{3x}{2} & \text{if } \dfrac{2\pi}{3} \le x \le \dfrac{4\pi}{3} \end{cases}$ on $\left[0, \dfrac{4\pi}{3}\right]$

10. $f(x) = \begin{cases} x^2 + 2x + 2 & \text{if } -1 \le x < 0 \\ 1 + \sin \dfrac{\pi x}{2} & \text{if } 0 \le x \le 2 \end{cases}$ on $[-1, 2]$

In Exercises 11–18, find the area of the shaded region.

11.

$y = (x-2)^2$

$y = 4$

12.

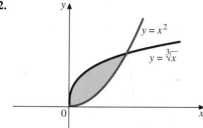

$y = x^2$

$y = \sqrt[3]{x}$

13.

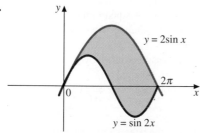

$y = 2\sin x$

2π

$y = \sin 2x$

14.

$y = \sqrt{x}$

$x + 2y = 3$

15.

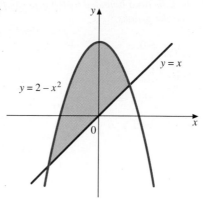

$y = x$

$y = 2 - x^2$

16.

$y = x^2$

$y^2 = x + 4$

17.

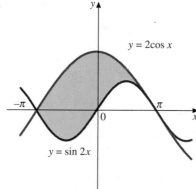

$y = 2\cos x$

$-\pi$ π

$y = \sin 2x$

18.

$2y = 3x^2$

$y = 1 + 3x^2 - x^4$

In Exercises 19–38, find the area bounded by the graphs of the given equations. Sketch the graphs.

19. $y = x^2 + 1$, $y = x$, $x = 0$, $x = 1$

20. $y = 4 - x^2$, $y = x^2 - 1$, $x = -1$, $x = 1$

21. $y = x^3 - 4x$, $y = 8 - x^2$, $x = -2$, $x = 2$

22. $x^2 + y = 0$, $x - y + 2 = 0$, $x = -1$, $x = 1$

23. $y = x^2$, $y = x + 2$

24. $x + y = 4$, $y = -\sqrt{x + 5}$, $x = -5$, $x = 4$

25. $x - y = 3$, $x^2 - 3x - y = 0$

26. $y = 8 - x^2$, $y = x^2$

27. $y = x^2$, $y^2 = x$

28. $y = x^2 - 3x + 4$, $y = -x^2 + 5x - 2$

29. $y = \sin \pi x$, $y = 4 + \cos 2\pi x$, $x = 0$, $x = 2$

30. $y = \sin \dfrac{\pi x}{2}$, $y = x^2 - 2x$

31. $y = \sin 2x$, $y = \cos x$, between $x = -\dfrac{\pi}{2}$ and $x = \dfrac{\pi}{6}$

32. $x = 4 - y^2$, $x = 0$

33. $x = y^2 - 2$, $x = y$

34. $y = \sqrt{x}$, $x - y = 2$, $y = 0$

35. $x = 2y + 1$, $x = y - 1$, $y = 2$

36. $x = y^3$, $y = x^2$ (Do this in two ways.)

37. $x = y^2 - 4y - 3$, $x = 1 - 2y^2$

38. $x = 2(y - 2)^2$, $x = y^2 - 4y + 8$

39. Find the area bounded by the graphs of the equations $y = 2x^2$, $x + y = 3$, and $y = 0$ (a) by using x as the independent variable and (b) by using y as the independent variable.

40. Follow the instructions for Exercise 39 for the area bounded by $x = \sqrt{y + 1}$, $x = 2 + \sqrt{3 - y}$, and $y = -1$.

41. Use integration to find the area of the triangle with vertices $(-1, 1)$, $(3, 2)$, and $(5, 4)$.

 In Exercises 42–44, use a graphing calculator or CAS to obtain the graphs.

42. Find the total area bounded by the curves $y = x^3 - 3x^2 - 2x + 4$ and $y = 1 - x$.

43. Find the area between the graphs of $f(x) = 1 + 2\sin x$ and $g(x) = \cos 2x$ on $[0, 2\pi]$.

44. Find the area between the graphs of $x = \sin \dfrac{\pi y}{2}$ and $y = x$.

45. Find the area of the region bounded on the left by the curves $x^2 - 2x - 4y + 9 = 0$ and $x + 2y = 5$ and on the right by $y^2 - 4y + x = 0$.

46. Find the limit of the ratio of the area of triangle AOB to the area of the parabolic region (in darker red) as $\alpha \to 0$.

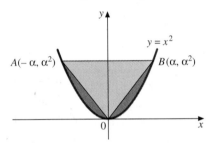

47. Suppose the area of the region bounded above by $y = f(x)$ and below by the x-axis, between the lines $x = 0$ and $x = 1$, equals A square units.
   (a) Show that the area bounded by $y = f(2x)$ on $[0, \frac{1}{2}]$ is $A/2$.
   (b) When are the areas bounded by $y = f(x)$ and $y = f(2x)$ on $[0, 1]$ equal?
   (c) Find the area bounded by $y = f(x)$ and $y = 2f(x)$ on $[0, 1]$.

48. Use a CAS to find the area of the region bounded by the two curves $f(x) = x^4 - 3x^2 + 2$ and $g(x) = x^3 + 7x^2 + x - 10$.

6.2 FINDING VOLUMES USING CROSS SECTIONS

We have seen how the integral is used in defining areas. In this section we will see that integrals also can be used to define volumes of certain solids. Consider the solid depicted in Figure 6.15. It is bounded on the left by a plane perpendicular to the x-axis at $x = a$ and bounded on the right by a plane perpendicular to the x-axis at $x = b$. If we pass a plane perpendicular to the x-axis at any point x between a and b, its intersection with the solid is a plane region called a **cross section** with respect to the x-axis. We denote the area of such a cross section by $A(x)$. *Throughout the remainder of this section we will consider only solids for which $A(x)$ exists and is a continuous function of x on the closed interval $[a, b]$.*

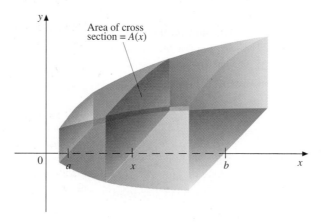

Area of cross section = $A(x)$

FIGURE 6.15

To find the volume of the solid, we partition the closed interval $[a, b]$ in the usual way. If we pass planes perpendicular to the x-axis at each of the partition points, we divide the solid into n parallel slices (like a sliced hard-boiled egg). In Figure 6.16 we have pictured the solid sliced in this way, highlighting a typical slice with faces that are the cross sections taken at x_{k-1} and x_k. The entire volume of the solid is the sum of the volumes of all the slices. In general, the cross section at x_{k-1} will be different from the one at x_k, but for Δx_k small, if we replace the kth slice with one that has constant cross-sectional area $A(c_k)$, as shown on the right in Figure 6.16, the volume will not differ much from that of the actual slice. We call this slice, which has width Δx_k and cross-sectional

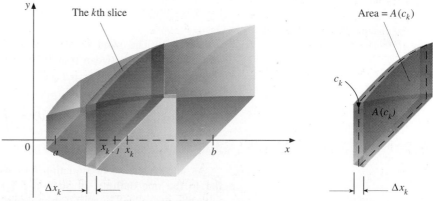

The kth slice

Area = $A(c_k)$

c_k

$A(c_k)$

Δx_k

Δx_k

FIGURE 6.16

area $A(c_k)$, a *typical element of volume* and denote its volume by ΔV_k. Its volume is defined as the *area of the cross section times the thickness*:

$$\Delta V_k = A(c_k)\Delta x_k$$

and the total volume is approximated by the sum of all the ΔV_k values:

$$V \approx \sum_{k=1}^{n} \Delta V_k = \sum_{k=1}^{n} A(c_k)\Delta x_k$$

The sum on the right is a Riemann sum that approaches the integral $\int_a^b A(x)dx$ as we take finer and finer partitions, which leads to the following definition.

Definition 6.1
Volumes Using
Cross Sections

If a solid is bounded on the left by a plane perpendicular to the x-axis at $x = a$ and on the right by a plane perpendicular to the x-axis at $x = b$, and if the area $A(x)$ of the cross section with respect to the x-axis of the solid at the point x is continuous for all x in the closed interval $[a, b]$, then the **volume** V of the solid is given by

$$V = \int_a^b A(x)dx \tag{6.6}$$

FIGURE 6.17

If the solid is bounded by planes perpendicular to the y-axis at $y = c$ and $y = d$, and if the cross-sectional area with respect to the y-axis is a continuous function $A(y)$ on $[c, d]$, then similar reasoning leads to the equation

$$V = \int_c^d A(y)dy \tag{6.7}$$

(See Figure 6.17.)

EXAMPLE 6.6   The base of a solid is the region in the xy-plane bounded by the graphs of $y = \sqrt{x}$, $y = 0$, and $x = 4$, and each cross section of the solid perpendicular to the x-axis is a square. Find the volume of the solid.

Solution   The solid is depicted in Figure 6.18. The cross section at an arbitrary point x between 0 and 4 is a square whose base has length $y = \sqrt{x}$. Thus, the area $A(x) = y^2 = (\sqrt{x})^2 = x$. So by Equation 6.6, the volume is

$$V = \int_0^4 A(x)dx = \int_0^4 x\,dx = \frac{x^2}{2}\Bigg]_0^4 = 8 \qquad \blacksquare$$

As an alternative to Equation 6.6, we can use a shorthand procedure similar to the one outlined in Section 6.1 for finding areas. As before, we must view the procedure as an aid to memory only and not a mathematically precise formulation. The final result, however, is correct.

FIGURE 6.18

Shorthand Technique for Finding Volumes Using Cross Sections

To find the volume of a solid with known cross-sectional area $A(x)$, between $x = a$ and $x = b$:

Step 1. Draw a figure.

Step 2. Find the integration limits a and b if they are not given.

Step 3. Show a typical element of volume and label its thickness dx and the area of its face $A(x)$.

Step 4. Write the volume of the typical element as $A(x)dx$.

Step 5. Affix the integral sign $\int_a^b$ to the volume in step 3 and perform the integration.

(a)

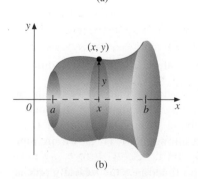

(b)

FIGURE 6.19

Obvious modifications in the procedure are needed when the cross-sectional areas are known with respect to the y-axis instead of the x-axis. We will illustrate the procedure after introducing a class of solids with simple cross sections.

Volumes of Revolution: Methods of "Disks" and "Washers"

The "Disk" Method

When a plane area is revolved about a line, the resulting solid is known as a **solid of revolution**. Cross sections with respect to the line are circular. For example, suppose the region R under the graph of a nonnegative continuous function f in Figure 6.19(a) is rotated about the x-axis. The result is the solid of revolution in part (b) of the figure. At any point x between a and b, the height of the curve is $y = f(x)$. Since the area of a circle of radius r is πr^2, the area of the circular cross section after rotation is

$$A(x) = \pi y^2 = \pi[f(x)]^2$$

The volume, by Equation 6.6, is therefore given by the following equation.

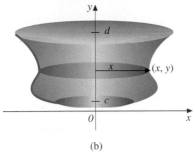

(a)

(b)

FIGURE 6.20

Volumes by the Disk Method: Rotation About the x-axis

$$V = \pi \int_a^b [f(x)]^2 dx \qquad (6.8)$$

For revolving about the y-axis instead of the x-axis, the analogous formula for a region bounded by $x = g(y)$, for $g(y) \geq 0$, between $y = c$ and $y = d$, is as follows.

Volumes by the Disk Method: Rotation About the y-axis

$$V = \pi \int_c^d [g(y)]^2 dy \qquad (6.9)$$

(See Figure 6.20.)

For a solid of revolution such as that pictured in Figure 6.19, the slicing technique produces elements of volume in the form of circular disks (like coins). We have shown a typical element in Figure 6.21. Because of the shape of such a typical element of volume, we often refer to the method described as the **disk method**.

FIGURE 6.21
Typical
element of volume
is a disk.

EXAMPLE 6.7   Find the volume of a cone of altitude h and base radius r.

Solution   For convenience, we orient the cone so that its vertex is at the origin and its axis is along the positive x-axis. The cone can be thought of as the solid of revolution formed by revolving the triangular region OAB in Figure 6.22(a) about the x-axis. The resulting cone is shown in Figure 6.22(b).

To apply Formula 6.8, we need to know the function $f(x)$ whose graph is the upper boundary of the region being rotated. In this case the graph is the

(a)

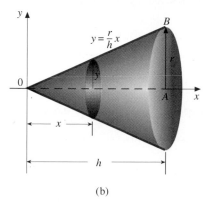

(b)

FIGURE 6.22

straight line through the origin, with slope $m = r/h$. Its equation is

$$y = \frac{r}{h}x$$

Thus, $f(x) = rx/h$, and from Formula 6.8 we obtain the volume as follows:

$$V = \pi \int_0^h [f(x)]^2\, dx = \pi \int_0^h \frac{r^2 x^2}{h^2}\, dx = \frac{\pi r^2}{h^2} \int_0^h x^2\, dx$$

$$= \frac{\pi r^2}{h^2} \left[\frac{x^3}{3}\right]_0^h = \frac{\pi r^2}{h^2}\left(\frac{h^3}{3}\right) = \frac{1}{3}\pi r^2 h$$

Since πr^2 is the area of the base of the cone, we can restate our result in the following words: *The volume of a right circular cone is one-third the area of the base times the altitude.* ■

EXAMPLE 6.8   Find the volume of the solid obtained by revolving the region under the graph of $y = \sqrt{x}$ from $x = 1$ to $x = 4$ about the x-axis.

Solution   This time we will illustrate the five-step shorthand technique. In Figure 6.23 we have shown in part (a) the region to be revolved, in part (b) a typical element of volume after rotation, and in part (c) the rotated solid. The typical volume element is a circular disk with radius $\sqrt{x}$, and we have labeled its thickness dx. Its volume is therefore $\pi(\sqrt{x})^2\, dx = \pi x\, dx$. Thus, the volume of the solid is

$$V = \int_1^4 \pi x\, dx = \pi \left[\frac{x^2}{2}\right]_1^4 = \pi \left[\frac{16}{2} - \frac{1}{2}\right] = \frac{15\pi}{2}$$ ■

(a)

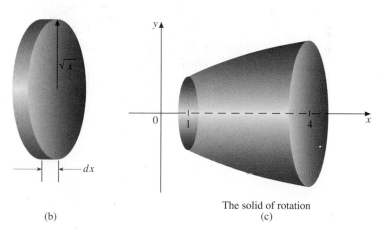

(b)

The solid of rotation
(c)

FIGURE 6.23

REMARK
■ Although up to now we have considered f (or g) to be nonnegative, this requirement is unnecessary, as the following reasoning shows. Suppose f is continuous on the closed interval $[a, b]$. Then the absolute value of f, $|f|$, is also continuous there (see Exercise 20 in Exercise Set 2.4) and is nonnegative. Figure 6.24 illustrates the graphs of f and $|f|$ for a typical function. When we rotate each of the shaded regions about the x-axis, we obtain identical solids,

(a)   (b)

FIGURE 6.24

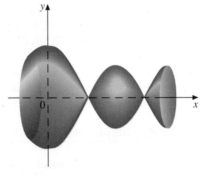

FIGURE 6.25
The same solid of revolution results from rotating either region in Figure 6.24.

shown in Figure 6.25. But $|f|$ satisfies the conditions for Equation 6.8 to hold. Thus,

$$V = \pi \int_a^b |f(x)|^2 dx$$

and since $|f|^2 = f^2$, we get

$$V = \pi \int_a^b [f(x)]^2 dx$$

That is, Equation 6.8 (and also 6.9) holds true even when we drop the requirement that f (or g) be nonnegative.

The "Washer" Method

If the region to be rotated lies between the graphs of two nonnegative continuous functions $y_1 = g(x)$ and $y_2 = f(x)$ and is entirely on one side of the axis of rotation, where $f(x)$ is always greater than or equal to $g(x)$ on the closed interval $[a, b]$, as shown in Figure 6.26, then the solid of revolution has a hole

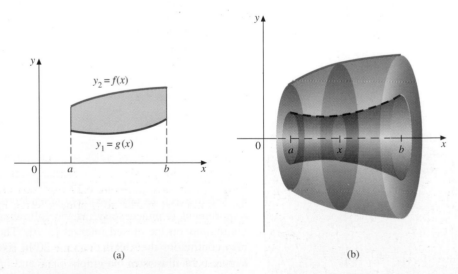

(a)   (b)

FIGURE 6.26

parallel to the axis in it. A cross section is the region between two concentric circles of radii y_1 and y_2, respectively. Thus, its area is

$$A(x) = \pi y_2^2 - \pi y_1^2 = \pi(y_2^2 - y_1^2) = \pi\{[f(x)]^2 - [g(x)]^2\}$$

The volume, by Equation 6.8, is therefore given by the following equation.

Volumes by the Washer Method: Rotation About the x-axis

$$V = \pi \int_a^b \{[f(x)]^2 - [g(x)]^2\}dx \qquad (6.10)$$

The analogous formula for a region bounded by the graphs of $x_1 = h(y)$ and $x_2 = g(y)$, where $g(y) \geq h(y) \geq 0$, between $y = c$ and $y = d$, revolved about the y-axis, is as follows.

Volumes by the Washer Method: Rotation About the y-axis

$$V = \pi \int_c^d \{[g(y)]^2 - [h(y)]^2\}dy \qquad (6.11)$$

The use of the dependent variable notations $y_1 = g(x)$ and $y_2 = f(x)$ for Equation 6.10 and $x_1 = h(y)$ and $x_2 = g(y)$ for Equation 6.11 simplifies the appearance of these formulas to

$$V = \pi \int_a^b (y_2^2 - y_1^2)dx \quad \text{about the } x\text{-axis}$$

and

$$V = \pi \int_c^d (x_2^2 - x_1^2)dy \quad \text{about the } y\text{-axis}$$

but you must remember that before integration can take place, the integrand must be expressed as a function of the variable of integration in each case.

Note that we can think of the volume in question as the volume within the outer surface minus the volume of the hole, just as though we drilled out a hole from the full volume.

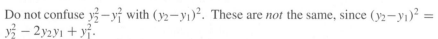

Do not confuse $y_2^2 - y_1^2$ with $(y_2 - y_1)^2$. These are *not* the same, since $(y_2 - y_1)^2 = y_2^2 - 2y_2y_1 + y_1^2$.

A typical element of volume in the slicing technique for solids of the type described is in the shape of a washer such as that used by carpenters or machinists, as shown in Figure 6.27 (for rotation about the x-axis). For this reason the method described here is often referred to as the **washer method.** You can think of a washer as a disk with a hole drilled through its center.

EXAMPLE 6.9   Find the volume of the solid obtained by revolving the region between the graphs of $y = x^2 + 1$ and $y = \sqrt{x}$ from $x = 0$ to $x = 1$ about the x-axis.

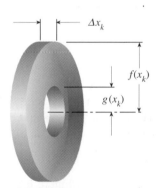

Δx_k

$f(x_k)$

$g(x_k)$

FIGURE 6.27
Typical element of volume is washer-shaped.

Solution   Let $f(x) = x^2 + 1$ and $g(x) = \sqrt{x}$. The region to be revolved is shown in Figure 6.28(a). By Equation 6.10, the volume is

$$V = \pi \int_0^1 [(x^2 + 1)^2 - (\sqrt{x})^2]dx$$

$$= \pi \int_0^1 (x^4 + 2x^2 + 1 - x)dx = \pi \left[\frac{x^5}{5} + \frac{2x^3}{3} + x - \frac{x^2}{2} \right]_0^1$$

$$= \pi \left[\frac{1}{5} + \frac{2}{3} + 1 - \frac{1}{2} \right] = \frac{41\pi}{30}$$ ■

(a)   (b)

FIGURE 6.28

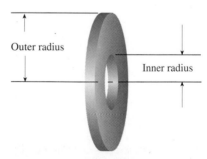

FIGURE 6.29
Area of cross section =
π [(outer radius)^2 − inner radius)^2]

As an aid in remembering Equations 6.10 and 6.11, observe that the area of a cross section for the washer method is

$$\pi[(\text{outer radius})^2 - (\text{inner radius})^2]$$

and this is the form of the integrand in each case. (See Figure 6.29.) As the next example shows, this formula for the cross-sectional area enables us to rotate areas about lines other than the x- and y-axes.

EXAMPLE 6.10   The region bounded by the graphs of $x = 2\sqrt{y}$ and $x = y$ is to be revolved about the line $x = 4$. Find the volume of the resulting solid of revolution.

Solution   Figure 6.30 shows the region in question, together with a typical element of volume of the revolved solid. Since the axis of revolution is vertical, we will use y as the variable of integration. By solving the two equations simultaneously, we find their points of intersection to be $(0, 0)$ and $(4, 4)$. (Verify.) Thus, the limits on y are 0 and 4. As shown in part (a) of the figure, we let x_1 and x_2 denote the distances from the y-axis to the leftmost and rightmost boundaries, respectively. Thus, for each y in $[0, 4]$,

$$x_1 = y \quad \text{and} \quad x_2 = 2\sqrt{y}$$

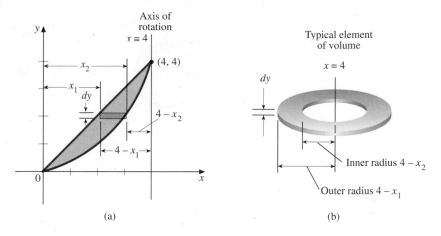

FIGURE 6.30

So the corresponding distances from the line $x = 4$ (the axis of rotation) are $4 - x_1$ and $4 - x_2$. As shown in Figure 6.30(b), these are the outer and inner radii, respectively, of the cross section of the solid of revolution taken at y:

$$\text{outer radius} = 4 - x_1 = 4 - y$$
$$\text{inner radius} = 4 - x_2 = 4 - 2\sqrt{y}$$

Thus, the area $A(y)$ of the cross section, by Equation 6.11, is

$$A(y) = \pi[(4 - y)^2 - (4 - 2\sqrt{y})^2]$$

and the volume of the solid is

$$V = \pi \int_0^4 A(y)dy = \pi \int_0^4 [(4 - y)^2 - (4 - 2\sqrt{y})^2]dy$$

$$= \pi \int_0^4 [16 - 8y + y^2 - 16 + 16\sqrt{y} - 4y]dy$$

$$= \pi \int_0^4 (y^2 - 12y + 16y^{1/2})dy = \pi \left[\frac{y^3}{3} - 6y^2 + \frac{32}{3}y^{3/2}\right]_0^4$$

$$= \pi \left[\frac{64}{3} - 96 + \frac{32}{3}(8)\right] = \frac{32\pi}{3} \qquad \blacksquare$$

EXAMPLE 6.11   A **torus**—a solid shaped like a doughnut—is generated by revolving a circle about an axis that does not intersect the circle. Figure 6.31(a) (on the following page) shows a region enclosed by a circle of radius r centered at $(0, R)$, with $R > r$. A torus is generated by revolving the region about the x-axis (Figure 6.31(b)). Use the washer method to find the volume of the torus. From your result, determine how many doughnuts can be made from a cube of dough 10 cm on a side, if the radius r of a cross section of the doughnut is 1 cm and the radius R is 4 cm.

Solution   The area of the circular shaded region in Figure 6.31(a) is πr^2, and since the center of the region is rotated a distance of $2\pi R$ (the circumference of a circle of radius R), intuitively we may guess that the volume of the torus

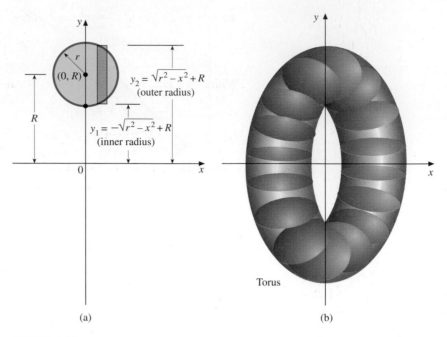

FIGURE 6.31

is

$$(\pi r^2)(2\pi R) = 2\pi^2 r^2 R$$

In Section 6.8 we will give a justification for this intuitive approach in the theorem called the First Theorem of Pappus. For now, we can verify the result using the washer method.

The circle in Figure 6.31 is simply the circle $x^2 + y^2 = r^2$ shifted vertically upward R units. The upper semicircle therefore has the equation

$$y_2 = \sqrt{r^2 - x^2} + R \qquad \text{Outer radius}$$

and the lower semicircle has the equation

$$y_1 = -\sqrt{r^2 - x^2} + R \qquad \text{Inner radius}$$

By Equation 6.10

$$V = \pi \int_{-r}^{r} (y_2^2 - y_1^2)\, dx = \pi \int_{-r}^{r} \left[\left(\sqrt{r^2 - x^2} + R \right)^2 - \left(-\sqrt{r^2 - x^2} + R \right)^2 \right] dx$$

$$= \pi \int_{-r}^{r} \left\{ \left[(r^2 - x^2) + 2R\sqrt{r^2 - x^2} + R^2 \right] \right.$$

$$\left. - \left[(r^2 - x^2) - 2R\sqrt{r^2 - x^2} + R^2 \right] \right\} dx$$

$$= \pi \int_{-r}^{r} 4R\sqrt{r^2 - x^2}\, dx$$

$$= 4\pi R \int_{-r}^{r} \sqrt{r^2 - x^2}\, dx$$

We have noted previously (see Example 5.8) that $\int_{-r}^{r} \sqrt{r^2 - x^2}\, dx$ is one-half the area of the circle $x^2 + y^2 = r^2$. Thus,

$$V = 4\pi R \left(\frac{\pi r^2}{2} \right) = 2\pi^2 r^2 R$$

With $r = 1$ cm and $R = 4$ cm, the volume of a doughnut is

$$V = 2\pi^2 \cdot 1^2 \cdot 4 = 8\pi^2 \approx 79 \text{ cm}^3$$

A cube of dough 10 cm on a side has a volume of $10^3 = 1,000$ cm^3. Since $1,000 \div 79 \approx 12.66$, we could make a dozen doughnuts—almost a baker's dozen—and have a little more than 50 cm^3 of dough left over.   ■

Exercise Set 6.2

In Exercises 1–10, find the volume of the solid obtained by revolving the region under the given curve between the specified limits about the x-axis.

1. $y = (x + 3)/2$ from $x = 0$ to $x = 1$

2. $y = x^2 + 1$ from $x = 0$ to $x = 2$

3. $y = 4 - x^2$ from $x = 0$ to $x = 2$

4. $y = 2x - x^2$ from $x = 0$ to $x = 2$

5. $y = \sqrt{x + 1}$ from $x = -1$ to $x = 3$

6. $y = \sqrt{2 - x}$ from $x = -2$ to $x = 2$

7. $y = x^3 + 1$ from $x = -1$ to $x = 1$

8. $y = \sqrt[3]{x}$ from $x = 1$ to $x = 8$

9. $y = \sqrt{25 - x^2}$ from $x = -4$ to $x = 3$

10. $y = 4x^2 - x^4$ from $x = -1$ to $x = 1$

In Exercises 11–20, find the volume of the solid obtained by rotating the region bounded by the given curves about the x-axis.

11. The x-axis and $y = x^2 - 1$ from $x = -2$ to $x = 2$

12. The x-axis and $y = \sqrt[3]{3x + 1}$ from $x = -3$ to $x = 0$

13. $y = x + 4$ and $y = 3 - \dfrac{x}{2}$ from $x = 1$ to $x = 4$

14. $y = x$ and $y = \sqrt{x}$

15. $y = x + 2$ and $y = x^2$

16. $y = \sqrt{2x - 3}$ and $y = \dfrac{x}{2}$

17. $y = x^2$ and $y = 2 - x^2$

18. $y = \sqrt{x}$ and $y = \sqrt{3x + 4}$ from $x = 0$ to $x = 4$

19. $y = 1$ and $y = \sqrt{5 - x^2}$

20. $y = 2/\sqrt[3]{x}$ and $y = \sqrt[3]{9 - x}$

In Exercises 21–30, find the volume of the solid obtained by rotating the region bounded by the given curves about the y-axis.

21. The y-axis and $x + 2y = 4$ from $y = 0$ to $y = 2$

22. The y-axis and $x = \sqrt{y}$ from $y = 4$ to $y = 9$

23. The y-axis and $x = y^{1/3}$ from $y = -1$ to $y = 8$

24. The y-axis and $x = y^2 - y - 2$

25. $y = 2x$, $x = 2y$, and $y = 2$

26. $y = 2x$, $y = \sqrt{3 - 4x}$, and the x-axis

27. $y = x$, $x = 6 - y^2$, and $x = 0$, for $x > 0$

28. $x = \sqrt{16 - y^2}$, $x = 10 - 2y$, $y = 0$, and $y = 4$

29. $y^3 + x = 0$, $y + 2 = 0$, and $x - 1 = 0$

30. $y^2 = x$ and $y^2 = 2x - 4$

31. Derive the formula for the volume of a sphere created by revolving the region under the semicircle $y = \sqrt{a^2 - x^2}$ about the x-axis.

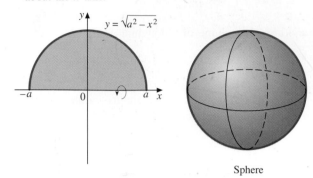

Sphere

32. If the region under the semiellipse $y = \frac{b}{a}\sqrt{a^2 - x^2}$ is revolved about the x-axis, the resulting solid is called an *ellipsoid of revolution*. Find its volume.

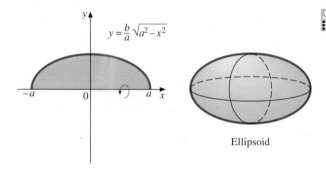

Ellipsoid

33. The region bounded by $y = x$, $x = 2$, and $y = 0$ is revolved about the line $y = 3$. Find the volume of the resulting solid.

34. Find the volume of the solid obtained by revolving the region under $y = \sqrt{x}$ from $x = 1$ to $x = 4$ about the line $y + 1 = 0$.

35. Find the volume of the solid obtained by revolving the region bounded by $x = \sqrt{y}$, $y = 8 - 2x$, and the x-axis about (a) the y-axis and (b) the line $x = 4$.

36. Find the volume of the solid obtained by revolving the region in the first quadrant bounded by $x^3 = 2y$ and $y = 2x$ about (a) the line $x = 2$ and (b) the line $y = 4$.

In Exercises 37–48, refer to the regions R_1, R_2, and R_3 in the figure. Find the volume of the solid of revolution formed by revolving the specified region about the line indicated.

37. R_1 about OA

38. R_1 about OC

39. R_1 about CB

40. R_1 about AB

41. R_2 about OA

42. R_2 about OC

43. R_2 about CB

44. R_2 about AB

45. R_3 about OA

46. R_3 about OC

47. R_3 about CB

48. R_3 about AB

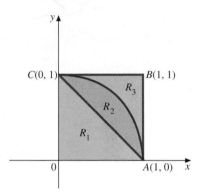

In Exercises 49–52, find the volume of the solid of revolution obtained by revolving the region between the graph of f and the x-axis from $x = a$ to $x = b$ about the x-axis. Use a graphing calculator or a CAS to sketch the regions.

49. $f(x) = \sin x + \cos x$; $a = -\dfrac{\pi}{3}$, $b = \dfrac{\pi}{6}$

50. $f(x) = \sec x$; $a = 0$, $b = \dfrac{\pi}{4}$

51. $f(x) = 2\csc x$; $a = \dfrac{\pi}{4}$, $b = \dfrac{3\pi}{4}$

52. $f(x) = \sqrt{2(1 - \sin x)}$; $a = -\dfrac{\pi}{2}$, $b = \dfrac{\pi}{6}$

In Exercises 53–56, find the volume using Definition 6.1.

53. The base of a solid is the triangular region bounded by $y = \frac{x}{2}$, $x = 4$, and $y = 0$, and each cross section perpendicular to the x-axis is a square. Find the volume.

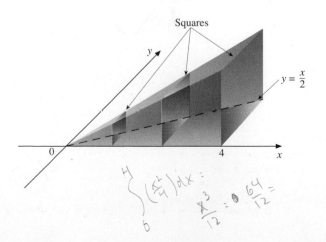

Squares

54. Find the volume of the pyramid shown in the figure, whose base is an isosceles right triangle with each leg 3 and altitude 5.

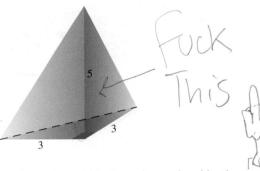

55. The base of a solid is the region enclosed by the parabola $y^2 = 2x$ and the line $x = 2$, and each cross section perpendicular to the x-axis is semicircular. Find the volume.

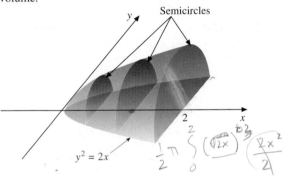

56. The base of a solid is the region enclosed by the graph of $y = 4 - x^2$ and the x-axis, and each cross section perpendicular to the y-axis is an isosceles triangle whose altitude is one-half its base. Find its volume.

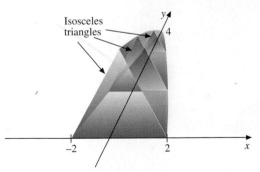

57. Find the volume of the solid obtained by revolving the leftmost region bounded by $y = 3x$, $y = x^2/2$, and $x + y = 4$ about the x-axis. (*Hint:* Divide the region into two parts.)

58. Find the volume of the solid obtained by revolving the region bounded by the x-axis and the curve $y = x^3 - x^2 - 2x$ about the x-axis.

59. An oil tank is in the shape of a sphere with radius 2 m,

and the oil in it is 1 m deep (see the figure). Find how many liters of oil are in the tank (1 m^3 = 1000 L).

60. In a sphere of radius a, at a point b units above the center on a vertical axis ($b < a$) a horizontal plane is passed. The portion of the sphere above the plane is called a *spherical segment*. Find its volume.

61. The base of a solid is the region inside the ellipse $4x^2 + 9y^2 = 36$, and each cross section perpendicular to the x-axis is an equilateral triangle. Find the volume of the solid.

62. The upper half of the region inside the *four-cusp hypocycloid* $x^{2/3} + y^{2/3} = a^{2/3}$ (see the figure) is revolved around the x-axis. Find the volume generated.

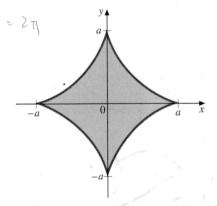

63. The region between the graphs of $y = \sqrt{x}$ and $y = 4 - \sqrt{x}$ from $x = 0$ to $x = 9$ is rotated about the line $y = -2$. Find the volume.

64. The region between the graphs of $y = \sin x$ and $y = \cos x$ from $x = \frac{\pi}{4}$ to $x = \pi$ is rotated about the line $y = 1$. Find the volume. (*Hint:* To perform the integration, make use of a trigonometric identity.)

65. The region bounded above by the graph of $f(x) = \sqrt[3]{x^3 + 2}$, below by the x-axis, and on each side by the vertical lines $x = 0$ and $x = 2$ is rotated about the x-axis. Approximate the volume of the solid of revolution using
(a) a right-hand Riemann sum with $n = 100$;
(b) Simpson's Rule with $n = 16$.

6.3 FINDING VOLUMES OF SOLIDS OF REVOLUTION USING CYLINDRICAL SHELLS

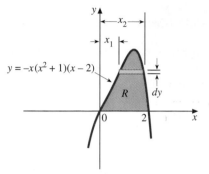

FIGURE 6.32

Using the method of disks or of washers to find the volume of a solid of revolution is sometimes difficult, or even impossible, because the resulting integral cannot always be evaluated easily. In other cases, the process is tedious because the shape of the region requires that it be divided into two or more parts in order to apply the method. To illustrate one such difficulty, consider the problem of rotating the region shown in Figure 6.32 about the y-axis. To use the washer method we would have to find the inner radius x_1 and the outer radius x_2 as functions of y. Thus, we would have to solve the equation

$$y = -x(x^2 + 1)(x - 2)$$

for x. (Try it!) We would also have to think of the graph in two parts—the part to the left of the maximum point and the part to the right. For each part x would have to be expressed as a function of y. Clearly, using the washer method for this problem is extremely difficult.

These difficulties can sometimes be eliminated by an alternative procedure for finding volumes of solids of revolution that we present in this section. The method involves the calculation of volumes of thin-walled hollow cylinders, called **cylindrical shells**. Pieces of pipe are such shells. More precisely, a cylindrical shell is the solid between two concentric right circular cylinders of fixed height. We show such a cylindrical shell in Figure 6.33. Our approach to finding the volume of a solid of revolution will be to use a collection of concentric cylindrical shells as an approximation to the volume. We will see that the sum of the volumes of the shells is a Riemann sum. Its limit is an integral that gives the total volume we are seeking.

FIGURE 6.33
A cylindrical shell

For a given cylindrical shell, we denote the inner radius by r_1, the outer radius by r_2, and the height by h. Then the volume V of the shell is the volume of the outer cylinder minus that of the inner cylinder:

$$V = \pi r_2^2 h - \pi r_1^2 h = \pi(r_2^2 - r_1^2)h$$

To write the sum of volumes of such cylindrical shells as a Riemann sum, we will find it convenient to write this formula in the following equivalent way:

$$V = \pi(r_2 + r_1)(r_2 - r_1)h = 2\pi\left(\frac{r_2 + r_1}{2}\right)(r_2 - r_1)h$$

Since $\frac{1}{2}(r_1 + r_2)$ is the average radius, multiplying it by 2π gives the average circumference. Also, $r_2 - r_1$ is the thickness of the shell. Thus, we can state the formula for its volume in words:

$$\text{volume of a shell} = (\text{average circumference}) \times (\text{height}) \times (\text{thickness}) \quad (6.12)$$

Now let us consider the region R between the x-axis and the graph of a continuous nonnegative function f between $x = a$ and $x = b$. Our objective is to rotate R about the y-axis and to obtain its volume as a limit of Riemann sums of volumes of shells. Let $P = \{x_0, x_1, x_2, \ldots, x_n\}$ be a partition of the closed interval $[a, b]$. Let c_k be the midpoint of the kth subinterval:

$$c_k = \frac{x_{k-1} + x_k}{2} \qquad k = 1, 2, \ldots, n$$

which matches the form we saw earlier for average radius.

Next, we construct a rectangle of height $f(c_k)$, with base equal to the subinterval $[x_{k-1}, x_k]$, for $k = 1, 2, 3, \ldots, n$. In Figure 6.34 we have shown a typical rectangle. If we rotate the kth rectangle about the y-axis, we get a cylindrical shell of average radius c_k, height $f(c_k)$, and thickness Δx_k, as shown in Figure 6.35. If this is done for $k = 1, 2, 3, \ldots, n$, we get n such concentric shells. Taken altogether, they approximate the solid of revolution that results from rotating the region R about the y-axis. By Equation 6.12 the volume of the kth cylindrical shell is

$$\Delta V_k = \underbrace{2\pi c_k}_{\substack{\text{average} \\ \text{circumference}}} \cdot \underbrace{f(c_k)}_{\text{height}} \cdot \underbrace{\Delta x_k}_{\text{thickness}}$$

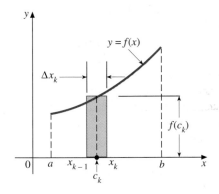

FIGURE 6.34

Thus, the volume V of the solid of revolution is approximated by

$$V \approx \sum_{k=1}^{n} \Delta V_k = \sum_{k=1}^{n} 2\pi c_k f(c_k) \Delta x_k$$

The sum on the right is the Riemann sum for the continuous function $2\pi x f(x)$ on $[a, b]$. We can see now the reason for choosing c_k as we did and for rewriting the original volume formula $V = \pi(r_2^2 - r_1^2)h$ in the form of Equation 6.12. By

FIGURE 6.35

doing so we obtained a Riemann sum. By letting max $\Delta x_k \to 0$, we obtain the following result:

Volume by Method of Cylindrical Shells

The volume of the solid obtained by revolving about the y-axis the area of the region under the graph of the continuous nonnegative function f from $x = a$ to $x = b$ is

$$V = 2\pi \int_a^b x f(x)\, dx \qquad (6.13)$$

Shorthand Technique for Finding Volumes Using Cylindrical Shells

FIGURE 6.36
Volume of shell $= 2\pi x f(x)\, dx$

As an aid to setting up the integral in Equation 6.13, we can label a typical element of volume (a cylindrical shell) as in Figure 6.36. Its volume is, by Equation 6.12,

$$\underbrace{2\pi x}_{\substack{\text{average} \\ \text{circumference}}} \cdot \underbrace{f(x)}_{\text{height}} \cdot \underbrace{dx}_{\text{thickness}}$$

Affixing the integral sign and the limits of integration is all that is needed to get the entire volume.

Analogous formulas hold for rotation about the x-axis and for regions between two graphs. Regions to be revolved along with typical cylindrical shells, are shown in Figures 6.37, 6.38, and 6.39. You can convince yourself of the validity of these formulas by finding the volume of a typical cylindrical shell. All functions involved are assumed to be continuous.

$$V = 2\pi \int_c^d y g(y)\, dy \qquad (6.14)$$

FIGURE 6.37
Rotation about the x-axis

$$V = 2\pi \int_a^b x[f(x) - g(x)]dx \qquad (6.15)$$

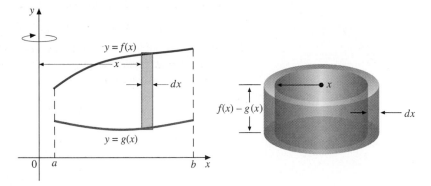

FIGURE 6.38
Rotation about the y-axis

$$V = 2\pi \int_c^d y[g(y) - h(y)]dy \qquad (6.16)$$

FIGURE 6.39
Rotation about the x-axis

REMARK ——————————————————————————————
■ Rather than memorize Equations 6.14–6.16, it is easier in each case to draw a typical rectangle and the resulting cylindrical shell. Calculating its volume leads to the appropriate integrand.
——————————————————————————————————————

For obvious reasons, calculating volumes in the way we have described is called the **method of cylindrical shells.** We illustrate the technique with several examples.

EXAMPLE 6.12   Use the method of cylindrical shells to find the volume of the solid obtained by revolving the region under the graph of $y = x^2$ from $x = 0$ to $x = 1$ about the y-axis.

Solution   We show the region to be rotated, together with the resulting solid of revolution, in Figure 6.40. Writing $f(x) = x^2$ and using Equation 6.13, we get

$$V = 2\pi \int_0^1 xf(x)dx = 2\pi \int_0^1 x(x^2)dx = 2\pi \int_0^1 x^3 \, dx$$

$$= 2\pi \left[\frac{x^4}{4} \right]_0^1 = \frac{\pi}{2}$$ ■

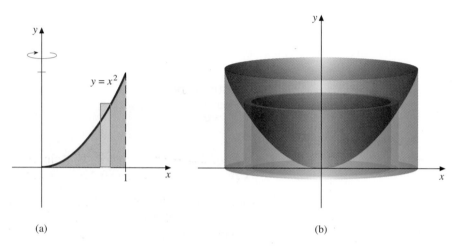

(a)   (b)

FIGURE 6.40

The volume in Example 6.12 can also be calculated by the method of washers. You should do it by the washer method, using Equation 6.11, with $g(y) = 1$ and $h(y) = \sqrt{y}$, and compare the difficulty with the cylindrical shell method.

EXAMPLE 6.13   Find the volume of the solid obtained by revolving about the x-axis the region bounded by $g(y) = 2y - y^2$ and the y-axis.

Solution   We show the region in Figure 6.41, along with the solid of revolution. We choose the method of cylindrical shells, since cross-sectional areas perpendicular to the x-axis are not easily calculated in this case. (It would be instructive for you to try to find the volume by the washer method to see the difficulties involved.) By Equation 6.14, we obtain

$$V = 2\pi \int_0^2 yg(y)dy = 2\pi \int_0^2 y(2y - y^2)dy$$

$$= 2\pi \int_0^2 (2y^2 - y^3)dy = 2\pi \left[\frac{2y^3}{3} - \frac{y^4}{4} \right]_0^2$$

$$= 2\pi \left[\frac{16}{3} - 4 \right] = \frac{8\pi}{3}$$ ■

(a)

(b)

FIGURE 6.41

(a)

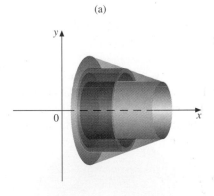

(b)

FIGURE 6.42

EXAMPLE 6.14   Find the volume of the solid obtained by revolving the region shown in Figure 6.42(a) about the x-axis.

Solution   We choose the method of cylindrical shells, since in this case the method avoids having to divide the region into two parts, as would be necessary with the washer method (since there are two different upper boundaries). The typical rectangle to be rotated is horizontal, so we need to express the left and right boundaries as functions of y:

$$x = \frac{y^2}{4} \qquad \text{Left boundary}$$

$$x = 5 - 2y \qquad \text{Right boundary}$$

We solve these two equations simultaneously to get their point of intersection. The y-coordinate gives the upper limit of integration. By setting the x values equal to each other, we obtain

$$\frac{y^2}{4} = 5 - 2y$$

$$y^2 + 8y - 20 = 0$$

$$(y - 2)(y + 10) = 0$$

$$y = 2 \text{ or } y = -10$$

but $y = -10$ is outside the domain of definition, so the upper limit is $y = 2$. Using Equation 6.16 with $g(y) = 5 - 2y$ and $h(y) = y^2/4$, we obtain the volume of the solid shown in Figure 6.42(b):

$$V = 2\pi \int_1^2 y \left(5 - 2y - \frac{y^2}{4} \right) dy = 2\pi \int_1^2 \left(5y - 2y^2 - \frac{y^3}{4} \right) dy$$

$$= 2\pi \left[\frac{5y^2}{2} - \frac{2y^3}{3} - \frac{y^4}{16} \right]_1^2$$

$$= 2\pi \left[\left(10 - \frac{16}{3} - 1 \right) - \left(\frac{5}{2} - \frac{2}{3} - \frac{1}{16} \right) \right] = \frac{91\pi}{24} \qquad \blacksquare$$

EXAMPLE 6.15   Find the volume of the solid of revolution obtained by revolving the region bounded by the graphs of $y = x^2/2$ and $y = x + 4$ about the line $x = 6$.

Solution   In Figure 6.43(a), we show a typical area element. Figure 6.43(b) shows the cylindrical shell that is generated when this element is revolved about the line $x = 6$. In Figure 6.43(c), we show the generated solid. We have used our simplified labeling technique, so the average radius of the shell is $6 - x$, the height is $f(x) - g(x)$, where $f(x) = x + 4$ and $g(x) = x^2/2$, and the thickness is dx. Thus, the volume of the shell is

$$\underbrace{2\pi(6 - x)}_{\substack{\text{average} \\ \text{circumference}}} \cdot \underbrace{\left[(x + 4) - \frac{x^2}{2} \right]}_{\text{height}} \cdot \underbrace{dx}_{\text{thickness}}$$

The total volume is therefore

$$V = 2\pi \int_{-2}^{4} (6 - x) \left(x + 4 - \frac{x^2}{2} \right) dx$$

The limits of integration were found by solving the two equations simultaneously. We simplify the integrand and perform the integration as follows:

$$V = 2\pi \int_{-2}^{4} \left(\frac{x^3}{2} - 4x^2 + 2x + 24 \right) dx$$

$$= 2\pi \left[\frac{x^4}{8} - \frac{4x^3}{3} + x^2 + 24x \right]_{-2}^{4}$$

$$= 2\pi \left[\left(32 - \frac{4 \cdot 64}{3} + 16 + 96 \right) - \left(2 + \frac{4 \cdot 8}{3} + 4 - 48 \right) \right]$$

$$= 2\pi(90) = 180\pi \qquad \blacksquare$$

(a)

(b)

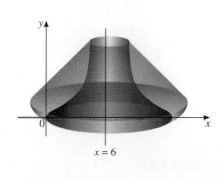

(c)

FIGURE 6.43

Exercise Set 6.3

In Exercises 1–10, find the volume obtained by revolving the region bounded by the graphs of the given equations about the axis specified. Use the method of cylindrical shells.

1. $y = x$, $y = 0$, $x = 2$; y-axis

2. $y = \sqrt{x}$, $y = 0$, $x = 4$; y-axis

3. $y = \sqrt{4 - x}$, $x = 0$, $y = 0$; x-axis

4. $x = 1/\sqrt{y}$, $y = 1$, $y = 4$, $x = 0$; x-axis

5. $x = \sqrt{y}$, $x = 0$, $x + y = 6$; y-axis

6. $y = \sqrt{6 - x}$, $y = x$, $y = 0$; x-axis

7. $y = (16 - x^2)/4$, $x - 2y + 6 = 0$, $x = 2$, $x = 4$; y-axis

8. $y = \sqrt{x + 5}$, $y = x - 1$, $y = 1$; x-axis

9. $y = x^3/2$, $y = 1/2$, $y = 4$, $x = 0$; x-axis

10. $y^2 = x^3$, $x = 1$; y-axis

In Exercises 11–16, find by two methods the volume of revolution obtained by revolving the region bounded by the graphs of the given equations about the axis specified.

11. $y = x^2$, $y = 2x$; y-axis

12. $y = \sqrt{2x - x^2}$, $y = x$; x-axis

13. $y^2 = 2x$, $x^2 = 2y$; x-axis

14. $x = 2\sqrt{4 - y^2}$, $x + 2y = 4$; y-axis

15. $y^2 = x^2 - 1$, $2x - y = 2$; x-axis

16. $y = \sqrt{8x}$, $y = x$, y-axis

In Exercises 17–20, use the method of cylindrical shells to find the volume of the solid generated by revolving the shaded region about the axis indicated.

17. y-axis

18. x-axis

19. x-axis

20. y-axis

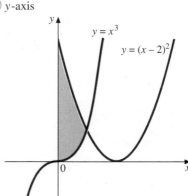

21. Find the volume of the solid obtained by revolving the region between the x-axis and the graph of $f(x) = 1 - x^2$ about the line $x = 2$.

22. Find the volume of the solid obtained by revolving the region between the graphs of $h(y) = y$ and $g(y) = 2 - y^2$ about the line $y = 1$.

23. Use the method of cylindrical shells to find the volume of a torus (see Example 6.11) generated by revolving a circle of radius a about an axis b units from the center of the circle, where $b > a$. (*Hint:* Orient the axes so that the center of the circle is at the origin.)

24. Find the volume of the solid obtained by revolving the region between the graphs of $y = 4x - x^2$ and $y = x^2 - 2x + 4$ about (a) the y-axis and (b) the x-axis.

25. Find the volume of the solid obtained by revolving the region between the graphs of $y = \frac{1}{3}(x - 4)(x^2 - 8x - 3)$ and $x + y = 4$ about the y-axis.

26. Let
$$f(x) = \begin{cases} x^2 & \text{if } 0 \le x < 2 \\ 6 - x & \text{if } 2 \le x \le 6 \end{cases}$$

Find the volume of the solid obtained by revolving the region bounded by the graph of f and the x-axis about the line $x = -1$. Use (a) the method of washers and (b) the method of cylindrical shells.

27. Find the volume generated by revolving the region bounded by the graphs of $y = 3x - x^2$ and $y = x^2 - 2$ about (a) the line $x = -1$ and (b) the line $y = -2$.

28. Maccabee the goldfish lived in a bowl that is in the shape of a sphere with its top and bottom cut off. (See the photo.) The radius of the sphere is 12 cm, and the top and bottom 3.5 cm of the sphere are cut off.

(a) When the bowl was half-filled with water, how much water was there, and how much did it weigh? (Water weighs 1 gm/cm³.)

(b) If water evaporates in 30 days so that the level is 2.5 cm below the middle, what volume of water evaporates per day?

 In Exercises 29 and 30, use a CAS to obtain the graph. Set up the integral for the volume of the solid described, and evaluate it using the CAS.

29. The solid obtained by revolving the region bounded by $y = \sin^2 x$ and the x-axis from $x = 0$ to $x = \pi$

(a) about the y-axis

(b) about the line $x + \pi = 0$

30. The solid obtained by revolving the region bounded by $y = x^2$ and $y = 2\cos\left(\dfrac{\pi x}{3}\right)$

(a) about the x-axis

(b) about the line $y = 2$

6.4 ARC LENGTH

In this section we show how integration can be used to find the length of a curve. The basic idea is very simple. As we show in Figure 6.44, we approximate the length from A to B by a series of straight line segments joining points on the curve. This collection of line segments is called a **polygonal path** from A to B. The length of the polygonal path is easily calculated, since the length of each segment is just the distance between two points. By taking the points on the curve closer and closer together, we get polygonal paths that more and more closely approximate the curve. We will show that the length of each polygonal path can be expressed as a Riemann sum, so in the limit we get an integral.

To carry out the process just described, let us assume that f is a continuous function on the closed interval $[a, b]$. For each k from 0 to n, let P_k be the point $(x_k, f(x_k))$ on the graph of f. Note that the initial point A is P_0 and

FIGURE 6.44
The length of the polygonal path approximates the length of the curve from a to b.

FIGURE 6.45

FIGURE 6.46

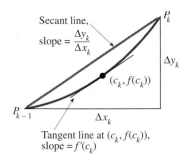

Secant line,
slope = $\dfrac{\Delta y_k}{\Delta x_k}$

$(c_k, f(c_k))$

Δy_k

Δx_k

Tangent line at $(c_k, f(c_k))$,
slope $= f'(c_k)$

FIGURE 6.47
The slope of the tangent line at
$(c_k, f(c_k))$ equals the slope of the
secant line $P_{k-1}P_k$.

The derivative of the function

$$f(x) = \begin{cases} x^2 \sin\left(\dfrac{1}{x}\right) & \text{if } x \neq 0 \\ 0 & \text{if } x = 0 \end{cases}$$

exists everywhere, but the derivative
is discontinuous at $x = 0$. (See
Exercise 76 in Exercise Set 3.7.)

the final point B is P_n. (See Figure 6.45.) The straight line segments joining the consecutive points $P_0, P_1, P_2, \ldots, P_n$ form a polygonal path from A to B corresponding to our partition. To find its length, consider the distance between two typical consecutive points P_{k-1} and P_k. As we indicate in Figure 6.46, we denote the horizontal distance between them by Δx_k and the vertical distance by Δy_k. So by the Pythagorean Theorem, the length of the segment joining P_{k-1} and P_k is

$$\sqrt{(\Delta x_k)^2 + (\Delta y_k)^2}$$

The length of the polygonal path is therefore

$$\sum_{k=1}^{n} \sqrt{(\Delta x_k)^2 + (\Delta y_k)^2} \tag{6.17}$$

To write the sum in Formula 6.17 in a form that looks like a Riemann sum, we apply the Mean-Value Theorem (Theorem 4.6) to f on each of the closed subintervals $[x_{k-1}, x_k]$. To do so, we require that $f'(x)$ exists, at least on each open interval (x_{k-1}, x_k). Assuming f satisfies this condition, the Mean-Value Theorem tells us that there is a point c_k in the open interval (x_{k-1}, x_k) such that

$$\frac{f(x_k) - f(x_{k-1})}{x_k - x_{k-1}} = f'(c_k)$$

or, equivalently,

$$\frac{\Delta y_k}{\Delta x_k} = f'(c_k) \qquad k = 1, 2, \ldots, n$$

Figure 6.47 illustrates this relationship. We solve the last equation for Δy_k to get

$$\Delta y_k = f'(c_k)\Delta x_k$$

When we substitute this value for Δy_k in the formula for the distance between P_{k-1} and P_k, we have

$$\sqrt{(\Delta x_k)^2 + (\Delta y_k)^2} = \sqrt{(\Delta x_k)^2 + [f'(c_k)\Delta x_k]^2} = \sqrt{1 + [f'(c_k)]^2}\,\Delta x_k$$

The length formula (6.17) for the polygonal path can now be written as

$$\sum_{k=1}^{n} \sqrt{1 + [f'(c_k)]^2}\,\Delta x_k$$

which we recognize as a Riemann sum for the function $\sqrt{1 + [f'(x)]^2}$. If this latter function is integrable on $[a, b]$, we know that its Riemann sums will approach its integral. To assure this integrability condition, we make the further assumption that $f'(x)$ be continuous on $[a, b]$. (See the marginal comment for a function that is differentiable everywhere but whose derivative is not continuous.) It follows that $\sqrt{1 + [f'(x)]^2}$ is also continuous there. A function whose derivative exists and is continuous on an interval is said to be a **smooth function**. Its graph has a continuously turning tangent line, with no sharp corners. We now have set the stage for the following definition.

Definition 6.2
The Length of a Curve
(Arc Length)

Let f be a smooth function on $[a, b]$. Then the length L of its graph from $A = (a, f(a))$ to $B = (b, f(b))$ is given by

$$L = \int_a^b \sqrt{1 + [f'(x)]^2}\, dx \qquad (6.18)$$

Similarly, when g is a smooth function of y on the closed interval $[c, d]$, the length of the graph of g from $A = (g(c), c)$ to $B = (g(d), d)$ is given by

$$L = \int_c^d \sqrt{1 + [g'(y)]^2}\, dy \qquad (6.19)$$

EXAMPLE 6.16   Find the length of the curve $y = 2(x + 1)^{3/2}$ from $x = -1$ to $x = 6$.

Solution   Let $f(x) = 2(x + 1)^{3/2}$. Then $f'(x) = 3(x + 1)^{1/2}$, so by Equation 6.18,

$$L = \int_{-1}^6 \sqrt{1 + 9(x + 1)}\, dx = \int_{-1}^6 \sqrt{9x + 10}\, dx \qquad \text{Mentally, let } u = 9x + 10, \text{ so that } du = 9\, dx.$$

$$= \frac{1}{9} \int_{-1}^6 (9x + 10)^{1/2} \cdot 9\, dx = \frac{1}{9} \cdot \frac{2}{3}(9x + 10)^{3/2} \Big]_{-1}^6$$

$$= \frac{2}{27}[(64)^{3/2} - (1)^{3/2}] = \frac{2}{27}(8^3 - 1) = \frac{2}{27}(511) = \frac{1022}{27} \qquad \blacksquare$$

EXAMPLE 6.17   Find the length of arc on the graph of $x = \frac{2}{3}(y^2 - 1)^{3/2}$ from $y = 1$ to $y = 3$.

Solution   The form in which the equation is written suggests using y as the independent variable, so Equation 6.19 is applicable:

$$g'(y) = \frac{dx}{dy} = \frac{2}{3} \cdot \frac{3}{2}(y^2 - 1)^{1/2} \cdot 2y = 2y(y^2 - 1)^{1/2}$$

Thus,

$$L = \int_1^3 \sqrt{1 + [2y(y^2 - 1)^{1/2}]^2}\, dy = \int_1^3 \sqrt{1 + 4y^4 - 4y^2}\, dy$$

$$= \int_1^3 \sqrt{(2y^2 - 1)^2}\, dy = \int_1^3 (2y^2 - 1)\, dy \qquad \text{Since } \sqrt{(2y^2 - 1)^2} = |2y^2 - 1| = 2y^2 - 1 \text{ for } 1 \le y \le 3$$

$$= \frac{2y^3}{3} - y \Big]_1^3 = (18 - 3) - \left(\frac{2}{3} - 1\right)$$

$$= \frac{46}{3} \qquad \blacksquare$$

Approximating Arc Length

We chose the functions in the two preceding examples carefully so that the integrals involved could be evaluated in exact form by means we have studied. When the integral does not turn out so nicely (which is usually the case), we can use one of our approximation methods. The next example illustrates such a case.

EXAMPLE 6.18   Use Simpson's Rule with $n = 10$ to approximate the arc of the parabola $y = x^2$ from $x = 0$ to $x = 2$.

Solution   Since $y = x^2$, we have $y' = 2x$, so

$$L = \int_0^2 \sqrt{1 + 4x^2}\, dx$$

In Chapter 8 we will show how to evaluate an integral of this type by means of a substitution. Here, however, the best we can do is to approximate its value. With $n = 10$, we have

$$\Delta x = \frac{2}{10} = 0.2$$

Writing $f(x) = \sqrt{1 + 4x^2}$, we have, by Simpson's Rule,

$$\int_0^2 \sqrt{1 + 4x^2}\, dx \approx \tfrac{0.2}{3}[f(0) + 4f(0.2) + 2f(0.4) + 4f(0.6) + 2f(0.8)$$

$$+ 4f(1.0) + 2f(1.2) + 4f(1.4) + 2f(1.6) + 4f(1.8)$$

$$+ f(2.0)]$$

By calculator, we obtain the answer

$$L \approx 4.65 \qquad\blacksquare$$

In the next example we consider the circumference of an ellipse. As in the previous example, we arrive at an integral of a function for which no antiderivative in the form of an elementary function is immediately apparent. In contrast to the previous example, however, there is no substitution, nor any other technique, by which the integral can be brought around to one we can evaluate in an exact form. We could use Simpson's Rule, or some other approximation technique. An alternative, however, is to consult a table. Not surprisingly, the integral in question is called an **elliptic integral**, and approximate values of such integrals have been widely tabulated. They also can be approximated using a CAS. Elliptic integrals occur in a variety of applications, particularly in physics.

EXAMPLE 6.19   Find an integral whose value gives the length of the ellipse

$$\frac{x^2}{a^2} + \frac{y^2}{b^2} = 1$$

Solution   We show a typical ellipse in Figure 6.48. The upper half of the ellipse has the equation

$$y = \frac{b}{a}\sqrt{a^2 - x^2}$$

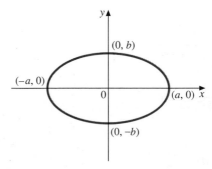

FIGURE 6.48

The ellipse $\dfrac{x^2}{a^2} + \dfrac{y^2}{b^2} = 1$

(obtained by solving the original equation for y and taking the positive square root). By symmetry, the entire length L of the ellipse is four times the arc in the first quadrant. Thus, by Equation 6.18,

$$L = 4 \int_0^a \sqrt{1 + (y')^2}\, dx$$

To find y', we can differentiate the equation

$$\frac{x^2}{a^2} + \frac{y^2}{b^2} = 1$$

implicitly to get

$$\frac{2x}{a^2} + \frac{2yy'}{b^2} = 0$$

Solving for y', we get

$$y' = -\frac{b^2 x}{a^2 y}$$

Thus,

$$L = 4 \int_0^a \sqrt{1 + \frac{b^4 x^2}{a^4 y^2}}\, dx$$

Since

$$y^2 = \frac{b^2}{a^2}(a^2 - x^2)$$

we finally obtain the result

$$L = 4 \int_0^a \sqrt{1 + \frac{b^2 x^2}{a^2(a^2 - x^2)}}\, dx$$

If we knew particular values of a and b, we could approximate the length by Simpson's Rule, or we could use a CAS or consult a table of elliptic integrals.

The special case in which $a = b$ is of interest. In this case the ellipse is a circle, and we know from previous experience (but have not yet proved) that the length (circumference) is $2\pi a$. So we have

$$2\pi a = 4 \int_0^a \sqrt{1 + \frac{x^2}{a^2 - x^2}}\, dx$$

$$= 4 \int_0^a \sqrt{\frac{a^2 - x^2 + x^2}{a^2 - x^2}}\, dx$$

$$= 4a \int_0^a \frac{1}{\sqrt{a^2 - x^2}}\, dx$$

Solving for the integral, we obtain

$$\int_0^a \frac{1}{\sqrt{a^2 - x^2}}\, dx = \frac{\pi}{2}$$

In Chapter 7 we will see another way of obtaining this result by making use of the Second Fundamental Theorem of Calculus. ∎

REMARK ─────────────────────────────

■ You might have noticed that each of the integrals we arrived at in Example 6.19 has an unbounded integrand on the interval $[0, a]$. The integrals are said to be *improper* in this case. We will study improper integrals in Chapter 7. We will also reconsider both integrals in Chapter 8 after we have studied trigonometric substitutions.

─────────────────────────────

Piecewise-Smooth Functions

Suppose f is continuous but f' is not continuous on the closed interval $[a, b]$ (so that f is not smooth), but we can divide the interval into a finite number of closed subintervals, on each of which f is smooth. Then we say that f is **piecewise smooth**. The length of its graph on $[a, b]$ is defined as the sum of the lengths on each of the subintervals. So if L_k denotes the arc length on the kth subinterval and there are m subintervals, then

$$L = \sum_{k=1}^{m} L_k$$

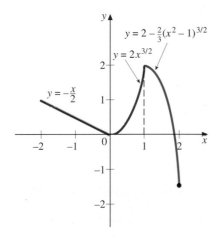

FIGURE 6.49

EXAMPLE 6.20   Find the length of the piecewise-smooth curve shown in Figure 6.49.

Solution   Let L_1 denote the length of the segment from $x = -2$ to $x = 0$, L_2 the length from $x = 0$ to $x = 1$, and L_3 the length from $x = 1$ to $x = 2$. We do not need calculus to find L_1, since it is the length of a straight line segment, but we will apply our theory anyway and show that the result agrees with what we would get from the Pythagorean Theorem.
Curve 1:

$$y = -\frac{x}{2} \qquad \text{on } [-2, 0]$$

$$y' = -\frac{1}{2}$$

$$L_1 = \int_{-2}^{0} \sqrt{1 + \frac{1}{4}}\, dx = \frac{\sqrt{5}}{2} x \Big]_{-2}^{0} = -\frac{\sqrt{5}}{2}(-2) = \sqrt{5}$$

Using the Pythagorean Theorem, we also get $L_1 = \sqrt{2^2 + 1^2} = \sqrt{5}$.
Curve 2:

$$y = 2x^{3/2} \qquad \text{on } [0, 1]$$

$$y' = 2 \cdot \frac{3}{2} x^{1/2} = 3x^{1/2}$$

$$L_2 = \int_{0}^{1} \sqrt{1 + 9x}\, dx = \frac{1}{9} \int_{0}^{1} (1 + 9x)^{1/2} (9\, dx) \qquad \text{Mentally, let } u = 1 + 9x, \text{ so that } du = 9\, dx.$$

$$= \frac{1}{9} \cdot \frac{2}{3} (1 + 9x)^{3/2} \Big]_{0}^{1} = \frac{2}{27} \left[(10)^{3/2} - 1 \right]$$

Curve 3:

$$y = 2 - \frac{2}{3}(x^2 - 1)^{3/2} \qquad \text{on } [1, 2]$$

$$y' = -\frac{2}{3} \cdot \frac{3}{2}(x^2 - 1)^{1/2}(2x) = -2x(x^2 - 1)^{1/2}$$

$$L_3 = \int_1^2 \sqrt{1 + 4x^2(x^2 - 1)}\, dx$$

$$= \int_1^2 \sqrt{4x^4 - 4x^2 + 1}\, dx$$

$$= \int_1^2 \sqrt{(2x^2 - 1)^2}\, dx$$

$$= \int_1^2 (2x^2 - 1)\, dx \quad \sqrt{(2x^2 - 1)^2} = |2x^2 - 1| = 2x^2 - 1 \text{ for } 1 \le x \le 2$$

$$= \frac{2x^3}{3} - x \Big]_1^2$$

$$= \left(\frac{16}{3} - 2\right) - \left(\frac{2}{3} - 1\right) = \frac{10}{3} - \left(-\frac{1}{3}\right) = \frac{11}{3}$$

The total length is therefore

$$L = L_1 + L_2 + L_3 = \sqrt{5} + \frac{2}{27}\left[(10)^{3/2} - 1\right] + \frac{11}{3} \approx 8.17 \qquad \blacksquare$$

The Arc-Length Function

It is instructive to consider the *arc-length function*, defined by

$$s(x) = \int_a^x \sqrt{1 + [f'(t)]^2}\, dt \qquad (x \in [a, b])$$

for the smooth function f on $[a, b]$. By the First Fundamental Theorem (Theorem 5.2),

$$s'(x) = \sqrt{1 + [f'(x)]^2}$$

Thus, the *differential of arc length*, $ds = s'(x)\, dx$, is given by

$$ds = \sqrt{1 + [f'(x)]^2}\, dx = \sqrt{1 + \left(\frac{dy}{dx}\right)^2}\, dx \qquad (6.20)$$

If $x = g(y)$, where g is smooth on $[c, d]$, then ds is given by

$$ds = \sqrt{1 + [g'(y)]^2}\, dy = \sqrt{1 + \left(\frac{dx}{dy}\right)^2}\, dy \qquad (6.21)$$

We now can express both Equations 6.18 and 6.19 in the briefer form

$$L = \int_{l_1}^{l_2} ds$$

where the correct limits l_1 and l_2 must be supplied. They will be x limits if ds is given by Equation 6.20, and they will be y limits if ds is given by Equation 6.21.

In the equation

$$ds = \sqrt{1 + \left(\frac{dy}{dx}\right)^2}\, dx$$

if we bring the final dx inside the radical and simplify, we get

$$ds = \sqrt{(dx)^2 + (dy)^2}$$

The same result can be obtained by bringing dy inside the radical in Equation 6.21. Squaring both sides gives

$$(ds)^2 = (dx)^2 + (dy)^2 \tag{6.22}$$

Equation 6.22 has an interesting geometric interpretation, as shown by Figure 6.50. The differential ds can be interpreted as the length of the hypotenuse of a right triangle with legs dx and dy. We let $dx = \Delta x$ be an arbitrary increment on x, which causes a corresponding increment Δy on y. From Section 3.9, we know that the distance dy to the tangent line approximates the distance Δy to the curve when Δx is small. Using Δs to denote the arc length from the point (x, y) to the point $(x + \Delta x, y + \Delta y)$, we see that $ds \approx \Delta s$. So we can think of the differential of arc as almost equal to a true section of arc when the change in x is small.

The use of this differential notation is illustrated in the following example.

FIGURE 6.50

EXAMPLE 6.21   An engineer plans to shape a length of fiber-optic cable according to the curve defined by $x^4 - 6xy + 3 = 0$ from $(1, \frac{2}{3})$ to $(3, \frac{14}{3})$, with distances in meters. The cable costs \$1 per meter. What is the total cost?

Solution   We show the computer-generated graph of the curve in Figure 6.51. Solving the equation for y gives

$$y = \frac{x^3}{6} + \frac{1}{2x}$$

For x in the interval $[1, 3]$ this equation defines a smooth function of x. (Why?)

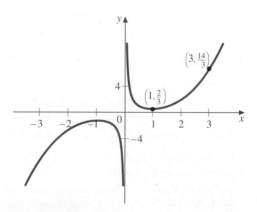

FIGURE 6.51

So we use Equation 6.20 to get ds:

$$ds = \sqrt{1 + \left(\frac{dy}{dx}\right)^2}\, dx = \sqrt{1 + \left(\frac{x^2}{2} - \frac{1}{2x^2}\right)^2}\, dx$$

$$= \sqrt{1 + \frac{x^4}{4} - \frac{1}{2} + \frac{1}{4x^4}}\, dx = \sqrt{\frac{x^4}{4} + \frac{1}{2} + \frac{1}{4x^4}}\, dx$$

$$= \sqrt{\left(\frac{x^2}{2} + \frac{1}{2x^2}\right)^2}\, dx = \left(\frac{x^2}{2} + \frac{1}{2x^2}\right) dx$$

Thus,

$$L = \int_{x=1}^{x=3} ds = \int_1^3 \left(\frac{x^2}{2} + \frac{1}{2x^2}\right) dx$$

$$= \frac{x^3}{6} - \frac{1}{2x} \Big]_1^3 = \left(\frac{27}{6} - \frac{1}{6}\right) - \left(\frac{1}{6} - \frac{1}{2}\right)$$

$$= \frac{14}{3}$$

The cable in question is therefore $4\frac{2}{3}$ m long, so its cost is \$4.67.   ■

FIGURE 6.52
A curve with infinite length

It may surprise you to learn that there are curves defined by continuous functions on finite intervals that are infinite in length. One such example is the "sawtooth" function shown in Figure 6.52. The peaks occur at the points $\left(\frac{1}{n}, \frac{1}{n}\right)$ for $n = 1, 2, 3, \ldots$, with the function defined as 0 at the origin. The sum of the lengths of the two rightmost segments that come together at the point $(1, 1)$ exceeds the vertical distance 1 from the x-axis to the peak. Similarly, the sum of the lengths of the two segments meeting at the point $\left(\frac{1}{2}, \frac{1}{2}\right)$ is greater than the vertical distance $\frac{1}{2}$ from the x-axis to the peak. In general, the sum of the lengths of the two segments meeting at $\left(\frac{1}{n}, \frac{1}{n}\right)$ is greater than $\frac{1}{n}$. Thus, if the total length L of the curve were finite, we would have

$$L > 1 + \frac{1}{2} + \frac{1}{3} + \frac{1}{4} + \cdots$$

In Chapter 10 we will show that by continuing to add terms on the right-hand side, the sum can be made arbitrarily large. That is, the sum is infinite. So L cannot be finite.

The function in Figure 6.52 does not have a simple formula. An example of a curve with infinite length that does have a simple formula is the graph of

$$f(x) = \begin{cases} x \sin \frac{1}{x} & \text{if } x \neq 0 \\ 0 & \text{if } x = 0 \end{cases}$$

The length over any interval containing the origin is infinite. We will ask you to find this result for the interval $0 \leq x \leq 1$ in Exercise 24 in Exercise Set 6.4.

Curves that have finite length are said to be **rectifiable**. The smoothness condition of a function f is sufficient (but not necessary) to guarantee that its graph is rectifiable.

Exercise Set 6.4

In Exercises 1–14, find the arc length of the graph of the given equation between the specified limits.

1. $y = 3x - 4$ from $x = -1$ to $x = 2$

2. $y = 3 - 2x$ from $x = 0$ to $x = 2$

3. $2x - 3y = 4$ from $(-1, -2)$ to $(2, 0)$

4. $4y - 5x = 7$ from $(-3, -2)$ to $(1, 3)$

5. $y = \dfrac{2}{3}x^{3/2}$ from $x = 0$ to $x = 3$

6. $x = (y - 1)^{3/2}$ from $y = 1$ to $y = 6$

7. $y^2 = x^3$ from $(0, 0)$ to $(5, 5\sqrt{5})$

8. $y = x^{2/3}$ from $x = 1$ to $x = 8$

9. $y = \dfrac{2}{3}(x^2 + 1)^{3/2}$ from $x = 0$ to $x = 2$

10. $y = 1 + \left(\dfrac{3x}{2}\right)^{2/3}$ from $(0, 1)$ to $(\frac{2}{3}, 2)$

11. $y = \left(\dfrac{1 - x}{2}\right)^{3/2} + 4$ from $x = -4$ to $x = 1$

12. $y = \dfrac{x^4}{8} + \dfrac{1}{4x^2}$ from $x = 1$ to $x = 2$

13. $4y^4 - 12xy + 3 = 0$ from $\left(\frac{7}{12}, 1\right)$ to $\left(\frac{109}{12}, 3\right)$

14. $9y^2 = (x^2 - 2)^3$ from $(\sqrt{2}, 0)$ to $\left(\sqrt{6}, \frac{8}{3}\right)$

15. Find the total length of the four-cusp hypocycloid $x^{2/3} + y^{2/3} = a^{2/3}$.

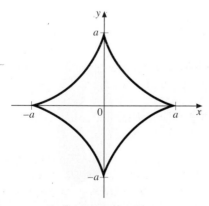

$x^{2/3} + y^{2/3} = a^{2/3}$

16. Find the length of the arc on the graph of $x = \frac{1}{3}\sqrt{y}(y - 3)$ from $y = 1$ to $y = 4$.

17. Find the fallacy in the following attempt to find the length of the curve $2y = 3(x + 1)^{2/3}$ from $(-2, \frac{3}{2})$ to $(7, 6)$, and correct it.

$$x = \left(\dfrac{2y}{3}\right)^{3/2} - 1, \qquad \text{so} \quad \dfrac{dx}{dy} = \left(\dfrac{2y}{3}\right)^{1/2}$$

$$L = \int_{3/2}^{6} \sqrt{1 + \dfrac{2y}{3}}\, dy = \left(1 + \dfrac{2y}{3}\right)^{3/2}\Bigg]_{3/2}^{6}$$

$$= 5\sqrt{5} - 2\sqrt{2}$$

18. Let
$$f(x) = \begin{cases} 1 - x & \text{if } 0 \le x < 1 \\ (x - 1)^{3/2} & \text{if } 1 \le x < 2 \\ 2 - (x - 1)^{2/3} & \text{if } 2 \le x \le 9 \end{cases}$$

   (a) Draw the graph of f.

   (b) Find the length of the graph of f from $x = 0$ to $x = 9$.

In Exercises 19 and 20, use ds to approximate the arc length in question.

19. Arc of $y = x^2$ from $x = 1.0$ to $x = 1.1$

20. Arc of $y = x/(x - 1)$ from $x = 2.05$ to $x = 2.07$

In Exercises 21–23, use Simpson's Rule with $n = 10$ to approximate the arc length.

21. $y = x^2$ from $x = 0$ to $x = 2$

22. $y = \sin x$ from $x = 0$ to $x = \pi$

23. $y = 4 - x^3$ from $x = -2$ to $x = 2$

24. Let
$$f(x) = \begin{cases} x \sin \dfrac{1}{x} & \text{if } x \ne 0 \\ 0 & \text{if } x = 0 \end{cases}, \qquad 0 \le x \le 1$$

Calculate $f(x)$ for $x = \dfrac{2}{\pi}, \dfrac{2}{3\pi}, \dfrac{2}{5\pi}, \dots$. Show how it follows that if the length of the graph of f is L, then

$$L > \dfrac{2}{\pi}\left(1 + \dfrac{1}{3} + \dfrac{1}{5} + \dfrac{1}{7} + \cdots\right)$$

In Chapter 10 we will show that the sum inside the parentheses is infinite. Thus the graph has infinite length.

6.5 FINDING AREAS OF SURFACES OF REVOLUTION

If an arc of a curve is revolved about an axis, a **surface of revolution** is generated. Our purpose in this section is to develop a means of finding the areas of such surfaces. First, we must define precisely what we mean by the area in this situation.

Our basic approach is to approximate the curve by a polygonal path and then rotate the polygonal path about the axis, to obtain an approximation to the actual surface area. We illustrate this idea in Figure 6.53. When a typical segment of the polygonal path is rotated about the axis, a band is formed. We highlight one such band in the figure. The band can be viewed as a section of a cone cut off by two parallel planes. The name given to such a portion of a cone is "frustum." So each segment of the polygonal path, when rotated, generates a frustum of a cone. As we indicate below, there is a simple formula for the area of a frustum of a cone. When we add up the areas of all the frustums (the bands), we obtain an approximation to the area of the surface formed by rotating the curve itself about the axis. We will show that our approximation is similar to a Riemann sum. As we take polygonal paths with points closer and closer together on the curve, we are led to an integral as the limit of Riemann sums.

As a starting point, we need to know the area of a frustum of a cone. Such a frustum is shown in Figure 6.54. An outline of how to find its area is presented in Exercise 22 at the end of this section. The result is that the surface area is given by

$$\text{area} = (\text{average circumference}) \times (\text{slant height})$$

or

$$\text{Area} = 2\pi r l \qquad (6.23)$$

where r is the average radius $= \frac{1}{2}(r_1 + r_2)$ and l is the slant height.

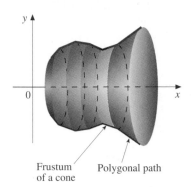

Now consider the arc of the graph of $y = f(x)$ between $x = a$ and $x = b$, where f is a smooth nonnegative function. Just as in defining arc length, we partition the closed interval $[a, b]$ and consider the chords that join successive points $P_k = (x_k, f(x_k))$ on the curve. (See Figure 6.55.) For each $k = 1, 2, \ldots, n$, we rotate the chord $P_{k-1}P_k$ about the x-axis. This process generates a band that is a frustum of a cone. In Figure 6.55 we show a typical band. We designate its area as ΔS_k. In Section 6.4 we found the slant height (the length of the line

Frustum   Polygonal path
of a cone

FIGURE 6.53
Each segment of the polygonal path generates a band, called a frustum of a cone, when rotated.

Average circumference $= 2\pi r$,
where $r = \frac{1}{2}(r_1 + r_2)$;
Area $= 2\pi r l$

FIGURE 6.54

(a)

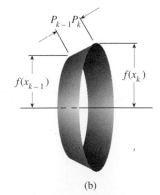

(b)

FIGURE 6.55

segment joining P_{k-1} and P_k) to be

$$\sqrt{1 + [f'(c_k)]^2}\, \Delta x_k$$

where c_k is between x_{k-1} and x_k. The average circumference is

$$2\pi \cdot \overbrace{\frac{1}{2}[f(x_{k-1}) + f(x_k)]}^{\text{average radius}}$$

and since $\frac{1}{2}[f(x_{k-1}) + f(x_k)]$ lies between $f(x_{k-1})$ and $f(x_k)$, we know by the Intermediate-Value Theorem (Theorem 2.8) that there is a number d_k in the interval $[x_{k-1}, x_k]$ such that

$$f(d_k) = \frac{1}{2}[f(x_{k-1}) + f(x_k)]$$

Thus, the surface area ΔS_k of the kth band is

$$\Delta S_k = \overbrace{2\pi f(d_k)}^{\substack{\text{average}\\\text{circumference}}} \cdot \overbrace{\sqrt{1 + [f'(c_k)]^2}\,\Delta x_k}^{\text{slant height}}$$

Summing all such areas, we arrive at what we would expect to be an approximation to the total surface area S:

$$S \approx \sum_{k=1}^{n} 2\pi f(d_k)\sqrt{1 + [f'(c_k)]^2}\, \Delta x_k \tag{6.24}$$

The sum on the right looks almost like a Riemann sum for the function $2\pi f(x)\sqrt{1 + [f'(x)]^2}$. The problem is that the points c_k and d_k are not necessarily the same. Fortunately, in the limit, it can be shown that this discrepancy does not matter, and the sum approaches the integral of this function. We therefore have the following definition.

Definition 6.3
Area of a Surface of Revolution

Let f be a nonnegative smooth function on the closed interval $[a, b]$. The area S of the surface of revolution generated by revolving the graph of f from $x = a$ to $x = b$ about the x-axis is given by

$$S = 2\pi \int_a^b f(x)\sqrt{1 + [f'(x)]^2}\, dx \tag{6.25}$$

The corresponding result for revolving the graph of the nonnegative smooth function g from $y = c$ to $y = d$ about the y-axis is

$$S = 2\pi \int_c^d g(y)\sqrt{1 + [g'(y)]^2}\, dy \tag{6.26}$$

In Section 6.4 we showed that the differential of arc length ds can be written as either

$$ds = \sqrt{1 + [f'(x)]^2}\, dx \quad \text{or} \quad ds = \sqrt{1 + [g'(y)]^2}\, dy$$

Using these equations, we can rewrite Equations 6.25 and 6.26 more simply as follows:

Area of a Surface of Revolution

rotation about x-axis: $S = 2\pi \displaystyle\int_{l_1}^{l_2} y\, ds$ (6.27)

rotation about y-axis: $S = 2\pi \displaystyle\int_{l_1}^{l_2} x\, ds$ (6.28)

The limits l_1 and l_2 of integration depend on the form of ds being used. When $ds = \sqrt{1 + [f'(x)]^2}\, dx$ is used, then x is the independent variable, and l_1 and l_2 are x-limits. When $ds = \sqrt{1 + [g'(y)]^2}\, dy$ is used, y is independent, so l_1 and l_2 are y-limits.

When the function f has an inverse (see Section 1.2), either Equation 6.27 or 6.28 may be used. In particular, when f is monotone increasing or monotone decreasing (still assuming f is smooth and nonnegative), either of the forms may be used. To be more explicit, suppose f is monotone increasing and that g is the inverse of f where both f and g are nonnegative. Then if $y = f(x)$, we have $x = g(y)$. If x satisfies $a \le x \le b$, then y satisfies $c \le y \le d$, where $c = f(a)$ and $d = f(b)$. Thus, Equation 6.27 can be written as

$$S = 2\pi \int_a^b f(x)\sqrt{1 + [f'(x)]^2}\, dx$$

or as

$$S = 2\pi \int_c^d y\sqrt{1 + [g'(y)]^2}\, dy$$

Similar remarks apply to Equation 6.28. We will illustrate this equivalence with the next example.

EXAMPLE 6.22   Find in two ways the area of the surface generated by revolving about the x-axis the arc of the curve $y = \sqrt{x}$ from $x = 1$ to $x = 4$.

Solution   Write $f(x) = \sqrt{x}$. Then for $1 \le x \le 4$ f is smooth, nonnegative, and monotone increasing (Figure 6.56). So we have a choice on how to write ds in Equation 6.27.

First, let x be the independent variable:

$$ds = \sqrt{1 + [f'(x)]^2}\, dx = \sqrt{1 + \left(\frac{1}{2\sqrt{x}}\right)^2}\, dx$$

$$= \sqrt{1 + \frac{1}{4x}}\, dx = \frac{\sqrt{4x + 1}}{2\sqrt{x}}\, dx$$

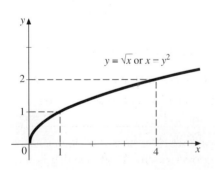

FIGURE 6.56

By Equation 6.27, with l_1 and l_2 as x-limits,

$$S = 2\pi \int_{l_1}^{l_2} y\, ds = 2\pi \int_1^4 \sqrt{x}\,\frac{\sqrt{4x + 1}}{2\sqrt{x}}\, dx$$

$$= \pi \int_1^4 \sqrt{4x + 1}\, dx$$

Now we mentally substitute $u = 4x + 1$, so that $du = 4\,dx$. We need a 4 to go with the dx, so we multiply by 4 inside the integral and by 1/4 outside. Thus,

$$S = \frac{\pi}{4} \int_1^4 \overbrace{(4x + 1)^{1/2}}^{u^{1/2}} \overbrace{(4\,dx)}^{du} = \frac{\pi}{4} \cdot \frac{2}{3}(4x + 1)^{3/2} \Big]_1^4$$

$$= \frac{\pi}{6}\left[(17)^{3/2} - (5)^{3/2}\right] \approx 30.85$$

Second, let us take y as the independent variable. We solve $y = \sqrt{x}$ for x to get $x = y^2$. When $x = 1$, $y = 1$, and when $x = 4$, $y = 2$, so the limits are from $y = 1$ to $y = 2$. Defining g by $g(y) = y^2$, we have

$$ds = \sqrt{1 + [g'(y)]^2}\,dy = \sqrt{1 + (2y)^2}\,dy = \sqrt{1 + 4y^2}\,dy$$

So by Equation 6.27, with l_1 and l_2 now given as y-limits,

$$S = 2\pi \int_{l_1}^{l_2} y\,ds = 2\pi \int_1^2 y\sqrt{1 + 4y^2}\,dy \quad u = 1 + 4y^2,\, du = 8y\,dy$$

$$= \frac{2\pi}{8} \int_1^2 y(1 + 4y^2)^{1/2}(8\,dy)$$

$$= \frac{\pi}{4} \cdot \frac{2}{3}(1 + 4y^2)^{3/2} \Big]_1^2$$

$$= \frac{\pi}{6}[(17)^{3/2} - (5)^{3/2}] \approx 30.85 \qquad \blacksquare$$

Shorthand Technique for Finding Areas of Surfaces of Revolution

Another advantage of Equations 6.27 and 6.28 is that they can be arrived at quite easily using the following technique. Referring to Figure 6.57, we treat ds as if it were a small arc on the curve. It may be thought of as being so small that it can be considered straight. Let (x, y) be its center. Then, when this arc is rotated about the x-axis, we get a frustum of a cone whose slant height is almost equal to the arc length. Its surface area is

$$(\text{average circumference}) \times (\text{``slant height''}) = 2\pi y\,ds$$

The total surface area is then the "sum" $2\pi \int_{l_1}^{l_2} y\,ds$. The same sort of reasoning leads to $2\pi \int_{l_1}^{l_2} x\,ds$ for the area of the surface of rotation about the y-axis.

FIGURE 6.57

REMARK —————————————————————————————

■ Just as with our other shorthand techniques, what we have described is simply a convenient device for arriving at a correct result. The reasoning is not mathematically rigorous.

—————————————————————————————

The next two examples further illustrate the use of Equations 6.27 and 6.28.

EXAMPLE 6.23   Find the surface area of the segment of a sphere of radius r cut off by a plane at a point a units from the center, where $0 < a < r$.

FIGURE 6.58

FIGURE 6.59

FIGURE 6.60

Solution   In Figure 6.58 we show the sphere with center at the origin and the segment cut off by a plane through $(a, 0)$ perpendicular to the x-axis. In Figure 6.59 we show an arc of the circle $x^2 + y^2 = r^2$ that generates the spherical segment when it is revolved about the x-axis. Treating y as a function of x, we have

$$x^2 + y^2 = r^2$$

Differentiating implicitly with respect to x, we obtain

$$2x + 2yy' = 0$$

Solving for y', we obtain

$$y' = -\frac{x}{y} \qquad y \neq 0$$

So

$$ds = \sqrt{1 + (y')^2}\, dx = \sqrt{1 + \frac{x^2}{y^2}}\, dx = \frac{\sqrt{x^2 + y^2}}{y}\, dx = \frac{r}{y}\, dx$$

Here, we have used the fact that $\sqrt{x^2 + y^2} = \sqrt{r^2} = r$ and also the fact that since $y \geq 0$ on the arc in question, $\sqrt{y^2} = y$. Thus,

$$S = 2\pi \int_{l_1}^{l_2} y\, ds = 2\pi \int_{a}^{r} y\left(\frac{r}{y}\right) dx = 2\pi r[x]_a^r$$

$$= 2\pi r(r - a) \qquad \blacksquare$$

EXAMPLE 6.24   A ceramic bowl is turned on a potter's wheel to approximate the surface obtained by revolving the curve $y = \frac{2}{3}(x^2 - 1)$ from $x = 1$ to $x = 2$ about the y-axis. (See Figure 6.60.) What is the approximate outside area of the bowl? Assume that each unit on the x- and y-axes represents 10 cm.

Solution   Letting $f(x) = \frac{2}{3}(x^2 - 1)$, we have $f'(x) = \frac{4}{3}x$, so

$$ds = \sqrt{1 + \frac{16}{9}x^2}\, dx = \frac{\sqrt{9 + 16x^2}}{3}\, dx$$

By Equation 6.28 the surface area S of the curved part of the bowl is given by

$$S = 2\pi \int_{l_1}^{l_2} x\, ds = 2\pi \int_{1}^{2} x\frac{\sqrt{9 + 16x^2}}{3}\, dx$$

If we think of u as the quantity $9 + 16x^2$ under the radical, we see that its differential is $32x\, dx$. So we multiply and divide by 32 to get

$$S = \frac{2\pi}{3} \cdot \frac{1}{32} \int_{1}^{2} (9 + 16x^2)^{1/2}(32x\, dx)$$

$$= \frac{\pi}{48} \cdot \frac{2}{3}(9 + 16x^2)^{3/2}\Big]_{1}^{2} = \frac{\pi}{72}\left[(73)^{3/2} - (25)^{3/2}\right] \approx 21.76$$

The circular bottom of the bowl has area $\pi(1^2) \approx 3.14$. Thus, the total outside area of the bowl is approximately 25.9 square units, or $(25.9)(10)^2 = 2{,}590\,\text{cm}^2$. $\blacksquare$

Exercise Set 6.5

In Exercises 1–10, find the area of the surface generated by revolving about the specified axis the arc of the curve whose equation is given, between the specified limits.

1. $y = 2x + 1$ from $x = 0$ to $x = 3$; x-axis

2. $2x + 3y = 6$ from $(0, 2)$ to $(3, 0)$; y-axis

3. $y = \sqrt{2x}$ from $x = 0$ to $x = 2$; x-axis

4. $y = x^2$ from $(1, 1)$ to $(2, 4)$; y-axis

5. $y = \sqrt{1 - x^2}$ from $(0, 1)$ to $(1, 0)$; y-axis

6. $y = x^3$ from $x = 0$ to $x = 1$; x-axis

7. $y = x^{1/3} + 1$ from $(1, 2)$ to $(8, 3)$; y-axis

8. $x^2 + y^2 = 9$ from $(0, 3)$ to $(2, \sqrt{5})$; x-axis

9. $x = \sqrt{5 - y}$ from $y = 1$ to $y = 4$; y-axis

10. $y^2 - 2x + 3 = 0$ from $(2, 1)$ to $(6, 3)$; x-axis

11. Derive the formula $S = 4\pi r^2$ for the surface area of a sphere of radius r.

12. Show that the surface area of a cone of height h and base radius r is $\pi r \sqrt{r^2 + h^2}$ by finding the area of the surface generated when the line segment $y = \dfrac{r}{h}x$ from $x = 0$ to $x = h$ is revolved about the x-axis.

13. Find the area of the surface that results from rotating the upper half of the four-cusp hypocycloid $x^{2/3} + y^{2/3} = a^{2/3}$ about the x-axis. (See Exercise 15 in Exercise Set 6.4.) Make use of implicit differentiation.

14. Find the area of the surface generated by revolving the segment of the line $2x + 3y = 6$ cut off by the x- and y-axes about (a) the line $x = -1$ and (b) the line $y = -2$.

15. Show that if f is smooth on $[a, b]$ but not necessarily nonnegative, the area of the surface generated by revolving the graph of f between $x = a$ and $x = b$ about the x-axis is

$$S = 2\pi \int_a^b |f(x)| \sqrt{1 + [f'(x)]^2} \, dx$$

16. Use the result of Exercise 15 to find the surface area generated by revolving the arc of the curve $y = x^3/6$ from $x = -1$ to $x = 2$ about the x-axis.

17. Find the area of the surface generated by revolving the arc of the curve

$$y = \frac{x}{2}\sqrt{\frac{1 - x^2}{2}}$$

from $x = -1$ to $x = 1$ about the x-axis. (See Exercise 15 and use the fact that $y = f(x)$ is odd in this case.)

18. Find the area of the surface generated by revolving the arc of the curve $y^4 + 3 = 6xy$ from $y = 1$ to $y = 2$ about the y-axis.

19. Find the area of the surface generated by revolving the arc of the curve $y = \sqrt{x}(x - 3)/3$ from $(1, -\frac{2}{3})$ to $(9, 6)$ about the x-axis.

20. A *zone* of a sphere is the band cut off by two parallel planes that intersect the sphere. Show that the surface area of a zone of a sphere of radius a cut off by two parallel planes h units apart ($h \le 2a$) is independent of the location of the planes (as long as both intersect the sphere). What is the area of such a zone?

21. (a) The earth is approximately a sphere of radius 6,374 km. The south-temperate zone extends from $-23\frac{1}{2}°$ (the Tropic of Capricorn) to $-67\frac{1}{2}°$ (the Antarctic Circle). What is the surface area of this south-temperate zone? (See Exercise 20.)

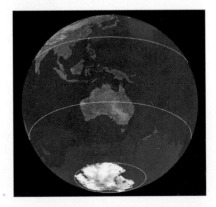

(b) A rain forest in Borneo covers a region 100 km by 200 km. What fraction of the area of the equatorial zone, from the Tropic of Cancer to the Tropic of Capricorn at $-23\frac{1}{2}°$, does the rain forest cover?

22. (a) Find a formula for the lateral surface area of a cone as follows. Imagine cutting a cone of base radius r along a line AB as shown in part (a) of the figure. Then flatten the cone so that it becomes a circular sector as in part (b). Calculate the area of this sector.

(b) Use the result of part (a) to find the area of a frustum of a cone with radii r_1 and r_2 and slant height l.

23. In each of the following use Simpson's Rule to approximate to two decimal places the area of the surface generated by revolving the curve about the x-axis.

(a) $y = \sin x$, $0 \le x \le \pi$

(b) $y = x^2$, $0 \le x \le 2$

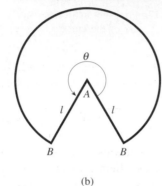

(a)   (b)

6.6 WORK

When a constant force F acts on an object, causing it to move a distance d in the direction of the force, physicists define the **work** W done by the force to be

$$W = Fd = (\text{force}) \times (\text{distance}) \tag{6.29}$$

Scientists now use units based on combinations of meters, kilograms, and seconds. In this *Système International d'Unités* (SI), force is measured in *newtons* (symbol: N), and the corresponding unit of work, a *newton-meter*, is called a *joule* (symbol: J). Energy is also measured in joules. Another set of metric units, now less used, involves centimeters, grams, and seconds (cgs). In this system, force is measured in *dynes* (1 newton = 10^5 dyne), so the unit of work is the *dyne-centimeter*, also called an *erg* (1 joule = 10^7 erg). In the English system of units, force is typically measured in pounds and distance in feet or inches, so the unit of work is the *foot-pound* or the *inch-pound*. These English units are hardly used anymore in science or internationally, though the British thermal unit (Btu), a unit of energy, is in common use in the United States.

Equation 6.29 applies only to constant forces that act in the direction of motion. There are many situations in which the acting force is variable. Either the magnitude or the direction of the force, or both, may change over time. For now, let us consider a variable force acting in a straight line. An example of a variable force is the force necessary to stretch a spring. The force is higher when the spring has been stretched more. As another example, the force of repulsion between two like electrostatic charges increases as the charges get closer to each other.

Let us now define the work done by a variable force F acting in a straight line over a fixed distance. For convenience, let us orient the x-axis so that the direction of force is along the axis, and let $F(x)$ denote the amount of the force at x. Suppose further that F is a continuous function on the closed interval from $x = a$ to $x = b$. As in Figure 6.61, we partition this closed interval

FIGURE 6.61

$[a, b]$, and let c_k denote an arbitrary point in the kth subinterval. If the width $\wedge x_k$ is sufficiently small, the error in replacing the variable force $F(x)$ with the constant force $F(c_k)$ on the kth subinterval will not be great. Using this constant force enables us to apply Equation 6.29. For this constant force, the work ΔW_k done across the kth subinterval is

$$\Delta W_k = \underbrace{F(c_k)}_{\text{force}} \cdot \underbrace{\Delta x_k}_{\text{distance}}$$

We can reasonably suppose that the total work W done by the force on the closed interval $[a, b]$ is approximated by the sum of the ΔW_k values:

$$W \approx \sum_{k=1}^{n} \Delta W_k = \sum_{k=1}^{n} F(c_k)\Delta x_k$$

As the widths of the subintervals all approach 0, the approximation becomes better and better. Since the Riemann sum on the right approaches the integral $\int_a^b F(x)\,dx$, we are led to the following definition.

Definition 6.4
Work

The **work** W done by a continuous force F acting along the x-axis from $x = a$ to $x = b$ is given by

$$W = \int_a^b F(x)dx \qquad (6.30)$$

EXAMPLE 6.25   A 100-kg mass is to be hoisted from the ground to a point 10 m high by a uniform chain passing over a pulley. If the chain has mass 3 kg/m, find the work done in raising the mass.

Solution   As shown in Figure 6.62, let x denote the distance from the pulley to the mass at any stage. Then x varies between 0 and 10. By Newton's Second Law the total force $F(x)$ is the total mass times the acceleration of gravity g (approximately 9.81 m/s^2 near the earth's surface). The total mass is the 100-kg mass plus the mass of the chain still to be raised. When a section of chain measuring x m is out, its mass is $3x$ kg. So $F(x) = (100 + 3x)g$ is the variable force. Thus, by Equation 6.30, the work W in joules is

$$W = \int_0^{10} (100 + 3x)g\,dx = g\left[100x + \frac{3x^2}{2}\right]_0^{10}$$

$$= (1,000 + 150)g = 1,150g$$

Using $g = 9.81$ m/s^2, we get $W \approx 11,280$ J. ∎

100 kg

FIGURE 6.62

Hooke's Law for Springs

The force necessary to stretch a spring x units beyond its natural length is given by $F(x) = kx$ (provided the force does not exceed what is called the *elastic limit* of the material from which the spring is made). The factor k is a constant, called the *spring constant*. This result is known as **Hooke's Law**, after the scientist Robert Hooke, a seventeenth-century contemporary (and rival) of Isaac Newton. It also holds true if the spring is compressed x units from its natural length.

EXAMPLE 6.26   A certain spring has a natural length of 50 cm, and a 5-N force is required to stretch it 0.15 m. Find the work done in stretching it from its natural length to a length of 60 cm. Find the additional work done in stretching it 10 cm more (from 60 cm to 70 cm).

Solution   By Hooke's Law, $F(x) = kx$, and we know that $F(x) = 5$ when $x = 0.15$. So we can find k:

$$5 \text{ kg} = k(0.15 \text{ m})$$

$$k = \frac{5}{0.15} = \frac{100}{3} \text{ N/m}$$

Thus, for the spring,

$$F(x) = \frac{100}{3}x$$

where $F(x)$ is in newtons and x is in meters. In stretching the spring from its natural length of 50 cm to a length of 60 cm, x varies from 0 to 0.1, since the spring is stretched 10 cm = 0.1 m. (Remember that x is the distance in meters that the spring is stretched *beyond its natural length*.) So we have

$$W = \int_0^{0.1} F(x)dx = \int_0^{0.1} \frac{100}{3}x\, dx = \frac{100}{3}\left[\frac{x^2}{2}\right]_0^{0.1}$$

$$= \frac{0.5}{3} \approx 0.17 \text{ J}$$

For the spring to be stretched an additional 10 cm, x will vary from 0.1 m to 0.2 m, so the additional work done is

$$W = \int_{0.1}^{0.2} \frac{100}{3}x\, dx = \frac{100}{3}\left[\frac{x^2}{2}\right]_{0.1}^{0.2} = \frac{50}{3}(0.04 - 0.01) = 0.5 \text{ J}$$

Note that the work required to stretch the spring the second 10 cm was almost three times that required to stretch it the first 10 cm. ∎

Pumping Liquid from Tanks

The next example is typical of a class of problems. In preparation, let us consider a tank of height b units filled with a liquid to a depth a units from the top. Figure 6.63 shows this situation. We want to find the work done in pumping the liquid over the top rim of the tank. We can imagine pumping (lifting) the liquid out of the tank one thin slab at a time and applying Equation 6.29 for the work done by a constant force (the weight of the slab) to each slab. The total work done is approximately equal to the sum that we get by adding up the work of lifting all the slabs. We will see that this sum is a Riemann sum whose limit, as the slab thicknesses approach zero, is an integral. The exact form of the integral will depend on the weight of the liquid and the cross-sectional area of the container.

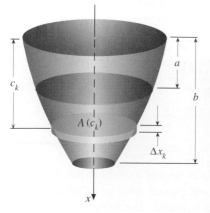

FIGURE 6.63

Suppose the density of the liquid is ρ, defined as mass per unit volume (for example, kilograms per cubic meter (kg/m^3)). As previously suggested, we can think of the liquid as being divided into thin layers by planes passed perpendicular to the vertical axis at points $\{x_0, x_1, x_2, \ldots, x_n\}$ that form a partition of the closed interval $[a, b]$. Suppose at the distance x units down from the top of the tank, the cross-sectional area $A(x)$ is known and is a continuous function of x on $[a, b]$. As in our study of volumes by the cross-sectional method, the kth layer is approximated by a disk of constant cross-sectional area $A(c_k)$ and thickness $\Delta x_k = x_k - x_{k-1}$, where c_k is any point in $[x_{k-1}, x_k]$. We have shown a typical disk in Figure 6.63. We imagine this "liquid disk" to be lifted to the top. The work required to do this lifting is the weight of the disk (a force) multiplied by the distance c_k through which it is lifted. The weight of the kth disk is its volume $A(c_k)\Delta x_k$ times its weight per unit volume, ρg (ρ is mass per unit volume and weight is mass times the acceleration g of gravity). So the work ΔW_k of lifting this weight to the top is

$$\Delta W_k = \overbrace{\rho g A(c_k)\Delta x_k}^{\text{force}} \cdot \overbrace{c_k}^{\text{distance}}$$

The total work is approximated by the sum

$$W \approx \sum_{k=1}^{n} \Delta W_k = \sum_{k=1}^{n} \rho g c_k A(c_k)\Delta x_k$$

On taking the limit as the thicknesses Δx_k all approach 0, we arrive at the result

$$W = \int_a^b \rho g x A(x)\,dx \qquad (6.31)$$

REMARK
■ As we have noted, the product ρg is weight per unit volume. For this reason it is sometimes called *weight density*. For water, in SI units, $\rho = 1,000$ kg/m^3. So, near the earth's surface, $\rho g \approx (1,000)(9.81) = 9,810$ N/m^3. In the English system, the weight density is given directly. For example, for water near the earth's surface, $\rho g \approx 62.4$ lb/ft^3.

EXAMPLE 6.27   A water tank 8 m high is in the form of a frustum of a cone, as shown in Figure 6.64. It is filled to a depth of 5 m. Find the work done in pumping all the water over the top of the tank.

Solution   As in the preceding discussion, we let x be the distance from the top of the tank to a typical element. To find the cross-sectional area $A(x)$, we need to find the radius r. By similar triangles (see Figure 6.65), we have

$$\frac{r - 2}{4} = \frac{8 - x}{8}$$

So

$$r - 2 = \frac{4}{8}(8 - x)$$

or

$$r = 2 + \frac{1}{2}(8 - x) = \frac{1}{2}(12 - x)$$

FIGURE 6.64

FIGURE 6.65

Thus,

$$A(x) = \pi r^2 = \frac{\pi}{4}(12 - x)^2$$

The density of water is 1000 kg/m³. So, by Equation 6.31, we have (taking g = 9.81 m/s²),

$$W = \int_a^b \rho g x A(x)\, dx = \int_3^8 9810 x \left(\frac{\pi}{4}\right)(12 - x)^2\, dx$$

$$= \frac{9810\pi}{4} \int_3^8 x(144 - 24x + x^2)\, dx$$

$$= \frac{9810\pi}{4} \left[72x^2 - 8x^3 + \frac{x^4}{4}\right]_3^8$$

$$\approx 8.350 \times 10^6 \text{ J}$$ ∎

Exercise Set 6.6

1. A variable force of $2x - x^2$ N acts in the direction of the positive x-axis from $x = 0$ to $x = 2$ (in meters). Find the work done by the force.

2. A weight of 200 lb is to be pushed a distance of 12 ft up a smooth ramp that makes an angle of 30° with the horizontal. Neglecting any friction, find the work done.

3. A force of 4 lb is required to stretch a certain spring 3 in. beyond its natural length of 10 in. Find the work done in stretching it from its natural length to a length of 16 in.

4. For the spring in Exercise 3, find the work done in stretching it from a length of 15 in. to a length of 18 in.

5. A spring of natural length 20 cm is stretched to a length of 28 cm by a force of 5 N. Find the work done in compressing the spring 6 cm from its natural length. What is the work done in stretching the spring from a length of 22 cm to a length of 30 cm?

6. A heavy-duty spring with spring constant 10,000 N/m is used to cushion the shock of a subway car at the end of the line in the event it does not come to a complete stop before bumping the end. Find the work done by the force of the subway car if in bumping the end, the spring is compressed a distance of 1 m.

7. The work done in stretching a spring from its natural length of 25 cm to a length of 30 cm is 0.5 J. Find the spring constant.

8. A spring of natural length 14 in. requires 30 ft-lb of work

to be compressed from a length of 12 in. to a length of 8 in. Find the spring constant. (Be careful with units.)

9. A uniform chain weighing 3 lb/ft is hanging over the edge of a building and is to be pulled up to the top. If a 30-ft length of chain initially hangs down, find the work required to raise it to the top.

10. A crane at a construction site is used to lift a bucket of cement from the ground to a point 10 m above the ground. The distance from the end of the crane arm to the ground is 25 m. If the bucket of cement has mass 250 kg and the cable has mass 7 kg/m, find the work required to lift the cement.

11. Water is to be raised from a well 15 m deep by means of a bucket attached to a rope. When the bucket is full of water its mass is 15 kg, but the bucket has a leak that causes it to lose water at the constant rate of $\frac{1}{4}$ kg for each meter that the bucket is raised. Neglecting the mass of the rope, find the work done in raising the bucket to the top.

12. A water tank is in the form of a vertical right circular cylinder with radius 2 m and height 8 m. If the tank is half full of water, find the work required to pump all the water over the top rim.

13. In Exercise 12 suppose the tank is initially full of water. Find the work required to pump all of it to a point 3 m above the top of the tank.

14. A storage tank for liquid ammonia is in the form of a sphere with radius 3 m. If the tank is full, find the approximate work required to pump half of the ammonia to a point 5 m above the top of the tank. The density of ammonia is $\rho = 891$ kg/m^3.

15. According to Coulomb's Law, the repulsive force between two like electrostatic charges Q_1 and Q_2 is given by

$$F = \frac{kq_1q_2}{x^2}$$

where q_1 and q_2 are the magnitudes of Q_1 and Q_2, respectively, x is the distance between them, and k is a constant of proportionality. Suppose $q_1 = 1$ C (coulomb) and $q_2 = 2$ C, and the charges are initially 6 cm apart. If Q_1 is stationary, find the work done in moving Q_2 on a line toward Q_1 to a point 3 cm from Q_1. (Leave the answer in terms of k.)

16. According to Newton's Universal Law of Gravitation, the gravitational force exerted by the earth on a body of mass m at a distance x above the earth's surface is

$$F(x) = G\frac{mM}{(x+R)^2}$$

where R is the radius of the earth, M is the earth's mass, and G is a constant called the universal gravitational constant. It can be shown that $G = gR^2/M$. Taking $g = 9.8$ m/s^2 and $R = 6.37 \times 10^6$ m, find the work done in lifting a satellite of mass 3.2×10^4 kg a distance of 300 km (300 km $= 3 \times 10^5$ m) above the earth's surface.

17. Hooke's Law can be extended to metal bars under tension or compression. The force necessary to elongate the bar, or compress it, by x units is given by

$$F(x) = \frac{EA}{L}x$$

where A is the cross-sectional area of the bar and L is its length. E is a constant called the *modulus of elasticity*

that is a property of the metal. Suppose a cylindrical aluminum bar of natural length 15 cm and cross-sectional radius 3 cm is stretched to a length of 15.5 cm. Find the work done. For aluminum, $E = 7.30 \times 10^{10}$ N/m^2.

18. A water tank is in the form of a frustum of a cone that has bottom radius 5 ft, top radius 8 ft, and height 12 ft. If the tank is initially full of water, find the work done in pumping half the water to a point 10 ft above the top. (Use $\rho g = 62.4$ lb/ft^3.)

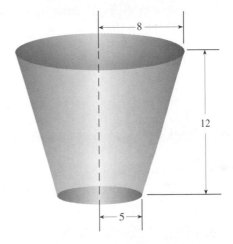

19. A swimming pool slopes from a depth of 1 m at the shallow end to a depth of 4 m at the deep end. It is 20 m long and 10 m wide. If it is full of water, find the work required to pump all the water over the top. (Use $\rho = 1000$ kg/m^3 and $g = 9.81$ m/s^2.)

20. A crane raises a bucket of sand weighing 500 lb from the ground to a height of 40 ft in 10 seconds. Sand spills out of the bucket at the rate of 10 lb/s as it is being raised. If the cable weighs 5 lb/ft, find the work done in raising the sand.

21. A tank has the shape of the surface generated by revolving the right half of the parabola $y = x^2/2$ from $x = 0$ to $x = 2$ about the y-axis. If x and y are measured in meters and the tank is initially half full of water (half of the volume), find the work done in pumping all the water over the top.

22. The force on a piston caused by the expansion of a gas in a cylinder (see the figure) is $F = pA$, where A is the area of the face of the piston and p is the pressure (force per unit area) of the gas. The volume v occupied by the gas, when the distance from the piston to the end of the cylinder is x, is given by $v = Ax$. For an ideal gas under an adiabatic process (in which no heat is transferred into or out of the cylinder), the relationship between pressure and volume is $pv^{1.4} = c$, where c is a constant. If when $x = a$ the volume is v_1 and when $x = b$ the volume is v_2, show that the work done by the gas in expanding from v_1 to v_2 is

$$W = \int_a^b F(x)\,dx = \int_{v_1}^{v_2} cv^{-1.4}\,dv$$

If the radius of the piston is 3 in., find the work done by the gas in moving the piston from $x = 5$ in. to $x = 8$ in. Express the answer in terms of c.

Area = A

Volume of gas = v

$\leftarrow$ x $\rightarrow$

6.7 FLUID PRESSURE

The Hoover Dam

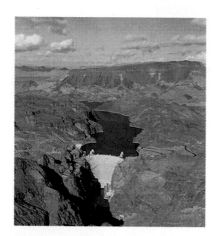

Density = ρ

h

Area = A

FIGURE 6.66
Force on plate = $\rho g A h$.

Our objective in this section is to show how integration can be used to calculate the force exerted by a liquid pressing against one side of a flat surface. For example, we can find the force of the water in a lake against the side of a dam. (We ignore the curvature of the dam's face.) We will consider only static forces rather than dynamic ones. This limitation means that we will suppose the liquid is not in motion.

Suppose the liquid in question has density ρ (mass per unit volume). If a flat plate of area A is submerged horizontally at a depth h (see Figure 6.66), then the total force on the plate is just the weight of the liquid directly above the plate. The volume occupied by this liquid is the area times the height, Ah. We multiply this volume by the weight per unit volume, ρg, to get the total weight of the liquid. So

$$F = \rho g A h$$

For example, the force exerted by the oil in an oil storage tank on the bottom of the tank is the area of the bottom times the depth of the oil times its density. Note that the force does not depend on the shape of the submerged plate, but only on its area and the depth.

The **pressure** exerted by the liquid is defined as the force on each square unit of area. So, taking $A = 1$, we get for the pressure p at depth h,

$$p = \rho g h \qquad (6.32)$$

As the depth increases, the pressure also increases, which explains why dams must be thicker at the bottom than at the top. It is an interesting fact, known as *Pascal's Principle*, that the pressure p at a given depth is the same in all directions. Thus, if a plate is submerged horizontally, vertically, or on a slant, the pressure at a fixed depth h units below the surface is given in all cases by Equation 6.32.

When a plate of known area is submerged horizontally, calculating the force on it is straightforward, as we have seen. When it is submerged vertically, however, finding the force is more involved. After all, the pressure varies continuously from the top of the plate to the bottom. We introduce a vertical axis that we show in Figure 6.67 as being directed positively downward. (In certain applications it will be more convenient to choose the positive direction upward.) The origin may be taken at any point, usually determined by the shape of the plate. In the examples we will see how certain choices can often simplify calculations. We assume that the vertical extremities of the plate are at $x = a$ and $x = b$, where $a < b$. We further assume that enough information is known about the shape of the plate that its width, $w(x)$, can be determined at each value of x in the interval. Finally, we assume that w is a continuous function of x on the closed interval $[a, b]$. We denote by $h(x)$ the depth at x.

FIGURE 6.67

We partition the interval $[a, b]$ in the usual way. Horizontal lines drawn at the partition points divide the plate into thin strips. The kth strip can be approximated by a rectangle that has dimensions $w(c_k)$ by Δx_k, where c_k is an intermediate point in the kth subinterval. A typical such rectangle is shown in Figure 6.67. Although the pressure varies from the top to the bottom of this rectangle, if Δx_k is very small, the error introduced by assuming the entire rectangle is at depth $h(c_k)$ will not be great. The pressure at this depth, by Equation 6.32, is

$$p = \rho g h(c_k)$$

We approximate the total force on the kth rectangle, which we designate by ΔF_k, by its area times this pressure:

$$\Delta F_k \approx \overbrace{\rho g}^{\substack{\text{weight} \\ \text{density}}} \cdot \overbrace{h(c_k)}^{\text{depth}} \cdot \overbrace{w(c_k) \Delta x_k}^{\text{area}}$$

It is useful to learn this result in words:

$$\text{force on a rectangular strip} \approx \text{weight density} \times \text{depth} \times \text{area}$$

We approximate the total force as the sum of all ΔF_k values:

$$F \approx \sum_{k=1}^{n} \Delta F_k \approx \sum_{k=1}^{n} \rho g h(c_k) w(c_k) \Delta x_k$$

As we take finer and finer partitions, the Riemann sum on the right approaches the integral $\int_a^b \rho g h(x) w(x) dx$. This discussion leads to the following definition.

Definition 6.5
Fluid Force

The total **force** F on one side of a flat plate submerged vertically in a liquid of density ρ is given by

$$F = \int_a^b \rho g h(x) w(x) dx \qquad (6.33)$$

where the vertical bounds of the plate are $x = a$ and $x = b$, respectively, $h(x)$ is the depth at x, and $w(x)$ is the width of the plate at x.

Shorthand Technique for Finding Force on a Submerged Plate

FIGURE 6.68

As a memory device, we can indicate the width of a typical rectangular element by dx, its length by $w(x)$, and its depth by $h(x)$, as shown in Figure 6.68. So the force on this element is

$$\text{weight density} \times \text{depth} \times \text{area} = \rho g h(x) w(x) dx$$

"Summing up" all such elements of force (in the sense of integration) we get Equation 6.33. Again, this technique is only a memory device, not a mathematically rigorous derivation.

EXAMPLE 6.28   A triangular plate with base 4 m and altitude 2 m is submerged vertically in a liquid of density ρ kg/m^3. It is oriented as in Figure 6.69, with its upper vertex 3 m below the surface of the liquid. Find the total force on one side of the plate.

FIGURE 6.69

Solution   We choose the origin for x as the upper vertex and direct x positively downward, as shown. The limits on x are then 0 and 2. To find the width $w(x)$, we use similar triangles:

$$\frac{w(x)}{4} = \frac{x}{2}$$

$$w(x) = 2x$$

The depth $h(x)$ is given by $h(x) = x + 3$. Thus, by Equation 6.33,

$$F = \int_0^2 \rho g (x + 3)(2x) dx$$

$$= 2\rho g \int_0^2 (x^2 + 3x) dx$$

$$= 2\rho g \left[\frac{x^3}{3} + \frac{3x^2}{2} \right]_0^2$$

$$= \frac{52\rho g}{3} \text{ N}$$

In the exercises you will be asked to rework this example, choosing the origin for x as the level of the fluid so that you can see the advantages of the choice we made. ∎

EXAMPLE 6.29   A dam across a river gorge is in the form of an inverted isosceles trapezoid, with upper base 100 ft, lower base 50 ft, and altitude 80 ft (Figure 6.70). Find the force of the water on the face of the dam (assumed vertical) when the water is within 10 ft of the top of the dam.

FIGURE 6.70

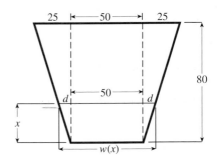

FIGURE 6.71

Solution   Because the trapezoid is inverted, it is convenient to choose as the origin for x the bottom of the dam and take the positive direction as upward. Doing so simplifies the calculation of the width $w(x)$, as we will show. The limits on x are then 0 and 70, since it is only this portion of the dam that has water against it. The depth from the water level to a point with coordinate x is $h(x) = 70 - x$. To find the width $w(x)$, we again make use of similar triangles, as shown in Figure 6.71. From this figure we see that $w(x) = 50 + 2d$, and using similar triangles, we have

$$\frac{d}{x} = \frac{25}{80}$$

So $d = 5x/16$ and

$$w(x) = 50 + 2\left(\frac{5x}{16}\right)$$

$$= 50 + \frac{5x}{8}$$

$$= \frac{5}{8}(80 + x)$$

Now, using Equation 6.33, we have

$$F = \int_0^{70} \rho g (70 - x) \cdot \frac{5}{8}(80 + x)dx$$

$$= \frac{5\rho g}{8} \int_0^{70} (5,600 - 10x - x^2)dx$$

$$= \frac{5\rho g}{8} \left[5,600x - 5x^2 - \frac{x^3}{3} \right]_0^{70}$$

$$= \frac{5\rho g}{8} \left(\frac{759,500}{3} \right)$$

Taking $\rho g = 62.4$ lb/ft^3 as the approximate density of water, we finally get $F = 9,873,500$ lb. At 2,000 lb/ton, this force is nearly 5,000 tons.   ■

Exercise Set 6.7

In Exercises 1–6, a plate is submerged vertically in a liquid of density ρ, as shown. Find the force of the liquid on one side of the plate.

1.

2.

3.

4.

5.

6.

7. The ends of a watering trough are in the form of inverted isosceles triangles, having base 3 ft and altitude 2 ft. Find the force of the water in the trough on one of the ends when the trough is full of water. What is the force when the water level is 1 ft?

8. If instead of a triangle, each end of the watering trough in Exercise 7 forms the lower half of a circle of radius 2 ft, find the force on one of the ends when the trough is full of water.

9. A swimming pool is 10 m wide and 15 m long. At the shallow end its depth is 1 m, and at the deep end it is 4 m; the depth increases linearly from the shallow to the deep end. When the pool is full of water find the force on (a) each of the ends and (b) each of the sides.

10. An oil drum is cylindrical, with diameter 1 m. If it is half full of oil that has density 920 kg/m³, and it is lying on its side, find the force on each end.

11. Rework Example 6.28, choosing the level of the fluid as the origin for x.

12. Rework Example 6.29, choosing the water level as the origin for x.

13. The plate in the accompanying figure is parabolic and is submerged as shown in a fluid of density ρ. Find the force on one side of the plate. (*Hint:* If the origin is taken at the vertex of the parabola, its equation is of the form $y = ax^2$. Find a, and use y instead of x as the variable of integration.)

14. Each end of a gasoline tank is elliptical, with equation $x^2 + 4y^2 = 1$ when referred to a coordinate system with origin at the center of the ellipse and y-axis vertical, where x and y are in feet. Find the force on one end of the tank when it is half full. Take the weight density of gasoline as 45 lb/ ft^3. (*Hint:* Use y as the variable of integration.)

15. Show that if a plate is submerged at an angle θ with the vertical, where $0 < \theta < \frac{\pi}{2}$, the force on it is given by

$$F = \int_a^b \rho g h(x) w(x) \sec \theta \, dx$$

with $h(x)$ and $w(x)$ having the same meaning as in Definition 6.5, with the x-axis directed vertically downward.

16. Find the force on the bottom of the swimming pool in Exercise 9. (Use the result of Exercise 15.)

17. A porthole in an undersea observation laboratory is circular, with a 0.3 m diameter. Assume the porthole lies in a vertical plane. If its center is at a depth of 400 m, find the force on the porthole. (The density of seawater is approximately 1040 kg/m^3.) (*Hint:* In carrying out the integration, use your knowledge of the area of the circle.)

18. Use Simpson's Rule with $n = 10$ to approximate the force on a vertical plate bounded by the x-axis and the top half of the hypocycloid $x^{2/3} + y^{2/3} = 4^{2/3}$, if the plate is submerged in water with its base 10 m beneath the surface.

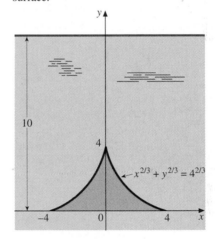

6.8 MOMENTS, CENTROIDS, AND CENTERS OF MASS

In this section we present some applications of the integral that are especially important in physics and engineering. The ideas are also used in statistics.

Children playing on a seesaw quickly learn that the heavier child must sit closer to the pivot point, called the *fulcrum*, to bring the seesaw into balance. Denoting the weights of the children by w_1 and w_2 and the corresponding distances from the fulcrum by d_1 and d_2, as in Figure 6.72, we must have $w_1 d_1 = w_2 d_2$ for the seesaw to be balanced. This relation is called the *law of the lever* and was known to Archimedes.

Now consider a finite number of weights, $w_1, w_2, \ldots, w_n$, placed at various points on an axis, with w_i at x-coordinate x_i, as in Figure 6.73. We wish to find the location of the balance point, also called the *center of gravity*, of the system. Denote its x-coordinate by $\bar{x}$. The product of a weight (or force) and its distance from a point is called the **moment** produced by that weight (or force)

FIGURE 6.72

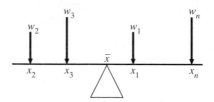

FIGURE 6.73

about that point. The moment thus is a measure of the tendency for the weight to cause a turning, or rotation, about the point. The directed distance of the weight from the point is called the **moment arm** of the weight. If the moment arm is positive, then the weight tends to produce a clockwise rotation, and if negative, a counterclockwise rotation. In Figure 6.73, for example, the moment arm of w_1 from the center of gravity is the positive distance $x_1 - \bar{x}$, and the moment about this point is $w_1(x_1 - \bar{x})$. The moment arm of w_3 is $x_3 - \bar{x}$, which is a negative number, and so w_3 produces the negative moment $w_3(x_3 - \bar{x})$. For balance, it must be true that the algebraic sum of all the moments is 0:

$$\sum_{i=1}^{n} w_i(x_i - \bar{x}) = 0$$

We can now solve for $\bar{x}$ from this equation:

$$\sum_{i=1}^{n} w_i x_i - \sum_{i=1}^{n} \bar{x} w_i = 0$$

Since $\bar{x}$ is a constant, we can factor it out of the second summation and then solve, getting

$$\bar{x} = \frac{\displaystyle\sum_{i=1}^{n} w_i x_i}{\displaystyle\sum_{i=1}^{n} w_i}$$

Since by Newton's Second Law the weight w of an object is its mass m times the acceleration of gravity g, we have for each i, $w_i = m_i g$, so that $\bar{x}$ can be rewritten in terms of masses as

$$\bar{x} = \frac{\displaystyle\sum_{i=1}^{n} m_i g x_i}{\displaystyle\sum_{i=1}^{n} m_i g} = \frac{\displaystyle\sum_{i=1}^{n} m_i x_i}{\displaystyle\sum_{i=1}^{n} m_i} \tag{6.34}$$

(assuming g is a constant and can therefore be divided out). Because of the shift in emphasis from weight to mass as given by Equation 6.34, it is customary to use the term **center of mass** instead of center of gravity for the balance point of the system. (Only if the force of gravity varies over the distance being studied is the center of gravity different from the center of mass.) The product $m_i x_i$ is the moment of mass m_i about the origin, since x_i is the directed distance of m_i from the origin. Thus, we can restate the result of Equation 6.34 in words as

$$\bar{x} = \frac{\text{sum of moments about the origin}}{\text{sum of masses}}$$

If we let $m = \sum_{i=1}^{n} m_i$ (the total mass m equals the sum of the individual masses m_i), then on clearing Equation 6.34 of fractions, we see that

$$m\bar{x} = \sum_{i=1}^{n} m_i x_i$$

From this result, we conclude that if the masses were all concentrated at $\bar{x}$— that is, at the center of mass—the moment about the origin of the total mass m would equal the sum of the moments about the origin of the individual masses.

EXAMPLE 6.30   Three masses, $m_1 = 35$ g, $m_2 = 56$ g, and $m_3 = 47$ g, are located on the x-axis at distances $x_1 = 15$, $x_2 = 36$, and $x_3 = -14$, respectively. Distances are in cm. Find the x-coordinate of the center of mass.

Solution   By Equation 6.34,

$$\bar{x} = \frac{(35)(15) + 56(36) + (47)(-14)}{35 + 56 + 47} = \frac{1883}{138} \approx 13.6 \text{ cm}$$ ■

Continuous One-Dimensional Mass

FIGURE 6.74

So far, we have been talking about a discrete system of masses. Now we consider a continuously distributed mass along a line, such as a thin rod or a wire. For definiteness, let us say it is a rod, and that we have introduced an x-axis as in Figure 6.74 such that the rod reaches from $x = a$ to $x = b$. Imagine that we cut the rod into n small segments by means of the partition points $a = x_0 < x_1 < x_2 < \cdots < x_n = b$. In Figure 6.74 we show a typical segment. Denote its center coordinate by c_k.

We will allow for the possibility that the density of the rod may vary with respect to distance along the axis, where density is defined as mass per unit length. (In general, density is mass per unit volume, but we are considering our rod to be essentially one-dimensional.) We denote the density by $\rho(x)$, and assume that ρ is continuous on the closed interval $[a, b]$. Let Δm_k denote the mass of the typical segment of the rod we have shown in Figure 6.74. If its length, Δx_k, is small, then the density over this short distance can be considered approximately the constant value $\rho(c_k)$. Thus,

$$\Delta m_k \approx \rho(c_k) \Delta x_k$$

We can consider our n small segments, $\Delta m_1, \Delta m_2, \ldots, \Delta m_n$, as point masses concentrated at the points $c_1, c_2, \ldots, c_n$, respectively. Thus, by Equation 6.34, the center of mass of the rod is approximately

$$\bar{x} \approx \frac{\displaystyle\sum_{k=1}^{n}(\Delta m_k)c_k}{\displaystyle\sum_{k=1}^{n}\Delta m_k} \approx \frac{\displaystyle\sum_{k=1}^{n}\rho(c_k)c_k \Delta x_k}{\displaystyle\sum_{k=1}^{n}\rho(c_k)\Delta x_k}$$

The numerator is a Riemann sum for the function $\rho(x)x$, and the denominator is a Riemann sum for the function $\rho(x)$. As we take finer and finer partitions, we get, from the limits on the numerator and denominator, the following formula for $\bar{x}$:

Center of Mass of a One-Dimensional Continuous Mass

$$\bar{x} = \frac{\int_a^b \rho(x)x \, dx}{\int_a^b \rho(x)dx} \qquad (6.35)$$

EXAMPLE 6.31   A rod 20 cm long has density given by $\rho(x) = (x+1)/2$. Find its center of mass.

Solution   By Equation 6.35,

$$\bar{x} = \frac{\displaystyle\int_0^{20} \left(\frac{x+1}{2}\right) x\, dx}{\displaystyle\int_0^{20} \left(\frac{x+1}{2}\right) dx} = \frac{\displaystyle\frac{1}{2}\int_0^{20} (x^2+x)\, dx}{\displaystyle\frac{1}{2}\int_0^{20} (x+1)\, dx} = \frac{\left.\dfrac{x^3}{3} + \dfrac{x^2}{2}\right]_0^{20}}{\left.\dfrac{x^2}{2} + x\right]_0^{20}}$$

$$= \frac{430}{33} \approx 13.03 \text{ cm}$$

Note that the center of mass is to the right of the geometric center, which makes sense because the density increases as we move from left to right.   ∎

Point Masses in the Plane

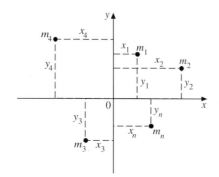

FIGURE 6.75

Now we extend these ideas to a system of masses in the xy-plane, as shown in Figure 6.75. We assume each mass to be concentrated at a point and refer to it as a *point mass*. The concept of a point mass is, of course, idealized. Let the point mass m_i be located at (x_i, y_i). If l is any line in the plane and d_i is the directed distance from l to (x_i, y_i), then the moment of m_i about l is $m_i d_i$, and the sum of the moments of all the masses about l is $\sum_{i=1}^{n} m_i d_i$. In particular, the sum of the moments about the x-axis is $\sum_{i=1}^{n} m_i y_i$ and about the y-axis is $\sum_{i=1}^{n} m_i x_i$. The point $(\bar{x}, \bar{y})$ for which the sum of the moments about the lines $x = \bar{x}$ and $y = \bar{y}$ each equals 0 is the center of mass of the system. By proceeding as in the one-dimensional case, we find that

$$\bar{x} = \frac{\displaystyle\sum_{i=1}^{n} m_i x_i}{\displaystyle\sum_{i=1}^{n} m_i} = \frac{\text{sum of moments about } y\text{-axis}}{\text{sum of masses}}$$

$$\text{(6.36)}$$

$$\bar{y} = \frac{\displaystyle\sum_{i=1}^{n} m_i y_i}{\displaystyle\sum_{i=1}^{n} m_i} = \frac{\text{sum of moments about } x\text{-axis}}{\text{sum of masses}}$$

EXAMPLE 6.32   Four point masses, of 12 g, 18 g, 24 g, and 30 g, are located in the plane as shown in Figure 6.76. Find the center of mass.

Solution   By Equations 6.36, we have

$$\bar{x} = \frac{(1)(12) + (5)(18) + (-1)(24) + (-4)(30)}{12 + 18 + 24 + 30} = -\frac{42}{84} = -\frac{1}{2}$$

$$\bar{y} = \frac{(3)(12) + (-4)(18) + (-7)(24) + 2(30)}{84} = -\frac{144}{84} = -\frac{12}{7}$$

So the center of mass is located at the point $(-\frac{1}{2}, -\frac{12}{7})$, as indicated in Figure 6.76.   ∎

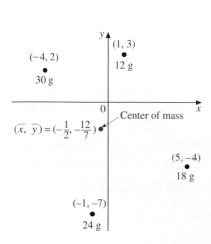

FIGURE 6.76

Continuous Mass in the Plane—A Lamina

FIGURE 6.77

If, instead of a system of discrete point masses in the plane, we have a continuously distributed mass, called a **lamina** (such as a thin sheet of material), we divide the lamina into vertical or horizontal strips (depending on its shape) and treat each strip as a point mass. For the present we will suppose the lamina has constant density ρ. When the density is constant, the lamina is said to be *homogeneous*. In multivariable calculus, using the concept of the double integral, we could consider a variable density. With the density constant, we can assume that each strip is a point mass concentrated at its geometric center.

In Figure 6.77 we show a lamina that occupies an area below the graph of a continuous nonnegative function, $y = f(x)$, between $x = a$ and $x = b$. We partition $[a, b]$ in the usual way, obtaining n rectangles. In the figure we show a typical rectangle, where we have taken c_k as the midpoint of the subinterval $[x_{k-1}, x_k]$. The "mass" of this rectangle is its area times the density ρ. Letting Δm_k be this mass, we have

$$\Delta m_k = \rho f(c_k) \Delta x_k$$

Since the density is constant, we can consider the rectangular element as a point mass concentrated at its geometric center,

$$\left(c_k, \frac{f(c_k)}{2} \right)$$

By the first of Equations 6.36, then, we have

$$\bar{x} \approx \frac{\displaystyle\sum_{k=1}^{n} c_k \Delta m_k}{\displaystyle\sum_{k=1}^{n} \Delta m_k}$$

and by the second,

$$\bar{y} \approx \frac{\displaystyle\sum_{k=1}^{n} \frac{f(c_k)}{2} \Delta m_k}{\displaystyle\sum_{k=1}^{n} \Delta m_k}$$

where $\Delta m_k = \rho f(c_k) \Delta x_k$. In each case, the numerator and denominator are Riemann sums. So as we let $n \to \infty$, with all $\Delta x_i \to 0$, the sums approach integrals. To simplify notation, we write

$$dm = \rho f(x) dx$$

and call dm the *differential of mass*. Then we can write $\bar{x}$ and $\bar{y}$ as follows.

> ### Center of Mass of a Homogeneous Lamina
> ### Under the Graph of $y = f(x)$ from $x = a$ to $x = b$
>
> $$\bar{x} = \frac{\displaystyle\int_a^b x\, dm}{\displaystyle\int_a^b dm} \qquad \text{and} \qquad \bar{y} = \frac{\displaystyle\int_a^b \frac{f(x)}{2}\, dm}{\displaystyle\int_a^b dm} \qquad (6.37)$$

By replacing dm by $\rho f(x)\, dx$ and noting that since ρ is a constant, we can divide numerator and denominator by ρ to get the equivalent equations

$$\bar{x} = \frac{\displaystyle\int_a^b x f(x)\, dx}{\displaystyle\int_a^b f(x)\, dx} \qquad \text{and} \qquad \bar{y} = \frac{\displaystyle\int_a^b \frac{[f(x)]^2}{2}\, dx}{\displaystyle\int_a^b f(x)\, dx}$$

REMARK ——————————————————————————————

■ If in the second of Equations 6.37 we replace $f(x)$ by y, we obtain

$$\bar{y} = \frac{\dfrac{1}{2}\displaystyle\int_a^b y\, dm}{\displaystyle\int_a^b dm}$$

(Of course, to apply this formula, we do have to replace y with its equivalent value in terms of x.) Since

$$\bar{x} = \frac{\displaystyle\int_a^b x\, dm}{\displaystyle\int_a^b dm}$$

there is a lack of symmetry in the equations for $\bar{x}$ and $\bar{y}$ because the factor $\frac{1}{2}$ appears in the formula for $\bar{y}$. From the preceding derivation, we can see why the factor $\frac{1}{2}$ is used for $\bar{y}$ but for $\bar{x}$. The center of the typical element in Figure 6.77, thought of as a point mass, is $\left(c_k, \frac{f(c_k)}{2}\right)$. The width of this rectangular element shrinks to 0 in the limit, but the height does not. So the y-coordinate continues to require the factor $\frac{1}{2}$.

——

EXAMPLE 6.33   Find the center of mass of a lamina of constant density ρ in the shape of a semicircular region of radius 2, shown in Figure 6.78.

Solution   Since the lamina is of constant density and the y-axis is a line of symmetry, it follows that $\bar{x} = 0$. That is, the center of mass lies on the y-axis. To find $\bar{y}$, let us first calculate the mass, m, given by

$$m = \int_{-2}^{2} dm = \int_{-2}^{2} \rho f(x)\, dx = \int_{-2}^{2} \rho\sqrt{4 - x^2}\, dx = \rho \int_{-2}^{2} \sqrt{4 - x^2}\, dx = \rho A$$

where A is the area of the region. Since the region is semicircular, we know its area A is $\frac{1}{2}\pi (2)^2 = 2\pi$. Thus, $m = 2\pi\rho$.

$y = \sqrt{4^2 - x^2}$

FIGURE 6.78

The numerator for $\bar{y}$ is

$$\int_{-2}^{2} \frac{f(x)}{2} dm = \frac{1}{2} \int_{-2}^{2} \sqrt{4 - x^2} \left(\rho \sqrt{4 - x^2} \, dx \right)$$

$$= \frac{\rho}{2} \int_{-2}^{2} (4 - x^2) dx$$

$$= \rho \int_{0}^{2} (4 - x^2) dx \qquad \text{Since the integrand is even}$$

$$= \rho \left[4x - \frac{x^3}{3} \right]_{0}^{2} = \frac{16\rho}{3}$$

Thus,

$$\bar{y} = \frac{16\rho/3}{2\pi\rho} = \frac{8}{3\pi}$$

The center of mass, therefore, is the point

$$(\bar{x}, \bar{y}) = \left(0, \frac{8}{3\pi} \right) \qquad \blacksquare$$

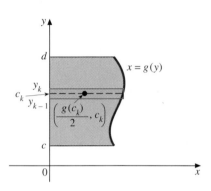

FIGURE 6.79

If the lamina occupies the region bounded by the nonnegative function $x = g(y)$, the lines $y = c$ and $y = d$, and the y-axis (see Figure 6.79), then the formulas for $\bar{x}$ and $\bar{y}$ become

$$\bar{x} = \frac{\displaystyle\int_{c}^{d} \frac{g(y)}{2} dm}{\displaystyle\int_{c}^{d} dm} \quad \text{and} \quad \bar{y} = \frac{\displaystyle\int_{c}^{d} y \, dm}{\displaystyle\int_{c}^{d} dm} \qquad (6.38)$$

where $dm = \rho g(y) dy$.

Shorthand Technique for Finding the Center of Mass of a Homogeneous Lamina

A useful device for obtaining the integrals in Equations 6.37 and 6.38, as well as those for the more general cases in which the region in question is bounded by two curves, is as follows. In Figure 6.80(a) we show the width of a typical vertical element as dx and in part (b) the width of a typical horizontal strip as dy. In either case let us call the area of the strip dA and its "mass"

$$dm = \rho \, dA$$

Call the moment arm from the y-axis to the center of the strip by the name x-arm, and call the moment arm from the x-axis to the center by the name y-arm. Then the moment of this elementary mass (the strip) about the y-axis is $(x\text{-arm}) \, dm$ and the moment about the x-axis is $(y\text{-arm}) \, dm$. "Summing" these moments (in the sense of integration), we get the total moments M_y and M_x of the lamina,

$$M_y = \int (x\text{-arm}) dm \quad \text{and} \quad M_x = \int (y\text{-arm}) dm$$

(a)

(b)

FIGURE 6.80

The appropriate limits of integration must be applied, depending on whether vertical strips (x independent) or horizontal strips (y independent) are being used. Also, x-arm and y-arm have to be calculated in each individual case and expressed in terms of the independent variable. We will illustrate how to do this shortly.

The total mass, m, of the lamina is

$$m = \int dm$$

where, again, the appropriate limits must be supplied. Now, using the earlier results, we have

$$\bar{x} = \frac{M_y}{m} \quad \text{and} \quad \bar{y} = \frac{M_x}{m}$$

We summarize our results as follows.

Center of Mass of a Homogeneous Lamina

$$\bar{x} = \frac{\int (x\text{-arm})dm}{\int dm} \quad \text{and} \quad \bar{y} = \frac{\int (y\text{-arm})dm}{\int dm} \tag{6.39}$$

where $dm = \rho\, dA$. The appropriate limits of integration must be supplied.

EXAMPLE 6.34   Find the center of mass of the homogeneous lamina bounded by the curves $y = 2 - x^2$, $y = x$, and $x = 0$.

Solution   We show the region in Figure 6.81. Solving $y = 2 - x^2$ and $y = x$ simultaneously, we find the point of intersection to be $(1, 1)$. Denote the upper curve by y_2 and the lower curve by y_1. The center of the vertical element shown is then

$$\left(x, \frac{y_1 + y_2}{2} \right)$$

so

$$x\text{-arm} = x \quad \text{and} \quad y\text{-arm} = \frac{y_1 + y_2}{2}$$

The area of the element is

$$dA = (y_2 - y_1)dx$$

and its mass is $dm = \rho\, dA$. The total mass is

$$m = \int dm = \int_0^1 \rho(y_2 - y_1)dx$$

$$= \int_0^1 \rho[(2 - x^2) - x]dx$$

$$= \rho\left[2x - \frac{x^3}{3} - \frac{x^2}{2} \right]_0^1$$

$$= \frac{7\rho}{6}$$

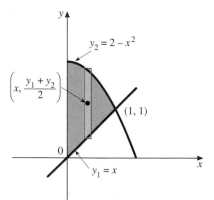

FIGURE 6.81

The denominator in each of Equations 6.39 is therefore $7\rho/6$. We calculate the numerators separately.

$$M_y = \int (x\text{-arm})dm = \int_0^1 x\rho(y_2 - y_1)dx$$

$$= \int_0^1 x\rho[(2 - x^2) - x]dx$$

$$= \rho \int_0^1 (2x - x^3 - x^2)dx$$

$$= \rho \left[x^2 - \frac{x^4}{4} - \frac{x^3}{3} \right]_0^1$$

$$= \frac{5\rho}{12}$$

Thus,

$$\bar{x} = \frac{M_y}{m} = \frac{5\rho/12}{7\rho/6} = \frac{5}{14}$$

For the moment about the x-axis, we have

$$M_x = \int (y\text{-arm})dm = \int_0^1 \left(\frac{y_1 + y_2}{2} \right) \rho(y_2 - y_1)dx$$

$$= \frac{\rho}{2} \int_0^1 (y_2^2 - y_1^2)dx$$

$$= \frac{\rho}{2} \int_0^1 [(2 - x^2)^2 - x^2]dx$$

$$= \frac{\rho}{2} \int_0^1 (4 - 5x^2 + x^4)dx$$

$$= \frac{\rho}{2} \left[4x - \frac{5x^3}{3} + \frac{x^5}{5} \right]_0^1$$

$$= \frac{19\rho}{15}$$

Thus,

$$\bar{y} = \frac{19\rho/15}{7\rho/6} = \frac{38}{35}$$

So the center of mass is $(\frac{5}{14}, \frac{38}{35})$. ∎

In Equations 6.39 if we replace dm by $\rho\, dA$ we get

$$\bar{x} = \frac{\int (x\text{-arm})\rho\, dA}{\int \rho\, dA} = \frac{\rho \int (x\text{-arm})dA}{\rho \int dA} = \frac{\int (x\text{-arm})dA}{\int dA}$$

$$\bar{y} = \frac{\int (y\text{-arm})\rho\, dA}{\int \rho\, dA} = \frac{\rho \int (y\text{-arm})dA}{\rho \int dA} = \frac{\int (y\text{-arm})dA}{\int dA}$$

(6.40)

That is, the density ρ cancels, since it is a constant factor. The results show that $\bar{x}$ and $\bar{y}$ are determined solely by the geometry of the region. If the density were variable, this conclusion would not hold true. To distinguish between the geometric center of a region and the center of mass of a physical lamina, we refer to the former as the **centroid** of the region. For a homogeneous lamina, then, the center of mass is the same as the centroid of the region occupied by the lamina. In summary, we have the following equations.

Centroid of a Plane Region

$$\bar{x} = \frac{\int (x\text{-arm})\, dA}{\int dA} = \frac{M_y}{A}$$

$$\bar{y} = \frac{\int (y\text{-arm})\, dA}{\int dA} = \frac{M_x}{A}$$

(6.41)

EXAMPLE 6.35   Find the centroid of the triangular region shown in Figure 6.82.

Solution   We choose to use horizontal elements this time to avoid having to divide the region into two parts. Denote the left boundary by x_1 and the right boundary by x_2. The center of a typical element of area is then

$$\left(\frac{x_1 + x_2}{2}, y \right)$$

so

$$x\text{-arm} = \frac{x_1 + x_2}{2} \qquad \text{and} \qquad y\text{-arm} = y$$

The area of the region is

$$A = \int dA = \int_0^2 (x_2 - x_1)\,dy = \int_0^2 (2y - y)\,dy = \left. \frac{y^2}{2} \right]_0^2 = 2$$

The moment with respect to the y-axis is

$$M_y = \int (x\text{-arm})\,dA = \int_0^2 \left(\frac{x_1 + x_2}{2} \right)(x_2 - x_1)\,dy$$

$$= \frac{1}{2} \int_0^2 (x_2^2 - x_1^2)\,dy$$

$$= \frac{1}{2} \int_0^2 [(2y)^2 - y^2]\,dy$$

$$= \frac{1}{2}[y^3]_0^2 = 4$$

and the moment with respect to the x-axis is

$$M_x = \int (y\text{-arm})\,dA = \int_0^2 y(x_2 - x_1)\,dy = \int_0^2 y[2y - y]\,dy = \left. \frac{y^3}{3} \right]_0^2 = \frac{8}{3}$$

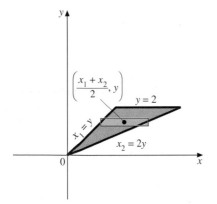

FIGURE 6.82

Thus, by Equations 6.41

$$\bar{x} = \frac{M_y}{A} = \frac{4}{2} = 2 \quad \text{and} \quad \bar{y} = \frac{M_x}{A} = \frac{8/3}{2} = \frac{4}{3}$$

So the centroid is

$$\left(2, \frac{4}{3}\right)$$

∎

REMARK ──

■ In the preceding example, if we had been seeking the center of mass of a homogeneous lamina occupying the same region, we would have replaced dA by dm, but since $dm = \rho\, dA$, the constant ρ would cancel, as we showed above. So the final result would be the same as the centroid.

──

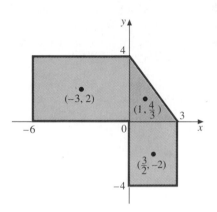

FIGURE 6.83

EXAMPLE 6.36   Find the centroid of the region shown in Figure 6.83.

Solution   Here we can make use of the fact that if the centroid $(\bar{x}, \bar{y})$ of a region is known, its moments with respect to the x- and y-axes, respectively, are

$$M_x = A\bar{y} \quad \text{and} \quad M_y = A\bar{x}$$

(In the case of a lamina of constant density ρ, the area A would be replaced by the mass $m = \rho A$.) Furthermore, we know that if a region consists of several parts, its total moment with respect to either axis is the sum of the moments of each individual part with respect to that axis.

In Exercise 22 of Exercise Set 6.8 you will be asked to show that the centroid of any triangular region is located one-third of the perpendicular distance from any side to the opposite vertex (at the intersection of the medians). Thus, the centroid of the triangular region in Figure 6.83 has coordinates $(1, \frac{4}{3})$. The area of the triangle is $\frac{1}{2}(3)(4) = 6$. The rectangular region to the left of the y-axis has centroid $(-3, 2)$ and area 24, and the rectangle below the x-axis has centroid $(\frac{3}{2}, -2)$ and area 12.

The total moment M_y is

$$M_y = (-3)(24) + (1)(6) + \left(\frac{3}{2}\right)(12) = -72 + 6 + 18 = -48$$

Similarly, the moment M_x is

$$M_x = 2(24) + \left(\frac{4}{3}\right)(6) + (-2)(12) = 48 + 8 - 24 = 32$$

The total area is $24 + 6 + 12 = 42$. Thus, the coordinates of the centroid of the total region are

$$\bar{x} = \frac{M_y}{A} = \frac{-48}{42} = -\frac{8}{7}$$

$$\bar{y} = \frac{M_x}{A} = \frac{32}{42} = \frac{16}{21}$$

∎

Centroids and Volumes of Solids of Revolution

We conclude this section by stating a theorem relating centroids and volumes of solids of revolution. You will be asked to supply the proof in Exercise 28 in Exercise Set 6.8. The theorem is attributed to Pappus of Alexandria, who flourished about A.D. 320. He was the last great Greek geometer.

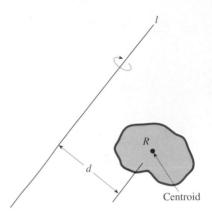

FIGURE 6.84
When R is rotated about l, the volume is $(2\pi d) \cdot$ (area of R).

The First Theorem of Pappus

If R is a region in a plane and l is a line in the plane that does not intersect R, then the volume of the solid formed by revolving R about l is the area of R times the distance traveled by the centroid of R.

Figure 6.84 illustrates the idea of this theorem.

In the next example we show how Pappus's Theorem enables us to find the volume of a torus quickly and easily. (Recall the work involved in using the methods of washers and cylindrical shells in Sections 6.2 and 6.3.)

EXAMPLE 6.37   Use the First Theorem of Pappus to find the volume of the torus formed by revolving the region inside the circle $(x - a)^2 + y^2 = r^2$ about the y-axis, where $a > r$.

Solution   In Figure 6.85, we show the region to be revolved. The centroid of the circle is its center, located at a distance a from the axis of rotation. The area of the circular region is πr^2. So, by Pappus's Theorem, the volume V of the torus generated after rotation (see Figure 6.86) is

$$V = 2\pi a(\pi r^2) = 2\pi^2 a r^2$$

In Example 6.11, we gave an intuitive argument for arriving at the volume of a torus as we have indicated in this example. The First Theorem of Pappus provides the needed justification for our earlier intuitive argument.   ■

FIGURE 6.85

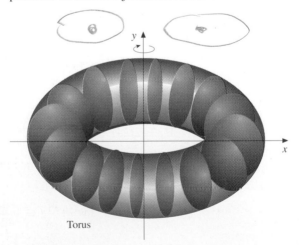

Torus

FIGURE 6.86

Exercise Set 6.8

In Exercises 1 and 2, point masses m_i are located on the x-axis, with x-coordinates x_i, for $i = 1, 2, 3, 4$. Find the center of mass.

1. $m_1 = 10;\ x_1 = 3$
$m_2 = 12;\ x_2 = -4$
$m_3 = 17;\ x_3 = 5$
$m_4 = 20;\ x_4 = -2$

2. $m_1 = 13;\ x_1 = -5$
$m_2 = 7;\ x_2 = -1$
$m_3 = 11;\ x_3 = 6$
$m_4 = 9;\ x_4 = 4$

In Exercises 3 and 4, point masses m_i are located in the plane, with coordinates as shown, for $i = 1, 2, 3, 4$. Find the center of mass.

3. $m_1 = 5$ at $(3, -2)$
$m_2 = 13$ at $(-4, 5)$
$m_3 = 6$ at $(2, 4)$
$m_4 = 8$ at $(-1, -2)$

4. $m_1 = 15$ at $(-4, 3)$
$m_2 = 18$ at $(2, 5)$
$m_3 = 11$ at $(-3, -4)$
$m_4 = 14$ at $(5, -6)$

In Exercises 5–8, a bar having density $\rho(x)$ is located on the x-axis between $x = a$ and $x = b$. Find the center of mass of the bar.

5. $\rho(x) = 2;\ a = 0,\ b = 3$

6. $\rho(x) = \frac{x}{2};\ a = 1,\ b = 5$

7. $\rho(x) = \sqrt{x};\ a = 4,\ b = 9$

8. $\rho(x) = 1 + x;\ a = 0,\ b = 7$

In Exercises 9–12, find M_x, M_y, and the center of mass of the lamina pictured, having the given density ρ.

9.

10.

11.

12.

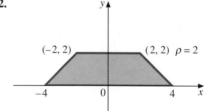

In Exercises 13–20, find the centroid of the region described.

13. Triangular region with vertices $(0, 0)$, $(4, 2)$, $(4, 8)$

14. Below $y = 4$ and above $y = |x|$

15. First-quadrant portion of circular region centered at the origin, with radius 2

16. Bounded by $y = \sqrt{x}$, $x = 4$, and the x-axis

17. Between $y = x$ and $y = 2 - x^2$

18. Under $y = 2x - x^2$ and above the x-axis

19. Between the y-axis and $y^2 = 4 - x$

20. Under $y = \cos x$ and above the x-axis, between $x = -\pi/2$ and $x = \pi/2$. (*Hint:* In one of the integrals you will need to use the identity $\cos^2 x = (1 + \cos 2x)/2$.)

21. Prove that the centroid of the parabolic region bounded by $y = ax^2$ and $y = b$, where $a > 0$ and $b > 0$, is independent of a. What is the value of $\bar{y}$?

22. Prove that the distance from any side of a triangular region to its centroid is one-third the perpendicular distance from that side to the opposite vertex. (*Hint:* Orient the axes so that one side is on the x-axis and the opposite vertex is on the y-axis. See the figure. Since any side can be placed on the x-axis in this way, you need only find $\bar{y}$ for the triangle shown.)

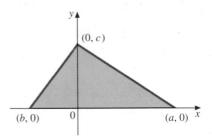

23. Prove that the distance from the base of a semicircular region of radius a to its centroid is $4a/3\pi$.

In Exercises 24–27, use the method of Example 6.36 to find the centroid of the region shown. Use the results of Exercises 22 and 23 where needed.

24.

25.

26.

27. (*Hint:* Treat the hole as having negative area.)

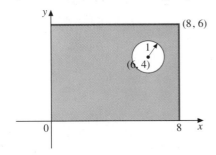

28. Prove the First Theorem of Pappus. (*Hint:* Orient the axes so that l coincides with the y-axis and take R to the right of the y-axis.)

In Exercises 29–31, use the First Theorem of Pappus to find the volume of the indicated solid.

29. Sphere of radius a (*Hint:* Use the result of Exercise 23.)

30. Cone of altitude h and base radius r

31. Solid obtained by revolving the region enclosed by the triangle with vertices $(2, 1)$, $(8, 1)$, and $(5, 4)$ about (a) the y-axis, (b) the x-axis, (c) the line $y = 6$

6.9 INTEGRATION IN PROBABILITY

Probability Distributions

If you were to flip a balanced coin 100 times, you would expect it to come up heads approximately 50 times. On any one toss of the coin there are two equally likely outcomes (heads or tails), so the likelihood it will be heads is 1 out of 2. We say that the **probability** of getting heads is 1/2. Similarly, if you toss a die a large number of times, you would expect that the number 3, say, would show up on the upper face about 1/6th of the time. That is, the probability of getting a 3 (or any other particular number from 1 to 6) is 1/6. In both of these cases we could reason in advance what the probability should be. In many cases, however, probabilities cannot be determined in advance. Suppose, for example, that as a quality-control measure, a manufacturer of light bulbs samples 100 of them periodically to see how many are defective. If, say, two of the 100 are found to be defective, then the manufacturer would assume that among the entire batch, approximately 2% are defective. That is, the probability that a bulb selected at random is defective is about 0.02.

In each of the cases cited, the number of possible outcomes is finite. In many cases, an investigator is interested in some characteristic, or attribute, that can take on any one of infinitely many possibilities. In particular, it may in some cases assume any real number on a given interval. Typically, these cases involve a measurement of some type, such as weight, height, or distance. For example, suppose the weights (in kilograms) of 100 college students were taken.* We could depict the results by means of a diagram such as that shown in Figure 6.87. In the diagram, points stacked vertically represent students who have approximately the same weight. The regions of approximately the same weight are called *bins*, and the procedure is called *binning*. The *density* of weights is greatest in the 50–100 kg range and diminishes to the right and left of this interval. As another example, a researcher might be interested in the cholesterol levels in the blood of persons in a certain age group. These values, too, can be any real numbers within some interval, say from 100 to 400.

FIGURE 6.87

---

*Strictly speaking, we should refer to the *masses* of the students in kilograms, but it is standard practice in this context to call these masses weights.

A Galton machine, on display in Boston's Museum of Science. Balls drop through posts and distribute themselves randomly in the approximate shape of the red curve, called a *normal curve* in probability.

It is interesting to note the similarity between the binning of students' weights and an experiment first introduced by the English scholar Sir Francis Galton (1822–1911). He devised an apparatus, subsequently known as a **Galton machine** like the one shown in the margin. In his experiment, many identical balls are dropped, one at a time, onto a triangular array of posts, as shown. Each ball hits the first post, bounces one way or the other and onto a post in the second row, then to the third row, and so on. In a random fashion the balls make their way into the bins at the bottom of the machine. Note the similarity between the typical distribution of balls shown in the Galton machine photograph and the distribution of students' weights in Figure 6.87.

Weighing a group of students, counting the number of defective light bulbs in a sample, or dropping balls in a Galton machine can be thought of as an *experiment* in which the outcomes are not predictable in advance. Such an experiment is sometimes called a *random* experiment, and the attribute or characteristic being measured is called a **random variable**. Some random variables can take on discrete values only—such as the number of defective light bulbs in a sample. We call these *discrete* random variables. Others can take on real values on some interval—examples include weights or cholesterol levels of people in a given population. These are called *continuous* random variables. We will be concerned with continuous random variables, since they lend themselves to the use of calculus.

It is often useful to be able to determine in advance the likelihood, or **probability**, that the value of such a continuous random variable will be in some specified interval. For example, we might want to know the probability that a college student chosen at random will weigh between 70 and 90 kg, or that the cholesterol level of a person in the age group under study will exceed 250. In general, these probabilities cannot be determined with absolute certainty, but they often can be approximated by the use of integration.

Suppose we indicate bins by rectangles as in Figure 6.88. We let the area of each rectangle represent the *proportion* of the total population falling within that bin. For example, if we are dealing with weights of 100 college students, and the area of the bin from 50 to 60 is 0.15, then 15% of the students weigh between 50 kg and 60 kg. The sum of the areas of all the rectangles must equal 1, since this sum is the proportion of students weighing between the lowest and highest weights, which includes all of them.

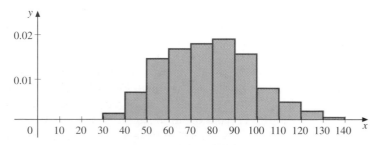

FIGURE 6.88

Figure 6.88 is an example of what is called a **histogram**. It looks suspiciously like the rectangles whose areas form a Riemann sum. In fact, if we were to consider narrower and narrower bins, the areas of the rectangles would approach more and more closely the area under a curve such as that shown in Figure 6.89.

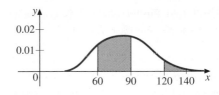

FIGURE 6.89

We can think of Figure 6.89 as an idealization of Figure 6.88. The shaded area between $x = 60$ and $x = 90$ represents the proportion of students weighing between 60 kg and 90 kg. Equivalently, we can say this area is the *probability* that the weight of a student selected at random will fall somewhere in this range. Similarly, the shaded area between $x = 120$ and $x = 140$ is the probability that the weight of a student selected at random is between 120 kg and 140 kg. The total area under the graph must be 1, since this area is the probability that a student's weight is between the lowest weight and the highest weight.

We can generalize the preceding discussion by considering any nonnegative integrable function such that the area under its graph on some specified interval is 1. (At present we consider only finite intervals. Later, we will remove this restriction.) In the following definition we follow the standard practice of using a capital letter, such as X, to designate a random variable, with the corresponding lowercase letter, such as x, designating one of its values.

Definition 6.6
Probability Density Function

Let f be a nonnegative integrable function on an interval I, such that the area under the graph of f on I is 1. If there exists a random variable X having I as the set of all of its possible values and such that for any two numbers a and b in I with $a < b$, the probability that X takes on a value between a and b is given by

$$P(a \leq X \leq b) = \int_a^b f(x)dx \qquad (6.42)$$

then f is called a **probability density function** for X.

REMARK
■ Since omitting one or both endpoints from the integral in Equation 6.42 will not change its value,

$$P(a \leq X \leq b) = P(a < X \leq b) = P(a \leq X < b) = P(a < X < b)$$

In actual practice the probabilities given by Equation 6.42 are usually only approximate, and the function f is an idealized mathematical model of the true probability density function for X. Many random variables have an apparent bell-shaped density function, similar to that shown in Figure 6.89. This type of graph is called a **normal curve** and the corresponding function a **normal density function**. We will come back to this type of function later. The choice of the particular density function to use depends on the nature of the random variable in question. Sometimes it can reasonably be predicted that the values of a random variable will be distributed in a certain way, and in other cases the choice of the model is based on experimental evidence.

A particularly simple probability density function is the **uniform probability density function**, defined by

$$f(x) = \begin{cases} \dfrac{1}{b-a} & \text{if } a \leq x \leq b \\ 0 & \text{otherwise} \end{cases} \qquad (6.43)$$

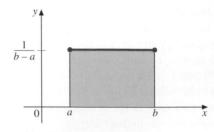

FIGURE 6.90
Uniform probability density function

FIGURE 6.91

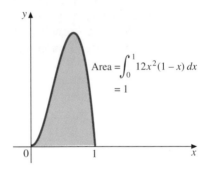

FIGURE 6.92
The beta density function with $\alpha = 3, \beta = 2$

The graph is shown in Figure 6.90. Observe that the function f is constant over the specified interval. The next example shows one case for which such a density function is an appropriate model.

EXAMPLE 6.38   Suppose a bus is known to arrive at a certain bus stop between 9:00 and 9:10 each weekday morning and that from observation over a period of time, it can be assumed that its arrival time during the 10-min interval is uniformly distributed; that is, the bus is just as likely to arrive at one time in the interval as at any other time. Find the probability that a person who arrives at 9:00 will have to wait more than 7 min for a bus.

Solution   The random variable X of interest in this case is the number of minutes after 9:00 that the bus arrives, and from the given information we know that the values have a uniform distribution on the interval $[0, 10]$, as shown in Figure 6.91. Although we can find the area in this case without integration, we choose to illustrate Equation 6.42. We want $P(7 < X \leq 10)$:

$$P(7 < X \leq 10) = \int_7^{10} f(x)\,dx = \int_7^{10} \frac{1}{10}\,dx$$

$$= \frac{1}{10}[x]_7^{10} = \frac{3}{10}$$

So if this person arrives at 9:00 each morning, we would expect that on average he or she will have to wait at least 7 min about 30% of the time.   ∎

The model for the next example is a special case of what is known as a **beta distribution**, with the general form

$$f(x) = Bx^{\alpha-1}(1-x)^{\beta-1}, \qquad 0 \leq x \leq 1 \qquad (6.44)$$

The Greek letters α (alpha) and β (beta) can be any positive numbers. The constant B is chosen so that the area under the graph is 1. In Figure 6.92 we show the graph of the beta distribution in which $\alpha = 3$ and $\beta = 2$. You can verify that $B = 12$ in this case. (See Exercise 28 in Exercise Set 6.9.)

EXAMPLE 6.39   A certain grass seed is advertised as a mixture containing at least 70% Kentucky bluegrass, with the remainder composed of red fescue, rye, and other extraneous seeds. As a means of quality control, a sample of seeds is tested periodically, and if the proportion of non–Kentucky-bluegrass seeds exceeds 30%, the production is halted so that corrective measures can be taken. By experiment, the probability density function for the proportion of undesirable seeds is found to be approximated by

$$f(x) = 10(1-x)^9, \qquad 0 \leq x \leq 1$$

Find the probability that on testing a random sample the production process will be halted.

Solution   We define the random variable X to be the proportion of undesirable grass seeds in the sample. From the given information, we see that this variable has a beta density function, Equation 6.44, with $\alpha = 1$ and $\beta = 10$ (so $B = 10$). Production will be halted if the value of X exceeds 0.3. So the probability of halting production is

$$P(0.3 < X \le 1) = \int_{0.3}^{1} 10(1-x)^9 dx$$

$$= -10 \left[\frac{(1-x)^{10}}{10} \right]_{0.3}^{1}$$

$$= -[0 - (0.7)^{10}] \approx 0.028$$

There is about a 3% chance of having to halt the production process.   ∎

The Mean, Expected Value, Variance, and Standard Deviation

An important characteristic of a probability density function is its **mean**, usually denoted by the Greek letter μ (mu), and defined by

$$\mu = \int_{a}^{b} x f(x) dx \tag{6.45}$$

It can be shown that if values of the random variable X, with probability density function $f(x)$, are repeatedly sampled, there is a high probability they will be concentrated near μ. In a sense, μ is the average value of X. Looked at geometrically, μ is the x-coordinate of the centroid of the region under the graph of f. For, as we know from the previous section,

$$\bar{x} = \frac{\displaystyle\int_{a}^{b} x f(x) dx}{\displaystyle\int_{a}^{b} f(x) dx}$$

But the denominator in this case is 1. Thus, $\bar{x} = \mu$.

Another way of viewing the mean is to compare it with the center of mass of a one-dimensional bar having variable density. We can take the density as $f(x)$ (it is even called a probability *density*), so according to Equation 6.35,

$$\bar{x} = \frac{\displaystyle\int_{a}^{b} x f(x) dx}{\displaystyle\int_{a}^{b} f(x) dx}$$

Again, the denominator is 1, so $\bar{x} = \mu$. Thus, the mean μ can be thought of as the center of mass, with "mass" (i.e., probabilities) distributed on a line from a to b, with density $f(x)$.

The mean of a distribution is a special case of what is called an **expected value**. In general, if X is a random variable with probability density $f(x)$ on the interval $[a, b]$, the **expected value of $g(X)$**, symbolized by $E[g(X)]$, is defined by

$$E[g(X)] = \int_{a}^{b} g(x) f(x) dx \tag{6.46}$$

We assume $g(x) f(x)$ is integrable, so that $E[g(X)]$ is defined. In the same sense that μ is an average value of X, the expected value of $g(X)$ is an average of the values taken on by $g(X)$. In repeated experiments in which $g(X)$ is measured, these measurements would be concentrated near $E[g(X)]$. By taking $g(X) = X$, we see that

$$\mu = E[X]$$

That is, the mean is the expected value of X.

Another expected value that is important in describing a distribution is obtained by taking $g(X) = (X - \mu)^2$ in Equation 6.46. We call this expected value the **variance** of X and symbolize it by σ^2 (σ is the lowercase Greek letter *sigma*). Thus,

$$\sigma^2 = E[(X - \mu)^2]$$

That is,

$$\sigma^2 = \int_a^b (x - \mu)^2 f(x)dx \qquad (6.47)$$

The number $x - \mu$ is called the *deviation* of x from the mean. By squaring the deviations, we eliminate the effect of their variations in sign (some positive, some negative). The variance is therefore a measure of how closely the values of X are clustered about the mean. The smaller the variance, the more closely they are clustered. The square root of the variance, denoted by σ, is called the **standard deviation** of X. In Figure 6.93 we show a normal probability density function with mean μ and standard deviation σ. The equation of the function is

$$f(x) = \frac{1}{\sqrt{2\pi}\sigma}e^{-(x-\mu)^2/\sigma^2}$$

We will study functions of this type (exponential functions) in Chapter 7. There you will be asked to show that for this normal curve, σ is the distance from the mean μ to each of the points of inflection to the right and left of the mean.

FIGURE 6.93

The mean μ and standard deviation σ for the normal curve

EXAMPLE 6.40   A certain random variable X has probability density function $f(x) = kx^2$ on $[0, 2]$. Find the value of k and calculate μ, σ^2, and σ. Also find the approximate probability that X is no more than one standard deviation from the mean.

Solution   The constant k must be chosen so that the total area under the graph of f is 1. Since

$$\int_0^2 kx^2 dx = k\left[\frac{x^3}{3}\right]_0^2 = \frac{8k}{3}$$

we must have $8k/3 = 1$ or $k = 3/8$. Thus, $f(x) = 3x^2/8$.

The mean, μ, is

$$\mu = E[X] = \int_0^2 xf(x)dx = \int_0^2 x\left(\frac{3}{8}x^2\right)dx = \frac{3}{8}\int_0^2 x^3 dx$$

$$= \frac{3}{8}\left[\frac{x^4}{4}\right]_0^2 = \frac{3}{2}$$

The variance, σ^2, is

$$\sigma^2 = E[(X - \mu)^2] = E\left[\left(X - \frac{3}{2}\right)^2\right] = \int_0^2 \left(x - \frac{3}{2}\right)^2 f(x)dx$$

$$= \int_0^2 \left(x^2 - 3x + \frac{9}{4} \right) \cdot \frac{3}{8} x^2 dx = \frac{3}{8} \int_0^2 \left(x^4 - 3x^3 + \frac{9x^2}{4} \right) dx$$

$$= \frac{3}{8} \left[\frac{x^5}{5} - \frac{3x^4}{4} + \frac{3x^3}{4} \right]_0^2 = \frac{3}{8} \left[\frac{32}{5} - 12 + 6 \right]$$

$$= \frac{3}{20}$$

Thus, the standard deviation, σ, is

$$\sigma = \sqrt{\frac{3}{20}} = \sqrt{\frac{3}{20} \cdot \frac{5}{5}} = \frac{\sqrt{15}}{10} \approx 0.3873$$

Finally, we want $P(\mu - \sigma \leq X \leq \mu + \sigma)$ where

$$\mu - \sigma \approx 1.5 - 0.3873 = 1.1127$$

and

$$\mu + \sigma \approx 1.5 + 0.3873 = 1.8873$$

So we have

$$P(\mu - \sigma \leq X \leq \mu + \sigma) \approx \int_{1.1127}^{1.8873} \frac{3}{8} x^2 \, dx = \frac{1}{8} x^3 \Big]_{1.1127}^{1.8873}$$

$$\approx 0.668$$

Notice that about two-thirds of the time we would expect X to fall within one standard deviation of the mean. ∎

Exercise Set 6.9

In Exercises 1–10, find the constant k that will make f a probability density function on the given interval.

1. $f(x) = k(2x + 3)$ on $[0, 4]$

2. $f(x) = k(4 - 2x)$ on $[0, 2]$

3. $f(x) = k(x^2 + x)$ on $[1, 3]$

4. $f(x) = kx(1 - x)$ on $[0, 1]$

5. $f(x) = k/x^2$ on $[1, 4]$

6. $f(x) = k\sqrt{x - 1}$ on $[1, 5]$

7. $f(x) = kx^3(1 - x)^2$ on $[0, 1]$

8. $f(x) = kx/(1 + x^2)^2$ on $[0, 1]$

9. $f(x) = kx/\sqrt{1 + 2x^2}$ on $[0, 2]$

10. $f(x) = kx^2/\sqrt{1 + x^3}$ on $[0, 2]$

In Exercises 11–16, verify that the given function is a probability density function. Also find the probability specified for a random variable X that has f as its probability density function.

11. $f(x) = \frac{1}{2}x$ on $[0, 2]$, $P(\frac{1}{2} \leq X \leq \frac{3}{2})$

12. $f(x) = \frac{1}{6}(2x + 3)$ on $[-1, 1]$; $P(0 \leq X < 1)$

13. $f(x) = \frac{1}{24}(x^2 + 1)$ on $[-3, 3]$; $P(-1 < X \leq 2)$

14. $f(x) = 12x^2(1 - x)$ on $[0, 1]$; $P(\frac{1}{2} < X < 1)$

15. $f(x) = \frac{3}{125}x\sqrt{25 - x^2}$ on $[0, 5]$; $P(3 \leq X \leq 4)$

16. $f(x) = x/\left(2\sqrt{x^2 + 9}\right)$ on $[0, 4]$; $P(0 < X \leq 1.25)$

In Exercises 17–22, find the mean and variance of the random variable that has the given probability density function.

17. $f(x) = \frac{2}{9}(x + 1)$ on $[-1, 2]$

18. $f(x) = \frac{1}{6}(3 - 2x)$ on $[-1, 1]$

19. $f(x) = 6x(1 - x)$ on $[0, 1]$

20. $f(x) = \frac{3}{16}\sqrt{x}$ on $[0, 4]$

21. $f(x) = 4x^3$ on $[0, 1]$

22. $f(x) = 3[1 - (x - 3)^2]/4$ on $[2, 4]$

23. Find the mean and variance of a random variable that has the uniform density function of Equation 6.43.

24. The number X of parts per million of a certain pollutant emitted from the smokestack of a cement plant is known to range from a low of 1 to a high of 27. Measurements of samples collected over a period of time suggest that an appropriate probability density model for X is $f(x) = cx^{-2/3}$, where $1 \le x \le 27$.
    (a) Find the value of c.
    (b) Find the expected value of X.
    (c) Find the standard deviation of X.

25. The income level of persons in a certain city whose incomes exceed a certain amount $x_0 > 0$ is approximately modeled by a **Pareto distribution**:

$$f(x) = \begin{cases} \dfrac{\alpha x_0^{\alpha}}{x^{\alpha+1}} & \text{if } x \ge x_0 \\ 0 & \text{otherwise} \end{cases}$$

where α is a positive constant. If x is measured in thousands of dollars, $\alpha = 2$, and $x_0 = 10$, find the probability that a person with an income that exceeds \$10,000 makes less than \$20,000.

26. A **truncated Pareto distribution** (see Exercise 25) is of the form

$$f(x) = \begin{cases} c\left(\dfrac{x_0}{x}\right)^{\alpha+1} & \text{if } x_0 \le x \le x_1 \\ 0 & \text{otherwise} \end{cases}$$

where c and α are positive constants. Find the constant c in terms of x_0, x_1, and α so that f will be a probability density function.

27. The truncated Pareto function of Exercise 26 is sometimes used to model the distribution of the size of oil fields in a certain region. Suppose that x is measured in hundreds of millions of barrels, $\alpha = 2$, $x_0 = 2$, and $x_1 = 10$ for a certain region.
    (a) Find the probability that a randomly selected oil field

in that region would produce more than 500 million barrels of oil.
    (b) Find the average (mean) number of barrels for oil fields in that region.

28. The proportion X of impurities in a product resulting from a certain chemical process has the beta distribution of Equation 6.44 with $\alpha = 3$ and $\beta = 2$.
    (a) Find the constant B.
    (b) Find the expected value of X.

29. The lifetime, in hundreds of hours, of a certain type of light bulb has been found empirically to have a probability density function approximated by

$$f(x) = \frac{5\sqrt{17}}{12(1 + x^2)^{3/2}}, \qquad 0 \le x \le 12$$

Find the expected lifetime of a bulb of this type chosen at random.

30. Let X be a random variable with probability density function f on $[a, b]$. Prove the following properties of expected value:
    (a) $E[c] = c$, where c is a constant.
    (b) $E[cg(x)] = cE[g(x)]$, where c is a constant.
    (c) $E[g_1(x) + g_2(x)] = E[g_1(x)] + E[g_2(x)]$

31. Prove that $\sigma^2 = E[X^2] - \mu^2$. (*Hint:* Use the results of Exercise 30, along with the definitions of μ and σ^2.)

32. Use the result of Exercise 31 to find σ^2 for a random variable with the probability density function given by

$$f(x) = \frac{3}{20}\left(\frac{x + 1}{\sqrt{x}}\right), \qquad 1 \le x \le 4$$

33. The random variable X with probability density function

$$f(x) = \frac{8}{(\pi + 2)(1 + x^2)^2} \quad \text{on } [0, 1]$$

is known to have variance

$$\sigma^2 = \frac{\pi^2 - 8}{(\pi + 2)^2}$$

Use this together with the result of Exercise 31 to evaluate the integral

$$\int_0^1 \frac{x^2 dx}{(1 + x^2)^2}$$

34. Show that if X is a random variable with finite mean and t can assume any real value, then $E[(x - t)^2]$ takes on its minimum value when $t = \mu$.

6.10 INTEGRATION IN ECONOMICS

Total Cost and Total Revenue

In Section 4.3 we introduced the notion of the **marginal cost** of a product, defined as the derivative of the cost function. We saw that the marginal cost evaluated at $x = n$ approximates the cost of producing the $(n + 1)$st unit. For example, the product in question might be pairs of shoes of a certain type. The marginal cost evaluated at $x = 100$, say, approximates the cost of producing the 101st pair, after having already produced 100 pairs. This marginal-cost concept is a key tool in the decision-making process.

In this section we are interested in the reverse problem to that of finding marginal cost when we know the cost function. That is, we now want to find the cost function when we know the marginal cost. We will refer to the cost function as the **total-cost function**, to emphasize that it gives the total cost (fixed cost plus variable cost) of producing a given number of units of the product in question.

Let $C(x)$ denote this total-cost function, and let $M_C(x)$ denote the marginal-cost function. Then

$$C'(x) = M_C(x)$$

We assume that $C(0) = C_0$, where C_0 is the fixed cost. We want to find $C(x)$, so we have an initial-value problem to solve. We can write the solution as

$$C(x) = \int_0^x M_C(t)dt + C_0 \tag{6.48}$$

You can check this solution by applying the First Fundamental Theorem of Calculus to get $C'(x) = M_C(x)$. It is also easily seen that $C(0) = C_0$.

REMARK ————————————————————————

■ We are assuming here that $M_C(x)$ is continuous for $x > 0$. Even though x might actually take on integer values only, we assume it can be any positive real number so that we can make use of calculus techniques.

————————————————————————

FIGURE 6.94

Suppose now that a units of the product in question are already being produced, and we want to know the *additional* cost of increasing the production level to b units, where $b > a$. This additional cost is

$$C(b) - C(a) = \int_a^b M_C(t)dt \tag{6.49}$$

as illustrated in Figure 6.94.

A similar analysis can be applied to the **total-revenue function**, $R(x)$. The **marginal revenue**, $M_R(x)$, is by definition the derivative of $R(x)$:

$$R'(x) = M_R(x)$$

We assume that $R(0) = 0$. (There is no income if no items are sold.) The solution of this initial-value problem is

$$R(x) = \int_0^x M_R(t)dt \tag{6.50}$$

As with the total cost, the change in revenue when production is increased from a units to b units is

$$R(b) - R(a) = \int_a^b M_R(t)\,dt \qquad (6.51)$$

Whether to increase production from one level to another clearly depends both on the change in cost and the change in revenue. So long as the increased revenue is greater than the increased cost, it is advantageous to increase production.

EXAMPLE 6.41   In the production of a new type of tennis racquet, the King Company determines that the fixed initial cost is \$65,000 and the marginal cost is

$$M_C(x) = -0.4x + 25 \qquad (0 \le x \le 60)$$

where x is in hundreds of racquets and $M_C(x)$ is in hundreds of dollars. Find

(a) the total-cost function

(b) the increased cost of producing 3,000 racquets instead of 2,000 racquets.

Solution

(a) By Equation 6.49, the total-cost function (in hundreds of dollars) is

$$C(x) = \int_0^x (-0.4t + 25)\,dt + 65,000 = -0.2x^2 + 25x + 65,000$$

(b) By Equation 6.48, the increased cost is

$$
\begin{aligned}
C(30) - C(20) &= \int_{20}^{30} (-0.4t + 25)\,dt \\
&= -0.2t^2 + 25t \big]_{20}^{30} \\
&= [-0.2(900) + 25(30)] - [-0.2(400) + 25(20)] \\
&= (-180 + 750) - (-80 + 500) \\
&= 150
\end{aligned}
$$

That is, the additional cost of producing an extra 1,000 racquets for a total of 3,000, is \$15,000. ∎

Consumers' Surplus and Producers' Surplus

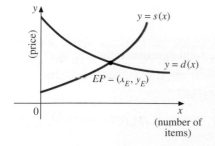

FIGURE 6.95

Two important functions in economics are the **demand function**, $d(x)$, giving the price per unit when there is a consumer demand of x units, and the **supply function**, $s(x)$, giving the price per unit at which the producer is willing to supply x units. In general, from a consumer's point of view, a lower price corresponds to a higher demand for the product in question. So the demand function decreases as x increases. Similarly, from a producer's point of view, a higher price will correspond to an increased supply; that is, a producer is willing to produce more items when the price goes up. Thus, the function s increases as x increases. In Figure 6.95 we show typical supply and demand curves. The point where the graphs of $y = s(x)$ and $y = d(x)$ intersect is called the **equilibrium point**, $EP = (x_E, y_E)$. Clearly, consumers want to buy as many items as possible at the lowest possible price, and producers want to sell as many items as possible at the highest possible price. The equilibrium point therefore represents the optimum quantity and price, taking into consideration the competing goals of consumers and producers.

Whatever the price, it is generally the case that some consumers would have been willing to pay more. So they saved money they were willing to spend by getting the item at a lower price. The total amount saved in this way is called the **consumers' surplus**. That is,

consumers' surplus = (total amount consumers were willing to pay)

− (total amount actually paid)

Similarly, whatever the actual price, some producers would likely have been willing to sell at a lower price. So these producers gained by selling at the higher price. This total amount gained is called the **producers' surplus**. That is,

producers' surplus = (total amount producers actually received)

− (total amount they were willing to accept)

To illustrate the concept of consumers' surplus, consider the following example.

EXAMPLE 6.42   Suppose that to attract new customers, a computer retailer slashes the price on its newest microcomputer system to $2,200. Further, assume that the sale attracts 50 customers who, before the sale, would have been willing to pay up to the amounts shown in the table. Find the consumers' surplus.

Number of customers	Maximum price willing to pay
10	$3,200
20	$2,700
20	$2,500

Solution   If the cost of the system had been $3,200, only 10 customers would have bought it; they would have spent a total of $10(\$3,200) = \$32,000$. If the price were $2,700, an additional 20 customers would have purchased the system, at an additional cost of $54,000. If the price had been $2,500, another 20 customers would have purchased the system, at an additional total cost of $50,000. By comparison, at the sale price of $2,200 we assume that all 50 customers purchased one of the systems, at a total cost of $50(\$2,200) = \$110,000$. Consequently, the total savings to the customers (the consumers' surplus) is

$$\underbrace{(\$32,000 + \$54,000 + \$50,000)}_{\substack{\text{amount customers} \\ \text{were willing to pay}}} - \underbrace{(\$110,000)}_{\text{amount paid}} = \$26,000$$

We can indicate the result graphically as shown in Figure 6.96.   ∎

FIGURE 6.96

FIGURE 6.97

If we assume continuous supply and demand functions for some product and that the product is actually sold at the equilibrium price, then the areas shown in Figure 6.97 represent the consumers' surplus and producers' surplus, as indicated. These areas are given by the integrals

$$\text{consumers' surplus} = \int_0^{x_E} [d(x) - y_E]\,dx \qquad (6.52)$$

$$\text{producers' surplus} = \int_0^{x_E} [y_E - s(x)]\,dx \qquad (6.53)$$

EXAMPLE 6.43   For a certain commodity, the demand function is

$$d(x) = 12 - 0.25x^2$$

and the supply function is

$$s(x) = 0.25x^2 + 0.50x + 2$$

where x is in thousands of units and both $d(x)$ and $s(x)$ are in hundreds of dollars. Find

(a) the equilibrium price, which is the y-coordinate of the equilibrium point;

(b) the consumers' surplus and the producers' surplus.

Show the results graphically.

Solution

(a) To find the equilibrium price we set $d(x) = s(x)$ and solve for x:

$$12 - 0.25x^2 = 0.25x^2 + 0.50x + 2$$

$$0.50x^2 + 0.50x - 10 = 0 \quad \text{Multiply by 2.}$$

$$x^2 + x - 20 = 0$$

$$(x + 5)(x - 4) = 0$$

$$x = -5 \quad \text{or} \quad x = 4$$

We reject $x = -5$, since x must be positive in the context of the problem. Thus, $x_E = 4$. To find y_E we can substitute 4 for x in $d(x)$ or in $s(x)$. Choosing $d(x)$, we get

$$y_E = d(4) = 12 - 0.25(4)^2 = 12 - 4 = 8$$

Since $d(x)$ is in hundreds of dollars, the equilibrium price is $800.

(b) By Equation 6.52,

$$\text{consumers' surplus} = \int_0^4 [(12 - 0.25x^2) - 8]dx$$

$$= 4x - 0.25\frac{x^3}{3}\Bigg]_0^4 = 16 - \frac{16}{3} = \frac{32}{3}$$

By Equation 6.53,

$$\text{producers' surplus} = \int_0^4 [8 - (0.25x^2 + 0.50x + 2)]dx$$

$$= 6x - 0.25\frac{x^3}{3} - 0.50\frac{x^2}{2}\Bigg]_0^4 = 24 - \frac{16}{3} - 4 = \frac{44}{3}$$

In dollars, the consumers' surplus is approximately $1,067 and the producers' surplus is approximately $1,467. The results are shown graphically in Figure 6.98.   ∎

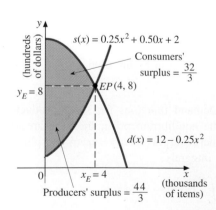

FIGURE 6.98

Exercise Set 6.10

1. A company determines that the marginal-cost function for a certain product is $M_C(x) = x^3 - 2x$. Find the total-cost function, assuming a fixed cost of $200.

2. The assembly line at the True-View Company produces 300 VCRs per day. The manager knows that the marginal-cost function (in dollars) is given by

$$M_C(x) = 600 - 1.5x \qquad (0 \le x \le 400)$$

where x is the number of VCRs made each day. Find the increase in total cost if the manager decides to have the assembly line increase production to 350 units daily.

3. The price per bushel of apples that a small apple grower receives depends on the size of the crop. From experience, the grower knows that if the size of the crop is x bushels, the marginal revenue is

$$M_R(x) = 10 - 0.004x + \frac{1}{\sqrt{x}} \qquad (500 \le x \le 2000)$$

Last year, the grower harvested 1,000 bushels. How much more revenue will the grower receive if the crop is 1,400 bushels? Sketch the marginal-revenue function, and shade the region whose area corresponds to the increased revenue.

4. Assume that in a market under pure competition (in which the actual price of an item is the equilibrium price, where supply equals demand), the supply function for a certain item is given by $s(x) = 2x + 1$, and the demand function for the same item is given by $d(x) = 12 - x^2$. Determine (a) the price at which the item will be sold, (b) the consumers' surplus, and (c) the producers' surplus.

5. The demand for a certain commodity in a local market is given by

$$d(x) = \frac{30x + 6000}{x + 100} \qquad (0 \le x \le 500)$$

where $d(x)$ is in hundreds of dollars. If the supply function is given by

$$s(x) = 8\sqrt{x+1} \qquad (0 \le x \le 500)$$

use a graphing calculator or CAS to approximate the equilibrium point, and then find the consumers' surplus and producers' surplus at market equilibrium.

6. The publisher of a successful magazine wants to sell more copies of each issue. Current circulation is 60,000 copies each month. When x is in thousands of copies circulated per month, the marginal revenue (in thousands of dollars) is given by

$$M_R(x) = 3.5 - 0.04x - \frac{50}{x}$$

where $40 \le x \le 80$. How much additional revenue would be produced by increasing the circulation to 70,000 copies?

7. The demand and supply functions for a certain commodity are, respectively, $d(x) = -x^2 - 20x + 1,020$ and $s(x) = x^2 + 60x + 20$, where $0 \le x \le 30$. The price is in dollars, and the quantity x is in thousands of units.
   (a) Find the equilibrium price and quantity.
   (b) Find the consumers' surplus.
   (c) Find the producers' surplus.
   Show your results graphically.

8. **Consumers' surplus when the supplier has a monopoly.** When a supplier has a monopoly on a commodity, market equilibrium will not exist (the product is not sold at the equilibrium price). In this situation the price will be set to maximize the supplier's profit. Suppose that a supplier has a monopoly on a certain product, with cost function $c(x) = x^2 + 920x + 3,000$ and demand function $d(x) = -x^2 - 2x + 2,000$ where $0 \le x \le 50$. It is assumed that $c(x)$ is in thousands of dollars, x is in thousands of units, and $d(x)$ is in dollars.
   (a) Find the number of units and the price per unit that will maximize profit. (*Hint:* profit = revenue − cost, and revenue is the number of units sold times the price per unit.)
   (b) Find the consumers' surplus if the product is sold at the price that maximizes the supplier's profit.

Chapter 6 Review Exercises

In Exercises 1–10, find the area of the region indicated. Sketch the region.

1. Under the graph of

$$f(x) = \begin{cases} 1 + \sqrt{2x} & \text{if } 0 \le x < 2 \\ x^2 - 4 & \text{if } 2 \le x \le 4 \end{cases} \quad \text{on } [0, 4]$$

2. Under the graph of

$$f(x) = \begin{cases} 1 - (x+2)^{-2} & \text{if } -1 \le x < 0 \\ 0 & \text{if } x = 0 \\ 4 - 2x & \text{if } 0 < x \le 2 \end{cases} \quad \text{on } [-1, 2]$$

3. Bounded by $y = 3x + 4$, $y = 2 - x$, $x = 0$, and $x = 3$

4. Between $y = x^3 - x$ and $y = 1 - x^2$

5. Between $y = x^2 - 2x$ and $y = 4\sqrt{x}$

6. Bounded by $y = \sqrt{x + 5}$, $4y = 3x$, and $y = 0$

7. Between $y = x^2 - x$ and $y = \sin \pi x$

8. Between $y^2 = x + 2$ and $y^2 + x - 2y - 2 = 0$

9. One of the regions bounded above by $y = \sqrt{3} \cos x$ and below by $y = \sin x$

10. Bounded by $y = 1/x^2$, $y = 1$, $2x - 4y = 3$, and $x = 0$

In Exercises 11–20, find the volume obtained by revolving the region about the specified axis. Sketch the region.

11. Bounded by $y = \frac{1}{x}$, $x = 1$, $x = 2$, and the x-axis; x-axis

12. Bounded by $y = \sqrt{2x + 3}$, $x = -1$, $x = 3$, and the x-axis; x-axis

13. Bounded by $x = 8 - 2y$, $x = 2$, $x = 6$, and the x-axis; y-axis

14. Bounded by $y = x$, $y = 2x$, $x = 1$, and $x = 4$; y-axis

15. Between $y = x^2$ and $y = x + 6$; x-axis

16. Between $y = x^2 + 1$ and $y = 3x - x^2$; y-axis

17. Bounded by $y = \sqrt{2x + 1}$, $y = x - 1$, and $y = 1$; x-axis

18. Between $x + \sqrt{2y} = 0$ and $x = y - y^2$; x-axis

19. Region of Exercise 15 revolved about $x + 2 = 0$

20. Region of Exercise 18 revolved about $x = 1$

21. The base of a wedge is the region enclosed by the ellipse

$x^2 + 2y^2 = 4$, and each cross section perpendicular to the x-axis is an isosceles right triangle with right angle on the upper half of the ellipse. Find the volume of the wedge.

22. In the accompanying figure, ABC and DEF are equilateral triangles of sides 4 and 1, respectively, in planes perpendicular to edge AD, which is of length 6, as shown. Find the volume of the solid.

In Exercises 23–25, find the arc length of the curve on the specified interval.

23. $x = 2 + (y - 1)^{3/2}$ from $(2, 1)$ to $(10, 5)$

24. $y = \frac{1}{20}x^5 + \frac{1}{3}x^{-3}$ from $x = \frac{1}{2}$ to $x = 2$

25. $x = y^{3/2} - \frac{1}{3}y^{1/2}$ from $(\frac{2}{3}, 1)$ to $(\frac{22}{3}, 4)$

In Exercises 26–29, find the area of the surface formed by revolving the given arc about the specified axis.

26. $y = 2\sqrt{3x - 2}$ from $x = 3$ to $x = 6$, about x-axis

27. $x^2 - 3y = 3$ from $(0, -1)$ to $(2, \frac{1}{3})$, about y-axis

28. $x = \left(\dfrac{y + 5}{4}\right)^3$ from $(1, -1)$ to $(8, 3)$, about y-axis

29. The arc of Exercise 25 about (a) the x-axis and (b) the y-axis

30. A watering trough 20 ft long has ends in the shape of inverted isosceles trapezoids 4 ft across the top, 2 ft across the bottom, and 3 ft high. If the trough has water in it to a depth of 2 ft, find the work required to pump all the water over the top of the trough.

31. A ship's anchor, weighing 1000 lb, is being raised. The anchor chain weighs 10 lb/ft, and it is being hoisted over a pulley that is 40 ft above the water. How much work is required to raise the anchor from the water level to a position 10 ft below the pulley?

32. A force of 12 lb is required to stretch a certain spring from its natural length of 20 in. to a length of 24 in. Find the work required to (a) stretch it from a length of 24 in. to a length of 30 in. and (b) compress it 4 in. from its natural length.

33. The ends of a long trough are parabolic segments with dimensions as shown in the figure (in feet). Find the total force on one end of the trough when it (a) is full of water and (b) has water in it to a depth of 1 ft.

34. A plate as shown in the figure is submerged vertically in water so that the top edge of the plate is at a depth of 9 ft. Find the total force on one side of the plate.

35. A bar with density $\rho(x) = \sqrt{x+1}$ is located on the x-axis between $x = -1$ and $x = 8$. Find its center of mass.

36. A homogeneous lamina with density $\rho = 3$ occupies the region in the plane bounded by the curves $y = 2 - x^2$, $y = x$, and $x = 0$. Find its center of mass.

In Exercises 37–40, find the centroid of the region described.

37. Between the graphs of $y = x^2$ and $y = x + 2$

38. The triangular region bounded by $y = x/2$, $y = 2x$, and $x = 4$

39. Between the y-axis and the parabola $y^2 - x - 4y = 0$

40. Bounded by $y = \sqrt{x}$, $x + y = 6$, and the x-axis

In Exercises 41 and 42, use the results of Exercises 22 and 23 of Exercise Set 6.8 to find the centroid.

41.

42.

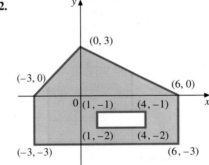

43. The probability density function for the random variable X, denoting the time for students in a certain class to complete a 3-hr final exam, is of the form $f(x) = \frac{x}{9}(1 + kx)$, where $0 \le x \le 3$. Find the following:
(a) The constant k
(b) The mean μ
(c) The variance σ^2
(d) $P(X < 1)$
(e) $P(2 \le X \le 3)$

44. The proportion X of defective items in a large batch has a probability density function that is approximated by a beta function with $\alpha = 1$ and $\beta = 8$. Find (a) the constant B and (b) the probability that more than 20% are defective.

45. Suppose that the marginal-cost and marginal-revenue functions for production of a certain product are

$$M_C(x) = 4 - \frac{x}{1000}$$

and

$$M_R(x) = 40 - \frac{x}{100}$$

The fixed cost is $5,400. Find the total-cost function $C(x)$ and the total-revenue function $R(x)$. Also find the profit function $P(x)$, and show that the maximum profit occurs when $M_C(x) = M_R(x)$. Find the value of x that yields the maximum profit. What is the maximum profit?

46. The supply and demand functions for a certain product are given by

$$s(x) = 5 - \frac{(x-8)^2}{16}$$

and

$$d(x) = 5 - \frac{x^2}{8}$$

respectively, where $0 \le x \le 6$. The quantity x is in thousands of units, and both $s(x)$ and $d(x)$ are in hundreds of dollars. Find
(a) the equilibrium price for the product;
(b) the consumers' surplus;
(c) the producers' surplus.
Show the results graphically.

Chapter 6 Concept Quiz

1. Let f be a nonnegative continuous function of x on $[a, b]$, where $b > a > 0$. Denote the graph of $y = f(x)$ by C, and let R be the region under C and above the x-axis from $x = a$ to $x = b$. Write integrals for the following:
(a) The area of R
(b) The volume of the solid obtained by revolving R about the x-axis
(c) The volume of the solid obtained by revolving R about the y-axis
(d) The coordinates $\bar{x}$ and $\bar{y}$ of the centroid of R
(e) The length of C (assuming f is smooth)
(f) The surface area obtained by revolving C about the x-axis (assuming f is smooth)
(g) The surface area obtained by revolving C about the y-axis (assuming f is smooth)

2. Let g be a nonnegative continuous function of y on $[c, d]$ where $d > c > 0$. Denote the graph of $x = g(y)$ by C, and let R be the region to the left of C and to the right of the y-axis from $y = c$ to $y = d$. Write integrals for parts (a)–(g) of Exercise 1, replacing f with g.

3. Repeat parts (a)–(d) of Exercise 1 for the region R

between the graphs of $y = f(x)$ and $y = g(x)$ for x on $[a, b]$, where $f(x) \ge g(x) \ge 0$ and $b > a > 0$.

4. Repeat parts (a)–(d) of Exercise 1 for the region R between the graphs of $x = g(y)$ and $x = h(y)$ for y on $[c, d]$, where $g(y) \ge h(y) \ge 0$ and $d > c > 0$.

5. Write integrals for the following.
(a) The center of mass of a thin rod that reaches from $x = a$ to $x = b$ if the density is $\rho(x)$
(b) The work done by the continuous force function $F(x)$ acting from $x = a$ to $x = b$
(c) The total force on a plate submerged vertically in a liquid of density $\rho(x)$ if the top of the plate is a units below the surface and the bottom is b units below the surface. Assume that the width of the plate at depth x ($a \le x \le b$) is $w(x)$.
(d) The probability that $a < X < b$ if the probability density function on $[a, b]$ is $f(x)$
(e) The mean μ of a random variable whose probability density function is $f(x)$ on $[a, b]$
(f) The variance σ^2 of a random variable with mean μ and probability density function $f(x)$ on $[a, b]$

APPLYING CALCULUS

1. *Income distribution—Lorenz curve and Gini coefficient.* When a high percentage of the population earns a low percentage of the total national income, there are income inequities. To measure the degree of inequity, economists use a Lorenz curve, which is the graph of a function showing the cumulative percentage, $I(x)$, of total income received, versus the cumulative percentage, x, of income recipients. The function I would be given by $y = x$ if all income recipients had equal shares (when $x = 0.2$, $y = 0.2$, so the poorest 20% earn 20% of the total income). When income is not equally distributed, $I(x)$ has some other form, giving a Lorenz curve that deviates from the 45° line $y = x$, *called the line of absolute income equality*. The accompanying figure illustrates such a situation. The Gini coefficient is defined as the area between the Lorenz curve and the line $y = x$ divided by the area of the right triangle with legs one unit long on the x and y axes and hypotenuse along the line $y = x$. Since the area of the right triangle is $(\frac{1}{2})(1)(1) = \frac{1}{2}$, the Gini coefficient is

$$G = 2 \int_0^1 [x - I(x)]dx$$

Suppose that

Unequally distributed income

$$I(x) = \frac{x^2 + 5x}{7} \qquad 0 \le x \le 1$$

(a) Sketch the Lorenz curve.

(b) What portion of the total income is earned by the lowest 15% of the population?

(c) Calculate the Gini coefficient.

2. *(Refer to Problem 1 above)* In 1990 the percent of aggregate income received by each fifth of families in the U.S. was as follows:

lowest fifth	5.1
second fifth	11.6
third fifth	17.5
fourth fifth	24.3
highest fifth	41.5

(a) Find a function I so that the Lorenz curve $y = I(x)$ closely fits the given data.

(b) Using the function you found in part (a), calculate the Gini coefficient.

3. *Marginal propensity to consume and to save.* The rate at which total national consumption changes with respect to change in total national disposable income is called the **marginal propensity to consume**. So if $C = f(I)$, where I is the total national disposable income and C is the total national consumption, the derivative dC/dI is the marginal propensity to consume. Similarly, if $S = g(I)$ represents the total national savings, then dS/dI represents the **marginal propensity to save**. Since the amount spent on consumption plus the amount saved equals the total income, we have $C + S = I$, and thus

$$\frac{dC}{dI} + \frac{dS}{dI} = 1$$

Suppose that the marginal propensity to save (in billions of dollars) is given by

$$\frac{dS}{dI} = 0.9 - \frac{0.4}{\sqrt{I+1}}$$

and consumption is $7.8 billion when disposable income is $15 billion. Find the consumption function. what is the consumption when disposable income is $20 billion?

4. Coulomb's Law states that the force of attraction or repulsion between two point charges is directly proportional to the product of the charges and inversely proportional to the square of the distance between them:

$$F = k\frac{q_1 q_2}{r^2}$$

where q_1 and q_2 are the magnitudes of the charges at the points P_1 and P_2, respectively, r is the distance between P_1 and P_2 and k is the constant of proportionality.

(a) Determine the rate of change of the force with respect to the distance between the points. Suppose the distance between the charges is changing with respect to time according to the equation

$$\frac{dr}{dt} = r^2 + 2r$$

What is the change in the force with respect to time?
Suppose the point charges are both negative and have equal magnitudes, $q_1 = q_2 = q$.

(b) Determine the constant k if the points P_1 and P_2 are 4 centimeters apart and the repulsion force is 10 dynes.

(c) How much work is required to bring the charge at P_2 to a distance of 2 centimeters from the charge at P_1?

5. Poiseuille discovered that the velocity v (centimeters per second) of a blood particle flowing through a blood vessel having a circular cross section of radius R is related to its distance r from the central axis by the equation

$$v(r) = \frac{P(R^2 - r^2)}{4\mu L}$$

where P is the blood pressure, μ (mu) is the coefficient of viscosity of the blood and L is the length of the vessel. Flow characterized by this equation is called laminar flow, since the blood particles are assumed to flow in cylindrical shells, or laminae, parallel to the walls of the vessel at a distance r from the center, with all the blood particles in a particular shell moving with the same velocity.

If the midpoint of the ith shell is at a distance r_i^* from the center of the vessel, then the volume of blood flowing through that shell is approximately

$$V_i \approx 2\pi r_i^* v(r_i^*)\Delta r_i$$

Thus, the total volume V of blood per second flowing through the cross section is given approximately by

$$\sum_{i=1}^{n} 2\pi r_i^* \mu(r_i^*)\Delta r_i$$

from which it follows that

$$V = \int_0^R 2\pi r v(r)dr$$

(a) Show that the volume of blood flowing through the blood vessel is proportional to the 4th power of the radius by evaluating the integral derived above. This result is known as Poiseuille's Law.

(b) Suppose that blood pressure is constant. Calculate the percentage decrease in the volume of blood that flows through the vessel per unit of time when the radius is constricted from R to $R/2$.

(c) The average velocity of blood flow through a vessel at a circular cross section of radius R and area A is defined to be

$$\bar{v} = \frac{L}{A} \int_0^R 2\pi r v(r) dr$$

Show that $V = PR^2/\delta\mu L$ and that $\bar{v}$ is 1/2 the maximum velocity of the blood flow.

6. *Volumes of barrels.*

(a) A barrel is formed by rotating an ellipse around its major axis and then cutting off equal caps from the top and bottom. Determine the volume of the barrel if the (maximum) radius at the center is R, its height is H and the top and bottom have radius r. (Problems of this type were considered by Kepler in his studies of the volumes of barrels.)

(b) Suppose a barrel has the shape of the solid of revolution formed by rotating the region between the y-axis and the parabola $x = R - ky^2$, $-h/2 \le y \le h/2$, R, $k > 0$, and $h < R/k$, around the y-axis. Show that the volume of the barrel is

$$V = \frac{1}{3}\pi h \left(2R^2 + r^2 - \frac{6}{15}\delta^2 \right)$$

where h is the height of the barrel, R is the radius at the center, r is the radius of the top and bottom, and $\delta = R - r$.

7. It can be shown that the shape of a flexible and inelastic cable supporting a load that is uniformly distributed horizontally is a parabola. As an application, consider the main cable in a suspension bridge. The dimensions L and h shown in the figure are called the span and the sag, respectively, of the bridge.

(a) Let the origin of an xy-coordinate system be at the lowest point of a parabolic cable of a suspension bridge whose span is L and whose sag is h. Determine the equation for the cable.

(b) Determine the length of the cable in part (a).

(c) The George Washington Bridge across the Hudson River connecting New York City to Fort Lee, New Jersey, has a span of 3500 feet and a sag of 316 feet. Determine the length of the main cables of the bridge.

8. According to *Newton's Law of Universal Gravitation*, the gravitational force on an object of mass m that has been projected vertically upward from the Earth's surface is

$$F = \frac{-mgR^2}{(x+R)^2}$$

where $x = x(t)$ is the object's distance above the surface at time t, R is the radius of the Earth and g is the acceleration due to gravity. (Assume all the force is exerted at the instant of launch.) Also, by *Newton's second law*, $F = ma = mdv/dt$ and so

$$m\frac{dv}{dt} = -\frac{mgR^2}{(x+R)^2}$$

(a) Suppose a space shuttle is fired vertically upward with an initial velocity v_0. Let h be the maximum height above the surface reached by the object. Show that

$$v_0 = \sqrt{\frac{2gRh}{R+h}}$$

(Hint: By the Chain Rule, $m\, dv/dt = mv\, dv/dx$.)

(b) Calculate $v_e = \lim_{h \to 0} v_0$. This limit is called the *escape velocity* for the Earth.

(c) Use $R = 3960$ mi and $g = 32$ ft/s^2 to calculate v_e in ft/s and mi/s.

9. A space shuttle of mass m is fired vertically upward from the surface of the Earth. The force F of the earth's gravitational field acting on the space shuttle is given by Newton's Law of Universal Gravitation.

(a) Determine the work that must be done against F to lift the space shuttle to a height h above the surface of the Earth.

(b) Determine the work that must be done to lift the space shuttle out of the Earth's gravitational field, that is, to lift the space shuttle to an infinite height.

(c) Suppose that the space shuttle is to be projected upward by imposing an initial kinetic energy $\frac{1}{2}mv_0^2$. Determine the initial velocity required to lift the space shuttle infinitely high. Compare with the result in Problem 9.

10. The sun's rate of travel across the sky is uneven, sometimes ahead and sometimes behind the average solar time shown on our watches. Part (a) below shows a photo on a single piece of film that shows the position of the Sun in the sky at 8:00 a.m. every 10 days for a year. The figure-8 is known as an analemma. Part (b) shows a sundial with an analemma rotated around its axis as the shadow stick (known as the *gnomon*). As the Sun rises higher in the sky as summer approaches, the bulge in the analemma compensates for the sun's divergence from its average rate of travel. We can thus read exact time as the position where the shadow of one side of the analemma crosses a circular bar at the bottom of the sundial. This analemma is a solid of revolution. If we assume that it is 30 cm long and 4 cm in its greatest diameter, and is approximated by $y = 2 \sin (2\pi t/365)$ rotated around the axis $y = -0.2$, calculate its volume. It is suspended by thin wires, so we have to know its weight. If it is made of aluminum, whose density is 2 g/cm^3, how much does this solid analemma weigh?

(a)

(b)

Keith Geddes

You're a professor of computer science in the faculty of mathematics at the University of Waterloo. Are you a computer scientist or a mathematician?

I see myself as a computer scientist and a mathematician. Computer science spans topics from mathematics through physics and engineering and more, but mathematics is even broader. In high school I found mathematics interesting but it never occurred to me that you could specialize in mathematics. I thought it was only a subject that supported "real" disciplines like engineering and physics. In second year, I switched my major from physics to mathematics.

What exactly is "computer algebra"?

The subject of computer algebra deals with mathematical computation in a general sense. This includes the manipulation of unevaluated symbols and unevaluated functions according to the rules of algebra and other branches of mathematics. In computer algebra programs, variables and expressions don't have to be evaluated to numerical values prior to being manipulated, but computer algebra programs can also work with arbitrary precision arithmetic. Computer algebra programs are like calculators that do algebra as well as arithmetic. They can also display your results graphically in a fast and sometimes spectacular way.

You and Gaston Gonnet are the original architects of the computer algebra program Maple. What motivated you to start the Maple project?

When we conceived of the project in November of 1980, the goal was to design a computer algebra system that would be useful to large numbers of researchers in mathematics, engineering, and the sciences—and also to assist students in their mathematical education. Existing computer algebra systems required large computers that were only usable by specialists. The other important motivation was to provide a focus for research into algorithms for algebraic computation. Maple has grown considerably since we started. There are now more than 2000 mathematical functions in Maple's libraries. But designers can't possibly anticipate all the needs of the users of mathematics. If Maple doesn't already know how to carry out a particular computation, you can program it to do so.

Maple's libraries are programs written in Maple's own language. How is the Maple language different from, say, Pascal or FORTRAN?

The fundamental difference is that the Maple language was designed specifically for algebraic computation. It incorporates commands for many high-level mathematical operations that are not built into other languages. Unlike FORTRAN or Pascal, the Maple language is designed to be used interactively. Specific directions such as "integrate $\cos x$ with respect to x" are given with one-line commands, which are easy to learn. Also, if you need to, it's not hard to program in Maple.

What types of research is Maple useful for?

Maple has proved useful in research that uses mathematics at almost any level. This includes pure mathematics, applied mathematics, mechanical engineering, economics, computer science, and so on. A few areas that come to mind are ongoing research in robotics, econometric modeling, general relativity, number theory, and computer-aided design. I recently heard about some models to improve voting systems and to study the AIDS epidemic that were developed and are taught using Maple. In the case of the AIDS model, students are exploring changes in parameters and using Maple to numerically integrate the differential equations involved.

How did you come to apply your own research in algebra to other areas of mathematics like those of calculus?

I came to the topic from the point of view of algebraic algorithms. I was attracted by the precision and beauty of algebraic computation and the very powerful algorithms that could be developed to factor multivariate polynomials, for instance. Many of the capabilities of computer algebra are straightforward, but indefinite integration is an exception. Many people view integration not as an algorithmic process, but as a collection of tricks that can solve a limited number of problems. Finding that the Risch algorithm applied algebraic algorithms to the indefinite integration problem of calculus is what led me deeper into computer algebra.

What is the Risch algorithm and why is it important?

In 1969 Robert Risch presented a decision procedure for the indefinite integration of the elementary functions. These are the functions typically encountered in calculus: algebraic, trigonometric, inverse trigonometric, exponential, logarithmic, and hyperbolic. By performing a finite number of computational steps, the algorithm can determine whether or not a given elementary function has an antiderivative that is also an elementary function. The Risch algorithm either determines a closed formula expressing the integral as an elementary function, or else it proves that expressing the integral as an elementary function is impossible. It has a clear advantage over standard integration techniques. If students learn traditional integration techniques only, and then fail to find an answer, they are sometimes left wondering if they simply did not find the right trick.

It's clear that Maple is a powerful research tool, but does it actually help students learn mathematics?

Yes. The greatest benefit of Maple to students is in making it possible to carry out "experimental mathematics." It allows mathematical topics that might seem distant and abstract to be played with on a level that is meaningful to students. It's been said that systems like Maple allow students and professors to devote more time to the ideas: the theory and the applications of the mathematics, and less to the routine computations. In the future I expect to see more emphasis on the problem-solving process with a significant decrease in the emphasis on teaching "tricks of the trade."

How would you contrast this approach with your own mathematical education?

My mathematical education is continuing, but it seems that today's students have more opportunity to explore "what if" scenarios related to core mathematical concepts. They can now learn more by "doing mathematics." Of course, they can also do many more problems than they can by hand. By removing some mathematical drudgery, there is greater potential for bringing some mathematical excitement to the learning process. It shouldn't just be drudgery. We should be teaching students how to take advantage of computers, not to compete with them in algorithmic computations.

O.K., but if you can now solve integrals in closed form with Maple, why should you learn to integrate by hand with pencil and paper?

People once worried that pocket calculators would make finding square roots by hand, and even knowing multiplication and division, less important. Of course, calculators didn't do this. The calculators made the concepts more important since more people were using mathematics because of them. In the same way, a small computer running a powerful computer algebra system is bound to change some aspects of using higher mathematics. I believe problem solving, as a way of mathematical thinking, will become more important. The real goal of a mathematical education is to be able to take a problem, stated in natural language, and express that problem in mathematical terms (a new language, if you will), to solve that mathematical problem, and finally to reinterpret the results in terms of the originally stated problem. The intermediate step [to] "solve the mathematical problem" will continue to require many skills and creativity, but the aspects that are pure calculation like "factor a polynomial" or "find an indefinite integral" can be relegated to the computer algebra calculator. The important thing, it seems to me, is to teach the understanding of mathematical concepts such as indefinite integration, or the meaning of a solution to a differential equation, rather than teaching how to "turn the crank." Sadly, there are generations of students who think, because of their education, that learning mathematics is learning how to turn the crank. Much of mathematics is really in trying to "find the crank" or, at a higher level, "is there a crank?" or "prove that there isn't a crank." Mathematics involves much more creativity than most people are aware.

Students seem to take well to Maple, after they learn the syntax and the computer keyboard. Is it true that students have been active in testing and developing Maple code?

Yes. Students have been very active with Maple from the beginning. At Waterloo most of our students are in the co-operative work/study program where they alternate four-month study terms with four-month work terms. We're lucky to have some good students who want to work on Maple itself. Many students have started with Maple in their calculus courses and have continued throughout their careers. There have always been several graduate students working on Maple at any given time. Waterloo has five separate departments in the Faculty of Mathematics. Faculty and students from all of these subdisciplines have had a hand in developing Maple. Maple is now an international research effort coordinated from Waterloo and other locations throughout the world. Many talented and hardworking people deserve credit for its success.

What are the limits to the type of mathematics Maple can help with? Will we ever be able to use Maple to set up a word problem or prove a theorem?

Well, I'm pretty sure that Maple can solve all of the limit problems in this text, for instance, but it's important to remember that Maple is just an algebraic calculator that does what it is programmed to do. While it is extremely useful to have such a calculator, calculation is only one aspect of doing mathematics. Maple will sometimes make mistakes because there are trade-offs between accuracy and efficiency in its design. Generally speaking, Maple doesn't "know calculus"; rather, it knows how to carry out the algebraic manipulations associated with calculus. I do not see Maple evolving into a system that could set up a word problem and solve it, or prove a theorem, although there is computer science research being done on these subjects. These aspects of mathematics require creative human mathematicians. Mathematics is very much a human endeavor.

Keith Geddes is Professor of Computer Science and Associate Dean for Graduate Studies and Research in the Faculty of Mathematics at the University of Waterloo. His research interests lie in the areas of algebraic algorithms, computer algebra systems, and scientific computation. Professor Geddes has coauthored a graduate-level textbook, *Algorithms for Computer Algebra* and continues to be involved in the research development of the Maple computer algebra system. He holds a B.A. in Mathematics and a Ph.D. in Computer Science from the University of Toronto.

7

TRANSCENDENTAL FUNCTIONS

In Chapter 1 we characterized the algebraic functions as those that can be obtained from polynomial functions by any finite combination of the operations of addition, subtraction, multiplication, raising to an integral power, and extracting integral roots. Stated precisely, an algebraic function is any function f such that $y = f(x)$ satisfies an equation of the form

$$P_n(x)y^n + P_{n-1}(x)y^{n-1} + \cdots + P_1(x)y + P_0(x) = 0$$

where each $P_i(x)$ is a polynomial with rational coefficients. For example, $f(x) = 1/\sqrt{x^2 + 1}$ is algebraic since $y = 1/\sqrt{x^2 + 1}$ satisfies the polynomial equation $(x^2 + 1)y^2 - 1 = 0$. The class of algebraic functions is very broad, but it by no means includes all possible functions.

In fact, we have already studied one class of nonalgebraic (or *transcendental*) functions, namely, the trigonometric functions. In this chapter we introduce the following additional transcendental functions: logarithmic, exponential, inverse trigonometric, hyperbolic, and inverse hyperbolic functions. These, together with the trigonometric and algebraic functions, constitute what are called the **elementary functions**. After introducing our new functions and studying their properties, we will concentrate on their derivatives and integrals. We will also show some important applications.

We begin with the natural logarithm function, $\ln(x)$, which we define as a special area function. This approach enables us to use properties of area functions that we studied in Chapter 5. In particular, we will see that the First Fundamental Theorem of Calculus plays an important role. Next, we define the natural exponential function, $\exp(x)$, as the inverse of $\ln(x)$. The natural logarithm and exponential functions comprise one of the most important function–inverse function pairs in mathematics and science. From these two basic functions we will obtain more general logarithm and exponential functions. (We will see, for example, how irrational powers such as $\pi^{\sqrt{2}}$ are defined.) We conclude the chapter by studying the inverses of each of the trigonometric functions (with suitably restricted domains) and then the so-called hyperbolic functions and their inverses.

7.1 THE NATURAL LOGARITHM FUNCTION

Logarithms were invented in about 1590 by John Napier (1550–1617), a Scottish mathematician who was interested in methods of rapid calculation for practical use in such fields as astronomy, navigation, and surveying. He developed his idea and calculated tables to 7 places of accuracy. Such "log tables" were in use well into the twentieth century. Though logarithms are easy to use, the idea had apparently not been thought of before Napier. A Swiss watchmaker, Jobst Bürgi, independently invented logarithms and published his results in 1620.

Word of Napier's discovery reached the Danish astronomer Tycho Brahe in the late 1590s, but Napier's tables were not ready in time for Brahe to use. But Brahe's successor, Johannes Kepler, did correspond with Napier and found logarithms important for astronomical use. Kepler's astronomical tables of 1620 made use of Napier's discovery.

You are probably familiar with what is known as the *common* logarithm of a number, usually designated $\log x$. This logarithm is by definition the power to which 10 must be raised to give the number x. For example, $\log 100 = 2$, since $10^2 = 100$. On your calculator you will find a $\boxed{\log x}$ key. Before scientific calculators were available, tables of common logarithms were used to obtain their values. These common logarithms were formerly used in simplifying arithmetical computations involving products, quotients, powers, and roots, based on the properties

$$\log ab = \log a + \log b$$

$$\log \frac{a}{b} = \log a - \log b$$

$$\log a^p = p \log a$$

Their use in computations is of little importance now, however, because of the calculator and computer.

Although common logarithms still have some uses, it is the *natural* logarithm that is of fundamental importance in calculus and its applications. The natural logarithm of x is denoted by $\ln(x)$, or more often by $\ln x$ (read "el en x"). It is possible to define $\ln x$ in a manner similar to the way $\log x$ is defined, with 10 being replaced by a number e whose value is approximately 2.71828. You may have seen this definition before in previous mathematics courses. Now that we have calculus at our disposal we will take a different approach, but we will see that this characterization of $\ln x$ as a power of e is valid.

We use the symbol $\ln$ to denote the **natural logarithm function**. We have previously (in Chapter 1) referred to the fact that your calculator has an $\boxed{\ln x}$ key. With it, you can obtain values of natural logarithms to a high degree of accuracy. But where these values come from is something of a mystery. One way of obtaining them is based on the following definition. (Another way will be given in Chapter 10.)

For any positive real number x, the value of $\ln x$ is defined as the integral of the function $f(t) = 1/t$ from $t = 1$ to $t = x$ (Figure 7.1). That is,

---

Definition 7.1
The Natural Logarithm Function

$$\ln x = \int_1^x \frac{1}{t} dt \qquad (x > 0) \tag{7.1}$$

If $x > 1$, we can interpret $\ln x$ as the area under the graph of $f(t) = 1/t$ from 1 to x. (See Figure 7.1(a).) If $0 < x < 1$, then

$$\ln x = \int_1^x \frac{1}{t} dt = -\int_x^1 \frac{1}{t} dt$$

which is the negative of the area under the graph of f from x to 1. (See Figure 7.1(b).)

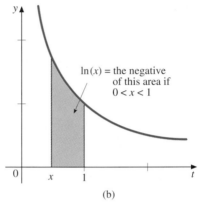

FIGURE 7.1

If $x = 1$, we see by Equation 7.1 that $\ln x = 0$. We therefore have

$$
\begin{aligned}
\ln(x) > 0 && \text{if } x > 1 \\
\ln 1 = 0 && \\
\ln x < 0 && \text{if } 0 < x < 1
\end{aligned}
$$

Note that $\ln x$ is not defined for $x \le 0$. That is, the domain of $\ln x$ is the set of all *positive* real numbers. We will see what the range is after investigating some of the properties of $\ln x$.

Properties of $\ln x$

Recall that by the First Fundamental Theorem of Calculus, when a function f is continuous on a closed interval, for a fixed number a and any x in this interval

$$
\frac{d}{dx} \int_a^x f(t)dt = f(x)
$$

Applying this theorem to the integral in Equation 7.1, we obtain the important result

$$
\frac{d}{dx}(\ln x) = \frac{1}{x} \qquad (x > 0) \tag{7.2}
$$

Since $x > 0$, we see that

$$
\frac{d}{dx}(\ln x) > 0
$$

for all x in the domain of $\ln x$. So $\ln x$ is an increasing function, and hence is one-to-one, throughout its domain. Differentiating again, we get

$$
\frac{d^2}{dx^2}(\ln x) = -\frac{1}{x^2}
$$

Since this second derivative is negative, we conclude that the graph of $y = \ln x$ is concave down throughout its domain.

We now know quite a bit about the graph of $y = \ln x$. Its x-intercept is 1, it is always rising as x increases, and it is always concave down. We are not sure, on the basis of what we have shown so far, whether there are any asymptotes. To investigate this question, we consider the two limits

$$
\lim_{x \to \infty} (\ln x) \qquad \text{and} \qquad \lim_{x \to 0^+} (\ln x)
$$

Before doing so, we need to establish some additional properties of the natural logarithm function.

Arithmetic Properties

By means of the following derivation, we will show that the three properties of the common logarithm, involving products, quotients, and powers, also hold true for the natural logarithm.

First, consider $\ln(ax)$, where a is a positive constant and x is any positive number. By Equation 7.2 and the Chain Rule, we have

$$\frac{d}{dx}\ln(ax) = \frac{1}{ax} \cdot a = \frac{1}{x}$$

But by Equation 7.2 we know that $1/x$ is also the derivative of $\ln x$. So

$$\frac{d}{dx}\ln(ax) = \frac{d}{dx}\ln x$$

By Corollary 4.9(b) we know that if two functions have the same derivative, they differ by a constant. Thus,

$$\ln(ax) - \ln x = C$$

for some constant C. We can determine C by substituting $x = 1$ and using the fact that $\ln 1 = 0$:

$$\ln a - \ln 1 = C$$

or, $C = \ln a$. So we have, for any $x > 0$ and $a > 0$,

$$\ln(ax) - \ln x = \ln a$$

In this equation we substitute $x = b$, where b is any positive number. After rearranging terms, we get

$$\ln(ab) = \ln a + \ln b \qquad (7.3)$$

In words, we can say *the natural logarithm of the product of two positive real numbers is the sum of their natural logarithms*; or more briefly, we might say "the log of a product is the sum of the logs," with the understanding that "log" here means "natural logarithm." (Later, we will see that the same result holds for other types of logarithms as well.)

We can obtain a formula for the natural logarithm of a quotient by the following trick. Let a and b again denote any two positive numbers. Then we can write a as the product of b and a/b, so

$$\ln a = \ln\left(b \cdot \frac{a}{b}\right) = \ln b + \ln\left(\frac{a}{b}\right) \qquad \text{By Equation 7.3}$$

Solving for $\ln(a/b)$ gives

$$\ln\left(\frac{a}{b}\right) - \ln a - \ln b \qquad (7.4)$$

So "the log of a quotient is the log of the numerator minus the log of the denominator."

A third property, involving the log of a power, can be derived as follows. Let r denote any rational number, and let x be positive. Then, by Equation 7.2 and the Chain Rule,

$$\frac{d}{dx}\ln(x^r) = \frac{1}{x^r} \cdot rx^{r-1} = \frac{r}{x}$$

Since

$$\frac{r}{x} = r \cdot \frac{1}{x} = r\frac{d}{dx}(\ln x) = \frac{d}{dx}(r \ln x) \quad \text{Since } r \text{ is a constant}$$

we see that $\ln x^r$ and $r \ln x$ have the same derivative for all x in $(0, \infty)$. So, again using Corollary 4.9(b), we conclude that these two functions differ only by some constant value C:

$$\ln(x^r) = r \ln x + C$$

Putting $x = 1$ gives $\ln 1 = r \ln 1 + C$, and since $\ln 1 = 0$, we find that $C = 0$. So we have the result

$$\ln(x^r) = r \ln x \qquad\qquad (7.5)$$

In words we can say "the log of x to a power is the exponent times the log of x."

REMARK

■ At this stage the exponent in Equation 7.5 is limited to being a rational number, since in our derivation we used the Power Rule for differentiating x^r, and we have proved the Power Rule only for rational r. In Section 7.3 we will show that this restriction can be removed and that r can be any real number, rational or irrational. (See Theorem 7.5 and the Remark following it.)

Formulas 7.3, 7.4, and 7.5 account for much of the utility of logarithms. We pull them together here for emphasis, referring to them as arithmetic properties.

Arithmetic Properties of the Natural Logarithm Function

For $a > 0$, $b > 0$, and r rational,

$$\ln(ab) = \ln a + \ln b$$

$$\ln\left(\frac{a}{b}\right) = \ln a - \ln b$$

$$\ln a^r = r \ln a$$

EXAMPLE 7.1   Use the arithmetic properties of $\ln x$ to write each of the following in terms of $\ln 2$ or $\ln 3$, or both.

(a) $\ln 6$ (b) $\ln 1.5$ (c) $\ln 8$

Solution

(a) Since $6 = 2 \cdot 3$, by Formula 7.3 we have $\ln 6 = \ln 2 + \ln 3$
(b) Since $1.5 = 3/2$, by Formula 7.4 we have $\ln 1.5 = \ln \frac{3}{2} = \ln 3 - \ln 2$
(c) Since $8 = 2^3$, by Formula 7.5 we have $\ln 8 = \ln 2^3 = 3 \ln 2$ ∎

The Graph of $y = \ln x$

We return now to our investigation of the graph of $y = \ln x$. First, let us show that $\ln x \to \infty$ as $x \to \infty$. Choose any number greater than 1, so that its logarithm is positive. For definiteness, let us choose 2. Our plan is to show that $\ln(2^n)$ can be made arbitrarily large by choosing n sufficiently large. By Equation 7.5 we know that for any positive integer n,

$$\ln(2^n) = n \ln 2$$

If M is any positive number whatsoever, no matter how large, we can make $n \ln 2$ even larger than M by choosing

$$n > \frac{M}{\ln 2}$$

(multiplying by the positive number $\ln 2$ gives $n \ln 2 > M$). For such an n, therefore,

$$\ln(2^n) > M$$

Since $\ln$ is an increasing function, it follows that for all $x > 2^n$, $\ln x > \ln(2^n)$, so

$$\ln x > M$$

What we have shown is that $\ln x$ can be made as large as we please (greater than any number M) provided x is sufficiently large (greater than 2^n, where $n > M/\ln 2$). It follows that

$$\lim_{x \to \infty} \ln x = \infty$$

So while the graph of $y = \ln x$ is concave down, it has no horizontal asymptote. It continues to rise as x increases (but ever so slowly, as we will soon see).

Now we use a similar argument to show that $\ln x \to -\infty$ as $x \to 0^+$ (i.e., as x approaches 0 from the right). For n a positive integer, we have, by Equation 7.5,

$$\ln(2^{-n}) = -n \ln 2$$

since $-n$ is rational. So if M is an arbitrarily large positive number, $-n \ln 2$ will be less than $-M$, provided

$$n > \frac{M}{\ln 2}$$

Remember that dividing an inequality by a negative number reverses the direction of the inequality.

For any such n, therefore,

$$\ln(2^{-n}) < -M$$

Now choose any $x < 2^{-n}$ and positive. Then

$$\ln x < -M$$

since ln is an increasing function on its domain. That is, by choosing x sufficiently close to zero (but positive), we can make its logarithm as large negatively as we wish. In other words,

$$\lim_{x \to 0^+} \ln x = -\infty$$

The graph of $\ln x$ is therefore asymptotic to the negative y-axis as x approaches 0 from the right.

Because $\ln x$ is continuous for all x in its domain (its derivative exists everywhere in the domain, which implies it is continuous for all x in the domain), and since $\ln x$ takes on arbitrarily large positive and negative values, as we have just shown, we can conclude by the Intermediate-Value Theorem (Theorem 2.7) that $\ln x$ takes on every real value. That is, the range of the natural logarithm function is $(-\infty, \infty)$.

We can now indicate the general shape of the graph of $y = \ln x$, as shown in Figure 7.2. Note that we arrived at this graph using the properties we have shown but knowing only one specific value of $\ln x$, namely, $\ln 1 = 0$. We will discuss how to approximate other values shortly. (Of course, the easiest way is to use your calculator.)

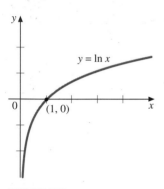

FIGURE 7.2

EXAMPLE 7.2   Use the graph of ln x shown in Figure 7.2, together with the arithmetic properties, to estimate the value of each of the following.

(a) ln 2   (b) ln 4096   (c) $\ln \dfrac{1}{64}$

Solution

(a) From the graph we find that $\ln 2 \approx 0.7$.

(b) Since $4096 = 2^{12}$ (you should verify this result) we have, by Formula 7.5

$$\ln 4096 = \ln 2^{12} = 12 \ln 2$$

Using our estimate for ln 2 from part (a), we see that $\ln 4096 \approx 12(0.7) = 8.4$. The fact that 4096 is a rather large number but its natural logarithm is only about 8.3 illustrates that the graph in Figure 7.2 rises very slowly.

(c) Since $1/64 = 1/2^6 = 2^{-6}$, we have, again by Formula 7.5 and the result of part (a),

$$\ln \frac{1}{64} = \ln 2^{-6} = -6 \ln 2 \approx -6(0.7) = -4.2$$

■

Let us summarize what we know so far about the natural logarithm function.

The Natural Logarithm Function

1. $\ln x = \displaystyle\int_1^x \frac{1}{t}\,dt,\ x > 0$
2. domain $= (0, \infty)$, range $= (-\infty, \infty)$
3. $\dfrac{d}{dx}\ln x = \dfrac{1}{x}$
4. $\ln x$ is an increasing function on its domain and its graph is concave down.
5. $\ln x > 0$ if $x > 1$
    $\ln x < 0$ if $0 < x < 1$
    $\ln 1 = 0$
6. $\displaystyle\lim_{x\to\infty} \ln x = \infty$ and $\displaystyle\lim_{x\to 0^+} \ln x = -\infty$
7. For $a > 0$, $b > 0$, and r rational,

$$\ln(ab) = \ln a + \ln b$$

$$\ln \frac{a}{b} = \ln a - \ln b$$

$$\ln a^r = r \ln a$$

EXAMPLE 7.3   Make use of the facts summarized above to do the following.

(a) Estimate the value of $\int_1^2 \frac{du}{u}$.
(b) Arrange in increasing order of magnitude:
     $\ln 22/7,\ \ln 2/3,\ \ln \pi,\ \ln 10^4,\ \ln 1 + \ln 0.02,\ \ln 2^{13},\ \ln 4 - \ln 5$
(c) Show that $y = \ln x$ is a solution of the differential equation

$$x^2 y'' + 2xy' = 1$$

for all $x > 0$.

Solution

(a) By the first item in the summary, we see that

$$\int_1^2 \frac{du}{u} = \ln 2$$

(The fact that the variable of integration here is u instead of t is of no consequence, since it is a "dummy" variable.) We saw in Example 7.2 that $\ln 2 \approx 0.7$. So

$$\int_1^2 \frac{du}{u} \approx 0.7$$

(b) First note that $\ln 1 + \ln 0.02 = \ln (0.02)$, since $\ln 1 = 0$. Also, $\ln 4 - \ln 5 = \ln (4/5)$ by the second arithmetic property in item 7. Since the natural logarithm function is monotone increasing, it suffices to arrange the numbers themselves in increasing order. That is, if we can show $x_1 < x_2$, it will follow that $\ln x_1 < \ln x_2$.

Clearly, $0.02 < 2/3 < 4/5$. In decimal form $22/7 = 3.1428\ldots$ and $\pi = 3.14159\ldots$, so $\pi < 22/7$. (It is interesting to note that $22/7$ is sometimes used as a rational approximation to π; in fact, a state legislature once passed a law declaring that π was *equal* to $22/7$!) We saw in Example 7.2 that $2^{12} = 4096$, so $2^{13} = 8192$. Since $10^4 = 10,000$, then, we see that $2^{13} < 10^4$.

We can now arrange the given logarithms in increasing order, as follows:

$$\ln 1 + \ln 0.02 < \ln \frac{2}{3} < \ln 4 - \ln 5 < \ln \pi < \ln \frac{22}{7} < \ln 2^{13} < \ln 10^4$$

(c) If $y = \ln x$, then we know by item 3 that $y' = 1/x$. Thus $y'' = -1/x^2$. So we have for $x > 0$

$$x^2 y'' + 2xy' = x^2 \left(-\frac{1}{x^2}\right) + 2x \left(\frac{1}{x}\right)$$
$$= -1 + 2 = 1$$

Thus, $y = \ln x$ is a solution to the given differential equation.   ∎

The Slow Growth Rate of $\ln x$

Although we have shown that $\ln x$ becomes arbitrarily large as $x \to \infty$, it is interesting to note how slowly the graph rises. For $x = 1$ billion, $\ln x$ is just somewhat more than 20. How large do you suppose x would need to be for $\ln x$ to equal 100? The answer is about $(2.7)^{100}$. As a way of getting some idea of the size of this number, we know by Definition 7.1 that for $\ln x$ to equal 100, we must have

$$\int_1^x \frac{1}{t} dt = 100$$

If we were to approximate the integral on the left by using a left-hand sum with $\Delta x = 1$, we would need to find n such that (see Figure 7.3)

$$\sum_{k=1}^n \frac{1}{k} = 100$$

(See Exercise 57 in Exercise Set 7.1.) That is, we would need to form the sum

$$1 + \frac{1}{2} + \frac{1}{3} + \frac{1}{4} + \cdots$$

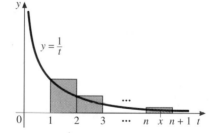

FIGURE 7.3
$\ln x = \int_1^x \frac{1}{t} dt \approx \sum_{k=1}^n \frac{1}{k}$, where $n \leq x < n+1$

and keep going until this sum first exceeds 100. How many terms are required? Well, as we said, it is about $(2.7)^{100}$. But if the fastest computer available today had started adding at the time of the Big Bang, about 20 billion years ago, it still would not have added enough terms!

 Do not confuse the graph of $f(t) = 1/t$, which we used to define $\ln x$, with the graph of $y = \ln x$ itself. The graph of $f(t) = 1/t$ was a means to an end.

The fact that $\ln x$ is defined in terms of the area under $f(t) = 1/t$ does enable us to approximate $\ln x$ for particular values of x, using an approximation technique such as Simpson's Rule. For example, it will be useful to obtain approximate values of ln 2 and ln 3. By Equation 7.1,

$$\ln 2 = \int_1^2 \frac{1}{t} dt$$

Let us use Simpson's Rule with $n = 10$ to approximate this value. We have $\Delta x = (2-1)/10 = 0.1$, so

$$\ln 2 \approx \frac{0.1}{3}\left[1 + 4\left(\frac{1}{1.1}\right) + 2\left(\frac{1}{1.2}\right) + 4\left(\frac{1}{1.3}\right) + 2\left(\frac{1}{1.4}\right) + \cdots \right.$$
$$\left. + 2\left(\frac{1}{1.8}\right) + 4\left(\frac{1}{1.9}\right) + \frac{1}{2}\right]$$
$$\approx 0.69315$$

Similarly, using $n = 20$, we find that $\ln 3 \approx 1.0986$. Using the error bounds we gave in Chapter 5, you can verify that these answers are correct to the number of decimal places given (see Exercise 72 in Exercise Set 7.1). In the next section we will have occasion to use the fact that $\ln 2 < 1$ and $\ln 3 > 1$.

Derivatives of Natural Logarithm Functions

If g is any positive differentiable function of x, then by Equation 7.2 and the Chain Rule,

$$\frac{d}{dx}\ln(g(x)) = \frac{1}{g(x)}g'(x)$$

Letting $u = g(x)$, we can write this result using Leibniz notation as

$$\frac{d}{dx}\ln u = \frac{1}{u}\frac{du}{dx} \qquad (u > 0) \tag{7.6}$$

We illustrate the use of this formula in the next example.

EXAMPLE 7.4   State the domain of each of the following, and find $f'(x)$.

(a) $f(x) = \ln(x\sqrt{1+x^2})$ (b) $f(x) = \dfrac{\ln x}{x}$

(c) $f(x) = (\ln\sqrt{2x-1})^3$ (d) $f(x) = \ln(\cos x)$

Solution

(a) The domain consists of all x for which $x\sqrt{1+x^2} > 0$, which is the set of all $x > 0$. To find the derivative, we first simplify, making use of Equations 7.3 and 7.5.

$$\ln(x\sqrt{1+x^2}) = \ln x + \ln(1+x^2)^{1/2} \qquad \text{By Equation 7.3}$$
$$= \ln x + \frac{1}{2}\ln(1+x^2) \qquad \text{By Equation 7.5}$$

Now we can differentiate each term, using Equation 7.6 on the second term, to get

$$f'(x) = \frac{1}{x} + \frac{1}{2}\frac{1}{1+x^2}\cdot 2x = \frac{1}{x} + \frac{x}{1+x^2}$$

Using the arithmetic properties of the natural logarithm function before taking the derivative greatly simplified our work. To see why, suppose that we had differentiated without applying these properties. Then, using Equation 7.6, we would have obtained

$$f'(x) = \frac{1}{x\sqrt{1+x^2}} \frac{d}{dx}\left(x\sqrt{1+x^2}\right)$$

$$= \frac{1}{x\sqrt{1+x^2}}\left[\sqrt{1+x^2} + x \cdot \frac{1}{2}\left(1+x^2\right)^{-1/2}(2x)\right]$$

$$= \frac{1}{x\sqrt{1+x^2}}\left[\sqrt{1+x^2} + \frac{x^2}{\sqrt{1+x^2}}\right]$$

$$= \frac{x}{1+x^2} + \frac{1}{x}$$

The result is the same as before, but getting there is harder. As a general rule, when taking the derivative of $\ln(g(x))$, where $g(x)$ involves products, quotients, or powers, make use of the arithmetic properties of the ln function *before* taking the derivative.

(b) The domain of the numerator is the set of all $x > 0$, so the requirement on the denominator that $x \neq 0$ is automatically satisfied. Thus, the domain is the set of all $x > 0$.

In this case, we can make use of the Quotient Rule to find the derivative.

$$f'(x) = \frac{\frac{1}{x} \cdot x - \ln x}{x^2} = \frac{1 - \ln x}{x^2}$$

(c) For $\ln\sqrt{2x-1}$ to be defined, we must have $2x - 1 > 0$, or equivalently, $x > \frac{1}{2}$. The domain is therefore the set of all $x > \frac{1}{2}$.

By Equation 7.5 we can write $f(x)$ as $[\frac{1}{2}\ln(2x-1)]^3$, or equivalently, $\frac{1}{8}[\ln(2x-1)]^3$. Now we use the Power Rule, together with Equation 7.6.

$$f'(x) = \frac{1}{8} \cdot 3[\ln(2x-1)]^2 \cdot \frac{1}{2x-1} \cdot 2 = \frac{3[\ln(2x-1)]^2}{4(2x-1)}$$

(d) The domain consists of all x values for which $\cos x > 0$ (the shaded regions in Figure 7.4), namely,

$$\left\{x : -\frac{\pi}{2} + 2n\pi < x < \frac{\pi}{2} + 2n\pi, \ n = 0, \pm 1, \pm 2, \ldots\right\}$$

Then, by Equation 7.6, we have

$$f'(x) = \frac{1}{\cos x}(-\sin x) = -\tan x \qquad \blacksquare$$

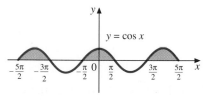

FIGURE 7.4

$\cos x$ is positive for
$-\frac{\pi}{2} + 2n\pi < x < \frac{\pi}{2} + 2n\pi$

We can generalize Equation 7.6 to include cases where $g(x) < 0$ by making use of absolute values. When $g(x) < 0$, $|g(x)| = -g(x)$, so

$$\frac{d}{dx}\ln|g(x)| = \frac{d}{dx}\ln[-g(x)] = \frac{1}{-g(x)} \cdot [-g'(x)] = \frac{g'(x)}{g(x)}$$

In terms of u, where $u = g(x)$, this result is

$$\frac{d}{dx} \ln |u| = \frac{1}{u} \frac{du}{dx} \qquad (u \neq 0) \qquad\qquad (7.7)$$

When $u > 0$, Equation 7.7 is equivalent to Equation 7.6. Thus, Equation 7.7 is true if $u > 0$ or if $u < 0$. The next example shows how we can use this result.

EXAMPLE 7.5   Find $f'(x)$ if

$$f(x) = \ln \left[\frac{x(x-1)}{x+2} \right]$$

FIGURE 7.5

Sign graph for $\dfrac{x(x-1)}{x+2}$

Solution   First, we find the domain of f. A sign graph (Figure 7.5) shows that the quotient

$$\frac{x(x-1)}{x+2}$$

is positive if x is in $(-2, 0) \cup (1, \infty)$. This set is therefore the domain of f. For x in this domain, we can write $f(x)$ as

$$f(x) = \ln \left[\frac{x(x-1)}{x+2} \right] = \ln \left| \frac{x(x-1)}{x+2} \right| \qquad \text{Since for } N > 0, \\ N = |N|$$

$$= \ln \frac{|x||x-1|}{|x+2|} \qquad \text{Using properties of absolute value}$$

$$= \ln |x| + \ln |x-1| - \ln |x+2| \qquad \text{By Equations 7.3 and 7.4}$$

Thus, by Equation 7.7,

$$f'(x) = \frac{1}{x} + \frac{1}{x-1} - \frac{1}{x+2} \qquad\qquad \blacksquare$$

REMARK ————————————————————————

■ In the preceding example, if we had ignored questions of the domain and forgotten about absolute values, we could have written, formally,

$$f(x) = \ln \left[\frac{x(x-1)}{x+2} \right] = \ln x + \ln(x-1) - \ln(x+2)$$

and on differentiation we would have obtained

$$f'(x) = \frac{1}{x} + \frac{1}{x-1} - \frac{1}{x+2}$$

which is exactly the answer we got before. So it is tempting to do the problem in this way. Unfortunately, it is implicit in this shortened procedure that each of the factors x, $x-1$, and $x+2$ is positive (otherwise their individual logarithms would not be defined). So the result would appear to be valid only on the interval $(1, \infty)$, whereas, as we showed in the example, it is actually valid on $(-2, 0)$ as well.

Before attempting to differentiate a function consisting of the natural logarithm of products, quotients, powers, or roots, *always* simplify it using the arithmetic properties 7.3, 7.4, and 7.5. Failure to do so is almost a guarantee that the problem will be more difficult algebraically.

Logarithmic Differentiation

In the next example we show how we can sometimes simplify the calculation of derivatives of complicated expressions involving products, quotients, and powers by first finding the logarithm of the expression and then differentiating. This technique is called **logarithmic differentiation**.

EXAMPLE 7.6   Find y' if

$$y = \frac{(x-1)^3}{(2x+3)\sqrt{1-2x}}$$

Solution   Of course, we could differentiate in a straightforward manner, using the Quotient Rule and Product Rule, along with the Chain Rule, but it would be messy. (Try it!) Instead, we will use logarithmic differentiation.

First, let us find the domain of the function. For $\sqrt{1-2x}$ to be a real number, we must have $1 - 2x \geq 0$, or equivalently, $x \leq \frac{1}{2}$. But we cannot allow $x = \frac{1}{2}$, since it causes the denominator to be 0. Similarly, we rule out $x = -\frac{3}{2}$, since we cannot allow $2x + 3$ to be 0. Thus, the domain is the set $D = \{x : x < \frac{1}{2}$ and $x \neq -\frac{3}{2}\}$.

To take the logarithm, we must have a positive function, since the logarithm is defined only for positive values. We can make the function positive by taking its absolute value:

$$|y| = \frac{|x-1|^3}{|2x+3||1-2x|^{1/2}} \qquad (x \in D)$$

Note that if $x = 1$ were in the domain D, we would have to restrict x to be different from 1 before taking the logarithm in order not to have ln of 0. However, $x = 1$ is not in the domain. So for all x in D, $|y| > 0$. Thus, we can take the logarithm and get

$$\ln|y| = \ln \frac{|x-1|^3}{|2x+3||1-2x|^{1/2}} = 3\ln|x-1| - \left[\ln|2x+3| + \frac{1}{2}\ln|1-2x|\right]$$

where we have made use of the arithmetic properties of the ln function. Now we differentiate both sides with respect to x, using Equation 7.7, bearing in mind that on the left-hand side y is a function of x:

$$\frac{1}{y}y' = \frac{3}{x-1} - \left[\frac{1}{2x+3}\cdot 2 + \frac{1}{2}\frac{1}{1-2x}(-2)\right]$$

$$= \frac{3}{x-1} - \frac{2}{2x+3} + \frac{1}{1-2x}$$

Finally, we solve for y' by multiplying both sides by y, and substitute its value in terms of x on the right:

$$y' = \frac{(x-1)^3}{(2x+3)\sqrt{1-2x}}\left[\frac{3}{x-1} - \frac{2}{2x+3} + \frac{1}{1-2x}\right]$$

Just as in the preceding example, we could have arrived at the same answer even if we had ignored absolute values. Had we done so, however, we would have been tacitly assuming $y > 0$, whereas our derivation is valid also if $y < 0$. ∎

Note the difference between Examples 7.5 and 7.6. In the former we were *given* the function in the form of a logarithm. No initiative on our part was required. We simply took the derivative. In the second, there was no logarithm. We began with the initial steps of *taking* the absolute value and then the logarithm of both sides, and then we differentiated.

An Integral Whose Value Is a Natural Logarithm

Each time we find a new derivative formula, we also have a new antiderivative formula. In particular, from Formula 7.7, we obtain

$$\int \frac{1}{u}\,du = \ln|u| + C \qquad (7.8)$$

This formula fills a gap in the Power Rule for antiderivatives. That is, the Power Rule, given by

$$\int u^n\,du = \frac{u^{n+1}}{n+1} + C$$

is valid *except* when $n = -1$. Formula 7.8 tells what the result is when $n = -1$. This new integration formula enables us to integrate many functions that we have been unable to integrate until now.

EXAMPLE 7.7   Evaluate the following indefinite integrals:

(a) $\int \frac{x}{x^2+1}\,dx$ (b) $\int \frac{\cos x}{1+2\sin x}\,dx$ (c) $\int \tan 3x\,dx$

Solution   In each case our strategy will be to make a substitution that brings the given integral around to the form of Equation 7.8.

(a) Let $u = x^2 + 1$. Then $du = 2x\,dx$. So we have

$$\int \frac{x}{x^2+1}\,dx = \frac{1}{2}\int \frac{2x\,dx}{x^2+1} = \frac{1}{2}\int \frac{du}{u} = \frac{1}{2}\ln|u| + C$$

$$= \frac{1}{2}\ln|x^2+1| + C$$

$$= \frac{1}{2}\ln(x^2+1) + C$$

In the last step we dropped the absolute values, since $x^2 + 1$ is always positive. We could also write the answer, using Equation 7.5, as $\ln\sqrt{x^2+1}+C$.

(b) Here we will do the substitution mentally and write

$$\int \frac{\cos x}{1 + 2\sin x}dx = \frac{1}{2}\int \frac{\overbrace{2\cos x\, dx}^{du}}{\underbrace{1 + 2\sin x}_{u}} = \frac{1}{2}\ln|1 + 2\sin x| + C$$

(c) $\displaystyle\int \tan 3x\, dx = \int \frac{\sin 3x}{\cos 3x}dx = -\frac{1}{3}\int \frac{(-3\sin 3x)dx}{\cos 3x}$

$$= -\tfrac{1}{3}\ln|\cos 3x| + C$$

(What mental substitution did we use?)   ∎

EXAMPLE 7.8   Find a formula for the general antiderivative of $\sec x$.

Solution   The solution involves a trick. We multiply the numerator and denominator by $\sec x + \tan x$:

$$\int \sec x\, dx = \int \frac{\sec x}{1}\cdot\frac{\sec x + \tan x}{\sec x + \tan x}dx = \int \frac{\sec^2 x + \sec x\tan x}{\sec x + \tan x}dx$$

Now recall from Section 3.6 that the derivative of $\tan x$ is $\sec^2 x$ and the derivative of $\sec x$ is $\sec x\tan x$. So if we let $u = \sec x + \tan x$, we have $du = (\sec x\tan x + \sec^2 x)dx$. Thus,

$$\int \sec x = \int \frac{du}{u} = \ln|u| = \ln|\sec x + \tan x| + C$$ ∎

Because we will have occasions later to refer to the integration formula from Example 7.8, we set it off here for emphasis.

$$\int \sec x\, dx = \ln|\sec x + \tan x| + C \qquad\qquad (7.9)$$

Graphs Involving Logarithms

We conclude this section with an example showing how to analyze and draw the graph of a function involving a natural logarithm.

EXAMPLE 7.9   Discuss and sketch the graph of $y = \ln|x^2 - 1|$.

Solution   Since $|x^2 - 1| > 0$ except when $x^2 = 1$, the domain consists of all real numbers except $x = \pm 1$.

(a) *Intercepts.* Setting $x = 0$, we get $y = \ln|-1| = \ln 1 = 0$. So the y-intercept is 0. Setting $y = 0$, we get $\ln|x^2 - 1| = 0$, which is true only when $|x^2 - 1| = 1$. (Why?) So $x^2 - 1 = \pm 1$. If $x^2 - 1 = 1$, then $x^2 = 2$, and $x = \pm\sqrt{2}$. If $x^2 - 1 = -1$, then $x^2 = 0$, and so $x = 0$ also. Thus, the x-intercepts are 0, $\sqrt{2}$, and $-\sqrt{2}$.

(b) *Symmetry.* Since no change results when x is replaced by $-x$, the graph is symmetric with respect to the y-axis.

(c) *Maxima and Minima.* By Equation 7.6,

$$y' = \frac{1}{x^2 - 1} \cdot 2x = \frac{2x}{x^2 - 1}$$

The only critical value is $x = 0$ (since $x = 1$ and $x = -1$ are outside the domain). By the sign graph for y' in Figure 7.6 we see that $(0, 0)$ is a local maximum point. We also see that the graph of our function is decreasing on $(-\infty, -1) \cup (0, 1)$ and is increasing on $(-1, 0) \cup (1, \infty)$.

(d) *Concavity and Inflection Points.* For the second derivative, we obtain

$$y'' = \frac{2 \cdot (x^2 - 1) - 2x(2x)}{(x^2 - 1)^2} = \frac{-2(x^2 + 1)}{(x^2 - 1)^2}$$

For all $x \neq \pm 1$, $y'' < 0$, so the graph is concave down at all points of the domain. There are no inflection points.

(e) *Asymptotes.* As $x \to 1$, $|x^2 - 1| \to 0^+$, so $\ln|x^2 - 1| \to -\infty$. Thus, $x = 1$ is a vertical asymptote. By symmetry, so is $x = -1$. As $|x| \to \infty$, $|x^2 - 1| \to \infty$, so $\ln|x^2 - 1| \to \infty$. Thus, there are no horizontal asymptotes. Combining all our information, we are able to draw the graph shown in Figure 7.7. ■

FIGURE 7.6

Sign graph for $y' = \dfrac{2x}{x^2 - 1}$

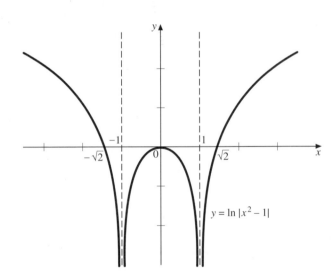

FIGURE 7.7

Exercise Set 7.1

In Exercises 1–20, give the domain of f and find $f'(x)$.

1. $f(x) = \ln\dfrac{1}{x}$

2. $f(x) = \ln\sqrt{x}$

3. $f(x) = \ln x^3$

4. $f(x) = (\ln x)^3$

5. $f(x) = \ln(x\sqrt{1 - x})$

6. $f(x) = \ln\left(\dfrac{x^2}{x - 1}\right)$

7. $f(x) = x \ln x$

8. $f(x) = \ln x^2$

9. $f(x) = \ln \sqrt{\dfrac{1-x}{1+x}}$

10. $f(x) = \sqrt{\ln x}$

11. $f(x) = \ln(\ln x)$

12. $f(x) = \ln \sin x$

13. $f(x) = \ln \sec x$

14. $f(x) = \sin(\ln x)$

15. $f(x) = \dfrac{1 - \ln x}{x}$

16. $f(x) = \ln \left(\dfrac{x-2}{x+1} \right)$

17. $f(x) = \ln \dfrac{\sqrt{x^2 - 1}}{2 - x}$

18. $f(x) = \ln \sqrt[3]{2x - x^2}$

19. $f(x) = \ln \left(\dfrac{x+3}{x^2 - 2x} \right)$

20. $f(x) = \ln \left(\dfrac{x\sqrt{1-x}}{1+x} \right)$

In Exercises 21–26, use L'Hôpital's Rule to find the limits.

21. $\lim\limits_{x \to 1} \dfrac{\ln x}{x - 1}$

22. $\lim\limits_{x \to \infty} \dfrac{\ln x}{x}$

23. $\lim\limits_{x \to 0^+} \dfrac{x^{-1/2}}{\ln x}$

24. $\lim\limits_{x \to 0^+} \dfrac{\cot x}{\ln x}$

25. $\lim\limits_{x \to \infty} \dfrac{1 + (\ln x)^3}{x \ln x}$

26. $\lim\limits_{x \to -1^-} \dfrac{x \ln |x|}{(x+1)^2}$

27. (a) Show that $\ln x < \sqrt{x}$ for $x > 0$.
(b) Show that $\ln x \leq x - 1$ for $x > 0$.

In Exercises 28–37, find the antiderivatives.

28. $\displaystyle \int \dfrac{2x \, dx}{x^2 + 4}$

29. $\displaystyle \int \dfrac{\sin x \, dx}{1 - \cos x}$

30. $\displaystyle \int \dfrac{x \, dx}{2 - 3x^2}$

31. $\displaystyle \int \dfrac{x^2 dx}{x^3 + 1}$

32. $\displaystyle \int \dfrac{(x-1) dx}{x^2 - 2x + 3}$

33. $\displaystyle \int \cot x \, dx$

34. $\displaystyle \int \dfrac{dx}{\sqrt{x}(\sqrt{x} + 1)}$

35. $\displaystyle \int \dfrac{dx}{x \ln x}$

36. $\displaystyle \int \dfrac{x \cos x}{x \sin x + \cos x} dx$

37. $\displaystyle \int \dfrac{\sqrt{x} - 1}{x(2\sqrt{x} - 3)} dx$

In Exercises 38–41, use logarithmic differentiation to find y'.

38. $y = \dfrac{(x-1)^2 \sqrt{3x + 4}}{x + 2}$

39. $y = \dfrac{x\sqrt{1-x}}{(x+2)(x-3)}$

40. $y = \sqrt{\dfrac{(x-1)(x+2)}{(x-3)^3}}$

41. $y = \dfrac{\sqrt[3]{(x+1)^2}}{x^4(3 - 4x)^2}$

In Exercises 42–47, find all local maxima and minima.

42. $y = x - \ln x$

43. $y = \ln \sqrt{x} - x^2$

44. $y = x^3 - \ln x^3$

45. $y = x + \ln \dfrac{x}{2x - 1}$

46. $y = \ln \dfrac{\sqrt{1-x}}{3 - 2x}$

47. $y = \ln \dfrac{x(2 - x)}{(x - 3)^2}$

In Exercises 48–55, discuss domain, intercepts, symmetry, asymptotes, maxima and minima, concavity, and points of inflection and sketch the graph.

48. $y = \ln 2x$

49. $y = \ln x^2$

50. $y = \ln 1/x$

51. $y = \ln(1 - x^2)$

52. $y = \ln \sqrt{x^2 - 1}$

53. $y = x - \ln x$

54. $y = \ln |x|$

55. $y = \ln(1 + x^2)$

56. Using the approximations $\ln 2 \approx 0.70$ and $\ln 3 \approx 1.1$, together with the arithmetic properties 7.3, 7.4, and 7.5, find approximations for the following:
(a) $\ln 6$ (b) $\ln 8$ (c) $\ln 81$ (d) $\ln 1.5$
(e) $\ln 36$

57. If $x > 0$ and n is an integer such that $n \leq x < n + 1$, show that

$$\int_1^x \frac{1}{t} dt \leq \sum_{k=1}^n \frac{1}{k}$$

Hence, conclude that for $\ln x$ to equal 100, x must be greater than the smallest integer n for which

$$\sum_{k=1}^n \frac{1}{k} \geq 100$$

(*Hint:* Use a left-hand Riemann sum on the interval $[1, n+1]$, taking $\Delta t = 1$.)

In Exercises 58–61, find $f'(x)$ and $f''(x)$.

58. $f(x) = \dfrac{\ln(x+1)}{x+1}$

59. $f(x) = x(\ln x)^2$

60. $f(x) = \dfrac{x}{\ln x}$

61. $f(x) = \sqrt{\ln x^2}$

In Exercises 62 and 63, use implicit differentiation to find y'.

62. $x \ln y - y \ln x = 1$ **63.** $\ln x^2 y = (\ln y)/x$

In Exercises 64–67, find the antiderivatives.

64. $\displaystyle \int \frac{\tan 2x}{\ln \cos 2x} dx$ **65.** $\displaystyle \int \frac{\sin 2x}{1 + \cos^2 x} dx$

66. $\displaystyle \int \frac{dx}{x + x^{1/3}}$ (*Hint:* Factor the denominator.)

67. $\displaystyle \int \left(\frac{x-2}{x-1} \right)^2 dx$ (*Hint:* First divide.)

68. Derive the formula

$$\int \csc x \, dx = \ln |\csc x - \cot x| + C$$

(*Hint:* Use a trick similar to the one used in Example 7.8 to get the answer $-\ln |\csc x + \cot x|$. Then show that this answer can be put in the form given.)

In Exercises 69–71, discuss domain, intercepts, symmetry, asymptotes, maxima and minima, concavity, and points of inflection and sketch the graph.

69. $y = \ln \left| \dfrac{x-1}{x+1} \right|$ **70.** $y = \ln \dfrac{x^2}{|x^2 - 2|}$

71. $y = \ln(x^2 - 3x) - \ln(x^2 + 3)$

 72. Verify the result stated in the text that $\ln 3 \approx 1.0986$, using Simpson's Rule with $n = 20$. Prove that this result is correct to the four decimal places given. Similarly, using Simpson's Rule with $n = 10$, verify that the result $\ln 2 \approx 0.69315$ obtained in the text is correct to five decimal places.

73. Show that for $n \geq 2$,

$$\frac{1}{2} + \frac{1}{3} + \cdots + \frac{1}{n} < \ln n < 1 + \frac{1}{2} + \frac{1}{3} + \cdots + \frac{1}{n-1}$$

 74. Let $f(x) = \ln x$ and $g(x) = \sqrt[n]{x}$, where n is a positive integer. Both $f(x)$ and $g(x)$ become infinite as $x \to \infty$. Use a graphing calculator or a CAS to graph f and g for several values of n. Conjecture which function grows more rapidly as $x \to \infty$. Prove your conjecture.

 75. Consider the family of curves $y = a \ln(bx)$, where a is any real number and $b > 0$. Use a CAS or a graphing calculator to graph this function for several different values of a and b. How does the parameter b affect the growth of the function? How does the parameter a affect the shape of the graph?

 76. Use a CAS to calculate each of the following to 12 decimal places.
(a) $\ln 2$ (b) $\ln 3$ (c) $\ln 100$ (d) $\ln 10^9$

7.2 THE NATURAL EXPONENTIAL FUNCTION

FIGURE 7.8
$f(x) = \ln x$ passes the Horizontal Line Test and thus is one-to-one.

We have seen that $f(x) = \ln x$ has domain $(0, \infty)$ and range $(-\infty, \infty)$. Furthermore, f is increasing for all x in its domain. Its graph (Figure 7.8) clearly shows it passes the horizontal line test. Hence, f is one-to-one and therefore has an inverse, which we designate by $f^{-1}(x) = \exp(x)$ and call the **exponential function** (with "natural" understood). The reason for the name "exponential" will soon become clear. Recall from Chapter 1 that the graph of the inverse of a function can be obtained by reflecting the graph of the function about the line $y = x$. So from the graph of $f(x) = \ln x$, we immediately obtain the graph of $f^{-1}(x) = \exp(x)$, shown in Figure 7.9. We know also that the domain and range of f^{-1} are the range and domain, respectively, of f. So the domain of exp is $(-\infty, \infty)$ and the range is $(0, \infty)$.

The Number e

Before analyzing the exponential function further, we single out for special emphasis the number whose natural logarithm is 1. We know that such a number exists, since $\ln x$ assumes every real value. This special number is designated by the letter e (after Leonhard Euler, the greatest mathematician of

FIGURE 7.9

FIGURE 7.10

FIGURE 7.11

the eighteenth century and one of the most prolific mathematicians of all time). Thus, e is defined by the equation

$$\ln e = 1$$

We show e graphically in Figure 7.10. Since, as we have seen, $\ln 2 < 1$ and $\ln 3 > 1$, it follows from the fact that ln is an increasing function that $2 < e < 3$. It can be shown that e is irrational and that its approximate value is

$$e \approx 2.71828$$

With a computer algebra system, you can easily obtain e to a large number of decimal places. For example, in Maple, to get e to 100 places, simply type

    evalf(E,100);

and you get the result

2.71828182845904523536028747135266249775724709369995957496696\
7627724076630353547594571382178525166427

We will see a way of approximating e to any desired degree of accuracy in Chapter 10.

Properties of e^x

We know from arithmetic property 7.5 that if r is any rational number

$$\ln(e^r) = r \ln e$$

and since $\ln e = 1$ (by definition of e),

$$\ln(e^r) = r$$

Let us apply the exponential function to each side of this equation:

$$\exp[\ln(e^r)] = \exp(r)$$

Since exp and ln are inverses, the left-hand side is just e^r. (Remember that $f^{-1}(f(x)) = x$.) So we have

$$e^r = \exp(r)$$

for all rational numbers r. Up to now we have not assigned any meaning to an irrational power of e (or of any other number). Since $\exp(x)$ is defined for *all* real numbers x (rational or irrational), we define e^x so as to be consistent with its meaning when x is rational. That is, if x is *any* real number, we define e^x to be $\exp(x)$.

$$e^x = \exp(x) \qquad (7.10)$$

The graph of $y = e^x$ is therefore identical to the graph of $y = \exp(x)$. We show this important graph again in Figure 7.11. In the future when we speak of the exponential function, we can write it as either $f(x) = e^x$ or $f(x) = \exp(x)$. They are the same. Just as $\ln x$ grows extremely slowly as x increases, e^x

grows extremely rapidly. The term "exponential growth," as applied, say, to world population growth, comes from this fact.

If we apply ln to both sides of Equation 7.10, and again use the fact that ln and exp are inverse functions, we get, for all x,

$$\ln(e^x) = x \qquad (7.11)$$

It is useful to learn the meaning of Equation 7.11 in words:

e^x is that number whose natural logarithm is x.

In Equation 7.10 if we replace x by $\ln x$ and again use the inverse relationship between exp and ln, so that $\exp(\ln(x)) = x$ for $x > 0$, we get

$$e^{\ln x} = x \qquad (7.12)$$

In particular, since $\ln 1 = 0$, we have $e^0 = 1$.

Equations 7.11 and 7.12 are of sufficient importance to warrant special attention. First, let us reiterate the fact that since e^x is the inverse of $\ln x$, its domain and range are the range and domain, respectively, of $\ln x$. Thus, since the range of $\ln x$ is the set of all real numbers, $(-\infty, \infty)$, e^x has domain equal to this set. Also, the domain of $\ln x$ is the set of all positive real numbers, $(0, \infty)$, so e^x has this set as its range. Thus, $e^x > 0$ for all x.

Inverse Properties of $\ln x$ and e^x

$$\ln(e^x) = x \quad \text{for all real } x$$
$$e^{\ln x} = x \quad \text{for all } x > 0$$

We may also express these relationships in the following alternative form.

$$y = e^x \text{ if and only if } x = \ln y$$

EXAMPLE 7.10   Evaluate each of the following.

(a) $\ln e^2$ (b) $\ln\left(\dfrac{1}{\sqrt{e}}\right)$ (c) $e^{\ln 3}$ (d) $e^{2\ln 3}$

Solution

(a) By Equation 7.11, $\ln e^2 = 2$.

(b) By Equation 7.11, $\ln\left(\dfrac{1}{\sqrt{e}}\right) = \ln(e^{-1/2}) = -\dfrac{1}{2}$.

(c) By Equation 7.12, $e^{\ln 3} = 3$.

(d) By Equations 7.5 and 7.12, $e^{2\ln 3} = e^{\ln(3^2)} = 3^2 = 9$.   ∎

EXAMPLE 7.11   Solve for x:

(a) $e^x = 3$ (b) $e^{(1-2x)} = 7$ (c) $\ln x = 5$ (d) $\ln(x-1) = -2$

Solution

(a) We take the natural logarithm of both sides, making use of Equation 7.11.

$$\ln(e^x) = \ln 3$$

$$x = \ln 3$$

(b) Proceeding as in part (a) we get

$$\ln(e^{1-2x}) = \ln 7$$

$$1 - 2x = \ln 7$$

$$2x = 1 - \ln 7$$

$$x = \frac{1 - \ln 7}{2}$$

(c) Here we "exponentiate." That is, we use the fact that when $a = b$, then $e^a = e^b$. So we have

$$e^{\ln x} = e^5$$

But by Equation 7.12, $e^{\ln x} = x$, so

$$x = e^5$$

(d) Again, we exponentiate.

$$e^{\ln(x-1)} = e^{-2}$$

$$x - 1 = \frac{1}{e^2}$$

$$x = 1 + \frac{1}{e^2} \qquad \blacksquare$$

Laws of Exponents

There is nothing in the definition of e^x (Equation 7.10) to suggest that x really behaves like an exponent, as we understand exponents. The next theorem, however, confirms that it does, and that we are therefore justified in thinking of e^x as "e raised to the power x." Also, since $\exp(x)$ and e^x are equal, the reason for the name "exponential function" will now be clear.

THEOREM 7.1

If x and y are any real numbers and r is any rational number,

1. $e^{x+y} = e^x \cdot e^y$
2. $e^{x-y} = \dfrac{e^x}{e^y}$
3. $(e^x)^r = e^{rx}$

Proof

1. Using Equation 7.11, we have

$$\ln e^{x+y} = x + y$$

Also, by Equation 7.3,

$$\ln(e^x \cdot e^y) = \ln e^x + \ln e^y = x + y$$

where the last equality again follows by Equation 7.11. Since the natural logarithm function is one-to-one, the equality of $\ln e^{x+y}$ and $\ln(e^x \cdot e^y)$ guarantees the equality of e^{x+y} and $e^x \cdot e^y$.

2. This formula is proved in a similar way, and we leave it as an exercise (Exercise 58 in Exercise Set 7.2).

3. Using the fact that $\ln a^r = r \ln a$, we have

$$\ln(e^x)^r = r \ln e^x = rx$$

Also,

$$\ln e^{rx} = rx$$

Again, the one-to-one nature of $\ln$ gives the desired result.   ∎

REMARK ───────────────────────────────────

■ At this stage, in part 3 of the preceding theorem we must limit r to be rational since we have so far proved $\ln a^r = r \ln a$ only for rational values of r. We will remove this restriction in the next section.

───────────────────────────────────

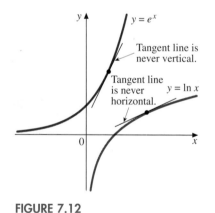

FIGURE 7.12

Derivatives and Integrals of Exponential Functions

The fact that $f(x) = \ln x$ has a nonzero derivative at each point of its domain suggests that the inverse function $f^{-1}(x) = \exp(x)$ (or equivalently, $f^{-1}(x) = e^x$) also has a derivative at each point in its domain. We can see this conclusion geometrically by reasoning that the graph of $y = \ln x$ has a nonhorizontal tangent line at all points on the graph, so the graph of $y = e^x$ has a nonvertical tangent line at all points on its graph. (See Figure 7.12.) There is a theorem, which we will not prove here, that confirms our conclusion without reference to geometry. It can be stated as follows. (See Appendix 5 for the proof.)

THEOREM 7.2

If f is differentiable on an open interval I, with range J, and if $f'(x) \neq 0$ for all x in I, then f^{-1} exists and is differentiable in J. Furthermore,

$$\frac{d}{dx} f^{-1}(x) = \frac{1}{f'(f^{-1}(x))} \qquad (7.13)$$

To find the derivative of e^x, we apply the result (Equation 7.13) to

$$y = e^x = f^{-1}(x)$$

with $f(x) = \ln x$. Since $f'(x) = 1/x$, we have $f'(f^{-1}(x)) = 1/e^x$. So by Equation 7.13,

$$\frac{d}{dx} e^x = \frac{1}{\dfrac{1}{e^x}} = e^x$$

That is,

$$\frac{d}{dx} e^x = e^x$$

We set off this important result for emphasis.

$$\frac{d}{dx} e^x = e^x \qquad (7.14)$$

This result is quite remarkable. The derivative of the exponential function is the function itself! So the second, third, fourth, ... derivatives all equal the original function. No other function has this self-replicating derivative property (except for constant multiples of e^x). Graphically, Equation 7.14 says that the slope of the tangent line at any point on the graph of $y = e^x$ is equal to the y-coordinate of that point.

If $u = g(x)$ is any differentiable function of x, we can use Equation 7.14 and the Chain Rule to obtain the formula

$$\frac{d}{dx} e^u = e^u \frac{du}{dx} \qquad (7.15)$$

Consequently, we also have the antidifferentiation formula

$$\int e^u \, du = e^u + C \qquad (7.16)$$

EXAMPLE 7.12

(a) Find dy/dx if $y = e^{\sin x}$.
(b) Evaluate the integral $\int x e^{x^2} \, dx$.

Solution

(a) By Equation 7.15,

$$\frac{dy}{dx} = e^{\sin x} \cos x$$

(b) Let $u = x^2$. Then $du/dx = 2x$. So we have

$$\int xe^{x^2}\,dx = \frac{1}{2}\int 2xe^{x^2}\,dx = \frac{1}{2}\int e^u\,du = \frac{1}{2}e^u + C = \frac{1}{2}e^{x^2} + C \qquad \blacksquare$$

EXAMPLE 7.13   Evaluate the following integrals:

(a) $\displaystyle\int \frac{e^x}{1+e^x}\,dx$ (b) $\displaystyle\int_0^{\ln 3} e^{-2x}\,dx$

Solution

(a) Let $u = 1 + e^x$. Then $du = e^x\,dx$. So we have

$$\int \frac{e^x}{1+e^x}\,dx = \int \frac{du}{u} = \ln|u| + C = \ln(1 + e^x) + C$$

Note that the absolute value symbol can be deleted, since $1 + e^x$ is always positive.

(b) $\displaystyle\int_0^{\ln 3} e^{-2x}\,dx = -\frac{1}{2}\int_0^{\ln 3} \underbrace{e^{-2x}}_{e^u}\underbrace{(-2\,dx)}_{du} = -\frac{1}{2}\left[e^{-2x}\right]_0^{\ln 3}$

$$= -\frac{1}{2}(e^{-2\ln 3} - e^0) = -\frac{1}{2}(e^{\ln(3^{-2})} - 1)$$

$$= -\frac{1}{2}[(3)^{-2} - 1] = -\frac{1}{2}\left[\frac{1}{3^2} - 1\right] = -\frac{1}{2}\left(\frac{1}{9} - 1\right) = \frac{4}{9}$$

Observe how we used each of the following facts:

$$-2\ln 3 = \ln(3)^{-2} \qquad \text{By Equation 7.5}$$

$$e^{\ln(3^{-2})} = 3^{-2} \qquad \text{By Equation 7.12}$$

$$e^0 = 1 \qquad \text{Since } \ln 1 = 0 \qquad \blacksquare$$

Graphs Involving the Exponential Function

The next example illustrates how to analyze and draw the graph of a function involving a power of e.

EXAMPLE 7.14   Analyze and draw the graph of $y = xe^{-x}$.

Solution   Since xe^{-x} is defined for all real values of x, the domain is all of $\mathbb{R}$.

(a) *Intercepts.* When $x = 0$, $y = 0$, and conversely. So both the x-intercept and y-intercept are 0.

(b) *Symmetry.* None, since replacing x by $-x$ does not leave y unchanged, nor does it give its negative.

(c) *Maxima and Minima.*

$$y' = e^{-x} + x(-e^{-x}) = e^{-x}(1 - x) \qquad \text{By the Product Rule}$$

$$y'' = (-e^{-x})(1 - x) + e^{-x}(-1) = e^{-x}(x - 2)$$

Since $e^{-x} \neq 0$ for all x, $y' = 0$ only when $x = 1$, so the only critical point is $x = 1$. Substituting $x = 1$ in y'', we get

$$y''\big|_{x=1} = e^{-1}(-1) = -\frac{1}{e} < 0$$

So by the Second-Derivative Test, the function has a local maximum at $x = 1$. The corresponding y value is $1(e^{-1}) = 1/e$. The maximum point is therefore $(1, 1/e)$.

(d) *Concavity.* From $y'' = e^{-x}(x-2)$ we see that $y'' > 0$ for $x > 2$ and $y'' < 0$ for $x < 2$. Thus, the graph is concave up on $(2, \infty)$ and concave down on $(-\infty, 2)$. The concavity changes at $x = 2$, so the point $(2, 2/e^2)$ is a point of inflection.

(e) *Asymptotes.* There is no vertical asymptote, since y is finite for all x values. For horizontal asymptotes, we consider the limits

$$\lim_{x \to -\infty} (xe^{-x}) \qquad \text{and} \qquad \lim_{x \to \infty} (xe^{-x})$$

For large negative values of x, e^{-x} is a large positive value (e.g., when $x = -100$, $e^{-x} = e^{100}$, which is a *huge* number). So the product of x and e^{-x} is a very large negative number. That is,

$$\lim_{x \to -\infty} (xe^{-x}) = -\infty$$

For the limit as $x \to \infty$, we have an indeterminate form of the type $0 \cdot \infty$, so, by L'Hôpital's Rule

$$\lim_{x \to \infty} xe^{-x} = \lim_{x \to \infty} \frac{x}{e^x} = \lim_{x \to \infty} \frac{1}{e^x} = 0$$

(It is also intuitively evident that the limit is 0, since e^x grows much faster than x, so that the fraction x/e^x becomes arbitrarily small.) We conclude that the line $y = 0$ (the x-axis) is an asymptote on the right.

When we combine our results, we get the graph in Figure 7.13.   ■

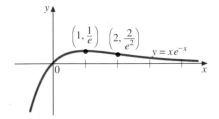

FIGURE 7.13

Relative Growth Rates of $\ln x$, e^x, and x^r

We conclude this section by considering the relative growth rates of the three functions $\ln x$, e^x, and x^r, where r is a positive rational number. (See also Exercise 74 in Exercise Set 7.1.) First, let us state precisely what we mean by one function growing faster than another.

Suppose

$$\lim_{x \to \infty} f(x) = \infty \qquad \text{and} \qquad \lim_{x \to \infty} g(x) = \infty$$

Then, we say that f **grows faster than** g as $x \to \infty$ if

$$\lim_{x \to \infty} \frac{f(x)}{g(x)} = \infty$$

or equivalently, if

$$\lim_{x \to \infty} \frac{g(x)}{f(x)} = 0$$

We also say in this case that g **grows more slowly than** f.

EXAMPLE 7.15　　Let $f(x) = e^x$, $g(x) = x^r$, and $h(x) = \ln x$, where r is a positive rational number. Prove that as $x \to \infty$, f grows faster than g, and g grows faster than h.

Solution　　Suppose n is an integer such that $n \le r < n + 1$ (i.e., $n = [\![r]\!]$ the greatest integer in r). Then, by repeated (if necessary) applications of L'Hôpital's Rule,

$$\lim_{x \to \infty} \frac{g(x)}{f(x)} = \lim_{x \to \infty} \frac{x^r}{e^x} = \lim_{x \to \infty} \frac{rx^{r-1}}{e^x} = \lim_{x \to \infty} \frac{r(r-1)x^{r-2}}{e^x}$$

$$= \cdots = \lim_{x \to \infty} \frac{r(r-1)\cdots(r-n+1)x^{r-n}}{e^x}$$

If $r = n$, we are through, since the numerator is a constant. In this case, the limit is 0. If $r > n$, we apply L'Hôpital's Rule one more time, and the exponent on x becomes $r - n - 1$, which is negative. Thus,

$$\lim_{x \to \infty} \frac{g(x)}{f(x)} = \frac{r(r-1)\cdots(r-n)}{e^x x^{n+1-r}} = 0$$

Therefore, e^x grows more rapidly than x^r, regardless of how large r is. For example, e^x grows more rapidly than x^{100}, or x^{1000}, or $x^{10,000}$.

Now consider the comparison of g and h.

$$\lim_{x \to \infty} \frac{h(x)}{g(x)} = \lim_{x \to \infty} \frac{\ln x}{x^r} = \lim_{x \to \infty} \frac{\dfrac{1}{x}}{rx^{r-1}} = \lim_{x \to \infty} \frac{1}{rx^r} = 0$$

Thus, x^r grows more rapidly than $\ln x$ regardless of how small the positive exponent r is. For example, even $x^{0.0001}$ grows more rapidly than $\ln x$. ∎

Exercise Set 7.2

In Exercises 1–8, solve for x.

1. (a) $e^{2\ln x} = 3$
 (b) $e^{-(1/2)\ln x} = 4$

2. (a) $\ln(e^{-2}) = x$
 (b) $\ln(e^{2x+3}) = 1$

3. (a) $e^{x+1} = 5$
 (b) $e^{1-x} = 2$

4. (a) $\dfrac{1}{e^{2x}} = 3$
 (b) $e^{-x^2} = 4$

5. (a) $\ln x = -1$
 (b) $\ln(x + 1) = 2$

6. (a) $\ln\left(\dfrac{x}{x+1}\right) = 3$
 (b) $\ln(2x - 7) = 0$

7. (a) $3\ln x + 1 = 0$
 (b) $2 - \ln(x - 1) = 5$

8. (a) $\ln x + \ln(2x - 1) = 0$
 (b) $2\ln x - \ln(3x + 4) = 0$

In Exercises 9–28, find $f'(x)$.

9. $f(x) = e^{x^2}$

10. $f(x) = e^{\cos x}$

11. $f(x) = e^{3\ln x}$

12. $f(x) = \ln e^{2x}$

13. $f(x) = e^{\sqrt{x}}$

14. $f(x) = \sqrt{e^x}$

15. $f(x) = e^{e^x}$

16. $f(x) = e^{1/x}$

17. $f(x) = e^{x + \ln x}$

18. $f(x) = e^{2x - 3\ln x}$

19. $f(x) = e^{2x} \sin 3x$

20. $f(x) = e^{\sqrt{1-x^2}}$

21. $f(x) = \dfrac{e^x - e^{-x}}{2}$

22. $f(x) = \dfrac{e^x + e^{-x}}{2}$

23. $f(x) = \dfrac{e^x - e^{-x}}{e^x + e^{-x}}$

24. $f(x) = e^{2x} \ln x$

25. $f(x) = \sin^2 e^x$

26. $f(x) = e^{-x} \cos 2x$

27. $f(x) = \dfrac{1}{1 - e^{-x}}$

28. $f(x) = e^{-x}(\sin x + \cos x)$

In Exercises 29–32, find y′ using implicit differentiation.

29. $e^{x+y} = xy$

30. $e^{xy} = x + y$

31. $e^{-x} \sin y = e^{-y} \cos x$

32. $y = 1 - e^{-(x+y)}$

In Exercises 33–48, evaluate the integrals.

33. $\displaystyle \int xe^{(1-x^2)}dx$

34. $\displaystyle \int \frac{x\,dx}{e^{x^2+1}}$

35. $\displaystyle \int \frac{\sin x}{e^{\cos x}}dx$

36. $\displaystyle \int \frac{e^x}{1 - e^x}dx$

37. $\displaystyle \int \frac{e^x - e^{-x}}{e^x + e^{-x}}dx$

38. $\displaystyle \int \frac{e^x}{\sqrt{e^x - 1}}dx$

39. $\displaystyle \int_0^{\ln 2} e^{3x}dx$

40. $\displaystyle \int_{\ln 2}^{\ln 5} \frac{e^x}{(1 + e^x)^2}dx$

41. $\displaystyle \int_0^3 e^{2x}\sqrt[3]{1 - e^{2x}}dx$

42. $\displaystyle \int_{2+1/e}^3 \frac{dx}{x - 2}$

43. $\displaystyle \int_0^{\ln 2} \frac{dx}{1 + e^{-x}}$

44. $\displaystyle \int \frac{dx}{1 + e^x}$ (*Hint:* Multiply numerator and denominator by e^{-x}.)

45. $\displaystyle \int_{1/\ln 3}^{2/\ln 3} \frac{e^{1/x}}{x^2}dx$

46. $\displaystyle \int_0^{\ln 2} \frac{e^{-x}}{4 - 3e^{-x}}dx$

47. $\displaystyle \int \frac{dx}{e^x - 2 + e^{-x}}$

48. $\displaystyle \int_{\ln \pi}^{\ln 4\pi/3} e^x \tan e^x dx$

In Exercises 49–57, analyze the function and draw its graph.

49. $y = e^{-x^2/4}$

50. $y = e^{1-|x|}$

51. $y = xe^{-x/2}$

52. $y = \dfrac{e^x + e^{-x}}{2}$

53. $y = \dfrac{1}{1 + e^{-x}}$

54. $y = \dfrac{e^x}{1 - e^x}$

55. $y = /x^3 e^{-x}$

56. $y = \dfrac{\ln x}{x^2}$

57. $f(x) = \begin{cases} e^{-1/x^2} & \text{if } x \neq 0 \\ 0 & \text{if } x = 0 \end{cases}$

58. Prove part 2 of Theorem 7.1.

59. Prove that $e^{-x} = 1/e^x$ for all real numbers x.

60. Prove the following, based on the definition of e^x:

(a) $e^0 = 1$ (b) $e^1 = e$.

61. Find the area of the region under the graph of $y = e^{-x}$ from x_1 to x_2, where $e^{x_1} = 1$ and $e^{x_2} = 2$.

62. Find the volume of the solid obtained by revolving the region under the graph of $y = \sqrt{1 + e^x}$ from $x = 0$ to $x = 1$ about the x-axis.

63. Find the length of the curve $y = (e^x + e^{-x})/2$ from $x = 0$ to $x = \ln 2$.

64. Find the volume of the solid obtained by revolving the region under the graph of $y = \sqrt{x}e^{(1-x^2)/2}$ from $x = 1$ to $x = 2$ about the x-axis.

65. Show that the function f defined by

$$f(x) = \frac{e^x - e^{-x}}{2}$$

is 1–1 on $\mathbb{R}$, and find f^{-1}.

66. Find $(f^{-1})'(x)$ for the function in Exercise 65.

In Exercises 67–69, analyze the function and draw its graph.

67. $f(x) = e^{1/x}$

68. $f(x) = \dfrac{e^{2x}}{x^2}$

69. $f(x) = xe^{(1-x^2)/2}$

70. A number of types of oscillatory motion, such as that of an oscillating spring, when the oscillation is occurring in a resisting medium, can be described by an equation of the type

$$y(t) = e^{-kt}(C_1 \cos \omega t + C_2 \sin \omega t)$$

where $y(t)$ is the displacement from the equilibrium position at time t. Suppose $k = 1$, $C_1 = 0$, $C_2 = 1$, and $\omega = 2$. Find the velocity and acceleration when $t = \pi$. Draw the graph of $y(t)$.

71. When limitations on size are considered for the growth of a colony of bacteria (because of physical limitations or limitations on nutrients, for example), the number $Q(t)$ present at time t is given with reasonable accuracy by

$$Q(t) = \frac{m}{1 + \left(\dfrac{m - Q_0}{Q_0}\right)e^{-kmt}}$$

where $Q_0 = Q(0)$, m is the maximum size, and k is a positive constant. If $Q_0 = 200$, $m = 1000$, and $Q(\frac{1}{2}) = 400$, find the constant k. Using this value of k, draw the graph of $Q(t)$. When does the maximum rate of growth occur?

72. The normal probability density function with mean μ and standard deviation σ is given by

$$f(x) = \frac{1}{\sqrt{2\pi}\sigma} e^{-(x-\mu)^2/2\sigma^2}$$

(a) Draw the graph of f.

(b) Show that inflection points occur at $x = \mu \pm \sigma$.

73. Let $f(x) = e^{ax}$, $g(x) = (\ln x)^b$, and $h(x) = x^c$ where a, b, and c are positive constants with b and c rational. Prove that as $x \to \infty$,

(a) f grows faster than h;

(b) h grows faster than g.

(Refer to Example 7.15.)

7.3 EXPONENTIAL AND LOGARITHMIC FUNCTIONS WITH OTHER BASES

Irrational Powers

We know that for rational numbers r and any real number $a > 0$,

$$a^r = e^{\ln a^r} = e^{r \ln a} \tag{7.17}$$

We still have not assigned a meaning to a raised to an irrational power. We now do so. Let x denote any real number, rational or irrational. We define a^x as follows:

Definition 7.2
The General Exponential Function

$$a^x = e^{x \ln a}, \qquad a > 0 \tag{7.18}$$

This definition agrees with Equation 7.17 when x is rational and extends the meaning in the natural way to irrational values. Note that this definition is meaningful, since $\ln a$ is a well-defined real number for each $a > 0$, so that $x \ln a$ is also a real number, and since the exponential function is defined on all of $\mathbb{R}$, $e^{x \ln a}$ is also well defined. For example, according to our definition,

$$\pi^{\sqrt{2}} = e^{\sqrt{2} \ln \pi}$$

With a calculator we obtain the approximation

$$\pi^{\sqrt{2}} \approx 5.0475$$

Observe that since $e^{g(x)}$ is always positive, it follows that $a^x > 0$ for all x.

Taking the natural logarithm of both sides of Equation 7.18 gives an extension of Property 7.5 of the natural logarithm function:

$$\ln a^x = x \ln a, \qquad a > 0 \tag{7.19}$$

Laws of Exponents

The next theorem shows that the laws of exponents for rational exponents extend to all real exponents.

THEOREM 7.3

Laws of Exponents

Let a and b denote positive real numbers. If x and y are any real numbers, then

1. $a^{x+y} = a^x \cdot a^y$
2. $a^{x-y} = \dfrac{a^x}{a^y}$
3. $(a^x)^y = a^{xy}$
4. $(ab)^x = a^x b^x$
5. $\left(\dfrac{a}{b}\right)^x = \dfrac{a^x}{b^x}$

Proof   We prove the first and third laws of exponents and leave the others as exercises.

$$
\begin{aligned}
\textbf{1.} \quad a^{x+y} &= e^{(x+y)\ln a} & &\text{By Equation 7.18} \\
&= e^{x\ln a + y\ln a} & &\text{Distributive Law} \\
&= e^{x\ln a} \cdot e^{y\ln a} & &\text{Theorem 7.1, part 1} \\
&= a^x \cdot a^y & &\text{By Equation 7.18}
\end{aligned}
$$

$$
\begin{aligned}
\textbf{3.} \quad (a^x)^y &= e^{y\ln a^x} & &\text{By Equation 7.18} \\
&= e^{y(x\ln a)} & &\text{By Equation 7.19} \\
&= e^{(xy)\ln a} & &\text{Associative and Commutative Laws} \\
&= a^{xy} & &\text{By Equation 7.18} \quad\blacksquare
\end{aligned}
$$

The next example illustrates how the natural logarithm can be used to solve certain types of equations for a variable that appears as an exponent.

EXAMPLE 7.16   Solve for x:

(a) $2^x = 3^{1-x}$ (b) $3 \cdot 4^{2x} = \dfrac{2}{\sqrt{5^x}}$

Solution   In each case we take the natural logarithm of both sides, using the fact that the natural logarithm is a one-to-one function.

(a)

$$
\ln 2^x = \ln 3^{1-x}
$$

$$
x \ln 2 = (1-x)\ln 3
$$

$$
x \ln 2 + x \ln 3 = \ln 3
$$

$$
x(\ln 2 + \ln 3) = \ln 3
$$

$$
x = \frac{\ln 3}{\ln 2 + \ln 3} = \frac{\ln 3}{\ln 6} \approx 0.6131
$$

(b) Rewrite the equation in the form

$$
3 \cdot 4^{2x} = \frac{2}{5^{x/2}}
$$

Then

$$\ln(3 \cdot 4^{2x}) = \ln\left(\frac{2}{5^{x/2}}\right)$$

$$\ln 3 + 2x \ln 4 = \ln 2 - \frac{x}{2}\ln 5$$

$$2\ln 3 + 4x \ln 4 = 2\ln 2 - x \ln 5$$

$$x(4\ln 4 + \ln 5) = 2\ln 2 - 2\ln 3$$

$$x = \frac{2\ln 2 - 2\ln 3}{4\ln 4 + \ln 5} = \frac{\ln 4 - \ln 9}{4\ln 4 + \ln 5} \approx -0.1133$$

Note that any further attempts to simplify the answer before using your calculator are not worthwhile in this case.   ∎

The Power Rule Extended to Real Exponents

We have often used the Power Rule,

$$\frac{d}{dx}x^r = rx^{r-1} \qquad (x > 0)$$

for *rational* numbers r. Now we can extend this rule to all real exponents.

THEOREM 7.4

Power Rule for Real Exponents

If $x > 0$ and α is any real number, then

$$\frac{d}{dx}x^\alpha = \alpha x^{\alpha-1}$$

Proof   By Equation 7.18 and the derivatives of e^u and $\ln u$,

$$\frac{d}{dx}x^\alpha = \frac{d}{dx}e^{\alpha \ln x} = \left(\frac{\alpha}{x}\right)e^{\alpha \ln x}$$

$$= \left(\frac{\alpha}{x}\right)x^\alpha = \alpha x^{\alpha-1}$$ ∎

For example, if $y = x^\pi$, then $dy/dx = \pi x^{\pi-1}$.

The Exponential Function a^x

If we take a as a fixed positive real number, then for each real number x, a^x is a well-defined real number, by Equation 7.18. Thus, $f(x) = a^x$ is a function from $\mathbb{R}$ to $\mathbb{R}$ (but its range is the set of positive real numbers). We call f the **exponential function with base a**. If the base $a = 1$, we have

$$1^x = e^{x \ln 1} = e^{x \cdot 0} = e^0 = 1$$

Thus, when $a = 1$, f is just the constant function $f(x) = 1$ for all x. Since the properties of constant functions are well known, *we will henceforth consider only those cases where $a > 0$ and $a \neq 1$.*

Derivatives of a^x

Since e^u is differentiable when u is differentiable and

$$\frac{d}{dx} e^u = e^u \frac{du}{dx}$$

it follows from Equation 7.18 that $a^x = e^{x \ln a}$ is differentiable, and the derivative is given by

$$\frac{d}{dx} a^x = \frac{d}{dx} e^{x \ln a} = e^{x \ln a} \cdot \ln a = a^x \ln a$$

Thus,

$$\frac{d}{dx} a^x = a^x \ln a \qquad\qquad (7.20)$$

Note that when $a = e$, this formula reduces to

$$\frac{d}{dx} e^x = e^x$$

since $\ln e = 1$. This derivative formula for e^x shows one reason why e is singled out as a special base. The derivative formula is simplified when e is the base.

The Graph of a^x

If $a > 1$, then $\ln a > 0$, and since a^x is always positive, it follows from Equation 7.20 that

$$\frac{d}{dx} a^x = a^x \ln a > 0 \qquad (a > 1)$$

If $0 < a < 1$, then $\ln a < 0$, so that

$$\frac{d}{dx} a^x < 0 \qquad (0 < a < 1)$$

Thus, the graph of $f(x) = a^x$ is increasing everywhere if $a > 1$ and decreasing everywhere if $0 < a < 1$. Consequently, there are no local (or absolute) maxima or minima. For f'' we have

$$f''(x) = \frac{d}{dx}(a^x \ln a) = \left(\frac{d}{dx} a^x\right) \ln a = a^x (\ln a)^2$$

which is always positive. So the graph is always concave up and there are no inflection points. Using the properties of e^x, we can now graph $f(x) = a^x$ as in Figure 7.14.

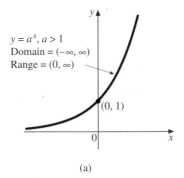

$y = a^x$, $a > 1$
Domain $= (-\infty, \infty)$
Range $= (0, \infty)$

$(0, 1)$

(a)

$y = a^x$, $0 < a < 1$
Domain $= (-\infty, \infty)$
Range $= (0, \infty)$

$(0, 1)$

(b)

FIGURE 7.14

Derivative of $a^{g(x)}$

Using the Chain Rule together with Equation 7.20, we obtain

$$\frac{d}{dx}a^u = a^u \ln a \cdot \frac{du}{dx} \tag{7.21}$$

where $u = g(x)$ is a differentiable function of x.

It is important to distinguish between a^x and x^a, where a is fixed (see Figure 7.15). We can contrast their derivatives as follows:

$$\begin{array}{cc} \underline{\quad y = a^x \quad} & \underline{\quad y = x^a \quad} \\[6pt] \dfrac{dy}{dx} = a^x \ln a & \dfrac{dy}{dx} = a x^{a-1} \end{array}$$

For example,

$$\frac{d}{dx}(3^x) = 3^x \ln 3, \quad \text{whereas} \quad \frac{d}{dx}x^3 = 3x^2$$

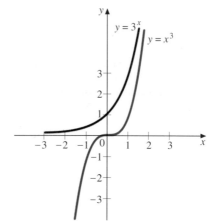

FIGURE 7.15

EXAMPLE 7.17   Find the derivative of each of the following:

(a) $y = 2^{x^2}$ (b) $y = x \cdot 3^{\sqrt{1-x}}$

Solution

(a) $\dfrac{dy}{dx} = 2^{x^2} \ln 2 \cdot 2x = (2 \ln 2)x \cdot 2^{x^2}$

(b) From the Product Rule and Equation 7.21,

$$\frac{dy}{dx} = 3^{\sqrt{1-x}} + x \cdot 3^{\sqrt{1-x}} \ln 3 \left(-\frac{1}{2\sqrt{1-x}} \right)$$

$$= 3^{\sqrt{1-x}} \left[1 - \frac{x \ln 3}{2\sqrt{1-x}} \right]$$ ∎

Integral of a^u

From Equation 7.21, we also obtain the antidifferentiation formula

$$\int a^u \, du = \frac{a^u}{\ln a} + C \tag{7.22}$$

for which we are continuing to require $a > 0$ and $a \neq 1$.

EXAMPLE 7.18   Evaluate the integral

$$\int \frac{3^{\sqrt{x}}}{\sqrt{x}} \, dx$$

Solution   Let $u = \sqrt{x}$. Then $du = dx/(2\sqrt{x})$. So by Equation 7.22,

$$\int \frac{3^{\sqrt{x}}}{\sqrt{x}} dx = 2 \int \underbrace{3^{\sqrt{x}}}_{3^u} \underbrace{\left(\frac{1}{2\sqrt{x}} dx \right)}_{du} = 2 \left(\frac{3^{\sqrt{x}}}{\ln 3} \right) + C$$ ∎

Differentiating $[f(x)]^{g(x)}$ (Variables to Variable Powers)

We now know how to differentiate constants to variable powers (such as 3^x) and variables to constant powers (such as x^3). The next example shows how to differentiate variables to variable powers.

EXAMPLE 7.19   Find y' if $y = (\sin x)^x$, with $0 < x < \pi$.
Note: We restrict x to be in the open interval $(0, \pi)$ so that $\sin x > 0$. If we allowed $\sin x$ to be negative, $(\sin x)^x$ would not be well defined.

Solution   We will find the derivative by two methods. In Method 1 we use Equation 7.18 to get $y = (\sin x)^x = e^{x \ln \sin x}$. So, by Equation 7.15,

$$y' = \frac{d}{dx} e^{x \ln \sin x} = e^{x \ln \sin x} \left(\ln \sin x + x \cdot \frac{1}{\sin x} \cdot \cos x \right) \quad \text{Using the Product Rule on } x \ln \sin x$$

$$= (\sin x)^x (\ln \sin x + x \cot x)$$

In Method 2 we use logarithmic differentiation:

$$\ln y = x \ln \sin x$$

$$\frac{y'}{y} = \ln \sin x + x \cdot \frac{1}{\sin x} \cdot \cos x \quad \text{Using the Product Rule}$$

$$= \ln \sin x + x \cot x$$

$$y' = y(\ln \sin x + x \cot x)$$

$$= (\sin x)^x (\ln \sin x + x \cot x)$$ ∎

The Indeterminate Forms 0^0, 1^∞, and ∞^0

By using logarithms and exponentials, we can often evaluate limits of the form

$$\lim_{x \to a} f(x)^{g(x)}$$

where either

(i) $\lim\limits_{x \to a} f(x) = 0$ and $\lim\limits_{x \to a} g(x) = 0$ (the form 0^0)

(ii) $\lim\limits_{x \to a} f(x) = 1$ and $\lim\limits_{x \to a} g(x) = \pm\infty$ (the form 1^∞)

or

(iii) $\lim\limits_{x \to a} f(x) = \pm\infty$ and $\lim\limits_{x \to a} g(x) = 0$ (the form ∞^0)

The basic approach is to set $y = f(x)^{g(x)}$ and then take the natural logarithm of both sides:

$$\ln y = \ln f(x)^{g(x)} = g(x) \ln f(x)$$

Then we write the product on the right as a quotient (in most cases writing $\ln f(x)$ as the numerator), and use L'Hôpital's Rule, if it is applicable (either 0/0 or ∞/∞). If we determine that

$$\lim_{x \to a} g(x) \ln f(x) = L$$

then we can conclude that

$$\lim_{x \to a} f(x)^{g(x)} = \lim_{x \to a} e^{g(x) \ln f(x)} = e^L$$

since $y = e^{\ln y}$. We used the continuity of the exponential function to get

$$\lim_{x \to a} e^{g(x) \ln f(x)} = e^L$$

We illustrate this approach in the next example.

EXAMPLE 7.20   Evaluate the following limits.

(a) $\displaystyle\lim_{x \to 0^+} (\sin x)^x$ (b) $\displaystyle\lim_{x \to 0} (1 - x)^{1/x}$

Solution

(a) As it stands, the limit is indeterminate, of the form 0^0. Proceeding as above, we set $y = (\sin x)^x$ and take logs and then write the product as a quotient.

$$\ln y = x \ln \sin x = \frac{\ln \sin x}{\dfrac{1}{x}}$$

As $x \to 0^+$, the fraction on the right is indeterminate, of the form ∞/∞, and since the conditions of L'Hôpital's Rule are met, we have

$$\lim_{x \to 0^+} \ln y = \lim_{x \to 0^+} \frac{\ln \sin x}{\dfrac{1}{x}}$$

$$= \lim_{x \to 0^+} \frac{\dfrac{1}{\sin x}(\cos x)}{-\dfrac{1}{x^2}}$$

$$= \lim_{x \to 0^+} \frac{-x^2 \cos x}{\sin x} \qquad \text{0/0 form; apply L'Hôpital's Rule again}$$

$$= \lim_{x \to 0^+} \frac{-2x \cos x + x^2 \sin x}{\cos x}$$

$$= \frac{0}{1} = 0$$

Thus,

$$\lim_{x \to 0^+} (\sin x)^x = \lim_{x \to 0^+} y = \lim_{x \to 0^+} e^{\ln y} = e^0 = 1$$

(b) As given, the limit is indeterminate, of the form 1^∞. We proceed as in part (a). Let

$$y = (1 - x)^{1/x}$$

Then

$$\ln y = \frac{1}{x} \ln(1 - x) = \frac{\ln(1 - x)}{x}$$

So

$$\lim_{x \to 0} \ln y = \lim_{x \to 0} \frac{\ln(1 - x)}{x}$$

Since we have a 0/0 indeterminate form, we apply L'Hôpital's Rule to get

$$\lim_{x \to 0} \ln y = \lim_{x \to 0} \frac{\dfrac{-1}{1 - x}}{1} = -1$$

Thus,

$$\lim_{x \to 0} (1 - x)^{1/x} = \lim_{x \to 0} y = \lim_{x \to 0} e^{\ln y} = e^{-1} = \frac{1}{e} \qquad \blacksquare$$

In limits of the type illustrated in the preceding example, it is tempting to stop after having found the limit of $\ln y$ by L'Hôpital's Rule. But it is essential that you find the limit of y itself, using

$$\lim_{x \to a} y = \lim_{x \to a} e^{\ln y} = e^{\lim_{x \to a} \ln y}$$

Otherwise, you haven't solved the problem that was posed.

In the next example we make use of L'Hôpital's Rule to aid in drawing the graph of a function.

EXAMPLE 7.21   Let $f(x) = x^x$ for $x > 0$.

(a) Find $\lim_{x \to 0^+} f(x)$.

(b) Find all maximum and minimum points on the graph of f.

(c) Draw the graph of f.

Solution

(a) Let $y = x^x$. Then $\ln y = x \ln x$. So

$$\lim_{x \to 0^+} \ln y - \lim_{x \to 0^+} \frac{\ln x}{1/x} \qquad \infty/\infty \text{ form}$$

$$= \lim_{x \to 0^+} \frac{1/x}{-1/x^2} = \lim_{x \to 0^+} (-x) = 0$$

So $\lim_{x \to 0^+} x^x = e^0 = 1$.

(b) From $\ln y = x \ln x$ we obtain by implicit differentiation

$$\frac{y'}{y} = \ln x + x \left(\frac{1}{x} \right)$$

$$y' = y(\ln x + 1) = x^x (\ln x + 1)$$

The only critical point occurs when $1 + \ln x = 0$ (since $x^x \neq 0$), or equivalently when $\ln x = -1$. Thus, $x = e^{-1} = 1/e \approx 0.37$. The corresponding

FIGURE 7.16

y value is $f(1/e) = (1/e)^{1/e} \approx 0.69$. Since $y' > 0$ when $x > 1/e$ and $y' < 0$ when $x < 1/e$ (you should verify these inequalities), the point $\left(1/e, (1/e)^{1/e}\right)$ is a minimum point.

(c) To check concavity we calculate y'':

$$y'' = \left[\frac{d}{dx}(x^x)\right] \cdot (\ln x + 1) + x^x \left(\frac{1}{x}\right) = x^x (\ln x + 1)^2 + x^{x-1}$$

(using the derivative of x^x found in part (b)). Thus, $y'' > 0$ for all $x > 0$, which tells us that the graph is always concave up. Clearly,

$$\lim_{x \to \infty} x^x = \infty$$

We can now draw the graph as in Figure 7.16.   ∎

The Logarithm Function with Base a

Since for $a > 0$ and $a \neq 1$, $f(x) = a^x$ is monotone (increasing if $a > 1$ and decreasing if $0 < a < 1$), f is one-to-one and therefore has an inverse. We denote its inverse by

$$f^{-1}(x) = \log_a x \qquad (a > 0, \ a \neq 1, \ x > 0)$$

and call f^{-1} the **logarithm function with base a**. Note that the domain of $f^{-1}(x)$ is the set of all $x > 0$, since the range of $f(x) = a^x$ is $(0, \infty)$. Remember that the domain of f^{-1} is the range of f. We call $\log_a x$ the **logarithm to the base a of x**. Thus, for $a > 0$, $a \neq 1$, we have

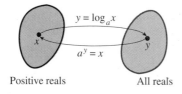

Positive reals   All reals

FIGURE 7.17

$$y = \log_a x \text{ if and only if } a^y = x \qquad (x > 0) \qquad (7.23)$$

This inverse relationship between a^y and $\log_a x$ is shown schematically in Figure 7.17. We can also state the same result as follows.

Inverse Properties

$$\log_a a^x = x \qquad \text{for all } x$$
$$a^{\log_a x} = x \qquad \text{for } x > 0$$

$$(7.24)$$

If $a = e$, then $\log_e x$ is the inverse of e^x. But $\ln x$ is the inverse of e^x, so we have

---

$$\log_e x = \ln x$$

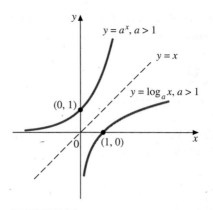

FIGURE 7.18

That is, the natural logarithm of x is the same as the logarithm to the base e of x.

We obtain the graph of $y = \log_a x$ by reflecting the graph of $y = a^x$ about the line $y = x$, with the result shown in Figure 7.18. We show the case $a > 1$ only.

Arithmetic Properties of $\log_a x$

The next theorem shows that the arithmetic properties for $\ln x$ also hold true for $\log_a x$. We leave the proof for the exercises.

THEOREM 7.5

> Assume x and y are in the domain of the logarithm function of base a, where $a > 0$ and $a \neq 1$. Then
>
> 1. $\log_a xy = \log_a x + \log_a y$
> 2. $\log_a \dfrac{x}{y} = \log_a x - \log_a y$
> 3. $\log_a x^\alpha = \alpha \log_a x$ for any real number α

REMARK

■ In particular, when $a = e$, Property 3 can be stated as

$$\ln x^\alpha = \alpha \ln x$$

This property extends Equation 7.5 for $\ln x$ to all real numbers. We previously had proved this property for rational α only.

The Derivative of $\log_a x$

To find the derivative of $y = \log_a x$, we use Equation 7.23 to convert to the equivalent form

$$a^y = x$$

and then differentiate implicitly to get

$$(a^y \ln a) y' = 1 \quad \text{By Equation 7.21}$$

Thus,

$$y' = \frac{1}{a^y \ln a}$$

But $a^y = x$, so we have the result

$$\frac{d}{dx} \log_a x = \frac{1}{x \ln a} \tag{7.25}$$

Here we see why e is the "natural" base for logarithms, for if $a = e$, the derivative becomes simply $1/x$. By the Chain Rule, from Equation 7.25 we obtain for $u = g(x)$ a differentiable function with $u > 0$,

$$\frac{d}{dx}\log_a u = \frac{1}{u\ln a}\cdot\frac{du}{dx} \tag{7.26}$$

We could use Equation 7.26 to obtain an antiderivative of $1/u$ in terms of $\log_a u$, but this is unnecessary, since we already know that an antiderivative is $\ln|u| + C$. That is, there is no need to write the integral formula coming from Equation 7.26.

Change of Base

Sometimes it is useful to express a logarithm to one base in terms of a logarithm to another base. Suppose a and b are positive and different from 1. For $x > 0$, let

$$y = \log_b x \qquad \text{so that } x = b^y$$

Now take the logarithm to the base a of both sides of the last equation:

$$\log_a x = \log_a b^y = y\log_a b$$

Thus, $y = (\log_a x)/(\log_a b)$, and since $y = \log_b x$, we have the **Change-of-Base Formula**:

$$\log_b x = \frac{\log_a x}{\log_a b} \tag{7.27}$$

Since in calculus the most convenient base to work with is e, we usually want to use the following special case of Equation 7.27:

$$\log_b x = \frac{\ln x}{\ln b} \tag{7.28}$$

For example, by Equation 7.28 we have

$$\log_2 7 = \frac{\ln 7}{\ln 2} \approx 2.8074 \quad \text{By calculator}$$

Integrals Involving $\log_a x$

The base e is by far the most commonly used base for logarithms and exponentials in calculus and its applications. The only other bases of any widespread use are 2 and 10. The base 2 is the foundation of binary arithmetic, which is very important in computer science. As we have previously noted, logarithms with base 10 are called common logarithms. In the next example we use the base 5, and in the exercises we will use a variety of bases to give practice in the theory of this section, but in the rest of the book we will concentrate on natural logarithms and exponentials.

EXAMPLE 7.22   Evaluate the indefinite integral

$$\int \frac{\log_5 x}{x}\, dx$$

Solution   By Formula 7.28,

$$\log_5 x = \frac{\ln x}{\ln 5}$$

So we can rewrite the integral as

$$\int \frac{\ln x}{x \ln 5}\, dx = \frac{1}{\ln 5} \int \frac{\ln x}{x}\, dx$$

Now we make the substitution $u = \ln x$. Then $du = \frac{1}{x}\, dx$. So we have

$$\frac{1}{\ln 5} \int \frac{\ln x}{x}\, dx = \frac{1}{\ln 5} \int u\, du = \frac{1}{\ln 5} \left(\frac{u^2}{2} + C \right) = \frac{1}{\ln 5} \left[\frac{(\ln x)^2}{2} + C \right]$$

If we let $C_1 = \frac{1}{\ln 5} C$, we can write the answer as

$$\int \frac{\log_5 x}{x}\, dx = \frac{(\ln x)^2}{2 \ln 5} + C_1$$ ∎

The Number e as a Limit

We conclude this section with a theorem that provides a means of estimating e, along with an application of the result.

THEOREM 7.6

The number e can be expressed in either of the following ways:

1. $e = \lim_{x \to 0} (1 + x)^{1/x}$

2. $e = \lim_{n \to \infty} \left(1 + \frac{1}{n} \right)^n$

Proof

1. Let $F(x) = \ln x$. We know that $F'(x) = 1/x$, so $F'(1) = 1$. From the definition of the derivative, we have

$$F'(1) = \lim_{h \to 0} \frac{F(1+h) - F(1)}{h}$$

$$= \lim_{h \to 0} \frac{\ln(1+h) - \ln 1}{h}$$

$$= \lim_{h \to 0} \frac{1}{h} \ln(1+h) \qquad \text{Since } \ln 1 = 0$$

$$= \lim_{h \to 0} \ln(1+h)^{1/h} \qquad \text{By Equation 7.5}$$

$$= \ln \left[\lim_{h \to 0} (1+h)^{1/h} \right] \qquad \text{By continuity of the ln function}$$

Thus,

$$\ln \left[\lim_{h \to 0} (1+h)^{1/h} \right] = 1 \quad \text{Since } F'(1) = 1$$

and since $\ln a = b$ implies $a = e^b$,

$$\lim_{h \to 0} (1+h)^{1/h} = e$$

By replacing h with x, we have proved part 1.

2. Let $1/n = x$. As $n \to \infty$, $x \to 0$, so we have

$$\lim_{n \to \infty} \left(1 + \frac{1}{n} \right)^n = \lim_{x \to 0} (1+x)^{1/x} = e$$

by part 1.   ∎

n	$(1 + \frac{1}{n})^n$
1	2.000000
10	2.59374
100	2.704814
1,000	2.716924
10,000	2.718146
100,000	2.718268
1,000,000	2.718280

With the aid of a calculator we get the approximate values for $(1 + 1/n)^n$ shown in the margin. This table gives evidence that $e \approx 2.71828$, and it can be proved that this approximation is correct to five decimal places.

Compound Interest

The limit in part 1 of Theorem 7.6 arises naturally in compound-interest problems. When an amount of money P (called the principal) is invested at an annual interest rate r (for example, if the interest rate is 8%, $r = 0.08$), and interest is compounded annually, the total amount A_1 in the account after t yr is

$$A_1(t) = P(1 + r)^t$$

(In the exercises, you will be asked to show why this formula is correct.) If interest is compounded semiannually, the amount after t yr becomes

$$A_2(t) = P \left(1 + \frac{r}{2} \right)^{2t}$$

and if it is compounded k times per year,

$$A_k(t) = P \left(1 + \frac{r}{k} \right)^{kt} \tag{7.29}$$

Some banks compound interest daily, using the formula

$$A_{365}(t) = P\left(1 + \frac{r}{365}\right)^{365t}$$

What we want to do now is to let $k \to \infty$ in Equation 7.29. The result will give the amount that results from *continuous* compounding of interest. If we write $x = r/k$, then $x \to 0$ as $k \to \infty$, since r is fixed. So we have

$$A(t) = \lim_{k\to\infty} A_k(t) = \lim_{k\to\infty} P\left(1 + \frac{r}{k}\right)^{kt}$$

$$= \lim_{k\to\infty} P\left(1 + \frac{r}{k}\right)^{(k/r)rt}$$

$$= \lim_{x\to 0} P(1 + x)^{(1/x)rt}$$

$$= P\left[\lim_{x\to 0}(1 + x)^{1/x}\right]^{rt} \qquad \text{Using continuity of the exponential function}$$

$$= Pe^{rt} \qquad\qquad\qquad \text{By Part 1 of Theorem 7.6}$$

So we have the following formula for continuously compounded interest.

Continuously Compounded Interest

The amount $A(t)$ after t yr is

$$A(t) = Pe^{rt}$$

where P is the principal and r is the annual interest rate.

Exercise Set 7.3

 In Exercises 1 and 2, express each number in terms of the base e (natural exponentials or logarithms) and then approximate its value correct to five decimal places using a calculator or CAS.

1. (a) $\pi^{\sqrt{3}}$
   (b) π^π
   (c) π^e
   (d) $(\sqrt{2})^{\sqrt{2}}$
   (e) $(\sin 2)^\pi$

2. (a) $\log_3 4$
   (b) $\log_{1/2} 5$
   (c) $\log_6 4$
   (d) $\log_7 3$
   (e) $\log_2 100$

In Exercises 3–18, find the derivative.

3. $y = 2^x$

4. $y = 3^{1-x^2}$

5. $y = x^2(4^{-x})$

6. $y = \log_2 x$

7. $y = \log_3 \dfrac{x}{1 + x^3}$

8. $y = 3^x \cdot \log_3 x$

9. $y = \log_{10} \dfrac{\sqrt{x^2 - 1}}{x}$

10. $y = (\log_3 \sqrt{4 - x})^2$

11. $y = \log_2(\log_3 x)$

12. $y = (3^x)(x^3)$

13. $y = x^x$

14. $y = x^{\sin x}$

15. $y = x^{1/(1-x)}$

16. $y = (\cos x)^{\sin x}$

17. $y = (1 - x^2)^{1/x}$

18. $y = (\ln x)^x$

In Exercises 19–25, evaluate the limits.

19. (a) $\displaystyle\lim_{x\to 0} \frac{x^2}{e^x - x - 1}$ (b) $\displaystyle\lim_{x\to\infty} \frac{e^{\sqrt{x}}}{x^2}$

20. (a) $\displaystyle\lim_{x\to\infty} x^2 e^{-x}$ (b) $\displaystyle\lim_{x\to 0^+} x^2 e^{1/x}$

21. (a) $\displaystyle\lim_{x\to 0} \left(\frac{1}{e^x - 1} - \frac{1}{x}\right)$ (b) $\displaystyle\lim_{x\to 0}\left(\frac{1}{xe^x} - \frac{1}{x}\right)$

22. (a) $\lim\limits_{x\to 0^+} x^x$ (b) $\lim\limits_{x\to\pi/2^-}(\tan x)^{\cos x}$

23. (a) $\lim\limits_{x\to 0}(1+\sin x)^{1/x}$ (b) $\lim\limits_{x\to 1} x^{1/(1-x)}$

24. (a) $\lim\limits_{x\to 0^-}\left(1-\dfrac{3}{x}\right)^x$ (b) $\lim\limits_{x\to\infty}\left(\dfrac{x+1}{x}\right)^{2x}$

25. (a) $\lim\limits_{x\to 1^+}(\ln x)^{1-x}$ (b) $\lim\limits_{x\to 0}(e^x+x)^{1/x}$

In Exercises 26–35, evaluate the integrals.

26. $\displaystyle\int 5^x dx$

27. $\displaystyle\int \dfrac{x}{2^{x^2}}dx$

28. $\displaystyle\int \dfrac{3^x}{1+3^x}dx$

29. $\displaystyle\int (\cos x)\cdot 2^{\sin x}dx$

30. $\displaystyle\int x^2\cdot 2^{x^3-1}dx$

31. $\displaystyle\int \dfrac{dt}{1+2^{-t}}$ (*Hint:* Multiply by $2^t/2^t$.)

32. $\displaystyle\int_1^4 \dfrac{2^{\sqrt{x}}}{\sqrt{x}}dx$

33. $\displaystyle\int_1^3 \dfrac{3^t}{\sqrt{3^t-2}}dt$

34. $\displaystyle\int \dfrac{dx}{3^x(1-3^{-x})^2}$

35. $\displaystyle\int \dfrac{dt}{(2^t-2^{-t})^2}$ (*Hint:* Multiply by $2^{2t}/2^{2t}$.)

In Exercises 36 and 37, draw the graphs of the functions in parts (a) through (d) on the same set of axes.

36. (a) $y=2^x$
(b) $y=e^x$
(c) $y=10^x$
(d) $y=(\tfrac{1}{2})^x$

37. (a) $y=\log_2 x$
(b) $y=\log_e x$
(c) $y=\log_{10} x$
(d) $y=\log_{1/2} x$

38. Prove that for $x\neq 0$, $a>0$, and $a\neq 1$,

$$\dfrac{d}{dx}\log_a|x|=\dfrac{1}{x\ln a}$$

39. Using the result of Exercise 38 along with the Chain Rule, differentiate each of the following:

(a) $y=\log_{10}\left|\dfrac{x-1}{x\mid 2}\right|$ (b) $y=\log_2\left|\dfrac{x\sqrt{1-x}}{(2x-3)^3}\right|$

In Exercises 40 and 41, analyze the function and sketch its graph.

40. $y=2^{1-x^2}$

41. $y=\log_{10}\sqrt{x^2-1}$

 In Exercises 42–45, solve for x in terms of natural logarithms and then approximate the answer using a calculator or CAS.

42. $3^{2x-1}=4^x$

43. $2\cdot 3^{x+1}=3\cdot 2^{2-x}$

44. $3e^{x+2}=4\cdot 5^x$

45. $2^x\cdot 3^{1-2x}=5\cdot 6^{-x}$

46. Complete the proof of Theorem 7.3 by proving Properties 2, 4, and 5.

47. Prove Theorem 7.5.

48. Prove the formula $A_1(t)=P(1+r)^t$ for the total amount of money $A_1(t)$ after t yr when a principal P is invested at a rate of interest r, compounded annually.

49. Suppose you invest \$10,000. Find the total amount in your account after 10 yr if the interest rate is 5% and is compounded (a) annually, (b) semiannually, (c) daily, (d) continuously.

50. Using the properties of logarithms, solve for x and check your answers:
(a) $\log_4(x+2)-\log_4(x-2)=\tfrac{1}{2}$
(b) $2\log_2|x-1|=2-\log_2(x+2)$

51. Find the volume of the solid obtained by revolving the region under the graph of $y=2^{1-x^2}$ from $x=0$ to $x=1$ about the y-axis.

52. Find the volume of the solid obtained by revolving the region under the graph of $y=\log_3 x$ from $x=1$ to $x=3$ about the y-axis.

53. Obtain an alternative proof of part 2 of Theorem 7.6 by the following steps:
(a) For $t\in[1,1+\tfrac{1}{n}]$ show that

$$\int_1^{1+1/n}\dfrac{1}{1+1/n}dt\le\int_1^{1+1/n}\dfrac{1}{t}dt\le\int_1^{1+1/n}1\,dt$$

(b) From part (a) show that

$$\dfrac{1}{n+1}\le\ln\left(1+\dfrac{1}{n}\right)\le\dfrac{1}{n}$$

(c) In part (b), multiply through by n and show that the result can be put in the form

$$\dfrac{1}{1+\dfrac{1}{n}}\le\ln\left(1+\dfrac{1}{n}\right)^n\le 1$$

(d) In part (c), let $n\to\infty$ and show how it follows that

$$\lim_{n\to\infty}\left(1+\dfrac{1}{n}\right)^n=e$$

54. (a) By graphing $f(x) = a^x$ for different values of $a > 0$, discuss the effect a has on the growth rate of a^x as $x \to \infty$.

  (b) Repeat part (a) with $f(x) = \log_a x$.

55. Compare the growth rates of $f(x) = x^n$ and $g(x) = a^x$ as $x \to \infty$. Consider various values of n and a.

56. Use a graphing calculator or a CAS to graph $f(x) = x^n$ and $g(x) = a^x$ by pairs for various values of n and a. In each case determine approximate points of intersection (e.g., graph x^2 and 2^x). (*Hint:* Choose your scales carefully.)

57. Carry out the following steps (supplying reasons) to determine the set of all positive x values for which $x^e < e^x$.

  (a) Show that $x^e < e^x$ if and only if $(\ln x)/x < 1/e$. (*Hint:* Take logs.)

  (b) Let $f(x) = (\ln x)/x$. Show that $f(e) = 1/e$ is the absolute maximum value of f on $(0, \infty)$.

  (c) For what positive x values is $f(x) < 1/e$?

58. Follow steps similar to those in Exercise 57 to determine the set of all positive x values for which $x^a < a^x$, where $a > 0$.

59. Investigate how the number of intersections of the graphs of $y = a^x$ and $y = x^b$ is affected by the values of a and b. (Consider $a > 0$ only.)

60. Use a CAS to compare the growth rates of the following functions as $x \to \infty$. Arrange in order according to increasing growth rates:

$$\ln x, \quad x^x, \quad e^{2x}, \quad x^{150}, \quad x^{-4}, \quad x^{10}e^{-x}, \quad x^{1/200}, \quad \frac{e^{4x}}{x^8}$$

61. (a) Redo Example 7.15 with r replaced by the positive real number α.

  (b) Redo Exercise 73 in Exercise Set 7.2 with b and c replaced by positive real numbers β and γ.

7.4 EXPONENTIAL GROWTH AND DECAY

The exponential function is used to model certain phenomena of nature. Often, the rate of change of some quantity at any given instant is approximately proportional to the amount of the quantity present at that instant. For example, under certain conditions, a culture of bacteria will grow at a rate proportional to the number present (the more bacteria there are, the faster the rate of growth of the culture).

Even in human populations, a crude model is that the rate of population increase is proportional to the size of the population at any given time. This model was used by the English demographer Thomas Malthus, in the late eighteenth century, to predict future world populations. It is thus referred to as the **Malthus model**. The Malthus model is reasonably accurate for relatively short time intervals but because it does not take into account such limiting factors as war, famine, epidemics, and limits on available food supply and available space, its growth predictions are much too large when applied to longer time intervals. We will indicate in Exercise 24 (in Exercise Set 7.4) a more realistic model.

Bacteria cultures and human (as well as other) populations are quantities that increase with time. Indeed, the observation that the rate of oxygen given off in a sample of Martian soil examined by NASA's *Viking* spacecraft started to grow exponentially gave hope that there might be life on Mars. But when the rate of growth dropped below the exponential rate, that hope was dashed.

For some quantities, rate of change is proportional to the amount present, but with a negative growth rate; that is, the quantity decreases with time. An example is the rate of change of a radioactive substance, called *radioactive decay*. One interesting application of radioactive decay is in determining the

A Viking spacecraft on the surface of Mars

age of certain plant or animal remains. Archaeologists in particular have used this technique. It is called *radiocarbon dating*, and we will describe how it works in Exercise 25 in Exercise Set 7.4.

We could cite many other examples, but for now we want to show how this class of problems can be solved using calculus. We let $Q(t)$ denote the quantity in question at time t, where t is measured in appropriate units starting from a convenient time origin. Thus, $Q(t)$ might denote the number of bacteria in a culture t days after the culture was prepared, or $Q(t)$ might be the population of the world t yr after 1900, or $Q(t)$ might be the amount of strontium-90 (a radioactive substance) that will be left t yr from now out of 200 g currently on hand. In some cases the actual quantity in question might change by only discrete amounts; population, for example, is always an integer. Nevertheless, in our model we will assume Q is a differentiable function, so that it varies in a continuous fashion over some interval (remember that differentiability implies continuity). In cases where the answer must be an integer, we round our answer to the nearest integer. We assume in all cases that $Q(t) > 0$.

The basic assumption is that *the rate of change of Q is proportional to Q itself.* Thus, we have the differential equation

$$Q'(t) = kQ(t) \tag{7.30}$$

where k is the constant of proportionality, sometimes called the *growth constant*. We want to solve Equation 7.30 to determine the explicit form of $Q(t)$. Since $Q(t) \neq 0$, we can write the equation in the equivalent form

$$\frac{Q'(t)}{Q(t)} = k$$

and since

$$\frac{d}{dt} \ln Q(t) = \frac{Q'(t)}{Q(t)}$$

we have

$$\frac{d}{dt} \ln Q(t) = k$$

Taking the antiderivative, we get

$$\int \frac{d}{dt} \ln Q(t)\, dt = \int k\, dt$$

$$\ln Q(t) = kt + C$$

Hence,

$$Q(t) = e^{kt+C} \qquad \text{Since } e^{\ln a} = a$$

Since $e^{kt+C} = e^{kt} \cdot e^{C}$, we can write this result in the form

$$Q(t) = C_1 e^{kt}$$

where $C_1 = e^{C}$. Now suppose that when $t = 0$, $Q = Q_0$; that is, $Q(0) = Q_0$ (the initial amount of the quantity is Q_0 units). Then we get, on setting $t = 0$,

$$Q(0) = C_1 e^{k \cdot 0} \quad \text{or} \quad Q_0 = C_1$$

Thus, we obtain the formula for $Q(t)$:

Exponential Growth ($k > 0$) and Decay ($k < 0$) Formula

$$Q(t) = Q_0 e^{kt} \qquad (7.31)$$

where $Q_0 = Q(0)$.

It is easy to show that the value of Q given by Equation 7.31 does actually satisfy the differential equation (Equation 7.30) and has the correct value when $t = 0$. So this value is the solution we were seeking. The growth constant k is determined in each individual instance from known data, usually based on observed values of Q at certain times. When $k > 0$ we say that Q **grows exponentially**, and when $k < 0$ we say that Q **decays exponentially**.

The equation $Q'(t) = kQ(t)$ is an example of what is called a *separable* first-order differential equation, and the method we used to solve the equation for $Q(t)$ is known as *separation of variables*. The reason for this terminology becomes clearer if we write Q for $Q(t)$ and dQ/dt for $Q'(t)$. Then Equation 7.30 becomes $dQ/dt = kQ$ or, equivalently,

$$dQ = kQ\,dt$$

On dividing by Q, we get

$$\frac{dQ}{Q} = k\,dt$$

Now the variables Q and t are *separated* on opposite sides of the equation. Taking antiderivatives, we get, as before,

$$\ln Q = kt + C$$

and then we solve for $Q = Q(t)$ as we did earlier.

EXAMPLE 7.23   A culture of HeLa cells is observed to triple in size in 2 days. Assuming it grows exponentially, how large will the culture be in 5 days?

Solution   We are not told the original size of the culture, so we let this initial value be denoted by Q_0. We let $Q(t)$ be the size of the culture t days after the initial observation. When $t = 2$ we know that the size has tripled, so $Q(2) = 3Q_0$. By Equation 7.31, we have

$$Q(t) = Q_0 e^{kt}$$

On setting $t = 2$, we get

$$3Q_0 = Q_0 e^{k \cdot 2}$$

$$e^{2k} = 3$$

We write this equation in the equivalent logarithmic form to find k:

$$2k = \ln 3$$

$$k = \frac{\ln 3}{2}$$

HeLa cells originated in 1951 in a tumor from a woman named Henrietta Lacks. They are now found worldwide in laboratories that do cell biology research. Cultures of HeLa cells grow exponentially and have been known to double in a single day.

Using a calculator, we could get a decimal approximation for k. However, it is often an advantage to leave k in an exact form at this stage. Substituting this value for k gives

$$Q(t) = Q_0 e^{t(\ln 3)/2} = Q_0 \left(e^{\ln 3}\right)^{t/2}$$

$$Q(t) = Q_0(3)^{t/2} \quad \text{Since } e^{\ln 3} = 3$$

(You can see now the advantage of leaving k in an exact form.) When $t = 5$ we have

$$Q(5) = Q_0(3)^{5/2} \approx 15.6 Q_0$$

So after 5 days the culture will be approximately 15.6 times its original size.

■

REMARK ——————————————————————————————

■ An alternative to solving for k in the preceding example is to solve for e^k, as follows. Since $e^{2k} = 3$, we have $(e^{2k})^{1/2} = (3)^{1/2}$ or $e^k = 3^{1/2}$. Now write

$$Q(t) = Q_0 e^{kt} = Q_0(e^k)^t = Q_0(3^{1/2})^t$$

or

$$Q(t) = Q_0(3)^{t/2}$$

as we had before. A similar procedure can be used in other problems of this type.

——

A useful measure of the rate of decomposition of a radioactive substance is its **half-life**, defined to be the time required for half the original amount to decompose. Thus, if 100 g of a radioactive substance with a half-life of 5 yr is on hand, then after 5 yr, 50 g will remain; after 10 yr, 25 g will remain; and so on. The next example makes use of this concept.

EXAMPLE 7.24   The radioactive isotope strontium-90 has a half-life of 25 yr. If 200 kg of the substance is on hand, how much will remain after 10 yr? How long will it be until 90% of the original amount has decomposed? Strontium-90 has not been added to the atmosphere since the nuclear test ban of 1963. What percentage of the amount present in 1963 remained in 1993?

Solution   Let $Q(t)$ be the amount, in kilograms, that remains after t yr. Then we know that $Q(0) = 200$, and so by Equation 7.31,

$$Q(t) = Q_0 e^{kt} = 200 e^{kt}$$

Since the half-life is 25 yr, one-half of the original 200 kg will remain after 25 yr. That is, $Q(25) = 100$. Thus, setting $t = 25$ gives

$$100 = 200 e^{k \cdot 25}$$

$$e^{25k} = \frac{1}{2}$$

$$25k = \ln \frac{1}{2} = -\ln 2$$

$$k = -\frac{1}{25} \ln 2$$

So we obtain

$$Q(t) = 200e^{-(\ln 2/25)t} = 200(e^{\ln 2})^{-t/25} = 200(2)^{-t/25}$$

When $t = 10$, we get

$$Q(10) = 200(2)^{-10/25} = 200(2)^{-2/5} \approx 151.6 \text{ kg}$$

To determine how long it will be until 90% has decomposed, we first observe that at this time 10%, or 20 kg, will remain. Thus, we want to find t for which $Q(t) = 20$; that is,

$$200(2)^{-t/25} = 20$$

$$2^{-t/25} = \frac{1}{10}$$

We take the natural logarithm of both sides to find t:

$$-\frac{t}{25} \ln 2 = \ln \frac{1}{10}$$

or

$$-\frac{t}{25} \ln 2 = -\ln 10$$

$$t = \frac{25 \ln 10}{\ln 2} \approx 83 \text{ yr}$$

To determine the percentage of strontium-90 remaining in 1993 of the amount present in 1963, we let Q_0 be the unspecified amount present in 1963, and we let $t = 30$ (30 yr after 1963). In our equation for $Q(t)$, we replace the initial amount 200 by the unspecified amount Q_0, giving

$$Q(t) = Q_0(2)^{-t/25}$$

Thus,

$$Q(30) = Q_0(2)^{-30/25} = Q_0(2)^{-1.2} \approx 0.435 Q_0$$

As a fractional part of Q_0, we have

$$\frac{Q(30)}{Q_0} \approx \frac{0.435 Q_0}{Q_0} = 0.435$$

That is, about 43.5% of the amount in 1963 remained in 1993. ∎

EXAMPLE 7.25   In 1930 the world population was approximately 2 billion and in 1960 approximately 3 billion. Using the Malthus model, estimate the population in the year 2000.

Solution   Let $Q(t)$ be the population in billions, with t in years measured from 1930. Then $Q(0) = 2$, $Q(30) = 3$, and, according to the Malthus model,

$$Q(t) = Q_0 e^{kt} = 2e^{kt}$$

Setting $t = 30$ and proceeding in the usual way, we get

$$3 = 2e^{30k}$$

$$e^{30k} = \frac{3}{2}$$

$$30k = \ln \frac{3}{2}$$

$$k = \frac{1}{30} \ln \frac{3}{2}$$

So

$$Q(t) = 2(e^{\ln 3/2})^{t/30} = 2 \cdot \left(\frac{3}{2}\right)^{t/30}$$

In the year 2000, $t = 70$. So, according to the model, the population should be

$$Q(70) = 2 \cdot \left(\frac{3}{2}\right)^{70/30} = 2 \cdot \left(\frac{3}{2}\right)^{7/3} \approx 5.15 \text{ billion}$$ ∎

EXAMPLE 7.26   **Newton's Law of Cooling** states that when a body at initial temperature T_0 is introduced into a medium of temperature T_m, where $T_m < T_0$, the rate at which the body cools at any given instant is proportional to the difference between its temperature at that instant and the temperature T_m of the surrounding medium.

(a) Find a formula for the temperature of the body at time t units after being introduced into the medium.

(b) Suppose a body at $120°C$ is placed in air at $20°C$ and it has cooled to a temperature of $80°C$ after $\frac{1}{2}$ hr. Find its temperature after 1 hr.

Solution

(a) Let $T(t)$ be the temperature of the body t units of time after it is introduced into the medium. Then, by Newton's Law of Cooling,

$$T'(t) = k[T(t) - T_m]$$

Writing $T = T(t)$ and $dT/dt = T'(t)$, we obtain the equivalent differential equation

$$dT = k(T - T_m)dt$$

Now we separate variables and take antiderivatives:

$$\int \frac{dT}{T - T_m} = \int k\,dt$$

$$\ln(T - T_m) = kt + C \qquad \text{Remember that } T - T_m > 0.$$

$$T - T_m = e^{kt+C} = e^{kt} \cdot e^C = C_1 e^{kt}$$

So, on solving for T, we get

$$T(t) = T_m + C_1 e^{kt}$$

To find the constant C_1 we set $t = 0$, observing that $T(0) = T_0$:

$$T_0 = T_m + C_1 e^0$$

$$C_1 = T_0 - T_m$$

Finally,

$$T(t) = T_m + C_1 e^{kt}$$
$$= T_m + (T_0 - T_m)e^{kt} \qquad (7.32)$$

(b) We are given that $T_0 = 120$, $T_m = 20$, and $T(\frac{1}{2}) = 80$. So from Equation 7.32,

$$T(t) = 20 + 100e^{kt} \qquad (7.33)$$

$$80 = 20 + 100e^{k(1/2)}$$

$$60 = 100e^{k/2}$$

$$e^{k/2} = 0.6$$

$$\frac{k}{2} = \ln(0.6)$$

$$k = 2\ln(0.6) = \ln(0.36)$$

Substituting for k in Equation 7.33 gives

$$T(t) = 20 + 100e^{(\ln 0.36)t}$$

$$= 20 + (100)(e^{\ln 0.36})^t$$

$$= 20 + 100(0.36)^t$$

Finally, we want $T(1)$:

$$T(1) = 20 + 100(0.36) = 56$$

So after 1 hr the temperature should be approximately 56°C.   ■

REMARK ─────────────────────────────────

■ Equation 7.32 is valid also if the body is introduced into a warmer medium (so that $T_m > T_0$), and you will be asked to show this result in Exercise 17 in Exercise Set 7.4.

Exercise Set 7.4

1. A culture of HeLa cells originally numbers 500. After 2 hr there are 1500 cells in the culture. Assuming exponential growth, find how many are present after 6 hr.

2. A radioactive substance has a half-life of 64 hr. If 200 g of the substance is present initially, how much will remain after 4 days?

3. A culture of 100 bacteria doubles after 2 hr. How long will it take for the number of bacteria to reach 3200?

4. How long will it take 100 g of a radioactive substance that has a half-life of 40 yr to be reduced to 12.5 g?

5. One hundred kilograms of a certain radioactive substance decays to 40 kg after 10 yr. Find how much will remain after 20 yr.

6. In the initial stages the growth of a HeLa cell culture is approximately exponential. If the number doubles after 3 hr, what proportion of the original number will be present after 6 hr? After 12 hr?

7. If a culture of bacteria doubles in size after 2 hr, how long will it take for it to triple in size? Assume exponential growth.

8. If 100 g of a radioactive material diminishes to 80 g in 2 yr, find the half-life of the substance.

9. Find the half-life of a radioactive substance that is reduced by 30% in 20 hr.

10. A radioactive substance has a half-life of 6 yr. If 20 lb of the substance is present initially, how much will remain after 2 yr?

11. A culture of bacteria increases from 1000 to 5000 in 8 hr. Assuming the growth is exponential, find how much will be present after 20 hr.

12. The population of a certain Florida city increased by 40% between 1970 and 1978. If the population in 1978 was 210,000, what is the expected population in the year 2000, according to the Malthus model?

13. A certain radioactive substance has a half-life of 8 yr. If 200 g of the substance is present initially, how much will remain at the end of 12 yr? How long will it be until 90% of the original amount has decayed?

14. A body heated to 120°F is brought into a room in which the temperature is 70°F. After 15 min the temperature of the body is 100°F. According to Newton's Law of Cooling, how long will it take for the temperature of the body to drop to 80°F?

15. A thermometer registering 20°C is placed in a freezer in which the temperature is −10°C. After 10 min the thermometer registers 5°C. According to Newton's Law of Cooling, what will it register after 30 min?

16. According to Newton's Law of Cooling, will the temperature of the body ever reach the temperature T_m of the surrounding medium? Show why or why not. Does this result agree with reality? If not, how do you explain the discrepancy?

17. Prove that the result in Equation 7.32 remains valid if $T_m > T_0$. Use this to find how long it will take a thermometer that registers 5°C when on the outside to rise to 18°C after being brought to the inside where the temperature is 20°C, if it takes 10 min to rise to 12°C.

18. The rate of change of air pressure p with respect to altitude h is approximately proportional to the pressure. If at sea level the pressure is p_0, find a formula for p as a function of h. If $p_0 = 1.01 \times 10^5$ pascals (Pa), and at altitude $h = 2$ km (6,560 ft) the pressure is 8.08×10^4 Pa, find the approximate pressure at the 4 km (13,800 ft) altitude of the Mauna Kea Observatory in Hawaii.

19. The atmospheric pressure at sea level is approximately 15 lb/in.^{2}, and at Denver, which is 1 mi above sea level, the pressure is approximately 12 lb/in.^{2}. What is the approximate pressure at the top of Vail Pass, which is 2 mi above sea level? (See Exercise 18.)

20. The rate of change of the intensity I of light passing through a translucent material, with respect to the depth x of penetration of the light ray, is approximately proportional to the intensity at depth x. If the sunlight that strikes water is reduced to half of its original intensity at a depth of 3 m, what will be the intensity (as a fractional part of the original intensity) at a depth of 9 m?

21. Use the data of Exercise 20 to find at what depth the intensity of illumination will be 10% of the original.

22. In a certain type of chemical reaction a substance is converted in such a way that the rate of conversion is always proportional to the amount of unconverted substance. If 20 g of the substance is present originally, and half of it has been converted in 2 min, find how long it will be until only 4 g remains unconverted.

23. The rate at which the number of items of a certain type of merchandise are sold is proportional to the number still on hand. Let $N(t)$ be the number of items still on hand t days after the items were first introduced, and let $N_1 = N(1)$.
    (a) Derive a formula for $N(t)$.
    (b) If 20 items are on hand on the first day and 10 on the fifth day, how many can be expected to be on hand on the ninth day?

24. A more realistic model than the Malthus model for population growth is the **Verhulst model**, which assumes that the rate of growth of the population is jointly proportional to the size of the population and the further growth capacity. Thus, if $Q(t)$ denotes the size of the population at time t and m is the ultimate limit on population, then

$$\frac{dQ}{dt} = kQ(m - Q)$$

    (a) Prove, by direct substitution into the differential equation, that

$$Q(t) = \frac{mQ_0}{Q_0 + (m - Q_0)e^{-mkt}}$$

    is a solution of the Verhulst model, where $Q_0 = Q(0)$. (A typical graph of $Q(t)$, called a **logistic curve**, is shown in the accompanying figure.)
    (b) Prove that for the solution in part (a), $\lim_{t \to \infty} Q(t) = m$.
    (c) Redo Example 7.22 using the Verhulst model with the assumption that the maximum supportable world population is 10 billion.

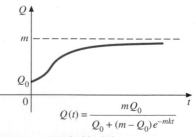

$$Q(t) = \frac{mQ_0}{Q_0 + (m - Q_0)e^{-mkt}}$$

A typical logistic curve
for population growth

25. A radioactive isotope of carbon, called carbon-14, is found in all living organisms. During the lifetime of an organism the amount that decays is replenished through the atmosphere, so that the amount in the

organism remains constant, but after the death of the organism the decayed carbon-14 is no longer replenished. Because the half-life of carbon-14 is very long (approximately 5730 yr), the remains of organisms that died thousands of years ago may still contain measurable amounts of carbon-14. By comparing the amount remaining with the amount known to be present in the particular organism when it was alive, the age of the remains can be determined. Suppose archaeologists find a human skull and determine that 10% of the original amount of carbon-14 remains. (The original amount is determined by measuring the ratio of carbon-14 to a stable isotope of carbon in a present-day object.) Find the approximate age of the skull.

26. The radioactive substance einsteinium-253 decays at a rate given approximately by $dQ/dt = -0.0939Q$, where t is measured in days, and $Q(t)$ is the amount present at time t.
    (a) Find the half-life.
    (b) Determine the proportion of an original amount that will remain after 8 days.
    (c) Find the number of days required for 80% of an original amount to decay.

27. Certain radioactive substances decay into a second radioactive substance. If $Q(t)$ denotes the amount of such a radioactive substance at time t and $R(t)$ denotes the amount of the second substance, then

    $$Q(t) = Q_0 e^{kt}, \qquad Q_0 = Q(0)$$

    and since the amount of substance that has decomposed is the original amount Q_0 minus the amount $Q(t)$ still present,

    $$R(t) = Q_0 - Q(t) = Q_0(1 - e^{kt})$$

    (a) Find a formula for the time required for the ratio of R to Q to equal a given value λ.
    (b) Rubidium-87, with a half-life of 5×10^{11} yr, decays exponentially into strontium-87. Find the time required for there to be 1% as much strontium-87 as rubidium-87.

28. Refer to Exercise 27, part (b). By measuring the ratio of strontium-87 to rubidium-87 in fossils, scientists can estimate the age of the fossil. Use this method to estimate the age of a fossil that contains 237 parts per million of rubidium and 1.85 parts per million of strontium.

7.5 INVERSE TRIGONOMETRIC FUNCTIONS

We have seen that the natural logarithm function and its inverse, the exponential function, are important in modeling many physical phenomena, especially those involving growth and decay. Other things in nature are modeled using trigonometric functions, especially the sine and cosine. The wave generated by the apparatus in the photo traces a sine and a cosine curve. Such things as ocean tides, electric impulses, human heartbeats, and many other periodic phenomena can be described using trigonometric functions. A natural question to ask is whether inverses of the trigonometric functions might also be useful. But there is a problem here. Recall that for a function to have an inverse, its graph must pass the horizontal line test. Clearly, none of the six trigonometric functions passes this test. For example, the graph of the sine function is shown in Figure 7.19, and it is clear that any horizontal line drawn between $y = -1$ and $y = 1$ intersects the graph in infinitely many points.

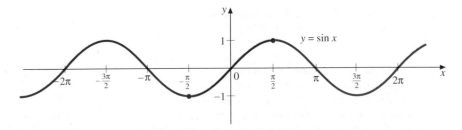

FIGURE 7.19
The part of the sine curve in red is used for the inverse.

A mechanical way of showing a sine and a cosine

The Inverse Sine Function

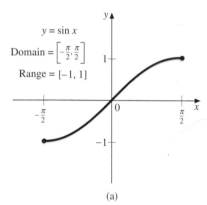

$y = \sin x$

Domain $= \left[-\frac{\pi}{2}, \frac{\pi}{2}\right]$

Range $= [-1, 1]$

(a)

We can remedy the problem of the failure of the sine function to have an inverse by considering a restricted domain on which the sine function is one-to-one. A convenient choice of such a restriction is the interval $[-\pi/2, \pi/2]$. We show the portion of the graph on this restricted domain in red in Figure 7.19. On that interval, the sine is increasing (since its derivative, $\cos x$, is positive between $-\pi/2$ and $\pi/2$), and it assumes all values in its range $[-1, 1]$. We denote the inverse of the sine restricted in this way by $\mathbf{\sin^{-1}}$ or by $\mathbf{arcsin}$. We will normally use $\sin^{-1}$, since it is consistent with the notation f^{-1} that we have been using for inverses in general. However, since both notations are in widespread use, you should be familiar with both. Relationships between the restricted sine function and its inverse are shown in Figure 7.20.

In summary, we have the following.

$y = \sin^{-1} x$

Domain $= [-1, 1]$

Range $= \left[-\frac{\pi}{2}, \frac{\pi}{2}\right]$

(b)

FIGURE 7.20

The Inverse Sine Function

$$y = \sin^{-1} x \text{ if and only if } \sin y = x$$

where $-1 \le x \le 1$ and $-\frac{\pi}{2} \le y \le \frac{\pi}{2}$

For example, we have

$$\sin^{-1} 1 = \frac{\pi}{2} \qquad\qquad \sin^{-1}\frac{1}{2} = \frac{\pi}{6}$$

$$\sin^{-1}(-1) = -\frac{\pi}{2} \qquad \sin^{-1}\left(-\frac{1}{2}\right) = -\frac{\pi}{6}$$

and so on.

Note carefully that $\sin^{-1} x$ *must* be a number in the interval $[-\pi/2, \pi/2]$.

It would be *incorrect*, for example, to say that $\sin^{-1}(-1/2) = (11\pi)/6$ even though $\sin(11\pi)/6 = -1/2$, because $(11\pi)/6$ does not lie in the interval $[-\pi/2, \pi/2]$.

The value of $\sin^{-1} x$, then, can be stated in words:

$\sin^{-1} x$ is that number in $[-\pi/2, \pi/2]$ whose sine is x.

For any function f for which an inverse f^{-1} exists, we know that

$$f^{-1}(f(x)) = x \qquad \text{for all } x \text{ in the domain of } f$$

and

$$f(f^{-1}(x)) = x \qquad \text{for all } x \text{ in the domain of } f^{-1}$$

Thus, we have the *cancellation laws*

$$\sin^{-1}(\sin x) = x \qquad \text{for } -\tfrac{\pi}{2} \le x \le \tfrac{\pi}{2}$$

and

$$\sin(\sin^{-1} x) = x \qquad \text{for } -1 \le x \le 1$$

 Be very careful in applying the first of these equations. For example,

$$\sin^{-1}\left(\sin \frac{2\pi}{3}\right) \ne \frac{2\pi}{3}$$

since $(2\pi)/3$ is not in the interval $[-\pi/2, \pi/2]$. Instead, we should write

$$\sin^{-1}\left(\sin \frac{2\pi}{3}\right) = \sin^{-1}\left(\frac{\sqrt{3}}{2}\right) = \frac{\pi}{3}$$

since $\pi/3$ is between $-\pi/2$ and $\pi/2$, and $\sin(\pi/3) = \sqrt{3}/2$.

The Derivative of $\sin^{-1} x$

To find the derivative of the inverse sine, we let $y = \sin^{-1} x$, so that $\sin y = x$. Then, differentiating the latter equation implicitly, we get

$$(\cos y)y' = 1$$

Thus, if $\cos y \ne 0$,

$$y' = \frac{1}{\cos y}$$

Since $\cos y > 0$ for $-\pi/2 < y < \pi/2$, we conclude that the derivative of $\sin^{-1} x$ exists on the open interval $(-1, 1)$. To obtain the answer in terms of x, observe that $\sin^2 y + \cos^2 y = 1$, so $\cos y = \sqrt{1 - \sin^2 y} = \sqrt{1 - x^2}$ (we choose the positive sign when taking the square root, since we know that $\cos y > 0$ when $-\pi/2 < y < \pi/2$).

An alternative way of finding $\cos y$ in terms of x is by the right triangle in Figure 7.21. There we label an acute angle as y, the side opposite as x, and the hypotenuse as 1. Labeling the triangle in this way expresses the fact that

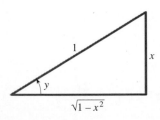

FIGURE 7.21
$y = \sin^{-1} x$

$y = \sin^{-1} x$; that is, $\sin y = x$. By the Pythagorean Theorem, the side adjacent to y is $\sqrt{1 - x^2}$. Thus, $\cos y = \sqrt{1 - x^2}$. We therefore have

$$\frac{d}{dx} \sin^{-1} x = \frac{1}{\sqrt{1 - x^2}} \qquad \text{if } x \in (-1, 1) \qquad (7.34)$$

Combining this formula with the Chain Rule in the usual way, we have the following formula for the derivative of $\sin^{-1} u$, where $u = g(x)$ is a differentiable function such that $g(x) \in (-1, 1)$:

$$\frac{d}{dx} \sin^{-1} u = \frac{1}{\sqrt{1 - u^2}} \frac{du}{dx} \qquad (7.35)$$

EXAMPLE 7.27   Find the derivative of each of the following:

(a) $y = \sin^{-1} 1/x$ (b) $y = x \sin^{-1} 2x$

Solution

(a) To apply Equation 7.35, we need to have $g(x) = 1/x$ in the interval $(-1, 1)$, which will be true if $x < -1$ or $x > 1$. With such values of x, we have, by Equation 7.35,

$$\frac{dy}{dx} = \frac{1}{\sqrt{1 - \dfrac{1}{x^2}}} \left(-\frac{1}{x^2} \right)$$

$$= \frac{1}{\dfrac{1}{|x|} \sqrt{x^2 - 1}} \left(-\frac{1}{x^2} \right) \qquad \text{Remember that } \sqrt{x^2} = |x|.$$

$$= -\frac{|x|}{x^2} \cdot \frac{1}{\sqrt{x^2 - 1}}$$

$$= \begin{cases} -\dfrac{1}{x\sqrt{x^2 - 1}} & \text{if } x > 1 \\[2mm] \dfrac{1}{x\sqrt{x^2 - 1}} & \text{if } x < -1 \end{cases}$$

(b) Using the Product Rule and Equation 7.35, we get

$$\frac{dy}{dx} = \sin^{-1} 2x + x \cdot \frac{1}{\sqrt{1 - 4x^2}} \cdot 2 = \sin^{-1} 2x + \frac{2x}{\sqrt{1 - 4x^2}}$$

This result is valid if $2x \in (-1, 1)$—that is, if $x \in (-\frac{1}{2}, \frac{1}{2})$.   ∎

Integral Formulas Involving the Inverse Sine

Formula 7.35 for the derivative of $\sin^{-1} u$ gives rise to the following antidifferentiation formula:

$$\int \frac{du}{\sqrt{1 - u^2}} = \sin^{-1} u + C, \qquad u \in (-1, 1) \qquad (7.36)$$

In Exercise 45 in Exercise Set 7.5 we will ask you to prove the more general formula given below, where $u = g(x)$ is a differentiable function on the open interval $(-a, a)$, for any positive constant a.

$$\int \frac{du}{\sqrt{a^2 - u^2}} = \sin^{-1} \frac{u}{a} + C, \qquad u \in (-a, a) \qquad (7.37)$$

EXAMPLE 7.28   Evaluate the following integrals:

(a) $\displaystyle \int_0^1 \frac{dx}{\sqrt{4 - x^2}}$ (b) $\displaystyle \int \frac{dx}{\sqrt{2x - x^2}}$

Solution

(a) By Equation 7.37,

$$\int_0^1 \frac{dx}{\sqrt{4 - x^2}} = \sin^{-1} \frac{x}{2} \Big]_0^1 = \sin^{-1} \frac{1}{2} - \sin^{-1} 0 = \frac{\pi}{6} - 0 = \frac{\pi}{6}$$

(b) We can make the integral conform to the pattern of Equation 7.36 by completing the square on $2x - x^2$:

$$\int \frac{dx}{\sqrt{2x - x^2}} = \int \frac{dx}{\sqrt{1 - (x^2 - 2x + 1)}} = \int \frac{dx}{\sqrt{1 - (x - 1)^2}}$$

Now let $u = x - 1$; then $du = dx$. So we have

$$\int \frac{dx}{\sqrt{1 - (x - 1)^2}} = \int \frac{du}{\sqrt{1 - u^2}} = \sin^{-1} u + C = \sin^{-1}(x - 1) + C$$

∎

EXAMPLE 7.29   **The Circumference of a Circle** Verify the formula $C = 2\pi r$ for the circumference of a circle of radius r.

Solution   Place the center of the circle at the origin, so that its equation is $x^2 + y^2 = r^2$. By symmetry, we can find the length of the upper semicircle (see Figure 7.22) and double it. The equation of the upper semicircle is

$$y = \sqrt{r^2 - x^2}$$

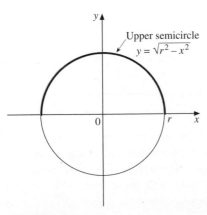

FIGURE 7.22

By Equation 6.18 for arc length, we know that the length of this semicircle is

$$\int_{-r}^{r} \sqrt{1 + \left(\frac{dy}{dx}\right)^2}\, dx = \int_{-r}^{r} \sqrt{1 + \left(\frac{-x}{\sqrt{r^2 - x^2}}\right)^2}\, dx$$

$$= \int_{-r}^{r} \sqrt{\frac{r^2 - x^2 + x^2}{r^2 - x^2}}\, dx$$

$$= \int_{-r}^{r} \frac{r}{\sqrt{r^2 - x^2}}\, dx \qquad \sqrt{r^2} = r \text{ since } r > 0$$

$$= r \int_{-r}^{r} \frac{dx}{\sqrt{r^2 - x^2}}$$

$$= r \left[\sin^{-1} \frac{x}{r}\right]_{-r}^{r} \qquad \text{By Equation 7.37}$$

$$= r[\sin^{-1} 1 - \sin^{-1}(-1)]$$

$$= r\left[\frac{\pi}{2} - \left(-\frac{\pi}{2}\right)\right] = \pi r$$

Therefore, the circumference is $2\pi r$.   ∎

The Other Inverse Trigonometric Functions

We can obtain inverses of the remaining trigonometric functions also by restricting their domains so that they become one-to-one functions. We show them (along with $\sin^{-1} x$ for completeness) in the following box.

The Inverse Trigonometric Functions

1. $y = \sin^{-1} x$ if and only if $\sin y = x$,
    where $-1 \le x \le 1$ and $-\pi/2 \le y \le \pi/2$
2. $y = \cos^{-1} x$ if and only if $\cos y = x$,
    where $-1 \le x \le 1$ and $0 \le y \le \pi$
3. $y = \tan^{-1} x$ if and only if $\tan y = x$,
    where $-\infty < x < \infty$ and $-\pi/2 < y < \pi/2$
4. $y = \cot^{-1} x$ if and only if $\cot y = x$,
    where $-\infty < x < \infty$ and $0 < y < \pi$
5. $y = \sec^{-1} x$ if and only if $\sec y = x$, where $1 \le x < \infty$ or
    $-\infty < x \le -1$ and $0 \le y < \pi/2$ or $\pi \le y < 3\pi/2$
6. $y = \csc^{-1} x$ if and only if $\csc y = x$, where $1 \le x < \infty$ or
    $-\infty < x \le -1$ and $0 < y \le \pi/2$ or $\pi < y \le 3\pi/2$

REMARK ───────────────────────────

■ There are other equally valid ways to restrict the domains of the secant and cosecant functions for forming the inverse. The choices we have made are motivated by the fact that the derivative formulas with these choices are simpler than with some of the other possible choices.

In Figures 7.23–7.27 we show the graphs of the restricted functions along with the graphs of their inverses.

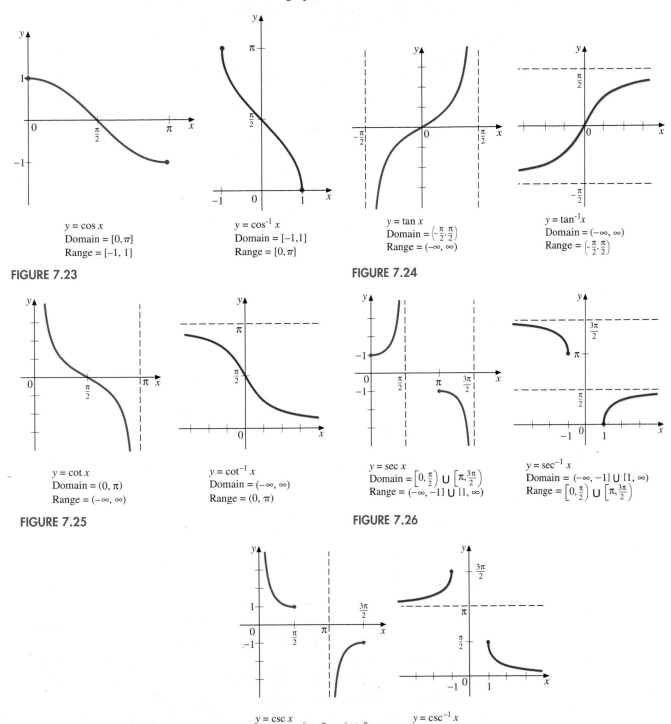

$y = \cos x$
Domain $= [0, \pi]$
Range $= [-1, 1]$

FIGURE 7.23

$y = \cos^{-1} x$
Domain $= [-1, 1]$
Range $= [0, \pi]$

$y = \tan x$
Domain $= \left(-\frac{\pi}{2}, \frac{\pi}{2}\right)$
Range $= (-\infty, \infty)$

FIGURE 7.24

$y = \tan^{-1} x$
Domain $= (-\infty, \infty)$
Range $= \left(-\frac{\pi}{2}, \frac{\pi}{2}\right)$

$y = \cot x$
Domain $= (0, \pi)$
Range $= (-\infty, \infty)$

FIGURE 7.25

$y = \cot^{-1} x$
Domain $= (-\infty, \infty)$
Range $= (0, \pi)$

$y = \sec x$
Domain $= \left[0, \frac{\pi}{2}\right) \cup \left[\pi, \frac{3\pi}{2}\right)$
Range $= (-\infty, -1] \cup [1, \infty)$

FIGURE 7.26

$y = \sec^{-1} x$
Domain $= (-\infty, -1] \cup [1, \infty)$
Range $= \left[0, \frac{\pi}{2}\right) \cup \left[\pi, \frac{3\pi}{2}\right)$

$y = \csc x$
Domain $= \left(0, \frac{\pi}{2}\right] \cup \left(\pi, \frac{3\pi}{2}\right]$
Range $= (-\infty, -1] \cup [1, \infty)$

$y = \csc^{-1} x$
Domain $= (-\infty, -1] \cup [1, \infty)$
Range $= \left(0, \frac{\pi}{2}\right] \cup \left(\pi, \frac{3\pi}{2}\right]$

FIGURE 7.27

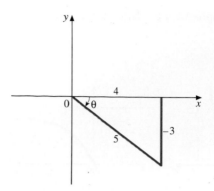

FIGURE 7.28

EXAMPLE 7.30   Evaluate each of the following:

(a) $\cos\left[\tan^{-1}\left(-\dfrac{3}{4}\right)\right]$ (b) $\sec^{-1}\left(\tan\dfrac{3\pi}{4}\right)$

Solution

(a) Let $\theta = \tan^{-1}(-3/4)$. Then $\tan\theta = -3/4$ and $-\pi/2 < \theta < \pi/2$. Since $\tan\theta$ is negative, θ is between $-\pi/2$ and 0. Viewed as an angle, θ is in the fourth quadrant, measured clockwise, as we show in Figure 7.28. By selecting the point $(4, -3)$ on the terminal side of θ we form the right triangle shown and see that the hypotenuse is 5. Thus,

$$\cos\left[\tan^{-1}\left(-\frac{3}{4}\right)\right] = \cos\theta = \frac{4}{5}$$

(b) Since $\tan 3\pi/4 = -1$, we want to find $\sec^{-1}(-1)$. Viewed as an angle, we want an angle that lies between π and $3\pi/2$ and whose secant is -1. The answer is π. Thus,

$$\sec^{-1}\left(\tan\frac{3\pi}{4}\right) = \pi$$

∎

Derivatives of the Other Inverse Trigonometric Functions

We will ask you to derive the following differentiation formulas in the exercises.

$$\frac{d}{dx}\cos^{-1}x = -\frac{1}{\sqrt{1-x^2}}, \qquad x \in (-1, 1) \qquad (7.38)$$

$$\frac{d}{dx}\tan^{-1}x = \frac{1}{1+x^2}, \qquad x \in (-\infty, \infty) \qquad (7.39)$$

$$\frac{d}{dx}\cot^{-1}x = -\frac{1}{1+x^2}, \qquad x \in (-\infty, \infty) \qquad (7.40)$$

$$\frac{d}{dx}\sec^{-1}x = \frac{1}{x\sqrt{x^2-1}}, \qquad |x| > 1 \qquad (7.41)$$

$$\frac{d}{dx}\csc^{-1}x = -\frac{1}{x\sqrt{x^2-1}}, \qquad |x| > 1 \qquad (7.42)$$

Formulas for the derivatives of $\cos^{-1}u$, $\tan^{-1}u$, $\cot^{-1}u$, $\sec^{-1}u$, and $\csc^{-1}u$, where u is a differentiable function of x that satisfies suitable restrictions, follow from these by the Chain Rule.

Integral Formulas Involving the Inverse Tangent and the Inverse Secant

From Formulas 7.39 and 7.41, we obtain

$$\int \frac{du}{1+u^2} = \tan^{-1} u + C \tag{7.43}$$

$$\int \frac{du}{u\sqrt{u^2-1}} = \sec^{-1} u + C, \qquad |u| > 1 \tag{7.44}$$

Observe that there is no need to write the integration formulas that arise from Equations 7.38, 7.40, and 7.42, since the negative sign can be factored out, leaving integrals for $\sin^{-1} x, \tan^{-1} x$, or $\sec^{-1} x$. For example, rather than write

$$\int -\frac{dx}{\sqrt{1-x^2}} = \cos^{-1} x + C$$

(which is correct), we write

$$\int -\frac{dx}{\sqrt{1-x^2}} = -\int \frac{dx}{\sqrt{1-x^2}} = -\sin^{-1} x + C$$

The following generalizations of Equations 7.43 and 7.44 are often useful.

For $a > 0$,

$$\int \frac{du}{a^2+u^2} = \frac{1}{a} \tan^{-1} \frac{u}{a} + C \tag{7.45}$$

$$\int \frac{du}{u\sqrt{u^2-a^2}} = \frac{1}{a} \sec^{-1} \frac{u}{a} + C, \qquad |u| > a \tag{7.46}$$

We will verify the first of these and leave the second for the exercises. The goal is to change the integrand so that it conforms to the form in Equation 7.43. To replace a^2 by 1, we divide numerator and denominator by a^2, getting

$$\int \frac{\dfrac{1}{a^2}}{1+\dfrac{u^2}{a^2}} \, du = \int \frac{\dfrac{1}{a^2}}{1+\left(\dfrac{u}{a}\right)^2} \, du$$

Now we make the substitution $v = u/a$. Then $dv = 1/a \, du$. Thus, we can write the integral as

$$\int \frac{\dfrac{1}{a}\left(\dfrac{1}{a}\right)}{1+\left(\dfrac{u}{a}\right)^2} \, du = \int \frac{\dfrac{1}{a}}{1+v^2} \, dv = \frac{1}{a} \int \frac{dv}{1+v^2}$$

By Equation 7.43,

$$\frac{1}{a} \int \frac{dv}{1+v^2} = \frac{1}{a} \left[\tan^{-1} v + C \right]$$

We therefore have, on replacing v by u/a,

$$\int \frac{du}{a^2 + u^2} = \frac{1}{a} \tan^{-1} \left(\frac{u}{a} \right) + C_1$$

where $C_1 = \frac{1}{a} C$. (Since C is an arbitrary constant, so is C_1.)

EXAMPLE 7.31   Differentiate each of the following:

(a) $y = \cos^{-1} \sqrt{x}$ $(0 < x < 1)$

(b) $y = \tan^{-1} e^x$

(c) $y = x \sec^{-1} x,$ $|x| > 1$

(d) $y = \cot^{-1}(\tan x)$

Solution

(a) $\dfrac{dy}{dx} = \dfrac{-1}{\sqrt{1-x}} \cdot \dfrac{1}{2\sqrt{x}} = \dfrac{-1}{2\sqrt{x - x^2}}$

(b) $\dfrac{dy}{dx} = \dfrac{1}{1 + e^{2x}} \cdot e^x = \dfrac{e^x}{1 + e^{2x}}$

(c) $\dfrac{dy}{dx} = \sec^{-1} x + x \cdot \dfrac{1}{x\sqrt{x^2 - 1}} = \sec^{-1} x + \dfrac{1}{\sqrt{x^2 - 1}}$

(d) $\dfrac{dy}{dx} = -\dfrac{1}{1 + \tan^2 x} \cdot \sec^2 x = -\dfrac{\sec^2 x}{\sec^2 x} = -1$ ∎

EXAMPLE 7.32   Evaluate the following integrals:

(a) $\displaystyle\int_{-\sqrt{3}}^{3\sqrt{3}} \frac{dx}{9 + x^2}$ (b) $\displaystyle\int \frac{dx}{x\sqrt{4x^2 - 9}}$

(c) $\displaystyle\int \frac{dx}{x^2 + 2x + 5}$ (d) $\displaystyle\int_0^{(\ln 3)/2} \frac{e^x}{e^{2x} + 1} dx$

Solution

(a) By Equation 7.45, with $a = 3$, we obtain

$$\int_{-\sqrt{3}}^{3\sqrt{3}} \frac{dx}{9 + x^2} = \frac{1}{3} \tan^{-1} \frac{x}{3} \Bigg]_{-\sqrt{3}}^{3\sqrt{3}} = \frac{1}{3} \left[\tan^{-1} \sqrt{3} - \tan^{-1} \left(-\frac{\sqrt{3}}{3} \right) \right]$$

$$= \frac{1}{3} \left[\frac{\pi}{3} - \left(-\frac{\pi}{6} \right) \right] = \frac{1}{3} \left(\frac{\pi}{2} \right) = \frac{\pi}{6}$$

As an alternative, we could divide the numerator and denominator by 9 to write the integrand in the form

$$\frac{1}{9} \left[\frac{1}{1 + \dfrac{x^2}{9}} \right] = \frac{1}{9} \left[\frac{1}{1 + \left(\dfrac{x}{3} \right)^2} \right]$$

Then we could substitute $u = x/3$ and proceed as in the derivation of Formula 7.45.

(b) To change this integral to the form of Equation 7.46, we let $u = 2x$. Then $du = 2\,dx$. So on multiplying numerator and denominator by 2, we have

$$\int \frac{dx}{x\sqrt{4x^2 - 9}} = \int \frac{2\,dx}{2x\sqrt{4x^2 - 9}} = \int \frac{du}{u\sqrt{u^2 - 9}} = \frac{1}{3}\sec^{-1}\frac{u}{3} + C$$

$$= \frac{1}{3}\sec^{-1}\frac{2x}{3} + C$$

(c) First, we complete the square in the denominator:

$$\int \frac{dx}{x^2 + 2x + 5} = \int \frac{dx}{(x^2 + 2x + 1) + 4} = \int \frac{dx}{(x + 1)^2 + 4}$$

Now if we mentally substitute $u = x + 1$, we see that this integral is in the form that occurs in Equation 7.45. So the answer is

$$\int \frac{dx}{x^2 + 2x + 5} = \frac{1}{2}\tan^{-1}\frac{x + 1}{2} + C$$

(d) This integral is in the form of Equation 7.45, with $u = e^x$ and $a = 1$. So we have

$$\int_0^{(\ln 3)/2} \frac{e^x}{e^{2x} + 1}\,dx = \tan^{-1} e^x\Big]_0^{(\ln 3)/2} = \tan^{-1}(e^{(\ln 3)/2}) - \tan^{-1} e^0$$

$$= \tan^{-1}\sqrt{3} - \tan^{-1} 1$$

$$= \frac{\pi}{3} - \frac{\pi}{4} = \frac{\pi}{12} \qquad\blacksquare$$

EXAMPLE 7.33   Find the area under the graph of

$$f(x) = \frac{1}{1 + x^2}$$

from $x = -1$ to $x = 1$.

Solution   The area A of the region (see Figure 7.29) is given by

$$A = \int_{-1}^1 \frac{dx}{1 + x^2} = \tan^{-1} x\Big]_{-1}^1$$

$$= \tan^{-1} 1 - \tan^{-1}(-1)$$

$$= \frac{\pi}{4} - \left(-\frac{\pi}{4}\right) = \frac{\pi}{2}$$

Note that we therefore have the following representation of π:

$$\pi = 2\int_{-1}^1 \frac{dx}{1 + x^2} \qquad\blacksquare$$

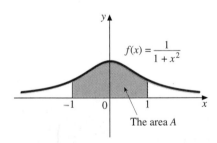

$$f(x) = \frac{1}{1 + x^2}$$

The area A

FIGURE 7.29

EXAMPLE 7.34   The TV broadcasting antennas for New York City are 10 m high and are on top of one of the towers of the World Trade Center, which is 330 m high. How far should a portable-television viewer at street level stand from the base of the building to get the best reception from the TV antennas, if the portable TV's antenna is about 2 m above the ground? (The reception is assumed to be best where the angle of the observer's eye subtended by the antenna is a maximum.)

Solution   The problem is to find x as shown in Figure 7.30 so as to maximize θ. We see that $\theta = \alpha - \beta$ and since $\cot\alpha = x/338$ and $\cot\beta = x/328$, we have

$$\alpha = \cot^{-1}\frac{x}{338} \quad \text{and} \quad \beta = \cot^{-1}\frac{x}{328}$$

FIGURE 7.30

(We could have used tangents instead, but the derivatives involved are somewhat simpler using cotangents.) So

$$\theta = \cot^{-1}\frac{x}{338} - \cot^{-1}\frac{x}{328} \qquad (x \geq 0)$$

Thus,

$$\frac{d\theta}{dx} = -\frac{1}{1+\left(\dfrac{x}{338}\right)^2}\cdot\frac{1}{338} + \frac{1}{1+\left(\dfrac{x}{328}\right)^2}\cdot\frac{1}{328}$$

$$= -\frac{338}{(338)^2 + x^2} + \frac{328}{(328)^2 + x^2}$$

Setting $d\theta/dx$ equal to 0 gives the critical points. After simplification we get

$$10x^2 = 328(338)^2 - 338(328)^2$$

$$= (328)(338)(338 - 328)$$

$$= (328)(338)(10)$$

$$x^2 = (328)(338)$$

$$x = \sqrt{110864} \approx 333 \text{ m} \qquad \text{\small Since } x > 0, \text{ ignore the negative square root.}$$

It is intuitively evident that as x varies from 0 to arbitrarily large values, θ attains a maximum value, and since we found only one critical point, it is the one we are seeking. We could confirm that this value of θ gives a maximum by using either the First- or Second-Derivative Test, but we will not do so. (See also the CAS-generated graph of $\theta(x)$ in Figure 7.31, showing the maximum value occurring at about $x = 333$.) ∎

$$\theta = \cot^{-1}\frac{x}{338} - \cot^{-1}\frac{x}{328}$$

FIGURE 7.31

Exercise Set 7.5

InExercises 1–12, give the exact value. (If answers involve π, leave them in terms of π.)

1. (a) $\sin^{-1} \frac{1}{2}$
   (b) $\cos^{-1} \frac{1}{2}$
   (c) $\arctan 1$
   (d) $\operatorname{arcsec} 2$
   (e) $\cot^{-1} 0$

2. (a) $\arccos(-\frac{1}{\sqrt{2}})$
   (b) $\tan^{-1}(-1)$
   (c) $\sin^{-1}(-\frac{\sqrt{3}}{2})$
   (d) $\operatorname{arccsc}(-2)$
   (e) $\sec^{-1}(-\frac{2}{\sqrt{3}})$

3. (a) $\arcsin(-1)$
   (b) $\cos^{-1}(-1)$
   (c) $\arctan 0$
   (d) $\cos^{-1}(-\frac{1}{2})$
   (e) $\cot^{-1}(-1)$

4. (a) $\sin[\cos^{-1}(-\frac{1}{3})]$
   (b) $\sec[\sin^{-1}(-\frac{1}{4})]$

5. (a) $\tan[\arccos \frac{3}{5}]$
   (b) $\cos[\tan^{-1}(-\frac{4}{3})]$

6. (a) $\sin[\arctan \frac{5}{12}]$
   (b) $\csc[\arccos(-\frac{2}{3})]$

7. (a) $\cos[2 \sin^{-1}(-\frac{1}{3})]$
   (b) $\sin[2 \cos^{-1}(-\frac{3}{5})]$

8. (a) $\sin^{-1}(\cos \frac{\pi}{3})$
   (b) $\cos^{-1}[\sin(-\frac{\pi}{6})]$

9. (a) $\cos^{-1}[\cos(-\frac{\pi}{5})]$
   (b) $\sin^{-1}(\sin \frac{9\pi}{8})$

10. (a) $\tan^{-1}(\tan \frac{3\pi}{4})$
    (b) $\cot^{-1}(\cot \frac{5\pi}{3})$

11. $\sin[\sin^{-1} \frac{3}{5} + \cos^{-1}(-\frac{5}{13})]$ (*Hint:* Use the identity for $\sin(\alpha + \beta)$.)

12. $\cos[\tan^{-1}(-\frac{1}{2}) - \sec^{-1} \frac{5}{3}]$ (*Hint:* Use the identity for $\cos(\alpha - \beta)$.)

 In Exercises 13 and 14, use a calculator or CAS to find the approximate radian measure of θ, where $0 \le \theta < 2\pi$.

13. (a) $\sin \theta = \frac{2}{3}$, $\cos \theta < 0$
    (b) $\cos \theta = \frac{1}{4}$, $\sin \theta < 0$

14. (a) $\tan \theta = 2$, $\sec \theta < 0$
    (b) $\cos \theta = -\frac{2}{5}$, $\cot \theta > 0$

In Exercises 15–24, find y'.

15. $y = \tan^{-1} \frac{2x}{3}$

16. $y = \sin^{-1}(2x - 1)$

17. $y = \arccos \sqrt{1 - x}$

18. $y = \sec^{-1} \frac{1}{x}$ $(0 < x < 1)$

19. $y = (1 + x^2) \tan^{-1} x$

20. $y = \sin^{-1} \sqrt{1 - x^2}$ $(0 < x < 1)$

21. $y = \dfrac{\cos^{-1} x}{\sqrt{1 - x^2}}$

22. $y = x^2 \sec^{-1} x^2$

23. $y - \sin^{-1} y = x^2$

24. $y = \tan^{-1} \dfrac{y}{x}$

In Exercises 25–40, evaluate the integrals.

25. $\displaystyle\int_{-1/2}^{\sqrt{3}/2} \frac{dx}{\sqrt{1 - x^2}}$

26. $\displaystyle\int_{-1}^{\sqrt{3}} \frac{dx}{1 + x^2}$

27. $\displaystyle\int_{\sqrt{2}}^{2} \frac{dx}{x\sqrt{x^2 - 1}}$

28. $\displaystyle\int_{-1}^{1} \frac{dx}{\sqrt{4 - x^2}}$

29. $\displaystyle\int_{0}^{4} \frac{dx}{16 + x^2}$

30. $\displaystyle\int_{4/\sqrt{3}}^{4} \frac{dx}{x\sqrt{x^2 - 4}}$

31. $\displaystyle\int \frac{dx}{\sqrt{16 - 9x^2}}$

32. $\displaystyle\int \frac{dx}{4x^2 + 25}$

33. $\displaystyle\int \frac{dx}{x\sqrt{16x^2 - 9}}$

34. $\displaystyle\int \frac{dx}{\sqrt{4x - 4x^2}}$

35. $\displaystyle\int \frac{dx}{9x^2 - 12x + 5}$

36. $\displaystyle\int_{0}^{4} \frac{dx}{x^2 - 2x + 4}$

37. $\displaystyle\int \frac{x \, dx}{\sqrt{1 - x^4}}$

38. $\displaystyle\int \frac{dx}{e^x + e^{-x}}$ (*Hint:* Multiply numerator and denominator by e^x.)

39. $\displaystyle\int_{-\sqrt{3}}^{-1} \frac{dx}{x\sqrt{4x^2 - 3}}$

40. $\displaystyle\int_{1}^{2} \frac{dx}{\sqrt{12x - 3x^2 - 8}}$

41. Find the area of the region under the graph of
$$y = \frac{1}{\sqrt{4 - x^2}}$$
from $x = -1$ to $x = 1$.

42. Find the area under the graph of
$$y = \frac{4}{4 + x^2}$$
from $x = -2$ to $x = 2$.

43. Find the volume of the solid generated by revolving the region under the graph of
$$y = \frac{1}{\sqrt{x^2 + 3}}$$
from $x = -1$ to $x = 3$ about the x-axis.

44. Find the volume of the solid generated by revolving the region under the graph of

$$y = \frac{1}{x^2\sqrt{x^2 - 9}}$$

from $x = 2\sqrt{3}$ to $x = 6$ about the y-axis.

In Exercises 45–52, prove Formulas 7.37–7.44, and in Exercise 53, prove Formula 7.46.

54. Give the exact value of $\tan^{-1}\frac{1}{3} + \tan^{-1}(-2)$. (*Hint:* Use $\tan(\alpha + \beta)$.)

55. Prove that if $0 < x < 1$,

$$\sin^{-1} x + \sin^{-1}\sqrt{1 - x^2} = \frac{\pi}{2}$$

(*Hint:* Use differentiation to prove that the left-hand side is a constant, and then substitute a particular value of x to show that the constant is $\pi/2$.)

56. Prove that if $x > 0$,

$$\tan^{-1} x + \tan^{-1}\frac{1}{x} = \frac{\pi}{2}$$

(See the hint for Exercise 55.)

In Exercises 57–60, evaluate the integrals.

57. $\displaystyle\int_1^3 \frac{dx}{\sqrt{x}(1 + x)}$

58. $\displaystyle\int \frac{dx}{\sqrt{e^{2x} - 1}}$

59. $\displaystyle\int_{\pi/2}^{\pi} \frac{\sin x}{1 + \cos^2 x}\,dx$

60. $\displaystyle\int \frac{dx}{x\sqrt{x^6 - 1}}$

61. Find the volume of the solid generated by rotating the region under the graph of $y = (1 + x^4)^{-1}$ from $x = 0$ to $x = 1$ about the y-axis.

62. A lighthouse is located on an island 2 km from the nearest point P on a straight shore. At a point on the shore 1 km from P the beam of light from the lighthouse beacon is moving at 40 km/min. Through how many radians per minute is the beacon turning? Through how many revolutions per minute is it turning?

63. A helicopter is initially 2 km above an observer on the ground, and it then moves horizontally at a speed of 30 m/s. What is the rate of change of the angle of elevation of the helicopter from the observer when the helicopter has gone 0.5 km?

 64. Use a calculator or CAS to approximate

$$\int_{-n}^{n} \frac{1}{1 + x^2}\,dx$$

for large values of n. As n increases, does the value of the integral get close to a familiar number? If so, which?

7.6 THE HYPERBOLIC FUNCTIONS

The combinations of exponential functions

$$\frac{e^x + e^{-x}}{2} \quad \text{and} \quad \frac{e^x - e^{-x}}{2}$$

have many interesting and useful properties. Because of their importance, they are given special names, as stated in the following definition.

Definition 7.3
The Hyperbolic Sine and Hyperbolic Cosine

The **hyperbolic sine**, abbreviated **sinh**, and the **hyperbolic cosine**, abbreviated **cosh**, are defined by

$$\sinh x = \frac{e^x - e^{-x}}{2}$$

$$\cosh x = \frac{e^x + e^{-x}}{2}$$

The domain of each of these functions is the set of all real numbers.

FIGURE 7.32

FIGURE 7.33

FIGURE 7.34

■ It is common practice to pronounce sinh like "cinch" and cosh to rhyme with "gosh."

We can obtain the graphs of the hyperbolic sine and cosine from their definitions in terms of the exponential function. Since

$$\sinh x = \frac{e^x - e^{-x}}{2} = \frac{1}{2}e^x + \left(-\frac{1}{2}e^{-x}\right)$$

we can draw the graphs of $y = \frac{1}{2}e^x$ and $y = -\frac{1}{2}e^{-x}$ and add the y values at different x values to find points on the graph $y = \sinh x$. In Figure 7.32 we show $y = \frac{1}{2}e^x$ and $y = -\frac{1}{2}e^{-x}$ (in dashed curves) and the graph of $y = \sinh x$ that results from this addition.

Similarly, for $\cosh x$, we write

$$\cosh x = \frac{e^x + e^{-x}}{2} = \frac{1}{2}e^x + \frac{1}{2}e^{-x}$$

So we can add the y values on the graphs of $y = \frac{1}{2}e^x$ and $y = \frac{1}{2}e^{-x}$ to get the graph of $y = \cosh x$, shown in Figure 7.33.

Since

$$\sinh(-x) = \frac{e^{-x} - e^x}{2} = -\sinh x$$

and

$$\cosh(-x) = \frac{e^{-x} + e^x}{2} = \cosh x$$

we see that $\sinh x$ is odd and $\cosh x$ is even, confirming the symmetries our graphs indicate. Also note that

$$\sinh 0 = \frac{e^0 - e^{-0}}{2} = \frac{1 - 1}{2} = 0$$

and

$$\cosh 0 = \frac{e^0 + e^{-0}}{2} = \frac{1 + 1}{2} = 1$$

Recall that for the trigonometric functions $\sin x$ and $\cos x$, we also know that $\sin x$ is odd, $\cos x$ is even, $\sin 0 = 0$, and $\cos 0 = 1$. So while the graphs of the hyperbolic sine and cosine do not resemble those of the trigonometric sine and cosine, they do have these even–odd characteristics in common, and their values at $x = 0$ are the same. These common features give the first hint of why the names *sine* and *cosine* are used with the hyperbolic functions. We will soon see that the analogy between the hyperbolic and trigonometric functions goes much further. We will also see why the name *hyperbolic* is used.

The hyperbolic cosine curve has an interesting application. When a uniform flexible cable hangs between two supports, if the only force acting on it is that caused by gravity, then it assumes a shape known as a **catenary**. For example, an electric power line hangs in this shape (see Figure 7.34). Any flexible object suspended between two points hangs in the shape of a catenary, even a spider web. If the axes are suitably chosen, the equation of the catenary can be shown to be of the form

$$y = a \cosh \frac{x}{a}$$

where a is the height of the catenary at its lowest point.

Another instance of the hyperbolic cosine is the Gateway Arch in St. Louis, Missouri. It is in the shape of an inverted hyperbolic cosine curve in which its height and span are the same: 630 feet.

The Gateway Arch in St Louis is in the form of a catenary.

Flexible objects suspended between two supports assume the shape of catenaries.

We define four other hyperbolic functions by analogy with the corresponding trigonometric functions. They are the **hyperbolic tangent** (tanh), **hyperbolic cotangent** (coth), **hyperbolic secant** (sech), and **hyperbolic cosecant** (csch).

Definition 7.4
Other Hyperbolic Functions

$$\tanh x = \frac{\sinh x}{\cosh x} \qquad\qquad \operatorname{sech} x = \frac{1}{\cosh x}$$

$$\coth x = \frac{\cosh x}{\sinh x}, \quad x \neq 0 \qquad\qquad \operatorname{csch} x = \frac{1}{\sinh x}, \quad x \neq 0$$

We show the graphs of these four functions in Figures 7.35, 7.36, 7.37, and 7.38.

$y = \tanh x$

FIGURE 7.35

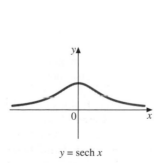

$y = \coth x$

FIGURE 7.36

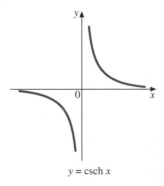

$y = \operatorname{sech} x$

FIGURE 7.37

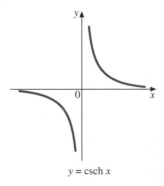

$y = \operatorname{csch} x$

FIGURE 7.38

Hyperbolic Identities

There are many relationships among the hyperbolic functions, called hyperbolic identities, that bear a striking resemblance to trigonometric identities. We list the most important ones here.

Basic Hyperbolic Identities

1. $\cosh^2 x - \sinh^2 x = 1$
2. $\sinh(x + y) = \sinh x \cosh y + \cosh x \sinh y$
3. $\cosh(x + y) = \cosh x \cosh y + \sinh x \sinh y$
4. $\sinh 2x = 2 \sinh x \cosh x$
5. $\cosh 2x = \cosh^2 x + \sinh^2 x = 2\cosh^2 x - 1 = 1 + 2\sinh^2 x$
6. $\sinh \dfrac{x}{2} = \pm\sqrt{\dfrac{\cosh x - 1}{2}}$
7. $\cosh \dfrac{x}{2} = \sqrt{\dfrac{\cosh x + 1}{2}}$
8. $1 - \tanh^2 x = \operatorname{sech}^2 x$
9. $\coth^2 x - 1 = \operatorname{csch}^2 x$
10. $\tanh 2x = \dfrac{2 \tanh x}{1 + \tanh^2 x}$
11. $\tanh \dfrac{x}{2} = \dfrac{\sinh x}{1 + \cosh x} = \dfrac{\cosh x - 1}{\sinh x}$

To verify Identity 1, we use the definitions of $\sinh x$ and $\cosh x$ to get

$$\cosh^2 x - \sinh^2 x = \left(\frac{e^x + e^{-x}}{2}\right)^2 - \left(\frac{e^x - e^{-x}}{2}\right)^2$$

$$= \frac{e^{2x} + 2 + e^{-2x}}{4} - \frac{e^{2x} - 2 + e^{-2x}}{4} = \frac{4}{4} = 1$$

For Identity 2, it is easier to show that the right-hand side equals the left:

$$\sinh x \cosh y + \cosh x \sinh y = \frac{e^x - e^{-x}}{2} \cdot \frac{e^y + e^{-y}}{2} + \frac{e^x + e^{-x}}{2} \cdot \frac{e^y - e^{-y}}{2}$$

After multiplying terms and simplifying, using the laws of exponents, we get

$$\frac{e^{x+y} - e^{-(x+y)}}{2}$$

which, by Definition 7.3, is $\sinh(x + y)$. Identity 3 can be verified similarly. The remaining identities can be verified using Identities 1, 2, and 3, along with Definition 7.4.

As we have noted, the hyperbolic identities are very similar to trigonometric identities. Some have exactly the same form; for example,

$$\sinh 2x = 2 \sinh x \cosh x \qquad \text{and} \qquad \sin 2x = 2 \sin x \cos x$$

Others differ only by a sign; for example,

$$\cosh^2 x - \sinh^2 x = 1 \qquad \text{and} \qquad \cos^2 x + \sin^2 x = 1$$

These similarities provide further explanation for the fact that the names of the hyperbolic functions involve the names of the trigonometric functions.

Derivatives of Hyperbolic Functions

To find the derivatives of $\sinh x$ and $\cosh x$, we use Definition 7.3 again. We have

$$\frac{d}{dx}(\sinh x) = \frac{d}{dx}\left(\frac{e^x - e^{-x}}{2}\right) = \frac{e^x + e^{-x}}{2} = \cosh x$$

and

$$\frac{d}{dx}(\cosh x) = \frac{d}{dx}\left(\frac{e^x + e^{-x}}{2}\right) = \frac{e^x - e^{-x}}{2} = \sinh x$$

Again, observe the similarities with the derivatives of $\sin x$ and $\cos x$. The derivatives of the other hyperbolic functions can now be obtained from the derivatives of $\sinh x$ and $\cosh x$, along with Definition 7.4. We ask you to derive these in the exercises. We list the results in the following box, where we have given their generalized forms, using the Chain Rule, with u a differentiable function of x.

Derivatives of Hyperbolic Functions

$$\frac{d}{dx}\sinh u = \cosh u \,\frac{du}{dx} \qquad (7.47)$$

$$\frac{d}{dx}\cosh u = \sinh u \,\frac{du}{dx} \qquad (7.48)$$

$$\frac{d}{dx}\tanh u = \operatorname{sech}^2 u \,\frac{du}{dx} \qquad (7.49)$$

$$\frac{d}{dx}\coth u = -\operatorname{csch}^2 u \,\frac{du}{dx} \qquad (7.50)$$

$$\frac{d}{dx}\operatorname{sech} u = -\operatorname{sech} u \tanh u \,\frac{du}{dx} \qquad (7.51)$$

$$\frac{d}{dx}\operatorname{csch} u = -\operatorname{csch} u \coth u \,\frac{du}{dx} \qquad (7.52)$$

EXAMPLE 7.35   Find dy/dx for the following:

(a) $y = \sinh(\ln x)$ (b) $y = x - \tanh x$

Solution

(a) By Formula 7.47,

$$\frac{dy}{dx} = \cosh(\ln x) \cdot \frac{d}{dx}(\ln x) = \frac{1}{x}\cosh(\ln x)$$

(b) By Formula 7.49 and Identity 8,

$$\frac{dy}{dx} = 1 - \ \text{sech}^2 x = \tanh^2 x$$

∎

Integrals of Hyperbolic Functions

From the derivative formulas, we immediately have the following antiderivative formulas (indefinite integrals):

Integrals of Hyperbolic Functions

$$\int \sinh u \, du = \cosh u + C \tag{7.53}$$

$$\int \cosh u \, du = \sinh u + C \tag{7.54}$$

$$\int \text{sech}^2 u \, du = \tanh u + C \tag{7.55}$$

$$\int \text{csch}^2 u \, du = -\coth u + C \tag{7.56}$$

$$\int \text{sech } u \tanh u \, du = -\text{sech } u + C \tag{7.57}$$

$$\int \text{csch } u \coth u \, du = -\text{csch } u + C \tag{7.58}$$

EXAMPLE 7.36   Evaluate the following integrals:

(a) $\displaystyle\int \frac{\sinh x}{1 + \cosh x} \, dx$ (b) $\displaystyle\int_0^{\ln 2} \frac{\text{sech}^2 x}{1 + \tanh^2 x} \, dx$ (c) $\displaystyle\int \cosh^2 x \, dx$

Solution

(a) If we let $u = 1 + \cosh x$, then $du = \sinh x \, dx$. So we have

$$\int \frac{\sinh x}{1 + \cosh x} \, dx = \int \frac{du}{u} = \ln |u| + C = \ln(1 + \cosh x) + C$$

Absolute values are not necessary since $1 + \cosh x$ is always positive (remember that $\cosh x \geq 1$).

(b) Since $\dfrac{d}{dx}(\tanh x) = \operatorname{sech}^2 x$, we observe that the integral is in the form

$$\int \frac{du}{1 + u^2}$$

with $u = \tanh x$. The antiderivative is therefore $\tan^{-1} u$ (by Equation 7.43). So we have

$$\int_0^{\ln 2} \frac{\operatorname{sech}^2 x \, dx}{1 + \tanh^2 x} = \tan^{-1}(\tanh x)\Big]_0^{\ln 2}$$

$$= \tan^{-1}[\tanh(\ln 2) - \tanh 0]$$

Since

$$\tanh x = \frac{\sinh x}{\cosh x} = \frac{e^x - e^{-x}}{e^x + e^{-x}}$$

we have $\tanh 0 = 0$, and

$$\tanh(\ln 2) = \frac{e^{\ln 2} - e^{-\ln 2}}{e^{\ln 2} + e^{-\ln 2}} = \frac{2 - \frac{1}{2}}{2 + \frac{1}{2}} = \frac{3}{5}$$

Thus,

$$\int_0^{\ln 2} \frac{\operatorname{sech}^2 x}{1 + \tanh^2 x} \, dx = \tan^{-1} \frac{3}{5} \approx 0.5404$$

(c) From Identity 5, we can write

$$\cosh^2 x = \frac{\cosh 2x + 1}{2}$$

So

$$\int \cosh^2 x \, dx = \frac{1}{2} \int (\cosh 2x + 1) dx$$

$$= \frac{1}{2} \cdot \frac{1}{2} \int \cosh 2x (2 \, dx) + \frac{1}{2} \int dx$$

$$= \frac{1}{4} \sinh 2x + \frac{1}{2}x + C \qquad \blacksquare$$

EXAMPLE 7.37   High-tension wires hang in the form of a catenary between poles that are 20 m apart. With the axes chosen as in Figure 7.39 (x-axis on the ground), the equation of the catenary is $y = 10 \cosh(x/10)$.

(a) Find the height of the wire at its lowest and highest points.

(b) Find the length of the wire between the two poles.

Solution

(a) The lowest point occurs at $x = 0$ and is given by

$$y_{\min} = 10 \cosh \frac{0}{10} = 10 \quad \text{Since } \cosh 0 = 1$$

The curve assumes its maximum height at each end. So

$$y_{\max} = 10 \cosh \frac{10}{10} = 10 \cosh 1 = 10 \cdot \frac{e + e^{-1}}{2} = 5\left(e + \frac{1}{e}\right) \approx 15.43$$

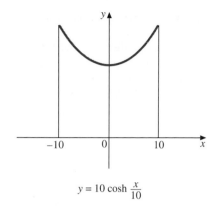

$$y = 10 \cosh \frac{x}{10}$$

FIGURE 7.39

(b) The total length is twice the length from $x = 0$ to $x = 10$. So we have

$$L = 2 \int_{x=0}^{x=10} ds = 2 \int_0^{10} \sqrt{1 + \left(\frac{dy}{dx}\right)^2}\, dx \qquad \text{By Formula 6.18}$$

$$= 2 \int_0^{10} \sqrt{1 + \sinh^2 \frac{x}{10}}\, dx$$

$$= 2 \int_0^{10} \cosh \frac{x}{10}\, dx \qquad \text{By Identity 1}$$

$$= 20 \int_0^{10} \cosh \frac{x}{10} \left(\frac{1}{10}\, dx\right)$$

$$= 20 \sinh \frac{x}{10} \Big]_0^{10} = 20(\sinh 1 - \sinh 0)$$

$$= 20 \sinh 1 = 20 \left(\frac{e - e^{-1}}{2}\right) = 10 \left(e - \frac{1}{e}\right) \approx 23.50 \qquad \blacksquare$$

Origin of the Name "Hyperbolic"

To understand why the functions we have been studying in this section are called *hyperbolic* functions, let us first recall how the trigonometric functions $\sin t$ and $\cos t$ are defined in terms of the unit circle $x^2 + y^2 = 1$. In Section 1.6 we saw that the point P that is t units along the circle from $A(1, 0)$ has coordinates $(\cos t, \sin t)$, as we indicate in Figure 7.40. These coordinates therefore satisfy the equation of the circle, giving the well-known identity

$$\cos^2 t + \sin^2 t = 1$$

Because of this relationship of $\sin t$ and $\cos t$ to the unit circle, these functions (as well as the other trigonometric functions) are sometimes called *circular* functions.

Since the hyperbolic functions $\sinh t$ and $\cosh t$ satisfy the identity

$$\cosh^2 t - \sinh^2 t = 1$$

it follows that the point P with coordinates $(\cosh t, \sinh t)$ lies on the "unit" hyperbola

$$x^2 - y^2 = 1$$

whose graph is shown in Figure 7.41. The point P is on the right-hand branch, since $\cosh t$ is always positive. It is above the x-axis when $t > 0$ and below when $t < 0$.

The geometric analogy between the relationship of trigonometric functions to the unit circle and hyperbolic functions to the unit hyperbola goes even further. The shaded areas in Figures 7.40 and 7.41 are called a *circular sector* and a *hyperbolic sector*, respectively. For $0 < t < 2\pi$, the area of each sector can be shown to be $t/2$; that is, in each case

$$\text{area of sector } OAP = \frac{t}{2}$$

(See Exercises 54 and 55 in Exercise Set 7.6.)

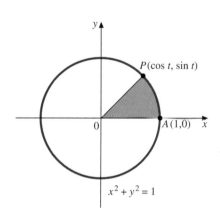

$$P(\cos t, \sin t)$$
$$A(1,0)$$
$$x^2 + y^2 = 1$$

FIGURE 7.40

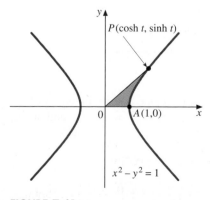

$$P(\cosh t, \sinh t)$$
$$A(1,0)$$
$$x^2 - y^2 = 1$$

FIGURE 7.41

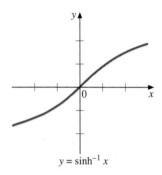

$y = \sinh x$

FIGURE 7.42

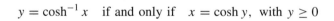

$y = \sinh^{-1} x$

FIGURE 7.43

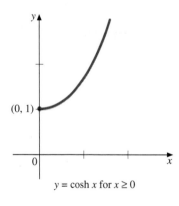

$y = \cosh x$ for $x \geq 0$

$(0, 1)$

$(1, 0)$

$y = \cosh^{-1} x$

FIGURE 7.44

Inverse Hyperbolic Functions

Just as we defined inverses of the trigonometric functions, after suitable restrictions on their domains, we can also define inverses of the hyperbolic functions. Restrictions on the domain are not necessary in all cases. Consider first $\sinh x$. Since $\frac{d}{dx}(\sinh x) = \cosh x$ and $\cosh x > 0$ for all x, it follows that $\sinh x$ is always increasing, so it is one-to-one. (See Figure 7.42.) It therefore has an inverse, which we designate by $\sinh^{-1}$. Thus,

$$y = \sinh^{-1} x \quad \text{if and only if} \quad x = \sinh y$$

The graph of $y = \sinh^{-1} x$ is shown in Figure 7.43.

To find the derivative of $\sinh^{-1} x$, we set $y = \sinh^{-1} x$ and solve for x to get

$$x = \sinh y$$

Now we differentiate implicitly.

$$1 = (\cosh y)y'$$

$$y' = \frac{1}{\cosh y}$$

By Identity 1, $\cosh^2 y = 1 + \sinh^2 y$. Also, $\sinh y = x$, so $\cosh^2 y = 1 + x^2$. We therefore have

$$\frac{d}{dx}(\sinh^{-1} x) = \frac{1}{\sqrt{1 + x^2}}$$

The hyperbolic cosine is not 1–1 (apply the horizontal line test to its graph in Figure 7.33). But when we restrict its domain to the interval $[0, \infty)$, it is then 1–1. So it has an inverse, designated by $\cosh^{-1}$. Thus,

$$y = \cosh^{-1} x \quad \text{if and only if} \quad x = \cosh y, \text{ with } y \geq 0$$

We show the graphs of the restricted hyperbolic cosine and its inverse in Figure 7.44. Differentiating both sides of the equation $x = \cosh y$ and using Identity 1 again, we obtain

$$1 = (\sinh y)y'$$

$$y' = \frac{1}{\sinh y} = \frac{1}{\sqrt{\cosh^2 y - 1}} \qquad y > 0$$

Since $\cosh y = x$, we have

$$\frac{d}{dx}\cosh^{-1}x = \frac{1}{\sqrt{x^2-1}} \qquad x > 1$$

Derivatives of the other inverse hyperbolic functions can be obtained in a similar way. We will ask you to draw their graphs and to derive formulas for their derivatives in the exercises. Just as with $\cosh x$, we restrict the domain of $\operatorname{sech} x$ to $[0, \infty)$ in order to define its inverse. Except for these two, the hyperbolic functions are 1–1 on their entire domains. We summarize all the formulas for derivatives of the inverse hyperbolic functions in the following box, replacing x by u, where u is a differentiable function of x.

Derivatives of Inverse Hyperbolic Functions

$$\frac{d}{dx}\sinh^{-1}u = \frac{1}{\sqrt{1+u^2}}\frac{du}{dx} \tag{7.59}$$

$$\frac{d}{dx}\cosh^{-1}u = \frac{1}{\sqrt{u^2-1}}\frac{du}{dx}, \qquad u > 1 \tag{7.60}$$

$$\frac{d}{dx}\tanh^{-1}u = \frac{1}{1-u^2}\frac{du}{dx}, \qquad |u| < 1 \tag{7.61}$$

$$\frac{d}{dx}\coth^{-1}u = \frac{1}{1-u^2}\frac{du}{dx}, \qquad |u| > 1 \tag{7.62}$$

$$\frac{d}{dx}\operatorname{sech}^{-1}u = -\frac{1}{u\sqrt{1-u^2}}\frac{du}{dx}, \qquad 0 < u < 1 \tag{7.63}$$

$$\frac{d}{dx}\operatorname{csch}^{-1}u = -\frac{1}{|u|\sqrt{1+u^2}}\frac{du}{dx}, \qquad u \neq 0 \tag{7.64}$$

EXAMPLE 7.38   Find the derivative of each of the following:

(a) $y = \sinh^{-1}(\tan x), \qquad -\pi/2 < x < \pi/2$

(b) $y = \tanh^{-1}\dfrac{1+x}{1-x}, \qquad x \neq 1$

Solution

(a) By Formula 7.59,

$$\frac{dy}{dx} = \frac{1}{\sqrt{1+\tan^2 x}}\cdot \sec^2 x = \frac{1}{\sec x}\cdot \sec^2 x = \sec x$$

Here we used the fact that $\sqrt{1+\tan^2 x} = \sqrt{\sec^2 x} = \sec x$, since $\sec x > 0$ for x in $(-\pi/2, \pi/2)$.

(b) By Formula 7.61,

$$\frac{dy}{dx} = \frac{1}{1 - \left(\dfrac{1+x}{1-x}\right)^2} \cdot \frac{(1-x) - (1+x)(-1)}{(1-x)^2}$$

$$= \frac{(1-x)^2}{(1-x)^2 - (1+x)^2} \cdot \frac{2}{(1-x)^2}$$

$$= \frac{2}{(1-x)^2 - (1+x)^2}$$

$$= \frac{2}{[(1-x) + (1+x)][(1-x) - (1+x)]}$$

$$= -\frac{1}{2x}$$

Note how we used the algebraic identity $a^2 - b^2 = (a+b)(a-b)$ to simplify the denominator in the next-to-last step.   ■

Integral Formulas Involving Inverse Hyperbolic Functions

One of the primary uses of inverse hyperbolic functions is in integration. From the derivative formulas we have the following antiderivative formulas.

Integral Formulas

$$\int \frac{du}{\sqrt{1+u^2}} = \sinh^{-1} u + C \qquad\qquad (7.65)$$

$$\int \frac{du}{\sqrt{u^2-1}} = \cosh^{-1} u + C, \qquad u > 1 \qquad (7.66)$$

$$\int \frac{du}{1-u^2} = \begin{cases} \tanh^{-1} u + C, & \text{if } |u| < 1 \\ \coth^{-1} u + C, & \text{if } |u| > 1 \end{cases} \qquad (7.67)$$

$$\int \frac{du}{u\sqrt{1-u^2}} = -\operatorname{sech}^{-1} u + C, \qquad 0 < u < 1 \qquad (7.68)$$

$$\int \frac{du}{u\sqrt{1+u^2}} = -\operatorname{csch}^{-1} u + C, \qquad u > 0 \qquad (7.69)$$

Before giving examples involving integration, we give alternative ways of writing the inverse hyperbolic functions, involving natural logarithms. These formulas will be useful in integration problems. Since the hyperbolic functions can be expressed in terms of exponentials, it is not surprising that their inverses can be expressed in terms of logarithms. The next theorem makes this representation explicit.

THEOREM 7.7

The inverse hyperbolic functions have the following explicit representations:

$$\sinh^{-1} x = \ln(x + \sqrt{x^2 + 1}) \qquad -\infty < x < \infty \qquad (7.70)$$

$$\cosh^{-1} x = \ln(x + \sqrt{x^2 - 1}) \qquad x \geq 1 \qquad (7.71)$$

$$\tanh^{-1} x = \frac{1}{2} \ln \frac{1 + x}{1 - x} \qquad |x| < 1 \qquad (7.72)$$

$$\coth^{-1} x = \frac{1}{2} \ln \frac{x + 1}{x - 1} \qquad |x| > 1 \qquad (7.73)$$

$$\operatorname{sech}^{-1} x = \ln \left(\frac{1 + \sqrt{1 - x^2}}{x} \right) \qquad 0 < x \leq 1 \qquad (7.74)$$

$$\operatorname{csch}^{-1} x = \ln \left(\frac{1}{x} + \frac{\sqrt{x^2 + 1}}{|x|} \right) \qquad x \neq 0 \qquad (7.75)$$

We will prove Equation 7.70 and leave the others as exercises.

Proof of Equation 7.70 Let $y = \sinh^{-1} x$. Then $x = \sinh y$, so by the definition of $\sinh y$,

$$x = \frac{e^y - e^{-y}}{2}$$

Clearing the equation of fractions and rearranging, we get

$$e^y - 2x - e^{-y} = 0$$

Now we multiply both sides by e^y:

$$e^{2y} - 2xe^y - 1 = 0$$

This equation is quadratic in e^y (with coefficients $a = 1$, $b = -2x$, and $c = -1$), so by the Quadratic Formula,

$$e^y = \frac{2x \pm \sqrt{4x^2 + 4}}{2} = x \pm \sqrt{x^2 + 1}$$

But $e^y > 0$ and $\sqrt{x^2 + 1} > x$, so the negative sign cannot hold. Thus,

$$e^y = x + \sqrt{x^2 + 1}$$

In the equivalent logarithmic form,

$$y = \ln(x + \sqrt{x^2 + 1}) \qquad \blacksquare$$

By Theorem 7.7, each of the integral formulas (7.65–7.69) can be written in an equivalent form involving logarithms. We call attention in particular to the fact that the two parts of Formula 7.67 can be written in the single form

$$\int \frac{du}{1-u^2} = \frac{1}{2}\ln\left|\frac{1+u}{1-u}\right| + C, \qquad u \neq 1 \tag{7.76}$$

You will be asked to verify this result in the exercises. We should note that because the integral formulas can be given in different ways, the answers given by computer algebra systems will vary, though they are equivalent.

EXAMPLE 7.39 Evaluate the following integrals:

(a) $\displaystyle\int \frac{x}{\sqrt{x^4+1}}\,dx$ (b) $\displaystyle\int_1^{7/2} \frac{dx}{\sqrt{4x^2-1}}$

Solution

(a) Let $u = x^2$. Then $du = 2x\,dx$, and we have, by Equation 7.65,

$$\int \frac{x}{\sqrt{x^4+1}}\,dx = \frac{1}{2}\int \frac{2x\,dx}{\sqrt{x^4+1}} = \frac{1}{2}\int \frac{du}{\sqrt{u^2+1}} = \frac{1}{2}\sinh^{-1}u + C$$

$$= \frac{1}{2}\sinh^{-1}x^2 + C$$

(b) Let $u = 2x$; then $du = 2\,dx$. When $x = 1$, $u = 2$, and when $x = 7/2$, $u = 7$. So we have

$$\int_1^{7/2} \frac{dx}{\sqrt{4x^2-1}} = \frac{1}{2}\int_2^7 \frac{du}{\sqrt{u^2-1}} = \frac{1}{2}\cosh^{-1}u\Big]_2^7 \qquad \text{By Equation 7.66}$$

$$= \frac{1}{2}[\cosh^{-1}7 - \cosh^{-1}2]$$

$$= \frac{1}{2}[\ln(7+\sqrt{48}) - \ln(2+\sqrt{3})] \qquad \text{By Equation 7.71}$$

$$= \frac{1}{2}[\ln(7+4\sqrt{3}) - \ln(2+\sqrt{3})] = \frac{1}{2}\ln\frac{7+4\sqrt{3}}{2+\sqrt{3}}$$

$$= \frac{1}{2}\ln\left[\frac{7+4\sqrt{3}}{2+\sqrt{3}} \cdot \frac{2-\sqrt{3}}{2-\sqrt{3}}\right] = \frac{1}{2}\ln\frac{14+\sqrt{3}-12}{4-3}$$

$$= \frac{1}{2}\ln(2+\sqrt{3}) \qquad\qquad\qquad\qquad \blacksquare$$

Exercise Set 7.6

1. Give the domain and range of each of the hyperbolic functions.

2. Express $\tanh x$, $\coth x$, $\operatorname{sech} x$, and $\operatorname{csch} x$ in terms of exponential functions, and evaluate the following limits:
   (a) $\lim_{x\to\infty} \tanh x$ (b) $\lim_{x\to 0^+} \coth x$
   (c) $\lim_{x\to\infty} \coth x$ (d) $\lim_{x\to\infty} \operatorname{sech} x$
   (e) $\lim_{x\to 0^+} \operatorname{csch} x$ (f) $\lim_{x\to\infty} \operatorname{csch} x$

3. Use a calculator or a CAS to approximate the value of each of the following for $x = 0.5$, 1, 2, and 4:
   (a) $\tanh x$ (b) $\coth x$
   (c) $\operatorname{sech} x$ (d) $\operatorname{csch} x$

4. Show that sech is an even function and that tanh, coth, and csch are odd functions.

In Exercises 5–12, verify the hyperbolic identities.

5. $\cosh(x + y) = \cosh x \cosh y + \sinh x \sinh y$

6. $\sinh 2x = 2 \sinh x \cosh x$

7. $\cosh 2x = \cosh^2 x + \sinh^2 x$

$$= 2 \cosh^2 x - 1$$

$$= 1 + 2 \sinh^2 x$$

8. $\sinh \dfrac{x}{2} = \pm \sqrt{\dfrac{\cosh x - 1}{2}}$

9. $\cosh \dfrac{x}{2} = \sqrt{\dfrac{\cosh x + 1}{2}}$

10. $1 - \tanh^2 x = \operatorname{sech}^2 x$

11. $\coth^2 x - 1 = \operatorname{csch}^2 x$

12. $\tanh 2x = \dfrac{2 \tanh x}{1 + \tanh^2 x}$

In Exercises 13–24, find dy/dx.

13. $y = \tanh^3 2x$

14. $y = \dfrac{\cosh x}{1 - \sinh x}$

15. $y = \ln(\sinh x)$

16. $y = \tan^{-1}(\sinh x)$

17. $y = \sec^{-1}(\cosh x)$

18. $y = x - \coth x$

19. $y = \sqrt{\operatorname{sech} x^2}$

20. $y = \dfrac{\tanh x}{1 + \operatorname{sech} x}$

21. $x \cosh y - y \sinh x = 1$

22. $y^2 = \ln(x \cosh y)$

23. $y = (\sinh x)^x$

24. $y = (\sinh x)^{\operatorname{sech} x}$

In Exercises 25–42, evaluate the integrals.

25. $\displaystyle\int \sinh x \cosh x \, dx$

26. $\displaystyle\int \dfrac{1 + \tanh x}{\cosh^2 x} dx$

27. $\displaystyle\int \dfrac{\sinh x \, dx}{\sqrt{1 + \sinh^2 x}}$

28. $\displaystyle\int \sinh^2 x \, dx$

29. $\displaystyle\int \dfrac{1 + \cosh x}{x + \sinh x} dx$

30. $\displaystyle\int e^{\sinh^2 x} \sinh 2x \, dx$

31. $\displaystyle\int \dfrac{\tanh x}{\ln \cosh x} dx$

32. $\displaystyle\int x \sinh x^2 \cosh^2 x^2 dx$

33. $\displaystyle\int \sinh^3 x \, dx$

34. $\displaystyle\int \tanh^2 x \, dx$

35. $\displaystyle\int \cosh^3 x \, dx$

36. $\displaystyle\int \operatorname{sech}^3 x \tanh x \, dx$

37. $\displaystyle\int \operatorname{sech}^4 x \tanh x \, dx$

38. $\displaystyle\int \cosh^2 x \sinh^3 x \, dx$

39. $\displaystyle\int_{\ln 2}^{\ln 3} \coth x \, dx$

40. $\displaystyle\int_{-1}^{1} (\sinh x + \cosh x)^2 dx$

41. $\displaystyle\int_{0}^{\ln 2} \dfrac{\cosh x}{\sqrt{9 - 4 \sinh^2 x}} dx$

42. $\displaystyle\int_{-\ln 3}^{\ln 3} \dfrac{\sinh x}{1 + \cosh^2 x} dx$

In Exercises 43–45, evaluate the limits.

43. (a) $\displaystyle\lim_{x \to 0} \dfrac{\cosh x - x^2 - 1}{\sinh^2 x}$ (b) $\displaystyle\lim_{x \to 1^-} \dfrac{\tanh^{-1} x}{\ln(1 - x)}$

44. (a) $\displaystyle\lim_{x \to \infty} x^2 \operatorname{sech} x$

(b) $\displaystyle\lim_{x \to 0} (\operatorname{csch} x - \coth x)$

45. (a) $\displaystyle\lim_{x \to 0^+} (\sinh x)^x$

(b) $\displaystyle\lim_{x \to \infty} (\tanh x)^{\cosh x}$

46. Find the volume of the solid obtained by revolving the region under the graph of $y = \operatorname{sech} x$ between $x = -1$ and $x = 1$ about the x-axis.

47. Find the volume of the solid obtained by revolving the region under the graph of $y = \tanh x$ between $x = 0$ and $x = 2$ about the x-axis.

48. Find the surface area obtained by revolving the arc of the curve $y = \cosh x$ from $x = 0$ to $x = 2$ about the x-axis.

In Exercises 49–52, verify that the equation is an identity for all admissible values of x by transforming the left-hand side into the right-hand side.

49. $\sinh x (\coth x - \tanh x) = \operatorname{sech} x$

50. $\dfrac{\sinh x}{\coth x - \operatorname{csch} x} = \cosh x + 1$

51. $\dfrac{2}{\coth x - \tanh x} = \sinh 2x$

52. $\dfrac{\coth x + \tanh x}{\coth x - \tanh x} = \cosh 2x$

53. Verify the identities

$$\tanh \frac{x}{2} = \frac{\sinh x}{1 + \cosh x}$$

and

$$\tanh \frac{x}{2} = \frac{\cosh x - 1}{\sinh x}$$

54. Prove that the area of the circular sector OAP shown in the figure is $t/2$, where $0 < t < \pi/2$. (*Hint:* Make use of your knowledge of the area of the circle.)

55. Prove that the area of the hyperbolic sector OAP shown in the figure is $t/2$, where $t > 0$, by completing the following steps.

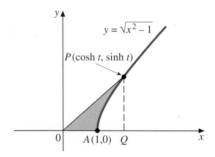

(a) Find the area of the triangle OPQ.
(b) Find the area under the arc AP by using Area $= \int_1^{\cosh t} y\,dx$. To evaluate the integral, substitute $x = \cosh u$.
(c) Subtract the area in part (b) from the area in part (a).

In Exercises 56–59, evaluate the integrals.

56. $\displaystyle\int \sinh^4 x\,dx$ **57.** $\displaystyle\int \tanh^5 x\,dx$

58. $\displaystyle\int \frac{dx}{\operatorname{sech}^2 x - \operatorname{sech}^4 x}$

59. $\displaystyle\int \frac{dx}{\cosh x - 1}$ (*Hint:* Multiply numerator and denominator by $\cosh x + 1$.)

In Exercises 60 and 61, derive the formulas.

60. $\displaystyle\int \operatorname{sech} x\,dx = \tanh^{-1} |\sinh u| + C$
   (*Hint:* Write $\operatorname{sech} x = \cosh x / \cosh^2 x$.)

61. $\displaystyle\int \operatorname{csch} x\,dx = \ln |\tanh \frac{x}{2}| + C$.   (*Hint:* Write $\operatorname{csch} x = \sinh x / \sinh^2 x$.)

In Exercises 62–64, verify that the given equation is an identity for all admissible values of x.

62. $\cosh^4 x = \dfrac{3}{8} + \dfrac{\cosh 2x}{2} + \dfrac{\cosh 4x}{8}$

63. $\dfrac{\sinh 3x}{\sinh x} - \dfrac{\cosh 3x}{\cosh x} = 2$

64. $2 \cosh^2 \dfrac{x}{2} \operatorname{csch} x = \coth \dfrac{x}{2}$

65. A uniform flexible cable suspended between two supporting towers 60 m apart hangs in the form of a catenary with the equation $y = 20 \cosh \frac{x}{20} + 10$, where y is the distance above the ground.
(a) Find the height of the cable at its lowest point.
(b) Find the height at each end.
(c) Determine the length of the cable.

66. (a) Show that $y = a \cosh \omega t + B \sinh \omega t$ satisfies the differential equation

$$\frac{d^2 y}{dt^2} - \omega^2 y = 0$$

(b) In part (a), find A and B such that when $t = 0$, $y = 3$ and $y' = 0$.

In Exercises 67–70, give the domain and range of f and draw its graph.

67. $f(x) = \tanh^{-1} x$ **68.** $f(x) = \coth^{-1} x$

69. $f(x) = \operatorname{sech}^{-1} x$ **70.** $f(x) = \operatorname{csch}^{-1} x$

71. (a) Prove $\operatorname{sech}^{-1} x = \cosh^{-1} \dfrac{1}{x}$, $x > 0$.
   (b) Prove $\operatorname{csch}^{-1} x = \sinh^{-1} \dfrac{1}{x}$, $x \neq 0$.

In Exercises 72–75, derive the formula for the derivative.

72. $\dfrac{d}{dx} \tanh^{-1} x$ **73.** $\dfrac{d}{dx} \coth^{-1} x$

74. $\dfrac{d}{dx} \operatorname{sech}^{-1} x$ **75.** $\dfrac{d}{dx} \operatorname{csch}^{-1} x$

In Exercises 76–85, find y'.

76. $y = \sinh^{-1} \sqrt{x}$

77. $y = \cosh^{-1} x^2$

78. $y = \tanh^{-1} \dfrac{1}{x}$

79. $y = \text{sech}^{-1}(\cos x), \quad 0 < x < \dfrac{\pi}{2}$

80. $y = \coth^{-1} \dfrac{x-1}{x+1}$

81. $y = x\,\text{csch}^{-1} x, \quad x > 0$

82. $y = \cosh^{-1}(\ln x)$

83. $y = \text{sech}^{-1} \sqrt{x+1}, \quad -1 < x < 0$

84. $y = (\sinh^{-1} 2x)^3$

85. $y = \tanh^{-1} \dfrac{e^x}{1+e^x}$

In Exercises 86–95, evaluate the integrals.

86. $\displaystyle \int \frac{dx}{\sqrt{1+9x^2}}$

87. $\displaystyle \int \frac{dx}{\sqrt{4x^2-1}}$

88. $\displaystyle \int_{\ln 1/4}^{\ln 1/2} \frac{e^x}{1-e^{2x}}\,dx$

89. $\displaystyle \int \frac{dx}{x\sqrt{1-4x^2}}$

90. $\displaystyle \int \frac{\sin x\,dx}{\sqrt{1+\cos^2 x}}$

91. $\displaystyle \int \frac{\sec^2 x\,dx}{\sqrt{\tan^2 x - 1}} \quad \left(\frac{\pi}{4} < x < \frac{\pi}{2}\right)$

92. $\displaystyle \int \frac{\sec^2 x}{\tan^2 x - 1}\,dx \quad \left(-\frac{\pi}{4} < x < \frac{\pi}{4}\right)$

93. $\displaystyle \int_0^{1/2} \frac{x\,dx}{1-x^4}$

94. $\displaystyle \int_0^5 \frac{dx}{\sqrt{x^2+4x+3}}$

95. $\displaystyle \int \frac{dx}{\sqrt{x^2+2x+2}}$

In Exercises 96–100, verify the given antidifferentiation formula, where $a > 0$.

96. $\displaystyle \int \frac{du}{\sqrt{a^2+u^2}} = \sinh^{-1}\frac{u}{a} + C$

97. $\displaystyle \int \frac{du}{\sqrt{u^2-a^2}} = \cosh^{-1}\frac{u}{a} + C, \ u > a$

98. $\displaystyle \int \frac{du}{a^2-u^2} = \begin{cases} \dfrac{1}{a}\tanh^{-1}\dfrac{u}{a} + C & \text{if } |u| < a \\[2mm] \dfrac{1}{a}\coth^{-1}\dfrac{u}{a} + C & \text{if } |u| > a \end{cases}$

99. $\displaystyle \int \frac{du}{u\sqrt{a^2-u^2}} = -\frac{1}{a}\,\text{sech}^{-1}\frac{u}{a} + C, \ 0 < u < a$

100. $\displaystyle \int \frac{du}{u\sqrt{a^2+u^2}} = -\frac{1}{a}\,\text{csch}^{-1}\frac{u}{a} + C, \ u > 0$

In Exercises 101–106, use the results of Exercises 96–100 to evaluate the integrals.

101. $\displaystyle \int \frac{dx}{\sqrt{4+9x^2}}$

102. $\displaystyle \int \frac{dx}{\sqrt{4x^2-9}}$

103. $\displaystyle \int_0^1 \frac{dx}{9-4x^2}$

104. $\displaystyle \int_3^4 \frac{dx}{x\sqrt{25-x^2}}$

105. $\displaystyle \int \frac{dx}{\sqrt{4+e^{2x}}}$

106. $\displaystyle \int \frac{dx}{x^2-6x+5}$

107. Prove Equations 7.71 and 7.72.

108. Prove Equations 7.73 and 7.74.

109. Prove Equations 7.75 and 7.76.

110. Prove $\dfrac{d}{dx}\,\text{sech}^{-1}|x| = -\dfrac{1}{x\sqrt{1-x^2}}$ if $|x| < 1$ and $x \neq 0$. Using this result, show that the formula of Exercise 99 can be generalized to

$$\int \frac{du}{u\sqrt{a^2-u^2}} = -\frac{1}{a}\,\text{sech}^{-1}\frac{|u|}{a} + C, \qquad 0 < |u| < a$$

111. Following the idea of Exercise 110, show that

$$\int \frac{du}{u\sqrt{a^2+u^2}} = -\frac{1}{u}\,\text{csch}^{-1}\frac{|u|}{a} + C, \qquad u \neq 0$$

112. Make use of inverse hyperbolic functions to obtain the result

$$\int \sec u\,du = \ln|\sec u + \tan u| + C$$

(Hint: $\sec u = \dfrac{1}{\cos u} = \dfrac{\cos u}{\cos^2 u} = \dfrac{\cos u}{1-\sin^2 u}$. Now use Equations 7.67 and 7.73.)

113. Consider a body falling under the influence of gravity, taking air resistance into consideration. The downward force is the weight of the object (mg, where m is its mass) and the resisting force is known experimentally to be approximately proportional to the square of the velocity. According to Newton's Second Law of Motion, the sum of forces acting on the body is its mass times its acceleration ($F = ma$). So we have

$$ma = mg - kv^2$$

Suppose that when $t = 0$, we have $v = 0$ and $s = 0$. (Measure distance positively downward from the initial position.) Using the fact that $a = dv/dt$ and $v = ds/dt$, find the following: (a) $v(t)$, (b) $s(t)$, and (c) the "terminal velocity" $\lim_{t \to \infty} v(t)$. (*Note:* We assume $mg > kv^2$.)

TRANSCENDENTAL FUNCTIONS USING COMPUTER ALGEBRA SYSTEMS

All the standard transcendental functions defined so far have definitions in every computer algebra system. The systems typically contain long lists of mathematical functions that can be accessed directly.

CAS 22

Let $f(x) = x^2 e^{1-x^2}$. Plot the graph and approximate $\int_0^4 x^2 e^{1-x^2} \, dx$.

To get a reasonable plot of the function, sketch the curve on the interval $[-4, 4]$ (see Figure 7.6.1).

Maple:

f:= x–>x^2*exp(1–x^2);
plot (f(x),x=–4..4,numpoints=500);

FIGURE 7.6.1
$y = f(x) = x^2 e^{1-x^2}$

Mathematica:

f[x_]=x^2*Exp[1–x^2]
Plot[f[x],{x,–4,4}]

DERIVE:

(At the □ symbol, go to the next step.)
a (author) □ x^2exp(1–x^2) □ p (plot window) □ p (plot)

The graph is shown in the margin. We see that the graph always lies above the x-axis, but that it approaches 0 very rapidly as x approaches plus or minus infinity. We can easily check this behavior using the computer algebra system.

Maple:

limit(f(x),x=infinity);   0
limit(f(x),x=–infinity);   0

Mathematica:

Limit[f[x], x–> Infinity]   0
Limit[f[x],x –> –Infinity]   0

DERIVE:

(At the □ symbol, go to the next step.)
c (calculus) □ l (limit) □ [choose expression] □ x (variable) □ inf (limit point) □ s (simplify) □ [choose expression]

Output: 0
c (calculus) □ l (limit) □ [choose expression] □ x (variable) □ [choose expression] □ –inf (limit point) □ s (simplify) □ [choose expression]
Output: 0

The computer algebra systems Maple, Mathematica, or DERIVE *cannot* find an antiderivative for the function f. In fact, the antiderivative cannot be expressed in terms of simple functions.

Maple:

int(f(x),x);

Output:

$$\left(-\frac{1}{2}\frac{x}{e^{x^2}} + \frac{1}{4}\sqrt{\pi}\,\mathrm{erf}(x) \right) e$$

Mathematica:

Integrate[f[x],x]

DERIVE:

int(x^2exp(1–x^2),x)
s (simplify) □
[choose expression]

In each case we see something of the form erf(x), which is called the error function and is another integral that cannot be evaluated.

The integral can be approximated by Maple, Mathematica, or DERIVE.

Maple:

evalf(int(f(x),x=0..4),6);

Mathematica:*

NIntegrate[f[x],{x,0,4}]

DERIVE:

int(x^2exp(1–x^2),x,0,4)
x (approximate)
[choose expression]

The approximate value for the integral is given as 1.20451 by Maple and Mathematica and as 1.20450 by DERIVE.

⌨ CAS 23

Approximate the volume of the solid of revolution obtained by rotating the region under $y = f(x) = \cosh(x)$ from $x = 0$ to $x = 2$ about the x axis.

Maple:

f := x–> cosh(x);
plot(f(x),x=–5..5,y=0..5);
evalf(int(Pi*f(x)^2,x=0..2));

The approximate volume is

24.57504350

Mathematica:

g[x_] = Cosh[x]
Plot[g[x],{x,–5,5},Plotrange–>{0,5}]
NIntegrate[Pi*(g[x])^2,
  {x,0,2}]

Output: 24.575

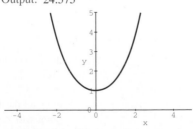

DERIVE:

(At the □ symbol, go to the next step.)
a (author) □ g(x):=cosh(x) □ p (plot
window) □ p (plot) □ int(Pi*(g(x))^2,x,0,2)
□ x (approximate) □ [choose expression]

Output: 24.5750

FIGURE 7.6.2
$y = f(x) = \cosh x$

---

*Mathematica's NIntegrate gives a numerical approximation of the definite integral.

Chapter 7 Review Exercises

In Exercises 1–7, find $f'(x)$ and give its domain.

1. (a) $f(x) = \ln(x\sqrt{x-2})$
    (b) $f(x) = \ln\left(\dfrac{x^2 - x}{x^2 - 2x - 8}\right)$

2. (a) $f(x) = \sqrt{\ln(\sec^2 x)}$
    (b) $f(x) = x \ln\dfrac{1}{x^2}$

3. (a) $f(x) = x^2 e^{-x}$
    (b) $f(x) = e^{\ln(e^{-x})}$

4. (a) $f(x) = e^{\sin x + \ln x}$
    (b) $f(x) = \ln\left(\dfrac{e^x - 1}{e^x + 1}\right)$

5. (a) $f(x) = \cos e^{-x} - \sin e^{-x}$
    (b) $f(x) = (\ln\sqrt{e^x + e^{-x}})^2$

6. (a) $f(x) = x \log_3 x$
    (b) $f(x) = \dfrac{2^x}{1 - 2^x}$

7. (a) $f(x) = x^{\ln x}$
    (b) $f(x) = (\sec x)^{\tan x}$

8. Use logarithmic differentiation to find y':
    (a) $y = \dfrac{x^3\sqrt{2x + 1}}{(x - 3)^2}$
    (b) $y = \dfrac{(x - 1)^5 (x^2 + 1)^{2/3}}{\sqrt{x^3 - 5x^2 + 6x}}$

9. Find y' using implicit differentiation:
    (a) $e^y - \ln\dfrac{x}{y} = 1$
    (b) $x = \ln\left(\dfrac{x + y}{x - y}\right)$

In Exercises 10–15, evaluate the given definite or indefinite integral.

10. (a) $\displaystyle\int \dfrac{\cos 2x}{1 + \sin 2x}\,dx$
    (b) $\displaystyle\int \dfrac{x + 1}{x^2 + 2x + 4}\,dx$

11. (a) $\displaystyle\int e^{-2\ln x}\,dx$
    (b) $\displaystyle\int \dfrac{e^x - 1}{e^{2x} - 2xe^x + x^2}\,dx$

12. (a) $\displaystyle\int \dfrac{\sqrt{x}}{\sqrt{x^3} + 4}\,dx$
    (b) $\displaystyle\int \dfrac{dx}{x(\ln 2x + 3)}$

13. (a) $\displaystyle\int_{\ln 2}^{\ln 3} \dfrac{e^{-2x}}{1 + e^{-2x}}\,dx$
    (b) $\displaystyle\int_{e^{-2}}^{e^2} \dfrac{dx}{x + \sqrt{x}}$

14. (a) $\displaystyle\int_{(1/2)\ln 7}^{2\ln 2} e^{2x}\sqrt{9 + e^{2x}}\,dx$
    (b) $\displaystyle\int_{\ln 2}^{\ln 3} \dfrac{e^{2x} - 1}{e^{2x} + 1}\,dx$ (*Hint:* Multiply the integrand by e^{-x}/e^{-x}.)

15. (a) $\displaystyle\int \dfrac{2^{2x}}{4^x + 1}\,dx$
    (b) $\displaystyle\int_0^2 \dfrac{\sqrt{1 - 2^{-t}}}{2^t}\,dt$

In Exercises 16–19, analyze the function and draw its graph.

16. $y = \dfrac{e^x - e^{-x}}{e^x + e^{-x}}$

17. $y = \ln|1 - x|$

18. $y = x \ln x^2$

19. $y = x^3 e^{-x^2/2}$

20. Find the volume of the solid obtained by revolving the region under the curve $y = e^{x^2/2}$ from $x = 0$ to $x = \sqrt{\ln 4}$ about the y-axis.

21. Find the surface area obtained by revolving the arc of the curve $y = e^{x/2} + e^{-x/2}$ from $x = -\ln 3$ to $x = \ln 3$ about the x-axis.

 22. Solve for x in terms of natural logarithms and approximate the answer using a calculator or CAS:
    (a) $2 \cdot 3^{-x} = 5 \cdot 4^{2x}$
    (b) $6e^{2x-1} - 3^x \cdot 2^{1-x} = 0$

23. Solve for x and check your answers:
    (a) $\ln x + \ln(x - 1) = \ln 2$
    (b) $2\log_3 x - \log_3(2x - 3) = 1$

 24. The radioactive isotope cobalt-60 has a half-life of approximately 5.26 yr. Of a given initial amount, how much will remain after 3 yr? How long will it take for 80% of the original amount to decay?

 25. A culture of bacteria is observed to double in size in 13 hr. How long will it take to triple in size? Assume exponential growth.

 26. A steel plate that has been heated to 40°C by exposure to direct sunlight is brought into an air-conditioned room where the temperature is 20°C. After 2 min it has cooled to 38°C. According to Newton's Law of Cooling, what will be the temperature of the plate after 5 min? How long will it take for the temperature to drop to 21°C?

27. A modification of the Malthus model for population growth is given by

$$Q'(t) = kQ \ln\dfrac{m}{Q}, \qquad Q < m$$

where k and m are positive constants. (This differential equation is known as the **Gompertz model**.) Find $Q(t)$ for this model. What is $\lim_{t \to \infty} Q(t)$?

In Exercises 28–31, give the exact value of the expression.

28. (a) $\sin^{-1} 1$ (b) $\cos^{-1}\left(-\dfrac{\sqrt{3}}{2}\right)$

     (c) $\tan^{-1}\left(-\dfrac{1}{\sqrt{3}}\right)$ (d) $\text{arcsec}(-\sqrt{2})$

     (e) $\text{arccot}(0)$ (f) $\text{arccsc}\left(-\dfrac{2}{\sqrt{3}}\right)$

29. (a) $\tan\left[\cos^{-1}\left(-\dfrac{3}{\sqrt{13}}\right)\right]$ (b) $\sin\left[2\tan^{-1}\left(-\dfrac{1}{2}\right)\right]$

30. (a) $\cos\left[2\sin^{-1}\left(-\dfrac{3}{5}\right)\right]$ (b) $\sin^{-1}\left(\sin\dfrac{8\pi}{9}\right)$

31. (a) $\sin\left[\tan^{-1}\dfrac{4}{3} + \sec^{-1}\left(-\dfrac{13}{5}\right)\right]$

     (b) $\cos\left[\cos^{-1}\left(-\dfrac{1}{3}\right) - \tan^{-1}(-2)\right]$

32. Use a calculator or CAS to find the approximate radian measure of θ, if $0 \le \theta < 2\pi$.

     (a) $\cos\theta = \dfrac{1}{3}$, $\tan\theta < 0$

     (b) $\sin\theta = -\dfrac{3}{4}$, $\tan\theta > 0$

In Exercises 33–39, find y'.

33. (a) $y = \dfrac{1}{x}\sin^{-1}x^2$, $0 < |x| < 1$

     (b) $y = \tan^{-1}\sqrt{x^2 - 1}$, $|x| > 1$

34. (a) $y = \cos^{-1}(e^{-x})$, $x \ge 0$

     (b) $y = x^2\sec^{-1}\dfrac{1}{x} + \sin^{-1}x$, $0 < x < 1$

35. (a) $y = (\sin^{-1}\sqrt{x})^2 + (\cos^{-1}\sqrt{x})^2$

     (b) $\tan^{-1}\dfrac{y}{x} + \ln\sqrt{x^2 + y^2} = 1$

36. (a) $y = \ln(\cosh^2 x)$

     (b) $y = \cos^{-1}(\text{sech}\,x)$, $x > 0$

37. (a) $y = \dfrac{\sinh x}{1 + \cosh x}$

     (b) $\tan^{-1}(\sinh y) + \cot^{-1}(\sinh x) = 1$

38. (a) $y = \sinh^{-1}\sqrt{x^2 - 1}$, $x > 1$

     (b) $y = \tanh^{-1}(\sin x)$, $|x| < \dfrac{\pi}{2}$

39. (a) $y = \coth^{-1}\dfrac{1+x}{1-x}$, $0 < x < 1$

     (b) $y = \sqrt{x}\cosh^{-1}(x + 1)$, $x > 0$

In Exercises 40–55, evaluate the given definite or indefinite integrals.

40. $\displaystyle\int \dfrac{dx}{\sqrt{9 - 4x^2}}$ **41.** $\displaystyle\int_{-3/2}^{3/2} \dfrac{dx}{4x^2 + 3}$

42. $\displaystyle\int_{-2\sqrt{2}}^{-2} \dfrac{dx}{x\sqrt{x^2 - 2}}$ **43.** $\displaystyle\int_{-\sqrt{2}/4}^{0} \dfrac{dx}{\sqrt{1 - 4x^2}}$

44. $\displaystyle\int_{1}^{e} \dfrac{dx}{x[1 + (\ln x)^2]}$ **45.** $\displaystyle\int_{1/\sqrt{3}}^{1} \dfrac{dx}{x\sqrt{4x^2 - 1}}$

46. $\displaystyle\int \dfrac{dx}{\sqrt{e^{2x} + 1}}$ **47.** $\displaystyle\int \dfrac{x}{1 + x^4}dx$

48. $\displaystyle\int \dfrac{\sinh 2x}{1 + \cosh 2x}dx$ **49.** $\displaystyle\int \dfrac{\sinh 2x}{e^{\sinh^2 x}}dx$

50. $\displaystyle\int x\tanh x^2\,dx$ **51.** $\displaystyle\int_{0}^{\ln(1+\sqrt{2})} \dfrac{\cosh x}{1 + \sinh^2 x}dx$

52. $\displaystyle\int_{0}^{\ln 2} \dfrac{\tanh x}{\cosh^2 x}dx$ **53.** $\displaystyle\int \dfrac{e^x dx}{\sqrt{e^{2x} - 1}}$

54. $\displaystyle\int_{-1}^{1} \dfrac{dx}{4 - x^2}$ **55.** $\displaystyle\int_{0}^{1} \dfrac{dx}{x^2 + 4x + 3}$

In Exercises 56–61, evaluate the limits.

56. (a) $\displaystyle\lim_{x\to 1^+} \dfrac{\ln(\ln x)}{\tan\frac{\pi x}{2}}$ (b) $\displaystyle\lim_{x\to 0^-} e^{1/x}\ln|x|$

57. (a) $\displaystyle\lim_{x\to 0} x\coth 2x$ (b) $\displaystyle\lim_{x\to 0^+} x\ln(\tanh x)$

58. (a) $\displaystyle\lim_{x\to 1^+} (\ln x)[\ln(\ln x)]$

     (b) $\displaystyle\lim_{x\to\infty} [\ln x - \ln(x + 2)]$

59. (a) $\displaystyle\lim_{x\to 0}\left[\dfrac{1}{e^x - 1} - \text{csch}\,x\right]$

     (b) $\displaystyle\lim_{x\to 1}\left[\dfrac{1}{x - 1} - \dfrac{1}{\ln x}\right]$

60. (a) $\displaystyle\lim_{x\to1^+}(x-1)^{\ln x}$ (b) $\displaystyle\lim_{x\to0^+}[\ln(x+1)]^x$

61. (a) $\displaystyle\lim_{x\to0^+}\left(\frac{1}{e^x+1}\right)^{\sqrt{x}}$ (b) $\displaystyle\lim_{x\to0}(\cosh x)^{\coth^2 x}$

62. Find the volume of the solid obtained by revolving the region under the graph of

$$y=\frac{4}{x\sqrt{4-x^2}}$$

from $x=1$ to $x=\sqrt{3}$ about the y-axis.

63. Find the area between the graphs of

$$y=\frac{2e^x}{1+e^x}\quad\text{and}\quad y=\frac{2e^x}{1+e^{2x}}$$

from $x=0$ to $x=\frac{1}{2}\ln 3$.

64. The bottom of a 5 m-high billboard is 2 m above an observer's eye level. At what distance from the billboard should the observer stand to get the best view of the billboard?

65. A helicopter is initially 120 m directly above an observer on the ground. Then the helicopter begins to ascend at a speed of 20 m/s on a 45° path with the vertical. How fast is the angle of elevation of the helicopter changing, as measured from the observer, after 4 seconds?

66. Derive the formula for $\coth(x+y)$ that involves only $\coth x$ and $\coth y$ and no other hyperbolic functions.

In Exercises 67–70, verify the identities.

67. $\dfrac{1}{1-\tanh x}-\dfrac{1}{1+\tanh x}=\sinh 2x$

68. $\dfrac{\coth^2 x+\cosh x\,\operatorname{csch}^2 x}{1+\cosh x}=\operatorname{csch} x\coth x$

69. $\dfrac{\cosh x-\sinh x}{\cosh 2x-\sinh 2x}=\cosh x+\sinh x$

70. $\dfrac{1}{\operatorname{csch}^2 x+\operatorname{csch}^4 x}=\sinh^2 x-\tanh^2 x$

71. Let $f(x)=2\cosh\dfrac{x}{2}$ on $[-1,1]$.
(a) Find the area under the graph of f.
(b) Find the volume of the solid obtained by revolving the region under the graph of f about the x-axis.
(c) Determine the length of the graph of f.
(d) Find the surface area obtained by revolving the graph of f about the x-axis.

72. Show that the function

$$y=e^{-kt}(A\cosh\omega t+B\sinh\omega t)$$

satisfies the differential equation

$$\frac{d^2y}{dt^2}+2k\frac{dy}{dt}+(k^2-\omega^2)y=0$$

where k and ω are positive constants. For $k=2$ and $\omega=1$, find A and B such that when $t=0$, $y=-1$ and $dy/dt=2$.

Chapter 7 Concept Quiz

1. Define each of the following:
(a) $\ln x$
(b) e^x
(c) $\log_a x$
(d) a^x
(e) $\sin^{-1} x$
(f) $\cos^{-1} x$
(g) $\tan^{-1} x$
(h) $\sinh x$
(i) $\cosh x$
(j) $\tanh x$

2. (a) State the three arithmetic properties of the natural logarithm.
(b) Write an equivalent equation to $\ln a=b$ involving the exponential function.
(c) Express $e^{a\ln b}$ in an equivalent form that does not involve e or $\ln$.
(d) What restriction must be placed on x for the equation $\sin^{-1}(\sin x)=x$ to be true?

3. Give the derivative formula for each of the following, where u is a differentiable function of x:
(a) $\ln u$
(b) e^u
(c) $\log_a u$
(d) a^u
(e) $\sin^{-1} u$
(f) $\tan^{-1} u$
(g) $\sec^{-1} u$
(h) $\sinh u$
(i) $\cosh u$
(j) $\tanh u$

4. Give the following indefinite integral formulas:

(a) $\int \dfrac{du}{u}$

(b) $\int e^u \, du$

(c) $\int \dfrac{du}{\sqrt{a^2 - u^2}}$

(d) $\int \dfrac{du}{a^2 + u^2}$

(e) $\int \dfrac{du}{u\sqrt{u^2 - a^2}}$

(f) $\int \sinh u \, du$

(g) $\int \cosh u \, du$

(h) $\int \tan u \, du$

5. Fill in the blanks:

(a) e is the number whose _____ is _____.

(b) The inverse of $f(x) = \ln x$ is $f^{-1}(x) = $ _____.

(c) The domain of $f(x) = \ln x$ is _____ and its range is _____.

(d) The domain of $f(x) = e^x$ is _____ and its range is _____.

(e) In order to define $\sin^{-1} x$, the domain of $f(x) = \sin x$ is restricted to the interval _____.

(f) In order to define $\cos^{-1} x$, the domain of $f(x) = \cos x$ is restricted to the interval _____.

(g) The domain of $f(x) = \tan^{-1} x$ is _____ and its range is _____.

6. Which of the following statements are true, and which are false?

(a) If $\ln a = \ln b$, then $a = b$.

(b) $\int \dfrac{dx}{x^2} = \ln x^2 + C$

(c) If $Q'(t) = kQ(t)$, then $Q(t) = Q(0) \cdot e^{kt}$.

(d) $\ln(a + b) = \ln a + \ln b$

(e) $e^{\ln a + \ln b} = ab$

(f) $\tan^{-1} x = \dfrac{1}{\tan x}$

APPLYING CALCULUS

1. Psychologists attempt to measure the reaction R of a subject to an amount of stimulus. For example, one might have a change of one unit in the reaction correspond to the smallest change in the stimulus that most individuals can distinguish. This is the idea behind the decibel scale—1 decibel is intended to be the smallest change in loudness that the human ear can detect. The *Weber-Fechner Law* states that if the stimulus is proportional to the fraction ds/s by which the stimulus increased by an amount ds, then the corresponding change in the reaction dR is increased. That is: $dR = k(ds/s)$, where k is a positive constant. This equation can be written in the form of a differential equation:

$$\frac{dR}{ds} = \frac{k}{s}$$

(a) Determine R as a function of s where s_0 is the smallest stimulus detectable by the subject.

(b) Suppose that a person can distinguish between two weights only if the heavier one weighs at least 5% more than the lighter one. Use the Weber-Fechner Law to devise a scale $R(s)$ of apparent weights such that a difference of 1 unit in R corresponds to the smallest detectable change in actual weight s. Let $R = 1$ correspond to a weight of $s = 1$ pound.

(c) A variation of the Weber-Fechner Law is the *Brentano-Stevens Law*, which states that

$$\frac{dR}{ds} = k\frac{R}{s}$$

Determine R as a function of s in this case.

2. A large picture measures a ft from top to bottom. Its bottom is b ft above the eye level of an observer who is standing x ft from the wall on which the picture is hung.

(a) Show that the angle θ subtended by the picture at the eye of the observer is given by

$$\theta = \tan^{-1}\left(\frac{a+b}{x}\right) - \tan^{-1}\left(\frac{b}{x}\right)$$

(b) A picture hanging in an art gallery measures 5 ft from top to bottom and its bottom is 7 ft above the floor. Suppose that an observer's eye level is 5 ft above the floor. How far should the observer stand from the wall to maximize her angle of vision?

3. The concentration of a drug in the blood after a single injection decreases with time as the drug is either absorbed or eliminated from the body. Clinical evidence supports the assumption that the rate of decrease of the concentration at time t is proportional to the concentration at time t. Thus, if $C = C(t)$ is the concentration at time t, then

$$C'(t) = -kC(t)$$

where k is a positive constant, called the *elimination constant*.

(a) Suppose that the initial concentration of a drug in the bloodstream is 20 milligrams per liter and 3 hours later the concentration is 12 milligrams per liter. Determine an expression for the concentration $C = C(t)$ at any time t. What is the half-life of the drug?

(b) Sodium phenobarbitol, which has a half-life of 5 hours, is to be used to anesthetize a dog for an operation. The dog is anesthetized when its bloodstream contains at least 30 milligrams of the drug for each kilogram of body weight. Assuming that the concentration of sodium phenobarbitol decreases at a rate proportional to the amount present, what single dose of the drug should be administered in order to anesthetize a 25 kilogram dog for 1 hour?

4. Another model for the concentration $C = C(t)$ of a drug in the bloodstream at time t after an injection is

$$C(t) = C_0(e^{-\alpha t} - e^{-\beta t}), \qquad t > 0$$

where C_0 is the initial concentration, and α and β are constants with $\beta > \alpha$.

(a) Determine the relative extrema of this function and determine the intervals on which it is increasing and decreasing.

(b) Determine where the graph is concave up and concave down; find the points of inflection.

(c) Sketch the graph, including $\lim_{t \to \infty} C(t)$.

5. Let $A = A(t)$ be the amount of a given natural resource that is consumed from time $t = 0$ to time $t > 0$. Let $c = c(t)$ be the rate at which the resource is being consumed. If $c(t) = C$, a constant, then the amount consumed from $t = 0$ to $t = T$ is $A(T) = CT$. On the other hand, since population and industry tend to grow exponentially, it is reasonable to assume that the rate of consumption will increase exponentially; that is $c(t) = Ce^{kt}$, where C and k are positive constants.

(a) Suppose that the rate of consumption is growing exponentially. Show that

$$A(t) = \frac{C(e^t - 1)}{k}$$

(b) Let R be the amount of the resource that exists at time $t = 0$. Show that if the consumption rate is a constant, C, then the resource will last R/C years. This value is called the *static index* of the resource. Show that if the consumption rate increases exponentially, then the resource will last

$$r = \frac{\ln(ks + 1)}{k} \qquad \text{years.}$$

This value is called the *exponential index* of the resource.

(c) Suppose the known global reserves of oil at this time are 500×10^8 barrels. Determine the exponential index of this resource if the static index is 31 years and the rate of consumption increases 3.9% per year. What is the exponential index if the known reserves are 5 times as large?

6. Suppose income is to be received at the rate of $f(t)$ dollars per year from time $t = 0$ to $t = T$. For the purposes of analysis, it is convenient to assume that the income is generated and received continuously. In this case, the continuous function f is called the *rate of flow* for a *continuous income stream*. The total income received from time $t = 0$ to time $t = T$ is

$$I(t) = \int_0^T f(t)dt$$

If f is the rate of flow of a continuous income stream, then the *present value*, V_P, at an interest rate r (expressed as a decimal) compounded continuously for T years is

$$V_P(T, r) = \int_0^T e^{-rt} f(t)dt$$

If the income is to be received for all future time T, then the present value of all future income depends only on the interest rate r and is given by

$$V_P(r) = \int_0^\infty e^{-rt} f(t)dt$$

(Note: This equation can be viewed as "transforming" the function f into the function V_P. This particular relation has many important applications; it is called the *Laplace transform*.)

(a) Suppose that the rate of flow f is a constant, that is $f(t) = c$, $c > 0$. Calculate the total income I and the present value V_P at the interest rate r over the time interval $[0,T]$. Also, calculate $\lim_{T \to \infty} I(t)$ and $\lim_{T \to \infty} V_P(T)$.

(b) In a "cyclical" company, the rate of flow fluctuates with time. Suppose that the rate of flow of a cyclical company is given by

$$f(t) = 2500 + 1000 \cos \pi t \text{ dollars per year.}$$

Calculate the total income and the present value at the interest rate $r = 0.10$ for $0 < t < 10$ years.

(c) Determine the present value of all future income for the company in part (b).

7. In the investigation of a homicide or accidental death, it is usually important to estimate the actual time of death. Newton's Law of Cooling (see Section 6.7), which states that rate of change of temperature of an object is proportional to the difference in temperature between the object and the surrounding medium, can be used for this purpose. The mathematical formulation of this law is

$$\frac{dy}{dt} = -k(y - T)$$

where $y = y(t)$ is the temperature of the object at time t, T is the (constant) temperature of the surrounding medium, and k is a positive constant.

(a) Suppose a human body has temperature y_0 at time $t = 0$, and temperature y_1 at time $t = t_1 > 0$. Use Newton's Law of Cooling to derive an expression for the temperature of the body at any time t.

(b) A corpse with a body temperature of $77°F$ has been found in a ditch at 8:00 a.m. One hour later its temperature is $68°F$. The (constant) air temperature is $T = 60°F$. Assuming that the person had a normal body temperature of $98.6°F$ at the moment of death, determine the time of death.

8. An electric circuit consists of a resistor whose resistance is R (ohms) connected in series to an inductor with inductance L (henries) and a battery which supplies a constant voltage V. The switch is closed at time $t = 0$.

Let $i = i(t)$ denote the current in the circuit at time $t > 0$. The voltage across the resistor is Ri and the voltage across the inductor is $L\, di/dt$. By *Kirchhoff's Law*, the sum of these two voltages must equal the applied voltage V. Thus,

$$L\frac{di}{dt} + Ri = V$$

The initial condition is $i(0) = 0$, since there is no current in the circuit until the switch is closed at time $t = 0$.

(a) Show that the solution of the initial value problem is

$$i = \frac{V}{R}\left(1 - e^{-Rt/L}\right) = \frac{V}{R} - \frac{V}{R}e^{-Rt/L}$$

(b) Sketch the graph of this solution, including $\lim_{t \to \infty} i(t)$.

(c) The term $i_0 = V/R$ is called the *steady-state current* and the term $i_{tr} = \frac{V}{R}e^{-Rt/L}$ is the *transient current*. The time $T = L/R$ is called the *time constant* of the circuit. Calculate the current at the time $T = L/R$. How long does it take for the current to reach 99% of its steady state current?

9. On a certain day, snow began to fall at a heavy and constant rate. A snowplow started out at noon, going 2 miles the first hour and 1 mile the second hour. What time did it start snowing? Assume the plow clears snow at a constant rate.

Deborah Tepper Haimo

When do you remember first getting interested in mathematics?

I did not find mathematics particularly interesting until after high school algebra. In sophomore year I studied geometry and a new and exciting world opened up for me. Here we weren't required to carry out calculations involving specialized formulas. Instead, we were given a set of axioms, self-evident truths, that we used to justify every step we took in establishing a result. Mathematical rules now made much more sense to me. We could add reasonable assumptions to the axioms that we accepted at the start, and only these could be used to advance and develop a geometric result. Until then I had no idea how fascinating mathematics could be. I spent much time trying to derive geometric properties beyond the routine classroom problems. I was thrilled to observe the dramatic consequences of our few hypotheses.

One particular class made an impression on me. We were learning about indirect proofs (as opposed to direct proofs) and were assigned a problem using this method. I tried to imagine circumstances in which only an indirect proof would work and I thought I'd found one involving geometric figures positioned in a certain way. I had forgotten an axiom that we'd never had occasion to use previously. It stated that geometric figures could be moved in space without affecting their properties. Either a direct or indirect proof could be used and both would be sound. This was wondrous to me. Mathematics was so plausible. You could get the same correct result in many ways.

How and when did you decide to go into mathematics professionally?

At Radcliffe I thought a lot about what I wanted to do. I knew that I loved mathematics, but I had no idea what I could do with it. I knew I didn't have the temperament to be a high school teacher. I finally decided that there might be opportunities in the physical sciences that use mathematics, so I enrolled in Astronomy and Physics courses. I obtained part-time employment at the Harvard Observatory, which had some appeal but the women astronomers there seemed to have really tedious and laborious jobs. Physics, on the other hand, seemed more promising initially, but another classroom situation helped me change my mind. We were doing a routine experiment using magnets to record the patterns made by iron filings around the magnetic poles. Everyone else did the experiment easily but my iron filings would not conform to the expected pattern. Substituting magnets didn't help. Everyone was mystified, including the lab instructor, until a student opened the drawer under the table. Of course, we found it full of magnets!

Although a trivial experience, it really did have an influence on my early attitudes to mathematics and science. Unlike my earlier revelation when studying geometry in which I failed to notice an assumption that was already there, in this case I had visions of developing a theory in physics that might be distorted by a drawerful of magnets! This thought helped reconfirm my earlier inclination toward mathematics. I was drawn to the control of hypotheses, and the intrinsic elegance of the mathematics when these were well chosen.

Did your calculus course further develop your interest in mathematics?

My calculus class at Harvard was taught by Hassler Whitney, a creative and imaginative mathematician who was acclaimed for his ground-breaking work in differential geometry. His lectures were hard to follow since he became very excited over results, stuttered a bit, and wrote illegibly. But I credit him with transmitting an air of excitement about mathematics that was inescapable. He recommended Courant's two volumes on Differential and Integral Calculus that became my "bible" for calculus.

Professor Whitney was the assigned advisor for our class, and when I consulted him I expected advice on what opportunities I would have if I decided to continue in mathematics. I was surprised to find that he was totally uninterested in this issue. Instead, he asked me why I bothered to come to class when I could clearly learn much more rapidly on my own. This question totally startled me. In my earlier academic experience, there was never any question that one was expected to attend class. When I explained that I was concerned about my grades and had to maintain my scholarship, he encouraged me to stop coming to class, skip the next required course, and take advanced calculus in my sophomore year. He then said that in my junior year he would arrange to enroll me in graduate courses at Harvard that were open to undergraduates and that he would be my tutor to direct me in my area of concentration and to supervise me in writing the required undergraduate thesis. All this was very unusual at the time, especially for Radcliffe students.

You're the only person I've heard of who was told to cut class by her own professor.

Yes. At the time I was both confused and also elated—because he confirmed what I already knew—that mathematics was my first love as a discipline. I had been holding back because of my concerns about the future. There and then I made up my mind to follow my interests whatever the ultimate outcome. I followed Whitney's advice even though my original Radcliffe tutor was appalled at what she considered totally unrealistic advice. She was sure I would fail and that it would reflect badly on Radcliffe. Nonetheless, in my junior year, as promised, Whitney became my tutor, and I enrolled in graduate-level courses.

How did you eventually decide on what research to pursue?

Whitney was known for his uncanny intuition, which led him to important conjectures, and he assigned me an undergraduate thesis topic involving what he thought was a simple version of a major theory he had just developed. As I discovered, the problem was far more intractable than he had surmised. I was not able to do more than raise doubts, through an example, of the validity of his initial conjecture.

During the war years, job vacancies were being created that had never before been available to women. I began my first teaching position and was appointed for one semester as head of both mathematics and physics, with a 22-contact-hour teaching schedule! I enjoyed the experience, but I was eager to return to Cambridge to marry the fellow student to whom I was engaged and to study for my Ph.D. My husband, Frank, was initially surprised to learn that I wanted to pursue a Ph.D., since we had both come from traditional family backgrounds where the women had devoted all their time to their homes. However, he was very happy with my decision and became my most ardent professional supporter. He continued with his doctoral work and I remained intrigued with the problem that I had failed to solve as an undergraduate. As time passed, and my efforts did not yield substantive results, Frank urged me to abandon the problem and move into analysis, the field for which calculus is a foundation, and in which I had shown early interest and aptitude. I was stubborn, though, and Whitney encouraged me to continue. He kept saying that he was sure that I was on the verge of a major result.

When we moved to St. Louis, I had done all my course work but had not written a thesis, nor did I seem close to doing so. At Washington University I was given an irregular position with a very heavy teaching load. I had to develop the lower division college algebra course, transform the course to the large lecture format, and plan help sessions and multiple-choice examinations. I was also assigned to teach the graduate course in linear programming to engineering students.

After 12 years at Washington University, I accepted an offer at Southern Illinois University. All that time, our family was growing, I continued teaching, and I worked on my thesis whenever I could. Whitney had moved to the Institute for Advanced Study and I communicated with him. I could not make headway on the problem, but I refused to give up. One summer I determined that I would devote all my time to the problem, but when fall came and I hadn't made substantial progress, I reluctantly decided to give up. I followed the advice I'd been given for several years and decided on a problem that seemed tractable. Professor Widder at Harvard told me that, despite the passage of two decades, I would be granted my Ph.D. if I did an acceptable thesis and passed the general oral doctoral examination. With the availability for consultation of an expert in my new field of interest, Professor Hirschman at Washington U., I succeeded in finishing my thesis in a few months. Defending my thesis at Harvard was unquestionably one of the most traumatic experiences of my life. Except for Widder, I faced a roomful of strangers who could ask me any question in mathematics that they liked, and they did. After what seemed an eternity, Widder left, indicating that he felt the session had continued long enough, but this only invited more questions in other areas from the other examiners. At long last, the session ended. The results were almost immediate as the office filled with smiling faces and, at long last, I had my Ph.D.

It was after my thesis adventures that I learned that, after I had abandoned my original problem, new mathematical tools had been developed that showed that the problem I had worked on for so long, the classification of certain singular points, was proven to be impossible!

Isn't it frustrating to know now that you spent all that time on an impossible problem? Is this a risk that all research mathematicians face?

Well, as far as research goes, the important thing is to know that a solution to a problem is impossible. My real frustration occurred when I realized that I should not have devoted so much time exclusively to a problem that didn't yield any useful material within a reasonable period of time. Indeed, when I finally decided to change, I found that there were plenty of sound and interesting problems that lent themselves to serious investigation and that could be expected to produce worthwhile results. I was surprised to find writing a thesis much less overwhelming than I had expected. After completing Ph.D. requirements, my research activities centered on generalizations of the classical heat equation. My field now is classical harmonic analysis.

Being a research mathematician is very satisfying and interesting. Useful work can be done wherever one happens to be because neither a laboratory nor a library is essential for mathematical research. The major requirement is to have good ideas; and good thinking can be carried out anywhere. In addition, there is the challenge of educating students and arousing their appreciation of the intrinsic beauty of mathematics. Opportunities exist to travel both in the States and abroad. Invitations often are issued to lecture on one's results at various institutions and to participate in speciality conferences and in regional, national, and international meetings.

In recent years, I've become increasingly concerned about our educational system. I have become directly involved at the national level in trying to address some of the major issues in mathematics education.

CHAPTER

8

TECHNIQUES OF INTEGRATION

By now you have learned formulas for antiderivatives of quite a few functions. That is, you know how to evaluate indefinite integrals for these functions. You have also seen how other integrals can be converted to one of the standard forms, using the technique of substitution. Although substitution is a powerful technique, there are simple-looking integrals that defy any substitution. For example, consider

$$\int x \cos x \, dx$$

Try as you may, you will not be able to find a substitution that will change this integral to a form you can integrate. Some other technique is needed.

Finding antiderivatives is important for two primary reasons: (1) we can use them to evaluate definite integrals by means of the Second Fundamental Theorem, and (2) we can use them to solve differential equations. Since definite integrals and differential equations arise in a wide variety of applications, it is worthwhile to learn further techniques for finding antiderivatives. In this chapter we introduce the most important of these techniques.

Differentiation is usually a straightforward process once you have learned the basic derivatives and a few rules. On the other hand, antidifferentiation (indefinite integration) often requires a good deal of ingenuity. You now know how to differentiate many of the elementary functions, but you don't know how to integrate all of them. In fact, as we have previously noted, there are elementary functions for which no antiderivative exists in the form of an elementary function. The function $f(x) = e^{-x^2}$ (see Figure 8.1) is one of these (a very important one, especially in probability and statistics). Although the First Fundamental Theorem of Calculus tells us that this function does have

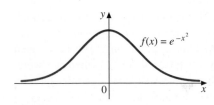

FIGURE 8.1
This simple function, important especially in probability and statistics, cannot be integrated in terms of elementary functions.

an antiderivative, there is no way of expressing it in terms of functions you know. In Chapter 10 we will see a way of approximating its antiderivative. We can, however, find antiderivatives of a great many functions without resorting to approximations. Learning the techniques in this chapter will greatly extend the class of functions you can integrate.

There are really only two significant techniques of integration: (1) **integration by parts** and (2) **substitution.** We will study the first of these in Section 8.1. It is based on the Product Rule for derivatives and is a powerful technique that enables us to integrate many functions we could not integrate by other means. We have used the method of substitution extensively already. In this chapter we show two further applications, both involving trigonometric functions. We also introduce a procedure called *partial-fraction decomposition*, which is not an integration technique per se but is an algebraic process by which complicated rational functions (quotients of polynomials) can be broken down into simpler ones. These simpler functions can either be integrated directly or by one of our other techniques.

The techniques of this chapter, together with those you have already learned, will provide you with the means of integrating most of the functions you are likely to encounter. Integration is an art, though, that can only be learned by practice. It would be nice if there were some definite step-by-step procedure (some *algorithm*) that you could apply to every integration problem and be sure of arriving at the answer. Unfortunately, there is no such procedure—the antiderivative may not even exist. With practice, you will learn which technique is appropriate for a given problem.

It was only in the late 1960s that it became possible to determine whether a function necessarily has an elementary antiderivative. The algorithm discovered by Robert Risch determines whether the elementary antiderivative exists and, if so, finds it. Modern computer algebra systems, such as Maple, use the same algorithm to accomplish this task.

Even when you know what technique to use on a particular integral, actually carrying out the steps can be tedious and time-consuming. Furthermore, if you need to evaluate a large number of integrals, each requiring a special technique, the time and effort needed may be more than you can afford to spend. In such situations, tables of integrals can be very helpful. Also, computer algebra systems can handle with ease many integrals that would be prohibitively difficult by other methods. We devote a section of this chapter, following the study of the various techniques, to an explanation of how to use integral tables and computer algebra systems to evaluate integrals.

We conclude the chapter by showing how the definition of the definite integral can be extended to two previously excluded cases: (1) the case in which the interval of integration is infinite, and (2) the case in which the integrand becomes infinite. Integrals of either type are referred to as *improper integrals*. Such integrals have both theoretical and practical applications. We have seen an example of an improper integral already, in Example 6.19, when we were deriving a formula for the length of an ellipse. We will consider this integral again in this chapter.

Before introducing the techniques, we summarize in Table 8.1 basic indefinite integral formulas that we have derived or have asked you to derive in the exercises. Our list is by no means exhaustive. A more extensive list is given on the front and back endpapers of this book. Tables of integrals that give hundreds of formulas are also available.

TABLE 8.1   Basic list of integrals

$$\int u^{\alpha}\,du = \frac{u^{\alpha+1}}{\alpha+1} + C \quad (\alpha \neq -1) \qquad \int \frac{du}{u} = \ln|u| + C$$

$$\int e^u\,du = e^u + C \qquad\qquad \int a^u\,du = \frac{a^u}{\ln a} + C \quad (a \neq 1)$$

$$\int \sin u\,du = -\cos u + C \qquad \int \cos u\,du = \sin u + C$$

$$\int \sec^2 u\,du = \tan u + C \qquad \int \csc^2 u\,du = -\cot u + C$$

$$\int \sec u \tan u\,du = \sec u + C \qquad \int \csc u \cot u\,du = -\csc u + C$$

$$\int \tan u\,du = -\ln|\cos u| + C \qquad \int \cot u\,du = \ln|\sin u| + C$$

$$\int \sec u\,du = \ln|\sec u + \tan u| + C$$

$$\int \csc u\,du = \ln|\csc u - \cot u| + C$$

$$\int \frac{du}{\sqrt{a^2-u^2}} = \sin^{-1}\frac{u}{a} + C \qquad \int \frac{du}{a^2+u^2} = \frac{1}{a}\tan^{-1}\frac{u}{a} + C$$

$$\int \frac{du}{u\sqrt{u^2-a^2}} = \frac{1}{a}\sec^{-1}\frac{u}{a} + C$$

$$\int \sinh u\,du = \cosh u + C \qquad \int \cosh u\,du = \sinh u + C$$

$$\int \text{sech}^2 u\,du = \tanh u + C \qquad \int \text{csch}^2 u\,du = -\coth u + C$$

$$\int \text{sech}\,u \tanh u\,du = -\text{sech}\,u + C \quad \int \tanh u\,du = \ln(\cosh u) + C$$

$$\int \text{csch}\,u \coth u\,du = -\text{csch}\,u + C \quad \int \coth u\,du = \ln|\sinh u| + C$$

$$\int \text{sech}\,u\,du = \tan^{-1}|\sinh u| + C$$

$$\int \text{csch}\,u\,du = \ln\left|\tanh\frac{u}{2}\right| + C$$

$$\int \frac{du}{\sqrt{a^2+u^2}} = \sinh^{-1}\frac{u}{a} + C = \ln(u + \sqrt{u^2+a^2}) + C$$

$$\int \frac{du}{\sqrt{u^2-a^2}} = \cosh^{-1}\frac{u}{a} + C = \ln(u + \sqrt{u^2-a^2}) + C$$

$$\int \frac{du}{a^2-u^2} = \left\{\begin{array}{ll} \dfrac{1}{a}\tanh^{-1}\dfrac{u}{a} + C, & |u| < a \\[2mm] \dfrac{1}{a}\coth^{-1}\dfrac{u}{a} + C, & |u| > a \end{array}\right\} = \frac{1}{2a}\ln\left|\frac{a+u}{a-u}\right| + C$$

$$\int \frac{du}{u\sqrt{a^2-u^2}} = -\frac{1}{a}\text{sech}^{-1}\frac{|u|}{a} + C = -\frac{1}{a}\ln\left(\frac{a+\sqrt{a^2-u^2}}{|u|}\right) + C$$

$$\int \frac{du}{u\sqrt{a^2+u^2}} = -\frac{1}{a}\text{csch}^{-1}\frac{|u|}{a} + C = -\frac{1}{a}\ln\left(\frac{a+\sqrt{a^2+u^2}}{|u|}\right) + C$$

8.1 INTEGRATION BY PARTS

One of the most important techniques of integration is derived from the formula for the derivative of a product of two functions. If $u = f(x)$ and $v = g(x)$ are differentiable functions on some common domain, then we know that

$$\frac{d}{dx}[f(x)g(x)] = f'(x)g(x) + f(x)g'(x)$$

Taking antiderivatives of both sides, we get

$$f(x)g(x) = \int f'(x)g(x)\,dx + \int f(x)g'(x)\,dx$$

which can also be written in the form

$$\int f(x)g'(x)\,dx = f(x)g(x) - \int g(x)f'(x)\,dx \qquad (8.1)$$

Formula 8.1 is called the **integration by parts** formula. It often enables us to replace an integral that we *can't* integrate directly (the integral on the left) with one we *can* integrate (the one on the right). In our examples we will illustrate how, by appropriate choices of $f(x)$ and $g(x)$, the integral on the right is often simpler to work with than the one on the left.

It is customary in Formula 8.1 to write $u = f(x)$ and $v = g(x)$. Then $du = f'(x)\,dx$ and $dv = g'(x)\,dx$, giving the following formula.

Integration by Parts Formula

$$\int u\,dv = uv - \int v\,du \qquad (8.2)$$

Integration by parts can be applied to a wide variety of integrals. In particular, it often works when the integrand is a product of a polynomial and a transcendental function, such as

$$\int x^3 e^x, \quad \int x \ln x\,dx, \quad \text{and} \quad \int (x^2 + 2x + 3)\cos x\,dx$$

We illustrate how to apply the technique in the following example.

EXAMPLE 8.1   Evaluate the integral $\int xe^x\,dx$.

Solution   We want to express the integral $\int xe^x\,dx$ in the form $\int u\,dv$. The secret lies in choosing the "parts" u and dv. Usually, we want to choose u so that du/dx is simpler than u, thereby obtaining a simpler integral on the right-hand side of Equation 8.2 than the one on the left. Also, dv *must* always be chosen such that $v = \int dv$ can be found; otherwise, we could not evaluate either term on the right. In the integral in this problem, these criteria suggest that we choose $u = x$ and $dv = e^x\,dx$. (Note that when a choice is made for u, the choice for dv is determined.) With these choices, we have $du/dx = 1$ (which is certainly simpler than u), and v can be evaluated as

$$v = \int dv = \int e^x\,dx = e^x$$

It is not necessary to add an arbitrary constant at this stage. (See the discussion following this example.) We can organize our work as follows:

$$u = x \qquad dv = e^x\,dx$$
$$du = dx \qquad v = \int e^x\,dx = e^x$$

Now we apply Equation 8.2:

$$\int \overbrace{x}^{u}\ \overbrace{e^x\,dx}^{dv} = \overbrace{x}^{u}\ \overbrace{e^x}^{v} - \int \overbrace{e^x}^{v}\ \overbrace{dx}^{du}$$

The integral on the right is one we can integrate: $\int e^x\,dx = e^x$. So we have

$$\int xe^x\,dx = xe^x - e^x + C \qquad \blacksquare$$

REMARK ————————————————————————————
■ In the preceding example, when we calculated v by integrating dv, we did not include an arbitrary constant of integration. Suppose we had included it. Then we would have had

$$v = \int e^x\,dx = e^x + C$$

By Equation 8.2 we then would have had

$$\int xe^x\,dx = x(e^x + C) - \int (e^x + C)\,dx$$
$$= xe^x + Cx - (e^x + Cx) + C_1$$
$$= xe^x - e^x + C_1$$

which is equivalent to our first answer. In general, as in this example, you may omit the arbitrary constant in calculating v from dv. That is, v can be taken as the simplest antiderivative.

————————————————————————————————

Wrong Choice of Parts

What if we had tried a different choice of u and dv in the preceding example? If we had chosen $u = e^x$ and $dv = x\,dx$, then we would have had

$$du = e^x\,dx \qquad \text{and} \qquad v = \frac{x^2}{2}$$

Thus, by Equation 8.2,

$$\int xe^x\,dx = \frac{x^2}{2}e^x - \int \frac{x^2}{2}e^x\,dx$$

Even though this is a correct equality, we would be worse off than when we started, since the integral on the right is more complicated than the original integral. The lesson to be learned here is that if you make a bad choice of parts, abandon the choice when it becomes evident you are not getting anywhere, and try another.

EXAMPLE 8.2   Evaluate the integral $\int x^2 \cos x\,dx$.

Solution   Guided by the criteria stated in Example 8.1, we make the following choice of parts:

$$u = x^2 \qquad dv = \cos x\,dx$$

Then

$$du = 2x\,dx \qquad v = \int \cos x\,dx = \sin x$$

So, by Equation 8.2,

$$\int x^2 \cos x \, dx = x^2 \sin x - \int \sin x \, (2x \, dx)$$

$$= x^2 \sin x - 2 \int x \sin x \, dx$$

We have made progress (since the power of x in the last integral is less than in the first) but we still cannot evaluate the final integral directly. To evaluate it, we use integration by parts again, with

$$u_1 = x \qquad dv_1 = \sin x \, dx$$

$$du_1 = dx \qquad v_1 = \int \sin x \, dx = -\cos x$$

So

$$\int x \sin x \, dx = -x \cos x + \int \cos x \, dx$$

$$= -x \cos x + \sin x + C_1$$

Thus,

$$\int x^2 \cos x \, dx = x^2 \sin x - 2 \int x \sin x \, dx$$

$$= x^2 \sin x - 2[-x \cos x + \sin x + C_1]$$

$$= x^2 \sin x + 2x \cos x - 2 \sin x - 2C_1$$

But since C_1 is arbitrary, so is $-2C_1$. Calling this constant C, we get the final result:

$$\int x^2 \cos x \, dx = x^2 \sin x + 2x \cos x - 2 \sin x + C \qquad \blacksquare$$

Repeated Applications of Integration by Parts

A useful formula can be derived to give the result of repeated integration by parts in problems such as Example 8.2. Suppose u is a polynomial of degree n, with $n \geq 2$. We assume also that v and its successive integrals can be obtained. Let u_0 and v_0 be the initial choices of u and v, and for $k = 1, 2, \ldots, n$, let

$$u_k = u'_{k-1} \quad \text{and} \quad v_k = \int v_{k-1} dx$$

Then the result of integrating by parts n times is as follows:

$$\int u \, dv = u_0 v_0 - u_1 v_1 + u_2 v_2 - \cdots + (-1)^n u_n v_n + C$$

$$= \sum_{k=0}^{n} (-1)^k u_k v_k + C \qquad\qquad (8.3)$$

This formula was first given by the German mathematician Leopold Kronecker (1823–1891); you will be asked to prove it in Exercise 52 of Exercise Set 8.1. (See also Exercise 42 in Exercise Set 8.2 for an unusual function that is named after Kronecker.)

To illustrate Equation 8.3, consider again the integral in Example 8.2, where we initially took $u = x^2$ and $dv = \cos x \, dx$. Thus, our first u and v are $u_0 = x^2$ and $v_0 = \int \cos x \, dx = \sin x$. The accompanying table is a convenient way to organize the calculations. The first entries are u_0 and v_0. Each subsequent u_k is obtained by differentiating the preceding one, whereas each subsequent v_k is obtained by integrating the previous one. This sequence of steps is continued until a constant occurs in the u_k column. (This constant is u_n.) We now multiply each u_k by the corresponding v_k, assign $+$ signs and $-$ signs alternately, and add the results, supplying a constant of integration at the end:

u_k	v_k	
x^2	$\sin x$	$(k = 0)$
$2x$	$-\cos x$	$(k = 1)$
2	$-\sin x$	$(k = 2)$

$$\int x^2 \cos x \, dx = x^2 \sin x - (-2x \cos x) + (-2 \sin x) + C$$

$$= x^2 \sin x + 2x \cos x - 2 \sin x + C$$

EXAMPLE 8.3   Evaluate the integral $\int x^3 e^{2x} \, dx$ using Equation 8.3.

Solution   Let $u_0 = x^3$ and $dv_0 = e^{2x} \, dx$, so that $v_0 = e^{2x}/2$. From the accompanying table we get

u_k	v_k
x^3	$\frac{1}{2}e^{2x}$
$3x^2$	$\frac{1}{4}e^{2x}$
$6x$	$\frac{1}{8}e^{2x}$
6	$\frac{1}{16}e^{2x}$

$$\int x^3 e^{2x} \, dx = \frac{x^3 e^{2x}}{2} - \frac{3x^2 e^{2x}}{4} + \frac{6x e^{2x}}{8} - \frac{6 e^{2x}}{16} + C$$

$$= \frac{e^{2x}}{8}(4x^3 - 6x^2 + 6x - 3) + C \qquad \blacksquare$$

EXAMPLE 8.4   Evaluate the integral $\int x \ln x \, dx$.

Solution   In view of the choices of u and dv in the preceding examples, it might be tempting to try $u = x$ and $dv = \ln x \, dx$. But this choice would not help, since we do not know an integral of $\ln x$. So we could not find v. Instead, we take

$$u = \ln x \qquad dv = x \, dx$$
$$du = \frac{1}{x} \, dx \qquad v = \int x \, dx = \frac{x^2}{2}$$

Then we have

$$\int x \ln x \, dx = (\ln x)\left(\frac{x^2}{2}\right) - \int \frac{x^2}{2} \cdot \frac{1}{x} \, dx$$

$$= \frac{x^2}{2} \ln x - \frac{1}{2} \int x \, dx = \frac{x^2}{2} \ln x - \frac{x^2}{4} + C \qquad \blacksquare$$

Definite Integrals

Integration by parts can be applied to a definite integral in two ways.
   Method 1: We include limits of integration in Equation 8.1, getting

$$\int_a^b f(x)g'(x) \, dx = f(x)g(x)\Big]_a^b - \int_a^b g(x)f'(x) \, dx$$

or, in the briefer notation of Equation 8.2,

$$\int_a^b u\,dv = uv\Big]_a^b - \int_a^b v\,du$$

It is understood in this last formula that a and b are limits for x and, as before, $u = f(x)$ and $v = g(x)$.

Method 2: First find the indefinite integral

$$\int u\,dv = uv - \int v\,du$$

and after the result is obtained, substitute upper and lower limits.

We illustrate both methods in the next example.

EXAMPLE 8.5   Evaluate the integral $\displaystyle\int_0^1 \tan^{-1} x\,dx$.

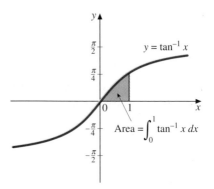

FIGURE 8.2

Solution   Note that we can interpret this integral geometrically as the area under the graph of $y = \tan^{-1} x$ from $x = 0$ to $x = 1$. (See Figure 8.2.)

Method 1: At first glance this integral does not seem suited to integration by parts, since there appears to be only one "part." But we can take dv as dx itself. (We wouldn't want to take $dv = \tan^{-1} x\,dx$. Why not?) Then we have

$$u = \tan^{-1} x \qquad dv = dx$$
$$du = \frac{1}{1+x^2}\,dx \qquad v = \int dx = x$$

Thus,

$$\int_0^1 \tan^{-1} x\,dx = x\tan^{-1} x\Big]_0^1 - \int_0^1 \frac{x}{1+x^2}\,dx \qquad \text{\small Let } w = 1+x^2\text{. Then}$$
$$\text{\small } dw = 2x\,dx.$$

$$= 1\tan^{-1} 1 - 0\tan^{-1} 0 - \frac{1}{2}\int_0^1 \frac{2x\,dx}{1+x^2}$$

$$= \frac{\pi}{4} - \frac{1}{2}\Big[\ln(1+x^2)\Big]_0^1$$

$$= \frac{\pi}{4} - \frac{1}{2}\ln 2$$

Method 2: Taking u and dv as above, we have

$$\int \tan^{-1} x\,dx = x\tan^{-1} x - \int \frac{x}{1+x^2}\,dx$$

$$= x\tan^{-1} x - \frac{1}{2}\int \frac{2x}{1+x^2}\,dx$$

$$= x\tan^{-1} x - \frac{1}{2}\ln(1+x^2) + C$$

Thus, the simplest antiderivative of $\tan^{-1} x$ is the one with $C = 0$:

$$x\tan^{-1} x - \frac{1}{2}\ln(1+x^2)$$

So, by the Second Fundamental Theorem,

$$\int_0^1 \tan^{-1} x \, dx = \left[x \tan^{-1} x - \frac{1}{2} \ln(1 + x^2) \right]_0^1$$

$$= \tan^{-1} 1 - \frac{1}{2} \ln 2$$

$$= \frac{\pi}{4} - \frac{1}{2} \ln 2$$

In general, the first method is preferable, since limits are substituted in the uv term as the work progresses, which sometimes simplifies the expression.

∎

REMARK ───────────────────────────────

■ Examples 8.4 and 8.5 suggest that when an integration involving a logarithm or an inverse trigonometric function (also an inverse hyperbolic function) is to be done by parts, u should be taken as the logarithm or the inverse function.

────────────────────────────────────

Recurrence of the Original Integral

The next two examples show how repeated integration by parts can result in the recurrence of the original integral and how it can then be evaluated.

EXAMPLE 8.6   Evaluate the integral $\int e^{-x} \cos 2x \, dx$.

Solution   There is no clear choice for u and dv in this case. Let us try

$$u = e^{-x} \qquad dv = \cos 2x \, dx$$

Then

$$du = -e^{-x} \, dx \qquad v = \int \cos 2x \, dx = \frac{1}{2} \sin 2x$$

Note that Equation 8.3 for repeated integration by parts is not applicable, since u is not a polynomial. We have

$$\int e^{-x} \cos 2x \, dx = \frac{1}{2} e^{-x} \sin 2x - \int \left(\frac{1}{2} \sin 2x \right) (-e^{-x} \, dx)$$

$$= \frac{1}{2} e^{-x} \sin 2x + \frac{1}{2} \int e^{-x} \sin 2x \, dx \qquad (8.4)$$

Although we did not achieve any simplification in this case, we can use integration by parts again on the last integral, with

$$u_1 = e^{-x} \qquad dv_1 = \sin 2x \, dx$$
$$du_1 = -e^{-x} \, dx \qquad v_1 = \int \sin 2x \, dx = -\frac{1}{2} \cos 2x$$

Thus,

$$\int e^{-x} \sin 2x \, dx = -\frac{1}{2} e^{-x} \cos 2x - \int \left(-\frac{1}{2} \cos 2x \right) (-e^{-x} \, dx)$$

$$= -\frac{1}{2} e^{-x} \cos 2x - \frac{1}{2} \int e^{-x} \cos 2x \, dx$$

Substituting this result in Equation 8.4 gives

$$\int e^{-x} \cos 2x \, dx = \frac{1}{2} e^{-x} \sin 2x + \frac{1}{2} \left[-\frac{1}{2} e^{-x} \cos 2x - \frac{1}{2} \int e^{-x} \cos 2x \, dx \right]$$

or

$$\int e^{-x} \cos 2x \, dx = \frac{1}{2} e^{-x} \sin 2x - \frac{1}{4} e^{-x} \cos 2x - \frac{1}{4} \int e^{-x} \cos 2x \, dx \quad (8.5)$$

It might seem that we are going in circles, since the integral we are attempting to evaluate occurs on both sides of Equation 8.5. But the way out of this dilemma is simple—we just solve the equation algebraically for the unknown integral. It is helpful to introduce a single letter—say, I—for the unknown integral:

$$I = \int e^{-x} \cos 2x \, dx$$

Then Equation 8.5 becomes

$$I = \frac{1}{2} e^{-x} \sin 2x - \frac{1}{4} e^{-x} \cos 2x - \frac{1}{4} I$$

To solve for I, we add $\frac{1}{4} I$ to both sides:

$$\frac{5I}{4} = \frac{e^{-x}}{4} (2 \sin 2x - \cos 2x)$$

$$I = \frac{e^{-x}}{5} (2 \sin 2x - \cos 2x)$$

In this process we did not introduce a constant of integration. To get the most general antiderivative, we must do so now:

$$\int e^{-x} \cos 2x \, dx = \frac{e^{-x}}{5} (2 \sin 2x - \cos 2x) + C \qquad \blacksquare$$

REMARK ───────────────────────────────
■ Example 8.6 is typical of integrals of the form

$$\int e^{ax} \cos bx \, dx \quad \text{or} \quad \int e^{ax} \sin bx \, dx$$

In such integrals it is immaterial whether we take u as the exponential or the trigonometric function. We will ask you to redo Example 8.6 in the exercises (Exercise 23), letting $u = \cos 2x$ and $dv = e^{-x} dx$ to show that this choice works just as well. We must emphasize, however, that whatever choice you make initially, you *must* make the same type of choice when you integrate by parts the second time. Otherwise, you really will go in circles, winding up with the true but unhelpful statement that the original integral equals itself!

─────────────────────────────────────

EXAMPLE 8.7   Evaluate the integral $\int \sec^3 x \, dx$.

Solution   We write the integral in the form $\int \sec x (\sec^2 x \, dx)$. The reason for writing it in this way is to get the factor $\sec^2 x$ with the dx, since we know its antiderivative is $\tan x$. Now we choose

$$u = \sec x \qquad dv = \sec^2 x \, dx$$

Then

$$du = \sec x \tan x \, dx \qquad v = \int \sec^2 x \, dx = \tan x$$

Then we have

$$\int \sec^3 x \, dx = \sec x \tan x - \int \tan x (\sec x \tan x \, dx)$$

$$= \sec x \tan x - \int \sec x \tan^2 x \, dx$$

Now we use the trigonometric identity $\tan^2 x = \sec^2 x - 1$ in the last integral to obtain

$$\int \sec^3 x \, dx = \sec x \tan x - \int \sec x (\sec^2 x - 1) \, dx$$

$$= \sec x \tan x - \int \sec^3 x \, dx + \int \sec x \, dx$$

$$= \sec x \tan x - \int \sec^3 x \, dx + \ln|\sec x + \tan x| \quad \text{From Equation 7.9}$$

Here again, the unknown integral appears on both sides. Calling this integral I, we get

$$I = \sec x \tan x - I + \ln|\sec x + \tan x|$$

$$2I = \sec x \tan x + \ln|\sec x + \tan x|$$

Finally, we divide by 2 and add a constant of integration:

$$\int \sec^3 x \, dx = \frac{1}{2}(\sec x \tan x + \ln|\sec x + \tan x|) + C \qquad \blacksquare$$

Exercise Set 8.1

Use integration by parts to evaluate the integrals in Exercises 1–29.

1. $\displaystyle\int xe^{-x} \, dx$

2. $\displaystyle\int x \sin x \, dx$

3. $\displaystyle\int_0^{\pi/4} x \cos 2x \, dx$

4. $\displaystyle\int x \sec^2 x \, dx$

5. $\displaystyle\int x^4 e^x \, dx$

6. $\displaystyle\int x^2 \cos^3 x \, dx$

7. $\displaystyle\int x \sec x \tan x \, dx$

8. $\displaystyle\int x \sinh 2x \, dx$

9. $\displaystyle\int_0^{\ln 2} x^2 \cosh x \, dx$

10. $\displaystyle\int \ln x \, dx$

11. $\displaystyle\int x \csc^2 x \, dx$

12. $\displaystyle\int_0^{\sqrt{3}/2} \sin^{-1} x \, dx$

13. $\displaystyle\int x \ln \frac{1}{x} \, dx$

14. $\displaystyle\int_1^4 \frac{\ln x}{\sqrt{x}} \, dx$

15. $\displaystyle\int \cos x \ln \sin x \, dx$

16. $\displaystyle\int x^3 (1-x)^9 \, dx$

17. $\displaystyle\int_1^2 \frac{x}{(2x-1)^2} \, dx$

18. $\displaystyle\int xe^{1-2x} \, dx$

19. $\displaystyle\int x^3 e^{-2x} \, dx$

20. $\displaystyle\int x^3 e^{x^2} \, dx$ (*Hint:* Take $u = x^2$.)

21. $\displaystyle\int x \tan^2 x \, dx$

22. $\displaystyle\int \sqrt{x} \ln x^3 \, dx$

23. Redo Example 8.6, taking $u = \cos 2x$ and $dv = e^{-x} \, dx$.

24. $\displaystyle\int e^x \sin x \, dx$ **25.** $\displaystyle\int e^x \cosh x \, dx$

26. $\displaystyle\int \csc^3 x \, dx$ **27.** $\displaystyle\int \operatorname{sech}^3 x \, dx$

28. $\int x \tan^{-1} x \, dx$ (*Hint:* In evaluating the integral that results from integration by parts, use long division.)

29. $\int x^2 \cot^{-1} x \, dx$ (See the hint for Exercise 28.)

30. Find the area under the graph of $y = \ln x^2$ from $x = 1$ to $x = 3$. Does $\ln x^2 = 2 \ln x$ for x between 1 and 3? Does $\ln x^2 = 2 \ln x$ for all $x \neq 0$? Explain.

31. Find the area of the region bounded by the graphs of $y = x \sin x$ and $y = x \cos x$ from $x = \pi/2$ to $x = \pi$.

32. Find the volume of the solid obtained by revolving the region under the graph of $y = e^x$ from $x = 0$ to $x = \ln 3$ about the y-axis.

33. Find the volume of the solid obtained by revolving the region under the graph of $y = \cos(x/2)$ from $x = 0$ to $x = \pi$ about the y-axis.

34. Find the surface area generated by revolving the arc of $y = \cosh x$ from $(0, 1)$ to $(2, \cosh 2)$ about the y-axis.

35. (a) Find the x-coordinate of the centroid of the region bounded by $y = \sin x$ and the x-axis from $x = 0$ to $x = \pi/2$.
    (b) Redo part (a) with $y = \cos x$ replacing $y = \sin x$.

36. Find the centroid of the region bounded by $y = \ln x$ and the x-axis from $x = 1$ to $x = e$.

37. Evaluate the integral $\int \sec^2 x \tan x \, dx$ by parts in two ways:
    (a) choosing $u = \sec x$
    (b) choosing $u = \tan x$

Show that the two answers are equivalent.

Use integration by parts to evaluate the integrals in Exercises 38–45.

38. $\displaystyle\int \sin \sqrt{x} \, dx$ (*Hint:* First make a substitution.)

39. $\displaystyle\int \cos(\ln x) \, dx$ (*Hint:* See the hint for Exercise 38.)

40. $\displaystyle\int x \sin^2 x \, dx$ **41.** $\displaystyle\int x \cos^2 x \, dx$

42. $\displaystyle\int \frac{x^3 \, dx}{\sqrt{1 - x^2}}$ **43.** $\displaystyle\int x^3 (x^2 - 4)^{3/2} \, dx$

44. $\displaystyle\int e^{ax} \sin bx \, dx$ **45.** $\displaystyle\int e^{ax} \cos bx \, dx$

In Exercises 46–48, derive the given reduction formulas, where n is a positive integer greater than 1.

46. $\displaystyle\int (\ln x)^n \, dx = x(\ln x)^n - n \int (\ln x)^{n-1} \, dx$

47. (a) $\displaystyle\int x^n \sin x \, dx = -x^n \cos x + n \int x^{n-1} \cos x \, dx$

    (b) $\displaystyle\int x^n \cos x \, dx = x^n \sin x - n \int x^{n-1} \sin x \, dx$

48. (a) $\displaystyle\int \sin^n x \, dx$
$$= -\frac{\sin^{n-1} x \cos x}{n} + \frac{n-1}{n} \int \sin^{n-2} x \, dx$$

    (b) $\displaystyle\int \cos^n x \, dx$
$$= \frac{\cos^{n-1} x \sin x}{n} + \frac{n-1}{n} \int \cos^{n-2} x \, dx$$

49. Prove **Wallis's Formula** for $n > 1$:
$$\int_0^{\pi/2} \sin^n x \, dx = \int_0^{\pi/2} \cos^n x \, dx$$
$$= \begin{cases} \dfrac{2 \cdot 4 \cdot 6 \cdot \,\cdots\, (n-1)}{1 \cdot 3 \cdot 5 \cdot \,\cdots\, n} & \text{if } n \text{ is odd} \\[2ex] \dfrac{1 \cdot 3 \cdot 5 \cdot \,\cdots\, (n-1)}{2 \cdot 4 \cdot 6 \cdot \,\cdots\, n} \cdot \dfrac{\pi}{2} & \text{if } n \text{ is even} \end{cases}$$
(*Hint:* Use the result of Exercise 48.)

50. Use Wallis's Formula from Exercise 49 to evaluate each of the following:
    (a) $\displaystyle\int_0^{\pi/2} \sin^5 x \, dx$
    (b) $\displaystyle\int_0^{\pi/2} \cos^6 x \, dx$
    (c) $\displaystyle\int_0^{\pi/2} \sin^{10} x \, dx$
    (d) $\displaystyle\int_0^{\pi/2} \cos^7 x \, dx$

51. Prove the reduction formulas
    (a) $\displaystyle\int \sec^n x \, dx$
$$= \frac{\sec^{n-2} x \tan x}{n-1} + \frac{n-2}{n-1} \int \sec^{n-2} x \, dx$$

    (b) $\displaystyle\int \csc^n x \, dx$
$$= -\frac{\csc^{n-2} x \cot x}{n-1} + \frac{n-2}{n-1} \int \csc^{n-2} x \, dx$$
where n is a positive integer greater than 1.

52. Prove Equation 8.3 for repeated integration by parts.

8.2 INTEGRATION OF TRIGONOMETRIC FORMS

In this section we consider integrals of the following types:

I. $\int \sin^m x \cos^n x \, dx$

II. $\int \sec^m x \tan^n x \, dx$

III. $\int \sin mx \cos nx \, dx, \quad \int \sin mx \sin nx \, dx, \quad \text{and} \quad \int \cos mx \cos nx \, dx$

Unless otherwise specified, we will assume that m and n are nonnegative integers.

Type I: $\int \sin^m x \cos^n x \, dx$

The approach to take with integrals of the form $\int \sin^m x \cos^n x \, dx$ depends on whether at least one of the exponents is odd or both are even. We begin with the case in which at least one of these exponents is odd. For definiteness, suppose n is odd. Our objective is to replace the given integral by one or more integrals of the form

$$\int \sin^k x \cos x \, dx$$

We can then make the substitution $u = \sin x$, so that $du = \cos x \, dx$, to obtain

$$\int \sin^k x \cos x \, dx = \int u^k \, du = \frac{u^{k+1}}{k+1} + C = \frac{\sin^{k+1} x}{k+1} + C$$

We can accomplish our objective by using the trigonometric identity $\cos^2 x = 1 - \sin^2 x$. We illustrate this procedure in the next example.

m or n Odd

EXAMPLE 8.8   Evaluate the integral

$$\int \sin^2 x \cos^3 x \, dx$$

Solution   We make use of the fact that the differential of $\sin x$ is $\cos x \, dx$. First we rewrite the integral in the form

$$\int (\sin^2 x)(\cos^2 x)(\cos x \, dx)$$

putting one of the $\cos x$ factors with the differential dx. We use the trigonometric identity $\cos^2 x = 1 - \sin^2 x$ to obtain

$$\int \sin^2 x (1 - \sin^2 x)(\cos x \, dx) = \int (\sin^2 x - \sin^4 x)(\cos x \, dx)$$

Now we see that the substitution $u = \sin x$, $du = \cos x \, dx$, gives

$$\int (\sin^2 x - \sin^4 x)(\cos x \, dx) = \int (u^2 - u^4) du = \frac{u^3}{3} - \frac{u^5}{5} + C$$

All that remains is to replace u by $\sin x$ to get the final result

$$\int \sin^2 x \cos^3 x \, dx = \frac{\sin^3 x}{3} - \frac{\sin^5 x}{5} + C$$

By now, you can probably handle substitutions of this sort mentally. The important step is to write the integrand in powers of $\sin x$, with the differential $\cos x \, dx$. ■

If the exponent m in the integral $\int \sin^m x \cos^n x \, dx$ is odd, we follow a similar procedure to that in Example 8.8, using the fact that the differential of $\cos x$ is $(-\sin x) \, dx$ and that $\sin^2 x = 1 - \cos^2 x$.

m and n Both Even

The next example shows the approach to take when both m and n are even. We will need the following trigonometric identities that are obtained from the *double-angle formulas* for $\cos 2x$. (See Appendix 2.)

$$\sin^2 x = \frac{1 - \cos 2x}{2} \tag{8.6}$$

$$\cos^2 x = \frac{1 + \cos 2x}{2} \tag{8.7}$$

When we use these identities, we can replace our integral with integrals involving powers of $\cos 2x$. For odd powers, we can use the technique of Example 8.8. For even powers, we apply Identity 8.7 again (with $2x$ in place of x), obtaining integrals involving powers of $\cos 4x$. We continue in this way until we are able to evaluate each of the integrals. We illustrate this technique in the next example.

EXAMPLE 8.9   Evaluate the integral

$$\int \sin^4 x \cos^2 x \, dx$$

Solution   Since $\sin^4 x = (\sin^2 x)^2$, we can use Identities 8.6 and 8.7 to get

$$\int \sin^4 x \cos^2 x \, dx = \int \left(\frac{1 - \cos 2x}{2} \right)^2 \left(\frac{1 + \cos 2x}{2} \right) dx$$

Now we expand the first factor on the right and then multiply the result by the second factor. The algebra involved is a bit tedious, and we omit the details. The result (which you should check) is

$$\frac{1}{8} \int (1 - \cos 2x - \cos^2 2x + \cos^3 2x) \, dx$$

which can be written as four integrals:

$$\frac{1}{8} \left[\int 1 \, dx - \int \cos 2x \, dx - \int \cos^2 2x \, dx + \int \cos^3 2x \, dx \right] \tag{8.8}$$

The first two of these integrals can easily be evaluated. The third integral again involves an even power, and we can evaluate it by applying Identity 8.7 again (with $2x$ replacing x), as follows.

$$\int \cos^2 2x\, dx = \int \frac{1 + \cos 4x}{2}\, dx = \frac{1}{2}\int (1 + \cos 4x)\, dx$$

$$= \frac{x}{2} + \frac{\sin 4x}{8}$$

(We will supply the constant of integration at the end.) The fourth integral in Formula 8.8 involves an odd power of $\cos 2x$, so we proceed as in Example 8.8. That is, we write

$$\int \cos^3 2x\, dx = \int \cos^2 2x (\cos 2x\, dx)$$

$$= \int (1 - \sin^2 2x)(\cos 2x\, dx)$$

Let $u = \sin 2x$. Then $du = 2\cos 2x\, dx$. So we have

$$\int \cos^3 2x\, dx = \frac{1}{2}\int (1 - \sin^2 2x)(2\cos 2x) = \frac{1}{2}\int (1 - u^2)\, du$$

$$= \frac{1}{2}\left(u - \frac{u^3}{3}\right) = \frac{1}{2}\left(\sin 2x - \frac{\sin^3 2x}{3}\right)$$

Now we can put everything together in Formula 8.8 to write the final answer (adding a constant of integration),

$$\int \sin^4 x \cos^2 x\, dx$$

$$= \frac{1}{8}\left[x - \frac{\sin 2x}{2} - \left(\frac{x}{2} + \frac{\sin 4x}{8}\right) + \frac{1}{2}\left(\sin 2x - \frac{\sin^3 2x}{3}\right)\right] + C$$

$$= \frac{x}{16} - \frac{\sin^3 2x}{48} - \frac{\sin 4x}{64} + C \qquad\blacksquare$$

We can summarize our approaches to integrals of Type I as follows:

Type I: $\quad \int \sin^m x \cos^n x\, dx$

Case (i): n is odd. Factor off $\cos x$ to go with dx. Then use the identity $\cos^2 x = 1 - \sin^2 x$ to write the rest of the integrand in powers of $\sin x$. Make the substitution $u = \sin x$.

Case (ii): m is odd. Factor off $\sin x$ to go with dx. Then use the identity $\sin^2 x = 1 - \cos^2 x$ to write the rest of the integrand in powers of $\cos x$. Make the substitution $u = \cos x$ and note that $du = -\sin x\, dx$. You will have to adjust the differential to obtain the minus sign.

Case (iii): m and n are both even. Use Identities 8.6 and 8.7 to reduce the powers. In the resulting integrals if an odd power of $\cos 2x$ occurs, use Case (ii). If an even power occurs, use Case (iii) again.

REMARKS

■ In Case (i), m need not be a nonnegative integer. It may be negative, or it may not be an integer. For example,

$$\int \frac{\cos^3 x}{\sin^2 x}\,dx \qquad \text{and} \qquad \int \frac{\cos^5 x}{\sqrt{\sin x}}\,dx$$

both can be handled as in Case (i). Similarly, in Case (ii) n need not be a nonnegative integer.

■ In Case (iii) m and n must both be nonnegative integers (so one or the other could be 0). For example, both

$$\int \sin^2 x\,dx \qquad \text{and} \qquad \int \cos^4 x\,dx$$

fall under Case (iii).

■ An alternative way to handle integrals of the form

$$\int \sin^m x\,dx \qquad \text{and} \qquad \int \cos^n x\,dx$$

is to use the reduction formulas in Exercise 48 of Exercise Set 8.1.

Type II: $\int \sec^m x \tan^n x\,dx$

For integrals of the form $\int \sec^m x \tan^n x\,dx$, the approach again depends on whether the exponents m and n are even or odd. The approach in each case is similar to that used in Type I. The trigonometric identities we will need are

$$\sec^2 x = 1 + \tan^2 x$$
$$\tan^2 x = \sec^2 x - 1$$

(You can obtain these identities from $\sin^2 x + \cos^2 x = 1$ by dividing both sides by $\cos^2 x$ and then rearranging terms.) We also need the differentials

$$d(\tan x) = \sec^2 x\,dx$$

$$d(\sec x) = \sec x \tan x\,dx$$

We illustrate the various cases with examples.

m Even

EXAMPLE 8.10   Evaluate the integral

$$\int \sec^4 x \tan^6 x\,dx$$

Solution   Since the differential of $\tan x$ is $\sec^2 x\,dx$, we split off the factor $\sec^2 x$ to go with the differential and write the rest of the integrand in terms of $\tan x$, using the identity $\sec^2 x = 1 + \tan^2 x$. We *must* have an even power of $\sec x$. Otherwise, when we factor off $\sec^2 x$, we will be left with an odd power of $\sec x$, and the substitution $\sec^2 x = 1 + \tan^2 x$ will lead to an irrational expression.

$$\int \sec^4 x \tan^6 x \, dx = \int \sec^2 x \tan^6 x (\sec^2 x \, dx)$$

$$= \int (1 + \tan^2 x) \tan^6 x (\sec^2 x \, dx)$$

$$= \int (\tan^6 x + \tan^8 x)(\sec^2 x \, dx)$$

Now we let $u = \tan x$. Then $du = \sec^2 x \, dx$. So our integral becomes

$$\int (u^6 + u^8) \, du = \frac{u^7}{7} + \frac{u^9}{9} + C$$

Replacing u by $\tan x$ gives the final result

$$\int \sec^4 x \tan^6 x \, dx = \frac{\tan^7 x}{7} + \frac{\tan^9 x}{9} + C$$

The key to this approach is that $\sec x$ is raised to a positive *even* power. That is, in the integral $\int \sec^m x \tan^n x \, dx$, the exponent m is even and positive. ∎

n Odd

EXAMPLE 8.11   Evaluate the integral

$$\int \sec^5 x \tan^3 x \, dx$$

Solution   This time we use the fact that the differential of $\sec x$ is $\sec x \tan x \, dx$, and split off the factor $\sec x \tan x$ to go with the differential. Then we are left with an even power of $\tan x$ (since the exponent on $\tan x$ was odd and we factored off one $\tan x$), so we can use the identity $\tan^2 x = \sec^2 x - 1$ to write the rest of the integrand in terms of $\sec x$:

$$\int \sec^5 x \tan^3 x \, dx = \int \sec^4 x \tan^2 x (\sec x \tan x \, dx)$$

$$= \int \sec^4 x (\sec^2 x - 1)(\sec x \tan x \, dx)$$

$$= \int (\sec^6 x - \sec^4 x)(\sec x \tan x \, dx)$$

Letting $u = \sec x$, we see that $du = \sec x \tan x \, dx$. Thus,

$$\int \sec^5 x \tan^3 x \, dx = \int (u^6 - u^4) \, du = \frac{u^7}{7} - \frac{u^5}{5} + C = \frac{\sec^7 x}{7} - \frac{\sec^5 x}{5} + C$$

The key to assuring success in this approach is that $\tan x$ is raised to a positive *odd* power. That is, in the integral $\int \sec^m x \tan^n x \, dx$, the exponent n is an odd integer greater than or equal to 1. ∎

REMARK ───

■ When $\tan x$ is raised to an odd power (and $\sec x$ is raised to *any* power, not necessarily an integral power), an alternative approach is to substitute

$$\sec x = \frac{1}{\cos x} \quad \text{and} \quad \tan x = \frac{\sin x}{\cos x}$$

The integral then is of Type I. In Exercise 41 of Exercise Set 8.2 we will ask you to redo Example 8.11, using this approach.

───

m **Odd,** n **Even**

There are four possible combinations for m and n:

$$m \text{ even, } n \text{ even} \qquad m \text{ even, } n \text{ odd}$$

$$m \text{ odd, } n \text{ even} \qquad m \text{ odd, } n \text{ odd}$$

We have already handled the first two cases by considering m even (n could be either even or odd) and the fourth case by considering n odd (m could be even or odd). The remaining case, in which m is odd and n is even (both nonnegative), can be changed to one or more integrals of the form $\int \sec^k x \, dx$ by replacing $\tan^2 x$ with $\sec^2 x - 1$. Then we can use integration by parts, as in Example 8.7, or alternatively, the reduction formula given in Exercise 51(a) of Exercise Set 8.1. For example, we can write

$$\int \sec^3 x \tan^2 x \, dx = \int \sec^3 x (\sec^2 x - 1) \, dx = \int (\sec^5 x - \sec^3 x) \, dx$$

$$= \int \sec^5 x \, dx - \int \sec^3 x \, dx$$

To complete the integration, we could use integration by parts, along with the result of Example 8.7 (see Exercise 38 in Exercise Set 8.2).

We can summarize the three cases for integrating $\int \sec^m x \tan^n x \, dx$ as follows:

───

Type II: $\displaystyle\int \sec^m x \tan^n x \, dx$

Case (i): m is even. Factor off $\sec^2 x$ to go with the differential. Then use the identity $\sec^2 x = 1 + \tan^2 x$ to write the rest of the integrand in terms of $\tan x$. Make the substitution $u = \tan x$.

Case (ii): n is odd. Factor off $\sec x \tan x$ to go with the differential. Then use the identity $\tan^2 x = \sec^2 x - 1$ to write the rest of the integrand in terms of $\sec x$. Make the substitution $u = \sec x$.

Case (iii): m is odd and n is even. Use the identity $\tan^2 x = \sec^2 x - 1$ to change the integral to one or more integrals of the form $\int \sec^k x \, dx$. Then use integration by parts, with $dv = \sec^2 x \, dx$, or use the reduction formula of Exercise 51(a) in Exercise Set 8.1.

REMARK

■ Integrals of the form $\int \csc^m x \cot^n x \, dx$ are handled as in Type II, using $\csc^2 x = 1 + \cot^2 x$, together with the differentials $d(\cot x) = -\csc^2 x \, dx$ and $d(\csc x) = -\csc x \cot x \, dx$.

Type III $\int \sin mx \cos nx \, dx$, $\int \sin mx \sin nx \, dx$, or $\int \cos mx \cos nx \, dx$ (m and n Positive Integers)

For integrals of one of these forms where $m \neq n$, we use one of the following trigonometric identities, known as *product formulas* (see Appendix 2):

$$\sin A \cos B = \frac{1}{2}[\sin(A + B) + \sin(A - B)] \tag{8.9}$$

$$\sin A \sin B = \frac{1}{2}[\cos(A - B) - \cos(A + B)] \tag{8.10}$$

$$\cos A \cos B = \frac{1}{2}[\cos(A + B) + \cos(A - B)] \tag{8.11}$$

We illustrate the technique with the following example, which is typical of Type III problems.

EXAMPLE 8.12   Evaluate the integral

$$\int \sin 5x \cos 2x \, dx$$

Solution   By Identity 8.9, we can write

$$\int \sin 5x \cos 2x \, dx = \frac{1}{2} \int (\sin 7x + \sin 3x) \, dx$$

$$= \frac{1}{2} \left[-\frac{\cos 7x}{7} - \frac{\cos 3x}{3} \right] + C$$

$$= -\frac{\cos 7x}{14} - \frac{\cos 3x}{6} + C$$

An alternative approach to this problem is to integrate by parts twice, solving the resulting equation for the original integral. We would obtain a different form of the answer, but the two results would be equivalent. ■

Exercise Set 8.2

In Exercises 1–40, evaluate the integrals.

1. $\displaystyle\int \cos^3 x \, dx$

2. $\displaystyle\int \sin^5 x \, dx$

3. $\displaystyle\int \sin^4 x \, dx$

4. $\displaystyle\int \cos^2 x \, dx$

5. $\displaystyle\int_0^{\pi/4} \sin^3 x \cos^2 x \, dx$

6. $\displaystyle\int \sin^2 x \cos^5 x \, dx$

7. $\displaystyle\int \sin^2 x \cos^2 x \, dx$

8. $\displaystyle\int \sin^2 x \cos^4 x \, dx$

9. $\displaystyle\int_0^{\pi/4} \sec^2 x \tan^3 x \, dx$

10. $\displaystyle\int \sec^4 x \tan^2 x \, dx$

11. $\displaystyle\int \sec x \tan^3 x \, dx$

12. $\displaystyle\int_0^{\pi/3} \sec^5 x \tan x \, dx$

13. $\displaystyle\int \sec x \tan^2 x \, dx$

14. $\displaystyle\int \sec^4 x \tan^3 x \, dx$ (Integrate by two methods.)

15. $\displaystyle\int_0^{\pi/3} \frac{\sin^3 x \, dx}{\cos x}$

16. $\displaystyle\int_{\pi/2}^{\pi} \sqrt{\sin x}\, \cos^3 x \, dx$

17. $\displaystyle\int \frac{\sec^4 x \, dx}{\tan^2 x}$ **18.** $\displaystyle\int \sec^{3/2} x \tan x \, dx$

19. $\displaystyle\int_0^{\pi/3} \tan^3 x \, dx$ **20.** $\displaystyle\int \tan^4 x \, dx$

21. $\displaystyle\int \sin^2 3x \cos^3 3x \, dx$

22. $\displaystyle\int_{\pi/12}^{\pi/4} \frac{\cos^5 2x}{\sin^4 2x} \, dx$

23. $\displaystyle\int_0^{\pi/8} \sqrt{\tan 2x}\, \sec^4 2x \, dx$

24. $\displaystyle\int \csc x \cot^2 x \, dx$

25. $\displaystyle\int x \sin^2(x^2) \cos^5(x^2) \, dx$ (*Hint:* Make the substitution $u = x^2$.)

26. $\displaystyle\int \frac{\tan^3 \sqrt{x}\, \sec^4 \sqrt{x}\, dx}{\sqrt{x}}$ (*Hint:* Make the substitution $u = \sqrt{x}$.)

27. $\displaystyle\int \csc^2 \frac{x}{2} \cot^2 \frac{x}{2} \, dx$ **28.** $\displaystyle\int \sin^4 x \cos^4 x \, dx$

29. $\displaystyle\int \sinh x \cosh^3 x \, dx$ **30.** $\displaystyle\int \sinh^2 x \cosh^2 x \, dx$

31. $\displaystyle\int \tanh^3 x \operatorname{sech}^4 x \, dx$

32. $\displaystyle\int \tanh^3 x \operatorname{sech}^3 x \, dx$

33. $\displaystyle\int \sin 3x \cos 2x \, dx$ **34.** $\displaystyle\int \sin 3x \sin 2x \, dx$

35. $\displaystyle\int \cos 3x \cos 2x \, dx$ **36.** $\displaystyle\int \sin x \cos 2x \, dx$

37. $\displaystyle\int \sin^6 x \, dx$ **38.** $\displaystyle\int \sec^3 x \tan^2 x \, dx$

39. $\displaystyle\int \frac{\tan^5 x \, dx}{\sec x}$

40. $\displaystyle\int (\sinh^3 x)(1 + \sinh^2 x)^{3/2} \, dx$

41. Redo Example 8.11 by writing the integrand in terms of $\sin x$ and $\cos x$.

42. For m and n positive integers, let

$$\delta_{mn} = \begin{cases} 1 & \text{if } m = n \\ 0 & \text{if } m \neq n \end{cases}$$

(δ_{mn} is known as the *Kronecker delta*.) Prove the following:
(a) $\displaystyle\int_{-\pi}^{\pi} \sin mx \sin nx \, dx = \int_{-\pi}^{\pi} \cos mx \cos nx \, dx = \pi \delta_{mn}$

(b) $\displaystyle\int_{-\pi}^{\pi} \sin mx \cos nx \, dx = 0$

43. A set $\{\phi_1(x), \phi_2(x), \ldots\}$ of continuous functions is said to be **orthogonal** on $[a, b]$ if

$$\int_a^b \phi_m(x)\phi_n(x) \, dx = 0 \qquad (m \neq n)$$

Prove that the set

$$\{1, \sin x, \cos x, \sin 2x, \cos 2x, \sin 3x, \cos 3x, \ldots\}$$

is orthogonal on $[-\pi, \pi]$. (*Hint:* See Exercise 42.)

44. Prove the following:
(a) $\int \sin^m x \cos^{2n+1} x \, dx = \int u^m (1 - u^2)^n \, du$,   where $u = \sin x$ and n is a nonnegative integer.
(b) $\int \sin^{2m+1} x \cos^n x \, dx = -\int u^n (1 - u^2)^m \, du$,   where $u = \cos x$ and m is a nonnegative integer.
(c) $\int \sec^{2m+2} x \tan^n x \, dx = \int u^n (1 + u^2)^m \, du$,   where $u = \tan x$ and m is a nonnegative integer.
(d) $\int \sec^m x \tan^{2n+1} x \, dx = \int u^{m-1} (u^2 - 1)^n \, du$,   where $u = \sec x$ and n is a nonnegative integer.

8.3 TRIGONOMETRIC SUBSTITUTIONS

We have previously encountered integrals of the form $\int_{-a}^{a} \sqrt{a^2 - x^2}\, dx$ and have used the fact that the graph of $y = \sqrt{a^2 - x^2}$ is the upper half of the circle $x^2 + y^2 = a^2$. Then, with our knowledge that the area of the entire circle is πa^2, we were able to arrive at the answer $\pi a^2/2$. In this section we will see another way of obtaining this result without reference to geometry.

More generally, we describe a method called *trigonometric substitution* that often works on integrals involving any one of the following expressions:

$$\sqrt{a^2 - u^2}, \quad \sqrt{a^2 + u^2}, \quad \text{or} \quad \sqrt{u^2 - a^2}$$

where $a > 0$. The appropriate substitutions are as follows:

If integral contains	Substitute
$\sqrt{a^2 - u^2}$	$u = a \sin\theta, \quad -\dfrac{\pi}{2} \le \theta \le \dfrac{\pi}{2}$
$\sqrt{a^2 + u^2}$	$u = a \tan\theta, \quad -\dfrac{\pi}{2} < \theta < \dfrac{\pi}{2}$
$\sqrt{u^2 - a^2}$	$u = a \sec\theta, \quad \begin{cases} 0 \le \theta < \dfrac{\pi}{2} & \text{if } u \ge a \\[2mm] \pi \le \theta < \dfrac{3\pi}{2} & \text{if } u \le -a \end{cases}$

(We are using the lowercase Greek letter θ as the substitution variable because it is suggestive of an angle, but other letters could be used.) These choices are based on the trigonometric identities

$$1 - \sin^2\theta = \cos^2\theta$$
$$1 + \tan^2\theta = \sec^2\theta$$
$$\sec^2\theta - 1 = \tan^2\theta$$

Using these identities, called the Pythagorean Identities, and the substitutions shown above, we eliminate the radicals:

$$\sqrt{a^2 - u^2} = \sqrt{a^2 - a^2 \sin^2\theta} = \sqrt{a^2(1 - \sin^2\theta)} = a\sqrt{\cos^2\theta} = a\cos\theta$$
$$\sqrt{a^2 + u^2} = \sqrt{a^2 + a^2 \tan^2\theta} = \sqrt{a^2(1 + \tan^2\theta)} = a\sqrt{\sec^2\theta} = a\sec\theta$$
$$\sqrt{u^2 - a^2} = \sqrt{a^2 \sec^2\theta - a^2} = \sqrt{a^2(\sec^2\theta - 1)} = a\sqrt{\tan^2\theta} = a\tan\theta$$

The final step in each case is justified as follows, by the restrictions we placed on θ. In the first case, since $\cos\theta \ge 0$ for $-\pi/2 \le \theta \le \pi/2$, we have $\sqrt{\cos^2\theta} = |\cos\theta| = \cos\theta$. Similarly, $\sec\theta > 0$ for $-\pi/2 < \theta < \pi/2$, so that $\sqrt{\sec^2\theta} = |\sec\theta| = \sec\theta$, and $\tan\theta > 0$ for θ satisfying either $0 \le \theta \le \pi/2$ or $\pi \le \theta < 3\pi/2$, so that $\sqrt{\tan^2\theta} = |\tan\theta| = \tan\theta$.

REMARK ———————————————————————————————

■ In the past when we have made a substitution in an integral, we have expressed the new variable in terms of the original one. For example, in the integral $\int \sin^3 x \cos x\, dx$, we would let $u = \sin x$. The new variable u is expressed in terms of the original variable x. Now we are going in reverse. We are saying, for example, that when an integral contains an expression of the form $\sqrt{a^2 - u^2}$

we let $u = a \sin \theta$, with $-\pi/2 \leq \theta \leq \pi/2$. That is, the original variable u is expressed in terms of the new variable θ. With the restrictions we have placed on θ, however, we can rewrite $u = a \sin \theta$ as $\theta = \sin^{-1}(u/a)$. Similarly, when we substitute $u = a \tan \theta$ with the specified restrictions, the substitution is equivalent to $\theta = \tan^{-1}(u/a)$, and when $u = a \sec \theta$, we have $\theta = \sec^{-1}(u/a)$. From now on, when we make the trigonometric substitutions $u = a \sin \theta$, $u = a \tan \theta$, or $u = a \sec \theta$ in the cases we have described, we will understand that these mean $\theta = \sin^{-1}(u/a)$, $\theta = \tan^{-1}(u/a)$, or $\theta = \sec^{-1}(u/a)$, respectively.

EXAMPLE 8.13   Evaluate the integral

$$\int \frac{x^3}{\sqrt{4 - x^2}}\, dx$$

Solution   We make the substitution $x = 2 \sin \theta$. Then $dx = 2 \cos \theta\, d\theta$ and $\theta = \sin^{-1}(x/2)$. So we have

$$\int \frac{x^3}{\sqrt{4 - x^2}}\, dx = \int \frac{(2 \sin \theta)^3 (2 \cos \theta\, d\theta)}{\sqrt{4 - 4 \sin^2 \theta}} = \int \frac{(8 \sin^3 \theta)(2 \cos \theta\, d\theta)}{2\sqrt{1 - \sin^2 \theta}}$$

$$= 8 \int \frac{\sin^3 \theta \cos \theta\, d\theta}{\sqrt{\cos^2 \theta}} = 8 \int \frac{\sin^3 \theta \, \cos \theta\, d\theta}{\cos \theta} = 8 \int \sin^3 \theta\, d\theta$$

Now we use ideas of the preceding section:

$$8 \int \sin^3 \theta\, d\theta = 8 \int \sin^2 \theta (\sin \theta\, d\theta) = 8 \int (1 - \cos^2 \theta) \sin \theta\, d\theta \qquad \text{Let } u = \cos \theta.$$

$$= 8 \left(-\cos \theta + \frac{\cos^3 \theta}{3} \right) + C$$

But we are not through, since this result must be expressed in terms of x. An easy way is to draw a right triangle as in Figure 8.3, showing θ as an acute angle with its opposite side equal to x and hypotenuse 2. Then the adjacent side is $\sqrt{4 - x^2}$, so that $\cos \theta = \sqrt{4 - x^2}/2$. (We could also find $\cos \theta$ without reference to the figure, using $\cos \theta = \sqrt{1 - \sin^2 \theta} = \sqrt{1 - (\frac{x}{2})^2} = \sqrt{4 - x^2}/2$.) Thus,

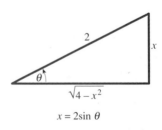

$x = 2\sin \theta$

FIGURE 8.3

$$\int \frac{x^3}{\sqrt{4 - x^2}}\, dx = 8 \left[-\frac{\sqrt{4 - x^2}}{2} + \frac{1}{3} \left(\frac{\sqrt{4 - x^2}}{2} \right)^3 \right] + C$$

$$= -4\sqrt{4 - x^2} + \frac{(4 - x^2)^{3/2}}{3} + C \qquad \blacksquare$$

When making a substitution, do not forget to substitute for dx.

EXAMPLE 8.14   Evaluate the integral

$$\int_0^4 \frac{dx}{(9 + x^2)^{3/2}}$$

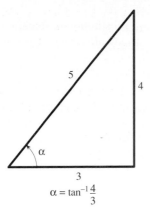

FIGURE 8.4

$\alpha = \tan^{-1}\dfrac{4}{3}$

Solution   We let $x = 3\tan\theta$. Then $dx = 3\sec^2\theta\,d\theta$, and $\theta = \tan^{-1}(x/3)$. We also change the limits of integration. When $x = 0$, we have $\theta = \tan^{-1} 0 = 0$, and when $x = 4$, $\theta = \tan^{-1}(4/3)$. Let us denote $\tan^{-1}(4/3)$ by α (that is, α is the angle whose tangent is $4/3$). Figure 8.4 shows a right triangle with α as one of its acute angles. Our integral now becomes

$$\int_0^\alpha \frac{3\sec^2\theta\,d\theta}{(9+9\tan^2\theta)^{3/2}} = \int_0^\alpha \frac{3\sec^2\theta\,d\theta}{27(1+\tan^2\theta)^{3/2}}$$

$$= \frac{1}{9}\int_0^\alpha \frac{\sec^2\theta}{\sec^3\theta}\,d\theta$$

$$= \frac{1}{9}\int_0^\alpha \cos\theta\,d\theta \quad \text{Since } \frac{1}{\sec\theta} = \cos\theta$$

$$= \frac{1}{9}[\sin\theta]_0^\alpha$$

$$= \frac{1}{9}\sin\alpha$$

From Figure 8.4 we see that $\sin\alpha = 4/5$. Thus, we have the final result

$$\int_0^4 \frac{dx}{(9+x^2)^{3/2}} = \frac{1}{9}\left(\frac{4}{5}\right) = \frac{4}{45} \qquad \blacksquare$$

EXAMPLE 8.15   Evaluate the integral

$$\int \frac{\sqrt{4x^2-9}}{x}\,dx$$

Solution   The radical is of the form $\sqrt{u^2-a^2}$, with $u = 2x$ and $a = 3$. Using $u = a\sec\theta$, we have $2x = 3\sec\theta$, or

$$x = \frac{3\sec\theta}{2}$$

Then $dx = \dfrac{3}{2}\sec\theta\tan\theta\,d\theta$ and $\theta = \sec^{-1}\dfrac{2x}{3}$. So the integral becomes

$$\int \frac{\sqrt{4x^2-9}}{x}\,dx = \int \frac{\sqrt{9\sec^2\theta-9}}{\frac{3}{2}\sec\theta}\cdot\left(\frac{3}{2}\sec\theta\tan\theta\,d\theta\right)$$

$$= \int 3\sqrt{\sec^2\theta-1}\,\tan\theta\,d\theta$$

$$= 3\int \tan^2\theta\,d\theta$$

$$= 3\int (\sec^2\theta-1)\,d\theta$$

$$= 3(\tan\theta-\theta)+C$$

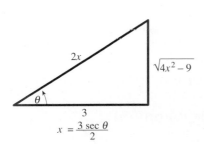

FIGURE 8.5

$x = \dfrac{3\sec\theta}{2}$

From the definition of θ, we get Figure 8.5, so

$$\int \frac{\sqrt{4x^2-9}}{x}\,dx = 3\left(\frac{\sqrt{4x^2-9}}{3} - \sec^{-1}\frac{2x}{3}\right)+C$$

$$= \sqrt{4x^2-9} - 3\sec^{-1}\frac{2x}{3}+C \qquad \blacksquare$$

The next example shows that a trigonometric substitution is sometimes helpful even when the integrand does not contain a radical.

EXAMPLE 8.16   Evaluate the integral

$$\int \frac{x^3 \, dx}{1 + 2x^2 + x^4}$$

Solution   The integral can be written in the form

$$\int \frac{x^3 \, dx}{(1 + x^2)^2}$$

Now let $x = \tan \theta$, so that $dx = \sec^2 \theta \, d\theta$ and $\theta = \tan^{-1} x$. This substitution gives

$$\int \frac{\tan^3 \theta \sec^2 \theta \, d\theta}{(1 + \tan^2 \theta)^2} = \int \frac{\tan^3 \theta \sec^2 \theta \, d\theta}{\sec^4 \theta}$$

$$= \int \frac{\tan^3 \theta \, d\theta}{\sec^2 \theta} = \int \frac{\sin^3 \theta}{\cos^3 \theta} \cdot \cos^2 \theta \, d\theta$$

$$= \int \frac{\sin^3 \theta}{\cos \theta} \, d\theta = \int \frac{1 - \cos^2 \theta}{\cos \theta} \sin \theta \, d\theta \qquad \text{Mentally, let } u = \cos \theta.$$

$$= \int \left(\frac{1}{\cos \theta} - \cos \theta \right) \sin \theta \, d\theta = -\ln |\cos \theta| + \frac{\cos^2 \theta}{2} + C$$

From Figure 8.6, we therefore have

$$\int \frac{x^3 \, dx}{1 + 2x^2 + x^4} = -\ln \frac{1}{\sqrt{1 + x^2}} + \frac{1}{2} \left(\frac{1}{1 + x^2} \right) + C$$

$$= -\ln(1 + x^2)^{-1/2} + \frac{1}{2} \left(\frac{1}{1 + x^2} \right) + C$$

$$= -\left(-\frac{1}{2} \right) \ln(1 + x^2) + \frac{1}{2} \left(\frac{1}{1 + x^2} \right) + C$$

$$= \frac{1}{2} \left[\ln(1 + x^2) + \frac{1}{1 + x^2} \right] + C \qquad \blacksquare$$

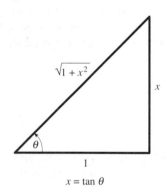

$x = \tan \theta$

FIGURE 8.6

EXAMPLE 8.17   Derive the integration formula

$$\int \sqrt{a^2 - u^2} \, du = \frac{1}{2} \left(a^2 \sin^{-1} \frac{u}{a} + u\sqrt{a^2 - u^2} \right) + C$$

Solution   Let $u = a \sin \theta$; then $du = a \cos \theta \, d\theta$ and $\theta = \sin^{-1} \frac{u}{a}$. So we have

$$\int \sqrt{a^2 - u^2} \, du = \int \sqrt{a^2 - a^2 \sin^2 \theta} \, (a \cos \theta \, d\theta)$$

$$= a^2 \int \sqrt{1 - \sin^2 \theta} \, (\cos \theta \, d\theta)$$

$$= a^2 \int \cos^2 \theta \, d\theta = \frac{a^2}{2} \int (1 + \cos 2\theta) \, d\theta$$

$$= \frac{a^2}{2} \left(\theta + \frac{\sin 2\theta}{2} \right) + C$$

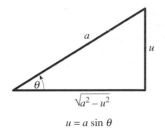

$$u = a \sin \theta$$

FIGURE 8.7

To express the answer in terms of u, use the trigonometric identity $\sin 2\theta = 2 \sin \theta \cos \theta$ and refer to Figure 8.7. This gives

$$\int \sqrt{a^2 - u^2}\, du = \frac{a^2}{2} \left(\sin^{-1} \frac{u}{a} + \frac{u}{a} \cdot \frac{\sqrt{a^2 - u^2}}{a} \right) + C$$

$$= \frac{1}{2} \left(a^2 \sin^{-1} \frac{u}{a} + u \sqrt{a^2 - u^2} \right) + C \qquad \blacksquare$$

REMARK
■ It is unlikely that you would want to commit the formula we just derived to memory. The important thing is to see how to derive it. In the future, when an integral of this type arises, you can use a trigonometric substitution, as we did in the example, or you can refer to the formula we derived. (It is listed in the tables on the endpapers.) You could also use a computer algebra system. We show one application of the formula in Example 8.18.

Area Enclosed by an Ellipse

In Example 5.25 we showed how to find the area enclosed by an ellipse by making use of our knowledge of the area of a circle. In the next example we calculate the area inside an ellipse again, without reference to the area of a circle. In fact, we will see that the area formula for a circle can be obtained as a special case of our new result.

EXAMPLE 8.18   Find the area inside the ellipse

$$\frac{x^2}{a^2} + \frac{y^2}{b^2} = 1 \qquad (a > 0, b > 0)$$

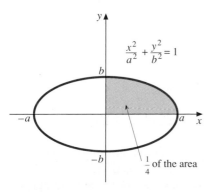

FIGURE 8.8

Solution   We show the ellipse in Figure 8.8. Taking symmetry into account, we see that it is sufficient to find the area in the first quadrant and multiply the result by 4.

The equation of the upper semiellipse is found by solving the given equation for y and taking the positive square root, since $y > 0$ for the upper half of the ellipse.

$$\frac{y^2}{b^2} = 1 - \frac{x^2}{a^2}$$

$$y^2 = b^2 \left(1 - \frac{x^2}{a^2} \right) = \frac{b^2}{a^2} (a^2 - x^2)$$

$$y = \frac{b}{a} \sqrt{a^2 - x^2} \qquad \text{For upper semiellipse}$$

The total area inside the ellipse is

$$A = 4 \int_0^a \frac{b}{a} \sqrt{a^2 - x^2}\, dx = \frac{4b}{a} \int_0^a \sqrt{a^2 - x^2}\, dx$$

To evaluate the integral, we could make the substitution $x = a \sin \theta$. But in the previous example we already evaluated the indefinite integral $\int \sqrt{a^2 - u^2}\, du$.

So we use that result and the Second Fundamental Theorem to get

$$A = \frac{4b}{a} \left[\frac{1}{2} \left(a^2 \sin^{-1} \frac{x}{a} + x\sqrt{a^2 - x^2} \right) \right]_0^a = \frac{4b}{a} \left(\frac{a^2}{2} \sin^{-1} 1 \right)$$

$$= 2ab \left(\frac{\pi}{2} \right) = \pi ab$$

In words, we can say that *the area of an ellipse is pi times the product of the semiaxes.*   ■

Area of a Circle

Since a circle is an ellipse in which $a = b$, we also have, from the formula πab for the area of an ellipse, that the area of a circle is πa^2, where a is the radius. Although this latter result is well known, and we have used it on several occasions, we now have verified it.

Trigonometric Versus Hyperbolic Functions

The next example illustrates further similarities between trigonometric and hyperbolic functions. Recall that we have already made some comparisons in Section 7.6.

EXAMPLE 8.19   Compare the area under the upper half of the unit circle $x^2 + y^2 = 1$ with the area under the upper half of the unit hyperbola $y^2 - x^2 = 1$ from $x = -1$ to $x = 1$.

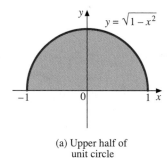

(a) Upper half of
unit circle

(b) Upper half of
unit hyperbola

FIGURE 8.9

Solution   We show the two regions in question in Figure 8.9. The graph of $y = \sqrt{1 - x^2}$ is the upper half of the unit circle $x^2 + y^2 = 1$, and the graph of $y = \sqrt{1 + x^2}$ is the upper half of the unit hyperbola $y^2 - x^2 = 1$.

In Example 8.17 we found a formula for the indefinite integral

$$\int \sqrt{a^2 - u^2} \, du$$

Taking $a = 1$ and replacing u by x in that formula gives

$$\int \sqrt{1 - x^2} \, dx = \frac{1}{2} \left(\sin^{-1} x + x\sqrt{1 - x^2} \right) + C$$

Recall that the result in Example 8.17 was obtained by the substitution $u = a \sin \theta$. In the special case we have, the substitution would therefore be $x = \sin \theta$. To get the area under the semicircle, we make use of the Second Fundamental Theorem:

$$A = \frac{1}{2} \left[\sin^{-1} x + x\sqrt{1 - x^2} \right]_{-1}^{1} = \left[\frac{1}{2} \sin^{-1} 1 + 0 \right] - \left[\frac{1}{2} \sin^{-1}(-1) + 0 \right]$$

$$= \frac{1}{2} \left(\frac{\pi}{2} \right) - \frac{1}{2} \left(-\frac{\pi}{2} \right) = \frac{\pi}{2}$$

Of course, we know this result already, since the area of a circle of radius 1 is $\pi (1)^2$, making half the area $\pi/2$. The emphasis we want to make in this example, however, is not on the numerical answer, but on the antiderivative used to get it, namely,

$$A = \int_{-1}^{1} \sqrt{1 - x^2} \, dx = \frac{1}{2} \left[\sin^{-1} x + x\sqrt{1 - x^2} \right]_{-1}^{1} \qquad (8.12)$$

Let us now consider the second curve, $y = \sqrt{1 + x^2}$. The area is given by

$$A = \int_{-1}^{1} \sqrt{1 + x^2} \, dx$$

To evaluate the integral, we will again use the method of finding the indefinite integral (antiderivative)

$$\int \sqrt{1 + x^2} \, dx$$

and then use the Second Fundamental Theorem.

A natural way to evaluate this indefinite integral is to make the substitution $x = \tan \theta$. However, to keep the parallelism with the method used for the area under the semicircle, we choose instead to use the hyperbolic sine (Section 7.6), letting

$$x = \sinh \theta$$

Then

$$dx = \cosh \theta \, d\theta$$

and

$$\sqrt{1 + x^2} = \sqrt{1 + \sinh^2 \theta} = \sqrt{\cosh^2 \theta} = |\cosh \theta| = \cosh \theta$$

Here we used the hyperbolic identity (derived in Section 7.6)

$$\cosh^2 \theta - \sinh^2 \theta = 1$$

giving $1 + \sinh^2 \theta = \cosh^2 \theta$. Thus, we have

$$\int \sqrt{1 + x^2} \, dx = \int (\cosh \theta)(\cosh \theta \, d\theta) = \int \cosh^2 \theta \, d\theta$$

Now from Hyperbolic Identity 5 (in Section 7.6) we find that $\cosh^2 \theta$ can be written as

$$\cosh^2 \theta = \frac{1 + \cosh 2\theta}{2}$$

Thus,

$$\int \sqrt{1 + x^2} \, dx = \frac{1}{2} \int (1 + \cosh 2\theta) \, d\theta = \frac{1}{2} \left[\theta + \frac{\sinh 2\theta}{2} \right] + C$$

By Hyperbolic Identity 4,

$$\sinh 2\theta = 2 \sinh \theta \cosh \theta$$

Thus,

$$\int \sqrt{1 + x^2} \, dx = \frac{1}{2} [\theta + \sinh \theta \cosh \theta] + C$$

Finally, since $x = \sinh\theta$, we have

$$\theta = \sinh^{-1} x$$

Also, $\cosh\theta = \sqrt{1 + \sinh^2\theta} = \sqrt{1 + x^2}$. Thus,

$$\int \sqrt{1 + x^2}\, dx = \frac{1}{2}\left[\sinh^{-1} x + x\sqrt{1 + x^2}\right] + C$$

By the Second Fundamental Theorem, the area is

$$A = \int_{-1}^{1} \sqrt{1 + x^2}\, dx = \frac{1}{2}\left[\sinh^{-1} x + x\sqrt{1 + x^2}\right]_{-1}^{1} \tag{8.13}$$

To get the final answer, we can use Formula 7.70 to write

$$\sinh^{-1} x = \ln\left(x + \sqrt{1 + x^2}\right)$$

Thus,

$$A = \frac{1}{2}\left[\left(\sinh^{-1} 1 + 1\sqrt{2}\right) - \left(\sinh^{-1}(-1) - 1\sqrt{2}\right)\right]$$

$$= \frac{1}{2}\left[\sinh^{-1} 1 - \sinh^{-1}(-1) + 2\sqrt{2}\right]$$

$$= \frac{1}{2}\left[\ln\left(1 + \sqrt{2}\right) - \ln\left(-1 + \sqrt{2}\right) + 2\sqrt{2}\right]$$

$$= \sqrt{2} + \frac{1}{2}\ln\frac{\sqrt{2} + 1}{\sqrt{2} - 1}$$

Rationalizing, we can write

$$\frac{\sqrt{2} + 1}{\sqrt{2} - 1} \cdot \frac{\sqrt{2} + 1}{\sqrt{2} + 1} = \frac{2 + 2\sqrt{2} + 1}{2 - 1} = 3 + 2\sqrt{2}$$

Thus,

$$A = \sqrt{2} + \frac{1}{2}\ln(3 + 2\sqrt{2}) \approx 3.18$$

Again, it is not the final answer we want to emphasize but the result in Equation 8.13. The analogy with Equation 8.12 is striking. Let us summarize what we have found:

Area Under Semicircle	**Area Under Hyperbola**
$$A = \int_{-1}^{1} \sqrt{1 - x^2}\, dx$$	$$A = \int_{-1}^{1} \sqrt{1 + x^2}\, dx$$
Trigonometric Substitution	Hyperbolic Substitution
$$x = \sin\theta$$	$$x = \sinh\theta$$
$$A = \frac{1}{2}\left[\sin^{-1} x + x\sqrt{1 - x^2}\right]_{-1}^{1}$$	$$A = \frac{1}{2}\left[\sinh^{-1} x + x\sqrt{1 + x^2}\right]_{-1}^{1}$$

Beyond this point, the analogy breaks down, since the values of $\sin^{-1} x$ and $\sinh^{-1} x$ are quite different. Nevertheless, this example provides further reinforcement of the similarities between trigonometric and hyperbolic functions.

■

Exercise Set 8.3

In Exercises 1–9, evaluate the integral by making a trigonometric substitution.

1. $\displaystyle\int \frac{dx}{\sqrt{x^2 - 4}}$

2. $\displaystyle\int \frac{\sqrt{9 - x^2}}{x}\, dx$

3. $\displaystyle\int_{-1}^{1} \frac{dx}{(2 - x^2)^{3/2}}$

4. $\displaystyle\int_{1}^{3} \frac{\sqrt{x^2 + 3}\, dx}{x^4}$

5. $\displaystyle\int_{\sqrt{3}}^{3} x^3 \sqrt{4x^2 - 9}\, dx$

6. $\displaystyle\int_{0}^{2} \frac{x^2\, dx}{(3x^2 + 4)^{3/2}}$

7. $\displaystyle\int_{-1/\sqrt{3}}^{1} \frac{x^3\, dx}{(4 - 3x^2)^{5/2}}$

8. $\displaystyle\int \frac{dx}{(x^2 - 1)^2}$

9. $\displaystyle\int \frac{x^5}{(1 - x^2)^3}\, dx$

In Exercises 10–16, evaluate the integrals by two methods: (a) trigonometric substitution and (b) algebraic substitution.

10. $\displaystyle\int \frac{x\, dx}{\sqrt{9 - 4x^2}}$

11. $\displaystyle\int_{0}^{1} x(3x^2 + 1)^{3/2}\, dx$

12. $\displaystyle\int_{1}^{\sqrt{3}} \frac{x}{\sqrt{4x^2 - 3}}\, dx$

13. $\displaystyle\int \frac{x}{1 + 2x^2 + x^4}\, dx$

14. $\displaystyle\int x^3 \sqrt{1 - x^2}\, dx$ (*Hint for (b)*: Let $u = 1 - x^2$. Then $du = -2x\, dx$ and $x^2 = 1 - u$.)

15. $\displaystyle\int \frac{x^3}{\sqrt{1 + x^2}}\, dx$ (See the hint for Exercise 14.)

16. $\displaystyle\int \frac{x^3}{\sqrt{x^2 - 1}}\, dx$ (See the hint for Exercise 14.)

In Exercises 17–23, verify the given integration formula by using a trigonometric substitution to evaluate the integral.

17. $\displaystyle\int \frac{du}{\sqrt{a^2 - u^2}} = \sin^{-1} \frac{u}{a} + C$

18. $\displaystyle\int \frac{du}{a^2 + u^2} = \frac{1}{a} \tan^{-1} \frac{u}{a} + C$

19. $\displaystyle\int \frac{du}{u\sqrt{u^2 - a^2}} = \frac{1}{a} \sec^{-1} \frac{u}{a} + C$

20. $\displaystyle\int \frac{du}{\sqrt{a^2 + u^2}} = \ln\left| u + \sqrt{a^2 + u^2} \right| + C$

21. $\displaystyle\int \frac{du}{u^2 - a^2} = \frac{1}{2a} \ln \left| \frac{u - a}{u + a} \right| + C$

22. $\displaystyle\int \frac{du}{u\sqrt{a^2 - u^2}} = -\frac{1}{a} \ln \left| \frac{a + \sqrt{a^2 - u^2}}{u} \right| + C$

23. $\displaystyle\int \frac{du}{u\sqrt{a^2 + u^2}} = -\frac{1}{a} \ln \left| \frac{a + \sqrt{a^2 + u^2}}{u} \right| + C$

24. Determine an appropriate rationalizing hyperbolic substitution for integrands involving the following:
(a) $\sqrt{a^2 - u^2}$
(b) $\sqrt{a^2 + u^2}$
(c) $\sqrt{u^2 - a^2}$

In Exercises 25–27, redo Exercises 2, 3, and 5, respectively, using the appropriate hyperbolic substitution determined in Exercise 24.

28. Find the length of the parabola $y = x^2/3$ from $x = 0$ to $x = 2$.

29. Find the area of the region under the curve $y = \sqrt{1 + x^2}$ from $x = 0$ to $x = 2\sqrt{2}$.

In Exercises 30–33, after completing the square, use a trigonometric substitution to evaluate the integral.

30. $\displaystyle\int \frac{dx}{\sqrt{x^2 + 2x + 2}}$

31. $\displaystyle\int (2x - x^2)^{3/2}\, dx$

32. $\displaystyle\int \frac{dx}{x^2 + 4x - 5}$

33. $\displaystyle\int \frac{(3x + 2)^2\, dx}{(5 - 12x - 9x^2)^{3/2}}$

Evaluate the integrals in Exercises 34–37.

34. $\displaystyle\int \frac{\cos t\, dt}{(1 + \sin^2 t)^{3/2}}$ (*Hint:* First let $u = \sin t$.)

35. $\displaystyle\int \frac{\sec^2 t \tan^3 t\, dt}{\sqrt{\tan^2 t - 1}}$

36. $\displaystyle\int \frac{\sin \theta\, d\theta}{(1 + \cos^2 \theta)^3}$

37. $\displaystyle\int \frac{e^{3x}\, dx}{\sqrt{1 - e^{2x}}}$

38. Derive the formula for the length of the circumference of a circle of radius a.

39. **The torus revisited.** Derive the formula for the volume of a torus by revolving the region under the upper half of the circle $x^2 + y^2 = r^2$ about the line $x = R$, where $R > r$, and doubling the result. Use the method of cylindrical shells, and use an appropriate trigonometric substitution to evaluate the integral.

40. The length of an ellipse revisited. In Example 6.19 of Section 6.4 we derived the formula

$$L = 4 \int_0^a \sqrt{1 + \frac{b^2 x^2}{a^2(a^2 - x^2)}} \, dx$$

for the length of the ellipse

$$\frac{x^2}{a^2} + \frac{y^2}{b^2} = 1$$

We noted that the integrand was unbounded at $x = a$ (the integral is improper).

(a) Use the trigonometric substitution $x = a \sin \theta$ to

convert the integral to

$$L = 4a \int_0^{\pi/2} \sqrt{1 - e^2 \sin^2 t} \, dt$$

where $e = \sqrt{a^2 - b^2}/a$ (e is called the *eccentricity* of the ellipse).

(b) The integral in part (a) is no longer improper, but there is no elementary antiderivative of the integrand, so the Second Fundamental Theorem of Calculus cannot be used. Use Simpson's Rule with $n = 20$ to approximate its value for $a = 4$ and $b = 3$.

8.4 PARTIAL FRACTIONS

We now turn to techniques for integrating rational functions—that is, functions of the form $P(x)/Q(x)$, where $P(x)$ and $Q(x)$ are polynomials. Actually, what we study is an algebraic technique for expressing $P(x)/Q(x)$ as a sum of simpler rational functions when the degree of Q is 2 or greater. The simpler functions can then be integrated by means already familiar to us. As a way of introducing the ideas, consider the following sum of two fractions:

$$\frac{2}{x - 1} + \frac{3}{x + 2} = \frac{5x + 1}{x^2 + x - 2}$$

Now suppose we are asked to perform the integration

$$\int \frac{5x + 1}{x^2 + x - 2} \, dx$$

By replacing the integrand with the sum of the two simpler fractions, whose integrals we can easily find, we get

$$\int \frac{5x + 1}{x^2 + x - 2} \, dx = \int \left(\frac{2}{x - 1} + \frac{3}{x + 2} \right) dx$$

$$= 2 \ln |x - 1| + 3 \ln |x + 2| + C$$

$$= \ln |(x - 1)^2 (x + 2)^3| + C$$

This procedure works, *provided* we know the two original fractions. The objective of this section is to introduce ways of determining the component fractions when we are given their sum; that is, we will introduce a procedure allowing us to do the reverse of adding fractions: given a rational function, we want to find the simplest fractions that add together to give that function. The process is called **decomposition into partial fractions** (or partial-fraction decomposition).

We begin with the assumption that $P(x)/Q(x)$ is in lowest terms and that the *degree of P is less than the degree of Q*—that is, that $P(x)/Q(x)$ is a *proper* fraction. If the fraction is not proper, then we divide to obtain

$$\frac{P(x)}{Q(x)} = P_1(x) + \frac{R(x)}{Q(x)}$$

$$
\begin{array}{r}
x - 3 \\
x^3 + x \overline{\smash{\big)}\ x^4 - 3x^3 + x^2 \qquad\ - 5} \\
\underline{x^4 \qquad\quad + x^2} \\
-3x^3 \qquad\qquad - 5 \\
\underline{-3x^3 \qquad - 3x} \\
3x - 5
\end{array}
$$

where the degree of R is less than the degree of Q. We then apply our theory to $R(x)/Q(x)$. For example, the long division in the margin shows that

$$
\frac{x^4 - 3x^3 + x^2 - 5}{x^3 + x} = x - 3 + \frac{3x - 5}{x^3 + x}
$$

Since the quotient $x - 3$ is easily integrated, we need only work with the fraction on the right.

It is proved in algebra courses that every polynomial with real coefficients can (in theory, at least) be factored into real linear and/or quadratic factors that are **irreducible** over the reals. A quadratic factor $ax^2 + bx + c$ is irreducible over the reals if it cannot be factored into linear factors with real coefficients, which is true when the *discriminant*, $b^2 - 4ac$, is negative. The factors $x^2 + 1$ and $x^2 + 2x + 4$ are examples of such irreducible quadratic factors. So we need to concern ourselves with only linear and irreducible quadratic factors for $Q(x)$. In particular, if $Q(x)$ is a cubic, or quartic, or a polynomial of higher degree, we first factor it into linear and/or quadratic factors. For example,

$$
x^3 - 2x^2 + x - 2 = (x - 2)(x^2 + 1)
$$

Computer algebra systems can do such factoring.

We illustrate in the next two examples how to handle situations in which $Q(x)$ has only linear factors. If a factor occurs to a power greater than 1, we say it is *repeated*. For example, we would say $x - 2$ is a repeated factor if it occurs as $(x - 2)^2$ or $(x - 2)^3$. The simplest case is the one in which no factor is repeated, as in the following example.

EXAMPLE 8.20   **Linear factors, none repeated**   Evaluate the integral

$$
\int \frac{5x + 1}{x^2 + x - 2}\, dx
$$

Solution   This integral is the one we considered at the beginning of this section. We know the answer, but we will do it again, this time showing how to decompose the integrand into its component fractions. First we factor the denominator:

$$
\frac{5x + 1}{x^2 + x - 2} = \frac{5x + 1}{(x - 1)(x + 2)}
$$

Now we would expect that the denominator $(x - 1)(x + 2)$ would arise from adding a fraction with $x - 1$ as its denominator to one with $x + 2$ as its denominator; that is, we expect that constants A and B exist such that

$$
\frac{5x + 1}{(x - 1)(x + 2)} = \frac{A}{x - 1} + \frac{B}{x + 2} \tag{8.14}
$$

Assume for the moment that this equation is true, and multiply both sides by the least common denominator $(x - 1)(x + 2)$ to obtain

$$
5x + 1 = A(x + 2) + B(x - 1) \tag{8.15}
$$

When we multiplied Equation 8.14 by the least common denominator, we were assuming that this common denominator was not 0. That is, we were assuming that $x \neq 1$ and $x \neq -2$. It would appear, then, that in Equation 8.15 we have to retain these restrictions. However, it can be proved that if two polynomials are equal for all but a finite number of values of the variable, they are equal for *all* values of the variable. Thus, whereas in Equation 8.14 we must restrict x to be different from 1 or -2, Equation 8.15 must be true for all values of x, including $x = 1$ and $x = -2$.

Equation 8.15 is particularly easy to solve for A or B if one of the terms on the right-hand side is 0. This situation occurs when $x = 1$ or when $x = -2$, respectively. By substituting these values for x, in turn, we will be able to solve for both A and B.

So we substitute first $x = 1$ and then $x = -2$:

$$
\begin{array}{ll}
\underline{x = 1:} \quad 5 + 1 = A(3) + B(0) & \underline{x = -2:} \quad 5(-2) + 1 = A(0) + B(-3) \\
\qquad\qquad 3A = 6 & \qquad\qquad\qquad -3B = -9 \\
\qquad\qquad\; A = 2 & \qquad\qquad\qquad\quad B = 3
\end{array}
$$

It is easy now to verify that these values of A and B do work—that is, that

$$
\frac{5x + 1}{(x - 1)(x + 2)} = \frac{2}{x - 1} + \frac{3}{x + 2}
$$

The integration is now straightforward and we get, as we have already seen,

$$
\int \frac{5x + 1}{x^2 + x - 2}\, dx = \ln |(x - 1)^2 (x + 2)^3| + C \qquad\qquad \blacksquare
$$

This example is typical of those integrands $P(x)/Q(x)$ (where the fraction is proper, that is, the degree of P is less than the degree of Q) in which $Q(x)$ factors into distinct linear factors, each appearing to the first power. Each such linear factor gives rise to a fraction in the decomposition that has a constant in the numerator. The constants can be determined by clearing of fractions and substituting, in turn, values of x that cause each factor to be 0. This procedure is called the **method of substitution.**

In the next example, one of the factors is repeated.

EXAMPLE 8.21   **Repeated linear factors**   Evaluate the integral

$$
\int \frac{x^2 - 13x + 20}{(x - 1)(x - 3)^2}\, dx
$$

Solution   The denominator again involves only linear factors, but $x - 3$ is repeated. The factor $x - 1$ is handled as in the preceding example, but for $(x - 3)^2$ we must allow for the denominator $x - 3$ as well as $(x - 3)^2$:

$$
\frac{x^2 - 13x + 20}{(x - 1)(x - 3)^2} = \frac{A}{x - 1} + \frac{B}{x - 3} + \frac{C}{(x - 3)^2}
$$

Note carefully the presence of the term $B/(x - 3)$ on the right-hand side. If the factor $(x - 3)^3$ had appeared, then we would need to allow for fractions with each of the denominators $x - 3$, $(x - 3)^2$, and $(x - 3)^3$.

As before, we clear the equation of fractions:

$$
x^2 - 13x + 20 = A(x - 3)^2 + B(x - 1)(x - 3) + C(x - 1) \qquad (8.16)
$$

Again we use the method of substitution, first setting $x = 1$ and then $x = 3$:

$$
\begin{array}{ll}
\underline{x = 1:} \quad\; 8 = A(-2)^2 & \underline{x = 3:} \quad -10 = 2C \\
\qquad\quad 4A = 8 & \qquad\qquad\quad C = -5 \\
\qquad\quad\; A = 2 &
\end{array}
$$

But now we have exhausted the substitutions that cause factors to be 0 and we still do not know B. One procedure is to substitute any other value for x and make use of the known values of A and C. An easy value to substitute is $x = 0$. When we do so, we get

$$x = 0: \qquad 20 = 9A + 3B - C$$

$$20 = 9(2) + 3B - (-5)$$

$$3B = -3$$

$$B = -1$$

An alternative method for finding B is known as **comparison of coefficients**. We will find this method is also useful when the denominator contains irreducible quadratic factors. With it, returning to Equation 8.16, we rewrite the right-hand side, expanding and collecting like powers of x. The result is (verify it)

$$x^2 - 13x + 20 = (A + B)x^2 + (-6A - 4B + C)x + (9A + 3B - C)$$

Now this equation is to be an *identity* in x; that is, the polynomial on the right must be the same polynomial as the one on the left. So coefficients of corresponding powers of x must be equal:

Coefficients of x^2: $\qquad 1 = A + B$

Coefficients of x: $\qquad -13 = -6A - 4B + C$

Constant terms: $\qquad 20 = 9A + 3B - C$

This system of three equations in three unknowns can be solved algebraically. However, we already know A and C, so we can get B using any one of the equations. From the first one, we have

$$1 = 2 + B$$

$$B = -1$$

We now know A, B, and C, so we can write

$$\int \frac{x^2 - 13x + 20}{(x - 1)(x - 3)^2}\, dx = \int \left[\frac{2}{x - 1} - \frac{1}{x - 3} - \frac{5}{(x - 3)^2} \right] dx$$

$$= 2 \int \frac{dx}{x - 1} - \int \frac{dx}{x - 3} - 5 \int \frac{dx}{(x - 3)^2}\, dx$$

$$= 2 \ln |x - 1| - \ln |x - 3| + 5(x - 3)^{-1} + C$$

$$= \ln \frac{(x - 1)^2}{|x - 3|} + \frac{5}{x - 3} + C \qquad \blacksquare$$

REMARK ────────────────────────────────────

■ The method of comparison of coefficients always works and can be used on all problems without first using substitution. However, it is usually easier to determine as many of the coefficients as possible by substitution (especially substitutions that cause a factor to be 0) and then go to comparison of coefficients to determine the remaining coefficients, as we did in the preceding example.

The next two examples show how to handle irreducible quadratic factors in the denominator.

EXAMPLE 8.22   Nonrepeated, irreducible quadratic factor
Evaluate the integral

$$\int \frac{7x-4}{x^3-2x^2+x-2}\,dx$$

Solution   We can factor the denominator by grouping as follows:

$$x^3-2x^2+x-2 = (x^3-2x^2)+(x-2)$$
$$= x^2(x-2)+(x-2)$$
$$= (x-2)(x^2+1)$$

We therefore want to find the partial fraction decomposition of

$$\frac{7x-4}{(x-2)(x^2+1)}$$

We expect two fractions, one with denominator $x-2$ and one with denominator x^2+1. The linear factor is handled as before, but for the irreducible quadratic factor x^2+1, we must assume a numerator that is a first-degree polynomial, rather than a constant. By allowing for a first-degree polynomial in the numerator, we are allowing for the worst case, in the sense that the fraction remains proper. Any higher degree in the numerator would cause an improper fraction. We therefore assume a decomposition of the form

$$\frac{7x-4}{(x-2)(x^2+1)} = \frac{A}{x-2}+\frac{Bx+C}{x^2+1}$$

When we clear of fractions, we get

$$7x-4 = A(x^2+1)+(Bx+C)(x-2)$$

Now we substitute $x=2$.

$$x=2: \qquad 10 = A(5)$$
$$A = 2$$

To find B and C, we compare coefficients of like powers of x. It is not actually necessary to multiply things out and collect terms on the right, since we can easily identify the terms of the same degree. It is particularly easy to read off the coefficients of x^2 (the highest degree) and the constant term. Note that there is no x^2 term on the left, so its coefficient is 0.

$$\text{Coefficient of } x^2: \qquad 0 = A+B$$
$$B = A = -2$$
$$\text{Constant term:} \qquad -4 = A-2C$$
$$2C = 6$$
$$C = 3$$

So we have

$$\int \frac{7x - 4}{(x - 2)(x^2 + 1)}\, dx$$

$$= \int \frac{2}{x - 2}\, dx + \int \frac{-2x + 3}{x^2 + 1}\, dx \qquad \text{Write the second integral as two integrals.}$$

$$= 2 \ln |x - 2| - \int \frac{2x}{x^2 + 1}\, dx + 3 \int \frac{dx}{x^2 + 1}$$

$$= 2 \ln |x - 2| - \ln(x^2 + 1) + 3 \tan^{-1} x + C \qquad \blacksquare$$

 Do not confuse a repeated linear factor in the denominator with an irreducible quadratic factor. For example,

$$x^2 - 2x + 1 = (x - 1)^2$$

is a repeated linear factor, but

$$x^2 - 2x + 2$$

is an irreducible quadratic factor, since $b^2 - 4ac = 4 - 8 = -4 < 0$. The first gives rise to the two fractions

$$\frac{A}{x - 1} + \frac{B}{(x - 1)^2}$$

whereas the second gives rise to the single fraction

$$\frac{Ax + B}{x^2 - 2x + 2}$$

The following example shows how to handle a repeated irreducible quadratic factor in the denominator.

EXAMPLE 8.23   **Repeated irreducible quadratic factor**   Evaluate the integral

$$\int \frac{2x - 3}{x^5 + 2x^3 + x}\, dx$$

Solution   The integrand can be written as

$$\frac{2x - 3}{x(x^4 + 2x^2 + 1)} = \frac{2x - 3}{x(x^2 + 1)^2}$$

Here the denominator has the linear factor x and the repeated irreducible quadratic factor $x^2 + 1$. The decomposition follows a pattern similar to that for repeated linear factors:

$$\frac{2x - 3}{x(x^2 + 1)^2} = \frac{A}{x} + \frac{Bx + C}{x^2 + 1} + \frac{Dx + E}{(x^2 + 1)^2}$$

As usual, we clear fractions and, because we will rely extensively on comparison of coefficients, we group the right-hand side in powers of x. We leave it to you to verify that the result is

$$2x - 3 = (A + B)x^4 + Cx^3 + (2A + B + D)x^2 + (C + E)x + A$$

Setting $x = 0$ gives $A = -3$. Comparing coefficients of like powers yields the following system. Note that powers of x on the left greater than 1 have a coefficient of 0.

$$A + B = 0$$
$$C = 0$$
$$2A + B + D = 0$$
$$C + E = 2$$

Since we have already found that $A = -3$, we get $B = 3$, $C = 0$, $D = 3$, and $E = 2$. Thus, the integral becomes

$$\int \frac{2x - 3}{x^5 + 2x^3 + x}\, dx$$

$$= \int \left[\frac{-3}{x} + \frac{3x}{x^2 + 1} + \frac{3x + 2}{(x^2 + 1)^2} \right] dx$$

$$= -3 \int \frac{dx}{x} + \frac{3}{2} \int \frac{2x}{x^2 + 1}\, dx + \int \frac{3x + 2}{(x^2 + 1)^2}\, dx \qquad \text{We mentally let } u = x^2 + 1 \text{ in the second integral.}$$

$$= -3 \ln |x| + \frac{3}{2} \ln(x^2 + 1) + \int \frac{3x + 2}{(x^2 + 1)^2}\, dx$$

To evaluate the last integral we make the substitution $x = \tan\theta$, so that $du = \sec^2\theta\, d\theta$. Then we get

$$\int \frac{3x + 2}{(x^2 + 1)^2}\, dx = \int \frac{3\tan\theta + 2}{\sec^4\theta} \sec^2\theta\, d\theta$$

$$= 3 \int \frac{\tan\theta}{\sec^2\theta}\, d\theta + 2 \int \frac{d\theta}{\sec^2\theta}$$

$$= 3 \int \frac{\sin\theta}{\cos\theta} \cdot \cos^2\theta\, d\theta + 2 \int \cos^2\theta\, d\theta \qquad \text{Since } \sec^2\theta = \frac{1}{\cos^2\theta}$$

$$= 3 \int \sin\theta \cos\theta\, d\theta + \int (1 + \cos 2\theta)\, d\theta \qquad \text{Since } \cos^2\theta = \frac{1 + \cos 2\theta}{2}$$

$$= \frac{3}{2} \sin^2\theta + \theta + \frac{\sin 2\theta}{2} + C \qquad \text{Since } d(\sin\theta) = \cos\theta\, d\theta$$

$$= \frac{3}{2} \sin^2\theta + \theta + \sin\theta \cos\theta + C \qquad \text{Since } \sin 2\theta = 2\sin\theta\cos\theta$$

$$= \frac{3}{2} \cdot \frac{x^2}{1 + x^2} + \tan^{-1} x + \frac{x}{1 + x^2} + C$$

$$= \frac{3x^2 + 2x}{2(1 + x^2)} + \tan^{-1} x + C$$

Observe how we made use of Figure 8.10 to express the answer in terms of x. So we have the final result:

$$\int \frac{2x - 3}{x^5 + 2x^3 + x}\, dx = \ln \frac{(x^2 + 1)^{3/2}}{|x|^3} + \frac{3x^2 + 2x}{2(1 + x^2)} + \tan^{-1} x + C \qquad \blacksquare$$

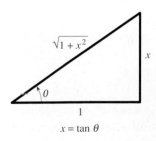

$$x = \tan\theta$$

FIGURE 8.10

We now summarize the procedures we have illustrated.

Partial-Fraction Decomposition

1. If $P(x)/Q(x)$ is improper (degree of $P \geq$ degree of Q), divide to obtain a quotient plus a proper fraction. The next steps apply to proper fractions only.

2. Factor the denominator $Q(x)$ into linear and irreducible quadratic factors.

3. For each factor of the form $(ax + b)^n$, write fractions

$$\frac{A_1}{ax + b} + \frac{A_2}{(ax + b)^2} + \cdots + \frac{A_n}{(ax + b)^n}$$

(If $n = 1$, only one fraction is present.)

4. For each factor of the form $(ax^2 + bx + c)^m$ where $b^2 - 4ac < 0$, write fractions of the form

$$\frac{B_1 x + C_1}{ax^2 + bx + c} + \frac{B_2 x + C_2}{(ax^2 + bx + c)^2} + \cdots + \frac{B_m x + C_m}{(ax^2 + bx + c)^m}$$

(If $m = 1$ only one fraction is present.)

5. Set $P(x)/Q(x)$ equal to the sum of all fractions obtained in steps 3 and 4 and clear of fractions.

6. Use substitution together with comparison of coefficients to find all unknown constants.

We will give one more example to illustrate how certain common complications can be handled.

EXAMPLE 8.24   Evaluate the integral

$$\int_0^1 \frac{x^3 - x^2 - 11x + 10}{x^3 - 2x + 4} \, dx$$

Solution   The first thing to observe is that the integrand is an improper fraction, so we divide:

$$
\begin{array}{r}
1 \\
x^3 - 2x + 4 \overline{)\, x^3 - x^2 - 11x + 10} \\
\underline{x^3 - 2x + 4} \\
- x^2 - 9x + 6
\end{array}
$$

We can therefore write

$$\frac{x^3 - x^2 - 11x + 10}{x^3 - 2x + 4} = 1 + \frac{-x^2 - 9x + 6}{x^3 - 2x + 4} = 1 - \frac{x^2 + 9x - 6}{x^3 - 2x + 4}$$

We concentrate on the last fraction, which is proper. To factor the denominator, we first try to determine any rational zeros. By trial and error, we find -2 to be a zero. Thus, $x + 2$ is a factor. By division (shown in the margin), we find

$$
\begin{array}{r}
x^2 - 2x + 2 \\
x + 2 \overline{)\, x^3 - 2x + 4} \\
\underline{x^3 + 2x^2} \\
- 2x^2 - 2x \\
\underline{- 2x^2 - 4x} \\
2x + 4 \\
\underline{2x + 4} \\
0
\end{array}
$$

the other factor to be the irreducible quadratic $x^2 - 2x + 2$. So we have

$$\frac{x^2 + 9x - 6}{x^3 - 2x + 4} = \frac{x^2 + 9x - 6}{(x + 2)(x^2 - 2x + 2)} = \frac{A}{x + 2} + \frac{Bx + C}{x^2 - 2x + 2}$$

$$x^2 + 9x - 6 = A(x^2 - 2x + 2) + (Bx + C)(x + 2) \qquad \text{Clear of fractions.}$$

$$\underline{x = -2}: \quad 4 - 18 - 6 = A(4 + 4 + 2) + 0 \qquad \text{Substitute } x = -2.$$

$$10A = -20$$

$$A = -2$$

$$\text{Coefficients of } x^2: \quad 1 = A + B \qquad\qquad \text{Equate coefficients.}$$

$$1 = -2 + B$$

$$B = 3$$

$$\text{Constants:} \quad -6 = 2A + 2C$$

$$-6 = -4 + 2C$$

$$2C = -2$$

$$C = -1$$

We therefore have the decomposition

$$\frac{x^2 + 9x - 6}{x^3 - 2x + 4} = \frac{-2}{x + 2} + \frac{3x - 1}{x^2 - 2x + 2}$$

Returning to the original fraction and the result of the first division,

$$\int_0^1 \frac{x^3 - x^2 - 11x + 10}{x^3 - 2x + 4}\, dx = \int_0^1 \left[1 - \frac{x^2 + 9x - 6}{x^3 - 2x + 4} \right] dx$$

$$= \int_0^1 \left[1 - \left(\frac{-2}{x + 2} + \frac{3x - 1}{x^2 - 2x + 2} \right) \right] dx$$

$$= \int_0^1 dx + 2\int_0^1 \frac{dx}{x + 2} - \int_0^1 \frac{3x - 1}{x^2 - 2x + 2}\, dx$$

$$= \left[x + 2\ln(x + 2) \right]_0^1 - \int_0^1 \frac{3x - 1}{x^2 - 2x + 2}\, dx$$

$$= 1 + 2\ln 3 - 2\ln 2 - \int_0^1 \frac{3x - 1}{x^2 - 2x + 2}\, dx$$

To evaluate the last integral we force the derivative of the denominator, $2x - 2$, into the numerator as follows (see the algebra in the margin):

$$\int \frac{3x - 1}{x^2 - 2x + 2}\, dx = \int \frac{\frac{3}{2}(2x - 2) + 2}{x^2 - 2x + 2}\, dx$$

$$= \frac{3}{2} \int \frac{2x - 2}{x^2 - 2x + 2}\, dx + 2 \int \frac{dx}{(x - 1)^2 + 1}$$

In the last integral we completed the square in the denominator to write

$$x^2 - 2x + 2 = (x^2 - 2x + 1) + 1$$

$$= (x - 1)^2 + 1$$

$$3x - 1 = 3\left(x - \frac{1}{3} \right)$$

$$= \frac{3}{2}\left(2x - \frac{2}{3} \right)$$

$$= \frac{3}{2}\left(2x - 2 + 2 - \frac{2}{3} \right)$$

$$= \frac{3}{2}(2x - 2) + \frac{3}{2}\left(2 - \frac{2}{3} \right)$$

$$= \frac{3}{2}(2x - 2) + \frac{3}{2}\left(\frac{4}{3} \right)$$

$$= \frac{3}{2}(2x - 2) + 2$$

The first integral is now in the form $\int du/u$ with $u = x^2 - 2x + 2$, and the second $\int du/(u^2 + 1)$ with $u = x - 1$. So we have

$$\int_0^1 \frac{3x - 1}{x^2 - 2x + 2}\, dx = \left[\frac{3}{2} \ln |x^2 - 2x + 2| + 2 \tan^{-1}(x - 1)\right]_0^1$$

$$= \frac{3}{2} \ln 1 + 2 \tan^{-1} 0 - \frac{3}{2} \ln 2 - 2 \tan^{-1}(-1)$$

$$= -\frac{3}{2} \ln 2 - 2 \left(-\frac{\pi}{4}\right) = -\frac{3}{2} \ln 2 + \frac{\pi}{2}$$

Combining this answer with the results already obtained, we have

$$\int_0^1 \frac{x^3 - x^2 - 11x + 10}{x^3 - 2x + 4}\, dx = 1 + 2 \ln 3 - 2 \ln 2 - \left(-\frac{3}{2} \ln 2 + \frac{\pi}{2}\right)$$

$$= 1 + 2 \ln 3 - \frac{1}{2} \ln 2 - \frac{\pi}{2} \qquad \blacksquare$$

Exercise Set 8.4

In Exercises 1–26, make use of partial fractions to evaluate the integrals.

1. $\displaystyle\int \frac{dx}{x^2 + 2x}$

2. $\displaystyle\int \frac{dx}{x^2 + 5x + 6}$

3. $\displaystyle\int \frac{x\, dx}{x^2 - x - 6}$

4. $\displaystyle\int_{-2}^0 \frac{5x + 7}{x^2 + 2x - 3}\, dx$

5. $\displaystyle\int_{-3}^2 \frac{x - 17}{x^2 + x - 12}\, dx$

6. $\displaystyle\int \frac{x + 9}{2x^2 + x - 6}\, dx$

7. $\displaystyle\int \frac{3x + 10}{x^2 + 5x + 6}\, dx$

8. $\displaystyle\int \frac{2}{x^2 - 9x + 20}\, dx$

9. $\displaystyle\int \frac{dx}{x^3 - x}$

10. $\displaystyle\int \frac{10x + 4}{4x - x^3}\, dx$

11. $\displaystyle\int \frac{x^2 + 4x - 2}{x^2 + x - 2}\, dx$

12. $\displaystyle\int_0^2 \frac{2x^2 - x - 20}{x^2 - x - 6}\, dx$

13. $\displaystyle\int \frac{3x^2 + 32x + 44}{(x + 3)(x^2 + 2x - 8)}\, dx$

14. $\displaystyle\int \frac{2x^2 + 19x - 45}{(x - 1)(x^2 - x - 6)}\, dx$

15. $\displaystyle\int \frac{2x^3 - x^2 - 4x + 5}{x^2 - 1}\, dx$

16. $\displaystyle\int \frac{5x + 1}{(x + 2)(x^2 - 2x + 1)}\, dx$

17. $\displaystyle\int_{-1}^1 \frac{x^2 - 6x + 23}{(x + 3)(x^2 - 4x + 4)}\, dx$

18. $\displaystyle\int_1^2 \frac{4 + 4x - x^2}{x^3 + 2x^2}\, dx$

19. $\displaystyle\int \frac{2x^2 - 6x + 12}{x^3 - 4x^2 + 4x}\, dx$

20. $\displaystyle\int \frac{6x - 4}{(x - 2)(x^2 - 4)}\, dx$

21. $\displaystyle\int \frac{5x + 6}{(x - 2)(x^2 + 4)}\, dx$

22. $\displaystyle\int_0^1 \frac{x^2 + 3}{(x + 1)(x^2 + 1)}\, dx$

23. $\displaystyle\int_1^3 \frac{3x - 6}{x^4 + 3x^2}\, dx$

24. $\displaystyle\int \frac{dx}{x^4 - 1}$

25. $\displaystyle\int \frac{dx}{x^3 - 1}$

26. $\displaystyle\int \frac{12}{x^3 + 8}\, dx$

27. (a) Find the area of the region under the graph of

$$y = \frac{1}{x^2 - x}$$

from $x = 2$ to $x = 3$.

(b) Find the volume of the solid obtained by revolving the region of part (a) about the x-axis.

28. Find the volume of the solid obtained by revolving the region under the graph of

$$y = \frac{1}{x\sqrt{x^2 + 4}}$$

from $x = 1$ to $x = 2$ about the x-axis.

In Exercises 29–33, derive the integration formulas for the given integrals, using partial fractions. For Exercises 29–32, compare your results with those given in the tables on the front and back endpapers.

29. $\displaystyle\int \frac{du}{u(au + b)}$ 30. $\displaystyle\int \frac{du}{u^2(au + b)}$

31. $\displaystyle\int \frac{du}{u(au + b)^2}$ 32. $\displaystyle\int \frac{du}{u^2 - a^2}$

33. $\displaystyle\int \frac{du}{u^3 - a^3}$

In Exercises 34–38, use partial fractions to evaluate the integrals.

34. $\displaystyle\int \frac{x^4 + 5x^3 + 3x}{(x - 1)(x^2 + 2)^2}\,dx$ 35. $\displaystyle\int \frac{4x}{x^3 - x^2 - x + 1}\,dx$

36. $\displaystyle\int \frac{x^2 + 3x + 6}{x^3 + x - 2}\,dx$ 37. $\displaystyle\int \frac{x^3 - 3x^2 - 5x}{x^4 + 5x^2 + 4}\,dx$

38. $\displaystyle\int \frac{20 - 5x^2}{4x^4 + 9x^2 - 11x + 3}\,dx$

39. In Exercise 24 of Exercise Set 7.4 we gave the Verhulst model for population growth in the form of the differential equation

$$\frac{dQ}{dt} = kQ(m - Q)$$

where $Q(t)$ is the size of the population at time t, k and

m are positive constants, and $m > Q$. Let $Q_0 = Q(0)$, and find Q as a function of t by separating variables and using partial fractions.

40. A second-order sodium-chloride chemical reaction involves the interaction of molecules of sodium with those of chloride ion to form molecules of sodium chloride. If the initial concentrations of sodium and chloride ion are a and b, respectively, and $x(t)$ is the concentration of sodium chloride at time t, then the rate at which the reaction occurs is given by

$$\frac{dx}{dt} = k(a - x)(b - x)$$

where k is a positive constant. If $x(0) = 0$, find $x(t)$ at any time t by separating variables and using partial fractions.

8.5 INTEGRAL TABLES AND COMPUTER ALGEBRA SYSTEMS

Tables containing long lists of integration formulas have been used for many years, often eliminating the need for evaluating a particular integral by pencil and paper. These tables are very useful but cannot possibly contain every integral you may encounter. Instead, many times an integral must be made to conform to one in the table, by using a substitution, integration by parts, resolving the integrand into partial fractions, or some other technique. Trying to fit an integral into the pattern of one in the table in this way can be a difficult task in itself.

An alternative is to use a computer algebra system (CAS). These powerful systems can do symbolic integration and so eliminate many of the algebraic difficulties encountered in using tables or in evaluating integrals by hand. However, not every CAS can integrate every function, even when the function is known to have an antiderivative that is a combination of simple functions. So, just as tables are not the final answer, computer algebra systems—although a significant improvement—also have limitations. Both tables and computer algebra systems are powerful aids in evaluating integrals. However, as we will see in the examples that follow, they cannot eliminate the need for understanding the underlying concepts.

Using Tables of Integrals

We illustrate with examples some of the techniques used to match integrals with forms found in the tables on the front and back endpapers of this book. Far more extensive tables are available, such as Burrington's *Handbook of Mathematical Tables and Formulas* (New York: McGraw-Hill) and the Chemical Rubber Company's *Standard Mathematical Tables* (Cleveland: Chemical Rubber Publishing Company).

EXAMPLE 8.25   Evaluate the following, using the integral tables on the endpapers.

(a) $\displaystyle\int \sqrt{2x^2 - 3}\,dx$ (b) $\displaystyle\int \frac{x\,dx}{\sqrt{3x - 4x^2}}$

Solution

(a) Formula 37 in the table, for $\int \sqrt{u^2 \pm a^2}\,du$, seems to be the appropriate one to use (with the negative sign). We take $u = \sqrt{2}\,x$ and $a = \sqrt{3}$. To have the proper form of du—namely, $du = \sqrt{2}\,dx$—we must multiply and divide by $\sqrt{2}$. So we have

$$\int \sqrt{2x^2 - 3}\,dx = \frac{1}{\sqrt{2}} \int \sqrt{\left(\sqrt{2}\,x\right)^2 - \left(\sqrt{3}\right)^2}\,\left(\sqrt{2}\,dx\right)$$

$$= \frac{1}{\sqrt{2}} \int \sqrt{u^2 - a^2}\,du$$

$$= \frac{1}{\sqrt{2}} \left[\frac{u}{2}\sqrt{u^2 - a^2} - \frac{a^2}{2} \ln\left|u + \sqrt{u^2 - a^2}\right| \right] + C$$

By Formula 37

$$= \frac{1}{\sqrt{2}} \left[\frac{\sqrt{2}\,x}{2}\sqrt{2x^2 - 3} - \frac{3}{2} \ln\left|\sqrt{2}\,x + \sqrt{2x^2 - 3}\right| \right] + C$$

(b) Formula 64 seems to be the one to use in this case. We let $u = 2x$ and $a = 3/4$. Then, adjusting the differential by multiplying and dividing by 2, we have

$$\int \frac{x\, dx}{\sqrt{3x - 4x^2}} = \frac{1}{2} \int \frac{2x\, dx}{\sqrt{3x - 4x^2}} = \frac{1}{2} \int \frac{u\, du}{\sqrt{2au - u^2}}$$

$$= \frac{1}{2} \left[-\sqrt{2au - u^2} + a\cos^{-1}\left(\frac{a - u}{a}\right) \right] + C$$

By Formula 64

$$= \frac{1}{2} \left[-\sqrt{3x - 4x^2} + \frac{3}{4}\cos^{-1}\left(\frac{\frac{3}{4} - 2x}{\frac{3}{4}}\right) \right] + C$$

$$= -\frac{\sqrt{3x - 4x^2}}{2} + \frac{3}{8}\cos^{-1}\left(\frac{3 - 8x}{3}\right) + C \qquad \blacksquare$$

EXAMPLE 8.26 Evaluate the integrals

(a) $\displaystyle \int \frac{x^2 dx}{2x^2 - 7x + 6}$ and (b) $\displaystyle \int x\sqrt{x^2 - 4x + 5}\, dx$

using the tables on the endpapers.

$$
\begin{array}{r}
\frac{1}{2} \\[2pt]
2x^2 - 7x + 6\ \overline{)\ x^2\phantom{-\frac{7}{2}x + 3}} \\
x^2 - \frac{7}{2}x + 3 \\ \hline
\frac{7}{2}x - 3
\end{array}
$$

Solution

(a) As the integral stands, there is no form in the table to match it. Since the integrand is an improper rational fraction, we divide (as shown in the margin) to get

$$\frac{x^2}{2x^2 - 7x + 6} = \frac{1}{2} + \frac{\frac{7}{2}x - 3}{2x^2 - 7x + 6} = \frac{1}{2} + \frac{\frac{7}{2}x - 3}{(2x - 3)(x - 2)}$$

We split the integral into three parts:

$$\int \frac{x^2 dx}{2x^2 - 7x + 6} = \int \frac{1}{2}\, dx + \frac{7}{2} \int \frac{x\, dx}{(2x - 3)(x - 2)} - 3 \int \frac{dx}{(2x - 3)(x - 2)}$$

The second integral matches Formula 27 and the third matches Formula 26, both with $u = x$, $a = -3$, $b = 2$, $c = -2$, and $d = 1$. So $ad - bc = -3 + 4 = 1 \neq 0$. Thus,

$$\int \frac{x\, dx}{(2x - 3)(x - 2)} = \int \frac{u\, du}{(a + bu)(c + du)}$$

$$= \frac{1}{ad - bc}\left(\frac{a}{b}\ln|a + bu| - \frac{c}{d}\ln|c + du|\right) + C$$

Formula 27

$$= -\frac{3}{2}\ln|2x - 3| + 2\ln|x - 2| + C$$

and

$$\int \frac{dx}{(2x-3)(x-2)} = \int \frac{du}{(a+bu)(c+du)} = \frac{1}{ad-bc}\ln\left|\frac{c+du}{a+bu}\right| + C$$

Formula 26

$$= \ln\left|\frac{x-2}{2x-3}\right| + C$$

Finally, then, from our original long division, we have (writing only one arbitrary constant)

$$\int \frac{x^2\,dx}{2x^2-7x+6} = \frac{x}{2} + \frac{7}{2}\left[-\frac{3}{2}\ln|2x-3| + 2\ln|x-2|\right] - 3\ln\left|\frac{x-2}{2x-3}\right| + C$$

Using properties of logarithms and some algebra, you can show that the answer can be written in the simpler form

$$\frac{x}{2} - \frac{9}{4}\ln|2x-3| + 4\ln|x-2| + C$$

In all probability we could have gotten the answer faster without using tables. All we would need to do, after the long division, is to resolve the proper fraction into partial fractions.

(b) Again, we do not find a form in the table that matches the integral, but by completing the square under the radical, we convert it to a standard form. We have

$$\int x\sqrt{x^2-4x+5}\,dx = \int x\sqrt{(x^2-4x+4)+1}\,dx$$

$$= \int x\sqrt{(x-2)^2+1}\,dx$$

Now if we let $u = x-2$, we have $x = u+2$, and $dx = du$. So our integral becomes

$$\int (u+2)\sqrt{u^2+1}\,du = \int u\sqrt{u^2+1}\,du + 2\int \sqrt{u^2+1}\,du$$

We can easily evaluate the first integral on the right without tables (using a substitution for $u^2 + 1$), after adjusting the differential. For the second integral we use Formula 37 again, with $a = 1$ (and with the plus sign). So we have

$$\int (u+2)\sqrt{u^2+1}\,du = \frac{1}{2}\int (u^2+1)^{1/2}(2u\,du) + 2\int \sqrt{u^2+1}\,du$$

$$= \frac{1}{2}(u^2+1)^{3/2}\cdot\frac{2}{3}$$

$$+ 2\left[\frac{u}{2}\sqrt{u^2+1} + \frac{1}{2}\ln\left|u + \sqrt{u^2+1}\right|\right] + C$$

or, on replacing u by $x-2$ and simplifying,

$$\int x\sqrt{x^2-4x+5}\,dx = \frac{1}{3}(x^2-4x+5)^{3/2} + (x-2)\sqrt{x^2-4x+5}$$

$$+ \ln\left|(x-2) + \sqrt{x^2-4x+5}\right| + C \qquad \blacksquare$$

INTEGRALS USING COMPUTER ALGEBRA SYSTEMS—TECHNIQUES OF INTEGRATION

We will illustrate how Maple, Mathematica, and DERIVE can be used to compute a wide range of integrals. Each of the examples duplicates an example that we previously have done by hand. In each case the actual input and output as they appear on the screen of the computer terminal are shown.

 CAS 24

Redo Example 8.3 using Maple, Mathematica, and DERIVE.

Maple:

int(x^3*exp(2*x),x);

Output:
$$\frac{1}{2}x^3 e^{2x} - \frac{3}{4}x^2 e^{2x} + \frac{3}{4}xe^{2x} - \frac{3}{8}e^{2x}$$

which can easily be shown to be equivalent to the answer obtained by hand. Note that the computer does not supply the constant of integration, which is true of other CAS as well.

Mathematica:

Integrate[x^3Exp[2x],x]

Note that in Mathematica, functions such as the exponential, as well as procedures such as integration, start with a capital letter. The x after the comma means that we are integrating with respect to x.
Output:

$$E^{2x}\left(-\left(\frac{3}{8}\right) + \frac{3x}{4} - \frac{3x^2}{4} + \frac{x^3}{2}\right)$$

Here E^{2x} means e^{2x}. We see, then, that the output is equivalent to the answer we found in Example 8.3.

DERIVE:

(At the □ symbol, go to the next step.)
a (author) □ x^3exp(2x) □ c (calculus)
□ i (integrate) □ [choose expression]
□ x (variable) □ enter (no limits) □ s
(simplify) □ [choose expression]
Output:
$$\frac{e^{2x}(4x^3 - 6x^2 + 6x - 3)}{8}$$

 CAS 25

Redo Example 8.7 using Maple, Mathematica, and DERIVE.

Maple:

int(sec(x)^3,x);

Output:
$$\frac{1}{2}\frac{\sin(x)}{\cos(x)^2} + \frac{1}{2}\ln(\sec(x) + \tan(x))$$

Mathematica:

Integrate[(Sec[x])^3,x]

Output: $\dfrac{-Log[Cos[\frac{x}{2}] - Sin[\frac{x}{2}]]}{2}$ +

$\dfrac{Log[Cos[\frac{x}{2}] + Sin[\frac{x}{2}]]}{2}$ +

$\dfrac{[Scc[x]\; Tan[x]]}{2}$

DERIVE:

(At the □ symbol, go to the next step.)
a (author) □ sec^3(x) □ c (calculus) □ i
(integrate) □ [choose expression] □ enter (no limits) □ s (simplify) □ [choose expression]

Output: $\dfrac{LN\left[\dfrac{SIN(x) + 1}{COS(x)}\right]}{2} + \dfrac{SIN(x)}{2COS(x)^2}$

In the preceding example the answers given by Maple and DERIVE are the same. It is not immediately evident that the answer given by Maple and DERIVE is equivalent to the answer given by Mathematica, nor that either answer is equivalent to the answer we got by hand in Example 8.7, namely $\frac{1}{2}(\sec x \tan x + \ln|\sec x + \tan x|) + C$. The answer given by Maple and DERIVE is essentially the same, however, since

$$\frac{1}{2}\frac{\sin x}{\cos^2 x} = \frac{1}{2}\left(\frac{\sin x}{\cos x}\right)\left(\frac{1}{\cos x}\right) = \frac{1}{2}\tan x \sec x$$

Note, however, that the absolute value signs on the logarithm terms are missing in Maple, Mathematica, and DERIVE. (In Mathematica "Log" means "ln.") You should be on the lookout for slight differences like this; otherwise, answers obtained could be interpreted incorrectly. The absolute value signs are essential in the solution of this problem.

In Exercise 25 of Section 8.5 we will ask you to show that the answer given by Mathematica in CAS 25 is equivalent to the one given by Maple and DERIVE.

📷 CAS 26

Redo Example 8.9 using Maple, Mathematica, and DERIVE.

Maple:

int(sin(x)^4*cos(x)^2,x);

Output: $-\frac{1}{6}\sin(x)^3\cos(x)^3$
$-\frac{1}{8}\sin(x)\cos(x)^3$
$+\frac{1}{16}\cos(x)\sin(x) + \frac{1}{16}x$

Mathematica:

Integrate[Sin[x]^4Cos[x]^2,x]

Output: $\dfrac{x}{16} - \dfrac{Cos[x]\,Sin[x]^3}{24}$

$+ \dfrac{Cos[x]\,Sin[x]^5}{6} - \dfrac{Sin[2\,x]}{32}$

DERIVE:

(At the □ symbol, go to the next step.)
a (author) □ sin^4(x)cos^2(x) □ c (calculus) □ i (integrate) □ [choose expression] □ x (variable) □ enter (no limits) □ s (simplify) □ [choose expression]

Output:

$$-COS(x)^3 \cdot \left[\frac{SIN(x)^3}{6} + \frac{SIN(x)}{8}\right]$$
$$+ \frac{SIN(x)COS(x)}{16} + \frac{x}{16}$$

Again, in this example Maple and DERIVE give answers essentially in the same form. Mathematica gives something quite different. Neither answer conforms to what we obtained in Example 8.9. Again, we will ask you to show the equivalence of the three answers in Exercise 26 of Section 8.5.

📷 CAS 27

Redo Example 8.15 using Maple, Mathematica, and DERIVE.

Maple:

int(sqrt(4*x^2–9)/x,x);

Output:

$$\sqrt{x^2 - \frac{9}{4}} - 3\,\text{arcsec}\left(\frac{2x}{3}\right)$$

Mathematica:

Integrate[(Sqrt[4x^2–9])/x,x]

Output: $Sqrt[-9 + 4\,x^2]$

$+ 3\,ArcTan\left[\dfrac{3}{Sqrt[-9 + 4\,x^2]}\right]$

DERIVE:

(At the □ symbol, go to the next step.)
a (author) □ sqrt(4x^2-9)/x □ c (calculus) □ i (integrate) □ [choose expression] □ x (variable) □ enter (no limits) □ s (simplify) □ [choose expression]
Output:

$$\sqrt{(4x^2 - 9)} - 3ATAN\left[\frac{\sqrt{(4x^2 - 9)}}{3}\right]$$

Observe that this time the result given by Maple is the same as the answer we got by hand in Example 8.15. This time Mathematica and DERIVE give essentially the same form for the answer. In Exercise 27 of Section 8.5 we will ask you to show that the result given by Mathematica is also equal to the result we got by hand.

 CAS 28

Redo Example 8.23 using Maple, Mathematica, and DERIVE.

Maple:

int((2*x–3)/(x^5–2*x^3+x),x);

Output: $-3\ln(x) + \dfrac{3}{2}\ln(x^2 + 1)$

$\qquad + \dfrac{1}{4}\dfrac{4x - 6}{x^2 + 1} + \arctan(x)$

Mathematica:

Integrate[(2x–3)/(x^5+2x^3+x),x]

Output: $\dfrac{-3 + 2\ x}{2\ (1 + x^2\ )} + \text{ArcTan}[x]$

$\qquad -3\ \text{Log}[x] + \dfrac{3\ \text{Log}[1 + x^2\ ]}{2}$

DERIVE:

(At the □ symbol, go to the next step.) a (author) □ (2x-3)/(x^5+2x^3+x) □ c (calculus) □ i (integrate) □ [choose expression] □ x (variable) □ enter (no limits) □ s (simplify) □ [choose expression]

Output: $\text{ATAN}\left[\dfrac{3\text{LN}(x^2 + 1)}{2}\right]$

$\qquad -3\text{LN}(x) + \dfrac{2x - 3}{2(x^2 + 1)}$

This time the three answers can easily be seen to be equivalent, but none of the three matches the answer we obtained by hand in Example 8.23. The answers do, however, differ by a constant, as we will ask you to show in Exercise 28 of Section 8.5. Again, notice the missing absolute values in $\ln|x|$.

Maple, Mathematica, and DERIVE certainly save work on this problem (we needed more than a page to do it by hand). Using a CAS, though, cannot replace an understanding of the techniques.

We give one final example to show the computational power of a CAS. To do this problem by hand would be prohibitively difficult.

 CAS 29

Evaluate the integral

$$\int \frac{1}{1 + x^5}dx$$

using Maple, Mathematica, and DERIVE.

Maple:

int(1/(1+x^5),x);

Output:

$$\frac{1}{5}\ln(x+1)-$$

$$\frac{1}{20}\ln(-2x^2+x+\sqrt{5}x-2)\sqrt{5}-$$

$$\frac{1}{20}\ln(-2x^2+x+\sqrt{5}x-2)$$

$$-\frac{\arctan\left(\frac{-4x+1+\sqrt{5}}{\sqrt{\%2}}\right)}{\sqrt{\%2}}$$

$$+\frac{1}{5}\frac{\arctan\left(\frac{-4x+1+\sqrt{5}}{\sqrt{\%2}}\right)\sqrt{5}}{\sqrt{\%2}}$$

$$+\frac{1}{20}\ln(2x^2-x+\sqrt{5}x+2)\sqrt{5}$$

$$-\frac{1}{20}\ln(2x^2-x+\sqrt{5}x+2)+$$

$$\frac{\arctan\left(\frac{4x-1+\sqrt{5}}{\sqrt{\%1}}\right)}{\sqrt{\%1}}+$$

$$\frac{1}{5}\frac{\arctan\left(\frac{4x-1+\sqrt{5}}{\sqrt{\%1}}\right)\sqrt{5}}{\sqrt{\%1}}$$

$$\%1:=10+2\sqrt{5}$$

$$\%2:=10-2\sqrt{5}$$

Mathematica:

Integrate[1/(1+x^5),x]

Output:

$$\frac{\text{Log}[1+x]}{5}-$$

$$\frac{\text{Cos}[\frac{\text{Pi}}{5}]\,\text{Log}[1+x^2-2\,x\,\text{Cos}[\frac{\text{Pi}}{5}]]}{5}-$$

$$\frac{\text{Cos}[\frac{3\,\text{Pi}}{5}]\,\text{Log}[1+x^2-2\,x\,\text{Cos}[\frac{3\,\text{Pi}}{5}]]}{5}+$$

$$\frac{2\,\text{ArcTan}[(x-\text{Cos}[\frac{\text{Pi}}{5}])\,\text{Csc}[\frac{\text{Pi}}{5}]]\,\text{Sin}[\frac{\text{Pi}}{5}]}{5}+$$

$$\frac{2\,\text{ArcTan}[(x-\text{Cos}[\frac{3\,\text{Pi}}{5}])\,\text{Csc}[\frac{3\,\text{Pi}}{5}]]\,\text{Sin}[\frac{3\,\text{Pi}}{5}]}{5}$$

DERIVE:

(At the □ symbol, go to the next step.)
a (author) □ 1/(1+x^5) □ c (calculus)
□ i (integrate) □ [choose expression]
□ x (variable) □ enter (no limits) □ s
(simplify) □ [choose expression]
Output:

$$\sqrt{\frac{\sqrt{5}}{50}+\frac{1}{10}}\,\text{ATAN}\left[\sqrt{\frac{1}{8}-\frac{\sqrt{5}}{40}}(4x+\sqrt{5}-1)\right]$$

$$+\sqrt{\frac{1}{10}-\frac{\sqrt{5}}{50}}\,\text{ATAN}\left[(4x-\sqrt{5}-1)\sqrt{\frac{\sqrt{5}}{40}+\frac{1}{8}}\right]$$

$$+\left[\frac{\sqrt{5}}{20}-\frac{1}{20}\right]\text{LN}(2x^2+x(\sqrt{5}-1)+2)$$

$$-\left[\frac{\sqrt{5}}{20}+\frac{1}{20}\right]\text{LN}(2x^2-x(\sqrt{5}+1)+2)$$

$$+\frac{\text{LN}(x+1)}{2}$$

The three answers certainly do not seem to bear much relation to one another. Trying to show they are equivalent would be a formidable task, so we will spare you time. (One easy way, however, to gain confidence that they are equivalent is to use Maple, Mathematica, or DERIVE to evaluate each answer at several values of x. The results should be the same, or else they should differ by a constant value.) It is a nontrivial problem.

Exercise Set 8.5

In Exercises 1–24, use the integral tables on the front and back endpapers to evaluate the integrals.

1. $\displaystyle\int \frac{dx}{x(2+3x)^2}$

2. $\displaystyle\int \frac{\sqrt{x^2-4}}{x}\,dx$

3. $\displaystyle\int x\tan^{-1}2x\,dx$

4. $\displaystyle\int x^2\sin 3x\,dx$

5. $\displaystyle\int \frac{dx}{x\sqrt{4x-x^2}}$

6. $\displaystyle\int \frac{x\,dx}{2x-3}$

7. $\displaystyle\int \frac{dx}{x^2(x-2)}$

8. $\displaystyle\int \frac{\sqrt{x^2-1}}{x^2}\,dx$

9. $\displaystyle\int \frac{dx}{(x^2+9)^{3/2}}$

10. $\displaystyle\int x^2\sqrt{9-4x^2}\,dx$

11. $\displaystyle\int \sqrt{2x-x^2}\,dx$

12. $\displaystyle\int (\ln x)^3\,dx$

13. $\displaystyle\int x^4 e^{3x}\,dx$

14. $\displaystyle\int \sin^6 x\,dx$

15. $\displaystyle\int \frac{dx}{x(3x+5)}$

16. $\displaystyle\int \frac{dx}{x^2\sqrt{3-2x^2}}$

17. $\displaystyle\int \frac{x\,dx}{(4x-3)^2}$

18. $\displaystyle\int \sec^5 x\,dx$

19. $\displaystyle\int e^{-2x}\cos 3x\,dx$

20. $\displaystyle\int \sqrt{x^2+2x+2}\,dx$

21. $\displaystyle\int \frac{x^2+1}{4+3x-x^2}\,dx$

22. $\displaystyle\int \frac{2x-3}{\sqrt{5+4x-x^2}}\,dx$

23. $\displaystyle\int x^2 \sin^{-1} x\,dx$

24. $\displaystyle\int \frac{\sqrt{3x^2-6x+5}}{x-1}\,dx$

25. Show that the answer given by Mathematica in CAS 25 is equivalent to the one given by Maple.

26. Show that the answer found in Example 8.9 can be changed to the form in CAS 26 given by (a) Mathematica and (b) Maple. (*Hint:* Use the identities $\sin 2\theta = 2\sin\theta\cos\theta$, $\cos 2\theta = 1-2\sin^2\theta$, and $\cos^2\theta = 1-\sin^2\theta$.)

27. Show that the answer given by Mathematica in CAS 27 is equivalent to the one given by Maple.

28. Show that the answer obtained in Example 8.23 can be put in a form that agrees with the output given by Mathematica and Maple in CAS 28, except that it differs by a constant (which can be incorporated into the constant of integration). (*Hint:* Write $3x^2 + 2x$ as $3(1+x^2) - 3 + 2x$.)

Do Exercises 29–32 only if you have access to a CAS other than Maple, Mathematica, or DERIVE. In each case redo the specified example using this CAS and show the answer is equivalent to the one obtained by either Mathematica, Maple, or DERIVE.

29. CAS 26

30. CAS 27

31. CAS 28

32. CAS 29

In Exercises 33–56, redo Exercises 1–24 using a CAS. Show that the answer is equivalent to the one found by using the integral table.

8.6 IMPROPER INTEGRALS

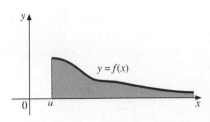

FIGURE 8.11
f is defined on the unbounded interval $[a, \infty)$.

In the definition of the definite integral $\int_a^b f(x)\,dx$, we understood that the interval of integration $[a, b]$ was *finite*; that is, a and b were finite real numbers. In this section we extend the definition of the integral to *unbounded* intervals of integration of the forms $[a, \infty)$, $(-\infty, a]$, and $(-\infty, \infty)$. In Figure 8.11 we illustrate the graph of a function defined on an unbounded interval of the form $[a, \infty)$.

Another requirement for the integral $\int_a^b f(x)\,dx$ was that the function f had to be *bounded*; that is, some positive constant M existed for which f had to satisfy an inequality of the form $-M \le f(x) \le M$. Now we extend the definition of the integral so as to remove this restriction also. Our extended definition will allow for f to have one or more infinite discontinuities (its graph

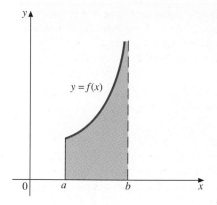

FIGURE 8.12
f is unbounded as $x \to b^-$.

will have vertical asymptotes) in $[a, b]$. In Figure 8.12 we show such a function with a vertical asymptote at $x = b$. Recall that we saw an integral with an unbounded integrand in calculating the length of an ellipse (Example 6.19).

Each of the two new types of integrals is called an **improper integral**. We consider the two types separately.

Unbounded Intervals of Integration

To motivate a plausible definition for the integral of a function on an infinite interval of integration, consider the function $f(x) = 1/x^2$. Suppose we wish to give some meaning to the area under the graph of f, to the right of $x = 1$. (See Figure 8.13.) On any finite interval $[1, t]$ with $t > 1$, the integral

$$\int_1^t \frac{1}{x^2}\, dx$$

is well defined (since f is continuous on any interval of this form). The value of this integral can be interpreted geometrically (Figure 8.14) as the area under the graph of $y = 1/x^2$ from $x = 1$ to $x = t$, namely,

$$\text{Area} = \int_1^t \frac{1}{x^2}\, dx = \int_1^t x^{-2}\, dx = -\frac{1}{x}\Big]_1^t = -\frac{1}{t} + 1$$

t	$\displaystyle\int_1^t \frac{1}{x^2}\, dx$
2	0.5
10	0.9
100	0.99
1000	0.999

Suppose we now assign larger and larger values to t. Some specific values of t and the corresponding values of the integral (the area) are shown in the table in the margin. Clearly, the areas seem to approach 1 as $t \to \infty$. We therefore can say that

$$\int_1^\infty \frac{1}{x^2}\, dx = \lim_{t \to \infty} \int_1^t \frac{1}{x^2}\, dx = 1$$

The preceding example suggests the following definition.

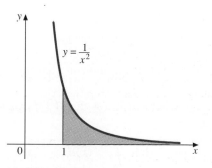

FIGURE 8.13
Unbounded region under the graph on $[1, \infty)$

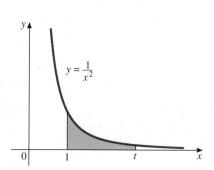

FIGURE 8.14

Definition 8.1
Improper Integrals with
Unbounded Intervals of
Integration

(a) If $\int_a^t f(x)dx$ exists for all $t \geq a$, then

$$\int_a^\infty f(x)dx = \lim_{t \to \infty} \int_a^t f(x)dx$$

provided the limit exists. (See Figure 8.15.)

(b) If $\int_t^a f(x)dx$ exists for all $t \leq a$, then

$$\int_{-\infty}^a f(x)dx = \lim_{t \to -\infty} \int_t^a f(x)dx$$

provided the limit exists. (See Figure 8.16.)

(c) If both $\int_{-\infty}^a f(x)dx$ and $\int_a^\infty f(x)dx$ exist for some real number a, then

$$\int_{-\infty}^\infty f(x)dx = \int_{-\infty}^a f(x)dx + \int_a^\infty f(x)dx$$

(See Figure 8.17).

FIGURE 8.15
$\int_a^\infty f(x)\,dx = \lim\limits_{t \to \infty} \int_a^t f(x)dx$

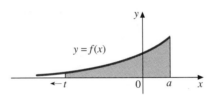

FIGURE 8.16
$\int_{-\infty}^a f(x)\,dx = \lim\limits_{t \to -\infty} \int_t^a f(x)\,dx$

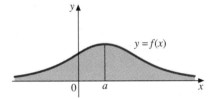

FIGURE 8.17
$\int_{-\infty}^\infty f(x)dx =$
$\int_{-\infty}^a f(x)\,dx + \int_a^\infty f(x)\,dx$

REMARKS ─────────────

■ When the limit in (a) or (b) exists, the corresponding integral is said to **converge**. Otherwise it is said to **diverge**. The integral in (c) converges if both integrals on the right converge, and it diverges if either integral on the right diverges.

■ In (c), any real number a can be used. It can be shown (see Exercise 54 in Exercise Set 8.6) that if both integrals on the right converge for any given real number a, they will converge for any other real number, and the sum will be the same.

■ If the function f is nonnegative on the specified infinite interval, then the improper integral can be interpreted as the area under the graph of f on that infinite interval, if the integral converges. If the integral diverges, we say the area does not exist.

■ The requirement in (c) that $\int_a^\infty f(x)dx$ and $\int_{-\infty}^a f(x)dx$ both exist means that the limits

$$\lim_{t_1 \to \infty} \int_a^{t_1} f(x)dx \qquad \text{and} \qquad \lim_{t_2 \to -\infty} \int_{t_2}^\infty f(x)dx$$

must exist (be a finite number) independently of one another.

EXAMPLE 8.27 Determine whether the integral $\int_1^\infty (1/x)\,dx$ converges or diverges.

Solution Since

$$\lim_{t\to\infty}\int_1^t \frac{1}{x}\,dx = \lim_{t\to\infty}\left[\ln x\right]_1^t = \lim_{t\to\infty}\left[\ln t - \ln 1\right]$$

$$= \lim_{t\to\infty}(\ln t) = \infty$$

it follows that the integral diverges. ∎

It is interesting to compare the integrals

$$\int_1^\infty \frac{1}{x^2}\,dx \qquad \text{and} \qquad \int_1^\infty \frac{1}{x}\,dx$$

When we were preparing for Definition 8.1, we showed that the first integral converges to the value 1, and in Example 8.27 we have shown that the second integral diverges to ∞. In Figure 8.18 we show the graphs of $y = 1/x^2$ and $y = 1/x$. Interpreted in terms of area, our results show that the area under $y = 1/x^2$ to the right of $x = 1$ is finite (it equals 1), whereas the area under $y = 1/x$ to the right of $x = 1$ is infinite. Although the graphs look very much alike, the key difference is that $1/x^2$ goes to 0 faster than $1/x$ as x increases. So although the graphs of $y = 1/x^2$ and $y = 1/x$ both have the horizontal asymptote $y = 0$, the first curve approaches the asymptote more rapidly than the second does, as x increases.

Both $\int_1^\infty (1/x^2)\,dx$ and $\int_1^\infty (1/x)\,dx$ are special cases of the family of integrals

$$\int_1^\infty \frac{1}{x^p}\,dx \qquad (p > 0)$$

We have seen that when $p = 2$ the integral converges and when $p = 1$ it diverges. As the next example shows, $p = 1$ is the dividing point between convergence and divergence, in the sense that the integral converges for any value of p greater than 1 and diverges for any value of p less than (or equal to) 1.

EXAMPLE 8.28 Show that the integral

$$\int_1^\infty \frac{1}{x^p}\,dx$$

converges if $p > 1$ and diverges if $p \le 1$.

Solution For $p \ne 1$, we have

$$\int_1^t \frac{1}{x^p}\,dx = \int_1^t x^{-p}\,dx = \frac{x^{-p+1}}{-p+1}\bigg]_1^t = \frac{1}{1-p}\left[\frac{1}{t^{p-1}} - 1\right]$$

Thus,

$$\int_1^\infty \frac{1}{x^p}\,dx = \lim_{t\to\infty}\int_1^t \frac{1}{x^p}\,dx = \lim_{t\to\infty}\frac{1}{1-p}\left[\frac{1}{t^{p-1}} - 1\right]$$

If $p > 1$ the exponent $p - 1$ is positive, so as $t \to \infty$, $1/t^{p-1} \to 0$, and the limit is finite, namely,

$$\int_1^\infty \frac{1}{x^p}\,dx = \frac{1}{p-1} \qquad (\text{if } p > 1)$$

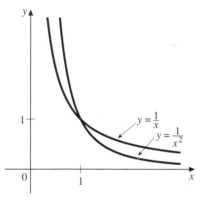

FIGURE 8.18

If $p < 1$, the exponent on t is negative, so

$$\frac{1}{t^{p-1}} = t^{1-p}$$

with $1 - p > 0$. Thus, as $t \to \infty$, $t^{1-p} \to \infty$, and the integral diverges. We saw in Example 8.27 that the integral diverges when $p = 1$. So we have completed the proof. ∎

Because of its importance, we summarize what we have just proved:

$$\int_1^\infty \frac{1}{x^p}\, dx \quad \begin{cases} \text{converges if } p > 1 \\ \text{diverges} \quad \text{if } p \leq 1 \end{cases}$$

REMARK ───────────────────────────────

■ In determining convergence or divergence, the lower limit does not need to be 1; it can be any positive number. So our conclusions continue to hold true for

$$\int_a^\infty \frac{1}{x^p}\, dx$$

where $a > 0$. Of course, when the integral converges, the *value* is affected by what a is, but whether the integral converges or diverges depends only on p.

───────────────────────────────

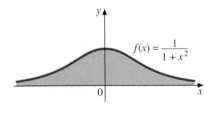

FIGURE 8.19

EXAMPLE 8.29   **An interesting representation of π**   Show that

$$\int_{-\infty}^{\infty} \frac{1}{1 + x^2}\, dx = \pi$$

Solution   We show the graph of $f(x) = 1/(1 + x^2)$ in Figure 8.19. By Definition 8.1(c), we must show that both of the integrals

$$\int_a^\infty \frac{1}{1 + x^2}\, dx \quad \text{and} \quad \int_{-\infty}^a \frac{1}{1 + x^2}\, dx$$

converge and that their sum is π. We are free to choose the number a. Let us take $a = 0$. Then we have

$$\int_0^\infty \frac{1}{1 + x^2}\, dx = \lim_{t \to \infty} \int_0^t \frac{1}{1 + x^2}\, dx = \lim_{t \to \infty} \left[\tan^{-1} x \right]_0^t$$

$$= \lim_{t \to \infty} \left[\tan^{-1} t - \tan^{-1} 0 \right] = \frac{\pi}{2}$$

since $\tan^{-1} t \to \pi/2$ as $t \to \infty$, and $\tan^{-1} 0 = 0$.
    Similarly,

$$\int_{-\infty}^0 \frac{1}{1 + x^2}\, dx = \lim_{t \to -\infty} \int_t^0 \frac{1}{1 + x^2}\, dx = \lim_{t \to -\infty} \left[\tan^{-1} x \right]_t^0$$

$$= \lim_{t \to -\infty} \left[\tan^{-1} 0 - \tan^{-1} t \right] = -\left(-\frac{\pi}{2} \right) = \frac{\pi}{2}$$

since $\tan^{-1} t \to -\pi/2$ as $t \to -\infty$.

Thus,

$$\int_{-\infty}^{\infty} \frac{1}{1+x^2}\,dx = \int_{0}^{\infty} \frac{1}{1+x^2}\,dx + \int_{-\infty}^{0} \frac{1}{1+x^2}\,dx = \frac{\pi}{2} + \frac{\pi}{2} = \pi \qquad \blacksquare$$

REMARK ——————————————————————————————

■ Since $f(x) = 1/(1+x^2)$ is even, we could have used symmetry in Example 8.29 to conclude from the fact that the integral on $[0, \infty)$ is $\pi/2$ that the integral on $(-\infty, 0]$ also has the value $\pi/2$.

———————————————————————————————————

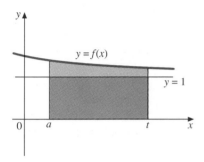

FIGURE 8.20
$\int_{a}^{t} f(x)\,dx > (t-a)$ and $t-a \to \infty$ as $t \to \infty$

In our examples the x-axis ($y = 0$) was a horizontal asymptote in each case. Suppose that the graph of f has some other horizontal asymptote, say $y = k$, where $k \neq 0$. Then the improper integral of f, whether on $[0, \infty)$ or on $(-\infty, a]$, diverges. Figure 8.20 illustrates this result for the case where the graph of f approaches the horizontal asymptote $y = 1$ from above, as $x \to \infty$. The area under the graph from a to t is greater than the area of the rectangle $(t - a)$ units long and 1 unit high. Thus,

$$\int_{a}^{t} f(x)\,dx > t - a \qquad (t > a)$$

and since $t - a \to \infty$ as $t \to \infty$, it follows that the integral $\int_{a}^{\infty} f(x)\,dx$ diverges. Although it is less evident in the case where the graph approaches a nonzero horizontal asymptote from below, the integral can again be shown to be divergent.

It is tempting to conclude from the preceding discussion that the only functions that have a chance of having convergent improper integrals over an unbounded interval of integration are those with the horizontal asymptote $y = 0$. Examples exist, however, to contradict this conclusion. We show the graph of one such function in Figure 8.21. The x-axis is not an asymptote, since $f(x) = 1$ throughout some small rectangle centered at each integer value. In Chapter 10 we will show that the areas, added together, give the value 1.

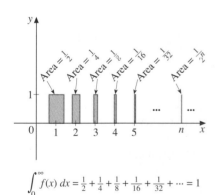

$\int_{0}^{\infty} f(x)\,dx = \frac{1}{2} + \frac{1}{4} + \frac{1}{8} + \frac{1}{16} + \frac{1}{32} + \cdots = 1$

FIGURE 8.21

A Comparison Test for Improper Integrals

Sometimes it may be difficult or impossible to find the exact value of an improper integral, but by comparing it with an improper integral whose value is known, we may be able to determine whether the integral in question converges or diverges. The next theorem shows how this comparison works.

THEOREM 8.1

> **A Comparison Test for Improper Integrals**
>
> Suppose that for all $x \geq a$, f and g are continuous and $0 \leq f(x) \leq g(x)$.
>
> (a) If $\int_{a}^{\infty} g(x)\,dx$ converges, then $\int_{a}^{\infty} f(x)\,dx$ converges.
> (b) If $\int_{a}^{\infty} f(x)\,dx$ diverges, then $\int_{a}^{\infty} g(x)\,dx$ diverges.

REMARK ——————————————————————————

■ A similar result holds for integrals on $(-\infty, a]$. We will not prove this theorem, but we will prove an analogous result in Chapter 10.

——————————————————————————————————

FIGURE 8.22

Intuitively, this result seems plausible. From Figure 8.22, the area under the graph of f is smaller than the area under the graph of g. So if the area under g is finite, we would expect the area under f to be finite also. On the other hand, if the area under f is infinite, the area under g, being greater, must also be infinite.

EXAMPLE 8.30   Test each of the following for convergence or divergence:

(a) $\displaystyle\int_1^\infty \frac{dx}{\sqrt{x^3 + 4}}$ (b) $\displaystyle\int_1^\infty \frac{dx}{\sqrt[3]{2x^2 - 1}}$

Solution

(a) Since for all $x \geq 1$,

$$0 \leq \frac{1}{\sqrt{x^3 + 4}} < \frac{1}{\sqrt{x^3}} = \frac{1}{x^{3/2}}$$

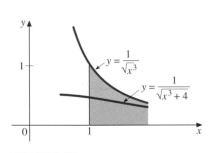

FIGURE 8.23

(see Figure 8.23), and since by Example 8.28,

$$\int_1^\infty \frac{dx}{x^{3/2}} \qquad p = \frac{3}{2} > 1$$

converges, it follows by the comparison test (Theorem 8.1) that

$$\int_1^\infty \frac{dx}{\sqrt{x^3 + 4}}$$

also converges.

(b) We observe that for all $x \geq 1$, $2x^2 - 1 < 2x^2$. So $\sqrt[3]{2x^2 - 1} < \sqrt[3]{2x^2}$. Thus, taking reciprocals, we get the reverse inequality:

$$\frac{1}{\sqrt[3]{2x^2 - 1}} > \frac{1}{\sqrt[3]{2x^2}} = \frac{1}{\sqrt[3]{2}} \cdot \frac{1}{x^{2/3}} > 0$$

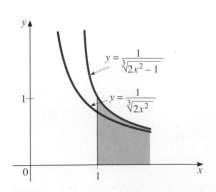

FIGURE 8.24

(see Figure 8.24.) Also, by Example 8.28,

$$\int_1^\infty \frac{dx}{x^{2/3}}\, dx \qquad p = \frac{2}{3} < 1$$

diverges, and multiplying by the constant $1/\sqrt[3]{2}$ does not alter this divergence. It follows by the comparison test (Theorem 8.1) that

$$\int_1^\infty \frac{dx}{\sqrt[3]{2x^2 - 1}}$$

also diverges.   ■

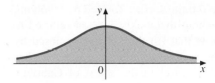

FIGURE 8.25

The standardized normal probability

density function $\phi(x) = \dfrac{1}{\sqrt{2\pi}} e^{-x^2/2}$

EXAMPLE 8.31   Show that the area under the standardized normal probability density function (see Figure 8.25)

$$\phi(x) = \frac{1}{\sqrt{2\pi}} e^{-x^2/2} \qquad (-\infty < x < \infty)$$

is finite. (*Note:* For $\phi(x)$ to be a probability density function, not only must the area be finite, but also its value must be 1. Here we are concerned only with showing convergence. It is possible to show its value is 1 using methods studied in multivariable calculus.)

Solution   Since ϕ is an even function, it suffices to show that the integral from 0 to ∞ is finite. First, we divide the interval of integration into the two intervals $[0, 1]$ and $[1, \infty)$, getting

$$\int_0^\infty e^{-x^2/2} dx = \int_0^1 e^{-x^2/2} dx + \int_1^\infty e^{-x^2/2} dx$$

The first integral on the right is finite, since $e^{-x^2/2}$ is continuous on $[0, 1]$. For the second integral, we use the comparison test, since we cannot integrate $e^{-x^2/2}$ directly. If, however, we could replace x^2 by x, then we could perform the integration. Because $x^2 \geq x$ for $x \geq 1$, we see that $e^{x^2/2} \geq e^{x/2}$ since the exponential function is increasing. Taking reciprocals gives the reverse inequality:

$$\frac{1}{e^{x^2/2}} \leq \frac{1}{e^{x/2}} \qquad (\text{for } x \geq 1)$$

or, equivalently, $e^{-x^2/2} \leq e^{-x/2}$. So if we can show that $\int_1^\infty e^{-x/2} dx$ converges, it will follow that $\int_1^\infty e^{-x^2/2}$ converges. For $t > 1$,

$$\int_1^t e^{-x/2} dx = -2 \int_1^t e^{-x/2} \left(-\frac{1}{2} dt\right) = -2 \left[e^{-x/2}\right]_1^t$$
$$= -2 \left[e^{-t/2} - e^{-1/2}\right]$$

So

$$\int_1^\infty e^{-x/2} dx = \lim_{t \to \infty} \left[-2e^{-t/2} + 2e^{-1/2}\right] = \frac{2}{\sqrt{e}}$$

Since this limit is finite, it follows by the comparison test (Theorem 8.1) that $\int_1^\infty e^{-x^2/2} dx$ converges.

We can now conclude that

$$\int_0^\infty e^{-x^2/2} dx = \int_0^1 e^{-x^2/2} dx + \int_1^\infty e^{-x^2/2} dx$$

exists. By symmetry, then, we see that

$$\int_{-\infty}^\infty e^{-x^2/2} dx$$

also exists. Finally, the convergence is unaltered if we multiply by the constant $1/\sqrt{2\pi}$. (The factor $1/\sqrt{2\pi}$ provides the "normalization" to make the integral have the value 1.) ∎

$y = \frac{1}{x}$

FIGURE 8.26

EXAMPLE 8.32   Gabriel's Horn   The infinite surface formed by revolving the graph of $f(x) = 1/x$ for $1 \leq x < \infty$ about the x-axis is often referred to as *Gabriel's Horn*. (See Figure 8.26.) Show that (a) the volume enclosed by the surface is finite, but (b) the area of the surface is infinite. (Someone has described this phenomenon by saying that a vessel in the shape of Gabriel's Horn would hold a finite amount of paint, but it would take an infinite amount of paint to cover its surface!)

Solution

(a) Using the disk method of Chapter 6, the volume V is

$$V = \pi \int_1^\infty \left(\frac{1}{x}\right)^2 dx = \pi \int_1^\infty \frac{1}{x^2}\, dx$$

From the result of Example 8.28, with $p = 2$, we know that this integral converges.

(b) Again, from Chapter 6 we know that the surface area S is given by

$$S = 2\pi \int_1^\infty y\, ds$$

where $y = \dfrac{1}{x}$ and

$$ds = \sqrt{1 + \left(\frac{dy}{dx}\right)^2}\, dx = \sqrt{1 + \left(-\frac{1}{x^2}\right)^2}\, dx = \frac{\sqrt{x^4 + 1}}{x^2}\, dx$$

Thus,

$$S = 2\pi \int_1^\infty \left(\frac{1}{x}\right) \frac{\sqrt{x^4 + 1}}{x^2}\, dx = 2\pi \int_1^\infty \frac{\sqrt{x^4 + 1}}{x^3}\, dx$$

Now, for $x \geq 1$,

$$\frac{\sqrt{x^4 + 1}}{x^3} > \frac{\sqrt{x^4}}{x^3} = \frac{x^2}{x^3} = \frac{1}{x}$$

and since $\int_1^\infty \frac{1}{x}\, dx$ diverges ($p = 1$), it follows by the comparison test that the integral for S also diverges. ∎

Unbounded Integrands

We consider first the case in which f is continuous for $a < x \leq b$ and $\lim_{x \to a^+} f(x) = \pm\infty$ (so the graph of f has a vertical asymptote at $x = a$). We illustrate such a function in Figure 8.27. As we indicate in that figure, we choose a number t in the open interval (a, b). Then, since f is continuous on the closed interval $[t, b]$, the integral $\int_t^b f(x)dx$ exists. After evaluating this

FIGURE 8.27

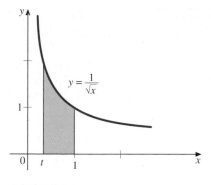

FIGURE 8.28

integral, we let $t \to a^+$ and see whether there is a finite limit. If so, this limit is the value we assign to the improper integral $\int_a^b f(x)dx$.

To illustrate what we have described, consider the integral

$$\int_0^1 \frac{dx}{\sqrt{x}}$$

We show the graph of $f(x) = 1/\sqrt{x}$ in Figure 8.28. Because the function f is unbounded in $(0, 1]$, the integral is improper. Following the procedure we outlined above, we let t be in the open interval $(0, 1)$ and evaluate the integral

$$\int_t^1 \frac{dx}{\sqrt{x}} = \int_t^1 x^{-1/2}dx = 2x^{1/2}\Big]_t^1 = 2 - 2\sqrt{t}$$

Now we let $t \to 0^+$ to get

$$\int_0^1 \frac{dx}{\sqrt{x}} = \lim_{t \to 0^+} (2 - 2\sqrt{t}) = 2$$

We can say that the area under the curve is 2.

If f is continuous for $a \le x < b$ and is unbounded as x approaches b from the left, we integrate from a to t and then let $t \to b^-$. (See Figure 8.29.)

A third possibility is that f becomes unbounded as x approaches some number c between the endpoints a and b of the interval of integration. We illustrate one such possibility in Figure 8.30. In this case we divide the interval into the two parts $[a, c]$ and $[c, b]$. For the integral over the entire interval $[a, b]$ to exist, the integral over each of these two subintervals (with c as an endpoint) must exist, as we have described above.

We are ready now for the formal definition of improper integrals with unbounded integrands.

FIGURE 8.29

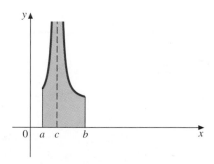

FIGURE 8.30
$\int_a^b f(x)dx =$
$\int_a^c f(x)dx + \int_c^b f(x)\,dx$

Definition 8.2
Improper Integrals with
Unbounded Integrands

(a) If f is continuous for $a < x \le b$ and $\lim_{x \to a^+} f(x) = \pm\infty$, then

$$\int_a^b f(x)\,dx = \lim_{t \to a^+} \int_t^b f(x)\,dx$$

provided the limit on the right exists.

(b) If f is continuous for $a \le x < b$ and $\lim_{x \to b^-} f(x) = \pm\infty$, then

$$\int_a^b f(x)\,dx = \lim_{t \to b^-} \int_a^t f(x)\,dx$$

provided the limit on the right exists.

(c) If f is continuous for $a \le x < c$ and $c < x \le b$ but has an infinite discontinuity at $x = c$, then

$$\int_a^b f(x)\,dx = \int_a^c f(x)\,dx + \int_c^b f(x)\,dx$$

provided both integrals on the right exist.

REMARKS ─────────────────────────────

■ As with improper integrals over unbounded intervals of integration, when the limit in (a) or (b) exists, we say the integral *converges*, and when both limits exist, we say the improper integral in (c) *converges*. Otherwise, these integrals *diverge*.

■ Definition 8.2 concerns integrals with unbounded integrands; that is, the integrands have infinite discontinuities. If f is continuous and *bounded* on the open interval (a, b), then it cannot have an infinite discontinuity at either $x = a$ or $x = b$, and even if $f(a)$ or $f(b)$ is undefined, it can be shown that $\int_a^b f(x)\,dx$ always exists. It is not an improper integral. For example, even though $(\sin x)/x$ is discontinuous at $x = 0$, it is bounded on $(0, 1)$, since

$$\lim_{x \to 0} \frac{\sin x}{x} = 1$$

Its integral,

$$\int_0^1 \frac{\sin x}{x}\,dx$$

therefore exists. The graph of $y = (\sin x)/x$ (Figure 8.31) clearly shows the boundedness property.

FIGURE 8.31

EXAMPLE 8.33   Evaluate the following integrals, or show that they are divergent:

(a) $\displaystyle\int_0^1 \frac{dx}{x}$ (b) $\displaystyle\int_0^1 \frac{dx}{\sqrt{1 - x^2}}$

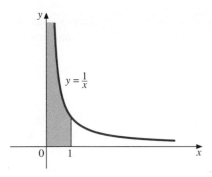

$$y = \frac{1}{x}$$

FIGURE 8.32

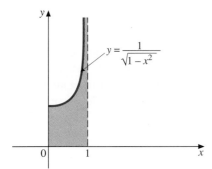

$$y = \frac{1}{\sqrt{1-x^2}}$$

FIGURE 8.33

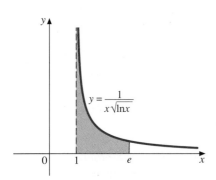

$$y = \frac{1}{x\sqrt{\ln x}}$$

FIGURE 8.34

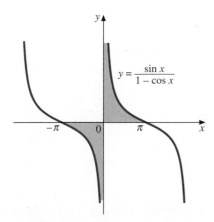

$$y = \frac{\sin x}{1 - \cos x}$$

FIGURE 8.35

Solution

(a) The integrand is unbounded as $x \to 0^+$ (see Figure 8.32), so by Definition 8.2 we have

$$\int_0^1 \frac{dx}{x} = \lim_{t \to 0^+} \int_t^1 \frac{dx}{x} = \lim_{t \to 0^+} [\ln x]_t^1 = \lim_{t \to 0^+} (-\ln t) = \infty$$

So the integral diverges.

(b) Since the denominator is 0 when $x = 1$, the point of discontinuity is $x = 1$. We show the graph of $y = 1/\sqrt{1-x^2}$ in Figure 8.33. Thus,

$$\int_0^1 \frac{dx}{\sqrt{1-x^2}} = \lim_{t \to 1^-} \int_0^t \frac{dx}{\sqrt{1-x^2}} = \lim_{t \to 1^-} \left[\sin^{-1} x\right]_0^t$$

$$= \lim_{t \to 1^-} (\sin^{-1} t - 0) = \sin^{-1} 1 = \frac{\pi}{2}$$

The integral converges to $\pi/2$. ∎

EXAMPLE 8.34   Evaluate the following integrals, or show that they are divergent:

(a) $\displaystyle\int_1^e \frac{dx}{x\sqrt{\ln x}}$ (b) $\displaystyle\int_{-\pi}^{\pi} \frac{\sin x}{1 - \cos x}\, dx$

Solution

(a) Since $\ln 1 = 0$, the integrand is unbounded near 1. We show its graph in Figure 8.34. Since $d(\ln x) = (1/x)\, dx$, we make the (mental) substitution $u = \ln x$ to get

$$\int_1^e \frac{dx}{x\sqrt{\ln x}} = \lim_{t \to 1^+} \int_t^e \frac{dx}{x\sqrt{\ln x}} = \lim_{t \to 1^+} \left[2\sqrt{\ln x}\right]_t^e$$

$$= \lim_{t \to 1^+} (2 - 2\sqrt{\ln t}) = 2$$

(b) To determine whether $\sin x/(1-\cos x)$ has a vertical asymptote on $[-\pi, \pi]$, we set the denominator equal to 0 and see that $x = 0$. However, the numerator is also 0 at this point. Thus, we have an indeterminate form of the type $0/0$. So we use L'Hôpital's Rule.

$$\lim_{x \to 0} \frac{\sin x}{1 - \cos x} = \lim_{x \to 0} \frac{\cos x}{\sin x}$$

When $x \to 0^+$ the limit is ∞, and when $x \to 0^-$ the limit is $-\infty$. In either case, the integrand is unbounded. We show the graph in Figure 8.35. Using Definition 8.2(c), we write

$$\int_{-\pi}^{\pi} \frac{\sin x}{1 - \cos x}\, dx = \int_{-\pi}^{0} \frac{\sin x}{1 - \cos x}\, dx + \int_{0}^{\pi} \frac{\sin x}{1 - \cos x}\, dx$$

For the last integral we have

$$\int_{0}^{\pi} \frac{\sin x}{1 - \cos x}\, dx = \lim_{t \to 0^+} \int_{t}^{\pi} \frac{\sin x}{1 - \cos x}\, dx \quad \text{Mentally, let } u = 1 - \cos x.$$

$$= \lim_{t \to 0^+} [\ln(1 - \cos x)]_t^{\pi}$$

$$= \lim_{t \to 0^+} [\ln 2 - \ln(1 - \cos t)] = \infty$$

(Note that since $1 - \cos x \geq 0$, we did not need absolute values for $\ln(1 - \cos x)$.) So this integral diverges, and hence the original integral also diverges. Observe that

$$f(x) = \frac{\sin x}{1 - \cos x}$$

is an odd function, so its graph is symmetric with respect to the origin. It is tempting, therefore, to say that the "net area" from $-\pi$ to π is 0. But remember that by Definition 8.2(c), each of the integrals $\int_0^\pi f(x)\,dx$ and $\int_{-\pi}^0 f(x)\,dx$ must exist independently of the other. In other words, we cannot say that ∞ and $-\infty$ add to give 0.   ■

REMARK ──

■ In contemporary physics, scientists doing quantum electrodynamics deal with equations in which "infinite terms" seem to cancel out. They use a process called "renormalization" to do so. The development of renormalization techniques helped lead to the Nobel Prize in physics for Richard Feynman, Julian Schwinger, and Sin-Itiro Tomonaga.

──

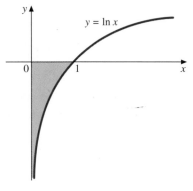

$y = \ln x$

FIGURE 8.36

EXAMPLE 8.35   Determine whether the integral

$$\int_0^1 \ln x \, dx$$

converges or diverges.

Solution   We show the graph of $y = \ln x$ in Figure 8.36. It is unbounded as $x \to 0^+$. Thus, by Definition 8.2(a),

$$\int_0^1 \ln x \, dx = \lim_{t \to 0^+} \int_t^1 \ln x \, dx$$

We need to find an antiderivative of $\ln x$ to complete the problem. We use integration by parts, with $u = \ln x$ and $dv = dx$. Then $du = (1/x)\,dx$ and $v = x$. So we have

$$\int \ln x \, dx = x \ln x - \int x \cdot \frac{1}{x} \, dx = x \ln x - x$$

Now we return to the evaluation of the improper integral.

$$\int_0^1 \ln x \, dx = \lim_{t \to 0^+} \int_t^1 \ln x \, dx = \lim_{t \to 0^+} [x \ln x - x]_t^1$$

$$= \lim_{t \to 0^+} [(1 \ln 1 - 1) - (t \ln t - t)]$$

$$= \lim_{t \to 0^+} [-1 - t \ln t + t] \qquad \text{Since } \ln 1 = 0$$

$$= -1 - \lim_{t \to 0^+} (t \ln t) \qquad \text{Since } \lim_{t \to 0^+} t = 0$$

provided this last limit exists. Now, we are confronted in $\lim_{t \to 0^+} t \ln t$ with an indeterminate form of the type $0 \cdot \infty$, since $\ln t \to -\infty$ as $t \to 0^+$. So we rewrite the product as a quotient and apply L'Hôpital's Rule.

$$\lim_{t \to 0^+} t \ln t = \lim_{t \to 0^+} \frac{\ln t}{\dfrac{1}{t}} = \lim_{t \to 0^+} \frac{\dfrac{1}{t}}{-\dfrac{1}{t^2}} = \lim_{t \to 0^+} (-t) = 0$$

Finally, we have the result

$$\int_0^1 \ln x \, dx = -1$$

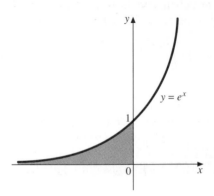

$y = e^x$

FIGURE 8.37

REMARK ————————————————————————————

■ It is interesting to note that the shaded region in Figure 8.36 is equivalent to the shaded region in Figure 8.37, obtained by reflecting through the line $y = x$. Thus, the integral $\int_0^1 \ln x \, dx$ in Example 8.35 can be replaced with

$$\int_{-\infty}^0 e^x \, dx$$

which is easier to analyze. (See Exercise 3 in Exercise Set 8.6.)

Special care must be taken when the integrand becomes unbounded at an interior point of the interval of integration, since failure to recognize the bad point can lead to incorrect results. For example, consider the integral

$$\int_0^2 \frac{dx}{(x-1)^2}$$

If we carelessly applied the Second Fundamental Theorem, we would get

$$\int_0^2 \frac{dx}{(x-1)^2} = -\frac{1}{x-1}\bigg]_0^2 = -1 - 1 = -2$$

But this result cannot be right, since the integrand is always positive, so that if the integral exists, it must be nonnegative. The trouble, of course, is that the integrand becomes unbounded in a neighborhood of $x = 1$, as the graph in Figure 8.38 shows. The correct procedure is to apply Definition 8.2(c) and write

$$\int_0^2 \frac{dx}{(x-1)^2} = \int_0^1 \frac{dx}{(x-1)^2} + \int_1^2 \frac{dx}{(x-1)^2}$$

For the first integral on the right, we have

$$\int_0^1 \frac{dx}{(x-1)^2} = \lim_{t \to 1^-} \int_0^t \frac{dx}{(x-1)^2} = \lim_{t \to 1^-} \left[-\frac{1}{x-1} \right]_0^t$$

$$= \lim_{t \to 1^-} \left[-\frac{1}{t-1} - 1 \right] = \infty$$

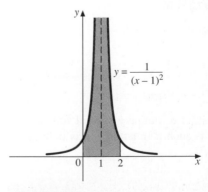

$y = \dfrac{1}{(x-1)^2}$

FIGURE 8.38

There is no need to evaluate the second integral, since the divergence of the first one implies the divergence of the original integral.

EXAMPLE 8.36   Evaluate the integral

$$\int_0^\infty \frac{dx}{x^2 + x}$$

or show that it diverges.

Solution   This integral is improper both because it has an unbounded interval of integration and because the integrand is unbounded in a neighborhood of $x = 0$. So that we can treat each problem separately, we write

$$\int_0^\infty \frac{dx}{x^2 + x} = \int_0^1 \frac{dx}{x^2 + x} + \int_1^\infty \frac{dx}{x^2 + x}$$

Without evaluating it, we can see by the comparison test that the second integral converges. (Do you see why?) So now we need consider only the first integral. By partial fractions, we find that

$$\frac{1}{x^2 + x} = \frac{1}{x(x + 1)} = \frac{1}{x} - \frac{1}{x + 1} \qquad \text{Verify.}$$

So

$$\int \frac{dx}{x^2 + x} = \int \left(\frac{1}{x} - \frac{1}{x + 1} \right) dx = \ln|x| - \ln|x + 1| = \ln\left| \frac{x}{x + 1} \right|$$

Thus,

$$\int_0^1 \frac{dx}{x^2 + x} = \lim_{t \to 0^+} \int_t^1 \frac{dx}{x^2 + x} = \lim_{t \to 0^+} \left[\ln \frac{x}{x + 1} \right]_t^1$$

$$= \lim_{t \to 0^+} \left[\ln \frac{1}{2} - \ln \frac{t}{t + 1} \right]$$

$$= \infty$$

since as $t \to 0^+$, $\ln[t/(t + 1)] \to -\infty$. Thus, this integral diverges, and hence the original integral also diverges.   ∎

EVALUATING IMPROPER INTEGRALS USING A COMPUTER ALGEBRA SYSTEM

CAS 30

The **Gamma Function:** Frequently it is useful to define a special function as an improper integral. The gamma function is such a function and is defined as

$$\Gamma(t) = \int_0^\infty x^{t-1} e^{-x} \, dx \quad \text{for all } t > 0$$

Investigate $\Gamma(n)$ for positive integers n.

In Maple we can enter:

G:= t–>int(x^(t–1)*exp(–x),x=0..a);

The output is:

$$G := t \rightarrow \int_0^a x^{(t-1)} \exp^{-x} dx$$

Enter: Output:
    limit (G(1),a=infinity);   1
    limit (G(2),a=infinity);   1
    limit (G(3),a=infinity);   2
    limit (G(4),a=infinity);   6
    limit (G(5),a=infinity);   24
    limit (G(6),a=infinity);   120

Do you recognize these values? They are factorials! That is

$$\Gamma(n+1) = n!$$

We can check this in Maple:

factorial(1);   1
factorial(2);   2
factorial(3);   6
factorial(4);   24
factorial(5);   120

The Gamma Function is such an important function in mathematics that Maple, Mathematica, and DERIVE have it built in. In Maple the function is called GAMMA (the function name must be specified in all caps), in Mathematica it is called Gamma, and it can be written as gamma in DERIVE. The Gamma Function is defined for all positive real numbers. For example,

Maple:

evalf(GAMMA(Pi));   2.288037796

Mathematica:

N[Gamma[Pi],10] 2.288037795

DERIVE:

a (author)
gamma(Pi)
x (approximate)
Output: 2.28803

Exercise Set 8.6

In Exercises 1–47, evaluate the integral or show that it diverges.

1. $\int_1^\infty \dfrac{dx}{x^3}$

2. $\int_1^\infty \dfrac{dx}{\sqrt{x^3}}$

3. $\int_{-\infty}^0 e^x \, dx$

4. $\int_0^\infty e^{-2x} dx$

5. $\int_{-\infty}^\infty \dfrac{dx}{x^2+4}$

6. $\int_2^\infty \dfrac{dx}{x\sqrt{x^2-1}}$

7. $\int_0^\infty \dfrac{e^{-x} dx}{1+e^{-x}}$

8. $\int_0^\infty \dfrac{x \, dx}{(1+x^2)^2}$

9. $\int_2^\infty \dfrac{x \, dx}{\sqrt{x^2+1}}$

10. $\int_1^\infty x^2 e^{-x} dx$

11. $\displaystyle\int_2^\infty \frac{dx}{x(\ln x)}$

12. $\displaystyle\int_2^\infty \frac{dx}{x(\ln x)^2}$

13. $\displaystyle\int_{-\infty}^\infty \frac{(x-1)dx}{(x^2-2x+4)^2}$

14. $\displaystyle\int_1^\infty \frac{dx}{x^2+x}$

15. $\displaystyle\int_2^\infty \frac{2\,dx}{x^2-1}$

16. $\displaystyle\int_3^\infty \frac{3}{x^2-x-2}\,dx$

17. $\displaystyle\int_0^\infty \tan^{-1} x\,dx$

18. $\displaystyle\int_1^\infty \frac{dx}{x\sqrt{1+x^2}}$

19. $\displaystyle\int_{-\infty}^{-3} \frac{5x\,dx}{x^2-x-6}$

20. $\displaystyle\int_1^\infty \frac{dx}{\sqrt{x}(1+\sqrt{x})^2}$

21. $\displaystyle\int_{-\infty}^\infty \frac{dx}{(x^2+1)^{3/2}}$

22. $\displaystyle\int_0^1 \frac{dx}{1-x}$

23. $\displaystyle\int_0^1 \frac{dx}{\sqrt{1-x}}$

24. $\displaystyle\int_0^1 \frac{x\,dx}{\sqrt[3]{1-x^2}}$

25. $\displaystyle\int_1^2 \frac{dx}{x\sqrt{x^2-1}}$

26. $\displaystyle\int_0^2 \frac{dx}{\sqrt{4-x^2}}$

27. $\displaystyle\int_2^3 \frac{dx}{(x-2)^{3/2}}$

28. $\displaystyle\int_2^3 \frac{dx}{(x-2)^{2/3}}$

29. $\displaystyle\int_0^3 \frac{dx}{(x-1)^2}$

30. $\displaystyle\int_{-2}^1 \frac{dx}{x+1}$

31. $\displaystyle\int_1^2 \frac{dx}{x\ln x}$

32. $\displaystyle\int_{-2}^{-1} \frac{dx}{x\sqrt{\ln|x|}}$

33. $\displaystyle\int_0^{\ln 2} \frac{e^x dx}{\sqrt{e^x-1}}$

34. $\displaystyle\int_0^{\pi/2} \frac{\cos x}{1-\sin x}\,dx$

35. $\displaystyle\int_0^\pi \frac{\sin x}{\sqrt[3]{\cos x}}\,dx$

36. $\displaystyle\int_0^\pi \tan^2 x \sec^2 x\,dx$

37. $\displaystyle\int_0^{\pi/3} \frac{\sec^2 x\,dx}{(\tan x-1)^2}$

38. $\displaystyle\int_{-2}^2 \frac{x^3 dx}{\sqrt{4-x^2}}$

39. $\displaystyle\int_0^2 \frac{dx}{x\sqrt{x^2+1}}$

40. $\displaystyle\int_0^2 \frac{dx}{x^2-2x}$

41. $\displaystyle\int_1^3 \frac{dx}{x^2-4x+3}$

42. $\displaystyle\int_{-1}^3 \frac{dx}{x^2+x-2}$

43. $\displaystyle\int_{-1}^1 \frac{dx}{x^3+x^2}$

44. $\displaystyle\int_{-1}^8 \frac{x^{1/3}dx}{x^{2/3}-4}$

45. $\displaystyle\int_0^\pi \frac{dx}{1-\sin x}$

46. $\displaystyle\int_0^2 \frac{dx}{\sqrt{2x-x^2}}$

47. $\displaystyle\int_a^{2a} \frac{dx}{\sqrt{x^2-a^2}}$

In Exercises 48–53, determine whether the integral is convergent or divergent, using Theorem 8.1.

48. $\displaystyle\int_1^\infty \frac{dx}{x^3+x+1}$

49. $\displaystyle\int_2^\infty \frac{dx}{x^{2/3}-1}$

50. $\displaystyle\int_3^\infty \frac{x\,dx}{\sqrt{x^3-1}}$

51. $\displaystyle\int_0^\infty \frac{x\,dx}{x^4+4}$

52. $\displaystyle\int_3^\infty \frac{dx}{x^2\ln x}$

53. $\displaystyle\int_1^\infty \frac{dx}{x^2+\sin^2 x}$

54. Prove that if $\int_a^\infty f(x)dx$ and $\int_{-\infty}^a f(x)dx$ both exist, then for any number $b \neq a$, $\int_b^\infty f(x)dx$ and $\int_{-\infty}^b f(x)dx$ also exist.

In Exercises 55–58, find the value of k so that f will be a probability density function for a random variable X on the indicated interval, and then find the specified probability.

55. $f(x) = ke^{-x/3}$ on $[0, \infty)$; $P(X \geq 1)$

56. $f(x) = kxe^{-x/2}$ on $[0, \infty)$; $P(X > 2)$

57. $f(x) = kxe^{-x^2}$ on $[0, \infty)$; $P(1 < X \leq 2)$

58. $f(x) = k[e^x/(1+e^x)^2]$ on $[0, \infty)$; $P(X \geq 1)$

59. Find the area of the region under the graph of $y = 1/(1+x^2)$ on $[1, \infty)$.

60. Find the volume of the solid obtained by revolving the region under the graph of $y = 1/(1+x)$ on $[0, \infty)$ about the x-axis.

61. Find the volume of the solid obtained by revolving the region under the graph of

$$y = \frac{1}{\sqrt{x}\,\ln x} \qquad (2 \leq x < \infty)$$

about the x-axis.

62. Show that

$$\lim_{t\to\infty} \int_{-t}^t \frac{x\,dx}{x^2+1} = 0$$

but that

$$\int_{-\infty}^\infty \frac{x\,dx}{x^2+1}$$

does not exist. Explain this result.

63. Find the area of the region under the graph of $y = 1/\sqrt{1-x}$ from 0 to 1, or show that it fails to exist.

64. Find the volume of the solid (if it exists) generated by rotating the region above $y = x$ and below $y = x^{-3/2}$ between $x = 0$ and $x = 1$ about the y-axis.

65. Prove each of the following.
(a) If f is even and $\int_0^\infty f(x)dx$ converges, then
$$\int_{-\infty}^\infty f(x)dx = 2\int_0^\infty f(x)dx$$
(b) If f is odd and $\int_0^\infty f(x)dx$ converges, then
$$\int_{-\infty}^\infty f(x)dx = 0$$

66. A random variable X has the **exponential** distribution if its probability density function is
$$f(x) = \frac{1}{\alpha} e^{-x/\alpha} \qquad (x \geq 0)$$
where α is a positive constant. Find the mean and variance of X. (See Section 6.9.) (This function is frequently used to model the distribution of times between successive events such as customers arriving at some facility.)

67. The lifetime X, in hours, of a certain type of light bulb is a random variable with an exponential probability density function having $\alpha = 1000$. (See Exercise 66.)
(a) Find the probability that a light bulb of this type selected at random will last at least 1500 hr.
(b) Of 10,000 light bulbs installed in an illuminated sign, how many will likely fail in less than 500 hr? in less than 100 hr?

68. In an experiment with mice in a maze, it is found that a reasonable model of the probability density function of the time X, in minutes, for a mouse to go through the maze is given by
$$f(x) = \frac{a}{x^2} \qquad (x \geq a)$$
where a is the minimum time required.
(a) Verify that f is a probability density function on $[a, \infty)$.
(b) Find the probability that a mouse selected at random will take longer than $2a$ minutes to go through the maze.

In Exercises 69–74, evaluate the integral or show that it diverges.

69. $\displaystyle\int_0^\infty \frac{dx}{\sqrt{x}(1+x)}$

70. $\displaystyle\int_1^\infty \frac{dx}{x\sqrt{x-1}}$

71. $\displaystyle\int_0^\infty \frac{dx}{e^x - 1}$

72. $\displaystyle\int_0^\infty \frac{dx}{\sqrt{x}+x^2}$

73. $\displaystyle\int_{-\infty}^\infty \frac{dx}{x(\ln|x|)^2}$

74. $\displaystyle\int_0^\pi \frac{\cot x \, dx}{\ln \sin x}$

75. Show that $\displaystyle\int_0^1 \cos\frac{1}{x}\, dx$ exists. (*Hint:* Let $u = \frac{1}{x}$.)

76. Evaluate the integral
$$\int_0^e \frac{dx}{x\ln x(\ln|\ln x|)^2}$$
or show that it diverges. (*Hint:* Look for all zeros of the denominator.)

77. Prove that the integral
$$\int_0^1 \frac{dx}{x^p}$$
converges if $p < 1$ and diverges if $p \geq 1$.

78. There are comparison tests for improper integrals with unbounded integrands analogous to those that have an unbounded interval of integration. For $f(x)$ unbounded in a neighborhood of a, one such test is as follows: If $0 \leq f(x) \leq g(x)$ for all x on $(a, b]$, and f and g are continuous on $(a, b]$, then
$$\int_a^b f(x)dx \text{ converges if } \int_a^b g(x)dx \text{ converges}$$
and
$$\int_a^b g(x)dx \text{ diverges if } \int_a^b f(x)dx \text{ diverges}$$
A similar test applies if f is unbounded near b. Use this test along with the result of Exercise 77 to determine the convergence or divergence of the following integrals.
(a) $\displaystyle\int_0^\pi \frac{\cos^2 x}{\sqrt{x}} dx$
(b) $\displaystyle\int_0^{\pi/3} \frac{\sec x}{x^{3/2}} dx$

79. Follow the instructions for Exercise 78 to determine convergence or divergence of the following integrals.
(a) $\displaystyle\int_0^{\pi/2} \frac{1 - \sin x}{x^{2/3} + x} dx$
(b) $\displaystyle\int_0^1 \frac{1 + e^{-x}}{x - x^2} dx$

80. Show that the formula for the length of an arc of the four-cusp hypocycloid $x^{2/3} + y^{2/3} = a^{2/3}$ (see the figure) results in an improper integral. Find the total length.

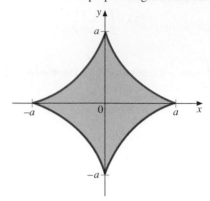

81. (a) Find the area of the region under the graph of $f(x) = 1/\sqrt{4 - x^2}$ from $x = 0$ to $x = 2$.

(b) Find the volume of the solid generated by revolving the region in part (a) about the line $x = 2$.

(c) Without finding its value, show that the surface obtained by revolving the graph of f in part (a) about the line $x = 2$ has a finite area.

82. Determine the values of p for which the integral

$$\int_{2}^{\infty} \frac{dx}{x(\ln x)^p}$$

converges and those for which it diverges.

83. Use the following outline to prove that

if $\int_{a}^{\infty} |f(x)|\,dx$ converges, then $\int_{a}^{\infty} f(x)\,dx$ converges.

(a) Explain why $0 \leq f(x) + |f(x)| \leq 2|f(x)|$.

(b) Why does it follow that $\int_{a}^{\infty}[f(x) + |f(x)|]\,dx$ converges?

(c) Write $f(x) = [f(x) + |f(x)|] - |f(x)|$, and explain why $\int_{a}^{\infty} f(x)\,dx$ must therefore converge.

84. Use the result to Exercise 83, together with the comparison test, to show that each of the following converges:

(a) $\int_{0}^{\infty} \frac{\sin x}{1 + x^2}\,dx$ (b) $\int_{0}^{\infty} e^{-x}\cos x\,dx$

85. Let $f(t)$ be continuous on $0 \leq t < \infty$. The **Laplace transform** of $f(t)$, denoted by $F(s)$, is defined by

$$F(s) = \int_{0}^{\infty} f(t)e^{-st}\,dt$$

provided the integral converges. Laplace transforms are useful in the study of differential equations. They also occur in various parts of physics, including, for example, the signal processing in compact discs and digital audiotapes. Find the Laplace transform of each of the following.

(a) $f(t) = 1$ (b) $f(t) = t$
(c) $f(t) = e^t$ (d) $f(t) = \sin t$

86. Refer to Exercise 85. Prove that if there are positive constants M and k such that

$$|f(t)| \leq Me^{kt} \quad \text{for } 0 \leq t < \infty$$

then $F(s)$ exists for $s > k$.

87. The normal probability density function is of the form

$$f(x) = \frac{1}{\beta\sqrt{2\pi}} e^{-(x-\alpha)^2/2\beta^2} \quad (-\infty < x < \infty)$$

Take it as given that $\int_{-\infty}^{\infty} f(x)\,dx = 1$, and prove that the mean $\mu = \alpha$ and the standard deviation $\sigma = \beta$. (Refer to Section 6.9.)

88. The probability density function for the random variable X representing the length of life of certain plants and animals is often modeled by the **Weibull** distribution:

$$f(x) = \frac{m}{\alpha} x^{m-1} e^{-x^m/\alpha} \quad (x > 0)$$

This function was first introduced by the Swedish physicist Waloddi Weibull in 1939. Let $m = 2$ and $\alpha = 4$ and find the following:

(a) $P(X \geq 2)$ (b) $E[X^2]$

Chapter 8 Review Exercises

In Exercises 1–30, evaluate the given definite or indefinite integral.

1. $\int x^2 e^{-2x}\,dx$

2. $\int \sqrt{x}\,\ln\sqrt{x}\,dx$

3. $\int_{0}^{1} x\tan^{-1} x^2\,dx$

4. $\int e^{-x}\cosh 2x\,dx$

5. $\int \sec^2\sqrt{x}\,dx$ *(Hint: Let $t = \sqrt{x}$.)*

6. $\displaystyle\int_{-\pi/6}^{\pi/3} \sin^2 x(\sin x - 1)dx$

7. $\displaystyle\int \sin^5 x \cos^3 x\, dx$ **8.** $\displaystyle\int \sec^3 x \tan^5 x\, dx$

9. $\displaystyle\int_{-\pi/3}^{\pi/4} \frac{\sin^2 x}{\cos^4 x}\, dx$ **10.** $\displaystyle\int \frac{\sin^2 x}{\cos^3 x}\, dx$

11. $\displaystyle\int x \sin x \cos^3 x\, dx$

12. $\displaystyle\int \sec^2 \sqrt{x}\, \tan \sqrt{x}\, dx$ (*Hint:* Let $t = \sqrt{x}$.)

13. $\displaystyle\int \sin 2x \cos 4x\, dx$

14. $\displaystyle\int \frac{\ln \cos x}{\sin^2 x}\, dx$ (*Hint:* Integrate by parts.)

15. $\displaystyle\int_{2}^{2\sqrt{3}} \frac{x^2}{(16 - x^2)^{3/2}}\, dx$

16. $\displaystyle\int_{1}^{2} x^3 \sqrt{x^2 + 1}\, dx$ (Do by two methods: integration by parts and trigonometric substitution.)

17. $\displaystyle\int_{1/\sqrt{3}}^{1} \frac{\sqrt{4x^2 - 1}}{x^3}\, dx$ **18.** $\displaystyle\int \frac{x\, dx}{\sqrt{x^2 - 2x + 5}}$

19. $\displaystyle\int \frac{e^{4x}\, dx}{\sqrt{1 + e^{2x}}}$ **20.** $\displaystyle\int \frac{\sinh x \cosh^3 x}{\sqrt{\cosh^2 x - 4}}\, dx$

21. $\displaystyle\int \frac{x^2\, dx}{\sqrt{4x - x^2}}$ **22.** $\displaystyle\int \frac{4x - 9}{2x^2 + 5x - 3}\, dx$

23. $\displaystyle\int_{1}^{2} \frac{x^2 - 3x + 2}{x^2 - 2x - 8}\, dx$ **24.** $\displaystyle\int \frac{5x + 1}{x^3 - x^2 - x + 1}\, dx$

25. $\displaystyle\int \frac{x^2 + x + 1}{x(x^2 + 1)^2}\, dx$ **26.** $\displaystyle\int \frac{x^4 + 20}{x^3 - 3x + 2}\, dx$

27. $\displaystyle\int_{-1}^{0} \frac{12\, dx}{x^3 - 8}$

28. $\displaystyle\int_{0}^{4} \frac{dx}{\sqrt{x} + 1}$ (*Hint:* Let $u = \sqrt{x}$.)

29. $\displaystyle\int \frac{x^3}{\sqrt{x^2 - 4}}\, dx$ (Do by two methods: trigonometric substitution and algebraic substitution.)

30. $\displaystyle\int_{0}^{\pi/2} \frac{1 + \cos x}{1 + \sin x}\, dx$ (*Hint:* Multiply numerator and denominator by $1 - \sin x$.)

31. Find the volume of the solid obtained by revolving the region under the graph of $y = \sin x$ from $x = 0$ to $x = \pi$ about (a) the x-axis and (b) the y-axis.

32. Find the length of the curve $y = x^2/4$ from $x = 0$ to $x = 2$.

33. Find the volume of the solid generated by revolving the region under the graph of

$$y = \frac{1}{\sqrt{5x - x^2}}$$

from $x = 1$ to $x = 4$ about (a) the x-axis and (b) the y-axis.

34. Find the area of the surface formed by revolving the arc of the curve $y = \cosh x$ from $x = 0$ to $x = \ln 3$ about (a) the x-axis and (b) the y-axis.

35. A model for a so-called *learning curve* in psychology is given by

$$\frac{dy}{dt} = k\sqrt{y^3(1 - y)^3} \qquad (0 \le y \le 1)$$

where k is a positive constant. Here $y = y(t)$ is the proportion of a task or of a body of knowledge learned by the subject at time t. For $0 < y < 1$, find y as a function of t by writing the equation in the form

$$\frac{dy}{y^{3/2}(1 - y)^{3/2}} = k\, dt$$

and integrating both sides. (*Hint:* To integrate the left-hand side, substitute $y = \sin^2 \theta$.)

In Exercises 36–51, evaluate the integral or show that it diverges.

36. $\displaystyle\int_{1}^{\infty} \frac{dx}{x^4 + x^2}$ **37.** $\displaystyle\int_{1}^{\infty} \frac{dx}{x^4 - x^2}$

38. $\displaystyle\int_{2}^{\infty} \frac{dx}{x\sqrt{x^2 - 4}}$ **39.** $\displaystyle\int_{-1}^{8} \frac{x + 1}{\sqrt[3]{x}}\, dx$

40. $\displaystyle\int_{2}^{4} \frac{x^3\, dx}{\sqrt{x^2 - 4}}$ **41.** $\displaystyle\int_{1}^{\infty} \frac{x\, dx}{\sqrt{x^4 + 1}}$

42. $\displaystyle\int_{0}^{1} \sqrt{\frac{1 - x}{x}}\, dx$ **43.** $\displaystyle\int_{3}^{\infty} \frac{3\, dx}{x^2 - x - 2}$

44. $\displaystyle\int_{-1}^{1} \frac{dx}{x^2\sqrt{1-x^2}}$

45. $\displaystyle\int_{0}^{1} x\ln x\, dx$

46. $\displaystyle\int_{-\infty}^{\infty} \frac{dx}{1+x^3}$

47. $\displaystyle\int_{1}^{e} \frac{dx}{x\sqrt{\ln x}}$

48. $\displaystyle\int_{-\ln 2}^{\ln 2} \coth x\, dx$

49. $\displaystyle\int_{0}^{\infty} x^3 e^{-x^2}\, dx$

50. $\displaystyle\int_{0}^{\infty} \frac{e^x\, dx}{e^{2x}+1}$

51. $\displaystyle\int_{\pi/3}^{\pi/2} \frac{\tan x}{(\ln\cos x)^2}\, dx$

52. Show that the volume of the solid generated by revolving the region under the graph of $y = xe^{-x}$ on $[0,\infty)$ about the x-axis is finite, and find its value.

53. Show that $\int_0^1 \sin(1/x^\alpha)dx$ converges for $\alpha > 0$.

54. Let

$$f(x) = \frac{1}{x\ln x[\ln(\ln x)]^p} \qquad (x > e)$$

Show that $\int_e^{e^2} f(x)dx$ converges if and only if $p < 1$, whereas $\int_{e^2}^{\infty} f(x)dx$ converges if and only if $p > 1$.

55. The **Pareto probability density function** is of the form

$$f(x) = \frac{k\theta^k}{x^{k+1}}$$

if $x \geq \theta$, and $f(x) = 0$ otherwise, where k and θ are positive constants.
(a) Show that f is a probability density function.
(b) For what values of k is the mean μ finite? Find μ for these values.
(c) For what values of k is the variance σ^2 finite? Find σ^2 for these values. (See Equation 6.47.)

56. The **gamma function,** defined by

$$\Gamma(\alpha) = \int_0^{\infty} x^{\alpha-1}e^{-x}\, dx \qquad (\alpha > 0)$$

has many applications in both pure and applied mathematics. Show the following:
(a) For $x \in (0, 1]$, $x^{\alpha-1}e^{-x} \leq x^{\alpha-1}$.
(b) For x sufficiently large, $x^{\alpha-1}e^{-x} \leq 1/x^2$.
(c) The integral that defines $\Gamma(\alpha)$ converges if $\alpha > 0$. (*Hint:* Use the results of parts (a) and (b).)

57. Refer to Exercise 56. Prove the following:
(a) $\Gamma(\alpha + 1) = \alpha\Gamma(\alpha)$
(b) For all natural numbers n,

$$\Gamma(\alpha + n) = (\alpha + n - 1)(\alpha + n - 2)\cdots(\alpha + 1)\Gamma(\alpha)$$

(c) For all natural numbers n, $\Gamma(n + 1) = n!$. Show why it is natural to define $0! = 1$.

58. The **gamma probability density function** is defined by

$$f(x) = \frac{1}{\beta^\alpha\Gamma(\alpha)} x^{\alpha-1}e^{-x/\beta} \qquad (x > 0)$$

and $f(x) = 0$ if $x \leq 0$, where α and β are positive constants. (See Exercise 57.) Prove that f is a probability density function, and find μ and σ^2. (The gamma distribution is often useful in modeling probability distributions that are skewed, or nonsymmetric.)

59. Analyze the function

$$f(x) = \begin{cases} e^{1-1/x} & \text{if } x \neq 0 \\ 0 & \text{if } x = 0 \end{cases}$$

and draw its graph. Pay particular attention to the right-hand derivative at 0.

Chapter 8 Concept Quiz

1. State the Integration by Parts Formula in terms of u and v. In each of the following apply the formula, indicating appropriate choices for u and dv. Do not complete the integration.
(a) $\int x^3 \sin x\, dx$
(b) $\int (2x^2 - 3x + 1)e^{-2x}\, dx$
(c) $\int x^2 \ln x\, dx$
(d) $\int \sin^{-1} x\, dx$
(e) $\int e^{-3x} \cos 2x\, dx$

2. Explain how you would evaluate the integral

$$\int \sin^m x \cos^n x\, dx \qquad (m, n \text{ nonnegative integers})$$

in each of the following cases:
(a) m is odd
(b) n is odd
(c) m and n are even

3. Explain how you would evaluate the integral

$$\int \sec^m x \tan^n x\, dx \qquad (m, n \text{ nonnegative integers})$$

in each of the following cases:
(a) m is even
(b) n is odd
(c) m is odd and n is even

4. Tell what substitution you would try in an integral involving the following:
   (a) $\sqrt{u^2 - a^2}$
   (b) $\sqrt{u^2 + a^2}$
   (c) $\sqrt{a^2 - u^2}$

   In each case indicate restrictions on the new variable, and sketch a right triangle that shows the relationship between u and the new variable.

5. Consider an integral of the form

$$\int \frac{P(x)}{Q(x)}\,dx$$

   where $P(x)$ and $Q(x)$ are polynomials. If $\deg P \geq \deg Q$, what do you do first? Suppose that $\deg P < \deg Q$ in each of the following cases. Show the appropriate form of the partial-fraction decomposition.
   (a) $Q(x) = x^2 - 4x - 5$
   (b) $Q(x) = (2x + 3)(x^2 - 6x + 9)$
   (c) $Q(x) = (x - 6)(x^2 + x + 1)$
   (d) $Q(x) = (x + 2)^3(x^2 + 1)$
   (e) $Q(x) = x^2(x^2 + 4)^2$

6. (a) How is $\int_a^\infty f(x)\,dx$ defined?
   (b) If f is continuous on $[a, b)$ and $\lim_{x \to b^-} f(x) = \pm\infty$, how is $\int_a^b f(x)\,dx$ defined?
   (c) If f is continuous on $(a, b]$ and $\lim_{x \to a^+} f(x) = \pm\infty$, how is $\int_a^b f(x)\,dx$ defined?
   (d) If f is continuous on $[a, c) \cup (c, b]$ and f is unbounded in a neighborhood of $x = c$, how is $\int_a^b f(x)\,dx$ defined?
   (e) For what values of p does the integral

$$\int_1^\infty \frac{1}{x^p}\,dx$$

   converge, and for what values does it diverge?

APPLYING CALCULUS

1. A catalyst for a chemical reaction is a substance that increases the speed of the reaction without itself undergoing change. An *autocatalytic reaction* is a reaction whose product acts as a catalyst for its own formation. Such a reaction proceeds slowly at first when the amount of catalyst present is small, and slowly at the end when most of the original substance is used up. In between, when both the product and the substance are present in some quantity, the reaction proceeds at a faster rate. One mathematical model for an autocatalytic reaction is based on the assumption that the rate $v = dx/dt$ of the reaction is proportional to both the amount of the original substance and the amount of the product; that is

$$v = \frac{dx}{dt} = kx(A - x)$$

where x is the amount of the product, A is the amount of the substance at the beginning of the reaction, and k, $k > 0$, is the constant of proportionality.

(a) At what value x is the rate of the reaction a maximum? What is the maximum value of v?

The differential equation above can be solved by rewriting it in the form

$$\frac{dx}{x(A - x)} = kdt$$

and integrating both sides

$$\int \frac{dx}{x(A - x)} = \int kdt = kt + C.$$

(b) Determine x as a function of t by calculating the integral derived above. Assume that $x = x_0$ when $t = 0$.

(c) Sketch the graph of the solution $x = x(t)$. Where does the curve have a point of inflection? Compare with the result in part (a).

2. A student who has just heard a new joke returns to his college campus whose enrollment is 10,000 students. Four hours later, 10 students have heard the joke. Let $x = x(t)$ be the number of students who have heard the joke at time t. A common model used by social scientists for the spread of a rumor through a population is that the rumor spreads at a rate $v = dx/dt$ which is proportional to the product of the number of people who have heard the rumor and the number of people who have not yet heard it. Applying the model in this case,

$$v = \frac{dx}{dt} = k(x + 1)(10,000 - x)$$

where k is a positive constant. To solve this differential equation, rewrite it in the form

$$\int \frac{dx}{(x + 1)(10,000 - x)} = \int kdt = kt + C$$

(a) Find x as a function of t by evaluating the integral and using $x(0) = 1$ and $x(4) = 10$ to determine C and k.

(b) Sketch the graph of $x = x(t)$, including $\lim_{t \to \infty} x(t)$, relative extrema (if any) and points of inflection.

(c) When is the joke spreading at its most rapid rate? How long will it take for 75% of the students to have heard the joke?

3. Consider the logistic differential equation

$$\frac{dy}{dt} = ky(M - y)$$

with initial condition $y(0) = y_0$. Show that the solution of the initial-value problem is

$$y(t) = \frac{y_0 M}{y_0 + (M - y_0)e^{-kMt}}$$

(a) Suppose $y_0 < M$. Show that the solution is increasing on $[0, \infty)$. Determine the point(s) of inflection and the intervals on which the graph is concave up and the intervals on which it is concave down. Use a CAS to draw an accurate sketch of the graph of the solution.

(b) Suppose $y_0 = M$. Draw the graph of the solution in this case.

(c) Suppose $y_0 > M$. Show that the solution is decreasing and concave up on $[0, \infty)$. Draw the graph of the solution, including $\lim_{t \to \infty} y(t)$. Note that, in all three cases, $\lim_{t \to \infty} y(t) = M$. This value is called *carrying capacity*.

4. In the logistic model of population growth

$$\frac{dy}{dt} = ky(M - y) = my - ky^2 \qquad (m = kM)$$

it is assumed that there is a "birth" rate that is proportional to the population size (the term "my") and a "death" rate which is proportional to the square of the population (the term "$-ky^2$"). In contrast, there are species in which the birth rate depends on chance encounters between males and females, and is therefore proportional to the product of the number of males and the number of females. If males and females are equally distributed throughout the population, then the birth rate is proportional to the square of the population. Suppose it is further assumed that the death rate is proportional to the population size. The differential equation corresponding to this model for population growth is

$$\frac{dy}{dt} = my^2 - ky = my(y - M) \qquad (M = k/m)$$

(a) Determine the solution of the differential equation which satisfies $y(0) = y_0$.

(b) Graph the solution in the three cases (i) $y_0 < M$, (ii) $y_0 = M$, and (iii) $y_0 > M$. How does this model compare with the logistic model?

5. A simplified model for the spread of a disease assumes that the rate at which the disease spreads is proportional to the product of the number of infected people and the number of susceptible people. Suppose that an influenza epidemic is spreading through a population of 10,000 people and that 500 people have it initially.

(a) Let $y = y(t)$ denote the number of people who have the flu at time t. Determine the differential equation which models the spread of the disease.

(b) Suppose that 1000 people have the disease after 10 days. Determine an expression for the number of infected people at any time t.

(c) How long will it take for 75% of the people to have the flu?

6. A spacecraft at a distance r above a planet of radius R can see less than half of the planet.

(a) For a rectangular coordinate system with origin at A, with $(g(y), y)$ giving the coordinates of a point on arc EC, explain why the surface area A_z is

$$A_z = 2\pi \int_{y_B}^{y_E} g(y)\sqrt{1 + [g'(y)]^2}\, dy$$

An astronaut in orbit

(Hint: Find $g(y)$, y_B, and y_E geometrically in terms of y and R.)

(b) Find A_z/A_p, where A_p is the planet's entire area.

(c) A space shuttle orbits at an altitude of about 300 km. What fraction of the Earth's surface can those astronauts see at one instant? The Earth's radius is 6378 km. What is the radius of a circle on the Earth's surface with this area? Give an example of a region of Earth of this size.

(d) How high would an astronaut have to go to see one-fourth of the Earth's surface?

(e) What fraction of the Earth could Neil Armstrong see when he became the first human to set foot on the Moon on July 20, 1969?

(f) Find $\lim\limits_{h \to \infty} \dfrac{A_z}{A_{\text{Earth}}}$ and compare with the answer for part (e).

7. On September 19, 1991, the frozen remains of a man were found encased in ice in the mountains on the Swiss-Italian border. (See the photo.) Using the technique of radiocarbon-dating (see Exercise 25 in Exercise Set 7.4), researchers determined that the man had been dead for 5,000 years (*Nature*, vol. 362, 4 March 1993). The half-life of carbon-14 is approximately 5,730 years. Assuming exponential decay, write and solve the differential equation for the amount of carbon-14 remaining in his body t years after his death. How much carbon-14 did the researchers find in his body, as a percentage of the amount present when he died?

Sheldon Glashow

How did you first get interested in mathematics and physics?

I got interested in science as a youngster. I remember my older brother explaining the kinematics of airplanes flying and dropping bombs and how the bombs would explode directly underneath the airplane unless the airplane took some evasive action. After a brief infatuation with biology, I realized that I had to know some chemistry. I learned some chemistry on my own and realized that it didn't make sense without physics. So I figured out by the time I graduated from high school that I pretty much had to know physics. I went to the Bronx High School of Science, where they didn't have a calculus course at that time. There formal mathematics ended with solid geometry, so I was forced to learn calculus in the lunchroom from a friend who had somehow picked some of it up. It was clear from the beginning that physics deals with things that change and that you can't deal with things that change unless you know calculus.

In view of new developments in mathematics, and discussions of discrete mathematics and computers, is it still useful to learn calculus?

I'm a computer illiterate but I'm pretty good at calculus, and physics is a continuous subject. It deals with things that change continuously, most of the time, and discrete mathematics is not usually useful for things that change like that. My father was a plumber, and despite all his knowledge and wisdom about plumbing, he couldn't work without a blowtorch and a wrench and the other tools of his trade. In exactly the same sense, the practicing physicist—the licensed quantum mechanic—has to have a set of tools, and primary among them is calculus.

How necessary is calculus for doing physics?

It is hard to explain, because you cannot isolate mathematics from physics. It is so much ingrained into all of physics so that you can't even think physics without knowing calculus. But I didn't win any prizes for doing integrals. Again and again, physicists had to invent the mathematics they needed to deal with the problems they encountered. Newton and Leibniz were forced to invent calculus. Einstein was forced to invent another form of mathematics. And today, string theorists [those working on a theory used to try to explain the basic structure of matter] are using fields of mathematics that have not yet been developed, but which depend at the very beginning on calculus.

In the discussion of polar coordinates, we use concepts of symmetry. How has symmetry proved of interest to you?

When we study mechanics, we learn that it doesn't matter where you put the origin of your coordinate system. That is, you can translate from one point to another. That is a symmetry of nature. You can also rotate the coordinate system—you can do physics while standing on your head and the laws come out the same—since the laws of physics are invariant under rotations in space. But from the point of view of calculus, you can show that the first symmetry—translational symmetry—is equivalent to the conservation of momentum and that the second symmetry—rotational symmetry—is equivalent to the rotation law for angular momentum. These are remarkable implications of symmetry but they can only be deduced within the framework of variational principles and calculus. Most of the progress in physics in the last few years, especially in the construction of the standard model of quarks and leptons and their interactions, involves a mathematical marriage between calculus and group theory (the theory of continuous groups). All of the forces of nature are described mathematically by things called Lie groups, which are the offspring of this marriage. A Lie group is a complicated set like the set of all possible rotations of a rigid body—that is, the rotation group. There are many other such groups. But the miracle is that the properties of finite rotations, which are very complicated things, can almost all be described in terms of very tiny rotations applied in sequence. Once again, you are building up finite objects by putting together infinitesimals, thereby describing symmetry groups in terms of calculus.

Your high school didn't teach calculus. How does that compare with the preparation of today's students?

There are two kinds of calculus students—those who know calculus and those who think they know calculus. The numbers who think they know calculus is very high, but it turns out that many don't know algebra and therefore cannot know calculus. Calculus is not a stand-alone subject. It depends crucially on a knowledge of geometry, trigonometry, and algebra. Without that foundation, there is no possibility of learning calculus. Many people want to learn about quarks and their behavior. But they don't know anything about calculus. They have no hope of understanding what they want to know without knowing some calculus.

Sheldon Glashow is a theoretical physicist at Harvard University. He shared the Nobel Prize for Physics in 1979 for his theory that unified two of the fundamental forces of nature—the weak force and the electromagnetic force. He is also known for predicting the existence of what is known as the charmed quark, one of the six kinds of quarks that make up the fundamental particles of nature.

Pierre de Fermat shares credit with René Descartes for the independent invention of analytic geometry, which is essentially coordinate geometry. The Cartesian coordinate system is so named to acknowledge Descartes's contribution. We have used aspects of analytic geometry without explicitly mentioning it throughout the text, every time we give graphs of equations. In the present chapter, we will use it even more, as we study graphs of parametric equations and of polar equations. An especially important use of analytic geometry is in the study of the conic sections, which we take up in relation to polar coordinates.

Pierre de Fermat (1601–1665) was a jurist in Toulouse, France. Mathematically, he contributed to many aspects of number theory, analytic geometry, and calculus. As the discoverer of methods for finding tangents to curves and the curves' maxima and minima, ideas later taken up by Isaac Newton, Fermat is considered one of the developers of differential calculus, although he used it without fully understanding it. He is also known to physicists for "Fermat's principle," which states that light traveling between two points seeks the path that is either a minimum or a maximum of time.

Fermat often wrote his mathematical results in letters to friends. Other results were kept to himself, and he sometimes wrote them as marginal comments in his books, such as the *Arithmetica of Diophantus* (c. A.D. 250), which was a then-recent translation. Fermat's notes were such important additions to this text that within 10 years after his death, a second edition of the translation had been published with Fermat's marginal notes included.

Fermat's most famous result, "Fermat's last theorem," can be stated as follows:

No positive integers x, y, z, and n exist for which

$$x^n + y^n = z^n \text{ for } n > 2$$

In the margin of his copy of *Arithmetica*, Fermat wrote, "It is impossible to divide a cube into two cubes, or a fourth power into two fourth powers, or, in general, any power from then on past the square into two like powers; of this I have discovered a very wonderful demonstration. This margin is too narrow to contain it."

"Fermat's last theorem" has tantalized mathematicians and been the subject of detailed investigation and debate for years. Euler (1707–1783) proved it for $n = 3$ and $n = 4$, and various other "proofs" have been advanced over the years, but all have turned out to contain flaws. Did Fermat himself really have a valid proof? We shall probably never know, but probably he did not. The theorem wasn't really his "last"; it may even have been written down 30 years before he died, and he was known to elaborate on other theorems.

The English mathematician Louis Mordell found in 1922 that equations such as Fermat's are linked to the numbers of holes in surfaces. Thus Fermat's ideas in number theory are important for topology, which deals with such surfaces.

Computers have been used to evaluate Fermat's conjecture, which has been tested for all $n < 3$ million! But it would take only one example to disprove it.

In 1993, the mathematical world was astounded by the proof of Fermat's last theorem by Andrew Wiles of Princeton University, using ideas of elliptic integrals and geometry. His proof is hundreds of pages long and is still being checked by his mathematical colleagues. If his proof is found to be correct, then a 250-year-old mystery will finally have been laid to rest.

C H A P T E R

9

PARAMETRIC EQUATIONS AND POLAR COORDINATES

Most of the curves we have considered so far have been graphs of functions—that is, graphs of equations of the form $y = f(x)$. One characteristic feature of such a curve is that each vertical line intersects it in at most one point. Many curves do not have this property, however. For example, none of the curves in Figure 9.1 passes the vertical line test, so none of these curves is the graph of a function of x.

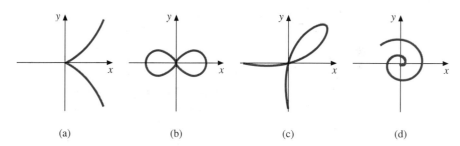

(a) (b) (c) (d)

FIGURE 9.1

In this chapter we introduce two new methods of representing curves in the plane. We will see that each of the curves in Figure 9.1, and many others, can be described by one of these methods. The first method makes use of an auxiliary variable, called a **parameter**. The coordinates x and y are each expressed in terms of the parameter, yielding what are called **parametric equations**. In the second method, we make use of an alternative coordinate system for locating points in the plane, called the **polar coordinate system**. We will see that certain curves that are difficult to describe by an equation in x and y (*rectangular* coordinates) can be conveniently represented by an equation in polar coordinates.

We conclude the chapter with a discussion of the curves known as conic sections—the circle, ellipse, parabola, and hyperbola—and show how their equations assume a particularly simple form using polar coordinates.

9.1 CURVES DEFINED PARAMETRICALLY

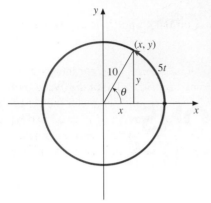

FIGURE 9.2

To introduce parametric equations, we consider the following problem. Suppose an object is moving at a constant speed of 5 m/s counterclockwise around a circle of radius 10 m, centered at the origin, as indicated in Figure 9.2. The path of the object is not the graph of a function. So we cannot express y as a (single) function of x (nor x as a (single) function of y). Instead, we describe the position of the object by expressing x and y separately as functions of time t. If at time $t = 0$ the object is at the point $(10, 0)$, what are the x- and y-coordinates of the object t seconds later?

As indicated in Figure 9.2, after t seconds the object has covered a distance of $5t$ meters on the circle. The angle θ, measured in radians, is therefore

$$\theta = \frac{5t}{10} = \frac{t}{2}$$

Here we used the fact that $\theta = s/r$, where s is the length of arc on a circle of radius r, subtended by a central angle of θ radians. (See Appendix 2.) Thus, $x = 10\cos\theta = 10\cos(t/2)$ and $y = 10\sin\theta = 10\sin(t/2)$. The two equations

$$\begin{cases} x = 10\cos\left(\dfrac{t}{2}\right) \\ y = 10\sin\left(\dfrac{t}{2}\right) \end{cases}$$

are called **parametric equations** for the path of the object. The variable t is called the **parameter**.

More generally, parametric equations are of the form

$$\begin{cases} x = f(t) \\ y = g(t) \end{cases}$$

where f and g are functions of the independent variable t. If no domain for t is specified, we will assume it to be the largest subset of the real numbers $\mathbb{R}$ for which both x and y are real numbers. As t varies over the domain, values of x and y are determined by the given parametric equations. In general, we will restrict the domain so that the curve is traced out only one time. The set of all pairs (x, y) determined in this way is the graph of the pair of parametric equations. In the example of the object moving on a circular path, we knew the graph at the outset, and our objective was to determine parametric equations for x and y. In the next example, we are given parametric equations and asked to determine the graph.

EXAMPLE 9.1   Make a table of values and draw the graph of the curve defined by the parametric equations

$$\begin{cases} x = t^2 \\ y = 2t - 1 \end{cases}$$

Solution   Since both x and y are defined for all real t, the domain is all of $\mathbb{R}$. We construct a table by assigning arbitrary values to t, thus determining the corresponding x and y values.

t	0	1	2	3	4	−1	−2	−3	−4
x	0	1	4	9	16	1	4	9	16
y	−1	1	3	5	7	−3	−5	−7	−9

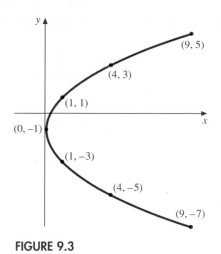

FIGURE 9.3

Now we plot the points (x, y) and connect them with a smooth curve, with the result shown in Figure 9.3. The curve appears to be a parabola with a horizontal axis. We will confirm this observation shortly.

Note that the parameter t does not appear on the graph. It is a means to an end only. The objective is to find x and y. ■

The letter t is customarily used as a parameter, because, as in our introductory example, parametric equations often result from a consideration of the location of a moving object at time t. Other letters can be used for the parameter, however. For example, parametric equations of the circle given by $x^2 + y^2 = a^2$ can be obtained as shown in Figure 9.4 by using the central angle θ as a parameter. Since the x-coordinate of a point on the circle is $a \cos \theta$ and the y-coordinate is $a \sin \theta$, the parametric equations are

$$\begin{cases} x = a \cos \theta \\ y = a \sin \theta \end{cases} \qquad 0 \le \theta \le 2\pi$$

The circle is traced out in a counterclockwise direction (as indicated by the arrows), starting at $(a, 0)$, as θ varies from 0 to 2π. (If θ is allowed to vary from 0 to 4π, the circle is traced two complete times.)

A curve defined parametrically is assigned a direction, or **orientation**, that is determined by the way points on the curve are traced out as the parameter increases.

Until now, we have relied on our intuitive understanding of what we mean by a curve. Now we give a precise definition.

Definition 9.1 A Plane Curve	A **curve** in the plane is a set of points (x, y) such that $x = f(t)$ and $y = g(t)$, where t is in an interval I and f and g are continuous on I.

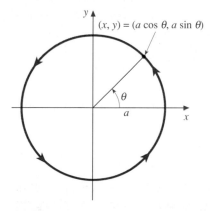

FIGURE 9.4
The circle given by $x^2 + y^2 = a^2$ has parametric equations

$$\begin{cases} x = a \cos \theta \\ y = a \sin \theta \end{cases} \qquad 0 \le \theta \le 2\pi$$

The key word in this definition is *continuous*. Its significance is that small changes in t produce small changes in x and y, thus assuring that there are no breaks in the graph.

Consider again the parametric equations

$$\begin{cases} x = t^2 \\ y = 2t - 1 \end{cases} \qquad -\infty < t < \infty$$

These equations define a curve, since both $f(t) = t^2$ and $g(t) = 2t - 1$ are continuous for all values of t.

Eliminating the Parameter

Sometimes we can eliminate the parameter between two parametric equations to find how x and y are related. We refer to the result as the **rectangular equation** of the curve. In the next two examples we illustrate some typical ways of eliminating the parameter.

EXAMPLE 9.2   Find the rectangular equation of the curve defined parametrically by the given equations:

(a) $\begin{cases} x = t^2 \\ y = 2t - 1 \end{cases}$ $-\infty < t < \infty$

(b) $\begin{cases} x = 2\cos\theta \\ y = 3\sin\theta \end{cases}$ $0 \le \theta \le 2\pi$

Solution

(a) We solve the second equation for t, getting

$$t = \frac{y + 1}{2}$$

Now we substitute this value of t in the first equation to get

$$x = \frac{(y + 1)^2}{4} \qquad \text{or} \qquad (y + 1)^2 = 4x$$

which is the equation of a parabola with vertex $(0, -1)$, opening to the right, confirming our observation about the graph in Figure 9.3.

(b) To eliminate θ, we make use of the identity $\sin^2\theta + \cos^2\theta = 1$. We divide the first equation by 2 and the second by 3, and then square and add:

$$\left(\frac{x}{2}\right)^2 = \cos^2\theta$$
$$\frac{\left(\frac{y}{3}\right)^2 = \sin^2\theta}{\frac{x^2}{4} + \frac{y^2}{9} = 1}$$

The graph of this equation is the ellipse shown in Figure 9.5, with x-intercepts ± 2 and y-intercepts ± 3. The curve is traced out in a counterclockwise direction, starting at $(2, 0)$ when $\theta = 0$ and terminating at the same point when $\theta = 2\pi$.   ∎

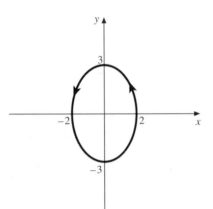

FIGURE 9.5
The graph of
$$\begin{cases} x = 2\cos\theta \\ y = 3\sin\theta \end{cases} \quad 0 \le \theta \le 2\pi$$
is the ellipse given by $x^2/4 + y^2/9 = 1$.

EXAMPLE 9.3   Find the rectangular equation of each of the following, and draw the graph:

(a) $\begin{cases} x = e^t \\ y = e^{-t} \end{cases}$

(b) $\begin{cases} x = t^2 \\ y = t^4 + 2 \end{cases}$

Solution

(a) Since $y = 1/e^t$ and $x = e^t$, we have

$$y = \frac{1}{x}$$

as the rectangular equation. Its graph is a hyperbola with asymptotes $x = 0$ and $y = 0$.

Note that the graph of this rectangular equation consists of two distinct branches, one in the first quadrant and one in the third (see Figure 9.6). However, the parametric equations require that both x and y be positive, since $e^t > 0$ and $e^{-t} > 0$ for all t. So the graph determined by the parametric equations is the right-hand branch only, shown in red in Figure 9.6. Since e^t is an increasing function, we see that x increases and y decreases with t, giving the orientation shown in the figure.

(b) Since $x = t^2$, we have $t^4 = x^2$, so on substituting x^2 for t^4 in the second equation, we get $y = x^2 + 2$ as the rectangular equation. Its graph is an upward-opening parabola with vertex $(0, 2)$. But here, again, the parametric equations impose a limitation not present in the rectangular equation, namely, that x be nonnegative. (After all, $x = t^2$, which is nonnegative.) We therefore have only the right half of the parabola $y = x^2 + 2$, as shown in red in Figure 9.7. ∎

FIGURE 9.6
The graph of

$$\begin{cases} x = e^t \\ y = e^{-t} \end{cases}$$

is the right-hand branch only of the hyperbola $y = 1/x$.

FIGURE 9.7
The graph of

$$\begin{cases} x = t^2 \\ y = t^4 + 2 \end{cases}$$

is the right half of the parabola $y = x^2 + 2$.

As the preceding example makes clear, when you eliminate the parameter between two parametric equations, it is essential that you check the original equations for any restrictions imposed on x and y, and that you take these into account when you draw the graph.

The Cycloid

We conclude this section by deriving parametric equations of a curve known as a **cycloid**. A cycloid can be described as the path traced out by a point on a circle as the circle rolls along a line. For example, if a point is marked on the outer edge of a bicycle tire and the bicycle moves along a straight path, the point traces out a cycloid. (See Figure 9.8.) The cycloid has many interesting properties, two of which we will mention after we derive the equations.

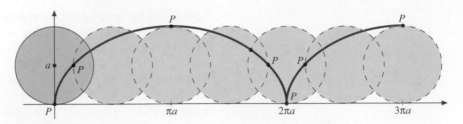

FIGURE 9.8
A cycloid generated by a circle rolling on a line.

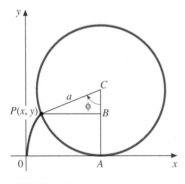

FIGURE 9.9

In Figure 9.9 we show the moving point P, originally at the origin, after the circle has rolled to the right. The angle ϕ (in radians) shown will be our parameter. Our objective is to express both x and y in terms of ϕ. Let the radius of the circle be a. From the figure we see that the x- and y-coordinates of P can be expressed as

$$x = \overline{OA} - \overline{PB} \qquad \text{and} \qquad y = \overline{AC} - \overline{BC}$$

Using the right triangle PBC, we have

$$\overline{PB} = a \sin \phi \qquad \text{and} \qquad \overline{BC} = a \cos \phi$$

Also, $\overline{AC} = a$. The distance $\overline{OA}$ is equal to the arc length $\widehat{AP}$. (Think of rolling the circle back to its starting position; the arc $\widehat{AP}$ would "unroll" as the distance $\overline{OA}$.) Since the arc length $\widehat{AP}$ equals $a\phi$ (because arc length equals the central angle in radians times the radius), we therefore have

$$\overline{OA} - \overline{PB} = a\phi - a \sin \phi \qquad \text{and} \qquad \overline{AC} - \overline{BC} = a - a \cos \phi$$

Thus, the parametric equations are

$$\begin{cases} x = a(\phi - \sin \phi) \\ y = a(1 - \cos \phi) \end{cases} \tag{9.1}$$

where ϕ can be any real number.

We can obtain the graph of the cycloid from Equations 9.1 by assigning values to ϕ. Finding the rectangular equation in this case is difficult (see Exercise 29 in Exercise Set 9.1), and it is not very helpful in drawing the graph. We show the graph in Figure 9.10. Note that the graph is periodic, with period equal to $2\pi a$, which is the circumference of the generating circle. Its maximum height is the diameter $2a$ of the circle.

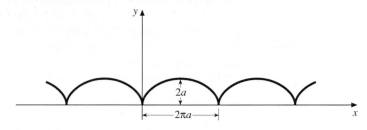

FIGURE 9.10
The cycloid

$$\begin{cases} x = a(\phi - \sin\phi) \\ y = a(1 - \cos\phi) \end{cases}$$

EXAMPLE 9.4   Find the area under one arch of the cycloid.

Solution   From Figure 9.10 we see that one arch is traced out as x goes from 0 to $2\pi a$. Thus, the area A is given by

$$A = \int_0^{2\pi a} y\, dx$$

To evaluate the integral, we use Equations 9.1, replacing y with $a(1 - \cos\phi)$ and dx with the differential of $a(\phi - \sin\phi)$, namely, $dx = a(1 - \cos\phi)\, d\phi$. The limits on ϕ are from $\phi = 0$ to $\phi = 2\pi$. So we have

$$A = \int_0^{2\pi} \overbrace{a(1 - \cos\phi)}^{y} \cdot \overbrace{a(1 - \cos\phi)\, d\phi}^{dx}$$

$$= a^2 \int_0^{2\pi} (1 - \cos\phi)^2\, d\phi$$

$$= a^2 \int_0^{2\pi} (1 - 2\cos\phi + \cos^2\phi)\, d\phi$$

$$= a^2 \int_0^{2\pi} [1 - 2\cos\phi + \tfrac{1}{2}(1 + \cos 2\phi)]\, d\phi$$

$$= a^2 \left[\frac{3\phi}{2} - 2\sin\phi + \frac{\sin 2\phi}{4} \right]_0^{2\pi} = 3\pi a^2$$

Thus, the area is three times the area inside the generating circle. Galileo guessed this result. It was first proved by Roberval around 1634 and was also proved independently by Torricelli in 1644. ■

Two famous problems in mechanics are known as the **Brachistochrone Problem** and the **Tautochrone Problem**. The brachistochrone is the "curve of quickest descent." Suppose, for example, a bead is allowed to slide (without

FIGURE 9.11

FIGURE 9.12

friction) along a wire from a point A to a lower point B, as in Figure 9.11. In what shape should the wire be bent so that the time required for the bead to go from A to B is a minimum? Surprisingly, the answer is not a straight line (the shortest distance) but is an inverted cycloid in which A is on the x-axis at a cusp of the cycloid and B is at the bottom of the arch. In effect, the bead picks up more speed by descending relatively quickly at first. This discovery was made independently by a number of seventeenth-century mathematicians (see the historical note).

The tautochrone is the "curve of equal time." Suppose beads are initially placed on a wire joining A and B, as in the Brachistochrone Problem, and then released. What should be the shape of the wire so that all the beads arrive at B at the same time, no matter how far down the wire they start? (See Figure 9.12.) Again, the answer is the inverted cycloid. This result is quite remarkable, for even if one of the beads starts very near to B and another starts at A, when the curve is a cycloid, they will arrive at the same instant. Dutch physicist and astronomer Christian Huygens (1629–1695) made this discovery in his effort to construct a pendulum clock in which the period of the pendulum is independent of the amplitude of its swing.

The brachistochrone problem represents the historical first of a new type of problem. For previous minimization problems, the object being minimized depended on only a few variables. In the brachistochrone problem, time is being minimized and it depends on the length of the whole curve. This approach introduced a whole new field of study that came to be known as the *calculus of variations*. This technique has many applications in physical science.

Historical Note: The Brachistochrone Problem

During the seventeenth and eighteenth centuries, mathematicians sometimes issued challenges to one another. In 1697, Johann Bernoulli had solved and published the question of finding the path by which a heavy body would descend most quickly from one position to another, given that the second position was not directly below the first. After the six-month deadline had passed, no answer had been given. Leibniz had written that he had solved the problem but suggested that the time be extended by a few months. Bernoulli added a second problem, sent the two problems for publication, and sent copies directly to both Newton and to one other person. Bernoulli and Leibniz apparently thought that the problem had stumped Newton and wanted to show their superiority over him.

Newton's recent biographer Richard Westfall concludes that Newton thought—probably correctly—that a direct challenge was being issued to his status as a mathematician. Newton wrote, "I do not love...to be...teezed by forreigners about Mathematical things...." But he had solved the problem by the next day. He wrote, "When the problem in 1697 was sent by Bernoulli-Sr I. N. was in the midst of the hurry of the great recoinage did not come home till four from the Tower very much tired, but did not sleep till he had solved it s^{ch} was by 4 in the morning."

Bernoulli received three solutions: one from Leibniz, one from L'Hôpital, and an anonymous one from England. Bernoulli immediately realized that this last solution must be from Newton. He said: "I recognize the lion from his claw" (*"tanquam ex ungue leonem"*).

Drawn from *Never at Rest*, by Richard S. Westfall, Cambridge University Press, 1980.

The Brachistochrone Problem and its solution in Newton's handwriting.

SKETCHING PARAMETRIC CURVES USING COMPUTER ALGEBRA SYSTEMS

Parametric representation of curves provides a method for describing a great variety of curves in the plane. The computer algebra systems Maple, Mathematica, and DERIVE have specific notations for easily plotting parametric curves.

 ### CAS 31

The Cycloid: Sketch the cycloid given by the parametric equations

$$x = 2(t - \sin t), \quad y = 2(1 - \cos t)$$

for $-10 \le t \le 10$.

FIGURE 9.1.1
$x = 2(t - \sin t), y = 2(1 - \cos t),$
$-10 \le t \le 10$

Maple:

plot([2*(t–sin(t)),
   2*(1–cos(t)),t=–10..10],
   numpoints=500);

Mathematica:

ParametricPlot[{2*(t–Sin[t]),
   2*(1–Cos[t])},
   {t,–10,10}]

DERIVE:

(At the □ symbol, go to the next step.)
a (author) □ [2(t–sin(t)),2(1–cos(t))]
□ p (plot window) □ s (scale) □ 3 (x-scale) □ 2 (y-scale) □ –10 (min) □ 10 (max)

Notice that in Maple and DERIVE, the parametric equations are enclosed in square brackets and that in Mathematica they are enclosed in curly brackets. Also, in Maple, the specification for the range of the parameter is given inside the square brackets.

CAS 32

Sketch the curve given by the parametric equations

$$x = \cos t(1 - 2\sin t), \quad y = \sin t(1 - 2\sin t)$$

for $0 \le t \le 2\pi$.

Maple:

x := t–> cos(t)*(1–2*sin(t));
y := t–> sin(t)*(1–2*sin(t));
plot([x(t),y(t),t=0..2*Pi]);

Mathematica:

x = Cos[t]*(1–2*Sin[t])
y = Sin[t]*(1–2*Sin[t])
ParametricPlot[{x,y},
  {t,0,2*Pi}]

DERIVE:

(At the □ symbol, go to the next step.)
(author) □ [cos(t)(1−2sin(t)), sin(t)
(1−2sin(t))] □ p (plot window) □ p
(plot) □ 0 (min) □ 2Pi (max)

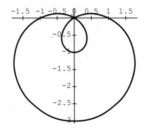

FIGURE 9.1.2
$x = \cos t(1 - 2\sin t)$,
$y = \sin t(1 - 2\sin t)$, $0 \le t \le \pi$

Exercise Set 9.1

In Exercises 1–6, draw the curve defined by the parametric equations by substituting values for the parameter. Indicate the orientation on the graph by the use of arrows.

1. $\begin{cases} x = 2t + 3 \\ y = 3t - 1 \end{cases}$

2. $\begin{cases} x = t^2 - 1 \\ y = t^2 + 1 \end{cases}$ $-3 \le t \le 0$

3. $\begin{cases} x = t - 2 \\ y = t^2 \end{cases}$ $-2 \le t \le 3$

4. $\begin{cases} x = u^2 \\ y = u^3 \end{cases}$

5. $\begin{cases} x = \dfrac{1}{v} \\ y = v \end{cases}$ $v > 0$

6. $\begin{cases} x = \sqrt{t} \\ y = t \end{cases}$ $t > 0$

In Exercises 7–20, eliminate the parameter and make use of the rectangular equation to draw the given curve. Indicate the orientation.

7. $\begin{cases} x = 3t - 2 \\ y = 1 - t \end{cases}$ $-1 \le t \le 4$

8. $\begin{cases} x = 4 - t^2 \\ y = t^2 + 2 \end{cases}$ $0 \le t \le 3$

9. $\begin{cases} x = \cos t \\ y = -\sin t \end{cases}$ $0 \le t \le \pi$

10. $\begin{cases} x = 1 - \cos\theta \\ y = \sin\theta \end{cases}$ $0 \le \theta \le 2\pi$

11. $\begin{cases} x = \cosh t \\ y = \sinh t \end{cases}$

12. $\begin{cases} x = 2t - 1 \\ y = 1 - t^2 \end{cases}$ $t \ge -1$

13. $\begin{cases} x = \sin^2\theta + 1 \\ y = \cos\theta \end{cases}$ $0 \le \theta \le \pi$

14. $\begin{cases} x = 3\sin\theta \\ y = 4\cos\theta \end{cases}$ $0 \le \theta \le 2\pi$

15. $\begin{cases} x = \sec\theta \\ y = \tan\theta \end{cases}$ $-\dfrac{\pi}{2} < \theta < \dfrac{\pi}{2}$

16. $\begin{cases} x = 2\cosh t \\ y = 3\sinh t \end{cases}$

17. $\begin{cases} x = 2 - \cos t \\ y = 1 + 3\sin t \end{cases}$ $0 \le t \le 2\pi$

18. $\begin{cases} x = \sqrt{4 - t} \\ y = \sqrt{t} \end{cases}$ $0 \le t \le 4$

19. $\begin{cases} x = \sin\theta \\ y = \cos 2\theta \end{cases}$ $0 \le \theta \le \dfrac{\pi}{2}$

20. $\begin{cases} x = 2\sec^2\theta - 3 \\ y = \tan\theta + 1 \end{cases}$ $0 \le \theta < \dfrac{\pi}{2}$

21. Compare the curves C_1, C_2, C_3, and C_4. Include a discussion of orientation.

C_1: $\begin{cases} x = 1 + 2t \\ y = -2 + 4t \end{cases}$ $0 \le t \le 1$

C_2: $\begin{cases} x = 3 - 2t \\ y = 2 - 4t \end{cases}$ $0 \le t \le 1$

C_3: $\begin{cases} x = t \\ y = 2t - 4 \end{cases}$ $1 \le t \le 3$

C_4: $\begin{cases} x = \tan^2 t \\ y = 2\sec^2 t - 6 \end{cases}$ $\dfrac{\pi}{4} \le t \le \dfrac{\pi}{3}$

22. If a curve crosses itself, the point where this occurs is called a **double point**. Show that the curve C defined by

$$\begin{cases} x = t^3 - 3t + 1 \\ y = t^2 + t - 1 \end{cases} \quad -\infty < t < \infty$$

has a double point, and find this point. What are the two distinct values of t that produce the double point?

23. If a curve has no double points (see Exercise 22), it is said to be a **simple curve.** Show that the curve C defined by

$$\begin{cases} x = t^3 + 1 \\ y = t^2 - 1 \end{cases} \quad 0 \le t \le 2$$

is simple. Draw its graph.

24. A curve C defined by $x = f(t)$ and $y = g(t)$, for $a \le t \le b$, is said to be a **simple closed curve** if it is simple for $a \le t < b$ (see Exercise 23) and $(f(a), g(a))$ is the same point as $(f(b), g(b))$. Show that the curve defined by

$$\begin{cases} x = t^2 - 3 \\ y = t^3 - 4t - 1 \end{cases} \quad -2 \le t \le 2$$

is a simple closed curve.

In Exercises 25–28, draw each curve, showing its orientation.

25. $\begin{cases} x = t^2 - 1 \\ y = t^3 - t \end{cases}$ $-\infty < t < \infty$

26. $\begin{cases} x = \cos^3\theta \\ y = \sin^3\theta \end{cases}$ $0 \le \theta \le 2\pi$

27. $\begin{cases} x = \cos^2\theta \\ y = \frac{1}{2}\sin 2\theta \end{cases}$ $0 \le \theta \le \pi$

28. $\begin{cases} x = (t^2 - 2)(t + 1) \\ y = (t^2 - 2)(t - 1) \end{cases}$ $-\infty < t < \infty$

29. Find the rectangular equation of the cycloid.

30. A **hypocycloid** is a curve traced out by a fixed point on a circle rolling on the inside of a fixed larger circle. (See the accompanying figure.) If the radius of the larger circle is a and the radius of the smaller circle is b, derive the parametric equations

$$\begin{cases} x = (a-b)\cos\theta + b\cos\dfrac{a-b}{b}\theta \\[2mm] y = (a-b)\sin\theta - b\sin\dfrac{a-b}{b}\theta \end{cases} \quad 0 \le \theta \le 2\pi$$

(*Hint:* Observe that $\overset{\frown}{AB} = \overset{\frown}{BP}$, so that $a\theta = b\phi$.)

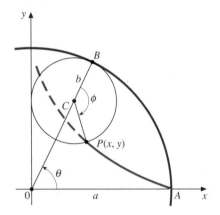

31. Use a CAS or a graphing calculator to obtain the graph of the hypocycloid (see Exercise 30) for each of the following values of a and b.
   (a) $a = 4$, $b = 1$
   (b) $a = 6$, $b = 2$
   (c) $a = 8$, $b = 1$

32. Refer to Exercise 30 and show that if $a = 4b$ the equations can be put in the form

$$\begin{cases} x = a\cos^3\theta \\ y = a\sin^3\theta \end{cases}$$

Also find the rectangular equation. (This is the four-cusp hypocycloid.) Use a CAS or graphing calculator to obtain its graph. How does changing a affect the graph?

33. If the smaller circle in Exercise 30 rolls on the outer circumference of the larger circle, the point traces out an **epicycloid**. (See the accompanying figure.) Derive the following parametric equations of the epicycloid:

$$\begin{cases} x = (a+b)\cos\theta - b\cos\left(\dfrac{a+b}{b}\right)\theta \\[2mm] y = (a+b)\sin\theta - b\sin\left(\dfrac{a+b}{b}\right)\theta \end{cases} \quad 0 \le \theta \le 2\pi$$

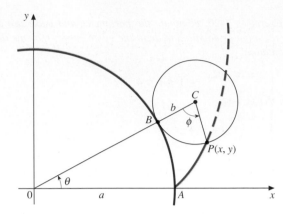

Use a CAS or graphing calculator to sketch the epicycloid for $a = 4$ and $b = 1$.

34. Suppose a point P is attached to an extended spoke of a wheel as in the figure and the wheel rolls along the x-axis. If the distance from the center of the wheel to P is b and the radius of the wheel is a (with $a < b$), find parametric equations of the path of P. This curve is called a **trochoid**. (See the accompanying figure.)

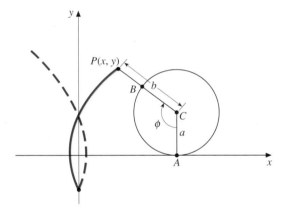

35. Find parametric equations for the x- and y-coordinates of a point that traces out the circle $x^2 + y^2 = r^2$ in the following manner:
   (a) once around, counterclockwise, starting at $(r, 0)$;
   (b) twice around, counterclockwise, starting at $(r, 0)$;
   (c) twice around, clockwise, starting at $(r, 0)$;
   (d) three times around, clockwise, starting at $(0, r)$;
   (e) three times around, counterclockwise, starting at $(-r, 0)$.

36. Find parametric equations for the x- and y-coordinates of a point that traces out the ellipse $x^2/a^2 + y^2/b^2 = 1$ in the following manner:
   (a) once around, counterclockwise, starting at $(a, 0)$;
   (b) twice around, counterclockwise, starting at $(a, 0)$;

(c) twice around, clockwise, starting at $(a, 0)$;

(d) three times around, clockwise, starting at $(0, b)$;

(e) three times around, counterclockwise, starting at $(-a, 0)$.

37. Verify that

$$\begin{cases} x = x_1 + (x_2 - x_1)t \\ y = y_1 + (y_2 - y_1)t \end{cases} \quad 0 \le t \le 1$$

are parametric equations for the straight line segment from (x_1, y_1) to (x_2, y_2).

38. Find parametric equations for the line passing through the points (x_1, y_1) and (x_2, y_2).

39. Describe the difference between the curves defined by $x = \cos t$, $y = \sin t$, where $-\pi \le t \le \pi$, and $x = \sin t$, $y = \cos t$, where $-\pi \le t \le \pi$.

 40. Use a CAS or graphing calculator to investigate the family of curves

$$\begin{cases} x = \cos t (a - b \sin t) \\ y = \sin t (a - b \sin t) \end{cases} \quad 0 \le t \le 2\pi$$

where a and b are real numbers. Consider cases where $a < b$, $a = b$, and $a > b$.

 41. Repeat Exercise 40 with

$$\begin{cases} x = \cos t (a - b \cos t) \\ y = \sin t (a - b \cos t) \end{cases} \quad 0 \le t \le 2\pi$$

42. Use a CAS or graphing calculator to investigate the family of curves given by

$$\begin{cases} x = \cos t (1 - \sin(at)) \\ y = \sin t (1 - \sin(at)) \end{cases} \quad 0 \le t \le 2\pi$$

where a is a real number.

43. Find parametric equations for a circle of radius r with a center at (h, k).

9.2 DERIVATIVES OF FUNCTIONS DEFINED PARAMETRICALLY

We begin this section by considering the question of how to find the slope (when it exists) of the tangent line to a curve given parametrically by

$$x = f(t), \quad y = g(t) \quad (a \le t \le b)$$

We saw in the previous section that it is sometimes possible, by eliminating the parameter, to write y as a function of x. Doing so explicitly is often difficult, however, and sometimes it is impossible. In many instances, it can be concluded that y is a differentiable function of x, even when the explicit function is unknown. In the remark following this discussion, we outline some general conditions that guarantee y is a differentiable function of x.

If we assume for the present discussion that y is a differentiable function of x, so that dy/dx exists, and if both dx/dt and dy/dt exist, then the Chain Rule gives

$$\frac{dy}{dt} = \frac{dy}{dx} \frac{dx}{dt}$$

If $dx/dt \ne 0$, we can solve for dy/dx to get

$$\frac{dy}{dx} = \frac{\dfrac{dy}{dt}}{\dfrac{dx}{dt}} \quad \left(\frac{dx}{dt} \ne 0\right) \tag{9.2}$$

Note that formally if we treat dy and dx as differentials, with $dx \neq 0$, we could get Formula 9.2 by dividing the numerator and denominator of dy/dx by dt.

If $dy/dt = 0$ and $dx/dt \neq 0$, then by Formula 9.2 the slope of the tangent line is 0, so the curve has a horizontal tangent. Reversing the roles of x and y, when $dx/dt = 0$ and $dy/dt \neq 0$, the curve has a vertical tangent. When both $dx/dt = 0$ and $dy/dt = 0$, the curve is said to have a **singular point**. The curve may or may not have a tangent line at a singular point. A curve is said to be **smooth** if dx/dt and dy/dt are continuous and there are no singular points. A smooth curve has a continuously turning tangent line. There are no sharp corners.

REMARK

■ It can be shown that if the curve defined by the parametric equations $x = f(t)$, $y = g(t)$, for $a \leq t \leq b$, is smooth and $f'(t) \neq 0$ for $a \leq t \leq b$, then y can be expressed as a differentiable function of x for $f(a) \leq x \leq f(b)$, say $y = F(x)$. Then, since $y = g(t)$ and $x = f(t)$, we have $g(t) = F(f(t))$. So by the Chain Rule,

$$g'(t) = F'(f(t)) f'(t) = F'(x) f'(t)$$

Thus, since $f'(t) \neq 0$,

$$F'(x) = \frac{g'(t)}{f'(t)}$$

or, equivalently,

$$\frac{dy}{dx} = \frac{\dfrac{dy}{dt}}{\dfrac{dx}{dt}}$$

That is, Formula 9.2 is true under the hypotheses stated.

EXAMPLE 9.5   Find dy/dx when $t = 2$ for the curve C defined parametrically by

$$\begin{cases} x = 2t^2 - 1 \\ y = t^3 - 3t \end{cases} \quad 1 \leq t \leq 4$$

Solution   Since $dx/dt = 4t$ and $dy/dt = 3t^2 - 3$, we see that these derivatives are continuous and not simultaneously 0 in the interval $[1, 4]$. So C is smooth. Also $dx/dt \neq 0$ for t in the interval $[1, 4]$. Thus, by Equation 9.2,

$$\frac{dy}{dx} = \frac{\dfrac{dy}{dt}}{\dfrac{dx}{dt}} = \frac{3t^2 - 3}{4t}$$

When $t = 2$, we obtain

$$\left. \frac{dy}{dx} \right|_{t=2} = \frac{9}{8}$$

EXAMPLE 9.6   Find the equation of the tangent line to the curve C defined by

$$\begin{cases} x = \dfrac{4}{t} \\ y = \sqrt{t} \end{cases} \quad 1 \le t \le 9$$

at the point $(1, 2)$.

Solution   We have, by Equation 9.2,

$$\frac{dy}{dx} = \frac{\dfrac{dy}{dt}}{\dfrac{dx}{dt}} = \frac{\dfrac{1}{2\sqrt{t}}}{-\dfrac{4}{t^2}} = -\frac{t^{3/2}}{8}$$

The value of t that produces the point $(1, 2)$ is $t = 4$. Thus, at this point,

$$\frac{dy}{dx}\bigg|_{t=4} = -\frac{(4)^{3/2}}{8} = -1$$

Using the point-slope form of the equation of a line, we obtain the tangent line

$$y - 2 = -(x - 1)$$

or

$$x + y = 3 \qquad \blacksquare$$

Point-slope form of the equation of a line:

$$y - y_1 = m(x - x_1)$$

The Second Derivative from Parametric Equations

As we have previously seen, second derivatives are useful in a variety of ways, such as in determining the concavity of a curve or the acceleration of a moving particle. Since $y'' = dy'/dx$, we can find the second derivative of y with respect to x, if it exists, by replacing y with y' in Equation 9.2. That is,

$$y'' = \frac{dy'}{dx} = \frac{\dfrac{dy'}{dt}}{\dfrac{dx}{dt}} \qquad \left(\frac{dx}{dt} \ne 0\right) \qquad (9.3)$$

We illustrate this way of calculating y'' in the next two examples.

EXAMPLE 9.7   Find y'' for the function defined in Example 9.6.

Solution   In Example 9.6 we found that

$$y' = \frac{dy}{dx} = -\frac{t^{3/2}}{8}$$

Then, since

$$\frac{dx}{dt} = -\frac{4}{t^2}$$

we have, by Equation 9.3,

$$y'' = \frac{dy'}{dx} = \frac{\dfrac{dy'}{dt}}{\dfrac{dx}{dt}} = \frac{-\dfrac{3t^{1/2}}{16}}{-\dfrac{4}{t^2}} = \frac{3t^{5/2}}{64}$$

■

EXAMPLE 9.8   For the cycloid

$$\begin{cases} x = a(t - \sin t) \\ y = a(1 - \cos t) \end{cases} \quad -\infty < t < \infty$$

do the following:

(a) Find $\dfrac{dy}{dx}$ and $\dfrac{d^2 y}{dx^2}$.

(b) Find all points where the tangent line is horizontal.

(c) Discuss concavity.

(d) Show that the origin is a singular point but that the curve has a vertical tangent line there.

Solution

(a) Since

$$\frac{dx}{dt} = a(1 - \cos t) \qquad \text{and} \qquad \frac{dy}{dt} = a \sin t$$

we have, by Equation 9.2,

$$y' = \frac{dy}{dx} = \frac{\dfrac{dy}{dt}}{\dfrac{dx}{dt}} = \frac{a \sin t}{a(1 - \cos t)} = \frac{\sin t}{1 - \cos t}$$

provided $\cos t \neq 1$. Similarly, by Equation 9.3,

$$\frac{d^2 y}{dx^2} = \frac{dy'}{dx} = \frac{\dfrac{dy'}{dt}}{\dfrac{dx}{dt}} = \frac{\dfrac{\cos t(1 - \cos t) - \sin t(\sin t)}{(1 - \cos t)^2}}{a(1 - \cos t)}$$

$$= \frac{\cos t - \cos^2 t - \sin^2 t}{a(1 - \cos t)^3}$$

$$= \frac{\cos t - 1}{a(1 - \cos t)^3} \qquad \text{Since } \cos^2 t + \sin^2 t = 1$$

$$= \frac{-1}{a(1 - \cos t)^2}$$

provided again that $\cos t \neq 1$.

(b) The tangent line is horizontal where $dy/dx = 0$, or equivalently where $dy/dt = 0$ and $dx/dt \neq 0$. Setting $dy/dt = 0$ gives

$$a \sin t = 0$$

$$\sin t = 0$$

This equation is true when t is any integral multiple of π. But we must also have $dx/dt \neq 0$, that is,

$$a(1 - \cos t) \neq 0$$

$$\cos t \neq 1$$

Thus, t cannot be an *even* multiple of π (i.e., $0, \pm 2\pi, \pm 4\pi, \ldots$). We conclude that the tangent line is horizontal when t is an odd multiple of π (i.e., $\pm \pi, \pm 3\pi, \pm 5\pi, \ldots$), which we can write as

$$t = (2n + 1)\pi \qquad n = 0, \pm 1, \pm 2, \ldots$$

At each such value of t, $\sin t = 0$ and $\cos t = -1$. Thus, the points on the curve are

$$(a(2n + 1)\pi, 2a) \qquad n = 0, \pm 1, \pm 2, \ldots$$

(c) Since

$$\frac{d^2 y}{dx^2} = \frac{-1}{a(1 - \cos t)^2}$$

is negative for all t except where $\cos t = 1$, the graph is concave down everywhere except for points corresponding to these values of t.

(d) When $t = 0$, we have $x = a(0 - \sin 0) = 0$ and $y = a(1 - \cos 0) = 0$. So the origin corresponds to $t = 0$. Also, at this point

$$\frac{dx}{dt} = a(1 - \cos 0) = 0 \qquad \text{and} \qquad \frac{dy}{dt} = a \sin 0 = 0$$

So the origin is a singular point. For t near 0 but not equal to 0, we have, by part (a),

$$\frac{dy}{dx} = \frac{\sin t}{1 - \cos t}$$

Using L'Hôpital's Rule, we have

$$\lim_{t \to 0^+} \frac{\sin t}{1 - \cos t} = \lim_{t \to 0^+} \frac{\cos t}{\sin t} = \infty$$

Similarly,

$$\lim_{t \to 0^-} \frac{\sin t}{1 - \cos t} = \lim_{t \to 0^-} \frac{\cos t}{\sin t} = -\infty$$

Thus, as t approaches 0 from the right, the slopes become larger and larger. That is, the tangent lines become more and more nearly vertical. As we approach from the left, the slopes are negative but large in absolute value, so again the tangent lines are nearly vertical. We illustrate this situation in Figure 9.13. The curve has a cusp at the origin. By periodicity it also has cusps at the points $(2n\pi a, 0)$ for $n = \pm 1, \pm 2, \ldots$.

It is easy to see that our conclusions are in agreement with the graph of the cycloid in Figure 9.10. ∎

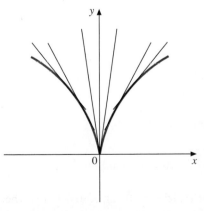

FIGURE 9.13
As $x \to 0^+$, slopes $\to \infty$. As $x \to 0^-$, slopes $\to -\infty$. When $x = 0$, tangent line is vertical.

Higher-Order Derivatives

To compute higher-order derivatives, we follow a procedure similar to the one used for y''. For example,

$$y''' = \frac{dy''}{dx} = \frac{\dfrac{dy''}{dt}}{\dfrac{dx}{dt}} \qquad \left(\frac{dx}{dt} \neq 0\right)$$

and in general, for $n \geq 1$,

$$y^{(n+1)} = \frac{dy^{(n)}}{dx} = \frac{\dfrac{dy^{(n)}}{dt}}{\dfrac{dx}{dt}} \qquad \left(\frac{dx}{dt} \neq 0\right)$$

provided the derivative exists. Note that in every case we divide by dx/dt and so must have this derivative nonzero.

Exercise Set 9.2

In Exercises 1–10, find dy/dx at the indicated point without eliminating the parameter.

1. $\begin{cases} x = t^2 - 1 \\ y = 2t + 3 \end{cases}$ at $t = 1$

2. $\begin{cases} x = \sqrt{t - 1} \\ y = t^2 \end{cases}$ at $t = 2$

3. $\begin{cases} x = t^3 \\ y = t^2 \end{cases}$ at $t = -2$

4. $\begin{cases} x = 2\sin t \\ y = 3\cos t \end{cases}$ at $t = \dfrac{\pi}{3}$

5. $\begin{cases} x = 4\sin^2 t - 3 \\ y = 2\cos t \end{cases}$ at $t = \dfrac{\pi}{3}$

6. $\begin{cases} x = 2\sinh t \\ y = \tanh t \end{cases}$ at $t = 0$

7. $\begin{cases} x = e^{-t} \\ y = 1 + e^t \end{cases}$ at $t = \ln 2$

8. $\begin{cases} x = \ln t \\ y = 2 - \ln\sqrt{t} \end{cases}$ at $t = e^2$

9. $\begin{cases} x = t^3 - 2t \\ y = \sqrt{t - 1} \end{cases}$ at $(4, 1)$

10. $\begin{cases} x = \dfrac{1}{t} \\ y = 4(1 - t^2) \end{cases}$ at $(-2, 3)$

In Exercises 11–14, find the equation of the tangent line to the curve defined by the parametric equations at the given point.

11. $\begin{cases} x = 3t^2 - 2t \\ y = t^2 + t - 1 \end{cases}$ at $t = -1$

12. $\begin{cases} x = \dfrac{1}{t - 1} \\ y = \dfrac{1}{t + 1} \end{cases}$ at $t = \frac{1}{2}$

13. $\begin{cases} x = t^3 + 3t^2 \\ y = 2t^2 - 3 \end{cases}$ at $(2, -1)$

14. $\begin{cases} x = t - \dfrac{2}{t} \\ y = t^2 \end{cases}$ at $(1, 4)$

15. Find all points on the curve C defined by

$$\begin{cases} x = t^3 - 4t^2 - 3t \\ y = 2t^2 + 3t - 5 \end{cases} \quad -\infty < t < \infty$$

where the tangent line is vertical and all points where the tangent line is horizontal.

16. (a) Show that the curve C defined by

$$\begin{cases} x = t^2 - 2t - 8 \\ y = t^2 - t - 2 \end{cases} \quad -\infty < t < \infty$$

is smooth.

(b) In what intervals do the parametric equations in part (a) define y as a differentiable function of x? Find dy/dx in terms of t for the intervals in which it exists.

17. Follow the instructions in Exercise 16 for the curve C defined by

$$\begin{cases} x = \dfrac{t^2 - 5}{t - 2} \\ y = \dfrac{1}{t - 2} \end{cases} \quad t \neq 2$$

In Exercises 18–21, find d^2y/dx^2 in terms of t without eliminating the parameter.

18. $\begin{cases} x = t^2 \\ y = 2t^3 \end{cases} \quad t > 0$

19. $\begin{cases} x = 2\cos t \\ y = 3\sin t \end{cases} \quad 0 < t < \pi$

20. $\begin{cases} x = \sinh t \\ y = \cosh^2 t \end{cases} \quad -\infty < t < \infty$

21. $\begin{cases} x = \sqrt{1 - t^2} \\ y = 1 + t^2 \end{cases} \quad 0 < |t| < 1$

22. A **double point** on a curve defined by $x = f(t)$ and $y = g(t)$ is a point for which $(f(t_1), g(t_1)) = (f(t_2), g(t_2))$ and $t_1 \neq t_2$. Show that the curve defined by

$$\begin{cases} x = t^2 + 1 \\ y = t^3 - t - 1 \end{cases} \quad -\infty < t < \infty$$

has a double point, and find equations of the two tangent lines at this point. Use a CAS or graphing calculator to obtain the graph of the curve.

23. When a smooth curve C is defined parametrically by $x = f(t)$ and $y = g(t)$ for t in some interval I, the notation $\dot{x} = dx/dt$, $\dot{y} = dy/dt$, $\ddot{x} = d^2x/dt^2$, and $\ddot{y} = d^2y/dt^2$ is sometimes used (it was introduced by Newton). Assuming the equations define y as a twice differentiable function of x, show that

$$\frac{d^2y}{dx^2} = \frac{\dot{x}\ddot{y} - \dot{y}\ddot{x}}{\dot{x}^3}$$

Similarly, find an expression for d^3y/dx^3.

24. Find the equation of the normal line (line perpendicular to the tangent line) to the cycloid at an arbitrary point $P(x, y)$ on the cycloid for which the parameter ϕ is in the open interval $(0, 2\pi)$. Show that the x-intercept of this line is the x-coordinate of the point of contact of the generating circle with the x-axis for the given value of ϕ.

9.3 ARC LENGTH AND SURFACE AREA

In Section 6.4, we showed that the differential of arc length, ds, can be written in the form

$$ds = \sqrt{(dx)^2 + (dy)^2} \tag{9.4}$$

If x and y are given parametrically by

$$\begin{cases} x = f(t) \\ y = g(t) \end{cases} \quad a \leq t \leq b$$

and f' and g' are continuous functions, we have

$$dx = f'(t)\, dt \quad \text{and} \quad dy = g'(t)\, dt$$

Thus, by Equation 9.4,

$$ds = \sqrt{[f'(t)\, dt]^2 + [g'(t)\, dt]^2} = \sqrt{[f'(t)]^2 + [g'(t)]^2}\; dt$$

or, equivalently,

$$ds = \sqrt{\left(\frac{dx}{dt}\right)^2 + \left(\frac{dy}{dt}\right)^2}\; dt \tag{9.5}$$

Thus, the length L of the curve is

Length of a Curve Defined Parametrically

$$L = \int_{t=a}^{t=b} ds = \int_a^b \sqrt{\left(\frac{dx}{dt}\right)^2 + \left(\frac{dy}{dt}\right)^2}\, dt \qquad (9.6)$$

In deriving Equation 9.4 for ds in Chapter 6, we assumed that y was a differentiable function of x. For Formula 9.6 to be valid, we no longer need this assumption. This formula remains true provided that f' and g' are continuous for $a \leq t \leq b$ and that the curve defined by $x = f(t)$, $y = g(t)$ is traced out exactly once as t increases from a to b. A formal proof can be constructed along the same lines as in the derivation of Formula 6.18 in Chapter 6, where we approximated the curve more and more closely with a series of smaller and smaller line segments (that is, by polygonal paths). We will omit the details.

It is easy to see that Formula 9.6 is equivalent to Formula 6.18 when y is a continuously differentiable function of x, say $y = F(x)$, with $a \leq x \leq b$. Here, we can parameterize the graph of F by the parametric equations

$$\begin{cases} x = t \\ y = F(t) \end{cases} \qquad a \leq t \leq b$$

So $dx/dt = 1$ and $dy/dt = F'(t)$. Thus, Equation 9.6 becomes

$$L = \int_a^b \sqrt{1 + [F'(t)]^2}\, dt$$

which is equivalent to Equation 6.18. A similar argument can be given when $x = G(y)$ to obtain Equation 6.19 as a special case of Formula 9.6.

EXAMPLE 9.9   A portion of a metal sculpture follows the curve C defined by

$$\begin{cases} x = t^3 - 1 \\ y = 2t^2 + 3 \end{cases} \qquad 0 \leq t \leq 1$$

where x and y are measured in meters. If the metal costs \$100 per meter, find the cost of the material.

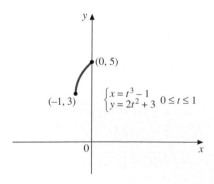

FIGURE 9.14

Solution   We show the graph of C in Figure 9.14. The derivatives

$$\frac{dx}{dt} = 3t^2 \qquad \text{and} \qquad \frac{dy}{dt} = 4t$$

are both continuous for $0 \leq t \leq 1$, so we can apply Equation 9.6.

$$L = \int_0^1 \sqrt{(3t^2)^2 + (4t)^2}\, dt = \int_0^1 \sqrt{9t^4 + 16t^2}\, dt$$

$$= \int_0^1 t\sqrt{9t^2 + 16}\, dt - \frac{1}{18}\left[(9t^2 + 16)^{3/2} \cdot \frac{2}{3}\right]_0^1 \qquad \text{Substitute } u = 9t^2 + 16.$$

$$= \frac{1}{27}[125 - 64] = \frac{61}{27}$$

So the length is $\dfrac{61}{27}$ m. At \$100/m, the cost is $\$100\left(\dfrac{61}{27}\right) \approx \$225.93$.   ∎

Ceiling of the Kimbell Art Museum

FIGURE 9.15

EXAMPLE 9.10   The ceiling of the Kimbell Art Museum in Forth Worth, Texas, consists of a series of arched sections, each in the shape of a cycloid, as shown in Figure 9.15. Find the parametric equations of the cycloid, and find the area of each section.

Solution   In Section 9.1 we showed that the period of the cycloid is $2\pi a$ and that the parametric equations are

$$\begin{cases} x = a(t - \sin t) \\ y = a(1 - \cos t) \end{cases}$$

The distance 20 ft in Figure 9.15 represents one period. So we have $2\pi a = 20$, or

$$a = \frac{10}{\pi}$$

For one arch we take t in the interval $[0, 2\pi]$. Thus, the equations are

$$\begin{cases} x = \dfrac{10}{\pi}(t - \sin t) \\ y = \dfrac{10}{\pi}(1 - \cos t) \end{cases} \qquad 0 \le t \le 2\pi$$

To find the area, we will first find the length L of the arch of the cycloid. Then, by imagining that we flatten the curved surface, we have a rectangle whose area is the length times the width, namely, $100L$ ft^2.

To find the length of the arch, we use Equation 9.6. To simplify calculations, we use $x = a(t - \sin t)$ and $y = a(1 - \cos t)$ and replace a by $10/\pi$ at the end. Thus,

$$\frac{dx}{dt} = a(1 - \cos t) \qquad \text{and} \qquad \frac{dy}{dt} = a \sin t$$

By Equation 9.6,

$$L = \int_0^{2\pi} \sqrt{a^2(1 - \cos t)^2 + a^2 \sin^2 t}\, dt$$

$$= a \int_0^{2\pi} \sqrt{1 - 2\cos t + \cos^2 t + \sin^2 t}\, dt$$

$$= a \int_0^{2\pi} \sqrt{2 - 2\cos t}\, dt$$

Now we use the half-angle formula (see Appendix 2)

$$\sin \tfrac{1}{2}t = \sqrt{\frac{1 - \cos t}{2}} \qquad (0 \le t \le 2\pi)$$

Thus,

$$\sqrt{2(1 - \cos t)} = \sqrt{4\left(\frac{1 - \cos t}{2}\right)} = 2 \sin \tfrac{1}{2}t$$

Making this substitution, we have

$$L = 2a \int_0^{2\pi} \sin \tfrac{1}{2}t\, dt = 2a(2) \int_0^{2\pi} (\sin \tfrac{1}{2}t)(\tfrac{1}{2}\, dt)$$

$$= 4a\left[-\cos \tfrac{1}{2}t\right]_0^{2\pi} = 8a$$

Thus, since $a = 10/\pi$ ft,

$$L = \frac{80}{\pi} \text{ ft}$$

The area is therefore

$$A = 100\left(\frac{80}{\pi}\right) = \frac{800}{\pi} \approx 254.65 \text{ ft}^2 \qquad \blacksquare$$

 In using Equation 9.6 to find arc lengths, it is important that the parameter range be such that the curve is traversed only once. Particular care must be exercised when x and y are expressed as periodic functions of t. For example, the equations $x = \cos 2t$ and $y = \sin 2t$ are parametric equations of the unit circle $x^2 + y^2 = 1$. The entire circle is traced out as t ranges from 0 to π. If we carelessly took t from 0 to 2π, then Equation 9.6 would give double the correct length, since the circle would have been traversed twice.

Surface Area

If C is defined parametrically by

$$\begin{cases} x = f(t) \\ y = g(t) \end{cases} \qquad a \le t \le b$$

we can derive formulas for the surface area formed by revolving C about the x-axis or the y-axis, analogous to Equations 6.25 and 6.26. We again assume f' and g' are continuous on $[a, b]$. For rotation about the x-axis, we require that $g(t)$ be nonnegative, so that C lies above the x-axis. Similarly, for rotation about the y-axis, we require that $f(t)$ be nonnegative. We omit the details of the derivation since it is similar to that in Section 6.5. For rotation about the x-axis, the result is

$$S = 2\pi \int_a^b g(t)\sqrt{[f'(t)]^2 + [g'(t)]^2}\, dt$$

or, equivalently,

$$S = 2\pi \int_a^b y\sqrt{\left(\frac{dx}{dt}\right)^2 + \left(\frac{dy}{dt}\right)^2}\, dt$$

We can put the result in a simpler form by using Equation 9.5:

$$S = 2\pi \int_{t=a}^{t=b} y\, ds \qquad \text{Rotation about } x\text{-axis} \qquad (9.7)$$

Similarly, for rotation about the y-axis, we have

$$S = 2\pi \int_{t=a}^{t-b} x\, ds \qquad \text{Rotation about } y\text{-axis} \qquad (9.8)$$

To carry out the integration in Equation 9.7, we must replace y with its value $g(t)$. In Equation 9.8, x must be replaced with $f(t)$.

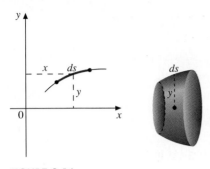

FIGURE 9.16
The rotated arc is an approximate frustum of a cone with area $2\pi y\,ds$.

Just as we indicated in Section 6.5, an easy way to remember the integrals in Equations 9.7 and 9.8 is to think of ds as a small element of arc, as in Figure 9.16. When this element is rotated about the x-axis we get (approximately) a frustum of a cone, with surface area $2\pi y\,ds$ (average circumference $2\pi y$ times "slant height" ds). Similarly, we get $2\pi x\,ds$ on rotation about the y-axis. When we "sum" each of these areas by integration, we get Equations 9.7 and 9.8, respectively.

EXAMPLE 9.11   Find the area of the surface generated by revolving the curve C defined by

$$\begin{cases} x = t^2 - 1 \\ y = 3t \end{cases} \quad 0 \le t \le 2$$

about the x-axis. (See Figure 9.17.)

Solution   We have

$$\frac{dx}{dt} = 2t \quad \text{and} \quad \frac{dy}{dt} = 3$$

So, by Equation 9.5,

$$ds = \sqrt{\left(\frac{dx}{dt}\right)^2 + \left(\frac{dy}{dt}\right)^2}\,dt = \sqrt{4t^2 + 9}\,dt$$

and by Equation 9.7, the surface area is

$$S = 2\pi \int_{t=0}^{t=2} y\,ds = 2\pi \int_0^2 3t\sqrt{4t^2 + 9}\,dt$$

$$= \frac{6\pi}{8} \int_0^2 8t\sqrt{4t^2 + 9}\,dt \quad \text{Substitute } u = 4t^2 + 9.$$

$$= \frac{3\pi}{4}(4t^2 + 9)^{3/2} \cdot \frac{2}{3}\Big]_0^2$$

$$= \frac{\pi}{2}\left[(25)^{3/2} - 9^{3/2}\right]$$

$$= \frac{\pi}{2}[125 - 27] = \frac{\pi}{2}(98) = 49\pi \qquad ■$$

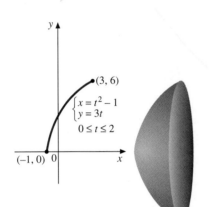

(3, 6)
$$\begin{cases} x = t^2 - 1 \\ y = 3t \end{cases}$$
$$0 \le t \le 2$$
(−1, 0)

The rotated surface

FIGURE 9.17

Exercise Set 9.3

In Exercises 1–16, find the length of the curve defined by the given parametric equations over the specified interval.

1. $\begin{cases} x = 2t - 1 \\ y = 3t + 4 \end{cases} \quad 0 \le t \le 4$

2. $\begin{cases} x = 2 - t \\ y = 2t - 1 \end{cases} \quad -2 \le t \le 2$

3. $\begin{cases} x = t \\ y = t^{3/2} \end{cases} \quad 0 \le t \le 5$

4. $\begin{cases} x = \sin t \\ y = \cos t \end{cases} \quad 0 \le t \le \pi$

5. $\begin{cases} x = 4t^2 \\ y = t^3 \end{cases} \quad 0 \le t \le 2$

6. $\begin{cases} x = t + \dfrac{1}{t} \\ y = 2\ln t \end{cases}$ $1 \le t \le 3$

7. $\begin{cases} x = 3e^t \\ y = 2e^{3t/2} \end{cases}$ $\ln 3 \le t \le 2\ln 2$

8. $\begin{cases} x = \sqrt{1 - t^2} \\ y = 1 - t \end{cases}$ $0 \le t \le \dfrac{1}{2}$

9. $\begin{cases} x = e^t \sin t \\ y = e^t \cos t \end{cases}$ $0 \le t \le \ln 3$

10. $\begin{cases} x = \cos^3 t \\ y = \sin^3 t \end{cases}$ $0 \le t \le \dfrac{\pi}{2}$

11. $\begin{cases} x = \sin t - t \cos t \\ y = \cos t + t \sin t \end{cases}$ $1 \le t \le 2$

12. $\begin{cases} x = \ln \cos t \\ y = t - \tan t \end{cases}$ $0 \le t \le \dfrac{\pi}{3}$

13. $\begin{cases} x = 4\sqrt{t} \\ y = t^2 + \dfrac{1}{2t} \end{cases}$ $\dfrac{1}{2} \le t \le 2$

14. $\begin{cases} x = \tanh t \\ y = \ln(\cosh^2 t) \end{cases}$ $-2 \le t \le 3$

15. $\begin{cases} x = 2\sin t \\ y = \cos 2t \end{cases}$ $0 \le t \le \dfrac{\pi}{3}$

16. $\begin{cases} x = 3\cos t \\ y = \cos 2t \end{cases}$ $0 \le t \le \dfrac{\pi}{2}$

17. Use Formula 9.6 to show that the circumference of a circle of radius r is $2\pi r$.

18. Use Formula 9.6 to find the length of the curve with the specified parametric equations.

(a) $\begin{cases} x = \cos t \\ y = \sin t \end{cases}$ $0 \le t \le 2\pi$

(b) $\begin{cases} x = \cos 3t \\ y = \sin 3t \end{cases}$ $0 \le t \le 2\pi$

Explain the difference in the results for (a) and (b).

In Exercises 19–26, find the area of the surface formed by revolving the given curve about the specified axis.

19. $\begin{cases} x = 3t^2 - 1 \\ y = 2t \end{cases}$ $0 \le t \le \sqrt{7}$; x-axis

20. $\begin{cases} x = 6t^2 \\ y = t^4 \end{cases}$ $0 \le t \le 2$; y-axis

21. $\begin{cases} x = \sin t + 2 \\ y = \cos t - 3 \end{cases}$ $0 \le t \le \pi/2$; y-axis

22. $\begin{cases} x = \cos 2t \\ y = 4\sin t \end{cases}$ $0 \le t \le \pi/2$; x-axis

23. $\begin{cases} x = \tan^2 t \\ y = 2\sec t \end{cases}$ $0 \le t \le \pi/3$; x-axis

24. $\begin{cases} x = 6t^2 \\ y = t^4 - 2 \end{cases}$ $0 \le t \le 2$; y-axis

25. $\begin{cases} x = 4e^t - 1 \\ y = 3e^t - 1 \end{cases}$ $0 \le t \le \ln 2$; the line $x + 2 = 0$

26. $\begin{cases} x = \cos^3 t \\ y = \sin^3 t \end{cases}$ $0 \le t \le \pi/2$; the line $y + 1 = 0$

27. Find the surface area generated by revolving the arch of the cycloid

$$\begin{cases} x = a(\theta - \sin\theta) \\ y = a(1 - \cos\theta) \end{cases} \quad 0 \le \theta \le 2\pi$$

about the x-axis.

28. Find the length of the epicycloid

$$\begin{cases} x = 2a\cos t - a\cos 2t \\ y = 2a\sin t - a\sin 2t \end{cases} \quad 0 \le t \le 2\pi$$

29. For the curve defined by

$$\begin{cases} x = 1/t \\ y = \ln t \end{cases} \quad 1 \le t \le \sqrt{3}$$

find (a) its length and (b) the surface area formed by revolving the curve about the y-axis.

30. Show that the graph of

$$\begin{cases} x = t^2 - 3 \\ y = \frac{1}{3}t^3 - t \end{cases} \quad -\infty < t < \infty$$

has a loop, and find the length of the loop. Draw the graph.

31. **Elliptic integrals**

(a) Show that the length of the ellipse given parametrically by $x = a\cos t$, $y = b\sin t$ for $0 \le t \le 2\pi$ and $a > b > 0$ is

$$4a \int_0^{\pi/2} \sqrt{1 - e^2 \sin^2 t}\, dt$$

where e is the eccentricity of the ellipse $e = \sqrt{a^2 - b^2}/a$. (See Example 6.19 and Exercise 40 in Exercise Set 8.3.)

(b) Approximate the length of the ellipse given by $x = 3\cos t$, $y = 2\sin t$ for $0 \le t \le 2\pi$ using Simpson's Rule with the number of subintervals $n = 10$.

32. Use a calculator or CAS to find the length of the **spiral of Archimedes** defined by

$$\begin{cases} x = t \cos t \\ y = t \sin t \end{cases} \quad 0 \le t \le 30$$

Sketch the spiral using a graphing calculator or a CAS.

In Exercises 33–48, sketch the graphs of the curves in Exercises 1–16 using a CAS or a graphing calculator.

9.4 THE POLAR COORDINATE SYSTEM

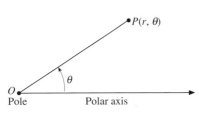

FIGURE 9.18

Although the rectangular system of coordinates we have used in all of our work so far is suitable for most graphing needs, other systems are sometimes more convenient to use. The most important alternative system is called the **polar coordinate system**. In it we take as references a fixed point O, called the **pole**, and a ray emanating from O, called the **polar axis**. It is customary to direct the polar axis horizontally to the right. If P is any point in the plane other than the pole, we indicate its location by means of the radial distance r from O to P and the measure θ of the angle from the polar axis to the ray from O through P, as shown in Figure 9.18. As usual, measures of angles resulting from a counterclockwise rotation are positive. The numbers r and θ are called the **polar coordinates** of P, and we indicate them as the ordered pair (r, θ). Although it is possible to take θ as either the degree or radian measure of the angle, we will always use radian measure. Coordinates of the pole are taken to be $(0, \theta)$ for any value of θ.

If a ray from the pole makes an angle α (alpha) with the polar axis, we refer to it as the α **ray**. For example, the polar axis is the 0 ray, and the $\pi/2$ ray is directed vertically upward. These correspond to the positive x- and y-axes, respectively, in a rectangular system. On a given ray, distances are measured positively in the direction of the ray. Figure 9.19 shows the points $A(3, \pi/4)$, $B(1, 2\pi/3)$, and $C(2, 3\pi/2)$. Also, we define the negative direction on a ray to be the opposite of the positive direction. More precisely, the negative direction of an α ray is in the direction of the $(\alpha + \pi)$ ray. With this understanding we can permit r to be negative, and to plot (r, θ) for r negative, we measure $|r|$ units in the negative direction of the θ ray. In Figure 9.20 we have plotted $P(-2, 0)$ and $Q(-3, 4\pi/3)$.

FIGURE 9.19   **FIGURE 9.20**

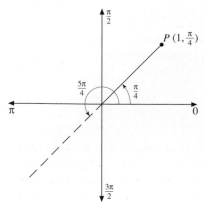

FIGURE 9.21

Point P has polar coordinates $(1, \frac{\pi}{4} + 2n\pi)$ and also $(-1, \frac{5\pi}{4} + 2n\pi)$ for $n = 0, \pm 1, \pm 2, \ldots$.

When we are given polar coordinates (r, θ), the point P represented by them is uniquely determined. Unfortunately, the converse is not true. If we are given a point P, there are infinitely many sets of polar coordinates for it, in contrast with the rectangular system, where there is a one-to-one correspondence between points in the plane and the set of all ordered pairs (x, y). You should convince yourself that $(1, \pi/4)$, $(1, 9\pi/4)$, $(1, -7\pi/4)$, $(-1, 5\pi/4)$, and $(-1, -3\pi/4)$ are all polar coordinates of the same point. In fact, $(1, \pi/4 + 2n\pi)$ and $(-1, 5\pi/4 + 2n\pi)$ for $n = 0, \pm 1, \pm 2, \ldots$ represent this point. (See Figure 9.21.) This lack of uniqueness of representation is not a serious drawback to the polar system, but it does mean that special care must be exercised when analyzing polar graphs or finding points of intersection of two curves, for example. Although two different ordered pairs may represent the same point, they are not equal as ordered pairs. For example, $(1, \pi/4)$ and $(-1, 5\pi/4)$ are coordinates of the same point, but $(1, \pi/4) \neq (-1, 5\pi/4)$, since by equality of ordered pairs we mean that first coordinates are equal and second coordinates are equal.

Change of Coordinates

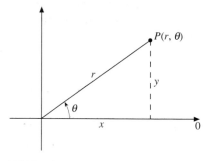

FIGURE 9.22

It is frequently useful to change from the polar coordinate system to the rectangular system, or vice versa. To see how to make these changes, superimpose a rectangular system onto a polar system so that the origin coincides with the pole, and the positive x-axis coincides with the polar axis. Let P be any point other than the origin. Then P has unique rectangular coordinates (x, y). As in Figure 9.22, let (r, θ) be polar coordinates of P for which $r > 0$ and θ is between 0 and 2π. Then from the definitions of $\sin\theta$ and $\cos\theta$, we have $\cos\theta = x/r$ and $\sin\theta = y/r$, or, equivalently, the following equations.

Polar Coordinates to Rectangular Coordinates

$$x = r \cos\theta$$
$$y = r \sin\theta$$

(9.9)

Adding a multiple of 2π to the angle θ does not alter $\sin\theta$ or $\cos\theta$, so when $r > 0$, any equivalent polar representation of the point P does not change x or y. The fact that we have shown P in the first quadrant is not significant. Equations 9.9 are true wherever P is located. It can also be shown that Equations 9.9 hold true when $r < 0$. (See Exercise 32 in Exercise Set 9.4.) They are true as well when $r = 0$, since for any value of θ, the point $(0, \theta)$ is the pole, or equivalently, the origin.

If we square both sides of Equations 9.9 and add, we get $r^2 = x^2 + y^2$, and if $x \neq 0$, on dividing the second equation by the first, we get $\tan\theta = y/x$. Thus, equations for r and θ in terms of x and y are as follows.

Rectangular Coordinates to Polar Coordinates

$$r^2 = x^2 + y^2$$

$$\tan \theta = \frac{y}{x} \qquad (x \neq 0) \tag{9.10}$$

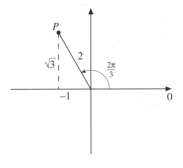

(a) If we choose $\theta = \frac{2\pi}{3}$ then $r = 2$.
Polar coordinates of P are $(2, \frac{2\pi}{3})$.

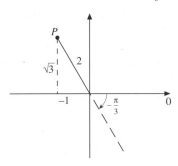

(b) If we choose $\theta = -\frac{\pi}{3}$ then $r = -2$.
Polar coordinates of P are $(-2, -\frac{\pi}{3})$.

FIGURE 9.23

FIGURE 9.24

To change from polar coordinates to rectangular coordinates, we use Equations 9.9 and get an unambiguous answer. To change from rectangular coordinates to polar coordinates, we use Equations 9.10, taking account of the quadrant in which the point lies and choosing $r = \sqrt{x^2 + y^2}$ or $-\sqrt{x^2 + y^2}$, depending on which value of θ is chosen. For example, if the rectangular coordinates of P are $(-1, \sqrt{3})$, then by Equations 9.10,

$$r^2 = (-1)^2 + (\sqrt{3})^2 = 4$$

and $\tan \theta = -\sqrt{3}$. If we take $\theta = 2\pi/3$, then $r = \sqrt{4} = 2$, whereas if we take $\theta = -\pi/3$, then we must choose $r = -\sqrt{4} = -2$. We illustrate this example in Figure 9.23. It will always work to take $r > 0$ if we choose θ as a positive angle from the polar axis to the ray from O through P, where P is the point with rectangular coordinates (x, y). There are times, however, when we might prefer some other polar representation of P, consistent with Equations 9.10. We illustrate these ideas in the next example.

EXAMPLE 9.12

(a) Change the polar coordinates $(-3, 4\pi/3)$ to rectangular coordinates. Plot the point.

(b) Give three sets of polar coordinates for the point P whose rectangular coordinates are $(\sqrt{3}, -1)$. For one set choose $r > 0$ and $\theta > 0$, and for the other two choose r and θ opposite in sign.

Solution

(a) The point P with polar coordinates $(-3, 4\pi/3)$ is plotted in Figure 9.24. Note that a simpler set of coordinates for P is $(3, \pi/3)$, but it is not necessary to make this change. By Equations 9.9, we have

$$x = r \cos \theta = (-3) \cos \frac{4\pi}{3} = (-3)\left(-\frac{1}{2}\right) = \frac{3}{2}$$

$$y = r \sin \theta = (-3) \sin \frac{4\pi}{3} = (-3)\left(-\frac{\sqrt{3}}{2}\right) = \frac{3\sqrt{3}}{2}$$

(b) The point P is plotted in Figure 9.25. From Equations 9.10 we have

$$r^2 = x^2 + y^2 = 3 + 1 = 4$$

From our discussion following Equations 9.10, we will choose $r = 2$ or $r = -2$, depending on which value of θ we choose. Since

$$\tan \theta = \frac{y}{x} = \frac{-1}{\sqrt{3}}$$

FIGURE 9.25

from our knowledge of special angles in trigonometry, we see that θ has a reference angle of $30°$, or $\pi/6$ radians. Since x is positive and y is negative, P is in the fourth quadrant. By choosing $\theta = 11\pi/6$ or $\theta = -\pi/6$, we take $r = 2$. By choosing $\theta = 5\pi/6$ (or any angle coterminal with it), we take $r = -2$. So three sets of polar coordinates for P can be taken as follows:

$$(2, 11\pi/6) \qquad (2, -\pi/6) \qquad (-2, 5\pi/6)$$
$$r > 0,\ \theta > 0 \quad r > 0,\ \theta < 0 \quad r < 0,\ \theta > 0$$

The second of these is probably the simplest form in this case.   ■

Equations in Polar Coordinates

A **polar equation** is an equation whose only variables are the polar variables r and θ. For example, each of the following is a polar equation.

$$r = 1 - 2\cos\theta \qquad r^2 = \sin 2\theta \qquad r = 2 \qquad \theta = \pi/4$$

By the **graph** of a polar equation, we mean the set of all points P that have *at least one* set of coordinates (r, θ) that satisfy the equation. We will consider general graphing techniques in the next section. For now, we observe that graphs of polar equations can sometimes be more easily obtained by finding a rectangular equation that has the same graph. On the other hand, complicated rectangular equations can sometimes be graphed more easily using an equivalent polar equation. Thus, it is desirable to be able to change an equation in one system to an equation in the other system that has the same graph. Equations 9.9 and 9.10 enable us to make such changes. (We should note, however, that a CAS or graphing calculator can handle both rectangular and polar equations in most cases.)

In the examples and exercises that follow, when we ask for a polar equation to be transformed to an equivalent rectangular equation, or vice versa, we mean that the resulting equation will have the same graph as the given equation.

EXAMPLE 9.13

(a) Transform $r = 2\sin\theta$ to an equivalent rectangular equation. Identify and draw the graph.

(b) Transform the equation $(x^2+y^2)^2 = x^2 - y^2$ to an equivalent polar equation.

Solution

(a) Since $r = \sqrt{x^2 + y^2}$, we could substitute this value for r, but we can avoid the square root by multiplying both sides of the original equation by r. We get $r^2 = 2r\sin\theta$. Then since $r^2 = x^2 + y^2$ and $r\sin\theta = y$, we have the rectangular equation

$$x^2 + y^2 = 2y$$

The graph of this equation is a circle. By completing the square in y, we can put it in the standard form $x^2 + (y-1)^2 = 1$. This form is the equation of a circle with center $(0, 1)$ and radius 1. (See Figure 9.26.)

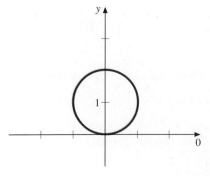

FIGURE 9.26

We should note that multiplication by r in this procedure did not change the graph, since the only possible new point it could have introduced would have been the pole, $r = 0$. We can see the reason for this result by writing $r^2 = 2r \sin\theta$ in the form

$$r^2 - 2r \sin\theta = 0$$

So

$$r(r - 2\sin\theta) = 0$$

That is, either $r = 0$ or $r = 2\sin\theta$. But the graph of $r = 0$ is just the pole, and the pole is on the graph of the original equation, $r = 2\sin\theta$, since when $\theta = 0$, $r = 0$.

(b) Substitute for x and y from Equations 9.9 and use $x^2 + y^2 = r^2$ to get

$$r^4 = r^2 \cos^2\theta - r^2 \sin^2\theta$$

or, since $\cos 2\theta = \cos^2\theta - \sin^2\theta$ (see Appendix 2),

$$r^4 = r^2 \cos 2\theta$$

Rewriting this equation as $r^4 - r^2 \cos 2\theta = 0$, we can factor to get $r^2(r^2 - \cos 2\theta) = 0$. So the graph consists of all points with coordinates that satisfy $r^2 = 0$ (the pole) or $r^2 = \cos 2\theta$. But the pole is on the graph of $r^2 = \cos 2\theta$, since when $\theta = \pi/4$, $r = 0$. Thus, the equation $r^2 = \cos 2\theta$ is equivalent to the original equation. We will see in the next section how to draw the graph using its polar equation. ∎

Exercise Set 9.4

In Exercises 1 and 2, plot the points in a polar coordinate system.

1. (a) $(2, \pi)$
   (b) $\left(3, -\dfrac{\pi}{2}\right)$
   (c) $\left(-1, \dfrac{\pi}{4}\right)$
   (d) $\left(0, \dfrac{5\pi}{9}\right)$

2. (a) $\left(-2, -\dfrac{\pi}{3}\right)$
   (b) $\left(4, \dfrac{7\pi}{2}\right)$
   (c) $(2, -5\pi)$
   (d) $\left(-3, -\dfrac{5\pi}{6}\right)$

3. Show the approximate location of the points that have the following polar coordinates (r, θ), where θ is in radians.
   (a) $(1, 2)$
   (b) $(2, 1)$
   (c) $(-5, 8)$
   (d) $(-3, -4)$

4. Let P be the point determined by the given polar coordinates. Determine two other pairs of polar coordinates for P with $r > 0$ and two with $r < 0$.

(a) $\left(2, \dfrac{\pi}{4}\right)$
(b) $\left(-1, -\dfrac{\pi}{2}\right)$
(c) $\left(-3, \dfrac{4\pi}{5}\right)$
(d) $\left(4, -\dfrac{2\pi}{9}\right)$

In Exercises 5 and 6, change the given polar coordinates to equivalent rectangular coordinates.

5. (a) $\left(2, \dfrac{5\pi}{6}\right)$
   (b) $\left(-3, \dfrac{3\pi}{2}\right)$
   (c) $\left(0, \dfrac{17\pi}{13}\right)$
   (d) $\left(-2, -\dfrac{3\pi}{4}\right)$

6. (a) $(3, 0)$
   (b) $\left(4, -\dfrac{11\pi}{6}\right)$
   (c) $(-2, 5\pi)$
   (d) $\left(7, \dfrac{4\pi}{3}\right)$

In Exercises 7 and 8, change the given rectangular coordinates to an equivalent pair of polar coordinates in which $r > 0$ and $0 \le \theta < 2\pi$.

7. (a) $(4, 0)$
    (b) $(0, 4)$
    (c) $(-4, 0)$
    (d) $(0, -4)$

8. (a) $(1, -\sqrt{3})$
    (b) $(-5, 5)$
    (c) $(-4\sqrt{3}, 4)$
    (d) $(-2\sqrt{2}, -2\sqrt{2})$

In Exercises 9–14, find an equivalent rectangular equation. Also identify and draw the graph.

9. (a) $r = 2$
    (b) $\theta = \dfrac{\pi}{4}$

10. (a) $r = -2$
    (b) $\theta = -\dfrac{\pi}{4}$

11. (a) $r \cos \theta = 3$
    (b) $r \sin \theta = -1$

12. (a) $r = -2 \csc \theta$
    (b) $r = 3 \sec \theta$

13. (a) $2r \cos \theta - 3r \sin \theta = 4$
    (b) $r(3 \cos \theta + 2 \sin \theta) = 5$

14. (a) $r = 2 \cos \theta$
    (b) $r = -2 \sin \theta$

In Exercises 15–22, find an equivalent rectangular equation.

15. $r = a(1 + \cos \theta)$

16. $r^2 = 2a \cos 2\theta$

17. $r = 1 - 2 \cos \theta$

18. $r = 2 + \sin \theta$

19. $r = a \cos 2\theta$

20. $r = a \sin 2\theta$

21. $r = \dfrac{2}{1 - \cos \theta}$

22. $r = \dfrac{1}{1 + 2 \sin \theta}$

In Exercises 23–31, find an equivalent polar equation.

23. (a) $x^2 + y^2 = a^2$
    (b) $y = \sqrt{3}\,x$

24. (a) $x + y = 0$
    (b) $2x - 3y = 4$

25. (a) $y + 1 = 0$
    (b) $x - 3 = 0$

26. $x^2 - y^2 = 4$

27. $y = x^2$

28. $x^2 + y^2 = 2x$

29. $x^2 - y^2 - 4x = 0$

30. $x^2 y^2 = 4(x^2 + y^2)^{3/2}$

31. $x^2 - y^2 = \dfrac{2xy}{x^2 + y^2}$

32. Prove that Equations 9.9 hold true when $r < 0$.

33. Prove that if P has rectangular coordinates (x, y) with $x \ne 0$, then as one set of coordinates (r, θ) for P we may take $\theta = \tan^{-1}(y/x)$ where $r = \sqrt{x^2 + y^2}$ if $x > 0$, and $r = -\sqrt{x^2 + y^2}$ if $x < 0$. (Recall that $-\pi/2 < \tan^{-1}(y/x) < \pi/2$.)

In Exercises 34–36, show that an equivalent polar equation is of the form $r = ed/(1 + e \cos \theta)$ for suitable values of d and e.

34. $y^2 = 1 - 2x$

35. $3x^2 + 4y^2 + 4x = 4$

36. $3x^2 - y^2 - 4x + 1 = 0$

37. By means of an example, show that a point P can lie on the graph of a polar equation even though a particular set of coordinates for P does not satisfy the equation.

9.5 GRAPHING POLAR EQUATIONS

Maple plot of two polar graphs

As we have seen, the graph of a polar equation can sometimes be obtained by changing to an equivalent rectangular equation. Many interesting curves, however, have rather complicated rectangular equations, so it is important to be able to draw the graph directly from the polar equation. By substituting convenient values of θ and solving the equation for r, we can obtain a table of values from which to draw a reasonable approximation of the graph. This "point plotting" method is not entirely satisfactory, though, since it is not always clear how to connect the points unless a great many points are used (whether in polar or in rectangular coordinates). Some additional analysis is usually needed. Computer algebra systems and graphing calculators can draw graphs of equations in polar form. At the end of this section we will describe how Maple, Mathematica, and DERIVE do such plotting.

The simplest polar equations to graph are of the form

$$\text{(a) } r = a \quad \text{and} \quad \text{(b) } \theta = \alpha$$

where a and α are constants. (a) The first equation, $r = a$, puts no restriction on θ, so its graph consists of points (a, θ) where θ takes on all real values.

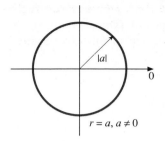

$r = a, a \neq 0$

(a)

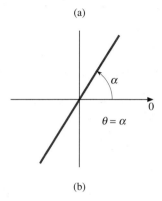

$\theta = \alpha$

(b)

FIGURE 9.27

Since a is a constant, the graph is a circle of radius $|a|$ if $a \neq 0$ and is the pole if $a = 0$. (b) The equation $\theta = \alpha$ places no restriction on r, and so its graph consists of all points of the form (r, α) where r takes on all real values. Thus, its graph is the line consisting of the α ray together with its extension in the opposite direction. Graphs of these equations are shown in Figure 9.27. For more complicated equations we want to consider several aids to graphing that can reduce reliance on plotting a large number of points.

Symmetry

We consider three types of symmetry: (a) symmetry with respect to the line $\theta = 0$ (x-axis, in rectangular coordinates), (b) symmetry with respect to the line $\theta = \pi/2$ (y-axis, in rectangular coordinates), and (c) symmetry with respect to the pole (origin). As Figure 9.28 indicates, in each case symmetry will exist if, whenever $P(r, \theta)$ is on the graph, the point P' as indicated is also on the graph. Since each point P' has many equivalent representations, devising a satisfactory test for symmetry is complicated. The following is perhaps the simplest test, but it should be emphasized that it provides *sufficient* conditions only (that is, if the conditions are satisfied, the curve will have the given type of symmetry, but the curve *may* have one or more of the symmetries even if the conditions are not satisfied). In Exercise 35 of Exercise Set 9.5 you will be asked to give a more general test, providing conditions that are both necessary and sufficient (by considering all possible polar representations of the points P' in Figure 9.28).

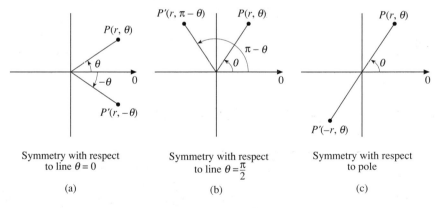

Symmetry with respect to line $\theta = 0$

(a)

Symmetry with respect to line $\theta = \frac{\pi}{2}$

(b)

Symmetry with respect to pole

(c)

FIGURE 9.28

Tests for Symmetry

(a) If replacing θ with $-\theta$ results in an equivalent equation, the graph is symmetric with respect to the line $\theta = 0$ (the x-axis).

(b) If replacing θ with $\pi - \theta$ results in an equivalent equation, the graph is symmetric with respect to the line $\theta = \pi/2$ (the y-axis).

(c) If replacing r with $-r$ results in an equivalent equation, the graph is symmetric with respect to the pole (the origin).

REMARK ───────────────────────────────

■ If a curve has any two of the three types of symmetry in (a), (b), or (c), it also has the third. You can convince yourself of this fact geometrically by combining any two of the types.

───────────────────────────────────────

To illustrate Test (a), consider the equation

$$r = \cos 2\theta - \cos \theta$$

If we replace θ with $-\theta$, we get

$$r = \cos(-2\theta) - \cos(-\theta)$$

But the cosine is an even function, so $\cos(-2\theta) = \cos 2\theta$ and $\cos(-\theta) = \cos \theta$. Thus, we obtain an equivalent equation, and we conclude that the graph is symmetric with respect to the line $\theta = 0$. The CAS-generated graph in Figure 9.29 illustrates this symmetry.

To illustrate Test (b), consider the equation

$$r = \sin^3 \theta$$

Replacing θ with $\pi - \theta$ gives $r = \sin^3(\pi - \theta)$. But $\sin(\pi - \theta) = \sin \theta$. So we get an equivalent equation and conclude that the graph is symmetric with respect to the line $\theta = \pi/2$. We show the CAS-generated graph in Figure 9.30.

To illustrate Test (c), consider the equation

$$r^2 = e^\theta$$

Clearly, replacing r with $-r$ yields an equivalent equation, so the graph (shown in Figure 9.31) is symmetric with respect to the pole.

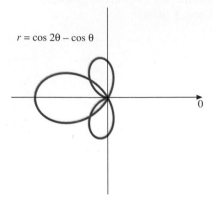

FIGURE 9.29

Symmetry with respect to the line $\theta = 0$

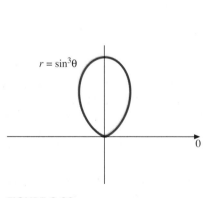

FIGURE 9.30

Symmetry with respect to the line $\theta = \frac{\pi}{2}$

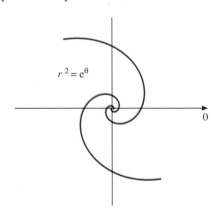

FIGURE 9.31

Symmetry with respect to the pole

To illustrate that the conditions stated in the test are sufficient only, and not necessary for symmetry, consider the equation

$$r = \sqrt{\sin \frac{\theta}{2}}$$

Since $\sin(-\theta/2) = -\sin(\theta/2)$, Test (a) fails. Also, $\sin((\pi - \theta)/2) = \sin((\pi/2) - (\theta/2)) = \cos(\theta/2)$, so Test (b) fails. Finally, replacing r with $-r$ gives $-r = \sqrt{\sin(\theta/2)}$, or $r = -\sqrt{\sin(\theta/2)}$. Thus, Test (c) fails. Yet the

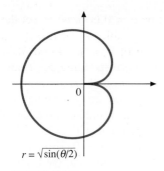

$r = \sqrt{\sin(\theta/2)}$

FIGURE 9.32
Symmetry with respect to the line $\theta = 0$.

CAS-generated graph in Figure 9.32 clearly shows symmetry with respect to the line $\theta = 0$. To see that we do, in fact, have this type of symmetry, we can use the coordinates $(r, 2\pi - \theta)$ for P' in Figure 9.28(a). Thus, if we can show that replacing θ with $2\pi - \theta$ leads to an equivalent equation, we will be able to conclude that we have symmetry with respect to the line $\theta = 0$. Since

$$\sin\left(\frac{2\pi - \theta}{2}\right) = \sin\left(\pi - \frac{\theta}{2}\right) = \sin\frac{\theta}{2}$$

we do get an equivalent equation.

Intercepts

The intercepts on what would correspond in a rectangular system to the x- and y-axes are almost always key points in drawing the graph. These are found by setting $\theta = 0$, $\pi/2$, π, and $3\pi/2$ and then solving for r.

Periodicity

Since many polar equations involve $\sin\theta$ or $\cos\theta$, the fact that each of these has period 2π can limit the domain for θ that needs to be considered. For example, in graphing $r = 1 + \cos\theta$ it is sufficient to consider values of θ that satisfy $-\pi < \theta \le \pi$. In fact, since the graph is symmetric with respect to the line $\theta = 0$, we need only consider θ in the range $0 \le \theta \le \pi$. The graph for $-\pi < \theta \le 0$ is then found by reflecting through the line $\theta = 0$.

Tangents at the Origin

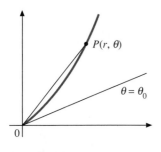

FIGURE 9.33

If the graph has one or more tangent lines at the origin, these can be found by setting $r = 0$ and solving for θ. We can reason as follows to see why. In Figure 9.33, the line OP that joins the pole O and a point $P(r, \theta)$ on the graph is a secant line. Now suppose that when $\theta = \theta_0$, $r = 0$. If we let $\theta \to \theta_0$, the secant line OP approaches the tangent line at O. The slope of OP is $\tan\theta$, which approaches $\tan\theta_0$ as $\theta \to \theta_0$. So the tangent line at O is the line with slope $\tan\theta_0$—namely, $\theta = \theta_0$. This reasoning will be made more precise in the next section, but we will use the result now, since it is often helpful in graphing.

To illustrate, consider the equation $r = 2 - 4\sin\theta$. Setting $r = 0$, we get $\sin\theta = 1/2$, so the lines $\theta = \pi/6$ and $\theta = 5\pi/6$ are tangent to the graph at the origin.

The next example uses these graphing aids.

EXAMPLE 9.14   Draw the graph of the equation $r = 2\cos\theta$.

Solution   We will consider the four aids to graphing discussed above.

Symmetry. Since $\cos(-\theta) = \cos\theta$, the equation is unchanged when θ is replaced with $-\theta$. So we have symmetry with respect to the line $\theta = 0$. Because of this symmetry and the periodicity of $\cos\theta$, we need only consider values of θ in the interval $[0, \pi]$. When we get the graph on this interval, we can flip it over (i.e., reflect it through) the line $\theta = 0$ (x-axis) to get the entire graph.

Intercepts. The intercepts on the rays $\theta = 0$, $\theta = \pi/2$, and $\theta = \pi$ are $r = 2$, $r = 0$, and $r = -2$, respectively. So the corresponding points are $(2, 0)$, $(0, \pi/2)$, and $(-2, \pi)$. Note that the points $(-2, \pi)$ and $(2, 0)$ are the same.

Periodicity. As already noted, $\cos \theta$ has period 2π, so taking values of θ between 0 and 2π is sufficient to give the entire graph. (We saw that by symmetry we can actually limit θ to be between 0 and π.)

Tangents at the Origin. Setting $r = 0$ yields $\cos \theta = 0$, so $\theta = \pi/2$. The curve is therefore tangent to the line $\theta = \pi/2$ (the y-axis) at the origin.

Finally, we make a table, choosing some convenient values of θ between 0 and π, and we plot the corresponding points.

θ	$\dfrac{\pi}{6}$	$\dfrac{\pi}{4}$	$\dfrac{\pi}{3}$	$\dfrac{2\pi}{3}$	$\dfrac{3\pi}{4}$	$\dfrac{5\pi}{6}$
r	$\sqrt{3}$	$\sqrt{2}$	1	-1	$-\sqrt{2}$	$-\sqrt{3}$

We show the graph in Figure 9.34. Notice that the entire graph is traced out for θ varying from 0 to π. The graph appears to be a circle. In fact, we can show that it is a circle by changing to rectangular coordinates as we did in Example 9.13(a). We first multiply by r, getting $r^2 = 2r \cos \theta$. Since $r^2 = x^2 + y^2$ and $r \cos \theta = x$, we have

$$x^2 + y^2 = 2x$$

and on completing the square in x,

$$(x - 1)^2 + y^2 = 1$$

We recognize the graph of this equation as a circle of radius 1 centered at $(1, 0)$.   ∎

FIGURE 9.34
The circle $r = 2 \cos \theta$

The equations in Examples 9.13(a) and 9.14 are special cases of the circles illustrated in Figure 9.35.

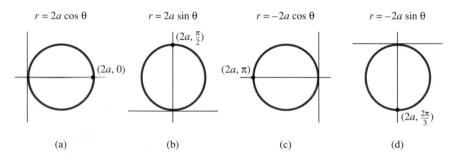

FIGURE 9.35
Family of circles $r = +2a \cos \theta$ and $r = \pm 2a \sin \theta$

In the next four examples we illustrate particular cases of some well-known polar graphs that have special names. These curves are more easily analyzed from their polar equations than from their rectangular equations.

EXAMPLE 9.15   **A cardioid**   Draw the graph of the equation $r = 1 + \sin\theta$.

Solution   Since $\sin(\pi - \theta) = \sin\theta$, the graph is symmetric with respect to the line $\theta = \pi/2$. Because of this symmetry and the periodicity of $\sin\theta$, it is sufficient to consider θ in the interval $[-\pi/2, \pi/2]$. The key intercepts in this interval are $(0, -\pi/2)$, $(1, 0)$, and $(2, \pi/2)$. Setting $r = 0$ yields $\sin\theta = -1$, so the tangent line at the origin is $\theta = -\pi/2$. Additional selected points are found as shown in the table.

θ	$-\pi/3$	$-\pi/4$	$-\pi/6$	$\pi/6$	$\pi/4$	$\pi/3$
r	$1 - \frac{\sqrt{3}}{2} \approx 0.13$	$1 - \frac{\sqrt{2}}{2} \approx 0.29$	$1 - \frac{1}{2} = 0.50$	$1 + \frac{1}{2} = 1.50$	$1 + \frac{\sqrt{2}}{2} \approx 1.71$	$1 + \frac{\sqrt{3}}{2} \approx 1.87$

The graph can now be drawn with reasonable accuracy, as shown in Figure 9.36. ∎

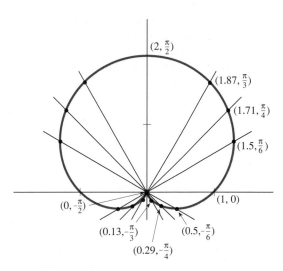

FIGURE 9.36
The cardioid $r = 1 + \sin\theta$

The graph in Figure 9.36 is called a **cardioid** because of its heartlike shape. It is a member of a family of curves called **limaçons** (pronounced lim-a-sons). The three types of limaçons are shown in Figure 9.37, along with their standard

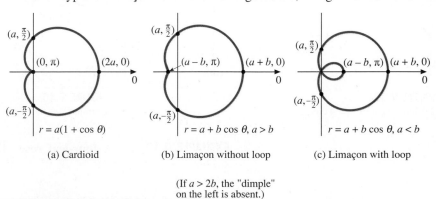

(a) Cardioid   (b) Limaçon without loop   (c) Limaçon with loop

(If $a > 2b$, the "dimple" on the left is absent.)

FIGURE 9.37
Family of limaçons $r = a + b\cos\theta$

equations when the axis of symmetry is horizontal. When the axis of symmetry is vertical, $\cos\theta$ is replaced with $\sin\theta$. The curves shown are for positive a and b. If b is negative and a is positive, the graph is rotated $180°$. If a is negative and b positive, the graph is unchanged, since when a point P with coordinates (r,θ) is on the graph of $r = a + b\cos\theta$, the same point with coordinates $(-r, \theta + \pi)$ is on the graph of $r = -a + b\cos\theta$. (See Exercise 25 in Exercise Set 9.5.)

EXAMPLE 9.16   **A lemniscate** Draw the graph of the equation $r^2 = 4\cos 2\theta$.

Solution   The graph is symmetric with respect to the line $\theta = 0$ and with respect to the pole, as can be seen by replacing, in turn, θ with $-\theta$ and r with $-r$. From these two symmetries, it follows that the graph is also symmetric with respect to the vertical line $\theta = \pi/2$. Since the period of $\cos 2\theta$ is π, it is sufficient to consider θ in the interval $[0, \pi]$. Solving for r yields $r = \pm 2\sqrt{\cos 2\theta}$, and we see that θ must be restricted to satisfy $\cos 2\theta \geq 0$ in order for the square root to be real. Thus, we must have $0 \leq 2\theta \leq \pi/2$, or equivalently, $0 \leq \theta \leq \pi/4$. (See Figure 9.38.) The tangent lines at the origin are solutions of $\cos 2\theta = 0$—namely, $\theta = \pi/4$ and $\theta = -\pi/4$. Using the intercepts $(2, 0)$ and $(0, \pi/4)$, together with the one additional point $(\sqrt{2}, \pi/6)$, we can now draw the portion of the graph shown in Figure 9.39(a). Then, by symmetry, we get the entire graph, shown in Figure 9.39(b). ∎

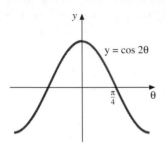

FIGURE 9.38
$\cos 2\theta \geq 0$ when
$0 \leq \theta \leq \pi/4$.

The "figure 8" curve in Figure 9.39 is called a **lemniscate**. These curves typically have equations of the form $r^2 = a^2\cos 2\theta$ or $r^2 = a^2\sin 2\theta$. The graphs of these two basic forms are shown in Figure 9.40. Note that when $\cos 2\theta$ is replaced by $\sin 2\theta$, the graph is rotated through an angle of $\pi/4$ radians. In Exercise 39 of Exercise Set 9.5 you will be asked to show in general that if $\cos n\theta$ is replaced by $\sin n\theta$, where the original equation is of the form $r = f(\cos n\theta)$, then the graph is rotated through an angle of $\pi/(2n)$ radians.

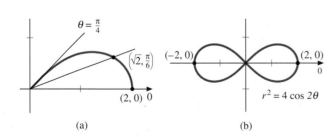

FIGURE 9.39
The lemniscate $r^2 = 4\cos 2\theta$

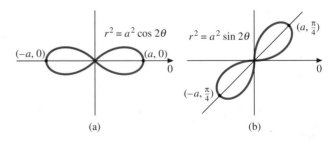

FIGURE 9.40
Lemniscates

EXAMPLE 9.17   **A four-leaf rose** Draw the graph of the equation $r = \sin 2\theta$.

Solution   To test for symmetry, we use alternative tests given in Exercise 36 of Exercise Set 9.5. If we replace r with $-r$ and θ with $-\theta$, we get $-r = \sin(-2\theta)$, or $-r = -\sin 2\theta$, which is equivalent to $r = \sin 2\theta$. Since the

θ	r
0	0
$\pi/6$	$\sqrt{3}/2$
$\pi/4$	1
$\pi/3$	$\sqrt{3}/2$
$\pi/2$	0

points $P(r, \theta)$ and $P'(-r, -\theta)$ are symmetrically placed with respect to the line $\theta = \pi/2$, it follows that the graph is symmetric with respect to this line. Also, $\sin 2(\pi + \theta) = \sin(2\pi + 2\theta) = \sin 2\theta$, and since the points $P(r, \theta)$ and $P'(r, \pi + \theta)$ are symmetrically placed with respect to the pole, it follows that the graph is symmetric with respect to the pole. Thus, it is also symmetric with respect to the line $\theta = 0$. The period of $\sin 2\theta$ is π and so, because of symmetry, we can limit our consideration of θ to values between 0 and $\pi/2$. Tangent lines at the origin are $\theta = 0$ and $\theta = \pi/2$, found by solving $\sin 2\theta = 0$. We list some points in the table and show the graph of the portion of the curve for θ in $[0, \pi/2]$ in Figure 9.41(a). We then use symmetry to get the complete graph, shown in Figure 9.41(b). ∎

The graph in Figure 9.41(b) is called a **four-leaf rose**. Standard equations of rose curves are $r = a \cos n\theta$ and $r = a \sin n\theta$. When n is even, there are $2n$ petals, as in Example 9.17, whereas when n is odd, there are n petals. These two cases are illustrated in Figure 9.42, using $r = a \cos 2\theta$ and $r = a \cos 3\theta$.

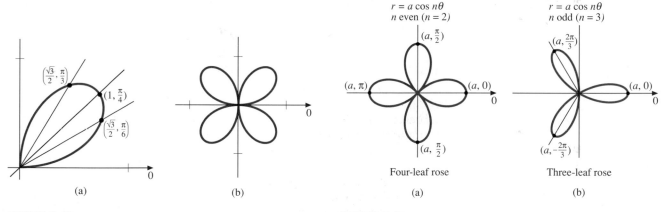

FIGURE 9.41
The four-leaf rose $r = \sin 2\theta$

FIGURE 9.42
Rose curves

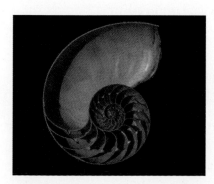

A nautilus shell has the shape of a logarithmic spiral

EXAMPLE 9.18 **A logarithmic spiral** Draw the graph of the equation $r = e^{\theta/4}$.

Solution The graph has no symmetry, and $e^{\theta/4}$ is not periodic. Nor can r ever be 0, so there are no tangent lines at the origin. We see, in fact, from the properties of the exponential function, that $r > 0$ for all values of θ, and that r increases as θ increases. Furthermore,

$$\lim_{\theta \to -\infty} r = 0 \quad \text{and} \quad \lim_{\theta \to \infty} r = \infty$$

There are infinitely many intercepts on every ray since when θ increases by 2π, the point is on the same ray, but the value of r is increased. We show the graph in Figure 9.43. Points can be found using a calculator. ∎

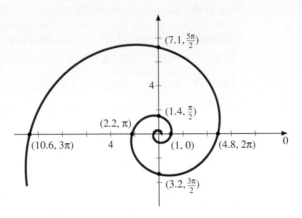

FIGURE 9.43
The logarithmic spiral $r = e^{\theta/4}$

The graph in Figure 9.43 is called a **logarithmic spiral**. There are other types of spirals as well. You will be asked to draw some of these in the exercises.

Polar Equations in Parametric Form

For the purpose of graphing a polar equation on some graphing calculators it is necessary to convert the polar equation to a pair of parametric equations. We will also see in Section 9.7 how the parametric representation is useful in deriving a formula for the length of a curve in polar coordinates.

Formulas 9.9, which involve polar to rectangular conversion,

$$x = r \cos \theta$$

$$y = r \sin \theta$$

provide a means of changing a polar equation

$$r = f(\theta), \qquad \alpha \leq \theta \leq \beta$$

to parametric form. We simply replace r with $f(\theta)$ to get

$$\begin{cases} x = f(\theta) \cos \theta \\ y = f(\theta) \sin \theta \end{cases} \qquad \alpha \leq \theta \leq \beta$$

To illustrate this conversion, parametric equations of the cardioid

$$r = 1 + \cos \theta, \qquad 0 \leq \theta \leq 2\pi$$

are

$$\begin{cases} x = (1 + \cos \theta) \cos \theta \\ y = (1 + \cos \theta) \sin \theta \end{cases} \qquad 0 \leq \theta \leq 2\pi$$

Intersections of Graphs in Polar Coordinates

In the next two examples we show how to find points of intersection of two polar graphs. The second example illustrates a problem that can arise because the coordinates of a given point are not unique.

EXAMPLE 9.19   Find the points of intersection of the curves

$$r = 2 + \sin\theta \qquad \text{and} \qquad r = 5\sin\theta$$

Show the graphs and label the points of intersection.

Solution   We set the two r values equal to one another and solve for θ.

$$2 + \sin\theta = 5\sin\theta$$
$$4\sin\theta = 2$$
$$\sin\theta = \frac{1}{2}$$

So $\theta = \pi/6$ or $5\pi/6$. Substituting these values into either of the original equations gives $r = 5/2$. The points of intersection are therefore $(5/2, \pi/6)$ and $(5/2, 5\pi/6)$.

We show the graphs in Figure 9.44. The first curve is a limaçon without a loop, and the second is a circle.   ∎

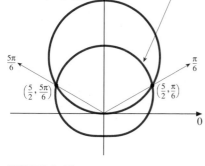

FIGURE 9.44

EXAMPLE 9.20   Find the points of intersection of the curves $r = 2 + 2\cos\theta$ and $r = -2\cos\theta$. Show the curves graphically.

Solution   Proceeding as in Example 9.19, we set the r values equal to one another and solve for θ:

$$2 + 2\cos\theta = -2\cos\theta$$
$$4\cos\theta = -2$$
$$\cos\theta = -\frac{1}{2}$$

Thus, $\theta = \pm 2\pi/3$ and $r = 1$. So we find the points of intersection $(1, 2\pi/3)$ and $(1, -2\pi/3)$. But as the graphs (Figure 9.45) clearly show, the pole is also a point of intersection.

Why did we fail to find this third point of intersection when we solved the equations simultaneously? The answer lies in the fact that the cardioid $r = 2 + 2\cos\theta$ goes through the pole when $\cos\theta = -1$, namely, when $\theta = \pi$. But the circle $r = -2\cos\theta$ goes through the pole when $\cos\theta = 0$, namely, when $\theta = \pi/2$. So the first curve goes through the pole because $(0, \pi)$ is on its graph, whereas the second curve goes through the pole because $(0, \pi/2)$ is on the graph. The points are the same (the pole) but the coordinates are different.   ∎

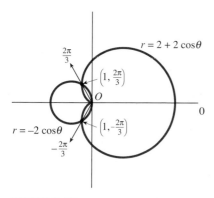

FIGURE 9.45
The pole is a point of intersection.

As the preceding example shows, there can be points of intersection of polar graphs that do not show up when we solve the equations simultaneously. One set of coordinates for such a point satisfies one of the equations and a different set of coordinates for the same point satisfies the other equation. Often, you can discover such "nonsimultaneous" solutions by drawing the graphs, as we did in Example 9.20. For a more general approach, see Exercise 54 in Exercise Set 9.5.

SKETCHING POLAR GRAPHS USING COMPUTER ALGEBRA SYSTEMS

As we have seen in this section, polar graphs can be very hard to sketch. Polar equations can also provide a wide range of beautiful and interesting curves. With the aid of a computer algebra system, many of these beautiful shapes that we could not otherwise envision become accessible to us.

In this section, we will illustrate how to use Maple, Mathematica, and DERIVE to sketch polar graphs. In the exercises, you will be asked to further experiment with graphing polar equations.

 CAS 33

A Cardioid: Sketch the polar curve $r = 2 + 2\cos t$.

Maple:

plot([2+2*cos(t),t,t=0..2*Pi],
    coords=polar);

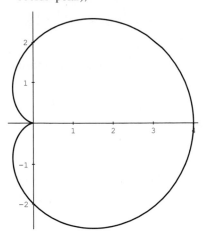

FIGURE 9.5.1
$r = 2 + 2\cos t,\ 0 \le t \le 2\pi$

Mathematica:

PolarPlot[2+2 Cos[t],
    {t,0,2 Pi}]

DERIVE:

(At the □ symbol, go to the next step.)
a (author) □ [2+2cos(t),t] □ p (plot window) □ o (options) □ p (polar) □ p (plot) □ 0 (min) □ 2Pi (max)

In Mathematica, you may first need to load a graphics library by executing the command <<Graphics`Graphics`.

In the examples that follow, assume that the plot mode in DERIVE has been changed to polar as we did in this example.

Notice the difference between the orientation of this graph and that of the cardioid in Example 9.15.

 CAS 34

A Limaçon with a Loop: Sketch the polar curve $r = 1 + 2\sin(t)$.

Maple:

plot([1+2*sin(t),t,t=0..2*Pi],
  coords=polar);

Mathematica:

PolarPlot[1+2 Sin[t],
  {t,0,2 Pi}]

DERIVE:

(At the □ symbol, go to the next step.)
a (author) □ [1+2sin(t),t] □ p (plot window) □ p (plot) □ 0 (min) □ 2Pi (max)

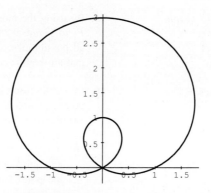

FIGURE 9.5.2
$r = 1 + 2\sin\ t,\ 0 \le t \le 2\pi$

CAS   **CAS 35**

A Rose: Sketch the polar curve $r = \cos 7t$.

Maple:

plot([cos(7*t),t,t=0..2*Pi],
  coords=polar,numpoints=200);

Mathematica:

PolarPlot[Cos[7 t],
  {t,0,2 Pi}]

DERIVE:

(At the □ symbol, go to the next step.)
a (author) □ [cos(7t),t] □ p (plot window) □ p (plot) □ 0 (min) □ 2Pi (max)

FIGURE 9.5.3
$r = \cos\ (7t),\ 0 \le t \le 2\pi$

Exercise Set 9.5

In Exercises 1–24, identify the curve and draw its graph.

1. (a) $r = 3$
   (b) $r = -3$

2. (a) $\theta = 3\pi/4$
   (b) $\theta = \pi/6$

3. (a) $r \cos\theta = 2$
   (b) $r \sin\theta = -3$

4. (a) $r = -3\csc\theta$
   (b) $r = 2\sec\theta$

5. $r = 3\cos\theta$

6. $r = -2\sin\theta$

7. $r = 2(1 - \cos\theta)$

8. $r = 1 - \sin\theta$

9. $r = -2(1 + \sin\theta)$

10. $r = \cos\theta - 1$

11. $r = 5 - 4\cos\theta$

12. $r = 4 - 5\sin\theta$

13. $r = 2 + 3\sin\theta$

14. $r = 3 + 2\cos\theta$

15. $r = 4 - 2\sin\theta$

16. $r = -2 + 4\cos\theta$

17. $r^2 = 4\sin 2\theta$

18. $r^2 = -2\cos 2\theta$

19. $r^2 = 6\sin\theta\cos\theta$

20. $r^2 = \cos^2\theta - \sin^2\theta$

21. $r = 2\cos 3\theta$

22. $r = 3\cos 2\theta$

23. $r = -2\sin 2\theta$

24. $r = \sin 3\theta$

25. Verify that the graphs of $r = a + b\cos\theta$ and $r = -a + b\cos\theta$, where $a > 0$ and $b > 0$, are the same.

In Exercises 26–29, draw the spirals.

26. $r = a\theta$, with $a = 1$ (**spiral of Archimedes**)

27. $r = a/\theta$, with $a = 2$ (**hyperbolic spiral**)

28. $r^2 = a\theta$, with $a = 1$ (**parabolic spiral**)

29. $r = e^{a\theta}$, with $a = \frac{1}{2}$ (**logarithmic spiral**)

 30. Use a CAS or graphing calculator to verify that your graphs in Exercises 26–29 are correct.

In Exercises 31–34, draw the graph by first changing to rectangular coordinates.

31. $r = \dfrac{6}{2\cos\theta - 3\sin\theta}$

32. $r^2 - r(2\cos\theta + 4\sin\theta) + 1 = 0$

33. $r = \dfrac{2}{1 - \cos\theta}$

34. $r = \dfrac{3}{1 + \cos\theta}$

35. By using the most general representations of the points P' in Figure 9.28, devise tests for symmetry that cover all possibilities; that is, devise tests that give conditions that are both necessary and sufficient.

36. Using the result of Exercise 35, show that the graph of a polar equation is symmetric with respect to the following:
    (a) $\theta = 0$ if replacing r by $-r$ and θ by $\pi - \theta$ results in an equivalent equation;
    (b) $\theta = \pi/2$ if replacing r by $-r$ and θ by $-\theta$ results in an equivalent equation;
    (c) the pole if replacing θ by $\pi + \theta$ results in an equivalent equation.

37. Use the results of Exercise 36 to show the specified type of symmetry for the following. Also show that the test as stated in this section fails in each case.
    (a) $r = \tan\theta$; symmetric with respect to $\theta = 0$, $\theta = \pi/2$, and the pole.
    (b) $r = \sin 2\theta\cos\theta$; symmetric with respect to $\theta = \pi/2$.
    (c) $r = 2\csc 2\theta - 1$; symmetric with respect to the pole.
    (d) $r = \sin 2\theta + \cos\theta$; symmetric with respect to $\theta = 0$.

38. Prove that if the polar axis is rotated through an angle α radians, a point P that has coordinates (r, θ) with respect to the original system has coordinates (r', θ') with respect to the new system, where $r' = r$ and $\theta' = \theta - \alpha$. Conclude from this that if the graph of a polar equation is known, the graph of the equation obtained by replacing θ by $\theta - \alpha$ is the original graph rotated through the angle α.

39. Use Exercise 38 to prove that if an equation can be expressed in the form $r = f(\cos\theta)$, then
    (a) when $\cos\theta$ is replaced by $\sin\theta$, the graph is rotated $\pi/2$ radians;
    (b) when $\cos\theta$ is replaced by $-\cos\theta$, the graph is rotated π radians;
    (c) when $\cos\theta$ is replaced by $-\sin\theta$, the graph is rotated $3\pi/2$ radians.

40. Prove that the graph of an equation of the form $r = f(\sin n\theta)$ is the same as that of $r = f(\cos n\theta)$ after rotating through an angle of $\pi/(2n)$ radians. (See Exercise 38.)

41. By multiplying and dividing the right-hand side of the equation $r = a\sin\theta + b\cos\theta$ by $\sqrt{a^2 + b^2}$, show that its graph is the circle $r = \sqrt{a^2 + b^2}\cos\theta$ rotated through the angle α for which $\sin\alpha = a/\sqrt{a^2 + b^2}$ and $\cos\alpha = b/\sqrt{a^2 + b^2}$. (See Exercise 38.)

42. Use the result of Exercise 41 to draw the graph of each of the following:

(a) $r = \sin\theta + \cos\theta$ (b) $r = 3\cos\theta - 4\sin\theta$

 In Exercises 43–47, draw the graph. Use a CAS or a graphing calculator to confirm your results.

43. $r = 2\cos\dfrac{\theta}{2}$ **44.** $r = \sqrt{4\sin\theta}$

45. $r = a\sin 2\theta\cos\theta$, for $a = 1$ **(bifolium)**

46. $r = a\sin\theta\tan\theta$, for $a = 2$ **(cissoid)**

47. $r = a\sec\theta + b$, for $a = 1$, $b = -2$ **(conchoid)**

In Exercises 48–53, find all points of intersection. Draw the graphs.

48. $r = 1 + \cos\theta$, $r = 1$

49. $r = 2(1 - \cos\theta)$, $r = 1$

50. $r^2 = 2\cos 2\theta$, $r = 1$

51. $r = 2\cos\theta$, $r = 2(1 - \cos\theta)$

52. $r = 2\sqrt{2} + \sin\theta$, $r = 5\sin\theta$

53. $r = 1 + 2\cos\theta$, $r = 1$

54. Show that a point P_0 is a point of intersection of the graphs of $r = f(\theta)$ and $r = g(\theta)$ if and only if (a) there exist numbers θ_1 and θ_2 such that $f(\theta_1) = 0$ and $g(\theta_2) = 0$, in which case P_0 is the pole, or (b) there exists a number θ_0 such that

$$f(\theta_0) = (-1)^k g(\theta_0 + k\pi)$$

for some integer k. In (b), P_0 has coordinates $(f(\theta_0), \theta_0)$ or, equivalently, $(g(\theta_0 + k\pi), \theta_0 + k\pi)$. (*Hint:* If P_0 has coordinates (r_0, θ_0) for $r_0 \neq 0$, then all other coordinates of P_0 have one of the forms $(r_0, \theta_0 + 2n\pi)$ or $(-r_0, \theta_0 + (2n + 1)\pi)$, where $n = 0, \pm 1, \pm 2, \ldots$.)

In Exercises 55 and 56, find all points of intersection, making use of Exercise 54.

55. $r = 2$, $r = 4\sin 2\theta$

56. $r = 25\cos\theta$, $r = 3 + 2\cos\theta$

 57. Family of butterflies. Use a CAS or graphing calculator to sketch several curves from the family

$$r = 1 + \sin(n\theta) + \cos^2(2n\theta), \qquad n \geq 1$$

 58. Family of hands. Repeat Exercise 57 with the family

$$r = 1 + \sin(2\theta) + \cos^2(n\theta), \qquad n \geq 5$$

 59. Use a CAS or graphing calculator to sketch several curves from each of the following families, where in each case m and n are positive integers.

(a) $r = \sin(m\theta) + \sin^2(n\theta)$
(b) $r = \cos(m\theta) + \cos^2(n\theta)$
(c) $r = \sin(m\theta) + \cos^2(n\theta)$
(d) $r = \cos(m\theta) + \sin^2(n\theta)$

 60. Use a CAS or a graphing calculator to sketch several curves from the family

$$r = [1 + \sin(m\theta) + \cos^2(n\theta)]^k$$

where m, n, and k are positive integers.

9.6 AREAS IN POLAR COORDINATES

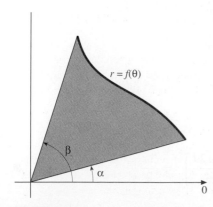

FIGURE 9.46

In Chapter 6 we derived formulas for areas between graphs whose equations were given in rectangular form. We cannot apply these formulas to regions between graphs of polar equations, however. In this section we develop formulas applicable to such regions. We begin with the simplest case, as shown in Figure 9.46. The region in question is bounded by the graph of $r = f(\theta)$ and the rays $\theta = \alpha$ and $\theta = \beta$, with $\beta \leq \alpha + 2\pi$. We assume f is continuous and nonnegative for $\alpha \leq \theta \leq \beta$. Our objective is to find a formula for the area of the region.

We partition the interval $[\alpha, \beta]$ by means of the numbers $\alpha = \theta_0 < \theta_1 < \theta_2 < \cdots < \theta_n = \beta$. For each k from 1 to n, let $\Delta\theta_k = \theta_k - \theta_{k-1}$. The θ_k rays divide the region in question into n subregions, as we show in Figure 9.47(a). Each of these subregions is shaped somewhat like a sector of a circle. In Figure

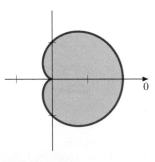

FIGURE 9.47

9.47(b) we show a typical subregion and a circular sector that approximates it. We find the radius r_k^* of the approximating sector by evaluating the function f at an arbitrary value of θ between θ_{k-1} and θ_k. (We have labeled this value θ_k^*.) If $\Delta\theta_k$ is small, we would expect that the area of the circular sector would differ very little from the actual area of the subregion. The area ΔA_k of the circular sector is (see Exercise 32 in Exercise Set 9.6)

$$\Delta A_k = \frac{1}{2}(r_k^*)^2 \Delta\theta_k = \frac{1}{2}\left[f(\theta_k^*)\right]^2 \Delta\theta_k$$

By summing all such ΔA_k values for $k = 1, 2, \ldots, n$, we obtain an approximation to the desired area A of the total region:

$$A \approx \sum_{k=1}^{n} \Delta A_k = \frac{1}{2}\sum_{k=1}^{n}\left[f(\theta_k^*)\right]^2 \Delta\theta_k$$

The sum on the right is a Riemann sum. Thus, by taking finer and finer partitions, such that each of the angles $\Delta\theta_k$ approaches 0, we get the exact value of A as the limit:

$$A = \lim_{\max \Delta\theta_k \to 0} \sum_{k=1}^{n} \Delta A_k = \lim_{\max \Delta\theta_k \to 0} \sum_{k=1}^{n} \frac{1}{2}\left[f(\theta_k^*)\right]^2 \Delta\theta_k = \int_{\alpha}^{\beta} \frac{1}{2}\left[f(\theta)\right]^2 d\theta$$

That is,

$$A = \int_{\alpha}^{\beta} \frac{1}{2}\left[f(\theta)\right]^2 d\theta$$

Since $r = f(\theta)$, we can write this result in the following briefer form:

Area Bounded by a Polar Curve $r = f(\theta)$

$$A = \frac{1}{2}\int_{\alpha}^{\beta} r^2 d\theta \qquad\qquad (9.11)$$

EXAMPLE 9.21   Find the area inside the cardioid $r = 1 + \cos\theta$.

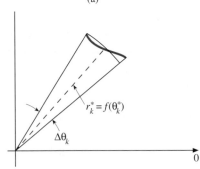

FIGURE 9.48

Solution   The graph is shown in Figure 9.48. By symmetry, it is sufficient to find the area between $\theta = 0$ and $\theta = \pi$ and then double the result. So we have, by Equation 9.11,

$$A = 2\cdot\frac{1}{2}\int_{0}^{\pi} r^2 d\theta = \int_{0}^{\pi} (1 + \cos\theta)^2 d\theta$$

$$= \int_{0}^{\pi} (1 + 2\cos\theta + \cos^2\theta)d\theta$$

$$= \int_{0}^{\pi} \left(1 + 2\cos\theta + \frac{1 + \cos 2\theta}{2}\right) d\theta$$

$$= \int_{0}^{\pi} \left(\frac{3}{2} + 2\cos\theta + \frac{\cos 2\theta}{2}\right) d\theta$$

$$= \frac{3\theta}{2} + 2\sin\theta + \frac{\sin 2\theta}{4}\bigg]_{0}^{\pi} = \frac{3\pi}{2} \qquad\blacksquare$$

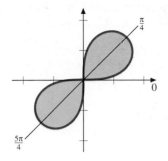

FIGURE 9.49

EXAMPLE 9.22   Find the area inside the lemniscate $r^2 = 4 \sin 2\theta$.

Solution   We show the graph in Figure 9.49. We again use symmetry (to the pole) by finding the area between $\theta = 0$ and $\theta = \frac{\pi}{2}$ and doubling the result:

$$A = 2 \cdot \frac{1}{2} \int_0^{\pi/2} r^2 d\theta = \int_0^{\pi/2} 4 \sin 2\theta \, d\theta = -2 \cos 2\theta]_0^{\pi/2}$$

$$= -2[-1 - 1] = 4 \qquad \blacksquare$$

EXAMPLE 9.23   Family of "Butterflies"   Consider the family

$$r = 1 + \sin(n\theta) + \cos^2(2n\theta), \qquad 0 \le \theta \le 2\pi$$

where n is a positive integer. Show that the area enclosed by the graph is the same for all positive integers n.

Solution   In Exercise 57 of Exercise Set 9.5, we asked you to use a CAS or graphing calculator to graph several members of this family. In Figure 9.50 we show the curves corresponding to $n = 2$ and $n = 3$.

For any positive integer n, the area enclosed is

$$A = \frac{1}{2} \int_0^{2\pi} r^2 d\theta = \frac{1}{2} \int_0^{2\pi} \left[1 + \sin(n\theta) + \cos^2(2n\theta) \right]^2 d\theta$$

$$= \frac{1}{2} \int_0^{2\pi} \left[1 + \sin^2(n\theta) + \cos^4(2n\theta) + 2\sin(n\theta) + 2\cos^2(2n\theta) \right.$$

$$\left. + 2\sin(n\theta)\cos^2(2n\theta) \right] d\theta$$

$$= \frac{1}{2} \int_0^{2\pi} \left[1 + \frac{1 - \cos(2n\theta)}{2} + \left(\frac{1 + \cos(4n\theta)}{2} \right)^2 + 2\sin(n\theta) \right.$$

$$\left. + (1 + \cos(4n\theta)) + \sin(n\theta)(1 + \cos(4n\theta)) \right] d\theta$$

Since

$$\left(\frac{1 + \cos(4n\theta)}{2} \right)^2 = \frac{1}{4} \left(1 + 2\cos(4n\theta) + \cos^2(4n\theta) \right)$$

$$= \frac{1}{4} \left[1 + 2\cos(4n\theta) + \frac{1 + \cos(8n\theta)}{2} \right]$$

$$= \frac{3}{8} + \frac{\cos(4n\theta)}{2} + \frac{\cos(8n\theta)}{8}$$

we have, on combining like terms in the last integral,

$$A = \frac{1}{2} \int_0^{2\pi} \left[\frac{23}{8} - \frac{\cos(2n\theta)}{2} + \frac{3\cos(4n\theta)}{2} + \frac{\cos(8n\theta)}{8} + 3\sin(n\theta) \right.$$

$$\left. + \sin(n\theta)\cos(4n\theta) \right] d\theta$$

It is not difficult to show that the only term of the integrand that does not give 0 when integrated over the interval $[0, 2\pi]$ is the constant term, 23/8. Thus,

$$A = \frac{1}{2} \int_0^{2\pi} \frac{23}{8} \, d\theta = \frac{1}{2} \cdot \frac{23}{8} \cdot 2\pi = \frac{23\pi}{8}$$

Since the area is the same for all positive integers n, each of the butterfly curves encloses the same area. $\qquad \blacksquare$

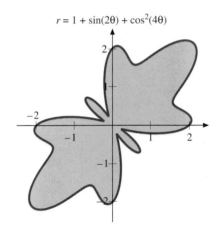

$r = 1 + \sin(2\theta) + \cos^2(4\theta)$

$r = 1 + \sin(3\theta) + \cos^2(6\theta)$

FIGURE 9.50
Two "butterflies"

Area Between Two Polar Graphs

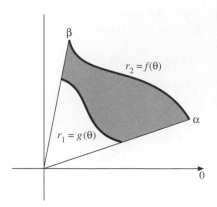

FIGURE 9.51

Suppose now that we want the area between the graphs of two continuous functions f and g for which $0 \le g(\theta) \le f(\theta)$, for θ between α and β, where again we have $\beta \le \alpha + 2\pi$. We illustrate a typical situation in Figure 9.51. Then we have

$$A = \frac{1}{2} \int_\alpha^\beta \left[f(\theta) \right]^2 d\theta - \frac{1}{2} \int_\alpha^\beta \left[g(\theta) \right]^2 d\theta$$

$$= \frac{1}{2} \int_\alpha^\beta \left\{ \left[f(\theta) \right]^2 - \left[g(\theta) \right]^2 \right\} d\theta$$

Writing $r_1 = g(\theta)$ and $r_2 = f(\theta)$, we can write the result in the following briefer form:

Area Between Two Polar Curves

$$A = \frac{1}{2} \int_\alpha^\beta (r_2^2 - r_1^2) d\theta \qquad (r_2 \ge r_1) \qquad (9.12)$$

EXAMPLE 9.24   Find the area outside the circle $r = 1$ and inside the circle $r = 2 \cos \theta$.

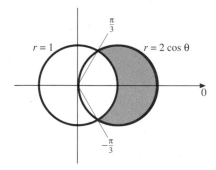

FIGURE 9.52

Solution   We show the two circles and the area in question in Figure 9.52. Our first task is to find the points of intersection so that we will know the limits on θ. Solving the equations simultaneously, we get

$$1 = 2 \cos \theta$$

$$\cos \theta = \frac{1}{2}$$

So $\theta = \pm \pi/3$. Because of symmetry we will find the area from 0 to $\pi/3$ and double the result. By Equation 9.12, we have

$$A = 2 \cdot \frac{1}{2} \int_0^{\pi/3} \left[(2 \cos \theta)^2 - 1^2 \right] d\theta$$

$$= \int_0^{\pi/3} (4 \cos^2 \theta - 1) d\theta = \int_0^{\pi/3} \left[2(1 + \cos 2\theta) - 1 \right] d\theta$$

$$= \int_0^{\pi/3} (2 \cos 2\theta + 1) d\theta = \sin 2\theta + \theta \Big]_0^{\pi/3} = \frac{\sqrt{3}}{2} + \frac{\pi}{3} \qquad ■$$

EXAMPLE 9.25   Find the area that is both inside the circle $r = 3 \sin \theta$ and outside the cardioid $r = 1 + \sin \theta$.

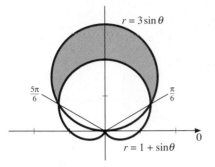

$r = 3 \sin \theta$

$\frac{5\pi}{6}$ $\frac{\pi}{6}$

$r = 1 + \sin \theta$

0

FIGURE 9.53

Solution   We show the graphs in Figure 9.53. Setting the two r values equal to each other gives

$$3 \sin \theta = 1 + \sin \theta$$

$$2 \sin \theta = 1$$

$$\sin \theta = \frac{1}{2}$$

$$\theta = \frac{\pi}{6}, \frac{5\pi}{6}$$

So we have, using symmetry,

$$A = 2 \cdot \frac{1}{2} \int_{\pi/6}^{\pi/2} \left[(3 \sin \theta)^2 - (1 + \sin \theta)^2 \right] d\theta$$

$$= \int_{\pi/6}^{\pi/2} (8 \sin^2 \theta - 2 \sin \theta - 1) d\theta$$

$$= \int_{\pi/6}^{\pi/2} [4(1 - \cos 2\theta) - 2 \sin \theta - 1] \, d\theta$$

$$= \int_{\pi/6}^{\pi/2} (-4 \cos 2\theta - 2 \sin \theta + 3) d\theta$$

$$= \left[-2 \sin 2\theta + 2 \cos \theta + 3\theta \right]_{\pi/6}^{\pi/2} = \frac{3\pi}{2} - \left(-\sqrt{3} + \sqrt{3} + \frac{\pi}{2} \right) = \pi \quad \blacksquare$$

Exercise Set 9.6

In Exercises 1–10, find the area enclosed by the graph of the given equation. Sketch the graph.

1. $r = 2(1 - \cos \theta)$

2. $r = 3 + 2 \sin \theta$

3. $r^2 = 9 \cos 2\theta$

4. $r = \cos 2\theta$

5. $r = \sin \theta - 1$

6. $r = 4 + 2 \cos \theta$

7. $r = 2 \sin 3\theta$

8. $r^2 = 2 \sin \theta \cos \theta$

9. $r = -4 \cos \theta$

10. $r = \sin \theta + \cos \theta$

In Exercises 11–20, find the area outside the graph of the first equation and inside the graph of the second. Sketch the graphs, showing the area in question.

11. $r = 2; r = 2(1 + \cos \theta)$

12. $r = \sqrt{3}; r = 2 \sin \theta$

13. $r = \sqrt{2} - \cos \theta; r = \cos \theta$

14. $r = \sqrt{2}; r^2 = 4 \cos 2\theta$

15. $r = 2 \cos \theta; r^2 = 2\sqrt{3} \sin 2\theta$

16. $r = 1; r = 2 \cos 2\theta$

17. $r = 3 + 2 \cos \theta; r = 1 - 2 \cos \theta$

18. $r = 2 - \sin \theta; r = 2(1 + \sin \theta)$

19. $r = 3(1 + \cos \theta); r = 1 - \cos \theta$

20. $r = \sin \theta; r = \sqrt{\sin 2\theta}$

21. Find the area enclosed by the small loop of the limaçon $r = 1 - 2 \cos \theta$.

22. Find the area inside the limaçon $r = 1 + 2 \cos \theta$ that is to the right of the line $r = \sec \theta$.

23. Find the area inside the limaçon $r = 3 - 2 \sin \theta$ that lies below the line $r = -2 \csc \theta$.

24. Find the area inside both the circle $r = 3 \sin \theta$ and the cardioid $r = 1 + \sin \theta$.

25. Find the area inside the cardioid $r = 2(1 + \sin\theta)$ and also inside the circle $r = 2\cos\theta$.

26. Find the smallest area bounded by $r = \ln\theta$, $\theta = 1$, and $\theta = e$. Sketch the graph.

27. Find the area inside the limaçon $r = 1 - \sqrt{2}\sin\theta$ that lies outside its small loop.

28. Find the area inside the graph of $r = 2\cos 3\theta$ and outside the graph of $r = -2\cos\theta$.

29. Find the area inside the circle $r = -2\sin\theta$ that lies outside the small loop of the limaçon $r = 1 - 2\sin\theta$.

30. (a) Find all points of intersection of the graphs of $r = 2\cos\theta/2$ and $r = -2\cos\theta$. Draw the graphs.
   (b) Find the area inside the graph of $r = -2\cos\theta$ that lies outside the two small loops of $r = 2\cos\theta/2$.

31. Find the shaded area shown in the figure at the top of the next column.

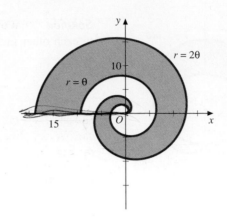

32. Derive the formula
$$A = \tfrac{1}{2}r^2\theta$$
for the area of a circular sector of radius r and central angle θ (in radians).

33. Show that for $n \geq 1$ the rose curves $r = \cos n\theta$ and $r = \sin n\theta$ all enclose the same area when
   (a) n is even
   (b) n is odd

34. Use a CAS or graphing calculator to show that
$$\int_0^{2\pi} \left[1 + \sin(m\theta) + \cos^2(n\theta)\right]^k \, d\theta = c_k$$
where for each positive integer k, c_k is a constant, independent of m and n ($m, n \geq 1$). In particular, show that $c_5 = \dfrac{8829}{128}\pi$.

9.7 TANGENT LINES TO POLAR GRAPHS; ARC LENGTH

FIGURE 9.54

In rectangular coordinates, if $y = f(x)$ and f is a differentiable function, then $f'(x)$ gives the slope of the tangent line to the graph. In polar coordinates, if $r = f(\theta)$ and f is a differentiable function, it might be tempting to assume that $f'(\theta)$ also gives the slope of the tangent line to the graph. But this apparent result clearly is not so, as the example $r = \theta$ shows. Here, $f(\theta) = \theta$, and so $f'(\theta) = 1$. But the graph (Figure 9.54) is the Spiral of Archimedes, which clearly does *not* always have slope 1. So we have to approach the slope of a polar graph in a different way.

The secret of success lies in changing the equation of a polar curve to parametric equations, as we did in Section 9.5. Recall that the equation $r = f(\theta)$ has the parametric representation

$$\begin{cases} x = f(\theta)\cos\theta \\ y = f(\theta)\sin\theta \end{cases} \qquad \alpha \leq \theta \leq \beta \qquad (9.13)$$

Now, by Equation 9.2 (with t replaced by θ),

$$\frac{dy}{dx} = \frac{\dfrac{dy}{d\theta}}{\dfrac{dx}{d\theta}} = \frac{f'(\theta)\sin\theta + f(\theta)\cos\theta}{f'(\theta)\cos\theta - f(\theta)\sin\theta} \qquad (9.14)$$

provided the denominator is nonzero. This result—though true—is not very satisfactory, since the formula is rather complicated. There is a more suitable measure than slope for describing the behavior of the tangent line in polar coordinates that we discuss presently. First, though, we can use Equation 9.14 to confirm our observation in Section 9.5 about tangent lines at the pole.

Tangent Lines at the Pole

Suppose that $r = f(\theta)$ and f is differentiable at the pole. Set $r = 0$ and let $\theta = \theta_0$ be a root of the equation $f(\theta) = 0$ (that is, $f(\theta_0) = 0$). Assume further that $f'(\theta_0) \neq 0$. Then, since $f(\theta_0) = 0$, we have, from Equation 9.14, the slope at the origin

$$\left.\frac{dy}{dx}\right|_{0,\theta_0} = \frac{f'(\theta_0)\sin\theta_0}{f'(\theta_0)\cos\theta_0} = \tan\theta_0$$

if $\cos\theta_0 \neq 0$. Since the slope of the tangent line is the tangent of its angle of inclination (see Figure 9.55), it follows that its angle of inclination is θ_0, or that it differs from this by a multiple of π. In any case, the line $\theta = \theta_0$ is tangent to the curve at the origin. If $\cos\theta_0 = 0$, then the tangent line is vertical, and again $\theta = \theta_0$ is this tangent line, since $\theta_0 = \pi/2$ or an odd multiple of $\pi/2$. Note that if $f'(\theta_0) = 0$, then both $dx/d\theta$ and $dy/d\theta$ are 0 at the origin, so the origin is a singular point.

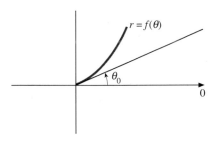

FIGURE 9.55
Slope of tangent line $= \tan\theta_0$

An Alternative to Slope of the Tangent Line

To find a way to describe the tangent line to a polar graph that is more convenient than its slope, consider the graph of the differentiable function $r = f(\theta)$. At an arbitrary point P on the curve, other than the pole, let ϕ (phi) be the angle of inclination of the tangent line at P. Then $\tan\phi$ is the slope of the tangent line. Its value must then be given by Equation 9.14, provided $\phi \neq \pi/2$ (if $\phi = \pi/2$, $\tan\phi$ is not defined):

$$\tan\phi = \frac{f'(\theta)\sin\theta + f(\theta)\cos\theta}{f'(\theta)\cos\theta - f(\theta)\sin\theta} \qquad (9.15)$$

As we show in Figure 9.56, we let ψ (psi) denote the smallest positive angle from the ray OP to the tangent line. Since the sum of the angles in a triangle is π radians (i.e., 180°), we must have

$$\theta + \psi + (\pi - \phi) = \pi$$

or $\psi = \phi - \theta$. Although this figure illustrates the special case in which $\phi > \theta$ and $0 \le \theta < 2\pi$, it can be shown that in all cases

$$\psi = \phi - \theta + n\pi \qquad (9.16)$$

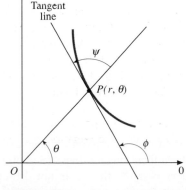

FIGURE 9.56

where n is an integer. Because the period of the tangent function is π, it follows that if $\psi \neq \pi/2$, $\tan \psi = \tan(\phi - \theta + n\pi) = \tan(\phi - \theta)$. By the trigonometric identity for $\tan(\phi - \theta)$ (see Appendix 2), therefore,

$$\tan \psi = \frac{\tan \phi - \tan \theta}{1 + \tan \phi \tan \theta}$$

In the exercises we ask you to show that when $\tan \phi$ is replaced by the right-hand side of Equation 9.15 and $\tan \theta$ is replaced by $\sin \theta / \cos \theta$, we get the simple formula

$$\tan \psi = \frac{f(\theta)}{f'(\theta)}$$

provided both $f(\theta)$ and $f'(\theta)$ are nonzero. Since $r = f(\theta)$, this equation can also be written in the form

$$\tan \psi = \frac{r}{dr/d\theta} \tag{9.17}$$

EXAMPLE 9.26   Find the angle ψ at the point $(3, \pi/3)$ on the graph of $r = 2(1 + \cos \theta)$. Show the result graphically.

Solution   By Equation 9.17 for $\theta = \pi/3$,

$$\tan \psi = \frac{r}{dr/d\theta} = \frac{2(1 + \cos \theta)}{-2 \sin \theta} = -\frac{1 + \cos \theta}{\sin \theta} = -\frac{\frac{3}{2}}{\frac{\sqrt{3}}{2}} = -\sqrt{3}$$

So $\psi = 2\pi/3$. We show the graph in Figure 9.57. Note that at the point in question the tangent line is horizontal.   ∎

From Equation 9.16, it follows that the angle of inclination ϕ of the tangent line is $\theta + \psi$ or differs by a multiple of π. In any case, if the tangent line is nonvertical,

$$\tan \phi = \tan(\theta + \psi) \tag{9.18}$$

This formula is an alternative to Equation 9.14 for calculating the slope, should it be needed. For example, suppose we want the slope of $r = \cos^2 \theta$ at $\theta = \pi/4$. First we calculate $\tan \psi$:

$$\tan \psi = \frac{r}{dr/d\theta} = \frac{\cos^2 \theta}{-2 \sin \theta \cos \theta} = -\frac{1}{2} \cot \theta$$

At $\theta = \pi/4$, $\tan \psi = -\frac{1}{2}$ and $\tan \theta = 1$. So the slope is

$$\tan \phi = \tan(\theta + \psi) = \frac{\tan \theta + \tan \psi}{1 - \tan \theta \tan \psi} = \frac{1 + \left(-\frac{1}{2}\right)}{1 - (1)\left(-\frac{1}{2}\right)} = \frac{1}{3}$$

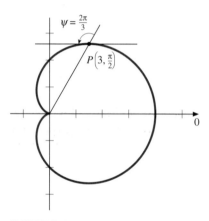

$\psi = \frac{2\pi}{3}$

$P\left(3, \frac{\pi}{2}\right)$

FIGURE 9.57

$r_2 = g(\theta)$

α

$r_1 = f(\theta)$

FIGURE 9.58
The angle α from $r_1 = f(\theta)$ to $r_2 = g(\theta)$ is the smallest positive angle from the tangent line of r_1 to the tangent line of r_2 at their point of intersection.

Angle Between Two Curves

When two curves intersect, the angle α between them at a point of intersection is defined as the smallest positive angle between their tangent lines. (See Figure 9.58.) If the equations are $r_1 = f(\theta)$ and $r_2 = g(\theta)$, then it is not difficult to show that the smallest positive angle α from the first curve to the second is

$\psi_2 - \psi_1$, or differs from that result by an integral multiple of π (see Exercise 47 in Exercise Set 9.7), where ψ_1 and ψ_2 are the angles defined above, for $r_1 = f(\theta)$ and $r_2 = g(\theta)$, respectively. So

$$\tan\alpha = \tan(\psi_2 - \psi_1) \qquad (9.19)$$

This result assumes that $\alpha \neq \pi/2$. When $\alpha = \pi/2$, the tangent lines are perpendicular, and the curves are said to be *orthogonal*. For such orthogonal curves, $\tan\alpha$ is undefined, but $\cot\alpha = 0$. You will be asked to show in Exercise 31 of Exercise Set 9.7 that setting $\cot\alpha = 0$ leads to the condition

$$\tan\psi_1 \tan\psi_2 = -1 \qquad (9.20)$$

for orthogonality, provided neither ψ_1 nor ψ_2 is $\pi/2$. Note the similarity to the perpendicularity condition $m_1 m_2 = -1$ for slopes m_1 and m_2.

EXAMPLE 9.27   Find the angle from $r_1 = 2 - \cos\theta$ to $r_2 = 2(1 + \cos\theta)$ at each of the points of intersection.

Solution   First we find the points of intersection. Setting $r_1 = r_2$ gives

$$2 - \cos\theta = 2 + 2\cos\theta$$

$$3\cos\theta = 0$$

$$\cos\theta = 0$$

$$\theta = \pm\frac{\pi}{2}$$

So the curves intersect at $(2, \pm\pi/2)$. An examination of the graphs (Figure 9.59) shows that there are no other points of intersection. By symmetry we need only consider the angle between the curves when $\theta = \pi/2$. To use Equation 9.19 we need $\tan\psi_1$ and $\tan\psi_2$:

$$\tan\psi_1 = \frac{r_1}{dr_1/d\theta} = \frac{2 - \cos\theta}{\sin\theta}$$

$$\tan\psi_2 = \frac{r_2}{dr_2/d\theta} = \frac{2(1 + \cos\theta)}{-2\sin\theta} = \frac{1 + \cos\theta}{-\sin\theta}$$

At $\theta = \pi/2$, $\tan\psi_1 = 2$ and $\tan\psi_2 = -1$. Thus,

$$\tan\alpha = \tan(\psi_2 - \psi_1) = \frac{\tan\psi_2 - \tan\psi_1}{1 + \tan\psi_1 \tan\psi_2}$$

$$= \frac{-1 - 2}{1 + 2(-1)} = 3$$

Using a calculator and the inverse tangent, we find that $\alpha \approx 1.25$ radians $= (1.25)(180/\pi)°$, or approximately $71.6°$.   ■

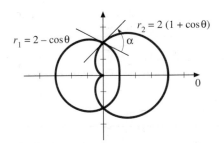

$r_1 = 2 - \cos\theta$

$r_2 = 2(1 + \cos\theta)$

α

FIGURE 9.59

Arc Length in Polar Coordinates

Equations 9.13 enable us to use the results from Section 9.3 to find the length of the curve $r = f(\theta)$, where f' is continuous for $\alpha \leq \theta \leq \beta$. We calculate $dx/d\theta$ and $dy/d\theta$ from Equations 9.13:

$$\frac{dx}{d\theta} = f'(\theta)\cos\theta - f(\theta)\sin\theta$$

and

$$\frac{dy}{d\theta} = f'(\theta)\sin\theta + f(\theta)\cos\theta$$

So, by Equation 9.4, with t replaced by θ,

$$ds = \sqrt{\left(\frac{dx}{d\theta}\right)^2 + \left(\frac{dy}{d\theta}\right)^2}\,d\theta$$

$$= \sqrt{[f'(\theta)\cos\theta - f(\theta)\sin\theta]^2 + [f'(\theta)\sin\theta + f(\theta)\cos\theta]^2}\,d\theta$$

We leave it for you in Exercise 48 of Exercise Set 9.7 to show that after simplification, this equation becomes

$$ds = \sqrt{[f(\theta)]^2 + [f'(\theta)]^2}\,d\theta \qquad (9.21)$$

We also note that if $f'(\theta)$ is continuous for θ in $[\alpha, \beta]$, both $dx/d\theta$ and $dy/d\theta$ are continuous there. It therefore follows from Equation 9.6 that arc length in polar coordinates is given by

$$L = \int_\alpha^\beta \sqrt{[f(\theta)]^2 + [f'(\theta)]^2}\,d\theta$$

provided f' is continuous in $[\alpha, \beta]$. Since $r = f(\theta)$, we can also write the equation in the following form:

Arc Length in Polar Coordinates

$$L = \int_\alpha^\beta \sqrt{r^2 + \left(\frac{dr}{d\theta}\right)^2}\,d\theta \qquad (9.22)$$

EXAMPLE 9.28   Find the length of the cardioid $r = 1 + \cos\theta$.

Solution   Because of symmetry it is sufficient to find the length for θ in $[0, \pi]$ and double the result. Since $dr/d\theta = -\sin\theta$, we have

$$L = 2\int_0^\pi \sqrt{r^2 + \left(\frac{dr}{d\theta}\right)^2}\,d\theta$$

$$= 2\int_0^\pi \sqrt{1 + 2\cos\theta + \cos^2\theta + \sin^2\theta}\,d\theta$$

$$= 2\int_0^\pi \sqrt{2(1+\cos\theta)}\,d\theta$$

To complete the integration, we need to eliminate the square root. To do so, we make use of the identity (see Appendix 2)

$$\cos\frac{\theta}{2} = \pm\sqrt{\frac{1+\cos\theta}{2}}$$

choosing the positive sign, since $\cos(\theta/2)$ is positive for θ in $[0, \pi]$. If we multiply both sides of this equation by 2, we get $\sqrt{2(1 + \cos \theta)} = 2 \cos(\theta/2)$. So we have

$$L = 2 \int_0^\pi 2 \cos \frac{\theta}{2} \, d\theta = 8 \sin \frac{\theta}{2} \Big]_0^\pi = 8 \qquad \blacksquare$$

Exercise Set 9.7

In Exercises 1–10, find $\tan \psi$ at the indicated value of θ.

1. $r = 2(1 - \sin \theta); \; \theta = \pi/6$

2. $r = 3 \cos 2\theta; \; \theta = \pi/6$

3. $r = \cos^2 \theta; \; \theta = \pi/4$

4. $r = 4 \sin \theta; \; \theta = \pi/3$

5. $r = 4 - 2 \sin \theta; \; \theta = 5\pi/6$

6. $r = \sqrt{\cos 2\theta}; \; \theta = -\pi/6$

7. $r = \dfrac{1}{1 - \cos \theta}; \; \theta = 2\pi/3$

8. $r = \dfrac{2}{2 - \sin \theta}; \; \theta = -\pi/6$

9. $r = \sin \theta \cos \theta; \; \theta = \pi/3$

10. $r = 1 - 2 \sin \dfrac{\theta}{2}; \; \theta = \pi/2$

In Exercises 11–20, make use of the fact that $\tan \phi = \tan(\theta + \psi)$, where ϕ is the angle of inclination of the tangent line, to find the slope of the curve at the given value of θ.

11. $r = 1 - \cos \theta; \; \theta = 2\pi/3$

12. $r = 2 + \sin \theta; \; \theta = -\pi/6$

13. $r = \dfrac{1}{1 - \cos \theta}; \; \theta = \pi/3$

14. $r = 2 \cos \theta; \; \theta = \pi/4$

15. $r = 2 \cos(\theta/2) - 1; \; \theta = \pi$

16. $r = \sqrt{\sin 2\theta}; \; \theta = \pi/3$

17. $r = 1 - 2 \sin \theta; \; \theta = \pi$

18. $r = \cos^2 \theta; \; \theta = \pi/4$

19. $r = \dfrac{2}{3 - 2 \sin \theta}; \; \theta = \pi$

20. $r = \sec \theta + \tan \theta; \; \theta = 2\pi/3$

In Exercises 21 and 22, show that the curve has a horizontal tangent line at the indicated value of θ, using $\tan \phi = \tan(\theta + \psi)$.

21. $r = 4 + 4 \cos \theta; \; \theta = -\pi/3$

22. $r = 2\sqrt{\cos 2\theta}; \; \theta = 5\pi/6$

In Exercises 23 and 24, show that the curve has a vertical tangent line at the indicated point, using the fact that $\cot \phi = \cot(\theta + \psi)$.

23. $r = 1 - \cos \theta; \; \theta = \pi/3$

24. $r = 3 + \tan^2 \theta; \; \theta = 3\pi/4$

25. Show that for the logarithmic spiral $r = ae^{b\theta}$, the angle ψ is constant. What is ψ when $b = 1$?

In Exercises 26–30, find the tangent of the angle from the first curve to the second at each of their points of intersection.

26. $r_1 = 2 - \cos \theta; \; r_2 = 3 \cos \theta$

27. $r_1 = \dfrac{1}{1 - \cos \theta}; \; r_2 = \dfrac{1}{2 + \cos \theta}$

28. $r_1 = \dfrac{3}{1 + \sin \theta}; \; r_2 = 4 \sin \theta$

29. $r_1 = \sqrt{3} \cos \theta; \; r_2 = \sin 2\theta$

30. $r_1 = \sqrt{2} + \sin \theta; \; r_2 = 3 \sin \theta$

31. Prove the orthogonality condition, Equation 9.20.

In Exercises 32–34, show that the graphs are orthogonal at all points of intersection.

32. $r_1 = \sin \theta; \; r_2 = \cos \theta$

33. $r_1 = 3 \cos \theta; \; r_2 = 2 - \cos \theta$

34. $r_1 = \dfrac{1}{1 - \sin \theta}; \; r_2 = \dfrac{1}{1 + \sin \theta}$

In Exercises 35–37, show that the curves intersect orthogonally for all nonzero values of a and b.

35. $r_1 = a \cos \theta$; $r_2 = b \sin \theta$

36. $r_1 = a\theta$; $r_2 = be^{-\theta^2/2}$

37. $r_1 = a(1 - \sin \theta)$; $r_2 = b(1 + \sin \theta)$, except at the pole

In Exercises 38–44, find the length of the curve on the given interval.

38. $r = 2a \cos \theta$; $0 \le \theta \le 2\pi$

39. $r = e^{3\theta}$; $0 \le \theta \le \ln 2$

40. $r = \cos \theta - 1$; $0 \le \theta \le \pi$

41. $r = 2 \cos^2(\theta/2)$; $0 \le \theta \le \pi$

42. $r = \sin^3(\theta/3)$; $0 \le \theta \le 3\pi$

43. $r = \theta^2 - 1$; $0 \le \theta \le 3$

44. $r = \sin \theta - \cos \theta$; $-\pi/2 \le \theta \le \pi$

45. Prove Equation 9.17.

46. Prove that the condition for orthogonality of the graphs of $r = f(\theta)$ and $r = g(\theta)$ can be put in the form $f(\theta)g(\theta) + f'(\theta)g'(\theta) = 0$.

47. Verify Equation 9.19. (*Hint:* Express $\phi_2 - \phi_1$ in terms of ψ_2 and ψ_1.)

48. Verify Equation 9.21.

49. Find the length of the curve $r = \sin^2 \theta$ from $\theta = 0$ to $\theta = \pi/2$.

50. Find the length of that portion of the curve $r = 2(1 - \sin \theta)$ that lies inside $r = -6 \sin \theta$.

51. (a) Suppose that f' is continuous and that $f(\theta) \sin \theta \ge 0$ for $\alpha \le \theta \le \beta$. Derive the following formula for the surface area obtained by rotating the graph of $r = f(\theta)$ about the line $\theta = 0$:

$$S = 2\pi \int_{\alpha}^{\beta} f(\theta) \sin \theta \sqrt{[f(\theta)]^2 + [f'(\theta)]^2} \, d\theta$$

(b) State appropriate conditions and derive an analogous formula for the surface area formed by rotation about the line $\theta = \pi/2$.

In Exercises 52–55, use the results of Exercise 51 to find the surface area formed when the graph of $r = f(\theta)$ is rotated about the specified axis.

52. $r = 1 - \cos \theta$, $0 \le \theta \le \pi$; about $\theta = 0$

53. $r = \sqrt{\sin 2\theta}$, $0 \le \theta \le \pi/2$; about $\theta = \pi/2$

54. $r = e^{2\theta}$, $0 \le \theta \le \pi$; about $\theta = 0$

55. $r = 2 \sin^2(\theta/2)$, $0 \le \theta \le \pi/2$; about (a) $\theta = 0$ and (b) $\theta = \pi/2$

9.8   POLAR EQUATIONS OF CONIC SECTIONS

Our goal in this section is to obtain polar equations of the curves known as **conic sections**—the circle, parabola, ellipse, and hyperbola. The name *conic sections* (also called **conics**) comes from the fact that when a plane passes through a right-circular cone at various angles, the curves of intersection are the conic sections, as we illustrate in the following photos.

Slicing a cone perpendicular to its axis gives a circle, and slicing it parallel to its axis gives a branch of a hyperbola.

Slicing through a cone at an angle gives an ellipse, and slicing it parallel to an edge gives a parabola.

We concentrate on the parabola, ellipse, and hyperbola, since we already know that a circle of radius a centered at the origin has the polar equation $r = a$.

Let us begin by reviewing some facts concerning the three types of curves in question, including their rectangular equations. (See also Supplementary Appendix 3.)

The Parabola

A parabola is defined as the set of points in the plane that are equidistant from a fixed point, called the **focus**, and a fixed line, called the **directrix**. If we take the focus as the point $(p, 0)$, in a rectangular coordinate system, and the directrix as the line $x = -p$ (see Figure 9.60), the definition requires that $\overline{PF} = \overline{PP'}$. That is,

$$\sqrt{(x - p)^2 + y^2} = x + p$$

If we square both sides and collect terms, we obtain the equation

$$y^2 = 4px \tag{9.23}$$

In an analogous way, by choosing the focus on the y-axis at $(0, p)$ and the directrix as the horizontal line $y = -p$, we would arrive at the equation

$$x^2 = 4py \tag{9.24}$$

with the graph shown in Figure 9.61.

If we replace x with $-x$ in Equation 9.23 we get $y^2 = -4px$, and the resulting graph is the reflection through the y-axis of the first graph (Figure 9.62). Similarly, the graph in Figure 9.63 is the reflection through the x-axis of the parabola $x^2 = 4py$, and its equation is $x^2 = -4py$.

FIGURE 9.60
The parabola $y^2 = 4px$

FIGURE 9.61
The parabola $x^2 = 4py$

FIGURE 9.62
The parabola $y^2 = -4px$

FIGURE 9.63
The parabola $x^2 = -4py$

Each of the parabolas $y^2 = \pm 4px$ and $x^2 = \pm 4py$ has its vertex at the origin. If we translate the curves h units horizontally and k units vertically, the new equations are obtained from the old with replacing x with $x - h$ and y by $y - k$. So we have the following equations.

Standard Forms for Equations of Parabolas

$(y - k)^2 = \pm 4p(x - h)$, vertex at (h, k), axis horizontal

$(x - h)^2 = \pm 4p(y - k)$, vertex at (h, k), axis vertical

Parabolas are important in optics, since light rays parallel to the axis of a parabola are reflected to the focus. (See Figure 9.64.) Thus, many telescope and searchlight mirrors are parabolic in shape. In the case of a searchlight, light emitted from the focus is reflected in parallel beams. Similarly, radio telescopes have parabolic shapes, since radio frequencies are reflected to the focus.

FIGURE 9.64
Light rays parallel to the axis of a parabola are reflected to the focus.

The Ellipse

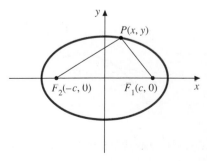

FIGURE 9.65

An ellipse is defined as the set of points in the plane such that the sum of the distances from each point in the set to two fixed points is constant. We call the two fixed points the **foci** (pronounced "fo sye"). Let us take the foci as $F_1(c, 0)$ and $F_2(-c, 0)$, and the constant as $2a$ (the final result is simpler when we use $2a$ rather than a). Then the definition requires of a point $P(x, y)$ on the ellipse that $\overline{PF_1} + \overline{PF_2} = 2a$ (see Figure 9.65). Using the distance formula, this requirement becomes

$$\sqrt{(x - c)^2 + y^2} + \sqrt{(x + c)^2 + y^2} = 2a$$

To simplify this equation, we can isolate one of the radicals and square, then

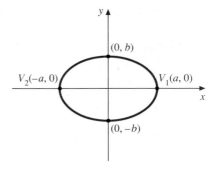

(a) Position I $\dfrac{x^2}{a^2} + \dfrac{y^2}{b^2} = 1$

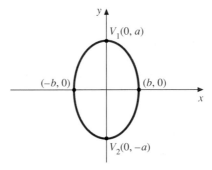

(b) Position II $\dfrac{x^2}{b^2} + \dfrac{y^2}{a^2} = 1$

FIGURE 9.66

isolate the remaining radical and square again. The result is

$$\frac{x^2}{a^2} + \frac{y^2}{b^2} = 1 \qquad \text{Position I} \qquad (9.25)$$

where $b^2 = a^2 - c^2$ (see Appendix 3 for the details).

If the foci are on the y-axis at $(0, c)$ and $(0, -c)$, the resulting equation is

$$\frac{x^2}{b^2} + \frac{y^2}{a^2} = 1 \qquad \text{Position II} \qquad (9.26)$$

Note that in both Equations 9.25 and 9.26, $a > b$.

In Figure 9.66 we show typical graphs of ellipses conforming to Equations 9.25 and 9.26. The points labeled V_1 and V_2 are called **vertices**. The line segment joining the vertices is called the **major axis** of the ellipse. Its length is $2a$. The **minor axis** is the line segment of length $2b$, through the center, perpendicular to the major axis and terminating on the ellipse.

When the center is translated from the origin to the point (h, k), the equations are obtained from Equations 9.25 and 9.26 by replacing x with $x - h$ and y with $y - k$.

Standard Forms for Equations of Ellipses

$$\frac{(x - h)^2}{a^2} + \frac{(y - k)^2}{b^2} = 1, \text{center } (h, k), \text{ major axis horizontal}$$

$$\frac{(x - h)^2}{b^2} + \frac{(y - k)^2}{a^2} = 1, \text{center } (h, k), \text{ major axis vertical}$$

Halley's Comet

Ellipses are particularly important because, as Johannes Kepler showed in 1609, the planets and other objects in our solar system all orbit the sun in ellipses. Figure 9.67 shows the orbit of Halley's Comet, which came closest to the sun most recently in 1986. It won't return to that point in its orbit (perihelion) until 2061.

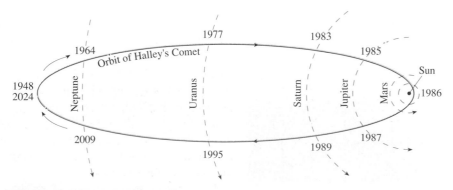

FIGURE 9.67
Orbits of Halley's Comet and the planets, showing the years when the comet crosses each planetary orbit.

The Hyperbola

A hyperbola is defined as the set of points in the plane for which the absolute value of the *difference* in their distances from two fixed points (the foci) is constant. As with the ellipse, the simplest equations result when the foci are either $(\pm c, 0)$ or $(0, \pm c)$, and the constant referred to in the definition is called $2a$. We leave it as an exercise for you to show that the two resulting equations can be put in the forms

$$\frac{x^2}{a^2} - \frac{y^2}{b^2} = 1 \quad \text{Position I} \tag{9.27}$$

and

$$\frac{y^2}{a^2} - \frac{x^2}{b^2} = 1 \quad \text{Position II} \tag{9.28}$$

where $b^2 = c^2 - a^2$. We show the graphs in Figure 9.68. The diagonals of the rectangles shown are asymptotes to the hyperbolas. Sketching these rectangles and their diagonals is an aid to drawing the graphs of the hyperbolas. For Position I the asymptotes have slopes $\pm b/a$, and for Position II they have slopes $\pm a/b$. The line through the vertices V_1 and V_2 is called the **transverse axis** and the line perpendicular to it through the center is called the **conjugate axis**.

When the center is translated to the point (h, k), we obtain the following:

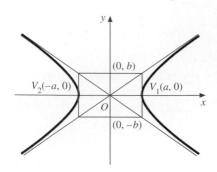

(a) Position I $\dfrac{x^2}{a^2} - \dfrac{y^2}{b^2} = 1$

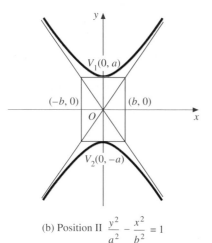

(b) Position II $\dfrac{y^2}{a^2} - \dfrac{x^2}{b^2} = 1$

FIGURE 9.68

Standard Forms of Equations of Hyperbolas

$\dfrac{(x-h)^2}{a^2} - \dfrac{(y-k)^2}{b^2} = 1$, center (h, k), transverse axis horizontal

$\dfrac{(y-k)^2}{a^2} - \dfrac{(x-h)^2}{b^2} = 1$, center (h, k), transverse axis vertical

---

Hyperbolas, too, have uses in optics. Mirrors for X-ray telescopes often use grazing rays (rays incident at very low angles) to bounce the X rays. The mirrors of NASA's Advanced X-Ray Facility will use combinations of hyperbolas and parabolas. In the photo at left we see one such mirror being constructed. The curvature of the inside of the cylinder, from top to bottom as we see it, is first a parabola and then a hyperbola. The combination is wrapped around to make a cylinder, and four such mirrors of different sizes will be nested.

Alternative Definition of Conic Sections

The following unites the parabola, ellipse, and hyperbola in a single definition:

Definition 9.2
Conics Defined by
Eccentricity

Let F be a fixed point and l a fixed line in the plane. The set of all points P in the plane such that

$$\frac{\text{distance between } P \text{ and } F}{\text{distance between } P \text{ and } l} = e \qquad e \text{ is a positive constant}$$

is a parabola if $e = 1$, an ellipse if $e < 1$, and a hyperbola if $e > 1$. The constant e is called the **eccentricity** of the conic, F is a **focus**, and l is a **directrix**.

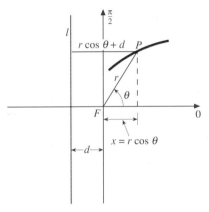

FIGURE 9.69

For the parabola this definition is equivalent to the one previously given. It is also equivalent in the case of the ellipse and the hyperbola, but we will not carry out the proof. We will show, however, that the equation that results when $e < 1$ is that of an ellipse, and the one for $e > 1$ is a hyperbola, as previously defined. The parabola has only one focus and one directrix, whereas the ellipse and the hyperbola have two foci and two directrices (plural of directrix).

Definition 9.2 provides a means of writing equations of the conic sections (other than the circle) in polar coordinates. To start with, we take the focus F at the pole and the directrix l perpendicular to the line $\theta = 0$, d units to the left of the pole, as shown in Figure 9.69. Then for a point P on the curve, the distance between P and F is r, and the distance between P and l is $r \cos \theta + d$, since $r \cos \theta$ is the x-coordinate of P. By Definition 9.2, we have

$$\frac{r}{r \cos \theta + d} = e$$

or, on solving for r,

$$r = \frac{ed}{1 - e \cos \theta} \qquad (9.29)$$

To verify that the graph of Equation 9.29 is one of the conic sections specified in Definition 9.2, let us obtain the corresponding rectangular equation. On clearing fractions and observing that $r \cos \theta = x$, we get

$$r = e(d + x)$$

If we square both sides and use the fact that $r^2 = x^2 + y^2$, we get

$$x^2 + y^2 = e^2(d^2 + 2xd + x^2) \qquad (9.30)$$

Now we consider three cases:

Case i: $e = 1$

Equation 9.30 becomes

$$y^2 = 2d\left(x + \frac{d}{2}\right)$$

which we recognize as a parabola with horizontal axis, vertex $(-d/2, 0)$, opening to the right. In fact, if we write $d = 2p$, the equation becomes

$$y^2 = 4p(x + p)$$

which is in the form $(y - k)^2 = 4p(x - h)$, with $h = -p$ and $k = 0$.

Case ii: $e < 1$

Equation 9.30 can be rearranged to give

$$x^2(1 - e^2) - 2e^2xd + y^2 = e^2d^2$$

If we divide by $1 - e^2$ (we know that $1 - e^2 > 0$ since $e < 1$) and complete the square in x, we get (you should verify it)

$$\left(x - \frac{e^2d}{1 - e^2}\right)^2 + \frac{y^2}{1 - e^2} = \frac{e^2d^2}{(1 - e^2)^2}$$

Finally, when we divide by the right-hand side, we get the standard form of the ellipse with center (h, k),

$$\frac{(x - h)^2}{a^2} + \frac{(y - k)^2}{b^2} = 1$$

in which

$$h = \frac{e^2d}{1 - e^2}, \quad k = 0, \quad a = \frac{ed}{1 - e^2}, \quad \text{and} \quad b = \frac{ed}{\sqrt{1 - e^2}}$$

(Note that $\sqrt{1 - e^2}$ is a real number, since $1 - e^2 > 0$.)

Case iii: $e > 1$

We leave it for you in Exercise 26 of Exercise Set 9.8 to show that the rectangular equation in this case can be put into the standard form of the hyperbola with center (h, k),

$$\frac{(x - h)^2}{a^2} - \frac{(y - k)^2}{b^2} = 1$$

in which

$$h = -\frac{e^2d}{e^2 - 1}, \quad k = 0, \quad a = \frac{ed}{e^2 - 1}, \quad \text{and} \quad b = \frac{ed}{\sqrt{e^2 - 1}}$$

From the three cases we have considered, we see that the polar equation

$$r = \frac{ed}{1 - e\cos\theta}$$

does have a parabola, an ellipse, or a hyperbola as its graph, according to whether $e = 1$, $e < 1$, or $e > 1$, respectively. To obtain this polar equation, we chose the directrix d units to the left of the focus (taken as the pole). If, instead, we take the directrix d units to the right, the resulting equation is

$$r = \frac{ed}{1 + e\cos\theta}$$

As a memory device, when the denominator $1 \pm e\cos\theta$ has a minus sign, the directrix in question is to the *left* of the pole, and when it has a plus sign, the directrix is to the *right* of the pole.

If we take the directrix as a horizontal line d units above or below the pole, we get

$$r = \frac{ed}{1 + e\sin\theta} \qquad \text{or} \qquad r = \frac{ed}{1 - e\sin\theta}$$

We will ask you to verify these forms in Exercise 27 of Exercise Set 9.8. Can you see how to determine from the sign in the denominator $1 \pm e \sin \theta$ when the directrix in question is above, and when it is below, the pole?

We summarize now our discussion of polar equations of conics having a focus at the pole and with directrix d units from the pole, directed vertically (when $\cos \theta$ is involved) or horizontally (when $\sin \theta$ is involved).

Polar Equations of Conic Sections

$$r = \frac{ed}{1 \pm e \cos \theta} \qquad \text{Directrix vertical} \tag{9.31}$$

$$r = \frac{ed}{1 \pm e \sin \theta} \qquad \text{Directrix horizontal} \tag{9.32}$$

The graph is a parabola if $e = 1$, an ellipse if $e < 1$, and a hyperbola if $e > 1$. In each case a focus is at the pole and the corresponding directrix is d units from the pole.

As the next two examples show, we can sketch the conic directly from its polar equation. It is not necessary to change to rectangular coordinates.

EXAMPLE 9.29   Identify the conic defined by

(a) $r = \dfrac{2}{1 - \cos \theta}$ (b) $r = \dfrac{3}{2 + \sin \theta}$

and draw its graph.

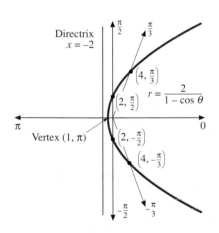

FIGURE 9.70

Solution

(a) The equation of this conic conforms to Equation 9.31, with $e = 1$, so the graph is a parabola with focus at the pole. The directrix is vertical and is $d = 2$ units to the left of the pole. (Its rectangular equation is $x = -2$.) When $\theta = \pm \pi/2$, $r = 2$, and when $\theta = \pi$, $r = 1$. The point $(1, \pi)$ is the vertex. Note that it is midway between the focus (the pole) and the directrix. Another easy pair of points to find are those corresponding to $\theta = \pi/3$ and $\theta = -\pi/3$. We find the corresponding r to be 4 in each case. With these points we can draw the graph as shown in Figure 9.70. We show the directrix, although it is not necessary for the graph.

REMARK ────────────────────────────────

■ If you had not remembered which way the parabola opened by looking at its equation, you could see that since r is undefined only when $\theta = 2n\pi$ ($n = 0$, ± 1, ± 2, ...), the parabola must open to the right.

────────────────────────────────

(b) To make the equation conform to Equation 9.32, we divide numerator and denominator by 2. The key is to get the 1 in the denominator. Thus,

$$r = \frac{3/2}{1 + \frac{1}{2} \sin \theta}$$

θ	r
0	3/2
$\pi/2$	1
π	3/2
$3\pi/2$	3

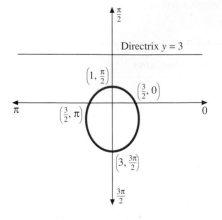

FIGURE 9.71

So we have $e = 1/2$. The curve is therefore an ellipse with major axis vertical. Since $ed = 3/2$ and $e = 1/2$, it follows that $d = 3$. So the directrix corresponding to the focus at the pole is 3 units from the pole, and is horizontal. Four key points on the ellipse can be obtained by taking $\theta = 0, \pi/2, \pi$, and $3\pi/2$ (corresponding to intercepts on the x- and y-axes). We show the corresponding values of r in the table in the margin. Using these four points and our knowledge of the symmetry of the ellipse, we obtain the graph in Figure 9.71.

Again, we did not need the directrix to draw the graph. Notice, however, that since the directrix corresponding to the focus at the pole is 3 units above the pole, if we consider the point $P(1, \pi/2)$, its distance from the pole (a focus) is 1, and its distance from the corresponding directrix is 2. Thus, the ratio of the first of these distances to the second is $\frac{1}{2}$, confirming that for this one point, the definition of e (Definition 9.2) is correct. (Of course, the ratio of the corresponding distances for any other point on the ellipse would also equal $\frac{1}{2}$.) ∎

EXAMPLE 9.30   Identify and sketch the graph of

$$r + 2r\cos\theta - 4 = 0$$

Solution   Solving for r, we obtain

$$r = \frac{4}{1 + 2\cos\theta} \tag{9.33}$$

θ	r
0	4/3
$\pm\pi/2$	4
π	-4

This equation conforms to Equation 9.31, with $e = 2$. Hence, the graph is a hyperbola with a vertical directrix. Consequently, its transverse axis is horizontal. Again, we make a table of values, using the four key angles $\theta = 0, \pm\pi/2$, and π. With these four points and our knowledge of symmetry of the hyperbola, we can make a sketch as in Figure 9.72. The accuracy can be improved by noting that

$$\lim_{\theta \to (2\pi/3)^+} \frac{4}{1 + 2\cos\theta} = -\infty$$

and

$$\lim_{\theta \to (2\pi/3)^-} \frac{4}{1 + 2\cos\theta} = \infty$$

These limits tell us that the asymptotes to the hyperbola, which go through its center, are parallel to the extended $2\pi/3$ ray and, by symmetry, the $-2\pi/3$ ray, shown in blue (see Exercise 28 in Exercise Set 9.8). We show these asymptotes in black in Figure 9.72.

If we superimpose a rectangular coordinate system on the polar system in the usual way, we can make further observations. We see, in particular, that the center is at $(\frac{8}{3}, 0)$, halfway between the two vertices. In rectangular coordinates the vertices are $(\frac{4}{3}, 0)$ and $(4, 0)$. The other focus (besides the pole) has rectangular coordinates $(\frac{16}{3}, 0)$, since it is $\frac{4}{3}$ units to the right of its corresponding vertex.

We can verify that $d = 2$, so the directrix corresponding to the focus at the pole is $x = 2$ (in rectangular coordinates). Using the point $P(\frac{4}{3}, 0)$, we can verify that the ratio defining e in Definition 9.32 is correct, namely,

$$\frac{4}{3} \div \frac{2}{3} = \frac{4}{3} \cdot \frac{3}{2} = 2$$

∎

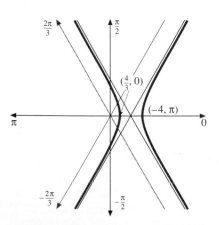

FIGURE 9.72
The graph of $r = 4/(1 + 2\cos\theta)$.

Exercise Set 9.8

In Exercises 1–10, identify the curve and draw its graph.

1. $r = \dfrac{1}{1 + \cos\theta}$

2. $r = \dfrac{2}{1 - \frac{1}{2}\sin\theta}$

3. $r = \dfrac{3}{1 - 2\sin\theta}$

4. $r = \dfrac{3}{2(1 - \sin\theta)}$

5. $r = \dfrac{6}{3 - 2\cos\theta}$

6. $r = \dfrac{3}{2 - \cos\theta}$

7. $r = \dfrac{9}{3 + 4\sin\theta}$

8. $r(4 + 3\cos\theta) = 8$

9. $2r = 12 - 3r\sin\theta$

10. $2r\sin\theta = 5 - 3r$

In Exercises 11–14, find the equation of the conic with the given eccentricity, having a focus at the pole, and with the corresponding directrix as specified.

11. $e = \frac{1}{3}$; directrix horizontal, 5 units above the pole.

12. $e = 1$; directrix vertical, 3 units to the right of the pole.

13. $e = \frac{3}{2}$; directrix vertical, 1 unit to the left of the pole.

14. $e = 0.3$; directrix horizontal, 1.2 units below the pole.

In Exercises 15–18, find the rectangular equation of the conic. If it is a parabola, give its vertex and its focus; if it is an ellipse, give its center, its vertices, and its foci; if it is a hyperbola, give its center, its vertices, its foci, and its asymptotes.

15. $r = \dfrac{2}{1 - \cos\theta}$

16. $r = \dfrac{4}{3 + \cos\theta}$

17. $r = \dfrac{6}{3 - 4\sin\theta}$

18. $r = \dfrac{3}{3 + \sin\theta}$

19. Denote the length of a semimajor axis of an ellipse by a. If its eccentricity is e and a focus is at the pole, verify the distances shown in the accompanying figure for the distances from the center to a focus and to a directrix. If c is the distance from a focus to the center, notice that $c = ae$, which implies that $e = c/a$ and also that $c < a$. (The same equations hold true for the hyperbola, but since $e > 1$, we see that $c > a$.)

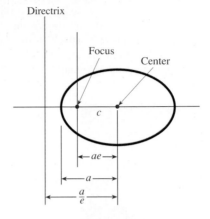

20. Use the result of Exercise 19 to show that the equation of the ellipse shown there can be written in the form

$$r = \frac{a(1 - e^2)}{1 - e\cos\theta}$$

Show that when $e = 0$, the ellipse becomes a circle.

21. Show that the polar form of the ellipse

$$\frac{x^2}{a^2} + \frac{y^2}{b^2} = 1$$

can be written as

$$r^2 = \frac{a^2(1 - e^2)}{1 - e^2\cos^2\theta}$$

22. The planets travel in elliptical orbits around the sun, with the sun at a focus. Let a polar coordinate system be introduced so that the sun is at the pole. Let r_a denote the greatest distance from the planet to the sun (called the *aphelion* distance) and r_p denote the minimum distance (called the *perihelion* distance).

(a) Prove that the eccentricity e of the orbit can be written in the form

$$e = \frac{r_a - r_p}{r_a + r_p}$$

(b) Prove that $r_a = a(1 + e)$ and $r_p = a(1 - e)$, where a is the semimajor axis of the ellipse.

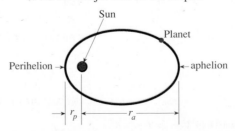

23. The planet Jupiter's semimajor axis is approximately 9.503 A.U. (astronomical units), and the eccentricity is 0.0484. Find the equation of its orbit (see Exercise 20). Find the aphelion and perihelion distances (see Exercise 22). *Note:* One A.U. is the semimajor axis of the earth's orbit, or approximately 1.496×10^{11} m. (Since the eccentricity of the earth's orbit is only 0.0167, its distance from the sun is always close to its semimajor axis, which we can think of as an average orbital radius.)

Jupiter, photographed with the
Hubble Space Telescope

24. Repeat Exercise 23 for Halley's Comet, whose semimajor axis is approximately 18 A.U. and whose orbit has eccentricity 0.9.

25. Repeat Exercise 23 for Asteroid (5100) Pasachoff, whose semimajor axis is approximately 2.47 A.U. and whose eccentricity is approximately 0.135.

26. Prove that the rectangular form of Equation 9.29 for $e > 1$ is

$$\frac{(x-h)^2}{a^2} - \frac{(y-k)^2}{b^2} = 1$$

where $h = -\dfrac{e^2 d}{e^2 - 1}$, $k = 0$, $a = \dfrac{ed}{e^2 - 1}$, and $b = \dfrac{ed}{\sqrt{e^2 - 1}}$.

27. Prove that if a focus of a conic section having eccentricity e is at the pole and its corresponding directrix is d units above the pole and is horizontal, then the equation of the conic is

$$r = \frac{ed}{1 + e \sin \theta}$$

Similarly, show that the equation is

$$r = \frac{ed}{1 - e \sin \theta}$$

if the directrix is d units below the pole.

28. Show that if $e > 1$, the asymptotes of the hyperbolas

$$r = \frac{ed}{1 \pm e \cos \theta} \qquad \text{and} \qquad r = \frac{ed}{1 \pm e \sin \theta}$$

are parallel to the lines obtained by setting the denominator equal to zero and solving for θ. (*Hint:* Use the formulas $a = ed/(e^2 - 1)$ and $b = ed/\sqrt{e^2 - 1}$, together with the slopes of the asymptotes $\pm b/a$ for Position I and $\pm a/b$ for Position II.)

Chapter 9 Review Exercises

In Exercises 1–6, eliminate the parameter and use the result to sketch the curve defined by the parametric equations.

1. $\begin{cases} x = 2t - 1 \\ y = \sqrt{1 - t} \end{cases} \quad -\infty < t \le 1$

2. $\begin{cases} x = \sin^2 t \\ y = \cos t \end{cases} \quad -\infty < t < \infty$

3. $\begin{cases} x = \ln t \\ y = 1 - \dfrac{1}{t} \end{cases} \quad 0 < t < \infty$

4. $\begin{cases} x = 1 + e^t \\ y = 1 - e^{-t} \end{cases} \quad -\infty < t < \infty$

5. $\begin{cases} x = \tan^2 \theta - 1 \\ y = \sec \theta + 1 \end{cases} \quad 0 \le \theta < \dfrac{\pi}{2}$

6. $\begin{cases} x = 2 \sinh t - 1 \\ y = 3 \cosh t + 2 \end{cases} \quad -\infty < t < \infty$

In Exercises 7–10, find dy/dx and d^2y/dx^2 without eliminating the parameter.

7. $\begin{cases} x = t^2 - 1 \\ y = 2t^3 \end{cases}$

8. $\begin{cases} x = \sqrt{t} \\ y = t^2 + 2 \end{cases}$

9. $\begin{cases} x = \sin^2 t \\ y = \ln \cos t \end{cases}$

10. $\begin{cases} x = (t - 1)/t \\ y = \ln \sqrt{t} \end{cases}$

11. Find the equations of the tangent line and normal line to the curve defined by $x = \ln(t^2 - 3)$ and $y = t/\sqrt{t^2 - 3}$ at $(0, 2)$.

12. Without eliminating the parameter, find all maximum and minimum points on the curve defined by

$$\begin{cases} x = \sqrt{t^2 - 1} \\ y = 10t^3 - 3t^5 \end{cases} \quad t \geq 1$$

Use the Second Derivative Test for testing critical values.

13. Find the length of the curve defined by

$$\begin{cases} x = \cosh^2 t \\ y = 2\sinh t \end{cases} \quad 0 \leq t \leq \ln 3$$

14. Find the area of the surface formed by revolving the curve of Exercise 13 about (a) the x-axis and (b) the y-axis.

15. For the curve defined by

$$\begin{cases} x = \ln t - 1 \\ y = 2\sqrt{t} \end{cases} \quad 1 \leq t \leq 3$$

find (a) its length and (b) the area of the surface formed by revolving the curve about the x-axis.

16. Let R denote the region enclosed by the x-axis and one arch of the cycloid

$$\begin{cases} x = a(t - \sin t) \\ y = a(1 - \cos t) \end{cases}$$

Find (a) the area of R and (b) the volume of the solid formed by revolving R about the x-axis.

In Exercises 17 and 18, change to an equivalent rectangular equation, identify the curve, and draw its graph.

17. (a) $r = 2(\sin\theta + \cos\theta)$
    (b) $r(2 + \sin\theta) = 1$

18. (a) $r = \dfrac{\sec\theta}{1 - \tan\theta}$
    (b) $r = \sec^2(\theta/2)$

In Exercises 19–21, discuss and graph the given equation.

19. (a) $r = 2\sin^2(\theta/2)$
    (b) $r^2 = 2\sin\theta\cos\theta$

20. (a) $r = -1 - 2\cos\theta$
    (b) $r = 5 + 3\sin\theta$

21. (a) $r = \sin^2\theta - \cos^2\theta$
    (b) $r\theta = 1, \ \theta > 0$

In Exercises 22–24, find the area of the specified region.

22. Inside both $r = \sqrt{3}\cos\theta$ and $r = \sin\theta$

23. Inside both $r = 1 - \cos\theta$ and $r = 2 + \cos\theta$

24. Outside $r = 2\cos 2\theta$ and inside $r^2 = 6\sin 2\theta$

25. Let C_1 be the curve $r = 2$ and C_2 the curve $r = 4\cos 3\theta$. Find the areas of the following regions:
    (a) outside C_1 and inside C_2
    (b) inside C_2 and inside C_1
    (c) outside C_2 and inside C_1

In Exercises 26 and 27, find the angle between the two curves at each of their points of intersection.

26. $r = 5\sin\theta, \ r = 2 + \sin\theta$

27. $r = \dfrac{1}{1 + \sin\theta}, \ r = 2(1 - \sin\theta)$

In Exercises 28 and 29, show that the curves intersect orthogonally.

28. $r = \tan(\theta/2), \ r = e^{\cos\theta}; \ 0 < \theta < \pi$

29. $r(1 + \cos\theta) = 3, \ r = 4\cos\theta; \ -\pi/2 < \theta < \pi/2$

In Exercises 30 and 31, find the length of the given curve.

30. $r = \sqrt{1 - \sin 2\theta}; \ -\pi/4 \leq \theta \leq \pi/4$

31. $r = 7/\theta; \ 1 \leq \theta \leq 7$

32. Find the area of the surface formed by revolving the arc of the curve $r = \sec^2(\theta/2)$ from $\theta = 0$ to $\theta = 2\pi/3$ about the polar axis.

33. The **witch of Agnesi** is a curve generated by the point P (as in the figure) moving so that QPR is always a right triangle with right angle at P. The point Q is on the circle of radius a centered at $(0, a)$ and R is on the line $y = 2a$. Find parametric equations for this curve, using θ as a parameter, and draw its graph. (*Hint:* Make use of the polar equation of the circle.)

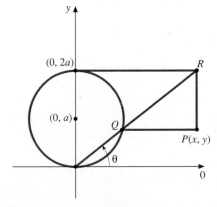

In Exercises 34–39, identify the curve defined by the given equation, and draw its graph.

34. $r = \dfrac{2}{\cos\theta - 1}$ **35.** $r = \dfrac{-2}{1 + \sin\theta}$

36. $r = \dfrac{1}{-1 + 2\cos\theta}$ **37.** $r = \dfrac{4}{5 - 3\cos\theta}$

38. $r(4\sin\theta - 3) = 8$ **39.** $r(\cos\theta - 2) = 5$

40. Show that the polar form of the hyperbola

$$\frac{x^2}{a^2} - \frac{y^2}{b^2} = 1$$

can be written as

$$r^2 = \frac{a^2(e^2 - 1)}{e^2 \cos^2\theta - 1}$$

41. Show that the parabolas

$$r = \frac{ed}{1 + \cos\theta} \quad \text{and} \quad r = \frac{ed}{1 - \cos\theta}$$

are orthogonal (intersect at right angles).

42. The planet Pluto's semimajor axis is approximately 39.529 A.U. and its eccentricity is 0.2481. Find each of the following:
(a) the equation of its orbit;
(b) its longest and shortest distances from the sun. (See Exercise 22 in Exercise Set 9.8.)

Chapter 9 Concept Quiz

1. Let a curve C be given parametrically by $x = f(t)$ and $y = g(t)$ for $a \leq t \leq b$. In the following, assume that f and g satisfy relevant differentiability conditions. Give formulas for each of the following:
(a) dy/dx and d^2y/dx^2 in terms of t;
(b) the length of C;
(c) the area of the surface formed by revolving C about the x-axis and about the y-axis. (Assume that C lies in the first quadrant.)

2. (a) Write equations expressing rectangular coordinates x and y in terms of polar coordinates r and θ.
(b) Suppose a point P_1 has polar coordinates (r_1, θ_1). Give three other sets of polar coordinates for P_1, one of which involves a negative angle.

3. Identify the graph of each of the following polar equations.
(a) $r = a$
(b) $\theta = \alpha$
(c) $r = 2a\cos\theta$
(d) $r = a(1 + \cos\theta)$
(e) $r = a + b\cos\theta, a > b > 0$
(f) $r = a + b\cos\theta, 0 < a < b$
(g) $r^2 = a^2 \cos 2\theta$
(h) $r = a\cos n\theta$
(i) $r = a\theta$
(j) $r \sec\theta = a$

4. (a) Let $r = f(\theta)$, where f is a nonnegative, continuous function for $\alpha \leq \theta \leq \beta$. Write the formula for the area of the region bounded by the graph of $r = f(\theta)$ and the rays $\theta = \alpha$ and $\theta = \beta$.
(b) Let $r = f(\theta)$ and $r = g(\theta)$ be continuous, nonnegative functions for $\alpha \leq \theta \leq \beta$, with $g(\theta) \leq f(\theta)$. Write the formula for the area of the region bounded by the graphs of $r = f(\theta)$, $r = g(\theta)$, and the rays $\theta = \alpha$ and $\theta = \beta$.

5. In each of the following, identify the conic and give the equation (you may use the rectangular equation) of the directrix corresponding to the focus at the pole:
(a) $r = \dfrac{3}{2 + \cos\theta}$
(b) $r = \dfrac{2}{1 - \sin\theta}$
(c) $r = \dfrac{4}{1 - 2\cos\theta}$
(d) $r = \dfrac{6}{3\sin\theta - 2}$
(e) $r = \dfrac{-4}{4 - 2\cos\theta}$

6. Indicate which of the following statements are true and which are false.
(a) If x and y are given parametrically and you eliminate the parameter, the graph of the resulting rectangular equation is the same as the graph defined by the parametric equations.
(b) All points of intersection of the graphs of $r = f(\theta)$ and $r = g(\theta)$ are found by solving the equations simultaneously.
(c) The graph of the equation $r = f(\theta)$ is symmetric with respect to the pole if $f(\theta - \pi) = f(\theta)$.
(d) The slope of the tangent line to the graph of $r = f(\theta)$ at a point (r, θ) on the graph is $f'(\theta)$.
(e) If $r = f(\theta)$, $\alpha \leq \theta \leq \beta$, then parametric equations for the graph are $x = f(t)\cos t$, $y = f(t)\sin t$, $\alpha \leq t \leq \beta$.

APPLYING CALCULUS

1. The electrostatic charge distribution consisting of a charge q at the point with polar coordinates, $(s, 0)$ and a charge $-q$ at the point (s, π) is called a *dipole*.

   (a) The lines of force (force field) are given by the family of curves $r = C \sin^2 \theta$, where each value of $C > 0$ determines a specific line of force. Sketch the lines of force corresponding to $C = 1$, $C = 2$ and $C = 3$.

   (b) The equipotential lines for the dipole are given by the family of curves $r^2 = B \cos \theta$, $B > 0$. Sketch the equipotential lines corresponding to $B = 1$, $B = 2$ and $B = 3$.

   (c) Show that the lines of force and the equipotential lines are orthogonal trajectories.

2. Consider orbits, which follow Kepler's Laws.

   (a) The orbit of the earth is approximately an ellipse with the sun at one focus. The length of the semimajor axis is 9.26×10^7 miles and the eccentricity of the orbit is 0.02. Determine the minimum and maximum distances between the sun and the earth (the earth's *perihelion* and *aphelion* distances, respectively).

   (b) Halley's comet travels in an elliptic orbit with the sun at one focus. In terms of astronomical units (1 AU $= 9.26 \times 10^7$ miles $=$ the semimajor axis of the earth's orbit), the major and minor axes of the comet's orbit are 36.18 AU and 5.12 AU, respectively. What is the eccentricity of the orbit? What are the perihelion and aphelion distances? Express your answers in terms of astronomical units.

3. If a point P is on a fixed radius and at a distance h from the center of a circle of radius r which is rolling without slipping around a fixed circle of radius R, then the path traced out by P is called an *epitrochoid*. The geometrical design of the Wankel rotary engine (used in the Mazda RX-7) involves an epitrochoid in which $R = 2r$ and $0 < h < r$. The orbit of the planet Mercury relative to the Earth (i.e., in a coordinate system with the Earth at the origin) is an epitrochoic. Consider the epitrochoid given parametrically by

   $$x = h \cos 3\theta + 3r \cos \theta, \qquad y = h \sin 3\theta + 3r \sin \theta$$

   This particular curve is generated by having a circle of radius r roll on a circle of radius $2r$.

    (a) Sketch the graph of this epitrochoid in the two cases: $0 < h < r$ and $r < h < 3r$.

   (b) Suppose that $0 < h < r$ and let $R = 3r$. Show that the arc length of the curve is given by

   $$L = \int_0^{2\pi} \sqrt{9h^2 + R^2 + 6hR \cos 2\theta} \, d\theta$$

   which can be written in the form

   $$L = \int_0^{2\pi} (3h + R)\sqrt{1 - k^2 \sin^2 \theta} \, d\theta$$

   where $k^2 = \dfrac{12hR}{(3h + R)^2} \leq 1$. There is no elementary antiderivative for the integrand. The integral belongs to a class of integrals known as *elliptic integrals* and tables of values are given in mathematical handbooks.

   (c) Let $b = 1/2$ and $r = 1$ in the integral in (b). Use Simpson's rule with $n = 8$ to find an approximation for the length of the corresponding epitrochoid.

4. Kepler's laws of planetary orbits can be derived from Newton's law of gravity. In particular, Kepler's First Law is that the planets orbit the Sun in ellipses with the Sun at one focus. From the law of conservation of energy, given Newton's law of gravity, we have

$$\left[\left(\frac{dr}{d\theta} \right)^2 + r^2 \right] \frac{J^2}{2\mu r^4} - \frac{k}{r} = W$$

where r is the distance between the Sun and a planet, J is the total angular momentum, μ is the mass function $m_1 m_2 / (m_1 + m_2)$, k is the gravitational constant $G m_1 m_2$ where G is the universal constant of gravitation, and W is the total energy. W and J are constant.

(a) Solve the equation of conservation of momentum for θ as an indefinite integral.

(b) Show that the integral can be evaluated as an inverse trigonometric function:

$$\theta \sin^{-1} \left(\frac{\mu k r - J^2}{r(\mu^2 k^2 + 2 J^2 \mu W)^{1/2}} \right) - \alpha$$

(c) Set the constant $\alpha = \pi/2$ and use $\sin(\theta + \alpha) = \cos\theta$, and then solve for r as a function of θ.

(d) Show that this equation can be put into the form of the general polar equation of an ellipse

$$r = \frac{a(1 - e^2)}{1 \pm e \cos\theta}$$

where a is the semimajor axis of the ellipse and e is the eccentricity.

(e) From the equations relating the numerators and the denominators, respectively, solve for W, the total energy, and show that it depends only on the semimajor axis of the orbit.

5. Kepler's third law of planetary motion, the harmonic law, states that the period squared of a planet is proportional to the cube of its semimajor axis. Derive this law from the law of conservation of angular momentum

$$r^2 (d\theta/dt) = J/\mu$$

where r is the distance between the planet and the Sun, J is the constant angular momentum, k is $G m_1 m_2$, and μ is a mass function $m_1 m_2 / (m_1 + m_2)$.

(a) Putting terms with r and θ on the left side of the equation and the constants and t on the right side, and integrating around the ellipse, evaluate the integral of the variables in terms of the area of an ellipse using its major and minor axes. Remember that the integral of time over the orbit is the period P and that the area element is $\frac{1}{2}(r^2)d\theta$. Show that $JP/2\mu = \pi ab$.

(b) Using the formula $b = a\sqrt{(1 - e^2)}$ for the semiminor axis of the ellipse in terms of the semimajor axis and eccentricity, and $J^2/\mu k = a(1 - e^2)$ for the angular momentum (see Problem 4(d)), solve for P in terms of a.

(c) Show that this result can be put in the usual form of Kepler's third law:

$$\frac{a^3}{P^2} = \frac{G}{4\pi^2}(m_1 + m_2)$$

6. A rope is anchored at a point on the ground at the base of a round silo that is 8 meters in diameter. The rope's length is half the circumference of the silo. A girl takes the free end of the rope and wraps the rope tightly around the silo until she is diametrically opposite the point where the rope is anchored. Then, while continuing to hold the rope taut, she starts walking away from the silo, so that the rope is always tangent to the circle. She continues until the rope stretches its full length in a straight line from the

anchor point. She then continues walking, holding the rope taut, now going in a circular path, until the rope again comes in contact with the silo on the opposite side from before. Finally, she comes back to the starting point with the rope wrapped all the way around the other half of the silo.

(a) Find parametric equations of the path of the girl. (*Hint:* The path will have to be divided into two parts.)

(b) Find the area of the region between the silo and the path of the girl.

Benoit B. Mandelbrot

Calculus is required as the first real mathematics course taught in most colleges. Is that still a reasonable requirement in view of the mathematics that is most applicable today?

It remains perfectly reasonable for many, or even most, students, but I would prefer to see the coexistence of several distinct introductory courses. At Yale, I work actively at promoting one that is based on fractal geometry. Calculus can be taught so as to leave little room for subjective grading, so it is tempting to use it as a kind of filter to select students among otherwise qualified candidates for professional schools. This may increase the number of jobs for math teachers, but mathematics as a profession is left the poorer. On the other hand, mathematicians find it hard to claim that their field must be supported simply because it is a noble and great adventure of humanity, like great music and drama. This is because music and drama are widely understood and appreciated in the general public, whereas mathematics is even less understood than the most extreme avant-garde in art.

Do fractals have a place in the teaching of calculus?

Yes, I think they have a very important place. The popular appeal of fractals shows, in my opinion, that almost everybody has some interest for some part of mathematics. Even when the overall presentation of calculus is very traditional and basic, fractals can help with a critically important lesson early on, when derivatives and tangents are introduced. My claim is that to understand tangents, it is best to start with curves that have no tangents, such as fractal curves. To the contrary, around 1977, an eminent international mathematical education committee recommended that teachers not confuse students by explaining that a curve may fail to have a tangent at any point. The only exceptions they allowed were those courses reserved for future professional mathematicians. Obviously, I think that this recommendation was badly timed and terribly out of touch. First of all, it happened to come out the same year my book *Fractals, Form, Chance and Dimension* was published. This book treated fractal geometry as a mathematical subspecialty. It effectively destroyed the well-entrenched belief that nondifferentiability was an abstract nonphysical notion, a kind of mathematical pathology. On the contrary, it is a very useful and broad notion in describing nature. In fact, the very same features that had been called "mathematical pathologies" are the very stuff of nature. Secondly, this committee seemed to be convinced that before starting calculus students' minds are blank slates. I take issue with this perception. I am convinced that every person, at a very early age, has a kind of vague but valid intuition based on their everyday experience that denies the physical existence of tangent lines or planes. Even a child knows that one cannot draw a tangent line to a coastline or a tangent plane to the bark of an oak tree. Even those who have never heard of fractals are at ease with the basic ideas of fractal geometry; yet textbooks throughout history have argued as if those intuitive ideas simply did not exist. Today, many high school students are exposed to fractal geometry through popular science books. They

already know examples of theoretical curves without tangents. Therefore, a calculus course would benefit from including some form of the following argument. "Let's talk about something that you probably already know, either intuitively or from reading or class work. Most curves you either see or can draw are best viewed as having no tangents. On the other hand, a circle has a tangent and so do other simple abstract shapes. You may feel that this is a difficult notion; but practice makes it intuitive." Calculus can serve to model natural phenomena as well as abstract phenomena. It is indispensable in science as well as higher mathematics, and if you pursue calculus far enough, you'll find it extraordinarily beautiful.

Your life has been very interesting and unconventional. How do you think your experiences influenced your decision to become a mathematician?

One of my uncles was a successful mathematician. I could have followed directly in his footsteps, but despite this family connection, my path into mathematics was convoluted. My schooling was always chaotic and thoroughly disrupted by the Second World War. My mother was a doctor and terrified of epidemics, so she made me skip early school grades and I studied privately with another of my uncles. He despised rote learning, including the alphabet and multiplication tables. This definitely made an impression; I'm still not good with either. Instead, he encouraged me to read and train my memory. Mostly we played chess. My father was a map nut, and I could read maps as early as I can remember. I've since thought that maybe the constant daily practice of playing chess and reading maps in my childhood helped develop my geometric intuition, which became my most important scientific skill.

Before World War II started we moved from Poland to Paris, then to the country outside Paris. From mid–1942 to the end of 1943, the German occupation of France was tightening, and my family decided that it was not safe for me to attend school. Periods of intense danger and fear—several times I narrowly escaped being executed or deported—were separated by periods when not much was happening. I had plenty of time for study on my own. Finally I landed in a large lycée (high school) that included two postgraduate grades designed to prepare good students for Ecole Normale and Ecole Polytechnique. You had to pass a series of notoriously tough "killer exams." Many students, some of whom later became great scientists, had to repeat the second grade. During the second week after I joined this postgraduate high school in Lyons, I found out that I had a talent I had never known or suspected.

Your geometric intuition?

Yes. Our math professor would describe a problem that he enjoyed, always stating it in terms of algebra or analytic geometry. But I didn't seem to hear the analytic problem; instead I heard a geometric problem. I soon realized that during my self-schooling, I had become intimately familiar with a large "zoo" of geometric shapes that I could instantly recognize and call upon, even when they were dressed up in analytic clothes.

The five months I spent in Lyons were among the most important of my life. I hardly ever left the school grounds because I couldn't afford to buy a meal elsewhere and we feared and avoided the Nazi boss in Lyons (I later found out his name was Klaus Barbie). I had a burning desire to do well. I worked at a rate that I could not have sustained indefinitely, learning the basics for the exams and polishing my geometric skills. But becoming an algebra whiz was not my goal. Anyhow, I was supposed to take the exams "for the practice." To my amazement and that of nearly everyone else, not only did I pass, I finished a very close second in all of France.

Did these exams contain calculus topics?

Yes, I remember a conversation with a professor who complained about a problem involving a triple integral that he couldn't solve in the time allowed on the exam. I told him that I had recognized that the integral was simply the volume of the sphere in certain coordinates suggested by the geometry. After I had explained my reasoning he went away repeating "of course, of course."

So you were mathematically precocious because of your gift?

Yes. My university schooling continued and I became certified in several disciplines but only because of my geometric gift. But I missed the main goal of certification, which is to train students for a profession by providing guidance and role models. This never happened to me, in any profession. Unlike most mathematicians I did my best known work later in life. I didn't devote substantial time to mathematical research until I was 40 and I discovered the Mandelbrot Set in my mid-fifties.

Your schooling may have adversely affected your early academic career, but it obviously gave you other opportunities.

There are many reasons why my particular story isn't a good model for academic success. Even though established fields find it hard to deal with diversity in training, diversity is just as indispensable to science as it is to society. There are those who believe that a given discipline at a given instant should have a single "best way" of training new workers in that field, and many believe that this best way is determined by some kind of inevitable tide of history. Were this true, those who depart from the one true way would deserve to be shunned by the mainstream. That is indeed how things happen, but this is an absolutely deplorable situation. It may not be corrected, but it must be mitigated. To mitigate it, society must intervene to allow at least a few exceptions to survive.

Could you say very briefly how you came to the ideas of fractal geometry?

It would be nice to say that fractal geometry was born in some kind of "Eureka moment," like when Archimedes ran from his overflowing bathtub. It would also be nice to respond by identifying the source in some early application, like records of prices, errors in telephone lines, or the shape of coastlines. But this

would be misleading. Fractal geometry didn't start with an over-reaching idea centered on an already formulated problem, much less a problem already recognized as significant enough to justify the effort needed to answer it. Let me point out that *problem solving* and *problem setting* seem to involve different talents. In the popular view—and in the restricted views of many mature disciplines—science is viewed as problem solving. I regret this, perhaps because my talent lies mainly in problem setting.

The study of fractals proceeded from bottom to top. It went from a mess of disregarded and disconnected odds and ends, on to ill-organized parts, and then toward increasing organization. Of the disconnected odds and ends with which I started, some were theoretical. In that sense, fractal geometry has incorporated many geometric shapes that were known long before I was born, but remained scattered with no common idea behind them. This is also why they were given no name, other than "mathematical monsters." My present name for them is *protofractals*. They were part of the zoo of shapes that played a central role in my early career. A second source of disconnected odds and ends were miscellaneous facts about nature. Most people don't register facts that do not yet fit into any established theory, but I was blessed with a sponge-like memory and could recall without conscious effort all these odd shapes and all these odd facts. I also read about mathematics a lot. On many occasions I was reading about some new mathematical shape and thought instantly and spontaneously of some real empirical phenomenon that somehow had the same "taste." On other occasions, I went from fact to formula. Each time, I dropped whatever I had been doing and followed up this new link between form and substance. So the "flavor" of my scientific activity has been oscillating back and forth. It was, at different times, dominated either by the search for more facts—to be identified in the literature or discovered in the laboratory, or by the search for new mathematics—again, to be identified in the literature or developed from scratch. My skills can best be described as "having a nose" for the subject.

Benoit B. Mandelbrot is best known for his *Fractal Geometry of Nature* (1982). "The father of fractal geometry," Professor Mandelbrot continues to work in this new field, with its concrete applications, and its growing impact on art and on high-school and college science teaching. Professor Mandelbrot worked for 35 years at the IBM T. J. Watson Research Center, where he is now IBM Fellow Emeritus. His contributions, first viewed as isolated and peripheral, are now widely recognized as central to the new ideas concerning chaos and complexity. Latest among his numerous awards is the 1993 Wolf Prize for Physics.

10

INFINITE SERIES

Leonhard Euler (1707–1783) began his research in mathematics and physics at age 18. He joined the two Bernoulli brothers in St. Petersburg, Russia, in 1725 at the newly organized St. Petersburg Academy of Sciences and stayed for 14 years.

Euler's mathematical work ranged from the foundations of mathematical physics to the theory of lunar and planetary motion to differential geometry of surfaces. He devoted much attention to the problem of logarithms of negative numbers, the equation of a vibrating string, and problems in optics.

Euler investigated many functions and extensively studied power series. He discovered, for example, that $\frac{\pi}{2} - \frac{x}{2} = \sin x + \sin \frac{2x}{2} + \sin \frac{3x}{3} + \cdots$, a type of Fourier series.

Euler provided many notations we use today: the symbol e for the base of natural logarithms, the use of f and of parentheses for a function, the modern signs for trigonometric functions, the Greek Sigma (Σ) for sum, and the letter i for the square root of -1. The use of the letters $a, b,$ and c for the sides of a triangle is also due to Euler. One of the most prolific mathematicians in history, Euler published more than 500 books and articles.

In 1736 the great Swiss mathematician Leonhard Euler (pronounced "oiler") discovered the remarkable formula

$$1 + \frac{1}{2^2} + \frac{1}{3^2} + \frac{1}{4^2} + \cdots = \frac{\pi^2}{6} \tag{10.1}$$

The sum on the left-hand side of the equation is an example of what is called an *infinite series* (or simply a *series*). The three dots mean that we are to continue adding terms indefinitely. The question immediately arises as to how we can find the sum of infinitely many terms. We will give a formal definition soon, but for now you can think of the sum as the value that is *approached* (if such a value exists) as we add more and more terms. The discovery of Formula 10.1 is a fascinating story in the history of mathematics.

We will not try to show how Euler found the sum $\pi^2/6$ for the series in Equation 10.1 but will indicate another way of obtaining the result in Exercise 8 of Exercise Set 10.10. To illustrate the idea of the sum of an infinite series, we consider the simpler series

$$1 + \frac{1}{2} + \frac{1}{4} + \frac{1}{8} + \frac{1}{16} + \cdots \tag{10.2}$$

in which each term after the first is one-half the preceding one. By starting with the first term, and listing the finite sums obtained by successively adding on one more term, we get the following:

1	$1\frac{1}{2}$	$1\frac{3}{4}$	$1\frac{7}{8}$	$1\frac{15}{16}$
first term	sum of first two terms	sum of first three terms	sum of first four terms	sum of first five terms

This list of finite sums is called the *sequence of partial sums* of the series. It appears that as we add more and more terms, the finite sums come arbitrarily close to 2. For this reason we say that the *infinite* sum *is* 2, and write

$$1 + \frac{1}{2} + \frac{1}{4} + \frac{1}{8} + \frac{1}{16} + \cdots = 2$$

We say that the series *converges* to 2.

Euler was one of the most prolific writers of mathematics in history.

Not all series converge. For example, the sum of the series $1+2+3+4+\cdots$ clearly becomes arbitrarily large as we add on more and more terms. As another example, the series $1-1+1-1+1-\cdots$ does not approach any one value as we add more and more terms, since the finite sums alternate between 1 and 0, depending on where we stop. When a series fails to converge to a unique finite number, we say that it *diverges*. So both of the series $1+2+3+4+\cdots$ and $1-1+1-1+\cdots$ diverge.

As one final example, consider the series

$$1 + \frac{1}{2} + \frac{1}{3} + \frac{1}{4} + \cdots$$

composed of the reciprocals of consecutive positive integers. Does this series converge? If so, can you find its sum? Or does it diverge? You should make a conjecture by considering the sums of the first 10, 20, or even 100 terms using a CAS, spreadsheet, or calculator. We will return to this series in Section 10.2 and answer the questions raised here.

Infinite series can also be made up of variable terms. The simplest ones involve integral powers of a variable. An example is

$$1 + x + x^2 + x^3 + \cdots \qquad (10.3)$$

We call such a series a *power series* in x, since each term is a power of x. We would expect the sum, if it exists, to depend on x; that is, the sum is a function of x. So, for the series 10.3, we can define a function f by

$$f(x) = 1 + x + x^2 + x^3 + \cdots$$

In general, a power series will converge for some values of x and diverge for other values (although it may converge for all values). For example, if we set $x = \frac{1}{2}$ in the series 10.3, we get the series 10.2, which converges to 2, as we have seen; but if we set $x = 1$, we get $1 + 1 + 1 + \cdots$, which clearly diverges.

If we consider only a finite number of terms of the series 10.3, we have a polynomial of the form

$$1 + x + x^2 + \cdots + x^n \qquad (n \geq 1)$$

which is an approximation to the infinite series and consequently to the function determined by the series. As we take polynomials of higher and higher degree (that is, as $n \to \infty$), the approximations become better and better, and in the limit we obtain the exact value of the function (assuming x is restricted to those values for which the series converges). Suppose now, instead of beginning with a series such as 10.3, we begin with a function $f(x)$. An important question is whether we can find an nth degree polynomial approximating $f(x)$ in such a way that as $n \to \infty$, the resulting "infinite polynomial" converges to $f(x)$ exactly.

For example, we will show in the next section that

$$e^x \approx 1 + x + \frac{x^2}{2!} + \frac{x^3}{3!} + \frac{x^4}{4!} + \cdots + \frac{x^n}{n!} \qquad (10.4)$$

and that the approximation becomes better and better as we let n increase. So we might expect that we can let $n \to \infty$ and obtain the exact equality

$$e^x = 1 + x + \frac{x^2}{2!} + \frac{x^3}{3!} + \frac{x^4}{4!} + \cdots \qquad (10.5)$$

Passing from the finite sum in Equation 10.4 to the infinite sum in Equation 10.5 and concluding that this latter sum converges to e^x requires justification. In this case the result is valid, as we will see later.

Our procedure in this chapter will be to begin with approximation of functions by polynomials. Next, we will discuss sequences and their limits, since, as we have indicated, the sum of an infinite series is determined from its sequence of partial sums. Then we will consider the general question of when an infinite series converges or diverges. In particular, we will study various convergence criteria for series of constants. Afterward, we will study power series and their properties. Then we will return to the crucial question of whether we can extend finite polynomial approximations of a function to an infinite power series that represents the function. We conclude the chapter by showing another way to represent functions by infinite series, one that involves sines and cosines instead of powers of x. This last kind of series, known as a *Fourier series,* is especially important in physics.

10.1 APPROXIMATION OF FUNCTIONS BY POLYNOMIALS; TAYLOR'S THEOREM

Polynomial functions are in many ways the simplest functions of all. First, they are the easiest to evaluate, requiring only addition and multiplication (since raising to positive integral powers is just repeated multiplication). Second, they possess derivatives of all orders, which are easily computed, for all real numbers. Third, they can be integrated with ease over any interval. In contrast, nonpolynomial functions can be difficult or impossible to evaluate, differentiate, or integrate. Even the exponential function $f(x) = e^x$ does not lend itself to easy evaluation. For example, try finding $e^{0.1}$ without the aid of a calculator, a computer, or tables. Similarly, $\ln 2$, $\sin 0.5$, and $\sqrt[3]{9.2}$ cannot readily be approximated as decimal quantities. We encounter the same problem in evaluating definite integrals. Although we can sometimes readily get the "answer," as in

$$\int_{\pi/6}^{\pi/2} \cot x \, dx = \ln(\sin x)\big]_{\pi/6}^{\pi/2} = \ln\left(\sin\frac{\pi}{2}\right) - \ln\left(\sin\frac{\pi}{6}\right)$$

$$= \ln 1 - \ln \frac{1}{2} = \ln 2$$

we are again confronted with the difficulty of giving this result as a decimal quantity. The situation is even worse with an integral such as

$$\int_0^1 e^{-x^2} dx$$

where no elementary antiderivative of the integrand exists. We could use Simpson's Rule (see Section 5.5) to approximate the integral in this case. If, however, we could find a polynomial approximation to e^{-x^2}, we could easily perform the integration on the polynomial, and the result would be an approximation to the given integral. We have seen that the function $f(x) = e^{-x^2}$ is especially important in probability theory.

Of course, calculators, computers, and tables *are* available, and with these a very large number of functions can be evaluated to a high degree of accuracy.

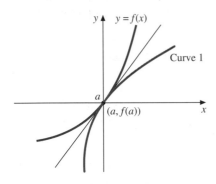

FIGURE 10.1
$P_1(x) = f(a) + f'(a)(x - a)$ is the linear approximation to $f(x)$ near $x = a$.

It is instructive, however, to learn how calculators and computers might arrive at the answers and how tables are constructed. Also, it is important to be able to determine just how accurate the approximations are. We will see answers to these questions in the procedures we develop in this section.

Linear and Quadratic Approximation

In Section 3.9 we discussed linear approximation of a function f near a point $x = a$. We found that if $f'(a)$ exists, then for x near a,

$$f(x) \approx f(a) + f'(a)(x - a)$$

If we let

$$P_1(x) = f(a) + f'(a)(x - a) \qquad (10.6)$$

then P_1 is the first-degree polynomial that best approximates f near a. The graph of P_1 is just the tangent line to the curve $y = f(x)$ at the point for which $x = a$. (See Figure 10.1.) The graph of P_1 has two things in common with the graph of f: both graphs go through the point $(a, f(a))$, and they both have the slope $f'(a)$ at that point.

If we also know the concavity of the curve $y = f(x)$ at $x = a$, it would make sense to approximate f with a polynomial that not only goes through the point $(a, f(a))$ and has slope $f'(a)$ there, but also has the same concavity as f at $x = a$. (See Figure 10.2.) That is, we want a quadratic polynomial $P_2(x)$ that satisfies the following three conditions:

$$P_2(a) = f(a) \qquad \text{Same value at } x = a$$

$$P_2'(a) = f'(a) \qquad \text{Same slope at } x = a$$

$$P_2''(a) = f''(a) \qquad \text{Same concavity at } x = a$$

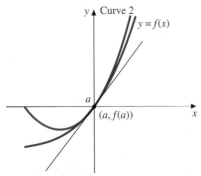

FIGURE 10.2
Curves 1 and 2 both go through the point $(a, f(a))$ with slope $f'(a)$, but Curve 2 better approximates the graph of f near that point, because its concavity agrees with that of f.

(We use the subscript 2 to indicate that P_2 is a polynomial of degree 2.) Let us write such a polynomial as

$$P_2(x) = c_0 + c_1 x + c_2 x^2$$

and determine what values the coefficients c_0, c_1, and c_2 must have so that our three conditions are satisfied. By differentiating $P_2(x)$ and substituting $x = a$ in $P_2(x)$, $P_2'(x)$, and $P_2''(x)$, our three conditions can be written as

$$P_2(a) = c_0 + c_1 a + c_2 a^2 = f(a)$$

$$P_2'(a) = c_1 + 2c_2 a \qquad\quad = f'(a)$$

$$P_2''(a) = 2c_2 \qquad\qquad\quad = f''(a)$$

From the third equation, we see that

$$c_2 = \frac{f''(a)}{2}$$

Thus, on substituting this value of c_2 in the second equation, we have

$$c_1 = f'(a) - 2\left(\frac{f''(a)}{2}\right)a = f'(a) - af''(a)$$

and from the first equation,

$$c_0 = f(a) - \frac{f''(a)}{2}a^2 - \left[f'(a) - af''(a)\right]a$$

$$= f(a) - af'(a) + \frac{a^2}{2}f''(a)$$

Our polynomial can now be written in the form

$$P_2(x) = \overbrace{f(a) - af'(a) + \frac{a^2}{2}f''(a)}^{c_0} + \overbrace{\left[f'(a) - af''(a)\right]}^{c_1} x + \overbrace{\frac{f''(a)}{2}}^{c_2} x^2$$

By rearranging the right-hand side and factoring out the derivatives $f'(a)$ and $f''(a)$, we obtain

$$P_2(x) = f(a) + f'(a)(x - a) + \frac{f''(a)}{2}(x^2 - 2ax + a^2)$$

or, finally,

$$P_2(x) = f(a) + f'(a)(x - a) + \frac{f''(a)}{2}(x - a)^2 \qquad (10.7)$$

EXAMPLE 10.1   Find the linear and quadratic polynomial approximations $P_1(x)$ and $P_2(x)$ for the function $f(x) = e^x$ near $x = 0$. Approximate $e^{0.1}$ using both $P_1(0.1)$ and $P_2(0.1)$.

Solution   Taking $a = 0$ in Equations 10.6 and 10.7, we get

$$P_1(x) = f(0) + f'(0)x$$

and

$$P_2(x) = f(0) + f'(0)x + \frac{f''(0)}{2}x^2$$

Since $f(x) = e^x$, both $f'(x)$ and $f''(x)$ also equal e^x. So $f(0) = 1$, $f'(0) = 1$, and $f''(0) = 1$. We therefore have

$$P_1(x) = 1 + x \qquad \text{and} \qquad P_2(x) = 1 + x + \frac{x^2}{2}$$

In Figure 10.3 we show the graph of $y = e^x$ along with the graphs of $y = P_1(x)$ and $y = P_2(x)$.

When $x = 0.1$, we have

$$P_1(0.1) = 1.1 \qquad \text{and} \qquad P_2(0.1) = 1.105$$

Using a calculator, we can find that

$$e^{0.1} \approx 1.10517092$$

to eight places. So $P_2(0.1)$ approximates $e^{0.1}$ to three places of accuracy.   ■

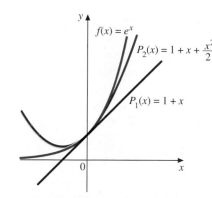

FIGURE 10.3

Taylor Polynomials of Degree n

Suppose now that f possesses derivatives up through the nth order at $x = a$. Our objective now is to find the polynomial of the nth degree, $P_n(x)$, whose first n derivatives at $x = a$ agree with the corresponding first n derivatives of

724 Chapter 10 Infinite Series

f at $x = a$ and whose value agrees with f at $x = a$. We know the result for $n = 1$ and $n = 2$. The pattern of the polynomials in these two cases suggests that we write $P_n(x)$ in the form

$$P_n(x) = c_0 + c_1(x - a) + c_2(x - a)^2 + \cdots + c_n(x - a)^n$$

Computing the successive derivatives, we obtain

$$P_n'(x) = c_1 + 2c_2(x - a) + 3c_3(x - a)^2 + \cdots + nc_n(x - a)^{n-1}$$
$$P_n''(x) = \qquad 2c_2 \qquad + 3 \cdot 2c_3(x - a) + \cdots + n(n - 1)c_n(x - a)^{n-2}$$
$$P_n'''(x) = \qquad\qquad\qquad 3 \cdot 2c_3 \qquad + \cdots + n(n - 1)(n - 2)c_n(x - a)^{n-3}$$
$$\vdots \qquad\qquad\qquad\qquad\qquad\qquad \vdots$$
$$P_n^{(n)}(x) = \qquad\qquad\qquad\qquad\qquad\qquad\qquad n!\,c_n$$

Setting $x = a$ and equating $P_n(a)$ with $f(a)$ and $P_n^{(k)}(a)$ with $f^{(k)}(a)$ for $k = 1, \ldots, n$, gives

$$P_n(a) = c_0 \qquad = f(a)$$
$$P_n'(a) = c_1 \qquad = f'(a)$$
$$P_n''(a) = 2c_2 \quad = f''(a), \quad \text{so } c_2 = \frac{f''(a)}{2!}$$
$$P_n'''(a) = 3 \cdot 2c_3 = f'''(a), \quad \text{so } c_3 = \frac{f'''(a)}{3!}$$
$$\vdots \qquad\quad \vdots \qquad\qquad\quad \vdots$$
$$P_n^{(n)}(a) = n!\,c_n \quad = f^{(n)}(a), \quad \text{so } c_n = \frac{f^{(n)}(a)}{n!}$$

We can write the kth coefficient c_k as

$$c_k = \frac{f^{(k)}(a)}{k!}, \qquad k = 0, 1, 2, \ldots, n$$

where $f^{(0)}(a)$ means $f(a)$. The polynomial $P_n(x)$ with these coefficients is named after the English mathematician Brook Taylor (1685–1731).

Definition 10.1
The nth Taylor Polynomial

Let f possess derivatives up through the nth order at $x = a$. The polynomial

$$P_n(x) = f(a) + f'(a)(x - a) + \frac{f''(a)}{2!}(x - a)^2 + \cdots + \frac{f^{(n)}(a)}{n!}(x - a)^n$$
(10.8)

is called the **nth Taylor polynomial for f about $x = a$**.

REMARK ───────────────────────────────

■ As our derivation of $P_n(x)$ shows, $P_n(x)$ is the polynomial of degree n in powers of $x - a$ that agrees with $f(x)$ at $x = a$ and whose first n derivatives agree with the corresponding derivatives of $f(x)$ at $x = a$.

───────────────────────────────

EXAMPLE 10.2   Find the Taylor polynomials $P_3(x)$, $P_4(x)$, and $P_5(x)$ about $x = 0$ for the function $f(x) = e^x$. Discuss the approximations to $e^{0.1}$ and $e^{1.0}$ by each of these polynomials.

Solution   Recall that in Example 10.1 we found $P_1(x)$ and $P_2(x)$ for this same function. To find $P_3(x)$, $P_4(x)$, and $P_5(x)$ we use Equation 10.8 with $a = 0$. Since all derivatives of $f(x) = e^x$ also equal e^x, their common value at $x = 0$ is 1. So we have

$$P_3(x) = 1 + x + \frac{x^2}{2!} + \frac{x^3}{3!}$$

$$P_4(x) = 1 + x + \frac{x^2}{2!} + \frac{x^3}{3!} + \frac{x^4}{4!}$$

$$P_5(x) = 1 + x + \frac{x^2}{2!} + \frac{x^3}{3!} + \frac{x^4}{4!} + \frac{x^5}{5!}$$

where $k! = 1 \cdot 2 \cdot 3 \cdot \cdots \cdot k$. The computer-generated graphs in Figure 10.4 show that P_3, P_4, and P_5 approximate f more and more closely as the degree of the Taylor polynomial increases. For x near 0, the polynomial graphs are almost indistinguishable from the graph of f. As we move farther from 0 the approximation is not as good. To confirm this observation, we compute values of each of the polynomials at $x = 0.1$ and at $x = 1.0$.

$$P_3(0.1) \approx 1.105167 \qquad P_3(1.0) = 2.6167$$

$$P_4(0.1) \approx 1.1051708 \qquad P_4(1.0) = 2.7083$$

$$P_5(0.1) \approx 1.105170917 \quad P_5(1.0) = 2.71667$$

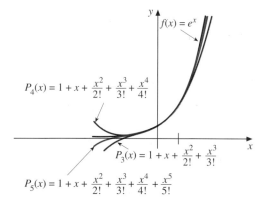

FIGURE 10.4

As we found with a calculator in Example 10.1, $f(0.1) = e^{0.1} \approx 1.10517092$ to eight places. Also, $f(1.0) = e^{1.0} \approx 2.71828183$ to eight places. We see, then, that for $x = 0.1$ each of the polynomials gives an answer close to the true value. In fact, $P_3(0.1)$, if rounded, is correct to five places, $P_4(0.1)$ is correct to six places, and $P_5(0.1)$ is correct to eight places.

In contrast, $P_3(1.0)$ is not even correct to one decimal place, $P_4(1.0)$ is correct to one place only, and $P_5(1.0)$ is correct only to two places. If we wanted more accuracy at $x = 1$, we would have to use a higher-degree Taylor polynomial. Also, to approximate e^x for values of x near $x = 1$, it would be better to use Taylor polynomials about $x = 1$ than about $x = 0$. ∎

As this example shows, the farther x is from the point $x = a$ (in our case $a = 0$), the less accurate is the Taylor polynomial approximation $P_n(x)$ for any fixed n. Also, for a fixed n we get better and better approximations by $P_n(x)$ as

n increases. Although these observations are valid for the function $f(x) = e^x$, it is not clear that they are true for other functions. We will have more to say about errors in approximating functions by their Taylor polynomials shortly. First, let us look at another example.

EXAMPLE 10.3   Find the Taylor polynomial $P_n(x)$ about $x = 0$ for the function $f(x) = \cos x$.

Solution   To apply Formula 10.8, we need to calculate successive derivatives of $f(x) = \cos x$ and evaluate them at $x = 0$.

$$
\begin{aligned}
f(x) &= \cos x & f(0) &= 1 \\
f'(x) &= -\sin x & f'(0) &= 0 \\
f''(x) &= -\cos x & f''(0) &= -1 \\
f'''(x) &= \sin x & f'''(0) &= 0 \\
f^{(4)}(x) &= \cos x & f^{(4)}(0) &= 1 \\
&\;\;\vdots & &\;\;\vdots
\end{aligned}
$$

The pattern is now clear: $f^{(n)}(0) = \pm 1$ when n is even, and $f^{(n)}(0) = 0$ when n is odd. Let $n = 2m$, where $m = 0, 1, 2, 3, \dots$. Then n is even. Since the signs of $f^{(2m)}(0)$ alternate from $+1$ to -1, we can write

$$ f^{(2m)}(0) = (-1)^m, \qquad m = 0, 1, 2, \dots $$

Thus, by Formula 10.8,

$$ P_{2m}(x) = 1 - \frac{x^2}{2!} + \frac{x^4}{4!} - \frac{x^6}{6!} + \cdots + \frac{(-1)^m x^{2m}}{(2m)!} = \sum_{k=0}^{m} \frac{(-1)^k x^{2k}}{(2k)!} $$

Note that since $f^{(2m+1)}(0) = 0$, it follows that $P_{2m+1}(x) = P_{2m}(x)$ for $f(x) = \cos x$.

In Figure 10.5 we show a computer-generated graph of $f(x)$ along with $P_0(x)$, $P_2(x)$, $P_4(x)$, $P_6(x)$, $P_8(x)$, and $P_{10}(x)$.   ∎

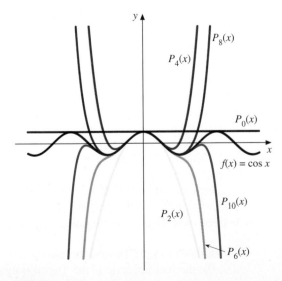

FIGURE 10.5

The Error in the Taylor Polynomial Approximation

We know that the nth Taylor polynomial $P_n(x)$ and its first n derivatives agree with $f(x)$ at $x = a$, and that for x near a, $P_n(x) \approx f(x)$. But just how good is this approximation? The following theorem provides the answer. We give its proof in Supplementary Appendix 5.

THEOREM 10.1

Taylor's Formula with Remainder

Let f have derivatives up through the $(n + 1)$st order in an open interval I centered at $x = a$. Then for each x in I there is a number c between a and x such that

$$f(x) = f(a) + f'(a)(x-a) + \frac{f''(a)}{2!}(x-a)^2 + \cdots + \frac{f^{(n)}(a)}{n!}(x-a)^n + R_n(x)$$

$$(10.9)$$

where

$$R_n(x) = \frac{f^{(n+1)}(c)}{(n+1)!}(x-a)^{n+1} \qquad (10.10)$$

REMARKS ———————————————————————————————

■ Formula 10.9 (Taylor's Formula) can be written more briefly as $f(x) = P_n(x) + R_n(x)$.

■ $R_n(x)$ is called the **Lagrange form of the remainder**, after Joseph Louis Lagrange (1736–1813), one of the greatest mathematicians of the eighteenth century. The term *remainder* is used here to mean what is left when the approximation $P_n(x)$ is subtracted from $f(x)$. It is the error in approximating $f(x)$ by $P_n(x)$.

■ The number c as stated in the theorem is dependent on x. It is between a and x, whether x is greater than a or less than a.

■ The interval I may be the entire real line, $(-\infty, \infty)$, or it may be finite, in which case it will be of the form $I = (a - r, a + r)$ for some $r > 0$. The size of the interval is determined by the point closest to a for which $f^{(n+1)}(x)$ fails to exist. If this point is r units from a, then $I = (a - r, a + r)$. If $f^{(n+1)}(x)$ exists everywhere, then $I = (-\infty, \infty)$.

■ Note that the remainder term $R_n(x)$ follows the same pattern as the terms in $P_n(x)$ *except* that the derivative is evaluated at the intermediate value c between a and x, rather than at a.

When $n = 0$ in Taylor's Formula and we replace x with b, where b is in the interval I and $b \neq a$, the result is

$$f(b) = f(a) + f'(c)(b - a)$$

or equivalently,

$$f'(c) = \frac{f(b) - f(a)}{b - a}$$

where c is between a and b. But this formula is just the statement of the conclusion in the Mean-Value Theorem (Theorem 4.9). For this reason Theorem 10.1 is sometimes called the *Extended Mean-Value Theorem*. (We should note that the Mean-Value Theorem is used in the proof of Theorem 10.1, so we cannot use this theorem as an independent way of getting the Mean-Value Theorem.)

EXAMPLE 10.4   Find $P_n(x)$ and $R_n(x)$ about $x = 0$ for $f(x) = e^x$. Use the result to find a bound on the error in estimating $e^{0.1}$ using $P_3(0.1)$.

Solution   Note first that $f(x) = e^x$ possesses derivatives of all orders everywhere, all equal to e^x, so that in this case Theorem 10.1 is valid on $(-\infty, \infty)$. Taking $a = 0$ in Theorem 10.1, we have

$$P_n(x) = 1 + x + \frac{x^2}{2!} + \frac{x^3}{3!} + \cdots + \frac{x^n}{n!}$$

and

$$R_n(x) = \frac{f^{(n+1)}(c)}{(n+1)!}x^{n+1} = \frac{e^c}{(n+1)!}x^{n+1}$$

where c is between 0 and x. Thus, the error in using $P_3(0.1)$ to estimate $e^{0.1}$ is

$$R_3(0.1) = \frac{e^c}{4!}(0.1)^4, \qquad 0 < c < 0.1$$

Although we do not know c, we can get an upper bound on e^c by the fact that

$$e^c < e^{0.1} < e^1 < 3$$

This is a very generous upper bound, since $e^{0.1}$ is much less than e, but since we presumably do not know $e^{0.1}$, we can take something larger, with a value that we do know. Thus,

$$R_3(0.1) < \frac{3}{4!}(0.1)^4 = \frac{3}{4 \cdot 3 \cdot 2 \cdot 1}(0.0001) = 0.0000125$$

Even using the generous upper bound on e^c, we see that the error is small, resulting in accuracy to at least four decimal places. Recall that in Example 10.2 we found that $P_3(0.1) \approx 1.10517$, which is correct to five decimal places. ∎

EXAMPLE 10.5   Find Taylor's Formula with remainder for arbitrary n for the function $f(x) = \ln x$ about $x = 1$, and state the interval in which it is valid.

Solution   In abbreviated form, Taylor's Formula is $f(x) = P_n(x) + R_n(x)$. For P_n we need the first n derivatives of f, and for R_n we need $f^{(n+1)}$. So we begin by calculating the successive derivatives:

$$\begin{aligned}
f(x) &= \ln x & f(1) &= 0 \\
f'(x) &= \tfrac{1}{x} = x^{-1} & f'(1) &= 1 \\
f''(x) &= -x^{-2} & f''(1) &= -1 \\
f'''(x) &= 2x^{-3} & f'''(1) &= 2 \\
f^{(4)}(x) &= -3 \cdot 2x^{-4} = -3!\,x^{-4} & f^{(4)}(1) &= -3! \\
f^{(5)}(x) &= 4 \cdot 3!\,x^{-5} = 4!\,x^{-5} & f^{(5)}(1) &= 4! \\
&\ \ \vdots & &\ \ \vdots \\
f^{(n)}(x) &= (-1)^{n-1}(n-1)!\,x^{-n} & f^{(n)}(1) &= (-1)^{n-1}(n-1)!
\end{aligned}$$

For $R_n(x)$ we need $f^{(n+1)}(x)$ evaluated at $x = c$:

$$f^{(n+1)}(x) = (-1)^n n! \, x^{-(n+1)} \qquad f^{(n+1)}(c) = (-1)^n n! \, c^{-(n+1)}$$

Observe carefully how we handled the coefficients of the successive derivatives. First, they alternate in sign—they are positive when the order is odd and negative when the order is even. The factor $(-1)^{n-1}$ in the nth derivative accomplishes the alternation of the signs. Second, we did not multiply out the factors as they accumulated but rather indicated the product as a factorial. Doing so enabled us to recognize the general pattern. We also used the fact that $k! = k(k-1)!$ for $k \geq 1$. You should convince yourself of this equality.

Since

$$f^{(n+1)}(x) = \frac{(-1)^n n!}{x^{n+1}}$$

is not defined when $x = 0$, the largest interval about $x = 1$ for which Taylor's Formula is valid is $(0, 2)$. (Later we will see that the formula remains valid at the right endpoint $x = 2$, but for now we will consider only the open interval.)
Now we can find $P_n(x)$ and $R_n(x)$. For $x \in (0, 2)$,

$$f(x) = P_n(x) + R_n(x)$$

$$= f(1) + f'(1)(x-1) + \frac{f''(1)}{2!}(x-1)^2 + \frac{f'''(1)}{3!}(x-1)^3$$

$$+ \frac{f^{(4)}(1)}{4!}(x-1)^4 + \cdots + \frac{f^{(n)}(1)}{n!}(x-1)^n + \frac{f^{(n+1)}(c)}{(n+1)!}(x-1)^{n+1}$$

$$= 0 + 1(x-1) + \frac{-1}{2!}(x-1)^2 + \frac{2}{3!}(x-1)^3 + \frac{-3!}{4!}(x-1)^4 + \cdots$$

$$+ \frac{(-1)^{n-1}(n-1)!}{n!}(x-1)^n + \frac{(-1)^n n!}{(n+1)! \, c^{n+1}}(x-1)^{n+1}$$

or

$$f(x) = (x-1) - \frac{(x-1)^2}{2} + \frac{(x-1)^3}{3} - \frac{(x-1)^4}{4} + \cdots$$

$$+ \frac{(-1)^{n-1}(x-1)^n}{n} + R_n(x)$$

where

$$R_n(x) = \frac{(-1)^n (x-1)^{n+1}}{(n+1)c^{n+1}}, \qquad c \text{ between } 1 \text{ and } x$$

In Figure 10.6 we show the graphs of $f(x) = \ln x$, along with the graphs of $P_n(x)$ for $n = 2, 5,$ and 10. Notice how all the graphs of $P_n(x)$ differ significantly from that of $\ln x$ for $x > 2$.   ∎

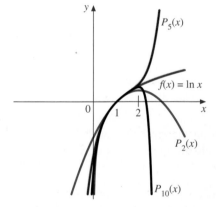

FIGURE 10.6

EXAMPLE 10.6   Using the results of Example 10.5,

(a) find $P_5(1.2)$ and an upper bound on the error in using $P_5(1.2)$ to approximate $\ln 1.2$, and

(b) find the smallest integer n that will guarantee that $P_n(1.2)$ estimates $\ln 1.2$ with eight-decimal-place accuracy.

Solution

(a) From Example 10.5,

$$P_5(x) = (x-1) - \frac{(x-1)^2}{2} + \frac{(x-1)^3}{3} - \frac{(x-1)^4}{4} + \frac{(x-1)^5}{5}$$

and

$$R_5(x) = \frac{(-1)^5(x-1)^6}{6c^6}, \qquad c \text{ between } 1 \text{ and } x$$

So

$$P_5(1.2) = 0.2 - \frac{(0.2)^2}{2} + \frac{(0.2)^3}{3} - \frac{(0.2)^4}{4} + \frac{(0.2)^5}{5}$$

$$\approx 0.182331$$

The error is given by

$$|R_5(1.2)| = \left| \frac{(-1)^5(0.2)^6}{6c^6} \right| = \frac{0.000064}{6c^6}$$

Since $1 < c < 1.2$, we get a larger fraction by replacing c by 1. Thus,

$$|R_5(1.2)| \leq \frac{0.000064}{6} \approx 0.000011$$

So the answer is correct to at least four decimal places.

(b) From Example 10.5,

$$R_n(x) = \frac{(-1)^n(x-1)^{n+1}}{(n+1)c^{n+1}}, \qquad c \text{ between } 1 \text{ and } x$$

With $x = 1.2$, an upper bound on $|R_n(1.2)|$ can be obtained by replacing c with 1. Then $c^{n+1} = 1^{n+1} = 1$, so we have

$$|R_n(1.2)| \leq \frac{(0.2)^{n+1}}{n+1}$$

To obtain eight decimal places of accuracy, we find the smallest integer n such that

$$\frac{(0.2)^{n+1}}{n+1} \leq 5 \times 10^{-9}$$

By trial and error, using a calculator, a CAS, or a spreadsheet, we find that $n + 1 = 11$, or $n = 10$, is the smallest value that works. So $P_{10}(1.2)$ estimates $\ln 1.2$ to at least eight decimal places of accuracy.   ∎

EXAMPLE 10.7   Find an upper bound on the error in using the nth Taylor polynomial $P_n(x)$ about $x = 0$ to approximate $f(x) = \cos x$. Use $P_6(x)$ to approximate $\cos 10°$ and estimate the error.

Solution   In Example 10.3 we found that

$$P_{2n}(x) = 1 - \frac{x^2}{2!} + \frac{x^4}{4!} + \cdots + \frac{(-1)^n}{(2n)!}x^{2n}$$

and that $P_{2n+1}(x) = P_{2n}(x)$. The remainder term in Taylor's Formula (Equation 10.9) can therefore be written

$$R_{2n}(x) = R_{2n+1}(x) = \frac{f^{(2n+2)}(c)}{(2n+2)!}x^{2n+2}$$

where c is between 0 and x. Since an even-order derivative of $\cos x$ is either $\cos x$ or $-\cos x$, we therefore have

$$|R_{2n}(x)| = \frac{|\cos c|}{(2n+2)!}|x|^{2n+2} \leq \frac{|x|^{2n+2}}{(2n+2)!}$$

since $|\cos c| \leq 1$. Thus, the error in using $P_{2n}(x)$ to approximate $\cos x$ for any real number x is at most $|x|^{2n+2}/(2n+2)!$.

To use $P_6(x)$ to approximate $\cos 10°$, we first express $10°$ in radians:

$$10° = 10° \left(\frac{\pi \text{ radians}}{180°} \right) = \frac{\pi}{18} \text{ radians}$$

Then we have

$$P_6 \left(\frac{\pi}{18} \right) = 1 - \frac{1}{2!} \left(\frac{\pi}{18} \right)^2 + \frac{1}{4!} \left(\frac{\pi}{18} \right)^4 - \frac{1}{6!} \left(\frac{\pi}{18} \right)^6 \approx 0.9848077530$$

The upper bound we found above on the error is

$$\left| R_6 \left(\frac{\pi}{18} \right) \right| \leq \frac{\left(\frac{\pi}{18} \right)^8}{8!} \approx 2.135 \times 10^{-11}$$

So our answer is correct to at least ten decimal places.   ∎

A Preview of Taylor Series

In Example 10.4 we found the nth Taylor polynomial about $x = 0$ for $f(x) = e^x$ to be

$$P_n(x) = 1 + x + \frac{x^2}{2!} + \frac{x^3}{3!} + \cdots + \frac{x^n}{n!}$$

So we have the approximation

$$e^x \approx 1 + x + \frac{x^2}{2!} + \frac{x^3}{3!} + \cdots + \frac{x^n}{n!}$$

and the approximation gets better and better as n increases. Later we will show that the approximation is valid for all real values of x. A natural question to ask is whether we can let n "go to infinity" and get the exact value of e^x. That is, can we say that

$$e^x = 1 + x + \frac{x^2}{2!} + \frac{x^3}{3!} + \frac{x^4}{4!} + \cdots \qquad (10.11)$$

for all real values of x? Recall from this chapter's introduction that we call a series such as this one a *power series in* x. If we set $x = 1$, for example, our question is whether it is valid to write

$$e = 1 + 1 + \frac{1}{2!} + \frac{1}{3!} + \frac{1}{4!} + \cdots$$

We are immediately confronted with another question: What do we mean by the sum of infinitely many numbers? Clearly, we must have some new definition of addition in such a case. We will give a definition in the next section and will explore its consequences in the sections that follow. We will find that Equation 10.11 is valid for all real values of x.

For now, we will say that the expression on the right-hand side of Equation 10.11 is the **Taylor Series of** e^x **about** $x = 0$ and that, under the meaning yet to be given, the series **converges** to e^x for all real x. More generally, by letting $n \to \infty$ in Taylor's Formula with Remainder, Equation 10.9, we obtain the **Taylor Series for the function** f **about** $x = a$:

$$f(a) + f'(a)(x - a) + \frac{f''(a)}{2!}(x - a)^2 + \frac{f'''(a)}{3!}(x - a)^3 + \cdots$$

As you might guess, the condition under which this series converges to $f(x)$ is that the error in the approximation of $f(x)$ by its nth Taylor polynomial gets smaller and smaller as $n \to \infty$; that is,

$$\lim_{n \to \infty} R_n(x) = 0$$

We will return to this important question of the representation of a function by its Taylor series after our discussion of infinite series in general.

COMPUTING TAYLOR POLYNOMIALS USING A COMPUTER ALGEBRA SYSTEM

Taylor's Theorem presents a method for using polynomials to approximate functions that cannot be explicitly evaluated. The theorem also provides a useful means of analyzing the error in such approximations. The degree of accuracy in using Taylor polynomials to approximate functions generally increases as the degree of the polynomial increases. The higher the degree of the polynomial desired, the greater the amount of computation necessary. For this reason, a computer algebra system becomes a valuable aid.

 CAS 36

Use Taylor polynomials to approximate the function $f(x) = e^{-x^2}$ close to 0.

We will begin by using Maple, Mathematica, and DERIVE to compute the Taylor polynomial of degree 8 for f near $x = 0$.

Maple:

```
f := x–>exp(–x^2);
taylor(f(x),x=0,9);
```

The output is:

$$1 - x^2 + \frac{1}{2}x^4 - \frac{1}{6}x^6 + \frac{1}{24}x^8 + O(x^9)$$

Notice the big-O term on the end of the output from Maple. This indicates that all higher degree terms have an exponent of at least 9. Because of this form of the output in Maple, if the Taylor polynomial of degree 8 is desired, the last parameter in the **taylor** command should be specified as 9.

Mathematica:

```
f[x_] = Exp[–x^2]
Normal[Series[f[x],{x,0,8}]]
```

DERIVE:

(At the □ symbol, go to the next step.)
a (author) □ exp(−x^2) □ c (calculus)
□ t (taylor) □ [choose expression] □
x (variable) □ 8 (degree) □ 0 (point) □
s (simplify)

Notice that Mathematica and DERIVE do not include a big-O term in the output. In Mathematica, the Normal command removes this big-O term.

To see the relative accuracy of the approximation near 0 a plot would be helpful. Maple does not directly plot the result of the command taylor(f(x),x=0,9);. We could enter a plot and retype the 8th degree polynomial as the function to plot. We can, however, avoid the problem by using a more general command in Maple, called mtaylor. To access this command, it must be read from a Maple library via readlib(‘mtaylor’):

Maple:

readlib(`mtaylor`):
P := mtaylor(f(x),[x],9);

The output is:

$$P := 1 - x^2 + \frac{1}{2}x^4 - \frac{1}{6}x^6 + \frac{1}{24}x^8$$

plot({f(x),P},x=−2..2);

We see from Figure 10.1.1 that near 0 the approximation looks very good.

Mathematica:

P[x_] = Normal[Series[f[x],
   {x,0,8}]]
Plot[{f[x],P[x]},{x,−2,2}]

DERIVE:

(At the □ symbol, go to the next step.)
[choose exp(−x^2)] □ p (plot window)
□ p (plot) □ a (algebra) □ [choose Taylor polynomial] □ p (plot window)
□ p (plot)

FIGURE 10.1.1

To see the relative accuracy obtained in approximating f by Taylor polynomials of different degrees, we could repeat what we have already done with different polynomials and compare the pictures obtained. This would be tedious. However, the computer algebra systems allow us to make this kind of comparison very quickly by using a loop structure.

Maple:

for n from 1 by 1 to 10 do
  P := mtaylor(f(x),[x],n);
od;

The output gives Taylor polynomials of degrees 0 to 9.

$$P := 1$$

$$P := 1$$

$$P := 1 - x^2$$

$$P := 1 - x^2$$

$$P := 1 - x^2 + \frac{1}{2}x^4$$

$$P := 1 - x^2 + \frac{1}{2}x^4$$

$$P := 1 - x^2 + \frac{1}{2}x^4 - \frac{1}{6}x^6$$

$$P := 1 - x^2 + \frac{1}{2}x^4 - \frac{1}{6}x^6$$

$$P := 1 - x^2 + \frac{1}{2}x^4 - \frac{1}{6}x^6 + \frac{1}{24}x^8$$

$$P := 1 - x^2 + \frac{1}{2}x^4 - \frac{1}{6}x^6 + \frac{1}{24}x^8$$

Notice the 1st and 2nd, 3rd and 4th, 5th and 6th, etc., Taylor polynomials are identical.

To get a series of plots in Maple, enter:

for n from 1 by 1 to 10 do
P[n] := mtaylor(f(x),[x],n);
plot({f(x),P[n]},x=−2..2);
od;

To plot f and the Taylor polynomials for f about $x = 0$, of orders 2, 4, 6, 8, and 10 on the same axes, use the Maple command:

plot({f(x),P[2],P[4],P[6],P[8],P[10]}, x=−5..5,y=−5..5);

Mathematica:

Do[p[x_,k] = Normal[Series[f[x],{x,0,k}]],
    {k,0,8}]
Do[Plot[{f[x],p[x,k]},
    {x,−2,2}],{k,0,8}]

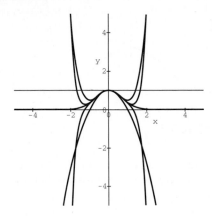

FIGURE 10.1.2
$f(x)$, $P_2(x)$, $P_4(x)$, $P_6(x)$, $P_8(x)$, and $P_{10}(x)$

The graphical analysis above gives an intuitive feel for where the Taylor polynomial is a good approximation to the function. For more precise numerical information, the error term should be analyzed.

Exercise Set 10.1

In Exercises 1–6, find the Taylor polynomial of the specified degree for f about $x = 0$.

1. $f(x) = e^{-x}$; $n = 5$

2. $f(x) = \sin x$; $n = 7$

3. $f(x) = 1/(x + 1)$; $n = 6$

4. $f(x) = \sqrt{x + 1}$; $n = 5$

5. $f(x) = \cosh x$; $n = 8$

6. $f(x) = \tan x$; $n = 5$

In Exercises 7–12, find the Taylor polynomial of degree n for f about the specified point a.

7. $f(x) = 1/x$; $a = 1$ **8.** $f(x) = e^{x-2}$; $a = 2$

9. $f(x) = \sin x$; $a = 0$

10. $f(x) = 1/(1 - x)^2$; $a = -1$

11. $f(x) = \ln(2 - x)$; $a = 1$

12. $f(x) = \sinh x$; $a = 0$

In Exercises 13–18, find $R_n(x)$ for the functions specified in Exercises 7–12 and give the interval about which $f(x) = P_n(x) + R_n(x)$.

In Exercises 19–32, estimate the specified function value using P_n for the given value of n and an appropriate value of a. Find an upper bound on the error, and determine the number of decimal places of accuracy. (When appropriate, you may use the results of preceding exercises or examples.)

19. e^{-1}; $n = 5$

20. $\sqrt{1.2}$; $n = 5$

21. $\sin 1.5$; $n = 7$

22. $\ln 1.5$; $n = 6$

23. $\cos 2$; $n = 8$

24. $\sinh 2$; $n = 7$

25. $\sqrt{5}$; $n = 5$ (*Hint:* Take $a = 4$.)

26. $\ln 4/5$; $n = 8$

27. e^3; $n = 12$

28. $\sqrt{e}$; $n = 4$

29. $\cosh 1$; $n = 6$

30. $\sin 32°$; $n = 5$ (*Hint:* Change to radians and use $a = \pi/6$.)

31. $\cos 58°$; $n = 4$ (*Hint:* Take $a = \pi/3$.)

32. $\sqrt[3]{7}$; $n = 4$

33. Use the result of Example 10.7 to find the degree of the Taylor polynomial for $\cos x$ about $x = 0$ that approximates $\cos 72°$ to six decimal places of accuracy. Find this approximation and compare it with the value given by a calculator.

34. Find $\sin 80°$ correct to five decimal places of accuracy using the appropriate Taylor polynomial about $x = 0$. (See Exercises 9 and 15.)

35. Find $\sinh 2$ to five decimal places of accuracy using the appropriate Taylor polynomial about $x = 0$. (See Exercises 12 and 18.)

36. Find $P_n(x)$ and $R_n(x)$ for $f(x) = \sqrt{x}$ about $x = 4$. In what interval is the result valid? Use the result to find $\sqrt{3}$ correct to four decimal places.

37. Let $f(x) = (1 + x)^\alpha$, where α is real.
   (a) Find $P_n(x)$ and $R_n(x)$ for f about $x = 0$.
   (b) Show that if $\alpha = m$, where m is a positive integer, then $f(x) = P_m(x)$ for all x.
   (c) If α is not a positive integer, show that $f(x) = P_n(x) + R_n(x)$ for $-1 < x < 1$.
   (d) Use the result of part (a) to estimate $\sqrt[3]{1.5}$, taking $n = 4$. Find an upper bound on the error.

38. (a) Show that if $f(x)$ is a polynomial of degree n, then the Taylor polynomial $P_n(x)$ for f about $x = a$ is identical to $f(x)$.

(b) Express the polynomial
$$f(x) = 2x^5 - 3x^4 + 7x^3 - 8x^2 + 2$$
in powers of $x - 1$.

39. For $-1 < x < 1$, let
$$f(x) = \ln \frac{1+x}{1-x}$$
   (a) Show that if u is any positive number, there is an x in $(-1, 1)$ such that $f(x) = \ln u$.
   (b) Find $P_n(x)$ and $R_n(x)$ for f about $x = 0$. (*Hint:* Rewrite $f(x)$ using properties of logarithms before taking derivatives.)
   (c) Use parts (a) and (b) to find $\ln 2$ correct to three decimal places.

40. (a) In Exercise 39(a), let $u = (N + 1)/N$, where N is a positive integer. What is x so that $f(x) = \ln u$?
   (b) Use the result of part (a), together with Exercise 39, to show that
$$\ln(N + 1) = \ln N + 2 \sum_{k=1}^{n} \frac{1}{(2k - 1)(2N + 1)^{2k-1}}$$
$$+ R_{2n}\left(\frac{1}{2N + 1}\right)$$

and that

$$\left| R_{2n}\left(\frac{1}{2N + 1}\right) \right| \le \frac{2}{(2n + 1)(2N)^{2n+1}}$$

(Note that $R_{2n-1}(x) = R_{2n}(x)$, since the Taylor polynomial consists of odd powers only.)
   (c) Find $\ln 3$ correct to three decimal places, making use of part (b) and Exercise 39(c).

41. Let $f(x) = \ln(1 + x)$.
   (a) Find the nth Taylor polynomial $P_n(x)$ for f about $x = 0$.
   (b) Use a CAS or graphing calculator to obtain the graphs of $f(x)$, $P_5(x)$, $P_{10}(x)$, and $P_{20}(x)$ on the same set of axes.
   (c) Describe the behavior of the graphs obtained in part (b). Pay particular attention to values of x in the interval $(-1, 1)$ as compared with values outside this interval.
   (d) Conjecture, based on your graphs, the values of x for which the remainder term $R_n(x)$ approaches 0 as $n \to \infty$.

42. Use the remainder term $R_n(x)$ in Taylor's Formula to determine the values of x for which the approximation
$$\sin x \approx x - \frac{x^3}{6}$$
is accurate to within 0.001 unit.

43. Repeat Exercise 42 with

$$\cos x \approx 1 - \frac{x^2}{2}$$

44. Repeat Exercise 42 with

$$\sqrt{1 + x^2} \approx 1 + \frac{x^2}{2}$$

45. Letting $f(x) = \sin x$,
   (a) find the nth Taylor polynomial for f about $x = 0$;
   (b) generate graphs of f and P_n on the same axes for $n = 1, 3, 5, 10$;
   (c) use the graphs to estimate on which interval $P_3(x)$ approximates $f(x)$ to within about 0.01 unit. (Zoom in.) Repeat for $P_5(x)$ and for $P_{10}(x)$.

46. Repeat Exercise 45 with $f(x) = e^{-x}$.

47. Repeat Exercise 45 with $f(x) = (1 - x)^{1/3}$.

48. Repeat Exercise 47 using the nth Taylor polynomial about $x = 2$.

49. Repeat Exercise 45 with $f(x) = \sqrt{x + 1}$, first using $P_n(x)$ about $x = 0$, and then using $P_n(x)$ about $x = 3$.

50. Let $f(x) = e^x$. Find a value of n so that $P(0.1)$ is within 10^{-6} units of $e^{0.1}$.

10.2 SEQUENCES

In the introduction to this chapter we indicated how the sum of an infinite *series* can be defined. The key lies in computing what we called the *sequence of partial sums* and determining if these partial sums approach some finite value as more and more terms are added. To illustrate this relationship between a series and its sequence of partial sums, we used the series

$$1 + \frac{1}{2} + \frac{1}{4} + \frac{1}{8} + \frac{1}{16} + \cdots \qquad (10.12)$$

Its sequence of partial sums is

$$1, \quad 1\frac{1}{2}, \quad 1\frac{3}{4}, \quad 1\frac{7}{8}, \quad 1\frac{15}{16}, \ldots \qquad (10.13)$$

which comes arbitrarily close to 2 as we add more and more terms. Thus, we concluded that the sum of series 10.12 is 2. Recall that we obtain the sequence 10.13 by writing, in order, the first term of the series 10.12, the sum of the first two terms, the sum of the first three terms, and so on.

Note carefully that in a series, such as 10.12, the terms are added together. In contrast, in a sequence, such as 10.13, the members (also called terms) are separated by commas. Though the words *sequence* and *series* are often used interchangeably in nonmathematical usage, in mathematics we are careful to distinguish between them.

We will explore more fully the relationship between a series and its sequence of partial sums in the next section. In this section we investigate general properties of sequences. Later, we will apply these properties to sequences of partial sums of a series. Since the convergence or divergence of a series is of central importance, one of our main goals is to define the notion of the *limit* of a sequence and to learn some methods of finding this limit when it exists.

Let us be more precise about the meaning of a sequence. One way of defining a sequence is to say it is an ordered collection of elements, called *terms*, formed according to some rule that gives a unique term for each positive integer. Since a rule that assigns a unique element to each positive integer is a function having the positive integers as its domain, we can state the definition more succinctly as follows:

Definition 10.2 A Sequence	A **sequence** is a function whose domain is the set of all positive integers.

REMARK ───────────────────────────────────

■ According to the definition, a sequence is a function, but it is customary to refer to the ordered list of function values (the range) as the sequence itself.

For most of the sequences we will be dealing with, the terms will be real numbers, but they could be other mathematical entities. Note that by our definition, there will always be infinitely many terms, one for each positive integer (although the terms may not be distinct—there can be repetitions). So when we speak of a sequence, we will always mean an *infinite sequence*. In Definition 10.2, if the domain is replaced by the set of the first n positive integers, the resulting function is called a *finite sequence*.

Notation

It is customary, in giving the rule for a sequence, to use subscript notation, such as a_n, rather than the usual functional notation $f(n)$. For example, if we write

$$a_n = \frac{1}{n}$$

then by substituting, in turn, $n = 1, 2, 3, \ldots$, we get the sequence

$$\frac{1}{1}, \frac{1}{2}, \frac{1}{3}, \frac{1}{4}, \cdots \qquad (10.14)$$

We refer to a_n as the **nth term**, or **general term**, of the sequence. Giving a formula for the nth term effectively defines the entire sequence.

When a formula for the nth term a_n of a sequence is known, we can designate the sequence with braces around a_n, such as $\{a_n\}$ or, more explicitly, $\{a_n\}_{n=1}^{\infty}$. The latter notation makes it explicit that n takes on the positive integers starting with 1. If the simple notation $\{a_n\}$ is used (as will usually be the case), we will understand it to mean the same as $\{a_n\}_{n=1}^{\infty}$. (There may, however, be times when we want to start with $n = 0$ or some other positive integer n_0 with $n_0 \neq 1$. If so, we will indicate the limits.) Using this notation, we can designate the sequence 10.14 by $\{1/n\}$.

In the next example we show several sequences using this notation.

EXAMPLE 10.8   Show the first five terms of each of the following sequences, indicating its continuation by three dots.

(a) $\{2n - 1\}$ (b) $\{(-1)^{n-1}\}$ (c) $\left\{\dfrac{n}{n+1}\right\}$

(d) $\left\{\dfrac{2^n}{n!}\right\}$ (e) $\{\sin n\}$

Solution

(a) $1, 3, 5, 7, 9, \ldots$ (the odd positive integers)

(b) $(-1)^0, (-1)^1, (-1)^2, (-1)^3, (-1)^4, \ldots$, or on simplifying, $1, -1, 1, -1, 1, \ldots$

(c) $\dfrac{1}{2}, \dfrac{2}{3}, \dfrac{3}{4}, \dfrac{4}{5}, \dfrac{5}{6}, \ldots$

(d) $\dfrac{2^1}{1!}, \dfrac{2^2}{2!}, \dfrac{2^3}{3!}, \dfrac{2^4}{4!}, \dfrac{2^5}{5!}, \ldots$ Since $1! = 1$, $2! = 2 \cdot 1 = 2$, $3! = 3 \cdot 2 \cdot 1 = 6$, $4! = 4 \cdot 3 \cdot 2 \cdot 1 = 24$, and $5! = 5 \cdot 4 \cdot 3 \cdot 2 \cdot 1 = 120$, we can rewrite the sequence as

$$\frac{2}{1}, \frac{4}{2}, \frac{8}{6}, \frac{16}{24}, \frac{32}{120}, \ldots$$

which becomes, after reducing the fractions,

$$2, 2, \frac{4}{3}, \frac{2}{3}, \frac{4}{15}, \ldots$$

(e) $\sin 1, \sin 2, \sin 3, \sin 4, \sin 5, \ldots$. Using a calculator (in radian mode), we get (rounding to four decimal places) approximately $0.8415, 0.9093, 0.1411, -0.7568, -0.9589, \ldots$. ∎

If the pattern of the terms is clear, so that we can infer the formula for the nth term, it is permissible to designate the sequence by merely showing a few of its terms. For example, the sequence 10.14 could be given simply by writing

$$1, \frac{1}{2}, \frac{1}{3}, \frac{1}{4}, \ldots$$

We could also show the sequence in parts (a), (b), and (c) of Example 10.8 in this way. However, showing the sequence in part (d) in its simplified form,

$$2, 2, \frac{4}{3}, \frac{2}{3}, \frac{4}{15}, \ldots$$

is *not* in itself a satisfactory way of designating the sequence, since it is not clear from this list alone how to form succeeding terms.

EXAMPLE 10.9   Assume that each of the following sequences continues in the pattern of the terms shown. Write the general term.

(a) $\dfrac{1}{2}, -\dfrac{1}{4}, \dfrac{1}{6}, -\dfrac{1}{8}, \ldots$ (b) $\dfrac{1}{2}, \dfrac{3}{4}, \dfrac{5}{8}, \dfrac{7}{16}, \dfrac{9}{32}, \ldots$

Solution

(a) The alternating signs can be indicated by means of the factor $(-1)^{n-1}$, as we saw in Example 10.8(b). Each denominator is twice the number of the term. So we can write

$$a_n = \frac{(-1)^{n-1}}{2n}$$

(b) The numerators are the odd positive integers, which are indicated by the factor $2n - 1$, as we saw in Example 10.8(a). The denominators appear to be powers of 2, with the nth term denominator equal to 2^n. So we can write

$$a_n = \frac{2n - 1}{2^n}$$ ∎

It is not always possible to give an explicit formula for the nth term. For example, consider the sequence $\{a_n\}$ where a_n is the nth prime number. Recall that a prime number is a positive integer greater than 1 whose only positive divisors are 1 and itself. Several terms of this sequence are

$$2, 3, 5, 7, 11, 13, 17, 19, 23, \ldots$$

It is known that there are infinitely many primes, but there is no known formula for the nth prime.

As another example, let a_n denote the digit in the nth decimal place of the decimal representation of π. The first few terms are

$$1, 4, 1, 5, 9, 2, 6, 5, \ldots$$

Incidentally, π has been calculated to more than two billion places! Again, no formula has been found for the nth decimal digit.

Recursive Definitions

Sometimes sequences are defined *recursively*. For example, suppose we are given that $a_1 = \sqrt{2}$ and that for all $n \geq 2$,

$$a_n = \sqrt{2 + a_{n-1}}$$

This formula giving a_n in terms of a_{n-1} is called a **recursion formula**. Since we know that $a_1 = 1$, we can find a_2 by the recursion formula as follows:

$$a_2 = \sqrt{2 + a_1} = \sqrt{2 + \sqrt{2}}$$

Now we use the recursion formula again to get a_3:

$$a_3 = \sqrt{2 + a_2} = \sqrt{2 + \sqrt{2 + \sqrt{2}}}$$

and again to get a_4:

$$a_4 = \sqrt{2 + a_3} = \sqrt{2 + \sqrt{2 + \sqrt{2 + \sqrt{2}}}}$$

and so on. We could approximate each term by using a calculator, but the pattern would no longer be evident.

Another example of a recursive definition is the sequence of approximations by Newton's Method to a root of an equation $f(x) = 0$ (see Section 4.5). After the initial approximation x_0 is made, we find subsequent approximations by the recursion formula:

$$x_{n+1} = x_n - \frac{f(x_n)}{f'(x_n)} \qquad (n = 0, 1, 2, \ldots)$$

Note that in this case we begin with $n = 0$.

The Fibonacci sequence is named after the Italian businessman and mathematician Leonardo Fibonacci (who lived about 1170–1250). He posed the following problem:

> How many pairs of rabbits can be produced from a single pair in a year if every month each pair begets a new pair that from the second month onward becomes productive?

If we assume the rabbits in the initial pair are newborn (so that they produce the first new pair after two months), then the number of pairs present in the nth month is f_n, the nth term of the Fibonacci sequence.

Perhaps the best-known sequence defined recursively is the **Fibonacci sequence** (see the note in the margin), defined by

$$f_1 = 1, \; f_2 = 1, \; \text{and} \; f_n = f_{n-1} + f_{n-2} \qquad \text{for } n \geq 2$$

In this case two initial terms are given, and the recursion formula gives the nth term as the sum of the two preceding terms. The first few terms are

$$1, 1, 2, 3, 5, 8, 13, \ldots$$

The numbers f_n are called *Fibonacci numbers*, and they occur in nature in various ways. For example, if you count the number of spirals on a pine cone or a pineapple, you will get a Fibonacci number. The same is true for the number of spirals formed by the seeds of a sunflower. Also, the number of petals on a daisy is normally equal to a Fibonacci number.

Graphical Representation of Sequences

There are two useful ways of showing sequences graphically. In the first we use a two-dimensional coordinate system just as with other functions. We plot the points (x, y), where $x = n$ and $y = a_n$ for $n = 1, 2, 3, \ldots$. The graph consists of isolated points that we do not connect. In the second method we show the numbers a_n, where $n = 1, 2, 3, \ldots$, as points on a number line. We illustrate both methods in the next example.

EXAMPLE 10.10   Show each of the following sequences graphically by two methods.

(a) $\left\{ \dfrac{1}{n} \right\}$ (b) $\left\{ \dfrac{n}{n + 1} \right\}$

Solution

(a) By the first method we plot the points $(1, 1)$, $(2, \frac{1}{2})$, $(3, \frac{1}{3})$, $(4, \frac{1}{4})$, ... We show the result in Figure 10.7(a). By the second method, we show the terms

$$1, \frac{1}{2}, \frac{1}{3}, \frac{1}{4}, \ldots$$

as points on a number line in Figure 10.7(b).

(b) In the first method we plot the points $(1, \frac{1}{2})$, $(2, \frac{2}{3})$, $(3, \frac{3}{4})$, $(4, \frac{4}{5})$, ..., as shown in Figure 10.8(a) at the top of page 741. In the second method we show the points

$$\frac{1}{2}, \frac{2}{3}, \frac{3}{4}, \frac{4}{5}, \ldots$$

as points on a number line in Figure 10.8(b). ∎

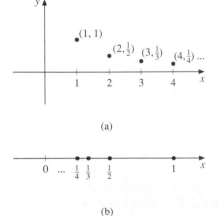

FIGURE 10.7

Limits of Sequences

(a)

(b)

FIGURE 10.8

In Figure 10.7(a) it appears that if we were to plot more and more points, they would come arbitrarily close to the x-axis, that is, to the line $y = 0$. Similarly, in Figure 10.7(b) the points seem to be coming arbitrarily close to 0. In fact, we can see from the nth term

$$a_n = \frac{1}{n}$$

that as $n \to \infty$, a_n comes arbitrarily close to 0.

In Figure 10.8(a) the points seem to be getting arbitrarily close to the horizontal line $y = 1$, and in Figure 10.8(b) the points on the number line seem to be getting close to 1. If we consider the nth term in this case,

$$a_n = \frac{n}{n + 1}$$

we can see intuitively that for large n, a_n is almost 1, since the numerator and denominator differ by only one unit. For example, with $n = 1,000$,

$$a_n = \frac{1000}{1001}$$

which is only slightly less than 1.

When the terms of a sequence come arbitrarily close to some number L as $n \to \infty$, we say that the sequence **converges** and that its **limit** is L. From our discussion in Example 10.10, we strongly suspect that both sequences $\{1/n\}$ and $\{n/(n + 1)\}$ converge, the first to the limit 0 and the second to the limit 1. In the case of the second sequence, we can use the "dominant term" approach, just as we did with rational functions of a continuous variable in Chapters 1 and 2. We reason that for large n, both numerator and denominator are approximately the same as their terms of highest degree. In this case, then, we could say that

$$\frac{n}{n + 1} \approx \frac{n}{n} = 1$$

for large n.

Another way to see this same result is to divide numerator and denominator by n:

$$\frac{n}{n + 1} = \frac{1}{1 + \frac{1}{n}}$$

As $n \to \infty$, $1/n \to 0$, so we conclude that the limit is 1.

The precise definition of the limit of a sequence is given below.

Definition 10.3
The Limit of a Sequence

We say that the **limit** of the sequence $\{a_n\}$ is L and write

$$\lim_{n \to \infty} a_n = L$$

provided that corresponding to each positive number ε, there is a positive integer N such that

$$|a_n - L| < \varepsilon$$

for all $n > N$. If the limit of $\{a_n\}$ is L, we say that the sequence **converges** to L. If the limit of $\{a_n\}$ does not exist, we say that the sequence **diverges**.

For $n > N$, the points (n, a_n) lie within an ε-band about L.

(a)

For $n > N$, a_n is the interval $(L - \varepsilon, L + \varepsilon)$

(b)

FIGURE 10.9

REMARKS

■ Since ε can be any positive number, by saying that $|a_n - L| < \varepsilon$ for all $n > N$, we are saying that the term a_n of the sequence will be *arbitrarily* close to L (closer than ε, however small ε may be) if we go far enough out in the sequence (beyond the Nth term).

■ A sequence can have at most one limit, since the terms cannot simultaneously all be arbitrarily close to two different values.

■ Definition 10.3 is the same as Definition 2.3 for the limit of a function $f(x)$ as $x \to \infty$, except that the continuous variable x is replaced by the integer variable n.

In Figure 10.9 we illustrate Definition 10.3 graphically by both methods.

EXAMPLE 10.11   Use Definition 10.3 to show that

$$\lim_{n \to \infty} \frac{n}{n+1} = 1$$

Solution   Consider the difference $|a_n - L|$, where $a_n = n/(n+1)$ and $L = 1$:

$$\left| \frac{n}{n+1} - 1 \right| = \left| \frac{n - n - 1}{n+1} \right| = \left| \frac{-1}{n+1} \right| = \frac{1}{n+1}$$

Now let ε denote any positive number. We want to see how large to make n so that $1/(n+1) < \varepsilon$. Since

$$\frac{1}{n+1} < \frac{1}{n}$$

we can simplify our choice by making $1/n < \varepsilon$. Solving this inequality for n, we get $n > 1/\varepsilon$. Thus, if we let N denote any integer larger than $1/\varepsilon$, we will have, for all $n > N$,

$$\left| \frac{n}{n+1} - 1 \right| < \varepsilon$$

So by Definition 10.3,

$$\lim_{n \to \infty} \frac{n}{n+1} = 1 \qquad \blacksquare$$

EXAMPLE 10.12   Show that each of the following sequences is divergent.
(a) $\{(-1)^{n-1}\}$ (b) $\{2n - 1\}$ (c) $\{\sin n\}$

FIGURE 10.10
Graph of $\{(-1)^{n-1}\}$

Solution

(a) The sequence is $1, -1, 1, -1, 1, \ldots$. There is no *single* number L that all terms come arbitrarily close to, no matter how far out in the sequence we go. More formally, if we choose $\varepsilon = \frac{1}{2}$, say, then regardless of the value of L, and regardless of how large N is, there are infinitely many terms a_n with $n > N$ such that $|a_n - L| \geq \frac{1}{2}$, contrary to the requirement on L given in Definition 10.3. The graph of $\{(-1)^{n-1}\}$ in Figure 10.10 clearly shows that the terms do not approach a limit.

(b) The sequence is $1, 3, 5, 7, 9, \ldots$, whose terms become arbitrarily large as $n \to \infty$, so the sequence diverges. We show its graph in Figure 10.11.

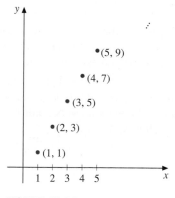

FIGURE 10.11
Graph of $\{2n - 1\}$

(c) In Figure 10.12 we show by means of the dashed line the graph of $f(x) = \sin x$. The points $(n, \sin n)$ lie on this graph. It is evident that as $n \to \infty$ these points do not come arbitrarily close to any horizontal line $y = L$. That is, $\lim_{n \to \infty} \sin n$ does not exist, so the sequence $\{\sin n\}$ diverges. ∎

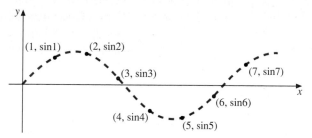

FIGURE 10.12
Graph of $\{\sin n\}$

EXAMPLE 10.13   Use Definition 10.3 to prove that if $|r| < 1$, then $\lim_{n \to \infty} r^n = 0$.

Solution   If $|r| < 1$, then its reciprocal $1/|r|$ is greater than 1, so we can write

$$\frac{1}{|r|} = 1 + p$$

for some positive number p. By the Binomial Theorem (see Appendix 4),

$$(1 + p)^n = 1 + np + \frac{n(n-1)}{2!}p^2 + \cdots + p^n > 1 + np > np$$

Thus, inverting, we get

$$|r|^n = \frac{1}{(1 + p)^n} < \frac{1}{np}$$

If ε denotes any positive integer, then we can make $1/np < \varepsilon$ by choosing $n > 1/\varepsilon p$. So if we choose N any integer greater than $1/\varepsilon p$, we have, for all $n > N$,

$$|r^n - 0| = |r^n| = |r|^n < \frac{1}{np} < \varepsilon$$

By Definition 10.3, it follows that

$$\lim_{n \to \infty} r^n = 0 \qquad\qquad ∎$$

n	$(0.9999)^n$
10	0.9990
100	0.9900
1000	0.9048
10,000	0.3679
100,000	0.000045

Even if we choose r only slightly less than 1—say, $r = 0.9999$—it is still true that r^n approaches 0 as $n \to \infty$. The table in the margin gives pretty convincing evidence for $r = 0.9999$.

If $r = 1$, then $r^n = 1$ for all n, and the sequence $\{r^n\}$ is the *constant* sequence

$$1, 1, 1, \ldots$$

whose limit is clearly 1. (After all, $|a_n - 1| = |1 - 1| = 0$, which is less than ε for any positive ε and for all n.)

If $r = -1$, the sequence $\{r^n\}$ is

$$-1, 1, -1, 1, -1, \ldots$$

which diverges, similar to the sequence in Example 10.12(a).

If $|r| > 1$, then, as we showed in Example 10.13, $|r| = 1 + p$ for some $p > 0$, and $|r|^n = (1 + p)^n > np$. Since $p > 0$, $np \to \infty$ as $n \to \infty$. Thus, $\{r^n\}$ diverges. For $r > 1$, r^n becomes arbitrarily large as $n \to \infty$. For $r < -1$, r^n oscillates in sign but becomes large in absolute value.

We now summarize our findings about the sequence $\{r^n\}$. (See also Figure 10.13.)

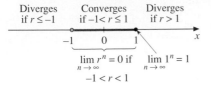

FIGURE 10.13
Convergence of $\{r^n\}$

The Sequence $\{r^n\}$

(a) If $|r| < 1$, $\lim\limits_{n\to\infty} r^n = 0$.

(b) If $r = 1$, $\lim\limits_{n\to\infty} r^n = 1$.

(c) For all other values of r, the sequence diverges.

Infinite Limits

As we saw with the sequence $\{2n - 1\}$ in Example 10.12(b) and again with the sequence $\{r^n\}$ when $r > 1$, some sequences diverge because their terms become arbitrarily large as $n \to \infty$. We indicate this type of divergence of a sequence $\{a_n\}$ by writing

$$\lim_{n\to\infty} a_n = \infty$$

Similarly, if the terms a_n are negative (at least for all sufficiently large n) but become arbitrarily large in absolute value as $n \to \infty$, we write

$$\lim_{n\to\infty} a_n = -\infty$$

In the first case we may say that "a_n approaches infinity" and in the second that "a_n approaches minus infinity," but in neither case does the sequence $\{a_n\}$ converge. We are simply saying it diverges in a particular way.

REMARKS

■ A sequence can diverge without approaching ∞ or $-\infty$, as we saw in Example 10.12, parts (a) and (c).

■ When we write $\lim_{n\to\infty} a_n = L$, we will always mean that L is finite.

Limit Properties of Sequences

The limit properties for functions given in Section 2.2 also hold true for sequences. After all, sequences are particular kinds of functions. We state these properties below in terms of sequences.

Limit Properties for Sequences

If $\lim\limits_{n \to \infty} a_n$ and $\lim\limits_{n \to \infty} b_n$ both exist, then the following properties hold true:

1. $\lim\limits_{n \to \infty} c a_n = c \left(\lim\limits_{n \to \infty} a_n \right)$ for any constant c

2. $\lim\limits_{n \to \infty} (a_n + b_n) = \lim\limits_{n \to \infty} a_n + \lim\limits_{n \to \infty} b_n$

3. $\lim\limits_{n \to \infty} (a_n - b_n) = \lim\limits_{n \to \infty} a_n - \lim\limits_{n \to \infty} b_n$

4. $\lim\limits_{n \to \infty} a_n b_n = \left(\lim\limits_{n \to \infty} a_n \right) \left(\lim\limits_{n \to \infty} b_n \right)$

5. $\lim\limits_{n \to \infty} \dfrac{a_n}{b_n} = \dfrac{\lim\limits_{n \to \infty} a_n}{\lim\limits_{n \to \infty} b_n}$ if $b_n \neq 0$ for all n and $\lim\limits_{n \to \infty} b_n \neq 0$

The next two theorems are often helpful in finding limits of sequences. We omit the proofs.

THEOREM 10.2

If $\lim\limits_{n \to \infty} a_n = L$, and f is a function whose domain includes L and a_n for $n \geq N$, and if f is continuous at $x = L$, then

$$\lim\limits_{n \to \infty} f(a_n) = f(L)$$

In particular, since $f(x) = x^k$, for k a positive integer, is continuous for all x, we have

$$\lim\limits_{n \to \infty} (a_n)^k = L^k$$

provided the sequence $\{a_n\}$ converges to L. Similarly,

$$\lim\limits_{n \to \infty} \sqrt[k]{a_n} = \sqrt[k]{L}$$

provided $a_n > 0$ and $L > 0$ for even-ordered kth roots.

THEOREM 10.3

Let $\{a_n\}$ be a sequence and f a function such that

$$f(n) = a_n, \qquad n = 1, 2, 3, \ldots$$

If

$$\lim\limits_{x \to \infty} f(x) = L$$

then also

$$\lim\limits_{n \to \infty} a_n = L$$

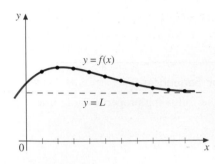

FIGURE 10.14
If $\lim\limits_{x\to\infty} f(x) = L$, then $\lim\limits_{n\to\infty} a_n = L$, where $a_n = f(n)$.

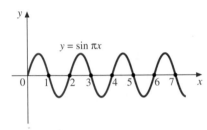

FIGURE 10.15
The sequence $\{\sin \pi n\} = \{0\}$, but $\lim\limits_{x\to\infty} \sin \pi x$ does not exist.

You will be asked to give a proof of Theorem 10.3 in Exercise 48 of Exercise Set 10.2, but the result is obvious intuitively by looking at the graph in Figure 10.14. The points (n, a_n) lie on the graph of $y = f(x)$, since $a_n = f(n)$. So if the graph of f is asymptotic to the line $y = L$, the a_n values come arbitrarily close to L.

REMARK ───
■ The converse of Theorem 10.3 is not true, as can be seen from the sequence $\{\sin \pi n\}$, which is constantly 0 and hence is convergent, but the corresponding function $f(x) = \sin \pi x$ does not approach a limit as $x \to \infty$. (See Figure 10.15.) So in this case $\lim\limits_{n\to\infty} f(n) = 0$, but $\lim\limits_{x\to\infty} f(x)$ does not exist.
───

In the next example we illustrate how we can make use of Theorem 10.3 to find limits of sequences.

EXAMPLE 10.14   Find the limit of each of the following sequences.

(a) $\left\{\dfrac{\ln n}{n}\right\}$ (b) $\{\sqrt[n]{n}\}$

Solution

(a) Let
$$f(x) = \frac{\ln x}{x}$$

As $x \to \infty$, the function is indeterminant of the type ∞/∞. Applying L'Hôpital's Rule, we get
$$\lim_{x\to\infty} \frac{\ln x}{x} = \lim_{x\to\infty} \frac{1/x}{1} = 0$$

Since $f(n) = (\ln n)/n$, we conclude by Theorem 10.3 that
$$\lim_{n\to\infty} \frac{\ln n}{n} = 0$$

(b) Let $f(x) = x^{1/x}$ for $x > 0$. To find the limit, write $y = x^{1/x}$ and take the natural logarithm of both sides:
$$\ln y = \frac{1}{x} \ln x = \frac{\ln x}{x}$$

As we have seen in part (a), $(\ln x)/x \to 0$ as $x \to \infty$. Thus, since $\ln y \to 0$, it follows that $y \to e^0 = 1$. By Theorem 10.3 we conclude that
$$\lim_{n\to\infty} n^{1/n} = \lim_{n\to\infty} \sqrt[n]{n} = 1 \qquad ■$$

The Squeeze Theorem (Theorem 2.3) also holds true for sequences. We restate it here in terms of sequences.

THEOREM 10.4

The Squeeze Theorem for Sequences

If $\lim_{n\to\infty} a_n = L$ and $\lim_{n\to\infty} b_n = L$ and if for all sufficiently large n the inequality
$$a_n \leq c_n \leq b_n$$
holds true, then $\lim_{n\to\infty} c_n = L$ also.

EXAMPLE 10.15   Find the limit of the sequence

$$\left\{ \frac{n \sin n}{1 + n^2} \right\}$$

or show that it diverges.

Solution   Since $|\sin n| \le 1$ for all n, we have

$$\left| \frac{n \sin n}{1 + n^2} \right| \le \frac{n}{1 + n^2} < \frac{n}{n^2} = \frac{1}{n}$$

Thus,

$$-\frac{1}{n} < \frac{n \sin n}{1 + n^2} < \frac{1}{n}$$

for all positive integers n. Since both $\{1/n\}$ and $\{-1/n\}$ converge to 0, it follows by the Squeeze Theorem that $\{n \sin n / (1 + n^2)\}$ converges to 0 also. That is,

$$\lim_{n \to \infty} \frac{n \sin n}{1 + n^2} = 0$$

■

Monotonicity and Boundedness

Sequences whose terms either always increase or always decrease (or else remain the same) play an especially important role in the study of infinite series. We use the following terminology to describe such sequences.

Definition 10.4
Monotone Sequences

A sequence $\{a_n\}$ is said to be

(a) **increasing** if $a_n \le a_{n+1}$, or

(b) **decreasing** if $a_n \ge a_{n+1}$

for all positive integers n. A sequence that is either always increasing or always decreasing is said to be **monotone**.

REMARKS ————————————————————

■ Frequently we will say *monotone increasing* or *monotone decreasing* to emphasize the monotonicity.

■ The terms *nondecreasing* and *nonincreasing* are sometimes used where we have used increasing and decreasing, respectively.

If it is necessary to distinguish between a sequence for which $a_n < a_{n+1}$ for all n and one for which $a_n \le a_{n+1}$, we will refer to the former as *strictly* increasing. By our definition, each of the following is increasing:

$$1, 2, 3, 4, 5, \dots$$

$$1, 2, 2, 3, 4, 4, 5, \dots$$

Only the first one is strictly increasing. Similarly, we can say a sequence in which $a_n > a_{n+1}$ is *strictly* decreasing.

Definition 10.5
Bounded Sequences

A sequence $\{a_n\}$ is said to be **bounded** if there is some positive constant M such that

$$|a_n| \leq M$$

for all positive integers n.

The sequences $\{(-1)^{n-1}\}$ and $\{\sin n\}$ considered earlier are bounded, since in each case $|a_n| \leq 1$. On the other hand, the sequence $\{2n - 1\}$ is unbounded.

Alone, neither monotonicity nor boundedness guarantees convergence. The sequence $\{(-1)^{n-1}\}$ is bounded but not convergent (also not monotone). The sequence $\{2n - 1\}$ is monotone but not convergent (also not bounded). We will shortly prove the important result that if a sequence is *both* monotone and bounded, then it must be convergent.

Before proving this result we need several more notions related to boundedness. A set S of real numbers is said to be **bounded above** if there is a real number M such that all elements of S are less than or equal to M. The number M is then said to be an **upper bound** of the set S. Similarly, S is **bounded below** if there is a real number m such that all elements of S are greater than or equal to m, in which case m is a **lower bound** of S. Clearly, if a sequence is bounded both above and below, then it is bounded, and the converse is also true. A fundamental property of the real-number system, called the *completeness property*, is that every nonempty set of real numbers that is bounded above has a **least upper bound**. To say L is the least upper bound of S means not only that L is an upper bound of S, but also that no number less than L is an upper bound of S. Similarly, if S is bounded below, it has a **greatest lower bound**. (These concepts are discussed further in Appendix 1.)

We are ready now for the main theorem on monotone bounded sequences.

THEOREM 10.5

Monotone Bounded Sequence Theorem

If $\{a_n\}$ is a sequence of real numbers that is both monotone and bounded, then it converges.

Proof   We will prove the theorem for a monotone increasing sequence only. The proof for the decreasing case is similar.

Let $\{a_n\}$ be monotone increasing and bounded. The terms of the sequence constitute a set of real numbers that is bounded above. By the completeness property of real numbers just discussed, there is a least upper bound, say L, for this set. That is, $a_n \leq L$ for all n, where L is the smallest number having this property.

Let ε denote any positive number. Then, since $L - \varepsilon < L$, the number $L - \varepsilon$ is *not* an upper bound to the sequence $\{a_n\}$. (Remember that L is the *least* upper bound, so $L - \varepsilon$ cannot be an upper bound.) Thus, there is some member

FIGURE 10.16

of the sequence, say a_N, such that $a_N > L - \varepsilon$. Since the sequence is increasing, it follows that $a_n \geq a_N$ for all $n > N$ (see Figure 10.16). Thus, for all $n > N$,

$$L - \varepsilon < u_N \leq a_n \leq L < L + \varepsilon$$

In particular, for $n > N$,

$$L - \varepsilon < a_n < L + \varepsilon$$

So, subtracting L from each member of this inequality,

$$-\varepsilon < a_n - L < \varepsilon$$

or equivalently,

$$|a_n - L| < \varepsilon \qquad \text{if } n > N$$

By Definition 10.3, we see that

$$\lim_{n \to \infty} a_n = L \qquad \blacksquare$$

Note that we have proved not only that the sequence converges but also that its limit is the least upper bound of the points in the sequence. (In many cases, however, finding this least upper bound is difficult.)

In applying Theorem 10.5 it is sufficient that the sequence be monotone *from some point onward*, since convergence or divergence depends only on the behavior of the sequence as n gets arbitrarily large. For example, the sequence

$$20, \quad 33, \quad -75, \quad 1, \quad \frac{1}{2}, \quad \frac{1}{4}, \quad \frac{1}{8}, \quad \frac{1}{16}, \ldots$$

converges since it is monotone decreasing starting with the fourth term and is bounded. In fact, in this case we see that the limit is 0, which is the greatest lower bound of the terms starting with the fourth term.

Showing Monotonicity

To apply Theorem 10.5, we need to know that the sequence is monotone. Sometimes monotonicity is obvious. For example, the sequence $\{1/n\}$ is clearly decreasing. Some sequences, however, we must analyze further to determine if they are monotone. The next example illustrates three ways to test for monotonicity.

EXAMPLE 10.16   Show that each of the following sequences is monotone.

(a) $\left\{\dfrac{2n + 3}{n}\right\}$ (b) $\left\{\dfrac{n}{\sqrt{1 + n^2}}\right\}$ (c) $\left\{\dfrac{n!}{n^n}\right\}$

Solution

(a) The first technique to test for monotonicity is to look at the difference $a_{n+1} - a_n$ between successive terms. If this difference is always greater than or equal to 0, the sequence is increasing; if it is always less than or equal to 0, the sequence is decreasing. For the sequence $\{(2n + 3)/n\}$, we have

$$a_{n+1} - a_n = \frac{2(n+1) + 3}{n + 1} - \frac{2n + 3}{n} = \frac{(2n + 5)n - (2n + 3)(n + 1)}{n(n + 1)}$$

$$= \frac{2n^2 + 5n - (2n^2 + 5n + 3)}{n(n + 1)} = \frac{-3}{n(n + 1)} < 0$$

Thus, for all n, $a_{n+1} - a_n < 0$, so $a_{n+1} < a_n$, and we conclude that the sequence is (strictly) decreasing.

(b) As a second technique for testing for monotonicity, we consider a function f for which $f(n) = a_n$. If we can show monotonicity of f on the interval $[1, \infty)$, then we can conclude that $\{a_n\}$ is also monotone for $n \geq 1$. If f is differentiable, we can test it for monotonicity by considering $f'(x)$.

For the sequence $\{a_n\} = \{n/\sqrt{1 + n^2}\}$, the natural choice for f is

$$f(x) = \frac{x}{\sqrt{1 + x^2}}$$

Taking its derivative, we have

$$f'(x) = \frac{\sqrt{1 + x^2} - x(\frac{1}{2})(1 + x^2)^{-1/2}(2x)}{1 + x^2}$$

$$= \frac{1 + x^2 - x^2}{(1 + x^2)^{3/2}} = \frac{1}{(1 + x^2)^{3/2}} > 0$$

(You should supply the missing algebra here.) Since $f'(x) > 0$ for all x, we know by Theorem 4.1 that f is an increasing function. Thus, since $a_n = f(n)$, we see that the sequence $\{a_n\}$ is also increasing.

(c) In the third approach to testing for monotonicity, we consider the ratio a_{n+1}/a_n of a given term (after the first) to the preceding term. If this ratio is always greater than or equal to 1, we conclude that $a_{n+1} \geq a_n$, and the sequence is increasing. If the ratio is less than or equal to 1, then $a_{n+1} \leq a_n$, and the sequence is decreasing. (In order to apply this technique, we must have $a_n > 0$ for all n.)

With $a_n = n!/n^n$, we have

$$\frac{a_{n+1}}{n} = \frac{\dfrac{(n + 1)!}{(n + 1)^{n+1}}}{\dfrac{n!}{n^n}} = \frac{(n + 1)!}{(n + 1)^{n+1}} \cdot \frac{n^n}{n!} = \frac{(n + 1)!}{n!} \cdot \frac{n^n}{(n + 1)^{n+1}}$$

$$= (n + 1) \cdot \frac{n^n}{(n + 1)^{n+1}} \qquad \text{Since } (n + 1)! = (n + 1)n!$$

$$= \frac{n^n}{(n + 1)^n} < \frac{n^n}{n^n} = 1 \qquad \text{Since } (n + 1)^n > n^n$$

Thus,

$$\frac{a_{n+1}}{a_n} < 1 \qquad \text{for all } n$$

Hence, $a_{n+1} < a_n$, and we conclude that $\{a_n\}$ is a decreasing sequence. ∎

Let us summarize the three ways of showing monotonicity illustrated in the preceding example.

Tests for Monotonicity

1. Calculate the difference $a_{n+1} - a_n$.

$$\text{If} \begin{cases} a_{n+1} - a_n \geq 0 & \text{for all } n, \text{ then } \{a_n\} \text{ is increasing;} \\ a_{n+1} - a_n \leq 0 & \text{for all } n, \text{ then } \{a_n\} \text{ is decreasing.} \end{cases}$$

2. Let $f(x)$ be such that $f(n) = a_n$. Calculate $f'(x)$ if it exists.

$$\text{If} \begin{cases} f'(x) \geq 0 & \text{on } [1, \infty), \text{ then } \{a_n\} \text{ is increasing;} \\ f'(x) \leq 0 & \text{on } [1, \infty), \text{ then } \{a_n\} \text{ is decreasing.} \end{cases}$$

3. If $a_n > 0$ for all n, calculate the ratio a_{n+1}/a_n.

$$\text{If} \begin{cases} \dfrac{a_{n+1}}{a_n} \geq 1 & \text{for all } n, \text{ then } \{a_n\} \text{ is increasing;} \\[2mm] \dfrac{a_{n+1}}{a_n} \leq 1 & \text{for all } n, \text{ then } \{a_n\} \text{ is decreasing.} \end{cases}$$

Again, for purposes of applying Theorem 10.5, it is sufficient that the sequence be monotone from some point onward, say, for $n \geq n_0$. Thus, in applying Tests 1 and 3 listed above, we can modify "for all n" by requiring only that the conditions be true "for all $n \geq n_0$." For Test 2, $f'(x)$ need be either positive or negative on $[a, \infty)$ for some $a > 0$.

If a sequence of nonnegative terms is found to be decreasing, then it must converge, since it is bounded (bounded below by 0 and above by its first term). Thus, we can conclude that the sequences in parts (a) and (c) of Example 10.16 both converge. In fact, for part (a) we can find the limit directly:

$$\lim_{n \to \infty} \frac{2n + 3}{n} = \lim_{n \to \infty} \left(2 + \frac{3}{n} \right) = 2$$

The limit in part (c) is 0, as you will be asked to show in the exercises (Exercise 50 of this section). In part (b) we showed that the sequence is increasing. It is bounded above by 1, since

$$\frac{n}{\sqrt{1 + n^2}} < \frac{n}{\sqrt{n^2}} = \frac{n}{n} = 1$$

(Of course, it is also bounded below by its first term, since subsequent terms are larger.) Thus, the sequence converges. The limit is actually 1, as we see from the dominant term approach:

$$\frac{n}{\sqrt{1 + n^2}} \approx \frac{n}{\sqrt{n^2}} = \frac{n}{n} = 1 \qquad \text{for large } n$$

So

$$\lim_{n \to \infty} \frac{n}{\sqrt{1 + n^2}} = 1$$

Exercise Set 10.2

In Exercises 1–3, write out the first five terms of the sequence $\{a_n\}$.

1. (a) $a_n = \dfrac{(-1)^{n-1}n}{2n-1}$ (b) $a_n = \dfrac{2^n}{n!}$

2. (a) $a_n = \dfrac{2 \cdot 4 \cdot 6 \cdot \cdots \cdot (2n)}{1 \cdot 3 \cdot 5 \cdot \cdots \cdot (2n-1)}$

   (b) $a_n = \dfrac{\sin[(2n+1)\pi/2]}{n(n+1)}$

3. (a) $a_1 = 1,\ a_n = -na_{n-1}$ if $n \geq 2$
   (b) $a_1 = 1,\ a_2 = 2,\ a_n = \dfrac{a_{n-1}}{a_{n-2}}$ if $n \geq 3$

In Exercises 4–6, determine a formula for the nth term of the sequence.

4. (a) $1, -\dfrac{1}{3}, \dfrac{1}{5}, -\dfrac{1}{7}, \dfrac{1}{9}, \ldots$

   (b) $\dfrac{1}{2}, \dfrac{2}{5}, \dfrac{3}{10}, \dfrac{4}{17}, \dfrac{5}{26}, \ldots$

5. (a) $\dfrac{1}{2}, \dfrac{3}{4}, \dfrac{7}{8}, \dfrac{15}{16}, \dfrac{31}{32}, \ldots$

   (b) $\dfrac{2}{5}, -\dfrac{4}{7}, \dfrac{6}{9}, -\dfrac{8}{11}, \dfrac{10}{13}, \ldots$

6. (a) $\dfrac{1}{7}, -\dfrac{3}{10}, \dfrac{5}{13}, -\dfrac{7}{16}, \dfrac{9}{19}, \ldots$

   (b) $-1, 3, \dfrac{5}{3}, \dfrac{7}{5}, \dfrac{9}{7}, \ldots$

In Exercises 7–30, find $\lim_{n\to\infty} a_n$ *or show that the sequence diverges.*

7. $a_n = \dfrac{n^2 - 2n}{3n^2 + n - 1}$ **8.** $a_n = \dfrac{n+1}{2n^2 - n + 1}$

9. $a_n = \dfrac{2n - n^3}{(n+1)(n+3)}$

10. $a_n = \dfrac{n}{\sqrt{1+n^2}}$ **11.** $a_n = \dfrac{\cos n\pi}{\sqrt{n}}$

12. $a_n = \dfrac{1 + \sin n}{\ln(n+1)}$ **13.** $a_n = \dfrac{1 - (-1)^n}{2}$

14. $a_n = n^3 e^{-n}$ **15.** $a_n = \tanh n$

16. $a_n = \dfrac{\ln n}{\sqrt{n}}$ **17.** $a_n = (-0.999)^n$

18. $a_n = \dfrac{1 - \cosh n}{\sinh n}$ **19.** $a_n = \ln\left(1 + \dfrac{1}{n}\right)$

20. $a_n = \dfrac{\ln n}{\ln(\ln n)}$ **21.** $a_n = \dfrac{2^n}{e^{n-1}}$

22. $a_n = \dfrac{(-1)^{n-1}(2\sin n + \cos n)}{n \sec n}$

23. $a_n = \sin n + \cos n$ **24.** $a_n = (1.00001)^n$

25. $a_n = \dfrac{n+1}{n}\left(\dfrac{2}{3}\right)^n$ **26.** $a_n = \dfrac{(-1)^n e^{2n}}{1 + 9^n}$

27. $a_n = \dfrac{n \ln n}{1 + n^2}$ **28.** $a_n = \dfrac{\sqrt{1+n^3}}{\sqrt[3]{1+n^4}}$

29. $a_n = n^2 \cdot 2^{-n}$ **30.** $a_n = n \sin \dfrac{1}{n}$

In Exercises 31–40, show that the sequence $\{a_n\}$ *is monotone for* $n \geq n_0$. *If* $n_0 \neq 1$, *give the smallest value of* n_0.

31. $a_n = \dfrac{2n-1}{n+1}$ **32.** $a_n = \dfrac{n}{2n-3}$

33. $a_n = n^2 e^{-n}$ **34.** $a_n = \dfrac{(\ln n)^2}{n}$

35. $a_n = \dfrac{2^n}{n!}$ **36.** $a_n = 2^{1/n}$

37. $a_n = \dfrac{n!}{1 \cdot 3 \cdot 5 \cdot \cdots \cdot (2n-1)}$

38. $a_n = \dfrac{1 \cdot 3 \cdot 5 \cdot \cdots \cdot (2n-1)}{2^n \cdot n!}$

39. $a_n = \ln n - \ln(n+1)$

40. $a_n = \tan^{-1} n$

In Exercises 41–46, prove that the sequence $\{a_n\}$ *converges.*

41. $a_n = \dfrac{e^n}{n!}$ **42.** $a_n = \dfrac{2^n n!}{(2n)!}$

43. $a_n = n^{1/n}$

44. $a_n = \dfrac{1 \cdot 3 \cdot 5 \cdot \cdots \cdot (2n-1)}{2 \cdot 4 \cdot 6 \cdot \cdots \cdot 2n}$

45. $a_1 = 1,\ a_n = 1 + \dfrac{a_{n-1}}{2}$ for $n \geq 2$

46. $a_1 = 2,\ a_n = \sqrt{a_{n-1}}$ for $n \geq 2$

47. Prove that a sequence can have at most one limit.

48. Prove Theorem 10.3.

49. Prove that $\lim_{n\to\infty} a_n = 0$ if and only if $\lim_{n\to\infty} |a_n| = 0$.

50. Prove that $\lim_{n\to\infty}(n!/n^n) = 0$. (*Hint:* Write

$$a_n = \frac{n!}{n^n} = \frac{1 \cdot 2 \cdot 3 \cdots n}{n \cdot n \cdot n \cdots n}$$

and show that $a_n < 1/n$.)

51. Prove that if

$$a_n = \frac{1}{n+1} + \frac{1}{n+2} + \cdots + \frac{1}{2n}$$

then $\{a_n\}$ is convergent. *Hint:* Use Theorem 10.4. To show boundedness, observe that

$$a_n \le n\left(\frac{1}{n+1}\right)$$

52. Let $a_1 = \sqrt{3}$ and for $n \ge 2$, $a_n = \sqrt{3a_{n-1}}$. Prove that $\{a_n\}$ converges, and find its limit. (*Hint:* Use mathematical induction (see Supplementary Appendix 4) to prove that $a_{n+1} > a_n$ and $a_n < 3$ for all n.)

53. For the Fibonacci sequence $\{f_n\}$, define

$$a_n = \frac{f_{n+1}}{f_n}, \qquad n = 1, 2, 3, \ldots$$

Use a spreadsheet, CAS, or calculator to find the first twenty terms of the sequence $\{a_n\}$. Assuming that $\lim_{n\to\infty} a_n = L$, find the exact value of L. (*Hint:* Show that $a_n = 1 + (1/a_{n-1})$ for $n \ge 2$ and take the limit as $n \to \infty$ to get $L = 1 + (1/L)$. Now solve for L.)

Note: The number L is the *golden mean*, defined as shown in the figure.

The golden mean L is the mean proportional between 1 and $1 - L$:
$$\frac{1}{L} = \frac{L}{1-L}$$

54. Use a spreadsheet, CAS, or calculator to do the following:
  (a) Find the first twenty terms of the Fibonacci sequence $\{f_n\}$.
  (b) Find the first twenty terms of the "Fibonacci-like" sequence $\{g_n\}$ in which $g_1 = 1$, $g_2 = 3$, and $g_n = g_{n-1} + g_{n-2}$ for $n \ge 2$.
  (c) Find the ratio g_n/f_n for $n = 1, 2, \ldots, 20$ for f_n and g_n in parts (a) and (b). Conjecture the limit

$$\lim_{n\to\infty} \frac{g_n}{f_n}$$

55. (a) Let x_1 be given and $x_{n+1} = \sqrt{x_n}$ for $n \ge 1$. Use a spreadsheet, CAS, or calculator to find the first twenty terms of the sequence $\{x_n\}$ for four different values of x_1. Conjecture the limit in each case. Prove your conjecture. Does the answer depend on x_1?
  (b) Repeat part (a) if $x_{n+1} = x_n^2 - 2$.

56. (1) Choose an arbitrary integer $A > 0$.
  (2) If $A = 1$, stop.
  (3) If A is even, replace A with $A/2$ and go to step 2.
  (4) If A is odd, replace A with $3A + 1$ and go to step 2.
  Let $A = 3$, $A = 34$, and $A = 75$, and in each case compute the terms generated. Is it true that the larger the value of A, the more steps are required before the algorithm stops?

57. Let $a_1 = 2$ and $a_n = \sqrt{2 + a_{n-1}}$ for $n \ge 2$.
  (a) Plot several points of the sequence $\{a_n\}$ on a number line. Does the sequence appear to be bounded above? If so, what appears to be an upper bound?
  (b) Use mathematical induction to prove that $a_n \le 2$ for all n.
  (c) Show that $a_{n+1}^2 - a_n^2 = (2 - a_n)(1 + a_n)$. Use this result to show that $\{a_n\}$ is monotone increasing.
  (d) Let $\lim_{n\to\infty} a_n = L$. By taking the limit of both sides of the equation $a_n = \sqrt{2 + a_{n-1}}$ as $n \to \infty$, find the value of L. (*Hint:* Note that $a_{n-1} \to L$ as $n \to \infty$.)

58. (a) Is it possible for a term of a convergent sequence to exceed the limit of the sequence? Is it possible for some terms to exceed and some terms to be less than the limit? Is it possible for some terms to equal the limit? Use examples in your answers to these questions.
  (b) If the terms of a sequence alternate in sign, can the sequence converge? If you think it is possible, give an example.

59. In each of the following give an example, whenever possible, satisfying the given condition. If it is not possible, explain why.
  (a) A sequence with limit 12 and no term equal to 12
  (b) An increasing sequence that is not bounded
  (c) A bounded sequence that does not converge
  (d) A bounded, convergent sequence that is not monotone
  (e) An increasing sequence that does not converge
  (f) A convergent sequence with infinitely many terms equal to 7, infinitely many terms not equal to 7, that converges to 7
  (g) A convergent sequence with infinitely many terms equal to 7 and infinitely many terms equal to 8
  (h) A convergent sequence that is not bounded
  (i) Two sequences with the same limit but with no common terms
  (j) A sequence with limit 2 that has infinitely many negative terms

10.3 INFINITE SERIES OF CONSTANTS

In the introduction to this chapter we discussed the question of how to define the sum of infinitely many terms. Now we want to explore this question in more detail.

Consider first the familiar decimal representation of the fraction 1/3. By repeated division, we get

$$\frac{1}{3} = 0.3333\ldots$$

We can write the repeating decimal on the right as the infinite sum

$$0.3 + 0.03 + 0.003 + 0.0003 + \cdots \tag{10.15}$$

We know, then, that the sum of this infinite series is 1/3.

As a second example, consider again the infinite series

$$1 + \frac{1}{2} + \frac{1}{4} + \frac{1}{8} + \cdots \tag{10.16}$$

in which each term after the first is half of the preceding term. We proceed as we did in the introduction to the chapter by recording its **partial sums**—that is, the successive finite sums:

First term $= 1$

Sum of first two terms $= 1\frac{1}{2}$

Sum of first three terms $= 1\frac{3}{4}$

Sum of first four terms $= 1\frac{7}{8}$

$\vdots$

FIGURE 10.17

In Figure 10.17 we show these partial sums as points on a number line. Notice that each successive point after the first is halfway between the previous point and 2. Clearly, as we add more and more terms, the partial sums come arbitrarily close to 2, so we say that the sum of the series is 2 and write

$$1 + \frac{1}{2} + \frac{1}{4} + \frac{1}{8} + \cdots = 2$$

Let us now generalize what we have illustrated with the two particular infinite series 10.15 and 10.16. Suppose that $a_1, a_2, a_3, \ldots$ are numbers (not necessarily all distinct). The sum

$$a_1 + a_2 + a_3 + \cdots \tag{10.17}$$

is called an **infinite series**, or simply a **series**. Writing the three dots at the end is important as an indication that the series continues indefinitely. For brevity, we will often write series 10.17 in summation notation as

$$\sum_{n=1}^{\infty} a_n$$

The particular letter used as the index of summation is not important, since the index is replaced by $1, 2, 3, \ldots$ when we write out the series. For example, we could write $\sum_{k=1}^{\infty} a_k$ for the series 10.17. Sometimes we will omit the range of the index and simply write $\sum a_n$, with the understanding that the index goes from 1 to ∞. There will be times when we want the index to begin with 0. Note, for example, that

$$\sum_{n=0}^{\infty} \frac{1}{2^n} \quad \text{and} \quad \sum_{n=1}^{\infty} \frac{1}{2^{n-1}}$$

are two ways of representing the same series. (Try writing the first few terms in each case.)

Just as we did with the series 10.16, we list the partial sums of the series 10.17:

First partial sum	$= a_1$
Second partial sum	$= a_1 + a_2$
Third partial sum	$= a_1 + a_2 + a_3$

$$\vdots$$

nth partial sum $\qquad = a_1 + a_2 + a_3 + \cdots + a_n$

$$\vdots$$

The question is: Do these partial sums approach more and more closely some limiting value as we let n increase indefinitely? If so, this limiting value is what we define as the sum of the series 10.17. In order to state this definition more succinctly, we introduce the notation

$$S_n = a_1 + a_2 + a_3 + \cdots + a_n$$

That is, $S_1 = a_1$, $S_2 = a_1 + a_2$, $S_3 = a_1 + a_2 + a_3$, and so on. For each positive integer n, S_n is the **nth partial sum** of the series 10.17. These partial sums, in order, $S_1, S_2, S_3, \ldots$, constitute a sequence $\{S_n\}$, called the **sequence of partial sums** of the series.

Note carefully that when we list the terms of a sequence, we separate them with commas, whereas the terms of a series are added together. For example, the *series* 10.16 is

$$1 + \frac{1}{2} + \frac{1}{4} + \frac{1}{8} + \cdots$$

and its *sequence* of partial sums is

$$1, 1\frac{1}{2}, 1\frac{3}{4}, 1\frac{7}{8}, \ldots$$

For this example, we can find a simple formula for S_n, namely,

$$S_n = 2 - \frac{1}{2^{n-1}}$$

since S_1 is 1 less than 2, S_2 is $\frac{1}{2}$ less than 2, S_3 is $\frac{1}{4}$ less than 2, and so on. As $n \to \infty$, we get the limit

$$\lim_{n \to \infty} S_n = 2$$

since $2 - (1/2^{n-1})$ can be made arbitrarily close to 2 by taking n sufficiently large. This result confirms again that series 10.16 converges to 2.

We now give the formal definition of the sum of an infinite series.

Definition 10.6
The Sum of an Infinite Series

The infinite series

$$\sum_{n=1}^{\infty} a_n = a_1 + a_2 + a_3 + \cdots$$

is said to **converge to the sum S** provided

$$\lim_{n \to \infty} S_n = S$$

where $S_n = a_1 + a_2 + \cdots + a_n$. In this case we write $\sum_{n=1}^{\infty} a_n = S$. If $\lim_{n \to \infty} S_n$ does not exist, we say that the series **diverges**.

Briefly, we can write

$$\sum_{n=1}^{\infty} a_n = \lim_{n \to \infty} (a_1 + a_2 + \cdots + a_n)$$

provided the limit on the right exists (that is, the limit is finite).

Recall from Section 10.2 that the precise meaning of

$$\lim_{n \to \infty} S_n = S$$

is that, given any positive number ε, there exists a positive number N such that for all $n > N$, $|S_n - S| < \varepsilon$.

In most cases it is not feasible to apply Definition 10.6 directly to determine whether a given series converges and if so, the value of its sum. In the next two sections we will develop means of testing for convergence or divergence without having to apply the definition directly. There are, however, certain series for which the definition can be used. The most important of these is called a *geometric series*, which we now describe.

Geometric Series

A **geometric series** is an infinite series of the form

$$\sum_{n=1}^{\infty} ar^{n-1} = a + ar + ar^2 + ar^3 + \cdots \tag{10.18}$$

in which a and r are fixed real numbers with $a \neq 0$. The number a is its first term, and the number r is called the **common ratio**. Note that the ratio of any term (after the first) to the preceding term is r if $r \neq 0$ (for example, $ar^3/ar^2 = r$), hence the name common ratio. Geometric series are particularly easy to work with not only because we can completely describe when they converge and diverge, but also because when they converge, we can easily compute the sum. To verify these facts, we investigate the sequence $\{S_n\}$ of partial sums, where

$$S_n = a + ar + ar^2 + \cdots + ar^{n-1}$$

First, suppose $r = 1$. Then

$$S_n = a + a + a + \cdots + a = na$$

Clearly, $\lim_{n\to\infty} S_n$ does not exist as a finite number in this case, since na will be arbitrarily large in absolute value if n is sufficiently large. That is, $\lim_{n\to\infty} na = \infty$ if $a > 0$ and $\lim_{n\to\infty} na = -\infty$ if $a < 0$. Thus, when $r = 1$, the geometric series 10.18 diverges.

If $r \neq 1$, we can write S_n in what is called a *closed form* by the following trick. We note that if we multiply S_n by r, we get many terms just like those in S_n. If we subtract rS_n from S_n, we get

$$S_n - rS_n = (a + ar + ar^2 + ar^3 + \cdots + ar^{n-1}) - (ar + ar^2 + ar^3 + \cdots + ar^n)$$

All the terms except a and ar^n cancel, so we have

$$S_n - rS_n = a - ar^n$$

or, equivalently,

$$S_n(1 - r) = a(1 - r^n)$$

Since with $r \neq 1$, the factor $1 - r$ is nonzero, we can solve for S_n:

$$S_n = \frac{a(1 - r^n)}{1 - r}$$

We know from Example 10.13 and the discussion following it that when $|r| < 1$,

$$\lim_{n\to\infty} r^n = 0$$

and when $r = -1$ or $|r| > 1$,

$$\lim_{n\to\infty} r^n \text{ does not exist}$$

Thus, when $|r| < 1$,

$$\lim_{n\to\infty} S_n = \lim_{n\to\infty} \frac{a(1 - r^n)}{1 - r}$$
$$= \frac{a}{1 - r}$$

and the geometric series converges to $a/(1 - r)$. If $|r| \geq 1$, $\{S_n\}$ does not converge, so the geometric series diverges.

Because of the importance of geometric series, we state our findings as a theorem.

THEOREM 10.6

If $|r| < 1$, the geometric series $\sum_{n=1}^{\infty} ar^{n-1}$ converges to the sum $a/(1-r)$. That is,

$$a + ar + ar^2 + ar^3 + \cdots = \frac{a}{1 - r} \qquad \text{if } |r| < 1 \qquad (10.19)$$

If $|r| \geq 1$, the series diverges.

Both of the series 10.15 and 10.16 are geometric. For 10.15 we see from the series

$$0.3 + 0.03 + 0.003 + \cdots$$

that $a = 0.3$ and $r = 0.1$. Since $|r| = 0.1 < 1$, we have, by Equation 10.19,

$$0.3 + 0.03 + 0.003 + \cdots = \frac{0.3}{1 - 0.1} = \frac{0.3}{0.9} = \frac{3}{9} = \frac{1}{3}$$

confirming what we already knew.

For the series 10.16,

$$1 + \frac{1}{2} + \frac{1}{4} + \frac{1}{8} + \cdots$$

$a = 1$ and $r = \frac{1}{2}$. Again, by Equation 10.19, the sum is

$$\frac{a}{1 - r} = \frac{1}{1 - \frac{1}{2}} = \frac{1}{\frac{1}{2}} = 2$$

The next four examples further illustrate Theorem 10.6.

EXAMPLE 10.17   Determine whether each of the following series is convergent or divergent. If convergent, find the sum.

(a) $1 - \dfrac{2}{3} + \dfrac{4}{9} - \dfrac{8}{27} + \cdots$

(b) $3 + 4 + \dfrac{16}{3} + \dfrac{64}{9} + \cdots$

Solution

(a) The series is geometric, with $a = 1$ and $r = -2/3$. One way to determine r is to divide any term after the first by the one before it. Another way is to ask by what factor you would multiply a given term to get the next term. In either case, be sure to check that the ratio (or the multiplicative factor) remains the same for all pairs of consecutive terms. Confirming that $r = -2/3$, we see that its absolute value is less than 1, so by Equation 10.19 the series converges to

$$\frac{a}{1 - r} = \frac{1}{1 - \left(-\frac{2}{3}\right)} = \frac{1}{\frac{5}{3}} = \frac{3}{5}$$

We can therefore write

$$1 - \frac{2}{3} + \frac{4}{9} - \frac{8}{27} + \cdots = \frac{3}{5}$$

(b) The series is geometric, with $a = 3$ and $r = 4/3$. Since $r > 1$, the series diverges.   ∎

EXAMPLE 10.18   Show that each of the following series is geometric and determine whether it is convergent or divergent. If convergent, find the sum.

(a) $\displaystyle\sum_{n=0}^{\infty} e^{-n}$ (b) $\displaystyle\sum_{k=1}^{\infty} \frac{(-1)^{k-1}}{2^k}$

Solution

(a) When a series is written in summation notation, it is sometimes helpful to write out the first few terms. In this case, we have

$$\sum_{n=0}^{\infty} e^{-n} = 1 + e^{-1} + e^{-2} + e^{-3} + \cdots = 1 + \frac{1}{e} + \frac{1}{e^2} + \frac{1}{e^3} + \cdots$$

Each term is obtained by multiplying the preceding one by $1/e$. So $r = 1/e$, and the series is geometric. Since $e \approx 2.7$, $|1/e| < 1$. The sum of the series is therefore

$$\frac{a}{1-r} = \frac{1}{1 - \dfrac{1}{e}} = \frac{e}{e-1}$$

(b) By writing out the first few terms,

$$\sum_{k=1}^{\infty} \frac{(-1)^{k-1}}{2^k} = \frac{1}{2} - \frac{1}{4} + \frac{1}{8} - \frac{1}{16} + \cdots$$

we see that the series is geometric, with $a = \frac{1}{2}$ and $r = -\frac{1}{2}$. Since $|-\frac{1}{2}| < 1$, the sum is

$$\frac{a}{1-r} = \frac{\frac{1}{2}}{1 - (-\frac{1}{2})} = \frac{\frac{1}{2}}{\frac{3}{2}} = \frac{1}{3} \qquad \blacksquare$$

EXAMPLE 10.19   Use a geometric series to express the repeating decimal $1.272727\ldots$ as the ratio m/n of two positive integers.

Solution   First, write the repeating decimal as the infinite series

$$1 + (0.27 + 0.0027 + 0.000027 + \cdots)$$

The series in parentheses is geometric, with $a = 0.27$ and $r = 0.01$. Thus, its sum is

$$\frac{a}{1-r} = \frac{0.27}{1 - 0.01} = \frac{0.27}{0.99} = \frac{27}{99} = \frac{3}{11}$$

The entire sum is therefore $1 + \frac{3}{11}$. That is,

$$1.272727\ldots = 1 + \frac{3}{11} = \frac{14}{11} \qquad \blacksquare$$

EXAMPLE 10.20   A ball is dropped from a height of 50 cm, and on each bounce it goes three-fourths as high as before. Approximate the total distance traveled by the ball in coming to rest.

Solution   Although the ball moves only along a vertical line, we can get a better picture by showing the path as in Figure 10.18 (on page 760). First it drops 50 cm, then it rises $37\frac{1}{2}$ cm and falls the same distance. Thereafter it rises and falls three-fourths of the up-and-down total for the previous bounce. To make the first motion analogous to the others, we can pretend it started from the ground, rising 50 cm and then falling 50 cm. We will subtract this imaginary initial 50-cm rise after getting the total. With this understanding, the total distance covered is

$$\left[100 + \frac{3}{4}(100) + \left(\frac{3}{4}\right)^2 (100) + \left(\frac{3}{4}\right)^3 (100) + \cdots \right] - 50$$

FIGURE 10.18

The expression in brackets is a geometric series, with $a = 100$ and $r = \frac{3}{4}$. So its sum is

$$\frac{a}{1-r} = \frac{100}{1 - \dfrac{3}{4}} = 400$$

Thus, on subtracting the initial imaginary 50-cm rise, we find the total distance the ball covers to be 350 cm.   ∎

Telescoping Series

The next example illustrates another type of series in which the limit of the nth partial sum can be found. For reasons that will become apparent, the series is called a **telescoping series**.

EXAMPLE 10.21   Show that the series

$$\sum_{k=1}^{\infty} \frac{1}{k^2 + k}$$

converges, and find its sum.

Solution   Using partial fractions, we can rewrite the general term in the form

$$\frac{1}{k^2 + k} = \frac{1}{k(k+1)} = \frac{1}{k} - \frac{1}{k+1}$$

Thus, the nth partial sum of the series is

$$S_n = \sum_{k=1}^{n} \left(\frac{1}{k} - \frac{1}{k+1} \right)$$

$$= \left(\frac{1}{1} - \frac{1}{2} \right) + \left(\frac{1}{2} - \frac{1}{3} \right) + \left(\frac{1}{3} - \frac{1}{4} \right) + \cdots + \left(\frac{1}{n} - \frac{1}{n+1} \right)$$

By regrouping the parentheses, we see that all terms except the first and last cancel out (this expression "telescopes," as in an old type of telescope that is collapsible), and so

$$S_n = 1 - \frac{1}{n+1}$$

As $n \to \infty$, $S_n \to 1$. This calculation shows that the original series converges, since the sequence $\{S_n\}$ converges, and also that its sum is 1. That is,

$$\sum_{k=1}^{\infty} \frac{1}{k^2 + k} = 1$$ ∎

A Necessary Condition for Convergence

Another very useful result can be obtained from Definition 10.6, which we state as a theorem.

THEOREM 10.7

If the series $\sum_{n=1}^{\infty} a_n$ converges, then $\lim_{n\to\infty} a_n = 0$.

Proof   Let S_n be the nth partial sum of $\sum_{n=1}^{\infty} a_n$; that is,

$$S_n = a_1 + a_2 + a_3 + \cdots + a_n$$

Then, if $n > 1$, we also have

$$S_{n-1} = a_1 + a_2 + a_3 + \cdots + a_{n-1}$$

On subtracting S_{n-1} from S_n, all terms drop out except a_n.

$$a_n = S_n - S_{n-1}$$

Now, since the series converges, we know that $\lim_{n\to\infty} S_n$ exists. Call its value S. But as $n \to \infty$, we also have $(n - 1) \to \infty$. So both $\lim_{n\to\infty} S_n = S$ and $\lim_{n\to\infty} S_{n-1} = S$. Thus,

$$\lim_{n\to\infty} a_n = \lim_{n\to\infty} (S_n - S_{n-1})$$
$$= \lim_{n\to\infty} S_n - \lim_{n\to\infty} S_{n-1}$$
$$= S - S = 0 \qquad \blacksquare$$

REMARK ————————————————————————————

■ Here we have made use of the limit property from Section 10.2 that says if $\lim_{n\to\infty} a_n$ and $\lim_{n\to\infty} b_n$ both exist, then $\lim_{n\to\infty}(a_n - b_n)$ exists and equals $\lim_{n\to\infty} a_n - \lim_{n\to\infty} b_n$.

An equivalent way of stating Theorem 10.7 (a logical equivalent called its *contrapositive*) is the following:

The nth-Term Test for Divergence

If $\lim_{n\to\infty} a_n \neq 0$, then $\sum_{n=1}^{\infty} a_n$ diverges.

Thus, if for a given series we can show that $\lim_{n\to\infty} a_n \neq 0$, we can conclude that the series $\sum a_n$ definitely diverges. For example, the series

$$\sum_{n=1}^{\infty} \frac{n}{2n + 1}$$

diverges, since

$$\lim_{n\to\infty} a_n = \lim_{n\to\infty} \frac{n}{2n + 1} = \frac{1}{2} \neq 0$$

 If you find merely that $\lim_{n\to\infty} a_n = 0$, you cannot draw any definite conclusion regarding the convergence or divergence of $\sum a_n$ except that the series at least has a chance to converge. It is a common mistake to assume that the converse of Theorem 10.7 is true; that is, that $\lim_{n\to\infty} a_n = 0$ implies the convergence of

$\sum a_n$. That this converse is *not true* can be seen by finding a divergent series whose terms approach 0, as we do now.

The Harmonic Series

The best-known example of a divergent series whose nth term tends to 0 is the so-called **harmonic series**—namely,

$$\sum_{n=1}^{\infty} \frac{1}{n} = 1 + \frac{1}{2} + \frac{1}{3} + \cdots + \frac{1}{n} + \cdots$$

Clearly, $\lim_{n \to \infty} 1/n = 0$, yet the series diverges, as the following analysis of certain of its partial sums shows:

$$S_1 = 1$$

$$S_2 = 1 + \frac{1}{2}$$

$$S_4 = 1 + \frac{1}{2} + \frac{1}{3} + \frac{1}{4} > 1 + \frac{1}{2} + \left(\frac{1}{4} + \frac{1}{4}\right) = 1 + \frac{1}{2} + \frac{1}{2}$$

$$S_8 = 1 + \frac{1}{2} + \frac{1}{3} + \frac{1}{4} + \frac{1}{5} + \frac{1}{6} + \frac{1}{7} + \frac{1}{8}$$

$$> 1 + \frac{1}{2} + \left(\frac{1}{4} + \frac{1}{4}\right) + \left(\frac{1}{8} + \frac{1}{8} + \frac{1}{8} + \frac{1}{8}\right) = 1 + \frac{1}{2} + \frac{1}{2} + \frac{1}{2}$$

Notice that the relationship between these partial sums and the number of times $\frac{1}{2}$ is added is as follows:

$$S_1 = S_{2^0} = 1 + 0\left(\frac{1}{2}\right)$$

$$S_2 = S_{2^1} > 1 + 1\left(\frac{1}{2}\right)$$

$$S_4 = S_{2^2} > 1 + 2\left(\frac{1}{2}\right)$$

$$S_8 = S_{2^3} > 1 + 3\left(\frac{1}{2}\right)$$

$$\vdots \qquad \vdots$$

In general, we have

$$S_{2^n} \geq 1 + n\left(\frac{1}{2}\right), \qquad n = 0, 1, 2, \ldots$$

(Mathematical induction can be used to give a formal proof.) Thus, $\lim_{n \to \infty} S_{2^n} = \infty$ and it follows that $\lim_{n \to \infty} S_n$ does not exist. Thus, the series diverges. Because we will have frequent occasions to refer to this result, we set it off for emphasis.

The Harmonic Series

The harmonic series

$$\sum_{n=1}^{\infty} \frac{1}{n} = 1 + \frac{1}{2} + \frac{1}{3} + \cdots + \frac{1}{n} + \cdots$$

diverges.

Since for the harmonic series $\lim_{n \to \infty} S_n = \infty$, we sometimes write $\sum_{n=1}^{\infty} 1/n = \infty$. It is interesting to note that the sum of the first billion terms of the harmonic series is only about 21. That is,

$$S_{10^9} = 1 + \frac{1}{2} + \frac{1}{3} + \cdots + \frac{1}{10^9} \approx 21$$

and each succeeding term is *very* small. Yet if we continue adding terms indefinitely, the cumulative effect is to cause the sum to become infinite! Sometimes infinite processes defy one's intuition.

Properties of Convergent Series

The next theorem is a direct consequence of Definition 10.6. We call for a proof in Exercise 38 of Exercise Set 10.3.

THEOREM 10.8

If $\sum_{n=1}^{\infty} a_n$ and $\sum_{n=1}^{\infty} b_n$ are convergent series, and if c is any real number, then each of the series $\sum_{n=1}^{\infty} c a_n$, $\sum_{n=1}^{\infty} (a_n + b_n)$, and $\sum_{n=1}^{\infty} (a_n - b_n)$ converges, and

1. $\displaystyle\sum_{n=1}^{\infty} c a_n = c \sum_{n=1}^{\infty} a_n$

2. $\displaystyle\sum_{n=1}^{\infty} (a_n + b_n) = \sum_{n=1}^{\infty} a_n + \sum_{n=1}^{\infty} b_n$

3. $\displaystyle\sum_{n=1}^{\infty} (a_n - b_n) = \sum_{n=1}^{\infty} a_n - \sum_{n=1}^{\infty} b_n$

EXAMPLE 10.22   Show that the series

$$\sum_{n=1}^{\infty} \left(\frac{1}{2^{n-1}} + \frac{1}{n^2 + n} \right)$$

converges, and find its sum.

Solution   The series

$$\sum_{n=1}^{\infty} \frac{1}{2^{n-1}}$$

is the geometric series with $a = 1$ and $r = \frac{1}{2}$, whose sum is

$$\frac{a}{1-r} = \frac{1}{1 - \dfrac{1}{2}} = 2$$

The series

$$\sum_{n=1}^{\infty} \frac{1}{n^2 + n}$$

is the telescoping series we considered in Example 10.21, where we found that it converges to 1. It follows by Property 2 of Theorem 10.8 that

$$\sum_{n=1}^{\infty} \left(\frac{1}{2^{n-1}} + \frac{1}{n^2 + n} \right) = 2 + 1 = 3 \qquad \blacksquare$$

If $\sum a_n$ converges and $\sum b_n$ diverges, then we can see from Theorem 10.8 that $\sum (a_n + b_n)$ and $\sum (a_n - b_n)$ both diverge. For suppose $\sum (a_n + b_n)$ converges. Then we can write $b_n = (a_n + b_n) - a_n$. Property 3 of Theorem 10.8 then gives

$$\sum b_n = \sum [(a_n + b_n) - a_n] = \sum (a_n + b_n) - \sum a_n$$

since by our assumption both series on the right converge. But we are given that $\sum b_n$ diverges. So $\sum (a_n + b_n)$ cannot converge. We can prove that $\sum (a_n - b_n)$ diverges in a similar way. If $\sum a_n$ and $\sum b_n$ both diverge, no definite conclusion can be drawn about convergence or divergence of $\sum (a_n + b_n)$ or $\sum (a_n - b_n)$. They may converge or diverge, depending on the particular series in question. (See Exercise 39 in Exercise Set 10.3.)

Another consequence of Theorem 10.8 is that if $\sum a_n$ diverges, so does $\sum ca_n$ for any $c \neq 0$. This result follows from the fact that we can write

$$a_n = \frac{1}{c}(ca_n)$$

So by Property 1 of Theorem 10.8 (with c replaced by $1/c$ and a_n replaced by ca_n), if $\sum ca_n$ converges, it would follow that $\sum a_n = \sum 1/c(ca_n)$ also would converge, contrary to what we are given.

We summarize these properties as a corollary to Theorem 10.8.

COROLLARY 10.8

1. If $\sum a_n$ converges and $\sum b_n$ diverges, then both $\sum (a_n + b_n)$ and $\sum (a_n - b_n)$ diverge.

2. If $\sum a_n$ diverges and $c \neq 0$, then $\sum ca_n$ diverges.

We mention one other property that will sometimes be useful.

The convergence or divergence of an infinite series is unaltered if we delete or add any finite number of terms.

Suppose, for example, we delete some or all of the first N terms of the series $\sum a_n$. For convenience of notation, let us replace each term we deleted with 0 and call the new series $\sum b_n$. Then some or all of the terms $b_1, b_2, \ldots, b_N$ equal 0, but for $n > N$, $b_n = a_n$. Let $\{A_n\}$ denote the sequence of partial sums of the series $\sum a_n$ and $\{B_n\}$ denote the sequence of partial sums of the series $\sum b_n$. Then for all $n > N$,

$$A_n - A_N = a_{N+1} + a_{N+2} + \cdots + a_n = b_{N+1} + b_{N+2} + \cdots + b_n = B_n - B_N$$

Since A_N and B_N are fixed constants (each being the sum of the first N terms of a series, where N is fixed), we see that if either A_n or B_n approaches a finite limit as $n \to \infty$, so does the other. It follows that $\sum a_n$ and $\sum b_n$ either both converge or both diverge. A similar argument can be given if a finite number of terms are added.

Note carefully that while altering a finite number of terms does not alter convergence or divergence, it *does* alter the sum when the series is convergent.

Fractals

The Mandelbrot Set

In 1904 the Swedish mathematician Helge von Koch gave an example of a curve without a tangent anywhere. To his contemporaries this was a "pathological" example, whereas now mathematicians are finding that such examples occur frequently in pure and applied mathematics. They are called **fractals**, geometric figures in which a set pattern repeats on an ever-decreasing scale. The Polish-born mathematician Benoit Mandelbrot, while working in this country at IBM during the late 1970s, did pioneering work in the theory of fractals. His findings have played a key role in the rapidly expanding field of fractal geometry. To construct the Koch curve, we start with the unit interval, remove the middle third and replace it with two pieces of equal length, forming the sides of an equilateral triangle. Then we repeat the process over and over. The first three stages of the construction are shown in Figure 10.19. A good approximation is given in Figure 10.20. The Koch curve is the "limit" of the sets shown. This limiting curve can never be drawn, existing only as a mathematical concept. In Figure 10.21, we can see that magnification of one-third of the edge of the curve by a factor of 3 results in the entire curve. If this property is preserved at every stage of any iterative construction, the curve is called self-similar. Fractals possess this property of self-similarity.

FIGURES 10.19–10.21

The construction of the **Koch island** is similar to the construction of the Koch curve described above. In the case of the island, we start with an equilateral triangle (Figure 10.22(a)). Now we remove the middle third of each of the three sides and replace each with two pieces of equal length, resulting in the star-shaped region in Figure 10.22(b). Now we repeat the process over and over. The next two approximations to the Koch island are given in Figure 10.23.

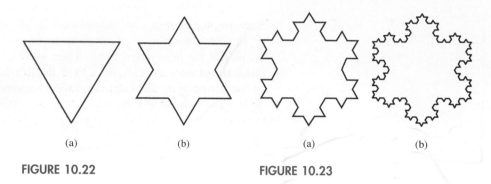

FIGURE 10.22   **FIGURE 10.23**

In Exercise 49 of Exercise Set 10.3 we will ask you to show that the perimeter of the Koch island is infinite and that its area is finite.

INFINITE SERIES USING COMPUTER ALGEBRA SYSTEMS

Computer algebra systems have commands for finding partial sums of infinite series and, in some cases, the sum of a convergent series.

CAS 37

Find several partial sums for the series

$$\sum_{n=1}^{\infty} \frac{1}{n^2}$$

If possible, determine the sum of the series.

Maple:

sum(1/k^2, k=1..4);
    $\frac{205}{144}$
evalf(sum(1/k^2, k=1..100));
        1.634983901
evalf(sum(1/k^2,
    k=1..1000));
        1.643934568

Mathematica:

Sum[1/n^2,{n,1,4}]
    $\frac{205}{144}$
N[Sum[1/n^2,{n,1,100}],10]
        1.6349839
N[Sum[1/n^2,{n,1,1000}],10]
        1.643934567

DERIVE:

(At the □ symbol, go to the next step.)
a (author) □ 1/k^2 □ c (calculus) □ s (sum) □ [choose expression] □ k (variable) □ 1 (lower limit) □ 4 (upper limit)

Output: $\frac{205}{144}$

Another approach:
a (author) □ 1/k^2 □ c (calculus) □ s (sum) □ [choose expression] □ k (variable) □ 1 (lower limit) □ n (upper limit) □ m (manage) □ s (substitute) □ k (variable) □ 100 (value for n) □ x (approximate) □ [choose expression]
Output: 1.63498

Maple can take an example like this one step further. In fact, the sum of this infinite series can be computed.

In Maple enter:

sum(1/k^2,k=1..infinity);

The output is the constant given by Euler, $\pi^2/6$.

Maple can also return formal infinite sums by using the Sum command (note the capital S). In Maple, sum and Sum are different commands. The command sum computes a sum, and Sum returns the formal sum in summation notation.

For example, in Maple enter:

Sum(1/k^2,k=1..infinity);

The output is:

$$\sum_{k=1}^{\infty} \frac{1}{k^2}$$

As one final example, in Maple enter:

sum(k^2, k=1..n);

The output is:

$$\frac{1}{3}(n+1)^3 - \frac{1}{2}(n+1)^2 + \frac{1}{6}n + \frac{1}{6}$$

If the output is factored, we see the formula given in Equation 5.5.

factor(");

The output is:

$$\frac{1}{6}n(n+1)(2n+1)$$

Exercise Set 10.3

In Exercises 1–4, write out several terms of the series and give the first four partial sums.

1. $\displaystyle\sum_{k=1}^{\infty} \frac{k}{k^2+1}$

2. $\displaystyle\sum_{n=1}^{\infty} \frac{(-1)^{n-1}}{\sqrt{2n-1}}$

3. $\displaystyle\sum_{m=1}^{\infty} \frac{\ln(m+1)}{m!}$

4. $\displaystyle\sum_{n=0}^{\infty} \frac{(-1)^n(n+1)}{2^n}$

In Exercises 5–8, write the series using summation notation. Also, give the nth partial sum.

5. $1 - \dfrac{1}{3} + \dfrac{1}{9} - \dfrac{1}{27} + \dfrac{1}{81} - \cdots$

6. $1 - \dfrac{1}{2!} + \dfrac{1}{4!} - \dfrac{1}{6!} + \dfrac{1}{8!} - \cdots$

7. $\dfrac{1}{2\ln 2} + \dfrac{1}{3\ln 3} + \dfrac{1}{4\ln 4} + \cdots$

8. $1 - \dfrac{2}{3^2} + \dfrac{3}{5^2} - \dfrac{4}{7^2} + \dfrac{5}{9^2} - \cdots$

In Exercises 9–12, the nth partial sum, S_n, of a series $\sum a_n$ is given. Determine whether the series converges and, if it does, give its sum.

9. $S_n = \dfrac{n}{n+1}$

10. $S_n = 1 - (-1)^n$

11. $S_n = 2 - \dfrac{\ln(n+1)}{n+1}$

12. $S_n = \dfrac{n^3 - 2n + 1}{\sqrt{n^4 + n^2 + 4}}$

In Exercises 13–20, show that the given series is geometric and determine whether it converges. If it converges, find its sum.

13. $2 - 1 + \dfrac{1}{2} - \dfrac{1}{4} + \dfrac{1}{8} - \cdots$

14. $\displaystyle\sum_{k=1}^{\infty} \dfrac{2}{3^{k-1}}$ **15.** $\displaystyle\sum_{n=0}^{\infty} 3 \cdot 2^{-n}$

16. $\dfrac{1}{\ln 3} + \dfrac{1}{(\ln 3)^2} + \dfrac{1}{(\ln 3)^3} + \cdots$

17. $\dfrac{2}{e} - \dfrac{4}{e^2} + \dfrac{8}{e^3} - \dfrac{16}{e^4} + \cdots$

18. $\displaystyle\sum_{n=1}^{\infty} \left(\dfrac{5}{4}\right)^n$ **19.** $\displaystyle\sum_{n=0}^{\infty} (0.99)^n$

20. $\displaystyle\sum_{k=1}^{\infty} \dfrac{(-1)^{k-1} 3^k}{4^{k-1}}$

In Exercises 21–24, express the repeating decimal in the form m/n, in lowest terms, where m and n are integers, making use of geometric series.

21. $0.151515\ldots$ **22.** $2.181818\ldots$

23. $0.148148148\ldots$ **24.** $1.135135135\ldots$

In Exercises 25–28, show that the series is telescoping, and find its sum.

25. $\displaystyle\sum_{k=1}^{\infty} \dfrac{1}{(k+1)(k+2)}$ **26.** $\displaystyle\sum_{n=1}^{\infty} \dfrac{2}{n(n+2)}$

27. $\displaystyle\sum_{n=1}^{\infty} \dfrac{2}{4n^2 - 1}$ **28.** $\displaystyle\sum_{k=1}^{\infty} \dfrac{1}{k^2 + 4k + 3}$

In Exercises 29–32, show that the series converges, and find its sum.

29. $\displaystyle\sum_{n=1}^{\infty} \left(\dfrac{1}{2^n} + \dfrac{2}{3^n}\right)$

30. $\displaystyle\sum_{k=1}^{\infty} \left[\left(\dfrac{2}{3}\right)^k - \left(\dfrac{3}{4}\right)^{k-1}\right]$

31. $\displaystyle\sum_{n=1}^{\infty} \left(\dfrac{1}{n^2 + n} - \dfrac{1}{3^{n-1}}\right)$

32. $\displaystyle\sum_{n=1}^{\infty} \left[\dfrac{(-1)^{n-1} 5}{2^n} + \dfrac{2}{n^2 + 2n}\right]$

33. A ball is dropped from a height of 10 m, and on each successive bounce it rises two-thirds as high as on the preceding bounce. Find the total distance the ball travels.

34. Determine whether the series $\sum_{k=100}^{\infty} 1/k$ converges or diverges. Justify your answer.

In Exercises 35 and 36, show that each series diverges.

35. (a) $\displaystyle\sum_{n=1}^{\infty} \dfrac{n}{100n + 1}$ (b) $\displaystyle\sum_{n=2}^{\infty} \dfrac{n}{(\ln n)^2}$

36. (a) $\displaystyle\sum_{n=1}^{\infty} \dfrac{2n^2 - 3n + 4}{3n^2 + n + 5}$

(b) $\displaystyle\sum_{n=1}^{\infty} \dfrac{(-1)^{n-1} n}{\sqrt{1 + n^2}}$

37. Indicate which of the following statements are true and which are false.
(a) If $\sum a_n$ diverges, then $\lim_{n\to\infty} a_n = 0$.
(b) If $\lim_{n\to\infty} a_n = 0$, then $\sum a_n$ converges.
(c) If $a_1 + a_2 + \cdots + a_n = 1/n$, then $\sum a_n$ converges to 0.
(d) If $\sum (a_n + b_n)$ converges, so do $\sum a_n$ and $\sum b_n$.
(e) If $c \neq 0$ and $\sum ca_n$ converges, then $\sum a_n$ also converges.

38. Prove Theorem 10.8, making use of the properties of limits of sequences from Section 10.2.

39. (a) Prove that if $\sum a_n$ converges and $\sum b_n$ diverges, then $\sum (a_n - b_n)$ diverges.
(b) Give examples to show that if $\sum a_n$ and $\sum b_n$ both diverge, then $\sum (a_n + b_n)$ and $\sum (a_n - b_n)$ may converge or may diverge.

40. Show that the series

$$\sum_{n=1}^{\infty} \ln \dfrac{n}{n+1}$$

diverges. (*Hint:* Show that it is a telescoping series and find S_n.)

41. Prove that if $a_n \geq 0$ and $a_1 + a_2 + a_3 + \cdots + a_n \leq k$ for all n, where k is a constant, then $\sum a_n$ converges.

In Exercises 42 and 43, show that the series is geometric. Find the values of x for which the series converges, and give the sum as a function of x.

42. (a) $\displaystyle\sum_{n=0}^{\infty} \left(\dfrac{x}{2}\right)^n$ (b) $\displaystyle\sum_{n=0}^{\infty} \dfrac{(x-1)^n}{3^{n+1}}$

43. (a) $\displaystyle\sum_{n=0}^{\infty} \frac{(-1)^n 3^n}{(x+2)^{n+1}}$ (b) $\displaystyle\sum_{n=0}^{\infty}(\ln x)^n$

44. A pendulum 1 m long is released from a position in which its angle with the vertical is 60°. On each swing after the first, it reaches a maximum angle with the vertical that is 0.9 times as large as the angle reached on the previous swing. Find the total distance covered by the bob of the pendulum in coming to rest.

45. Regrouping of terms. Consider the divergent series $\sum_{n=1}^{\infty}(-1)^{n+1}$. If we group the terms by pairs starting with the first two terms, we get

$$(1-1)+(1-1)+(1-1)+\cdots = 0+0+0+\cdots = 0$$

If, instead, we group by pairs starting with the second and third terms, we can write the result as

$$1-(1-1)-(1-1)-(1-1)-\cdots = 1-0-0-0-\cdots = 1$$

Prove that this phenomenon cannot occur in a convergent series. That is, prove that the terms of a convergent series

can be regrouped, preserving order, in any way and the resulting series has the same sum as the original one.

46. Use a CAS or a spreadsheet to approximate the value of $\sum_{n=1}^{\infty} 1/n^2$ by the partial sum S_{100}. Compare your answer with the known result $\pi^2/6$.

47. Use a CAS or a spreadsheet to approximate S_{100} for the harmonic series $\sum_{n=1}^{\infty} 1/n$.

48. Find the smallest integer n for which the partial sum S_n of the series $\sum_{n=0}^{\infty} 1/3^n$ agrees with the exact sum to 10 decimal places.

49. For the Koch island, let the initial equilateral triangle have sides each of length a.
(a) Find formulas for the number of sides, the length of each side, and the perimeter of the nth approximation to the Koch island. Denote these by s_n, l_n, and P_n, respectively.
(b) Show that the perimeter of the Koch island is infinite. That is,

$$\lim_{n\to\infty} P_n = \infty$$

(c) Express the area of the Koch island as an infinite geometric series and show that the area equals

$$\left(\frac{2\sqrt{3}}{5}\right)a^2$$

(*Hint:* The initial triangle has area $(\sqrt{3}/4)a^2$, and three triangles are added at the second stage, each with area $(\sqrt{3}/4)(a/3)^2$. So the combined area at the second stage is $(\sqrt{3}/4)a^2 + 3(\sqrt{3}/4)(a/3)^2$. Continue in this manner.)

10.4 SERIES OF POSITIVE TERMS; THE INTEGRAL TEST AND COMPARISON TESTS

Given an infinite series of constants, we usually want to answer two primary questions: (1) Does the series converge? and (2) If it does converge, what is its sum? The second question generally is much harder to answer than the first. (Geometric series and telescoping series are exceptions.) However, if we know that a series converges, we can at least approximate its sum by using a partial sum S_n for sufficiently large n. So it is very useful just to be able to answer the first question. In this section and the next we give certain tests for convergence that are applicable when the terms of the series are all positive or zero.

The basis for our tests is the following theorem.

THEOREM 10.9

If $a_n \geq 0$ for all n, then $\sum a_n$ converges if and only if its sequence of partial sums is bounded.

Proof   Let $\{S_n\}$ be the sequence of partial sums. Then, since $S_1 = a_1$, $S_2 = a_1 + a_2$, $S_3 = a_1 + a_2 + a_3, \ldots$, and all of the a_n's are nonnegative, we see that $S_1 \leq S_2 \leq S_3 \leq \cdots$. That is, $\{S_n\}$ is a monotone increasing sequence. If $\{S_n\}$ is bounded, we know from Theorem 10.5 that the sequence converges. Thus, the series $\sum a_n$ converges.

Suppose now we are given that $\sum a_n$ converges. Since $\{S_n\}$ is an increasing sequence, it is either bounded or it diverges to ∞. But if $\lim_{n \to \infty} S_n = \infty$, the series would diverge, contrary to our hypothesis. Thus $\{S_n\}$ must be bounded. Our proof is therefore complete.   ∎

The Integral Test

As our first application of Theorem 10.9, we prove a test for convergence of a series based on our knowledge of improper integrals (see Section 8.7).

THEOREM 10.10

The Integral Test

Let f be a continuous, positive, monotone decreasing function for $x \geq 1$, and let $a_n = f(n)$ for $n = 1, 2, 3, \ldots$. Then $\sum_{n=1}^{\infty} a_n$ is convergent if and only if the improper integral $\int_1^{\infty} f(x)dx$ is convergent. That is:

(a) If $\int_1^{\infty} f(x)dx$ converges, then $\sum_{n=1}^{\infty} a_n$ converges.
(b) If $\int_1^{\infty} f(x)dx$ diverges, then $\sum_{n=1}^{\infty} a_n$ diverges.

(a)

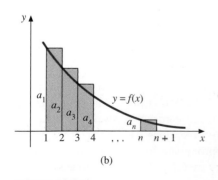

(b)

FIGURE 10.24

Proof   Suppose first that $\int_1^{\infty} f(x) \, dx$ converges. As Figure 10.24(a) shows, if we partition the interval $[1, n]$ at the integer points, the sum of the areas of the inscribed rectangles is

$$f(2) \cdot 1 + f(3) \cdot 1 + f(4) \cdot 1 + \cdots + f(n) \cdot 1$$

Since $a_n = f(n)$, the sum can be written as

$$a_2 + a_3 + a_4 + \cdots + a_n = S_n - a_1$$

where S_n is the nth partial sum of the series. The areas of the inscribed rectangles cannot exceed the area under the graph of f. So we have

$$S_n - a_1 \leq \int_1^n f(x)dx \leq \int_1^{\infty} f(x)dx$$

Thus,

$$S_n \leq a_1 + \int_1^{\infty} f(x)dx$$

By our hypothesis, the integral on the right converges. Therefore, the sequence $\{S_n\}$ is bounded. We also know that $a_n \geq 0$, since $a_n = f(n)$ and f is a positive function. It now follows from Theorem 10.9 that $\sum a_n$ converges.

Suppose now that $\int_1^\infty f(x)dx$ diverges. From Figure 10.24(b), by partitioning the interval $[1, n+1]$ and using circumscribed rectangles, we see that

$$S_n \geq \int_1^{n+1} f(x)dx$$

As $n \to \infty$, the integral on the right becomes arbitrarily large (otherwise $\int_1^\infty f(x)dx$ would be finite). Thus, $\lim_{n\to\infty} S_n = \infty$, and $\sum a_n$ diverges. ■

REMARK ─────────────────────────────────

■ Since deleting or adding any finite number of terms of a series does not affect its convergence or divergence, it is sufficient that the conditions of the Integral Test hold for $x \geq m$, where m is some positive integer.

──

EXAMPLE 10.23   Test each of the following series for convergence or divergence. (a) $\sum_{n=1}^\infty \dfrac{1}{1+n^2}$ (b) $\sum_{n=2}^\infty \dfrac{1}{n \ln n}$

Solution

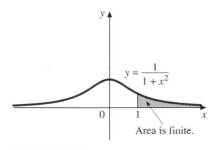

$y = \dfrac{1}{1+x^2}$

Area is finite.

FIGURE 10.25

(a) Let $f(x) = 1/(1+x^2)$ (see Figure 10.25). Then f satisfies the hypotheses of Theorem 10.10. So we consider the improper integral

$$\int_1^\infty \frac{dx}{1+x^2} = \lim_{t\to\infty} \int_1^t \frac{dx}{1+x^2} = \lim_{t\to\infty} \left[\tan^{-1} t - \tan^{-1} 1 \right] = \frac{\pi}{2} - \frac{\pi}{4} = \frac{\pi}{4}$$

Since the integral converges, we conclude from Theorem 10.10 that the series

$$\sum_{n=1}^\infty \frac{1}{1+n^2}$$

also converges.

Note that while we know the value of the improper integral is $\pi/4$, we cannot conclude that the series converges to this value. All we know is that the series does converge.

(b) The function $f(x) = 1/(x \ln x)$ (see Figure 10.26) satisfies the hypotheses of Theorem 10.10 for $x \geq 2$. So we consider

$$\int_2^\infty \frac{1}{x \ln x} dx = \lim_{t\to\infty} \int_2^t \frac{1}{x \ln x} dx$$

Since $d(\ln x) = (1/x)dx$, the integrand is in the form du/u with $u = \ln x$. So an antiderivative is $\ln |u| = \ln |\ln x|$. Thus,

$$\lim_{t\to\infty} \int_2^t \frac{1}{x \ln x} dx = \lim_{t\to\infty} \left[\ln |\ln t| - \ln |\ln 2| \right] = \infty$$

since $\ln t \to \infty$, and hence, $\ln |\ln t| \to \infty$ as $t \to \infty$. Thus, by Theorem 10.10, the series

$$\sum_{n=2}^\infty \frac{1}{n \ln n}$$

diverges. ■

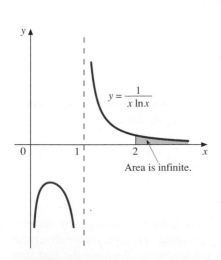

$y = \dfrac{1}{x \ln x}$

Area is infinite.

FIGURE 10.26

EXAMPLE 10.24   Determine all values of p for which the series $\sum 1/n^p$ converges.

Solution   If $p > 0$, let $f(x) = 1/x^p$. Then f is a positive, continuous, monotone decreasing function for $x \geq 1$. Furthermore, $f(n) = 1/n^p$. So the conditions of Theorem 10.10 are met. We found in Example 8.28 that

$$\int_1^\infty \frac{1}{x^p} dx$$

converges if $p > 1$ and diverges if $p \leq 1$. Thus, by Theorem 10.10, $\sum 1/n^p$ also converges if $p > 1$ and diverges if $p \leq 1$.

If $p \leq 0$, write $p = -\alpha$, where $\alpha \geq 0$. Then $1/n^p = n^\alpha \geq 1$. That is, each term of the series is at least 1, so the partial sums become arbitrarily large, and the series diverges. ∎

The series in the preceding example is referred to as the **p-series**. We give special prominence to what we have found.

The p-Series

The p-series

$$\sum_{n=1}^\infty \frac{1}{n^p}$$

converges if $p > 1$ and diverges if $p \leq 1$.

Note that by taking $p = 1$, we see again that the harmonic series diverges. Also, we confirm the fact that the series $\sum_{n=1}^\infty 1/n^2$ converges. This series is the one Euler proved converges to $\pi^2/6$. In fact, Euler also proved that $\sum_{n=1}^\infty 1/n^4$ converges to $\pi^4/90$, and he found a general formula for the sum of $\sum_{n=1}^\infty 1/n^{2k}$, but he was unable to find the sum of the reciprocals of odd powers (greater than or equal to 3) of the positive integers. In particular, he did not find the sum of the series

$$\sum_{n=1}^\infty \frac{1}{n^3}$$

Nor has anyone else ever been able to find this sum. It is one of the famous unsolved problems in mathematics. You will become famous overnight if you find the sum.

Comparison Tests

We can frequently determine whether a series converges or diverges by comparing its terms with those of some known convergent series or known divergent series. There are two ways to make such comparisons. We state the first way in the following theorem.

THEOREM 10.11

> **The Comparison Test**
>
> Let $\sum a_n$ and $\sum b_n$ be series of nonnegative terms, with $a_n \le b_n$ for all n.
>
> (a) If $\sum b_n$ converges, $\sum a_n$ converges.
> (b) If $\sum a_n$ diverges, $\sum b_n$ diverges.
>
> In other words, if each term of a positive-term series is smaller than the corresponding term of a known convergent series, the original series also converges, and if each term is larger than the corresponding term of a known divergent series of positive terms, the original series also diverges.

Proof   Let A_n and B_n denote the nth partial sums of $\sum a_n$ and $\sum b_n$, respectively. Suppose first that $\sum b_n$ converges. Then, by Theorem 10.9, $B_n \le M$ for some positive constant M. Since $a_n \le b_n$, it follows that

$$A_n = a_1 + a_2 + \cdots + a_n \le b_1 + b_2 + \cdots + b_n = B_n$$

So $A_n \le B_n \le M$. Using Theorem 10.9 again, we see that $\sum a_n$ converges, since $\{A_n\}$ is bounded.

Suppose now that $\sum a_n$ diverges. Then the sequence $\{A_n\}$ is unbounded (if it were bounded, $\sum a_n$ would be convergent). Since $B_n \ge A_n$, it follows that $\{B_n\}$ also is unbounded. Thus, by Theorem 10.9, $\sum b_n$ diverges.   ■

REMARKS

■ It is sufficient that $a_n \le b_n$ from some n onward, say for $n \ge N$, since convergence or divergence is unaffected by neglecting any finite number of terms.

■ When a_n and b_n are nonnegative and $a_n \le b_n$ for all $n \ge N$, we say that the series $\sum b_n$ *dominates* the series $\sum a_n$. Equivalently, we may say that $\sum a_n$ is *dominated by* $\sum b_n$. Theorem 10.11 can therefore be rephrased by saying that for series of nonnegative terms, when a series to be tested is dominated by a known convergent series, the series converges. Similarly, when a series to be tested dominates a known divergent series, the series diverges.

If $a_n \le b_n$ and $\sum a_n$ converges, we can draw no conclusion from the Comparison Test about convergence or divergence of $\sum b_n$. Similarly, if $\sum b_n$ diverges, we can draw no conclusion about $\sum a_n$. For example,

$$\frac{1}{n^2} \le \frac{1}{n}$$

but $\sum 1/n^2$ converges and $\sum 1/n$ diverges. Figure 10.27 shows the situation graphically. In the figure, C represents a known convergent series and D a known divergent series. Any series with terms less than those of C converges, and any series with terms greater than those of D diverges. No conclusion can be drawn about series between C and D, however.

Convergence	No conclusion	Divergence
C		D

FIGURE 10.27

EXAMPLE 10.25   Test each of the following series for convergence or divergence.

(a) $\displaystyle\sum_{n=1}^{\infty} \frac{1}{\sqrt{n^3 + 3}}$ (b) $\displaystyle\sum_{n=1}^{\infty} \frac{1}{2n - 1}$

Solution

(a) By ignoring the 3 under the radical in the denominator, we see that the general term behaves like $1/n^{3/2}$. But $\sum 1/n^{3/2}$ is a convergent p-series, and we might suspect, by the Comparison Test, that the given series also converges. Indeed, since for $n \geq 1$,

$$\frac{1}{\sqrt{n^3 + 3}} < \frac{1}{\sqrt{n^3}} = \frac{1}{n^{3/2}}$$

it follows by the Comparison Test that

$$\sum_{n=1}^{\infty} \frac{1}{\sqrt{n^3 + 3}}$$

also converges.

(b) Here we might suspect that the terms $1/(2n - 1)$ behave similarly to $1/n$. In fact,

$$\frac{1}{2n - 1} > \frac{1}{2n} \qquad \text{for } n \geq 1$$

since a smaller denominator results in a larger fraction. Now, we know that the harmonic series $\sum 1/n$ diverges. It follows that

$$\sum \frac{1}{2n}$$

also diverges (see Corollary 10.8). Thus, by the Comparison Test,

$$\sum \frac{1}{2n - 1}$$

diverges.   ■

The next test is a variation on the Comparison Test that often is easier to apply than the Comparison Test itself. In it we consider the *ratio* of the terms of a series we are testing to those of a series that we know to be either convergent or divergent. If this ratio approaches some positive (finite) limit as $n \to \infty$, we can conclude that the two series are similar with regard to convergence or divergence. That is, both series converge or else both series diverge.

THEOREM 10.12

The Limit Comparison Test

If $\sum a_n$ and $\sum b_n$ are series of positive terms and

$$\lim_{n \to \infty} \frac{a_n}{b_n} = L$$

where $0 < L < \infty$, then $\sum a_n$ and $\sum b_n$ both converge or else both diverge.

Proof   Since a_n/b_n approaches L as a limit, it follows that a_n/b_n will lie in the interval $(L/2, 3L/2)$ for all sufficiently large n, say for $n \geq N$. Thus, for $n \geq N$, since $a_n/b_n < 3L/2$, we have

$$a_n < \frac{3L}{2} b_n$$

If we know that $\sum b_n$ converges, then so does $\sum (3L/2)b_n$ (Theorem 10.8). Thus, by the Comparison Test, $\sum a_n$ also converges.

Since for $n \geq N$, the ratio a_n/b_n is in the interval $(L/2, 3L/2)$, we also have $a_n/b_n > L/2$, or

$$a_n > \frac{L}{2} b_n$$

when $n \geq N$. Thus, if we know that $\sum b_n$ diverges, so does $\sum (L/2)b_n$, by Corollary 10.8. Thus, again by the Comparison Test, $\sum a_n$ diverges.   ∎

REMARK —————————————————————————————————

■ When the conditions of Theorem 10.12 are met (i.e., $\lim_{n\to\infty} a_n/b_n = L$, with $0 < L < \infty$), we say that a_n and b_n are *of the same order of magnitude* as $n \to \infty$. You can often tell the order of magnitude of the nth term of a series by neglecting all but the highest powers in the numerator and denominator. You can also ignore any coefficients. (See also Exercise 50 in Exercise Set 10.4.)

EXAMPLE 10.26   Test each of the following for convergence or divergence.

(a) $\displaystyle\sum_{n=1}^{\infty} \frac{n^2 - n}{2n^3 + 3n - 4}$ (b) $\displaystyle\sum_{n=1}^{\infty} \frac{1}{\sqrt{3n^4 - 2n}}$

Solution   Observe that in each case it would be difficult to apply the Integral Test. Also, finding a suitable series for comparison and showing that the appropriate inequality in Theorem 10.11 is satisfied would be difficult. So we try the Limit Comparison Test in each case.

(a) Neglecting the coefficients and all but the highest powers, we see that

$$\frac{n^2 - n}{2n^3 + 3n - 4} \sim \frac{n^2}{n^3} = \frac{1}{n}$$

where we are using "$\sim$" to mean "is of the same order of magnitude." To confirm that the orders of magnitude are the same, we take the limit of the quotient:

$$\lim_{n\to\infty} \left[\frac{n^2 - n}{2n^3 + 3n - 4} \div \frac{1}{n} \right] = \lim_{n\to\infty} \frac{n^3 - n^2}{2n^3 + 3n - 4} = \frac{1}{2}$$

Since we know that the harmonic series $\sum 1/n$ diverges, we conclude by Theorem 10.12 that the given series also diverges.

(b) Again using the dominant-term approach, we have

$$\frac{1}{\sqrt{3n^4 - 2n}} \sim \frac{1}{\sqrt{n^4}} = \frac{1}{n^2}$$

To confirm that the orders of magnitude are the same, we divide and take the limit as $n \to \infty$:

$$\lim_{n \to \infty} \left[\frac{1}{\sqrt{3n^4 - 2n}} \div \frac{1}{n^2} \right] = \lim_{n \to \infty} \frac{n^2}{\sqrt{3n^4 - 2n}}$$

$$= \lim_{n \to \infty} \sqrt{\frac{n^4}{3n^4 - 2n}} = \frac{1}{\sqrt{3}}$$

We know that the p-series $\sum 1/n^2$ converges (since $p = 2 > 1$), so by Theorem 10.12,

$$\sum_{n=1}^{\infty} \frac{1}{\sqrt{3n^4 - 2n}}$$

also converges.   ∎

Exercise Set 10.4

In Exercises 1–10, use the Integral Test to determine convergence or divergence.

1. $\displaystyle\sum_{n=1}^{\infty} \frac{1}{\sqrt{2n-1}}$

2. $\displaystyle\sum_{n=2}^{\infty} \frac{1}{n(\ln n)^2}$

3. $\displaystyle\sum_{n=1}^{\infty} \frac{1}{n^2 + 4}$

4. $\displaystyle\sum_{n=1}^{\infty} ne^{-n^2}$

5. $\displaystyle\sum_{n=3}^{\infty} \frac{1}{n\sqrt{\ln n}}$

6. $\displaystyle\sum_{n=1}^{\infty} \frac{n}{(n^2 + 1)^{3/2}}$

7. $\displaystyle\sum_{n=1}^{\infty} \frac{1}{3n + 2}$

8. $\displaystyle\sum_{n=1}^{\infty} \frac{\ln n}{n}$

9. $\displaystyle\sum_{n=1}^{\infty} \frac{n}{n^2 + 1}$

10. $\displaystyle\sum_{n=1}^{\infty} ne^{-n}$

In Exercises 11–20, use the Comparison Test to determine convergence or divergence.

11. $\displaystyle\sum_{n=1}^{\infty} \frac{1}{n^2 + 1}$

12. $\displaystyle\sum_{n=1}^{\infty} \frac{1}{\sqrt{n^3 + 2n}}$

13. $\displaystyle\sum_{n=1}^{\infty} \frac{2}{2n - 1}$

14. $\displaystyle\sum_{n=1}^{\infty} \frac{n + 1}{n^2}$

15. $\displaystyle\sum_{n=1}^{\infty} \sqrt{\frac{n + 1}{n}}$

16. $\displaystyle\sum_{n=1}^{\infty} \frac{n}{2n^3 + 1}$

17. $\displaystyle\sum_{n=1}^{\infty} \frac{2}{3^n + 1}$

18. $\displaystyle\sum_{n=1}^{\infty} \frac{3^n}{2^n - 1}$

19. $\displaystyle\sum_{n=1}^{\infty} \frac{1 + \sin^2 n}{n\sqrt{n}}$

20. $\displaystyle\sum_{n=1}^{\infty} \frac{n + 1}{\sqrt{2n^3 - 1}}$

In Exercises 21–30, use the Limit Comparison Test to determine convergence or divergence.

21. $\displaystyle\sum_{n=1}^{\infty} \frac{2n - 1}{3n^2 + 4n - 2}$

22. $\displaystyle\sum_{n=1}^{\infty} \frac{3n}{n^4 - 2}$

23. $\displaystyle\sum_{n=1}^{\infty} \frac{\sqrt{2n + 3}}{n^3}$

24. $\displaystyle\sum_{n=1}^{\infty} \frac{n}{\sqrt{n^3 - 2n + 4}}$

25. $\displaystyle\sum_{n=1}^{\infty} \frac{1 + 2n}{(n^2 + 1)^{3/2}}$

26. $\displaystyle\sum_{n=1}^{\infty} \frac{n^2 - 2}{\sqrt{3n^5 + 1}}$

27. $\displaystyle\sum_{n=1}^{\infty} \frac{2n^3 - 3n^2 + 4}{5n^4 + 2n^3 - 1}$

28. $\displaystyle\sum_{n=1}^{\infty} \sqrt{\frac{n}{n^5 + 2}}$

29. $\displaystyle\sum_{n=1}^{\infty} \frac{\sqrt{n^2 + 3n}}{n^3 + 1}$

30. $\displaystyle\sum_{n=2}^{\infty} \frac{1}{\sqrt{n^3 - n}}$

In Exercises 31–48, test for convergence or divergence by any appropriate means.

31. $\displaystyle\sum_{n=1}^{\infty} \frac{n \cos^2 n}{1 + n^3}$

32. $\displaystyle\sum_{k=2}^{\infty} \frac{\sqrt{\ln k}}{k}$

33. $\displaystyle\sum_{n=1}^{\infty} \frac{1}{n \cdot 2^n}$

34. $\displaystyle\sum_{i=1}^{\infty} \frac{2i^2}{3i^2 - 2}$

35. $\displaystyle\sum_{n=0}^{\infty} \frac{\sqrt{n+1}}{n^2+1}$

36. $\displaystyle\sum_{j=0}^{\infty} \frac{j+1}{j^3+2}$

37. $\displaystyle\sum_{m=0}^{\infty} \frac{3^m}{4^m+3}$

38. $\displaystyle\sum_{n=0}^{\infty} \frac{1}{\cosh n}$

39. $\displaystyle\sum_{k=0}^{\infty} \frac{e^k}{1+e^{2k}}$

40. $\displaystyle\sum_{k=2}^{\infty} \frac{2k^2-3k+1}{k^5+2k^2+1}$

41. $\displaystyle\sum_{n=1}^{\infty} \frac{3n-2}{(n+1)(n+2)}$

42. $\displaystyle\sum_{m=1}^{\infty} \frac{m\sec^2 m}{1+m^2}$

43. $\displaystyle\sum_{n=1}^{\infty} \frac{1+\cos^2 n}{n^{3/2}}$

44. $\displaystyle\sum_{k=1}^{\infty} \frac{k}{\sqrt{100k^2+1}}$

45. $\displaystyle\sum_{n=3}^{\infty} \frac{1}{n(\ln n)[(\ln(\ln n)]}$

46. $\displaystyle\sum_{n=1}^{\infty} \frac{\sinh n}{\cosh 2n}$

47. $\displaystyle\sum_{n=1}^{\infty} \frac{n\ln n}{1+n^3}$

48. $\displaystyle\sum_{n=1}^{\infty} \frac{\sqrt{n+1}-\sqrt{n}}{n}$

49. Determine all values of p for which the series
$$\sum_{n=2}^{\infty} \frac{1}{n(\ln n)^p} \quad \text{converges.}$$

50. Prove the following extensions of the Limit Comparison Test for series $\sum a_n$ and $\sum b_n$ of positive terms:
(a) If $\lim_{n\to\infty} a_n/b_n = 0$ and $\sum b_n$ converges, then $\sum a_n$ also converges.
(b) If $\lim_{n\to\infty} a_n/b_n = \infty$ and $\sum b_n$ diverges, then $\sum a_n$ also diverges.

51. Use the results of Exercise 50 to test each of the following series for convergence or divergence.

(a) $\displaystyle\sum_{n=2}^{\infty} \frac{1}{\ln n}$ (b) $\displaystyle\sum_{n=2}^{\infty} \frac{\ln n}{n}$ (c) $\displaystyle\sum_{n=2}^{\infty} \frac{\ln n}{n^2}$

52. Prove that if $a_n \geq 0$ and $\sum a_n$ converges, then $\sum a_n^2$ also converges.

53. Prove that if $a_n \geq 0$ and $\lim_{n\to\infty} na_n$ exists and is positive, then $\sum a_n$ diverges.

54. Prove that if $\sum_{k=1}^{\infty} a_k$ converges, then $\lim_{n\to\infty} \sum_{k=n}^{\infty} a_k = 0$.

 55. Use a CAS or a spreadsheet to find the first 100 terms of each of the series in Example 10.26. In part (a) compare your result with the first 100 terms of the harmonic series, and in part (b) compare your result with the first 100 terms of the series $\sum 1/n^2$. Interpret your findings.

10.5 SERIES OF POSITIVE TERMS; THE RATIO TEST AND THE ROOT TEST

The next test is particularly well suited to series whose nth terms involve powers or products, especially factorials. In it, we consider the ratio of each term (after the first) to the preceding one.

THEOREM 10.13

The Ratio Test

Let $\sum a_n$ be a series of positive terms such that
$$\lim_{n\to\infty} \frac{a_{n+1}}{a_n} = L$$

(a) If $L < 1$, then $\sum a_n$ converges.
(b) If $L > 1$, then $\sum a_n$ diverges.
(c) If $L = 1$, then the test is inconclusive.

For $n \geq N$, the ratios $\dfrac{a_{n+1}}{a_n}$ are all to the left of r.

$\dfrac{a_{n+1}}{a_n}$ is in the shaded interval for $n \geq N$.

FIGURE 10.28

Proof Suppose first that $L < 1$. Let r denote any number such that $L < r < 1$. Since the ratios a_{n+1}/a_n eventually come arbitrarily close to L, they all lie to the left of r for n sufficiently large, say for $n \geq N$. (See Figure 10.28.)

Then we have

$$\frac{a_{n+1}}{a_n} < r \qquad \text{for all } n \geq N$$

or equivalently, $a_{n+1} < ra_n$ for all $n \geq N$. So

$$a_{N+1} < ra_N$$
$$a_{N+2} < ra_{N+1} < r^2 a_N$$
$$a_{N+3} < ra_{N+2} < r^3 a_N$$
$$\vdots$$

Thus, the terms of the series $\sum_{n=N+1}^{\infty} a_n$ are dominated by the terms of the series $\sum_{n=1}^{\infty} a_N r^n$. This latter series is a convergent geometric series, since the common ratio r is positive and less than 1. It follows by the Comparison Test that $\sum_{n=N+1}^{\infty} a_n$ converges. When we add the first N terms, the resulting series is also convergent; that is, the entire series $\sum_{n=1}^{\infty} a_n$ also converges.

If $L > 1$, then for all sufficiently large n,

$$\frac{a_{n+1}}{a_n} > 1, \qquad \text{or equivalently,} \qquad a_{n+1} > a_n$$

so the terms increase and thus cannot approach 0. Thus, $\sum a_n$ diverges (the nth term test for divergence).

Finally, the series $\sum 1/n$ and $\sum 1/n^2$ both satisfy

$$\lim_{n \to \infty} \frac{a_{n+1}}{a_n} = 1$$

(see Exercise 27 in Exercise Set 10.5), yet the first series diverges and the second converges. So when $L = 1$, the Ratio Test does not give us any conclusive information, and some other test is required.   ■

In applying the Ratio Test to show convergence, it is not enough to show $a_{n+1}/a_n < 1$ for all n. For example, consider the harmonic series

$$\sum \frac{1}{n} = 1 + \frac{1}{2} + \frac{1}{3} + \frac{1}{4} + \cdots$$

which we know to be divergent. Yet

$$\frac{a_{n+1}}{a_n} = \frac{1}{n+1} \div \frac{1}{n} = \frac{n}{n+1}$$

which is less than 1 for all n. It is the *limit* of this ratio as $n \to \infty$ that must be less than 1 to ensure convergence. In the case of the harmonic series, the limit of the ratio is 1. So the Ratio Test is not applicable.

EXAMPLE 10.27   Test each of the following series for convergence.

(a) $\displaystyle\sum_{n=1}^{\infty} \frac{2^n}{n!}$ (b) $\displaystyle\sum_{n=1}^{\infty} \frac{3^{n-1}}{n^2 \cdot 2^n}$

Solution   Since products and powers are involved, the Ratio Test appears to be the test to use.

(a) $\displaystyle \lim_{n \to \infty} \frac{a_{n+1}}{a_n} = \lim_{n \to \infty} \frac{2^{n+1}}{(n+1)!} \cdot \frac{n!}{2^n} = \lim_{n \to \infty} \frac{2^{n+1}}{2^n} \cdot \frac{n!}{(n+1)!} = \lim_{n \to \infty} \frac{2}{n+1} = 0$

Note that instead of writing the quotient of a_{n+1} over a_n, we inverted the denominator and multiplied. Note also that $(n+1)! = (n+1)n!$, so we canceled the $n!$ that appeared in both numerator and denominator. Since $L = 0$ and $0 < 1$, the series converges.

(b)

$$\lim_{n\to\infty} \frac{a_{n+1}}{a_n} = \lim_{n\to\infty} \frac{3^n}{(n+1)^2\, 2^{n+1}} \cdot \frac{n^2 2^n}{3^{n-1}}$$

$$= \lim_{n\to\infty} \frac{n^2}{(n+1)^2} \cdot \frac{3^n}{3^{n-1}} \cdot \frac{2^n}{2^{n+1}}$$

$$= \lim_{n\to\infty} \left(\frac{n}{n+1}\right)^2 \cdot \frac{3}{2} = \frac{3}{2}$$

Since $L = \frac{3}{2} > 1$, the series diverges.   ∎

EXAMPLE 10.28   Test the following series for convergence or divergence.

$$\sum_{n=1}^{\infty} \frac{1 \cdot 3 \cdot 5 \cdot \cdots \cdot (2n-1)}{3^n \cdot n!}$$

Solution   Before applying the Ratio Test, we simplify the quotient a_{n+1}/a_n. The numerator of the nth term is the product of the first n odd positive integers, so the numerator of the $(n+1)$st term is the product of the first $n+1$ odd positive integers, namely,

$$1 \cdot 3 \cdot 5 \cdot \cdots \cdot [2(n+1) - 1] = 1 \cdot 3 \cdot 5 \cdot \cdots \cdot (2n+1)$$

This product has all of the factors of the numerator of the nth term, together with the additional factor $2n + 1$. Thus,

$$\frac{a_{n+1}}{a_n} = \frac{1 \cdot 3 \cdot 5 \cdot \cdots \cdot (2n+1)}{3^{n+1}(n+1)!} \cdot \frac{3^n n!}{1 \cdot 3 \cdot 5 \cdot \cdots \cdot (2n-1)} = \frac{2n+1}{3(n+1)}$$

Now we take the limit:

$$\lim_{n\to\infty} \frac{a_{n+1}}{a_n} = \lim_{n\to\infty} \frac{2n+1}{3n+3} = \frac{2}{3}$$

Since $\frac{2}{3} < 1$, we conclude that the series converges.   ∎

We conclude the tests of positive term series with the **Root Test**. Its proof is similar to that of the Ratio Test, and we leave it for the exercises (Exercise 29 in Exercise Set 10.5).

THEOREM 10.14

The Root Test

Let $\sum a_n$ be a series of nonnegative terms such that

$$\lim_{n\to\infty} \sqrt[n]{a_n} = L$$

(a) If $L < 1$, then $\sum a_n$ converges.

(b) If $L > 1$, then $\sum a_n$ diverges.

(c) If $L = 1$, then the test is inconclusive.

The Root Test is more powerful than the Ratio Test in the sense that whenever the Ratio Test gives a definite conclusion concerning convergence or divergence, the same will be true of the Root Test. But there are series for which the Ratio Test is inconclusive and the Root Test gives definite information. (See Exercise 31 in Exercise Set 10.5.) However, the Ratio Test is usually easier to apply.

EXAMPLE 10.29   Test the series

$$\sum_{n=1}^{\infty} \left(\frac{n}{2n+1} \right)^n$$

Solution   Here, the Root Test seems appropriate, since the general term involves the nth power. When we take the nth root, we get a simple expression whose limit we can find:

$$\lim_{n \to \infty} \sqrt[n]{a_n} = \lim_{n \to \infty} \frac{n}{2n+1} = \frac{1}{2} < 1$$

So the series converges.   ∎

REMARK ────────────────────────────────────

■ In applying the Root Test you may sometimes need the following limits:

$$\lim_{n \to \infty} \sqrt[n]{a} = 1 \qquad \text{for } a > 0$$

$$\lim_{n \to \infty} \sqrt[n]{n} = 1$$

We proved the second of these limits in Example 10.14. You will be asked to prove the first in Exercise 28 of this section. You can also convince yourself that the results are reasonable with a CAS or a spreadsheet (Exercise 32 of this section).

────────────────────────────────────

You may have wondered why there are so many tests for convergence (there are still more, but we have given the main ones). The answer is that no single test works on all positive-term series. Even when more than one test *could* be used, one may be much easier to apply than the others. With a little practice you will probably develop a feel for which test to apply in a given situation.

Exercise Set 10.5

In Exercises 1–26, use the Ratio Test or Root Test to determine convergence or divergence.

1. $\displaystyle\sum_{n=1}^{\infty} ne^{-n}$

2. $\displaystyle\sum_{n=1}^{\infty} \frac{n^2 \cdot 2^n}{3^{n-1}}$

3. $\displaystyle\sum_{n=1}^{\infty} \frac{n!}{10^n}$

4. $\displaystyle\sum_{n=1}^{\infty} \left(\frac{n-1}{2n+1} \right)^n$

5. $\displaystyle\sum_{n=1}^{\infty} \frac{2^{3n}}{3^{2n}}$

6. $\displaystyle\sum_{n=1}^{\infty} \frac{n^2}{2^n}$

7. $\displaystyle\sum_{n=1}^{\infty} \frac{2^{n^2}}{n!}$

8. $\displaystyle\sum_{n=1}^{\infty} n^3 3^{-n}$

9. $\displaystyle\sum_{n=1}^{\infty} \frac{n}{\sqrt{e^n}}$

10. $\displaystyle\sum_{k=2}^{\infty} \left(\frac{4k^2+3}{3k^2-4} \right)^k$

11. $\displaystyle\sum_{m=1}^{\infty} \frac{(m+1)^{2m}}{3^{m^2}}$

12. $\displaystyle\sum_{k=1}^{\infty} k^2 \left(\frac{2}{3}\right)^k$

13. $\displaystyle\sum_{n=0}^{\infty} \frac{(100)^{2n+1}}{(2n+1)!}$

14. $\displaystyle\sum_{n=1}^{\infty} \frac{(n+1)^{n/2}}{2 \cdot 2^{n+1}}$

15. $\displaystyle\sum_{n=1}^{\infty} \frac{\sqrt{3^n}}{2^n}$

16. $\displaystyle\sum_{n=1}^{\infty} \frac{2^{3n+1}}{e^{2n}}$

17. $\displaystyle\sum_{n=1}^{\infty} \frac{n^n}{2^{n^2}}$

18. $\displaystyle\sum_{n=1}^{\infty} \frac{4^n}{n \cdot 3^n}$

19. $\displaystyle\sum_{n=1}^{\infty} \frac{2^n \cdot n!}{(2n)!}$

20. $\displaystyle\sum_{n=1}^{\infty} \frac{(n!)^2}{(2n)!}$

21. $\displaystyle\sum_{n=1}^{\infty} \frac{n!\,(2n+1)!}{(2n+3)!}$

22. $\displaystyle\sum_{n=1}^{\infty} \frac{n^n}{n!}$

23. $\displaystyle\sum_{n=1}^{\infty} \frac{n!}{1 \cdot 3 \cdot 5 \cdot \,\cdots\, \cdot (2n-1)}$

24. $\displaystyle\sum_{n=1}^{\infty} \frac{n!}{2 \cdot 4 \cdot 6 \cdot \,\cdots\, \cdot (2n)}$

25. $\displaystyle\sum_{n=1}^{\infty} \frac{1 \cdot 4 \cdot 7 \cdot \,\cdots\, \cdot (3n-2)}{3 \cdot 5 \cdot 7 \cdot \,\cdots\, \cdot (2n+1)}$

26. $\displaystyle\sum_{n=1}^{\infty} \frac{2^n \cdot n!}{1 \cdot 3 \cdot 5 \cdot \,\cdots\, \cdot (2n-1)}$

27. Prove that

$$\lim_{n\to\infty} \frac{a_{n+1}}{a_n} = 1$$

where (a) $a_n = 1/n$ and (b) $a_n = 1/n^2$. What conclusion can you draw?

28. Prove that

$$\lim_{n\to\infty} a^{1/n} = 1 \qquad \text{for } a > 0$$

(*Hint:* Let $y = a^{1/x}$. Take logs and use L'Hôpital's Rule. Then apply Theorem 10.2.)

29. Prove the Root Test (Theorem 10.14).

30. Prove the following version of the Ratio Test for the positive-term series $\sum a_n$.
(a) If there is a number r such that $0 < r < 1$ for which

$$\frac{a_{n+1}}{a_n} \le r \qquad \text{for all } n \ge N$$

where N is some positive integer, then $\sum a_n$ converges.
(b) If

$$\frac{a_{n+1}}{a_n} \ge 1 \qquad \text{for all } n \ge N$$

where N is some positive integer, then $\sum a_n$ diverges.

31. Consider the series

$$\sum_{n=1}^{\infty} 2^{(-1)^n - n} = \frac{1}{2^2} + \frac{1}{2} + \frac{1}{2^4} + \frac{1}{2^3} + \frac{1}{2^6} + \frac{1}{2^5} + \cdots$$

(a) Use the Root Test to show that the series converges.
(b) Show that

$$\frac{a_{n+1}}{a_n} = 2^{2(-1)^{n+1} - 1}$$

and $\lim_{n\to\infty}(a_{n+1}/a_n)$ does not exist, so that the Ratio Test does not apply.

32. Use a CAS or a spreadsheet to find
(a) $\sqrt[n]{2}$ for $n = 1$ to $n = 20$
(b) $\sqrt[n]{100}$ for $n = 1$ to $n = 20$
What do you observe in each case?

33. Use a CAS or a spreadsheet to find $\sqrt[n]{n}$ for $n = 1$ to $n = 20$. What do you observe?

10.6 SERIES WITH TERMS OF VARIABLE SIGNS; ABSOLUTE CONVERGENCE

In general, when a series consists of both positive and negative terms, tests for convergence are more complex than those for series consisting only of positive terms. There is one important test, however, that is relatively simple, and that is applicable to series in which the terms are alternately positive and negative. Such series are called **alternating series**. Though alternating series in general could begin with either a positive or a negative first term, it is sufficient to consider such series in which the first term is positive, writing, for example,

$$\sum_{n=1}^{\infty}(-1)^{n-1}a_n = a_1 - a_2 + a_3 - a_4 + \cdots$$

where for all n, $a_n > 0$. If the first term were negative, we could write

$$-a_1 + a_2 - a_3 + a_4 - \cdots = -(a_1 - a_2 + a_3 - a_4 + \cdots)$$

and determine convergence or divergence by studying the series in parentheses.

An example of an alternating series is

$$\sum_{n=1}^{\infty}\frac{(-1)^{n-1}}{n} = 1 - \frac{1}{2} + \frac{1}{3} - \frac{1}{4} + \cdots$$

This series is called the **alternating harmonic series**. Though the harmonic series itself, with all terms positive, diverges, we will soon see (Example 10.30) that the alternating harmonic series converges.

The following test shows that if the absolute values of the terms of an alternating series decrease monotonically and approach 0 as $n \to \infty$, then the series converges.

THEOREM 10.15

Alternating Series Test

If $a_n > 0$ for all n and the following two conditions are satisfied for all n,

(i) $a_{n+1} \le a_n$

(ii) $\lim_{n\to\infty} a_n = 0$

then the alternating series

$$\sum_{n=1}^{\infty}(-1)^{n-1}a_n$$

converges.

FIGURE 10.29

Proof   The idea of the proof is indicated in Figure 10.29. We start with the first partial sum S_1, which is just a_1. For the purpose of this illustration, let us suppose that the strict inequality $a_{n+1} < a_n$ holds true. Then $S_2 = a_1 - a_2$ is to the left of S_1, and $S_3 = a_1 - a_2 + a_3 = S_2 + a_3$ is to the right of S_2. Also, since $a_3 < a_2$, S_3 is less than S_1. Similarly, $S_4 = S_3 - a_4$ is to the left of S_3 but to the right of S_2, and so on. We see that the even-ordered partial sums $S_2, S_4, S_6, \ldots$ increase and are all to the left of the odd-ordered partial sums $S_1, S_3, S_5, \ldots$, which decrease. It seems reasonable to suppose that both even-ordered and odd-ordered partial sums converge to the same limit S, from which we would conclude that S is the sum of the alternating series. We now proceed to the proof.

First, consider the partial sums of even order:

$$S_2 = a_1 - a_2$$
$$S_4 = (a_1 - a_2) + (a_3 - a_4)$$
$$\vdots$$
$$S_{2n} = (a_1 - a_2) + (a_3 - a_4) + \cdots + (a_{2n-1} - a_{2n})$$

By condition (i), the pairs in parentheses are all nonnegative. So $\{S_{2n}\}$ is a sequence of nonnegative terms that is monotone increasing. Furthermore, this sequence is bounded above, since we can also write S_{2n} in the form

$$S_{2n} = a_1 - (a_2 - a_3) - (a_4 - a_5) - \cdots - (a_{2n-2} - a_{2n-1}) - a_{2n}$$

The pairs in parentheses are all positive or zero, and a_{2n} is positive. So $S_{2n} \leq a_1$. As we saw in the proof of Theorem 10.5, such a bounded, monotone increasing sequence must converge. Let the limit be S. That is,

$$\lim_{n \to \infty} S_{2n} = S$$

Now let us consider an odd-ordered partial sum. Except for S_1, we can write a typical odd-ordered partial sum as

$$S_{2n+1} = S_{2n} + a_{2n+1}$$

For example,

$$S_3 = (a_1 - a_2) + a_3 = S_2 + a_3$$

$$S_5 = (a_1 - a_2 + a_3 - a_4) + a_5 = S_4 + a_5$$

and so on. By condition (ii), $\lim_{n \to \infty} a_{2n+1} = 0$. So we have

$$\lim_{n \to \infty} S_{2n+1} = \lim_{n \to \infty} S_{2n} + \lim_{n \to \infty} a_{2n+1} = S + 0 = S$$

That is, both even- and odd-ordered partial sums approach the same limit S. Thus,

$$\lim_{n \to \infty} S_n = S$$

So the series $\sum_{n=1}^{\infty} (-1)^{n-1} a_n$ converges to S. ∎

REMARK ─────────────────────────────────

■ As with our other tests for convergence, it is sufficient that the conditions of Theorem 10.15 hold true from some point onward. That is, the series converges if the signs alternate and $a_{n+1} \leq a_n$ from some n onward (not necessarily starting with $n = 1$), provided $\lim_{n \to \infty} a_n = 0$.

───

Error in Using S_n to Approximate S for Alternating Series

When a series satisfies the hypotheses of the Alternating Series Test, there is an easy way of getting an upper bound on the error when using any given partial sum, S_n, to approximate the sum S. The following corollary shows how.

COROLLARY 10.15

Under the conditions of Theorem 10.15,

$$|S_n - S| \leq a_{n+1}$$

That is, the error in approximating the sum S by the nth partial sum S_n does not exceed the absolute value of the $(n + 1)$st term of the series.

Proof   In the proof of Theorem 10.15 we showed that the even-ordered partial sums were bounded above by a_1, the first term of the series. It follows that $\lim_{n \to \infty} S_{2n} = S$ must also not exceed a_1. Figure 10.29 clearly shows the fact that $S \leq a_1$.

Thus, for any alternating series whose terms in absolute value decrease monotonically and approach 0 as $n \to \infty$, the sum does not exceed the first term of the series (which we assumed to be positive). But

$$|S_n - S| = a_{n+1} - a_{n+2} + a_{n+3} - a_{n+4} + \cdots$$

is itself an alternating series whose terms in absolute value decrease monotonically and have 0 as a limit. By what we have just shown, the sum of the alternating series on the right does not exceed its first term, a_{n+1}. That is,

$$|S_n - S| \leq a_{n+1} \qquad \blacksquare$$

EXAMPLE 10.30   The Alternating Harmonic Series   Show that the alternating harmonic series

$$\sum_{n=1}^{\infty} \frac{(-1)^{n-1}}{n} = 1 - \frac{1}{2} + \frac{1}{3} - \frac{1}{4} + \cdots$$

converges. Determine an upper bound on the error in estimating the sum using S_{100}.

Solution   The conditions of Theorem 10.15 are clearly satisfied. That is, the series is an alternating series in which (i) $1/(n+1) < 1/n$, and (ii) $\lim_{n \to \infty} 1/n = 0$. The series therefore converges. Call its sum S. Then, by Corollary 10.15,

$$|S_{100} - S| \leq \frac{1}{101} \approx 0.009$$

So if we take the first 100 terms of the series to estimate the sum, we can be assured of accuracy to only one decimal place. This series converges very slowly. $\blacksquare$

EXAMPLE 10.31   Determine the convergence or divergence of each of the following series: (a) $\displaystyle\sum_{n=1}^{\infty}(-1)^{n-1} \ln\left(\frac{n+1}{n}\right)$ (b) $\displaystyle\sum_{n=1}^{\infty}(-1)^{n-1}\frac{n}{2n+1}$

Solution

(a) To apply Theorem 10.15, we must show both that $a_{n+1} \leq a_n$ and that $a_n \to 0$ as $n \to \infty$. To show that the terms $a_n = \ln((n+1)/n)$ decrease, we will use Method 2 of the tests for monotonicity given in Section 10.2. Replacing n with x in the formula for a_n and differentiating, we have

$$\frac{d}{dx}\ln\left(\frac{x+1}{x}\right) = \frac{d}{dx}[\ln(x+1) - \ln x]$$

$$= \frac{1}{x+1} - \frac{1}{x} = \frac{x-(x+1)}{x(x+1)} = \frac{-1}{x(x+1)}$$

Since the derivative is negative for all positive values of x, the function $\ln((x+1)/x)$ decreases. It therefore follows that for $n \geq 1$

$$\ln\left(\frac{n+2}{n+1}\right) \leq \ln\left(\frac{n+1}{n}\right)$$

That is, $a_{n+1} \leq a_n$ for all $n \geq 1$.

To see if $a_n \to 0$ as $n \to \infty$, we calculate the limit

$$\lim_{n\to\infty} \ln\left(\frac{n+1}{n}\right) = \ln\left[\lim_{n\to\infty}\left(\frac{n+1}{n}\right)\right] = \ln 1 = 0$$

(Note that we used the continuity of $\ln x$ here.)

Both conditions of the Alternating Series Test are met, so the series converges.

(b) Since

$$\lim_{n\to\infty} \frac{n}{2n+1} = \frac{1}{2} \neq 0$$

condition (ii) of the Alternating Series Test is not satisfied. We know, in fact, by the nth-Term Test for Divergence given in Section 10.3 that the series diverges. ∎

Absolute Convergence

Some series with variable signs have the property that when the signs are changed to be all positive, the resulting series converges. In this case we say that the original series is *absolutely convergent*. We can state the definition more succinctly as follows.

Definition 10.7 Absolute Convergence	The series $\sum a_n$ is said to be **absolutely convergent** if $\sum |a_n|$ converges.

REMARKS ————————————————————————————

■ Here we are allowing a_n to be either positive, negative, or zero, and we are not explicitly showing the signs, as we did with alternating series. No restriction is placed on the signs of the terms. They need not alternate but may do so.

■ If $a_n \geq 0$ for all n, then absolute convergence is equivalent to ordinary convergence. The concept of absolute convergence is useful only when the original series has variable signs (which may or may not alternate).

EXAMPLE 10.32   Test each of the following for absolute convergence.

(a) $\displaystyle\sum_{n=1}^{\infty} \frac{(-1)^{n-1}}{n}$ (b) $\displaystyle\sum_{n=1}^{\infty} \frac{(-1)^{n-1}}{n^2}$

Solution

(a) This series is the alternating harmonic series, which we have shown to be convergent, using the Alternating Series Test. However, when we take absolute values, we get the harmonic series $\sum 1/n$ itself, which we know to be divergent. So the alternating harmonic series is not absolutely convergent.

(b) Taking absolute values, we get the p-series $\sum 1/n^2$, with $p = 2$. Since $p > 1$, the series converges. Thus, the original series converges absolutely.

∎

Conditional Convergence

The alternating harmonic series in part (a) of Example 10.32 converges, but it is not absolutely convergent. This type of convergence is called *conditional* convergence.

Definition 10.8 Conditional Convergence	If $\sum a_n$ converges but $\sum |a_n|$ diverges, then $\sum a_n$ is said to be **conditionally convergent**.

Thus, of the two series in Example 10.32, the first is conditionally convergent and the second is absolutely convergent.

We now know that a series may be convergent without being absolutely convergent (in which case it is conditionally convergent). However, if a series is absolutely convergent, then it must also be convergent, as the next theorem shows.

THEOREM 10.16

If $\sum a_n$ is absolutely convergent, then it is convergent.

Proof   Since $|a_n|$ is either a_n or $-a_n$, it follows that

$$0 \le a_n + |a_n| \le 2|a_n|$$

(For example, if $a_n = -2$, then $|a_n| = 2$, and $0 = -2 + 2 < 2(2)$. Similarly, if $a_n = 4$, then $|a_n| = 4$, and $0 < 4 + 4 = 2(4)$.) Thus, since $\sum |a_n|$ (and hence also $\sum 2|a_n|$) converges, it follows by the Comparison Test that $\sum (a_n + |a_n|)$ converges. Finally, $a_n = (a_n + |a_n|) - |a_n|$, so by Theorem 10.8, part 3,

$$\sum a_n = \sum (a_n + |a_n|) - \sum |a_n|$$

Since each series on the right converges, we conclude that $\sum a_n$ converges.

∎

The next theorem, which we state without proof, shows that absolute convergence is a much stronger condition than conditional convergence. It deals with *rearrangements* of terms—that is, altering the order in which the terms occur. For example,

$$1 + \frac{1}{3} - \frac{1}{2} + \frac{1}{5} - \frac{1}{4} + \frac{1}{7} - \frac{1}{6} + \cdots$$

is a rearrangement of the alternating harmonic series.

THEOREM 10.17

> If $\sum a_n$ is absolutely convergent, then every rearrangement of it converges to the same sum. If $\sum a_n$ is conditionally convergent, then if S is any real number, there exists a rearrangement of the series that converges to S. Also, there are rearrangements that diverge to ∞ or to $-\infty$.

(See Exercises 39–43 in Exercise Set 10.6 for the ideas involved in the proof.)

REMARK ─────────────────────────────────

■ According to this theorem, no amount of rearranging of terms disturbs the convergence or the value of the sum of an absolutely convergent series. On the other hand, a conditionally convergent series is delicately balanced. Any change in the order may cause it to diverge or to converge to another sum. In fact, by a suitable rearrangement, the series can be made to converge to any sum we want. (See Exercise 43 in Exercise Set 10.6.) The sum is therefore *conditional* on the particular way the terms are written.

A Strategy for Testing Series with Variable Signs

When you are confronted with a series of variable signs, a good way to proceed is to test first to see that $a_n \to 0$ (so that the series has a chance to converge). If $a_n \to 0$, then test for absolute convergence, using one of the tests for positive-term series. If the given series is absolutely convergent, then we know by Theorem 10.16 that it is convergent. If it is not absolutely convergent and the signs alternate, then the Alternating Series Test can be applied. If the signs do not alternate, there are more delicate tests that can be used, but we will not study them here.

Exercise Set 10.6

In Exercises 1–10, use the Alternating Series Test to determine convergence or divergence.

1. $\displaystyle\sum_{n=1}^{\infty} \frac{(-1)^{n-1}}{\sqrt{n}}$

2. $\displaystyle\sum_{n=1}^{\infty} \frac{(-1)^{n-1}}{\ln(n+1)}$

3. $\displaystyle\sum_{n=1}^{\infty} \frac{(-1)^{n-1} \ln n}{n}$

4. $\displaystyle\sum_{n=1}^{\infty} \frac{(-1)^n n}{\sqrt{1+n^2}}$

5. $\displaystyle\sum_{n=2}^{\infty} \frac{(-1)^n}{n \ln n}$

6. $\displaystyle\sum_{n=1}^{\infty} (-1)^n \ln\left(1 + \frac{1}{n}\right)$

7. $\displaystyle\sum_{n=1}^{\infty} \frac{(-1)^n n}{100n + 1}$

8. $\displaystyle\sum_{n=1}^{\infty} \frac{\cos n\pi}{n}$

9. $\displaystyle\sum_{n=1}^{\infty} \frac{(-1)^n (n+1)}{n^2 - 2}$

10. $\displaystyle\sum_{n=1}^{\infty} (-1)^n \ln \frac{2n+3}{2n+1}$

In Exercises 11–14, show that the given series converges, and determine an upper bound on the error in using S_n to approximate the sum, for the specified value of n.

11. $\displaystyle\sum_{k=1}^{\infty} \frac{(-1)^{k-1}}{k^3}$; $n = 9$

12. $\displaystyle\sum_{k=1}^{\infty} \frac{(-1)^{k-1}}{k!}$; $n = 7$

13. $\displaystyle\sum_{k=1}^{\infty} \frac{(-1)^{k-1}}{k^k}$; $n = 5$

14. $\displaystyle\sum_{k=1}^{\infty} \frac{(-1)^k \sqrt{k}}{k + 1}$; $n = 9999$

In Exercises 15–18, show that the series converges, and deter-mine the smallest value of n for which the error in using S_n to approximate the sum does not exceed 0.005.

15. $\displaystyle\sum_{k=1}^{\infty} \frac{(-1)^k}{\sqrt{k+1}}$ **16.** $\displaystyle\sum_{k=2}^{\infty} \frac{(-1)^k}{k(\ln k)^2}$

17. $\displaystyle\sum_{k=1}^{\infty} \frac{(-1)^k(k+1)}{2k^2-3}$ **18.** $\displaystyle\sum_{k=1}^{\infty} \frac{(-1)^{k-1}(\ln k)^3}{k^2}$

19. Find an estimate, with an error of at most 0.0005, of the sum of the series

$$\sum_{n=1}^{\infty} \frac{(-1)^{n-1}(0.7)^n}{n}$$

In Exercises 20–31, test for absolute convergence, conditional convergence, or divergence.

20. $\displaystyle\sum_{n=1}^{\infty} \frac{(-1)^{n-1}}{\sqrt{n^3+1}}$ **21.** $\displaystyle\sum_{n=1}^{\infty} \frac{(-1)^{n-1}\sqrt{n+1}}{n}$

22. $\displaystyle\sum_{n=2}^{\infty} \frac{(-1)^n}{n\ln\sqrt{n}}$ **23.** $\displaystyle\sum_{n=1}^{\infty} \frac{(-1)^{n-1}n}{2^n}$

24. $\displaystyle\sum_{n=2}^{\infty} \frac{(-1)^n}{\ln n}$ **25.** $\displaystyle\sum_{n=0}^{\infty} (-1)^n \frac{e^n}{e^{2n}+1}$

26. $\displaystyle\sum_{n=1}^{\infty} \frac{(-1)^{n-1}n}{\sqrt{2n^2+3}}$ **27.** $\displaystyle\sum_{n=1}^{\infty} \frac{(-1)^{n-1}}{n(n+2)}$

28. $\displaystyle\sum_{n=1}^{\infty} \frac{\cos n}{n^2}$ **29.** $\displaystyle\sum_{n=1}^{\infty} \frac{n\sin n}{1+n^3}$

30. $\displaystyle\sum_{n=0}^{\infty} ne^{-n}\cos n$ **31.** $\displaystyle\sum_{n=0}^{\infty} \frac{(1-\pi)^n}{2^{n+1}}$

32. Determine the fallacy in the following "proof" that $2=1$. Let S denote the sum of the alternating harmonic series:

$$S = 1 - \frac{1}{2} + \frac{1}{3} - \frac{1}{4} + \frac{1}{5} - \frac{1}{6} + \frac{1}{7} - \frac{1}{8}\cdots$$

Then, on multiplying both sides by 2, we get

$$2S = 2 - 1 + \frac{2}{3} - \frac{1}{2} + \frac{2}{5} - \frac{1}{3} + \frac{2}{7} - \frac{1}{4} + \cdots$$

$$= 1 - \frac{1}{2} + \left(\frac{2}{3} - \frac{1}{3}\right) - \frac{1}{4} + \left(\frac{2}{5} - \frac{1}{5}\right) - \frac{1}{6} + \cdots$$

$$= 1 - \frac{1}{2} + \frac{1}{3} - \frac{1}{4} + \frac{1}{5} - \frac{1}{6} + \cdots = S$$

So $2S = S$, and since $S \neq 0$, $2 = 1$.

33. Let

$$a_n = \frac{2^n n!}{3 \cdot 5 \cdot 7 \cdot \cdots \cdot (2n+1)}$$

Prove that $\sum_{n=0}^{\infty}(-1)^n a_n$ is conditionally convergent by carrying out the following steps: Show that a_n can be written in the form

$$a_n = \frac{2 \cdot 4 \cdot 6 \cdot \cdots \cdot (2n)}{3 \cdot 5 \cdot 7 \cdot \cdots \cdot (2n+1)}$$

To show the series is not absolutely convergent, write

$$a_n = \frac{2}{1} \cdot \frac{4}{3} \cdot \frac{6}{5} \cdots \frac{2n}{2n-1} \cdot \frac{1}{2n+1}$$

Explain how it follows that

$$a_n > \frac{1}{2n+1}$$

To show $\{a_n\}$ is monotone decreasing, consider a_{n+1}/a_n. To show that $a_n \to 0$, verify the following:

$$a_n < \frac{3 \cdot 5 \cdot 7 \cdot \cdots \cdot (2n+1)}{4 \cdot 6 \cdot 8 \cdot \cdots \cdot (2n+2)} = \frac{1}{a_n(n+1)}$$

so

$$a_n^2 < \frac{1}{n+1}, \qquad \text{or equivalently,} \qquad a_n < \frac{1}{\sqrt{n+1}}$$

Now show that $a_n \to 0$ as $n \to \infty$.

34. Prove that

$$\sum_{n=1}^{\infty}(-1)^{n-1}\frac{1 \cdot 3 \cdot 5 \cdot \cdots \cdot (2n-1)}{2 \cdot 4 \cdot 6 \cdot \cdots \cdot (2n)}$$

is conditionally convergent. (*Hint:* Use steps similar to those in Exercise 33.)

35. Under the conditions of Theorem 10.15 show that the partial sums S_{2n+1} of odd order form a monotone decreasing, bounded sequence.

36. It can be shown (see Exercise 43 in Exercise Set 10.3) that in any convergent series the associative law holds unrestrictedly; that is, we may group terms by introducing parentheses (but not rearranging the order of the terms). Use this fact to show that

$$\sum_{n=1}^{\infty} \frac{1}{(2n-1)2n} = \frac{1}{1 \cdot 2} + \frac{1}{3 \cdot 4} + \frac{1}{5 \cdot 6} + \cdots$$

converges to the same sum as the alternating harmonic series. (*Hint:* In the alternating harmonic series, group terms by pairs.)

37. Let $\sum a_n$ be conditionally convergent and write

$$p_n = \begin{cases} a_n & \text{if } a_n \geq 0 \\ 0 & \text{if } a_n < 0 \end{cases} \quad \text{and} \quad q_n = \begin{cases} 0 & \text{if } a_n \geq 0 \\ -a_n & \text{if } a_n < 0 \end{cases}$$

(a) Show that $\sum a_n = \sum (p_n - q_n)$ and $\sum |a_n| = \sum (p_n + q_n)$.

(b) Show that $\sum p_n$ and $\sum q_n$ both diverge.

38. Show that in an absolutely convergent series the series $\sum p_n$ and $\sum q_n$ both converge, where p_n and q_n are defined as in Exercise 37. (*Hint:* Use part (a) of Exercise 37.)

39. Let p_n and q_n be defined as in Exercise 37 for the alternating harmonic series. Use a CAS or a spreadsheet to do the following. Add terms of $\sum p_n$ until the sum first exceeds 2. Then add one or more terms of $\sum(-q_n)$ until the sum is less than 2. Then continue with terms of $\sum p_n$ until the sum again exceeds 2. Continue in

this way until you have added a total of 50 nonzero terms. If this process were continued, what could you conclude about the convergence of this rearrangement of the alternating harmonic series? Justify your conclusion.

40. Use the idea of Exercise 39 with the roles of $\sum p_n$ and $\sum(-q_n)$ reversed in order to obtain a rearrangement of the alternating harmonic series with partial sums that can be made arbitrarily close to -1.

41. Rearrange the alternating harmonic series so that the resulting series diverges.

42. Rearrange the alternating harmonic series so that the resulting series converges to 4.

43. Use the result of Exercise 37 to explain how a suitable rearrangement of any conditionally convergent series can be made to converge to any sum S we choose.

10.7 POWER SERIES

In Section 10.1 we indicated that we would be considering "infinite polynomials" that result from letting $n \to \infty$ in the Taylor polynomial representation of a function. Such infinite series differ from those we have been studying in the previous two sections in that the terms involve powers of a variable x, or more generally of $(x - a)$. In fact, our motivation for studying infinite series of constants was that such series arise when we substitute a particular value of x into a series involving variables.

In the next section we will return to the question of representing a function by its extended Taylor polynomial. In this section we consider properties of series of powers of x, or of $(x - a)$. Our emphasis will be on the series itself, rather than on what function may have given rise to the series.

Definition 10.9
Power Series

A series of the form

$$\sum_{n=0}^{\infty} a_n x^n = a_0 + a_1 x + a_2 x^2 + \cdots \qquad (10.20)$$

is called a **power series in x**, and a series of the form

$$\sum_{n=0}^{\infty} a_n (x - a)^n = a_0 + a_1 (x - a) + a_2 (x - a)^2 + \cdots \qquad (10.21)$$

is called a **power series in $(x - a)$**.

REMARK ───

■ If we let $x = 0$ on the left-hand side of Equation 10.20, the first term of the summation is $a_0(0)^0$, which is not defined. We will adopt the convention, however, that the first term in this case is a_0, in agreement with the right-hand side. Similar remarks apply to the summation in Equation 10.21 when $x = a$ and $n = 0$.

───

The power series 10.20 is clearly the special case of 10.21, in which $a = 0$. Although we could develop our theory for the general case of power series in $(x - a)$, we will concentrate instead primarily on the simpler case of power series in x. All our results can be extended in obvious ways to power series in $(x - a)$.

An appropriate question to ask for a power series is not "Does the series converge?" but rather "For what values of x does the series converge?" This change arises because convergence is defined in terms of series of constants, and a power series becomes a constant series only when x is given a value. All power series in x converge in a trivial way for $x = 0$, as can be seen by putting $x = 0$ in Equation 10.20. Similarly, the series in Equation 10.21 converges for $x = a$. Some series converge for all real values of x, and others converge on an interval and diverge outside that interval. We will see examples of each of these types shortly. The nature of the set of x values where convergence occurs can be determined from the following theorem.

THEOREM 10.18

> If the power series $\sum a_n x^n$ converges at $x_0 \neq 0$, then it converges absolutely for all x such that $|x| < |x_0|$. If the series diverges at $x_1 \neq 0$, then it diverges for all x such that $|x| > |x_1|$.

Simply stated, if a power series in x converges for any nonzero value of x, it converges absolutely for all x that are smaller in absolute value. If it diverges for a given value of x, it diverges for all x that are greater in absolute value.

REMARK ───

■ For simplicity, we will sometimes drop the range of the index n (from 0 to ∞) in $\sum_{n=0}^{\infty} a_n x^n$ and write simply $\sum a_n x^n$.

Proof   Since $\sum a_n x_0^n$ converges, we know by Theorem 10.7 that $\lim_{n \to \infty} a_n x_0^n = 0$. So for n sufficiently large, say $n > N$, $|a_n x_0^n| < 1$. Now let x be any number for which $|x| < |x_0|$, and denote the ratio $|x/x_0|$ by r, so that $0 \leq r < 1$. Then we have for $n > N$

$$|a_n x^n| = |a_n x_0^n| \cdot \left| \frac{x}{x_0} \right|^n = |a_n x_0^n| \cdot r^n < r^n$$

Since $\sum r^n$ is a convergent geometric series, it follows by the Comparison Test that $\sum |a_n x^n|$ converges; that is, $\sum a_n x^n$ converges absolutely.

For the second part, if $\sum a_n x_1^n$ diverges and $|x| > |x_1|$, the series could not converge at x. If it did, since $|x_1| < |x|$, it would converge absolutely at x_1

(and thus would converge), by what we have just proved. But this conclusion contradicts the given fact that the series is assumed to diverge at x_1. So $\sum a_n x^n$ diverges when $|x| > |x_1|$. ∎

Radius of Convergence and Interval of Convergence

As a consequence of Theorem 10.18 the following result can be proved. (See Exercise 40 in Exercise Set 10.7.)

THEOREM 10.19

For the power series $\sum a_n x^n$, one and only one of the following cases holds true:

1. The series converges only for $x = 0$.
2. The series converges absolutely for all x.
3. There is a positive number R such that the series converges absolutely for $|x| < R$ and diverges for $|x| > R$.

The number R in Case 3 is called the **radius of convergence** of the series. For convenience, if Case 1 holds, we agree to call the radius of convergence 0, and if Case 2 holds, we say the radius of convergence is ∞. For Case 3, in which $0 < R < \infty$, nothing is said about convergence at $x = R$ or at $x = -R$. The series may or may not converge at these points, and if it does converge, the convergence may be absolute or conditional. These endpoints of the interval $(-R, R)$ must be individually tested using means studied in the previous three sections. Depending on whether the series converges or diverges at these endpoint values, the series will converge in an interval of one of the types $(-R, R)$, $[-R, R]$, $[-R, R)$, or $(-R, R]$, and it will diverge elsewhere. Do you see that if only one endpoint is included, the convergence is necessarily conditional there? The appropriate interval for a given series is called its **interval of convergence**. When $R = 0$, the interval of convergence degenerates to the single point $x = 0$, and if $R = \infty$, it is the entire real line $(-\infty, \infty)$.

Theorem 10.19 as well as the notions of radius of convergence and interval of convergence extend, with obvious modifications, to power series of the form $\sum a_n (x - a)^n$. If the radius of convergence is R, for example, where $0 < R < \infty$, the interval of convergence is of the form $(a - R, a + R)$, or this interval together with one or both of its endpoints.

Using the Ratio Test to Find the Radius of Convergence

When $\lim_{n \to \infty} |a_{n+1}/a_n|$ exists, the radius of convergence can be found using the Ratio Test. We illustrate this procedure in the next three examples.

EXAMPLE 10.33   Find the interval of convergence of the series

$$\sum_{n=0}^{\infty} \frac{x^n}{2n + 1}$$

Solution   We use the Ratio Test on the series of absolute values. (Remember that the Ratio Test is applicable only to series of positive terms, and since x can be either positive or negative, we must consider absolute values.)

$$\lim_{n \to \infty} \left| \frac{x^{n+1}}{2(n+1)+1} \cdot \frac{2n+1}{x^n} \right| = \lim_{n \to \infty} \frac{2n+1}{2n+3} |x| = |x|$$

(Note that for this limit x is held fixed and $n \to \infty$.) The series therefore converges absolutely when $|x| < 1$ and diverges when $|x| > 1$. The radius of convergence is then $R = 1$. Now we must test the endpoint values $x = \pm 1$. We test them by substituting in the original series:

$$x = -1: \qquad \sum_{n=0}^{\infty} \frac{(-1)^n}{2n+1} = 1 - \frac{1}{3} + \frac{1}{5} - \frac{1}{7} + \cdots$$

This series is an alternating series, and $1/(2n+1)$ decreases monotonically to 0. Thus, the series converges.

$$x = 1: \qquad \sum_{n=0}^{\infty} \frac{1}{2n+1} = 1 + \frac{1}{3} + \frac{1}{5} + \frac{1}{7} + \cdots$$

The term $1/(2n+1)$ appears to be of the same order of magnitude as $1/n$, so we can try the Limit Comparison Test:

$$\lim_{n \to \infty} \frac{1}{2n+1} \div \frac{1}{n} = \lim_{n \to \infty} \frac{n}{2n+1} = \frac{1}{2}$$

This limit confirms that our series and the harmonic series $\sum 1/n$ behave in the same way. Thus, at $x = 1$ our series diverges.

The complete interval of convergence of the original series is therefore $-1 \leq x < 1$. Note that the convergence is conditional at $x = -1$.   ■

EXAMPLE 10.34   Find the interval of convergence of the series

$$\sum_{n=0}^{\infty} \frac{(-1)^n x^{2n}}{(2n)!}$$

Solution   We will again apply the Ratio Test to the series of absolute values. Since $|(-1)^n| = 1$ for any n, we can ignore the factor $(-1)^n$. We then have

$$\lim_{n \to \infty} \left| \frac{x^{2(n+1)}}{[2(n+1)]!} \div \frac{x^{2n}}{(2n)!} \right| = \lim_{n \to \infty} \left| \frac{x^{2n+2}}{(2n+2)!} \cdot \frac{(2n)!}{x^{2n}} \right|$$

$$= \lim_{n \to \infty} \frac{x^2}{(2n+2)(2n+1)} = 0$$

for all fixed values of x. Since $0 < 1$, we conclude that the series converges absolutely for all real x; that is, $R = \infty$ and the interval of convergence is $(-\infty, \infty)$.   ■

EXAMPLE 10.35   Find the interval of convergence of the series $\sum_{n=0}^{\infty} n!\, x^n$.

Solution   To apply the Ratio Test we consider the following limit for $x \neq 0$,

$$\lim_{n \to \infty} \left| \frac{(n+1)!\, x^{n+1}}{n!\, x^n} \right| = \lim_{n \to \infty} (n+1)|x|$$

Thus, if $x \neq 0$, the limit is ∞, and so by the Ratio Test the series diverges. If $x = 0$, only the first term of the series is different from 0, so the series is finite, hence convergent. Thus, $R = 0$ and the interval of convergence degenerates to the single point $x = 0$.   ∎

We will now see how the same technique can be applied to power series in $(x - a)$ to find the interval of convergence.

EXAMPLE 10.36   Find the interval of convergence of the series

$$\sum_{n=0}^{\infty} \frac{(-1)^n (x - 2)^n}{(n + 1)^2 \cdot 3^n}$$

Solution   Consider the limit

$$\lim_{n \to \infty} \left| \frac{(x - 2)^{n+1}}{(n + 2)^2 3^{n+1}} \cdot \frac{(n + 1)^2 3^n}{(x - 2)^n} \right| = \lim_{n \to \infty} \frac{1}{3} \left(\frac{n + 1}{n + 2} \right)^2 |x - 2| = \frac{|x - 2|}{3}$$

Thus, by the Ratio Test the series converges absolutely if $|x - 2|/3 < 1$ or, equivalently, if $|x - 2| < 3$, and it diverges if $|x - 2| > 3$. Now we test the values $(x - 2) = \pm 3$, which correspond to the endpoint values $x = 5$ and $x = -1$.

$$\underline{(x - 2) = 3}: \qquad \sum_{n=0}^{\infty} \frac{(-1)^n 3^n}{(n + 1)^2 \cdot 3^n} = \sum_{n=0}^{\infty} \frac{(-1)^n}{(n + 1)^2}$$

and this series converges absolutely, since the series of absolute values is of the same order of magnitude as a p-series with $p = 2$.

$$\underline{(x - 2) = -3}: \qquad \sum_{n=0}^{\infty} \frac{(-1)^n (-3)^n}{(n + 1)^2 \cdot 3^n} = \sum_{n=0}^{\infty} \frac{(-1)^n \cdot (-1)^n \cdot 3^n}{(n + 1)^2 \cdot 3^n}$$

$$= \sum_{n=0}^{\infty} \frac{1}{(n + 1)^2}$$

and again we have a convergent p-series. So the complete interval of convergence is defined by $|x - 2| \le 3$, and the convergence is absolute for all values of x satisfying this inequality. We can determine the interval of convergence as follows:

$$|x - 2| \le 3$$

$$-3 \le x - 2 \le 3$$

$$-1 \le x \le 5$$

Thus, the interval of convergence is $[-1, 5]$. Notice that the center of the interval is $x = 2$ and the radius is $R = 3$.   ∎

Exercise Set 10.7

In Exercises 1–37, find the interval of convergence.

1. $\displaystyle\sum_{n=0}^{\infty} x^n$

2. $\displaystyle\sum_{n=1}^{\infty} \frac{x^n}{n}$

3. $\displaystyle\sum_{n=0}^{\infty} \frac{x^n}{n!}$

4. $\displaystyle\sum_{n=0}^{\infty} \frac{(-1)^n n x^n}{n + 1}$

5. $\displaystyle\sum_{n=0}^{\infty} \frac{(-1)^n x^n}{1 + n^2}$

6. $\displaystyle\sum_{n=0}^{\infty} \frac{x^{2n+1}}{(2n + 1)!}$

7. $\displaystyle\sum_{n=0}^{\infty} \frac{(x-1)^n}{\sqrt{n^2+1}}$

8. $\displaystyle\sum_{n=0}^{\infty} \frac{(x+1)^n}{2^n}$

9. $\displaystyle\sum_{n=0}^{\infty} \frac{(x+2)^n}{3^n(n+1)}$

10. $\displaystyle\sum_{n=0}^{\infty} \frac{(-1)^n 2^{n+1} x^n}{3^n}$

11. $\displaystyle\sum_{n=2}^{\infty} \frac{(-1)^n x^{n-2}}{n \ln n}$

12. $\displaystyle\sum_{n=2}^{\infty} \frac{\ln n}{n} x^{n-2}$

13. $\displaystyle\sum_{n=1}^{\infty} \frac{2^n x^{n-1}}{n^2}$

14. $\displaystyle\sum_{n=0}^{\infty} (-1)^n \sqrt{\frac{n+1}{n+2}}\, x^n$

15. $\displaystyle\sum_{n=0}^{\infty} \frac{n+1}{n^2+1} x^n$

16. $\displaystyle\sum_{n=0}^{\infty} \frac{n!\,(x-1)^n}{2^n}$

17. $\displaystyle\sum_{n=0}^{\infty} \frac{n!}{(2n)!} x^n$

18. $\displaystyle\sum_{n=0}^{\infty} ne^{-n} x^n$

19. $\displaystyle\sum_{n=1}^{\infty} \frac{(-1)^{n-1} 2^n (x-3)^{n-1}}{n^2}$

20. $\displaystyle\sum_{n=0}^{\infty} \frac{(-1)^n \sqrt{n}}{n+1} (x+2)^n$

21. $\displaystyle\sum_{n=2}^{\infty} \left(\frac{3}{4}\right)^{n-2} (x+1)^n$

22. $\displaystyle\sum_{n=2}^{\infty} \frac{(-2x)^{n-2}}{\ln n}$

23. $\displaystyle\sum_{n=0}^{\infty} \frac{n(-3x)^n}{2n+1}$

24. $\displaystyle\sum_{n=0}^{\infty} \frac{(-1)^n n^4 x^n}{e^n}$

25. $\displaystyle\sum_{n=0}^{\infty} \frac{\sqrt{n}(x-5)^n}{1+n^2}$

26. $\displaystyle\sum_{n=0}^{\infty} \frac{n^2-1}{n^2+1} x^{2n}$

27. $\displaystyle\sum_{n=0}^{\infty} \frac{(-1)^n x^{2n+1}}{(2n+1)!}$

28. $\displaystyle\sum_{n=1}^{\infty} \frac{(2x-1)^n}{n\sqrt{n+1}}$

29. $\displaystyle\sum_{n=0}^{\infty} \frac{(3x-2)^n}{(n+1)^{2/3}}$

30. $\displaystyle\sum_{n=0}^{\infty} \frac{(1-x)^n}{2^{n+1}}$

31. $\displaystyle\sum_{n=0}^{\infty} \frac{n!\,(x+4)^{2n}}{3^n}$

32. $\displaystyle\sum_{n=0}^{\infty} \frac{n!\,x^n}{1\cdot 3\cdot 5\cdot\,\cdots\,\cdot(2n+1)}$ (See Exercise 33 in Exercise Set 10.6.)

33. $\displaystyle\sum_{n=1}^{\infty} \frac{1\cdot 3\cdot 5\cdot\,\cdots\,\cdot(2n-1)}{2\cdot 4\cdot 6\cdot\,\cdots\,\cdot(2n)} x^{n-1}$ (See Exercise 34 in Exercise Set 10.6.)

34. $\displaystyle\sum_{n=1}^{\infty} \frac{n^n x^n}{2^{n+1}}$

35. $\displaystyle\sum_{n=0}^{\infty} \frac{(n!)^2}{(2n)!} x^{2n}$

36. $\displaystyle\sum_{n=0}^{\infty} \frac{2^n+n}{3^n+2} x^n$

37. $\displaystyle\sum_{n=1}^{\infty} \frac{x^n}{(\sqrt{n})^n}$

38. If $\sum a_n x^n$ has radius of convergence R, where $0 < R < \infty$, prove that the radius of convergence of $\sum a_n x^{2n}$ is $\sqrt{R}$.

39. Find the radius of convergence of the series $\sum_{n=0}^{\infty} \binom{\alpha}{n} x^n$, where α is a fixed real number and
$$\binom{\alpha}{n} = \frac{\alpha(\alpha-1)(\alpha-2)\cdots(\alpha-n+1)}{n!}$$

40. Prove Case 3 of Theorem 10.19 as follows.
   (a) Let R denote the least upper bound of the set of x values for which $\sum a_n x^n$ is absolutely convergent. If $x_0 \in (-R, R)$, choose x_1 such that $|x_0| < |x_1| \le R$, and explain why $\sum a_n x_1^n$ is absolutely convergent. Based on the definition of R, why can such an x_1 be chosen? (*Hint:* Otherwise $|x_0|$ would be an upper bound, smaller than R, of the set of points where $\sum a_n x^n$ is absolutely convergent.)
   (b) Now use Theorem 10.18. How do you conclude that $\sum a_n x^n$ converges absolutely for *all* x in $(-R, R)$?
   (c) To prove divergence for $|x| > R$, assume to the contrary that $\sum a_n x^n$ converges for some x such that $|x| > R$, and arrive at a contradiction.

10.8 DIFFERENTIATION AND INTEGRATION OF POWER SERIES

Within its interval of convergence, a power series defines a function. So it is appropriate to write, for example,

$$f(x) = \sum_{n=0}^{\infty} a_n x^n$$

where x is in the interval I of convergence. For any x_0 in I, $f(x_0)$ is equal to the sum of the convergent series of constants, $\sum a_n x_0^n$. It is appropriate to ask, then, what properties such a function has. For example, we may want to know about continuity, differentiability, or integrability. The following rather remarkable theorem answers these questions. We do not give its proof, which is quite technical. (A proof can be found in most textbooks on advanced calculus.) The theorem says that a power series can essentially be treated as a polynomial within its interval of convergence.

THEOREM 10.20

Let $\sum a_n x^n$ have nonzero radius of convergence R, and for $-R < x < R$ write

$$f(x) = \sum_{n=0}^{\infty} a_n x^n$$

Then

1. f is continuous on the interval $(-R, R)$.
2. f is differentiable on $(-R, R)$ and

$$f'(x) = \sum_{n=0}^{\infty} \frac{d}{dx}(a_n x^n) = \sum_{n=1}^{\infty} n a_n x^{n-1}$$

   The series on the right also has radius of convergence R.
3. f is integrable over any interval $[a, b]$ contained in $(-R, R)$, and

$$\int_a^b f(x)\,dx = \sum_{n=0}^{\infty} \int_a^b a_n x^n\,dx$$

   Furthermore, f has an antiderivative in $(-R, R)$ given by

$$\int f(x)\,dx = \sum_{n=0}^{\infty} \int a_n x^n\,dx = \sum_{n=0}^{\infty} \frac{a_n x^{n+1}}{n+1} + C$$

   The series on the right also has radius of convergence R.

REMARKS
■ Property 2 says that a power series may be differentiated term by term, and the resulting power series converges to the derivative of the function and has the same radius of convergence as the original series. Since the differentiated series is itself a power series with radius of convergence R, it too can be differentiated term by term to give $f''(x)$. Continuing in this way gives the very powerful result that within its interval of convergence (excluding endpoints) *a power series has derivatives of all orders*.

■ We can paraphrase Properties 2 and 3 by saying that a power series can be differentiated or integrated term by term within $(-R, R)$. If the integral is a definite integral over $[a, b]$, then we must have $-R < a < b < R$. If it is an indefinite integral (an antiderivative), the integration is valid for any x in $(-R, R)$.

■ None of the properties can be assumed to be true at the endpoints $x = \pm R$, even if the original series converges at one or both of these points. Actually, it can be proved that continuity *does* extend to an endpoint if the series converges there. So a power series is continuous on its entire interval of convergence.

■ The theorem extends in an obvious way to $f(x) = \sum a_n(x - a)^n$.

The following examples illustrate some of the consequences of Theorem 10.20. In the examples we will make repeated use of the geometric series

$$a + ar + ar^2 + \cdots = \frac{a}{1-r}, \qquad |r| < 1$$

By setting $r = x$ and $a = 1$ and reversing the sides of the equation, we obtain the following result.

$$\frac{1}{1-x} = 1 + x + x^2 + x^3 + \cdots = \sum_{n=0}^{\infty} x^n, \qquad |x| < 1 \qquad (10.22)$$

EXAMPLE 10.37   Find the sum of the series

$$\sum_{n=1}^{\infty} nx^n = x + 2x^2 + 3x^3 + 4x^4 + \cdots$$

and state the domain of validity. Use the result to find the sum of the series $\sum_{n=1}^{\infty} n(1/2)^n$.

Solution   On factoring out an x we get

$$x(1 + 2x + 3x^2 + 4x^3 + \cdots)$$

and we observe that the series in parentheses is the derivative of

$$x + x^2 + x^3 + x^4 + \cdots$$

Factoring out an x, we can write this series as $x(1 + x + x^2 + x^3 + \cdots)$. The series in parentheses is the geometric series in Equation 10.22. So if $|x| < 1$,

$$x + x^2 + x^3 + x^4 + \cdots = x(1 + x + x^2 + x^3 + \cdots) = x\left(\frac{1}{1-x}\right) = \frac{x}{1-x}$$

Thus, by Property 2 of Theorem 10.20, we have

$$\sum_{n=1}^{\infty} nx^n = x\frac{d}{dx}\left(x + x^2 + x^3 + x^4 + \cdots\right)$$

$$= x\left(\frac{d}{dx}\frac{x}{1-x}\right)$$

$$= \frac{x}{(1-x)^2}$$

and this result is valid for $|x| < 1$. In particular, for $x = 1/2$, we have

$$\sum_{n=1}^{\infty} n\left(\frac{1}{2}\right)^n = \frac{\frac{1}{2}}{\left(1 - \frac{1}{2}\right)^2} = 2 \qquad \blacksquare$$

EXAMPLE 10.38   Find a power series whose sum is $\ln(1 + x)$. Use the result to find a series that converges to $\ln(1/2)$.

Solution   We begin with the fact that

$$\ln(1 + x) = \int \frac{dx}{1 + x}$$

If in Equation 10.22 we replace x with $-x$, we obtain

$$\frac{1}{1 + x} = \sum_{n=0}^{\infty} (-1)^n x^n, \qquad |x| < 1$$

Thus, by Theorem 10.20,

$$\int \frac{dx}{1 + x} = \int \sum_{n=0}^{\infty} (-1)^n x^n \, dx$$

$$= \sum_{n=0}^{\infty} \int (-1)^n x^n \, dx$$

We therefore have

$$\ln(1 + x) = \sum_{n=0}^{\infty} \frac{(-1)^n x^{n+1}}{n + 1} + C, \qquad |x| < 1$$

To find C, we can substitute $x = 0$:

$$\ln 1 = 0 + C$$

Thus,

$$C = 0$$

So we have

$$\ln(1 + x) = \sum_{n=0}^{\infty} \frac{(-1)^n x^{n+1}}{n + 1} = x - \frac{x^2}{2} + \frac{x^3}{3} - \frac{x^4}{4} + \cdots$$

and this result is valid for $|x| < 1$. In particular, setting $x = -1/2$, we have

$$\ln\left(\frac{1}{2}\right) = -\frac{1}{2} - \frac{1}{2 \cdot 2^2} - \frac{1}{3 \cdot 2^3} - \frac{1}{4 \cdot 2^4} - \cdots$$

since $\ln(1/2) = -\ln 2$, we can also conclude that

$$\ln 2 = \frac{1}{2} + \frac{1}{2 \cdot 2^2} + \frac{1}{3 \cdot 2^3} + \frac{1}{4 \cdot 2^4} + \cdots$$

$$= \sum_{n=1}^{\infty} \frac{1}{n \cdot 2^n} \qquad \blacksquare$$

Note that when $x = 1$, the series for $\ln(1 + x)$ that we obtained in Example 10.38 is convergent (verify), and as stated in the third remark after Theorem 10.20, the continuity of the series thus extends to include the endpoint $x = 1$. We can therefore say that

$$\lim_{x \to 1} \ln(1 + x) = \lim_{x \to 1} \sum_{n=0}^{\infty} \frac{(-1)^n x^{n+1}}{n + 1}$$

and since the natural logarithm function also is continuous, we get the following result.

The Sum of the Alternating Harmonic Series

$$\ln 2 = 1 - \frac{1}{2} + \frac{1}{3} - \frac{1}{4} + \cdots$$

So we now know that the sum of the alternating harmonic series is $\ln 2$.

EXAMPLE 10.39   Find a power series representation of $\tan^{-1} x$ valid near $x = 0$.

Solution   We use the idea of the previous example. Since

$$\tan^{-1} x = \int \frac{1}{1 + x^2} dx$$

we can solve the problem provided we can determine a power series representation of $f(x) = 1/(1 + x^2)$. All we need to do is to replace x with $-x^2$ in Equation 10.22, giving

$$\frac{1}{1 + x^2} = 1 - x^2 + x^4 - x^6 + \cdots = \sum_{n=0}^{\infty} (-1)^n x^{2n}$$

where $x^2 < 1$, or equivalently, $|x| < 1$. So

$$\int \frac{1}{1 + x^2} dx = \int \sum_{n=0}^{\infty} (-1)^n x^{2n} dx$$

$$\tan^{-1} x = \sum_{n=0}^{\infty} \int (-1)^n x^{2n} dx \qquad \text{By Theorem 10.20}$$

$$= \sum_{n=0}^{\infty} \frac{(-1)^n x^{2n+1}}{2n + 1} + C$$

For $x = 0$ we get $\tan^{-1} 0 = 0 + C$. Thus, $C = 0$. So

$$\tan^{-1} x = \sum_{n=0}^{\infty} \frac{(-1)^n x^{2n+1}}{2n + 1}$$

$$= x - \frac{x^3}{3} + \frac{x^5}{5} - \frac{x^7}{7} + \cdots$$

valid so long as we have $|x| < 1$.

We can readily see that the series on the right converges at both endpoints. (Check this.) So, as we have indicated, the series represents a continuous function on the closed interval $[-1, 1]$. Since $\tan^{-1} x$ is also continuous at $x = \pm 1$, it follows as in the preceding example that

$$\tan^{-1} 1 = 1 - \frac{1}{3} + \frac{1}{5} - \frac{1}{7} + \cdots$$

and

$$\tan^{-1}(-1) = -1 + \frac{1}{3} - \frac{1}{5} + \frac{1}{7} - \cdots$$

Using the first of these equations we find that

$$\frac{\pi}{4} = 1 - \frac{1}{3} + \frac{1}{5} - \frac{1}{7} + \cdots$$

or

$$\pi = 4 \left[1 - \frac{1}{3} + \frac{1}{5} - \frac{1}{7} + \cdots \right]$$

We can view this result in two ways: first, it gives a means of calculating π to any degree of accuracy (although it is not efficient, since the series converges very slowly), and second, it is a formula for the sum of the series on the right.

∎

In the next example we give one way to verify that the nth Taylor polynomial for e^x about $x = 0$ can be extended to an infinite power series (the Taylor series) that converges to e^x.

EXAMPLE 10.40   Show that the series

$$\sum_{n=0}^{\infty} \frac{x^n}{n!} = 1 + x + \frac{x^2}{2!} + \frac{x^3}{3!} + \cdots + \frac{x^n}{n!} + \cdots$$

converges to e^x for all real x.

Solution   Let

$$f(x) = \sum_{n=0}^{\infty} \frac{x^n}{n!}$$

We first show that this series converges for all real x. Then we will show that $f'(x) = f(x)$, suggesting the possibility that $f(x) = e^x$. Applying the Ratio Test, we have

$$\lim_{n \to \infty} \left| \frac{x^{n+1}}{(n+1)!} \cdot \frac{n!}{x^n} \right| = \lim_{n \to \infty} \frac{|x|}{n+1} = 0$$

Since this limit is always less than 1, the given series converges for all values of x. Its derivative is, by Theorem 10.20,

$$f'(x) = \sum_{n=1}^{\infty} \frac{n x^{n-1}}{n!} = \sum_{n=1}^{\infty} \frac{x^{n-1}}{(n-1)!} \qquad \text{Since } n!/n = (n-1)!$$

$$= 1 + x + \frac{x^2}{2!} + \frac{x^3}{3!} + \cdots = f(x)$$

That is, $f'(x) = f(x)$ for all values of x. We know that e^x is a function with the property that its derivative equals itself. To show that our $f(x)$ and e^x are identical, consider the derivative

$$\frac{d}{dx} \left(\frac{f(x)}{e^x} \right) = \frac{f(x) \cdot e^x - e^x \cdot f'(x)}{e^{2x}}$$

Since $f'(x) = f(x)$, the numerator is 0 and it follows that

$$\frac{d}{dx} \left(\frac{f(x)}{e^x} \right) = 0$$

Thus,

$$\frac{f(x)}{e^x} = C$$

for some constant C. Setting $x = 0$, we get $C = 1$, since both $f(0)$ and e^0 equal 1. Thus, $f(x) = e^x$.

∎

The result of the preceding example deserves special emphasis.

A Power Series for e^x

$$e^x = \sum_{n=0}^{\infty} \frac{x^n}{n!} = 1 + x + \frac{x^2}{2!} + \frac{x^3}{3!} + \cdots + \frac{x^n}{n!} + \cdots \qquad (10.23)$$

valid for $-\infty < x < \infty$.

A consequence of the convergence of the series in Equation 10.23 is that the nth term goes to 0 as $n \to \infty$, and this result, too, is one we will need later.

A Special Limit

$$\lim_{n \to \infty} \frac{x^n}{n!} = 0 \qquad \text{for all real } x \qquad (10.24)$$

Exercise Set 10.8

In Exercises 1–6, differentiate the given series term by term, and determine in what interval the resulting series is the derivative of the function defined by the given series.

1. $1 + x + \dfrac{x^2}{2} + \dfrac{x^3}{3} + \cdots$

2. $x - \dfrac{x^3}{3} + \dfrac{x^5}{5} - \dfrac{x^7}{7} + \cdots$

3. $\displaystyle\sum_{n=0}^{\infty} \frac{(-1)^n x^{2n}}{(2n)!}$

4. $\displaystyle\sum_{n=0}^{\infty} \frac{(-1)^n x^{2n+1}}{(2n+1)!}$

5. $\displaystyle\sum_{n=0}^{\infty} \frac{x^n}{2^{n+1}}$

6. $\displaystyle\sum_{n=1}^{\infty} \frac{x^n}{n^2 + n}$

In Exercises 7–12, find the indicated antiderivatives and give the domain of validity.

7. $\displaystyle\int \sum_{n=0}^{\infty} \frac{x^n}{n+1} \, dx$

8. $\displaystyle\int \sum_{n=1}^{\infty} (-1)^{n-1} n x^n \, dx$

9. $\displaystyle\int \sum_{n=0}^{\infty} \frac{(-1)^n x^{2n}}{(2n)!} \, dx$

10. $\displaystyle\int \sum_{n=1}^{\infty} \frac{x^{2n}}{1 \cdot 3 \cdot 5 \cdot \, \cdots \, \cdot (2n-1)} \, dx$

11. $\displaystyle\int \sum_{n=0}^{\infty} \frac{(-1)^n (2n+1) x^{2n}}{2^n} \, dx$

12. $\displaystyle\int \sum_{n=1}^{\infty} \frac{(-1)^{n-1} x^{2n-1}}{(2n-1)!} \, dx$

In Exercises 13–18, show that term-by-term integration is valid over the given interval, and carry out the integration. Where possible, express the answer in closed form.

13. $\displaystyle\int_0^{1/2} \sum_{n=0}^{\infty} (n+1) x^n \, dx$

14. $\displaystyle\int_{-1}^{1} \sum_{n=0}^{\infty} \frac{x^n}{2^n} \, dx$

15. $\displaystyle\int_1^2 \sum_{n=0}^{\infty} \frac{x^n}{n!} \, dx$

16. $\displaystyle\int_{-1}^{2}\sum_{n=0}^{\infty}\frac{(-1)^{n}x^{n}}{3^{n+1}}\,dx$

17. $\displaystyle\int_{-1/2}^{1/2}\sum_{n=0}^{\infty}\frac{(-1)^{n}x^{2n}}{2n+1}\,dx$

18. $\displaystyle\int_{-1}^{3}\sum_{n=0}^{\infty}\frac{(-1)^{n}x^{2n}}{(2n)!}\,dx$

In Exercises 19–31, use differentiation or integration of an appropriate geometric series to find a series representation in powers of x of the given function, and give its radius of convergence.

19. $\dfrac{1}{(1+x)^{2}}$

20. $\dfrac{2}{(2-x)^{2}}$

21. $\ln(1-x)$

22. $\tanh^{-1}x$

23. $\dfrac{2x}{(1-x^{2})^{2}}$

24. $\dfrac{2x}{(1-2x)^{2}}$

25. $\ln\sqrt{1+x}$

26. $\ln\dfrac{1+x}{1-x}$

27. $\tan^{-1}\dfrac{x}{2}$

28. $6\left(\dfrac{x}{3-2x}\right)^{2}$

29. $\dfrac{1}{(1-x)^{3}}$

30. $\ln(3+2x)$

31. $\dfrac{x^{2}}{(1-x)^{2}}$

32. Show, by integrating the series in Equation 10.23, that $\int e^{x}\,dx = e^{x}+C$.

33. Let $f(x)=e^{2x}$. In Equation 10.23 replace x with $2x$ to find a series representation of f. Use this result to calculate $\int_{0}^{1}f(x)\,dx$. What conclusion can you draw concerning the sum of the following series?

$$1+\frac{2}{2!}+\frac{2^{2}}{3!}+\frac{2^{3}}{4!}+\cdots+\frac{2^{n}}{(n+1)!}+\cdots$$

34. Find the sum of the series $\sum_{n=1}^{\infty}(-1)^{n-1}nx^{n}$ for $|x|<1$. (*Hint:* Differentiate an appropriate geometric series and then multiply by x.)

35. Find the sum of the series $\sum_{n=0}^{\infty}n^{2}x^{n}$. (*Hint:* Write $n^{2}=n+n(n-1)$ and use the idea of Example 10.37.)

36. Find the sum of the series

$$\sum_{n=1}^{\infty}\frac{n}{2^{n}}$$

(*Hint:* See Example 10.37.)

37. Find the sum of the series

$$\sum_{n=1}^{\infty}\frac{n^{2}}{3^{n}}$$

(*Hint:* See Exercise 35.)

38. Let $f(x)=xe^{x}$. Make use of Equation 10.23 and $f'(1)$ to find the sum of the series

$$\sum_{n=0}^{\infty}\frac{n+1}{n!}$$

39. Make use of Equation 10.23 to find a series for $(e^{x}-1)/x$. By differentiating the result, show that

$$\frac{1}{2!}+\frac{2}{3!}+\frac{3}{4!}+\cdots=1$$

40. (a) Use partial fractions and geometric series to find a power series in x for the function

$$f(x)=\frac{1}{2+x-x^{2}}$$

and give its radius of convergence.

(b) Using part (a), find a power series representation of

$$\frac{2x-1}{(2+x-x^{2})^{2}}$$

What is the radius of convergence?

41. Use Equation 10.23 to evaluate

$$\int_{0}^{1}e^{-x^{2}}\,dx$$

correct to eight decimal places.

42. Define f and g on $(-\infty,\infty)$ by

$$f(x)=\sum_{n=0}^{\infty}\frac{(-1)^{n}x^{2n}}{(2n)!}\qquad g(x)=\sum_{n=0}^{\infty}\frac{(-1)^{n}x^{2n+1}}{(2n+1)!}$$

Show the following:
(a) $f''(x)+f(x)=0$ and $g''(x)+g(x)=0$
(b) $f(0)=1$, $f'(0)=0$, $g(0)=0$, and $g'(0)=1$
(c) $f'(x)=-g(x)$ and $g'(x)=f(x)$
Do the results in part (b) and part (c) remind you of functions you have seen before? If so, what are they?

43. Use a CAS, a spreadsheet, or a calculator to verify the plausibility of Equation 10.24 with the following values of x and n.
(a) $x=10$, $n=50$ (b) $x=-20$, $n=100$
(c) $x=100$, $n=1000$

10.9 TAYLOR SERIES

Recall (from Section 10.1) Taylor's Formula, which states that if f has derivatives up through the $(n + 1)$st order in an open interval I centered at $x = a$, then

$$f(x) = P_n(x) + R_n(x) \tag{10.25}$$

for all x in I. Here $P_n(x)$ is the Taylor polynomial

$$P_n(x) = f(a) + f'(a)(x - a) + \frac{f''(a)}{2!}(x - a)^2 + \cdots + \frac{f^{(n)}(a)}{n!}(x - a)^n \tag{10.26}$$

and $R_n(x)$ is the remainder term, given by

$$R_n(x) = \frac{f^{(n+1)}(c)}{(n + 1)!}(x - a)^{n+1} \tag{10.27}$$

where c is some number between a and x.

We want to consider now what happens when we let $n \to \infty$. First of all, we must require that f have derivatives of *all* orders in the interval I. By letting $n \to \infty$ in Equation 10.26 we get the infinite series shown below.

Taylor Series for f about $x = a$

$$\sum_{n=0}^{\infty} \frac{f^{(n)}(a)}{n!}(x - a)^n = f(a) + f'(a)(x - a) + \frac{f''(a)}{2!}(x - a)^2$$
$$+ \frac{f'''(a)}{3!}(x - a)^3 + \cdots \tag{10.28}$$

This series is called the **Taylor series for f about $x = a$**. In particular, if $a = 0$, we get the **Taylor series for f about $x = $ o**:

Taylor Series for f about $x = 0$: Maclaurin Series

$$\sum_{n=0}^{\infty} \frac{f^{(n)}(0)}{n!}x^n = f(0) + f'(0)x + \frac{f''(0)}{2!}x^2 + \frac{f'''(0)}{3!}x^3 + \cdots \tag{10.29}$$

The Taylor series for f about $x = 0$ in Equation 10.29 is also called the **Maclaurin series for f**, after the Scottish mathematician Colin Maclaurin (1698–1746).

The crucial question is whether the Taylor series for a function f converges to that function in some interval. If so, then we say that f is *represented* by its Taylor series in the interval where the convergence is valid. The answer to this question of representation hinges on the remainder term $R_n(x)$ given by Equation 10.27. Note that the Taylor polynomial $P_n(x)$ is the nth partial sum (starting with $n = 0$) of the Taylor series in Equation 10.28. By Equation 10.25, we have

$$|f(x) - P_n(x)| = |R_n(x)|$$

It follows that $P_n(x)$ will come arbitrarily close to $f(x)$ if and only if $|R_n(x)|$ comes arbitrarily close to 0 as $n \to \infty$. That is, $\lim_{n \to \infty} P_n(x) = f(x)$ if and only if $\lim_{n \to \infty} R_n(x) = 0$. We summarize our findings as a theorem.

THEOREM 10.21

Taylor's Theorem

Let f have derivatives of all orders in an open interval I centered at $x = a$. Then the Taylor series for f about $x = a$ converges to $f(x)$ for x in I if and only if for all x in I,

$$\lim_{n \to \infty} R_n(x) = 0$$

where $R_n(x)$ is the remainder term in Taylor's Formula, given in Equation 10.27.

EXAMPLE 10.41   Show that the Taylor series for $f(x) = e^x$ about $x = 0$ converges to e^x for all real x.

Solution   In Example 10.4 we found that the nth Taylor polynomial for e^x is

$$P_n(x) = 1 + x + \frac{x^2}{2!} + \frac{x^3}{3!} + \cdots + \frac{x^n}{n!}$$

and that

$$R_n(x) = \frac{e^c}{(n+1)!} x^{n+1}$$

where c is between 0 and x. If $0 < c < x$, then $e^c < e^x$ since $f(x) = e^x$ is an increasing function. Thus, for $x > 0$,

$$|R_n(x)| \le \frac{e^x}{(n+1)!} x^{n+1}$$

From the special limit given in Equation 10.24 we know that $x^{n+1}/(n+1)! \to 0$ as $n \to \infty$. So for fixed $x > 0$, e^x is constant and it follows that

$$\lim_{n \to \infty} R_n(x) = 0$$

If $x < c < 0$, then $e^c < e^0 = 1$. Thus,

$$|R_n(x)| \le \left| \frac{x^{n+1}}{(n+1)!} \right|$$

Again, by the special limit in Equation 10.24 the right-hand side goes to 0 as $n \to \infty$. Thus, for all $x < 0$,

$$\lim_{n \to \infty} R_n(x) = 0$$

By Theorem 10.21, we conclude that the Taylor series for e^x about $x = 0$ converges to e^x for all real x. That is,

$$e^x = 1 + x + \frac{x^2}{2!} + \frac{x^3}{3!} + \cdots = \sum_{n=0}^{\infty} \frac{x^n}{n!}$$

∎

In the preceding section (see Example 10.40) we arrived at the result of Example 10.41 in a different way. We began with the power series $\sum x^n/n!$ and proved that its radius of convergence was infinite. Then we proved that the function represented by the series was, in fact, identical to e^x. As you might suspect, it is no accident that we got the same series for e^x by the two different approaches. The next theorem explains why.

THEOREM 10.22

> If f can be represented by a power series in $x - a$ in an open interval I centered at $x = a$, then that power series is the Taylor series for f about $x = a$.

Proof   For simplicity, we prove the theorem for $a = 0$. The proof for $a \neq 0$ is similar. Suppose, then, that $f(x)$ is represented by the power series

$$f(x) = a_0 + a_1 x + a_2 x^2 + a_3 x^3 + a_4 x^4 + a_5 x^5 + \cdots$$

for all x in an interval I centered at $x = 0$. Then we have, on successive differentiation,

$$f'(x) = a_1 + 2a_2 x + 3a_3 x^2 + 4a_4 x^3 + 5a_5 x^4 + \cdots$$

$$f''(x) = 2a_2 + (3 \cdot 2)a_3 x + (4 \cdot 3)a_4 x^2 + (5 \cdot 4)a_5 x^3 + \cdots$$

$$f'''(x) = (3 \cdot 2)a_3 + (4 \cdot 3 \cdot 2)a_4 x + (5 \cdot 4 \cdot 3)a_5 x^2 + \cdots$$

$$f^{(4)}(x) = (4 \cdot 3 \cdot 2)a_4 + (5 \cdot 4 \cdot 3 \cdot 2)a_5 x + \cdots$$

$$\vdots$$

On setting $x = 0$ in each equation, we get

$$f(0) = a_0, \quad f'(0) = a_1, \quad f''(0) = 2a_2, \quad f'''(0) = 3!\, a_3,$$

$$f^{(4)}(0) = 4!\, a_4, \quad \dots$$

and in general,

$$f^{(n)}(0) = n!\, a_n$$

If we solve for a_n, we get

$$a_n = \frac{f^{(n)}(0)}{n!}, \qquad n = 0, 1, 2, \dots$$

Thus, the coefficients in the given power series representation of f are precisely those of the Taylor series for f about $x = 0$. The given series therefore is the Taylor series. ∎

An important consequence of Theorem 10.22 is that by whatever valid means we arrive at a power series representation of a function (such as by differentiation or integration of a geometric series, for example), the series will be the Taylor series. In particular, then, the series for $\ln(1 + x)$ and for $\tan^{-1} x$ that we found in Examples 10.37 and 10.38 are in fact the Taylor series about $x = 0$ for these functions.

It is important to emphasize in Theorem 10.22 the hypothesis that f *can* be represented by a power series. That is, we begin with the assumption that there is a power series in $x - a$ that *does converge to f* throughout some interval centered at $x = a$. Then we can conclude that f has derivatives of all orders at a, and that the given power series is, in fact, the Taylor series for f about $x = a$. If, on the other hand, we knew only that f had derivatives of all orders at $x = a$, we would know that it has a Taylor series about $x = a$, *but we would not know that f is represented by its series.* That is, we would not know that the series converges to $f(x)$. Theorem 10.21 gives the necessary and sufficient condition for the function to be represented by its Taylor series, namely, that $\lim_{n \to \infty} R_n(x) = 0$. In Exercise 29 of Exercise Set 10.9 we will ask you to show that the function f defined by

$$f(x) = \begin{cases} e^{-1/x^2} & \text{if } x \neq 0 \\ 0 & \text{if } x = 0 \end{cases}$$

has a Taylor series about $x = 0$ but that, except for $x = 0$, the series does not converge to $f(x)$.

In the next example we show two ways of finding the Taylor series about $x = 0$ of $f(x) = (1 - x)^{-2}$.

EXAMPLE 10.42   Find the Maclaurin series for $f(x) = 1/(1 - x)^2$ by two methods. (*Note:* Recall that the Maclaurin series is the Taylor series about $x = 0$.)

Solution   We first use the direct method to calculate the coefficients in the Taylor series about $x = 0$. We calculate the successive derivatives of f and evaluate them at $x = 0$.

$$f(x) = \frac{1}{(1 - x)^2} = (1 - x)^{-2} \qquad\qquad f(0) = 1!$$
$$f'(x) = -2(1 - x)^{-3}(-1) = 2!(1 - x)^{-3} \qquad\qquad f'(0) = 2!$$
$$f''(x) = -3 \cdot 2(1 - x)^{-4}(-1) = 3!\,(1 - x)^{-4} \qquad\qquad f''(0) = 3!$$
$$f'''(x) = -4 \cdot 3!\,(1 - x)^{-5}(-1) = 4!\,(1 - x)^{-5} \qquad\qquad f'''(0) = 4!$$
$$\vdots \qquad\qquad\qquad\qquad\qquad\qquad\qquad \vdots$$
$$f^{(n)}(x) = (n + 1)!\,(1 - x)^{-(n+2)} \qquad\qquad f^{(n)}(0) = (n + 1)!$$

The general term of the Maclaurin series in Equation 10.29 is therefore

$$\frac{f^{(n)}(0)}{n!}x^n = \frac{(n + 1)!}{n!}x^n = (n + 1)x^n$$

So the Maclaurin series for f is

$$\sum_{n=0}^{\infty}(n + 1)x^n = 1 + 2x + 3x^2 + 4x^3 + \cdots$$

Note that we have not shown that this series represents f in any interval. To do so, we would have to show that the remainder term in Taylor's Formula, $R_n(x)$, approaches 0 as $n \to \infty$. Rather than show this limit, we defer the question of whether the series represents f to our second method, where the representation is an automatic consequence of the method.

For the second method, we note that

$$\frac{1}{(1 - x)^2} = \frac{d}{dx}\left(\frac{1}{1 - x}\right)$$

From Equation 10.22 we know that

$$\frac{1}{1-x} = 1 + x + x^2 + x^3 + \cdots = \sum_{n=0}^{\infty} x^n, \qquad |x| < 1$$

By Theorem 10.20, therefore,

$$\frac{1}{(1-x)^2} = \frac{d}{dx}(1 + x + x^2 + x^3 + \cdots)$$

$$= 1 + 2x + 3x^2 + \cdots = \sum_{n=0}^{\infty} (n+1)x^n, \qquad |x| < 1$$

This result is the same as we found by the direct method, but we actually have the added information that the series does represent the function in the interval $-1 < x < 1$.   ∎

EXAMPLE 10.43

(a) Find the Maclaurin series for $f(x) = \cos x$.

(b) Use the result of part (a) to approximate the integral

$$\int_0^1 \frac{\sin^2 x}{x^2}\,dx$$

correct to three decimal places.

Solution

(a) In Example 10.3 we found the nth Taylor polynomial for $\cos x$ about $x = 0$ to be

$$P_n(x) = 1 - \frac{x^2}{2!} + \frac{x^4}{4!} - \frac{x^6}{6!} + \cdots + \frac{(-1)^n}{(2n)!}x^{2n}$$

and in Example 10.7, we found the remainder term

$$R_{2n}(x) = \frac{(-1)^{n+1}\cos c}{(2n+2)!}x^{2n+2}, \qquad c \text{ between } 0 \text{ and } x$$

To show where the Taylor series converges to $\cos x$, we need to find the values of x for which

$$\lim_{n \to \infty} R_{2n}(x) = 0$$

For any real value of x,

$$|R_{2n}(x)| = \frac{|\cos c|}{(2n+2)!}|x|^{2n+2} \le \frac{|x|^{2n+2}}{(2n+2)!}$$

since $|\cos c| \le 1$. By the special limit in Equation 10.24, the last expression approaches 0 as $n \to \infty$. Therefore, $\lim_{n \to \infty} R_{2n}(x) = 0$ for all real x. By Theorem 10.21 it follows that

$$\cos x = 1 - \frac{x^2}{2!} + \frac{x^4}{4!} - \frac{x^6}{6!} + \cdots = \sum_{n=0}^{\infty} \frac{(-1)^n}{(2n)!}x^{2n}$$

for all real x.

(b) To use the result of part (a) to evaluate the integral, we need to relate $\sin^2 x$ to the first power of the cosine. Such a relationship is provided by the trigonometric identity

$$\sin^2 x = \frac{1}{2}(1 - \cos 2x) \qquad \text{See Supplementary Appendix 2.}$$

Now we replace x with $2x$ in the result of part (a) to get

$$\sin^2 x = \frac{1}{2}\left[1 - \left(1 - \frac{(2x)^2}{2!} + \frac{(2x)^4}{4!} - \frac{(2x)^6}{6!} + \cdots\right)\right]$$

$$= \frac{2x^2}{2!} - \frac{2^3 x^4}{4!} + \frac{2^5 x^6}{6!} - \cdots, \qquad -\infty < x < \infty$$

Dividing by x^2 gives

$$\frac{\sin^2 x}{x^2} = 1 - \frac{2^3 x^2}{4!} + \frac{2^5 x^4}{6!} - \cdots \qquad (x \neq 0)$$

Thus, by Theorem 10.20,

$$\int_0^1 \frac{\sin^2 x}{x^2}\,dx = \int_0^1 \left(1 - \frac{2^3 x^2}{4!} + \frac{2^5 x^4}{6!} - \frac{2^7 x^6}{8!} + \cdots\right) dx$$

$$= x - \frac{(2x)^3}{3 \cdot 4!} + \frac{(2x)^5}{5 \cdot 6!} - \frac{(2x)^7}{7 \cdot 8!} + \cdots\Bigg]_0^1$$

$$= 1 - \frac{2^3}{3 \cdot 4!} + \frac{2^5}{5 \cdot 6!} - \frac{2^7}{7 \cdot 8!} + \cdots$$

Using the first three terms, we get

$$\int_0^1 \frac{\sin^2 x}{x^2}\,dx \approx 0.898$$

Since the series alternates, we know from Corollary 10.15 that the error does not exceed the fourth term:

$$\text{Error} \leq \frac{2^7}{7 \cdot 8!} \approx 0.000454$$

So our approximation is correct to three places. ∎

REMARK ————————————————————————

■ Although $(\sin^2 x)/x^2$ is not defined when $x = 0$, it has a removable discontinuity there. In fact, we see from its series representation that its limit is 1 as $x \to 0$.

Binomial Series

We wish to consider finally an important series known as the **binomial series**, which is the Maclaurin series for $f(x) = (1 + x)^\alpha$ for an arbitrary real number α. If $\alpha = 0$, then $f(x) = 1$. If α is a positive integer, say $\alpha = n$, we know from the **Binomial Formula** that

$$(1 + x)^\alpha = (1 + x)^n = 1 + nx + \frac{n(n-1)}{2!}x^2 + \frac{n(n-1)(n-2)}{3!}x^3 + \cdots + x^n$$

$$(10.30)$$

which is a finite series—that is, a polynomial—so there is no question of convergence. For all other values of α we proceed as usual to find the Maclaurin series:

$$
\begin{aligned}
f(x) &= (1+x)^{\alpha} & f(0) &= 1 \\
f'(x) &= \alpha(1+x)^{\alpha-1} & f'(0) &= \alpha \\
f''(x) &= \alpha(\alpha-1)(1+x)^{\alpha-2} & f''(0) &= \alpha(\alpha-1) \\
f'''(x) &= \alpha(\alpha-1)(\alpha-2)(1+x)^{\alpha-3} & f'''(0) &= \alpha(\alpha-1)(\alpha-2) \\
&\ \ \vdots & &\ \ \vdots \\
f^{(n)}(x) &= \alpha(\alpha-1)\cdots & f^{(n)}(0) &= \alpha(\alpha-1)\cdots \\
&\quad \cdots(\alpha-n+1)(1+x)^{\alpha-n} & &\quad \cdots(\alpha-n+1)
\end{aligned}
$$

So the binomial series is

$$
\begin{aligned}
1 + \alpha x &+ \frac{\alpha(\alpha-1)}{2!}x^2 + \frac{\alpha(\alpha-1)(\alpha-2)}{3!}x^3 + \cdots \\
&+ \frac{\alpha(\alpha-1)(\alpha-2)\cdots(\alpha-n+1)}{n!}x^n + \cdots
\end{aligned}
\tag{10.31}
$$

Note that if $\alpha = n$, then the series terminates with the term of nth degree since the factor $(\alpha - n)$ appears in all successive terms and is 0 when $\alpha = n$. Thus, series 10.31 reduces to the finite series 10.30, that is, to the Binomial Formula. It is convenient to introduce the notation

$$
\binom{\alpha}{n} = \frac{\alpha(\alpha-1)(\alpha-2)\cdots(\alpha-n+1)}{n!}
$$

with $\binom{\alpha}{0}$ defined as 1. Then we can write the series in the compact form

$$
\sum_{n=0}^{\infty} \binom{\alpha}{n} x^n
$$

It is not difficult to show that the radius of convergence for this series is $R = 1$ (see Exercise 27 in Exercise Set 10.9), so that the series converges in $(-1, 1)$, with only convergence at the endpoints remaining in question. This convergence in itself, however, does not guarantee that the function represented by the series in this interval is what we hope it will be—namely, $(1 + x)^{\alpha}$. We can show that the series does converge to $(1 + x)^{\alpha}$ by proving that $\lim_{n\to\infty} R_n(x) = 0$. The details are tedious, however. We simply state the result in the following theorem.

THEOREM 10.23

If α is any real number other than a nonnegative integer, the binomial series

$$
\sum_{n=0}^{\infty} \binom{\alpha}{n} x^n = 1 + \alpha x + \frac{\alpha(\alpha-1)}{2!}x^2 + \frac{\alpha(\alpha-1)(\alpha-2)}{3!}x^3 + \cdots
$$

converges to $(1 + x)^{\alpha}$ in the interval $(-1, 1)$ and diverges if $|x| > 1$. If α is a nonnegative integer, the series is finite and represents $(1 + x)^{\alpha}$ for all real values of x.

REMARK

■ When endpoint values are taken into consideration, it can be shown that the series represents the function precisely in the following intervals, depending on the size of α:

$$\begin{cases} -1 \le x \le 1 & \text{if } \alpha > 0 \\ -1 < x < 1 & \text{if } \alpha \le -1 \\ -1 < x \le 1 & \text{if } -1 < \alpha < 0 \end{cases}$$

EXAMPLE 10.44   Find the Maclaurin series for $f(x) = 1/\sqrt{1-x^2}$.

Solution   Since $f(x) = (1-x^2)^{-1/2}$, we can use the binomial series in which $\alpha = -\frac{1}{2}$ and x is replaced by $-x^2$. So we have

$$\frac{1}{\sqrt{1-x^2}} = \sum_{n=0}^{\infty} \binom{-\frac{1}{2}}{n}(-x^2)^n = \sum_{n=0}^{\infty}(-1)^n\binom{-\frac{1}{2}}{n}x^{2n}$$

$$= 1 - \left(-\frac{1}{2}\right)x^2 + \frac{\left(-\frac{1}{2}\right)\left(-\frac{3}{2}\right)}{2!}x^4$$

$$- \frac{\left(-\frac{1}{2}\right)\left(-\frac{3}{2}\right)\left(-\frac{5}{2}\right)}{3!}x^6 + \cdots$$

$$= 1 + \frac{1}{2}x^2 + \frac{1\cdot 3}{2^2\cdot 2!}x^4 + \frac{1\cdot 3\cdot 5}{2^3\cdot 3!}x^6 + \cdots$$

$$= 1 + \sum_{n=1}^{\infty}\frac{1\cdot 3\cdot 5\cdot \cdots \cdot(2n-1)}{2^n\cdot n!}x^{2n}$$

The series converges to the given function for $|-x^2| < 1$ or, equivalently, $-1 < x < 1$.   ■

EXAMPLE 10.45   Find the Maclaurin series for $\sin^{-1}x$ and give its radius of convergence.

Solution   Since

$$\sin^{-1}x = \int \frac{dx}{\sqrt{1-x^2}} + C$$

we can integrate the series obtained in the preceding example to get

$$\sin^{-1}x = \int\left[1 + \sum_{n=1}^{\infty}\frac{1\cdot 3\cdot 5\cdot \cdots \cdot(2n-1)}{2^n\cdot n!}x^{2n}\right]dx + C$$

$$= x + \sum_{n=1}^{\infty}\int \frac{1\cdot 3\cdot 5\cdot \cdots \cdot(2n-1)}{2^n\cdot n!}x^{2n}dx + C$$

$$= x + \sum_{n=1}^{\infty}\frac{1\cdot 3\cdot 5\cdot \cdots \cdot(2n-1)}{2^n\cdot n!}\cdot\frac{x^{2n+1}}{2n+1} + C$$

Putting $x = 0$ on each side, we see that $C = 0$. So we can write

$$\sin^{-1} x = x + \frac{1}{2} \cdot \frac{x^3}{3} + \frac{1 \cdot 3}{2^2 \cdot 2!} \cdot \frac{x^5}{5} + \frac{1 \cdot 3 \cdot 5}{2^3 \cdot 3!} \cdot \frac{x^7}{7} + \cdots$$

Since the radius of convergence of the series for $(1 - x^2)^{-1/2}$ is $R = 1$, it follows from Theorem 10.20 that the series for $\sin^{-1} x$ has the same radius of convergence.   ∎

Summary of Some Important Maclaurin Series

For reference, we list some of the more frequently used Maclaurin series. The first three, in particular, are especially useful to know.

Frequently Used Maclaurin Series

$$e^x \; = \sum_{n=0}^{\infty} \frac{x^n}{n!} = 1 + x + \frac{x^2}{2!} + \frac{x^3}{3!} + \cdots, \; -\infty < x < \infty$$

$$\sin x = \sum_{n=0}^{\infty} \frac{(-1)^n x^{2n+1}}{(2n+1)!} = x - \frac{x^3}{3!} + \frac{x^5}{5!} - \frac{x^7}{7!} + \cdots, \; -\infty < x < \infty$$

$$\cos x = \sum_{n=0}^{\infty} \frac{(-1)^n x^{2n}}{(2n)!} = 1 - \frac{x^2}{2!} + \frac{x^4}{4!} - \frac{x^6}{6!} + \cdots, \; -\infty < x < \infty$$

$$\ln(1 + x) = \sum_{n=1}^{\infty} \frac{(-1)^{n-1} x^n}{n} = x - \frac{x^2}{2} + \frac{x^3}{3} - \frac{x^4}{4} + \cdots, \; -1 < x \le 1$$

$$\tan^{-1} x = \sum_{n=1}^{\infty} \frac{(-1)^{n-1} x^{2n-1}}{2n - 1} = x - \frac{x^3}{3} + \frac{x^5}{5} - \frac{x^7}{7} + \cdots, \; -1 \le x \le 1$$

$$\sinh x = \sum_{n=0}^{\infty} \frac{x^{2n+1}}{(2n+1)!} = x + \frac{x^3}{3!} + \frac{x^5}{5!} + \frac{x^7}{7!} + \cdots, \; -\infty < x < \infty$$

$$\cosh x = \sum_{n=0}^{\infty} \frac{x^{2n}}{(2n)!} = 1 + \frac{x^2}{2!} + \frac{x^4}{4!} + \frac{x^6}{6!} + \cdots, \; -\infty < x < \infty$$

$$(1 + x)^\alpha = \sum_{n=0}^{\infty} \binom{\alpha}{n} x^n = 1 + \binom{\alpha}{1} x + \binom{\alpha}{2} x^2 + \binom{\alpha}{3} x^3 + \cdots, \; -1 < x < 1,$$

$$\text{where} \quad \binom{\alpha}{k} = \frac{\alpha(\alpha - 1) \cdots (\alpha - k + 1)}{k!}$$

COMPUTATIONS WITH POWER SERIES AND COMPUTER ALGEBRA SYSTEMS

When a function can be approximated via the Taylor series representation, then the function can be treated essentially as an infinite polynomial. In particular, differentiation and integration can be performed on the series representation term by term.

 CAS 38

Approximate $\int_0^1 e^{-x^2}\, dx$.

We will use the representation

$$e^{-x^2} = \sum_{n=1}^{\infty} (-1)^n \frac{x^{2n}}{n!}$$

and to approximate the integral, we replace the integrand with a 10th degree Taylor polynomial, which is a partial sum of the series representation for the function.

Maple:

f := x–>exp(–x^2);
readlib(`mtaylor`):   *Access mtaylor*
P := mtaylor(f(x),[x],12);

Output:

$$P := 1 - x^2 + \frac{1}{2}x^4 - \frac{1}{6}x^6 + \frac{1}{24}x^8 - \frac{1}{120}x^{10}$$

int(P,x);

Output:

$$x - \frac{1}{3}x^3 + \frac{1}{10}x^5 - \frac{1}{42}x^7 + \frac{1}{216}x^9 - \frac{1}{1320}x^{11}$$

int(P,x=0..1); $\frac{31049}{41580}$

evalf(");   .7467291967

Mathematica:

f[x_] = Exp[–x^2]
P[x_] = Normal[Series[f[x],
    {x,0,10}]]
Integrate[P[x],x]
Integrate[P[x],{x,0,1}]

DERIVE:

(At the □ symbol, go to the next step.)
a (author) □ exp(–x^2) □ c (calculus) □ t (taylor) □ [choose expression] □ x (variable) □ 10 (degree) □ 0 (point) □ s (simplify) □ [choose expression] □ c (calculus) □ i (integrate) □ [choose expression] □ x (variable) □ enter (no limits) □ s (simplify) □ [choose expression] □ c (calculus) □ i (integrate) □ [choose expression] □ x (variable) □ 0 (lower limit) □ 1 (upper limit) □ x (approximate) □ [choose expression]

Finally, we compare this result with that obtained from Simpson's Rule.

Maple:

with(student):
evalf(value(simpson(f(x),
    x=0..1,100)));

Output: .7468241331

Exercise Set 10.9

In Exercises 1–18, find the Maclaurin series for the given function and determine its interval of convergence.

1. $\sin x$

2. $\sinh x$

3. $\cosh x$

4. $\ln(1 - x)$

5. $\tanh^{-1} x$

6. $\sqrt{1 - x}$ (Do not test endpoints.)

7. $\sin^2 x$ (Hint: $\cos 2x = 1 - 2\sin^2 x$.)

8. $\cos^2 x$ (Hint: $\cos 2x = 2\cos^2 x - 1$.)

9. $\dfrac{1}{(1 - x)^3}$

10. $\sin x \cos x$ (Hint: $\sin 2x = 2 \sin x \cos x$.)

11. 2^x

12. xe^{-x}

13. $\dfrac{1 - \cos x}{x^2}$ (defined as its limiting value at 0)

14. $\sin x^2$

15. $\sqrt[3]{8 - x}$ (Do not test endpoints.)

16. $\dfrac{\sin x}{x}$ (defined at 0 to be its limiting value as $x \to 0$)

17. $\cos \sqrt{x}$

18. $x \ln \sqrt{1 + x}$

In Exercises 19–24, find the Taylor series for the given function about $x = a$ for the specified value of a, and determine the interval of convergence of the series.

19. e^x; $a = 1$

20. $\cos x$; $a = \dfrac{\pi}{3}$

21. $\dfrac{1}{x}$; $a = -1$

22. $\sqrt{x + 3}$; $a = 1$

23. $\sin \dfrac{\pi x}{6}$; $a = 1$

24. $1/(x - 1)^2$; $a = 2$

25. For each of the following, show that the Maclaurin series of the given function represents the function for all values of x.
    (a) $\sin x$ (b) $\sinh x$ (c) $\cosh x$

26. Find the Taylor series about $x = 1$ for $f(x) = \ln x$ by replacing x with $x - 1$ in the Maclaurin series for $\ln(1 + x)$. In what interval does the series represent $\ln x$?

27. Show that the radius of convergence of the binomial series is 1.

28. Prove that if $f^{(n)}(x)$ exists for all n and x and there exists a constant M such that $|f^{(n)}(x)| \le M$ for all n and x, then the Taylor series of f about $x = a$ exists and converges to $f(x)$ everywhere.

29. For $x \ne 0$, define $f(x) = e^{-1/x^2}$ and let $f(0) = 0$. Prove that the Maclaurin series for f exists and converges everywhere but that it represents f only at $x = 0$.

[CAS] In Exercises 30–33, approximate the integrals, making use of three terms of the Maclaurin series of the integrand. Estimate the error.

30. $\displaystyle\int_{-1}^{1} \dfrac{1 - \cos x}{x^2} dx$

31. $\displaystyle\int_{0}^{0.5} \sqrt{1 + x^3} dx$

32. $\displaystyle\int_{0}^{0.4} \dfrac{\tan^{-1} x}{x} dx$

33. $\displaystyle\int_{-0.2}^{0} e^{-x^2} dx$

34. Find the first three nonzero terms of the Maclaurin series for $\tan x$.

35. Find the first three nonzero terms of the Maclaurin series for $\sec^2 x$. (Hint: Use the results of Exercise 34.)

[CAS] 36. Use a CAS to find the first eight terms of the Maclaurin series for
    (a) $\tan x$;   (b) $\sec^2 x$.
    (Compare Exercises 34 and 35.)

37. The **error function** is defined by

$$\text{erf}(x) = \frac{2}{\sqrt{x}} \int_{0}^{x} e^{-t^2} dt, \qquad -\infty < x < \infty$$

Find the Maclaurin series for erf(x), and use it to approximate erf(1) correct to three decimal places.

[CAS] 38. Approximate the integral

$$\int_{0}^{2} xe^x dx$$

(a) using Simpson's Rule with 10 subintervals;
(b) using the first 10 nonzero terms of the Taylor series about $x = 0$.
Compare the results.

39. (a) Use a CAS to obtain the graph of

$$f(x) = |\sin^{-1}(\sin x)|$$

Explain the appearance of the graph.

(b) Define the function

$$g(x) = \sum_{n=0}^{\infty} \left(\frac{3}{4}\right)^n f(4^n x)$$

where f is the function of part (a). Approximate the graph of $g(x)$ by several partial sums of the series. *Note:* $g(x)$ is a nowhere-differentiable function first given by Karl Weierstrass (1815–1897).

10.10 FOURIER SERIES

The representation of functions by Taylor series is of fundamental importance. Yet there are many functions that cannot be expanded in Taylor series. For one thing, only functions that are infinitely differentiable have such expansions. Another type of series, whose terms consist of sines and cosines, can be used to represent a much broader class of functions. Series of this type are called Fourier (pronounced "foor-ee-yay") series, after the French mathematician and physicist Jean Baptiste Joseph Fourier (1768–1830), who studied these series in connection with his investigation of heat conduction.

Fourier series are especially important in studying all sorts of wave phenomena, such as light waves, sound waves, radio waves, and signals in electronic instruments. A concept closely related to a Fourier series, called a Fourier integral, has been used in such diverse ways as analyzing the sunspot cycle and investigating X rays in computer-assisted tomography (CAT) scanners.

We will give only a brief introduction here to Fourier series. There is a vast literature on this subject, and research related to Fourier series and Fourier integrals is ongoing. In this discussion we can only scratch the surface.

Let us begin by looking at the series

$$\cos x - \frac{\cos 3x}{3} + \frac{\cos 5x}{5} - \cdots = \sum_{k=1}^{\infty} \frac{(-1)^{k-1}\cos(2k-1)x}{2k-1} \qquad (10.32)$$

This series is an example of a Fourier series involving cosine terms only. (Others involve only sine terms or combine sines and cosines.) In Figure 10.30 we show graphs of the first three partial sums,

$$S_1(x) = \cos x, \quad S_2(x) = \cos x - \frac{\cos 3x}{3}, \quad S_3(x) = \cos x - \frac{\cos 3x}{3} + \frac{\cos 5x}{5}$$

Notice how adding the second term modifies the first by adding more "ripples" with smaller amplitude. The tops and bottoms seem to be flattening out. Adding the third term (getting $S_3(x)$) further flattens the tops and bottoms and adds "ripples" of still smaller amplitude. To see that this trend continues, we show the graph of the first ten terms,

$$S_{10}(x) = \sum_{k=1}^{10} \frac{(-1)^{k-1}\cos(2k-1)x}{2k-1}$$

in Figure 10.31. It appears that the partial sums are converging to the "square wave" shown in Figure 10.32. In Exercise 9 of Exercise Set 10.10 we will ask you to verify that series 10.32 is the Fourier series for the function whose graph is given in Figure 10.32.

(a)

(b)

(c)

FIGURE 10.30

FIGURE 10.31

FIGURE 10.32
The square wave
$$f(x) = \begin{cases} \frac{\pi}{4} & \text{if } -\frac{\pi}{2} < x < \frac{\pi}{2} \\ -\frac{\pi}{4} & \text{if } \frac{\pi}{2} < x < \frac{3\pi}{2} \end{cases}$$
$f(x + 2\pi) = f(x)$.

In our example above, we began with a series and conjectured the function to which it converged. Now we want to reverse this procedure. We will begin with a function and attempt to find a series of sines and cosines converging to the function. This process will lead us to the definition of a Fourier series.

Deriving the Fourier Series

We begin with a function f that is integrable on the interval $[-\pi, \pi]$. (Later, we will indicate how we can use a more general interval of the form $[-L, L]$ for any $L > 0$.) At present we place no other restriction on f. It may not even be continuous. Let us assume for the moment that $f(x)$ can be represented by the following series of sines and cosines:

$$f(x) = \frac{a_0}{2} + \sum_{n=1}^{\infty} (a_n \cos nx + b_n \sin nx) \tag{10.33}$$

We will see later why it is convenient to write the constant term as $a_0/2$ rather than a_0. (By doing so, it will turn out that the formula we develop for a_n when $n \geq 1$ will be true for $n = 0$ as well.)

To find the coefficients in Equation 10.33, we will need the following results, which can be obtained using the integration techniques of Section 8.3 (see Exercise 42 in Exercise Set 8.3).

For m and n positive integers,

$$\int_{-\pi}^{\pi} \cos mx \cos nx \, dx = \begin{cases} 0 & \text{if } m \neq n \\ \pi & \text{if } m = n \end{cases} \tag{10.34}$$

$$\int_{-\pi}^{\pi} \sin mx \sin nx \, dx = \begin{cases} 0 & \text{if } m \neq n \\ \pi & \text{if } m = n \end{cases} \tag{10.35}$$

$$\int_{-\pi}^{\pi} \sin mx \cos nx \, dx = 0 \tag{10.36}$$

By direct integration, you can also show that

$$\int_{-\pi}^{\pi} \cos nx \, dx = \int_{-\pi}^{\pi} \sin nx \, dx = 0 \tag{10.37}$$

Let us return now to Equation 10.33 and integrate both sides from $-\pi$ to π. We will assume at this stage that term-by-term integration on the right-hand side is valid.

$$\int_{-\pi}^{\pi} f(x) \, dx = \int_{-\pi}^{\pi} \frac{a_0}{2} \, dx + \sum_{n=1}^{\infty} \left[\int_{-\pi}^{\pi} a_n \cos nx \, dx + \int_{-\pi}^{\pi} b_n \sin nx \, dx \right]$$

By Equation 10.37 the integrals in the summation both equal 0 for all n. Thus, we get

$$\int_{-\pi}^{\pi} f(x) \, dx = \frac{a_0}{2} (2\pi)$$

Solving for a_0 gives

$$a_0 = \frac{1}{\pi} \int_{-\pi}^{\pi} f(x) \, dx \tag{10.38}$$

We now have found the value of a_0. To find a_m for $m \geq 1$, we multiply both sides of Equation 10.33 by $\cos mx$, where m is a fixed, but unspecified, positive integer. Then, we again integrate over the interval $[-\pi, \pi]$:

$$\int_{-\pi}^{\pi} f(x) \cos mx \, dx = \frac{a_0}{2} \int_{-\pi}^{\pi} \cos mx \, dx + \sum_{n=1}^{\infty} \left[a_n \int_{-\pi}^{\pi} \cos mx \cos nx \, dx \right.$$
$$\left. + b_n \int_{-\pi}^{\pi} \cos mx \sin nx \, dx \right]$$

By Equation 10.37 the first integral on the right-hand side is 0 if $m \geq 1$. By Equation 10.34 the only nonzero term in the summation is the term for which $n = m$. Thus,

$$\int_{-\pi}^{\pi} f(x) \cos mx \, dx = a_m \int_{-\pi}^{\pi} \cos^2 mx \, dx$$

By Equation 10.34 the integral on the right equals π. So, on solving for a_m, we have

$$a_m = \frac{1}{\pi} \int_{-\pi}^{\pi} f(x) \cos mx \, dx, \quad m = 1, 2, \ldots \tag{10.39}$$

Notice that if we let $m = 0$ in this equation, the result agrees with the value of a_0 in Equation 10.38.

In order to find b_m, we proceed in a similar way, multiplying both sides of Equation 10.33 by $\sin mx$ and integrating. The result is (see Exercise 3 in Exercise Set 10.10)

$$b_m = \frac{1}{\pi} \int_{-\pi}^{\pi} f(x) \sin mx \, dx, \quad m = 1, 2, \ldots \tag{10.40}$$

REMARK ─────────────────────────────

■ We used the letter m as a subscript in deriving Equations 10.39 and 10.40, since n was used as the index of summation. Now that we have the formulas, we can replace m with n.

─────────────────────────────────────

Although we have made a number of assumptions in deriving the formulas for the coefficients, the formulas themselves have meaning provided only that $f(x)$ is integrable on $[-\pi, \pi]$. We therefore can make the following definition.

Definition 10.10
The Fourier Series for f

Let f be integrable on $[-\pi, \pi]$. Then the series

$$\frac{a_0}{2} + \sum_{n=1}^{\infty} (a_n \cos nx + b_n \sin nx)$$

where

$$a_n = \frac{1}{\pi} \int_{-\pi}^{\pi} f(x) \cos nx \, dx, \qquad n = 0, 1, 2, \ldots$$

and

$$b_n = \frac{1}{\pi} \int_{-\pi}^{\pi} f(x) \sin nx \, dx, \qquad n = 1, 2, 3, \ldots$$

is called the **Fourier series for f** on $[-\pi, \pi]$.

EXAMPLE 10.46   Find the Fourier series for the function f defined by $f(x) = x$ on $[-\pi, \pi]$.

Solution   By Equation 10.39,

$$a_n = \frac{1}{\pi} \int_{-\pi}^{\pi} x \cos nx \, dx$$

Since $x \cos nx$ is the product of an odd function and an even function, the result is odd. Thus, as we showed in Section 5.4, the integral over a symmetric interval is 0. That is,

$$a_n = 0 \qquad \text{for } n = 0, 1, 2, \dots$$

Also, by Equation 10.40 and by what we showed in Section 5.4,

$$b_n = \frac{1}{\pi} \int_{-\pi}^{\pi} x \sin nx \, dx = \frac{2}{\pi} \int_{0}^{\pi} x \sin nx \, dx$$

since $x \sin nx$ is an even function (the product of two odd functions is even). Now we integrate by parts, with $u = x$ and $dv = \sin nx$.

$$b_n = \frac{2}{\pi} \left[-\frac{x \cos nx}{n} \Big|_{0}^{\pi} + \frac{1}{n} \int_{0}^{\pi} \cos nx \, dx \right]$$

$$= \frac{2}{\pi} \left[-\frac{\pi \cos n\pi}{n} \right] + \frac{2}{\pi n^2} \left[\sin nx \right]_{0}^{\pi}$$

$$= -\frac{2}{n} \cos n\pi$$

Since $\cos n\pi = 1$ when n is even and $\cos n\pi = -1$ when n is odd, we can write $\cos n\pi = (-1)^n$. Thus,

$$b_n = \frac{2(-1)^{n+1}}{n} \qquad n = 1, 2, 3, \dots$$

By Definition 10.10 the Fourier series for f is

$$2 \left[\sin x - \frac{\sin 2x}{2} + \frac{\sin 3x}{3} - \cdots \right] \tag{10.41}$$

We will graph some of the partial sums shortly.   ∎

A Convergence Theorem

The question of convergence of the Fourier series for a function is a deep one and is better left to more advanced courses. We will state below one set of criteria that will ensure that the series converges to the function, at least for most values of x. First, though, observe that since $\sin nx$ and $\cos nx$ have period 2π for all values of n, if the Fourier series converges in the interval $[-\pi, \pi]$, then it also converges outside that interval in a periodic manner. Thus, if $f(x)$ is to be represented by its Fourier series for all x, it must be periodic, of period 2π. That is, $f(x)$ must satisfy

$$f(x + 2\pi) = f(x)$$

If f is not periodic, its Fourier series still may converge to $f(x)$ on $[-\pi, \pi]$. Outside that interval it would then converge to the *periodic extension* of f.

The following theorem was first stated by the German mathematician P. G. Lejeune Dirichlet (1805–1859).

THEOREM 10.24

Dirichlet's Theorem

Let f be bounded, piecewise-continuous, and have a finite number of maxima and minima on the interval $-\pi < x \leq \pi$. For x outside this interval let f be defined periodically by $f(x + 2\pi) = f(x)$. Then the Fourier series for f converges to $f(x)$ for all points of continuity of f. At each point of discontinuity the series converges to the average of the right-hand and left-hand limits of f at that point.

REMARK ——————————————————————

■ A function is piecewise continuous on an interval if it has at most finitely many points of discontinuity, and each such discontinuity is a simple (jump) discontinuity. (See Section 2.3.)

Graphing the Partial Sums

Let us illustrate the manner in which the partial sums of a Fourier series converge by looking again at some graphs. We will use series 10.41 for the function $f(x) = x$. In Figure 10.33 we show the graph of f on $(-\pi, \pi]$ and extended periodically outside that interval. Because of the appearance of its graph, this function is often referred to as a "sawtooth" function.

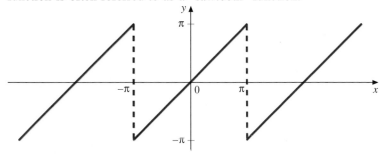

FIGURE 10.33
The "sawtooth" function $f(x) = x$ on $(-\pi, \pi]$, $f(x + 2\pi) = f(x)$

The hypotheses of Theorem 10.24 are satisfied for our function in Figure 10.33, so its Fourier series converges in the manner specified by the theorem. The points of discontinuity are the integral multiples of π. At each such point the average of the right-hand and left-hand limits of f is 0. If we redefine $f(x)$ to be 0 at these points, then we can write

$$f(x) = 2\left[\sin x - \frac{\sin 2x}{2} + \frac{\sin 3x}{3} - \cdots\right] = 2\sum_{k=1}^{\infty} \frac{(-1)^{k+1} \sin kx}{k}$$

In Figure 10.34 we show the graphs of the three partial sums $S_3(x)$, $S_6(x)$, and $S_9(x)$. Notice how they increasingly approach the sawtooth shape of the graph of f as n increases. Also, observe that each partial sum graph crosses the x-axis (the value of the partial sum is 0) at each integral multiple of π.

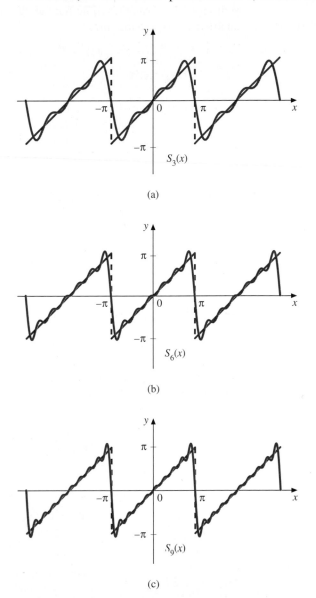

(a)

(b)

(c)

FIGURE 10.34

Harmonics

In the Fourier series for a function f,

$$\frac{a_0}{2} + \sum_{n=1}^{\infty}(a_n \cos nx + b_n \sin nx)$$

the constant term $a_0/2$ is known as *the fundamental*. In Exercise 1 of this section you will be asked to show that it is the average value of f over $[-\pi, \pi]$. The term $a_1 \cos x + b_1 \sin x$ is called the *first harmonic*, the term $a_2 \cos 2x + b_2 \sin 2x$ the *second harmonic*, and so on. When we listen to music, we are hearing the sum of many harmonics. The human ear and brain make a wonderful Fourier analyzer of these harmonics.

Other Intervals

By means of a horizontal stretching or shrinking of the type we studied in Section 1.2, it is possible to obtain the Fourier series for a function f defined on $[-L, L]$ for an arbitrary positive number L instead of on $[-\pi, \pi]$. The result, which we ask you to show in Exercise 16 of Exercise Set 10.10, is

$$f(x) = \frac{a_0}{2} + \sum_{n=1}^{\infty} \left(a_n \cos \frac{n\pi x}{L} + b_n \sin \frac{n\pi x}{L} \right) \qquad (10.42)$$

where

$$a_n = \frac{1}{L} \int_{-L}^{L} f(x) \cos \frac{n\pi x}{L} \, dx \qquad (10.43)$$

and

$$b_n = \frac{1}{L} \int_{-L}^{L} f(x) \sin \frac{n\pi x}{L} \, dx \qquad (10.44)$$

Exercise Set 10.10

1. Show that the constant term $a_0/2$ in the Fourier series for a function f is the average value of f on $[-\pi, \pi]$. (See Definition 5.3.)

2. (a) Show that if f is even, then $b_n = 0$ for $n = 1, 2, 3, \ldots$, and the Fourier series for f consists of cosine terms only.
   (b) Show that if f is odd, then $a_n = 0$ for $n = 0, 1, 2, \ldots$, and the Fourier series for f consists of sine terms only.

3. Carry out the details of deriving Formula 10.40 for b_m.

In Exercises 4–7, find the Fourier series for f.

4. $f(x) = |x|$ on $[-\pi, \pi]$

5. $f(x) = \begin{cases} 0 & \text{if } -\pi < x < 0 \\ x & \text{if } 0 \le x \le \pi \end{cases}$ and $f(x + 2\pi) = f(x)$

6. $f(x) = x^2$ on $[-\pi, \pi]$

7. $f(x) = \cos 2x$ on $[-\pi, \pi]$ (How could you have determined the answer without calculation?)

8. By substituting $x = \pi$ in the Fourier series for $f(x) = x^2$ (Exercise 6), obtain the sum

$$\sum_{n=1}^{\infty} \frac{1}{n^2} = \frac{\pi^2}{6}$$

(Recall that this sum was referred to in the introduction to this chapter, and it was first discovered by Euler. He used a completely different approach from the one used here.)

9. Show that the series 10.32 is the Fourier series for the square wave function shown in Figure 10.32.

10. Find the Fourier series for $f(x) = \cos^2 x$ for $-\pi \le x \le \pi$.

 In Exercises 11–15, use a CAS to graph the partial sums $S_n(x)$ for the specified series and the specified values of n.

11. The series of Example 10.46; $n = 2, 5, 7, 10$

12. The series of Exercise 4; $n = 2, 5, 8, 11$

13. The series of Exercise 5; $n = 1, 3, 5, 7$

14. The series of Exercise 6; $n = 2, 4, 6, 8$

15. The series of Exercise 10; $n = 2, 5, 8, 11$

16. Derive Equations 10.42, 10.43, and 10.44 as follows:
(a) Let f satisfy the conditions given in Theorem 10.24 on the interval $[-L, L]$. Make the change of

variables $u = \pi x/L$, and let

$$g(u) = f\left(\frac{Lu}{\pi}\right)$$

(b) Show that $g(u)$ is defined for $-\pi \le u \le \pi$.
(c) Write the Fourier series for $g(u)$ and the formulas for a_n and b_n in terms of u.
(d) In the results of part (c) replace u with $\pi x/L$. Simplify the formulas for a_n and b_n.

Chapter 10 Review Exercises

1. Find the Taylor polynomial of degree 5 about $x = 0$ for $f(x) = \ln(1 - x)$, and determine how accurately it approximates $f(x)$ for $|x| \le 0.2$.

2. Compute $\cos 62°$ with six decimal places of accuracy using Taylor's Formula with an appropriate value of n and a.

3. Find the Taylor polynomial of degree 5 about $x = \pi/4$ for $f(x) = \tan x$.

4. Find the nth Taylor polynomial for $f(x) = \cos^2 x$ about $x = 0$. What is $R_n(x)$ for this function? (*Hint:* Use a trigonometric identity.)

5. Find the nth Taylor polynomial about $x = \pi/4$ for $f(x) = \sin x - \cos x$. (*Hint:* Show that $f(x) = \sqrt{2} \sin(x - \pi/4)$.)

6. Use a cubic polynomial to approximate $\sqrt[3]{25}$. Estimate the error. (*Hint:* $\sqrt[3]{25} = (27 - 2)^{1/3} = (27)^{1/3}(1 - \frac{2}{27})^{1/3} = 3(1 - \frac{2}{27})^{1/3}$.)

7. Find $P_3(x)$ and $R_3(x)$ about $x = 1$ for $f(x) = \tan^{-1} x$.

8. Show that the Taylor polynomial

$$1 - \frac{x^2}{2!} + \frac{x^4}{4!} - \frac{x^6}{6!} + \frac{x^8}{8!}$$

for $\cos x$ gives accuracy to at least seven decimal places for x in $[-\pi/4, \pi/4]$.

9. Let $P_n(x)$ be the nth Taylor polynomial about $x = 0$ for $f(x) = e^x$. Use a calculator to make a table showing each of the following for $x = \pm 0.25$, $x = \pm 0.50$, $x = \pm 0.75$, $x = \pm 1$: e^x, $P_1(x)$, $P_2(x)$, $P_3(x)$, and $P_4(x)$.

10. Write out Taylor's Formula with Remainder with $a = 0$ for $f(x) = xe^{-x}$.

In Exercises 11 and 12, find $\lim_{n \to \infty} a_n$, or show the limit fails

to exist.

11. (a) $a_n = \dfrac{2n^3 - 3n + 1}{n^3 + 2n + 3}$ (b) $a_n = \dfrac{\sqrt{n + 10}}{n + 2}$

12. (a) $a_n = \dfrac{\sqrt{n^5 + 1}}{10n^2 + 3n}$ (b) $a_n = \dfrac{n \cos n\pi}{n + 1}$

In Exercises 13 and 14, show that each sequence is monotone. Then determine whether it converges by investigating boundedness.

13. (a) $\{e^{1 - (1/n)}\}$ (b) $\left\{\dfrac{2n - 3}{n + 3}\right\}$

14. (a) $\left\{\dfrac{e^{2n} n!}{(2n - 1)!}\right\}$
(b) $\{n \ln(n + 1) - (n + 1) \ln n\}$

15. Prove that if $\lim_{n \to \infty} a_n = L$, then $\lim_{n \to \infty} a_n^2 = L^2$. Is the converse also true? Prove or disprove.

16. Let

$$S_n = \frac{1}{2n + 1} + \frac{1}{2n + 2} + \frac{1}{2n + 3} + \cdots + \frac{1}{3n}$$

Prove that $\{S_n\}$ is a convergent sequence.

17. A sequence $\{S_n\}$ is defined recursively by $S_1 = 1$, and for $n \ge 2$, $S_n = S_{n-1} + 2/(n^2 + n)$. Find an explicit formula for S_n. Then show that $\{S_n\}$ converges, and find its limit.

18. What can you conclude about convergence or divergence of the series $\sum_{k=1}^{\infty} a_k$ under the following conditions?
(a) $a_k = \dfrac{2k^2 - 1}{k^2 + 1}$
(b) $a_1 + a_2 + \cdots + a_n = \dfrac{2n^2 - 1}{n^2 + 1}$
Give reasons.

19. Find the sum of each of the following series:

(a) $\displaystyle\sum_{n=0}^{\infty} \frac{(-1)^n 2^{n-1}}{3^n}$ (b) $\displaystyle\sum_{k=2}^{\infty} \frac{1}{k^2 - 1}$

20. Use geometric series to express each of the following repeating decimals as the ratio of two integers:

(a) $1.297297297\ldots$ (b) $3.2454545\ldots$

In Exercises 21–44, test for convergence or divergence. If the series has variable signs, test also for absolute convergence.

21. $\displaystyle\sum_{n=1}^{\infty} \frac{n}{\sqrt{n^3 + 2}}$ **22.** $\displaystyle\sum_{n=1}^{\infty} \frac{\sin n + \cos n}{n\sqrt{n + 1}}$

23. $\displaystyle\sum_{n=0}^{\infty} \frac{e^{-n}}{1 + e^{-n}}$ **24.** $\displaystyle\sum_{n=2}^{\infty} \frac{\cos n\pi}{\ln n}$

25. $\displaystyle\sum_{n=1}^{\infty} \frac{\tanh n}{n}$ **26.** $\displaystyle\sum_{n=1}^{\infty} ne^{-2n}$

27. $\displaystyle\sum_{k=1}^{\infty} \frac{\ln k}{k^2}$ **28.** $\displaystyle\sum_{n=1}^{\infty} \left(\frac{n}{3n + 4}\right)^n$

29. $\displaystyle\sum_{k=1}^{\infty} \frac{(-1)^{k-1} \ln k}{\sqrt{k}}$ **30.** $\displaystyle\sum_{k=1}^{\infty} \frac{3k^2 + 2k - 1}{k^5 + 2}$

31. $\displaystyle\sum_{k=2}^{\infty} \frac{\sin k}{k(\ln k)^2}$ **32.** $\displaystyle\sum_{n=1}^{\infty} \frac{n - 1}{(n + 1)^2}$

33. $\displaystyle\sum_{n=0}^{\infty} \frac{2^n n!}{(2n)!}$ **34.** $\displaystyle\sum_{k=0}^{\infty} (-1)^k k^2 e^{-k}$

35. $\displaystyle\sum_{k=1}^{\infty} \frac{k^k}{3^k k!}$ **36.** $\displaystyle\sum_{n=2}^{\infty} \frac{\ln n}{n^3 + 4}$

37. $\displaystyle\sum_{n=2}^{\infty} (-1)^n \ln\left(\frac{n + 1}{n - 1}\right)$

38. $\displaystyle\sum_{k=0}^{\infty} \frac{\sin(k + \frac{1}{2})\pi}{k + \frac{1}{2}}$ **39.** $\displaystyle\sum_{n=1}^{\infty} \frac{n^{2n}}{(n^3 + 1)^n}$

40. $\displaystyle\sum_{n=1}^{\infty} \left(\frac{1}{n} - \frac{1}{n + 2}\right)$ **41.** $\displaystyle\sum_{n=1}^{\infty} \frac{2 + \cos n}{\sqrt{2n^2 - 1}}$

42. $\displaystyle\sum_{n=0}^{\infty} \frac{\operatorname{sech}^2 n}{1 + \tanh n}$

43. $\displaystyle\sum_{n=0}^{\infty} (-1)^n (\sqrt{n + 1} - \sqrt{n})$

44. $\displaystyle\sum_{n=0}^{\infty} \frac{(-1)^n (n - 1)}{2n + 1}$

In Exercises 45–52, find the interval of convergence.

45. $\displaystyle\sum_{n=0}^{\infty} \frac{2^n x^n}{3^{n+1}}$ **46.** $\displaystyle\sum_{n=0}^{\infty} \frac{(-1)^n n x^n}{2n + 1}$

47. $\displaystyle\sum_{k=0}^{\infty} \frac{(x - 2)^k}{\sqrt{2k + 1}}$ **48.** $\displaystyle\sum_{k=0}^{\infty} \frac{(-1)^k (x + 1)^{2k}}{3^k}$

49. $\displaystyle\sum_{k=1}^{\infty} \frac{k x^k}{\ln(k + 1)}$

50. $\displaystyle\sum_{n=1}^{\infty} \frac{(nx)^n}{n! \, e^n}$ (Do not check endpoints.)

51. $\displaystyle\sum_{n=0}^{\infty} \frac{(-1)^n (2x - 1)^n n!}{1 \cdot 3 \cdot 5 \cdot \ \cdots \ \cdot (2n + 1)}$ (*Hint:* In testing endpoints, see Exercise 33 in Exercise Set 10.6.)

52. $\displaystyle\sum_{n=1}^{\infty} \frac{\cosh n}{n^2} (x - 1)^n$

In Exercises 53–56, use integration or differentiation of an appropriate series to find the Maclaurin series of f. Give the radius of convergence.

53. $f(x) = \ln(1 - x^2)$ **54.** $f(x) = x/(2 + x)^2$

55. $f(x) = \sinh^{-1} x$ **56.** $f(x) = \dfrac{1}{a} \tan^{-1} \dfrac{x}{a}$

57. Use infinite series to verify the following limits:

(a) $\displaystyle\lim_{x \to 0} \frac{\sin x}{x} = 1$

(b) $\displaystyle\lim_{x \to 0} \frac{1 - \cos x}{x^2} = \frac{1}{2}$

(c) $\displaystyle\lim_{x \to 0} \frac{e^x - 1}{x} = 1$

In Exercises 58–60, find the Taylor series about $x = a$.

58. $f(x) = 1/(2 - x); \ a = 1$

59. $f(x) = \ln \sqrt{x + 2}; \ a = -1$

60. $f(x) = \sin x; \ a = \dfrac{\pi}{4}$. Show that the series represents the function for all values of x.

[CAS] *In Exercises 61 and 62, approximate the integral using the first five nonzero terms of the Maclaurin series of the integrand. Give an upper bound on the error.*

61. $\displaystyle\int_0^{1/2} \frac{1 - e^{-x}}{x}\, dx$ **62.** $\displaystyle\int_0^{1/2} \frac{dx}{\sqrt{1 + x^3}}$

63. Find the Taylor series for $f(x) = \cos^2 x - \sin^2 x$ about $x = \pi/4$, and show that it converges to f for all x. (*Hint:* First use a trigonometric identity.)

64. Find the Fourier series for each of the functions
(a) $f(x) = \begin{cases} 0 & \text{if } -\pi < x < 0 \\ 1 & \text{if } 0 \le x < \pi \end{cases}$
(b) $f(x) = |\sin x|$ on $[-\pi, \pi]$

Chapter 10 Concept Quiz

1. Define each of the following:
(a) the limit of a sequence $\{a_n\}$
(b) the sum of a series $\sum_{n=1}^{\infty} a_n$
(c) a geometric series
(d) an absolutely convergent series
(e) the nth Taylor polynomial for a function f about $x = a$
(f) the Fourier series for a function f on $[-\pi, \pi]$

2. State each of the following:
(a) the nth-Term Divergence Test
(b) the Integral Test
(c) the Comparison Test
(d) the Limit Comparison Test
(e) the Ratio Test
(f) the Alternating Series Test
(g) Taylor's Formula with Remainder
(h) a necessary and sufficient condition for the Taylor series for a function f about $x = a$ to converge to $f(x)$
(i) conditions that guarantee the Fourier series for a function will converge to the function

3. (a) Suppose that $a_n \le a_{n+1} \le M$ for all n, where M is a constant. What can you conclude about convergence or divergence of the sequence $\{a_n\}$, and why?
(b) Suppose the power series $\sum a_n x^n$ converges when $x = -1$ and diverges when $x = 2$. State everything you can about where else the series converges and where else it diverges.
(c) Suppose that $a_n \ge 0$, $b_n \ge 0$, and $c_n \ge 0$ for all n, and that $\sum a_n$ converges and $\sum b_n$ diverges. State everything you can about convergence or divergence of $\sum c_n$ if for all n:
(i) $c_n \le a_n$
(ii) $a_n \le c_n \le b_n$
(iii) $c_n \ge b_n$
(d) Let $f^{(n)}(x)$ exist for all n for $-R < x < R$, and

suppose $\sum_{n=0}^{\infty} a_n x^n$ converges to $f(x)$ for all x in $(-R, R)$. What can you conclude about a_n?
(e) Explain under what conditions on a and b the equation

$$\int_a^b \left(\sum_{n=0}^{\infty} a_n x^n \right) dx = \sum_{n=0}^{\infty} \int_a^b a_n x^n \, dx$$

is valid.

4. Fill in the blanks.
(a) The series $\sum_{n=1}^{\infty} a_n$ converges to the sum S if and only if its _____ converges to S.
(b) The p-series $\sum_{n=1}^{\infty} 1/n^p$ converges if _____ and diverges if _____.
(c) The geometric series $\sum_{n=1}^{\infty} ar^{n-1}$ converges to _____ if _____ and diverges if _____.
(d) A sequence is a function whose domain is _____.
(e) If the power series $\sum_{n=0}^{\infty} a_n x^n$ has positive radius of convergence R and its sum is $f(x)$, then $\sum_{n=1}^{\infty} n a_n x^{n-1}$ converges to _____ for each x in the interval _____.

5. Indicate which of the following statements are true and which are false.
(a) If $\lim_{n \to \infty} a_n = 0$, then $\sum_{n=1}^{\infty} a_n$ converges.
(b) If $\sum_{n=1}^{\infty} a_n$ diverges, then $\lim_{n \to \infty} a_n \ne 0$.
(c) If $\lim_{n \to \infty} a_n \ne 0$, then $\sum_{n=1}^{\infty} a_n$ does not converge.
(d) If $\sum_{n=1}^{\infty} a_n$ is a series of nonnegative terms, and $a_1 + a_2 + a_3 + \cdots + a_n \le 10$ for all n, then $\sum_{n=1}^{\infty} a_n$ converges.
(e) If $\{a_n\}$ is a monotone decreasing sequence of positive numbers, then the sequence converges.
(f) If $\sum_{n=1}^{\infty} a_n$ is a series of positive terms and $a_{n+1}/a_n < 1$ for all n, then $\sum_{n=1}^{\infty} a_n$ converges.

APPLYING CALCULUS

1. When money is spent on goods and services, those that receive money also spend some of it. The people receiving some of the twice-spent money will spend some of that, and so on. Economists call this chain reaction the *multiplier effect*. In a hypothetical isolated community, the local government begins the process by spending $$D$. Suppose that each recipient of spent money spends 100 c% and saves 100 s% of the money that he or she receives. The values c and s are called the *marginal propensity to consume* and the *marginal propensity to save*, and, of course, $c + s = 1$.

    (a) Let S be the total spending that has been generated after n transactions. Determine an equation for S.

    (b) Show that $\lim_{n \to \infty} S = kD$, where $k = 1/s$. The number k is called the *multiplier*. What is the multiplier if the marginal propensity to consume is 80%?
    Note: The Federal Government uses this principle to justify deficit spending. Banks use this principle to justify lending out a large percentage of the money that they receive in deposits.

2. A certain ball has the property that each time it falls from height h onto a hard level surface, it rebounds to a height rh, where $0 < r < 1$. Suppose the ball is dropped from an initial height of H meters.

    (a) Assuming that the ball continues to bounce indefinitely, determine the total distance that it travels.

    (b) Calculate the total time that the ball is traveling. (Hint: If the ball is dropped from a height h, then the distance s it travels in t seconds is $s = 4.9t^2$).

    (c) Suppose that each time the ball strikes the surface with velocity v, it rebounds with velocity kv, where $0 < k < 1$. How long will it take for the ball to come to rest?

3. Let $S(x)$ and $C(x)$ be defined by the power series

    $$S(x) = \sum_{k=0}^{\infty} \frac{(-1)^k (\omega x)^{2k+1}}{(2k+1)!}$$

    $$C(x) = \sum_{k=0}^{\infty} \frac{(-1)^k (\omega x)^{2k}}{(2k)!}$$

    where ω is a constant.

    (a) Determine the domain of S and of C.

    (b) Show that

    $$\int_0^x C(t)\,dt = \frac{1}{\omega} S(x) \qquad \text{and} \qquad C'(x) = -\omega S(x)$$

    (c) Show that S and C are solutions of the differential equation

    $$y'' + \omega^2 y = 0$$

    satisfying the initial conditions: $y(0) = 0$, $y'(0) = 1$ and $y(0) = 1$, $y'(0) = 0$, respectively.

    (d) Show that $S^2(x) + C^2(x) = 1$. (Hint: Put $U(x) = S^2(x) + C^2(x)$ and calculate U'.)

4. Consider the initial value problem

    $$y' - 2y = e^x; \qquad y(0) = 0$$

Assume that the problem has a solution $y = y(x)$ that can be expressed as a power series expansion

$$y(x) = \sum_{n=0}^{\infty} a_n x^n = a_0 + a_1 x + a_2 x^2 \dots$$

with radius of convergence $R > 0$.

(a) Express y' as a power series and calculate the power series expansion for $y' - 2y$.

(b) Since two power series are equal if and only if the coefficients of the corresponding powers of x are equal, equate the power series found in part (a) to the MacLaurin series for e^x and solve for the coefficients. The result will be a recursion formula that gives a_{n+1} in terms of a_n, $n = 0, 1, 2, \dots$.

(c) Show that the initial condition $y(0) = 0$ implies $a_0 = 0$ and determine the solution of the initial value problem.

(d) Show that the power series found in (c) is equal to $y(x) = e^x - e^{2x}$.

5. What if you watch a TV showing a TV camera watching it?

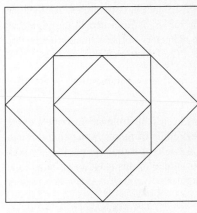

(a) The "nested" sequence of squares shown in the margin is formed as follows: join the midpoints of the sides of the outermost square S_1 to form the square S_2; join the midpoints of the sides of S_2 to form S_3; and so on.
Determine the sum of the areas of all the squares if the first square has area A.

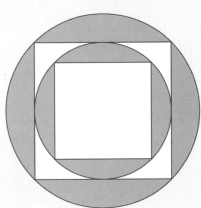

(b) The "nested" sequence of circles and squares shown in the margin is formed as follows: Inscribe a square in the circle of radius r; inscribe a circle in the square; inscribe a square in the circle; and so on.
Determine the area of the shaded region.

6. The curve defined parametrically by the equations

$$x = a\cos t, \qquad y = b\sin t, \qquad 0 < t < 2$$

where a and b are positive numbers, is an ellipse whose center is at the origin and whose axes are on the coordinate axes. Assuming that $a > b$, the major axis is on the x-axis. The arc length of this ellipse is

$$L = \int_0^{2\pi} \sqrt{a^2 \sin^2 t + b^2 \cos^2 t}\, dt \qquad\qquad (*)$$

(a) Show that the formula $(*)$ can be simplified to

$$L = 4a \int_0^{\pi/2} \sqrt{1 - e^2 \cos^2 t}\, dt$$

where e is the eccentricity of the ellipse. This form is an elliptic integral; the integrand does not have an elementary antiderivative.

(b) Use the binomial series for $\sqrt{1-x}$ to show that

$$L = 4a \int_0^{\pi/2} \left[1 - \frac{1}{2} e^2 \cos^2 t - \frac{1}{8} e^4 \cos^4 t - \frac{1}{16} e^6 \cos^6 t - \dots \right] dt$$

(c) Show that the arc length of the ellipse is given by

$$L = 2\pi a \left[1 - \frac{1}{4} e^2 - \frac{3}{64} e^4 - \frac{5}{256} e^6 - \dots \right]$$

by evaluating the integral in part (b). Hint: Integrate termwise using the identity

$$\int_0^{\pi/2} \cos^{2n} t\, dt = \frac{1 \cdot 3 \cdot 5 \cdots (2n-1)}{2 \cdot 4 \cdot 6 \cdots (2n)} \cdot \frac{\pi}{2}$$

See Exercise 49 in Exercise Set 8.1. This formula gives the arc length of an ellipse in terms of its semimajor axis and its eccentricity. Note that this formula reduces to $2\pi a$, the circumference of a circle of radius a when $e = 0$.

(d) The orbit of the moon about the Earth is (almost) a perfect ellipse with the Earth at one focus. Assume that $a = 400,000$ km and $e = 0.055$. Use the first three terms of the series in (c) to determine the length of the moon's path around the earth.

7. In the accompanying figure is a region bounded by two circles of radius 1 that are tangent to one another and by a straight line which is tangent in both circles. A sequence of smaller circles, each having the largest possible radius, is inscribed in the region as shown. Show geometrically that the lengths of the diameters of these smaller circles are the terms of a convergent series whose sum is one. Then show that this series is

$$\sum_{n=1}^{\infty} \frac{1}{n(n+1)}.$$

8. Construct a sequence of circles with radii $r_1, r_2, r_3, \ldots$, as follows. Inscribe a circle of radius $r_1 = 1$ in an equilateral triangle. Inscribe a circle of radius r_2, passing through the vertices of the triangle, in a square. Inscribe a circle of radius r_3, passing through the vertices of the square, in a regular pentagon. Continue in this way: inscribe a circle of radius r_{n-1}, passing through the vertices of a regular polygon of n sides, in a regular polygon of $n + 1$ sides.
(a) Sketch a diagram of what is happening.

(b) Show that $r_2 = r_1 \sec(\pi/3)$ and $r_3 = r_2 \sec(\pi/4)$. Find a general formula for r_n.

(c) Show that

$$\ln r_n = \ln r_1 + \ln \sec \frac{\pi}{3} + \ln \sec \frac{\pi}{4} + \cdots + \ln \sec \frac{\pi}{n+1}$$

(d) Does r_n have a limit as n tends to infinity? (Hint: Use the Limit Comparison Test with $\sum [\ln \sec(\pi/n)]$ and $\sum (1/n^2)$.)

9. An indicator of future population growth is the fertility rate, which is the average number of children a woman has during her childbearing years. If this rate exceeds the theoretical replacement level of 2 (it is slightly greater than 2 in the more developed countries and even higher in the less developed countries), the size of the population will continue to increase. One factor that affects a country's fertility rate is the type of family considered desirable. In each of the following assume that the probability of bearing male and female children is the same.

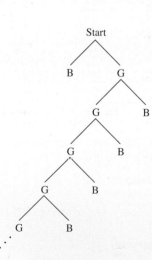

(a) Suppose each family will have children until they have a son. What is the fertility rate for the culture? (See the tree diagram.)

(b) Suppose each family will have children until they have at least one male and one female. Draw a tree diagram showing the possible outcomes. What is the fertility rate for this culture?

(c) Suppose each family has children until they have at least two males. Draw a tree diagram showing the possible outcomes. What is the fertility rate for this culture?

(d) A more general case is to suppose each family has children until they have at least m males and n females. Draw a tree diagram showing the possible outcomes. What is the fertility rate for this culture?

ANSWERS

CHAPTER 1

Exercise Set 1.1

1. (a) 0   (b) 0   (c) $-\frac{1}{2}$ (d) $\frac{a^2-1}{a+2}$ (e) $\frac{x^2+2x}{x+3}$ **3.** (a) 1   (b) 2   (c) 3   (d) $|t|$ (e) x^2 since $x^2 \geq 0$

5. (a) domain of $f = \mathbb{R}$, range of $f = -2 \cup (2, \infty)$.   (b) domain of $f = (-\infty, 3)$, range of $f = (0, \infty)$

7. domain of $f = \mathbb{R}$, range of $f = [-4, \infty)$ **9.** domain $f = (-\infty, -1]$, range $f = [0, \infty)$.   **11.** domain $f = \mathbb{R}$, range $f = \mathbb{R}$.

13.   **15.**   **17.**   **19.**   **21.**

23. Assume that $0 < w \leq 12$. The postage is 0.29 for the first ounce or fraction thereof plus 0.23 for each additional ounce or fraction thereof. This is equivalent to

$$P(w) = \begin{cases} 0.29 + 0.23\,[w] & \text{if } w \text{ is not an integer} \\ 0.29 + 0.23\,[w-1] & \text{if } w \text{ is an integer} \end{cases}$$

If $w = 1.5$, for example, $P(w) = 0.29 + 0.23\,[1.5] = 0.52$.
Similarly, if $w = 2.0$, $P(w) = 0.29 + 0.23\,[w-1] = 0.52$

25. $-\frac{1}{2x}$ **27.** $-\frac{3}{1+h}$ **29.** $f(x) = x^n$ is even if n is even and odd if n is odd.

31. (a) odd   (b) odd   (c) neither even nor odd   (d) even   (e) even

33. (a) odd   (b) even   (c) even   (d) neither even nor odd   (e) odd

35. (a) **(b)**

37. (a) $f(x) = 1000x - 25x^2$

(b) domain $= [x: 0 \leq x \leq 40]$

39. (a) $V = 2\pi r^3$
(b) $L = 4\pi r^2$
(c) $S = 6\pi r^2$

$h = 2r$

41. $\dfrac{12b - 2b^2}{3}$

43. domain $f = [-1, 3]$

45. domain $g = \{x: x \geq 1\}$. In Exercise 44, the square root is applied to the entire fraction, not the numerator and denominator individually.

47. domain $h = (-3, 2)$ range $h = \mathbb{R}$.

49. $V = 2\pi r^2 \sqrt{a^2 - r^2}$ **51.** Area $= \frac{4}{25}(10 - h)^2$ **53.** $f(x) = \begin{cases} 0.015x & \text{if } x \leq 1000 \\ 5 + 0.01x & \text{if } x > 1000 \end{cases}$

55. range $= \{y: |y| \leq 1\}$.   **57.** The parameter a is the slope of the line. If $a > 0$, the slope is positive, and $x_2 > x_1 \Rightarrow y_2 > y_1$. If $a > 0$, the slope is positive, and $x_2 > x_1 \Rightarrow y_2 > y_1$. If $a < 0$, the slope is negative, and $x_2 > x_1 \Rightarrow y_2 < y_1$. The parameter b represents the y-intercept, i.e., $y(0)$.

59. (a) the graph is shifted vertically upward a units.   **(b)** the graph is shifted horizontally left a units.   **(c)** $a = 1$ gives the original graph, $a = -1$ reflects that graph about the x-axis, $a = 2$ stretches the original graph vertically, and $a = -2$ stretches the graph and reflects it about the x-axis.
(d) $a = -1$ reflects the original graph about the y-axis, $a = 2$ shrinks the graph horizontally to half its original width (for a particular value of y), and $a = -2$ shrinks the graph and reflects it about the y-axis.

61. $\lim\limits_{x \to -3/2} f(x) = 0.2$

63. $\lim\limits_{x \to 8} f(x) = -12$

Exercise Set 1.2

1. (a) $y = f(x) + 2$ **(b)** $y = f(x - 3)$ **(c)** $y = 2f(x)$ **(d)** $y = f(2x)$ **(e)** $y = -f(x)$

(f) $y = f(-x)$ **3. (a)** $y = f(x) + 2$ **(b)** $y = f(x - 3)$ **(c)** $y = 2f(x)$ **(d)** $y = f(2x)$

(e) $y = -f(x)$ **(f)** $y = f(-x)$ **5. (a)** $y = f(x) + 2$ **(b)** $y = f(x - 3)$ **(c)** $y = 2f(x)$

(d) $y = f(2x)$ **(e)** $y = -f(x)$ **(f)** $y = f(-x)$ **7.** $y = \frac{1}{2}(x - 1)^3 + 2$
Move the graph 1
unit to the right, 2
units upward, and
then compress it by a
factor of $\frac{1}{2}$ in the
y-direction.

9. $y = g(x) = 3 - 2f(\frac{1}{2}x - 1)$

11. domain $f = \mathbb{R}$, domain $g = \mathbb{R}$
domain $f + g = \mathbb{R}$
domain $f - g = \mathbb{R}$
domain $fg = \mathbb{R}$
domain $f/g = \{x : x \neq -1\}$

13. domain $f = [-4, \infty)$
domain $g = (-\infty, 4]$
domain $f + g = [-4, 4]$
domain $f - g = [-4, 4]$
domain $fg = [-4, 4]$
domain $f/g = [-4, 4)$

15. domain $f = \{x : x \neq 0\}$
domain $g = \{x : x \neq 3\}$
domain $f + g = \{x : x \neq 0, 3\}$
domain $f - g = \{x : x \neq 0, 3\}$
domain $fg = \{x : x \neq 0, 3\}$
domain $f/g = \{x : x \neq 0, 3\}$

17. Domain of f and $g = \mathbb{R}$
domain $f + g = \mathbb{R}$
domain $f - g = \mathbb{R}$
domain $fg = \mathbb{R}$
domain $f/g = \{x : x < 0\}$

19.

21. (a) g is even. h is odd   **(b)** $g(x) + h(x) = \frac{1}{2}[f(x) + f(-x) + f(x) - f(-x)] = f(x)$

Exercise Set 1.3

1. domain f and domain $g = \mathbb{R}$, domain $f \circ g = \mathbb{R}$, domain $g \circ f = \mathbb{R}$,   **3.** domain $f = \{x : x \neq 1\}$, domain $g = \{x : x \neq 0\}$, domain
$f \circ g = \{x : x \neq 0, 3\}$, domain $g \circ f = \{x : x \neq 1, 0\}$ **5.** $F(x) = (4 - 3x^2)^6 = (f \circ g)(x)$, where $f(x) = x^6$ and $g(x) = 4 - 3x^2$
7. Let $f(x) = x^{3/2}$, $g(x) = x^2 - 1$.   **11.** $[f \circ (g \circ h)](x) = f \circ (g \circ h)(x) = f \circ (g(h(x)) = f(g(h(x)))$
$= (f \circ g)(h(x)) = [(f \circ g) \circ h](x)$. Hence, $f \circ (g \circ h) = (f \circ g) \circ h$ whenever both sides are defined.   **13.** $f^{-1}(x) = (3x + 2)/5$
domain $f^{-1} = \mathbb{R}$ **15.** $F^{-1}(x) = (x^2 + 3)/2$, domain F^{-1} must be restricted to range $F = \{x : x \geq 0\}$.
17. $\varphi^{-1}(t) = \sqrt{t + 1}$; domain $\varphi^{-1} = \{t : t \geq -1\}$ **19. (a)**   **(b)**

21. To make f 1 – 1, restrict the domain of f to $\{x : x \geq 0\}$. $f^{-1}(x) = \sqrt{\frac{1 - 2x}{x}}$; domain $f^{-1} = $ range $f = \left(0, \frac{1}{2}\right]$.   **23.** To make F 1 – 1,
restrict its domain to $\left[\frac{3}{2}, \infty\right)$. $F^{-1}(x) = \frac{3 + x}{2}$; domain $F^{-1} = $ range $F = [0, \infty)$. Alternatively, one can restrict domain F to $\left(-\infty, \frac{3}{2}\right]$ to
get $F^{-1}(x) = \frac{3 - x}{2}$ and domain $F^{-1} = [0, \infty)$.

25. $f^{-1}(x) = \frac{a-bx}{x-1}$; domain $f^{-1} = \{x : x \neq 1\}$ range $f^{-1} = \{y : y \neq -b\}$ **27.**

Exercise Set 1.4

1. $f(x) = -x + 2$ $f(10) = -10 + 2 = -8$ **3.** A: \$21 + 0.21/mile \$21 + 0.21 (320) = \$88.20, B: \$32 + 0.18/mile \$32 + 0.18 (320) = \$89.60. The better rate is given by agency A. For equal charges, $x \approx 367$ miles   If the mileage is greater than 367, agency B has the lower rate.   **5. (a)** $f(x) = 3x + 4$ or $f(x) = -3x - 8$ **(b)** $f(x) = d, d \in \mathbb{R}$ or $f(x) = x$

7. (a)

(b)

(c)

(d)

(e)

(f)

9. $y = -\frac{1}{2}(x+2)^2 + 3$ **11.** $P_{max} = \$44,000$ at $x = 800$

13. $R_{max} = \$3,698$ at $x = 7$ **15. (a)** $(f \circ g)(x) = a(cx + d) + b = acx + (ad + b)$, which is linear.   **(b)** $(f \circ g)(x) = a_2(b_2 x^2 + b_1 x + b_0)^2 + a_1(b_2 x^2 + b_1 x + b_0) + a_0$ which is a quartic function. Therefore, the composition of two quadratic functions is not quadratic.   **(c)** $p \circ q$ is of degree mn.   **17.** If $f_n(x) = x^n$ and n is an even positive integer, then increasing n causes the graph to lie closer to the x-axis on $(0,1)$ and to rise more steeply for $x > 1$. The functions are even. If n is an odd positive integer, the function is odd. As n increases, the graph lies closer to the x-axis on $(0,1)$ and rises more steeply for $x > 1$. **(a)**

(b)

19. $f(x) = 3x^3 - 9x^2 + 12$

21. lowest possible value of $n = 5$, since the concavity changes at three points

23.

$y = |2x - 3|$ $y = |x^2 + x - 2|$ $y = |2x - x^3|$

25. (a) $a + b + c = 1.$

(b) vertex is at $\left(-\frac{b}{2a}, \frac{4ac-b^2}{4a}\right)$.

(c) $c = 6.$ **(d)** $c = 6.$ $a = 5, b = -10,$ domain $f = \mathbb{R}$; range $f = [1, \infty)$

27. (a) $f_1(x) = x^3 + b_1 x^2 + c_1 x + d_1$, where $b_1 = \frac{b}{a}, c_1 = \frac{c}{a}$, and $d_1 = \frac{d}{a}$. The graph of f is a contraction of the graph of f_1 if $|a| < 1$ and an expansion if $|a| > 1$.   (b) $f_2(x) = x^3 + c_2 x + d_2$, where $c_2 = c_1 - \frac{b_1^2}{3}$ and $d_2 = \frac{2b_1^3}{27} - \frac{b_1 c_1}{3} + d_1$. The graph of f_1 is the graph of f_2 shifted $\frac{b_1}{3}$ units to the left.   (c) $f_3(x) = f_2(x) - d_2 = x^3 + c_2 x$ The graph of f_2 is displaced d_2 units upward from the graph of f_3.   (d) Let $k = c_2$, and write $g(x) = x^3 + kx$. Given the graph of g, find the graph of f_2 by displacing the graph d_2 units upward. The graph of f_1 is obtained by shifting the graph of f_2 to the left by $\frac{b_1}{3}$ units. The graph of f is now a contraction or expansion of the graph of f_1, as explained in part (a).   (e) If $k = 0$, the graph of g is simply that of the cubic x^3. If $k < 0$, the graph has zeros at $0, \sqrt{k}$, and $-\sqrt{k}$. It has a local maximum at $x = -\sqrt{k/3}$ and a local minimum at $x = \sqrt{k/3}$. If $k > 0$, the graph has no local maxima or minima. It rises more rapidly than the graph of x^3.

29. (a) local maxima at $\left(-3, \frac{72}{5}\right)$ and $\left(1, \frac{88}{15}\right)$. Local minima at $\left(-1, -\frac{88}{15}\right)$ and $\left(3, -\frac{72}{5}\right)$.   (b) increasing on $(-\infty, -3) \cup (-1, 1) \cup (3, \infty)$. decreasing on $(-3, -1) \cup (1, -3)$.   (c) concave up on $(-\sqrt{5}, 0) \cup (\sqrt{5}, \infty)$. concave down on $(-\infty, -\sqrt{5}) \cup (0, \sqrt{5})$.

Exercise Set 1.5

1. (a) Vertical asymptote: $x = 2$. Horizontal asymptote: $y = 2$ (b) Vertical asymptotes: $x = -2, 2$. Horizontal asymptote: $y = 0$

3. Vertical asymptote:
$x = 3$
Horizontal asymptote:
$y = 0$
y-intercept: $-\frac{1}{3}$

5. Vertical asymptotes:
$x = -1, 1$
Horizontal asymptote:
$y = 1$
x-intercepts: $-2, 2$
y-intercept: 4

7.

11. (a) oblique asymptote is $y = x$.   (b) oblique asymptote is $y = x - 4$.

13. (a) shift the graph of $y = \sqrt{x}$ two units to the left

$y = \sqrt{x + 2}$

(b) shift the graph of $y = \sqrt{x}$ one unit to the right.

$y = \sqrt{x - 1}$

15. (a) $y = \sqrt{3}\sqrt{-(x - 4/3)} - 2$; shift the graph of $y = \sqrt{x}$ to the right $\frac{4}{3}$ unit. Reverse the direction of the graph along the x-axis, and expand the graph in the y-direction by a factor of $\sqrt{3}$. Drop the resulting graph down 2 units.

$y = \sqrt{4 - 3x} - 2$

(b) expand the graph of $y = \sqrt{x}$ in the y-direction by a factor of 3. Raise the graph 1 unit.

$y = 1 + 3\sqrt{x}$

17. (a) upper half of an ellipse

$y = \sqrt{9 - 4x^2}$

(b) upper half of an ellipse

$y = \frac{3}{2}\sqrt{4 - x^2}$

19. (a) the graph opens upward like that of $y = x^2$, but does not rise so rapidly. The function is even.

$y = x^{3/2}$

(b) the graph is a displacement two units to the right of the graph of $y = x^{2/3}$.

$y = (x - 2)^{2/3}$

21. domain $f = \{x: -2 < x \le 2\}$
x-intercept: $x = 2$
y-intercept: $y = 1$
no symmetry
vertical asymptote: $x = -2$

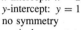

23. domain $f = \{x : |x| < 2\}$, x-intercept: $x = 0$, y-intercept: $y = 0$, symmetry: f is symmetric about the origin, vertical asymptotes: $x = \pm 2$

$y = \frac{x}{\sqrt{4 - x^2}}$

25. $y = \sqrt{\frac{x+2}{x-1}}$

27. $y = \sqrt{\frac{x^2-1}{x^2+1}}$

29. $y = \frac{2\sqrt{x^2-4}}{x-1}$

Exercise Set 1.6

1.

x	0	$\pi/6$	$\pi/4$	$\pi/3$	$\pi/2$
$\sin x$	0	$1/2$	$\sqrt{2}/2$	$\sqrt{3}/2$	1
$\cos x$	1	$\sqrt{3}/2$	$\sqrt{2}/2$	$1/2$	0
$\tan x$	0	$\sqrt{3}/3$	1	$\sqrt{3}$	DNE
$\cot x$	DNE	$\sqrt{3}$	1	$\sqrt{3}/3$	0
$\sec x$	1	$2\sqrt{3}/3$	$\sqrt{2}$	2	DNE
$\csc x$	DNE	2	$\sqrt{2}$	$2\sqrt{3}/3$	1

x	$2\pi/3$	$3\pi/4$	$5\pi/6$	π
$\sin x$	$\sqrt{3}/2$	$\sqrt{2}/2$	$1/2$	0
$\cos x$	$-1/2$	$-\sqrt{2}/2$	$-\sqrt{3}/2$	-1
$\tan x$	$-\sqrt{3}$	-1	$-\sqrt{3}/3$	0
$\cot x$	$-\sqrt{3}/3$	-1	$-\sqrt{3}$	DNE
$\sec x$	-2	$-\sqrt{2}$	$-2\sqrt{3}/3$	-1
$\csc x$	$2\sqrt{3}/3$	$\sqrt{2}$	2	DNE

x	$7\pi/6$	$5\pi/4$	$4\pi/3$	$3\pi/2$	$5\pi/3$	$7\pi/4$	$11\pi/6$
$\sin x$	$-1/2$	$-\sqrt{2}/2$	$-\sqrt{3}/2$	-1	$-\sqrt{3}/2$	$-\sqrt{2}/2$	$-1/2$
$\cos x$	$-\sqrt{3}/2$	$-\sqrt{2}/2$	$-1/2$	0	$1/2$	$\sqrt{2}/2$	$\sqrt{3}/2$
$\tan x$	$\sqrt{3}/3$	1	$\sqrt{3}$	DNE	$-\sqrt{3}$	-1	$-\sqrt{3}/3$
$\cot x$	$\sqrt{3}$	1	$\sqrt{3}/3$	0	$-\sqrt{3}/3$	-1	$-\sqrt{3}$
$\sec x$	$-2\sqrt{3}/3$	$-\sqrt{2}$	-2	DNE	2	$\sqrt{2}$	$2\sqrt{3}/3$
$\csc x$	-2	$-\sqrt{2}$	$-2\sqrt{3}/3$	-1	$-2\sqrt{3}/3$	$-\sqrt{2}$	-2

3. amplitude $= 2$,
period $= 2\pi$

$y = 2\cos x$

5. amplitude $= 3$,
period $= 4\pi$

$y = 3\sin x/2$

7. amplitude $= 1$,
period $= \frac{2\pi}{\pi} = 2$

$y = 2 + \sin \pi x$

9. amplitude $= \frac{1}{2}$,
period $= \frac{2\pi}{2} = \pi$

$y = \frac{1}{2}\sin 2x + \frac{\pi}{2}$

11. amplitude $= 3$,
period $= \frac{2\pi}{2} = \pi$

$y = 3\cos(2x - \pi/2)$

13. amplitude $= \frac{3}{2}$,
period $= \frac{2\pi}{2} = \pi$

$y = -\frac{3}{2}\sin(2x + 1) + 2$

15.

Period $= 4$. The graph
of $\cos x$ has been
expanded by 2 and
shifted down 1
unit. Thus,

$$y = 2\cos\left(\frac{\pi x}{2}\right) - 1.$$

17.

Period $= 4\left(\frac{\pi}{3}\right)$;

graph of $\cos x$ is
reversed about
the x-axis and
has a contraction
of $\frac{1}{2}$. Thus,

$$y = -\frac{1}{2}\cos\left(\frac{3}{2}x\right).$$

19. Period $= (4 - 1)(2) = 6$
(y attains minimum and
maximum values at $x = 1$ and
$x = 4$, respectively).
$ax = 2\pi$ when $x = 6$; $6a = 2\pi$; $a = \frac{\pi}{3}$
The graph of $\cos x$ has
been shifted 4 units to the
right and raised 2 units
on the y-axis. Thus,

$$y = \cos\left[\frac{\pi}{3}(x - 4)\right] + 2.$$

21. (a) Smallest positive value is $\frac{\pi}{3}$.
(b) $x = \frac{7\pi}{12}$ (c) $x = \frac{\pi}{12}$

23.

$y = 2\tan 3x$

25.

$y = \cot \pi x + 1$

27. Asymptotes $x = \frac{1}{b}(2n+1)\frac{\pi}{2} - \frac{c}{b}$.
The factor a causes expansion of
the graph in the y-direction if
$|a| > 1$ and contraction if
$0 < |a| < 1$. If $a < 0$, the graph
is reversed about the x-axis.

29.

The graph of
$y = \sin x + \cos x$ is periodic
with period 2π. maximum
of $\sqrt{2}$ at $x = \frac{\pi}{4}$; minimum
of $\sqrt{2}$ at $x = \frac{5\pi}{4}$; zeros
occur at $\frac{3\pi}{4}$ and $\frac{7\pi}{4}$. In
terms of the sine function
alone, $y = \sqrt{2}\sin\left(x + \frac{\pi}{4}\right)$.

31.

(a) The graph
appears to approach 1
as $x \to 0$. The
function is periodic
with period 2π. Its
amplitude $\to 0$ as
$x \to \pm\infty$.

(b)

$y = \sin\left(\frac{1}{x}\right) (x \neq 0)$

The graph oscillates
more and more
rapidly as $x \to 0$.
$f(x) \to 0$ as
$x \to \pm\infty$.

(c)

$y = \sin^2 x$
The function is periodic with period π. Its range is $[0,1]$.

(d)

$y = |\sin x|$
The graph is the same as that of $y = \sin x$ wherever $\sin x \geq 0$ and is its reflection across the x-axis wherever $\sin x < 0$.

(e)

$y = \sin x^2$
The graph oscillates more and more rapidly as $x \to \pm\infty$. Since $x^2 \geq 0$, the function is even.

(f)

$y = x \sin \frac{1}{x}$ $(x \neq 0)$
The graph appears to approach 1 as $x \to \pm\infty$. It appears to approach 0 as $x \to 0$. The function is even.

33. (a) $x = \frac{2}{(2n+1)\pi}$, $n = 0, \pm1, \pm2, \ldots$

(b) $x = \frac{2}{(4n+3)\pi}$, $n = 0, \pm1, \pm2, \ldots$

35.

Exercise Set 1.7

1. (a) domain $f =$
$\left(-\infty, -\frac{7}{3}\right] \cup [5, \infty)$

(b) domain $g =$
$(-\infty, -1) \cup (-1, 1] \cup (3, \infty)$

3. $(f \circ g)(t) = f(g(t)) = -t - 2$
domain $f \circ g = (-\infty, 1]$

$(g \circ f)(t) = \sqrt{4 - t^2}$
domain $g \circ f = [-2, 2]$

5. (a) $f(-3) = 3$
(b) $f(0) = \frac{1}{2}$
(c) $f(1) = 1$
(d) $f(1.8) = 5$
(e) vertical asymptote:
$x = 2$

7. $(g \circ h)(t) = -\frac{t+2}{t+1}$
domain $g \circ h = \{t : t \neq -2, -1\}$
$(h \circ g)(t) = \frac{t-1}{2t-1}$

domain $h \circ g = \left\{t : t \neq \frac{1}{2}, 1\right\}$

9. (a) $(f - g)(3) = 11$

(b) $\left(\frac{f}{g}\right)(-1) = -2$

(c) $(f \circ g)(x) =$
$\sec^2 \frac{\pi x}{4}$

11. (a) $y = f(x) + 1$

(b) $y = f(x + 1)$

(c) $y = 2f(x)$

(d) $y = f(-x)$

(e) $y = f\left(\frac{x}{2}\right)$

13. (a) $f^{-1}(x) = \frac{1}{(1-x)^2}$

(b) $f^{-1}(x) = \frac{3x+2}{x-1}$

15. zeros of f are $0, 0, -2, 4$

(a) increasing on $(-\infty, -1.4) \cup (0, 2.9)$
(b) decreasing on $(-1.4, 0) \cup (2.9, \infty)$
(c) concave up on $(-0.76, 1.76)$
(d) concave down on $(-\infty, -0.76) \cup (1.76, \infty)$

17. (a) The graph of $y = x^{2/3}$ is shifted right $\frac{1}{2}$ unit and stretched vertically by a factor of $2^{2/3} \approx 1.587$.

(b) domain $f = [-1, 1)$
vertical asymptote:
$x = 1$

19. (a) period $= 9 - 1 = 8$,
amplitude $= 2$,
horizontal shift $= 3$,
vertical shift $= -2$

(b) period $= 2\pi$
amplitude $= 2$

horizontal shift $= -\frac{5}{6\pi}$,
vertical shift $= 1$
$y = 2\sin\left(x + \frac{5\pi}{6}\right) + 1$

CHAPTER 2

Exercise Set 2.1

1. 3

3. does not exist.

5. does not exist.

7. $\lim\limits_{x \to 2^+} f(x) = \lim\limits_{x \to 2^-} f(x) = 2$; $\lim\limits_{x \to 2} f(x) = 2$

9. $\lim\limits_{x \to 2^+} f(x)$ does not exist; $\lim\limits_{x \to 2^-} f(x) = 0$; $\lim\limits_{x \to 2} f(x)$ does not exist

11. $\lim\limits_{x \to 2^+} f(x)$ does not exist; $\lim\limits_{x \to 2^-} f(x)$ does not exist; $\lim\limits_{x \to 2} f(x)$ does not exist.

13. $\lim\limits_{x \to 2^+} f(x) = 4$; $\lim\limits_{x \to 2^-} f(x) = 4$; $\lim\limits_{x \to 2} f(x)$ exists; its value is 4.

15. $\lim\limits_{x \to 1} f(x) = 2$; $\lim\limits_{x \to -1^-} f(x) = 2$; $\lim\limits_{x \to 1} f(x)$ exists; its value is 2.

17. $\lim\limits_{x \to 1^+} f(x) = 1$; $\lim\limits_{x \to 1^-} f(x) = -1$; $\lim\limits_{x \to 1} f(x)$ does not exist.

19.

x	-2.1	-2.01	-2.001	-1.9	-1.99	-1.999
$f(x)$	12.61	12.06	12.006	11.41	11.94	11.994

Conjecture: $\lim\limits_{x \to -2} f(x) = 12$

21.

t	1.1	1.01	1.001	0.9	0.99	0.999
$f(t)$	-1.281	-1.328	-1.333	-1.393	-1.339	-1.334

Conjecture: $\lim\limits_{t \to 1} f(t) = -\frac{4}{3}$

23.

x	4.1	4.01	4.001	3.9	3.99	3.999
$f(x)$	0.3315	0.3331	0.3333	0.3352	0.3335	0.3334

Conjecture: $\lim\limits_{x \to 4} f(t) = \frac{1}{3}$

25.

x	0.1	0.01	0.001
$f(x)$	0.4996	0.499995	0.50000

Conjecture: $\lim\limits_{x \to 0} f(x) = \frac{1}{2}$

27.

x	$\frac{\pi}{2} - 0.1$	$\frac{\pi}{2} - 0.01$	$\frac{\pi}{2} - 0.001$	$\frac{\pi}{2} - 0.0001$
$f(x)$	1.2580	1.0471	1.00693	1.00092

Conjecture: $\lim\limits_{x \to (\pi/2)^-} f(x) = 1$

29. $\lim\limits_{x \to 2} (3x - 1) = 5,\ x \in \left(\frac{5.99}{3}, \frac{6.01}{3} \right),\ x \in \left(\frac{5.999}{3}, \frac{6.001}{3} \right)$

31. $\lim\limits_{x \to 0} x \sin\left(\frac{1}{x} \right) = 0,\ x \in (-0.01, 0.01),\ x \in (-0.001, 0.001)$.

Exercise Set 2.2

1. (a) 0   (b) -2 **3.** (a) 1   (b) $\sqrt{2}$ **5.** -4 **7.** 5   **9.** -7 **11.** $\frac{1}{2}$ **13.** $\frac{3}{8}$ **15.** $\frac{1}{2}$

17. $f(x) = x - 2$
domain
$f = \{x : x \neq -1\}$
$\lim\limits_{x \to -1^+} f(x) = -3$
$\lim\limits_{x \to -1^-} f(x) = -3$
$\lim\limits_{x \to -1} f(x) = -3$

19. $f(x) = 2x - 1$
domain $f = \{x : x \neq \frac{1}{2}\}$
$\lim\limits_{x \to -\frac{2}{3}^+} f(x) = -\frac{7}{3}$
$\lim\limits_{x \to -\frac{2}{3}^-} f(x) = -\frac{7}{3}$
$\lim\limits_{x \to -\frac{2}{3}} f(x) = -\frac{7}{3}$

21. $f(x) = x^2$
domain $f = \{x : x \neq -2\}$
$\lim\limits_{x \to 2^+} f(x) = \lim\limits_{x \to -2^-} f(x) = 4$
$\lim\limits_{x \to -2} f(x) = 4$

23. $f(x) = \frac{1}{x-1}$
domain $f = \{x : x \neq -1, 1\}$
$\lim\limits_{x \to -1^+} f(x) = \lim\limits_{x \to -1^-} f(x) = -\frac{1}{2}$
$\lim\limits_{x \to -1} f(x) = -\frac{1}{2}$

25. 2   **27.** -1 **29.** 1   **31.** (a) Let $f(x) = \begin{cases} 1 & \text{if } x \geq 0 \\ 0 & \text{if } x < 0 \end{cases}$ $g(x) = \begin{cases} 0 & \text{if } x \geq 0 \\ -1 & \text{if } x < 0 \end{cases}$ (b) Let $f(x) = \begin{cases} 1 & \text{if } x \geq 0 \\ 0 & \text{if } x < 0 \end{cases}$

$g(x) = \begin{cases} 0 & \text{if } x \geq 0 \\ -1 & \text{if } x < 0 \end{cases}$ **33.** for $-\frac{\pi}{4} < x < \frac{\pi}{4},\ 0 \leq \lim\limits_{x \to 0} x^2 \cos 2x \leq \lim\limits_{x \to 0} x^2 = 0$ therefore $\lim\limits_{x \to 0} x^2 \cos 2x = 0$

Exercise Set 2.3

1. continuous except at $x = 0$ **3.** continuous except at $x = 2$ **5.** continuous except at $x = 0$ **7.** continuous except at $x = 3$
9. continuous except at $x = (2n + 1)\frac{\pi}{2}$ **11.** continuous at $x = 2$ **13.** continuous at $x = 3$ **15.** not continuous at $x = -1$
17. continuous at $x = 2$ **19.** continuous at $x = 0$ **21.** not continuous at $x = -3$ **23.** not continuous at $x = 0$
25. continuous on $(-\infty, \infty)$.   **27.** continuous on $(-\infty, 0] \cup (0, \infty)$.   **29.** G is continuous on $(-\infty, 0) \cup [0, \infty)$
31. continuous on all of $\mathbb{R}$.   **33.** discontinuous at $x = 2$ only.   **35.** define $f(3) = \frac{6}{5}$ **37.** ϕ is discontinuous at all points

39. $f(x) = \frac{1}{x}$ on $(0, 1)$ **41.** Yes, $\phi(x)$ of Exercise 37 is discontinuous at every point. Yes, $\psi(x)$ of Exercise 38 is continuous only at

$x = 0$.   **43.** (a) False. Let $f(x) = \begin{cases} 1 & \text{if } x \geq 0 \\ 0 & \text{if } x < 0 \end{cases}$ $g(x) = \begin{cases} 0 & \text{if } x \geq 0 \\ 1 & \text{if } x < 0 \end{cases}$ (b) False. Use same two functions as in part (a).

45. Show $P_n(a)$ and $P_n(-a)$ have opposite signs for a large. Then use the intermediate value theorem.   **47.** continuous for all real x when x is not an integer and is discontinuous when x is an integer value.

Exercise Set 2.4

1. (a) $-\infty$ (b) ∞ **3.** (a) ∞ (b) $-\infty$ **5.** (a) $-\infty$ (b) ∞ **7.** 0   **9.** 1   **11.** 2   **13.** does not exist.
15. (a) 2   (b) -2 **17.** (a) $\lim\limits_{x \to +\infty} f(x) = 1$ (b) $\lim\limits_{x \to +\infty} f(x) = 0$ **19.** (a) $\lim\limits_{x \to \infty} f(x) = 0$ (b) $\lim\limits_{x \to \infty} f(x)$ does not exist
$\lim\limits_{x \to -\infty} f(x) = -1$ $\lim\limits_{x \to -\infty} f(x) = 0$ $\lim\limits_{x \to -\infty} f(x) = 0$ $\lim\limits_{x \to -\infty} f(x) = -1$

21. vertical: $x = 2, x = -2$ **23.** vertical: $x = 3, x = -2$ **25.** vertical: $x = -1$ no horizontal asymptote
horizontal: $y = 0$ horizontal: $y = 2$

27. vertical: $x = 3$, horizontal: $y = 0$ **29.** vertical: $x = -1$, horizontal: $y = 2$ **31.** horizontal: $y = 0$, vertical: $x = 1$

33. horizontal: $y = 0$ (on right), no vertical   **35.** vertical: $\begin{cases} x = \pm\frac{\pi}{6} + 2n\pi \\ x = \pm\frac{5\pi}{6} + 2n\pi \end{cases}$ $n = 0, \pm 1, \pm 2, \ldots$, no horizontal

37. Conjecture: $\lim\limits_{x \to +\infty} \left(1 + \frac{1}{x}\right)^x \approx 2.7182$ **39.** Property 2: Consider $\lim\limits_{x \to -1} \left[\frac{1}{x+1} + \frac{x}{x+1} \right] = \lim\limits_{x \to -1} \frac{1+x}{x+1} = 1$.

Property 3: Consider $\lim\limits_{x \to 1} \left[\frac{1}{x-1} - \frac{x}{x-1} \right] = \lim\limits_{x \to 1} \frac{1-x}{x-1} = -1$.

Property 4: Consider $\lim\limits_{x \to 0} \left[x \cdot \frac{1}{x} \right]$.

Property 5: Consider $\lim\limits_{x \to 0} \frac{x}{1/x}$.

41. (a) Find $P(1)$, $P(2)$, and extend to $P(n)$.

(b)

How Compounded	Number of Times Per Year, n	$P(n)$
Annually	1	$1100.00
Semiannually	2	$1102.50
Quarterly	4	$1103.81
Monthly	12	$1104.71
Weekly	52	$1105.06
Daily	365	$1105.16
Hourly	8760	$1105.17

(c) Conjecture: $\lim\limits_{n\to\infty} P(n) = 1000e^{0.10} = \$1105.17$

43. 20,000   **45.** $\approx 2.39 \times 10^3$ m/sec

Exercise Set 2.5

1. take $\delta = \frac{\epsilon}{2}$.   **3.** Take $\delta = 3\epsilon$.   **5.** Take $\delta = \frac{\epsilon}{4}$.   **7.** Take $\delta = \frac{\epsilon}{2}$.   **9.** Take $\delta = 5\epsilon$.   **11.** Choose $\delta = \epsilon$.
13. Choose $\delta = \frac{\epsilon}{2}$.   **15.** Choose $\delta = \epsilon$.   **17.** There is a $\delta_1 > 0$ such that if $0 < |x - a| < \delta_1, |f(x) - 0| < 1$.   **19.** By
Problem 15 and Problem 18, we know that $\lim\limits_{x\to a} |x| = |a|$. $|x|$ is continuous at a, hence on all of $\mathbb{R}$.   **21.** Choose $\delta^+ = \sqrt{\frac{2}{M}}$.
Choose $\delta^- = \frac{3}{M}$, and let $\delta = \frac{3}{M}$.   **23.** Choose $\delta < \frac{2}{M}$.   **25.** $\frac{1}{2}$ Choose $N = \frac{3-2\epsilon}{4\epsilon}$ **27.** 0   Choose $N = \frac{2}{\epsilon}$
29. (a) $\lim\limits_{x\to\infty} f(x) = \infty$ if $f(x) > M$ for all $x > N$ in the domain of f where M is an arbitrary positive constant.   **(b)** $\lim\limits_{x\to\infty} f(x) = -\infty$
if $f(x) < -M$ for all $x > N$ in the domain of f where M is an arbitrary positive constant.   **(c)** $\lim\limits_{x\to-\infty} f(x) = \infty$ if $f(x) > M$ for
all $x < -N$ in the domain of f where M is an arbitrary positive constant and $N > 0$.   **(d)** $\lim\limits_{x\to-\infty} f(x) = -\infty$ if $f(x) < -M$ for
all $x < -N$ in the domain of f where M is an arbitrary positive constant and $N > 0$.   **31.** Choose $\delta = \frac{\epsilon}{2a}$ **33.** Choose
$\delta = \min\left\{\frac{1}{2}, \frac{\epsilon}{2}\right\}$.   **35.** Choose $\delta = \min\left\{1, \frac{\epsilon}{3}\right\}$.   **37.** For $a = 0$, let $\delta = \epsilon$. For $a \neq 0$, assuming continuity leads to a
contradiction.   **39.** By continuity of f at a, $f(a) = \lim\limits_{x\to a} f(x)$; so let $f(a) = L > 0$ in Exercise 38. This holds for $x = a$ also.
If $f(a) = L < 0$, the proof is analogous.

Chapter Review

1. (a) ∞ **(b)** $-\infty$ **(c)** $-\infty$ **(d)** ∞ **(e)** 1   **(f)** 1   **(g)** 0 and 1.   **3. (a)** 0   **(b)** $\frac{1}{4}$ **5. (a)** does not exist.   **(b)** does not
exist.   **7. (a)** $-\frac{1}{4}$ **(b)** -1 **9. (a)** 0   **(b)** 0   **11.** $\frac{1}{\sqrt{2x+3}}, \frac{1}{\sqrt{x}}$ **13. (a)** Discontinuous at $x = 2$.   **(b)** Discontinuous at
$x = \pm\sqrt{2}$ **15.** Take $\delta = \frac{\epsilon}{2}$.   **17.** Take $\delta = \min\{2, \epsilon\}$.   **19.** Choose $\delta = \frac{\epsilon}{|m|}$.   **21.** Choose $M = \frac{3}{\sqrt{\epsilon}}$

CHAPTER 3

Exercise Set 3.1

1. (a)

(b)
$h = 1$:	slope = 2.5
$h = 0.5$:	slope = 2.25
$h = 0.1$:	slope = 2.05

(c) 2   **(d)** $y = 2x - 2$

3. $2x + y = 2$

5. $x + 2y = -3$

7. $2x - y = 5$ **9.** $4x + 2y + 1 = 0$

11. (a) 6; 5; 4.2; 4.02
(b) $v_{av} = 4 + 2h$
(c) $v(1) = 4$
(d)

13. (a) $v(t) > 0$
(b) $v(t) < 0$
(c) $v(t)$ increasing
(d) $v(t)$ decreasing
(e) $v(t_0) = 0$

15. (a) -8 m/sec   **(b)** $v(5) = 0$ **17.** 2   **19.** $\frac{1}{2}$ **21.** 5   **23.** $-\frac{4}{9}$ **25.** $\frac{9}{2}$ **27.** $2(x - 1)$

29. $\frac{1}{(x+1)^2}$ **31.** $x - 3y + 5 = 0$ **33.** Increasing on $(-\infty, 1) \cup (1, \infty)$ and decreasing on $(-1, 1)$.

35. (a) Increasing on $(0.45, 3.72) \cup (5.42, 6.86) \cup (5.42 + n\pi, 6.86 + n\pi), n = 1, 2, \ldots$ Also on
$(-0.45, -3.72) \cup (-5.42, -6.86) \cup (-5.42 - n\pi, -6.86 - n\pi), n = 1, 2, \ldots$ Decreasing on $(-0.45, 0.45) \cup (3.72, 5.42)$
$\cup(3.72 + n\pi, 5.42 + n\pi), n = 1, 2, \ldots$ Also on $(-5.42, -3.72) \cup (-5.42 - n\pi, -3.72 - n\pi), n = 1, 2, \ldots$ **(b)** ≈ -0.10 m/s

Exercise Set 3.2

1. $3; 3x - y + 4 = 0$ **3.** $-4; 4x + y + 8 = 0$ **5.** $1; x - y - 1 = 0$ **7.** $\frac{1}{6}; \frac{1}{6}x - 6y + 9 = 0$ **9.** $-2; 2x + y + 3 = 0$

11. $3; 3x - y + 4 = 0$ **13.** $-4; 4x + y + 8 = 0$ **15.** $1; x - y - 1 = 0$ **17.** $\frac{1}{6}; \frac{1}{6}x - 6y + 9 = 0$ **19.** $-2; 2x + y + 3 = 0$

21. 2   **23.** $\frac{-2}{x^2}$ **25.** $\frac{1}{(x+1)^2}$ **27. (a)**   **(b)**   **29. (a)**

(b)   **31. (a)**   **(b)**   **33.** $\frac{-1}{(2x-3)^{3/2}}$

domain $f' = \left(\frac{3}{2}, \infty\right)$

35. $\frac{3}{2}x^{1/2}$ domain $f' = [0, \infty)$ **37.** 0   **39.**

$y = f(x)$

$y = f'(x)$

41. $\frac{1}{(t+1)^2}$

domain $f' = (-\infty, 0) \cup (0, \infty)$
f' and g are the same.

43. $-2\frac{Gm_1 m_2}{r^3}$ **45. (a)** The graph of $y = f'(x)$ appears to be that of $y = \cos x$.   **(b)** The graph of $y = g'(x)$ appears to be that of $y = -\sin x$.   **47.** Conjecture: $g'(x) = e^x$

$y = f'(x)$ $y = \sin x$

$y = \cos x$ $y = g'(x)$

Exercise Set 3.3

1. 0   **3.** $4x^3$ **5.** -2 **7.** a **9.** $12x^2 - 14x + 8$ **11.** $5x^4 - 24x^3$ **13.** $2ax + b$ **15.** $4x + 3$ **17.** $18x + 24$
19. $15x^2 - 6x - 6$ **21.** $y = 5x - 4$ **23.** $k = \frac{4}{3}$ **25.** tangent line: $2x - y - 9 = 0$ **27.** 20
normal line: $x + 2y + 3 = 0$

29. -10 **31.** $a = \frac{5}{4}, b = -4$ **33.** $(2, 4), (-2, 4)$
(1) $y = 4x - 4$
(2) $y = -4x - 4$

$(x_0, f(x_0))$ $(0, -4)$

35. $v(t) = 48t^5 - 144t^3 + 108t$;
$a(t) = 240t^4 - 432t^2 + 108$

Exercise Set 3.4

1. $f'(x) = 12x - 1$ **3.** $-20x^3 + 6x$ **5.** $f'(x) = 14x^6 + 4x^3 - 1$ **7.** $-12x^3 + 18x^2 - 10x - 8$
9. $7x^6 - 15x^4 + 12x^3 + 9x^2 - 12x - 1$ **11.** $70x^6 + 72x^5 - 75x^4 - 52x^3 + 60x^2 - 21$ **13.** $\frac{-2}{(x-2)^2}$ **15.** $\frac{4x}{(x^2+1)^2}$ **17.** $\frac{-8x}{(x^2-4)^2}$

19. $\frac{-2x^2 + 6x + 6}{(x^2 + 2x)^2}$ **21.** $\frac{-27x^2}{(x^3-1)^2}$ **23.** $-6x^{-4}$ **25.** $2x - 2x^{-3}$ **27.** $1 - 4x^{-2}$ **29.** $\frac{8}{x^3}$ **31.** $4x + y = 13$ **33.** $\frac{27}{16}$

35. If g is continuous at $x = a$, then $\lim\limits_{x \to a} g(x) = g(a)$. Let $x = a + h$ and consider $\lim\limits_{h \to 0} g(x + h)$.   **37.** $-\frac{2(9x^4 - 18x^3 + 14x^2 - 5x + 5)}{[(3x^2 + 5)(x^2 - 2x)]^2}$

39. $1 - \frac{2x}{(x^2+1)^2}$ **41.** $2x(2x+1)(3x^2 - 4x) + (2x+1)(6x-4)(x^2+4) + 2(3x^2 - 4x)(x^2 + 4)$

Exercise Set 3.5

1. $y' = 8x^3, y'' = 24x^2, y''' = 48x$ **3.** $y' = 12x^2 - 4x, y'' = 24x - 4, y''' = 24$ **5.** $y' = 2x + 2x^{-3}, y'' = 2 - 6x^{-4},$
$y''' = 24x^{-5}$ **7.** $12x + 7; 12$ **9.** $4x^3 - 6x - \frac{5}{x^2}; 12x^2 - 6 + \frac{10}{x^3}$ **11.** $\frac{x^2 - 3}{x^2}; \frac{6}{x^3}$ **13.** $3x^2 - 6x$ **15.** $\frac{5}{(x+3)^2}$

17. $6x - 6x^{-4}$ **19.** $2 + \frac{3}{2}x^{-4}$ **21.** $n! a_n; 0$ **23.** $f'(x) = \frac{1}{2}x^{-1/2}; f''(x) = -\frac{1}{2^2}x^{-3/2},$
$f'''(x) = \frac{1 \cdot 3}{2^3}x^{-5/2}; f^{(4)}(x) = -\frac{1 \cdot 1 \cdot 5}{2^4}x^{-7/2}, f^{(n)}(x) = (-1)^n \cdot \frac{(-1)1 \cdot 3 \cdot 5 \cdots (2n-3)}{2^n}x^{-(2n-1)/2}, (n = 1, 2, \ldots)$

25. $v(t) = 4t - \frac{t^2 + 6t + 1}{t^2 + 6t + 9}, a(t) = 4 - \frac{16}{(t+3)^3}, v(1) = \frac{7}{2}, a(1) = \frac{15}{4}$

27. $y' = 2 - \frac{1}{x^2}; y'' = \frac{2}{x^3}$ Substitute into the differential equation   **29. (a)** $6(3x - 2)$ **(b)** $8(x + 2)$ **(c)** $4(x - 1)^3$ **(d)** $6x^2(x^3 + 1)$

Exercise Set 3.6

1. 2   **3.** 1   **5.** 0   **7.** $\frac{3}{2}$ **9.** $\frac{2}{5}$ **11.** $2x \sin x + x^2 \cos x$ **13.** $\frac{-2 \sin x \cos x}{(1+\sin x)^2}$ **15.** $\frac{x \cos x - \sin x}{x^2}$

17. $-2 \sin x + \cos x$ **19.** $\frac{1}{1 - \sin 2x}$ **21.** (a) $2 \sin x \cos x$ (b) $-2 \sin x \cos x$ **23.** $y = \frac{\pi}{2}$ **25.** $-x \cos x - \sin x$

27. $\frac{x(1+\sin^2 x)+2\sin x \, \cos x}{\cos^3 x}$ **29.** $v(t) = (t+1)\cos t + \sin t$ **31.** $y' = x(-2 \sin x + \cos x) + (2\cos x + \sin x)$
$\qquad\qquad\qquad\qquad\qquad a(t) = -(t+1)\sin t + 2\cos t \qquad\qquad y'' = -x(2\cos x + \sin x) + 2(\cos x - 2\sin x)$
$\qquad\qquad\qquad\qquad\qquad\qquad\qquad\qquad\qquad\qquad\qquad$ Substitute into equation.

33. Let $\sec x = \frac{1}{\cos x}$ and use quotient rule.   **35.** (a) $s(0) = 3$; $v(0) = 2$ (b) $s\left(\frac{2\pi}{3}\right) \approx 0.2321$, $v\left(\frac{2\pi}{3}\right) \approx -3.5981$, $a\left(\frac{2\pi}{3}\right) \approx -0.2321$,

moving to the left and speeding up.   (c) $t \approx 2.1588$ **37.** No $\lim\limits_{x \to 0} \frac{\sin x}{x} = \frac{\pi}{180} \approx 0.01745$.

39. $\frac{d}{dx}(2\cos^2 x - 1) = -4\sin x \cos x = -2 \sin 2x$ **41.** (a) odd   (b) $\frac{1+\cos x + x + \sin x}{\cos x}$ (c) $y \approx 6.826x - 3.322$

Exercise Set 3.7

1. $8x(x^2 + 2)^3$ **3.** $-20x(1 - 2x^2)^4$ **5.** $120x^3(3x^4 - 2)^9$ **7.** $3(x^2 - 3x + 4)^2(2x - 3)$ **9.** $-8x(3x^2 - 1)(3x^4 - 2x^2 + 1)^{-3}$

11. $2(2x + 1)(x^2 + 2)^2(8x^2 + 3x + 4)$ **13.** $\frac{-12x}{(x^2+1)^4}$ **15.** $\frac{(3x-4)^2(6x+13)}{(x+1)^2}$ **17.** $\frac{-2x^3 - 6x}{(x^2-1)^3}$ **19.** $\frac{6(x-1)^2}{(x+1)^4}$ **21.** $3\cos 3x$

23. $2\sin(1 - 2x)$ **25.** $6x \sec^2(3x^2)$ **27.** $-6 \csc 3x \cot 3x$ **29.** $10 \sin x \cos x$ **31.** $4\tan(2x + 3)\sec^2(2x + 3)$
33. $6x \sec^2 x^2 \tan^2 x^2$ **35.** $-2\cos 2x$ **37.** $-3(3x + 2)\csc 3x \cot 3x + 3\csc 3x$ **39.** $-\sin x \cos(\cos x)$ **41.** $6(2x + 3)^2$

43. $-2(x^2 + \sin x)^{-3}(2x + \cos x)$ **45.** $18x^2(x^3 - 8)$ **47.** $12x(x^2 - 3)^2$ **49.** $-\frac{18x^2}{(x^3+4)^3}$ **51.** $2(\sin x + \cos x - \cos 2x)$

53. $8(x^2 + 1)^2(7x^2 + 1)$ **55.** $\frac{2}{(1-x)^3}$ **57.** $2[(1 - 2x^2)\cos 2x - 4x \sin 2x]$ **59.** $48\sec^5 2x - 36\sec^3 2x$

61. $(h \circ g)'(t) = \frac{-4t}{(t^2-1)^2}$, $(g \circ h)'(t) = \frac{-4(t+1)}{(t-1)^3}$ **63.** $3(y^2 + 3y - 1)(y^2 + 3y - 3)(2y + 3)$ **65.** $\frac{4(6x^3-5x^2-6x+1)}{(x^2+1)^4}$

67. $36x^2(2x^3 - 3)[(2x^3 - 3)^2 + 4]^2$ $f(x) = x^3, g(x) = x^2 + 4, h(x) = 2x^3 - 3$ **69.** $-2\sqrt{3} + 80\pi \frac{\sqrt{3}}{3}, 32 + 16\sqrt{3} - \frac{1280\pi}{3}$

71. $\left[-\sin(y^2 + 2xy)\right]\left[2y\frac{dy}{dx} + 2x\frac{dy}{dx} + 2y\right]$ **73.** $f'(f_{n-1}(x))f'(f_{n-2}(x))\ldots f'(f_2(x))f'(f(x))f'(x)$ **75.** (a) $-\frac{1}{x}\cos\frac{1}{x} + \sin\frac{1}{x}$

(b) $f'(0) = \lim\limits_{h \to 0} \sin\frac{1}{h}$, which does not exist.   (c) If $x \neq 0$, f is the product of two continuous functions. If $x = 0$,

$\lim\limits_{x \to 0} x \sin\frac{1}{x} = f(0) = 0$

Exercise Set 3.8

1. $\frac{x}{y}$ **3.** $\frac{2(1-xy^3)}{1+3x^2y^2}$ **5.** $\frac{3x^2}{1+3y^2}$ **7.** $\frac{2xy}{4y-x^2}$ **9.** $\frac{1}{2y-1}$ **11.** $-\frac{\sin y + y\cos x}{\sin x + x\cos y}$ **13.** $\frac{3 - y\sec(xy)\tan(xy)}{x\sec(xy)\tan(xy)}$ **15.** $\frac{1-\cos x \cos y}{1-\sin x \, \sin y}$

17. $\frac{x(2-x^2-y^2)}{y(2+x^2+y^2)}$ **19.** $\frac{x}{\sqrt{x^2+1}}$ **21.** $\frac{x}{(4-x^2)^{3/2}}$ **23.** $\frac{4x^2}{\sqrt[3]{2x^3+5}}$ **25.** $\frac{1}{(1-x^2)^{3/2}}$ **27.** $\frac{\cos x}{2\sqrt{\sin x}}$ **29.** $-\frac{1}{2x^{3/2}}\sec^2\left(\frac{1}{\sqrt{x}}\right)$

31. $-\frac{2\sec(2x+3)}{\tan^2(2x+3)} = -2\csc(2x+3)\cot(2x+3)$ **33.** $2\cos x$ **35.** $\frac{-4}{x^2\sqrt{4-x^2}}$ **37.** $-\frac{1}{y^3}$ **39.** $\frac{-4}{x^{1/2}y^2}$ **41.** $-\frac{1}{(1-x^2)^{3/2}}$

43. $-\left[\frac{1}{2\sqrt{x}}\right]^2 \sin\sqrt{x} - \frac{1}{4x^{3/2}}\cos\sqrt{x}$ **45.** $y = y_0 - \frac{b^2 x_0}{a^2 y_0}(x - x_0)$ **47.** $-\sin^2(x + y); -2\sin(x + y)\cos^3(x + y)$

49. (a) $f_1(x) = x + \sqrt{4 - x^2}$; $f_2(x) = x - \sqrt{4 - x^2}$ **51.** $v(t) = \frac{4t}{3(t^2-1)^{1/3}}$ **53.** Product of the derivatives is -1

(b) $f_1'(x) = \frac{\sqrt{4-x^2}-x}{\sqrt{4-x^2}}$; $f_2'(x) = \frac{\sqrt{4-x^2}+x}{\sqrt{4-x^2}}$ $a(t) = \frac{4(t^2-3)}{9(t^2-1)^{4/3}}$

(c) $y' = \frac{y-2x}{y-x}$ $v(3) = 2$; $a(3) = \frac{1}{6}$

Exercise Set 3.9

1. $(2x + 2)dx$ **3.** $-\dfrac{x\,dx}{\sqrt{4-x^2}}$ **5.** $\dfrac{2-x}{2(1-x)^{3/2}}\,dx$ **7.** $(x\sin x + \sin x)dx$ **9.** $(\sec x\,\tan x + \sec^2 x)dx$ **11.** $\dfrac{4x^5\,dx}{(x^6+1)^{1/3}}$

13. $\left(\dfrac{x\sec^2 x - \tan x}{x^2}\right)dx$ **15.** $(2x\sin 2x^2)dx$ **17.** $\dfrac{3x^2-2y^2}{4xy}$ **19.** $\dfrac{3y-4x}{8y-3x}$ **21.** $-\dfrac{y^{1/3}}{x^{1/3}}$ **23.** $\dfrac{y\cos x - \sin y}{x\cos y - \sin x}$

25. (a) $1.8 \text{ cm}^2 \approx \Delta A$ **27.** $0.001; 0.1\%;$ **29.** $\approx 1.958; 0.289\%$ **31.** $0.099; 0.015\%$
    (b) $0.2 \text{ cm} \approx \Delta P$ $0.0005; 0.05\%$

33. $1 + 2\left(x - \dfrac{\pi}{4}\right)$ **35.** $3x - 4$ **37.** $4x - \dfrac{25}{3}$ **39.** $1 + 3x$ **41.** $1 - 2x$ **43.** $1 - 3x$

45. $V_{\max} \approx 16\frac{1}{4}$
$V_{\min} \approx 15\frac{3}{4}$
Actual $V_{\max} \approx 16.2092$
Actual $V_{\min} \approx 15.7925$

47. $\$3.$

49. (a) $4\cos 2t\,dt$ **51.** $\dfrac{z(1+z^2)\sec^2 z + \tan z}{(1+z^2)^{3/2}}dz$

53. ≈ 26.74

Chapter Review

1. 9 **3.** $\dfrac{1}{9}$ **5.** $60x(3x^2 - 2)^9$ **7.** $\dfrac{1-x}{(1-2x)^{3/2}}$ **9.** $\dfrac{1}{1+\cos x}$ **11.** $-\dfrac{(a^{2/3}-x^{2/3})^{1/2}}{x^{1/3}}$ **13.** $\sqrt{2}\sec^2 x$

15. tangent line: **17.** $v(3) = 12$ **19.** $\dfrac{4x^3+2y^2}{3y^2-4xy}$ **21.** $\dfrac{\sec y - y\sec^2 x}{\tan x - x\sec y\,\tan y}$ **23.** $1 - \dfrac{a^{1/2}}{x^{1/2}}, \quad y'' = \dfrac{a^{1/2}}{x^{3/2}}$
    $3x + y = 4$ $a(3) = -\dfrac{52}{5}$
    normal line:
    $x - 3y = 8$

25. (a) $= 33.9 \text{ cm} \approx 50\%$ **27.** Substitute $y, y' = x^{-1/2}\cos x - \frac{1}{2}x^{-3/2}(1 + \sin x)$
    (b) $\approx 11.92 \text{ cm}^2 \approx 3.32\%$
    Uncertainty in calorie count $y'' = -x^{-1/2}\sin x - \frac{1}{2}x^{-3/2}\cos x - \dfrac{1}{2x^{-3/2}}\cos x + \frac{3}{4}x^{-5/2}(1 + \sin x)$ into the
    $\approx 5.0\%$ differential equation.

CHAPTER 4

Exercise Set 4.1

1. increasing on $(-\infty, -1)$,
decreasing on $(-1, \infty)$

3. increasing on
$\left(-\infty, -\frac{4}{3}\right) \cup (2, \infty)$
decreasing on $\left(-\frac{4}{3}, 2\right)$

5. increasing on
$\left(-\infty, -\frac{2}{3}\right) \cup (2, \infty)$
decreasing on $\left(-\frac{2}{3}, 2\right)$

7. decreasing on
$(-\infty, 0) \cup (0, 1)$
increasing on $(1, \infty)$

9. increasing on
$(-\infty, 2)$, decreasing
on $(2, \infty)$

11. increasing on
$(-\infty, -1) \cup (3, \infty)$
decreasing on $(-1, 3)$

13. minimum at $(4, -9)$

15. maximum at $(-1, 8)$,
minimum at $(1, 4)$

17. $(4, -76)$ is a minimum point; $(-2, 32)$ is a maximum point.

19. $(0, 12)$ is a maximum point; $(-2, -4)$ and $(2, -4)$ are minimum points.

21. $(3, -19)$ is a minimum point.

23. $(3, 1)$ is a minimum point.

25. local minimum at $f(2) = -1$
local maxima:
$f(0) = 3$, $f(5) = 8$
absolute minimum $= -1$ at $x = 2$;
absolute maximum $= 8$ at $x = 5$

27. local maximum at $\left(-\frac{2}{3}, \frac{256}{27}\right)$, local minimum at $(2, 0)$
absolute maximum at $\left(-\frac{2}{3}, \frac{256}{27}\right)$, absolute minimum at $(-2, 0)$ and $(2, 0)$

29. local minima at $\left(-\frac{3}{2}, \frac{297}{16}\right)$, $(3, -27)$
local maximum at $(0, 27)$
absolute minimum at $(0, -27)$
absolute maximum at $(0, 27)$

31. local and absolute maximum at $\left(\frac{\pi}{4}, \sqrt{2}\right)$;
absolute minimum at $(\pi, -1)$

33. local and absolute maximum at $\left(-\frac{\pi}{4}, \frac{\pi}{2}\right)$
local and absolute minimum at $\left(\frac{\pi}{4}, 2 - \frac{\pi}{2}\right)$

35. local maximum at $\left(\frac{\pi}{6}, \frac{5\sqrt{3}}{2} + \frac{\pi}{6} - 1\right)$
local minimum at $\left(\frac{5\pi}{6} - \frac{5\sqrt{3}}{2} + \frac{5\pi}{6} - 1\right)$
absolute max at 2π; 9.28 $\left(\frac{5\pi}{6}, -2.71\right)$

37. absolute maximum $= 24$ at $x = 3$
absolute minimum $= -12$ at $x = -3$

39. absolute maximum $= 32$ at $x = -2$
absolute minimum $= -76$ at $x = 4$

41. absolute maximum $= 21$ at $x = 3$ and $x = -3$
absolute minimum $= -4$ at $x = 2$ and $x = -2$

43. absolute maximum $= 8$ at $x = -4$
absolute minimum $= -19$ at $x = 3$

45. absolute maximum $= 9$ at $x = -5$
absolute minimum $= 1$ at $x = 3$

47. only critical number is $x = 0$. $f'(x)$ does not change sign at $x = 0$. If n is an odd positive integer, then f does not have a local extremum at $x = 0$. $f(x) = (x - a)^n + b$ changes sign at $x = a$ only if n is an even positive integer   **49.** If $f(a)$ is a local maximum, then $f(a) > f(a - h)$ and $f(a) > f(a + h)$. If $g(x) = -f(x)$, then $-f(a) < -f(a - h)$ and $g(a) < g(a - h)$. Similarly, $g(a) < g(a + h)$.

51. $f(x) = 0$ can have two solutions or no solutions.   **53.** g is increasing on $(0,1)$.

55.

57. Absolute minimum $= 0$ at $x = 1$; no absolute maximum.

59. on $\left(-\infty, \frac{1}{2}\right]$
absolute maximum $= 2$ at $x = \frac{1}{2}$
local and absolute minimum $= -\frac{1}{4}$ at $x = -1$
On $(-\infty, 1)$ f has no maximum. Absolute minimum remains $-\frac{1}{4}$ at $x = -1$. On $(1, \infty)$ f has no maximum or minimum.

61. local minimum at $\left(4, \frac{1}{3}\right)$;
absolute maximum $= \frac{1}{3}$ at $x = 3$ and $x = 12$
absolute minimum $= 0$ at $x = 4$

63. local minimum at $(-3, -177)$
absolute max $= 255$ at $x = -5$; absolute min $= -177$ at $x = -3$

65. absolute max $= 13$ at $x = 3$; local and absolute min $= -19$ at $x = -1$

67. local and absolute max $= 51.3$ at -1.92; local and absolute min $= 25.6$ at 0.97 (All results are approximate.)

69. Assume that $\lim_{x \to a} f(x) = L < 0$. Show that this contradicts the given information. Consider $f(x) = x^2$. In every deleted neighborhood of 0, $f(x) > 0$. But $\lim_{x \to 0} f(x) = 0$.

71. As $h \to 0$, $\sin \frac{1}{h}$, used in the definition, does not approach a limit.

73. Let $f(x) = \begin{cases} x & \text{if } x \epsilon (0, 1) \\ \frac{1}{2} & \text{if } x = 0 \text{ or } 1 \end{cases}$

75. max $(f + g) \le$ max f +max g.
If max $f \ge 0$ and max $g \ge 0$, then max $(fg) \le$ (max f)(max g).
If max $f \ge 0$ and max $g \le 0$, then max $(fg) \le 0$.
If max $f \le 0$ and max $g \le 0$, then we can draw no conclusion about max fg, unless we know that min f and min g exist, in which case max $fg \le$ (min f)(min g).

Exercise Set 4.2

1. f is always concave up.

3. f is always concave up if $a > 0$ and always concave down if $a < 0$.

5. f is concave up for $x > -2$ and concave down for $x < -2$.

7. G is concave up on $(-\infty, 0) \cup (1, \infty)$ and concave down on $(0,1)$.

```
  +       −       +
 ─────┼───────┼─────
      0       1
```

9. f is concave up on $(-\infty, -1) \cup (1, \infty)$ and concave down on $(-1, 1)$.

```
  +       −       +
 ────┼───────┼────
    -1       1
```

11. (a) $f'(x) > 0$ at points B and C
(b) $f''(x) > 0$ at points A and B

13. $(-2, 13)$ is a local maximum
$\left(\frac{2}{3}, \frac{95}{27}\right)$ is a local minimum

15. $(3, 11)$ is a local maximum
$\left(\frac{1}{3}, \frac{41}{27}\right)$ is a local minimum

17. $(-2, 16)$ and $(1, 38)$ are local maximum points $(-1, -38)$ and $(2, 16)$ are local minimum points

19. $(-1, 2)$ is a local minimum point

21. $(-2, 71)$ is a local maximum point
$(2, -57)$ is a local minimum point

```
       −       −
 ────┼───┼───┼────
    -2   0   2
```

23. decreasing on $(-2, 2)$
increasing on $(-\infty, -2) \cup (2, \infty)$
concave down on $(-\infty, 0)$
concave up on $(0, \infty)$
point of inflection: $(0, 0)$
intercepts: $x = 0$, $x = \pm 2\sqrt{3}$
local maximum at $x = -2$
$f(-2) = 16$
local minimum at $x = 2$
$f(2) = -16$

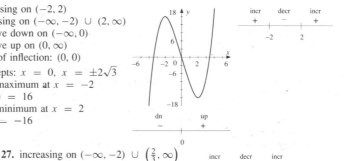

```
 incr   decr   incr
  +      −      +
 ───┼──────┼───
    -2     2

 dn        up
  −         +
 ─────┼──────
      0
```

25. increasing on $\left(-\infty, -\frac{1}{3}\right) \cup (1, \infty)$
decreasing on $\left(-\frac{1}{3}, 1\right)$
concave down on $\left(-\infty, \frac{1}{3}\right)$
concave up on $\left(\frac{1}{3}, \infty\right)$
point of inflection $\left(\frac{1}{3}, \frac{16}{27}\right)$
intercepts: $(-1, 0)$, $(1, 0)$ and $(0, 1)$
local maximum at $x = -\frac{1}{3}$
$f\left(-\frac{1}{3}\right) = \frac{32}{27}$
local minimum at $x = 1$
$f(1) = 0$

```
  +       −       +
 ────┼───────┼────
    −1/3     1
```

```
 dn        up
  −         +
 ─────┼──────
      1/3
```

27. increasing on $(-\infty, -2) \cup \left(\frac{2}{3}, \infty\right)$
decreasing on $\left(-2, \frac{2}{3}\right)$
concave down on $\left(-\infty, -\frac{2}{3}\right)$
concave up on $\left(-\frac{2}{3}, \infty\right)$
point of inflection: $\left(-\frac{2}{3}, \frac{230}{27}\right)$
local maximum at $x = -2$
$f(-2) = 18$
local minimum at $x = \frac{2}{3}$
$f\left(\frac{2}{3}\right) = -\frac{26}{27}$

```
 incr   decr   incr
  +      −      +
 ───┼──────┼───
    -2     2/3

 dn        up
  −         +
 ─────┼──────
      −2/3
```

29. increasing on $(-\infty, 2) \cup (2, \infty)$
concave down on $(-\infty, 2)$
concave up on $(2, \infty)$
point of inflection $(2, 5)$
no local extrema
horizontal tangent at $(2, 5)$

incr	incr
+	+

2

dn	up
−	+

2

31. increasing on $(-\infty, 1)$
decreasing on $(1, \infty)$
no point of inflection
concave down on
$(-\infty, 0) \cup (0, \infty)$
local maximum at $x = 1$
$f(1) = 6$

incr	decr
+	−

1

−	−

0

33. increasing on $\left(-\infty, -\frac{2}{3}\right) \cup (2, \infty)$

decreasing on $\left(-\frac{2}{3}, 2\right)$

concave down on $\left(-\infty, \frac{2}{3}\right)$

concave up on $\left(\frac{2}{3}, \infty\right)$

point of inflection: $\left(\frac{2}{3}, \frac{128}{27}\right)$

intercept: $x = 2$ or $x = -2$

local maximum at $x = -\frac{2}{3}$

$f\left(-\frac{2}{3}\right) = \frac{256}{27}$

local minimum at $x = 2$
$f(2) = 0$

incr	decr	incr
+	−	+

$-\frac{2}{3}$ 2

dn	up
−	+

$\frac{2}{3}$

35. increasing on $(-\infty, -1) \cup (2, \infty)$
decreasing on $(-1, 2)$

concave down on $\left(-\infty, \frac{1}{2}\right)$

concave up on $\left(\frac{1}{2}, \infty\right)$

point of inflection: $\left(\frac{1}{2}, \frac{27}{2}\right)$

intercepts: $x = 2$ or $x = -\frac{5}{2}$
local maximum $(-1, 27)$
local minimum $(2, 0)$

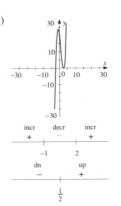

incr	decr	incr
+	−	+

−1   2

dn	up
−	+

$\frac{1}{2}$

37. increasing on $\left(0, \frac{\pi}{4}\right) \cup \left(\frac{5\pi}{4}, 2\pi\right)$

decreasing on $\left(\frac{\pi}{4}, \frac{5\pi}{4}\right)$

concave down on $\left(0, \frac{3\pi}{4}\right) \cup \left(\frac{7\pi}{4}, 2\pi\right)$

concave up on $\left(3\frac{\pi}{4}, \frac{7\pi}{4}\right)$

point of inflection: $\left(\frac{3\pi}{4}, 0\right), \left(\frac{7\pi}{4}, 0\right)$

local maximum $\left(\frac{\pi}{4}, \sqrt{2}\right)$

local minimum $\left(\frac{5\pi}{4}, -\sqrt{2}\right)$

incr	decr	incr
+	−	+

0    $\frac{\pi}{4}$ $\frac{5\pi}{4}$ 2π

−	+	−

0    $\frac{3\pi}{4}$ $\frac{7\pi}{4}$ 2π

39. decreasing on $\left(-\frac{\pi}{3}, \frac{2\pi}{3}\right)$,

increasing on $\left(\frac{2\pi}{3}, \frac{5\pi}{3}\right)$

concave down on $\left(-\frac{\pi}{3}, \frac{\pi}{6}\right) \cup \left(\frac{7\pi}{6}, \frac{5\pi}{3}\right)$,

concave up on $\left(\frac{\pi}{6}, \frac{7\pi}{6}\right)$

local minimum at $\left(\frac{2\pi}{3}, -2\right)$

endpoint maximum at $\left(-\frac{\pi}{3}, 2\right)$ and $\left(\frac{5\pi}{3}, 2\right)$

points of inflection: $\left(\frac{\pi}{6}, 0\right)$ and $\left(\frac{7\pi}{6}, 0\right)$

−	+

$-\frac{\pi}{3}$ $\frac{2\pi}{3}$ $\frac{5\pi}{3}$

−	+	−

$-\frac{\pi}{3}$ $\frac{\pi}{6}$ $\frac{7\pi}{6}$ $\frac{5\pi}{3}$

41. increasing on $\left(-\pi, -\frac{7\pi}{8}\right) \cup \left(-\frac{3\pi}{8}, \frac{\pi}{8}\right) \cup \left(\frac{5\pi}{8}, \pi\right)$

decreasing on $\left(-\frac{7\pi}{8}, -\frac{3\pi}{8}\right) \cup \left(\frac{\pi}{8}, \frac{5\pi}{8}\right)$

concave up on $\left(-\frac{5\pi}{8}, -\frac{\pi}{8}\right) \cup \left(\frac{3\pi}{8}, \frac{7\pi}{8}\right)$

concave down on $\left(-\pi, -\frac{5\pi}{8}\right) \cup \left(-\frac{\pi}{8}, \frac{3\pi}{8}\right) \cup \left(\frac{7\pi}{8}, \pi\right)$

local minima at $x = -\frac{3\pi}{8}, \frac{5\pi}{8}$

local maxima at $x = -\frac{7\pi}{8}, \frac{\pi}{8}$

$f\left(-\frac{3\pi}{8}\right) = -1 - \sqrt{2} = f\left(\frac{5\pi}{8}\right)$;

$f\left(-\frac{7\pi}{8}\right) = f\left(\frac{\pi}{8}\right) = \sqrt{2} - 1$

absolute maximum $= \sqrt{2} - 1$ at $\frac{\pi}{8}$

absolute minimum $= -1 - \sqrt{2}$ at $x = -\frac{3\pi}{8}$

points of inflection:

$\left(-\frac{\pi}{8}, -1\right), \left(\frac{3\pi}{8}, -1\right), \left(\frac{7\pi}{8}, -1\right), \left(-\frac{5\pi}{8}, -1\right)$

43. intercept: (0, 0)
symmetric with respect to the origin
vertical asymptotes: $x = -1, x = 1$
horizontal asymptote: $y = 0$
inflection point: (0,0)
f decreasing on $(-\infty, -1) \cup (-1, 1) \cup (1, \infty)$
f concave down on $(-\infty, -1) \cup (0, 1)$
f concave up on $(-1, 0) \cup (1, \infty)$

45. intercept: (0,0)
symmetric with respect to the origin
vertical asymptotes: none
horizontal asymptote: $y = 0$
f decreasing on $(-\infty, -1) \cup (1, \infty)$
f increasing on $(-1, 1)$
f concave down on $(-\infty, -\sqrt{3}) \cup (0, \sqrt{3})$
f concave up on $(-\sqrt{3}, 0) \cup (\sqrt{3}, \infty)$
inflection points: $\left(-\sqrt{3}, \frac{-\sqrt{3}}{4}\right)$, (0, 0),

$\left(\sqrt{3}, \frac{\sqrt{3}}{4}\right)$

local minimum at $\left(-1, -\frac{1}{2}\right)$

local maximum at $\left(1, \frac{1}{2}\right)$

47. intercepts: $(-2, 0), (2, 0), (0, 4)$
symmetric with respect to the y-axis
vertical asymptotes: $x = -1, x = 1$
horizontal asymptote: $y = 1$
f decreasing on $(-\infty, -1) \cup (-1, 0)$
f increasing on $(0, 1) \cup (1, \infty)$
f concave down on $(-\infty, -1) \cup (1, \infty)$
f concave up on $(-1, 1)$
local minimum at $(0, 4)$
inflection points: none

49. intercepts: $(-1, 0), (1, 0), \left(0, -\frac{1}{4}\right)$
symmetric with respect to the y-axis
vertical asymptotes: $x = -2, x = 2$
horizontal asymptote: $y = -1$
f decreasing on $(-\infty, -2) \cup (-2, 0)$
f increasing on $(0, 2) \cup (2, \infty)$
f concave down on $(-\infty, -2) \cup (2, \infty)$;
f concave up on $(-2, 2)$
local minimum at $\left(0, -\frac{1}{4}\right)$;
inflection points: none

51. intercepts: none, vertical asymptotes: $x = -1, x = 0$, horizontal asymptote: $y = 0$,

symmetry: none (Actually, symmetric about $x = -\frac{1}{2}$.) f decreasing on $\left(-\frac{1}{2}, 0\right) \cup (0, \infty)$,

f increasing on $(-\infty, -1) \cup \left(-1, -\frac{1}{2}\right)$, f concave down on $(-1, 0)$; f concave up on

$(-\infty, -1) \cup (0, \infty)$, local maximum at $\left(-\frac{1}{2}, -4\right)$; inflection points: none

53. $\lim\limits_{x \to 0^-} f'(x) = -\infty$,

$\lim\limits_{x \to 0^+} f'(x) = +\infty$

f has a vertical tangent line at $x = 0$.

55. $\lim\limits_{x \to 1^-} f'(x) = -\infty$,

$\lim\limits_{x \to 1^+} f'(x) = -\infty$

point of inflection at $(1, 0)$, with vertical tangent

57. $\lim\limits_{x \to 0^-} f'(x) = +\infty$,

$\lim\limits_{x \to 0^+} f'(x) = -\infty$

f has a vertical tangent line at $x = 0$.

59. $\lim\limits_{x \to 0^+} f'(x) = -\infty$

$f(0) = 1$

f has a vertical tangent line at $x = 0$.

61. $\left(-2, \frac{-19}{3}\right)$ is a local minimum

$\left(1, \frac{59}{12}\right)$ is a local maximum

$\left(2, \frac{13}{3}\right)$ is a local maximum

63. $\left(-1, \frac{7}{2}\right)$ is a local minimum

65. $(3, 11)$ is a local minimum

67. decreasing on $(-\infty, -2)$,
increasing on $(-2, 2) \cup (2, \infty)$
concave up on

$\left(-\infty, -\frac{2}{3}\right) \cup (2, \infty)$

concave down on $\left(-\frac{2}{3}, 2\right)$

points of inflection:

$\left(-\frac{2}{3}, -\frac{121}{27}\right)$ and $(2, 5)$

local minimum at $(-2, -11)$
horizontal tangent line at $(2, 5)$

69. increasing on

$(-\infty, -\sqrt{2}) \cup (-1, 1) \cup (\sqrt{2}, \infty)$
decreasing on $(-\sqrt{2}, -1) \cup (1, \sqrt{2})$

concave up on $\left(-\sqrt{\frac{3}{2}}, 0\right) \cup \left(\sqrt{\frac{3}{2}}, \infty\right)$

concave down on $\left(-\infty, -\sqrt{\frac{3}{2}}\right) \cup \left(0, \sqrt{\frac{3}{2}}\right)$
points of inflection:

$\left(-\sqrt{\frac{3}{2}}, \left(-\frac{19}{4}\right)\sqrt{\frac{3}{2}} - 3\right)$, $(0, -3)$,

$\left(\sqrt{\frac{3}{2}}, \left(\frac{19}{4}\right)\sqrt{\frac{3}{2}} - 3\right)$

maximum points: $(-\sqrt{2}, -4\sqrt{2} - 3)$,
$(1, 3) \approx (-1.41, -8.66)$, $(1, 3)$
minimum points: $(-1, -9)$, $(\sqrt{2}, 4\sqrt{2} - 3)$
$\approx (-1, -9)$, $(1.41, 2.66)$

71. increasing on $(-1, 2) \cup (2, \infty)$
decreasing on $(-\infty, -1)$
concave up on $(-\infty, 0) \cup (2, \infty)$
concave down on $(0, 2)$
points of inflection: $(0, -4)$, $(2, 12)$
local minimum at $(-1, -15)$
horizontal tangent at $(2, 12)$

73. increasing on $\left(-\pi, \frac{\pi}{6}\right) \cup \left(\frac{5\pi}{6}, \pi\right)$

decreasing on $\left(\frac{\pi}{6}, \frac{5\pi}{6}\right)$

concave up on $\left(-\pi, -\frac{\pi}{2}\right) \cup \left(\frac{\pi}{2}, \pi\right)$

concave down on $\left(-\frac{\pi}{2}, \frac{\pi}{2}\right)$

points of inflection: $\left(-\frac{\pi}{2}, 0\right)$, $\left(\frac{\pi}{2}, 0\right)$

local maximum: $\left(\frac{\pi}{6}, \sqrt{3}\right)$

local minimum: $\left(\frac{5\pi}{6}, -\sqrt{3}\right)$

endpoints: $\left(-\pi, -\frac{3}{2}\right)$, $\left(\pi, -\frac{3}{2}\right)$

75. symmetry about y-axis

increasing on $\left(-\frac{3\pi}{2}, -\pi\right) \cup (0, \pi)$

decreasing on $(-\pi, 0) \cup \left(\pi, \frac{3\pi}{2}\right)$

concave down on
$\left(-\frac{3\pi}{2}, -\frac{4\pi}{3}\right) \cup \left(-\frac{4\pi}{3}, -\frac{\pi}{2}\right) \cup \left(\frac{\pi}{2}, \frac{4\pi}{3}\right) \cup \left(\frac{4\pi}{3}, \frac{3\pi}{2}\right)$

concave up on $\left(-\frac{\pi}{2}, 0\right) \cup \left(0, \frac{\pi}{2}\right)$

points of inflection: $\left(-\frac{\pi}{2}, 3\right)$, $\left(\frac{\pi}{2}, 3\right)$

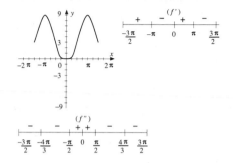

77. intercepts: $(\pm\sqrt{3}, 0)$
symmetry: $f(-x) = -f(x)$;
symmetry about origin
asymptotes: $x = 0$, $y = 0$
maxima, minima: $f'(x) = 0$ if
$x = \pm 3$
increasing on $(-\infty, -3) \cup (3, \infty)$,
decreasing on $(-3, 0) \cup (0, 3)$

local maximum at $\left(-3, \frac{2}{9}\right)$,

local minimum at $\left(3, -\frac{2}{9}\right)$

concave down on
$(-\infty, -3\sqrt{2}) \cup (3\sqrt{2}, \infty)$
concave up on $(-3\sqrt{2}, 3\sqrt{2})$
points of inflection:

$\left(-3\sqrt{2}, \frac{5}{18\sqrt{2}}\right)$, $\left(3\sqrt{2}, -\frac{5}{18\sqrt{2}}\right)$

79. intercepts: $(0, 0)$
symmetry: none
asymptotes: $x = 2$, $y = 0$
local minimum at $(-1, -1)$
vertical tangent at $x = 0$.
concave down on
$(-\infty, -2.3416) \cup (0, 0.3416) , \cup (2, \infty)$
concave up on
$(-2.3416, 0) \cup (0.3416, 2)$
points of inflection:
$(-2.3416, -0.9176)$, $(0, 0)$, $(0.3416, 1.2646)$

81. intercepts: $(-1, 0)$
symmetry: none
asymptotes: $x = 0$, $y = x$

increasing on $(-\infty, 0) \cup (\sqrt[3]{2}, \infty)$

decreasing on $(0, \sqrt[3]{2})$

local minimum at $\left(\sqrt[3]{2}, \frac{3}{\sqrt[3]{4}}\right)$

point of inflection: none. Always concave up.

83. $a = 0$: $f(x) = x^3$ is symmetric about the origin, concave down on $(-\infty, 0)$, concave up on $(0, \infty)$, has one critical point at $(0, 0)$, and is increasing on $(-\infty, \infty)$.

$a > 0$: The graph is symmetric about the origin, concave down on $(-\infty, 0)$, concave up on $(0, \infty)$, has no critical points, and is increasing on $(-\infty, \infty)$.

$a < 0$: The graph is symmetric about the origin, concave down on $(-\infty, 0)$, concave up on $(0, \infty)$, has two critical points, has a local maximum at $\left[-\sqrt{\frac{-a}{3}}, \ f\left(-\sqrt{\frac{-a}{3}}\right)\right]$ and a local minimum at $\left[\sqrt{\frac{-a}{3}}, \ f\left(\sqrt{\frac{-a}{3}}\right)\right]$. The general cubic equation, $g(x) = ax^3 + bx^2 + cx + d$, can be converted to the simpler form $f(x) = a_1 x^3 + c_1 x$. The graph of the general equation is related to the graph of f by horizontal and vertical shifts and compression or expansion in the y-direction.

85. **(a)** $|c| > 0.648$ (approximate).   **(b)** inflection points at $x = \pm\sqrt{\frac{1}{3}}$; $(-\sqrt{\frac{1}{3}}, \frac{3}{4} - c\sqrt{\frac{1}{3}}), (\sqrt{\frac{1}{3}}, \frac{3}{4} + c\sqrt{\frac{1}{3}})$ **(c)** $c \approx -0.6495, 0.6495$

Exercise Set 4.3

1. 12.5 ft × 25 ft.

3. Dimensions are 100 m × 120 m. Minimum cost is $3,600.

5. 1.5 ft × 3 ft × 2 ft.

7. Dimensions are 2 × 3.

9. Construct a rectangle of sides x and y and hypotenuse $2a$. Show that $x = y$ for maximum area.

11. $r = \frac{a\sqrt{6}}{3}, \ h = \frac{2a\sqrt{3}}{3}$

13. 80 **15.** 210 **17.** 70 **19.** 5 **21.** dimensions are $\sqrt{3}$ ft × $\sqrt{6}$ ft.

23. dimensions are 10 in × 12 in.

25. $12\frac{1}{2}$ in wide and 10 in high.

27. base radius $= \frac{2a\sqrt{2}}{3}$, height $= \frac{4a}{3}$

29. $x = 8\sqrt{5}$.

31. $x = \frac{24}{\pi + 4}, \ y = \frac{12}{\pi + 4}$

33. 30 kilometers

35. $\left(3^{2/3} + 5^{2/3}\right)^{3/2} \approx 11.19$ feet

37. **(a)** $x = 4$ **(b)** $x = \frac{9}{2}$

Exercise Set 4.4

1. 5 m/sec

3. 1 ft/sec

5. 10π m^2sec

7. $\frac{2,000\pi}{3}$ m/sec

9. $\frac{48}{125}$ m/sec

11. $\frac{2}{5}$ ft/min

13. $-\frac{5}{2}$ m/sec

15. -0.0114 m³/s

17. -20 N/sec

19. $72,000$ π cm³/min

21. $\frac{17}{4}$ m/s

23. $\frac{63\pi}{10}$ mm³/day

25. 10 cm/sec

27. $\frac{5}{3}$ cm/min

29. ≈ 0.1548 m/min

Exercise Set 4.5

1. 2.2247

3. -0.8850

5. 0.4429

7. $r_1 \approx 0.7140$; $r_2 \approx 1.6449$

9. 1.4142

11. -2.0801

13. $x_0 = 6.5$

15. $x_0 = 1.2$

17. $f'(x) > 0$ for all x and $f''(x) > 0$ for $x > \frac{1}{2} f(x_0)$ agrees in sign with $f''(x)$ on I. On $I = (1.7, 1.8)$, take $M = 18$ and $m = 4$.
$E_3 \leq 0.0000029$

19. f is monotone in I and is concave down. $f(2.1) \approx -0.4786 < 0$. Let $I = (1.9, 2.1)$, $m \geq \frac{\pi}{2}$, $M \leq 3$. $E_3 \leq 0.000002$

21. $z_3 = 1.703928$, $x_3 = 1.7039397$ **23.** $z_3 = 1.9333131$, $x_3 = 1.933754$ **25.** 2.44

27. (i) $f(0.2594) \approx -19.42$, local maximum
(ii) $f(2.2994) \approx = -50.63$
(iii) $f(-0.5588) \approx -22.11$

29. Use Eq. 4.7 with $E_n \leq 5 \times 10^{-(k+1)}$

Exercise Set 4.6

1. -6 **3.** $\frac{1}{2}$ **5.** 1   **7.** limit does not exist   **9.** 6   **11.** 4   **13.** $\frac{1}{4}$ **15.** $\frac{1}{2}$ **17.** 1   **19.** 2

21. 0   **23.** $-\frac{1}{3}$ **25.** $\frac{1}{2}$ **27.** (a)

x	$f(x)$
0.1	-0.05449
0.01	-0.005064
0.001	-0.000827
0.0001	-0.0000306

The limit appears to be 0.

(b)

x	$f(x)$
0.1	0.9161
0.01	1.0086
0.001	1.0006
0.0001	0.9999

The limit appears to be 1.

29. (a) $y = 0$ (b) $x \approx 0.4534$

(c)

(d) local maximum at $(0, -1)$
local minimum at
$(-0.9180, -1.2526)$

Exercise Set 4.7

1. $c = \frac{3}{2}$ **3.** $c = \frac{3}{4}$ **5.** $c = \sqrt{3}$ **7.** $c = 1$ **9.** $c = \frac{1}{2}$ **11.** $c = \frac{7}{3}$ **13.** $c = \frac{2}{3}$ **15.** $c = \frac{3}{2}$

17. f' does not exist at 0.   **19.** f is discontinuous at $x = \frac{\pi}{2}$ **21.** f is discontinuous at $x = 0$.   **23.** f' does not exist at $x = 1$

25. $c = \frac{2}{3}$ satisfies conclusion of Rolle's Theorem. No contradiction, since theorem gives sufficient conditions only.

27. $\left(-\frac{1}{3}, \frac{37}{27}\right)$, $(2, -9)$

29. Use the result of Ex. 28 and let $h = x - a$ and $\theta = \frac{(c-a)}{h}$

31. Use $\frac{f(x) - f(0)}{x - 0} = \frac{\tan x}{x} = \sec^2 c > 0$

33. Suppose $f'(a) > 0$ and $f'(b) < 0$. Since f is continuous on $[a, b]$, it must attain a maximum at some number c in (a, b). By Theorem 4.2, $f'(c) = 0$. A similar argument applies if $f'(a) < 0$ and $f'(b) > 0$.

Exercise Set 4.8

1. $\frac{x^6}{6} + C$ **3.** $\frac{2x^3}{3} - 2x^2 + 5x + C$ **5.** $\frac{3x^4}{4} - \frac{5x^3}{3} + x^2 - 3x + C$ **7.** $-\frac{1}{2}\left[\frac{1}{x} + \frac{x^3}{3}\right] + C$ **9.** $\frac{4}{3}x^{3/2} - 6x^{1/2} + C$

11. $\frac{3x^2}{2} - \frac{8}{3}x^{3/2} + 7x + C$ **13.** $\frac{2}{5}x^{5/2} + \frac{3}{2}x^2 + C$ **15.** $-2\cos x - 3\sin x + C$ **17.** $\frac{x^2}{2} + \cot x + C$ **19.** $x + \csc x + C$

21. $\frac{2}{5}t^{5/2} - \frac{2}{3}t^{3/2} + C$ **23.** $-\frac{1}{2x^2} + \frac{1}{x} + C$ **25.** $\sec x + \tan x + C$ **27.** $v = t^2 - 3t + 2$; $v(t) = \frac{1}{3}t^3 - \frac{3}{2}t^2 + 2t + C$

29. $f(x) = x^2 - 3x + 5$ **31.** $g(x) = x - \frac{x^2}{2} - \frac{4}{3}x^3 + \frac{79}{2}$ **33.** $y = x^2 - \frac{x^3}{3} - \frac{8}{3}$ **35.** $f(x) = \frac{1}{2}x^3 + 2$

37. $f(x) = \frac{x^4}{12} - \frac{x^2}{2} - \frac{2x}{3} + 2$ **39.** $y = \frac{x^2}{2} - \cos x + 4$ **41.** 64 feet = maximum height
Strikes ground when $t = 4$ sec
$v(4) = -64$ ft/sec

43. Maximum height = 78.4 meters above canyon floor, or 19.6 meters above top of cliff. Rock strikes canyon floor at $t = 6$ sec

Exercise Set 4.9

1. (a) increasing on
$(-1, 1) \cup (3, \infty)$
decreasing on
$(-\infty, -1) \cup (1, 3)$

(b) increasing on
$(-\infty, -2) \cup$
$\left(-2, -\frac{1}{2}\right) \cup \left(\frac{1}{2}, \infty\right)$
decreasing on
$\left(-\frac{1}{2}, \frac{1}{2}\right)$

3. local maximum at $(2, 13)$
local minimum at $\left(-\frac{2}{3}, \frac{95}{27}\right)$

5. local maximum at $(0, 0)$
local minimum at $(2, 4)$

7. local minimum at $(2, -19)$
absolute maximum at
$(-2, 13)$
absolute minimum at
$(2, -19)$

9. local and absolute maximum at $\left(\frac{\pi}{12}, \frac{\sqrt{3}}{2} + \frac{\pi}{12}\right)$
local minimum at $\left(\frac{5\pi}{12}, -\frac{\sqrt{3}}{2} + \frac{5\pi}{12}\right)$
absolute minimum at $\left(-\frac{\pi}{2}, -1 - \frac{\pi}{2}\right)$

11. (a) concave up on
$\left(-\infty, -\frac{1}{2}\right) \cup (1, \infty)$ concave down
on $\left(-\frac{1}{2}, 1\right)$

(b) concave up on $(3, \infty) \cup (-1, 1)$
concave down on $(-\infty, -1) \cup (1, 3)$

13. intercepts: $(0, -1), (1, 0), (-1, 0)$
symmetry: none
asymptotes: none
local maximum at $(-1, 0)$
local minimum at $\left(\frac{1}{3}, -\frac{32}{27}\right)$
point of inflection: $x = -\frac{1}{3}$

15. intercept: $(0, 60)$
$\approx (\pm 3.7, 0), (\pm 2.1, 0)$
symmetry: y-axis
asymptotes: none
maxima, minima:
$(\pm 3, -21)$ local minimum
$(0, 60)$ local maximum
points of inflection: $(\pm\sqrt{3}, 15)$

17. intercept: $(0, 0)$
symmetry: origin
asymptote: $y = 0$
maxima, minima: max $(2, 2)$, min
$(-2, -2)$
points of inflection:
$(-2\sqrt{3}, -\sqrt{3}), (0, 0), (2\sqrt{3}, \sqrt{3})$

19. intercepts: $(0, 15)$
x intercepts between 1 and 2
and between 3 and 4
symmetry: none
asymptotes: none
maxima, minima: $(3, -12)$ local
minimum
point of inflection: $(2, -1)$;
horizontal tangent at $(0, 15)$

21. intercepts: $(2, 0)$, $(-1, 0)$, $(0, 2)$
symmetry: none
asymptotes: $x = 1$, $y = x$
maxima, minima: none
points of inflection: none

23. intercepts: $(0, -1)$
x-intercepts between 0 and $\frac{1}{3}$, between $\frac{1}{3}$ and 1, and between 8 and 9
symmetry: none
asymptotes: none
maxima, minima: $(3, -1)$ local minimum
$\left(\frac{1}{3}, \frac{1}{3}\right)$ local maximum,
cusp points of inflection: none

25. intercept: $(0, 0)$
symmetry: none
asymptotes: $x = 1$, $y = 0$
maximum, minimum: $(-1, -1)$
is a local min
point of inflection: $\left(-2, -\frac{8}{9}\right)$

27. intercepts: $\left(\pm \frac{1}{\sqrt{3}}, 0\right)$; symmetry: origin; asymptotes:
$x = 0$, $y = 0$; maxima, minima: $(1, 2)$ local
maximum; $(-1, -2)$ local minimum; points of
inflection: $\left(\sqrt{2}, \frac{5}{2\sqrt{2}}\right)$, $\left(-\sqrt{2}, -\frac{5}{2\sqrt{2}}\right)$

29. intercept: $(0, -1)$; symmetry: y-axis; asymptotes: $x = \pm \frac{\pi}{4}$
in $\left[-\frac{\pi}{2}, \frac{\pi}{2}\right]$; maxima, minima: $f(0) = -1$ (local
maximum); $f\left(\pm \frac{\pi}{2}\right) = 3$ (local minimum); no inflection
points; f is periodic with period π

31. 6 ft $\times$ 6 ft $\times$ 9 ft.   **33.** $x = 18$, $y = 36$ **35.** $h = 10$ in, $r = 5$ **37.** $x = \frac{m}{2}$

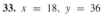

39. $5\sqrt{5}$ ft   **41.** 17 mph   **43.** (a) $\frac{2}{\pi}$ ft/min   (b) $\frac{1}{\pi \sqrt[3]{2}}$ ft/min

45. $r \approx 0.7430$ **47.** $r_1 \approx 2.1149$; $r_2 \approx -0.2541$; $r_3 \approx -1.8608$ **49.** $f(0.8603336) \approx 0.561096$ **51.** 4

53. $f\left(\frac{\pi}{6}\right) \approx 0.142$, $f\left(\frac{\pi}{3}\right) \approx -0.523$, f' and f'' do not change sign on $\left(\frac{\pi}{6}, \frac{\pi}{3}\right)$, $E_{n+1} \leq 0.113 E_n$ **55.** (a) 1   (b) 0

57. (a) 0   (b) $\frac{3}{2}$ **59.** (a) f and f' are continuous everywhere $c = 1$ (b) f & f' are continuous everywhere. $c = 0$

61. (a) f and f' are continuous everywhere $c = \frac{1}{2}$ (b) f and f' are continuous everywhere   **63.** (a) 7   (b) $g(x) = x^2 + 1$

65. (a) $-4\cos x - 3\sin x + C$ **(b)** $\tan x + 2\sec x + C$ **67. (a)** $f(x) = x^3 + \frac{2}{x} + 1$ **(b)** $y = -2\cos x - \frac{x^2}{2} + 7$

69. (a) $f(x) =$ **71. (a)** $v(t) = 3t^2 - 12$ **(b)** $s(0) = 0$ when   **(c)** increasing speed for
$-\frac{1}{6x} - \frac{x^2}{6} + \frac{x}{2} + \frac{5}{2}$ $s(t) = t^3 - 12t$ $t = 2$, 16 units to $t > 2$ decreasing
                             the left of the origin.   speed for $0 < t < 2$

(b) $y = \cos x + x^2 - x + 1$

(d)

CHAPTER 5

Exercise Set 5.1

1. $L_5 = 12.5$, $U_5 = 15.0$, $L_{10} = 13.125$, $U_{10} = 14.375$, $L_{20} = 13.4375$, $U_{20} = 14.0625$, $L_{100} = 13.6875$, $U_{100} = 13.8125$.
3. $L_5 = 4.48$, $U_5 = 6.08$, $L_{10} = 4.92$, $U_{10} = 5.72$, $L_{20} = 5.13$, $U_{20} = 5.53$, $L_{100} = 5.2932$, $U_{100} = 5.3732$ **5.** $L_5 = 4.704$, $U_5 = 7.656$, $L_{10} = 5.298$, $U_{10} = 6.792$, $L_{20} = 5.635125$, $U_{20} = 6.385875$, $L_{100} = 5.925453$, $U_{100} = 6.075447$

7. $L = 141.5$, $U = 181.5$ **9.** Add $1 + n, 2 + (n - 1), \ldots, n + 1$. Sum and divide by 2.   **11.** $P(1)\ 1^2 = \frac{1(1+1)(2+1)}{6}$. Assume $P(k)$ is true, and add $(k + 1)^2$ to both sides.   **13.** $P(1)\ 1^3 = \left[\frac{1(1+1)}{2}\right]^2$. Assume $P(k)$ is true and add $(k + 1)^3$ to both sides.

15. $L_n = \frac{b-a}{n} \sum_{k=1}^{n} m_k$, $U_n = \frac{b-a}{n} \sum_{k=1}^{n} M_k$, error $\leq |U_n - L_n| = \frac{b-a}{n} \left| \sum_{k=1}^{n} [f(x_{k-1}) - f(x_k)] \right| = \frac{b-a}{n} |f(x_n) - f(x_0)| = \frac{b-a}{n} |f(b) - f(a)|$

17. (a) 240   **(b)** 640, with $\Delta x = 0.0125$ **19. (a)** Area $= \frac{1}{2} \left(2r\tan\left(\frac{\pi}{n}\right) \right)(r)$ **(b)** Let $n = \frac{\pi}{t}$, $\lim_{t \to 0} \pi r^2 \frac{\tan t}{t} = \lim_{t \to 0} \frac{\pi r^2}{\cos t} \frac{\sin t}{t}$

Exercise Set 5.2

1. -19 **3.** -5.025 **5.** $\sum_{k=1}^{n} \frac{1}{n+k}$ **7.** $\frac{3}{m} \sum_{k=1}^{m} \sqrt[3]{3\left(1 + \frac{3k}{m}\right)} - 4$

9. $\sum_{k=1}^{n} \frac{1}{n+k-1}$ **11.** $\frac{3}{m} \sum_{k=1}^{m} \sqrt[3]{3\left(1 + \frac{3(k-1)}{m}\right)} - 4$ **13.** $\int_0^2 \sqrt{1 + x^3}\,dx$ **15.** $\int_0^2 x^2\,dx$

17. $\int_{-2}^{3} \sin x\,dx$ **19.** $\int_0^3 (1 + x)\,dx$ **21.** $\int_0^2 x\,dx + \int_2^4 (4 - x)\,dx$ **23.** 9   **25.** 2π **27.** $6 + \pi$

29. $9\left(\frac{\pi}{4} + 1\right)$ **31.** 10   **33.** 10   **35.** $3 < \int_1^4 f(x)\,dx < 7$, $11 < \int_{-3}^{2} f(x)\,dx < 17$

37.

n	2	8	16	32	100	200	
LHS	0.125	0.2734	0.3027	0.3179	0.3284	0.3308	≈ 0.3333
RHS	0.625	0.3984	0.3652	0.3491	0.3384	0.3358	

n	2	8	16	32	100	200	
LHS	-0.523	1.221	1.418	1.479	1.515	1.523	≈ 1.532
RHS	1.277	1.672	1.643	1.592	1.551	1.541	

39. (see table above)

41. $f'(x) = 2x \geq 0$
$n \geq 80$
Integral ≈ 2.7

43. 0.41

45. 4.97   **47. (a)**

(b) $\int_0^{2\pi} \cos^5 x\,dx = 0$ by symmetry with respect to $x = \pi$ and $\int_0^{2\pi} \sin^5 x\,dx = 0$ by symmetry about the point $x = \pi$.

49. $\lim\limits_{n\to\infty} \Delta x \sum\limits_{k=1}^{n} (c) = \lim\limits_{n\to\infty} \frac{b-a}{n} \cdot nc = c(b-a)$ **51.** $\lim\limits_{n\to\infty} \Delta x \sum\limits_{k=1}^{n} [f(x_k) + g(x_k)] = \lim\limits_{n\to\infty} \Delta x \sum\limits_{k=1}^{n} f(x_k) + \lim\limits_{n\to\infty} \Delta x \sum\limits_{k=1}^{n} g(x_k)$

$= \int_a^b f(x)dx + \int_a^b g(x)dx$ **53.** Let $h(x) = g(x) - f(x)$, $\int_a^b h(x)\,dx = \int_a^b g(x)\,dx - \int_a^b f(x)\,dx \geq 0$, or $\int_a^b f(x)\,dx \leq \int_a^b g(x)\,dx$

55. $P(1)$ is true for $m = 2$ by Property 4. Assume for $n = m$ and add $\int_a^b f_{m+1}(x)dx$ to both sides.   **57.** $c_k = \frac{kx}{n}$, $\frac{x}{n} \sum\limits_{k=1}^{n} c_k = \frac{x^2}{n^2} \sum\limits_{k=1}^{n} k$

$= \frac{x^2}{n^2}\frac{n(n+1)}{2} = \frac{x^2}{2} + \frac{x^2}{2n} = \frac{x^2}{2}$ as $n \to \infty$ **59.** $c_k = \frac{kx}{n}$, $\frac{x}{n} \sum\limits_{k=1}^{n} c_k^3 = \frac{x^4}{n^4} \sum\limits_{k=1}^{n} k^3 = \frac{x^4}{n^4}\left[\frac{n(n+1)}{2}\right]^2 = \frac{x^4}{4} + \frac{x^4}{2n} + \frac{x^4}{4n^2} = \frac{x^4}{4}$ as

$n \to \infty$

Exercise Set 5.3

1. $\frac{44}{3}$ **3.** 9   **5.** 16   **7.** $\frac{17}{6}$ **9.** $\frac{84}{5}$ **11.** -2 **13.** 4   **15.** $\sqrt{3}$ **17.** 2   **19.** $\frac{961}{15}$ **21.** $\frac{48}{5}$ **23.** $\frac{1}{3}\pi^2 + \sqrt{3}$

25. 1   **27.** $\sqrt{3} - 1$ **29.** $\frac{28}{3}$ **31.** -1 **33.** $-\frac{15}{4}$ **35.** (a) $V_{ave} = \frac{s(b)-s(a)}{b-a}$ (b) $V_{ave} = \frac{\int_a^b s'(t)dt}{b-a}$ **37.** 2

39. $1 - \sqrt{3}$ **41.** $\frac{x}{x-1}$ for all $x > 1$ **43.** $\frac{\sin x}{2+\cos x}$ for all x **45.**   **47.** $\approx 5,692,100$ kg; $\approx 58,706,700$ kg

49. $\approx 116,000,000$ kwh.   **51.** $-\frac{2}{3}$ **53.** $-\frac{\sin x}{x}$ **55.** $2(x-1)\sin^3(x^2-2x)$ **57.** $2 - \frac{\sqrt{2}}{2}$ **59.** $\frac{4}{3}$ **61.** $\frac{1}{6}$

63. 3   **65.** $\frac{8}{\sqrt{3}}$ **67.** $2(\sqrt{3} + \sqrt{2} - 1) - \frac{7\pi}{12}$ **69.** $\int_a^{x+h} f(t)\,dt - \int_a^x f(t)\,dt = -\int_{x+h}^x f(t)\,dt = -f(c)[x - (x+h)] = f(c)h$

Thus, $F'(x) = \lim\limits_{h\to 0} \frac{f(c)h}{h} = f(x)$, by the Squeezing Theorem, Property 3, and the mean value theorem for integrals.

71. If $x = 0$, $f'(0) = \lim\limits_{h\to 0} \frac{h^2 \sin\frac{1}{h^2}}{h} = \lim\limits_{h\to 0} h\left(\sin\frac{1}{h^2}\right) = 0$. For all other x, $f'(x) = \frac{-2}{x}\cos\frac{1}{x^2} + 2x\sin\frac{1}{x^2}$ which is unbounded as $x \to 0$.

f' is not integrable.   **73.** Let $F' = f$. Then, $\int_a^b f(x)\,dx = \int_a^b F'(x)\,dx = F(b) - F(a) = F'(c)(b-a)$ **75.** Let x

and $x_0 \in [a, b]$, with x_0 fixed. Since f is integrable, it is bounded, so there is an $M > 0$ such that $|f(t)| \leq M$ for all $t \in [a, b]$.

$|F(x) - F(x_0)| = \left| \int_x^b f(t)\,dt - \int_{x_0}^b f(t)\,dt \right| = \left| \int_x^{x_0} f(t)\,dt \right| = \int_x^{x_0} |f(t)|dt \leq M\int_x^{x_0} dt = M(x_0 - x)$ when $x < x_0$,

and when $x > x_0 |F(x) - F(x_0)| = |\int_{x_0}^x f(t)\,dt| \leq \int_{x_0}^x |f(t)|dt \leq M(x - x_0)$. Let $\delta = \frac{\epsilon}{M}$.   **77.** Define $g(x) = f(x)$ on

(a,b) and $g(a) = \lim\limits_{x\to a^+} f(x)$, $g(b) = \lim\limits_{x\to b^-} f(x)$. Then g is continuous on $[a,b]$. Let c be any number in (a,b). Then, by

Exercise 74, $\int_c^b f(x)\,dx$ exists and equals $\int_c^b g(x)\,dx$ and by Exercise 76, $\int_a^c f(x)\,dx$ exists and equals $\int_a^c g(x)\,dx$. Therefore,

$\int_a^b f(x)\,dx = \int_a^c f(x)\,dx + \int_c^b f(x)\,dx$ exists and equals $\int_a^b g(x)\,dx$.

Exercise Set 5.4

1. $\frac{1}{8}(1 + x^2)^4 + C$ **3.** $\frac{1}{18}(x^3 - 1)^6 + C$ **5.** $\frac{1}{9}(3x^2 + 4)^{3/2} + C$ **7.** $\frac{2}{9}(1 + 3x)^{3/2} + C$ **9.** $\frac{1}{3}(x^2 - 2x + 3)^{3/2} + C$

11. $-\frac{1}{2}\cos 2x + C$ **13.** $-\frac{1}{k}\cos kx + C$ **15.** $\sin^2 x + C$ or $-\cos^2 x + C_1$ or $-\frac{1}{2}\cos 2x + C_2$ **17.** $\frac{1}{2}\tan x^2 + C$

19. $\frac{2}{3}(\tan x)^{3/2} + C$ **21.** $-\frac{1}{1-\cos x} + C$ **23.** $-2\cot\sqrt{x} + C$ **25.** $\frac{1}{6}\left(\frac{x-1}{x}\right)^6 + C$ **27.** $\frac{2}{5}(x^{3/2} - 1)^{5/3} + C$ **29.** $\frac{121}{5}$

31. $\frac{33}{10}$ **33.** $\frac{254}{7}$ **35.** 60   **37.** $\frac{1}{3}$ **39.** $\frac{1}{8}$ **41.** -1 **43.** $\frac{7}{3}$ **45.** $\frac{2}{9}$ **47.** (a) $\frac{1}{2}\tan^2 x + C$ (b) $\frac{1}{2}\sec^2 x + C_1$

Both are correct; they differ in the constant.   **49.** $g(x) = \frac{1}{6}(4x + 2)^{3/2} - \frac{1}{3}$ **51.** \$90,990   **53.** $\sqrt{x^2 + 1} + C$

55. $\frac{1}{2}(x^2 + \sin 2x) + C$ **57.** $\frac{2}{5}(1 + \tan t)^{5/2} + C$ **59.** $y = \frac{2}{3}\sqrt{x^3 + 1} + 3$ **61.** $y = \frac{2}{3}\left[(1 + \sin^2 x)^{3/2} + 2\right]$

63. $y = -\frac{1}{2}\left(\frac{1}{x^2+1} + 1\right)$ **65.** (a) ≈ 413.1 ft   (b) ≈ 128.6 m   **67.** (a) $-9.8t - 39.2$ (b) $s(t) = -4.9t^2 - 39.2t + 1176$; 12 sec

69. $-\frac{1}{\tan x+1} + C$ **71.** $-4\sqrt{1 - \sqrt{x}} + C$ **73.** $\sin x - \frac{2}{3}\sin^3 x + C$ **75.** $\frac{d}{dx} - 2\left[\frac{1}{\sqrt{x}}\sin\frac{1}{\sqrt{x}} + \cos\frac{1}{\sqrt{x}}\right] + C = \frac{1}{x^2}\cos\frac{1}{\sqrt{x}}$

77. Property 1: $\frac{d}{dx}[F(x) + C] = F'(x) = f(x)$. Property 2: If f is differentiable on I, f is then an antiderivative of f'. Property 3: Let $u(x) = cF(x)$ where F is an antiderivative of f. Then, $du = cF'(x)\,dx = cf(x)\,dx$. $\int cf(x)\,dx = \int du = u + C = cF(x) + C = c\int f(x)\,dx$. Property 4: Let $u(x) = F(x) \pm G(x)$ Then, $du = [F'(x) \pm G'(x)]\,dx = [f(x) \pm g(x)]\,dx$. $\int[f(x) + -g(x)]\,dx = \int du = u + C = F(x) + -G(x) + C = \int f(x)\,dx + -\int g(x)\,dx$.

Exercise Set 5.5

1. (a) ≈ 0.50899, $E_4 \leq 0.03125$ **(b)** ≈ 0.4955, $E_4 \leq 0.015625$ **(c)** ≈ 0.5004, $E_4 \leq 0.002604$, $\int_0^1 \frac{dx}{(x+1)^2} = \frac{1}{2}$ **3. (a)** ≈ 0.62955, $E_8 \leq 1.3125$.   **(b)** ≈ 0.77489, $E_8 \leq 0.65625$ **(c)** ≈ 0.73782, $E_8 \leq 0.2703$, $\int_1^3 x\sin x^2\,dx = -\frac{1}{2}(\cos 9 - \cos 1) \approx 0.72572$

5. (a) ≈ -0.9375, $E_6 \leq 0.75$ **(b)** ≈ -0.6563, $E_6 \leq 0.375$ **(c)** -0.7500, $E_6 = 0$, $\int_{-2}^1 (1 + x^3)\,dx = -0.7500$ **7. (a)** $n = 30$
   (b) $n = 22$ **(c)** $n = 5$ **9. (a)** $n = 116$ **(b)** $n = 82$ **(c)** $n = 11$ **11. (a)** $n = 142$ **(b)** $n = 100$ **(c)** $n = 14$
13. (a) ≈ 2.29 **(b)** 2.30 **15.** ≈ 243 cu meters   **17. (a)** ≈ 0.6592, $E_{10} \leq 0.00833\ldots$ **(b)** ≈ 0.6594, $E_{10} \leq 0.00417$
   (c) ≈ 0.6593, $E_{10} \leq 0.0000933$ **19.** 2.6368, $M_2 = 2.00$. $E_n \leq 0.00667$ **21.** $n = 10$: ≈ 0.785398153, $n = 20$: ≈ 0.785398163,
   $n = 100$: ≈ 0.785398163, $n = 1000$: ≈ 0.785398163, $\pi/4 \approx 0.785398163$,

Chapter Review

1. $U_n = \frac{14}{3} + \frac{4}{n} + \frac{4}{3n^2}$, $L_n = \frac{14}{3} - \frac{4}{n} + \frac{4}{3n^2}$, $\lim_{n \to \infty} U_n = \lim_{n \to \infty} L_n = \frac{14}{3}$ **3. (a)** $-\frac{8}{3}$ **5.** Overestimate by Trapezoidal rule: $= 17.8$
   liters. Underestimate by Midpoint rule: $= 17.2$ liters   **7. (a)** 10.47 **(b)** 28.07 **9.** $-\frac{55}{6}$ **11.** $\sqrt{3} - 1$ **13.** $\frac{1}{16}$

15. $\frac{9}{2}$ **17.** 21 **19.** $\frac{3072}{5}$ **21.** $\frac{\tan^4 \theta}{4} - \sec \theta + C$ **23.** $\frac{(x^2+2)^6}{12} + C$ **25.** $-\cot x^2 - \frac{x^2}{2} + C$ **27.** $-\frac{(1+\cos^2 t)^4}{4} + C$
29. 28 **31. (a)** 0.   **(b)** $-\frac{3}{\pi}$ **33.** $\frac{1}{2}$ **35.** $y = x^2 + \frac{1}{2}\cos 2x + C$ **37.** $\frac{1}{3}(t^2 + 4)^{3/2} + C$ **39.** $\frac{1}{5}(t^2 + 4)^{5/2} - \frac{22}{5}$
41. $\frac{1}{2}(\sqrt{x} + 2)^4 - \frac{71}{2}$ **43. (a)** ≈ 0.4137, $E_{10} \leq 0.000716$ **(b)** ≈ 0.4145, $E_{10} \leq 0.000358$ **(c)** ≈ 0.41422, $E_{10} \leq 4.48 \times 10^{-6}$
$\int_0^1 \frac{x}{\sqrt{x^2+1}}\,dx = \sqrt{2} - 1 \approx 0.414214$ **45. (a)** ≈ 15.7536, $E_8 \leq 0.0992$ **(b)** ≈ 15.7484, $E_4 \leq 0.0496$ **(c)** ≈ 15.7511,
$E_8 \leq 0.0563$, $\int_1^8 (x^{1/3} + x^{-1/3})\,dx = 15.75$ **47. (a)** $n = 260$ **(b)** $n = 184$ **(c)** $n = 24$ **49.** 0; $E_{20} \leq 0.00889$
51. ≈ 34.2 m   **53. (a)** ≈ -0.0073115 **(b)** 0 **(c)** ≈ 46.7097

CHAPTER 6

Exercise Set 6.1

1. $\frac{14}{3}$ **3.** $\frac{22}{3}$ **5.** $\frac{20}{3}$ **7.** 3 **9.** $\frac{8\pi}{3}$

11. $\frac{32}{3}$ **13.** 4 **15.** $\frac{9}{2}$ **17.** 4 **19.** $\frac{5}{6}$ **21.** $\frac{80}{3}$ **23.** $4\frac{1}{2}$

25. $\frac{4}{3}$

27. $\frac{1}{3}$

29. 8

31. $\frac{9}{4}$

33. $\frac{9}{2}$

35. 8

37. $\frac{256}{27}$

39. $\frac{8}{3}$

41. 3

43. 8

45. 3

47. **(a)** For $\int_0^{1/2} f(2x)\,dx$, let $u = 2x$, $\frac{1}{2}\int_0^1 f(u)\,du = \frac{1}{2}A$ **(b)** If $f(x) = c$, $f(2x) = c$

then $\int_0^1 f(x)\,dx = \int_0^1 f(2x)\,dx = c$. If f is periodic, with period 1, then

$\int_0^1 f(x)\,dx = \int_0^1 f(2x)\,dx = 0$.   **(c)** A

Exercise Set 6.2

1. $\frac{37\pi}{12}$

3. $\frac{256\pi}{15}$

5. 8π

7. $\frac{16\pi}{7}$

9. $\frac{434\pi}{3}$

11. $\frac{92\pi}{15}$

13. $\frac{477\pi}{4}$

15. $\frac{72\pi}{5}$

17. $\frac{16\pi}{3}$

19. $\frac{32\pi}{3}$

21. $\frac{32\pi}{3}$

23. $\frac{99\pi}{5}$

25. 10π

27. $\frac{277\pi}{3}$

29. $\frac{120\pi}{7}$

31. $\frac{4}{3}\pi a^3$

33. $\frac{28\pi}{3}$

35.

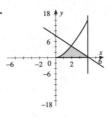

(a) $\frac{88\pi}{3}$
(b) 24π

37. $\frac{\pi}{3}$ **39.** $\frac{2}{3}\pi$ **41.** $\frac{\pi}{3}$

43. $\frac{\pi}{6}(3\pi - 8)$ **45.** $\frac{\pi}{3}$ **47.** $\frac{\pi}{6}(10 - 3\pi)$ **49.** $\frac{\pi}{2}(\pi - 1)$ **51.** 8π **53.** $\frac{16}{3}$

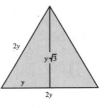

55. 2π **57.** $\frac{1612\pi}{15}$ **59.** $\frac{5000\pi}{3}$ liters   **61.** $16\sqrt{3}$

63. $\frac{256\pi}{3}$ **65. (a)** ≈ 15.4282
(b) ≈ 15.3319

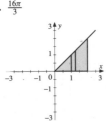

Exercise Set 6.3

1. $\frac{16\pi}{3}$ **3.** 8π **5.** $\frac{32\pi}{3}$ **7.** $\frac{110\pi}{3}$ **9.** $\frac{381\pi}{14}$

11. $\frac{8\pi}{3}$ **13.** $\frac{12\pi}{5}$ **15.** $\frac{4\pi}{27}$ **17.** $\frac{16\pi}{5}$ **19.** $\frac{5\pi}{6}$ **21.** $\frac{16\pi}{3}$

23. $2\pi^2 a^2 b$

25. $\frac{1024\pi}{3}$

27.

(a) $\frac{875\pi}{48}$ (b) $\frac{2125\pi}{96}$

29.

(a) $2\pi \int_0^\pi x \sin^2 x \, dx$; $\frac{\pi^3}{2} \approx 15.503$

(b) $2\pi \int_0^\pi (x + \pi) \sin^2 x \, dx$; $\frac{3\pi^3}{2} \approx 46.509$

Exercise Set 6.4

1. $3\sqrt{10}$ **3.** $\sqrt{13}$ **5.** $\frac{14}{3}$ **7.** $\frac{335}{27}$ **9.** $\frac{22}{3}$ **11.** $\frac{1}{54\sqrt{2}}\left[(77)^{3/2} - (32)^{3/2}\right]$ **13.** $\frac{53}{6}$ **15.** $6a$

17. $x = \pm\left(\frac{2y}{3}\right)^{3/2} - 1$

The point $\left(-2, \frac{3}{2}\right)$ is on $x = -\left(\frac{2y}{3}\right)^{3/2} - 1$

and $(7, 6)$ is on $\left(\frac{2y}{3}\right)^{3/2} - 1$. Thus, there are

two functions of y. Each must be dealt with

independently of the other. $L = 5\sqrt{5} + 2\sqrt{2} - 2$

19. ≈ 0.22 **21.** ≈ 4.6468 **23.** ≈ 17.261

Exercise Set 6.5

1. $24\pi\sqrt{5}$ **3.** $\frac{2\pi}{3}(5\sqrt{5} - 1)$ **5.** 2π **7.** $\frac{\pi}{27}[(145)^{3/2} - (10)^{3/2}]$ **9.** $\frac{\pi}{6}[(17)^{3/2} - (5)^{3/2}]$

11. $4\pi a^2$

13. $\frac{12}{5}\pi a^2$

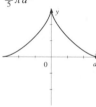

15. Partition [a,b] so that the zeros of f occur at partition points. For those intervals on which $f(x) < 0$, the band obtained has radius $|f(x_k^{**})|$, where $x_k^{**} \in (x_{k-1}, x_k)$, and so its area is

$2\pi |f(x_k^{**})|\sqrt{1 + [f'(x_k^*)]^2}\Delta x_k$. If

$f(x) > 0$, the same result holds. Sum and let $\|P\| \to 0$ to get the result.

17. $\frac{\pi}{2}$ **19.** $\frac{448\pi}{9}$

21.

(a) $\approx 1.341 \times 10^8 \text{ km}^2$
(b) $\approx 9.824 \times 10^{-5}$

23. (a) 14.42
(b) 53.23

Exercise Set 6.6

1. $\frac{4}{3}$ joules   **3.** 24 inch-pounds   **5.** $\frac{45}{4}$ newton-cm; 30 newton-cm   **7.** $k = 0.04$ N/m   **9.** 1350 ft-lb   **11.** 1930 joules

13. $\approx 6.904 \times 10^6$ joules   **15.** $\frac{k}{3}$ dyne-cm   **17.** $\approx 17,200$ joules   **19.** $\approx 6.87 \times 10^6$ joules   **21.** $\approx 17,020$ joules

Exercise Set 6.7

1. $56\rho g$

3. $135\rho g$

$\frac{w}{x} = \frac{10}{8}$
$w = \frac{5x}{4}$

5. $\frac{184\rho g}{3}$

7. ≈ 124.8 lb; ≈ 62.4 lb

9.

(a) $\approx 49,050$N;
$\approx 784,800$N
(b) $\approx 515,000$N

11. $\frac{52\rho g}{3}$

13. $\frac{304\rho g}{15}$

15. The kth strip, cut off by horizontal planes $x = k_{k-1}$ and $x = x_k$ is inclined and so its dimensions are $w(x*)$ by $\Delta x \sec\theta$ (see sketch).

$$F \approx \sum_{k=1}^{n} \rho g_{\text{depth}} h(x*) w(x*) \Delta x_{\text{area}} \sec\theta$$

Take limits to get the integral.

17. $\approx 285,700$N

Exercise Set 6.8

1. $\frac{27}{59}$ **3.** $\left(-\frac{33}{32}, \frac{63}{32}\right)$ **5.** $\frac{3}{2}$ **7.** $\frac{633}{95} \approx 6.663$ **9.** $\left(\frac{4}{3}, \frac{4}{3}\right)$ **11.** $\left(-\frac{2}{3(4+\pi)}, 0\right)$

13. $\left(\frac{8}{3}, \frac{10}{3}\right)$

15. $\left(\frac{8}{3\pi}, \frac{8}{3\pi}\right)$

17. $\left(-\frac{1}{2}, \frac{2}{5}\right)$

19. $\left(\frac{8}{5}, 0\right)$

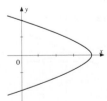

21. $(\bar{x}, \bar{y}) = \left(0, \frac{3}{5}b\right)$, which is independent of a.

23. $\bar{y} = \frac{2}{\pi a^2} \int_{-a}^{a} \left(\frac{1}{2}\sqrt{a^2 - x^2}\right)\left(\sqrt{a^2 - x^2}\right) dx = \frac{4a}{3\pi}$

25. $\left(\frac{23}{8}, \frac{5}{8}\right)$

27. $\left(\frac{192 - 7\pi}{48 - \pi}, \frac{144 - 4\pi}{48 - \pi}\right)$

29. Rotate a semicircular region of radius a about the y-axis. Centroid of region is $\left(\frac{4a}{3\pi}, a\right)$. Area $= \frac{\pi a^2}{2}$. Volume $= \frac{\pi a^2}{2} \cdot 2\pi\left(\frac{4a}{3\pi}\right) = \frac{4}{3}\pi a^3$

31.

(a) 90π
(b) 36π
(c) 72π

Exercise Set 6.9

1. $\frac{1}{28}$ **3.** $\frac{3}{38}$ **5.** $\frac{4}{3}$ **7.** 60   **9.** 1   **11.** $\frac{1}{2}$ **13.** $\frac{1}{4}$ **15.** $\frac{37}{125}$ **17.** $\mu = 1$, $\sigma^2 = \frac{1}{2}$

19. $\mu = \frac{1}{2}$, $\sigma^2 = \frac{1}{20}$ **21.** $\mu = \frac{4}{5}$, $\sigma^2 = \frac{2}{75}$ **23.** $\mu = \frac{1}{2}(a + b)$, $\sigma^2 = \frac{1}{12}(a^2 - 2ab + b^2)$ **25.** $\frac{3}{4}$ **27. (a)** $\frac{1}{8}$

(b) μ = about 333,000,000 barrels   **29.** ≈ 1.58 hundred hours   **31.** $\sigma^2 = E[(x - \mu)^2] = E[x^2 - 2\mu x + \mu^2] = E[x^2] - 2\mu E[x] + E[\mu^2] = E[x^2] - 2\mu^2 + \mu^2 = E[x^2] - \mu^2$ **33.** $\frac{\pi - 2}{8}$

Exercise Set 6.10

1. $\frac{x^4}{4} - x^2 + 200$ **3.** $\approx \$2092$

5. Consumers' surplus = $\$15,340$ Producers' surplus = $\$66,120$

7. (a) equilibrium price = $\$700$; equilibrium quantity = 10,000 units
(b) $\approx \$1867$
(c) $\approx \$3467$
(d)

9. $C = 0.1I + 0.8\sqrt{I + 1} + 3.1$; $C(20) \approx \$8.77$ billion

Chapter Review

1. $\frac{46}{3}$ **3.** 24   **5.** 16   **7.** $\frac{2}{\pi} + \frac{1}{6}$ **9.** 4

11. $\frac{\pi}{2}$ **13.** $\frac{176\pi}{3}$ **15.** $\frac{500\pi}{3}$ **17.** $\frac{28\pi}{3}$ **19.** $\frac{625\pi}{6}$

21. $\frac{32}{3}$ **23.** $\frac{1}{27}\left[(40)^{3/2}-8\right]$ **25.** $\frac{22}{3}$ **27.** $\frac{49\pi}{9}$ **29. (a)** $\frac{1744}{45}\pi$ **31.** 37,500 ft-lb

(b) $\frac{173\pi}{3}$

33.   **(a)** ≈ 266 lb   **35.** $\frac{22}{5}$ **37.** $\left(\frac{1}{2},\frac{8}{5}\right)$ **39.** $\left(-\frac{8}{5},2\right)$ **41.** $\left(\frac{19}{12(3+2\pi)},\frac{1+8\pi}{6+4\pi}\right)$

(b) ≈ 165 lb

43. (a) $\frac{1}{2}$ **(d)** $\frac{2}{27}$ **(e)** $\frac{17}{27}$ **45.** $C(x)=4x-\frac{x^2}{2000}+5400,\ R(x)=40x-\frac{x^2}{200},$

(b) $\frac{17}{8}$

(c) $\frac{139}{320}$ $P(x)=36x-\frac{9x^2}{2000}-5400;\ P'(x)=0\rightarrow x=4000,$ where

$M_C(x)=M_R(x)$

Maximum profit $\approx \$66,600$

CHAPTER 7

Exercise Set 7.1

1. $\{x: x>0\};\ -\frac{1}{x}$ **3.** $\{x: x>0\};\ \frac{3}{x}$ **5.** $\{x: 0<x<1\};\ \frac{1}{x}-\frac{1}{2(1-x)}$ **7.** $\{x: x>0\};\ 1+\ln x$

9. $\{x:-1<x<1\};\ \frac{1}{x^2-1}$ **11.** $\{x: x>1\};\ \frac{1}{x\ln x}$ **13.** $\{x:\ -\frac{\pi}{2}+2n\pi<x<\frac{\pi}{2}+2n\pi\};\ \tan x$

15. $\{x: x>0\};\ \ln\frac{x-2}{x^2}$ **17.** $\{x: x<-1\}\cup$ **19.** $\{x:-3<x<0\}\cup\{x:2<$ **21.** 1   **23.** $-\infty$

$\{x: 1<x<2\};\ \frac{x}{x^2-1}+\frac{1}{2-x}$ $x\};\ \frac{1}{x+3}-\frac{1}{x}-\frac{1}{x-2}$

25. 0   **27. (a)** Let $f(x)=\frac{\ln x}{\sqrt{x}}$ **(b)** Let $f(x)=x-1-\ln x,$

For $0<x\le 1,\ \ln x<0$ and   For $x>1,\ f'(x)=1-\frac{1}{x}>0$

$\ln x<\sqrt{x};\ f'(x)=\frac{2-\ln x}{2x^{3/2}}$ has critical   Therefore, f is an increasing

function, and $f(x)>0$ for

number: $x=e^2\ f(e^2)\approx 0.736,$ and $x>1.$

$\ln x<\sqrt{x},\ \lim\limits_{x\to+\infty}\frac{\ln x}{\sqrt{x}}=0$

Thus, $\ln x<\sqrt{x}$ for $x>0.$

29. $\ln(1-\cos x)+C,\ x\ne 2n\pi$ **31.** $\frac{1}{3}\ln|x^3+1|+C,\ x\ne -1$ **33.** $\ln|\sin x|+C,\ x\ne n\pi$

35. $\ln|\ln x|+C,\ x>0,\ x\ne 1$ **37.** $\frac{1}{3}\ln|x(2\sqrt{x}-3)|+C,\ x>0,\ x\ne\frac{9}{4}$ **39.** $\frac{x\sqrt{1-x}}{(x+2)(x-3)}\left[\frac{1}{x}-\frac{1}{2(1-x)}-\frac{1}{x+2}-\frac{1}{x-3}\right]$

41. $\dfrac{\sqrt[3]{(x+1)^2}}{x^4(3-4x)^2}\left[\dfrac{2}{3(x+1)}-\dfrac{4}{x}+\dfrac{8}{3-4x}\right]$ **43.** $\left(\dfrac{1}{2}, -\dfrac{1}{2}\ln 2 - \dfrac{1}{4}\right)$ is a local maximum.

45. $(1,1)$ is a local minimum, $\left(-\dfrac{1}{2}, -\dfrac{1}{2}-\ln 4\right)$ is a local maximum.

47. $\left(\dfrac{3}{2}, -\ln 3\right)$ is a local maximum.

49. domain $= \{x : x \neq 0\}$
intercepts: $(\pm 1, 0)$
symmetry: to y-axis
asymptote: $x = 0$
maxima, minima: none
concavity: always down
inflection points: none

51. domain $= (-1, 1)$
intercepts: $x = 0$, $y = 0$
symmetry: y-axis
asymptotes: $x = \pm 1$
maxima, minima: $(0,0)$
local maximum.
concavity: always down
inflection points: none

53. domain $= \{x: x > 0\}$
intercepts: none
symmetry: none
asymptote: $x = 0$
maxima, minima: $(1, 1)$
local minimum
concavity: always up
inflection points: none

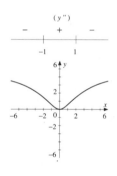

55. domain $= \mathbb{R}$
intercepts: $(0, 0)$
symmetry: y-axis
asymptotes: none
maxima, minima: $(0, 0)$
local minimum
concavity: down on $(-\infty, -1) \cup (1, \infty)$
up on $(-1, 1)$
inflection points: $(-1, \ln 2), (1, \ln 2)$

57. Let $x = n + 1$. Left-hand Riemann sum is $\displaystyle\sum_{k=1}^{n}\dfrac{1}{k}$. For a decreasing function, the area of the kth rectangle $\geq$ area under the graph, and $\displaystyle\sum_{k=1}^{n}\dfrac{1}{k} \geq \int_1^x \dfrac{1}{t}\,dt = \ln x$. If $\ln x = 100$, then $100 \leq \displaystyle\sum_{k=1}^{n}\dfrac{1}{k}$, or n is the smallest integer satisfying the given inequality. x must be greater than m.

59. $2\ln x + (\ln x)^2$; $\dfrac{2}{x}(1 + \ln x)$ **61.** $\dfrac{1}{x\sqrt{\ln x^2}}$; $-\dfrac{1+\ln x^2}{x^2(\ln x^2)^{3/2}}$ **63.** $\dfrac{y(2x+\ln y)}{x(1-x)}$ **65.** $-\ln(1 + \cos^2 x) + C$

67. $x - 2\ln|x - 1| - \dfrac{1}{x-1} + C$

69. domain: $x \neq 1, -1$
intercept: $(0, 0)$
asymptotes: $y = 0$, $x = \pm 1$
symmetry: origin
maxima, minima: none
concavity: down if $x > 0$
up if $x < 0$
inflection points: $(0, 0)$

71. domain $= (-\infty, 0) \cup (3, \infty)$
$\left(-3, \ln\dfrac{3}{2}\right)$ is a local maximum.
intercept: $(-1, 0)$
symmetry: none
asymptotes: $x = 0$, $x = 3$
$y = 0$
Inflection point: $(-4.91, .36)$
Concave up on: $(-\infty, -4.91)$
Concave down on: $(-4.91, 0) \cup (0, \infty)$

73. Let $\ln n = \int_1^n \frac{1}{x}\,dx$, and

consider the Riemann sums L_{n-1} and U_{n-1} with $\Delta x = 1$. Then, $L_{n-1} \le \ln n \le U_{n-1}$.

$$\sum_{k=1}^{n-1} \frac{1}{k+1} < \ln n < \sum_{k=1}^{n-1} \frac{1}{k}.$$

Hence, $\frac{1}{2} + \frac{1}{3} + \ldots + \frac{1}{n} < \ln n$

$< 1 + \frac{1}{2} + \frac{1}{3} + \ldots + \frac{1}{n-1}$.

75. If $0 < a < 1$, it shrinks the graph of $\ln (bx)$ in the vertical direction. If $1 < a < \infty$, it stretches the graph. If $a < 0$, it reflects the graph about the x-axis and shrinks or stretches the graph depending on $|a|$. Let $y = a(\ln b + \ln x) = a \ln b + a \ln x$. For $b > 1$, the graph is moved up if $a > 0$ and down if $a < 0$. For $0 < b < 1$, the graph is moved down if $a > 0$ and up if $a < 0$.

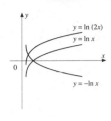

Exercise Set 7.2

1. (a) $\sqrt{3}$ (b) $\frac{1}{16}$ **3.** (a) $\ln 5 - 1$ (b) $1 - \ln 2$ **5.** (a) $\frac{1}{e}$ (b) $e^2 - 1$ **7.** (a) $\frac{1}{\sqrt[3]{e}}$ (b) $1 + \frac{1}{e^3}$ **9.** $2xe^{x^2}$

11. $3x^2$ **13.** $\frac{1}{2\sqrt{x}} e^{\sqrt{x}}$ **15.** $e^x e^{e^x} = e^{x + e^x}$ **17.** $e^x(x+1)$ **19.** $e^{2x}(3\cos 3x + 2\sin 3x)$ **21.** $\frac{e^x + e^{-x}}{2}$ **23.** $\frac{4}{(e^x + e^{-x})^2}$

25. $2e^x(\sin e^x \cos e^x) = e^x \sin 2e^x$ **27.** $\frac{-e^x}{(e^x-1)^2}$ **29.** $\frac{y - e^{x+y}}{e^{x+y} - x}$ **31.** $\frac{e^{-x}\sin y - e^{-y}\sin x}{e^{-x}\cos y + e^{-y}\cos x}$ **33.** $-\frac{1}{2}e^{(1-x^2)} + C$

35. $e^{-\cos x} + C$ **37.** $\ln(e^x + e^{-x}) + C$ **39.** $\frac{7}{3}$ **41.** $-\frac{3}{8}(1 - e^6)^{4/3}$ **43.** $\ln \frac{3}{2}$ **45.** $3 - \sqrt{3}$ **47.** $\frac{1}{1 - e^x} + C$

49. intercept: $(0, 1)$
symmetry: to y-axis
asymptote: $y = 0$
maxima, minima: local max $(0, 1)$
inflection points:
$\left(2, \frac{1}{e}\right)$, $\left(-2, \frac{1}{e}\right)$
concave up on $(-\infty, -2) \cup (2, \infty)$
down on $(-2, 2)$

51. intercept: $(0, 0)$
symmetry: none
asymptote: $y = 0$ is an asymptote on the right
maxima, minima: $\left(2, \frac{2}{e}\right)$ is a local max
inflection points: $\left(4, \frac{4}{e^2}\right)$ is an inflection point
concave up on: $(4, \infty)$
concave down on: $(-\infty, 4)$

53. intercept: $\left(0, \frac{1}{2}\right)$
symmetry: none
asymptotes: $y = 1$ on right
$y = 0$ on left
maxima, minima: none
inflection point: $\left(0, \frac{1}{2}\right)$
concave up on: $(-\infty, 0)$
concave down on: $(0, \infty)$

55. intercept: $(0, 0)$
symmetry: none
asymptote: $y = 0$
maxima, minima: $(3, 27e^{-3})$ is a local maximum
inflection points: $(0, 0)$,
$(3 - \sqrt{3}, 0.574)$,
$(3 + 3\sqrt{3}, 0.9334)$
concave down on:
$(-\infty, 0) \cup (3 - \sqrt{3}, 3 + \sqrt{3})$
concave up on:
$(0, 3 - \sqrt{3}) \cup (3 + \sqrt{3}, \infty)$

57. intercept: $(0, 0)$
symmetry: about y-axis
asymptote: $y = 1$
maxima, minima: $(0, 0)$ is a local minimum
inflection points:

$$\left(-\sqrt{\tfrac{2}{3}}, e^{-3/2}\right) \left(\sqrt{\tfrac{2}{3}}, e^{-3/2}\right)$$

concave down on:

$$\left(-\infty, -\sqrt{\tfrac{2}{3}}\right) \cup \left(\sqrt{\tfrac{2}{3}}, \infty\right)$$

concave up on: $\left(-\sqrt{\tfrac{2}{3}}, \sqrt{\tfrac{2}{3}}\right)$

(f'')

f'

59. $\ln e^{-x} = -x$ and
$\ln \frac{1}{e^x} = \ln 1 -$
$\ln e^x = 0 - x = -x.$ So
$\ln e^{-x} = \ln \frac{1}{e^x}$ and,
therefore, $e^{-x} = \frac{1}{e^x}$.

61. $\frac{1}{2}$ **63.** $\frac{3}{4}$

65. $f'(x) = \frac{e^x + e^{-x}}{2} > 0$ for all x. Therefore,
monotone increasing; hence $1 - 1$ on $\mathbb{R}$.
$f^{-1}(x) = \ln\left(x + \sqrt{x^2 + 1}\right)$

67. intercepts: none
symmetry: none
asymptotes: $x = 0$, $y = 1$
maxima, minima: no local extrema
inflection point: $\left(-\frac{1}{2}, e^{-2}\right)$
concave up on: $\left(-\frac{1}{2}, 0\right) \cup (0, \infty)$
concave down on: $\left(-\infty, -\frac{1}{2}\right)$

69. intercept: $(0, 0)$
symmetry: origin
asymptote: $y = 0$
maxima, minima: $(1,1)$ local max,
$(-1, -1)$ local min
inflection points:
$(0, 0)$, $\left(\sqrt{3}, \frac{1}{e}\right)$, $\left(-\sqrt{3}, -\frac{1}{e}\right)$
concave up on: $(-\sqrt{3}, 0) \cup (\sqrt{3}, \infty)$
concave down on: $(-\infty, -\sqrt{3}) \cup (0, \sqrt{3})$

71. $k = -\frac{1}{500} \ln \frac{3}{8}$; $\left(\frac{\ln 2}{\ln \frac{8}{3}}, 500\right)$ is
where rate of growth is fastest

(Q'')

.7067
max for Q'

73. (a) $\lim\limits_{x \to \infty} \frac{h(x)}{f(x)} = \lim\limits_{x \to \infty} \frac{x^c}{e^{ax}} = 0.$ Use L'Hôpital's rule until the exponent of x in the numerator ≤ 0. **(b)** $\lim\limits_{x \to \infty} \frac{g(x)}{h(x)} = \lim\limits_{x \to \infty} \frac{(\ln x)^b}{x^c} = 0.$
Use L'Hôpital's rule until the exponent of $\ln x \leq 0$.

Exercise Set 7.3

1. (a) $e^{\sqrt{3} \ln \pi} \approx 7.26255$ **(b)** $e^{\pi \ln \pi} \approx 36.46216$ **(c)** $e^{e \ln \pi} \approx 22.45916$ **(d)** $e^{\sqrt{2} \ln \sqrt{2}} \approx 1.63253$
(e) $e^{\pi \ln(\sin 2)} \approx 0.74177$ **3.** $2^x \ln 2$ **5.** $4^{-x} x (2 - x \ln 4)$ **7.** $\frac{1}{\ln 3}\left[\frac{1}{x} - \frac{3x^2}{1+x^3}\right]$ **9.** $\frac{1}{x(x^2-1)\ln 10}$ **11.** $\frac{1}{x(\ln 2)(\ln x)}$

13. $x^x(1 + \ln x)$ **15.** $x^{1/(1-x)}\left[\frac{1}{x(1-x)} - \frac{\ln x}{(1-x)^2}\right]$ **17.** $(1 - x^2)^{1/x}\left[\frac{2}{x^2 - 1} - \frac{1}{x^2}\ln(1 - x^2)\right]$ **19. (a)** 2 **(b)** ∞

21. (a) $-\frac{1}{2}$ **(b)** -1 **23. (a)** e **(b)** $\frac{1}{e}$ **25. (a)** 1 **(b)** e^2 **27.** $-\frac{1}{2^{x^2+1}\ln 2} + C$ **29.** $\frac{2^{\sin x}}{\ln 2} + C$

31. $\frac{1}{\ln 2} \ln(2^t + 1) + C$ **33.** $\frac{8}{\ln 3}$ **35.** $-\frac{1}{2 \ln 2} \frac{1}{(2^{2t}-1)} + C = \frac{1}{(\ln 4)(1-4^t)} + C$ **37.**

39. (a) $\left(\frac{1}{x-1} - \frac{1}{x+2}\right) \frac{1}{\ln 10} = \frac{3}{(x-1)(x+2)\ln 10}$

(b) $\frac{1}{\ln 2} \left[\frac{1}{x} - \frac{1}{2(1-x)} - \frac{6}{2x-3}\right]$

41. intercepts: $(\pm\sqrt{2}, 0)$
symmetry: y-axis
asymptotes: $x = \pm 1$ for negative y
maximum, minimum: none
inflection points: none
concave down on
$(-\infty, -1), \cup (1, \infty)$

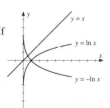

43. $\frac{\ln 2}{\ln 6} \approx 0.3869$ **45.** $\frac{\ln \frac{5}{3}}{\ln \frac{4}{3}} \approx 1.7757$ **47. (a)** Let $u = \log_a x$ and $v = \log_a y$. Then

$x = a^u$ and $y = a^v$; $xy = a^u \cdot a^v = a^{u+v}$ $\log_a xy = u + v = \log_a x + \log_a y$

(b) Let $u = \log_a x$ and $v = \log_a y$ $x = a^u$, $y = a^v$; $\frac{x}{y} = \frac{a^u}{a^v} = a^{u-v}$ $\log_a \frac{x}{y} = u - v = \log_a x - \log_a y$

(c) $\log_a x^\alpha = \alpha \log_a x$ Let $u = \log_a x$ $x = a^u$; $x^\alpha = (a^u)^\alpha = a^{\alpha u}$ $\log_a x^\alpha = \alpha u = \alpha \log_a x$

49. (a) $\approx \$16,288.95$ **(b)** $\approx \$16,386.16$ **(c)** $\approx \$16,486.65$ **(d)** $\approx \$16,487.21$ **51.** $\frac{\pi}{\ln 2}$ **53. (a)** Since for

$1 \le t \le 1 + \frac{1}{n}$, $\frac{1}{1+\frac{1}{n}} \le \frac{1}{t} \le \frac{1}{1}$, it follows that $\int_1^{1+1/n} \frac{1}{1+\frac{1}{n}} dt \le \int_1^{1+1/n} \frac{1}{t} dt \le \int_1^{1+1/n} 1 dt$ **(b)** Left integral:

$\int_1^{1+1/n} \frac{1}{1+\frac{1}{n}} dt = \frac{1}{n+1}$. Middle integral: $\int_1^{1+1/n} \frac{1}{t} dt = \ln\left(1 + \frac{1}{n}\right)$. Right integral: $\int_1^{1+1/n} 1 dt = \frac{1}{n}$. So $\frac{1}{n+1} \le \ln\left(1 + \frac{1}{n}\right) \le \frac{1}{n}$

(c) $\frac{n}{n+1} \le n \ln\left(1 + \frac{1}{n}\right) \le 1$ or $\frac{1}{1+\frac{1}{n}} \le \ln\left(1 + \frac{1}{n}\right)^n \le 1$ **(d)** Let $n \to \infty$, $1 \le \lim_{n \to \infty} \ln\left(1 + \frac{1}{n}\right)^n \le 1$, and thus,

$\lim_{n \to \infty} \ln\left(1 + \frac{1}{n}\right)^n = 1$. Since the logarithm is continuous, $\lim(\ln) = \ln(\lim)$, and so $\ln\left[\lim_{n \to \infty} \left(1 + \frac{1}{n}\right)^n\right] = 1$. Thus,

$\lim_{n \to \infty} \left(1 + \frac{1}{n}\right)^n = e$ **55.** Let $h(x) = \frac{f(x)}{g(x)} = \frac{x^n}{a^x}$; if $n > 0$ and $a > 1$, $\lim_{x \to \infty} \frac{x^n}{a^x} = \lim_{x \to \infty} \frac{nx^{n-1}}{a^x \ln a} = 0$. f grows more slowly than g.

If $n < 0$ and $a > 0$, $f(x)$ decreases and $g(x)$ increases as $x \to \infty$. If $n > 0$ and $0 < a < 1$, $f(x)$ increases and $g(x)$ decreases as $x \to \infty$. If $n < 0$ and $0 < a < 1$, both $f(x)$ and $g(x) \to 0$ as $x \to \infty$. Let $h(x) = a^x x^n$; $h'(x) = na^x x^{n-1} + a^x (\ln a) x^n$. $h'(x) < 0$, since $n < 0$ and $\ln a < 0$. Hence, h is a decreasing function. Because $\lim_{x \to \infty} h(x) = 0$, we can conclude that a^x decreases

more rapidly than x^n.   **57. (a)** If $x^e < e^x$, then $\ln x^e < e^x$, since $\ln x$ is an increasing function. Or, $e \ln x < x$, and

$\frac{\ln x}{x} < \frac{1}{e}$, since $x > 0$. If $\frac{\ln x}{x} < \frac{1}{e}$, then $e \ln x < x$, $\ln x^e < x$, and $e^{\ln x^e} = x^e < e^x$, since the exponential function is

increasing.   **(b)** $f'(x) = \frac{1-\ln x}{x^2}$. The only critical value is at $\ln x = 1$, or $x = e$. $f''(e) = -\frac{1}{e^3} < 0 \Rightarrow$ local maximum

$\lim_{x \to 0^+} f(x) = -\infty$; $\lim_{x \to \infty} f(x) = 0$. Therefore, $\left(e, \frac{1}{e}\right)$ is the absolute maximum of f.   **(c)** $x \in (0, e) \cup (e, \infty)$

59. $a^x = x^b \Rightarrow x \ln a = b \ln x, a > 0$; $x = \frac{b}{\ln a} \ln x$. We are looking for

intersections of $y = x$ and $y = \frac{b}{\ln a} \ln x$. There can be 0, 1, or 2 intersections. If

$\frac{b}{\ln a} > 0$, the graph will intersect the line $y = x$ zero or two times. If $\frac{b}{\ln a} < 0$,
the graph of $\ln x$ will intersect the graph of $y = x$ at one point.

61. (a) $f(x) = e^x$, $g(x) = x^\alpha$, $h(x) = \ln x$; $\alpha > 0$ and real. Suppose that n is an integer such that $n \le \alpha \le n + 1$ $n = [\alpha]$, the

greatest integer in α. $\lim_{x \to \infty} \frac{g(x)}{f(x)} = \lim_{x \to \infty} \frac{x^\alpha}{e^x} = \lim_{x \to \infty} \frac{\alpha(\alpha-1)...(\alpha-n+1)x^{\alpha-n}}{e^x}$. If $\alpha = n$, we are through, since the numerator is a constant,

and in this case, the limit is 0. If $\alpha > n$, apply L'Hôpital's rule once more, and the limit is 0. e^x grows more rapidly than x^α, regardless

of how large α is. Similarly, $\lim_{x \to \infty} \frac{h(x)}{g(x)} = \lim_{x \to \infty} \frac{\ln x}{x^\alpha} = \lim_{x \to \infty} \frac{1/x}{\alpha x^{\alpha-1}} = \lim_{x \to \infty} \frac{1}{\alpha x^\alpha} = 0$. Thus, x^α grows more rapidly than $\ln x$,

regardless of how small the positive exponent α is.   **(b)** Let $f(x) = e^{ax}$, $g(x) = (\ln x)^\beta$, and $h(x) = x^\gamma$; $a > 0, \beta, \gamma > 0$ and

real. $\lim_{x \to \infty} \frac{h(x)}{f(x)} = \lim_{x \to \infty} \frac{x^\gamma}{e^{ax}} = \lim_{x \to \infty} \frac{\gamma(\gamma-1)...(\gamma-n+1)x^{\gamma-n}}{a^n e^{ax}}$, where $n = [\gamma]$. If $\gamma = n$, we are through, since the numerator is a constant.

If $\gamma > n$, apply L'Hôpital's rule once more. In either case, the limit of the quotient is 0. Therefore, e^{ax} grows more rapidly than x^{γ}. Next consider, $\lim_{x\to\infty} \frac{g(x)}{h(x)} = \lim_{x\to\infty} \frac{(\ln x)^{\beta}}{x^{\gamma}} = \lim_{x\to\infty} \frac{(\beta/x)(\ln x)^{\beta-1}}{\gamma x^{\gamma-1}}$. Continue applying L'Hôpital's rule until the exponent on $\ln x$ is ≤ 0. The limit will then be zero. Thus, x^{α} grows more rapidly than $(\ln x)^{\beta}$.

Exercise Set 7.4

1. 13,500   **3.** 10 hours   **5.** 16 kilograms   **7.** $\frac{\ln 9}{\ln 2} \approx 3.17$ hours   **9.** ≈ 33.87 hours   **11.** ≈ 55, 902

13. ≈ 70.7 grams; ≈ 26.6 years   **15.** $-6.25°$ **17.** Suppose $T_m > T_0$, and assume $\frac{dT}{dt} = k(T - T_m)$. Then, $\frac{dT}{T-T_m} = kdt$, $\ln|T - T_m| = kt + C$; $|T - T_m| = e^{kt+C} = C_1 e^{kt}$, $T - T_m < 0$, so $|T - T_m| = T_m - T$. $T = T_m - C_1 e^{kt}$. When $t = 0, T = T_0; C_1 = T_m - T_0$ and $T = T_m - (T_m - T_0)e^{kt}$ or $T(t) = T_m + (T_0 - T_m)e^{kt}$ $t \approx 32.1$ minutes   **19.** 9.6 psi

21. ≈ 9.97 m   **23. (a)** $N(t) = N_1 e^{k(t-1)}$; $N(1) = N_1$.   **(b)** 5   **25.** $\approx 19,035$ years   **27. (a)** $t = -\frac{1}{k}\ln(1+\lambda)$

(b) $\approx 7.18 \times 10^9$ years

Exercise Set 7.5

1. (a) $\frac{\pi}{6}$ **(b)** $\frac{\pi}{3}$ **(c)** $\frac{\pi}{4}$ **(d)** $\frac{\pi}{3}$ **(e)** $\frac{\pi}{2}$ **3. (a)** $-\frac{\pi}{2}$ **(b)** π **(c)** 0   **(d)** $\frac{2\pi}{3}$ **(e)** $\frac{3\pi}{4}$

5. (a) $\frac{4}{3}$ **(b)** $\frac{3}{5}$ **7. (a)** $\frac{7}{9}$ **(b)** $-\frac{24}{25}$

9. (a) $\frac{\pi}{5}$ **11.** $\frac{33}{65}$ **13. (a)** 2.412   **(b)** 4.965   **15.** $\frac{6}{4x^2+9}$ **17.** $\frac{1}{2\sqrt{x-x^2}}$

(b) $-\frac{\pi}{8}$

19. $1 + 2x\tan^{-1}x$ **21.** $\frac{x\cos^{-1}x - \sqrt{1-x^2}}{(1-x^2)^{3/2}}$ **23.** $\frac{2x\sqrt{1-y^2}}{\sqrt{1-y^2}-1}$ **25.** $\frac{\pi}{2}$ **27.** $\frac{\pi}{12}$ **29.** $\frac{\pi}{16}$ **31.** $\frac{1}{3}\sin^{-1}\frac{3x}{4} + C$

33. $\frac{1}{3}\sec^{-1}\frac{4x}{3} + C$ **35.** $\frac{1}{3}\tan^{-1}(3x-2) + C$ **37.** $\frac{1}{2}\sin^{-1}x^2 + C$ **39.** $-\frac{\pi}{6\sqrt{3}}$ **41.** $\frac{\pi}{3}$ **43.** $\frac{\pi^2}{2\sqrt{3}}$

45. $\int \frac{du}{\sqrt{a^2-u^2}} = \int \frac{du}{a\sqrt{1-\frac{u^2}{a^2}}} = \int \frac{dv}{\sqrt{1-v^2}} = \sin^{-1}v + C$ Let $v = \frac{u}{a}$, $dv = \frac{1}{a}du = \sin^{-1}\frac{u}{a} + C$

47. Let $y = \tan^{-1}x$, $-\infty < x < \infty$, $-\frac{\pi}{2} < y < \frac{\pi}{2}$, $x = \tan y$; $1 = (\sec^2 y)y'$, $y' = \frac{1}{\sec^2 y} = \frac{1}{1+\tan^2 y} = \frac{1}{1+x^2}$

49. Let $y = \sec^{-1}x$, $|x| > 1$ $0 < y < \frac{\pi}{2}$ if $x > 1$, $\pi < y < \frac{3\pi}{2}$ if $x < -1$, $x = \sec y$; $1 = (\sec y \tan y)y'$, $y' = \frac{1}{\sec y \tan y} = \frac{1}{\sec y \sqrt{\sec^2 y-1}}$ (since $\tan y > 0$ for y in given range) $= \frac{1}{x\sqrt{x^2-1}}$

51. $\int \frac{du}{1+u^2} = \tan^{-1}u + C$, since $\frac{d}{du}(\tan^{-1}u + C) = \frac{1}{1+u^2}$ **53.** $\int \frac{du}{u\sqrt{u^2-a^2}} = \frac{1}{a}\int \frac{du}{u\sqrt{\frac{u^2}{a^2}-1}} = \frac{1}{a}\int \frac{\frac{1}{a}du}{\frac{u}{a}\sqrt{\frac{u^2}{a^2}-1}} = \frac{1}{a}\int \frac{dv}{v\sqrt{v^2-1}}$

Let $v = \frac{u}{a}$ $= \frac{1}{a}\sec^{-1}v + C$
$dv = \frac{du}{a}$ $= \frac{1}{a}\sec^{-1}\frac{u}{a} + C$

55. $\frac{d}{dx}\left[\sin^{-1}x + \sin^{-1}\sqrt{1-x^2}\right] = \frac{1}{\sqrt{1-x^2}} - \frac{x}{\sqrt{x^2}\sqrt{1-x^2}} = 0$. Therefore, $\sin^{-1}x + \sin^{-1}\sqrt{1-x^2} = C$. Let $x = \frac{1}{2}$ and

$\sin^{-1}x + \sin^{-1}\sqrt{1-x^2} = \frac{\pi}{2}$ which holds for all $x \in (0,1)$.   **57.** $\frac{\pi}{6}$ **59.** $\frac{\pi}{4}$ **61.** $\frac{\pi^2}{4}$ **63.** $-\frac{6}{425}$ rad/sec

Exercise Set 7.6

1. $\sinh x$; Domain $= \mathbb{R}$, Range $= \mathbb{R}$, $\cosh x$; Domain $= \mathbb{R}$, Range $= [1, \infty)$, $\tanh x$; Domain $= \mathbb{R}$, Range $= (-1, 1)$, $\coth x$; Domain $= \{x : x \neq 0\}$, Range $= (-\infty, -1) \cup (1, \infty)$, $\operatorname{sech} x$; Domain $= \mathbb{R}$, Range $= (0, 1]$, $\operatorname{csch} x$; Domain $= \{x : x \neq 0\}$, Range $= (-\infty, 0) \cup (0, \infty)$

3. (a)

x	0.5	1	2	4
$\tanh x$	0.462	0.762	0.964	0.9993

(b)

x	0.5	1	2	4
$\coth x$	2.16	1.31	1.04	1.00

(c)

x	0.5	1	2	4
$\operatorname{sech} x$	0.887	0.648	0.266	0.0366

(d)

x	0.5	1	2	4
$\operatorname{csch} x$	1.92	0.851	0.276	0.0366

5. $\cosh x \cosh y + \sinh x \sinh y$

$= \left(\frac{e^x + e^{-x}}{2}\right)\left(\frac{e^y + e^{-y}}{2}\right) + \left(\frac{e^x - e^{-x}}{2}\right)\left(\frac{e^y - e^{-y}}{2}\right) = \frac{1}{4}[e^{x+y} + e^{-x+y} + e^{x-y} + e^{-(x+y)} + e^{(x+y)} - e^{-x+y} - e^{x-y} + e^{-(x+y)}]$

$= \frac{1}{4}[2e^{x+y} + 2e^{-(x+y)}] = \frac{e^{x+y} + e^{-(x+y)}}{2} = \cosh(x+y)$

7. $\cosh(x+x) = \cosh 2x = \cosh x \cosh x + \sinh x \sinh x = \cosh^2 x + \sinh^2 x$, which $= \cosh^2 x + \cosh^2 x - 1 = 2\cosh^2 x - 1$ or $= \sinh^2 x + 1 + \sinh^2 x = 2\sinh^2 x + 1$

9. In $\cosh 2x = 2\cosh^2 x - 1$, replace x by $\frac{x}{2}$ and solve for $\cosh \frac{x}{2}$: $\cosh x = 2\cosh^2 \frac{x}{2} - 1$; $\cosh^2 \frac{x}{2} = \frac{\cosh x + 1}{2}$ $\cosh \frac{x}{2} = \frac{\sqrt{\cosh x + 1}}{2}$ since $\cosh \frac{x}{2} > 0$

11. $\cosh^2 x - \sinh^2 x = 1$. Divide through by $\sinh^2 x$: $\coth^2 x - 1 = \operatorname{csch}^2 x$

13. $6 \tanh^2 2x \operatorname{sech}^2 2x$

15. $\coth x$

17. $\frac{1}{\cosh x \sqrt{\cosh^2 x - 1}} = \frac{1}{\cosh x \, |\sinh x|}$

19. $-x \sqrt{\operatorname{sech} x^2} \, \tanh x^2$

21. $\frac{y \cosh x - \cosh y}{x \sinh y - \sinh x}$

23. $(\sinh x)^x [x \coth x + \ln \sinh x]$

25. $\frac{\sinh^2 x}{2} + C$

27. $\ln(\cosh x) + C$

29. $\ln|x + \sinh x| + C$

31. $\ln(\ln \cosh x) + C$

33. $\frac{\cosh^3 x}{3} - \cosh x + C$

35. $\sinh x + \frac{\sinh^3 x}{3} + C$

37. $\frac{\tanh^2 x}{2} - \frac{\tanh^4 x}{4} + C$

39. $\ln \frac{16}{9}$

41. $\frac{\pi}{12}$

43. (a) $-\frac{1}{2}$

(b) $-\frac{1}{2}$

45. (a) 1 **(b)** 1

47. $\pi[2 - \tanh 2]$

49. $\sinh x(\coth x - \tanh x) = \sinh x \left(\frac{\cosh x}{\sinh x} - \frac{\sinh x}{\cosh x}\right)$

$= \cosh x - \frac{\sinh^2 x}{\cosh x} = \frac{\cosh^2 x - \sinh^2 x}{\cosh x} = \frac{1}{\cosh x} = \operatorname{sech} x$

51. $\frac{2}{\coth x - \tanh x} = \frac{2}{\frac{\cosh x}{\sinh x} - \frac{\sinh x}{\cosh x}} = \frac{2 \sinh x \cosh x}{\cosh^2 x - \sinh^2 x} = \frac{\sinh 2x}{1} = \sinh 2x$

53. $\tanh \frac{x}{2} = \frac{\sinh \frac{x}{2}}{\cosh \frac{x}{2}} \cdot \frac{2 \cosh \frac{x}{2}}{2 \cosh \frac{x}{2}} = \frac{2 \sinh \frac{x}{2} \cosh \frac{x}{2}}{2 \cosh^2 \frac{x}{2}} = \frac{\sinh x}{2\left(\frac{1 + \cosh x}{2}\right)} = \frac{\sinh x}{1 + \cosh x}$ Also, $\frac{\sinh x}{\cosh x + 1} \cdot \frac{\cosh x - 1}{\cosh x - 1} = \frac{\sinh x(\cosh x - 1)}{\cosh^2 x - 1} = \frac{\sinh x(\cosh x - 1)}{\sinh^2 x}$

$= \frac{\cosh x - 1}{\sinh x}$

55. (a) $\frac{1}{2}(\cosh t)(\sinh t)$

(b) $\frac{1}{2}\sinh t \cosh t - \frac{t}{2}$ **(c)** $\frac{t}{2}$

57. $\ln \cosh x - \frac{\tanh^2 x}{2} - \frac{\tanh^4 x}{4} + C$

59. $-\operatorname{csch} x - \coth x + C$

61. $\int \frac{\sinh x}{\sinh^2 x} dx = \int \frac{\sinh x}{\cosh^2 x - 1} dx$ Let $u = \cosh x$, $du = \sinh x \, dx$ $\int \frac{du}{u^2 - 1} = \frac{1}{2}\int \left[\frac{1}{u-1} - \frac{1}{u+1}\right] du$

$= \frac{1}{2} \ln \frac{u-1}{u+1} + C = \frac{1}{2} \ln \left|\frac{\cosh x - 1}{\cosh x + 1}\right| + C$ and $\sqrt{\frac{\cosh x - 1}{\cosh x + 1} \cdot \frac{\cosh x + 1}{\cosh x + 1}} = \left|\frac{\sinh x}{\cosh x + 1}\right| = \left|\tanh \frac{x}{2}\right|$

63. $\frac{\sinh 3x}{\sinh x} - \frac{\cosh 3x}{\cosh x} = \frac{\sinh 3x \cosh x - \cosh 3x \sinh x}{\sinh x \cosh x} = \frac{\sinh(3x - x)}{\sinh x \cosh x} = \frac{2 \sinh x \cosh x}{\sinh x \cosh x} = 2$

65. (a) 30 m
(b) ≈ 57.05 m
(c) ≈ 85.17 m

67. domain $= (-1, 1)$
range $= \mathbb{R}$

69. domain $= (0, 1]$
range $= [0, \infty)$

71. (a) Let $y = \operatorname{sech}^{-1} x$, $x > 0$.
Then $x = \operatorname{sech} y$. So

$\cosh^{-1}\left(\frac{1}{x}\right) = \cosh^{-1}\left(\frac{1}{\operatorname{sech} y}\right) = \cosh^{-1}(\cosh y) = $

$y = \operatorname{sech}^{-1} x$ **(b)** Let
$y = \operatorname{csch} x$, $x \neq 0$; $\operatorname{csch} y = x$. So,

$\sinh^{-1} \frac{1}{x} = \sinh^{-1}\left(\frac{1}{\operatorname{csch} y}\right) = \sinh^{-1}(\sinh y) = $
$y = \operatorname{csch} x$

73. Let $y = \coth^{-1} x$, $|x| > 1$, $x = \coth y$ $y' = -\frac{1}{\operatorname{csch}^2 y} = -\frac{1}{\coth^2 y - 1}$, $1 = (-\operatorname{csch}^2 y)y' = \frac{1}{1 - x^2}$

75. Let $y = \operatorname{csch}^{-1} x$, $x \neq 0$; $x = \operatorname{csch} y$ $1 = -(\operatorname{csch} y \coth y)y'$; $y' = -\frac{1}{\operatorname{csch} y \coth y}$. Since $\coth^2 y = 1 + \operatorname{csch}^2 y$,

$\coth y = \begin{cases} \sqrt{1 + \operatorname{csch}^2 y} & \text{if } \coth y > 0 \\ -\sqrt{1 + \operatorname{csch}^2 y} & \text{if } \coth y < 0 \end{cases}$. That is, $\coth y = \begin{cases} \sqrt{1 + x^2} & \text{if } x > 0 \\ -\sqrt{1 + x^2} & \text{if } x < 0 \end{cases}$. Thus, $y' = -\frac{1}{|x|\sqrt{1+x^2}}$

77. $\frac{2x}{\sqrt{x^4 - 1}}$ **79.** $\sec x$ **81.** $\operatorname{csch}^{-1} x - \frac{1}{\sqrt{1+x^2}}$ **83.** $-\frac{1}{\sqrt{-(x+x^2)}}$ **85.** $\frac{e^x}{1 + 2e^x}$ **87.** $\frac{1}{2}\cosh^{-1} 2x + C$

89. $-\operatorname{sech}^{-1} 2x + C$ **91.** $\cosh^{-1}(\tan x) + C$ **93.** $\frac{1}{2}\tanh^{-1}\frac{1}{4} = \frac{1}{4}\ln\frac{5}{3}$ **95.** $\sinh^{-1}(x+1) + C$

97. $\int \frac{du}{a\sqrt{(\frac{u}{a})^2-1}} = \int \frac{\frac{1}{a}du}{\sqrt{(\frac{u}{a})^2-1}} = \cosh^{-1}\frac{u}{a} + C, \frac{u}{a} > 1$ **99.** $\int \frac{du}{au\sqrt{1-(\frac{u}{a})^2}} = \frac{1}{a}\int \frac{\frac{1}{a}du}{\frac{u}{a}\sqrt{1-(\frac{u}{a})^2}} = -\frac{1}{a}\operatorname{sech}^{-1}\frac{u}{a} + C$ if $0 < \frac{u}{a} < 1$

101. $\frac{1}{3}\sinh^{-1}\frac{3x}{2} + C$ **103.** $\frac{1}{2} \cdot \frac{1}{3}\tanh^{-1}\frac{2}{3} = \frac{1}{12}\ln 5$ **105.** $-\frac{1}{2}\operatorname{csch}^{-1}\frac{e^x}{2} + C$

107. (a) If $\cosh^{-1}x = y$, $x \geq 1$, $y \geq 0$ then $x = \cosh y = \frac{e^y + e^{-y}}{2}$; $e^y - 2x + e^{-y} = 0$; $e^{2y} - 2xe^y + 1 = 0$. $e^y = x \pm \sqrt{x^2-1}$.

$e^y > 1$. The solution $x - \sqrt{x^2-1}$ is not possible, therefore, $y = \ln(x + \sqrt{x^2-1})$ **(b)** Let $y = \tanh^{-1}x$. $x = \tanh y = \frac{e^y - e^{-y}}{e^y + e^{-y}}$.

$e^{2y}(1-x) = 1+x$; $y = \frac{1}{2}\ln\left(\frac{1+x}{1-x}\right)$ **109. (a)** Let $y = \operatorname{csch}^{-1}x$. $x = \operatorname{csch} y = \frac{2}{e^y - e^{-y}}$; $xe^{2y} - 2e^y - x = 0$;

$e^y = \frac{1+\sqrt{1+x^2}}{x}$ if $x > 0$ and $e^y = \frac{1-\sqrt{1-x^2}}{x}$ if $x < 0$. Thus $y = \ln\left[\frac{1}{x} + \frac{\sqrt{1+x^2}}{|x|}\right]$.   **(b)** $\int \frac{du}{(1+u)(1-u)} = \frac{1}{2}\int \left(\frac{1}{1+u} + \frac{1}{1-u}\right) du$

$= \frac{1}{2}(\ln|1+u| - \ln|1-u|) + C = \frac{1}{2}\ln\left|\frac{1+u}{1-u}\right| + C$

111. If $x > 0$, $\frac{d}{dx}\operatorname{csch}^{-1}|x| = -\frac{1}{x\sqrt{1+x^2}}$ If $x < 0$, $\frac{d}{dx}\operatorname{csch}^{-1}|x| = \frac{d}{dx}\operatorname{csch}^{-1}(-x) = -\frac{1}{(-x)\sqrt{1+x^2}}(-1) = -\frac{1}{x\sqrt{1+x^2}}$. So, for all

$x \neq 0$, $\frac{d}{dx}\operatorname{csch}^{-1}|x| = -\frac{1}{x\sqrt{1+x^2}}$ Thus, $\int \frac{du}{u\sqrt{a^2+u^2}} = \frac{1}{a}\int \frac{\frac{1}{a}du}{\frac{u}{a}\sqrt{1+(\frac{u}{a})^2}} = -\frac{1}{a}\operatorname{csch}^{-1}\frac{|u|}{a} + C$, $u \neq 0$, $a > 0$.

113. (a) $v(t) = \sqrt{\frac{mg}{k}}\tanh\left(\sqrt{\frac{kg}{m}}t\right)$ **(b)** $s(t) = \frac{m}{k}\ln\cosh\sqrt{\frac{kg}{m}}t$ **(c)** $\lim_{t\to\infty} v(t) = \sqrt{\frac{mg}{k}}$

Chapter Review

1. (a) $f'(x) = \frac{1}{x} + \frac{1}{2(x-2)}$, domain: $x > 2$ **(b)** $f'(x) = \frac{1}{x} + \frac{1}{x-1} - \frac{1}{x-4} - \frac{1}{x+2}$, domain: $(-\infty, -2) \cup (0, 1) \cup (4, \infty)$

3. (a) $f'(x) = xe^{-x}(2-x)$, domain: $\mathbb{R}$
(b) $f'(x) = -e^{-x}$, domain: $\mathbb{R}$ **5. (a)** $f'(x) = e^{-x}(\sin e^{-x} + \cos e^{-x})$,

domain: $\mathbb{R}$ **(b)** $f'(x) = \frac{1}{2}\frac{e^x - e^{-x}}{e^x + e^{-x}} \cdot \ln(e^x + e^{-x})$,

domain: $\mathbb{R}$ **7. (a)** $f'(x) = 2(\ln x)x^{\ln x - 1}$,
domain: $x > 0$ **(b)** $f'(x) = (\sec x)^{\tan x}(\tan^2 x + \sec^2 x \ln \sec x)$,

domain: $\left(-\frac{\pi}{2}, \frac{\pi}{2}\right)$, $\left(\frac{3\pi}{2}, \frac{5\pi}{2}\right)$, ... or $\{x : \sec x > 0\}$ **9. (a)** $\frac{y}{xye^y + x}$ **(b)** $\frac{x^2 - y^2 + 2y}{2x}$ **11. (a)** $-\frac{1}{x} + C$ **(b)** $\frac{1}{x - e^x} + C$

13. (a) $\frac{1}{2}\ln\frac{9}{8}$ **(b)** 2   **15. (a)** $\frac{1}{\ln 4}\ln(4^x + 1) + C$ **(b)** $\frac{\sqrt{3}}{4\ln 2}$

17. intercepts: $(2, 0)$, $(0, 0)$, symmetry: to $x = 1$, asymptote: $x = 1$, maxima, minima: none, inflection points: none, concave down on: $(-\infty, 1) \cup (1, \infty)$

no max nor min

19. intercept: $(0, 0)$, symmetry: origin, asymptote: $y = 0$, maxima, minima: $(\sqrt{3}, 3\sqrt{3}e^{-3/2})$ local maximum, $(-\sqrt{3}, -3\sqrt{3}e^{-3/2})$ local minimum, inflection points: $(0, 0)$, $(\pm 1, \pm e^{-1/2})$, $(\pm\sqrt{6}, \pm 6\sqrt{6}e^{-3})$, concave up on: $(-\sqrt{6}, -1) \cup (0, 1) \cup (\sqrt{6}, \infty)$, concave down on: $(-\infty, -\sqrt{6}) \cup (-1, 0) \cup (1, \sqrt{6})$

21. $\pi\left(\frac{16}{3} + 4\ln 3\right)$ **23. (a)** $x = 2$ **(b)** $x = 3$ **25.** ≈ 20.6 hours

27. $Q(t) = m\left(\frac{Q_0}{m}\right)^{e^{-kt}}$. As $t \to \infty$, $Q(t) \to m$. Thus, m is the limiting size of the population.   **29. (a)** $-\frac{2}{3}$ **(b)** $-\frac{4}{5}$

31. (a) $-\frac{56}{65}$ **(b)** $-\frac{1+4\sqrt{2}}{3\sqrt{5}}$

33. (a) $\frac{2}{\sqrt{1-x^4}} - \frac{1}{x^2}\sin^{-1}x^2$ **(b)** $\frac{1}{x\sqrt{x^2-1}}$ **35. (a)** $\frac{\sin^{-1}\sqrt{x} - \cos^{-1}\sqrt{x}}{\sqrt{x-x^2}}$ **(b)** $\frac{y-x}{y+x}$ **37. (a)** $\frac{1}{1+\cosh x}$ **(b)** $-\frac{\cosh y}{\cosh x}$

39. (a) $-\frac{1}{2x}$ **(b)** $\frac{1}{\sqrt{x+2}} + \frac{1}{2\sqrt{x}}\cosh^{-1}(x+1)$ **41.** $\frac{\pi}{3\sqrt{3}}$ **43.** $\frac{\pi}{8}$ **45.** $\frac{\pi}{6}$ **47.** $\frac{1}{2}\tan^{-1}x^2 + C$ **49.** $-\frac{1}{e^{\sinh^2 x}} + C$

51. $\frac{\pi}{4}$ **53.** $\cosh^{-1} e^x + C$ **55.** $\frac{1}{2} \ln \frac{3}{2}$ **57. (a)** $\frac{1}{2}$ **(b)** 0   **59. (a)** $-\frac{1}{2}$ **(b)** $-\frac{1}{2}$ **61. (a)** 1   **(b)** $\sqrt{e}$

63. $2 \ln \frac{(1+\sqrt{3})}{2} - \frac{\pi}{6}$ **65.** ≈ -0.04937 radian/sec

67. $\frac{1}{1-\tanh x} - \frac{1}{1+\tanh x}$

$= \frac{1+\tanh x - 1 + \tanh x}{1 - \tanh^2 x}$

$= \frac{2 \tanh x}{\mathrm{sech}^2 x} = \frac{2 \sinh x}{\cosh x}$

• $\cosh^2 x = 2 \sinh x \cosh x = \sinh 2x$

69. $\frac{\cosh x - \sinh x}{\cosh 2x - \sinh 2x} = \frac{\cosh x - \sinh x}{\cosh^2 x + \sinh^2 x - 2 \sinh x \cosh x}$

$= \frac{\cosh x - \sinh x}{(\cosh x - \sinh x)^2} = \frac{1}{\cosh x - \sinh x}$

$= \frac{1}{\cosh x - \sinh x} \cdot \frac{\cosh x + \sinh x}{\cosh x + \sinh x}$

$= \frac{\cosh x + \sinh x}{\cosh^2 x - \sinh^2 x} = \cosh x + \sinh x$

71. (a) $8 \sinh \frac{1}{2}$

CHAPTER 8

Exercise Set 8.1

1. $-xe^{-x} - e^{-x} + C$ **3.** $\frac{\pi}{8} - \frac{1}{4}$ **5.** $e^x(x^4 - 4x^3 + 12x^2 - 24x + 24) + C$ **7.** $x \sec x - \ln|\sec x + \tan x| + C$

9. $\frac{3}{4}\left[(\ln 2)^2 + 2\right] - \frac{5}{2} \ln 2$ **11.** $-x \cot x + \ln|\sin x| + C$ **13.** $-\frac{x^2 \ln x}{2} + \frac{x^2}{4} + C$ **15.** $\sin x \ln(\sin x) - \sin x + C$

17. $\frac{1}{6} + \frac{1}{4} \ln 3$ **19.** $\frac{-e^{-2x}}{8}\left[4x^3 + 6x^2 + 6x + 3\right] + C$ **21.** $x \tan x + \ln|\cos x| - \frac{x^2}{2} + C$ **23.** $\frac{e^{-x}}{5}(2 \sin 2x - \cos 2x) + C$

25. $\frac{1}{4}(e^{2x} + 2x) + C$ **27.** $\frac{1}{2}\left[\mathrm{sech}\, x \, \tanh x + \tan^{-1}(\sinh x)\right] + C$ **29.** $\frac{x^3}{3} \cot^{-1} x + \frac{x^2}{6} - \frac{1}{6} \ln(1 + x^2) + C$ **31.** $\frac{3\pi}{2}$

33. $4\pi(\pi - 2)$ **35. (a)** $\left(1, \frac{\pi}{8}\right)$ **(b)** $\left(\frac{\pi-2}{2}, \frac{\pi}{8}\right)$ **37. (a)** $\frac{1}{2} \sec^2 x + C$ **(b)** $\frac{1}{2} \tan^2 x + C_1$ **39.** $\frac{x}{2}[\sin(\ln x) + \cos(\ln x)] + C$

41. $\frac{x^2}{4} + \frac{x \sin 2x}{4} + \frac{\cos 2x}{8} + C$ **43.** $\frac{(x^2-4)^{5/2}}{35}(5x^2 + 8) + C$ **45.** $\frac{e^{ax}(a \cos bx + b \sin bx)}{a^2 + b^2} + C$ **47. (a)** Let $u = x^n$ $du = nx^{n-1}\, dx$
$dv = \sin x \, dx$ $v = -\cos x$

(b) Let $u = x^n$ $du = nx^{n-1}\, dx$
$dv = \cos x \, dx$ $v = \sin x$

49. By Exercise 48, $\int_0^{\pi/2} \sin^n x \, dx = \left. -\frac{\sin^{n-1} x \, \cos x}{n} \right]_0^{\pi/2} + \frac{n-1}{n}\int_0^{\pi/2} \sin^{n-2} x \, dx$

$= \frac{n-1}{n}\int_0^{\pi/2} \sin^{n-2} x \, dx$; apply this same result again so that: $\int_0^{\pi/2} \sin^n x \, dx = \frac{n-1}{n} \cdot \frac{n-3}{n-2}\int_0^{\pi/2} \sin^{n-4} x \, dx$. Continue. If n is odd,

$\int_0^{\pi/2} \sin^n x \, dx = \frac{n-1}{n} \cdot \frac{n-3}{n-2} \cdot \frac{n-5}{n-4} \cdots \frac{2}{3}\int_0^{\pi/2} \sin x \, dx$ and $\int_0^{\pi/2} \cos^n x \, dx = \frac{n-1}{n} \cdot \frac{n-3}{n-2} \cdot \frac{n-5}{n-4} \cdots \frac{2}{3}\int_0^{\pi/2} \cos x \, dx$. Since

$\int_0^{\pi/2} \sin x \, dx = 1$ and $\int_0^{\pi/2} \cos x \, dx = 1$, $\int_0^{\pi/2} \sin^n x \, dx = \int_0^{\pi/2} \cos^n x \, dx = \frac{2 \cdot 4 \cdot 6 \cdots (n-1)}{1 \cdot 3 \cdot 5 \cdots n}$. If n is *even* the result in each case

is: $\frac{n-1}{n} \cdot \frac{n-3}{n-2} \cdot \frac{n-5}{n-4} \cdots \frac{1}{2}\int_0^{\pi/2} dx = \frac{1}{2} \cdot \frac{3}{4} \cdot \frac{5}{6} \cdots \frac{n-1}{n} \frac{\pi}{2}$

51. (a) Let $u = \sec^{n-2} x$ $\Big|$ $du = (n-2)\sec^{n-3} x \, \sec x \, \tan x \, dx$
$dv = \sec^2 x \, dx$ $v = \tan x$.

Then $I = \sec^{n-2} x \, \tan x - (n-2)\int \sec^{n-2} x(\sec^2 x - 1)dx = \sec^{n-2} x \, \tan x - (n-2)\left[I - \int \sec^{n-2} x \, dx\right]$.

$I(n-1) = \sec^{n-2} x \, \tan x + (n-2)\int \sec^{n-2} x \, dx$; and $I = \frac{\sec^{n-2} x \, \tan x}{n-1} + \frac{n-2}{n-1}\int \sec^{n-2} x \, dx$.

(b) Let $u = \csc^{n-2} x$ $\Big|$ $du = -(n-2)\csc^{n-3} x \, \csc x \, \cot x \, dx$
$dv = \csc^2 x \, dx$ $v = -\cot x$.

Then $I = -\csc^{n-2} x \, \cot x - (n-2)\int \csc^{n-2} x(\csc^2 x - 1) \, dx = -\csc^{n-2} x \, \cot x - (n-2)\left[I - \int \csc^{n-2} x \, dx\right]$;

$I(n-1) = -\csc^{n-2} x \, \cot x + (n-2)\int \csc^{n-2} x \, dx$ and $I = \frac{-\csc^{n-2} x \, \cot x}{n-1} + \frac{n-2}{n-1}\int \csc^{n-2} x \, dx$

Exercise Set 8.2

1. $\sin x - \frac{\sin^3 x}{3} + C$ **3.** $\frac{3x}{8} - \frac{\sin 2x}{4} + \frac{\sin 4x}{32} + C$ **5.** $\frac{16 - 7\sqrt{2}}{120}$ **7.** $\frac{x}{8} - \frac{\sin 4x}{32} + C$ **9.** $\frac{1}{4}$ **11.** $\frac{\sec^3 x}{3} - \sec x + C$

13. $\frac{1}{2}(\sec x \, \tan x - \ln|\sec x + \tan x|) + C$ **15.** $\ln 2 - \frac{3}{8}$ **17.** $\tan x - \cot x + C$ **19.** $\frac{3}{2} - \ln 2$ **21.** $\frac{\sin^3 3x}{9} - \frac{\sin^5 3x}{15} + C$

23. $\frac{10}{21}$ **25.** $\frac{1}{2}\left[\frac{\sin^3 x^2}{3} - \frac{2 \sin^5 x^2}{5} + \frac{\sin^7 x^2}{7}\right] + C$ **27.** $-\frac{2 \cot^3(x/2)}{3} + C$ **29.** $\frac{\sinh^2 x}{2} + \frac{\sinh^4 x}{4} + C = \frac{\cosh^4 x}{4} + C_1$

31. $\frac{\tanh^4 x}{4} - \frac{\tanh^6 x}{6} + C$ **33.** $-\frac{\cos 5x}{10} - \frac{\cos x}{2} + C$ **35.** $\frac{\sin 5x}{10} + \frac{\sin x}{2} + C$ **37.** $\frac{1}{8}\left[\frac{5x}{2} - 2 \sin 2x + \frac{3 \sin 4x}{8} + \frac{\sin^3 2x}{6}\right] + C$

39. $\frac{\sec^3 x}{3} + 2 \ln|\cos x| + \cos x + C$ **41.** $\int \frac{\sin^3 x}{\cos^8 x} dx = \frac{1}{7} \sec^7 x - \frac{1}{5} \sec^5 x + C$

43. $\int_{-\pi}^{\pi} \sin mx \, \sin nx \, dx = \frac{1}{2}\int_{-\pi}^{\pi} [\cos(m-n)x - \cos(m+n)x]\, dx = \frac{1}{2}\left[\frac{\sin(m-n)x}{m-n} - \frac{\sin(m+n)x}{m+n}\right]_{-\pi}^{\pi} = 0 \quad$ if $\; m \neq n;$

$\int_{-\pi}^{\pi} \cos mx \, \cos nx \, dx = \frac{1}{2}\int_{-\pi}^{\pi} [\cos(m+n)x + \cos(m-n)x]\, dx = \frac{1}{2}\left[\frac{\sin(m+n)x}{m+n} + \frac{\sin(m-n)x}{m-n}\right]_{-\pi}^{\pi} = 0 \quad$ if $\; m \neq n$ and

$\int_{-\pi}^{\pi} \sin mx \, \cos nx \, dx = \frac{1}{2}\int_{-\pi}^{\pi}[\sin(m+n)x + \sin(m-n)\,x]\,dx = 0.$ Therefore, the set is orthogonal on $[-\pi, \pi]$

Exercise Set 8.3

1. $\ln|\sec\theta + \tan\theta| + C$ **3.** 1 **5.** $\frac{2511}{20\sqrt{3}}$ **7.** $\frac{1}{27} + \frac{5}{81\sqrt{3}}$ **9.** $\frac{1}{4(1-x^2)^2} - \frac{1}{1-x^2} - \ln\sqrt{1-x^2} + C$ **11.** $\frac{31}{15}$

13. (a) $\frac{x^2}{2(1+x^2)} + C$ **(b)** $-\frac{1}{2(1+x^2)} + C$ **15.** $\frac{\sqrt{1+x^2}}{3}(x^2 - 2) + C$ **17.** Let $u = a\sin\theta, \theta = \sin^{-1}\frac{u}{a}, du = a\cos\theta\, d\theta$

19. Let $u = a\sec\theta, du = a\sec\theta\tan\theta\, d\theta, \theta = \sec^{-1}\frac{u}{a}$ **21.** Let $u = a\sec\theta, du = a\sec\theta\tan\theta\, d\theta, \theta = \sec^{-1}\frac{u}{a},$

$\int \frac{a\sec\theta\tan\theta\, d\theta}{a^2\tan^2\theta} = \frac{1}{a}\int\csc\theta\, d\theta, = \frac{1}{a}\left[\ln|u - a| - \frac{1}{2}\ln|u + a| - \frac{1}{2}\ln|u - a|\right] + C, = \frac{1}{2}a \ln\left|\frac{u-a}{u+a}\right| + C$ **23.** Let

$u = a\tan\theta, du = a\sec^2\theta\, d\theta, \theta = \tan^{-1}\frac{u}{a}, \int\frac{a\sec^2\theta\, d\theta}{a\tan\theta\, a\sec\theta} = \frac{1}{a}\int\csc\theta\, d\theta, = \frac{1}{a} \ln|\csc\theta - \cot\theta| + C,$

$= \frac{1}{a} \ln\left|\frac{\sqrt{a^2+u^2}-a}{u} \cdot \frac{\sqrt{a^2+u^2}+a}{\sqrt{a^2+u^2}+a}\right| + C$ **25.** $3 \ln\left|\frac{3-\sqrt{9-x^2}}{x}\right| + \sqrt{9-x^2} + C$ **27.** $\frac{2511}{20\sqrt{3}}$ **29.** $3\sqrt{2} + \frac{1}{2} \ln(3 + 2\sqrt{2})$

31. $\frac{3}{8}\sin^{-1}(x - 1) + \frac{3}{8}(x - 1)\sqrt{2x - x^2} + \frac{1}{4}(x - 1)(2x - x^2)^{3/2} + C$ **33.** $\frac{1}{3}\left[\frac{3x+2}{\sqrt{5-12x-9x^2}} - \sin^{-1}\frac{3x+2}{3}\right] + C$

35. $\frac{\sqrt{\tan^2 t - 1}}{3}(\tan^2 t + 2) + C$ **37.** $\frac{1}{2}\sin^{-1}e^x - \frac{e^x\sqrt{1-e^{2x}}}{2} + C$ **39.** $V = 2(2\pi)\int_{-r}^{r}(R - x)\sqrt{r^2 - x^2}\, dx.$ Let

$x = r\sin\theta, dx = r\cos\theta\, d\theta.$ Then $4\pi R\int_{-\pi/2}^{\pi/2}r^2\sqrt{1 - \sin^2\theta}\cos\theta\, d\theta = 4\pi r^2 R\int_{-\pi/2}^{\pi/2}\cos^2\theta\, d\theta = 2\pi^2 r^2 R$

Exercise Set 8.4

1. $\frac{1}{2}\left[\frac{1}{x} - \frac{1}{x+2}\right]$ **3.** $\frac{2}{5} \ln|x + 2| + \frac{3}{5} \ln|x - 3| + C$ **5.** $5 \ln 6$ **7.** $\ln\frac{(x+2)^4}{|x+3|} + C$ **9.** $\ln\frac{\sqrt{|x^2-1|}}{|x|} + C$

11. $x + 2 \ln|x + 2| + \ln|x - 1| + C$ **13.** $5 \ln|x + 3| - 6 \ln|x + 4| + 4 \ln|x - 2| + C$

15. $x^2 - x - 3 \ln|x + 1| + \ln|x - 1| + C$ **17.** $\ln 12 + 2$ **19.** $\ln\left|\frac{x^3}{(x-2)}\right| - \frac{4}{x-2} + C$ **21.** $\ln\frac{(x-2)^2}{x^2+4} + \frac{1}{2}\tan^{-1}\frac{x}{2} + C$

23. $\frac{\pi}{3\sqrt{3}} - \frac{4}{3} + \ln\sqrt{3}$ **25.** $\frac{1}{6} \ln\frac{(x-1)^2}{x^2+x+1} - \frac{1}{\sqrt{3}}\tan^{-1}\frac{2x+1}{\sqrt{3}} + C$ **27. (a)** $\ln\frac{4}{3}$ **(b)** $\pi\left(\frac{2}{3} + \ln\frac{9}{16}\right)$

29. $\int\frac{du}{u(au+b)} = \frac{1}{b}\int\left(\frac{1}{u} - \frac{a}{au+b}\right)du = \frac{1}{b} \ln\left|\frac{u}{au+b}\right| + C$ **31.** $\int\frac{du}{u(au+b)^2} = \frac{1}{b^2}\int\left(\frac{1}{u} - \frac{a}{au+b}\right)du - \frac{1}{b}\int\frac{a}{(au+b)^2}du$

$= \frac{1}{b^2} \ln\left|\frac{u}{au+b}\right| + \frac{1}{b(au+b)} + C$ **33.** $\int\frac{du}{u^3-a^3} = \frac{1}{3a^2}\int\left[\frac{1}{u-a} - \frac{u+2a}{u^2+au+a^2}\right]du = \frac{1}{3a^2}\left[\ln|u - a| - \frac{1}{2}\int\frac{(2u+a)+3a}{u^2+au+a^2}du\right]$

$= \frac{1}{3a^2}\left[\ln|u - a| - \frac{1}{2} \ln(u^2 + au + a^2) - \frac{3a}{2}\int\frac{du}{(u+a/2)^2+3a^2/4}\right] = \frac{1}{3a^2}\left[\ln\frac{|u-a|}{\sqrt{u^2+au+a^2}} - \sqrt{3}\tan^{-1}\frac{2u+a}{\sqrt{3}a}\right] + C$

35. $\ln\left|\frac{x-1}{x+1}\right| - \frac{2}{x-1} + C$ **37.** $-\ln(x^2 + 1) + \tan^{-1}x + \frac{3}{2} \ln(x^2 + 4) - 2\tan^{-1}\frac{x}{2} + C$ **39.** $Q(t) = \frac{mQ_0}{Q_0+(m-Q_0)e^{-mkt}}$

Exercise Set 8.5

1. $\frac{1}{2(2+3x)} - \frac{1}{4} \ln\left|\frac{2+3x}{x}\right| + C$ (#22) **3.** $\frac{4x^2+1}{8}\tan^{-1}2x - \frac{x}{4} + C$ (#97) **5.** $-\frac{\sqrt{4x-x^2}}{2x} + C$ (#66) **7.** $\frac{1}{2x} + \frac{1}{4} \ln\left|\frac{-2+x}{x}\right| + C$ (#23)

9. $\frac{x}{9\sqrt{x^2+9}} + C$ (#49) **11.** $\frac{x-1}{2}\sqrt{2x - x^2} + \frac{1}{2}\cos^{-1}(1 - x) + C$ (#59) **13.** $\frac{1}{3}x^4e^{3x} - \frac{4}{9}x^3e^{3x} + \frac{4}{9}x^2e^{3x} - \frac{8}{27}xe^{3x} + \frac{8}{81}e^{3x} + C$ (#102)

15. $\frac{1}{5} \ln\left|\frac{x}{3x+5}\right| + C$ (#26) **17.** $\frac{-3}{16(-3+4x)} + \frac{1}{16} \ln|-3 + 4x| + C$ (#20) **19.** $\frac{e^{-2x}}{13}(-2\cos 3x + 3\sin 3x) + C$ (#112)

21. $-x - \frac{17}{5} \ln|x - 4| + \frac{2}{5} \ln|x + 1| + C$ (#26, 27) **23.** $\frac{1}{3}\left[x^3\sin^{-1}x + \sqrt{1-x^2} - \frac{(1-x^2)^{3/2}}{3}\right] + C$ (#98)

25. $-\frac{1}{2}\left[\ln\left(\cos\frac{x}{2} - \sin\frac{x}{2}\right)\right] + \frac{1}{2}\left[\ln\left(\cos\frac{x}{2} + \sin\frac{x}{2}\right)\right] + \frac{\sec x\tan x}{2}$ (Mathematica result) $= \frac{1}{2} \ln\left[\frac{\cos^2\left(\frac{x}{2}\right)+2\sin\left(\frac{x}{2}\right)\cos\left(\frac{x}{2}\right)+\sin^2\left(\frac{x}{2}\right)}{\cos^2\left(\frac{x}{2}\right)-\sin^2\left(\frac{x}{2}\right)}\right] + \frac{1}{2}\frac{\sin x}{\cos^2 x}$

$= \frac{1}{2} \ln\left[\frac{1+\sin x}{\cos x}\right] + \frac{1}{2}\frac{\sin x}{\cos^2 x}$ (Maple result) **27.** $\sqrt{-9 + 4x^2} + 3\tan^{-1}\frac{3}{\sqrt{-9+4x^2}}$ (Mathematica result) $= 2\sqrt{x^2 - \frac{9}{4}} + 3\left(\frac{\pi}{2} - \sec^{-1}\frac{2x}{3}\right)$

(Maple result, except for a constant)

Exercise Set 8.6

1. $\frac{1}{2}$ **3.** 1   **5.** $\frac{\pi}{2}$ **7.** ln 2   **9.** ∞ diverges   **11.** ∞ diverges   **13.** 0   **15.** ln 3   **17.** does not exist
19. diverges   **21.** 2   **23.** 2   **25.** $\frac{\pi}{3}$ **27.** diverges   **29.** diverges   **31.** diverges   **33.** 2   **35.** 0

37. diverges   **39.** diverges   **41.** diverges   **43.** diverges   **45.** diverges   **47.** $\ln(2+\sqrt{3})$ **49.** diverges   **51.** converges

53. converges   **55.** $k = \frac{1}{3}; \frac{1}{\sqrt[3]{e}} \approx 0.72$ **57.** $k = 2; \frac{e^3-1}{e^4} \approx 0.35$ **59.** $\frac{\pi}{4}$ **61.** $\frac{\pi}{\ln 2}$ **63.** 2

65. (a) f even:, Consider $\lim\limits_{t\to-\infty}\int_t^0 f(x)dx$. Let $u = -x$, $x = -u$, $dx = -du$. $\int_t^0 f(x)dx = \int_{-t}^0 f(-u)(-du) = \int_0^{-t} f(u)\,du$,

since $f(u) = f(-u)$. Write $\tau = -t$. Then $\lim\limits_{t\to-\infty}\int_t^0 f(x)\,dx = \lim\limits_{\tau\to-\infty}\int_0^\tau f(u)\,du = \int_0^\infty f(u)\,du$, and this converges,

by hypothesis. Since, $\int_0^\infty f(u)\,du = \int_0^\infty f(x)dx$, $\int_{-\infty}^\infty = \int_0^\infty + \int_{-\infty}^0 = \int_0^\infty + \int_0^\infty = 2\int_0^\infty f(x)dx$,   **(b)** f odd:

Make substitutions as above. Then, since $f(-u) = -f(u)$, $\int_{-\infty}^0 f(x)\,dx = \lim\limits_{t\to-\infty}\int_t^0 f(x)\,dx = \lim\limits_{t\to-\infty}\int_0^{-t} -f(u)\,du$

$= -\lim\limits_{\tau\to\infty}\int_0^\tau f(u)\,du = -\int_0^\infty f(u)\,du = -\int_0^\infty f(x)\,dx$; $\int_{-\infty}^\infty f(x)\,dx = 0$ **67. (a)** $0 + e^{-1.5} \approx 0.223$ **(b)** 3930; 950

69. π **71.** diverges   **73.** diverges   **75.** Let $u = \frac{1}{x}$; $\int_0^1 \cos\frac{1}{x}dx = \int_\infty^1 \cos u \left(-\frac{1}{u^2}\right)du =$

$\int_1^\infty \frac{\cos u}{u^2}du$; $\left|\cos\frac{u}{u^2}\right| \le \frac{1}{u^2}$

and $\int_1^\infty \frac{du}{u^2}$ converges. Thus,

$\int_1^\infty \cos\frac{u}{u^2}du$, converges.

77. $\int_0^1 \frac{dx}{x^p} = \lim\limits_{t\to0^+}\int_t^1 \frac{dx}{x^p} = \lim\limits_{t\to0^+}\frac{x^{1-p}}{1-p}\Big]_t^1$

$= \frac{1}{1-p}\left(1 - \lim\limits_{t\to0^+} t^{1-p}\right) = \begin{cases} 1/(1-p) & \text{if } p < 1 \\ \infty & \text{if } p > 1 \end{cases}$

If $p = 1$, $\int_0^1 \frac{dx}{x} = \ln|x|\Big]_0^1$, which diverges.

79. (a) Converges; compare with $\int_0^{\pi/2} \frac{dx}{x^{2/3}}$

(b) Diverges; compare with $\int_0^1 \frac{dx}{x}$

81. (a) $\frac{\pi}{2}$

(b) $2\pi^2 - 4\pi$

(c) $f'(x) = \frac{x}{(4-x^2)^{3/2}}$, $ds = \sqrt{1 + \frac{x^2}{(4-x^2)^3}}\,dx$;

$S = \int_0^2 2\pi \frac{\sqrt{(4-x^2)^3 + x^2}}{(2+x)^{3/2}(2-x)^{1/2}}\,dx$; $0 < \frac{2\pi\sqrt{(4-x^2)^3+x^2}}{(2+x)^{3/2}(2-x)^{1/2}} \le \frac{2\pi\sqrt{4^3+2^2}}{2^{3/2}(2-x)^{1/2}}$

$= \pi\sqrt{34}\frac{1}{(2-x)^{1/2}}$ on $[0,2)$, $\int_0^2 \frac{dx}{(2-x)^{1/2}} = 2\sqrt{2}$. Thus, by the

comparison test, S is finite.

83. (a) $-|f(x)| \le f(x) \le |f(x)|$.   So, $0 \le f(x) + |f(x)| \le 2|f(x)|$ **(b)** By comparison test, $\int_1^\infty (f(x) + |f(x)|)\,dx$

converges, since $2\int_a^\infty |f(x)|\,dx$ converges, by hypothesis.   **(c)** $\int_0^t f(x)\,dx = \int_0^t [f(x) + |f(x)|]\,dx - \int_0^t |f(x)|\,dx$ Now let

$t \to \infty$. Both integrals on the right approach finite limits. The integral on the left approaches a finite limit also (the difference of

the limits on the right). That is, $\int_a^\infty f(x)\,dx$ converges.   **85. (a)** $\frac{1}{s}$, $s > 0$ **(b)** $\frac{1}{s^2}$, $s > 0$ **(c)** $-\frac{1}{1-s}$, $(s > 1)$

(d) $\frac{1}{1+s^2}$, $(s > 0)$ **87.** $\mu = E[x] = \int_{-\infty}^\infty x f(x)\,dx = \int_{-\infty}^\infty x\frac{1}{\beta\sqrt{2\pi}}e^{-(x-\alpha)^2/2\beta^2}\,dx$. Write $x = (x - \alpha) + \alpha$,

and get $\mu = \frac{1}{\beta\sqrt{2\pi}}\int_{-\infty}^\infty (x-\alpha)e^{-(x-\alpha)^2/2\beta^2}\,dx + \alpha\int_{-\infty}^\infty \frac{1}{\beta\sqrt{2\pi}}e^{-(x-\alpha)^2/2\beta^2}\,dx = \left[\frac{-\beta}{\sqrt{2\pi}}e^{-(x-\alpha)^2/2\beta^2}\right]_{-\infty}^\infty + \alpha = \alpha$,

$\sigma^2 = E[(x-\mu)^2] = \int_{-\infty}^\infty (x-\alpha)^2\frac{1}{\beta\sqrt{2\pi}}e^{-(x-\alpha)^2/2\beta^2}\,dx = \frac{-\beta^2}{\beta\sqrt{2\pi}}(x-\alpha)e^{-(x-\alpha)^2/2\beta^2}\Big]_{-\infty}^\infty + \beta^2\underbrace{\int_{-\infty}^\infty \frac{1}{\beta\sqrt{2\pi}}e^{-(x-\alpha)^2/2\beta^2}\,dx}_{=1} = \beta^2$.

So, $\sigma = \beta$

Chapter Review

1. $-\frac{x^2 e^{-2x}}{2} - \frac{xe^{-2x}}{2} - \frac{1}{4}e^{-2x} + C$ **3.** $\frac{\pi}{8} - \frac{\ln 2}{4}$ **5.** $2\left[\sqrt{x}\tan\sqrt{x} + \ln|\cos\sqrt{x}|\right] + C$ **7.** $\frac{\sin^6 x}{6} - \frac{\sin^8 x}{8} + C$

9. $\frac{1}{3}(1 + 3\sqrt{3})$ **11.** $-\frac{1}{4}x\cos^4 x + \frac{3x}{32} + \frac{\sin 2x}{16} + \frac{\sin 4x}{128} + C$ **13.** $\frac{1}{2}\left[\frac{\cos 2x}{2} - \frac{\cos 6x}{6}\right] + C$ **15.** $\frac{2\sqrt{3}}{3} - \frac{\pi}{6}$

17. $\frac{\pi}{3}$ **19.** $\frac{\sqrt{1+e^{2x}}}{3}(e^{2x} - 2) + C$ **21.** $6\sin^{-1}\frac{x-2}{2} - 4\sqrt{4x - x^2} - \frac{1}{2}(x-2)\sqrt{4x - x^2} + C$ **23.** $1 + \ln\frac{3}{8}$

25. $\ln\frac{|x|}{\sqrt{x^2+1}} + \frac{1}{2}\left[\tan^{-1}x + \frac{x}{x^2+1}\right] + C$ **27.** $\left(-\frac{\pi}{2\sqrt{3}} - \frac{\ln 3}{2}\right)$ **29.** $\frac{\sqrt{x^2-4}}{3}(x^2 + 8) + C$ **31. (a)** $\frac{\pi^2}{2}$ **(b)** $2\pi^2$

33. (a) $\frac{2\pi}{5}\ln 4$ **(b)** $10\pi\sin^{-1}\frac{3}{5}$ **35.** $y = \frac{1}{2}\left[1 - \frac{kt+C}{\sqrt{16+(kt+C)^2}}\right]$ **37.** diverges   **39.** $\frac{243}{10}$

41. diverges   **43.** $\ln 4$ **45.** $-\frac{1}{4}$ **47.** 2   **49.** $\frac{1}{2}$ **51.** $\frac{1}{\ln 2}$ **53.** Let $u = \frac{1}{x^\alpha}$, $x^\alpha = \frac{1}{u}$, $x = \frac{1}{u^{1/\alpha}}$,

$dx = -\frac{1}{\alpha}(u^{-(1/\alpha)-1})\,du$. $\int_0^1 \sin\frac{1}{x^\alpha}\,dx = \int_\infty^1 (\sin u)\left(-\frac{1}{\alpha}u^{-(1/\alpha)-1}\,du\right) = -\frac{1}{\alpha}\int_1^\infty \frac{\sin u}{u^{1+1/\alpha}}\,du \left|\frac{\sin u}{u^{1+1/\alpha}}\right| \le \frac{1}{u^{1+1/\alpha}}$, and

$\int_1^\infty \frac{du}{u^{1+1/\alpha}}$ converges. $\left(p = 1 + \frac{1}{\alpha} > 1\right)$. Therefore, $\int_2^\infty \frac{\sin u}{u^{1+1/\alpha}}\,du$ converges.   **55. (a)** $\int_e^\infty f(x)\,dx = -\frac{k\theta^k}{kx^k}\Big]_\theta^\infty = 1$

(b) $k > 1; \mu = \frac{k\theta}{k-1}$.   **(c)** $k > 2; \sigma^2 = \frac{k\theta^2}{(k-1)^2(k-2)}$

57. (a) $\Gamma(\alpha + 1) = \int_0^\infty x^\alpha e^{-x}dx = -x^\alpha e^{-x}\Big]_0^\infty + \alpha\int_0^\infty x^{\alpha-1}e^{-x}\,dx$, and $\lim_{x\to\infty}\frac{x^\alpha}{e^x} = 0$. Hence,

$\Gamma(\alpha + 1) = \alpha\int_0^\infty x^{\alpha-1}e^{-x}dx = \alpha\Gamma(\alpha)$.

(b) By repeated applications of the result in (a) we get: $\Gamma(\alpha + n) = (\alpha + n - 1)\Gamma(\alpha + n - 1) = (\alpha + n - 2)\Gamma(\alpha + n - 2) = \ldots = (\alpha + n - 1)(\alpha + n - 2)\ldots(\alpha + 1)\Gamma(\alpha)$ **(c)** In (b), let $\alpha = 1$. $\Gamma(n + 1) = n(n - 1)\ldots(2)\Gamma(1)$. But we have shown $\Gamma(1) = 1$. So, $\Gamma(n + 1) = n!$ If we take $n = 0$, it is natural to define $0! = \Gamma(1) = 1$.

59. Right-hand derivative at 0 equals 0. $x = 0$ is a vertical asymptote. Left-hand derivative does not exist at 0. $y = e$ is a horizontal asymptote. minimum at (0, 0) f is always increasing for $x \ne 0$.

Inflection point: $\left(\frac{1}{2}, \frac{1}{e}\right)$ concave up on:

$(-\infty, 0) \cup (0, \frac{1}{2})$ concave down on: $(\frac{1}{2}, \infty)$

CHAPTER 9

Exercise Set 9.1

1.

3.

5.

7.

9.

11.

13.

15.

17.

19.

21.

All represent the same curve, a straight line segment with endpoints $(1, -2)$ and $(3,2)$. C_1, C_3, and C_4 have the same orientation; C_2 has the reverse orientation. If t represents time, we can say that the point (x,y) covers the same path in each case, but it reaches points along the way at different times.

23.

If t_1 and t_2 produce the same point (x,y), then $t_1^3 + 1 = t_2^3 + 1$ and $t_1 = t_2$. Thus there is no double point.

25.

27.

$C\left(\frac{1}{2}, 0\right)$

29.

$x = r\left[\cos^{-1}\left(\frac{r-y}{r}\right) - \frac{\sqrt{2ry-y^2}}{r}\right]$

for $0 \le x \le \pi r$

$x = r\left[2\pi - \cos^{-1}\left(\frac{r-y}{r}\right) + \frac{\sqrt{2ry-y^2}}{r}\right]$

for $\pi r < x \le 2\pi r$ The graph is periodic of period $2\pi r$.

31.

33.

$x = (a+b)\cos\theta + b\sin\left(\phi+\theta-\frac{\pi}{2}\right) = (a+b)\cos\theta - b\cos(\phi+\theta)$

$a\theta = B\phi$, so $\phi+\theta = \left(\frac{a+b}{b}\right)\theta$

$\therefore x = (a+b)\cos\theta - b\cos\left(\frac{a+b}{b}\right)\theta$

$y = (a+b)\sin\theta - b\cos\left(\phi+\theta-\frac{\pi}{2}\right)$

$= (a+b)\sin\theta - b\sin\left(\frac{a+b}{b}\right)\theta$

35. In each case, parameter θ is restricted to the interval $[0, 2\pi]$ and θ increases. **(a)** $x = r\cos\theta$, $y = r\sin\theta$ **(b)** $x = r\cos 2\theta$, $y = r\sin 2\theta$ **(c)** $x = r\cos 2\theta$, $y = -r\sin 2\theta$ **(d)** $x = r\sin 3\theta$, $y = r\cos 3\theta$ **(e)** $x = -r\cos 3\theta$, $y = -r\sin 3\theta$

37. When $t = 0$, $x = x_1$, and $y = y_1$. When $t \ne 0$, $\frac{x-x_1}{y-y_1} = \frac{(x_2-x_1)t}{(y_2-y_1)t}$, is the equation of a line passing through (x_1, y_1) and (x_2, y_2).

When $t = 1, x = x_2$, and $y = y_2$. **39.** Both are circles, but the first starts at $(-1, 0)$ and is traced counterclockwise, while the second starts at $(0, -1)$ and is traced clockwise. **41.**

If $a < b$, the curves are limaçons with loops. If $a = b$, the inner loop disappears, and the figure is called a cardioid.

If $a > b$, the figure is that of a limaçon without loop. If $a > 2b$, the dimple on the right is absent.

43. $x = h + r\cos t$, $y = k + r\sin t$, $0 \le t \le 2\pi$.

Exercise Set 9.2

1. 1 **3.** $-\frac{1}{3}$ **5.** $-\frac{1}{2}$ **7.** $-\frac{1}{4}$ **9.** $\frac{1}{20}$ **11.** $x - 8y = 13$. **13.** $4x - 3y = 11$. **15.** horizontal tangent at $\left(\frac{-27}{64}, \frac{-49}{8}\right)$, vertical tangents at $\left(\frac{14}{27}, -\frac{52}{9}\right)$ and $(-18, 22)$ **17. (a)** Both $x'(t)$ and $y'(t)$ are rational functions with nonzero denominators in their domain; hence both are continuous. Neither is ever zero. So C is smooth, and the equations define y as a differentiable function of x in $(-\infty, 2)$ and $(2, \infty)$. **(b)** $\frac{-1}{t^2-4t+5}$ **19.** $-\frac{3}{4}\csc^3 t$ **21.** -2 **23.** $\frac{d^2y}{dx^2} = \frac{d}{dt}\left(\frac{\dot{y}}{\dot{x}}\right)/\frac{dx}{dt} = \frac{\dot{x}\ddot{y}-\dot{y}\ddot{x}}{\dot{x}^3}$ $\frac{d^3y}{dx^3} = \frac{\dot{x}^2\dddot{y}-\dot{x}\dot{y}\dddot{x}-3\dot{x}\ddot{x}\ddot{y}+3\dot{y}\ddot{x}^2}{\dot{x}^5}$

Exercise Set 9.3

1. $4\sqrt{13}$ **3.** $\frac{335}{27}$ **5.** $\frac{488}{27}$ **7.** $2(5^{3/2} - 8) \approx 6.36$ **9.** $2\sqrt{2}$ **11.** $\frac{3}{2}$ **13.** $\frac{9}{2}$

15. $\sqrt{3} + \frac{1}{2}\ln(\sqrt{3}+2)$ **17.** $ds = \sqrt{r^2(\sin^2 t + \cos^2 t)}\, dt = r\, dt$; $L = \int_0^{2\pi} r\, dt = 2\pi r$

19. $\frac{4088\pi}{27}$ **21.** $2\pi(\pi+1)$ **23.** $\frac{8\pi}{3}(5\sqrt{5} - 2\sqrt{2})$ **25.** 70π **27.** $\frac{64}{3}\pi a^2$ **29. (a)** $\frac{\sqrt{6}-2}{\sqrt{3}} + \ln\frac{\sqrt{3}+2}{\sqrt{2}+1}$

(b) $\pi \left[\sqrt{2} - \frac{2}{3} - \ln(\sqrt{6} - \sqrt{3}) \right]$ **31. (a)** $L = 4\int_0^{\pi/2} \sqrt{(a^2 - b^2)\sin^2 t + b^2}\, dt = 4a\int_0^{\pi/2} \sqrt{e^2 \sin^2 t - e^2 + 1}\, dt$

$= 4a\int_0^{\pi/2} \sqrt{1 - e^2(1 - \sin^2 t)}\, dt = 4a\int_0^{\pi/2} \sqrt{1 - e^2 \sin^2 t}\, dt$, by symmetry.   **(b)** ≈ 15.87

33.   **35.**   **37.**   **39.**

41.   **43.**   **45.**   **47.**

Exercise Set 9.4

1. (a)   **(b)**   **(c)**   **(d)**

3. (a)   **(b)**   **(c)**   **(d)**

5. (a) $(-\sqrt{3}, 1)$ **7. (a)** $(4, 0)$ **9. (a)** $x^2 + y^2 = 4$ **(b)** $y = x$
(b) $(0, 3)$ **(b)** $\left(4, \frac{\pi}{2}\right)$
(c) $(0, 0)$ **(c)** $(4, \pi)$
(d) $(\sqrt{2}, \sqrt{2})$ **(d)** $\left(4, \frac{3\pi}{2}\right)$

11. (a) $x = 3$ **(b)** $y = -1$ **13. (a)** $2x - 3y = 4$ **(b)** $3x + 2y = 5$

 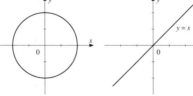

15. $(x^2 + y^2)^2 - 2ax^3 - ay^2(2x + a) = 0$ **17.** $(x^2 + y^2)^2 + 4x^3 + y^2(4x - 1) + 3x^2 = 0$ **19.** $(x^2 + y^2)^{3/2} = a(x^2 - y^2)$
21. $y^2 = 4(x + 1)$ **23. (a)** $r = a$ **(b)** $\theta = \frac{\pi}{3}$ **25. (a)** $r = -\csc\theta$ **(b)** $r = 3\sec\theta$ **27.** $r = \sec\theta \, \tan\theta$
29. $r = \frac{4\cos\theta}{\cos 2\theta}$ **31.** $r^2 = \tan 2\theta$ **33.** If P is in quadrant I or IV, $\theta = \tan^{-1}\frac{y}{x}$ and $r = \sqrt{x^2 + y^2}$. If P is in quadrant II or III,
$\theta = \tan^{-1}\frac{y}{x}$ if we choose $r = -\sqrt{x^2 + y^2}$.   **35.** $e = 1/2, \ d = 2$

37. One example $r = 2\cos\theta$. The pole lies on the graph, since the equation is satisfied by $\left(0, \frac{\pi}{2}\right)$. But the equation is not satisfied by $(0,0)$; that is $r = 0, \theta = 0$.

Exercise Set 9.5

1. (a) circle

(b) circle

3. (a) line

(b) line

5. circle

(3, 0)

7. cardioid

$\left(3, \frac{2\pi}{3}\right)$ $\left(2, \frac{\pi}{2}\right)$ $\left(1, \frac{\pi}{3}\right)$ $(4, \pi)$

9. cardioid

$(-2, 0)$ $(-2, \pi)$ $\left(0, \frac{3\pi}{2}\right)$ $\left(-4, \frac{\pi}{2}\right)$

11. Limaçon without loop.

$\left(5, \frac{\pi}{2}\right)$ $(9, \pi)$ $(1, 0)$ $\left(5, -\frac{\pi}{2}\right)$

13. Limaçon with loop.

$\left(5, \frac{\pi}{2}\right)$ $\left(-1, \frac{3\pi}{2}\right)$ $(2, \pi)$ $(2, 0)$

15. Limaçon without loop, no dimple.

$\left(2, \frac{\pi}{2}\right)$ $(4, \pi)$ $(4, 0)$ $\left(6, \frac{3\pi}{2}\right)$

17. Lemniscate.

$\left(2, \frac{\pi}{4}\right)$ $\left(0, \frac{\pi}{2}\right)$ $(0, 0)$ $\left(-2, \frac{\pi}{4}\right)$

19. Lemniscate.

$\left(\sqrt{3}, \frac{\pi}{4}\right)$ $\left(0, \frac{\pi}{2}\right)$ $(0, 0)$ $\left(-\sqrt{3}, \frac{\pi}{4}\right)$

21. Three-Leaf Rose.

$\left(2, \frac{2\pi}{3}\right)$ $(2, 0)$ $\left(-2, \frac{\pi}{3}\right)$

23. Four-Leaf Rose.

$\left(2, \frac{3\pi}{4}\right)$ $\left(-2, \frac{5\pi}{4}\right)$ $\left(-2, \frac{\pi}{4}\right)$ $\left(2, \frac{7\pi}{4}\right)$

25. $r_1 = a + b\cos\theta$ and $r_2 = -a + b\cos\theta$, where $a > 0$ and $b > 0$, have the same graphs, although θ is shifted by π radians for a given value of the radius. $r_2(\theta) = -r_1(\theta + \pi)$.

27.

$\left(\frac{12}{\pi}, \frac{\pi}{6}\right)$ $\left(\frac{2}{\pi}, \pi\right)$ $\left(\frac{-2}{\pi}, -\pi\right)$

29.

$(\sqrt{e}, 1)$ $(1, 0)$

31.

33.

35. (a) The graph is symmetric to the line $\theta = 0$ if, and only if, when (r, θ) satisfies the equation, either $(r, 2n\pi - \theta)$ or $(-r, (2n+1)\pi - \theta)$ satisfies the equation for some integer n.   **(b)** The graph is symmetric to the line $\theta = \pi/2$ if, and only if, when (r, θ) satisfies the equation, either $(r, (2n+1)\pi - \theta)$ or $(-r, 2n\pi - \theta)$ satisfies the equation for some integer n.   **(c)** The graph is symmetric to the pole if, and only if, when (r, θ) satisfies the equation, either $(r, (2n+1)\pi + \theta)$ or $(-r, 2n\pi + \theta)$ satisfies the equation for some integer n.   **37. (a)** Let $F(r, \theta) = r - \tan\theta$. Then, $F(-r, \pi - \theta) = -F(r, \theta)$. $\therefore$ the graph is symmetric to the polar axis. $F(r, -\theta) = r + \tan\theta$, however, and this does not necessarily equal 0 when $r - \tan\theta = 0$. Thus, the test given in the text fails. Similarly, $F(-r, -\theta) = -F(r, \theta) = 0$ if $F(r, \theta) = 0$. But, $F(r, \pi - \theta) = r + \tan\theta$, and so the test for symmetry to $\theta = \pi/2$ fails. The graph is, nevertheless, symmetric to $\theta = \pi/2$.   **(b)** Let $F(r, \theta) = r - \sin 2\theta \cos\theta$ Then $F(-r, -\theta) = -F(r, \theta)$. But $F(r, \pi - \theta) \neq \pm F(r, \theta)$, and so the test for symmetry to $\theta = \pi/2$, as given in the text, fails.   **(c)** Let $F(r, \theta) = r - 2\csc 2\theta + 1$ Then $F(r, \pi + \theta) = F(r, \theta)$. But $F(-r, \theta) \neq \pm F(r, \theta)$, and so the test for symmetry to the pole fails.   **(d)** Let $F(r, \theta) = r - \sin 2\theta - \cos\theta$. But $F(r, \theta)$ Then

$F(-r, \pi - \theta) = -F(r, \theta) \neq \pm F(r, \theta)$ and so the test for symmetry to $\theta = 0$ fails.   **39. (a)** $\sin \theta = \cos \left(\theta - \frac{\pi}{2} \right)$. Hence,

$F(r, \sin \theta) = F \left(r, \cos \left(\theta - \frac{\pi}{2} \right) \right)$, or the graph of $F(r, \cos \theta) = 0$ is rotated $\frac{\pi}{2}$ radians.   **(b)** $F(r, -\cos \theta) = F(r, \cos(\theta - \pi))$

(c) $F(r, -\sin \theta) = F \left(r, \cos \left(\theta - \frac{3\pi}{2} \right) \right)$ **41.** Let $\sin \alpha = \frac{a}{\sqrt{a^2 + b^2}}$ and $\cos \alpha = \frac{b}{\sqrt{a^2 + b^2}}$. Then, $r = \sqrt{a^2 + b^2} \left[\cos(\theta - \alpha) \right]$, which

is the equation of the circle $r = \sqrt{a^2 + b^2} \cos \alpha$, rotated through the angle α.

43.

45.

47.

49. $\left(1, \frac{\pi}{3} \right), \left(1, \frac{5\pi}{3} \right)$

51. $\left(1, \frac{\pi}{3} \right), \left(1, \frac{5\pi}{3} \right)$, and $(0, 0)$

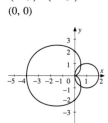

53. $\left(1, \frac{\pi}{2} \right), \left(1, \frac{3\pi}{2} \right), (-1, \pi)$

55. $\left(2, \frac{\pi}{12} \right), \left(2, \frac{5\pi}{12} \right), \left(2, \frac{13\pi}{12} \right)$, and $\left(2, \frac{17\pi}{12} \right)$

57.

(a)

(b)

(c)

(d)

(e)

(f)

(g)

(h)

(i)

(j)

(k)

(l)

(m)   **(n)**   **(o)**   **(p)**

Exercise Set 9.6

1. 6π **3.** 9 **5.** $\frac{3\pi}{2}$ **7.** π

9. 4π **11.** $8+\pi$ **13.** $2-\frac{\pi}{2}$ **15.** $\sqrt{3}-\frac{\pi}{3}$

17. $8\sqrt{3}-\frac{8\pi}{3}$ **19.** $9\sqrt{3}-4\pi$ **21.** $\pi-\frac{3\sqrt{3}}{2}$ **23.** $\frac{11\pi}{3}+\frac{5\sqrt{3}}{2}$

25. $2\pi-4$ **27.** $\pi+3$ **29.** $1+\frac{3\sqrt{3}}{2}-\pi$ **31.** $\frac{27\pi^3}{2}$ **33.** $r=\cos n\theta$ n odd:
$A=\frac{1}{4}\int_0^\pi (1+\cos 2n\theta)d\theta = \frac{\pi}{4}$
n even:
$A=\frac{1}{2}\int_0^\pi (1+\cos 2n\theta)d\theta = \frac{\pi}{2}$
$r=\sin n\theta$; n odd:
$A=\frac{1}{4}\int_0^\pi (1-\cos 2n\theta)d\theta = \frac{\pi}{2}$
n even:
$A=\frac{1}{2}\int_0^\pi (1-\cos 2n\theta)d\theta = \frac{\pi}{2}$

Exercise Set 9.7

1. $-\frac{1}{\sqrt{3}}$ **3.** $-\frac{1}{2}$ **5.** $\sqrt{3}$ **7.** $-\sqrt{3}$ **9.** $-\frac{\sqrt{3}}{2}$ **11.** 0 **13.** $\frac{1}{\sqrt{3}}$ **15.** 1 **17.** $\frac{1}{2}$ **19.** $-\frac{3}{2}$ **21.** $\tan\phi=0$

23. $\cot\phi=0$ **25.** $\tan\psi=\frac{1}{b}=\frac{\pi}{4}$ **27.** $\tan\alpha=-\sqrt{3}$ at $\theta=\frac{2\pi}{3}$ and $\sqrt{3}$ at $\theta=-\frac{2\pi}{3}$ **29.** At $\theta=\frac{\pi}{2}$, $\tan\alpha=0$. At

$\theta=\frac{3\pi}{2}$, $\tan\alpha=0$. At $\theta=\frac{\pi}{3}$, $\tan\alpha=-\frac{\sqrt{3}}{9}$. At $\theta=\frac{2\pi}{3}$, $\tan\alpha=\frac{\sqrt{3}}{9}$ **31.** $\cot\alpha=\frac{1}{\tan(\psi_2-\psi_1)}=\frac{1+\tan\psi_2\psi_1}{\tan\psi_2-\tan\psi_1}=0$ if and only

if $\tan\psi_1\tan\psi_2=-1$ for ψ_1 and $\psi_2\neq\frac{\pi}{2}$. This condition is satisfied if and only if $\alpha=\frac{\pi}{2}$.   **33.** For $\theta=\frac{\pi}{3}$, $\tan\psi_1=-\frac{1}{\sqrt{3}}$, and

$\tan\psi_2=\frac{2-1/2}{\sqrt{3}/2}=\sqrt{3}$ $\tan\psi_1\tan\psi_2=-1$. For $\theta=-\frac{\pi}{3}$, $\tan\psi_1=\frac{1}{\sqrt{3}}$, and $\tan\psi_2=\frac{2-1/2}{-\sqrt{3}/2}=-\sqrt{3}$ $\tan\psi_1\tan\psi_2=-1$

35. $\tan\psi_1=-\cot\theta$, $\tan\psi_2=\tan\theta$, $\tan\psi_2=-1$ **37.** $\tan\psi_1=\frac{1-\sin\theta}{-\cos\theta}$, $\tan\psi_2=\frac{1+\sin\theta}{\cos\theta}$, $\tan\psi_1\tan\psi_2=-1$ **39.** $\frac{7\sqrt{10}}{3}$

41. 4   **43.** 12   **45.** $\dfrac{\tan\phi - \tan\theta}{1 + \tan\phi\ \tan\theta}$ $\dfrac{\frac{f'(\theta)\sin\theta + f(\theta)\cos\theta}{f'(\theta)\cos\theta - f(\theta)\sin\theta} - \tan\theta}{1 + \left(\frac{f'(\theta)\sin\theta + f(\theta)\cos\theta}{f'(\theta)\cos\theta - f(\theta)\sin\theta}\right)\tan\theta}$ $= \dfrac{f(\theta)[\sin^2\theta + \cos^2\theta]}{f'(\theta)[\sin^2\theta + \cos^2\theta]} = \dfrac{f(\theta)}{f'(\theta)}$

47. $\alpha = \phi_2 - \phi_1 = [(\theta + \psi_2) + n\pi] - [(\theta + \psi_1) + m\pi] = (\psi_2 - \psi_1) + (n - m)\pi$ $(m, n$ integers). Hence, $\tan\alpha = \tan(\psi_2 - \psi_1)$, since $\tan$ has period π.   **49.** $1 + \frac{1}{2\sqrt{3}}\ln(2 + \sqrt{3})$

51. (a) $S = 2\pi\int y\,ds = 2\pi\int_\alpha^\beta r\sin\theta\,\sqrt{r^2 + (dr/d\theta)^2}\,d\theta = 2\pi\int_\alpha^\beta f(\theta)\sin\theta\sqrt{[f(\theta)]^2 + [f'(\theta)]^2}\,d\theta$ **(b)** Assume that $\cos\theta \geq 0$

for $\alpha \leq \theta \leq \beta$ and that f' is continuous. Then, about $\theta = \frac{\pi}{2}$, $S = 2\pi\int x\,ds = 2\pi\int_\alpha^\beta r\cos\theta\,\sqrt{r^2 + (dr/d\theta)^2}\,d\theta$

$= 2\pi\int_\alpha^\beta f(\theta)\cos\theta\,\sqrt{[f(\theta)]^2 + [f'(\theta)]^2}\,d\theta$ **53.** 2π **55. (a)** $\frac{4\pi\sqrt{2}}{5}$ **(b)** $\frac{8\pi}{5}(3\sqrt{2} - 4)$

Exercise Set 9.8

1. parabola   **3.** hyperbola   **5.** ellipse   **7.** hyperbola

9. hyperbola   **11.** $r = \frac{5}{3 + \sin\theta}$ **13.** $r = \frac{3}{2 - 3\cos\theta}$ **15.** $y^2 = 4(x + 1)$ vertex: $(-1, 0)$; focus: $(0, 0)$

17. $\dfrac{(y + 24/7)^2}{(18/7)^2} - \dfrac{x^2}{\left(6/\sqrt{7}\right)^2} = 1$, center: $\left(0, -\frac{24}{7}\right)$; vertices: $\left(0, -\frac{6}{7}\right)$, $(0, -6)$; foci: $(0, 0)$, $\left(0, -\frac{48}{7}\right)$; asymptotes: $y = \frac{3}{\sqrt{7}}x - \frac{24}{7}$ and

$y = -\frac{3}{\sqrt{7}}x - \frac{24}{7}$ **19.** Let $r = \frac{de}{1 - e\,\cos\theta}$, $e < 1$; center $= \frac{de^2}{1 - e^2}$; $a = \frac{de}{1 - e^2}$. Thus, $c = ae$. The distance from the directrix to the

center $= \frac{a}{e}$.   **21.** Let $b^2 = a^2 - c^2$; $(a^2 - c^2)x^2 + a^2y^2 = a^2(a^2 - c^2)$. Let $e = c/a$; $a^2(1 - e^2)x^2 + a^2y^2 = a^4(1 - e^2)$ and

$(1 - e^2)r^2\cos^2\theta + r^2\sin^2\theta = a^2(1 - e^2)$. Thus $r^2 = \frac{a^2(1 - e^2)}{(1 - e^2\cos^2\theta)}$.

23. $r \approx \frac{9.481}{1 - 0.0484\cos\theta}$; $r_a \approx 9.963$ A.U. $\approx 1.490 \times 10^{12}m$; $r_p \approx 9.043$ A.U. $\approx 1.353 \times 10^{12}m$ **25.** $r \approx \frac{2.42}{1 - 0.135\ \cos\theta}$, $r_a \approx 2.80$ A.U.,

$r_p \approx 2.14$ A.U.   **27.** From Definition 9.3, $\frac{\text{distance between } P \text{ and } F}{\text{distance between } P \text{ and } \ell} = e$,

where P is a point on the graph, F is a focus, and ℓ is the corresponding directrix. Let $F =$ the pole

$= (0, 0)$. The ratio becomes $\frac{\sqrt{x^2 + y^2}}{d - y} = e$ if the

directrix is d units above the pole and is horizontal.

Thus, $\frac{r}{d - r\sin\theta} = e$; $r = \frac{ed}{1 + e\sin\theta}$. If the directrix is d units below the pole and is horizontal, then:

$r(1 - e\sin\theta) = ed$; and $r = \frac{ed}{1 - e\sin\theta}$

Exercise Set 9.9

1. $y^2 = \frac{1}{2}(1-x)$,
$y \geq 0$

3. $y = 1 - e^{-x}$,
$-\infty < x < \infty$

5. $(y-1)^2 = x+2$,
$-1 \leq x \leq \infty$

7. $\frac{dy}{dx} = 3t$; $\frac{d^2y}{dx^2} = \frac{3}{2t}$

9. $\frac{dy}{dx} = -\frac{1}{2}\sec^2 t$; $\frac{d^2y}{dx^2} = -\frac{1}{2}\sec^4 t$

11. tangent line:
$3x + 4y = 8$
normal line:
$4x - 3y + 6 = 0$

13. $\frac{20}{9} + \ln 3$

15. (a) $4 - 2\sqrt{2} - \ln(9 - 6\sqrt{2})$
(b) $8\pi(\sqrt{3} - \frac{\sqrt{2}}{2} +$
$\frac{1}{2}\ln\frac{\sqrt{3}+2}{\sqrt{2}+1})$

17. (a) $(x-1)^2 + (y-1)^2$
$= 2$, circle

(b) $\frac{x^3}{1/3} + \frac{(y+1/3)^2}{4/9} = 1$,
ellipse

19. (a) cardioid

(b) lemniscate

21. (a) four-leaf rose

(b) hyperbolic spiral

23. $\frac{5\pi}{2} - 3\sqrt{3}$

25.

(a) $\frac{4}{3}\pi + 2\sqrt{3}$
(b) $\frac{8\pi}{3} - 2\sqrt{3}$
(c) $\frac{4\pi}{3} + 2\sqrt{3}$

27. At $\theta = \frac{\pi}{4}, \alpha = \frac{\pi}{4}$
At $\theta = -\frac{\pi}{4}, \alpha = \frac{3\pi}{4}$
At $\theta = -\frac{3\pi}{4}, \alpha = \frac{\pi}{4}$

29. At $\theta = \pm\frac{\pi}{3}$, $\tan\psi_1\tan\psi_2 =$
$-\frac{(1/2)}{(1-1/2)} = -1$

31. $2\sqrt{2} + 7\ln(3 + 2\sqrt{2})$

33. $x = 2a\cot\theta$, $y = 2a\sin^2\theta$, $y > 0$

35. parabola

37. ellipse

39. ellipse

41. $\tan\psi_1 = \frac{1+\cos\theta}{\sin\theta}$;
$\tan\psi_2 = -\frac{(1-\cos\theta)}{\sin\theta}$;
$\tan\psi_1\tan\psi_2 = -1$

CHAPTER 10

Exercise Set 10.1

1. $1 - x + \frac{x^2}{2!} - \frac{x^3}{3!} + \frac{x^4}{4!} - \frac{x^5}{5!}$ **3.** $1 - x + x^2 - x^3 + x^4 - x^5 + x^6$ **5.** $1 + \frac{x^2}{2!} + \frac{x^4}{4!} + \frac{x^6}{6!} + \frac{x^8}{8!}$

7. $1 - (x-1) + (x-1)^2 - (x-1)^3 + (x-1)^4 - \ldots + (-1)^n(x-1)^n$.   **9.** $x - \frac{x^3}{3!} + \frac{x^5}{5!} - \ldots + \frac{(-1)^{k+1}x^{2k-1}}{(2k-1)!}$, $n = 2k-1$ or $n = 2k$

$k = 1, 2, 3, \ldots$ **11.** $-(x-1) - \frac{(x-1)^2}{2} - \frac{(x-1)^3}{3} - \frac{(x-1)^4}{4} - \ldots - \frac{(x-1)^n}{n}$ **13.** $R_n(x) = \frac{(-1)^{n+1}(x-1)^{n+1}}{c^{n+2}}$; c is between 1 and x,

where $x \in (0,2)$ **15.** $R_{2k}(x) = \frac{(-1)^k \cos c}{(2k+1)!} x^{2k+1}$, $k = 1, 2, \ldots$; c is between 0 and x; $x \in (-\infty, \infty)$

17. $R_n(x) = \frac{-(x-1)^{n+1}}{(2-c)^{n+1}(n+1)}$, c is between 0 and x; $x \in (0, 2)$ **19.** $P_5(1) \approx 0.36667$, $|R_5(1)| \le \frac{1}{6!} \approx 0.0013889$, $P_5(1)$ has at least a

2-decimal place accuracy. The most we can say, then, is $e^{-1} \approx 0.37$.   **21.** $P_7(1.5) \approx 0.99739$, $|R_8(1.5)| \le \frac{(1.5)^9}{9!} \approx 0.0001059$,

$\sin 1.5 \approx 0.997$, correct to 3 places.   **23.** $P_8(2) \approx -0.415873$, $|R_9(2)| \le \frac{2^{10}}{(10)!} \approx 0.0002822$, $\cos 2 \approx -0.416$, correct to 3 places.

25. $P_5(5) \approx 2.236076$, $|R_5(5)| \le \frac{1 \cdot 3 \cdot 5 \cdot 7 \cdot 9}{2^6 \cdot 2^{11} \cdot 6!} \approx 0.000010014$. $\sqrt{5} \approx 2.2361$ to 4 places.

27. $P_{12}(3) \approx 20.0852$; $|R_{12}(3)| < \frac{3^{16}}{(13)!} \approx 0.0069$, $e^3 \approx 20.1$, with accuracy assured only to 1 place.   **29.** $P_6(1) \approx 1.54306$,

$R_7(1) < \frac{2}{8!} \approx 0.0000496$, $\cosh 1 \approx 1.5431$ to 4 places.   **31.** $P_4\left(\frac{\pi}{3} - \frac{\pi}{90}\right) \approx 0.5299193$; $\left|R_4\left(\frac{\pi}{3} - \frac{\pi}{90}\right)\right| \le \frac{\left(\frac{\pi}{90}\right)^5}{5!} \approx 4.3 \times 10^{-10}$,

$\cos 58° = 0.5299193$ is accurate to 7 places, and we would have accuracy to 9 places if the calculator went that far.

33. $k = 4$, $\cos 72° = \cos \frac{2\pi}{5} \approx 0.309017$. By calculator, $\cos 72° = 0.309016994$.   **35.** Use $k = 7$. $\sinh 2 \approx 3.626860$

37. (a) $P_n(x) = 1 + \alpha x + \frac{\alpha(\alpha-1)}{2!}x^2 + \frac{\alpha(\alpha-1)(\alpha-2)}{3!}x^3 + \ldots + \frac{\alpha(\alpha-1)(\alpha-2)\ldots(\alpha-n+1)}{n!}x^n$; $R_n(x) = \frac{\alpha(\alpha-1)(\alpha-2)\ldots(\alpha-n)}{(n+1)!}\frac{1}{(1+c)^{n+1-\alpha}}x^{n+1}$; c is between 0 and x

(b) If $\alpha = m$, then $R_m(x) = 0$, since $(\alpha - m) = m - m = 0$. Then, $f(x) = P_m(x) + R_m(x) = P_m(x)$ **(c)** $f^{(n+1)}(x)$ fails

to exist for $x = -1$; the neighborhood of 0 in which $f(x) = P_n(x) + R_n(x)$ is $(-1, 1)$.   **(d)** Let $x = \frac{1}{2}$, $P_4\left(\frac{1}{2}\right) \approx 1.1440$;

$\left|R_4\left(\frac{1}{2}\right)\right| \le 0.000943$; $\sqrt[3]{1.5} \approx 1.14$, is correct to 2 places   **39. (a)** If $\frac{1+x}{1-x} = u$, then $x = \frac{u-1}{u+1} = 1 - \frac{2}{u+1} < 1$. Also

$1 - \frac{2}{u+1} > -1$. For any positive number u, there is an x in $(-1, 1)$ such that $\ln u = \ln \frac{1+x}{1-x}$.   **(b)** $P_{2k-1}(x) = P_{2k}(x) = $

$2\left[x + \frac{x^3}{3} + \frac{x^5}{5} + \frac{x^7}{7} + \ldots + \frac{x^{2k-1}}{(2k-1)}\right]$, $k = 1, 2, 3, \ldots$, $R_{2k}(x) = \left[\frac{1}{(1+c)^{2k+1}} + \frac{1}{(1-c)^{2k+1}}\right]\frac{x^{2k+1}}{2k+1}$, $k = 1, 2, \ldots$ **(c)** Let $u = 2$.

Then, $x = \frac{1}{3}$, $\left|R_8\left(\frac{1}{3}\right)\right| \le 0.000223$, $P_8\left(\frac{1}{3}\right) = 2\left[\frac{1}{3} + \frac{1}{3 \cdot 3^3} + \frac{1}{5 \cdot 3^5} + \frac{1}{7 \cdot 3^7}\right] \approx 0.693$.

41. (a) $P_n(x) = x - \frac{x^2}{2} + \frac{x^3}{3} - \frac{x^4}{4}$ **(b)**   **(c)** The graphs of $P_n(x)$ approach the graph of $\ln(1 + x)$

$+ \ldots + (-1)^{n-1}\frac{x^n}{n}$

on $(-1, 1)$, but diverge outside that interval.   **(d)** Conjecture: $R_n(x) \to 0$ as $n \to \infty$ for $x \in (-1, 1)$. The actual interval is

$(-1, 1]$.   **43.** $-0.33098 \le x \le 0.33098$.   **45. (a)** $P_n(x) = P_{2k}(x) = x - \frac{x^3}{3!} + $ **(b)**

$\frac{x^5}{5!} + \ldots + (-1)^k \frac{x^{2k+1}}{(2k+1)!}$

(c) $P_3(x): (-1.04, 1.04)$.   **47. (a)** $P_n(x) = 1 - \frac{1}{3}x - \frac{2}{9}\frac{x^2}{2!} - \frac{10}{27}\frac{x^3}{3!} - \ldots - \frac{1 \cdot 2 \cdot 5 \ldots (3n-4)}{3^n}\frac{x^n}{n!}$ **(b)**
$P_5(x): (-1.75, 1.75)$.
$P_{10}(x): (-3.25, 3.25)$.

(c) $P_3(x) : (-0.78, 0.60)$.
$P_5(x) : (-0.94, 0.74)$.
$P_{10}(x) : (-0.995, 0.875)$.

49. (a) About $x = 0$: $P_n(x) = 1 + \frac{1}{2}x - \frac{1}{4}\frac{x^2}{2!} + \frac{3}{8}\frac{x^3}{3!} - \frac{15}{16}\frac{x^4}{4!} +$
$\cdots + (-1)^{n+1} \frac{1 \cdot 3 \cdot 5 \cdots (2n-3)}{2^n} \frac{x^n}{n!}$

(b)

(c) $P_3(x): (-0.611, 0.792)$.
$P_5(x): (-0.755, 0.974)$.
$P_{10}(x): (-0.883, 1.083)$.

(a) About $x = 2$: $P_n(x) = \sqrt{3} + \frac{1}{2\sqrt{3}}(x - 2) -$
$\frac{1}{4 \cdot 3^{3/2}} \frac{(x-2)^2}{2!} + \frac{3}{8 \cdot 3^{5/2}} \frac{(x-2)^3}{3!}$
$+ \frac{\cdots + (-1)^{n+1} 1 \cdot 3 \cdot 5 \cdots (2n-3)}{2^n 3^{(2n-1)/2}} \frac{(x-2)^n}{n!}$

(b)

(c) $P_3(x) : (0.361, 4.046)$.
$P_5(x) : (-0.107, 4.647)$.
$P_{10}(x) : (-0.554, 5.084)$.

51. (a) ≈ 15.865

Exercise Set 10.2

1. (a) $1, \frac{-2}{3}, \frac{3}{5}, \frac{-4}{7}, \frac{5}{9}$ (b) $2, \frac{4}{2!}, \frac{8}{3!}, \frac{16}{4!}, \frac{32}{5!}$ 3. (a) $1, -2, 6, -24, 120$ (b) $1, 2, 2, 1, \frac{1}{2}$ 5. (a) $a_n = \frac{2^n - 1}{2^n}$

(b) $a_n = \frac{(-1)^{n-1}(2n)}{2n+3}$ 7. $\frac{1}{3}$ 9. diverges 11. 0 13. Does not exist. $a_{2k} = 0$, $a_{2k+1} = 1$ 15. 1 17. 0
19. 0 21. 0 23. Does not exist. 25. 0 27. 0 29. 0 31. monotone increasing. 33. monotone decreasing for
$n \geq n_0 = 2$. 35. monotone decreasing. 37. monotone decreasing. 39. monotone increasing.

41. $\frac{a_{n+1}+1}{a_n} = \frac{e}{n+1} < 1$ for $n \geq 2$ ∴ $\{a_n\}$ is decreasing. Also, $a_n \geq 0$. ∴ convergent. 43. Let $y = x^{1/x}$.

$y' < 0$ for $x \geq e$. ∴ $x^{1/x}$ is decreasing for $x \geq e$, and $\{n^{1/n}\}$ is decreasing. Since it is bounded below by 0, the sequence converges.
45. Show that $\{a_n\}$ is monotone increasing and bounded above by 2. 47. Suppose that L_1 and L_2 are both limits of $\{a_n\}$, $L_1 \neq L_2$,

say $L_2 > L_1$. Let $\varepsilon = \frac{L_2 - L_1}{2}$. Then for all n large enough, $|a_n - L_1| < \varepsilon$, so that $a_n < L_1 + \varepsilon = \frac{L_1 + L_2}{2}$. Similarly,

$a_n > L_2 - \varepsilon = \frac{L_1 + L_2}{2}$. The two inequalities cannot both be true. 49. Since $|a_n - 0| = ||a_n| - 0|$, it follows that for any $\varepsilon > 0$,
there is an N such that, for all $n > N$, $|a_n - 0| < \varepsilon$ if, and only if, there is an N such that, for all $n > N$, $||a_n| - 0| < \varepsilon$. Or

$\lim_{n \to \infty} a_n = 0$ if, and only if, $\lim_{n \to \infty} |a_n| = 0$. 51. $a_{n+1} - a_n > 0$. $a_n < n\left(\frac{1}{n+1}\right) < 1$. ∴ $\{a_n\}$ is monotone increasing and

bounded above; it is convergent. 53. $\frac{1+\sqrt{5}}{2}$ 55. (a) The limit is 1 if $0 < x_1 < \infty$ and 0 if $x_1 = 0$. Proof: Let $\lim_{n \to \infty} x_n = L$.

Then, $L = \sqrt{L}; L^2 - L = 0; L(L - 1) = 0; L = 0, 1$. The answer depends on x_1 if x_1 may be zero; otherwise, it does not.
(b) $x_{x+1} = x_n^2 - 2$. Let $\lim_{x \to \infty} x_n = L$. Then $L = L^2 - 2; L^2 - L - 2 = 0; L = -1, 2$. The numbers -1 and 2 are fixed points of the

function $f(x) = x^2 - 2$. If some other number is substituted into the function, the interation may or may not converge. If we let
$x_1 = -1 + \delta$ or $x_1 = 2 + \delta$, both sequences diverge.

57.

(a) upper bound of 2 (b) Assume that $a_1 = 1$, which is ≤ 2. Assume that $a_n \leq 2$. Then
$a_{n+1} = \sqrt{2 + a_n} = \sqrt{b}$, where $b \leq 4$. Hence, $a_{n+1} \leq 2$. By induction, $a_n \leq 2$ for all n.
(c) $a_{n+1}^2 - a_n^2 = 2 + a_n - (2 + a_{n+1}) = 2 + a_n - a_n^2 = (2 - a_n)(1 + a_n) > 0$, since $2 - a_n$ and
$1 + a_n$ are > 0. Then, $a_{n+1}^2 - a_n^2 > 0$, which shows that $\{a_n\}$ is monotone increasing $(a_n > 0)$.
(d) $L = 2$.

59. (a) $\left\{11 + \frac{n-1}{n}\right\}$; (b) $\{n\}$ (c) $\{\sin n\}$ (d) $\left\{\frac{\sin n}{n}\right\}$ (e) $\{n\}$ (f) $\left\{7\left(1 + \frac{\sin(n\pi/2)}{n}\right)\right\}$ (g) Such a sequence is not possible because
if it converges to 7, there exists a finite number N such that $|a_n - 7| < \epsilon$ for $n > N$. (h) Not possible, since every convergent
sequence is bounded. (i) $\left\{1 + \frac{1}{n}\right\}$ and $\left\{1 - \frac{1}{n}\right\}$ (j) Not possible, by the same reasoning as in part (g), with 7 replaced by 2.

Exercise Set 10.3

1. $\frac{1}{2} + \frac{2}{5} + \frac{3}{10} + \frac{4}{17} + \cdots + \frac{n}{n^2+1} + \cdots$ $S_1 = \frac{1}{2}$, $S_2 = \frac{9}{10}$, $S_3 = \frac{6}{5}$, $S_4 = \frac{122}{85}$ 3. $\frac{\ln 2}{1!} + \frac{\ln 3}{2!} + \frac{\ln 4}{3!} + \frac{\ln 5}{4!} + \cdots + \frac{\ln(n+1)}{n!} + \cdots$

$S_1 = \ln 2$, $S_2 = \ln 2 + \frac{\ln 3}{2}$, $S_3 = \ln 2 + \frac{\ln 3}{2} + \frac{\ln 4}{6}$, $S_4 = \ln 2 + \frac{\ln 3}{2} + \frac{\ln 4}{6} + \frac{\ln 5}{24}$ 5. $\sum_{n=1}^{\infty} \frac{(-1)^{n-1}}{3^{n-1}}$; $S_n = \sum_{k=1}^{n} \frac{(-1)^{k-1}}{3^{k-1}}$

7. $\sum\limits_{k=1}^{\infty} \frac{1}{(k+1)\ln(k+1)}$; $S_n = \sum\limits_{k=1}^{n} \frac{1}{(k+1)\ln(k+1)}$ **9.** converges to 1.   **11.** converges to 2.   **13.** $a = 2$, $r = -\frac{1}{2}$; $S = \frac{4}{3}$

15. $a = 3$, $r = \frac{1}{2}$; $S = 6$ **17.** $a = \frac{2}{e}$, $r = \frac{-2}{e}$; $S = \frac{2}{e+2}$ **19.** $a = 1, r = 0.99; S = 100$ **21.** $\frac{5}{33}$ **23.** $\frac{4}{27}$

25. $\frac{1}{2}$ **27.** 1   **29.** 2   **31.** $-\frac{1}{2}$ **33.** 50 m   **35. (a)** $\lim\limits_{n\to\infty} \frac{n}{100n+1} = \frac{1}{100} \neq 0$ **(b)** $\lim\limits_{x\to\infty} \frac{x}{(\ln x)^2} = \infty \neq 0$

37. (a) False   **(b)** False   **(c)** True   **(d)** False   **(e)** True   **39. (a)** Suppose that $\sum (a_n - b_n)$ converges. Then,

$-\sum b_n = \sum -b_n = \sum [(a_n - b_n) - a_n] = \sum (a_n - b_n) - \sum a_n$ converges also, contrary to the statement that $\sum b_n$ diverges.

Thus, $\sum (a_n - b_n)$ diverges.   **(b)** Let $A = \sum\limits_{n=1}^{\infty} \frac{1}{2m-1}$ and $B = \sum\limits_{n=1}^{\infty} \frac{1}{2n}$; both diverge but $A - B$ converges.   **41.** Since

$a_n \geq 0$ and $S_{n+1} = S_n + a_n$, it follows that $\{S_n\}$ is monotone increasing. Also, $S_n = a_1 + a_2 + \ldots + a_n \leq k$ for all n. Thus $\{S_n\}$ is

bounded above. $\therefore$ the sequence $\{S_n\}$ converges. Call its limit S. Then, by definition, $\sum\limits_{n=1}^{\infty} a_n = S$.   **43. (a)** $x < -5$ or $x > 1$

(b) $e^{-1} < x < e$; $\frac{1}{1-\ln x}$; $\frac{1}{x+5}$ **45.** For a convergent series, $|S_n - L| < \epsilon$ whenever $n > N$, a finite number. The partial sum S_n is the sum of the first N terms, and since addition is associative, these terms may be arranged in any order whatsoever.   **47.** ≈ 5.1873775

49. (a) $s_n = 3 \cdot 4^{n-1}$, $l_n = \frac{a}{3^{n-1}}$, $P_n = 3a\left(\frac{4}{3}\right)^{n-1}$ **(b)** $\lim\limits_{n\to\infty} P_n = \infty$, since $\lim\limits_{n\to\infty} r^n = \infty$ if $r > 1$, and $r = \frac{4}{3}$ **(c)** After the

first term the series is geometric, with $r = \frac{4}{9}$. Area $= \frac{2\sqrt{3}}{5}a^2$

Exercise Set 10.4

1. diverges   **3.** converges   **5.** diverges   **7.** diverges   **9.** diverges   **11.** converges   **13.** diverges   **15.** diverges
17. converges   **19.** converges   **21.** diverges   **23.** converges   **25.** converges   **27.** diverges   **29.** converges
31. converges   **33.** converges   **35.** converges   **37.** converges   **39.** converges   **41.** diverges   **43.** converges
45. diverges   **47.** converges   **49.** converges if, and only if, $p > 1$ **51. (a)** diverges   **(b)** diverges   **(c)** converges   **53.** Let

$\lim\limits_{n\to\infty} na_n = L > 0$. Then there is a natural number N such that, for all $n > N$, $na_n > \frac{L}{2}$; that is, $a_n > \frac{L}{2} \cdot \frac{1}{n}$ for all $n > N$. But

$\sum\limits_{n=N+1}^{\infty} \frac{L}{2} \cdot \frac{1}{n} = \frac{L}{2} \sum\limits_{n=N+1}^{\infty} \frac{1}{n}$ diverges, since $\sum\limits_{n=1}^{\infty} \frac{1}{n}$ diverges. $\therefore \sum a_n$ diverges.   **55. (a)** The first series behaves like $\sum \frac{1}{n}$ and

diverges   **(b)** The second series behaves like $\sum \frac{1}{n^2}$ and converges

Exercise Set 10.5

1. converges   **3.** diverges   **5.** converges   **7.** diverges   **9.** converges   **11.** converges   **13.** converges   **15.** converges
17. converges   **19.** converges   **21.** diverges   **23.** converges   **25.** diverges   **27. (a)** $\lim\limits_{n\to\infty} \frac{a_{n+1}}{a_n} = \lim\limits_{n\to\infty} \frac{n}{n+1} = 1$, No

conclusion   **(b)** $\lim\limits_{n\to\infty} \frac{n^2}{(n+1)^2} = 1$, No conclusion   **29.** If $0 \leq L < 1$, choose r such that $L < r < 1$. Then, $\sqrt[n]{a_n} < ra_n < r^n$;

$\sum\limits_{k=N}^{\infty} a_k$ converges as a geometric series. If $1 < L$, choose r such that $1 < r < L$ and $\sum\limits_{k=N}^{\infty} a_k$ diverges. If $L = 1$, no conclusion

can be drawn.   **31. (a)** $\lim\limits_{n\to\infty} \sqrt[n]{a_n} = \frac{1}{2} < 1$ **(b)** n even: 2^{-3}; n odd: 2^1 $\lim\limits_{n\to\infty} \frac{a_{n+1}}{a_n}$ does not exist   **33.** It appears that $n^{1/n}$

passes through a maximum at $n = 3$ and then decreases toward its limit of 1.

Exercise Set 10.6

1. converges   **3.** converges   **5.** converges   **7.** diverges   **9.** converges   **11.** Error $\leq a_{10} = \frac{1}{10^3} \approx 0.001$ **13.** Error

$\leq a_6 = \frac{1}{6^6} \approx 0.0000214$.   **15.** $n \geq 39,998$ **17.** $n \geq 101$ **19.** ≈ 0.531 **21.** converges conditionally   **23.** converges

absolutely   **25.** converges absolutely   **27.** converges absolutely   **29.** converges absolutely   **31.** diverges.   **33.** Alternate

the 2's from 2^n with the elements of $n!$; group factors to get form for a_n; compare to $\sum\limits_{n=0}^{\infty} \frac{1}{2n+1}$ which diverges. $0 < a_n < \frac{1}{\sqrt{n+1}}$ implies

$\lim\limits_{n\to\infty} a_n = 0$ and the alternating series converges conditionally.   **35.** $S_{2n+1} = (a_1 - a_2) + (a_3 - a_4) + \ldots + (a_{2n-1} - a_{2n}) + a_{2n+1} \geq 0$

for all n. Also, $S_{2n+1} - S_{2n-1} = a_{2n+1} - a_{2n} \leq 0$, since $a_{2n+1} \leq a_{2n}$. Hence, $\{S_{2n+1}\}$ is a decreasing sequence and is bounded
below by 0.   **37. (a)** $p_n - q_n = a_n$, since when $a_n \geq 0$, $p_n = a_n$ and $q_n = 0$, and when $a_n < 0$, $-q_n = a_n$ and

$p_n = 0$. Also, $p_n + q_n = \begin{cases} a_n & \text{if } a_n \geq 0 \\ -a_n & \text{if } a_n < 0 \end{cases}$ or $p_n + q_n = |a_n|$.   **(b)** The series could not both converge, since if they did,

$\sum (p_n + q_n) = \sum |a_n|$ would also converge and $\sum a_n$ would be absolutely convergent. Assume one series converges and compare S_n of each to get a contradiction.   **39.** If the process is continued, the sum can be made arbitrarily close to 2. At any stage the maximum difference between the partial sum and 2 is q_k, for the last q_k subtracted. Since $\lim_{k \to \infty} Q_k = 0$, the partial sums approach 2.

41. $1 - \frac{1}{2} + \left(\frac{1}{3} + \frac{1}{5} + \frac{1}{7}\right) - \frac{1}{4} + \left(\frac{1}{9} + \frac{1}{11} + \ldots + \frac{1}{19}\right) - \frac{1}{6} + \ldots$ **43.** Add terms of $\sum p_k$ until the sum first exceeds S. Then subtract terms of $\sum q_k$ until the sum less than S.

Exercise Set 10.7

1. $-1 < x < 1$ **3.** $-\infty < x < \infty$ **5.** $-1 \leq x \leq 1$ **7.** $0 \leq x < 2$ **9.** $-5 \leq x < 1$ **11.** $-1 < x \leq 1$
13. $-\frac{1}{2} \leq x \leq \frac{1}{2}$ **15.** $-1 \leq x < 1$ **17.** $-\infty < x < \infty$ **19.** $\frac{5}{2} \leq x \leq \frac{7}{2}$ **21.** $-\frac{7}{3} < x < \frac{1}{3}$ **23.** $-\frac{1}{3} < x < \frac{1}{3}$

25. $4 \leq x \leq 6$ **27.** $-\infty < x < \infty$ **29.** $\frac{1}{3} \leq x < 1$ **31.** converges only if $x = -4$.   **33.** $-1 \leq x < 1$

35. $-2 < x < 2$ **37.** $-\infty < x < \infty$ **39.** $R = 1$ **41. (a)** Choose x_1 such that $|x_0| < |x_1| \leq R$ and such that $\sum_{a^n} |x_1|^n$

converges. If no such x_1 existed, then $|x_0|$ would be an upper bound to the set for which the series converges absolutely, contradicting the definition of R.   **(b)** Thus, by Theorem 10.18, $\sum a_n x^n$ converges absolutely for all x such that $|x| < |x_0|$.   **(c)** If $|x| > R$ and $\sum a_n x^n$ converged, then we could find an x_1 such that $R < |x_1| < |x|$ and conclude by Theorem 10.18 that $\sum a_n x_1^n$ converges absolutely, contrary to the definition of R. Hence, $\sum a_n x^n$ diverges.

Exercise Set 10.8

1. $\displaystyle\sum_{n=1}^{\infty} x^{n-1}, \; |x| < 1.$ **3.** $\displaystyle\sum_{n=1}^{\infty} \frac{(-1)^n x^{2n-1}}{(2n-1)!}, \; |x| < \infty.$ **5.** $\displaystyle\sum_{n=1}^{\infty} \frac{n x^{n-1}}{2^{n+1}}, \; |x| < 2$ **7.** $\displaystyle\sum_{n=0}^{\infty} \frac{x^{n+1}}{(n+1)^2} + C, \; |x| < 1.$

9. $\displaystyle\sum_{n=0}^{\infty} \frac{(-1)^n x^{2n+1}}{(2n+1)!} + C, \; |x| < \infty.$ **11.** $\displaystyle\sum_{n=0}^{\infty} \frac{(-1)^n x^{2n+1}}{2^n} + C, \; |x| < \sqrt{2}.$ **13.** $\displaystyle\sum_{n=0}^{\infty} \left(\frac{1}{2}\right)^{n+1} = 1$

15. $\displaystyle\sum_{n=0}^{\infty} \frac{2^{n+1}}{(n+1)!} - \sum_{n=0}^{\infty} \frac{1}{(n+1)!}, e^2 - e.$ **17.** $\displaystyle\sum_{n=0}^{\infty} \frac{(-1)^n}{2^n(2n+1)^2}$ **19.** $\displaystyle\sum_{n=0}^{\infty} (-1)^{n-1} n x^{n-1}, R = 1.$ **21.** $-\displaystyle\sum_{n=0}^{\infty} \frac{x^{n+1}}{n+1}, R = 1$

23. $2\displaystyle\sum_{n=1}^{\infty} n x^{2n-1}, R = 1,$ **25.** $\frac{1}{2}\displaystyle\sum_{n=0}^{\infty} \frac{(-1)^n x^{n+1}}{n+1}, R = 1.$ **27.** $\displaystyle\sum_{n=0}^{\infty} \frac{(-1)^n x^{2n+1}}{2^{2n+1}(2n+1)}, R = \sqrt{2}$

29. $\frac{1}{2}\displaystyle\sum_{n=2}^{\infty} n(n-1)x^{n-2}, R = 1$ **31.** $\displaystyle\sum_{k=1}^{\infty} k x^{k+1}, R = 1$ **33.** $\int_0^1 f(x)dx = \frac{1}{2}(e^2 - 1) = \displaystyle\sum_{n=0}^{\infty} \frac{2^n}{(n+1)!}$ **35.** $\frac{x^2+x}{(1-x)^3}$

37. $\frac{3}{2}$ **39.** $\displaystyle\sum_{n=1}^{\infty} \frac{x^{n-1}}{n!}$ **41.** ≈ 0.74682413 **43. (a)** $\frac{10^{50}}{50!} \approx 3.288 \times 10^{-15}$ **(b)** $\frac{20^{100}}{100!} \approx 1.3583 \times 10^{-28}$

(c) $\frac{100^{1000}}{1000!} \approx 10^{-567.6}$

Exercise Set 10.9

1. $\displaystyle\sum_{k=0}^{\infty} \frac{(-1)^k x^{2k+1}}{(2k+1)!}, -\infty < x < \infty.$ **3.** $\displaystyle\sum_{k=0}^{\infty} \frac{x^{2k}}{(2k)!}, -\infty < x < \infty$ **5.** $\displaystyle\sum_{k=0}^{\infty} \frac{x^{2k+1}}{2k+1}, -1 < x < 1$

7. $\displaystyle\sum_{k=1}^{\infty} \frac{(-1)^{k-1} 2^{2k-1} x^{2k}}{(2k)!}, -\infty < x < \infty.$ **9.** $\displaystyle\sum_{k=0}^{\infty} \left(\binom{-3}{k}\right)(-1)^k x^k, -1 < x < 1$ **11.** $\displaystyle\sum_{k=0}^{\infty} \frac{(\ln 2)^k x^k}{k!}, -\infty < x < \infty.$

13. $\displaystyle\sum_{k=1}^{\infty} \frac{(-1)^{k+1} x^{2k-2}}{(2k)!}, -\infty < x < \infty$ **15.** $\displaystyle\sum_{k=0}^{\infty} (-1)^k \left(\binom{1/3}{k}\right)\frac{x^k}{2^{3k+1}}, -8 < x < 8$ **17.** $\displaystyle\sum_{k=0}^{\infty} \frac{(-1)^k x^k}{(2k)!}, -\infty < x < \infty$

19. $e\displaystyle\sum_{k=0}^{\infty} \frac{(x-1)^k}{k!}, -\infty < x < \infty$ **21.** $-\displaystyle\sum_{k=0}^{\infty} (x+1)^k, -2 < x < 0.$

23. $\frac{1}{2}\left[1 + \sqrt{3}\left(\frac{\pi}{6}\right)(x-1) - \left(\frac{\pi}{6}\right)^2 \frac{(x-1)^2}{2!} - \sqrt{3}\left(\frac{\pi}{6}\right)^3 \frac{(x-1)}{3!} + \left(\frac{\pi}{6}\right)^4 \frac{(x-1)^4}{4!} + - - + + \ldots\right], -\infty < x < \infty$

25. (a) $|R_{2k+1}(x)| = |R_{2k+2}(x)| = \frac{|\cos c| |x|^{2k+2}}{(2k+2)!} \leq \frac{|x|^{2k+2}}{(2k+2)!},$ and $\lim_{k \to \infty} |R_{2k+2}(x)| = 0$ **(b)** $|R_{2k+1}(x)| = |R_{2k+2}(x)| = \frac{(\cosh c) |x|^{2k+2}}{(2k+2)!},$

and $\lim_{k \to \infty} R_{2k+2}(x) = 0.$ **(c)** $|R_{2k}(x)| = |R_{2k+1}(x)| = \frac{(\cosh c) |x|^{2k+1}}{(2k+1)!} \leq \frac{(\cosh x) |x|^{2k+1}}{(2k+1)!}$ and $\lim_{k \to \infty} R_{2k+1}(x) = 0.$

27. $R = \lim_{n\to\infty}\left|\frac{a_n}{a_{n+1}}\right| = \lim_{n\to\infty}\left|\frac{\alpha(\alpha-1)(\alpha-2)...(\alpha-n+1)}{n!} \cdot \frac{(n+1)!}{\alpha(\alpha-1)(\alpha-2)...(\alpha-n)}\right| = \lim_{n\to\infty}\left|\frac{n+1}{\alpha-n}\right| = 1$ **29.** The series converges to 0 for all x

and so represents $f(x)$ only at $x = 0$. By using the definition of the derivative $f'(0) = 0$. Also $f^{(n)}(0) = 0$ for all n.

31. ≈ 0.5077. error ≤ 0.000006.   **33.** ≈ 0.19737, error ≤ 0.0000003.   **35.** $1 + x^2 + \frac{2}{3}x^4$

37. $\frac{2}{\sqrt{\pi}}\sum_{n=0}^{\infty}\frac{(-1)^n}{(2n+1)n!}$ erf $(1) \approx 0.843$ **39. (a)** $|\sin^{-1}(\sin x)| = |x|$; but with a period of π **(b)** The graph of the sum of the first three terms is

Exercise Set 10.10

1. $\frac{a_0}{2} = \frac{\int_{-\pi}^{\pi} f(x)dx}{\pi-(-\pi)} = f_{ave}$ on $[-\pi, \pi]$ **3.** $\int_{-\pi}^{\pi} f(x)\sin mx\ dx = \frac{a_0}{2}\int_{-\pi}^{\pi}\sin mx\ dx + \sum_{n=1}^{\infty}\left[a_n\int_{-\pi}^{\pi}\cos nx\sin mx\ dx\right.$

$\left. +b_n\int_{-\pi}^{\pi}\sin nx\ \sin mx\ dx\right]$. Apply equations 10.37, 10.34 and 10.35. $\int_{-\pi}^{\pi} f(x)\sin mx\ dx = b_m\int_{-\pi}^{\pi}\sin^2 mx\ dx = b_m\pi$ (equation

10.26). Hence, $b_m = \frac{1}{\pi}\int_{-\pi}^{\pi} f(x)\sin mx\ dx$.   **5.** $\frac{\pi}{4} - \frac{2}{\pi}\left(\cos x + \frac{\cos 3x}{9} + \frac{\cos 5x}{25} + ...\right) + \sin x - \frac{\sin 2x}{2} + \frac{\sin 3x}{3} - ...$

7. $a_n = 0$, except for $a_2 = 1$, and $b_n = 0$.   **9.** $a_0 = 0$, $a_n = \frac{1}{n}\sin\frac{n\pi}{2}$, $b_n = 0$, So $f(x) = \cos x - \frac{\cos 3x}{3} + \frac{\cos 5x}{5} - ...$

11. For 10 terms   **13.** Graphs for $n = 1$ and 3   The graph for $n = 7$ **15.** The graphs are identical for $n = 2, 5, 8, 11$.

Chapter Review

1. $P_5(x) = -x - \frac{x^2}{2} - \frac{x^3}{3} - \frac{x^4}{4} - \frac{x^5}{5}$; error ≤ 0.0000407; actual error ≈ 0.000013

3. $P_5(x) = 1 + 2\left(x - \frac{\pi}{4}\right) + 2\left(x - \frac{\pi}{4}\right)^2 + \frac{8}{3}\left(x - \frac{\pi}{4}\right)^3 + \frac{10}{3}\left(x - \frac{\pi}{4}\right)^4 + \frac{64}{15}\left(x - \frac{\pi}{4}\right)^5$

5. $\sqrt{2}\sin\left(x - \frac{\pi}{4}\right) = \sqrt{2}\left[x - \frac{\pi}{4} - \frac{(x-\pi/4)^3}{3!} + \frac{(x-\pi/4)^5}{5!} - ... + (-1)^{2n+1}\frac{(x-\pi/4)^{2n+1}}{(2n+1)!} + ...\right]$

7. $P_3(x) = \frac{\pi}{4} + \frac{1}{2}(x - 1) - \frac{1}{4}(x - 1)^2 + \frac{1}{12}(x - 1)^3$, $R_3(x) = \frac{c(1-c)^2}{(1+c^2)^4}(x - 1)^4$

9.

x	$P_1(x)$	$P_2(x)$	$P_3(x)$	$P_4(x)$	e^x
0.25	1.25	1.2813	1.2839	1.2840	1.2840
-0.25	0.75	0.7813	0.7786	0.7788	0.7788
0.50	1.50	1.6250	1.6458	1.6484	1.6487
-0.50	0.50	0.6250	0.6042	0.6068	0.6065
0.75	1.75	2.0313	2.1016	2.1147	2.1170
-0.75	0.25	0.5313	0.4609	0.4741	0.4724
1.00	2.00	2.5000	2.6667	2.7083	2.7183
-1.00	0.00	0.5000	0.3333	0.3750	0.3779

11. (a) 2   **(b)** 0

13. (a) Increasing, bounded above by e, limit is e.   **(b)** Increasing, 2   **15.** For large n, $|a_n^2 - L^2| = |a_n + L|\,|a_n - L| \le (|a_n| + |L|)\,|a_n - L| < (2|L| + 1)\,|a_n - L|$. For $\epsilon > 0$ and large n, $|a_n - L| < \frac{\epsilon}{2|L|+1}$, and

$|a_n^2 - L^2| < \epsilon$. The converse is not true. Consider $a_n = \frac{(-1)^n n}{n+1}$.

17. $S_n = 2 - \frac{2}{n+1}$; $\lim_{n\to\infty} S_n = 2$ **19. (a)** $\frac{3}{10}$ **(b)** $\frac{3}{4}$ **21.** diverges   **23.** converges   **25.** diverges.

27. converges.   **29.** converges conditionally   **31.** converges.   **33.** converges   **35.** converges   **37.** conditionally convergent.

39. converges.   **41.** diverges   **43.** converges conditionally.   **45.** $-\frac{3}{2} < x < \frac{3}{2}$ **47.** $1 \le x < 3$.

49. $-1 < x < 1$ **51.** $-\frac{1}{2} < x \le \frac{3}{2}$.   **53.** $-2 \displaystyle\sum_{n=0}^{\infty} \frac{x^{2n+2}}{2n+2}$, $R = 1$.   **55.** $\displaystyle\sum_{n=0}^{\infty} \binom{-1/2}{n} \frac{x^{2n+1}}{2n+1}$, $R = 1$.

57. (a) $\displaystyle\lim_{x\to 0} \left(1 - \frac{x^2}{3!} + \frac{x^4}{5!} - \ldots\right) = 1$ (b) $\displaystyle\lim_{x\to 0} \left(\frac{1}{2!} - \frac{x^2}{4!} + \frac{x^4}{6!} - \ldots\right) = \frac{1}{2}$ (c) $\displaystyle\lim_{x\to 0} \left(1 + \frac{x}{2!} + \frac{x^2}{3!} + \ldots\right) = 1$

59. $\frac{1}{2} \displaystyle\sum_{n=1}^{\infty} \frac{(-1)^{n-1}(x+1)^n}{n}$ **61.** ≈ 0.444, $|\text{error}| \le 0.000052$ **63.** $\cos^2 x - \sin^2 x = \cos 2x$; $\displaystyle\lim_{n\to\infty} R_n(x) = 0$;

$\displaystyle\sum_{k=0}^{\infty} \frac{(-1)^{k+1} 2^{2k+1} (x-\pi/4)^{2k+1}}{(2k+1)!}$

Photo Credits

PROLOGUE: xii, top, Kathleen Olson; bottom, courtesy of Texas Instruments.
CHAPTER 1: 4, Fig. 1.1, Keeling, C.D. et al., in *Aspects of Climate Variability in the Pacific and the Western Americas*, Part 1: Atmospheric CO_2 Data and Analysis, Geophysical Monograph **55**, pp. 165–236, American Geophysical Union, Washington, D.C., 1989 (updated, private communication); **33**, Kathleen Olson; **51**, Ken Whitmore/Tony Stone Images; **67**, TIBWest/Jay Brousseau © 1994; **69**, Loren M. Winters; **79**, courtesy, Central Scientific Company. **80**, top, Art Resource, NY; center, Erich Lessing/Art Resource, NY; bottom, Bettmann Archive.
CHAPTER 2: 82, photo by Jay M. Pasachoff.
CHAPTER 3: 139; © Chad Slattery/Tony Stone Images; **142**, © Loubat// Vandystadt/Photo Researchers; **173**, courtesy, Central Scientific Company; **176**, © Gary Buss 1990/FPG Selects; **192**, © Jim Goodwin/Photo Researchers; **203**, © 1990 Smithsonian Institution and Lockheed Corporation; **204**, Mick Roessler/ Superstock.
CHAPTER 4: 207, Ralph Cowan/Tony Stone Worldwide; **242**, photo by Jay M. Pasachoff; **247**, © 1992, 1993, PhotoDisc, Inc.; **251**, left, © 1992, 1993, PhotoDisc, Inc.; right, © Day Williams/Photo Researchers; **257**, top left, Uniphoto/ Pictor; top right, William Strode/Superstock; bottom, © 1992, 1993, PhotoDisc, Inc.; **259**, Westlight/Orion; **289**, Exercise 4, reprinted by permission from Mathematical Association of America Notes, Vol. 4 (by John Ramsay, College of Wooster); **298**, courtesy, Central Scientific Company.
CHAPTER 5: 302, photo by Jay M. Pasachoff; **304**, Steve Lipofsky; **311**, Globus Brothers/The StockMarket; **319**, photo by Jay M. Pasachoff; **339**, ARS-USDA; **373**, TIB West/Harald Sund © 1994; **383**, Exercise 3, reprinted by permission of Dr. Judith H. Morrel, Department of Mathematics and Computer Science, Butler University; **384**, top, © J. Fields/Science Source/Photo Researchers; below, courtesy, Central Scientific Company.
CHAPTER 6: 388, photo by Jay M. Pasachoff; **399**, photo by Jay M. Pasachoff; **403**, Kathleen Olson; **405**, Kathleen Olson; **412**, © 1992, 1993, PhotoDisc, Inc.; **420**, photo by Jay M. Pasachoff; **427**, Robert Llewellyn/Superstock; **430**, Charles Orrica/Superstock; **435**, © 1990 Tom Van Sant/The GeoSphere™ Project, Santa Monica, CA; **438**, courtesy, Central Scientific Company; **441**, NASA Photo/Lyndon B. Johnson Space Center; **442**, Grant Heilman/Grant Heilman Photography; **447**, Billy E. Barnes/PhotoEdit; **462**, photo by Jay M. Pasachoff; **479**, photo by Jay M. Pasachoff; **480**, top photo, Eloise Pasachoff; left, Dennis Di Cicco/Peter Arnold, Inc; right, photo by Jay M. Pasachoff.
CHAPTER 7: 525, Jet Propulsion Laboratory; **527**, M.G. Giles/Pix* Elation/Fran Heyl Associates; **530**, © 1992, 1993, PhotoDisc, Inc.; **534**, courtesy, Central Scientific Company; **548**, left, © Garry D. McMichael/Photo Researchers; center, Jeremy Hayhurst; right, © 1992, 1993, PhotoDisc, Inc.; **552**, Larry Chiger/ Superstock; **568**, photo by Jay M. Pasachoff/courtesy the Sterling and Francine Clark Art Institute; **570**, © Dr. R. Clark and M. Goff/Science Photo Library/Photo Researchers.
CHAPTER 8: 572, photo by Jay M. Pasachoff; **612**, © Yoav Levy/ Phototake NYC; **644**. © Paul Hanny/Gamma Liaison; **644**, courtesy Jay M. Pasachoff.
CHAPTER 9: 646, Musées de Narbonne; **652**, Fundamental Photographs, NY; **655**, The Bettmann Archive; **667**, courtesy, Kimbell Art Museum; **676**, Kathleen Olson; **683**, photo by Jay M. Pasachoff; **700**, Lee Hocker; **701**, Lee Hocker;

INDEX

Page numbers followed by a d refer to definitions on those pages.

Forms Containing $\sqrt{u^2 \pm a^2}$ (continued)

41. $\displaystyle\int \frac{\sqrt{u^2 - a^2}}{u}\,du = \sqrt{u^2 - a^2} - a\sec^{-1}\left|\frac{u}{a}\right| + C$

42. $\displaystyle\int \frac{\sqrt{u^2 \pm a^2}}{u^2}\,du = -\frac{\sqrt{u^2 \pm a^2}}{u} + \ln\left|u + \sqrt{u^2 \pm a^2}\right| + C$

43. $\displaystyle\int \frac{du}{\sqrt{u^2 \pm a^2}} = \ln\left|u + \sqrt{u^2 \pm a^2}\right| + C$

44. $\displaystyle\int \frac{u^2\,du}{\sqrt{u^2 \pm a^2}} = \frac{u}{2}\sqrt{u^2 \pm a^2} \mp \frac{a^2}{2}\ln\left|u + \sqrt{u^2 \pm a^2}\right| + C$

45. $\displaystyle\int \frac{du}{u\sqrt{u^2 + a^2}} = -\frac{1}{a}\ln\left|\frac{a + \sqrt{u^2 + a^2}}{u}\right| + C$

46. $\displaystyle\int \frac{du}{u\sqrt{u^2 - a^2}} = \frac{1}{a}\sec^{-1}\left|\frac{u}{a}\right| + C$

47. $\displaystyle\int \frac{du}{u^2\sqrt{u^2 \pm a^2}} = \mp\frac{\sqrt{u^2 \pm a^2}}{a^2 u} + C$

48. $\displaystyle\int (u^2 \pm a^2)^{3/2}\,du = \frac{u}{8}(2u^2 \pm 5a^2)\sqrt{u^2 \pm a^2} + \frac{3a^4}{8}\ln\left|u + \sqrt{u^2 \pm a^2}\right| + C$

49. $\displaystyle\int \frac{du}{(u^2 \pm a^2)^{3/2}} = \pm\frac{u}{a^2\sqrt{u^2 \pm a^2}} + C$

Forms Containing $\sqrt{a^2 - u^2}$

50. $\displaystyle\int \sqrt{a^2 - u^2}\,du = \frac{u}{2}\sqrt{a^2 - u^2} + \frac{a^2}{2}\sin^{-1}\frac{u}{a} + C$

51. $\displaystyle\int u^2\sqrt{a^2 - u^2}\,du = \frac{u}{8}(2u^2 - a^2)\sqrt{a^2 - u^2} + \frac{a^4}{8}\sin^{-1}\left(\frac{u}{a}\right) + C$

52. $\displaystyle\int \frac{\sqrt{a^2 - u^2}}{u}\,du = \sqrt{a^2 - u^2} - a\ln\left|\frac{a + \sqrt{a^2 - u^2}}{u}\right| + C$

53. $\displaystyle\int \frac{\sqrt{a^2 - u^2}}{u^2}\,du = -\frac{\sqrt{a^2 - u^2}}{u} - \sin^{-1}\frac{u}{a} + C$

54. $\displaystyle\int \frac{u^2}{\sqrt{a^2 - u^2}}\,du = -\frac{u}{2}\sqrt{a^2 - u^2} + \frac{a^2}{2}\sin^{-1}\left(\frac{u}{a}\right) + C$

55. $\displaystyle\int \frac{du}{u\sqrt{a^2 - u^2}} = -\frac{1}{a}\ln\left|\frac{a + \sqrt{a^2 - u^2}}{u}\right| + C$

56. $\displaystyle\int \frac{du}{u^2\sqrt{a^2 - u^2}} = -\frac{\sqrt{a^2 - u^2}}{a^2 u} + C$

57. $\displaystyle\int (a^2 - u^2)^{3/2}\,du = \frac{u}{4}\left(a^2 - u^2\right)^{3/2} + \frac{3a^2 u}{8}\sqrt{a^2 - u^2} + \frac{3a^4}{8}\sin^{-1}\frac{u}{a} + C$

58. $\displaystyle\int \frac{du}{(a^2 - u^2)^{3/2}} = \frac{u}{a^2\sqrt{a^2 - u^2}} + C$

Forms Involving $\sqrt{2au - u^2}$

59. $\displaystyle\int \sqrt{2au - u^2}\,du = \frac{u - a}{2}\sqrt{2au - u^2} + \frac{a^2}{2}\cos^{-1}\left(\frac{a - u}{a}\right) + C$

60. $\displaystyle\int u\sqrt{2au - u^2}\,du = \frac{2u^2 - au - 3a^2}{6}\sqrt{2au - u^2} + \frac{a^3}{2}\cos^{-1}\left(\frac{a - u}{a}\right) +$

61. $\displaystyle\int \frac{\sqrt{2au - u^2}}{u}\,du = \sqrt{2au - u^2} + a\cos^{-1}\left(\frac{a - u}{a}\right) + C$

62. $\displaystyle\int \frac{\sqrt{2au - u^2}}{u^2}\,du = -\frac{2\sqrt{2au - u^2}}{u} - \cos^{-1}\left(\frac{a - u}{a}\right) + C$

63. $\displaystyle\int \frac{du}{\sqrt{2au - u^2}} = \cos^{-1}\left(\frac{a - u}{a}\right) + C$

64. $\displaystyle\int \frac{u\,du}{\sqrt{2au - u^2}} = -\sqrt{2au - u^2} + a\cos^{-1}\left(\frac{a - u}{a}\right) + C$

65. $\displaystyle\int \frac{u^2\,du}{\sqrt{2au - u^2}} = -\frac{(u + 3a)}{2}\sqrt{2au - u^2} + \frac{3a^2}{2}\cos^{-1}\left(\frac{a - u}{a}\right) + C$

66. $\displaystyle\int \frac{du}{u\sqrt{2au - u^2}} = -\frac{\sqrt{2au - u^2}}{au} + C$

67. $\displaystyle\int \frac{\sqrt{2au - u^2}}{u^n}\,du = \frac{(2au - u^2)^{3/2}}{(3 - 2n)au^n} + \frac{n - 3}{(2n - 3)a}\int \frac{\sqrt{2au - u^2}}{u^{n-1}}\,du,\ \ n \neq \frac{3}{2}$

68. $\displaystyle\int \frac{u^n\,du}{\sqrt{2au - u^2}} = -\frac{u^{n-1}\sqrt{2au - u^2}}{n} + \frac{a(2n - 1)}{n}\int \frac{u^{n-1}}{\sqrt{2au - u^2}}\,du$

69. $\displaystyle\int \frac{du}{u^n\sqrt{2au - u^2}} = \frac{\sqrt{2au - u^2}}{a(1 - 2n)u^n} + \frac{n - 1}{(2n - 1)a}\int \frac{du}{u^{n-1}\sqrt{2au - u^2}}$

70. $\displaystyle\int \frac{du}{(2au - u^2)^{3/2}} = \frac{u - a}{a^2\sqrt{2au - u^2}} + C$

71. $\displaystyle\int \frac{u\,du}{(2au - u^2)^{3/2}} = \frac{u}{a\sqrt{2au - u^2}} + C$

Forms Containing Trigonometric Functions

72. $\displaystyle\int \sin^2 u\,du = \frac{u}{2} - \frac{\sin 2u}{4} + C$

73. $\displaystyle\int \cos^2 u\,du = \frac{u}{2} + \frac{\sin 2u}{4} + C$